Biology

A Journey Into Life

"A Fish Tale," painting by Sandy Delahanty

SECOND EDITION

Karen Arms

Pamela S. Camp

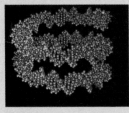

Biology

A
Journey
Into
Life

SAUNDERS COLLEGE PUBLISHING

Philadelphia   Fort Worth   Chicago   San Francisco   Montreal   Toronto
London   Sydney   Tokyo

Text Typeface: Garamond
Compositor: York Graphic Services
Acquisitions Editor: Julie Levin Alexander
Developmental Editor: Richard Koreto
Managing Editor: Carol Field
Project Editor: Margaret Mary Anderson
Copy Editor: Teresa L. Danielsen
Manager of Art and Design: Carol Bleistine
Art and Design Coordinator: Doris Bruey
Text Designer: Tracy Baldwin
Cover Designer: Lawrence R. Didona
Text Artwork: J & R Technical Services
Layout Artist: Dorothy Chattin
Director of EDP: Tim Frelick
Production Manager: Bob Butler
Marketing Manager: Marjorie Waldron
Cover Credit: Clown Fish, Coral Sea, Australia © 1989 Carl Roessler/FPG International

Printed in the United States of America

Biology: A Journey Into Life, Second Edition

0-03-072617-4

Library of Congress Catalog Card Number: 90-052774

1234   061   987654321

The first edition of *Biology: A Journey Into Life* was very well received and we are happy to have had the opportunity to revise and update it. Our aim continues to be to produce a concise text emphasizing the important concepts in biology. Instructors can easily use this text in courses taught to nonmajors, courses with a mix of majors and nonmajors, and even in brief courses for majors.

One of our major goals has been to revise first edition art and create new art to make the illustrations even more useful. Of course, any new edition must also be updated to reflect new discoveries in this rapidly moving field. In addition, we have fine-tuned the book's coverage to the needs of its users. We scrutinized every sentence to be sure it was as clear and understandable as possible, and we reorganized some chapters for a better flow of thought. We also rewrote most of the chapter introductions to make sure readers will have a good overview of the chapter's significant points before they plunge into its details. All of these changes have been aided by the suggestions of teachers who used the first edition.

School of jacks

Improved Illustrations

The illustrations have been thoroughly revised. Many line drawings are new, and many others have been redrawn to make the concepts more vivid for the students. We have also chosen dozens of new photos. Notable among these are the beautiful and interesting selections for the chapter and part openings, which set the stage by illustrating various themes in each area.

Environmental Emphasis

Only time will tell whether the 1990s will really be the "environmental decade," but biologists are under pressure to introduce their students to the scientific background they need to understand debates in this area. Accordingly, much of the new material in the second edition of *Biology: A Journey Into Life* emphasizes areas that are important to understanding environmental questions, including new boxes on solar energy (Chapter 6), environmental roles of social insects, including Africanized bees (16), prokaryotes and pollution (19), noise pollution and hearing loss (29), soil destruction (37), and declining productivity (38).

Content Changes

First edition users will also find several changes in the content of this edition:

1. The chapter on techniques in genetic engineering has been omitted because a survey of users showed that few teachers assigned it. The important concepts in that chapter have been incorporated into other parts of the book, especially boxes in Chapters 9, 10, and 36. The treatment of cancer has been completely rewritten to reflect current understanding and moved to Chapter 10 (on protein synthesis and gene expression). We have omitted the chemical details of genetic engineering techniques, which tend to bog down most students at this introductory level of biology.
2. Human evolution has been moved from the last chapter of the book into the chapter on animal diversity and evolution (21). Most users felt

it belonged in the evolution part of the book, and this reorganization allowed us to expand the ecology sections of the book with such important new topics as soil loss, deforestation, and ozone depletion.

3. Instead of the first edition's "Essays" we now have features called "A Journey Into Life." Several new ones have been added, mostly on health or environmental topics. In addition to those already mentioned, there is a look at membrane specializations in the function of intestinal epithelial cells (Chapter 4), genetic and environmental threats facing cheetahs (16), insects in the environment (21), smoking (23), drugs and neurons (28), and osteoporosis (30). Others have been extensively rewritten, including those on Gregor Mendel (Chapter 12), Alzheimer's disease (28), and the tragedy of the commons (40).

4. Chapters 4 (Cells and Their Membranes) and 10 (RNA and Protein Synthesis) have been reorganized, rewritten, and updated.

5. To make room for all the new material on recent findings and on environmental concerns, the coverage of chemistry has been tightened and the most advanced details eliminated. Our aim here has been to keep the level of detail low enough to avoid turning off the many nonmajors who use this book, without shortchanging students who will go on to major in biology or another science.

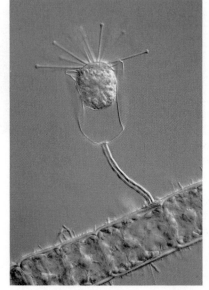

Suctorian attached to an alga

Organization

The book is designed to be flexible. Although it is organized in the traditional cell-to-ecology order, many instructors will want to rearrange the chapters to suit their own curriculums, which may start with evolution, ecology, or physiology. Each chapter can be read independently, and important terms are defined in more than one chapter. Cross references to sections in other chapters point to background material for the subject under discussion. Many satisfied users of the first edition have found that *A Journey Into Life* can work in many different arrangements.

Part 1 covers the life of cells, from their structure to their use and production of energy sources. Part 2 is on the transmission of genetic information from cell to cell and from generation to generation. Part 3 introduces evolution and the diversity of life on earth. Parts 4 and 5 cover animal and plant physiology, respectively. Part 6 covers ecology and the relationship of our own species to its environment.

Features to Aid Learning

In addition to its flexible organization, *Journey* contains a number of features designed to make it easier to study. Three of these are new:

PART PREVIEWS, with a selection of photos representing the part's chapters, survey the common ground to be covered by each group of chapters. These will be especially helpful to students assigned chapters in the part in an order other than that presented in the book.

CHAPTER OUTLINES next to the list of Objectives at the beginning of each chapter tell readers what they will be studying.

SUGGESTED READINGS have been moved from the ends of Parts to the end of each chapter. This makes it easier for readers who wish to explore the subject of the chapter further to locate suitable sources.

In addition to these new features, the book retains all those from the first edition:

OBJECTIVES, at the beginning of each chapter, tell students what they should expect to get out of the chapter and what they may expect on a test. Instructors will modify the objectives to suit their own purposes.

KEY CONCEPTS, near the beginning of each chapter, state the main ideas explored in the chapter. These are the signposts that guide students through the wealth of details, clarifying the main points of each chapter. Ideally, students will ask themselves how each part of the chapter illustrates one of its key concepts.

TAKE-HOME MESSAGES in this edition are placed in the margin for emphasis. Studies have shown that learners carry away from the average study session no more than a few ideas, perhaps illustrated by a few examples. These are the messages that we think students should carry away from their study of each chapter.

BOLDFACE within the text calls attention to terms being defined.

A SUMMARY at the end of each chapter provides an overview of the material covered.

A SELF-QUIZ provides a brief test of the objectives. It is intended to reassure students that they have mastered most of the material in the chapter or warn them that they need to spend more time on it. The answers are at the back of the book.

QUESTIONS FOR DISCUSSION stimulate thinking. They are ideas to be batted about in discussion groups or over lunch. The student should be able to work out the answers to some of them from the information in the chapter. However, many of them are unanswerable—questions research workers are attempting to answer, ethical and scientific problems that confront our society, or questions raised but not yet answered by the modern explosion of biological knowledge.

A GLOSSARY at the end of the book gives readers one source where they can look up many terms. The glossary will be particularly useful when the chapters are used in an order different from that in which they are presented in the book.

A JOURNEY INTO LIFE, the new name for the kind of material called "Essay" in the first edition, contains interesting anecdotes, unusual examples, historical background, environmental and human concerns, which are related to the chapter but are so peripheral to the main story, or so speculative, that we would not normally expect students to learn them in detail. Consequently, objectives and self-quizzes do not cover material mentioned only in these boxes.

This list of features has been carefully thought out to present students new to biology with everything they need to learn the material successfully. *A Journey Into Life* is the only book to offer such a complete learning system. As a result, we hope that students using the book will find it easier than ever to learn and remember biology. Learning is, in the last analysis, something students must do for themselves. All that we, as teachers and authors, can do is to try to whet their curiosities and select and organize material in a way that is interesting, attractive, and easy to master.

Grizzly cub

Supplements

All of the ancillaries that came with the first edition are available in revised editions: The Ohio State team of Russell V. Skavaril, Mary K. Finnen, and Steven M. Lawton has written a **study guide** with the same popular features: concept trees, which display each chapter's key ideas and show how they interrelate; vocabulary flashcards to help students master the language of biology; and unlabeled illustrations (Bio Art) from the text that test comprehension of the visual aspects of biology. Most Bio Art pieces are also available in color transparencies.

Florence C. Ricciuti of Albertus Magnus College has written a new **test bank** with over 1,500 questions, many of them new to this edition. Questions are designed to test not only the student's ability to recall facts, but also to use these facts to solve problems. A **computerized test bank** is available for both the Macintosh and IBM PC.

Also by Florence Ricciuti is a new edition of the **instructor's manual** that provides information and instruction to those who teach the course. Included are course organization suggestions, lecture preparation, lists of sources for audiovisual materials and software, essay questions keyed to the chapter's learning objectives, and a lecture outline on disk. This last item allows instructors to combine their own lecture notes with those prepared by Professor Ricciuti.

Visual supplements include approximately 250 **overhead transparencies,** also available as **35-mm slides,** and 70 **transparencies of electron micrographs.** Both sets of overheads feature printed captions for the instructor's use in explaining each transparency to the students. A special feature of the line art overheads is their extra-large labels, designed to be seen easily from the back of a large classroom.

Donald A. Keefer of Loyola College has revised his **Bio/XL tutorial program** to reflect the changes in the text. This program allows students to test themselves, review material, and keep track of topics they need to work on. It is available in both IBM and Macintosh formats.

Carolyn Eberhard of Cornell University has written a new **laboratory manual.** This manual is suitable for use with *A Journey Into Life* and other Saunders biology texts. Complete guides to its use with *Journey* are provided in the manual itself and in the instructor's guide to the manual.

New to this edition is a **laser disk** with many images and brief films.

Attempting to cover the main features of biology in a "short" book has proved an interesting, and at times frustrating, challenge. We hope that we have produced a book with which many newcomers will take off on the journey into biology that has fascinated and entertained us for many years.

Karen Arms
Savannah, Georgia

Pamela S. Camp
Ithaca, New York

January, 1991

ACKNOWLEDGMENTS

Dozens of people have helped us with this book. They have read chapters, housed us and fed us, argued with us about grammar, biology, and teaching, found photographs, drawings, reprints, and information for us. Our grateful thanks to: Alzheimer's Association, Paul Feeny, Virginia Fry, Eric Greene, Carol Johnson, Allan Larson, Michael Lelage, Marian and John McGrath, Fred McLafferty, Verne Rockcastle, Katherine and Walter Sharp, Thom Smith, and Milton Zaitlin. Bill Camp has again given invaluable help with computer, camera, and criticism.

As usual, Gordon Leedale and Helena Cmiech of Biophoto Associates have provided most of the lovely photographs, assisted by a small battalion of photographers. Tricia Smith, of Savannah College of Art and Design, brought back fine photographs from her trip to China. Karen Roeder, of the University of Georgia Marine Extension Service, toured the Southeast for photographs of wildlife and marshes. Edith Schmidt of the same institution combed the family photo collection. Matt Gilligan, of Savannah State College, contributed shots of deserts and fish. Steve Webster, now at the Monterey Aquarium, provided our favorite Caribbean shots. Paul Feeny, of Cornell University, turned up fine photographs of city life and eroding Himalayan hillsides after a trip to India. Cindy Vines Bright, of Carolina Biological Supply Company, shipped and catalogued hundreds of photographs for us. The Audiovisual Department of Carolina Biological Supply Company has been hospitable and enormously helpful. To all these competent, helpful people, our thanks for all your hard work.

We also thank everyone at Saunders College Publishing for their hard work coping with the problems that inevitably beset a book.

This book's official reviewers are teachers. They have taken time from their busy schedules to read the manuscript and to suggest numerous improvements. For your many thoughtful contributions, our thanks to:

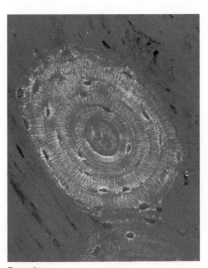

Bone tissue

Second Edition Survey Participants

Maurice Clark, Our Lady of the Lake University

Brad Henry, Pan American University

Cliff Knight, East Carolina University

Chuck Kugler, Radford Univesity

Terry Morrow, Clarion University

Bernie Morse, Norwalk Community College

Steven Murray, California State University at Fullerton

Brian Myres, Cypress College

Beverly Shue, Los Angeles Harbor College

Larry St. Clair, Brigham Young University

Michael Tansey, Indiana University

Paul Wright, Western Carolina University

Second Edition Reviewers

Maurice Clark, Our Lady of the Lake University

Stanley Cornish, Mohawk Valley C.C.

John Curtis, University of Wisconsin—Stevens Point

Lawrence DeFilippi, Lurleen B. Wallace State Junior College

James Forbes, Hampton University

Bernard Frye, University of Texas at Arlington
Bill Gaddis, University of Arizona
John Harley, Eastern Kentucky University
Fred Hinson, Western Carolina University
Charles Leavell, Fullerton College
Neil Miller, Memphis State University
Sallie Noel, Austin Peay State University
Thomas Terry, University of Connecticut
David Whitenberg, Southwest Texas State University
Bernard Woodhouse, Savannah State College

First Edition Reviewers

Donald Collins, Orange Coast College
Loren Denney, Southwest Missouri State University
Stephen J. Dina, St. Louis University
Patrick J. Doyle, Middle Tennessee State University
Nathan Dubowsky, Westchester Community College
Charles Duggins, Jr., University of South Carolina
Michael S. Gaines, The University of Kansas
Holt Harner, Broward Community College
George A. Hudock, Indiana University
Jerry L. Kaster, The University of Wisconsin—Milwaukee
James A. Morrow, Allan Hancock College
Robert L. Neill, The University of Texas at Arlington
Florence C. Ricciuti, Albertus Magnus College
David M. Senseman, The University of Texas at San Antonio
Russell V. Skavaril, The Ohio State University
Gerald Summers, University of Missouri—Columbia
Barry D. Tanowitz, University of California, Santa Barbara
Richard R. Tolman, Brigham Young University
R. W. Van Norman, The University of Utah
Nancy K. Webster, Prince George's Community College
George West, Northern Virginia Community College
Tommy Elmer Wynn, North Carolina State University

The Unity of Life: Cells

PART

1

*T*he star of Part 1 of this book is the cell—a tiny mass of living matter, often called the unit of life. In Chapter 1 we listed cells as the first characteristic of living things: all organisms consist of one or more cells, which show all the other features common to life. The chapters in Part 1 focus on the cell itself and two other features of life: orderliness and the use of energy.

A high degree of order occurs in crystals, which are not alive. It is possible to predict the precise, fixed position of every atom in a crystal. Some cells produce structures with even more complex but regular order, such as hair or other fibers, or intricately patterned external walls. The orderliness of these static structures is impressive. However, living cells also show a more interesting sort of order: the dynamic relationships among their working parts. For instance, a thin, fatty membrane covers the outside of every living cell and exchanges substances with the external environment. Inside the cell are chemical components such as enzymes, which carry on orderly sequences of chemical reactions, using the substances taken in through the membrane as raw materials. Here, too, is the cell's genetic material, which determines what enzymes and other chemical components the cell makes. By the coordinated interaction of these and other structures, the cell continuously processes energy and materials taken from its environment and uses them to maintain its orderly structure and activity, and eventually to reproduce.

A crystal or fiber has order because its parts have settled into the lowest state of energy possible (under prevailing conditions). A cell, by contrast, maintains its order by constantly acquiring energy from its environment and using this energy to work against the tendency toward lower-energy states.

To understand how the cell carries out all the activities of life, we need to take it apart and examine its chemical components to see how their properties contribute to the life of the cell (Chapters 2 and 3). Then we shall see how these components are assembled into the parts of a cell and how these parts work together to carry on the business of living (Chapters 4 and 5). Finally, Chapters 6 to 8 look at the rules of energy transactions that cells must obey and see how cells capture and process the energy they need to carry out the activities that keep them going and growing.

Throughout our study of cells, we shall see a theme of the unity of life: all cells contain the same basic kinds of chemicals; these chemicals form certain universal cellular structures and are organized in a limited number of ways that permit cells to obtain, release, and use energy.

The study of cells is one of the most important fields of modern biology. As the units of life, cells also harbor its secrets. However, most cells are so small that it is hard to tell what is going on in there. Our knowledge of these secrets is growing rapidly, thanks to the development of new instruments and techniques for viewing, analyzing, and manipulating the tiny entities involved.

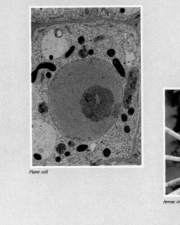

An energy transfer

18

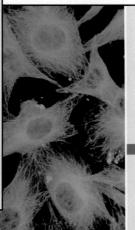

Microtubules

Lipid (wax) and carbohydrate (chitin)

Plant cell

Arrow crab feeding

19

PART OPENER
Overview of chapters relates diverse ideas and points to unifying theme.

PART PHOTO PREVIEW
Selected photographs from chapter opening provide a visual preview of upcoming material.

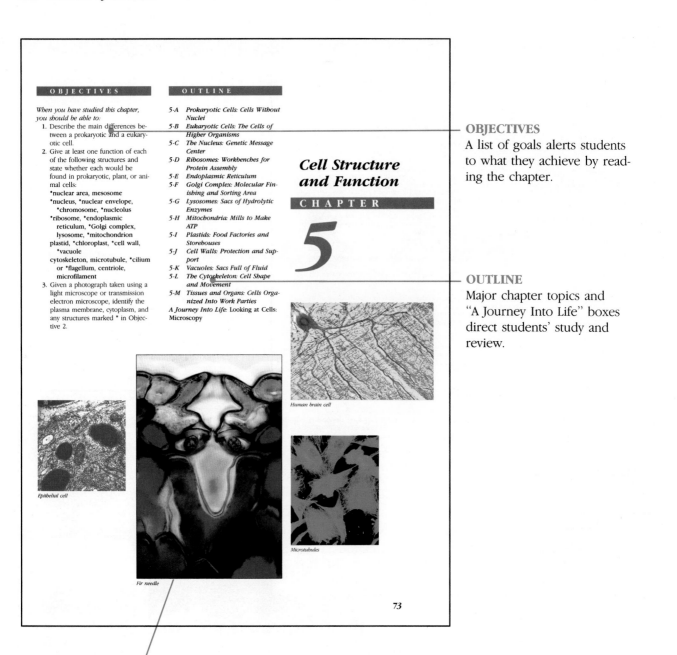

Cell Structure and Function

CHAPTER

5

Human brain cell

Microtubules

Epithelial cell

Fir needle

73

OBJECTIVES
A list of goals alerts students to what they achieve by reading the chapter.

OUTLINE
Major chapter topics and "A Journey Into Life" boxes direct students' study and review.

CHAPTER PHOTO PREVIEW
All-new photos visually summarize the chapter.

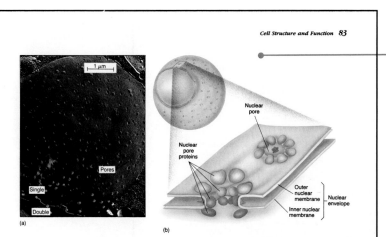

LINE ART AND MICROGRAPHS
Side-by-side micrographs and line art help students visualize structures.

Figure 5-6
The nuclear envelope. (a) The surface of a nucleus as seen with a scanning electron microscope. The specimen has fractured in such a way that the double layer of the nuclear envelope shows at the bottom, but only the single inner layer remains in most of the photograph. Nuclear pores are visible, some of them apparently clogged with material. (b) Diagram of two nuclear pores. The pores are partly blocked by proteins of the nuclear pore complex. The particle in the center of the far pore is often seen in electron micrographs. It is not known whether it is part of the pore's structure or merely a large particle caught during transit. (a, Biophoto Associates)

though, the nucleus is not dividing, and the chromosomes are uncoiled in a loose, indistinct tangle called **chromatin** (Figure 5-5).

DNA in the chromatin determines what RNA is made in the nucleus. The RNA travels to the cytoplasm, where it directs protein synthesis. Therefore, DNA also determines what proteins the cell makes. This, in turn, affects most of the cell's structures and activities. Hence the nucleus has been called the cell's "control center." However, control is a two-way street. Substances in the cytoplasm enter the nucleus and influence the DNA in such a way that it changes the particular mix of RNA (and hence proteins) produced. So the kinds and amounts of proteins made in the cell change depending on its needs.

With the light microscope, the most obvious features in the nuclei of nondividing cells are the nucleoli. **Nucleoli** (singular: **nucleolus**) are areas where ribosomes are made; they appear as one or more dense areas, which disappear when the cell divides (Figure 5-5).

The nucleus is surrounded by a double membrane, the **nuclear envelope** or **nuclear membrane,** which is perforated by pores (Figure 5-6). RNA leaves the nucleus via these pores. (RNA molecules are very large as well as hydrophilic, so they could not pass through the lipid layers of the nuclear envelope.)

The inner and outer membranes of the nuclear envelope connect with each other through the nuclear pores, and the outer membrane, in turn, is continuous with a system of membranes in the cytoplasm called the endoplasmic reticulum (Section 5-E). These interconnections allow the nuclear envelope to shrink or grow rapidly by losing material to, or gaining it from, the endoplasmic reticulum. In most eukaryotes, the nuclear envelope disappears completely during cell division and re-forms afterwards. The envelope may expand very rapidly when a dormant cell gears up to produce DNA or RNA.

5-D Ribosomes: Workbenches for Protein Assembly

Ribosomes are necessary components of cells because they are the sites of protein synthesis. In eukaryotes, ribosomes are made in the nucleolus area(s) of the nucleus and then travel to the cytoplasm by way of the nuclear pores. Eukaryotic

built of, and run by, thousands of different molecule has distinctive properties that have been selected during the course of evolution. skeletons" of carbon atoms, bonded together se carbons are also bonded to one or more ecules are made chiefly by living organisms, **s.** In addition to carbon and hydrogen, most and some also contain nitrogen, sulfur, or

all organic molecules, which play important of these small organic molecules also serve lding units that are joined together to make er molecules contain only a few monomers. ge number of monomers and so are called nown as **macromolecules** (macro = big).

Living organisms can assemble an enormous array of different polymers from a relatively small number of monomers. Natural organic polymers include wool, silk, rubber, starch, and cotton. The artificial organic polymers known as "plastics" are made by joining small organic monomers in various ways—a case of art, or at least industry, imitating nature.

There are four main classes of biologically important organic compounds:

1. **Lipids:** nonpolar substances, which do not dissolve in water—including fats, oils, waxes, and steroids.
2. **Carbohydrates:** sugars, starches, cellulose, and related compounds.
3. **Nucleic acids:** the genetic material (containing instructions for making proteins) and molecules that help assemble proteins.
4. **Proteins:** molecules that make up silk, hair, tendons, and so on; carry out cell movements; act as hormones; transport substances in the blood; and fight infections. One important group of proteins is the enzymes, which carry out the cell's hundreds of biochemical reactions.

This chapter introduces all four groups and considers enzymes in some detail. In each group we find different chemical features, which give members of the group properties useful to living organisms in various ways.

KEY CONCEPTS
Summary of main ideas provides a framework for studying the chapter.

KEY CONCEPTS

• Living organisms have a distinctive chemistry based on large molecules with skeletons made up of strings and rings of carbon atoms. These organic molecules also contain hydrogen and oxygen, and often nitrogen, sulfur, or phosphorus.
• Organisms make and use thousands of kinds of organic molecules, which fall into four main classes: lipids, carbohydrates, nucleic acids, and proteins.
• Biological macromolecules are composed of many similar or identical monomer subunits joined together.
• A cell's enzymes convert organic molecules into different forms, step by step, in complex, controlled pathways.

TAKE-HOME MESSAGES
Messages in the margins highlight concepts students need to remember.

3-A Structure of Organic Molecules

This chapter contains many diagrams of molecular structure. We show them, not expecting you to learn each molecule's name and structure, but so you can see a sample of biological molecules. At first, even small organic monomers may seem complicated. However, they start to look familiar once you know how to approach them. When you meet a new molecule, you should first get an idea of its size and shape by examining its carbon skeleton. Then look at the other atoms or groups of atoms present—the molecule's functional groups.

■ *Chains and rings of carbon are the "skeletons" of organic monomers. Attached to the carbon skeleton are atoms, or groups of atoms, of other elements: hydrogen (H), oxygen (O), and sometimes also nitrogen (N), sulfur (S), or phosphorus (P).*

35

Sounds Make Silence

A JOURNEY INTO LIFE

Calling noise a nuisance is like calling smog an inconvenience. Noise must be considered a hazard to the health of people everywhere. (William H. Stewart, former U.S. Surgeon General)

One hazard of environmental noise is hearing loss. A baby is born with up to 20,000 hair cells in each cochlea—not very many when you consider that this is the total lifetime supply; damaged or killed hair cells are never replaced. Loud noise damages the hair cells' microvilli by disrupting their internal skeletal framework, so that they flop like wet noodles. Noise also causes the microvilli surfaces to develop blister-like vesicles, which rupture as the noise continues. Once the damage starts, it takes *less* sound pressure to do an equal amount of additional damage.

Hearing loss caused by noise is insidious because it is painless and so gradual that it is not noticed. At first, sounds must simply be louder (or closer) in order to be heard. Since the first hair cells to die detect high pitches, above the range of the human voice, victims can still hear conversations. Then the higher-pitched speech sounds fade out, and these people can no longer distinguish between similar words (this often gives rise to complaints that speakers are mumbling). Hearing loss is so widespread that it is regarded as a natural part of aging. However, much of the blame should be placed on noises of modern civilization. The elders of a Sudanese tribe living in a quiet Stone Age culture hear better at 80 than most Americans do at 30.

Tests of over 4000 students entering college in the early 1980s showed that more than one in three already had measurable loss of hearing for high-pitched sounds. These were mostly not serious yet, but today's young people can expect to suffer greater hearing loss by the time they reach their 50s and 60s than we see in people now in that age group.

Noise is the most widespread occupational hazard. The federal government regulates noise in the workplace, and employers have found it cheaper to provide workers with ear protectors than to pay damage awards from lawsuits for work-related hearing loss (Figure 29-A). However, many workers do not realize the damage noise can do and do not use the equipment properly. Everyone is exposed to dangerous levels of noise off the job. The worst offenders are gunfire (from military duty, target shooting, or hunting), vehicles (aircraft, subways, heavy city traffic, snowmobiles, and speedboats), and loud music. The noise level in front of a speaker at a rock band concert is equivalent to that of a thunderclap, and efficient stereo headphones, turned to maximum volume, can deliver sound louder than a jet taking off at the other end of a football field.

Hearing loss is not the only biological damage done by noise. It also causes symptoms of stress—high blood pressure, increases in the heartbeat and breathing rates and in muscle tension, and changes in the digestive tract that can lead to ulcers. The brain can learn to tune out noise, so that it no longer reaches the level of conscious awareness, but the body cannot adapt in this way. Prolonged exposure to loud noise keeps the body in a constant state of stress.

Figure 29-A
Ear protection. Employers are required to supply employees in noisy jobs, such as airport baggage handlers, with equipment to protect their hearing. (Karen Roeder)

which detect light as opposed to darkness. **Eyes** detect more detail. In an eye, light from the visual field projects an image onto light-sensitive receptor cells. True eyes occur only among vertebrates, arthropods, and molluscs.

In the eyes of vertebrates, the curved **cornea** focuses light entering the eye (Figure 29-10). The focus is adjusted by the **lens,** which is somewhat elastic and changes shape when it is pulled by muscles around its edges. The **iris** is a diaphragm that controls the size of its central opening, the **pupil,** and so regulates the amount of light entering the eye. The **retina** is a delicate layer contain-

A JOURNEY INTO LIFE
Boxed essays apply biological concepts to current human and environmental topics and other interesting subjects.

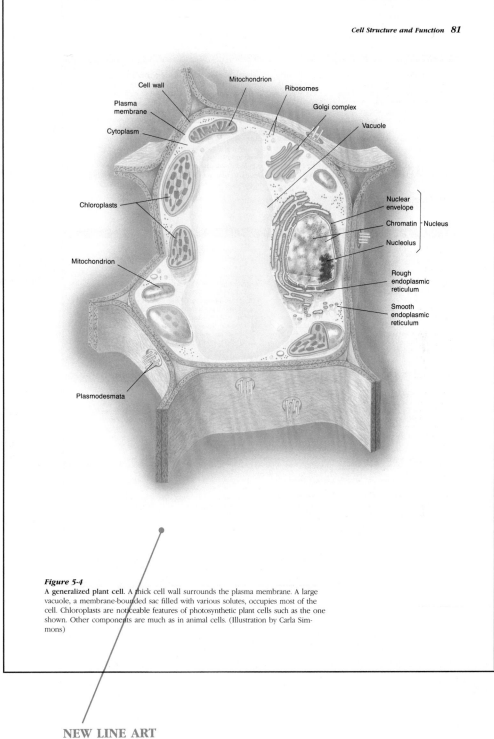

Cell wall

Mitochondrion

Ribosomes

Plasma membrane

Golgi complex

Cytoplasm

Vacuole

Chloroplasts

Nuclear envelope

Chromatin — Nucleus

Nucleolus

Mitochondrion

Rough endoplasmic reticulum

Smooth endoplasmic reticulum

Plasmodesmata

Figure 5-4
A generalized plant cell. A thick cell wall surrounds the plasma membrane. A large vacuole, a membrane-bounded sac filled with various solutes, occupies most of the cell. Chloroplasts are noticeable features of photosynthetic plant cells such as the one shown. Other components are much as in animal cells. (Illustration by Carla Simmons)

NEW LINE ART

New line art with larger, easier-to-read labels illustrate important concepts.

SUMMARY

Living organisms require energy in order to maintain their chemical composition, move, repair damage, grow, and reproduce. Energy cannot be created nor destroyed. However, each time energy is converted from one form to another, some useful energy is degraded into an unusable form. To remain alive, organisms must constantly acquire fresh supplies of energy.

The central energy-processing pathways of life are photosynthesis and respiration. In photosynthesis, the sun's energy is captured and stored in the chemical bonds of food molecules, which can later be broken down during respiration to release the trapped energy:

photosynthesis: $CO_2 + H_2O + energy \longrightarrow (CH_2O)_n + O_2$
respiration: $(CH_2O)_n + O_2 \longrightarrow CO_2 + H_2O + energy$

Both photosynthesis and respiration are essentially oxidation-reduction reactions: many steps in each process involve transferring electrons (or hydrogen atoms, which contain electrons). Many of these redox reaction steps release a great deal of useful energy. The ultimate task of both photosynthesis and respiration is to trap this energy and use it to produce energy intermediates, which can then drive endergonic reactions.

The most common energy intermediate is ATP. Gradients of ions across membranes also serve as energy intermediates for some energy-requiring processes. In fact, most ATP is formed by using the energy of such a gradient to join phosphate groups to ADP.

SELF-QUIZ

1. Does the following equation describe photosynthesis or respiration?

$$CO_2 + H_2O \longrightarrow C_6H_{12}O_6 + O_2$$

2. In the above equation, is carbon oxidized or reduced as a result of the reaction?
3. True or False? The cells of plants contain mitochondria and carry on cellular respiration.
4. ATP is made by joining _____ and _____. This reaction is (endergonic, exergonic). Hence it requires _____ in addition to the reactants.
5. The electron transport system:
   a. makes ATP
   b. contains ATP synthetase
   c. is responsible for separating hydrogen atoms into their components
   d. can work against the second law of thermodynamics
   e. cannot perform redox transfers if the membrane is broken
6. The inner membranes of mitochondria and chloroplasts:
   a. contain channels for movement of H^+
   b. have ATP synthetase enzymes attached
   c. contain molecules of the electron transport system
   d. form closed compartments
   e. all of the above

QUESTIONS FOR DISCUSSION

1. List as many energy-requiring activities carried out by living organisms as you can.
2. Why must photosynthetic plants carry on respiration?
3. Of the light energy reaching the earth from the sun, the earth's plants are believed to convert less than 1% into the form of potential energy stored in the chemical bonds of food molecules. What happens to the rest of the energy?
4. Organisms cannot use heat energy to drive their energy-requiring processes. Does this mean that the heat released by metabolism is of no use to them? Why or why not?

SUGGESTED READINGS

Atkins, P. W. *The Second Law*. Scientific American Books, 1984.
Hinkle, P. C., and R. E. McCarty. "How cells make ATP." *Scientific American*, March 1978. Compares the chemiosmotic mechanism for ATP synthesis in bacteria, mitochondria, and chloroplasts. [Note: to understand this article, you must know that biochemists ignore the outer mitochondrial membrane. When they say, "outside the mitochondrion," they mean "outside the inner membrane."]

SUMMARY
A recap of major concepts aids study and review.

SELF-QUIZ
Varied question formats help students test their understanding.

QUESTIONS FOR DISCUSSION
Problems stimulate and challenge students to use their new knowledge and extend their understanding.

SUGGESTED READINGS
Encourages further reading about chapter topics in books and articles.

CONTENTS OVERVIEW

Succulent plant

Earth Day 1990: solar collector running water pump

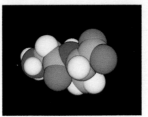

A dipeptide

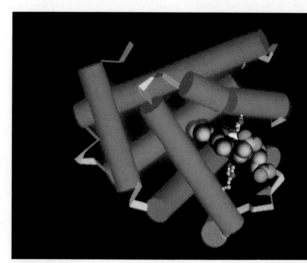

Protein (myoglobin)

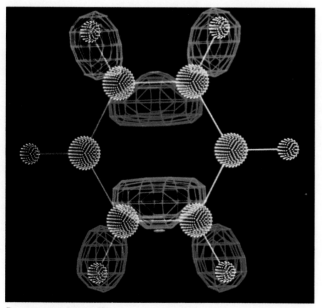

Benzene, computer graphic

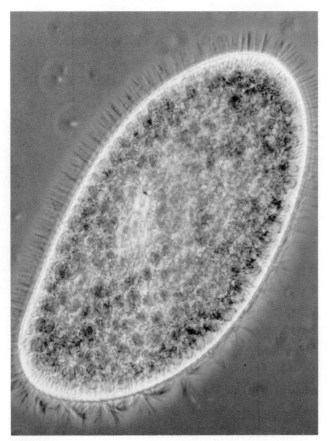

Paramecium

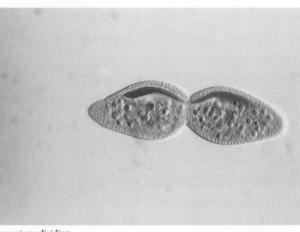

Paramecium *dividing*

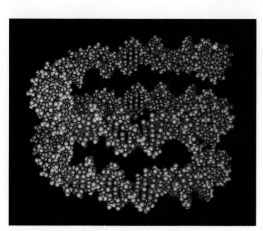

DNA molecule (computer graphic)

Flamingo

Snapdragons

Charles Darwin

Phantasmal dart poison frog (Epipedobates tricolor)

SEM, penicillin fungus

Hawiian lobster

Impala herd

Red blood cells

Manatee

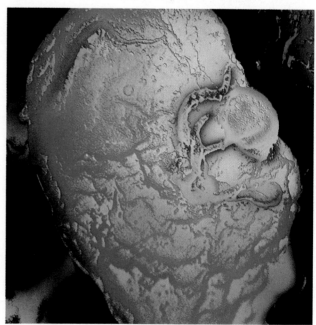

AIDS virus

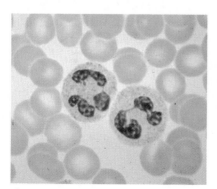

Phagocytes

Galago ("bush baby")

Tulip

Monterey cypress

Pear blossoms

Vegetation reproduction

Mt. Ranier

Cholla and ocotillo cacti

Wheat harvest

Biology

A Journey Into Life

Introduction

CHAPTER

1

Maple leaves, autumn

Hyacinth macaw

Mountain gorilla

Silver Dry Lake

*E*very autumn, we prepare to go back to school: buying notebooks and pens, a bookbag, some new shirts, maybe a sweater. Soon the natural world changes too. Goldenrod blooms. Trees blaze with brilliant colors. Birds fly south, and summer visitors to the bird feeder are replaced by winter residents, arriving from their northern breeding grounds.

Why does goldenrod bloom in the fall? Why do the leaves turn color? How do birds find their way south to their wintering grounds, and back in the spring? And why are there more robins in some years than in others?

All of these questions—and more—are the subject of biology, the study of living organisms. An **organism** is an individual living thing—animal, plant, fungus, bacterium, or one of the single-celled creatures called **protists.** Every one of the organisms sharing our earth today is descended from a long line of ancestors stretching back to the first life on earth, more than 3.5 billion years ago. Because modern-day organisms trace their ancestry back to these primitive forms of life, they all share certain similarities. But over these billions of years, the descendants of the first organisms have diversified into an astounding variety of species, including the millions alive today and many, like dinosaurs and dodos, that became extinct along the way.

The job of biology is to describe the immense diversity of life, outline the history of life on earth from ancient to modern forms, and learn more about the organisms around us. How are organisms alike? How do they differ? How do they work, what do they do, how are they made? And how do they affect their physical surroundings, the other organisms around them, and the human race? In this chapter, we consider how scientists go about asking and answering questions like these. Finally, we summarize the most important concepts in biology— which is really a brief synopsis of what is covered in this book.

1-A   Scientific Method

The **scientific method** is a way of answering questions about cause and effect. It is a logical way to try to solve problems. Similar reasoning is used by business people, athletes, and each of us in everyday life. Once we understand the reasoning and procedures scientists use, we are better able to decide for ourselves whether the conclusions of the latest scientific study are justified from the data presented. We can ask for further tests if a claim does not appear to be well supported by the evidence, and we can agree or disagree with predictions based on such claims.

Experimental Method

When we speak of scientific method, we usually mean a method of doing science that involves experiments. This experimental method has three main steps (although in practice scientists work in many different ways). The first, and key, step is to collect **observations,** not only by sight, but perhaps using other senses too (hearing, smell, taste, and touch). Scientists often use instruments to extend human senses or to detect things our senses cannot; examples are microscopes, radar, voltmeters, or Geiger counters. Second, the scientist thinks of several alternative **hypotheses** (singular: hypothesis), proposed answers to questions about what has been observed. The third step is **experimentation,** performing tests designed to show that one or more of the hypotheses is more or less likely to be incorrect. As a result of these experiments, the scientist should be able to draw some conclusions about *why* the original observed events occur. Let us see how this works in practice.

■ *The experimental method of doing science has three main steps:*
1. *Collecting observations*
2. *Forming hypotheses*
3. *Performing experiments*

Scientists usually start with observations that stimulate questions. Some years ago, one of your authors was part of a group of biologists discussing the clusters of butterflies that seemed to be everywhere that June.

"Today," said one, "I saw about 20 yellow sulfur butterflies by a stream and some black swallowtails on a manure heap. What are they doing?"

"It's called 'puddling behavior,'" replied another. "You find puddling butterflies in groups in open places such as the edges of drying puddles, or sandbars (Figure 1-1). I don't think anyone knows what they are doing. Another odd thing is that in many species only the males puddle."

These observations of puddling led us to ask what the butterflies were doing and why. To answer these questions, we had to think of some hypotheses that would account for the observations. That evening, the hypotheses came thick and fast from our armchair scientists.

"An article I read suggested it was a method of population control. Coming together permits the males to count each other. A newcomer can see if there is likely to be enough land for him to set up a territory in the area. Puddling saves them having to fight over territories."

"That sounds wrong to me," replied one of the company. "How can a butterfly figure out the density of males in the area from a group like that? Besides, swallowtails do fight for territories—I've seen them."

"I think it is more likely they're feeding," another contributed. "It was called 'puddling' in the first place because the butterflies often have their proboscises [tongues] out and seem to be sucking something up from the ground."

"I wonder if they are feeding on substances that contain nitrogen. In our lab we've shown that butterfly caterpillars grow faster if you feed them extra nitrogen, and there is lots of nitrogen in a manure pile."

Figure 1-1
Sulfur butterflies puddling on a sandbank. (Keith Brown)

"But not in sand," came the objection. "And if they are after nitrogen, you'd expect females to puddle, not males. The females lay the eggs that hatch into caterpillars, and extra nitrogen in the egg might be very useful, but it's not the females that puddle."

"It sounds to me," chipped in another, "as if they're after salts—perhaps salts containing sodium. All the puddling places contain quite a lot of salts: manure piles have salts from urine, and puddles have salts at the edges, left behind by evaporation of water. Lots of animals that feed on plants are short of sodium because plants contain so little of it. We put out salt blocks for cows and horses and end up attracting deer and rabbits as well. Perhaps male butterflies need more sodium than females do."

We could test these alternative hypotheses only by doing experiments. Some hypotheses are of no use because they cannot be tested. For instance, the hypothesis "puddling butterflies count each other" is probably untestable because it is hard to imagine an experiment that could show us whether or not an animal has counted its neighbors. Even a testable hypothesis usually cannot be tested directly. We must first develop a testable **prediction** from it. From the hypothesis that butterflies sucked up sodium when they puddled, we predicted that if we put out trays containing sodium, butterflies would be attracted to puddle on them. The hypothesis that puddling butterflies suck up nitrogen generated the prediction that butterflies would puddle on trays of amino acids, substances that contain nitrogen. These predictions can be tested and, in this case, both can be tested at the same time, in the same experiment.

We must design experiments to make their results as clear-cut as possible. For this reason, experiments have to include **control treatments** as well as **experimental treatments.** The two differ only by the factor(s) being investigated. For instance, to test our hypotheses, we had to show that butterflies would puddle on an experimental tray containing amino acids or one containing sodium but would not puddle on control trays that were identical except that they did not contain the amino acids or sodium.

Suppose we put out three trays—one containing sodium, another containing amino acids, and a third containing something butterflies are most unlikely to eat, such as plain sand or sand and water (the control). We would predict that if butterflies are attracted to puddle on sodium, they would come to puddle on the sodium tray but not on the other two. If they are attracted to amino acids, they would puddle only on that tray. If they are attracted to both, they would puddle on both of these but not on the control tray, and if they are attracted neither to amino acids nor to sodium, they would not puddle on the trays at all. Note that there are dozens of other possible reasons for the last result. If no butterflies turned up to puddle on our trays, we would have learned nothing. Butterflies might not puddle on trays because they won't come near trays for some reason, or because they never see the trays, or because they avoid the human watchers nearby, or for any one of a number of other reasons.

So that our experiment would not fail for lack of butterflies, we put our trays on a sandbank by a lake where tiger swallowtail butterflies often puddled in large numbers. We filled the trays with clean sand for the butterflies to stand on, and in each tray we pinned a dead male tiger swallowtail as a decoy, because we thought butterflies might be attracted to puddling places by seeing other butterflies there. We put out ten trays of sand and poured the same volume of solution (substances dissolved in water) into each one. Then we sat nearby, with binoculars, notebooks, and watches, to see what would happen.

Soon dozens of tiger swallowtails were hovering over the trays. Whenever a butterfly landed on a tray, it stuck its proboscis into the sand. At times, as many as 30 butterflies were on a tray together. Most of the butterflies spent a few seconds on every tray, but they puddled (which we defined as staying for more than 15

Figure 1-2

Arrangement of trays on one day of the puddling experiment. Each tray contained the same volume of sand. Each of eight trays also contained 1.5 litres of water or solution. Different solutions were placed in different trays on subsequent days. (Sugar was tested because swallowtail butterflies eat sugar-filled nectar from flowers, and therefore we wondered if they might be attracted to puddle on sugar.) The black number on each tray shows the number of "sampling" visits (lasting less than 15 seconds) by butterflies. Colored numbers show the number of butterfly-minutes spent puddling on the tray in visits lasting more than 15 seconds. The numbers make it obvious that butterflies puddled on the trays containing sodium and those containing amino acids but not on any of the other trays.

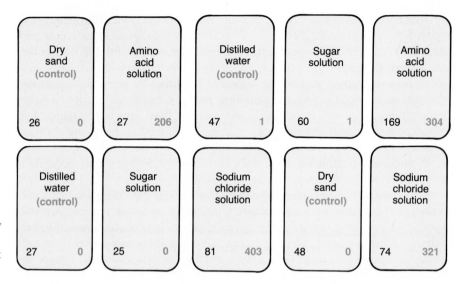

| Dry sand (control) 26 0 | Amino acid solution 27 206 | Distilled water (control) 47 1 | Sugar solution 60 1 | Amino acid solution 169 304 |
| Distilled water (control) 27 0 | Sugar solution 25 0 | Sodium chloride solution 81 403 | Dry sand (control) 48 0 | Sodium chloride solution 74 321 |

seconds) on only a few trays: all those containing sodium in any form and those containing amino acids (Figure 1-2).

We were satisfied that these results were accurate because we had taken another precaution: the people recording the butterflies' visits did not know which tray contained which solution (Figure 1-3). Making an experiment "blind" in this way is important. Psychologists have shown that, even in a carefully controlled experiment, experimenters tend to find the results they want to find. This is also why scientists try to form many hypotheses to explain their observations: it's too easy to bend the truth, without even realizing it, to support the only available hypothesis.

Those of us who favored the hypothesis that butterflies puddle in response to sodium were disappointed that they also puddled on amino acids. But prejudice can sometimes be useful, even in science! Not only were we disappointed by the results, we were inclined to think they were wrong. Back we went to our bottle of amino acids. We now made an observation that should have preceded the experiment: the label said, "Prepared in sodium citrate." According to popular myth, scientists are calm and objective, but we were very excited as a technician analyzed our amino acids: they were chock full of sodium! There followed

Figure 1-3

The puddling experiment. (a) Observers watching puddling trays. (b) Butterflies visiting the trays. In the middle of each tray is pinned a dead butterfly to serve as a decoy. (Paul Feeny)

(a)

(b)

frantic phone calls and special deliveries to obtain amino acids free from sodium. At last came a suspenseful experiment, which showed that butterflies did not puddle on our new, sodium-free amino acids.

We had now conducted a well-controlled scientific experiment. What conclusions could we draw? Had we proved the hypothesis that butterflies puddle so as to obtain sodium? No. We had not even shown that the butterflies actually drank the sodium solution. All we had shown was that male tiger swallowtail butterflies would puddle on sand containing sodium salts but not on sand containing various other solutions. Many more hypotheses and experiments were needed if we were to learn more.

One peculiarity of the scientific method is that a hypothesis can never formally be proved but can only be disproved. A correct hypothesis leads to predictions that are borne out by experiments, but an incorrect hypothesis may also produce correct predictions (that is, the prediction was right, but for the wrong reason). Therefore, if the results of an experiment agree with the prediction, we are still not sure that the hypothesis is correct. For instance, the hypothesis that butterflies puddle to obtain sodium for food is not proven by the experimental finding that butterflies puddle on sodium. They might puddle because wherever there is sodium in nature there is also nitrogen and they really obtain nitrogen from puddling. We have not even disproved the hypothesis that puddling is a means for the butterflies to "count" each other. They might puddle on sodium merely as a convenient rendezvous (although the fact that the butterflies appear to feed when they puddle makes this hypothesis unlikely). The more alternative hypotheses we disprove or cast doubt on, however, the greater the likelihood that the remaining hypothesis is correct.

Scientists also hesitate to accept the results of an experiment until they have tested its repeatability. Repeating an experiment guards against two kinds of errors. The first is **human error** (a polite term for mistakes); we might have inadvertently switched the solutions, written our results in the wrong column of our data notebook, or alarmed the butterflies. (Even in this simple experiment, the possibilities are endless.) Second, any experiment is subject to **sampling error,** error due to using a relatively small number of subjects. Organisms are notoriously variable. Our experiment sampled only a few dozen butterflies on six days. These butterflies might not have been representative of all tiger swallowtails. We could be more confident of our results if we were to repeat the experiment, using more butterflies (that is, a larger sample) and following precisely the same procedure. How many butterflies do we need? The more the better, but we could not possibly test all the butterflies in the world. In practice, we can use statistical tests to tell how "sure" we are of our results with a given sample size.

A hypothesis supported by many different lines of evidence from repeated experiments is generally regarded as a **theory,** and after even further testing it comes to be generally accepted.

■ *Scientists tentatively accept hypotheses that are consistent with experimental results or observations and reject those that are not.*

Correlation Studies

Many scientific questions cannot be studied by the experimental method. In some cases, experiments would be unethical (for example, it would be unethical to test a vaccine against AIDS by injecting live AIDS viruses into vaccinated and unvaccinated people). Geologists who study the earth's crust, or paleontologists who study dinosaur fossils, work with processes that occurred millions of years ago: it is too late for experiments. We also cannot study long-term changes in the earth's climate by doing experiments because we have no "control" planet, and because the experiment is too big to perform.

To test their hypotheses, scientists in such fields depend upon **correlations,** reliable associations between two events. We may observe many occa-

Figure 1-4

Tropical forest goes up in smoke. This view from the space shuttle *Discovery* shows smoke (not clouds) completely covering an area about three times the size of Texas in the Amazon River Basin of Brazil in September, 1988. Settlers cut down the forest to clear land for farming and ranching, and they burn the felled trees. The plume in the middle of the photograph is about the same size as the one formed by all the fires in and around Yellowstone National Park in 1988. This addition to the greenhouse gases is highly visible, but the greatest contribution of carbon dioxide comes from the burning of fossil fuels in vehicles and industry. (NASA)

sions on which one event always accompanies another. Also, we may find that if one of these increases, the other increases (or decreases) in a way we can predict. The purpose of establishing a correlation in this way is to assert that one of the events causes the other. If we can explain why this might be so, we may propose the hypothesis that one causes the other. As in the experimental method, we have made observations and formed hypotheses, but we cannot perform an experimental test. Instead, we test our hypotheses by subjecting them to more observations of as many kinds as possible. The trouble with correlation studies is that events may be linked even when they do not have a causal relationship, and observation is less conclusive than experimentation as a way to show causality.

For instance, the earth is kept warm by the "greenhouse effect" of gases in the atmosphere, including water vapor, carbon dioxide, and methane. These gases form a layer that traps heat from the earth and prevents it from escaping into space. Since about 1965, scientists have argued that the earth is warming up, largely because burning fossil fuels, such as coal and oil, releases carbon dioxide into the atmosphere and increases the greenhouse effect.

This hypothesis is extraordinarily difficult to test. There is good evidence that the amount of carbon dioxide in the atmosphere has increased every year since 1958, when records were first kept. Also, Swiss workers found air 1000 years old trapped in bubbles in the ice of Antarctica and Greenland. It contained considerably less carbon dioxide than modern air. This and similar studies have established that the carbon dioxide content of the air has increased steadily since the industrial revolution, about 200 years ago, when people first began to burn large amounts of fossil fuel. But is the temperature on earth increasing? And can we show a connection between greenhouse gases and temperature?

Temperature records from old newspapers and books show that temperatures on earth vary considerably over time, but the average temperature has indeed increased steadily for the last 100 years. Some people see no reason to blame this increase on the greenhouse effect. They argue that evidence from fossils and rocks shows that the earth's average temperature has changed every few thousand years during much of the earth's 4-billion-year history.

Laboratory experiments have been used to establish how much extra heat is trapped by various concentrations of carbon dioxide or methane. All the available data are then fed into a computer model, which shows a correlation between increases in greenhouse gases and temperature during the last 100 years. When the model is used to predict climate in the future, it shows that average temperatures may be from 1 to 5°C warmer by the year 2030 (which would seriously disrupt life on earth).

Since the burning of fossil fuel is the main factor increasing the carbon dioxide content of the air over its natural levels, many scientists are convinced that this practice is increasing the temperature on earth. They are convinced for two reasons. First, the correlation between temperature change and greenhouse gases is strong. Second, in theory we would expect increasing levels of greenhouse gases to raise the temperature on earth. These are the two main parts of a scientific conclusion based on correlation studies: a hypothesis that makes sense and correlations that support the hypothesis.

1-B   It's a Fact?

"It's a scientific fact" is often presented as the clincher to an argument. Most scientists, however, would argue that any scientific finding is open to question (Figure 1-5). As we have just seen, the doubts and uncertainties inherent in scientific method make it impossible to be 100% sure that a scientific discovery is "right."

"Frogs breed in the spring" and "spiders have eight legs" look like facts at first glance, but they are really predictions about what will happen in the future, based on past experience. "This is a table" may also seem like a fact, but it is really a statement resulting from an agreement: all have agreed to call that sort of object a table.

"Facts" are also less sure than they seem because they depend on our faith in our senses. Suppose several people look at two photographs, one of a table and one of an object floating in a lake. Everyone may agree that the first photo clearly shows a table. When they look at the second one, however, some may say, "That is a Loch Ness monster," but the others may disagree. When technology, in the form of a camera, microscope, or oscilloscope, intervenes between our senses and an object, as it often must in scientific research, the problem of interpreting what we see or hear or smell becomes even more difficult. Thus, a "fact" is really a piece of information that we believe in strongly or that seems highly likely to be repeated without change.

The history of science abounds with dogmas that turned out to be wrong, although for a time they were widely accepted. Indeed, many statements in this book will undoubtedly prove untrue in the future. This is one reason why the cautious person or society will not place too much faith or invest too heavily in a new scientific discovery until it has been well tested.

Although scientific findings are less reliable than most people realize, scientists do believe that their methods discover useful information, and that careful study increases the probability that science's generalizations about nature come close to reality. Public support for science rests on the belief that a better understanding of the natural world increases our ability to promote human well-being.

Figure 1-5

When is a fact not a fact? Nineteenth century doctors were taught that men and women breathed differently: men used their diaphragms (the sheet of muscle below the rib cage) to expand their chests, whereas women raised the ribs near the top of the chest. Finally, a woman doctor found that women breathed in this way because their clothes were so fashionably tight that the diaphragm could not move far enough to admit air into the lungs. Some women, like the one in this drawing of 1870 styles, even had their lower ribs removed surgically so that they could lace their waists more tightly.

1-C   Fundamental Concepts of Biology

Biology is the branch of science that studies living things: their structure, function, reproduction, and interactions with one another and with the nonliving environment. We can identify several basic concepts in biology, which we shall explore in this book:

1. **Living things are organized into units called cells, which are the units of structure, function, and reproduction in organisms.** Most cells are so small that we must use a microscope to see them. Many small organisms, such as bacteria and protists, consist of one cell each. Larger organisms, such as grasses and humans, contain up to hundreds of millions of

Figure 1-6
Cells and energy. This transparent suctorian is a protist, an organism consisting of a single cell. It obtains energy by capturing prey with its tentacles. Its slender stalk is attached to a filament of an alga, a simple plant. The alga is made up of a row of cells, each containing many green chloroplasts, which capture the sun's energy and use it to make food. (P. W. Johnson and J. McN. Sieburth, Univ. of Rhode Island/BPS)

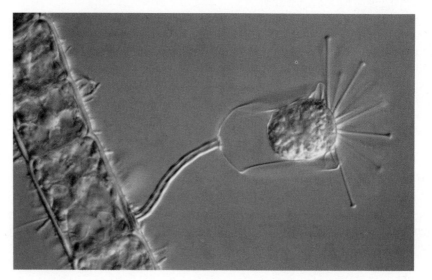

cells. Each cell is a discrete packet of living material, a biochemical factory that shows all the features of life listed here.

2. **Living things are highly ordered.** The chemicals that make up living organisms are much more complex and highly ordered than are the chemicals that make up most nonliving things. This chemical organization is reflected in the organized structure and function of the organism's body. All organisms contain very similar kinds of chemicals, and the proportions of various chemical elements in a living body differ markedly from those of its nonliving environment.

3. **Living things obtain and use energy from their environments to maintain and increase the high degree of orderliness of their bodies, to grow, and to reproduce.** Most organisms depend, directly or indirectly, on energy from the sun. Green plants use solar energy to make food, which supports the plants themselves. This food is also used by all organisms that eat plants, and eventually by those that eat the plant-eaters (Figure 1-6).

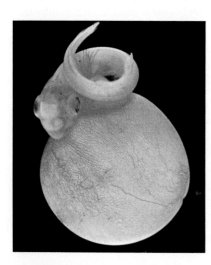

Figure 1-7
Living things develop. This dogfish embryo, lying on top of the yolk sac from which it draws nourishment, will develop into one of the smaller sharks. (Biophoto Associates)

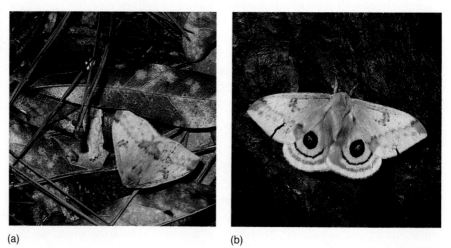

(a) (b)

Figure 1-8
Adaptation and response to the environment. (a) The color pattern on this moth's wings camouflages it among the dead leaves of its forest home. (b) The moth responds to disturbances by opening its forewings, revealing a pair of owl-like eyes sure to startle small insect-eating birds or animals.

Figure 1-9
Living things reproduce. These mushrooms, the reproductive structures of a fungus, produce spores that scatter and grow in new homes.

4. **Living organisms respond to stimuli from their environments.** Most animals respond rapidly to environmental changes by making some sort of movement—exploring, fleeing, or even rolling into a ball. Plants respond more slowly but still actively: stems and leaves bend toward light, and roots grow downward. The capacity to respond to environmental stimuli is universal among living things.

5. **Living things develop.** Everything changes with time, but living organisms change in particularly complex ways called development. A nonliving crystal grows by addition of identical or similar units, but a plant or animal develops new structures, such as leaves or teeth, that differ in chemistry and organization from the structures that produced them (Figure 1-7).

6. **Living things reproduce themselves.** New organisms arise only from the reproduction of other, similar organisms. New cells arise only from the division of other cells (Figure 1-9).

7. **The information each organism needs to survive, develop, and reproduce is segregated within the organism and passed from each organism to its offspring.** This information is contained in the organism's **genetic material**—its chromosomes and genes—which specifies the possible range of the organism's development, structure, function, and response to its environment. Each cell contains its own copy of the organism's genetic information. An organism passes genetic information to its offspring, and this is why offspring are similar to their parents. Genetic information does vary somewhat, though, so parents and offspring are usually similar but not identical (Figure 1-10).

8. **Living things evolve and are adapted to their environments.** Today's organisms have arisen by evolution, the descent and modification of organisms from more ancient forms of life. Evolution proceeds in such a way that living things and their components are well suited to their ways of life. Fish, earthworms, and frogs are all so constructed that we can predict roughly how they live merely by examining them. The adaptation of organisms to their environments is one result of evolution.

Figure 1-10
Genetic information. The information an organism needs is passed from each organism to its offspring. These flowers are similar because they have all inherited similar genetic information from their parents, but they do differ somewhat because of variations in the genetic material.

1-D *Evolution and Natural Selection*

A living organism is the product of interactions between its genetic information and its environment. This interaction is the basis for the most important concept in biology, that organisms evolve by means of natural selection—point 8 in our preceding list.

The theory of evolution states that today's organisms have arisen by descent and modification from more ancient forms of life. For instance, most biologists believe that human beings evolved from now-extinct animals which looked something like apes, and that this happened through accumulation of changes from generation to generation. In more modern terms, we can say that evolution is the process by which the members of a population of organisms come to differ from their ancestors.

Like many other great ideas in science, the theory of evolution by means of natural selection presents a simple explanation that makes sense of a great many observations of the natural world. Soon after Charles Darwin proposed this theory, his champion, Thomas Huxley, remarked, "How extremely stupid not to have thought of that!"

The theory is based on three familiar observations:

1. **Organisms are variable.** Even the most closely related individuals differ in some respects.
2. **Some of the differences among organisms are inherited.** Inherited differences between individuals result from differences in the genetic material that they inherited from their parents. Because parents and offspring have very similar genetic material, they tend to resemble each other more closely than they resemble organisms to which they are less closely related.
3. **More organisms are produced than live to grow up and reproduce.** Fish and birds may produce hundreds of eggs, oak trees thousands of acorns, but only a few of these survive to reproduce in their turn.

Some of the inherited variations among organisms are bound to affect the chances that an individual will live to reproduce. Individuals with some genetic variations produce more offspring (which inherit this genetic material) than do others. This is called **natural selection,** and it produces **evolution:** a change in the proportions of different genes from one generation of a population to the next.

To take an example of natural selection producing evolution, the length and thickness of an animal's hair is largely determined by its genes. A very cold winter may kill many individuals with short, sparse hair. Individuals with longer, thicker fur are more likely to survive the winter and reproduce in the following spring (Figure 1-11). Because more animals with thicker fur breed and pass on the genetic material that dictates the growth of thick hair, a larger proportion of individuals in the next generation of the population will have genes for thick fur. The genetic makeup of the population has changed somewhat from one generation to the next, and that is evolution. The agent of natural selection in this case is low temperature, which acts as a **selective pressure** against those individuals with short, sparse hair.

The result of natural selection is that populations undergo **adaptation,** a process of accumulating changes appropriate to their environments, over the course of many generations. The selective pressures acting on a population "select" those genetic characteristics that are adapted, or well suited, to the environment. For instance, through selection, populations living in cold areas evolve so as to become better adapted to withstand the cold.

In this discussion of evolution, we use "environment" as a catchall word meaning much more than merely whether an organism lives in a forest rather

Figure 1-11
Shaggy survivors. These highland cattle belong to a breed adapted to cold, windy winters on high moorland pastures in Scotland.

than a desert and whether or not it can obtain enough food. Environment includes all the external factors that affect the number of offspring the organism produces.

An organism's environment includes its external environment as an embryo, juvenile, and adult. Let us, for example, consider a frog. Whether it successfully meets the pressure of its environment depends on the speed and normality of its embryonic development, whether bacteria penetrate the jelly coat of the egg and destroy it during development, whether as a tadpole it can find enough food and avoid being eaten by a predator, whether the pond in which it lives as a tadpole dries up before it becomes a frog, and whether as a small frog it avoids death by disease or predation. To make things more complicated, environmental pressures are often contradictory. For instance, a hot summer benefits our frog in one way because frog embryos develop faster at higher temperatures, but a hot summer also increases the chance that the tadpole's pond will dry up before it is ready to live on land (Figure 1-12). Even worse, environmental pressures frequently change. The frog must have features that allow it to withstand both the heat of summer and the cold of winter; it should remain still to be safe from some predators and move quickly to escape from others; and so forth.

Figure 1-12
A tadpole, which will develop into an adult frog. (Biophoto Associates)

Figure 1-13
Armed for life in the desert. This euphorb's leaves are modified as sharp spines that deter animals from using its thick, water-storing stem as a source of food and water.

Figure 1-14
Always in hot water. The pastel bands of color in the runoff from this geyser are colonies of various species of bacteria.

So the frog's genetic makeup is a compromise brought about by selection for a number of opposing characteristics.

Although natural selection is probably the most far-reaching agent of evolution, it is not the only one. We shall discuss other mechanisms that change a population's genetic makeup in Chapter 15.

Adaptations

Figure 1-15
Deceptive behavior. A female *Photuris* firefly eating a *Photinus* male, which she has attracted by mimicking the pattern of light flashes produced by females of his species. (James E. Lloyd)

We have just defined adaptation as the process by which populations become better and better suited to their environments as a result of natural selection. Biologists also use the word in a second way. An **adaptation** is any genetically determined trait that has been selected for and that occurs in a large part of the population because it increases an individual's chance of reproducing successfully. (Note that this may include the capacity to change depending on environmental conditions, as with the ability to learn.) Much of biology involves the study of the adaptations of organisms.

Adaptations may be broadly classed as anatomical, physiological, or behavioral. **Anatomical** adaptations involve the organism's physical structure. For instance, many plants produce thorns or spines that protect them from being eaten by animals (Figure 1-13). An organism's **physiology** is all of its body's internal workings: the biochemistry of its cells and the processes that allow it to digest food, exchange gases, excrete wastes, reproduce, move, and sense and respond to the outside world. An example of an extreme physiological adaptation to temperature is seen in the ability of some bacteria to live in hot springs at temperatures up to 80°C (175°F), which would destroy all biochemical activity in most other organisms (Figure 1-14). An example of an impressive **behavioral** adaptation is the ability of a female *Photuris* firefly to mimic the pattern of light flashes by which female *Photinus* fireflies attract their mates. The female *Photuris* devours the male *Photinus* she has lured by this ruse (Figure 1-15). When she is ready for romance, she flashes in her own species' courtship pattern and attracts a male to mate with her.

The Limitations of Science

Science is only one way of exploring the world; history, religion, and philosophy are others. Science deals only with things that can be experienced directly or indirectly through the senses. By definition, science has nothing to say about the spiritual or supernatural. As biologist George Gaylord Simpson noted, "This is not to say that science necessarily denies the existence of immaterial or supernatural relationships, but only that, whether or not they exist, they are not the business of science." By the same token, many religious leaders argue that religion should not censor science. A Vatican Council report put it this way: "research performed in a truly scientific manner can never be in contrast with faith because both profane [nonreligious] and religious realities have their origin in the same God."

Although science does not deal with the supernatural or with the illogical, scientists themselves may be just as emotional, political, or illogical as anyone else, and this may affect their scientific efforts. Furthermore, scientists' social environment strongly influences how they think and what projects they undertake. For example, today many scientists study the biology of embryonic development or of differences between the sexes, or the physics of new weapons systems. People are interested in these topics, and governments and foundations are more apt to grant research funds for them than for studies of how to build pyramids or how to preserve mummies—activities of little interest today although the ancient Egyptians put a lot of effort into them.

Much public support goes to "applied science," investigating problems of immediate concern such as cancer, alternative energy sources, or food production. However, there is still a great deal of basic, or "pure," research to be done to discover the principles underlying the behavior of objects and organisms in the world around us. Although such research may not benefit humankind immediately, it adds to our understanding of the world and will almost inevitably be put to use sooner or later. And even knowledge that does not find an application may be as intellectually satisfying as painting a picture or writing a play. It is interesting that people may accept "art for art's sake" but often will not grant the same privilege to science, which must work for its keep.

Scientists often say that science is neither good nor bad; only the use of science has moral consequences. From a purist's point of view, this is true. The discovery that the atom could be split was merely a scientific discovery with no moral implications. It was the decision to use this knowledge to build an atom bomb that produced the moral dilemma of whether it was ever right to use such a weapon.

In the past, many scientists took an ostrich-like approach to the moral implications of their work. However, more and more people now feel that scientists must become involved in society's moral decisions about science and must take more moral responsibility for the consequences of their research.

This makes sense. However, scientists cannot always foresee all the practical implications of their work. In the early 1940s, a graduate student found that a chemical called TIBA could be applied to soybeans to improve their yield. Farmers began to use TIBA on the increasingly important soybean crop. The Army Chemical Corps, interested by the student's finding that higher levels of TIBA made the plants lose their leaves, began to develop chemical defoliants. TIBA itself was never used in warfare, but it paved the way for the defoliant spray (containing the notorious Agent Orange) used on over 6 million acres of Vietnamese forests and farms during the war there. Who would be rash enough to engage in science with the expectation of being held legally and morally responsible for such unforeseeable consequences? During the Dark Ages, the Western world experienced about 500 years in which very few scientific discoveries were made because of religious bans; few of us want to return to such times.

There are no simple solutions to these dilemmas. The successful partnership of science with society depends on citizens who understand what science can and cannot do, and who do not confuse scientific with moral, economic, or political values.

Figure 1-A
Science often becomes embroiled in political controversy. We have awakened to the fascinating intelligence of whales at a time when many whale species are in danger of extinction because so many individuals have been caught by whaling ships like this one. Dead whales are turned into many products, such as pet food and oil, all of which can be obtained easily from other sources. (Biophoto Associates)

Energy and Natural Selection

We have seen that all living things must take in and use energy to maintain their bodies, to grow, to obtain more energy, and to reproduce. The evolutionarily successful individual is one that leaves descendants, bearing its genetic information, in future generations of the population. Therefore, natural selection favors those individuals that can channel the most energy into producing offspring. The use of energy in other activities such as feeding, fighting, or growing is selectively advantageous only insofar as these activities result in the organism's producing more offspring.

Each individual has an "energy income," all of the energy that it acquires during its lifetime. It also has an "energy budget," its allotment of different amounts of energy to various activities. The most evolutionarily successful organisms are those most effective at converting energy to offspring. This does not mean that organisms use all their energy directly to produce offspring. For example, suppose that a tree converts some of its energy into growing a large root system. The energy thus spent cannot be used to produce offspring. Its large root system may enable the tree to obtain a great deal of water and minerals from the soil and so to produce more leaves, another diversion of energy away from the production of offspring. However, its many leaves may enable the tree to make more food than it would have otherwise and so allow it to recoup some of its previous energy expenditure by producing more offspring in the end. Thus organisms make energy investments that may ultimately yield energy gains which can be reinvested in the production of offspring. Sometimes these investments will turn out to be selectively disadvantageous because they postpone production of offspring. If the organism meets an early death, it will never get a chance to reproduce. So any item in an organism's energy budget must have the potential to confer an ultimate reproductive gain that makes it worth taking the risks of diverting energy away from the immediate production of offspring.

SUMMARY

Scientific knowledge is developed by subjecting problems to the scientific method. First, scientists make observations. Then they formulate alternative hypotheses that might explain the observations, and, when possible, they test the hypotheses by experiments designed to disprove one or more of the hypotheses and therefore to strengthen the evidence for those that remain.

Scientific discoveries and theories are useful, but they are always open to question; in science there is no such thing as "proof positive." Time and again in the history of science, widely accepted dogmas have turned out to be wrong, and even today scientists are busily discarding or remodeling some of the cherished "truths" presented in this book. Scientists must maintain a healthy skepticism toward scientific findings, both old and new.

Biology is the science that studies living things. We can group the fundamental concepts of biology under

three headings: cellular organization, biological information, and evolution:

1. All living things consist of one or more cells. Cells take energy from their environments and use it to maintain a high degree of chemical and structural orderliness, and for such activities as maintenance, growth, development, and ultimately, reproduction to produce more cells.
2. Living things contain information in the form of their genetic material. This information dictates how organisms develop, survive, and reproduce, and determines the characteristics they can pass on to their offspring.
3. The chief agent of evolution is natural selection, the phenomenon by which individuals with certain traits are more likely to survive and reproduce, thereby increasing the proportion of their own genetic information in future generations of the popu-

lation. This ensures that a population of organisms will become increasingly well adapted to its environment. An adaptation that increases the ability to survive and reproduce in the population's environment becomes more common in the population as natural selection eliminates members lacking the adaptation. Natural selection ensures that those individuals most effective in converting energy to offspring will be evolutionarily successful.

QUESTIONS FOR DISCUSSION

1. After every hard rain you find dead earthworms lying on the sidewalk. What experiments would you perform to show the cause of death?
2. To what extent should scientists be held responsible for the social and moral consequences of their discoveries?
3. Many professional scientific societies have adopted ethical conduct guidelines for their members and have pledged legal aid to members who "blow the whistle" on employers who make dangerous products or dispose of hazardous materials unsafely. Nevertheless, employees who bring valid protests often find themselves out of a job (management can always find an excuse to eliminate a person's position, or a way to make an employee so uncomfortable that he or she resigns). Why do company managers act this way? What might our society do to ensure its own safety by guaranteeing security to these whistle-blowers?
4. Is some scientific information too dangerous to know?
5. Many characteristics of life can be found in some nonliving things. Can you think of examples of these?
6. What might you expect was the selective pressure that resulted in each of the following adaptations?

   an elephant's trunk   the scent of honeysuckle
   a leopard's spots   the bark of a tree
   human language

SUGGESTED READINGS

Arms, K., P. Feeny, and R. C. Lederhouse. "Sodium: Stimulus for puddling behavior by tiger swallowtail butterflies, *Papilio glaucus.*" *Science* 185:372, 1974. The story of the puddling experiments described in this chapter.

Mayr, E. *The Growth of Biological Thought.* Boston: Belknap Press of Harvard University Press, 1982. A historical perspective from an eminent evolutionary geneticist.

Roszak, T. *Where the Wasteland Ends.* Garden City, N.Y.: Doubleday, 1973. Critique of modern science by a man who believes science dominates Western society and causes much of its malaise.

The Unity of
Life: Cells

*T*he star of Part 1 of this book is the cell—a tiny mass of living matter, often called the unit of life. In Chapter 1 we listed cells as the first characteristic of living things: all organisms consist of one or more cells, which show all the other features common to life. The chapters in Part 1 focus on the cell itself and two other features of life: orderliness and the use of energy.

A high degree of order occurs in crystals, which are not alive. It is possible to predict the precise, fixed position of every atom in a crystal. Some cells produce structures with even more complex but regular order, such as hair or other fibers, or intricately patterned external walls. The orderliness of these static structures is impressive. However, living cells also show a more interesting sort of order: the dynamic relationships among their working parts. For instance, a thin, fatty membrane covers the outside of every living cell and exchanges substances with the external environment. Inside the cell are chemical components such as enzymes, which carry on orderly sequences of chemical reactions, using the substances taken in through the membrane as raw materials. Here, too, is the cell's genetic material, which determines what enzymes and other chemical components the cell makes. By the coordinated interaction of these and other structures, the cell continuously processes energy and materials taken from its environment and uses them to maintain its orderly structure and activity, and eventually to reproduce.

A crystal or fiber has order because its parts have settled into the lowest state of energy possible (under prevailing conditions). A cell, by contrast, maintains its order by constantly acquiring energy from its environment and using this energy to work against the tendency toward lower-energy states.

To understand how the cell carries out all the activities of life, we need to take it apart and examine its chemical components to see how their properties contribute to the life of the cell (Chapters 2 and 3). Then we shall see how these components are assembled into the parts of a cell and how these parts work together to carry on the business of living (Chapters 4 and 5). Finally, Chapters 6 to 8 look at the rules of energy transactions that cells must obey and see how cells capture and process the energy they need to carry out the activities that keep them going and growing.

Throughout our study of cells, we shall see a theme of the unity of life: all cells contain the same basic kinds of chemicals; these chemicals form certain universal cellular structures and are organized in a limited number of ways that permit cells to obtain, release, and use energy.

The study of cells is one of the most important fields of modern biology. As the units of life, cells also harbor its secrets. However, most cells are so small that it is hard to tell what is going on in there. Our knowledge of these secrets is growing rapidly, thanks to the development of new instruments and techniques for viewing, analyzing, and manipulating the tiny entities involved.

An energy transfer

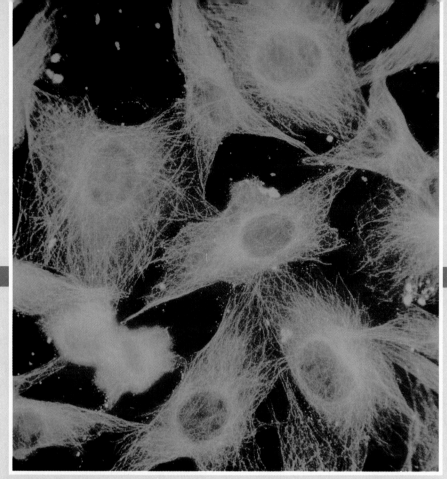

Microtubules

Lipid (wax) and carbohydrate (chitin)

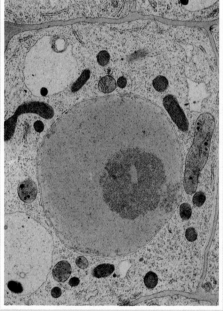

Plant cell

Arrow crab feeding

Some Basic Chemistry

CHAPTER

2

Atom

Water

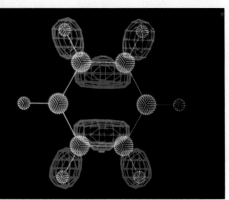

Benzene (computer graphic)

Grunts obtain oxygen from water

A human being has been defined as "20 gallons of water and $5 worth of assorted chemicals." This definition omits many human attributes, but it does emphasize an important chemical point: the material in living bodies, including our own, ultimately comes from some of the same commonplace substances that make up the nonliving environment. However, living and nonliving things differ in the structure and organization of their chemical components. The $5 price tag applies to chemicals in a fairly simple form. In a living body, these materials are assembled into large, highly organized chemical components based on carbon: enzymes, hormones, the DNA of the genes, and so on. To fill an order for as much of these complex chemicals as a human body contains, a chemical supply house would charge an estimated $6 million!

In this chapter we shall look at some basic properties of matter, and especially of water, as background for our study of biology. Chapter 3 covers the carbon-containing chemicals made by living organisms.

- The chemistry of living and nonliving things obeys the same rules.
- The chemistry of life has two notable features:

  1. Living things are composed mainly of water.
  2. The large chemical components characteristic of living things have structures based on "skeletons" of carbon.

2-A Chemical Elements and Atoms

Chemical elements are substances that cannot be broken down into other kinds of substances (except by radioactive decay). Each of the 100 known elements has a unique set of chemical properties. Living organisms use only about 20 elements (Table 2-1). These are not the most common elements, but those with properties that have made them useful to organisms.

Table 2-1 Chemical Elements Found in Animals, Their Approximate Abundance by Weight, and Their Atomic Weights

Major elements (composing most of the body)				Trace elements (present in very small amounts)		
Element	Symbol*	Weight, percent	Atomic weight†	Element	Symbol*	Atomic weight†
Oxygen	O	62	16.0	Copper	Cu	63.5
Carbon	C	20	12.0	Manganese	Mn	55.0
Hydrogen	H	10	1.0	Molybdenum	Mo	96.0
Nitrogen	N	3.3	14.0	Cobalt	Co	59.0
Calcium	Ca	2.5	40.0	Boron	B	11.0
Phosphorus	P	1.0	31.0	Zinc	Zn	65.5
Sulfur	S	0.25	32.0	Fluorine	F	19.0
Potassium	K	0.25	39.0	Selenium	Se	79.0
Chlorine	Cl	0.2	35.5	Chromium	Cr	52.0
Sodium	Na	0.10	23.0			
Magnesium	Mg	0.07	24.5			
Iodine	I	0.01	127.0			
Iron	Fe	0.01	56.0			
		99.69				

* Each element is assigned a one- or two-letter symbol that is used as a chemical "shorthand" in writing chemical formulas and equations.

† Where atomic weight is not an integer, the weight given is the average of commonly occurring isotopes (atoms of the element with different weights; see *A Journey Into Life* for this chapter).

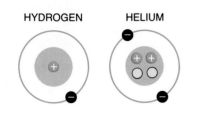

HYDROGEN   HELIUM

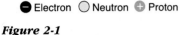

● Electron ○ Neutron ⊕ Proton

Figure 2-1
Models of hydrogen and helium
atoms. (Not drawn to scale.)

The element carbon occurs in two pure forms, graphite (part of the "lead" in pencils) and diamond. More often, carbon is combined with other elements. Thousands of different carbon-containing substances occur in living organisms or in their remains, such as coal and oil. Carbon's properties make it peculiarly suitable to form this variety of chemical structures.

Imagine dividing a piece of graphite into ever-smaller pieces until you had separated it into carbon atoms. An **atom** is the smallest unit of an element that retains all of the element's properties. Atoms are the units of matter.

Atoms can be split even further, into three main kinds of particles:

Proton, with electric charge of +1 and weight of 1.
Neutron, with electric charge of 0 and weight of 1.
Electron, with electric charge of −1 and weight of 0.

Picture an atom the size of a football field, with a grape in the center. The grape represents the atom's **nucleus,** composed of its protons and neutrons clustered together. The much tinier electrons whiz around the nucleus so fast that they occupy most of the atom's space. This space is called the **electron orbitals** or **electron shells.** Since electrons have negative charges, they are attracted to the positively charged protons in the nucleus, and this attraction holds the atom together. An atom contains equal numbers of electrons and protons, and so its net electric charge is zero.

The smallest and simplest atoms are those of the element hydrogen, only 0.1 nanometre (nm) in diameter. (1 nm is one billionth of a metre, or one millionth of a millimetre.)* A hydrogen atom contains one positively charged proton, with one negatively charged electron zipping around it. Most hydrogen atoms do not have neutrons, but all other atoms do have neutrons. For example, a helium atom's nucleus contains two protons and two neutrons (Figure 2-1).

The number of protons in an atom determines what element the atom is. This is the element's **atomic number;** each of its atoms has this many protons. Each element has a different atomic number: hydrogen 1, helium 2, and so on.

An atom's **atomic weight** is the sum of its numbers of protons and neutrons. (Protons and neutrons both weigh one atomic unit, and the weight of electrons is so low that we can ignore it.) For instance, all carbon atoms have six protons, and most have six neutrons, for an atomic weight of twelve.

In 1913, physicist Niels Bohr imagined electrons as moving in spherical spaces, called electron shells or **energy levels** (Figures 2-1, 2-2). Electrons move to the lowest possible energy level. The first shell, nearest the nucleus, must be filled before electrons begin to occupy the second shell. The first shell can hold only two electrons, the second shell can hold eight, and the third can also hold eight when it is the outermost shell. For example, carbon has six protons, six neutrons, and six electrons. It has two electrons in its first energy level, but only four in the second, which is its outer shell. This energy level has room for four more electrons (Figure 2-2).

A more realistic atomic model treats electrons as particles in constant, random motion. The electron's orbital is defined by an imaginary sheath around the space where the electron passes most of its time. Different electrons have different amounts of energy, which determine the orbitals they occupy. An electron in the first and lowest energy level, closest to the nucleus, has a spherical orbital. This orbital can hold two electrons and no more.

The first two electrons in the second energy level (and in all higher energy levels) also occupy a spherical orbital. However, the second energy level can also hold up to six more electrons. These electrons occupy three dumbbell-shaped orbitals, each of which can hold two electrons (Figure 2-3).

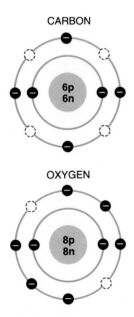

CARBON

6p
6n

OXYGEN

8p
8n

Figure 2-2
Bohr models for carbon and oxygen. The letters *p* and *n* in the nuclei stand for protons and neutrons. Carbon has room for four more electrons in its outer shell (empty dotted circles). Oxygen can take two more electrons before its outer shell has a full set of eight.

---

* As requested by the United States Metric Association, this book uses the international spellings "metre" and "litre" (instead of "meter" and "liter"). This spelling is used by scientists of all other nations and by international businesses.

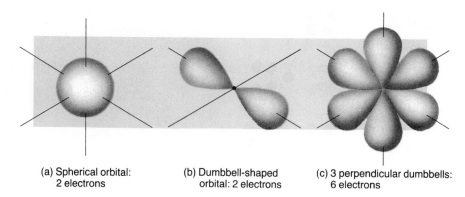

(a) Spherical orbital:
2 electrons

(b) Dumbbell-shaped
orbital: 2 electrons

(c) 3 perpendicular dumbbells:
6 electrons

Figure 2-3
Electron orbitals. (a, b) Spherical and dumbbell-shaped orbitals can each hold two electrons. (c) The second energy level has three dumbbell-shaped orbitals, at right angles to each other, and one spherical orbital, and so it holds a maximum of eight electrons.

2-B   Bonds Between Atoms

Atoms are most stable when their outermost electron shells are filled. A helium atom has two electrons (see Figure 2-1). Since its electron shell is filled, the helium atom is stable and inert. This is why helium gas is used in blimps and balloons instead of lighter, but explosive, hydrogen gas.

Atoms that do not have exactly enough electrons to fill their outermost shells take part in chemical reactions in which atoms join, or **bond,** with one or more other atoms, ending up with stable (filled) outer electron shells.

Three types of bonds between atoms are important in living things:

1. Covalent bonds. A **covalent bond** is a link between two atoms that share a pair of electrons—one electron from each atom—so that each atom has a stable, complete outer electron shell. For instance, two hydrogen atoms, with one electron each, may share their electrons. Each atom ends up with a filled shell of two electrons. By combining in this way, two hydrogen atoms form a molecule of hydrogen gas (Figure 2-4). A **molecule** is a unit made up of two or more atoms which are joined by covalent bonds and so have stable, filled outer electron shells.

In a double covalent bond, each atom contributes two electrons, for a total of two pairs of shared electrons. Oxygen molecules in the air all consist of two oxygen atoms linked by a double covalent bond. Each atom thereby fills its outer shell with a total of eight electrons (Figure 2-5).

When two atoms of the same element bond covalently, they attract the shared pair of electrons equally, and so the average position of the shared

HYDROGEN GAS (H₂)

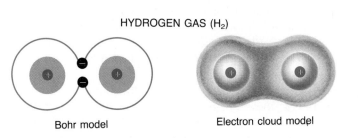

Bohr model

Electron cloud model

Figure 2-4
Two ways to show a covalent bond between two hydrogen atoms. Each hydrogen atom has one electron and tends to gain another, completing its shell of two. By sharing electrons, both hydrogen atoms fill their electron shells. In the electron cloud model, density of shading shows the proportion of time spent by the rapidly moving electrons in various areas of the atoms' electron shells. The electrons in a covalent bond spend most of their time between the nuclei that share them.

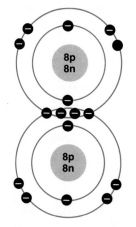

Figure 2-5
A double covalent bond between two oxygen atoms. The atoms share two pairs of electrons, each filling its outer shell with eight electrons.

Figure 2-6
Two ways to show the polar bond in hydrogen chloride (HCl). The electron pair shared in the covalent bond is attracted more strongly by the chlorine nucleus than by the hydrogen nucleus. As a result, the chlorine has a partial negative charge, and the hydrogen atom has partly lost its electron, leaving it with a partial positive charge.

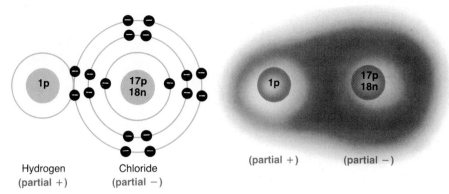

Hydrogen
(partial +)

Chloride
(partial −)

(partial +)   (partial −)

■ **Atoms get stable, filled outer electron shells by:**
1. **sharing electrons (forming covalent bonds), or**
2. **gaining or losing electrons (forming ionic bonds).**

electrons is midway between them (see Figure 2-4). If the two atoms are of different elements, one of them is usually more **electronegative;** that is, it attracts electrons more strongly than the other. Hence the shared electrons spend more time near this atom, giving it a partial negative charge. Meanwhile, the other atom bears a partial positive charge. Such an electrically lopsided covalent bond is said to be **polar** (Figure 2-6).

Oxygen and nitrogen are much more electronegative than hydrogen is. So, when oxygen (or nitrogen) bonds with hydrogen, the bond is polar. The oxygen bears a partial negative charge, the hydrogen a partial positive charge. On the other hand, carbon and hydrogen are about equally electronegative. Therefore, a carbon-to-hydrogen bond is **nonpolar,** with the average position of the shared electrons about midway between the two atomic nuclei, and no difference in electrical charge between them. Both polar and nonpolar covalent bonds play vital roles in the chemistry of life.

2. Ionic bonds. An **ionic bond** is a strong electrical attraction between ions that bear opposite electrical charges. These **ions** are electrically charged particles formed when an atom (or a molecule) gives up one or more of its outermost electrons to another atom. Atoms that give up negatively charged electrons end up as ions with net positive charges, whereas electron recipients become negatively charged ions.

As a result of this atomic give and take, the newly formed ions end up with stable outermost energy shells. For instance, a sodium atom has one electron in its outer shell. If this electron leaves, the resulting sodium ion will have a stable outer shell of eight electrons (Figure 2-7). A sodium ion has 11 protons and 10 electrons, for a net charge of +1. In contrast, a chlorine atom has seven electrons in its outermost shell. If it takes an electron from sodium, it will have a stable outer shell of eight. The ionic form of chlorine is called a chlor<u>ide</u> ion; it has 18 electrons but only 17 protons, for a net charge of −1. Oppositely charged sodium (Na^+) and chloride (Cl^-) ions are attracted to each other and form crystals of sodium chloride (NaCl), also called table salt. [Note that an ion is represented by its chemical symbol (see Table 2-1) followed by a superscript showing its charge.]

We can arrange bonds in a series according to the distribution of electrons. In nonpolar covalent bonds, the average position of the shared electrons is symmetrical. In polar covalent bonds, the electrons spend more time near one end of the bond. The bond is lopsided, with opposite partial electrical charges at

SODIUM ATOM

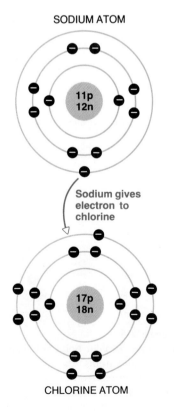

Sodium gives electron to chlorine

CHLORINE ATOM

Figure 2-7
Formation of an ionic bond. Sodium becomes stable by giving up the lone electron in its outer shell (red arrow). Chlorine adds one electron to its outer shell of seven and ends up with a filled shell of eight. After losing a negatively charged electron, sodium is a positively charged ion, and chlorine, by accepting the electron, becomes negatively charged.

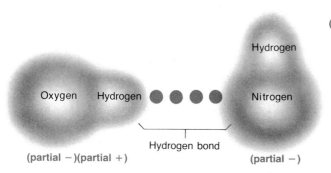

(partial +)

Hydrogen

Oxygen Hydrogen ● ● ● ● Nitrogen

Hydrogen bond

(partial −)(partial +) (partial −)

Figure 2-8
Hydrogen bonding. A hydrogen bond (red dots) is a weak attraction between a polar-bonded hydrogen with a partial positive charge and a polar-bonded atom of nitrogen or oxygen with a partial negative charge.

its ends. Ionic bonding is the extreme case. Here, one atom gives up one or more electrons to another atom, resulting in two separate particles, each with one or more full electrical charges.

3. Hydrogen bonds. A **hydrogen bond** is a weak and often fleeting electrical attraction between two atoms with opposite partial electrical charges. One partner is a hydrogen atom, bearing a partial positive charge because it is linked to oxygen or nitrogen by a polar bond. Because of its partial positive charge, the hydrogen is attracted to a third atom, with a partial negative charge (again usually oxygen or nitrogen), and this forms the hydrogen bond (Figure 2-8). Hydrogen bonds can form between atoms in different molecules or between atoms on different parts of a large molecule.

Although we apply the term "bond" to all three of these atomic interactions, they differ greatly in strength. An ionic or covalent bond may be ten or more times as strong as a hydrogen bond. But even though single hydrogen bonds are very weak and easily broken, the countless hydrogen bonds in an organism's body add up to exert forces that literally hold life together.

The atoms of each element gain, lose, or share a particular number of electrons when they form stable energy levels. Hence the numbers and types of bonds any atom can form are predictable (Table 2-2).

2-C Molecules and Compounds

Some molecules contain atoms of only one element, as in hydrogen and oxygen gases. However, many molecules contain atoms of different elements.

A **compound** is a substance made up of atoms of two or more different elements, in specific proportions, and with a specific pattern of bonds. A compound's properties differ from those of its component elements. A molecule of a compound is the smallest unit that retains all the compound's properties, just as an atom is the smallest unit that retains all the element's properties. Ionically bonded compounds are said to consist of ions instead of molecules.

A **molecular formula** is a shorthand way to show the kinds and numbers of atoms in a molecule, using the symbols for elements (see Table 2-1). The formula for sodium chloride, NaCl, says that table salt contains sodium and chloride ions in a 1:1 ratio. Water is H_2O; the subscript 2 shows that a water molecule has two hydrogen atoms (H) as well as an oxygen atom (O). Likewise, CO_2—carbon dioxide—has two oxygen atoms and one carbon (C) atom. A molecule of oxygen gas, also called molecular oxygen, contains two oxygen atoms: O_2.

Structural formulas take more space than molecular formulas but show the arrangement of atoms and bonds as well as the numbers and kinds of atoms. For instance, the structural formula for water, H—O—H, shows that each hydrogen atom is separately attached to the oxygen atom; the lines between atoms represent covalent bonds. In carbon dioxide, each oxygen is double-bonded to the carbon atom: O=C=O. When two different compounds have the same

■ *Hydrogen bonds are weak electrical attractions between two atoms that both have polar covalent bonds to other atoms.*

Table 2-2	Numbers of Bonds Formed by the Most Common Elements in Living Organisms	
Element		**Number of bonds**
Carbon	(C)	4
Hydrogen	(H)	1
Oxygen	(O)	2
Nitrogen	(N)	3

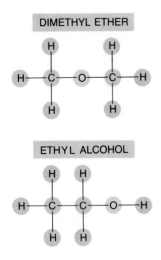

Figure 2-9
Both dimethyl ether and ethyl alcohol have the same molecular formula, C_2H_6O. However, their structural formulas differ, since the same atoms can be arranged in different ways. Ethyl alcohol, also called ethanol, is the type of alcohol in alcoholic beverages.

molecular formula, only a structural formula will distinguish between them (Figure 2-9).

The **molecular weight (MW)** of a molecule is the sum of the atomic weights of all its atoms. Using the atomic weights from Table 2-1, we can determine that the molecular weight of water (H_2O) is 18: 2×1 for the two hydrogens, $+ 16$ for the oxygen.

The **gram molecular weight** (1 **mole**) of a substance is the molecular weight of the substance in grams. For example, a mole of water weighs 18 grams. A mole of any substance contains 6.023×10^{23} molecules. The mole is a useful quantity since it is based on the number of molecules; a mole of table sugar (342 grams) and a mole of ethanol (46 grams) contain the same number of molecules, although one gram of ethanol contains more than seven times as many molecules as a gram of sugar.

Biologists often measure the concentration of molecules or ions in a solution in moles per litre. A **solution** consists of a **solvent,** most commonly water, plus the substances dissolved in it, called **solutes.** The **concentration** of a solution is a measure of the proportion of solutes it contains.

2-D Movement of Molecules

All molecules are in constant, random motion. In solids, the molecules occupy fixed positions, and each vibrates in its own space, much like passengers in a crowded bus knocking into each other. In a liquid the molecules are still so close that they constantly jostle one another, but they can slide past each other and so change places. A gas consists mostly of space, and the scattered molecules move quickly and freely, occasionally colliding with one another.

In any substance, some molecules move faster than others. The faster a particle moves, the greater its **kinetic energy,** or energy of motion. **Temperature** is a measure of the average kinetic energy of molecules: the faster the average speed, the higher the temperature. Heating a substance increases the energy of its molecules, and hence their average speed and their temperature. If we add enough heat to a solid such as a block of baking chocolate, the molecules will begin to move so fast that the solid melts into a liquid. The fastest molecules will even reach escape velocity and vaporize into the gaseous state, entering the air. It is in this gaseous state that we smell them.

2-E Chemical Reactions

When molecules bump into each other, they usually remain intact but bounce off in new directions. However, if molecules with high internal energy collide forcefully at a specific angle, they may undergo a change. The energy of the impact distorts the electron orbitals, raising the molecules into an unstable, high-energy transition state. Next, one of two things can happen. Either the molecules settle back to their original state, or the electrons rearrange themselves further, forming a new set of bonds and therefore making new substances. This is called a **chemical reaction.** The energy needed to raise the molecules to the transition state is the **activation energy.** At normal temperatures on earth, few molecules have enough energy, and so few collisions produce reactions.

Reactions can be written as equations, like this one for the burning of marsh gas (methane):

$$CH_4 + 2\,O_2 \longrightarrow CO_2 + 2\,H_2O$$

methane   oxygen   carbon dioxide   water

reactants   products

The **reactants** (starting materials) are shown before the reaction arrow, the **products** after it. This equation says that two molecules of oxygen combine with one of methane, and that for each carbon dioxide molecule produced, two water molecules are also formed. The equation is *balanced:* the products contain all the atoms of each element from the reactants, now combined into different molecules. The number of molecules denotes the proportions of reactants and products: complete combustion of one mole of methane takes two moles of oxygen, and two moles of water are produced for each mole of carbon dioxide.

The arrows in a chemical equation may point in both directions:

$$CO_2 + H_2O \rightleftharpoons H_2CO_3$$

carbon water carbonic
dioxide acid

This means that the reaction can go either from left to right (forward) or from right to left (backward), depending on the conditions. Such a reaction is said to be reversible.

■ *Chemical reactions rearrange atoms, forming new molecules from old.*

2-F Water

Living cells perform a continuous series of chemical reactions, most of which must take place in aqueous (watery) solution. The properties of water make it a suitable environment for these reactions, and for living cells. In fact, biologists list the abundant supply of water on earth as one of the major factors that made the evolution of life possible (Figure 2-10).

Water's unique properties result from the structure of the water molecule. An atom of oxygen is covalently bonded to two atoms of hydrogen (Figure 2-11). The molecule is polar because it is bent at an angle, and the electronegative oxygen attracts the electrons to the point of the angle. This gives the oxygen a partial negative charge and each hydrogen a partial positive charge.

A water molecule's partially negative oxygen is attracted to the partially positive hydrogens of other molecules, including other water molecules, and so

Figure 2-10
Water. Swirling cloud formations and blue oceans dominate this photograph of earth. Earth's organisms depend on this abundant water for their existence. (NASA)

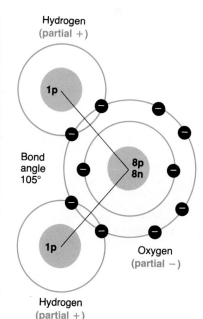

Hydrogen
(partial +)

1p

Bond
angle
105°

8p
8n

Oxygen
(partial −)

1p

Hydrogen
(partial +)

Figure 2-11
A water molecule. One atom of oxygen is covalently bonded to two atoms of hydrogen, forming an angle.

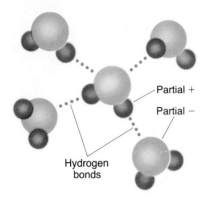

Figure 2-12
Hydrogen bonds (red dots) between **water molecules.** Oxygen atoms (blue) bear partial negative charges, and so they hydrogen-bond to hydrogen atoms (gray), bearing partial positive charges, on other water molecules. In liquid water these hydrogen bonds last only about a hundred billionth of a second before they break, and new ones form to other molecules, as the molecules tumble about.

■ *The polar molecules of water form extensive networks of hydrogen bonds.*

Figure 2-13
Walking on water. Surface tension, a result of the cohesion of water molecules, permits small animals such as this water strider to stand on the water's surface without sinking.

water molecules attach to one another by hydrogen bonds (Figure 2-12). Each water molecule can form four hydrogen bonds at a time. These weak hydrogen bonds form and break rapidly as the molecules tumble past each other.

The water molecule's structure, polar electrical nature, and ability to form hydrogen bonds give water several properties important to life:

1. Water is cohesive and adhesive. Cohesion is the holding together of like substances. **Adhesion** is the attachment of different substances to each other. You can fill a glass of water slightly above its brim; some aquatic insects can stand on the surface of a pond (Figure 2-13). Such feats are possible because of water's **surface tension,** which makes the surface appear to be covered by a "skin." Surface tension results from the cohesion of water molecules to one another by their hydrogen bonds. The polar molecules of water also adhere strongly to electrically charged surfaces, such as cell surfaces.

2. Water has a high specific heat. That is, it takes a lot of heat to raise the temperature of water, and much heat must be lost to lower its temperature. Compared with the air above it, a body of water warms more slowly in spring and cools more slowly in autumn. For aquatic organisms, this means more gradual changes in the environmental temperature. Since living bodies are largely water, they also gain and lose heat relatively slowly. An organism's chemical reactions may produce a lot of heat, but the surrounding water absorbs heat and keeps the organism's body temperature from rising too far.

Water also has a high **thermal conductivity:** heat applied to one part of a body of water rapidly spreads to all the rest. This keeps the heat produced in an organism's body from generating destructive local "hot spots."

3. Water has a high boiling point. It takes a lot of heat energy to break all of the hydrogen bonds between water molecules and so change water from a liquid to a gas, in which each molecule is separate. Only in volcanic and thermal areas do surface temperatures on earth reach the boiling point of water (100°C at sea level). Hence organisms do not face the prospect of boiling away.

4. Water is a good evaporative coolant. It takes a lot of heat to vaporize water molecules to the gaseous state of water vapor. Those that reach escape velocity and leave the body carry away the heat they have absorbed. Sweating in humans and panting in dogs are examples of evaporative cooling.

5. Water has a high freezing point and is less dense as a solid than as a liquid. These properties have both advantages and disadvantages for living things. Many climates often reach the freezing point of water, 0°C. As warm water cools, it contracts and becomes more dense. However, water has the peculiar property that it is most dense at 4°C. As water cools from 4° to 0°C, it expands again, becoming less dense as the molecules form ice crystals.

Ice is a regular latticework with each molecule hydrogen-bonded to four others (Figure 2-14). Ice is less dense than liquid water because its molecules are packed less closely; therefore an ice crystal is larger than the volume of water it replaces, and ice floats in water. The low density of ice has an advantage for aquatic organisms: in winter, floating ice forms a blanket of insulation between the water and the cold air above. This slows the formation of more ice from the remaining water, and so protects organisms below the ice from freezing. In spring, the sun shines directly on the ice and melts it.

Because water expands when it freezes, the formation of ice crystals within an organism may destroy its delicate internal structure and cause death. Some organisms have adaptations that allow them to avoid freezing, such as natural antifreezes like glycerol, which is also used in automobile antifreeze (see Figure 3-5). Other organisms have tissues that are not damaged by ice crystals. Organisms without such adaptations, such as tomato and dahlia plants, are killed by freezing and must complete a generation of growth and reproduction in the summer months between frosts.

6. Water is a solvent. More substances dissolve in water than in any other known liquid. When a substance dissolves, its individual molecules or ions sepa-

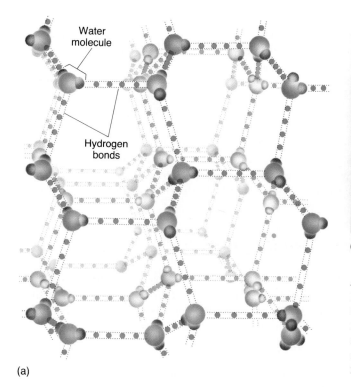

Water molecule

Hydrogen bonds

(a)

(b)

Figure 2-14
Ice. (a) The arrangement of water molecules in an ice crystal. Each molecule becomes hydrogen-bonded to four others in the crystal structure. (Blue spheres = oxygen atoms; gray spheres = hydrogen atoms.) (b) Because the molecules in ice crystals are farther apart than those in liquid water, ice forms on top of water, as shown by this stream in winter.

rate from one another and mingle with molecules of the solvent (in this case, water). The partial electrical charges of polar water molecules are attracted to charged ions and to partially charged polar molecules. So water readily surrounds and dissolves these solutes (Figure 2-15).

Nonpolar molecules, such as those made up mainly of carbon and hydrogen, do not dissolve in water because they lack electrical charge to interact with water molecules. While water and its solutes form one big, "friendly" crowd of molecules all connected by many electrical attractions, these "shy" nonpolar molecules are elbowed aside. Here the nonpolar molecules form groups, not by mutual attraction but by default—they hang around together because they are

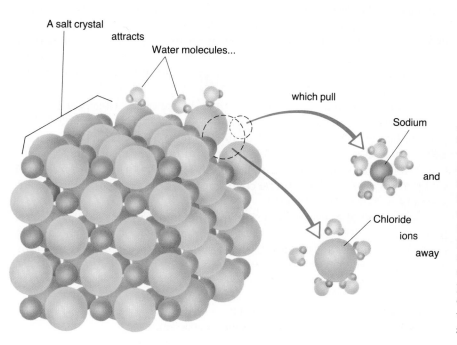

A salt crystal
attracts
Water molecules...
which pull
Sodium
and
Chloride ions
away

Figure 2-15
Sodium chloride (NaCl) dissolving in water. The opposite electrical charges of sodium (Na^+) and chloride (Cl^-) ions are attracted to one another, forming the ionic bonds that hold the crystal together. When NaCl is placed in water, sodium ions (+) attract the partial negative charges of oxygen atoms in water molecules. Likewise, chloride ions (−) attract the partially positive hydrogens of water molecules. All the tiny electrical tugs from water molecules pull Na^+ and Cl^- away from each other.

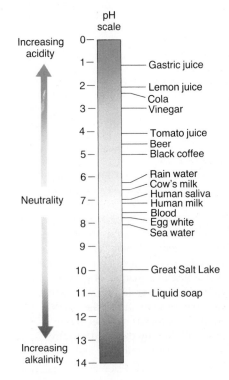

Figure 2-16

An interface. These petals produce a waxy surface coating. The wax is nonpolar, and so it will not mix with water—that is, it is waterproof. When it rains, wax and water "keep to themselves"—the water stands up in round droplets on the petals. Each drop is held in shape by the surface tension of water, which results from the attraction among water molecules.

all excluded from the mass of water molecules. So instead of dissolving in water, nonpolar molecules form interfaces with it, such as the interface in salad dressing between polar water and nonpolar oil. Similar interfaces are the basis for membranes in cells. Hence, water's inability to dissolve nonpolar substances is also necessary to life (Figure 2-16).

2-G Dissociation and the pH Scale

Many substances come apart, or **dissociate,** into ions when they dissolve in water. Some compounds dissociate completely, others only partially, so that some molecules are intact and some ionized. Water itself dissociates partially, most commonly into hydrogen ions (H^+) and hydroxide ions (OH^-):

$$H_2O \rightleftharpoons H^+ + OH^-$$

Because water molecules carry both partial negative and partial positive charges, they can assist dissociation by forming "shells" around ions. The watery shells shield the ions from the attraction of oppositely charged ions in the solution and allow them to move independently (see Figure 2-15).

Substances are classified by the particles they yield when they dissociate in water. An **acid** releases H^+ when it dissociates in water. For example, when hydrogen chloride gas (HCl) dissolves in water and forms hydrochloric acid, it yields hydrogen ions (H^+) and chloride ions (Cl^-) in solution.

A **base** (also called an **alkali**) is a substance that releases hydroxide ions (OH^-) in water, or that accepts H^+. The base sodium hydroxide (NaOH), used in drain cleaners, dissociates into sodium and hydroxide ions (Na^+ and OH^-) in solution. (Note that water is both an acid and a base!)

A **salt** is a substance in which the H^+ of an acid has been replaced by another positive ion. A salt dissociates into oppositely charged ions, as when sodium chloride (NaCl) separates into sodium and chloride ions (Na^+ and Cl^-).

A solution's acidity or alkalinity is indicated by its **pH,** a measure of the concentration of H^+. The pH scale goes from 0 to 14. A pH of 7 is neutral, neither acidic nor basic. Pure water is neutral because it gives off equal numbers of H^+ and OH^- ions when it dissociates. Values of pH below 7 are acidic, and those above 7 are alkaline (Figure 2-17).

The numbers on the pH scale come from the exponents (logarithms) of the H^+ concentrations in solutions. For example, a solution with 10^{-5} moles of H^+ per litre has a pH of 5, a solution with 10^{-6} moles of H^+ per litre has a pH of 6, and so on. Also, a solution of pH 5 is ten times more acidic than a solution of pH 6, and 100 times more acidic than a solution of pH 7. (The fact that *lower* pH values mean *greater* acidity can be confusing. It is worth repeating this to yourself until it comes naturally.)

Most chemical reactions of life occur most rapidly at a pH near the neutral point. In our own bodies, the pH of blood and most other fluids is about 7.4. A notable exception is the stomach contents during digestion of a meal: the stomach lining secretes hydrochloric acid, with a pH of 1 or less.

■ *In any group of water molecules, some are always dissociated into H^+ and OH^-.*

pH scale

Increasing acidity

0 —
1 — Gastric juice
2 — Lemon juice
— Cola
3 — Vinegar
4 — Tomato juice
— Beer
5 — Black coffee
6 — Rain water
— Cow's milk
— Human saliva
Neutrality 7 — Human milk
— Blood
8 — Egg white
— Sea water
9 —
10 — Great Salt Lake
11 — Liquid soap
12 —
13 —
Increasing alkalinity 14 —

Figure 2-17

The pH scale, with the pH readings of some familiar substances.

Isotopes: Chemical Tools for Biology

A JOURNEY INTO LIFE

Isotopes are atoms of the same element that have different numbers of neutrons, and hence different atomic weights. For example, all carbon atoms have six protons (this is what makes them carbon atoms) and most have six neutrons, for an atomic weight of 12. However, one in every trillion carbon atoms has eight neutrons, for a weight of 14. These isotopes are named carbon 12 and carbon 14 (the numbers refer to their atomic weights).

Like many other rare isotopes, carbon 14 is **radioactive;** that is, its atomic nuclei are more or less unsta-ble and will eventually decompose to form atoms of other elements, emit-ting radioactive energy in the pro-cess. In terms of chemical bonding and chemical reactions, however, all isotopes of a given element behave alike.

Isotopes come in handy for trac-ing the fate of chemicals in living organisms. For instance, an experi-mental organism can be fed some compound prepared so that it con-tains a much higher than normal per-centage of the rare isotope carbon 14. The carbon 14 "labels" the newly added carbon, distinguishing it from the many carbon atoms (mostly car-bon 12) already in the organism's body. The fate of radioactive isotopes can be followed by using a Geiger counter, photographic film, or medi-cal imaging equipment to detect the energy emitted during radioactive decay (see Figure 34-25). A nonradio-active isotope used as a label can be traced by its different weight, using very sensitive analytical instruments.

Radioactive isotopes can also be used for dating fossils or archeologi-cal finds. Each atom of a particular radioactive isotope has the same probability of undergoing radioactive decay per unit of time. An isotope's **half-life** is the time it takes for half of its atoms to undergo radioactive decay. After one half-life, half the original atoms remain; after two half-lives, only a quarter remain; and so on. We could measure the amounts of uranium 238 and of its decay prod-uct, lead 206, in a rock. Using the known half-life of uranium 238 (4.5 billion years), we could calculate when the rock was formed.

Since all living things contain carbon, carbon dating ought to be a good way to determine the age of fossils. However, carbon 14 has such a short half-life (5570 years) that dat-ing by radioactive carbon is reliable only for objects less than 30,000 years old. In terms of the long history of life on earth (more than 3.5 billion years), such organisms disappeared but a moment ago.

This last 30,000 years, however, does include the emergence of human societies and cultures. Carbon dating is very useful in studying the remains of human beings, the timbers of buildings, charcoal from campfires, or bones of slain animals.

In 1988, the archbishop of Turin requested carbon dating to settle a 600-year debate within the Roman Catholic church on the authenticity of the Shroud of Turin, the supposed burial sheet of Jesus Christ. The shroud's linen fabric came from fibers of flax plants. Like other living organ-isms, these plants cycled carbon through their bodies, and they con-tained the same, constant ratio of car-bon 12 to carbon 14. At death, the carbon atoms were fixed in place, and as carbon 14 decayed, its ratio to carbon 12 decreased.

The shroud's age was deter-mined by taking a sample of the cloth and measuring the amount of carbon 14 still left, and the amount of car-bon 12, which has not changed. Using these data, the half-life of car-bon 14, and the ratio of carbon 14 to carbon 12 known to have existed when the plants were alive, it is possible to calculate the age of the sample.

A 1- × 7-cm strip of the shroud was divided among three laboratories for carbon analysis by extremely sen-sitive techniques. The result: the shroud's fibers dated from closer to 1356, the year the shroud was "dis-covered," than to the first century A.D., when Jesus Christ died.

Figure 2-A
Excavating the remains of a building timber. Carbon dating reveals that the pine timbers of this building in a Guale Indian village in Georgia are about 6,500 years old. (Karen Roeder)

■ *The pH scale measures the H$^+$ concentration of a solution and indicates its acidity or alkalinity. The units of the pH scale are logarithmic.*

If we add drops of acid or base to water, its pH changes rapidly. If we add the acid or base to blood instead of to water, the pH remains steady until we have added a great excess of acid or base. Blood and other body fluids contain **buffers,** substances that tend to keep the pH constant by absorbing or releasing H$^+$ or OH$^-$ as needed. One of the most important buffers in many body fluids is the bicarbonate ion, HCO$_3^-$, which takes up excesses of either H$^+$ or OH$^-$:

$$H^+ + HCO_3^- \rightleftharpoons H_2CO_3 \rightleftharpoons H_2O + CO_2$$

bicarbonate ion   carbonic acid   water   carbon dioxide

$$OH^- + HCO_3^- \rightleftharpoons CO_3^{2-} + H_2O$$

bicarbonate ion   carbonate ion   water

In each case, the bicarbonate ion removes the added ion from the solution; hence the pH does not change. Both equations are reversible, providing H$^+$ or OH$^-$ to make up any deficit if some other reaction removes H$^+$ or OH$^-$.

The first equation shows that the carbonic acid formed from bicarbonate and hydrogen ions can break down to water and carbon dioxide. Carbon dioxide is a major waste product of respiration in living organisms, and we usually think of it as such, forgetting its role as an important buffer.

Many of the body's chemical reactions produce acids or bases, but buffers prevent wide swings in pH. Buffers are vital because the chemical reactions of living organisms work best at particular pH values. This is partly because enzymes, the molecules that carry out chemical reactions in living organisms, usually work best in a particular narrow range of pH (Section 3-F).

SUMMARY

Living organisms are subject to the same rules that govern nonliving systems. Like nonliving matter, organisms are made up of atoms, which bond in various ways, forming compounds. Covalent bonds form when atoms share electron pairs. Covalent bonds may be polar or nonpolar, depending on the average position of the shared electrons between the ends of the bond. Ionic bonds form when one atom takes one or more electrons from another atom, and the resulting ions are attracted to each other by their opposite electrical charges. Hydrogen bonds are weak electrical attractions between partial positive and partial negative charges on polarly bonded atoms of different molecules.

Chemical reactions rearrange the bonding of atoms, ions, and molecules and so form different compounds. Living organisms constantly carry out a variety of chemical reactions, forming different compounds as required.

Water is the most abundant substance in living things, and is necessary for life as we know it. The water molecule's structure and hydrogen-bonding ability give water a unique set of properties that make it essential to life: water dissolves polar and ionic substances; it forms interfaces with nonpolar substances; it absorbs heat and disperses it throughout the body; it carries away body heat when it vaporizes from the body surface; and it is denser as a liquid than as a solid.

Many substances dissociate when they dissolve in water. The pH of a solution is a measure of its hydrogen ion concentration; the pH value indicates whether a solution is acidic or alkaline. Buffers, chiefly bicarbonate ion, keep the body fluids of living organisms at a nearly constant pH.

SELF-QUIZ

In questions 1, 5, and 6, choose the correct alternative from the words in parentheses. For multiple choice questions, choose the one *best* answer.

1. If the pH of a solution changes from 2 to 5, it has become more (acidic, alkaline); its hydrogen ion concentration has (increased, decreased, remained constant).

2. A positively charged ion has:
   a. more protons than electrons
   b. more electrons than protons
   c. equal numbers of neutrons and electrons
   d. equal numbers of protons and electrons
   e. more neutrons than electrons
3. Write the chemical formulas for:
   a. water _____   c. carbon dioxide _____
   b. table salt _____   d. oxygen gas _____
4. Compute the weight of a mole of hydrogen chloride (HCl) and of carbon dioxide.
5. What kind of bond involves the sharing of electrons between atoms such that each atom completes its outer electron shell? (covalent, ionic)
6. In a water molecule, the hydrogen atoms are joined to the oxygen atom by (ionic, covalent, hydrogen) bonds.
7. In the dissociation of NaCl in water:
   a. water exerts forces that induce dissociation
   b. water is a passive solvent, accepting particles that dissociate because of their own internal forces

   c. water molecules lose hydrogen ions
   d. twice as many H^+ ions are formed as Na^+ ions
   e. equal numbers of H^+ and Na^+ ions are formed

Matching: for each event listed below, select the property of water responsible from the list (a–h) at the bottom of this column.

_____ 8. Heat applied to the bottom of a kettle spreads evenly through the water in the kettle
_____ 9. Ionic and polar substances dissolve in water
_____ 10. Some insects can stand on the surface of water
_____ 11. Freezing kills begonia plants
_____ 12. Lake water remains warm in the autumn after the air above it cools

   a. adhesion
   b. high boiling point
   c. cohesion
   d. denser as liquid than as solid

   e. evaporative coolant
   f. polar molecules
   g. specific heat
   h. thermal conductivity

QUESTIONS FOR DISCUSSION

1. Water makes up the bulk of an organism's body. A water molecule contains two atoms of hydrogen but only one of oxygen. Why, then, does oxygen account for 62% of an organism's weight and hydrogen only 10%?
2. Do oxygen molecules move faster in air or in water?
3. What effect does wind have on the movement of molecules? Using your knowledge of molecular movement and evaporation, explain the phenomenon of the "wind chill factor."
4. Fabrics are sometimes made "water repellent" by coating them with substances that cause water to form beads instead of spreading out on the fabric surface. What do you suppose happens on a molecular level when a surface repels water in this way?
5. Imagine that water, like most other substances, was denser as a solid than as a liquid. How would the freezing of water in winter, and melting of ice in spring, be different? How would these differences affect organisms living in lakes?

SUGGESTED READINGS

Henderson, L. S. *The Fitness of the Environment.* Boston: Beacon Press, 1958. How the earth's physical and chemical conditions support life. Chapter 3, on water and its relationship to life, is especially good.

Hill, J. W. *Chemistry for Changing Times,* 5th ed. New York: Macmillan, 1988. An excellent and entertaining "chemistry for poets," useful also for Chapter 3.

Biological Chemistry

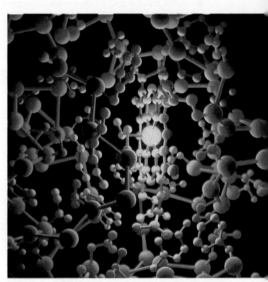

Protein (cytochrome c)

Lipid (wax) and carbohydrate (chitin)

Structural proteins in hair and horns

A living organism's body is built of, and run by, thousands of different kinds of molecules. Each kind of molecule has distinctive properties that have proven advantageous and so have been selected during the course of evolution. These molecules of life all have "skeletons" of carbon atoms, bonded together to form chains or rings. Most of these carbons are also bonded to one or more hydrogen atoms. Because such molecules are made chiefly by living organisms, they are called **organic molecules.** In addition to carbon and hydrogen, most organic molecules contain oxygen, and some also contain nitrogen, sulfur, or phosphorus.

Organisms make a variety of small organic molecules, which play important roles in the chemistry of life. Many of these small organic molecules also serve as **monomers** (mono = one), building units that are joined together to make larger molecules. Some of the larger molecules contain only a few monomers. However, many of them have a large number of monomers and so are called **polymers** (poly = many), also known as **macromolecules** (macro = big).

Living organisms can assemble an enormous array of different polymers from a relatively small number of monomers. Natural organic polymers include wool, silk, rubber, starch, and cotton. The artificial organic polymers known as "plastics" are made by joining small organic monomers in various ways—a case of art, or at least industry, imitating nature.

There are four main classes of biologically important organic compounds:

1. **Lipids:** nonpolar substances, which do not dissolve in water—including fats, oils, waxes, and steroids.
2. **Carbohydrates:** sugars, starches, cellulose, and related compounds.
3. **Nucleic acids:** the genetic material (containing instructions for making proteins) and molecules that help assemble proteins.
4. **Proteins:** molecules that make up silk, hair, tendons, and so on; carry out cell movements; act as hormones; transport substances in the blood; and fight infections. One important group of proteins is the enzymes, which carry out the cell's hundreds of biochemical reactions.

This chapter introduces all four groups and considers enzymes in some detail. In each group we find different chemical features, which give members of the group properties useful to living organisms in various ways.

- Living organisms have a distinctive chemistry based on large molecules with skeletons made up of strings and rings of carbon atoms. These organic molecules also contain hydrogen and oxygen, and often nitrogen, sulfur, or phosphorus.
- Organisms make and use thousands of kinds of organic molecules, which fall into four main classes: lipids, carbohydrates, nucleic acids, and proteins.
- Biological macromolecules are composed of many similar or identical monomer subunits joined together.
- A cell's enzymes convert organic molecules into different forms, step by step, in complex, controlled pathways.

3-A Structure of Organic Molecules

This chapter contains many diagrams of molecular structure. We show them, not expecting you to learn each molecule's name and structure, but so you can see a sample of biological molecules. At first, even small organic monomers may seem complicated. However, they start to look familiar once you know how to approach them. When you meet a new molecule, you should first get an idea of its size and shape by examining its carbon skeleton. Then look at the other atoms or groups of atoms present—the molecule's functional groups.

■ *Chains and rings of carbon are the "skeletons" of organic monomers. Attached to the carbon skeleton are atoms, or groups of atoms, of other elements: hydrogen (H), oxygen (O), and sometimes also nitrogen (N), sulfur (S), or phosphorus (P).*

Figure 3-1
Carbon and its bonds. (a) A carbon atom forms four covalent bonds, which point to the four corners of a tetrahedron. This maximizes the distance between the atoms bonded to the carbon atom. (b) Methane (CH_4). Four hydrogens are bonded to carbon. (c) Carbon dioxide (CO_2). Each oxygen atom is joined to the carbon atom by a double bond. The bonds move in such a way that each oxygen has as much room as possible, and the molecule is linear.

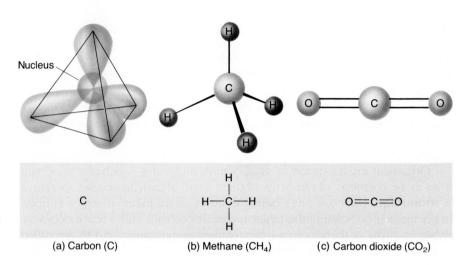

(a) Carbon (C)   (b) Methane (CH_4)   (c) Carbon dioxide (CO_2)

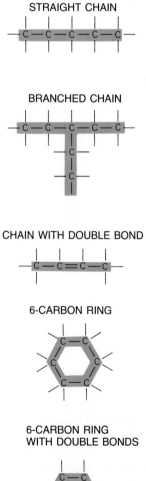

STRAIGHT CHAIN

BRANCHED CHAIN

CHAIN WITH DOUBLE BOND

6-CARBON RING

6-CARBON RING
WITH DOUBLE BONDS

Carbon Skeletons

Carbon is the only element that can form enough different, complex, stable compounds to make up the variety of molecules found in living organisms. A carbon atom can form four covalent bonds with other atoms (Section 2-B). These bonds point toward the corners of a tetrahedron with the carbon nucleus at its center (Figure 3-1). Carbon atoms can be joined together to form long chains, sometimes with shorter carbon chains as side branches. Sometimes the ends of carbon chains are joined, forming ring structures (Figure 3-2).

Functional Groups

To simplify the great variety of biological molecules, we can sort them by their functional groups. **Functional groups** are clusters of atoms that behave in particular ways no matter what the rest of the molecule is like (Table 3-1). For instance, compounds containing **carboxyl groups** (—COOH), such as fatty acids and amino acids, are acidic because they release hydrogen ions in solution: —COOH $\longrightarrow$ —COO$^-$ + H$^+$ (see Section 2-G).

As we saw in Section 2-B, bonds between carbon and hydrogen (C—H bonds) are nonpolar (electrically symmetrical). A part of a molecule containing only C and H is also nonpolar and tends to be insoluble in water.

Other functional groups contain polar bonds, particularly O—H, N—H, and C=O. In each case, the O or N bears a partial negative charge, the H or C a partial positive charge. Polar functional groups interact with other polar groups or with charged ions. In particular, they attract water molecules, giving the molecule in which they occur a tendency to dissolve in water (Section 2-F). Hence, functional groups contribute to a molecule's solubility properties and determine what kinds of chemical reactions it will take part in.

Figure 3-2
Carbon atoms bond together in many ways, forming the "carbon skeletons" of organic molecules (color). The unconnected lines sticking out from each carbon atom (C) can bond to atoms of other elements, commonly hydrogen, oxygen, nitrogen, or sulfur. Because a carbon atom forms tetrahedral bonds, the "straight" chain actually forms a zigzag in space.

Building Biological Polymers

Organic monomers, such as amino acids and glucose, are joined to form larger molecules. This occurs by a type of reaction called a **condensation reaction:** two molecules become joined as one loses an —H and the other an —OH, which join one another to form a water molecule (the term "condensation" refers to this loss of water).

Condensation:

Monomer—H + HO—Monomer $\longrightarrow$ Monomer—Monomer + H_2O

Condensation is readily reversible. Our two-monomer product can be split into its component monomers by adding a water molecule into the bond linking them. This is a **hydrolysis reaction** (hydro = water; lysis = loosening):

Hydrolysis:

Monomer—Monomer + $H_2O \longrightarrow$ Monomer—H + HO—Monomer

When we eat, our digestive systems hydrolyze the polymers in our food into monomers, which are distributed throughout the body. Our cells may then assemble these monomers into our own macromolecules by condensation reactions. Sooner or later, these too are hydrolyzed, as part of the body's continuous turnover of molecules, and the monomers are either broken down or used to build still other macromolecules.

3-B Lipids

Lipids are organic compounds that vary in structure but share one distinguishing property: they are nonpolar and so do not dissolve appreciably in water. This is because lipids contain a high proportion of carbon-hydrogen bonds. They therefore dissolve in nonpolar organic solvents, such as ether, chloroform, and benzene, but not in water, which is polar. Because they repel water, lipids such as oils and waxes are often found as waterproof coatings on the outer surfaces of many organisms, such as leaves, wool, or feathers.

Lipids contain mostly carbon and hydrogen, with a very small proportion of oxygen. Some lipids also contain the elements phosphorus and nitrogen.

Because lipids are insoluble in water, they are vital components of the membranes that separate aqueous (watery) compartments from one another in living organisms. They also offer an excellent way to store energy, for two reasons. First, lipids contain a high proportion of carbon-hydrogen bonds, which are rich in stored energy. Weight for weight, lipids yield more than twice the energy of carbohydrates, the other group generally used to supply energy. (Carbohydrates contain a lot of oxygen, which adds to the molecule's weight but not to its energy content.) Second, lipids do not attract water, and so they can be stored in concentrated form. In contrast, polar carbohydrates must inevitably be surrounded by water. When carbohydrate plus its storage water is compared with an equal weight of lipid, the lipid contains six times as much energy! This is undoubtedly why lipids have become increasingly important food reserves in the bodies of animals during the course of evolution, especially in birds and mammals, which use a great deal of energy. Without lipid reserves, for instance, the annual nonstop migrations of many birds that winter in warmer climates would be impossible. A small bird may almost double its body weight in the fall as it stores fat reserves for the long flight. To carry the same energy reserve in carbohydrate, its weight would have to increase so much that it could not fly.

Small lipid molecules can combine to form somewhat larger ones, but these do not contain enough subunits to qualify as polymers or macromolecules.

■ *Organic monomers are linked together by condensation: removal of the components of a water molecule between the two units. They are broken apart by hydrolysis: addition of a water molecule into the bond joining them.*

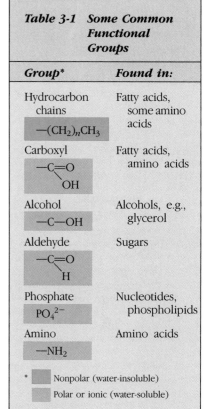

Table 3-1 Some Common Functional Groups	
Group*	**Found in:**
Hydrocarbon chains —$(CH_2)_n CH_3$	Fatty acids, some amino acids
Carboxyl —C=O OH	Fatty acids, amino acids
Alcohol —C—OH	Alcohols, e.g., glycerol
Aldehyde —C=O H	Sugars
Phosphate PO_4^{2-}	Nucleotides, phospholipids
Amino —NH_2	Amino acids

* ▨ Nonpolar (water-insoluble)
  ▨ Polar or ionic (water-soluble)

■ *Lipids are assorted organic molecules containing many carbon-hydrogen bonds and little oxygen. Hence they are nonpolar and tend not to dissolve in water. Lipids store energy and form biological membranes.*

A FATTY ACID ($C_6H_{12}O_2$)

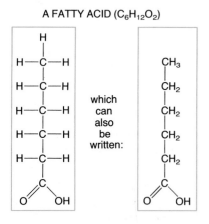

Figure 3-3
Two ways of showing the structure of a six-carbon fatty acid.

Fatty Acids

Fatty acids are the simplest lipids. To show you what to look for when meeting a new molecule, let us examine a small fatty acid containing six carbons (Figure 3-3). We can simplify this assembly of 20 atoms by looking at the molecule's features one at a time:

1. Each carbon atom (C) forms four covalent bonds to other atoms.
2. The fatty acid's six carbon atoms are linked to one another, forming a chain down the center of the molecule.
3. Most of the carbons are attached only to hydrogen (H) atoms, an arrangement called a **hydrocarbon chain.** Since carbon-hydrogen bonds are nonpolar, this five-carbon-long hydrocarbon chain is nonpolar and will not easily dissolve in water.
4. At the bottom end of the molecule is a carboxyl group (—COOH). It contains the sixth carbon, double-bonded to an oxygen, and also bonded to —OH. This carboxyl group is the acidic part of the molecule: it can ionize to give —COO⁻ and H⁺. Because the carboxyl group contains the polar C=O and O—H groups, it tends to dissolve in water even though the rest of the molecule tends not to. This results in the unique behavior of lipid molecules.

In summary, a **fatty acid** is a simple lipid molecule consisting of a long hydrocarbon chain, with a carboxyl group at one end. If all carbon atoms of the hydrocarbon chain are "filled" with as many hydrogens as they can possibly hold, the fatty acid is said to be **saturated.** A fatty acid with one or more double bonds in its hydrocarbon chain is said to be **unsaturated** because it could hold more hydrogens if one of the two bonds were broken and two hydrogen atoms were attached to the carbons instead (Figure 3-4).

The bonds in a carboxyl group are polar, and so this end of a fatty acid is **hydrophilic** (hydro = water; philic = loving), that is, it attracts water molecules and forms hydrogen bonds with them. However, the carbon-hydrogen bonds in the hydrocarbon chain are nonpolar; therefore the chain is **hydrophobic** ("water-fearing"). A fatty acid's hydrophilic end will dissolve in aqueous solutions, and the hydrophobic end will dissolve in nonpolar organic compounds. This behavior makes some lipids containing fatty acids important parts of the membranes that divide living systems into compartments (Section 4-A).

Fatty acids seldom occur free, but are usually combined with other molecules to form substances such as **glycolipids** (carbohydrate + lipid) or **lipoproteins** (lipid + protein). They are also parts of many larger lipids.

Fats and Oils

Fats and **oils** are lipids used to store energy reserves. They are formed by condensation reactions joining three fatty acids to a molecule of the alcohol glycerol (Figure 3-5). This structure gives them another name, **triglycerides.**

Fats, such as butter, lard, or suet, are solid at room temperature, whereas oils are liquid, like olive, corn, and peanut oils. Fats are solid because they contain mostly saturated fatty acids. Oils generally contain more unsaturated fatty acids than fats do. Oils occur more commonly in plants than in animals. However, animals living in cold areas, such as Arctic and Antarctic fish, usually

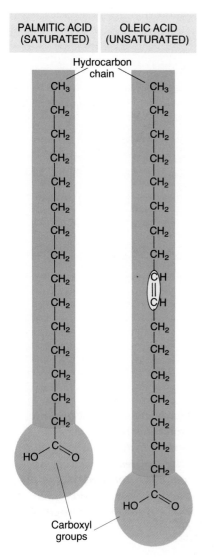

Figure 3-4
Fatty acids. Most of a fatty acid molecule is a long, nonpolar hydrocarbon chain. The acidic carboxyl group is shown in a purple circle at one end. Most fatty acids have even numbers of carbon atoms in chains 14 to 22 carbon atoms long. The yellow oval highlights the double bond that makes oleic acid unsaturated.

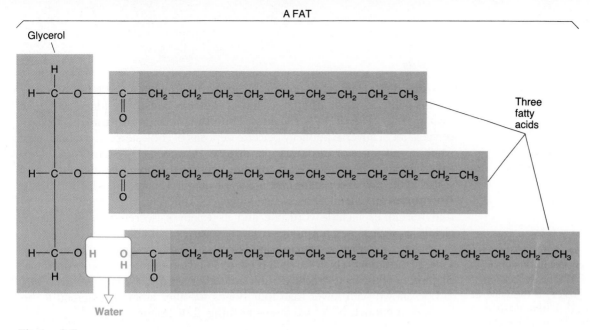

Figure 3-5

A fat. Three fatty acids (alike or different) are attached to a molecule of glycerol, an alcohol with three alcohol groups (C—OH). Here, the fatty acid shown at the bottom is about to become attached to glycerol by a condensation reaction, with the removal of the components of a water molecule between them (blue). The two other fatty acids are already linked to glycerol. This also happened by condensation reactions, not shown here.

keep their bodies flexible by producing a relatively high proportion of unsaturated fatty acids.

Phospholipids

Phospholipids are similar to fats, except that one (or two) of the fatty acids is replaced by a phosphate group, which in turn is usually linked to a nitrogen-containing group (Figure 3-6). Phospholipids are **structural** molecules, that is, building materials that contribute to the shape of the body. They are the chief lipid components of biological membranes, with their polar phosphate and nitrogenous groups facing aqueous areas and their fatty acid tails buried in the membrane's nonpolar interior (Section 4-A).

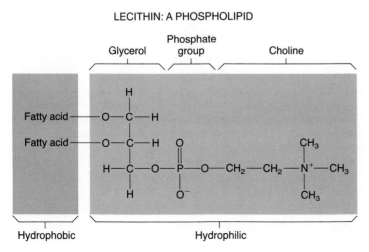

Figure 3-6

Lecithin, a phospholipid found in the membranes of all cells. This molecule contains glycerol and two fatty acids. It differs from a fat in having a phosphate group, joined to choline, in place of a third fatty acid.

39

CHOLESTEROL

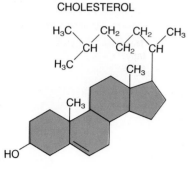

Figure 3-7
A steroid: cholesterol. Cholesterol is an important component of animal cell membranes. The basic skeleton of carbon rings (color) is the same in all steroids, but other steroids have other groups attached to this skeleton.

Steroids

Steroids differ from other lipids in structure, but qualify as lipids because they are insoluble in water. The basic steroid skeleton consists of four contiguous carbon rings (Figure 3-7). The most abundant steroid, **cholesterol,** has received a lot of adverse publicity because deposits of cholesterol in the body may result in gallstones or cardiovascular disease. However, cholesterol is an essential component of animal cell membranes and of the brain and nerves. It also serves as the raw material for production of vitamin D and of steroid hormones. **Hormones** are chemical messengers between different parts of the body. Steroid hormones include cortisone and its relatives, secreted by the adrenal glands, as well as sex hormones from the ovaries and testes.

Figure 3-7 shows a standard simplified way of drawing molecules, omitting some of the atoms, which we shall use often. Carbon atoms are assumed to lie at all corners of the ring structures, unless another atom is shown there. Also, any carbon shown with fewer than four bonds to other atoms is assumed to be bonded to enough hydrogen atoms to fill its four bonds.

3-C *Carbohydrates*

Carbohydrates are the sugars, starches, and related compounds. Sugars and starches are energy-storage molecules. Other carbohydrates are structural molecules, such as cellulose in plant cell walls.

The simplest carbohydrates are the **monosaccharides** or "simple sugars," such as glucose and fructose. These may occur singly, but they are also the

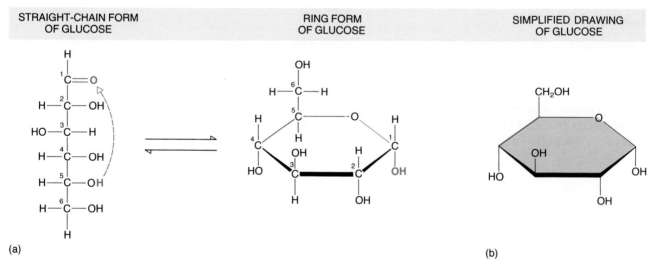

Figure 3-8
Forms of glucose. (a) When straight-chain glucose dissolves in water, the molecule bends back on itself, bringing the —OH group on carbon #5 next to the =O on carbon #1. The hydrogen atom "hops" from one oxygen to the other, and the leftover bonds of C #1 and O #5 join, so that the molecule becomes a ring with O at one corner. The bond lines in the lower side of the ring are drawn heavier to show that this edge of the ring projects out of the page toward you, while the upper half of the ring is behind the page. (b) Simplified structures take less space. Again, carbon atoms lie at all corners except the upper right one, occupied by oxygen. Also, hydrogen atoms filling out the four bonds of carbon atoms are omitted.

monomers used to build larger carbohydrates. A monosaccharide contains carbon, hydrogen, and oxygen in about a 1:2:1 ratio—represented by the general formula $(CH_2O)_n$. In this formula, n is any number from 3 to 9 (usually 5 or 6). (Note that H and O atoms occur in the same ratio as in water; carbohydrate means "water of carbon.") The proportion of oxygen is much higher in monosaccharides than in lipids, where the typical unit in a molecule is (CH_2), not (CH_2O). One of the oxygens in a monosaccharide is double-bonded to carbon, and the others occur in —OH groups (Figure 3-8). All of these functional groups are polar, and so sugars dissolve readily in water.

When a monosaccharide with five or more carbons is dissolved in water, as it always is in a living system, the bonds rearrange and the molecule takes on the shape of a ring (Figure 3-8).

Monosaccharides play an important role by providing ready energy. Sugary foods are quick-energy foods because sugars are easily digested into a form that the body can use for energy. Carbohydrates are transported in our blood mainly in the form of the six-carbon monosaccharide glucose, which cells take up and use for energy. Certain five-carbon sugars are also important parts of nucleic acids (Section 3-D).

Two monosaccharides (mono = one) may be joined together by a condensation reaction, forming a **disaccharide** (di = two). This reaction forms a molecule of water by removing —H from one sugar and —OH from the other. The disaccharide can be hydrolyzed back to its component monosaccharides by adding a molecule of water into the oxygen bridge linking them (Figure 3-9).

Sucrose (table sugar), maltose (malt sugar), and lactose (milk sugar) are familiar disaccharides. It is interesting that several disaccharides seem to be used mainly as means of transporting carbohydrate. Sucrose is the main carbohydrate transported in the bodies of plants, lactose in milk.

By a series of condensation reactions, many monosaccharides can be joined to form a polymer called a **polysaccharide.** Three important polysaccharides made of glucose monomers occur in living things: glycogen, starch, and cellulose. They differ in the arrangement of the bonds between the glucose subunits, in the branching patterns of the polymer, and in the total number of glucose subunits per chain.

Animals store energy in the form of the polysaccharide **glycogen.** The liver and muscles remove glucose from the blood and assemble it into glycogen, which is later broken back down into glucose as it is needed for energy.

The energy-storage polysaccharide in plants is **starch.** Starch consists of two kinds of glucose polymers: amylose, a long, unbranched chain of glucose subunits, and amylopectin, a branched polymer like glycogen but with less frequent

> ■ *Carbohydrates supply energy and provide structural support. Monosaccharides $[(CH_2O)_n]$ are the monomers of carbohydrates.*

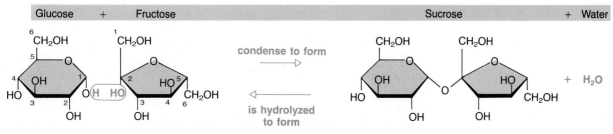

Figure 3-9

Formation of sucrose, a common disaccharide. A condensation reaction between the monosaccharides glucose and fructose produces the disaccharide sucrose (table sugar) and water. This reaction is reversible by hydrolysis.

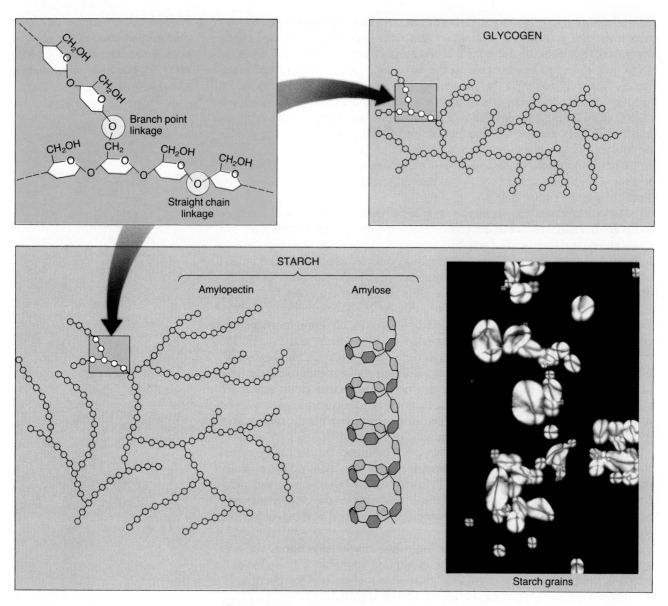

Figure 3-10
Storage polysaccharides. Both glycogen and starch (which consists of amylopectin plus amylose) contain chains of glucose monomers. Glycogen and amylopectin have branched structures. The unbranched amylose chain assumes a spiral (helix) shape. The photograph shows starch grains from a potato, viewed with polarized light. (Biophoto Associates)

branch points (Figure 3-10). When you boil potatoes, the water becomes cloudy as amylose dissolves in it. Amylopectin stays in the potatoes and is later digested to glucose subunits in your intestine. Left to itself, a living potato would eventually break down its starch to glucose and use it for energy, growth, and reproduction.

The major building material made from glucose in plants is the structural polysaccharide **cellulose**—probably the most abundant organic material on earth. Cellulose is made up of long, straight chains of glucose monomers, linked alternately right-side-up and upside-down (Figure 3-11). The —OH groups on the glucose subunits form hydrogen bonds between different cellulose chains.

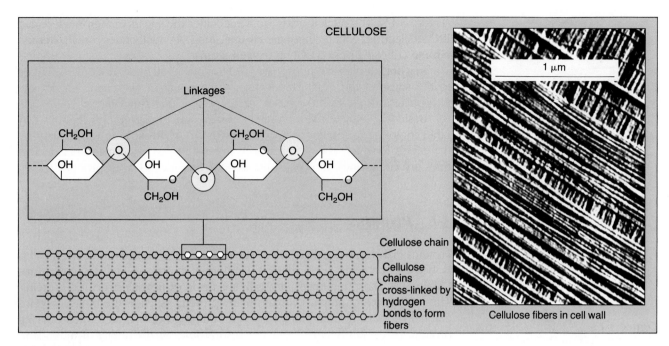

CELLULOSE

Linkages

Cellulose chain

Cellulose chains cross-linked by hydrogen bonds to form fibers

Cellulose fibers in cell wall

Figure 3-11
Cellulose. Glucose subunits are linked into long, unbranched chains (note that each glucose is "upside down" from its neighbors). Hydrogen bonding (red dots) between cellulose chains produces fibers containing many cellulose molecules. Cellulose fibers are deposited in layers to form the protective cell wall outside a plant cell (photo) (μm = 1/1000 mm). (Biophoto Associates)

This cross-links many chains together, forming cellulose fibers. Cotton fibers are almost entirely cellulose.

In plants, each cell surrounds itself with a tough external cell wall made up of several layers of cellulose fibers, often reinforced with other substances. The cell walls help to stiffen and support the plant.

Humans cannot digest cellulose, and so cellulose in our food is not broken down into glucose to provide energy. However, indigestible cellulose is important in our diet because it provides bulk (often called "fiber" or "roughage"), which stimulates the intestines to keep things moving along.

Another important structural polysaccharide is **chitin.** Its subunit is glucose with a nitrogen-containing group attached. Chitin is a major component of the external skeletons of arthropods (crabs, insects, spiders) (Figure 3-12).

3-D Nucleic Acids

Nucleic acids include the largest biological molecules, and possibly the most fascinating. Their story takes up most of Chapters 9 and 10, so here we only summarize what you need to know before then.

Nucleic acids come in two kinds. **Deoxyribonucleic acid (DNA)** is the genetic material. It contains the organism's genetic information, including instructions for how to make proteins. DNA also carries the information needed to make the other nucleic acid, **ribonucleic acid (RNA).** RNA then directs the building of proteins (Chapter 10).

Nucleotides, the monomers of nucleic acids, have three distinct parts. In the center is a five-carbon sugar (a carbohydrate), with one to three phosphate groups attached on one side, and a ring-shaped nitrogenous base on the other

Figure 3-12
Structural polysaccharides. Chitin is a major component of this cotton stainer's "armor plating." Cellulose makes up the cotton fibers it is sitting on. (Biophoto Associates)

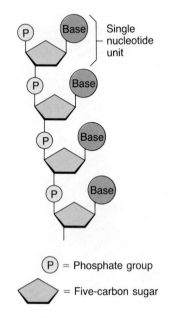

(P) = Phosphate group

= Five-carbon sugar

Figure 3-13
Nucleotides and nucleic acids. A nucleotide (top) consists of a sugar, a phosphate group, and a base. Nucleic acids are made up of nucleotides, with the phosphate group of one attached to the sugar of the next.

■ *From the 20 common amino acids, living things build thousands of different kinds of proteins.*

(Figure 3-13). The nucleic acids are named after the sugars in their nucleotides: RNA nucleotides contain the sugar **ribose,** and DNA nucleotides contain **deoxyribose** (ribose stripped of one oxygen atom).

In a nucleic acid, the phosphate group of one nucleotide monomer is linked to the sugar of the next. This forms a long string of alternating sugars and phosphate groups, with the bases sticking out at one side (Figure 3-13).

Besides serving as the monomers of nucleic acids, several nucleotides play other important roles. For example, **adenosine triphosphate (ATP)** supplies the energy for many chemical reactions in most living organisms. Other nucleotides are coenzymes, molecules needed for enzymes to work properly.

3-E Proteins

Proteins make up more than 50% of the dry weight of animals and bacteria. This is because proteins perform many important functions in living organisms (Table 3-2). Hair, fingernails, and silk are made of fibrous structural proteins. The body fluids of animals contain many proteins, such as the blood proteins that help combat disease. Hemoglobin, in red blood cells, is an oxygen-carrying protein. Other kinds of proteins are responsible for the ability of muscles to contract. Insulin is one of a number of small protein hormones. Proteins are also components of biological membranes, and they help to regulate the passage of many substances through membranes. But the most numerous class of proteins is the enzymes, which speed up chemical reactions (Section 3-F).

Proteins contain the elements carbon, oxygen, hydrogen, nitrogen, and usually some sulfur.

The monomers of proteins are **amino acids.** Each amino acid has a carboxyl group (—COOH) and an **amino group** (—NH$_2$), both attached to the same carbon atom. Also attached to this carbon is one of 20 possible side chains, collectively called R groups (Figure 3-14). The proteins of all organisms contain the same 20 common amino acids, distinguished from each other by their different R groups.

Like other biological polymers, proteins are made by condensation reactions, which join amino acid monomers. A **peptide bond** forms between the

Table 3-2 Some Functions of Proteins

Type	Example	Function
Enzymes	Amylase	Converts starch to glucose
Structural proteins	Keratin	Hair, wool, nails, horns, hoofs
	Collagen	Tendons, cartilage
Hormones	Insulin, glucagon	Regulate glucose metabolism
Contractile proteins	Actin, myosin	Contractile filaments in muscle
Storage proteins	Ferritin	Stores iron in spleen and egg yolk
Transport proteins	Hemoglobin	Carries O$_2$ in blood
	Serum albumin	Carries fatty acids in blood
Immunological proteins	Antibodies	Form complexes with foreign proteins
Toxins	Neurotoxin	Cobra venom blocker of nerve function

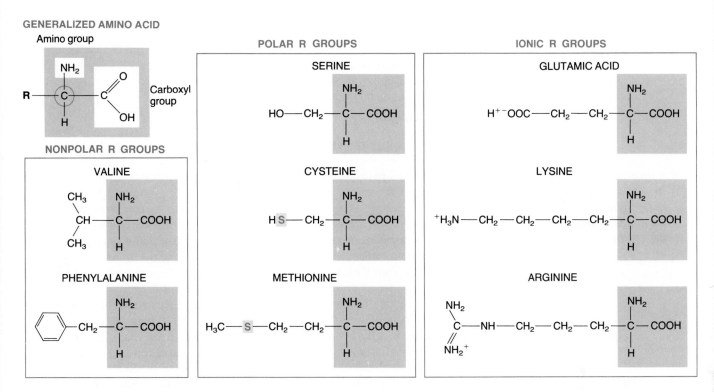

GENERALIZED AMINO ACID

Figure 3-14
Amino acids. (top left) The general structure of amino acids consists of an amino
and a carboxyl group, both attached to the same carbon atom (circled), which is also
bonded to a hydrogen atom. This carbon's fourth bond is attached to the R group. In
the representative amino acids shown in red boxes, the parts shown on blue backgrounds are the same for all amino acids. The parts on white backgrounds are the R
groups, which are different for each kind of amino acid. The R groups may be nonpolar, polar, or ionic (carrying an electric charge). Some R groups contain sulfur (S),
nitrogen (N), or ring structures.

carboxyl carbon of one amino acid and the amino nitrogen of another. The
condensation of two amino acids forms a **dipeptide** (Figure 3-15). **Polypeptides** are long strings of amino acids; most contain 100 to 300 amino acids.

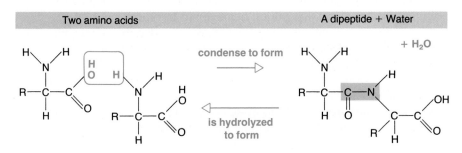

Figure 3-15
Formation of a peptide bond by a condensation reaction between two amino acids.
The atoms that go to make up the molecule of water released are colored blue. A
peptide bond (shaded) is one between the carboxyl carbon of one amino acid and
the amino nitrogen of another.

A **protein** is a functional unit composed of one or more polypeptides. For instance, the protein insulin (a hormone that stimulates removal of glucose from the bloodstream) is made up of two linked polypeptide chains (Figure 3-16). With only 51 amino acids, insulin is small as proteins go.

Protein Structure

Proteins are long, unbranched chains of amino acids, but they may fold up into complex shapes. Protein structure is analyzed by considering four aspects.

A protein's **primary structure** is the unique sequence of its amino acids, which are arranged in a long chain (Figure 3-16).

The **secondary structure** is the shape imparted to local areas of the chain by the amino acids there. The most common type of secondary structure is the alpha helix, a coil assumed by parts of some polypeptide chains. An alpha helix results from a regular pattern of hydrogen bonds among the amino acids in its area of the chain. Each hydrogen bond joins the —NH group in one peptide bond and the C=O group of another nearby (Figure 3-17a).

Another type of secondary structure is the beta pleated sheet. Again, hydrogen bonds form between —NH and C=O groups next to peptide bonds. However, the amino acids involved are not near neighbors in the chain. Instead, they lie close enough to form hydrogen bonds because of the bending of the chain (Figure 3-17b). Beta sheets occur in the fibrous protein of silk.

A protein's **tertiary structure** is the characteristic three-dimensional shape assumed by the folding of its polypeptide chain(s). This tertiary structure is strongly influenced by four types of interactions that occur between R groups in different parts of the chain:

1. Ionic bonds (Section 2-B) between R groups with positive charges and those with negative charges (see Figure 3-14).
2. Hydrogen bonds (Section 2-B) between polar R groups bearing partial positive and partial negative charges.
3. **Disulfide bonds,** covalent bonds linking the sulfur atoms of two molecules of the amino acid cysteine (see Figure 3-14). Such sulfur bridges may link two parts of the same polypeptide chain or may join one chain to another (Figures 3-16 and 3-18).
4. **Hydrophobic interactions,** due to the tendency of nonpolar R groups to stay close together because they are pushed aside when water, polar groups, and ionic groups interact electrically with one another.

Many proteins are made up of two or more polypeptide chains. These proteins also have a **quaternary structure:** the specific arrangement in which the different chains fit together to form a complete, functional protein. As we have seen, insulin has two polypeptides; hemoglobin has four (Figure 3-19).

Once the amino acids are linked in the proper order, the polypeptide chain automatically coils, loops, and folds into its correct secondary and tertiary structures. Likewise, the different polypeptides of a complex protein have shapes and forces of attraction that automatically fit them together. Polypeptide molecules can be made to lose their shape by gentle heating or by certain chemical treatments. When they are returned to more normal surroundings, they fold or coil up again and join together, re-forming their original structure. However, with

■ *A protein's three-dimensional structure is dictated by its primary structure—that is, by the sequence of amino acids in the chain.*

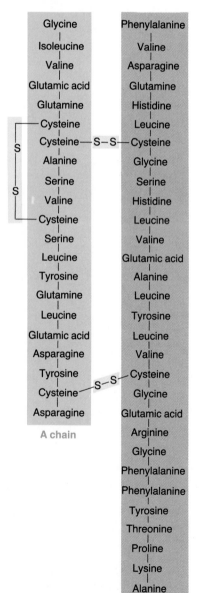

A chain

B chain

Figure 3-16
The amino acid sequence of the hormone insulin from cattle. This small protein is made up of two short polypeptide chains, A and B, joined by (yellow) "sulfur bridges," attachments between sulfur-containing cysteine amino acids (see Figure 3-14).

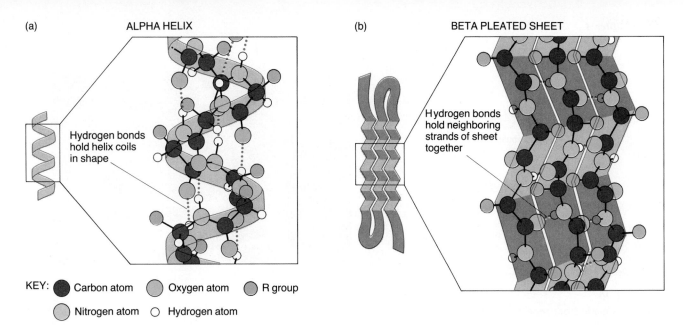

(a) ALPHA HELIX

Hydrogen bonds hold helix coils in shape

(b) BETA PLEATED SHEET

Hydrogen bonds hold neighboring strands of sheet together

KEY:
- Carbon atom
- Oxygen atom
- R group
- Nitrogen atom
- Hydrogen atom

Figure 3-17
Secondary structure of proteins. Hydrogen bonds (red dots) form between the hydrogen and oxygen atoms that lie next to peptide bonds (see Figure 3-15). (a) An alpha helix. (b) A beta pleated sheet.

harsher treatment (stronger chemicals or higher temperature), polypeptides lose their shape permanently and are said to be **denatured.**

Having certain amino acids in certain positions is crucial to the protein's overall shape. For example, a change of just one amino acid makes the difference between normal hemoglobin and the hemoglobin of sickle cell anemia (Section 13-B). Proteins are masterpieces of molecular engineering, tailored to their functions by hundreds of millions of years of natural selection.

Many structures in living organisms are made up of repeating protein subunits—identical or very similar pieces. For example, globular molecules of the protein actin fit together in a helix resembling a twisted double string of beads (Figure 3-20). These actin filaments play vital roles in muscle contraction. The tendons holding muscles to bones are made up of fibers of another protein, collagen. The basic structure of collagen consists of three polypeptides wound

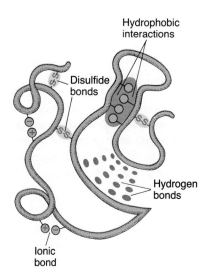

Figure 3-18
Tertiary structure of proteins. The three-dimensional folding pattern of this hypothetical protein results from interactions between R groups of various amino acids. For example, covalent disulfide bonds (—S—S—) between two molecules of cysteine join parts of the structure firmly together. Hydrophobic interactions keep nonpolar R groups together, forming nonpolar areas (orange) separate from water and the polar or ionic R groups of other amino acids. Hydrogen bonds (red dots) form between polar R groups, and ionic attractions occur between oppositely charged R groups (+ and −).

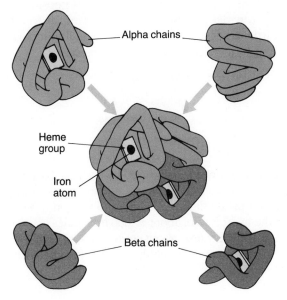

Figure 3-19
Hemoglobin. The complete protein consists of four intricately folded polypeptide chains—two alpha chains (shown from different sides) and two beta chains. The heme groups are nonprotein structures attached to the polypeptide chains. They contain iron atoms (black dots), which bind the oxygen molecules transported by hemoglobin.

47

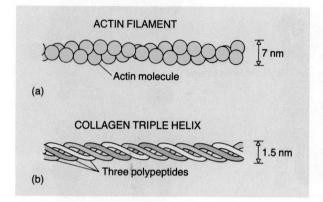

(c) Collagen fibers

Figure 3-20
Structures made from repeating protein subunits. (a) An actin filament consists of many globular proteins lined up in a formation that looks like a tightly wound double helix. (b) A collagen helix consists of three intertwined polypeptide molecules. (c) Collagen fibers. Each fiber is made up of thousands of triple helices like the one in (b). (c, Jerome Gross)

■ *An enzyme is like a robot on an assembly line, performing the same task over and over on a particular set of parts— the enzyme's substrates.*

around each other into a triple helix (Figure 3-20). Covalent cross-linkages tie collagen molecules to their neighbors, forming tremendously strong fibers.

3-F Enzymes

Many of an organism's proteins are **enzymes,** proteins that act as catalysts. A **catalyst** is a substance that increases the rate of a chemical reaction. The catalyst is not permanently changed by the reaction. The reactants in an enzyme-catalyzed reaction are called the enzyme's **substrates.**

Each of the 2000 known enzymes catalyzes a particular reaction. Some enzymes carry out condensation or hydrolysis reactions—joining monomers into larger molecules, or vice versa. Other enzymes transfer groups, such as amino groups, from one molecule to another. Oxidation-reduction reactions transfer electrons or hydrogen atoms from one molecule to another. These are especially important energy-transfer reactions, and oxidation-reduction reactions play leading roles in the manufacture and breakdown of food (photosynthesis and respiration, Chapters 6, 7, 8).

Enzymes are named according to their substrates and the reactions they catalyze, and all official enzyme names end with the suffix "-ase." For example, RNA polymerase links nucleotides to form RNA polymers, and sucrase hydrolyzes sucrose. However, some enzymes are still usually called by their old-fashioned names. Examples are digestive enzymes such as trypsin and pepsin.

As an example of enzyme activity, let us look at a reaction familiar to cat owners, the hydrolysis of urea from urine into carbon dioxide and ammonia, which gives its characteristic odor to a litterbox in need of cleaning:

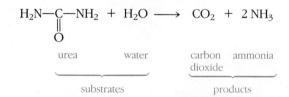

This reaction is catalyzed by the enzyme **urease,** produced by bacteria. Bacteria are always floating around in the air, and some of them settle and grow in the litterbox. At room temperature and pH 8 (slightly alkaline; see Section 2-G), a molecule of urease can catalyze the hydrolysis of about 30,000 molecules of urea per second. Without a catalyst, this would take about three million years. So urease makes the reaction go a trillion times faster than it would otherwise. Some enzymes work faster than urease, some more slowly.

It's Delightful,
It's Delicious,
It's Denatured!

A JOURNEY INTO LIFE

A protein's overall three-dimensional shape is determined by interactions between its amino acids. The protein can be denatured, that is, made to lose its characteristic shape, by various treatments that disrupt these interactions. For example, a human hair contains hundreds of strands of the protein keratin. The strength of a hair comes from abundant sulfur bridges linking molecules of the amino acid cysteine in neighboring keratin strands. When hair is given a permanent wave, the first solution used breaks the sulfur bridges. This allows the hair structure to be distorted as the hair is wound around curlers. Then another solution is applied, allowing the formation of new sets of sulfur bridges, which hold the keratin strands in the configuration imparted by the curlers.

Perhaps the most common way to denature a protein is heat, and the most familiar place this happens is the kitchen. Many proteins in food fold up into roughly globular shapes, with water molecules surrounding the outside of the protein and also lying among some of its internal loops and folds. At high temperatures, the protein's atoms, and the associated water molecules, have so much energy that their motion disrupts the hydrophobic, hydrogen, and ionic bonds that give the protein its normal shape. The protein unfolds and the loose ends form new bonds to other protein molecules, which have also been denatured. As the proteins form a meshwork with one another, there is less room available for water molecules, so water is squeezed out, and some of it is lost by evaporation.

When you roast meat, much of the water is squeezed out into the tissue spaces. If you carve a roast fresh from the oven, this juice runs out as the knife slices through, and the slices of meat are quite dry. However, if you let the roast sit for 15 to 20 minutes, the cooling proteins undergo a partial reversal of their denaturation, allowing water to move back among them. The result: moister meat.

The same thing happens when you cook eggs. The unfolding of globular proteins allows them to interact with each other and eventually form one big, tangled, solid network, moist with infiltrating water molecules. But if cooking is not stopped at this crucial moment, the mesh tightens further and squeezes the water out. Overcooking eggs has two possible outcomes: the proteins coagulate in lumps, floating in the squeezed-out liquid, or they form a single, rubbery mass, with the water either separated and floating on top or simply evaporated off altogether.

Another interesting form of denaturation occurs with some of the globular proteins of egg whites. When you whip egg whites, you force air in next to the proteins, exposing them to air on one side and water on the other. The proteins' hydrophilic regions are attracted to the liquid, while the hydrophobic areas are not, but tend to associate with the air pockets instead. So the proteins un-fold, and the hydrophilic stretches of different proteins bond to each other. They form a lacy network, reinforcing the liquid wall of the bubble around the trapped air. This structure is not strong enough to withstand baking, however. In the heat of the oven, the air bubbles expand, and they can burst the protein bonds in their walls, collapsing the whole structure. This does not happen in a properly cooked meringue, soufflé, or angel cake, thanks to still other proteins, which did not participate in forming the foam itself. These proteins are denatured by heat, and they form a stronger meshwork that stabilizes the bubble walls before they can be ruptured by escaping hot air.

Gelatin also consists of a loose protein meshwork, with water held in the spaces between proteins. The directions on gelatin packages warn cooks against adding fresh or frozen pineapple. Pineapple contains a protein-digesting enzyme, which chops the gelatin proteins into pieces too short to form a gelled meshwork. Gelatin made with fresh pineapple never sets but remains a soupy liquid. Canned pineapple is fine to use because heat applied in the canning process denatures the enzyme, so it cannot attack the gelatin molecules (Figure 3-A).

Figure 3-A
It's digested! A culinary disaster: enzymes in fresh pineapple have digested the proteins in the gelatin in the center, so that they are too short to form the meshwork that makes gelatin set. Heat used during the canning process denatures these enzymes, and so gelatin containing canned pineapple (right) sets as well as the fruit-free control (left).

How do enzymes speed up reactions? First, let us review how a reaction occurs without a catalyst. For urea and water to react by themselves, they must collide at the proper angle, and both molecules must have enough internal energy to reach a reactive transition state. The energy needed to reach the transition state is called the **activation energy,** and the higher the temperature, the more molecules have enough energy (Section 2-D). At room temperature, few molecules have enough energy, and so few reactions occur. An enzyme actually combines with its substrates (urea and water) and holds them at the correct angle for the reaction. Also, the enzyme pulls on the substrates' bonds and "loosens" them. This lowers the activation energy needed to reach the transition state, and so the reaction occurs readily, even at much lower temperatures.

■ *Enzymes are protein catalysts that permit organisms to carry out chemical reactions rapidly, at the relatively low temperatures found in their bodies.*

Enzyme-Substrate Complexes

Each enzyme catalyzes only particular reactions of one or a few kinds of molecules. Enzymes are specific because an enzyme actually binds its substrate(s) to form an enzyme-substrate complex. The substrate binds to an area called the enzyme's **active site,** a small groove formed as the protein folds up (Figure 3-21). The size, shape, and electrical charge of amino acid R groups at the active site determine which substrates can fit there. The active site forms a space complementary to the substrate, and the two bind by specific point-to-point interactions. This means that each enzyme can bind only one or a few very similar kinds of substrates, and that it always orients them in the same way. The close correlation of enzyme and substrate shape has been compared to a "lock and key" fit. However, this analogy conjures up a misleading image of unyielding hardware. Actually, both enzyme and substrate(s) change shape slightly when they combine, a phenomenon known as "induced fit" (Figure 3-22). When the reaction is over, the enzyme releases the products and emerges exactly as it started, ready to catalyze another reaction.

■ *An enzyme is specific because only a few kinds of substrate molecules fit its active site.*

Some enzymes do not bind their substrates unless their active sites contain additional substances called **cofactors.** Some cofactors are inorganic ions, such as zinc or iron. Others are nonprotein organic molecules called **coenzymes,** such as some nucleotides. Many vitamins are coenzymes or parts of coenzymes. We need very little of them because coenzymes are not destroyed in a chemical reaction. Like enzymes, coenzymes can be used over and over again.

Factors That Affect Enzyme Activity

The rate of an enzyme-mediated reaction depends partly on the concentration of substrate: when substrate molecules are scarce, the enzyme spends some time empty, waiting for substrate to arrive. As substrate concentrations increase, the

Figure 3-21
The enzyme lysozyme. This enzyme, found in egg white, protects the developing chick by hydrolyzing polysaccharides in the walls of some bacteria. The active site is the area where the polysaccharide is bound and then hydrolyzed.

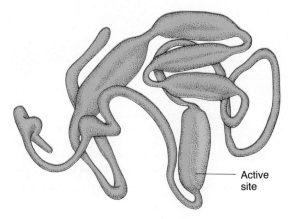

Active
site

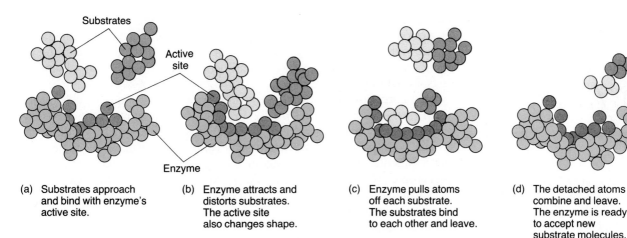

(a) Substrates approach and bind with enzyme's active site.

(b) Enzyme attracts and distorts substrates. The active site also changes shape.

(c) Enzyme pulls atoms off each substrate. The substrates bind to each other and leave.

(d) The detached atoms combine and leave. The enzyme is ready to accept new substrate molecules.

Figure 3-22
How an enzyme-catalyzed reaction between two substrates might look. Only part of the enzyme is shown. Note that both the enzyme and substrates change shape somewhat when the substrates bind to the enzyme's active site. The enzyme probably loosens bonds in the substrate molecules, allowing them to release some of their atoms more easily and re-bond to each other. (Thom Smith)

reaction rate does too. However, the rate has an upper limit because an enzyme molecule, working at top speed, can catalyze only a certain number of reactions per unit time. This number depends on the kind of enzyme. A urease molecule can catalyze 30,000 reactions per second. The enzyme carbonic anhydrase, which combines carbon dioxide and water to form carbonic acid (H_2CO_3), works much faster: 600,000 reactions per second. A very slow enzyme is lysozyme (see Figure 3-21), which catalyzes only one reaction every two seconds.

Inhibitors are substances that decrease an enzyme's reaction rate. Some inhibitors bind to the active site and prevent substrates from binding and reacting there. Others disrupt the enzyme's three-dimensional structure and so destroy its function.

Allosteric Interactions **Allosteric enzymes** (allos = other; stereos = space) are enzymes that can exist in two or more different shapes. This, in turn, affects their activity.

In addition to the active site, most allosteric enzymes have at least one **regulatory site,** which also has a specific shape enabling it to bind certain molecules. The binding of molecules at the regulatory site alters the enzyme's shape and therefore changes its activity.

Allosteric inhibitors are molecules that, when bound to the regulatory site, cause distortion of the active site so that it cannot function properly (Figure 3-23). On the other hand, stimulatory molecules cause inactive allosteric enzymes to change shape, bringing the active site into the proper form to catalyze the enzymatic reaction.

pH The activity of enzymes is affected by the pH of the surrounding solution. Enzymes carry electrical charges on the ionic R groups of some of their amino acids. However, the pH of the solution may change these charges. For example,

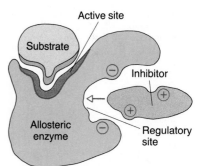

(a) Active form of the enzyme. Substrate can bind to active site.

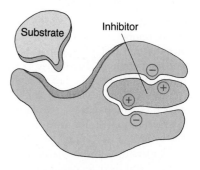

(b) When inhibitor binds, it distorts the enzyme's active site. Substrate can no longer bind.

Figure 3-23
How the activity of a hypothetical allosteric enzyme is controlled. (a) When the enzyme is in the active form, substrate can bind to its active site. (b) When an inhibitor binds to the enzyme's regulatory site, the enzyme's active site is changed so that substrate can no longer bind there.

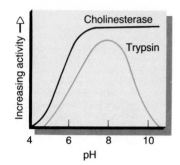

Figure 3-24
Effect of pH on the activity of two enzymes of mice. Trypsin is a digestive enzyme that hydrolyzes proteins in the small intestine. It works best (that is, its activity is highest) at the pH of 8 found in the intestine. Cholinesterase hydrolyzes substances that are important in the nervous system. The graph line shows that it works well at any basic pH but quickly loses activity when the pH is acidic.

the H^+ ions in an acid solution tend to combine with negatively charged R groups on the enzyme, making the R groups neutral and hence disrupting ionic bonds in the enzyme's folding pattern (see Figure 3-18). This changes the enzyme's three-dimensional structure and hence its function. This is why pH affects the rate of an enzyme-mediated reaction (Figure 3-24).

Most proteins work best in a surrounding medium of approximately neutral pH. However, some organisms live in acidic or basic environments. For instance, organisms living in the mineral springs of Yellowstone National Park (where both acidic and basic waters occur) have enzymes that work at the pH of their particular environment. Digestive enzymes of the human stomach work best at an extremely acidic pH, around 1.5 to 2.0. When these enzymes pass into the small intestine with the food, they become relatively inactive: sodium bicarbonate secreted into the small intestine neutralizes stomach acid and raises the pH to about 8.

Temperature   Temperature also affects the rate of enzyme reactions. When the temperature is warm, molecules move around faster, collide harder and more often, and so are more likely to react than when it is cool. (This is true whether the reaction is catalyzed by an enzyme or not.) On the other hand, high temperatures (usually above about 60°C) denature proteins, permanently destroying their three-dimensional structure so that they can no longer function. Cooking preserves food by destroying the enzyme activity of organisms that cause decay.

At the other extreme, chemical reactions proceed slowly at low temperatures because molecules move so slowly that few collisions occur between enzyme and substrate molecules. Refrigeration also preserves food, in this case by slowing the activity of enzymes in organisms that cause decay, or enzymes in the food itself.

Evolution has produced enzymes adapted to function in a particular range of temperatures. This range depends on the temperatures in the organism's environment (Figure 3-25). For example, some of the organisms known as cyanobacteria live on the surface of glacial ice, and they are adapted to temperatures close to the freezing point of water (0°C). Other members of this group inhabit the hot springs of Yellowstone, which, besides having unusually high or low pH values, may be at temperatures of 80 to 85°C (see Figure 1-14). In a more familiar example, the characteristic color pattern of Siamese cats is due to a temperature-sensitive enzyme (Figure 3-26).

3-G   Metabolism

We have met members of all four main classes of biological molecules: lipids, carbohydrates, nucleic acids, and proteins. Now let's put them all together—in a living cell. Table 3-3 shows the chemical content of a common bacterium living in your large intestine. In keeping with their many and crucial functions, proteins are usually the most abundant organic compounds inside cells.

A living cell is a busy biochemical factory. Here organic monomers such as sugars, amino acids, and nucleotides are converted from one form to another, new macromolecules are built and old ones dismantled, and the energy required to run all this is extracted from food molecules. Each reaction is carried out by an enzyme.

An organism's **metabolism** is the total of all its biochemical reactions. These reactions are organized into sequences called **metabolic pathways.** Each pathway consists of several enzyme-mediated steps such that one enzyme's product is the next one's substrate. Substrates enter the pathway and are converted, step by step, into a final product. For example, in animals one pathway

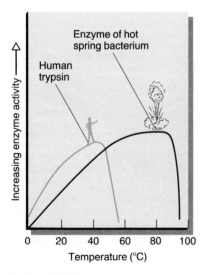

Figure 3-25
Effect of temperature on two enzymes that normally function at different temperatures. The human body is normally at 37.6°C; the hot spring bacterium lives at about 80°C. Note that each enzyme is most active in the temperature range it normally encounters. The curves fall off steeply at high temperatures as the enzymes are denatured and their function completely destroyed.

Figure 3-26
A Siamese cat. The enzyme that makes the dark pigment melanin is unstable at body temperature, and so fur on the cat's body is light in color. The cat's ears, nose, feet, and tail are cooler. At these lower temperatures, the enzyme can work and make its product. Hence, the fur in these areas is dark.

converts the six-carbon sugar glucose into the five-carbon sugar ribose, which is needed to make nucleotides. This pathway contains four different enzymes. Other pathways have more or fewer steps.

The various metabolic pathways interconnect in a staggeringly complex pattern. Furthermore, each substrate occurs in very low concentrations. These tiny amounts of so many different substrates must all be channelled into the appropriate pathways to meet the organism's needs, with no shortages or surpluses. How can these intricate biochemical traffic patterns be controlled?

To bring order into this chaos, different metabolic pathways may be confined to specific locations. For instance, the enzymes for a series of related reactions may be arranged together in large complexes. This increases efficiency because one enzyme's product is its neighbor's substrate. As the first enzyme releases the molecule, the second binds it before it can move away. Membranes provide a second way to organize metabolic pathways for greater efficiency. Many membranes partition off compartments where particular kinds of molecules are confined and concentrated. Sometimes all the molecules needed to complete a task are embedded in a membrane or attached to its surface.

Metabolic pathways are often controlled by **negative feedback:** the regulation of the rate of a process by the concentration of its product (Figure 3-27). Often the enzyme at the beginning of a metabolic pathway is allosteric. When the pathway works too fast, surplus product accumulates. Some of the product molecules attach to molecules of the pathway's first (allosteric) enzyme and "turn them off." This slows activity in the entire pathway. Later, as the product is used up by other pathways, this process reverses. Some product molecules unbind from the start-up enzymes, which resume their active forms and go back to work.

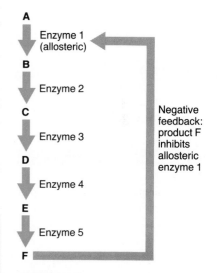

Figure 3-27
Regulation of a hypothetical metabolic pathway. The raw material for the pathway is substrate A, which is converted to B, C, D, E, and finally end product F, by a series of enzymatic reactions. The first enzyme in the pathway, which binds A, is allosteric. Product F acts as an inhibitor of allosteric enzyme 1. Hence, when excess F accumulates, some of it feeds back and combines with molecules of enzyme 1, thereby slowing the operation of the entire pathway.

Table 3-3	Chemical Composition (Excluding Water) of a Common Bacterium	
Type of molecule	**Percent of total dry weight**	**Comments**
Small molecules	10	Inorganic ions, monomers, coenzymes
Polysaccharides and lipids	16	Protective outer wall and membrane; some glycogen stored inside bacterium
DNA	4	One or two molecules per bacterium; each molecule is about 1 millimetre (mm) long and highly folded; the bacterium itself is only about 0.002 mm long
RNA	20	About 3000 different kinds
Proteins	50	About 2500 different kinds: about $\frac{1}{3}$ structural protein, $\frac{2}{3}$ enzymes

SUMMARY

Aside from water, the main chemical components of living organisms are organic molecules. Organic molecules are based on carbon, a versatile element able to form molecular skeletons of myriad sizes and shapes. Most of the organic molecules in a living body are macromolecules. These polymers are made up of many monomeric subunits. A polymer contains identical or similar monomers.

Monomers are joined together to form polymers by condensation reactions, in which a bond is formed by removing the components of a water molecule from the subunits. Macromolecules are broken down by hydrolysis, the addition of water molecules between the subunits, which thus become separated.

Biological molecules fall into four main groups: lipids, carbohydrates, nucleic acids, and proteins. Lipids do not form molecules large enough to be called polymers, but the other three groups contain polymers formed from monomers as just described:

Group	Monomers	joining together $\rightleftharpoons$ breaking apart	Polymers
Carbohydrates	Monosaccharides	$-H_2O$ $\rightleftharpoons$ $+H_2O$	Polysaccharides
Proteins	Amino acids	$-H_2O$ $\rightleftharpoons$ $+H_2O$	Polypeptides
Nucleic acids	Nucleotides	$-H_2O$ $\rightleftharpoons$ $+H_2O$	DNA, RNA

Lipids and carbohydrates are composed mainly of carbon, hydrogen, and oxygen. Some lipids and carbohydrates are important energy-storage compounds that may be broken down to release energy. Unlike members of the other three groups, lipids are nonpolar and so do not dissolve in water; they are vital components of all biological membranes. Some are also important hormones.

Structural polysaccharides include cellulose in plants, and chitin in arthropods.

Nucleic acids and proteins play vital roles in directing an organism's growth, activity, and reproduction. Nucleic acids contain the elements carbon, hydrogen, oxygen, nitrogen, and phosphorus. Proteins contain carbon, hydrogen, oxygen, nitrogen, and some sulfur. Important proteins include enzymes, structural and transport proteins, and hormones.

Enzymes are protein catalysts. About 2000 kinds of enzymes have been named. Each is adapted to facilitate reactions between specific substrates. Enzymes enable organisms to carry out chemical reactions quickly at the relatively low temperatures of their bodies. Enzymatic reactions are organized into the various metabolic pathways that convert one kind of molecule to another, build up or break down polymers, and break down food to release energy. The channelling of substrates into different metabolic pathways is under negative feedback control: allosteric enzymes at the beginnings of metabolic pathways are activated or inactivated according to the organism's metabolic needs. The activity of metabolic enzymes is also affected by substrate concentration, cofactors, inhibitors, pH, and temperature.

SELF-QUIZ

Make a summary table of this chapter for yourself by filling in the blanks numbered 1–13. (For example, in #1, fill in the class that contains the elements C, H, O, N, P; in #2, name that class's monomer subunits, etc.)

Summary of the Major Classes of Biological Compounds

Class	Chemical elements	Monomer subunits	Main roles
1. _____	C, H, O, N, P	2. _____	3. a. _____ b. _____
4. _____	5. _____	Fatty acids, glycerol, etc.	6. a. _____ b. _____ c. _____
Proteins	7. _____	8. _____	9. a. _____ b. _____ c. _____
10. _____	11. _____	12. _____	13. a. _____ b. _____

14. The molecules shown below:
    a. both contain amino groups
    b. both contain carboxyl groups
    c. both belong to the same major class of organic compounds
    d. could both serve as monomer subunits of polymers
    e. all of the above

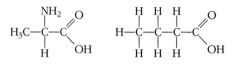

15. The molecule shown below is a:
    a. fatty acid
    b. dipeptide
    c. disaccharide
    d. nucleotide
    e. steroid

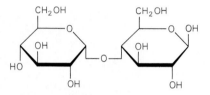

16. Which of the following is *not* made up of monosaccharides?
    a. sucrose
    b. starch
    c. glycogen
    d. insulin
    e. cellulose

17. You have a solution of an enzyme. You put half of it into each of two beakers containing identical substrate at equal concentrations. After waiting awhile, you test both solutions and find that the substrate in beaker A has been changed but the substrate in beaker B has not been acted on by the enzyme. Suddenly you notice that beaker B has been sitting on a hot plate with the switch turned to "high." The enzyme in beaker B probably did not work because it had been:
    a. hydrolyzed
    b. denatured
    c. condensed
    d. catalyzed
    e. dehydrated

18. Which of the following statements about enzymes is *false?* Enzymes:
    a. catalyze only a particular reaction of specific substrates
    b. usually work only in a particular pH range
    c. increase the energy of the reactant molecules
    d. work better at moderate temperatures than at very high or low ones
    e. bind their substrates and hold them in a particular orientation

QUESTIONS FOR DISCUSSION

1. Science fiction tales sometimes feature life forms based on silicon rather than carbon. Silicon is much more abundant on earth than carbon, and like carbon its atoms can bond to four other atoms. Bonds between two silicon atoms are unstable in the presence of O_2, but bonds between silicon and oxygen atoms are extremely stable and difficult to break. What implications would these properties have for silicon-based life forms?

2. An organism produces different amounts of the various proteins it can synthesize. Based on your own understanding of natural selection from Chapter 1, explain how this variation may have arisen.

3. You go on a journey, taking your cat along but leaving its litterbox at home. Draw a graph to show the rate of hydrolysis of the urea in the litterbox to ammonia and carbon dioxide during your absence (see Section 3-F).

4. During the winter, you are too lazy to take your cat's box outside to empty it, until finally the stench is so overpowering you take the box outside and leave it in the snow. What happens to the rate of hydrolysis of urea by the enzyme urease, and why?

5. When you cut an apple or banana, phenol oxidase enzymes in the injured areas quickly begin a "wound reaction," which results in the cut surfaces turning brown. Good cooks sprinkle lemon juice on sliced fruit to prevent it from discoloring in this way. Why does this work?

SUGGESTED READINGS

Dickerson, R. E., and I. Geis. *The Structure and Action of Proteins.* Menlo Park, Calif.: Benjamin-Cummings, 1984. Excellent illustrations, highly readable.

Lehninger, A. L. *Principles of Biochemistry.* New York: Worth Publishers, 1982. A readable text, useful background for many chapters in this part of the book.

McGee, H. *On Food and Cooking: The Science and Lore of the Kitchen.* New York: Charles Scribner's Sons, 1984. Great for browsing.

Scientific American, October 1985 issue, *The Molecules of Life.*

Cells and Their Membranes

CHAPTER

4

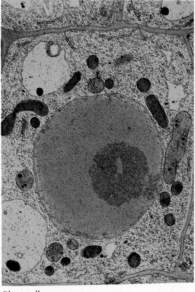

Plant cell

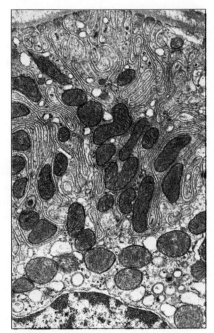

Rat kidney tubule cells

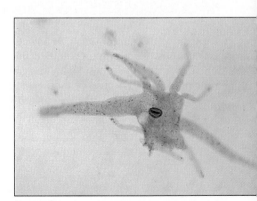

An amoeba

I took a good clear piece of Cork and with a Pen-knife sharpen'd as keen as a razor . . . cut off . . . an exceeding thin piece of it . . . I could exceedingly plainly perceive it to be all perforated and porous . . . these pores, or cells, were not very deep, but consisted of a great many little Boxes . . . Nor is this kind of texture peculiar to Cork only; for upon examination with my Microscope, I have found that the pith of an Elder, or almost any other Tree, the inner pulp or pith of the Cany hollow stalks of several other Vegetables: as of Fennel, Carrets, Daucus, Bur-docks, Teasels, Fearn . . . & c. have much such a kind of Schematisme, as I have lately shewn that of Cork.

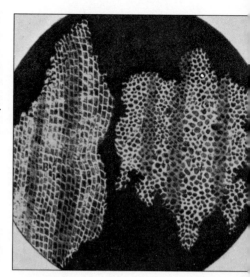

Figure 4-1
Cell portrait. Robert Hooke drew this illustration, published in 1665 in his book *Micrographia*. It shows the structure of thin slices of cork as seen through his microscope. The tiny dark pores reminded him of the cells where monks slept in their monastery dormitories.

With these words, written in 1665, the Englishman Robert Hooke first reported the existence of cells. What Hooke really saw in the bark of the cork oak tree were empty, dead **cell walls,** without the living matter they once contained (Figure 4-1). Other early microscopists soon observed cells in all kinds of plants. They found similar structures in animals too, but animal cells were harder to distinguish because they lack the thick cell walls that surround plant cells. Observers also reported the existence of many tiny organisms consisting of only one cell each.

Eventually, biologists recognized the main features of the **cell theory:**

1. All organisms consist of one or more cells.
2. Cells are the fundamental units of life—the smallest entities that can be called "living."
3. Cells arise only by division of existing cells.

Why do organisms need cells? The metabolic reactions of a living organism can take place only in a delicately balanced environment different from any found in the nonliving world. Cells are the life-support chambers that contain this special environment. A living cell keeps its chemical composition steady within narrow limits, a condition known as **homeostasis** (homeo = same; stasis = standing). In the controlled environment of a cell, all the activities of life occur: acquiring energy; using this energy to maintain the cell's internal chemical environment, to build organic molecules, and to grow; and reproducing by division into two new cells.

Many organisms are **unicellular** (consisting of only one cell). However, a cell's size is limited. The biochemical reactions of its metabolism require raw materials from outside the cell and generate waste products that must be expelled. Hence the cell must keep up a lively trade in chemicals with its environment. This occurs through the cell's outer surface. As a cell grows, its surface area, which imports and exports materials, increases more slowly than its volume, which uses the materials. Also, the surface lies farther from the innermost areas that need materials, so it takes longer for things to go where they are needed. At some point, the cell divides into two new cells, each with a lower (and more favorable) ratio of surface area to volume (Figure 4-2). Large organisms such as animals and most plants are **multicellular,** composed of many cells derived by repeated division from one original cell.

All cells must perform certain basic tasks. In addition, each cell of a multicellular organism makes a specialized contribution to the body as a whole. For example, a muscle cell in the heart is specialized to contract and help pump blood. Since it is deep inside the body, it cannot capture its own food nor obtain oxygen from the air, but must rely on other specialized cells, such as those of the lungs and blood, to provide the food and oxygen it needs. Thus there is division of labor among the cells of a multicellular organism.

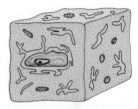

(a) One 2-cm cube

(b) Eight 1-cm cubes

Figure 4-2
Surface area and volume. A cube-shaped cell two centimetres (cm) on a side has a surface area of 24 cm^2, and volume of 8 cm^3. If it were divided into eight 1-cm cubes, the same volume of cell contents would be served by twice as much surface area (48 cm^3). Also, the center of each cell would be only 0.5 cm, rather than 1 cm, from the nearest points on the surface.

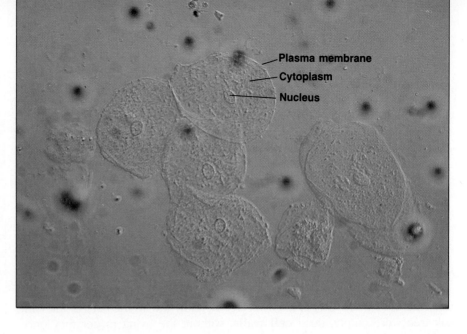

Plasma membrane
Cytoplasm
Nucleus

Figure 4-3
Human cells obtained by scraping the inside of the cheek with a toothpick. The cells were then placed in a drop of salt solution on a glass slide and viewed through a light microscope, using phase contrast techniques to bring out depth and detail. Each cell is surrounded by a plasma membrane and contains a large, oval cell nucleus. The rest of the cell is filled with cytoplasm. (Biophoto Associates)

Most cells have three main parts (Figure 4-3):

1. The **plasma membrane,** covering the outside of the cell and controlling what enters and leaves. (In plants, this is surrounded by a tough cell wall.)
2. The **cytoplasm** ("cell fluid"), containing water, various salts, and organic molecules, including many metabolic enzymes. The cytoplasm also contains a variety of larger structures, collectively called **organelles,** which perform various tasks. Many of these "little organs" are surrounded by membranes very similar to the plasma membrane.
3. The **cell nucleus** (in bacteria, the **nuclear area**), housing the cell's genetic material (DNA and associated RNA and proteins). The genetic material contains directions for making the cell's proteins.

In this chapter, we examine the structure and function of the plasma membrane. In the next chapter we discuss the cell's other components and look at the differences among the cells of plants, animals, and bacteria.

- Cells are the basic structural, functional, and reproductive units of life.
- To remain alive, a cell must maintain homeostasis: keeping its internal chemical composition fairly constant within the narrow limits suitable for life.
- Every cell is surrounded by a membrane, which helps control what enters and leaves the cell.

4-A *Structure of Biological Membranes*

The plasma membrane is too thin to be seen with a regular light microscope, such as you might use in the laboratory. In photographs made with more powerful electron microscopes, the membrane appears as a continuous double line around the cell (Figure 4-4). However, long before electron microscopes were invented, biologists had deduced the plasma membrane's existence by observing cell behavior, and they had collected and studied membrane material.

All biological membranes have similar structures and functions, whether they are plasma membranes or membranes of organelles inside the cell. These membranes consist mainly of lipids and proteins, which vary from one type of membrane to another.

The main lipids in biological membranes are phospholipids (see Figure 3-6). In addition, plasma membranes usually contain small amounts of glycolipids (lipid + carbohydrate). Membranes of animal cells also contain a lot of cho-

Figure 4-4
The two-layered structure of a plasma membrane. This photograph was taken with an electron microscope (see *A Journey Into Life,* Chapter 5). Hydrophilic areas at each surface of the membrane show as two dark lines, with the membrane's hydrophobic interior visible as a light area between them. (Biophoto Associates)

lesterol. All of these are long, asymmetrical molecules with one hydrophilic (polar) end and one hydrophobic (nonpolar) end (Section 3-B). When surrounded by water, these molecules tend to form groups with their hydrophilic heads exposed to the water and their hydrophobic tails huddled together as far from the water as possible. They can do this either by forming spheres or by forming **bilayers** (two layers, each one molecule thick), with the hydrophobic tails sandwiched between the hydrophilic heads (Figure 4-5). It is this bilayer arrangement that occurs in biological membranes.

■ *All biological membranes consist of lipids and proteins.*

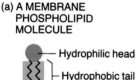

(a) A MEMBRANE PHOSPHOLIPID MOLECULE

— Hydrophilic head

— Hydrophobic tail

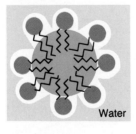

(b) SPHERE

Water

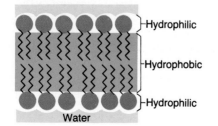

(c) BILAYER

—Hydrophilic

—Hydrophobic

—Hydrophilic

Water

Figure 4-5
How phospholipid molecules arrange themselves in water. (a) A phospholipid molecule has a hydrophilic head and two hydrophobic fatty acid tails. In water, phospholipids become oriented with their hydrophobic tails as far from the water as possible. They may (b) form a sphere, or (c) form a double layer with their hydrophilic heads outward and their hydrophobic tails buried in the center.

Lipid bilayers are fluid: individual lipid molecules can move about, changing places with their neighbors. This fluidity allows the membrane to stretch under stress and to reseal itself if it is disrupted. The membrane's fluidity also permits some of the proteins to move within it.

Figure 4-6 summarizes the current model of biological membrane structure. This is called the **fluid mosaic model** because the lipid layers form a two-dimensional fluid that acts as a solvent for various proteins. Some proteins move around in the membrane; others are relatively immobile, some because they are attached to structures in the cell's cytoplasm.

A membrane's inner and outer surfaces contain different molecules and have different functions. Some lipids are more common in one layer or the other, and particular proteins associate with only one surface. Some proteins that span the entire membrane have distinct inner and outer ends, and must be aligned right to work properly. In addition, many membrane lipids and proteins are combined with carbohydrates to form glycolipids and glycoproteins. The attached carbohydrates are short chains of sugars (**oligosaccharides:** oligo = few) found only on the outside surface of the plasma membrane (Figure 4-6).

Glycoproteins on the cell surface act as **receptors,** structures that recognize and bind specific molecules in the cell's surroundings. The molecules to be recognized may be part of another cell's membrane, or hormones or other chemical signals. For example, cell-surface receptors on eggs and sperm bind specifically to each other during sexual reproduction. These receptors are very precise: sperm recognize and fertilize eggs of only their own species (or of closely related species, which produce very similar receptor molecules).

Now that we have seen how molecules are organized into membranes, we can go on and see how the plasma membrane's structure is related to its function.

Figure 4-6

Structure of a plasma membrane. Phospholipid molecules form a double layer, with their hydrophilic heads facing the outside and the inside of the cell, their hydrophobic tails hidden in the middle of the membrane. Proteins extend through the membrane or sit on one surface or the other. Some phospholipids and some proteins have oligosaccharides attached on the outside surface of the membrane.

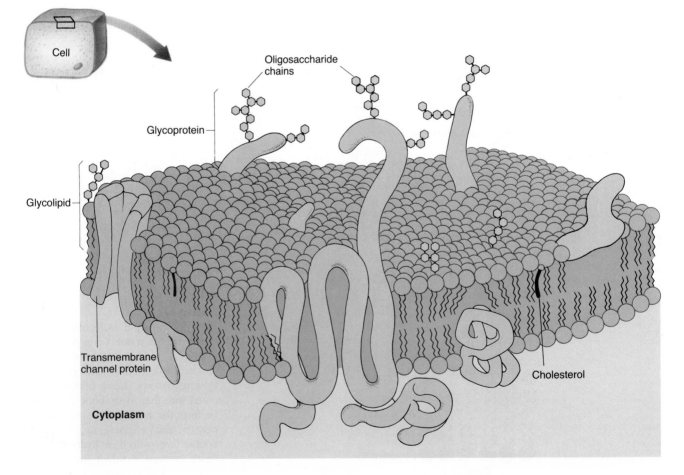

4-B   Roles of the Plasma Membrane

The plasma membrane lies at the frontier between the living cell and its environment. In this strategic location, the membrane performs many vital roles. First, it forms a continuous, closed covering that keeps the cell's contents separate from the external environment. These roles of *containment* and *separation* are performed by the lipid bilayer: its hydrophobic interior excludes the aqueous solutions of the cytoplasm and the external environment.

Some membrane proteins are enzymes that catalyze chemical reactions as part of the cell's *metabolism*. Others mediate the *exchange* of various substances between the cytoplasm and the external environment and so help to control the cell's chemical homeostasis. The membrane's various receptor molecules carry out the task of *recognition* of other cells or of substances in the environment. Some receptors also participate in another membrane function, *irritability,* that is, response to stimuli impinging on the cell. For example, in nerve and muscle cells, stimulation of certain receptors results in a brief change in the rate at which sodium and potassium ions are permitted to pass through the membrane. This leads to other changes in the cells, causing nerve cells to transmit messages or muscles to contract (Chapters 28 and 30).

We now know that substances cross biological membranes in three distinct ways: some pass straight through the membrane; some are transported by proteins in the membrane; and some move within a sac formed from part of the membrane. We shall consider each of these methods in turn.

4-C   How Small Uncharged Molecules Cross Membranes

Diffusion

All molecules are in constant, random motion (Section 2-D). On average, each different kind of molecule or ion tends to move from areas where it is more abundant to those where it is scarcer. It is said to move down its **concentration gradient,** which is a gradual decrease in its concentration over a distance.

Consider a sugar cube dissolving in a beaker of water (Figure 4-7). A sugar molecule moves in one direction until it bumps into another molecule, either another sugar molecule or a water molecule. Both molecules bounce off in new directions. The sugar molecule may bounce back toward the sugar cube, but most of the possible new directions will carry it still further from the original cube. So, on the whole, sugar molecules tend to move down their concentration gradient, toward areas where they are less concentrated, until they disperse throughout the solution. At the same time, water molecules also move and spread evenly throughout the beaker, including the part once occupied by the sugar cube. This process, whereby molecules of two or more substances move about at random and become evenly mixed, is called **diffusion.**

The structure of the plasma membrane does not permit molecules to diffuse through it freely. The membrane is **selectively permeable,** meaning that some

■ *The plasma membrane forms a physical barrier around a cell, controls the passage of substances into or out of the cell, and recognizes and responds to external stimuli.*

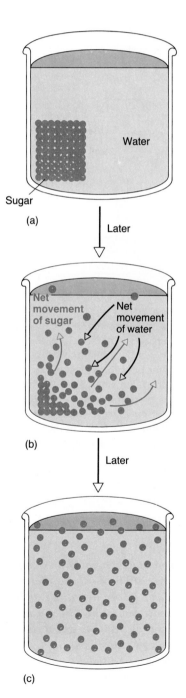

Figure 4-7
Diffusion. (a) A sugar cube is placed in a beaker of water. (b) Both water and sugar move down their concentration gradients: sugar molecules gradually diffuse away from the cube, and water molecules diffuse into the corner where the cube was. (c) Eventually, the molecules of sugar and water are spread evenly throughout the solution. They continue to move randomly, maintaining this even distribution.

substances can pass through it more readily than others. The membrane even prevents the passage of certain kinds of molecules, to which it is **impermeable.** A lipid bilayer's hydrophobic interior makes it relatively impermeable to ions and to many polar molecules. As a result, the plasma membrane prevents most of the water-soluble cell contents from escaping.

Small uncharged molecules cross the plasma membrane by diffusion, each moving down its own concentration gradient. These substances can slip between the hydrophilic heads of the membrane lipids and pass through the bilayer. In essence, these small molecules dissolve in the lipid on one side of the membrane and emerge at the opposite face. For instance, oxygen molecules (O_2) are usually more concentrated outside a cell than in the cytoplasm because the cell constantly uses oxygen. So, on the whole, oxygen molecules dissolve in the outside of the plasma membrane and eventually reach the cell's interior, where O_2 is less concentrated. Carbon dioxide, produced when the cell uses oxygen, leaves the cell by the reverse route.

How fast a substance can diffuse through the lipid bilayer depends on its solubility in lipids and its molecular size. Small, nonpolar molecules such as oxygen, nitrogen (N_2), and ether (see Figure 2-9) cross membranes rapidly. Uncharged polar molecules also cross the lipid bilayer rapidly if they are small enough. For example, urea (the main waste product in human urine) and ethyl alcohol (from alcoholic beverages) cross rapidly. Glycerol (an antifreeze), which is also uncharged but larger, crosses much more slowly, and the sugar glucose, twice the size of glycerol, hardly crosses a lipid bilayer at all. The bilayer is also virtually impermeable to ions, even such small ones as hydrogen, sodium, and potassium. This is partly because of the ions' electric charge, and partly because ions are surrounded by a layer of water molecules, which in effect makes them much larger (see Figure 2-15).

Osmosis

The most abundant substance in a typical cell is water, the solvent in which most of the cell's other molecules are dissolved. Because water does not dissolve readily in lipid, it is somewhat surprising that water crosses lipid bilayers quite rapidly. This is partly because of the water molecule's small size, but it may also be that the molecule's unique bipolar structure (see Figure 2-11) somehow permits it to pass the bilayer's hydrophilic outer layers especially easily.

Osmosis is the process by which water moves through a selectively permeable membrane. Osmosis is a special case of diffusion because it involves the movement of a *solvent* (water) rather than a solute, and because the water is moving *through a membrane.* In osmosis, the net diffusion of water through the membrane is down its concentration gradient: from a weak, or dilute, solution, into a strong, or concentrated solution—that is, from higher to lower concentration of *water.*

A simple way to demonstrate osmosis is to separate distilled (pure) water from an aqueous solution with a membrane permeable to water but not to the solute (Figure 4-8). As time passes, the volume of the solution increases and the volume of the distilled water decreases. Some of the water molecules are moving by osmosis from the pure water, through the membrane, and into the solution.

The **osmotic potential** of a solution is its tendency, or potential, to gain water by osmosis when separated from pure water by a selectively permeable membrane. The osmotic potential depends on the concentration of particles (ions or molecules) in the solution. The higher the concentration, the greater the solution's tendency to gain water by osmosis. The osmotic potential of pure water is zero. The osmotic potential of any solution is less than zero—that is, negative.

■ *Small uncharged molecules enter or leave a cell by diffusion through the plasma membrane's lipid bilayer, each moving down its own concentration gradient.*

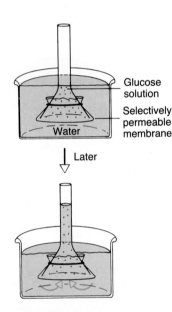

Later

Figure 4-8
An osmotic system. The tube is open at both ends. The bottom of the tube is covered by a membrane permeable to water but not to glucose. The tube is partly filled with glucose solution and immersed in pure water to the level of the glucose solution. Later, the solution has risen in the tube because water has moved from the container into the tube by osmosis. This continues until the weight of solution in the tube pushes water molecules back out as fast as they enter.

■ *The higher a solution's concentration of solutes, the lower (more negative) its osmotic potential. The net movement of water is down its own concentration gradient, into the area of lowest (most negative) osmotic potential.*

Cells as Osmotic Systems

Cells are osmotic systems. A living cell has a selectively permeable plasma membrane, which encloses the cell's internal solution of various particles dissolved in water. To remain alive, the cell must have a covering of water, which also contains solutes. If the internal and external solutions are in osmotic balance, no net exchange of water occurs between them, and the cell is said to be living in an **isotonic** (iso = same; tonus = tension) solution. Solutions that are isotonic with the body fluids are used for such purposes as washing contact lenses and injecting drugs into the bloodstream.

If the solution outside the cell is made more concentrated, so that the cell loses water to its environment, this external solution is said to be **hypertonic** (hyper = exceeding) to the cell contents. And if the cell is placed in a solution dilute enough for the cell to gain water from outside, this environment is said to be **hypotonic** (hypo = lower) to the cell. Plant cells in a hypotonic solution can gain only a limited amount of water before their rigid cell walls prevent them from expanding further. After this, the cell wall squeezes water back out as fast as it enters. In the same hypotonic solution, an animal cell may swell so far that it bursts (Figure 4-9).

Many animals live in fresh water, which is hypotonic to their cells' cytoplasm. Why don't these animals take up so much water by osmosis that they

Figure 4-9
Osmotic behavior of plant and animal cells. (left) Cells in an isotonic solution are in osmotic balance with their environment and lose as much water as they gain by osmosis. Most of the plant cell's water is stored in a large central vacuole, surrounded by its own membrane. (center) In a hypertonic solution, cells lose more water by osmosis than they gain. In the plant cell, the vacuole loses fluid, the cytoplasm shrinks away from the cell wall, and the wall sags. This is what happens when a plant wilts. (right) In a hypotonic solution, cells gain water by osmosis faster than they lose it. In the plant cell, the pressure of water entering the cell is opposed by the cell wall, so the cell gains only a limited amount of water. However, an animal cell may swell so much that it bursts.

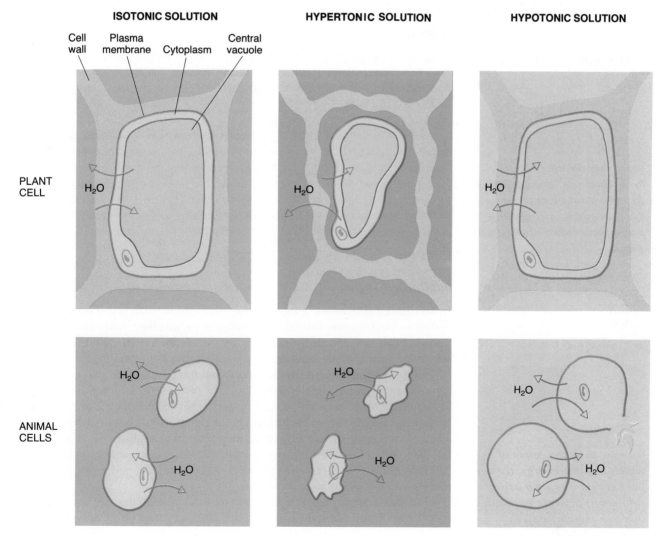

ISOTONIC SOLUTION **HYPERTONIC SOLUTION** **HYPOTONIC SOLUTION**

Cell wall Plasma membrane Cytoplasm Central vacuole

PLANT CELL

H_2O

ANIMAL CELLS

H_2O

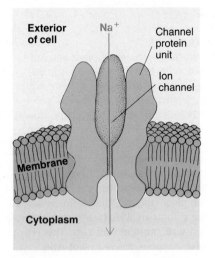

Figure 4-10
Channel proteins. The channel shown here passes through the center of a group of five proteins, two of which have been removed to reveal the channel. This is a gated channel from muscle. It opens when it binds chemicals released by a nearby nerve cell. This briefly changes the shapes of the channel proteins such that sodium ions can rush in through the narrow channel, signalling the muscle to contract.

■ *Some membrane proteins form channels through which specific substances can cross the membrane by diffusing down their concentration gradients.*

■ *Carrier proteins, some of which require energy, transport substances such as ions and small organic molecules through membranes.*

swell up and burst? Most of a freshwater animal's body surface is covered by a layer of rather impermeable material, which retards water uptake. Such layers include the mucus of fish and worms or the chitinous armor of aquatic insects and spiders. In addition, these organisms have well-developed excretory systems that form large volumes of very dilute urine, thereby ridding their bodies of excess water (Chapter 26).

Because water moves freely across the plasma membrane by osmosis, a cell can control its water content only indirectly. The cell can create a difference in osmotic potential across its membrane by moving solutes from one side of the membrane to the other (as we shall see in the next section). Water then enters or leaves the cell, moving by osmosis toward the side of the membrane with the lower osmotic potential.

4-D   Transport by Membrane Proteins

Many molecules cross biological membranes rapidly even though they are not very soluble in lipid. Examples include various small ions, glucose, and amino acids. These substances are transported by **membrane carrier proteins.** Here we consider three important types of carrier proteins, which have some features in common. First, all carrier proteins extend entirely through the lipid bilayer (or are part of protein complexes that do). Each carrier is also specific: it transports only one or a few chemically similar substances.

Channel Proteins

Perhaps the simplest case occurs where proteins form channels through the lipid membrane. These channels permit molecules that are soluble in water to pass through the membrane by simple diffusion, avoiding the membrane's hydrophobic interior.

While some channels are open all the time, others behave as if they have "gates" that open and close. These **gated channels** let certain substances through only when the gates are open. Hence the membrane's permeability changes from time to time, depending on whether the gates are open or not. The opening and closing of gates are thought to involve changes in the shape of the channel proteins in response to some signal (Figure 4-10). This feature is a vital part of the membrane's irritability—its ability to react to stimuli. It plays a key role, among other things, in the working of nerves and muscles (Chapters 28 and 30).

Facilitated Diffusion

In **facilitated diffusion,** a carrier protein combines with a specific substance and moves it from one side of the membrane to the other, down its concentration gradient. This effectively increases the membrane's permeability to the substance, by speeding the substance's passage through the membrane.

An example is the system that speeds up the diffusion of glucose into the cells of some tissues. In the liver, the lens of the eye, and red blood cells, facilitated diffusion moves glucose through the plasma membrane in both directions by means of a carrier molecule. The carrier molecule is more likely to bind a glucose molecule on that side of the membrane where glucose is more plentiful. When the cell is using glucose quickly, the glucose concentration inside the cell falls. Glucose is then more plentiful outside the cell, and more glucose is moved into the cell than is moved out.

Facilitated diffusion is just as important in increasing the rate at which glucose leaves a cell. Cells in the liver, for instance, remove glucose from the bloodstream when the blood glucose level is high, after a meal, but also replenish blood glucose later, when its level drops.

Active Transport

Active transport differs in two ways from the two passive transport mechanisms discussed above. First, active transport can move substances against their concentration gradients. Second, active transport requires energy, usually provided by ATP or by a concentration gradient of ions.

The plasma membranes of many cells contain "calcium pumps," which actively transport calcium ions (Ca^{2+}) out of the cell and so keep the Ca^{2+} concentration much lower inside than outside the cell. Some cells in the stomach wall secrete stomach acid via another active transport pump. It uses energy to export hydrogen ions (H^+) from the cell into the stomach fluid against a concentration gradient of about a million to one!

Still another active transport pump occurs in some cancerous cells that are not harmed by chemotherapy (Figure 4-11). Drug-resistant malaria parasites have a similar pump that rids them of the antimalaria medicine chloroquine.

By far the most important active transport mechanism, though, is the **sodium-potassium pump,** often called the **sodium pump.** It uses energy from ATP to expel sodium (Na^+) from the cell and bring potassium (K^+) in, moving both ions against their concentration gradients (Table 4-1). Transport of these ions is linked: the pump moves three Na^+ ions out of the cell for every two K^+ ions that it moves in.

The sodium-potassium pump is enormously important to cells. An estimated one third of all our energy goes to power this pump! The ability of nerves to conduct electrical impulses, of muscles to contract, of the digestive tract to absorb sugars and amino acids from food, and of the kidneys to form urine, all depend on the working of this ionic pump. Cells also use the sodium-potassium pump to control their water content. The pump adjusts the concentration of sodium and potassium ions inside and outside the cell by active transport, and water then moves passively through the plasma membrane by osmosis (Section 4-C).

All of the active transport pumps just mentioned use energy from ATP. Other active transport mechanisms use a different source of energy: the movement of ions down a steep gradient across the membrane. In our own cells, the sodium-potassium pump builds a steep gradient of Na^+ across the plasma membrane. The Na^+ gradient, in turn, powers the active transport of glucose and amino acids into cells. For example, glucose is transported into cells of the intestine

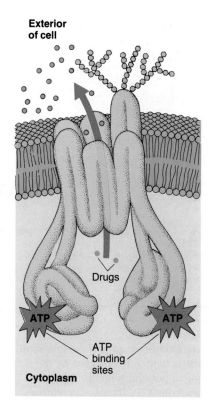

Figure 4-11
The active transport protein that ejects drugs from cancer cells resistant to chemotherapy. The protein chain folds in such a way that parts of it weave through the membrane 12 times, forming a ring around a central tunnel. ATP attaches to sites where the protein juts into the cytoplasm and provides energy needed to move drugs out through the central tunnel. Unlike most other pumps, this one is nonspecific. It exports various drugs that are not chemically related to each other, making these cells resistant to many drugs.

■ *The sodium-potassium pump drives the active transport of sugars and amino acids, and controls the cell's water content.*

Table 4-1	Comparison of Concentrations of Various Substances Inside and Outside a Mammalian Muscle Cell		
	Substance	Concentration inside cell (millimoles)	Concentration outside cell (millimoles)
Positively charged ions	Na^+	12	145
	K^+	150	5
	Mg^{2+}	30	1
	Ca^{2+}	1	4
Negatively charged ions	Cl^-	4	120
	HCO_3^-	27	8
	Proteins, etc.	155	7

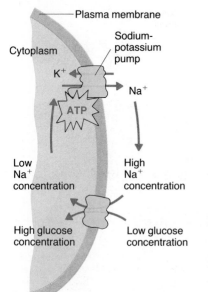

Figure 4-12
Active transport of glucose into some cells is indirectly powered by the sodium-potassium pump, using energy from ATP. The pump builds up a high concentration of sodium outside the cell by active transport. The steep gradient of sodium is then used for the active transport of glucose into the cell, accompanied by sodium ions moving into the cell down their own gradient.

and kidney when sodium and glucose outside the cell bind to sites on a membrane carrier protein. This binding causes the carrier to change shape, pushing the sodium and glucose into the cell. The rate of glucose transport depends on the sodium gradient, which in turn depends on how much sodium the sodium-potassium pump has pumped out of the cell (Figure 4-12).

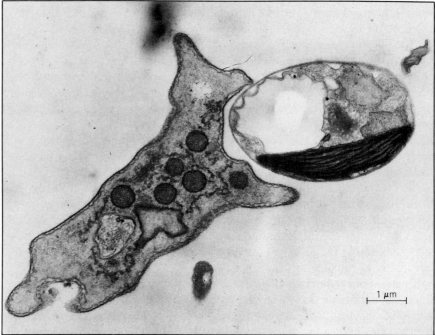

Figure 4-13
"Hello, lunch." An *Amoeba* extends pseudopods around a unicellular green alga, a small photosynthetic organism, the first step in phagocytosis. (Biophoto Associates)

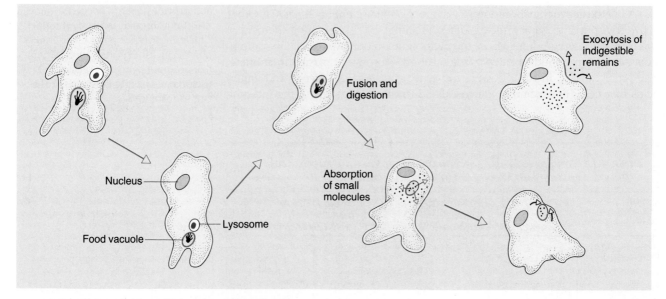

Figure 4-14
Phagocytosis, a form of endocytosis. An amoeba engulfs its prey, and part of the plasma membrane pinches off to form a food vacuole inside the cell. The vacuole fuses with a lysosome, a membrane-enclosed sac full of enzymes that digest the prey. The resulting small food molecules are absorbed into the cytoplasm before the indigestible remains are expelled by exocytosis.

The sodium-potassium pump also maintains the cell's **membrane potential,** the difference in electrical charge across the plasma membrane. The inside of the membrane is negative compared to the outside. Most of the negative charge inside the cell comes from proteins and other organic molecules too large to escape through the membrane. Most of the positive charge outside the membrane comes from Na^+ pushed out of the cell by the sodium-potassium pump.

4-E Membrane Transport of Large Particles

We have now seen how small molecules and ions can cross biological membranes. Sometimes a cell must take in or expel very large molecules, or even bigger particles of matter. Items of this size cannot pass through membranes either by penetrating the lipid bilayer or via membrane proteins. However, the transfer of such large particles is possible because of the membrane's fluid nature: the membrane can change shape and fuse with, or pinch off, small membrane-enclosed sacs. When this happens, the membrane automatically seals itself.

In the process of **endocytosis,** cells take in material from their surroundings. The unicellular organism *Amoeba* feeds by extending projections called pseudopods from the cell body and surrounding the food (Figure 4-13). The pseudopods then fuse so that the food ends up inside the *Amoeba* in a membranous sac called a **vacuole.** This kind of endocytosis is called **phagocytosis** ("cell eating")—the engulfing of large particles, such as an entire bacterium or a fragment of a disintegrating cell, into a vacuole. Phagocytosis is a major feeding method of many unicellular organisms and simple multicellular animals (Figure 4-14).

In most animals, phagocytosis rids the body of debris such as dead cells and also plays a part in defense against disease. For example, some human white blood cells are phagocytes, which engulf and digest invading bacteria. Some of the body's cells normally undergo "programmed death," and their remains are also cleaned up by phagocytes. In your body, phagocytes remove 100 billion worn-out red blood cells per day.

Phagocytosis of, say, a bacterium involves specific binding of protein receptors on the phagocyte's plasma membrane to complementary molecules on the bacterium's surface. Some strains of *Streptococcus* ("strep") bacteria surround themselves with a capsule of carbohydrates that inhibits binding and ingestion by phagocytes; these strains are apt to cause illness, whereas phagocytes easily dispose of strains lacking the capsule, before they can cause disease. In contrast, the bacteria that cause leprosy and tuberculosis are easily engulfed but have adaptations making them resistant to being digested by phagocytes.

Many cells take in some of the external fluid, and whatever solutes it contains, by a kind of endocytosis called **pinocytosis** ("cell drinking").

In contrast, **receptor-mediated endocytosis** takes up small particles selectively. The particles first bind to specific protein receptors on the plasma membrane. Next, the membrane forms a depression around many of these receptors and the molecules they have bound (Figure 4-15). The membrane then pinches off a small sac, containing the loaded receptors, into the cytoplasm.

Materials can be released from cells as well as engulfed. In **exocytosis,** the membrane of an internal sac or vacuole fuses with the plasma membrane, which then opens and allows the sac's contents to escape from the cell. Substances released in this way may be indigestible food particles, or secretions such as hormones.

Figure 4-14 shows how a patch of plasma membrane becomes a vacuolar membrane in the cytoplasm by endocytosis and eventually returns to the plasma membrane during exocytosis. Likewise, membrane material that merges with

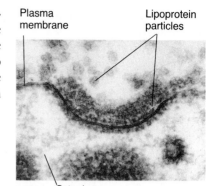

Plasma membrane   Lipoprotein particles

(a)   Cytoplasm

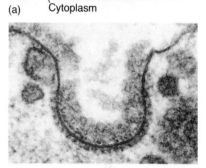

(b)

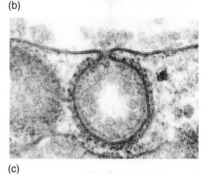

(c)

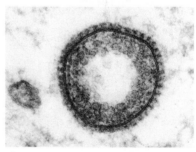

(d)

Figure 4-15
Receptor-mediated endocytosis. This sequence of photographs is thought to show the stages by which the yolk of a forming chicken egg takes in lipoprotein material. (a) Lipoprotein particles bind to receptors in the plasma membrane, which (b) folds inward and (c) pinches off, (d) becoming a separate membrane-enclosed sac (a vesicle) in the cytoplasm. (M. M. Perry and A. B. Gilbert, *J. Cell Sci.* 39:257–272, 1979)

■ *Membrane material cycles between the plasma membrane and membranous compartments in the cytoplasm.*

the plasma membrane during exocytosis is reclaimed by endocytosis and recycled into more vacuoles. This membrane-recycling process is called the endocytic cycle.

4-F Membrane Attachments Between Cells

The many cells of an animal's body are held together in various ways. Some of these involve attachments between the plasma membranes of adjacent cells. The most common is a glue containing a polysaccharide, hyaluronic acid. This glue is often reinforced by several kinds of attachments.

Tight junctions are areas where plasma membranes of animal cells are sealed together, forming such a tight barrier that even small molecules cannot move through the spaces between the cells (Figure 4-16).

Even stronger attachments between cells are provided by **desmosomes,** which occur in regions of high mechanical stress. In these areas, circular patches of the membranes are held together by the interaction of proteins that extend through each membrane into the space between the cells. On the cytoplasmic side of each membrane is a dense plate of proteins that provides mechanical support (Figure 4-16).

One place where tight junctions and desmosomes occur is between the cells lining the small intestine. These attachments keep food from seeping indiscriminately into the body by slipping between cells. Instead, the cells of the lining selectively absorb food and pass it to the bloodstream (see *A Journey Into Life* for this chapter).

Figure 4-16
Tight junctions and desmosomes. (left) Tight junctions occur where proteins extend through the lipid bilayer and attach to their counterparts in the membrane of a neighboring cell. In effect, this stitches the two cells' plasma membranes together like layers of fabric. (right) In desmosomes, the attachment between two cells is reinforced by proteins between their plasma membranes and just inside each cell.

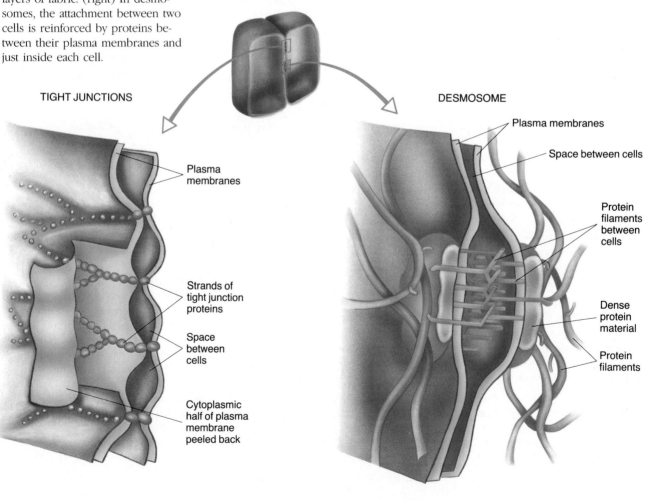

TIGHT JUNCTIONS

Plasma membranes

Strands of tight junction proteins

Space between cells

Cytoplasmic half of plasma membrane peeled back

DESMOSOME

Plasma membranes

Space between cells

Protein filaments between cells

Dense protein material

Protein filaments

The Inner Seal

A JOURNEY INTO LIFE

A cell's plasma membrane separates the living cytoplasm from the nonliving environment and controls what enters the cell. In the cells lining the small intestine, the plasma membranes perform these roles not only for the cells, but also for the entire body: they form a barrier between the living body and the nonliving "soup" of digested food in the intestinal canal, and control what food molecules enter the body proper. In these specialized cells, the features of the plasma membranes are arranged in a way that enhances the cells' role of absorbing food.

First, the membrane surface area available to absorb food is increased about 20-fold. On side of the cell facing the intestinal canal, the plasma membrane projects out in finger-shaped extensions. These **microvilli** are each about one micrometre (one thousandth of a millimetre) long. The inner surface of the intestine has about 200,000 microvilli per square millimetre.

At the edge of the microvilli area, where neighboring cells meet, their plasma membranes form tight junctions, sealed together so that the entire layer of cells acts as a single sheet. Nothing can enter the body from the intestinal canal without passing through a plasma membrane. Below the tight junction area, desmosomes strengthen the attachment between cells.

The membrane covering the microvilli contains various enzymes that help digest many kinds of food molecules. It also contains proteins for active transport of glucose and amino acids into the cell, powered by the Na^+ gradient, which is "downhill" from the intestinal canal into the cell.

As Na^+ enters the cell with food molecules, it must be pumped out again to maintain this gradient. Contrary to what one might expect, Na^+ is not pumped back out to the intestinal canal. Instead, sodium pumps in the plasma membrane on the body side of the tight junctions push it out to become part of the body fluids. Thus the body conserves Na^+, a nutrient that is often scarce in the envi-

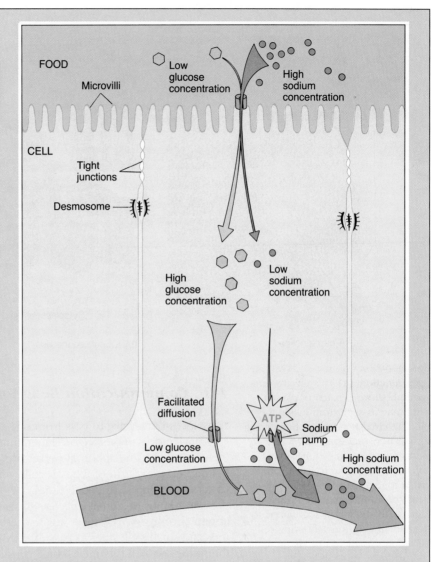

Figure 4-A
Cells in the epithelium of the small intestine absorb digested food and pass it to the bloodstream.

ronment, rather than eventually losing it as the food passes on down the intestine and is eliminated in the feces. (Where does all the Na^+ in the intestinal canal come from? Some comes in with the food, but most is from secretions of the digestive system.)

The plasma membrane on the body side of the tight junctions also contains proteins for facilitated diffusion of glucose. Glucose leaves the cell passively via these proteins, moving down its concentration gradient. (Remember, active transport in the microvilli area has built up the cell's glucose concentration.) Once outside the cell, glucose enters the bloodstream, which whisks it away and so keeps the concentration of glucose on this side of the cell low. Hence glucose (and other food molecules) can continue to leave the intestinal

lining cells by diffusing down their concentration gradients.

In effect, both Na^+ and glucose are pumped through the intestinal lining cell from the intestinal canal to the body fluids. For this to work, the various membrane transport proteins must stay in the proper area of the plasma membrane (facing the intestinal canal or facing the body fluids). These proteins stay in the correct membrane area because they cannot diffuse past the very snug tight junctions.

The large surface of the microvilli, and the placement of tight junctions and membrane transport proteins, makes the cells of the intestinal lining well suited for their role of passing nutrients from the digestive tract to the rest of the body.

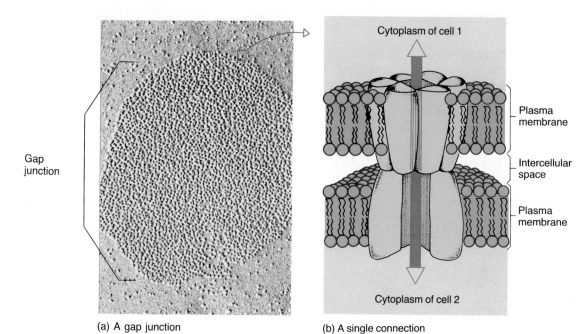

(a) A gap junction

(b) A single connection

Figure 4-17

A gap junction. (a) This electron micrograph shows a patch of plasma membrane as seen from inside the cell. The circular group of bumps is a gap junction, and each bump is one end of a protein "pipe" that forms a direct cytoplasm-to-cytoplasm connection to the adjacent cell. (b) Drawing of one pipe, enlarged and turned sideways to show how it connects to a similar pipe in a neighboring cell. Each pipe is made up of six protein units arranged in a circle with a tunnel in the middle. Materials can pass through the tunnel from one cell to the other but cannot leak into the intercellular space between the two cells. Here, some proteins in the pipe in cell 2 have been removed to show the tunnel's interior. (a, E. Anderson, *J. Morphol.* 156:339–366, 1978)

4-G Communication Between Cells

Substances often have to pass from one cell to another. This happens when a substance moves out through the membrane of one cell and then in through the membrane of its neighbor. However, it is much faster to have a direct bridge linking the cytoplasm of the two cells, with no membranes to cross. In many cases we do find such connections.

Many kinds of animal cells have direct cytoplasm-to-cytoplasm connections at **gap junctions.** These are areas where an array of protein "pipes" permits ions and small molecules to pass directly from cell to cell without leaking into the space between them. Each pipe consists of a ring of six roughly cylindrical proteins that sticks through the plasma membrane and butts against a similar pipe in the adjacent cell's membrane (Figure 4-17).

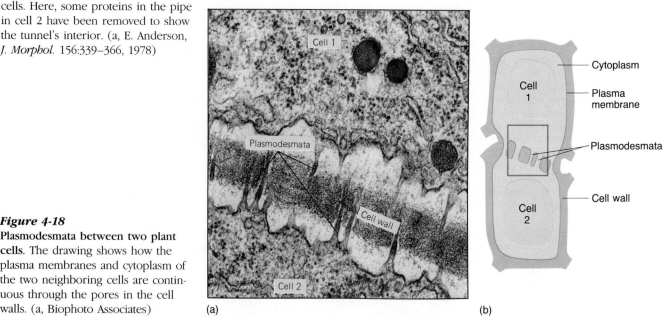

Figure 4-18

Plasmodesmata between two plant cells. The drawing shows how the plasma membranes and cytoplasm of the two neighboring cells are continuous through the pores in the cell walls. (a, Biophoto Associates)

(a)

(b)

In plants, neighboring cells are often connected by strands of cytoplasm called **plasmodesmata** (singular: **plasmodesma**). These cytoplasmic bridges pass through openings in the cell walls between the two cells, and both the cytoplasm and the plasma membranes of these cells are essentially continuous with each other (Figure 4-18).

■ *Cells may communicate with their neighbors by exchanging substances through direct cytoplasm-to-cytoplasm connections.*

SUMMARY

Organisms are composed of one or more cells, the units of life. Under the general direction of its genetic material, a cell maintains chemical homeostasis, carries on its metabolism, and eventually perhaps reproduces by dividing in two. All the time a cell is conducting these internal affairs, it must also interact with its environment.

The plasma membrane is responsible for these interactions between the cell and its environment. It regulates what enters or leaves the cell and detects and responds to changes in the environment. Paradoxically, it also serves as a barrier between the two.

A biological membrane consists of a fluid lipid bilayer, with various proteins embedded and floating in it. This basic structure has two properties crucial to membrane function. First, lipid bilayers spontaneously form closed compartments, thereby keeping the solutions inside and outside the membrane separate. Second, the membrane has different lipid and protein components on each side of the bilayer. Hence the membrane has distinct "cytoplasmic" and "exterior" sides, with distinct functions.

Biological membranes are selectively permeable. Most are freely permeable to water and to small, lipid-soluble molecules, which diffuse through the lipid layers down their concentration gradients. Most ions and polar molecules can cross a membrane only with the aid of the membrane's transport proteins. Channel proteins form aqueous channels through the membrane. Some are gated so that they open in response to specific stimuli. The channel proteins, and protein carriers for facilitated diffusion, provide for the diffusion of specific polar molecules and ions down their concentration gradients. Other proteins carry out active transport, which can move a solute against its concentration gradient. Active transport requires energy, provided either by ATP or by a steep gradient of ions. The sodium-potassium pump, powered by ATP, pumps Na^+ out of a cell and K^+ in. The Na^+ gradient then powers active transport of sugars and amino acids. The sodium pump also regulates the cell's water content by actively transporting these ions through the plasma membrane. Water follows the solutes passively by osmosis, moving toward the area with lower osmotic potential.

A cell takes in macromolecules or larger particles by endocytosis. The membrane surrounds the particles and pinches off to become a vacuole inside the cell. Substances can be discharged from many cells by the opposite process of exocytosis. Membrane material cycles between the plasma membrane and some membranous compartments in the cytoplasm by endocytosis and exocytosis.

The plasma membranes of adjacent cells may interact. Tight junctions between some animal cells seal membranes together and prevent seepage of substances between cells. Desmosomes provide mechanical strength by attaching the membranes of adjacent cells. Gap junctions contain an array of "pipes" through adjacent plasma membranes, providing for direct transfer of ions from cell to cell. In plants, direct transfer between cytoplasm of adjacent cells occurs by way of plasmodesmata.

SELF-QUIZ

1. The U-tube in the figure below is divided by a membrane that is impermeable to starch but permeable to water. A 10% starch solution is put into the right-hand half of the tube and an equal amount of 6% starch solution is put into the left-hand half of the tube.
   In this situation, which of the following occurs?
   a. water will move from the right to the left
   b. water will move from the left to the right
   c. starch will move from the right to the left
   d. water will move in both directions, but more from left to right than right to left
   e. water will move in both directions, but more from right to left than left to right

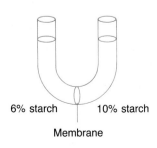

6% starch   10% starch

Membrane

2. *Cambarus* is an animal that excretes a very dilute urine. Therefore you would expect that *Cambarus* lives in an environment that is (hypertonic, isotonic, hypotonic) to

its body fluids. Compared to its body fluids, this environment has a (higher, lower) osmotic potential, and the net movement of water is (from the animal into the environment, from the environment into the animal). The habitat of *Cambarus* is most likely which of the following?

a. a freshwater pond (contains little salt)
b. the ocean (saltier)
c. Great Salt Lake (even saltier)

3. A nerve cell sends messages to other cells by means of a special transmitter substance. Membrane-enclosed sacs containing transmitter molecules fuse with the nerve cell's plasma membrane and then open, releasing the transmitter outside the cell. This is an example of:

a. exocytosis
b. endocytosis
c. active transport
d. facilitated diffusion
e. phagocytosis

4. An area where the plasma membranes of neighboring cells are pressed directly to each other is a:

a. tight junction
b. gap junction
c. desmosome
d. plasmodesma
e. bilayer

Match the substances from the list (a–d) to the way(s) they cross membranes.

____ **5.** Active transport		**a.** ions
____ **6.** Diffusion through bilayer		**b.** macromolecules
____ **7.** Diffusion through channels		**c.** small uncharged molecules
____ **8.** Endocytosis		**d.** water
____ **9.** Osmosis		

Which process(es) in questions 5–9 depend(s) on:

____ **10.** The membrane's fluidity
____ **11.** Membrane proteins

QUESTIONS FOR DISCUSSION

1. Why is it important for cells to maintain chemical homeostasis?

2. Hydrogen cyanide (HCN) and carbon monoxide (CO) are poisons that penetrate cell membranes readily. By what route do these molecules probably cross the plasma membrane into cells? Can you think of an explanation for the fact that no cells have evolved adaptations to keep these molecules out?

3. If cells act as osmotic systems, why don't we swell up and burst when we go swimming in fresh water, which is hypotonic to our blood and to the fluids outside and inside our cells?

4. Figure 4-3 shows cells as seen using a light microscope, with the plasma membrane labelled. Yet light microscopes are not powerful enough to show the very thin plasma membrane. Why does the edge bearing this label appear in the photo?

SUGGESTED READINGS

Bretscher, M. S. "The molecules of the cell membrane." *Scientific American,* October 1985.

Bretscher, M. S. "How animal cells move." *Scientific American,* December 1987.

Dautry-Varsat, A., and H. F. Lodish. "How receptors bring proteins and particles into cells." *Scientific American,* May 1984.

Lodish, H. F., and J. E. Rothman. "The assembly of cell membranes." *Scientific American,* January 1979.

Kartner, N., and V. Ling. "Multidrug resistance in cancer." *Scientific American,* March 1989.

Staehelin, L. A., and B. E. Hull. "Junctions between living cells." *Scientific American,* May 1978.

Cell Structure and Function

CHAPTER

5

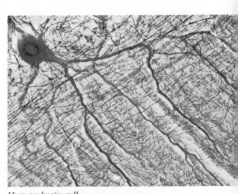

Human brain cell

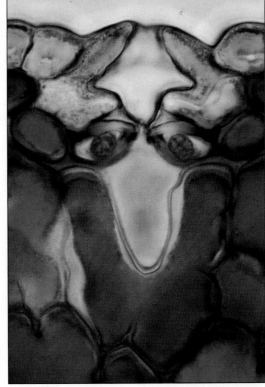

Fir needle

Microtubules

Epithelial cell

*I*n Chapter 4, we saw that cells are the structural, functional, and reproductive units of life. We also examined the plasma membrane, the thin, flexible covering of the cell, and saw how it controls the vital exchange of substances between the cell and its environment. Now it's time to delve further and see what's going on inside the cell.

The viscous, fluid cytoplasm forms the ground substance in a living cell. It contains food molecules, metabolic enzymes, and so on. In the cytoplasm of virtually all cells we find a nucleus or nuclear area containing the cell's genetic material (DNA). We also see many ribosomes, structures that make the cell's proteins according to instructions from the DNA.

What else we find inside the cell depends on which of two basic cell types we have chosen: prokaryotic or eukaryotic. **Prokaryotic cells** (pro = before; karyon = nucleus) have a simpler structure, in particular lacking a discrete nucleus separated from the cytoplasm by a membrane. In addition, they are usually smaller than eukaryotic cells, and they usually have a rigid cell wall outside the plasma membrane. Most are also unicellular—in function if not in structure, for when a cell divides, its offspring may stick together and form clumps or strings of attached but independent cells. For these reasons, prokaryotic cells are believed to have evolved earlier than eukaryotic cells. Organisms with prokaryotic cells are the bacteria (including the photosynthetic cyanobacteria), often simply called **prokaryotes.**

A **eukaryotic cell** (eu = good) contains a nucleus bounded by membrane, as well as other membrane-bounded organelles (Figure 5-1). Eukaryotic cells make up the bodies of all organisms other than bacteria. These **eukaryotes** include the unicellular **protists** and the multicellular fungi, plants, and animals.

Figure 5-1 shows a human white blood cell called a lymphocyte, a small eukaryotic cell. It has all the machinery it needs to live. Its plasma membrane takes in food, such as sugars and amino acids, and expels wastes. Some of its cytoplasmic enzymes, and its mitochondria, break down sugars and use their energy to make ATP, the energy source for many biochemical reactions. The

Figure 5-1

A **eukaryotic cell.** This is a human lymphocyte, a white blood cell. It contains a large nucleus, and smaller mitochondria, enclosed by membranes. Such membrane-bounded organelles are characteristic of eukaryotic cells. This is a transmission electron micrograph, a photograph taken with a transmission electron microscope (see *A Journey Into Life,* this chapter). Electron microscopes produce black and white photographs. This is a "false-color" micrograph: it has been colored to make it easier to see the difference between the cytoplasm (green) and the nucleus and mitochondria (orange). The cell's diameter is about 8 micrometres (μm) (Table 5-1). (CNRI/Science Photo Library/Photo Researchers)

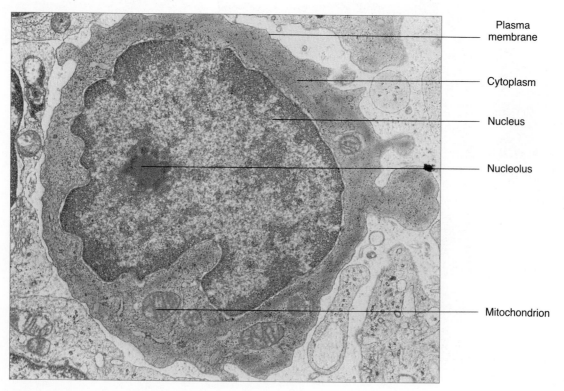

Plasma membrane

Cytoplasm

Nucleus

Nucleolus

Mitochondrion

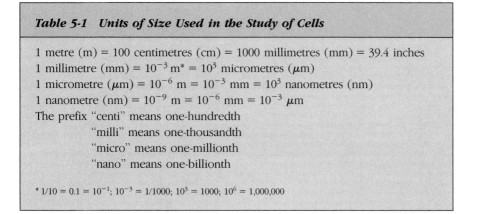

Table 5-1   Units of Size Used in the Study of Cells

1 metre (m) = 100 centimetres (cm) = 1000 millimetres (mm) = 39.4 inches
1 millimetre (mm) = 10^{-3} m* = 10^3 micrometres (μm)
1 micrometre (μm) = 10^{-6} m = 10^{-3} mm = 10^3 nanometres (nm)
1 nanometre (nm) = 10^{-9} m = 10^{-6} mm = 10^{-3} μm
The prefix "centi" means one-hundredth
            "milli" means one-thousandth
            "micro" means one-millionth
            "nano" means one-billionth

* 1/10 = 0.1 = 10^{-1}; 10^{-3} = 1/1000; 10^3 = 1000; 10^6 = 1,000,000

lymphocyte's ribosomes use amino acids taken in by the plasma membrane, ATP from mitochondria, and message molecules sent out from the genetic information in the nucleus, to make its proteins. Some of these proteins perform the "housekeeping" activities needed in every cell. Others are made especially by this cell as its contribution to the whole body. These special proteins are sent to the surface of the plasma membrane. Here they act as sentinels, ready to detect invasion of the body by "foes" such as disease-causing organisms.

Because our lymphocyte's job is patrolling the body, it has no fixed address, although it spends a lot of time hanging out in lymph nodes (such as those in the neck and armpits that swell up during illness). Most other kinds of cells are attached to their neighbors or to a framework of material outside the cell, forming an organized structure of many similar cells, called a tissue. Various kinds of tissues arranged in certain ways form organs, such as the intestine. Blood is also a tissue, but one in which the material outside the cells is a fluid.

KEY CONCEPTS

- Cells can be divided into two groups—prokaryotic and eukaryotic—based on fundamental differences in their organization and size.
- All cells contain components essential for life: a plasma membrane, genetic material (DNA), ribosomes, and cytoplasm. In addition, most prokaryotic cells have a cell wall.
- In addition to the basic cell components, eukaryotic cells have a membrane-bounded nucleus, containing the genetic material, and many other membrane-bounded organelles.
- Eukaryotic organisms may be unicellular or multicellular. Each cell of a multicellular organism must carry on its own life processes and in addition perform some specialized task that contributes to the body as a whole.
- In most multicellular organisms, cells are organized into tissues, tissues into organs, and organs into the various organ systems of the body.

5-A   Prokaryotic Cells: Cells Without Nuclei

Prokaryotic cells are so small that a light microscope, such as you might use in the laboratory, shows little of their structure. The fact that bacterial cell structure differs from that of other organisms was revealed after the invention of the more powerful transmission electron microscope (see *A Journey Into Life* for this chapter). One major difference is prokaryotes' lack of a membrane-bounded nucleus and most other organelles.

Looking at Cells: Microscopy

Most cells are too small to be seen without magnification. Much of our knowledge of cells has depended upon the gradual improvement of microscopes since 1590, when Dutch lens grinders Hans and Zacharias Janssen mounted two lenses in a tube to produce the first microscope.

The **compound light microscopes** used today contain two main lenses. The **objective lens,** close to the object being viewed, forms a magnified image of the specimen. The image is further magnified by the **ocular lens,** near the viewer's eye (Figure 5-A).

Surprisingly, the most important factor determining how small an object may be viewed with a microscope is not its magnifying power but its **resolving power,** its ability to distinguish the separateness of two objects that are close together. Without good resolving power, a microscope produces a fuzzy image, and more magnification only produces a larger fuzzy image. The resolving power of a lens system is limited by diffraction (scattering) of light as it passes through the lens opening. Since diffraction at the objective lens enlarges the image of the specimen, small objects close together will have overlapping images that cannot be resolved as separate.

The most important way to increase the resolving power of microscopes is by reducing the wavelength of the light used. Light of short wavelength (such as violet light) is diffracted less than light with a long wavelength (such as red light).

Entire cells and their larger components can be seen with a light microscope, but many smaller cell parts cannot be seen with visible light. Electron microscopes overcome this problem by using beams of electrons, which have shorter wavelengths than visible light. Hence electron microscopes have higher resolving powers than light microscopes.

In **transmission electron microscopes,** invented in the 1930s, electrons pass through the specimen, just as light does in a light microscope. An electron microscope's working parts are much like those of a light microscope assembled upside down (Figure 5-A). The lenses are not glass but electromagnets, which can deflect the negatively charged electrons. A beam of electrons is produced by heating a tungsten filament in the electron gun. The beam is accelerated through the **condenser lens,** which focuses it, and the focused beam then passes through the specimen and the objective lens. The **projector lens** in an electron microscope is the equivalent of the ocular lens in a light microscope. Since our eyes cannot detect electrons, the projector lens focuses the final electron beam onto a photographic film or fluorescent screen, where it produces a visible image.

One disadvantage of transmission electron microscopes is that electrons are easily deflected or absorbed by air molecules or by the specimen itself. For this reason, specimens must be observed in an almost complete vacuum and must also be sliced very thin.

Scanning electron microscopes were invented during the 1930s and 40s and first manufactured in 1963. Here electrons do not pass through the specimen. Instead, the electron beam hitting the object causes atoms at its surface to emit lower-energy **secondary electrons,** which are collected and used to vary the intensity of a spot on a television screen that scans in synchrony with the electron beam. The resolving power is less than with a transmission electron microscope, but the scanning electron microscope has other advantages. First, the specimen needs less preparation. Second, since the microscope has an extraordinary depth of focus, the three-dimensional surface of an intact specimen can be observed in great detail. In addition, some living organisms, such as hardy insects, can withstand the high vacuum of a scanning electron microscope and can be viewed alive.

Electron microscopes certainly do not make light microscopes obsolete. Light microscopes are much better for examining larger biological specimens and can also be used to study live organisms and tissues, which is rarely possible with an electron microscope. Perhaps the most important advantage of a light microscope, though, is that it produces a colored image. Photographs taken through microscopes are called **micrographs.** Light micrographs are usually colored, but electron micrographs are always black and white because electron beams have no color. The colored electron micrographs in this book were produced by coloring a black and white photograph to emphasize certain parts.

A microscope produces an image with light or dark areas because light or electrons pass through some parts of the specimen but are absorbed or deflected by others. Light microscopes also show the specimen's colors. To increase the contrast in the image, most specimens are specially prepared and stained for microscopy. Specimens are first fixed (killed) and then embedded in wax or resin so that they can be sectioned into thin slices with a glass or metal knife.

Stains for microscopy give contrast to the image by absorbing light or electrons. The chemicals used for staining react specifically with certain cell components. For instance, the alkaline proteins attached to DNA in eukaryotic cells react with a blue stain that does not affect the rest of the cell. In this way, genetic material can be stained blue for the light microscope. For transmission electron microscopy, structures are stained with heavy metal ions, which absorb electrons and so produce dark areas in the final image. A specimen might be stained with lead solution, which reacts with acid structures to leave a deposit of electron-absorbing lead. For scanning electron microscopy, the whole surface of the dried specimen is coated with a thin layer of gold, platinum, or some other good emitter of secondary electrons. In effect, the viewer "sees" the metal coating, not the specimen itself.

Several types of **scanning probe microscopes** were invented in the 1980s. Instead of aiming light or electrons at a sample, these microscopes feel their way across an object's surface with an extremely fine needle, much as a blind person may explore the surface of the ground with a cane. In many cases, the resolution of these microscopes is good enough to reveal individual atoms. More important, they permit us to look at specimens in new ways, measuring such features as magnetic fields, ion flow, electrical charge, or temperature of the object.

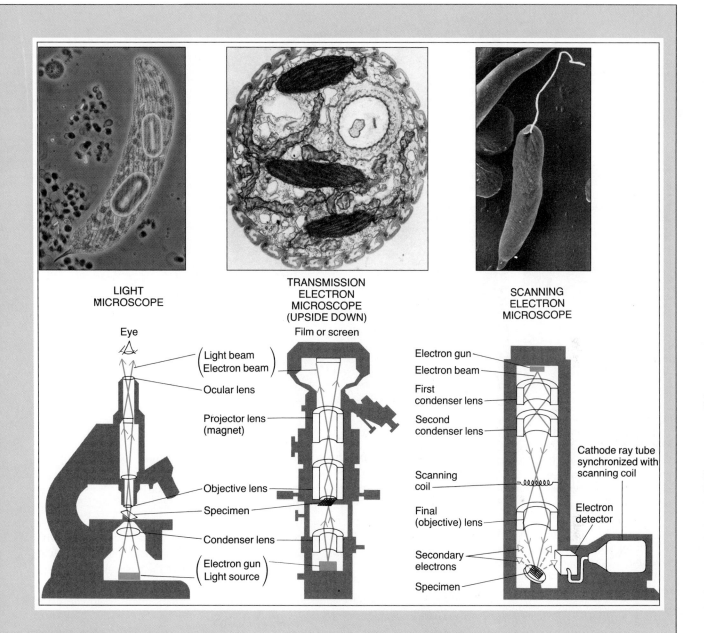

LIGHT MICROSCOPE

TRANSMISSION ELECTRON MICROSCOPE (UPSIDE DOWN)

SCANNING ELECTRON MICROSCOPE

Figure 5-A

Microscopes. (left and center drawings) Comparison of a light microscope with a transmission electron microscope (TEM). The electron microscope is shown upside down so that its parts correspond with those of the light microscope. The EM uses electrons instead of light, and its lenses are electromagnets rather than curved pieces of glass. Since the human eye cannot detect electrons, the image is projected onto a screen or photographic film.

(Right) the scanning electron microscope (SEM) also uses electrons and electromagnets. The scanning coils deflect a fine beam of electrons so that it travels rapidly across the specimen, in synchrony with the spot on a cathode ray tube. The user views the specimen as a "television" picture on the screen of the cathode ray tube.

The photographs show views of the protist *Euglena* taken with each kind of microscope. The light micrograph shows a whole cell; the TEM shows a thin cross section taken by slicing the cell in the light micrograph from left to right; and the SEM shows the cell's outer surface, with its pattern of fine ridges and its flagellum (the white line at the top). (Biophoto Associates)

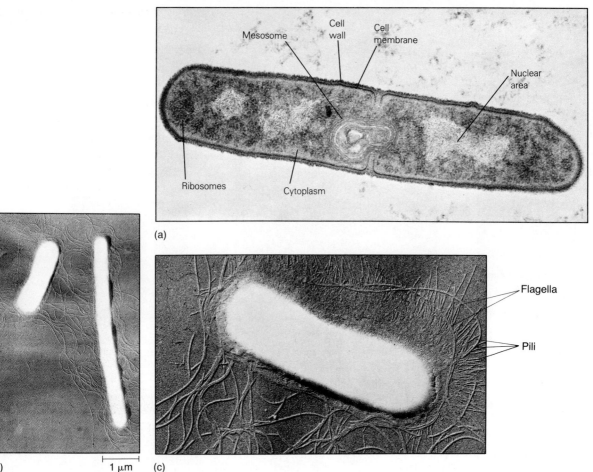

Mesosome
Cell wall
Cell membrane
Nuclear area
Ribosomes
Cytoplasm

(a)

(b)   1 μm   (c)

Flagella
Pili

Figure 5-2

Prokaryotic cells. (a) Electron micrograph of a thin section of a bacterium. This shows the basic components of all prokaryotic cells: a rigid cell wall outside the plasma membrane, and cytoplasm containing the nuclear area and many ribosomes. Note the mesosome, the only internal membranous structure. The ingrowth of the cell wall in the center of the cell indicates that the cell is in the process of dividing. (b) Two bacteria as seen with the scanning electron microscope (SEM). The long, wavy lines are flagella. (c) Close-up of the smaller cell from (b), showing many short, straight pili. (Biophoto Associates)

The DNA of a prokaryote occurs as a large, circular molecule folded up in a **nuclear area** (Figure 5-2). The cytoplasm contains many ribosomes, clusters of RNA and protein that carry on protein synthesis. A prokaryote's ribosomes are smaller than those of eukaryotic cells.

The plasma membrane, present in all cells, is the only membrane found in many prokaryotes. However, some prokaryotes also contain various internal membranes. A membrane known as the **mesosome** appears continuous with the plasma membrane in some electron micrographs of bacteria, although some biologists now doubt that this membrane exists in living bacteria. Most photosynthetic bacteria have a system of internal membranes involved in capturing the light energy needed for this process (see Figure 19-2).

The cytoplasm of a prospering bacterial cell contains granules of storage polymers. These include glycogen or other food (energy) reserves, and polyphosphate granules, which store phosphorus reserves.

A prokaryotic cell is usually surrounded by a thick cell wall, which performs essentially the same functions as a plant cell wall: it protects the cell, gives it shape, and keeps it from bursting in hypotonic media (Section 4-C). Prokaryote cell walls contain unique polymers of amino sugars (sugars with amino groups) and amino acids. Penicillin and related drugs interfere with the building of these walls and therefore inhibit the growth of bacteria but not of their host organisms, which neither need nor make these kinds of polymers. The walls of some bacteria also contain toxic substances. Many diseases caused by bacteria can be duplicated by injecting only these toxic cell-wall chemicals into an animal.

Some bacteria produce a polysaccharide or polypeptide **capsule** outside the cell wall. In nature, a dense felt-like mat of capsule polysaccharides enables bacteria to stick to surfaces—soil particles, rocks in streams, or cells of host animals, for example. *Streptococcus mutans,* the chief agent of tooth decay, glues itself to teeth in this way, using sucrose (but not other sugars) as a raw material. This is why candy and other sugary foods are so bad for our teeth.

Some bacteria produce hundreds of hollow protein strands called **pili,** which serve as means of attachment. In some species, pili are used to attach to another cell and transfer DNA to it during mating. In *Neisseria gonorrhoeae,* only strains that produce pili can attach to host cells and cause gonorrhea. Bacteria of many species move by the rotation of one or more long, thin, helix-shaped **flagella** (singular: **flagellum**) (Figure 5-2).

Because of the small size and simple structure of prokaryotic cells, most biologists believe that they arose earlier in evolution than eukaryotic cells. Chapter 19 discusses how eukaryotes may have originated.

■ *The prokaryotic cells of bacteria contain DNA and ribosomes but not the membrane-bounded organelles found in eukaryotic cells. Most importantly, they do not have a nucleus surrounded by a nuclear envelope.*

5-B Eukaryotic Cells: The Cells of Higher Organisms

Most eukaryotic cells are much larger than prokaryotic cells and contain a greater variety of components (Table 5-2). In addition to the plasma membrane, cytoplasm, nucleus, and ribosomes, eukaryotic cells contain many membrane-

Table 5-2   Comparison of Eukaryotic and Prokaryotic Cells

Feature	Function	Eukaryotic cells		Prokaryotic cells
		Animal	Plant	
Average size		10–20 μm	30–50 μm	1–10 μm
Cell wall	External support and protection	−	+ (mainly cellulose)	+ (amino sugar, amino acid polymers)
Plasma membrane	Containment; exchange of substances with environment	+	+	+
Nucleus (nuclear area)	Housing of genetic material	+	+	Nuclear area
Nuclear envelope	Enclosure of chromosomes	+	+	No membranes around DNA
Chromosomes	Storage of genetic information	Many, linear	Many, linear	One, circular
Nucleoli	Production of ribosomes	+	+	−
Ribosomes	Protein synthesis	+	+	+ (smaller, different)
Endoplasmic reticulum	Segregation of proteins to be secreted; site of new membrane synthesis	+	+	−
Golgi complex	Modification, sorting, packaging of cell products	+	+	−
Lysosomes	Digestion of food and worn-out cell components	+ (in many cells)	+ (some cells)	−
Mitochondria	Respiration of food to produce ATP for energy	+	+	−
Plastids	Photosynthesis; food storage	−	+	−
Vacuoles	Storage of fluid, food, pigments	+ (some)	+ (most)	−
Cytoskeleton	Provision of cell shape; movement	+	+	−
Microtubules	Cell shape, spindle for chromosome separation during cell division	+	+	−
Centrioles	Organization of microtubules and basal bodies of cilia	+	Only in lower plants	−
Cilia, flagella	Locomotion of cell or movement of fluid past cell	+	Only in lower plants	+ (in some, but different type)
Intermediate filaments	Strengthening of cytoskeleton	+	+	−
Microfilaments	Cell motion and changes in shape	+	+	−

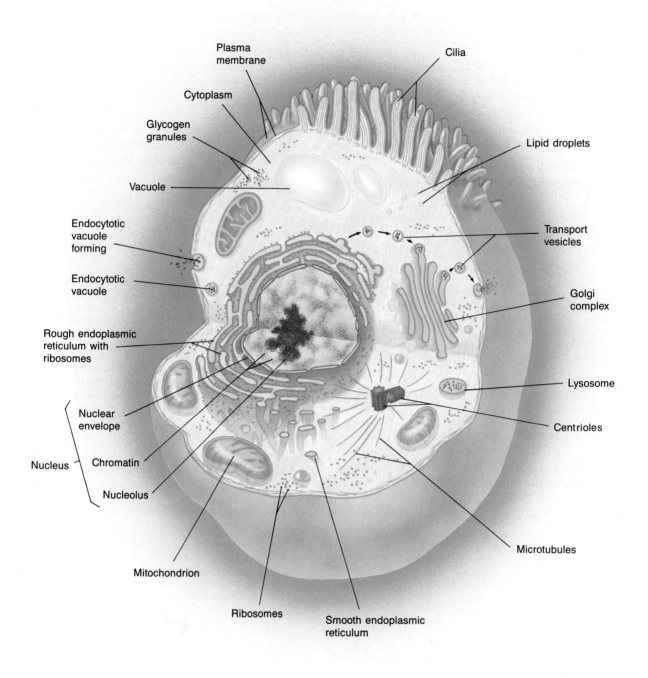

Figure 5-3

A generalized animal cell. This cell has components not usually found in the same cell. For instance, cilia occur on the surfaces of cells in the respiratory passages and in the oviducts of females, and glycogen granules are found in liver cells. The arrows show the relationships among cellular membranes, discussed in Sections 5-E and 5-F. The endoplasmic reticulum forms part of the nuclear envelope, the membrane system that surrounds the nucleus. Transport vesicles bud off from the endoplasmic reticulum and carry substances to the membranous sacs of Golgi complexes. Vesicles from Golgi complexes travel to lysosomes or expel materials by exocytosis. (Illustration by Carla Simmons)

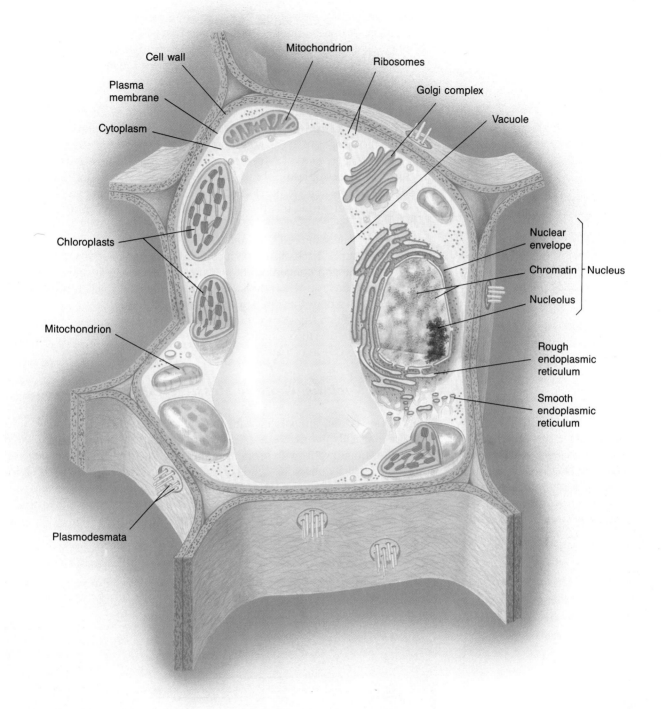

Figure 5-4
A generalized plant cell. A thick cell wall surrounds the plasma membrane. A large vacuole, a membrane-bounded sac filled with various solutes, occupies most of the cell. Chloroplasts are noticeable features of photosynthetic plant cells such as the one shown. Other components are much as in animal cells. (Illustration by Carla Simmons)

bounded organelles and various food storage inclusions such as lipid droplets and glycogen granules (Figures 5-3 and 5-4). There is also a framework of protein fibers, the so-called cytoskeleton, running through the cytoplasm, giving shape and support to the cell, and serving as tracks for the movement of various organelles to different parts of the cell. All these components can be seen with the microscope if the cell is prepared suitably.

Cells also contain other components too small to be seen even with the most powerful electron microscope. These make up the **cytosol,** the soluble portion of the cytoplasm. It is a watery suspension that contains ions, molecules, and molecular aggregates of submicroscopic size, and that surrounds the visible structures in the cytoplasm. About half the cell's volume is occupied by the cytosol. Most of the cell's general metabolism and protein synthesis occur here, and so it is not surprising that the cytosol is about 20 percent protein—mostly metabolic enzymes.

The other half of the cell's volume is split up among the various membrane-bounded organelles. Each of these is a compartment separated from the cytosol by its membrane and containing the proteins and other molecules needed to perform a specific task. When the task is complete, its products move to other parts of the cell where they are used or processed further.

In the rest of this chapter we shall consider the most common components of eukaryotic cells and the elements of the cytoskeleton, and finally consider the organization of cells into tissues.

5-C *The Nucleus: Genetic Message Center*

■ *A eukaryotic cell's nucleus houses its chromosomes. Chromosomal DNA directs the synthesis of RNA molecules, which leave the nucleus and participate in protein synthesis in the cytoplasm.*

When we look at an animal cell with the light microscope, the most obvious structure is usually the **nucleus** (see Figure 5-1). A plant cell nucleus is often less obvious but is visible with proper staining (Figure 5-5). The genetic material in the nucleus of a eukaryotic cell is organized into chromosomes. Each **chromosome** consists of a long, linear DNA molecule and associated RNA and proteins. The chromosomes coil up into short, thread-like structures just before and during division of the nucleus, which precedes cell division. Most of the time,

Figure 5-5
Nuclei and chromosomes. These bean root tip cells were photographed using a light microscope. The large, bluish circular areas in the centers of most of these cells are cell nuclei. In these nuclei, the purple chromatin is spread out in an indistinct mass, but the chromosomes appear as distinct purple threads in the dividing cell in the center. The nucleoli are large dark purple blobs in the nondividing nuclei. (Biophoto Associates)

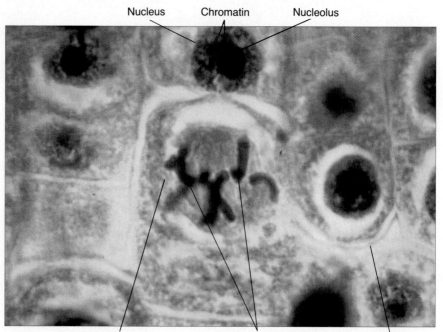

Nucleus   Chromatin   Nucleolus

Dividing cell   Condensed chromosomes   Cell wall

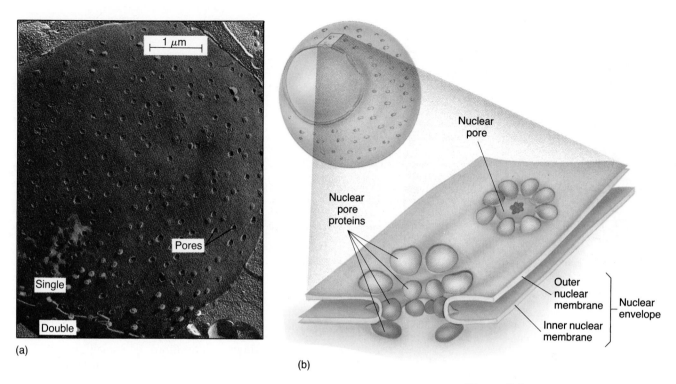

(a)

(b)

Figure 5-6
The nuclear envelope. (a) The surface of a nucleus as seen with a scanning electron microscope. The specimen has fractured in such a way that the double layer of the nuclear envelope shows at the bottom, but only the single inner layer remains in most of the photograph. Nuclear pores are visible, some of them apparently clogged with material. (b) Diagram of two nuclear pores. The pores are partly blocked by proteins of the nuclear pore complex. The particle in the center of the far pore is often seen in electron micrographs. It is not known whether it is part of the pore's structure or merely a large particle caught during transit. (a, Biophoto Associates)

though, the nucleus is not dividing, and the chromosomes are uncoiled in a loose, indistinct tangle called **chromatin** (Figure 5-5).

DNA in the chromatin determines what RNA is made in the nucleus. The RNA travels to the cytoplasm, where it directs protein synthesis. Therefore, DNA also determines what proteins the cell makes. This, in turn, affects most of the cell's structures and activities. Hence the nucleus has been called the cell's "control center." However, control is a two-way street. Substances in the cytoplasm enter the nucleus and influence the DNA in such a way that it changes the particular mix of RNA (and hence proteins) produced. So the kinds and amounts of proteins made in the cell change depending on its needs.

With the light microscope, the most obvious features in the nuclei of nondividing cells are the nucleoli. **Nucleoli** (singular: **nucleolus**) are areas where ribosomes are made; they appear as one or more dense areas, which disappear when the cell divides (Figure 5-5).

The nucleus is surrounded by a double membrane, the **nuclear envelope** or **nuclear membrane,** which is perforated by pores (Figure 5-6). RNA leaves the nucleus via these pores. (RNA molecules are very large as well as hydrophilic, so they could not pass through the lipid layers of the nuclear envelope.)

The inner and outer membranes of the nuclear envelope connect with each other through the nuclear pores, and the outer membrane, in turn, is continuous with a system of membranes in the cytoplasm called the endoplasmic reticulum (Section 5-E). These interconnections allow the nuclear envelope to shrink or grow rapidly by losing material to, or gaining it from, the endoplasmic reticulum. In most eukaryotes, the nuclear envelope disappears completely during cell division and re-forms afterwards. The envelope may expand very rapidly when a dormant cell gears up to produce DNA or RNA.

5-D   Ribosomes: Workbenches for Protein Assembly

Ribosomes are necessary components of cells because they are the sites of protein synthesis. In eukaryotes, ribosomes are made in the nucleolus area(s) of the nucleus and then travel to the cytoplasm by way of the nuclear pores. Eukaryotic

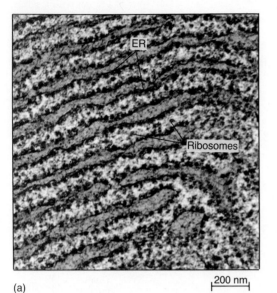

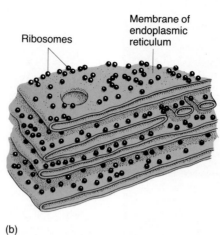

Ribosomes

Membrane of endoplasmic reticulum

(a)

|200 nm|

(b)

Figure 5-7

Endoplasmic reticulum. (a) Rough endoplasmic reticulum, with many ribosomes attached. This cell is from the pancreas of a mouse. (b) Diagram showing the three-dimensional structure of a segment of endoplasmic reticulum.

ribosomes are larger than prokaryotic ones; however, they are still so tiny that even electron micrographs show them as little more than small granules (see Figure 5-7). Most of our knowledge of ribosomes comes from laboratory research using ribosomes that have been isolated from cells. The function of ribosomes in protein synthesis is discussed in Chapter 10.

A cell may contain up to half a million ribosomes, the number varying with how much protein the cell makes. Some ribosomes appear to be attached to part of the cytoskeleton (Section 5-L), and some are attached to membranes in the cell, especially those of the endoplasmic reticulum, discussed next.

5-E *Endoplasmic Reticulum*

The **endoplasmic reticulum (ER)** is a system of membranous tunnels and sacs found in most eukaryotic cells. Often it lies just outside the nucleus, since it is continuous with the outer membrane of the nuclear envelope. The endoplasmic reticulum usually accounts for half or more of the cell's total membrane. In electron micrographs, the endoplasmic reticulum appears as piles and sacs of membrane, but it is believed to consist of one continuous sac surrounded by a single, highly convoluted membrane (Figure 5-7). The interior of this sac, the **lumen** of the ER, provides the cell with a compartment to contain substances that must be kept separate from the cytosol.

Some of the ER is called **rough endoplasmic reticulum** because ribosomes attached to the outer (cytoplasmic) surface give it a bumpy appearance in electron micrographs (Figure 5-7). Rough endoplasmic reticulum is especially abundant in cells that make proteins which will be secreted from the cell (for example, cells of the pancreas that make digestive enzymes for export to the small intestine). This is because the endoplasmic reticulum and ribosomes co-operate in solving the problem of getting these large protein molecules out through the plasma membrane. As the proteins are made, they are pushed through the ER membrane into the lumen. Here, enzymes modify the proteins. Next, the proteins move on into a transitional area, where patches of the ER membrane bud off to form membranous sacs called **vesicles,** in this case transport vesicles carrying the proteins (Figure 5-8). These vesicles move to a Golgi complex (Section 5-F, next), where the proteins are modified further before being packaged into other vesicles bound for exocytosis at the plasma membrane.

■ *The endoplasmic reticulum provides a large surface area where ribosomes can attach and produce proteins to be exported from the cell. New membrane is also made in the ER. Transport vesicles carry proteins and membrane material from the ER to a Golgi complex.*

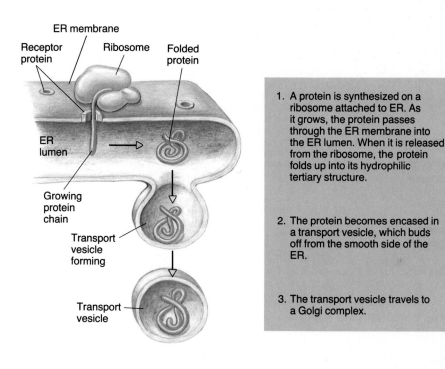

1. A protein is synthesized on a ribosome attached to ER. As it grows, the protein passes through the ER membrane into the ER lumen. When it is released from the ribosome, the protein folds up into its hydrophilic tertiary structure.

2. The protein becomes encased in a transport vesicle, which buds off from the smooth side of the ER.

3. The transport vesicle travels to a Golgi complex.

Figure 5-8
How protein synthesis on rough endoplasmic reticulum permits a protein to cross a membrane. Receptor proteins provide sites for ribosome attachment and pores through which growing protein chains reach the lumen of the endoplasmic reticulum as they leave the ribosome. Proteins leave the endoplasmic reticulum inside transport vesicles pinched off from the ER membrane. Transport vesicles from the endoplasmic reticulum carry proteins to Golgi complexes.

Most of the cell's new membrane is produced in the endoplasmic reticulum. The ER membrane contains enzymes that use substrates from the cytosol to make new lipids. These become inserted into the ER membrane, and this is how the membrane grows. We have seen that the ER membranes are continuous with the nuclear envelope, and that parts of the ER can pinch off as vesicles, which may fuse with other membranous structures such as Golgi complexes. New membrane material can thus be added to existing membranes either by direct transfer from the ER or by the integration of vesicles into existing membranes (see Figure 5-3).

Smooth endoplasmic reticulum is ER without attached ribosomes. Most cells contain little smooth ER, but it is abundant in cells responsible for lipid metabolism, such as those making cholesterol and steroid hormones.

5-F Golgi Complex: Molecular Finishing and Sorting Area

Most transport vesicles pinched off from the ER soon fuse with larger sacs that are part of a **Golgi complex**—a stack of flattened, membranous sacs like a pile of pita bread. Around the edges of the stack, swarms of small, round transport vesicles carry molecules to or from these large sacs (Figure 5-9). Like the ER, the Golgi often lies near the nucleus. A cell may have one large Golgi complex or up to hundreds of much smaller ones.

The Golgi's overall role appears to be to modify, sort, and package molecules made in other parts of the cell, before they travel to their final destinations in other organelles or outside the cell. Molecules being processed in the Golgi move from one of the large sacs to another in sequence, carried by transport vesicles. In each sac, the molecules are modified by enzymes and then enclosed in another transport vesicle, which fuses with a sac containing enzymes for the next biochemical steps. The last sac finishes, sorts, and packages the molecules into transport vesicles according to "address labels" that have been attached in the Golgi. For example, proteins en route to lysosomes are labeled with phosphate groups at their first stop in the Golgi. In the last sac these proteins bind to phosphate-specific membrane receptors and become organized into a transport

■ *The Golgi puts the finishing biochemical touches to molecules destined for other organelles or for export from the cell.*

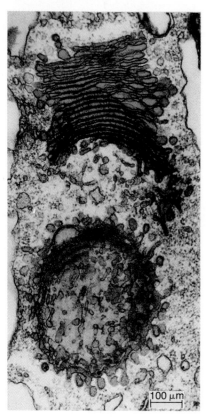

(a)

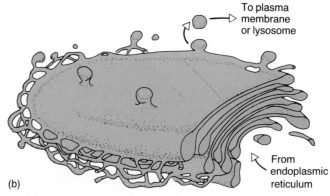

(b)

Figure 5-9

Golgi complex. (a) Two Golgi complexes at right angles to each other. The membranes that look like a pile of pita bread (top) are a stack of Golgi vesicles cut from top to bottom of the stack. The circular structure at the bottom of the photo is one vesicle seen from above. (b) Drawing of a Golgi complex, sliced at the right to show the characteristic structure seen in electron micrographs such as (a). (a, Biophoto Associates)

vesicle that will fuse with a lysosome. Molecules to be exported from the cell are enclosed in vesicles that pinch off from the Golgi sac membranes, move to the plasma membrane, and discharge their contents by exocytosis. These molecules include things like mucus and digestive enzymes, secreted by cells lining the digestive tract, or hormones from gland cells. Other vesicles leaving the Golgi carry new protein and lipid material to be added to the plasma membrane itself.

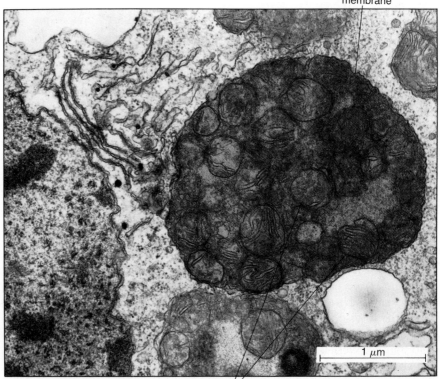

Figure 5-10

Lysosomes in a starved *Euglena* (the same kind of unicellular organism shown in *A Journey Into Life* Figure 5-A). The lysosomes contain many of the protist's own mitochondria, which it is digesting as a source of food. (Biophoto Associates)

5-G Lysosomes: Sacs of Hydrolytic Enzymes

A **lysosome** is a membrane-bounded sac containing hydrolytic (digestive) enzymes (Figure 5-10). These enzymes are made in the endoplasmic reticulum and then modified and packaged in the Golgi. Vesicles filled with lysosomal enzymes bud off from the Golgi and fuse with vesicles containing material to be digested.

Lysosomes digest a variety of things: food molecules, disease-causing viruses brought into the cell by endocytosis, damaged organelles, or macromolecules in the cell. The entire cell may even die and then be digested by its lysosomes. For example, as a tadpole develops into a frog, lysosomes in the tadpole's tail digest the tail, and the molecules released are absorbed back into the body and reused by other cells.

■ *Lysosomes break down food, unneeded macromolecules, or damaged cell components into their component monomers, which the cell can use to meet its present needs.*

5-H Mitochondria: Mills to Make ATP

Mitochondria (singular: **mitochondrion**) are large organelles that produce most of a cell's adenosine triphosphate (ATP), the energy supply for many of its metabolic chemical reactions (Section 6-E). Mitochondria get the energy they need to make ATP by **cellular respiration,** a series of reactions that use oxygen in the breakdown of small food molecules to carbon dioxide and water. Cells that use a lot of energy, such as muscle cells in animals, or cells in the growing root tips of plants, have many mitochondria. A liver cell, which is a busy biochemical factory, may contain about 2500 mitochondria.

A mitochondrion has two membranes, an outer membrane separating the mitochondrion from the cytoplasm, and a highly folded inner membrane (Figure 5-11). We shall see how mitochondrial structure contributes to respiration in Chapter 7.

Mitochondria contain their own genetic machinery (DNA, RNA, and ribosomes) and make some of their own proteins and membrane material. They also reproduce themselves: new mitochondria arise only by division of existing ones, and cells cannot make them from raw materials. Many biologists think that mitochondria evolved from prokaryotic cells that came to live inside larger cells (Section 19-E).

■ *Mitochondria are the organelles that carry out cellular respiration and produce most of a cell's ATP. ATP leaves the mitochondrion and provides energy to power chemical reactions in the cytoplasm.*

Figure 5-11
Mitochondria. (a) Electron micrograph of a mitochondrion from the pancreas of a bat. The inner membrane forms narrow folds extending deep into the matrix inside the organelle; the outer membrane is much smoother. (b) Interpretive drawing of mitochondrial structure. (a, Keith Porter)

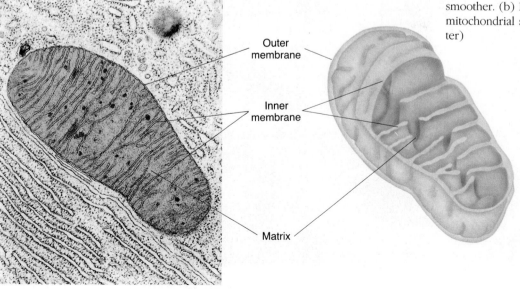

Outer membrane

Inner membrane

Matrix

(a)   (b)

5-I   *Plastids: Food Factories and Storebouses*

The organelles mentioned so far in this chapter can be found in both plants and animals because they perform functions essential to nearly all cells. However, if we were to examine a cell under the microscope, we could tell that it came from a plant rather than an animal by the presence of three features: plastids, a cell wall, and a large vacuole—the subjects of the next three sections.

Like mitochondria and bacteria, plastids contain DNA, RNA, and ribosomes, and reproduce themselves. Plastids are bounded by two outer membranes separating them from the cytoplasm, and they also contain a separate system of internal membranes. There is good evidence that plastids may be descendants of free-living bacteria that set up housekeeping inside larger cells (Section 19-E). While virtually all eukaryotic cells contain mitochondria, plastids are found only in photosynthetic eukaryotes: plants and some protists.

Chloroplasts are the green plastids that carry out photosynthesis—the manufacture of food using carbon dioxide, water, and light energy. The internal membrane of a chloroplast is highly folded (Figure 5-12). Chloroplasts are green because this membrane contains the green pigment chlorophyll, which traps the light used in photosynthesis. Leaves are green because of the green chloroplasts in some of their cells. The details of chloroplast structure are covered in Chapter 8.

Chromoplasts are plastids that make and store the yellow and orange pigments which give their colors to many flowers, fruits, or roots. Other common plastids are **amyloplasts,** which store starch as a plant's reserve food supply. Amyloplasts are abundant in the cells of potatoes and in roots (Figure 5-13).

Figure 5-12
Plant cells. (a) A cell from a leaf of timothy as seen using the electron microscope. Note the characteristic features of plant cells: the cell walls, large vacuoles, and plastids (in this case, several green chloroplasts). (b) Drawing of a chloroplast, showing the two outer membranes and the highly folded green photosynthetic membrane inside. (a, E. H. Newcomb, University of Wisconsin-Madison/BPS)

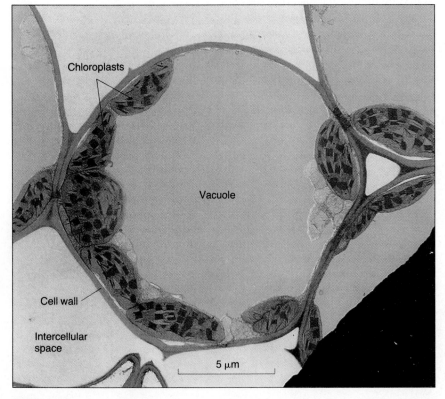

Chloroplasts

Vacuole

Cell wall

Intercellular space

5 μm

(a) Plant cell

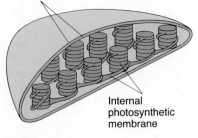

Two outer membranes

Internal photosynthetic membrane

(b) Chloroplast

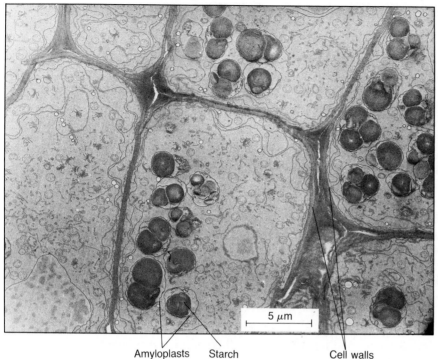

Amyloplasts   Starch   Cell walls

Figure 5-13
Amyloplasts. These cells from a root tip of a corn plant contain many starch-storing amyloplasts. The cell walls are made up of layers of cellulose fibers (see Figure 3-11) and other substances. (Barrie Juniper)

5-J   Cell Walls: Protection and Support

A plant cell is surrounded by a thick but porous **cell wall,** lying just outside the plasma membrane (Figure 5-14).

Composed of cellulose and other fibers, the cell wall is porous enough to allow water and dissolved substances to pass through it freely, tough enough to give the plant body structure and support, and flexible enough to permit the plant to bend in the wind instead of breaking.

Because every plant cell is irrevocably cemented to its neighbors, the cells and organs of plants cannot move much with respect to one another. This, along with the relative rigidity of cell walls, accounts for many of the special characteristics of plants and plant cells. Cell walls contribute largely to the structural support of a plant's cells and of the entire plant body (stems, leaves, and roots). The cell walls of soft-bodied plants are thinner and more flexible than those of woody plants, and the cell walls alone cannot support the plant body. The cells must also be well filled with water, so that they exert an outward pressure on their walls (see Figure 4-9). If the plant loses too much water, the cells do not fill their walls completely, and the plant wilts.

■ *The cell wall holds the enclosed cell in shape, and the cell walls of adjacent plant cells are cemented firmly to one another.*

5-K   Vacuoles: Sacs Full of Fluid

A vacuole is a sac of fluid surrounded by a membrane. Vacuoles occur in many cells but are particularly prominent in plant cells. Typically, most of a plant cell's volume is occupied by a single, large vacuole, and as a result the nucleus, plastids, mitochondria, and other organelles in the cytoplasm are crowded around the edges (see Figure 5-12).

The vacuole holds stored food, pigments, and other substances. It is also convenient as a "storage locker" for toxic substances. For instance, some acacia trees produce and store cyanides (which make them poisonous to animals that eat plants) inside their vacuoles. If the cyanides were in the cytoplasm, they would poison the rest of the cell.

Figure 5-14
Cell walls in the stem of a squash plant. The cells have died and only their walls remain. (Carolina Biological Supply Company)

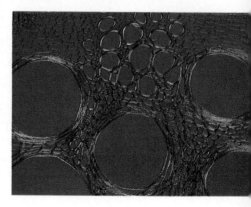

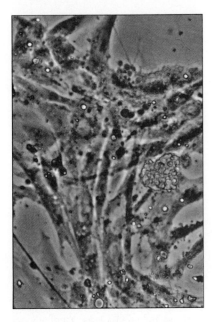

Figure 5-15
Heart muscle cells from a chick embryo growing in cell culture. The cells have put out long filaments by which they move around in the culture dish.

5-L   *The Cytoskeleton: Cell Shape and Movement*

The world of living cells is filled with constant motion. Membrane material is exchanged among the plasma membrane, vesicles, Golgi complexes, endoplasmic reticulum, and nuclear envelope. In the cells of a pondweed, the green chloroplasts circulate around the central vacuole, moved by **cytoplasmic streaming.** Heart cells, whether in an intact heart or isolated in culture dishes, contract rhythmically several times a minute. Cells growing in a culture dish may change shape, putting out long filaments and moving around (Figure 5-15). And nerve cells in an embryo's brain and spinal cord grow long, thin processes to other nerve cells and to the fingers, toes, and other distant parts of the body.

Such movements and changes of shape are still poorly understood. However, we can now use the electron microscope to see some of the delicate structures responsible. The cytoplasm of all eukaryotic cells contains a network of assorted protein filaments attached to the plasma membrane and to various organelles. This has been called the "cytoskeleton" because it provides a framework for cell shape and movement. However, unlike our own skeletons, the structure is not permanent. It seems more like a scaffolding: its components can be disassembled, moved to new locations, and used to erect new structures as needed.

The cytoskeleton is made up of three types of fibers: microtubules, intermediate filaments, and microfilaments, all composed of proteins.

Microtubules

The thickest filaments in the cytoskeleton are **microtubules** (Figure 5-16). They are hollow tubes with walls made up of thousands of protein subunits, and they can be assembled and dismantled rapidly from subunits that are always present in the cytoplasm.

Microtubules serve both as a skeletal framework and as tracks for the movement of organelles. Proteins associated with the microtubules use energy from ATP to move the organelle along.

Microtubules play an important part in cell division (Chapter 11). A great many eukaryotic cells also have other structures composed of complex arrangements of microtubules: cilia or flagella, and centrioles.

Figure 5-16
Microtubules. Lengthwise section (top) and cross section (bottom). (Biophoto Associates)

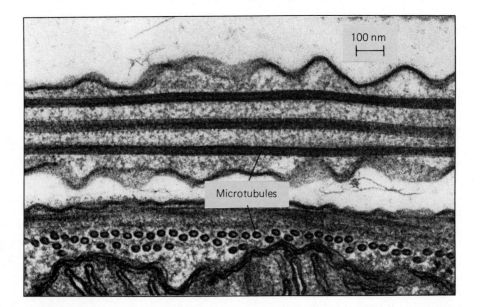

Cilia and Flagella **Cilia** (singular: **cilium**) and **flagella** are thread-like organelles present on the surfaces of many eukaryotic cells. Cilia are generally shorter and more numerous than flagella, but both have the same basic structure.

Cilia and flagella are organelles of locomotion, serving either to propel the cell through its environment or to move something past the cell. For example, many protists move by the beating of their cilia or flagella, and a human sperm moves by lashing the single flagellum that forms its tail. Cilia also move mucus and debris up and out of the human air passages, thereby helping to keep the lungs clear. Cilia and flagella do not occur in higher plants.

Electron micrographs of eukaryotic cilia and flagella show that each contains a circle of nine pairs of microtubules, with two single microtubules in the center. All this is covered by an extension of the plasma membrane. A cilium or flagellum grows from its **basal body,** found where the organelle joins the cell body. The basal body consists of a circle of nine microtubule triplets (instead of pairs), and there are no microtubules in the center of the circle (Figure 5-17).

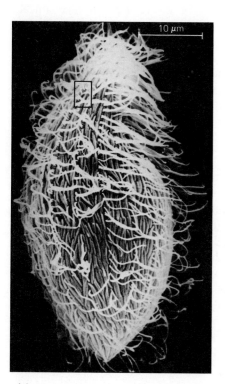

(a)

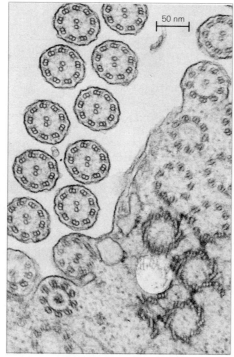

(b)

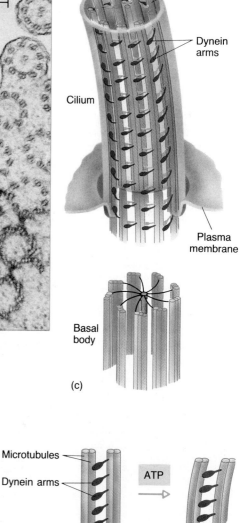

(c)

(d)

Figure 5-17
Cilia. (a) Scanning electron micrograph of the ciliated protist *Colpidium.* The cell moves by the beating of its many cilia. (b) Cross section of a field of cilia such as those shown in the box in part (a). The section has cut straight across several cilia outside the cell (top left) and across basal bodies inside the cell (bottom right). (c) Drawing of a cilium and of its basal body. The cilium has nine double microtubules around two single tubules. The basal body has nine triple microtubules arranged in a circle. The structure in the middle of the basal body is not a microtubule; its makeup is unknown. (d) The dynein arms on each microtubule doublet attach to the next doublet and use energy from ATP to produce a sliding force. This bends the cilium, causing it to move through the surrounding fluid. (a, b, Biophoto Associates)

A cilium (or flagellum) moves by the action of "arms" of the protein dynein that extend from one microtubule of each pair. These dynein arms attach briefly to a microtubule in the next pair and use energy from ATP to produce a sliding force. This bends the whole bundle of microtubules in the cilium, and the cilium pushes against the fluid outside the cell (Figure 5-17).

Centrioles   Eukaryotic cells (except cells of higher plants) contain a pair of **centrioles,** oriented at right angles to each other. Each centriole has the same arrangement of microtubules as the basal body of a cilium (Figure 5-18).

Before cell division, the two centrioles move apart and each somehow directs the formation of a new partner at right angles to itself. When the cell divides, the two pairs of centrioles separate and each pair ends up in one of the two new cells (Chapter 11). Centrioles also give rise to the basal bodies of cilia.

Intermediate Filaments

Intermediate filaments are protein fibers intermediate in thickness between microtubules and microfilaments. They consist of rope-like assemblies of fibrous polypeptides, with the apparent role of giving the cell mechanical strength and thereby maintaining its shape. This is important because cells are often stretched or squeezed when parts of the body move. In animals, cells in different types of tissue, such as epithelial, nervous, and muscle tissue (Section 5-M), have different kinds of proteins in their intermediate filaments. The various proteins are thought to be adapted to withstand the sorts of mechanical stress experienced by each kind of tissue.

Microfilaments

Microfilaments are even thinner than intermediate filaments. They are also called **actin filaments** because they are made up of the globular protein actin (see Figure 3-20). Actin was familiar as one of the contractile proteins in muscle cells long before its role in the protein framework of eukaryotic cells was recognized. We now know that it is the most abundant cytoskeletal protein, and also the most abundant protein inside many eukaryotic cells. As with microtubules, free actin molecules in the cytoplasm can be assembled into filaments at need and later disassembled into actin subunits, which can be reused. Actin filaments can also serve as tracks for the movement of organelles.

Microfilaments are especially plentiful just under the plasma membrane (Figure 5-19). Here they are attached to the membrane and can strengthen, pull, or push the membrane during changes in cell shape. Associated with microfilaments are molecules of **myosin,** the protein that interacts with actin in muscle

Figure 5-18
Centrioles. These structures take part in cell division and are composed of microtubules. (a) Centrioles lie at right angles to one another. This section has cut lengthwise through one centriole (left) and crosswise through the other (right). (b) Three-dimensional drawing of centrioles such as those shown in part (a). Parts that actually appear in the photograph are drawn in color. Parts that lie behind the plane of the photograph are shown in black and gray. (a, Biophoto Associates)

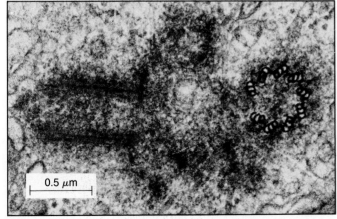

0.5 μm

(a)

(b)

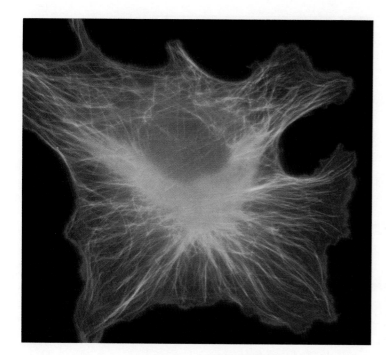

Figure 5-19
The cytoskeleton. This cell has been stained so that actin filaments are red, microtubules green. Note the differences in distribution of each type of element. The oval area outlines the nucleus, which does not contain either element. (Dr. M. Schliwa/Visuals Unlimited)

cells to produce contraction. Actin filaments and myosin are responsible for many cellular motions, such as cytoplasmic streaming in plant cells, endocytosis, and exocytosis, as well as for muscle contraction. Microfilaments also play an active role in ameboid movement and other changes in cell shape. When an animal cell divides, bundles of microfilaments constrict the cell around the middle and divide it into two. Actin filaments can also give shape to protrusions of the cell surface. For example, in cells of the small intestine, bundles of actin filaments form the cores of microvilli (see *A Journey Into Life,* Chapter 4).

Organization of the Cytoskeleton

Most of what we know about the cytoskeleton has been discovered since 1970, and we do not yet have a complete picture of how it is built and how its components interact. With this in mind, we can make a tentative outline to encompass the evidence now available.

Fibers of the cytoskeleton extend throughout the cytoplasm of eukaryotic cells, and can be broken down and reassembled as required. Microtubules give the cell its general shape, act as tracks on which various organelles move, and make up the framework of cilia and flagella. Next smaller are the intermediate filaments, which are thought to provide mechanical strength. Even thinner are the microfilaments, which interact with myosin as the main contractile elements of the cytoskeleton. These are the fibers responsible for much of the actual movement within cells and changes in cell shape, and they can also serve as tracks for transport of organelles.

5-M   Tissues and Organs: Cells Organized Into Work Parties

All but the simplest multicellular organisms contain various different types of cells, many of them arranged in groups specialized for different functions. A **tissue** is a group of cells of one or a few types, held together in a characteristic pattern and performing a particular function. **Organs** are functional units of the body made up of more than one type of tissue; examples are eyes, kidneys, muscles, leaves, and roots.

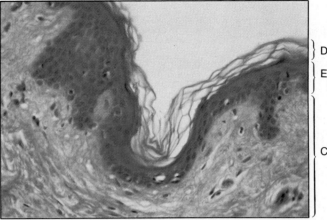

Figure 5-20
Epithelial and connective tissue: human skin. The dark, close-packed cells of the epithelial tissue divide constantly, replacing cells that die and wear off from the skin surface. Beneath this dark layer of epithelium lies connective tissue containing many fibers. (Ed Reschke)

Dead cells

Epithelial tissue

Connective tissue

Animal Tissues

In animals, every cell is surrounded by a space containing **extracellular fluid**—the cell's immediate environment. It is the source of nearly all the substances the cell takes in, as well as being the immediate sink for the cell's wastes.

Animal tissues are divided into four main types:

1. **Epithelial tissues** form coverings and linings (Figure 5-20). Epithelia cover the outside of the body and line the cavities of tubes such as the digestive tract, lungs, vagina, and mouth. In keeping with its function, this type of tissue forms sheets one to several cells thick, with the cells tightly packed together. Tight junctions and desmosomes often strengthen the attachments between epithelial cells (Section 4-F).
2. **Connective tissue** is probably the most abundant type of animal tissue. Adipose tissue (fat), cartilage, and bone are familiar examples. Connective tissue has a great deal of material between the cells, usually in the form of fibers or gelatinous substances secreted by the cells.
3. **Nervous tissue** contains nerve cells, which conduct electrical impulses.
4. **Muscle tissue** is made up of cells that can both conduct electrical impulses and contract (Chapter 30).

Plant Tissues

Plant cells surround themselves with thick but porous cell walls, which are firmly cemented to the walls of neighboring cells. Cells exchange substances with the fluid that fills the spaces between cells and seeps into the pores of the cell walls.

There are four main types of plant tissues:

1. **Epidermis,** the equivalent of epithelium in animals, covers the outsides of leaves and of young stems and roots (Figure 5-21).
2. **Vascular tissue** transports water, food, hormones, and so on between different parts of the plant. It is familiar as the veins in leaves and the wood of trees.
3. **Ground tissue** fills the spaces between the epidermis and vascular tissue inside leaves and in nonwoody stems and roots. It consists mostly of **parenchyma** cells, which are often loosely packed, with many spaces between the cells. Some parenchyma cells contain chloroplasts, a specialization for photosynthesis.
4. **Meristems** are tissues made up of cells that are ready to divide and develop into the other three types of tissue whenever the plant grows new parts.

Epidermis   Epidermis   Parenchyma

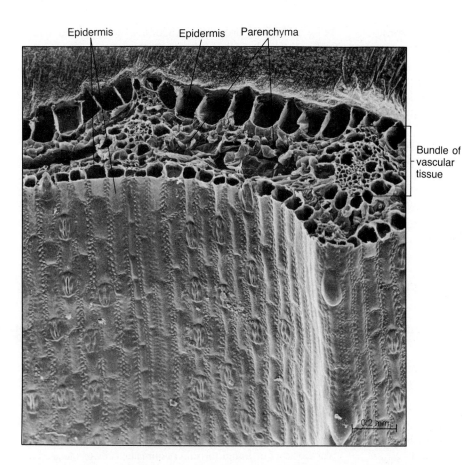

Bundle of vascular tissue

0.2 mm

Figure 5-21

Plant tissues in a leaf. This scanning electron micrograph shows a leaf cut through to expose the cells in the interior. The leaf's outer surface can be seen at the bottom of the photograph. The epidermis, composed of brick-like cells, forms a protective covering over this outer surface. The many slits in the surface are pores called stomata. They can be opened to admit air into the spaces between the loosely arranged parenchyma cells in the leaf interior. The parenchyma cells use carbon dioxide from the air in their photosynthesis. Vascular bundles, in the veins of the leaf, contain pipe-like cells that transport food and water from one part of the plant to another. (Biophoto Associates)

SUMMARY

The prokaryotic cells of bacteria contain everything necessary for life. The plasma membrane, a double layer of lipid and protein, separates the cell contents from the environment and controls the passage of substances into and out of the cell. The cytoplasm contains small molecules and ions, food storage deposits, metabolic enzymes, and ribosomes, where the cell's proteins are made following instructions from the DNA in the cell's nuclear area. Most prokaryotes also have a cell wall, which gives the cell shape and protection.

Eukaryotic cells have all these features (except that animal cells and some protists lack cell walls). In addition, they contain a number of separate interior compartments in the form of membrane-bounded organelles. They also have various proteinaceous structures that make up the cytoskeleton. Table 5-2 summarizes the important structures found in eukaryotic cells.

The hallmark of eukaryotic cells is a nucleus surrounded by a nuclear envelope, consisting of two membranes, and pierced by pores. The nucleus contains genetic material in the form of the DNA of the chromosomes. RNA is made in the nucleus, ribosomes in the nucleolus.

RNA and ribosomes leave the nucleus and participate in protein synthesis in the cytoplasm. Ribosomes making proteins destined for lysosomes, the plasma membrane, or export from the cell associate with the membrane of the endoplasmic reticulum. The ER then sends these proteins on to the flattened membranous sacs of the Golgi complex, where they are finished, sorted, and packaged into transport vesicles for their final destinations.

Many membranous compartments in the cell are connected, either physically or by way of the membranes of transport vesicles that bud off from one, move to another, and fuse to become part of it. The outer part of the nuclear envelope is continuous with the endoplasmic reticulum. The far side of the ER buds off transport vesicles containing substances produced in the ER. These merge with the membranous sacs of Golgi complexes, which in turn bud off new transport vesicles to nearby Golgi sacs or to other areas. Some of the vesicles from Golgi complexes expel materials by exocytosis, and the vesicle membrane becomes part of the plasma membrane for a time. As we saw in Chapter 4, membrane material is returned to structures inside the cell via endocytosis.

Some endocytotic vesicles become part of lysosomes, which digest their contents. Other vesicles, coming from the Golgi to lysosomes, bring the enzymes needed to digest items such as food, foreign matter, or worn-out cellular structures. In all these cases, the contents of the membrane-bounded compartments are kept separate from the cytosol.

Mitochondria are membrane-bounded organelles that do not take part in the membrane exchanges just described. They produce most of the cell's energy supply of ATP.

Eukaryotic plant cells, including photosynthetic protists, have three features not found in animal cells: plastids, such as photosynthetic chloroplasts and starch-storing amyloplasts; a cell wall, a porous but fairly rigid protective and supportive structure outside the plasma membrane, made largely of cellulose fi-bers; and large, prominent vacuoles, storing fluid and various soluble cell products.

The eukaryotic cell's cytoskeleton is responsible for shape, strength, and movement of the cell or of organelles inside the cell. It is attached to the plasma membrane and probably to some internal organelles. It is composed of a variety of protein fibers: microfilaments, intermediate filaments, and microtubules. Microtubules also form the framework of cilia and flagella, thread-like projections at the cell surface that move the cell itself or move substances past the cell surface.

In multicellular organisms, cells are organized into tissues, such as epithelium and connective tissue. Several types of tissue may be organized to form organs, such as kidneys or roots, each with its own particular function.

SELF-QUIZ

Questions 1 to 11 describe cell components. From the list of structures (a to q) below, choose one *or more* that matches each description.

a. cell walls
b. centrioles
c. chromosomes
d. cilia
e. endoplasmic reticulum
f. flagella
g. Golgi complexes
h. intermediate filaments
i. lysosomes
j. mesosome
k. microfilaments
l. microtubules
m. mitochondria
n. nucleoli
o. plastids
p. ribosomes
q. vacuoles

_____ 1. Sites of protein synthesis.
_____ 2. Used to propel a cell through a fluid, or to move a fluid past the surface of a cell.
_____ 3. Rigid coverings of some cells.
_____ 4. Carry hereditary information of cell.
_____ 5. Finish and package products for export from cell.
_____ 6. Contain digestive enzymes of cell.
_____ 7. Impart color to leaves.
_____ 8. Storage compartments in plant cells.
_____ 9. Sites of ribosome synthesis.
_____ 10. Make most of the cells' ATP by cellular respiration.
_____ 11. Involved in cell movement.

For the items listed below, place check marks indicating whether each is found in cells of animals, cells of plants, and/or prokaryotic cells.

	Animal	Plant	Prokaryote
12. ribosome	____	____	____
13. flagellum	____	____	____
14. cell wall	____	____	____
15. chromosome	____	____	____
16. mitochondrion	____	____	____

17. Which of the following is *not* found in the cells of higher plants?
    a. plasma membrane
    b. cell wall
    c. chloroplast
    d. ribosome
    e. centriole

QUESTIONS FOR DISCUSSION

1. Would you expect cells that produce hair to contain more ribosomes than cells that store fat? Why?
2. It has been said that animals, as we know them, could not exist if they had cell walls. Why not?
3. What is the advantage to cells of keeping microtubule protein subunits on hand in the cytoplasm, rather than making them anew from amino acids each time they are needed?
4. Tobacco smoke reduces the activity of cilia in the air passages between the throat and lungs. How does this contribute to "smoker's cough" and lung disease?

SUGGESTED READINGS

Alberts, B., D. Bray, J. Lewis, M. Raff, K. Roberts, and J. D. Watson. *Molecular Biology of the Cell,* 2d ed. New York: Garland Publishing, 1989. A comprehensive but readable text.

Allen, R. D. "The microtubule as an intracellular engine." *Scientific American,* February 1987.

DeDuve, C. *A Guided Tour of the Living Cell,* student edition. New York: Scientific American Books, 1984.

Orci, L., J. Vassalli, and A. Perrelet. "The insulin factory." *Scientific American,* September 1988. Traces the path taken by this important protein hormone from the production in the nucleus of RNA instructions for making it to its export from the cell by exocytosis.

Rothman, J. E. "The compartmental organization of the Golgi apparatus." *Scientific American,* September 1985.

Weber, K., and M. Osborn. "The molecules of the cell matrix." *Scientific American,* October 1985. The cytoskeleton and methods for studying it.

Wickramasinghe, H. K. "Scanned-probe microscopes." *Scientific American,* October 1989. How scanning probe microscopes produce images.

Energy and Living Cells

CHAPTER

6

When you have studied this chapter, you should be able to:

1. Use or interpret the following terms correctly:
   potential energy
   entropy
   autotroph, heterotroph
   oxidation, reduction
   ADP, P$_i$
2. State the first and second laws of thermodynamics, and explain how living organisms carry out endergonic chemical reactions.
3. List the basic starting materials and end products of photosynthesis and respiration, and describe the roles of photosynthesis and respiration in the energy economy of the living world.
4. Discuss the role of ATP in the energy economy of living organisms.
5. Describe chemiosmotic ATP synthesis, including the role of the electron transport system, the membrane housing this system, the H$^+$ gradient, and ATP synthetase.

Nerve endings on muscle fibers

An energy transfer

The sun: energy for life

A one-celled plant

A cartoon character once said that he disliked bombs because they put everything too everywhere (and in little bits, too). He may not have realized that bombs are only a faster way to arrive at an inevitable end result. Left to its own devices, everything will eventually go everywhere anyway. The intrinsic, random movement of molecules carries them into a state of increasing disorder, as everything proceeds to the lowest possible level of energy.

At first glance, living organisms appear to run counter to this universal tendency to increasing disorder, in that their bodies maintain molecular orderliness. A living body has a chemical composition very different from that of its environment, and it keeps this internal chemistry within narrow limits. An organism also increases orderliness by joining simple molecules together to form more complex ones, and by growing, thereby increasing the amount of matter that is organized into living cytoplasm.

However, if we step back and look at the organism *plus its surroundings,* we see an increase in disorder that more than makes up for any increased orderliness of the organism itself. The organism achieves order by using energy (from its surroundings) to push molecules against their natural tendencies to become randomly ordered (Figure 6-1).

Organisms obtain the energy to perform this feat mainly from food. The living world depends on two main energy-processing pathways to meet its energy needs: photosynthesis and respiration. In photosynthesis, green plants capture solar energy and store it in the form of chemical bonds in food molecules, usually carbohydrates. The food's stored energy can later be used by the plant (or by an animal that has eaten the plant in the meantime). During cellular respiration, food molecules are broken down and their energy is released to drive energy-requiring processes.

In this chapter we introduce the rules of energy transactions and show how they apply to the chemical reactions—including the steps of respiration and

Figure 6-1
Maintaining order. These seaweeds, resembling tiny palm trees, maintain the orderliness of their bodies despite the constant pounding of the surf. They use the sun's energy in photosynthesis and make their own food, which they can then use to supply energy needed to maintain their bodies.

photosynthesis—carried out by cells. The next two chapters cover respiration and photosynthesis in greater detail.

- Energy enters the living world when green plants capture solar energy during photosynthesis. Plants use this energy to build food molecules from carbon dioxide and water.
- Energy is released and made available to do useful things during respiration as food is broken back down to carbon dioxide and water.
- Organisms use energy to combat the universal tendency toward increasing disorder. They preserve their organization by using energy to make molecules needed for growth, repair, and reproduction; to move substances within the body; and often to move the body itself.

6-A *Energy Transformations*

We can define energy as the capacity to do work, that is, to cause a change. Energy occurs in many familiar forms, such as heat, light, and chemical and electrical energy. Energy can be **transformed,** or changed from one form into another. For example, the filament of a light bulb converts the energy of an electric current into light energy and heat energy.

■ *Energy cannot be created nor destroyed, but it can be transformed from one kind to another. In every energy transaction, useful energy decreases and some energy escapes in the form of useless heat.*

Energy transformations are governed by physical laws called the **laws of thermodynamics.** The first law states that energy can be neither created nor destroyed; it can only be transformed from one form to another. In other words, "you can't win." This is also called the law of conservation of energy.

The second law of thermodynamics says, in short, "you can't even break even." In any change of energy from one form to another, some useful energy is inevitably converted to useless heat energy, and **entropy,** the degree of disorder, increases. Therefore the energy available to do useful work decreases. So useful energy always proceeds one way—"downhill."

The laws of thermodynamics apply to living organisms. As an example of the second law, we saw in Chapter 4 that molecules tend to diffuse and become evenly mixed—an increase in entropy and decrease in orderliness. To combat this tendency, by which molecules enter or leave at random, cells continuously need more energy, which they obtain from the environment. They can then transform this energy to forms they can use to maintain orderliness.

Plants obtain fresh energy from the sun, animals from eating other organisms. In good times, organisms capture enough energy to store some in reserve for times of energy shortage. So organisms must constantly process energy—obtaining it, storing it, releasing it from food, and using the released energy to drive energy-requiring activities.

■ *Energy flows continuously through living things.*

Energy can exist in kinetic or potential form. **Kinetic energy** is the energy of things in *motion,* such as a rolling stone, running water, or a vibrating molecule. Only kinetic energy can actually perform work. **Potential energy** is stored energy, and it is associated with *position;* it is the possible energy that can be released to do work if the position of something is permitted to change. A rock perched on a cliff has potential energy, which can be changed into kinetic energy if the rock is pushed over the edge. Chemical potential energy is stored in bonds between atoms in a molecule. Electrical potential energy is present when oppositely charged particles are held apart by some barrier, which counteracts the electrical force of attraction that pulls opposite charges together. Each of these forms of potential energy can do work if its components are permitted to move into new positions—that is, if the stored potential energy is transformed into kinetic energy (Figure 6-2).

■ *A cell can use energy only as it is released and "runs downhill" to a lower energy level. The cell can never trap and use all of the energy released during an energy transaction.*

The driving force of an energy transaction is the decrease in useful energy during the energy transformation. Conditions permitting, things will proceed to

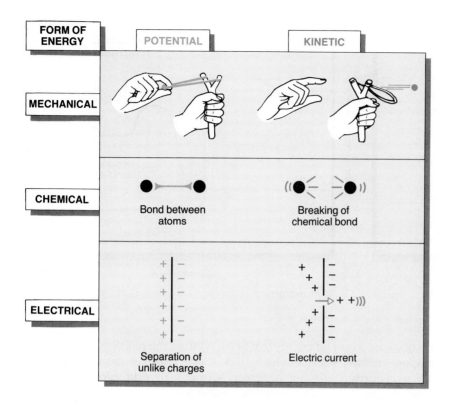

FORM OF ENERGY

POTENTIAL   KINETIC

MECHANICAL

CHEMICAL

Bond between atoms   Breaking of chemical bond

ELECTRICAL

Separation of unlike charges   Electric current

Figure 6-2
Some important forms of energy found in biological systems. Both potential and kinetic examples of mechanical, chemical, and electrical energy are shown. The potential energy is transformed to kinetic energy if the marble is released from the slingshot, the chemical bond is broken, or the electric charge is permitted to move under the influence of the force of attraction between opposite electrical charges.

a lower level of useful energy, as in the familiar case of water running downhill. As this energy decrease occurs, useful energy is released. We can tap and use some of this released energy if we put a paddle wheel or turbine in the water's path (Figure 6-3). The same idea applies to other systems, including living cells.

Living cells use energy in all the forms shown in Figure 6-2. For example, nerve cells perform electrical work as they transmit an electrical nerve impulse, perhaps from the brain to a muscle, telling the muscle to contract. A nerve cell sets up an electrical potential, similar to the one shown in Figure 6-2, by separating positively and negatively charged ions across the barrier of its plasma membrane. This membrane potential is transformed into the kinetic electrical energy of a nerve impulse when some of the ions are permitted to move through the membrane. The force moving the ions is the attraction between oppositely charged particles. However, the useful kinetic energy of the moving ions is dissipated as the ions come together, leaving only the intrinsic, random kinetic energy (heat) of the ions themselves. Also, the entropy has increased, because the ions are now mixed more randomly. To restore its membrane potential, the cell must expend energy on the active transport of ions against their electrical and chemical gradients.

Muscles perform mechanical work, using potential chemical energy stored in molecules of ATP (adenosine triphosphate). Breaking a chemical bond in ATP releases energy, which is used to "cock" certain protein molecules in the muscle. The proteins then spring back to their original position, causing the muscle to contract. Again, the cell must continually expend energy to keep the muscle going. Fresh supplies of ATP are needed constantly, and making new ATP is one of the many kinds of chemical work the cell must perform.

6-B Chemical Reactions and Energy

All cells perform a lot of chemical work in the many reactions of their metabolism. Each reaction is an energy transaction and so obeys the laws of thermodynamics. As in other energy transactions, the driving force of a chemical reaction

Figure 6-3
Energy transformation. In this small power plant, the kinetic energy of water running downhill is transformed to electrical energy, which supplies some of the needs of the Cornell University campus.

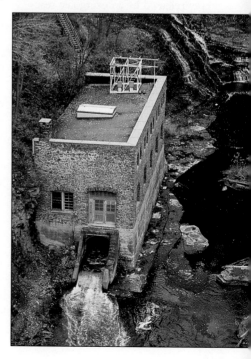

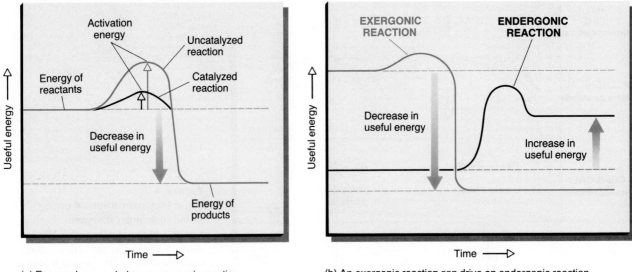

(a) Energy changes during an exergonic reaction

(b) An exergonic reaction can drive an endergonic reaction

Figure 6-4

Changes in useful energy during chemical reactions. (a) A chemical reaction occurs spontaneously only if the useful energy of the products is less than that of the reactants. However, it may not actually happen unless additional activation energy is supplied to bring the reactants to a transition state from which they can form products. An enzyme lowers the amount of activation energy needed to make the reaction occur (black curve). The reaction shown is exergonic because the useful energy has decreased. Conversion of the products back into the reactants would be an endergonic reaction. (b) The energy released by an exergonic reaction can be used to drive an endergonic one. The exergonic reaction must release more useful energy than the endergonic one requires to raise the useful energy content of the products above that of the reactants (red arrows).

is the decrease in useful energy that occurs during the reaction. A reaction that goes spontaneously, with the release (that is, a decrease) of useful energy, is said to be **exergonic** (ex = out of; ergon = work). In **endergonic reactions** (endon = within), the useful energy increases as the reactants are converted to products (Figure 6-4). Endergonic reactions will not go without an input of additional energy from some other source, usually an exergonic reaction.

Even though a reaction is exergonic, it still needs **activation energy** to start (Section 2-E). At the temperatures found in living organisms, few molecular collisions have enough energy to reach this activation energy threshold, and so chemical reactions occur very slowly. Living cells can increase the rate of reactions because their enzymes act as catalysts, lowering the activation energy. An enzyme binds the reactants, called its **substrates,** holds them in the correct position, and exerts forces that loosen the bonds to be broken (Section 3-F). So enzymes enable organisms to carry out their metabolic reactions at the relatively low temperatures of their bodies.

We can compare what happens in nonliving systems versus living cells when they carry out the same chemical reaction—the breakdown of glucose, in the presence of oxygen, to carbon dioxide and water:

$$C_6H_{12}O_6 + 6\,O_2 \longrightarrow 6\,CO_2 + 6\,H_2O + heat$$

To perform this reaction in the laboratory, we must apply a healthy dose of activation energy, perhaps by lighting the glucose with a match. We can measure the heat given off as the glucose burns: 673 kilocalories per mole of glucose. (1 kilocalorie = 1 Calorie = 1000 calories. A calorie is the amount of heat needed to raise one gram of water 1°C. The energy values of foods are expressed in Calories.) The entropy has also increased: atoms from the single, highly ordered glucose molecule are now bouncing around in 12 low-energy CO_2 and H_2O molecules.

When living cells perform the same overall reaction, the energy follows a different path but eventually reaches the same final state (Figure 6-5). Cells, too, must provide some energy (in the form of ATP) to begin a series of enzyme-catalyzed reactions that release even more energy. In living cells, some of the useful energy released by breaking down glucose is stored as chemical potential energy in ATP molecules. This energy can later be released to do work such as active transport, assembly of monomers into polymers, or muscle contraction. However, cells cannot convert all the useful energy from glucose into potential

(a)

(b)

energy stored in ATP. Some energy escapes in the form of useless heat. When energy is later released from ATP itself, some of it may be stored temporarily in the form of new chemical bonds or a membrane potential, but again some heat is released. Eventually all the useful energy originally released from glucose will be degraded to heat. For energy-requiring processes to continue, the organism must constantly acquire new supplies of energy.

Strongly exergonic reactions, such as the oxidation of glucose to carbon dioxide and water, proceed until almost all the reactants are used up. Where the energy change in a reaction is not great, however, the reaction has a tendency to go in both directions, as in the conversion of glucose-phosphate to another sugar phosphate, fructose-phosphate:

$$\text{glucose-phosphate} \rightleftharpoons \text{fructose-phosphate}$$

Under standard laboratory reaction conditions, this reaction is exergonic in the direction of the longer arrow; the opposite direction is endergonic. How much useful energy a reaction releases—or uses—depends on reaction conditions: the concentrations of reactants and products, temperature, and pH. As a reaction proceeds, the concentration of reactants decreases, and that of products increases. Also, the energy released by the reaction declines until it reaches zero. At this point, the rates of the forward and backward reactions are equal, and so there is no net change in the proportions of reactants and products. This state is called **equilibrium.**

A reaction at equilibrium releases no energy and so cannot be used to do work. We could get the reaction going again, and doing work, by adding more reactants or by removing products. This is what happens in a cell's metabolism: reactions are arranged in series, so that one reaction's products are the next one's reactants, and so on. The cell can get energy from some of its reactions because they never reach equilibrium.

The role of enzymes is to make all of this happen faster. An enzyme lowers the energy barriers to both the forward and reverse reactions, and so it does not alter a reaction's equilibrium point.

■ *To extract energy from reactions, a cell must keep substrates and products in disequilibrium by obtaining new energy sources and expelling end products.*

6-C Photosynthesis and Respiration

The energy to run most cellular processes comes from organic **food molecules,** which contain chemical potential energy in the form of the bonds between atoms. This energy can be released by breaking the bonds, and the useful energy so obtained can be used by the cell.

Food is made by the process of **photosynthesis,** which captures solar energy and stores it in chemical bonds of organic molecules, usually carbohydrates: $(CH_2O)_n$. The raw materials of photosynthesis are the simple, low-energy

Figure 6-6
Photosynthesis and respiration. Flow of matter (gray arrows) and energy (red arrows) in living things. Green plants capture the sun's radiant energy and use it to make organic molecules, such as glucose, from water and carbon dioxide. Oxygen is released as a by-product of photosynthesis. All organisms break down organic molecules to release energy during the process of cellular respiration. Respiration uses oxygen and produces carbon dioxide and water, which can be reused in photosynthesis. Although matter cycles indefinitely between photosynthesis and respiration, energy does not. The sun must constantly supply fresh energy.

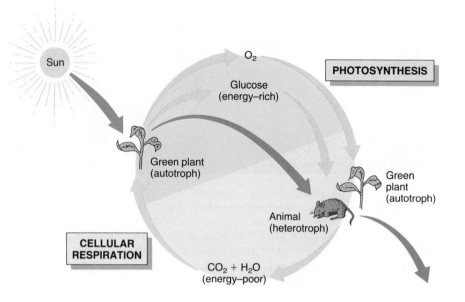

■ *Life on earth depends on the sun to power photosynthesis, by which green plants make food molecules. Food molecules are broken down during respiration, releasing energy that powers the cell's endergonic reactions.*

inorganic molecules of carbon dioxide and water, which are assembled to form more complex, high-energy food molecules. Oxygen is given off as a by-product. Virtually all the O_2 in the air today came from photosynthesis.

All of the photosynthetic organisms—plants, some bacteria, and some protists—are called **autotrophs** (auto = self; trophe = food) because they do not require food molecules from other organisms to meet their energy needs. The food made by these photosynthetic organisms is used by almost all the other life on earth. (There are also some autotrophic bacteria that make food using sources other than the sun for energy [Section 19-B], but their contribution to the world's food supply is negligible.)

Most organisms, including photosynthetic ones, break down food molecules by **cellular respiration,** releasing energy that their cells can use. Respiration uses oxygen in the complete breakdown of food molecules, usually carbohydrates. It gives off carbon dioxide and water (plus the released energy) as its end products. Thus respiration has the overall effect of undoing photosynthesis. The carbon dioxide and water that went into photosynthesis are returned to the environment, and the energy released is used to drive the cell's energy-requiring processes. Autotrophic organisms use respiration to break down the food they have made as its stored energy is needed. Animals and other organisms that cannot make their own food molecules are called **heterotrophs** (hetero = other). They must obtain their food from other organisms (Figure 6-6).

The overall equations for photosynthesis and respiration are:

$$\textbf{photosynthesis:} \quad CO_2 + H_2O + energy \longrightarrow (CH_2O)_n + O_2$$

$$\textbf{respiration:} \quad (CH_2O)_n + O_2 \longrightarrow CO_2 + H_2O + energy$$

These equations appear to be direct opposites: the raw materials of each are the end products of the other; and photosynthesis uses energy, whereas respiration releases it. However, the two processes are not simply the opposite of one another. They are similar in some ways and different in others.

6-D *Oxidation-Reduction Reactions*

One major similarity between the metabolic pathways of photosynthesis and respiration is the large number of oxidation-reduction reactions. Some of these reactions release a lot of energy, enough to drive endergonic reactions that require considerable energy input.

Oxidation-reduction (nicknamed **redox**) reactions involve the transfer of one or more electrons (e^-) from an electron donor molecule (or ion) to an electron acceptor. The molecule that loses the electron is **oxidized,** and the one that gains the electron is **reduced.** Oxidation and reduction are complementary: for every oxidation, there is a corresponding reduction, because electrons cannot float around on their own.

$$Ae^- + B \longrightarrow A + Be^-$$

| electron donor | electron acceptor | has been oxidized | has been reduced |

The simplest case of oxidation is loss of an electron, as from an ion of iron: Fe^{2+} may be oxidized to $Fe^{3+} + e^-$. Substances are also oxidized if they lose an entire hydrogen atom, which contains an electron. Oxidation takes its name from oxygen, which is often the electron acceptor. Because oxygen attracts electrons strongly, it can remove electrons from a wide variety of substances. Conversely, reduction is the gain of one or more electrons or entire hydrogen atoms (Figure 6-7).

Looking again at the overall equations of respiration and photosynthesis, we see that they are redox equations. Respiration is the oxidation of small organic molecules, such as carbohydrates:

$$(CH_2O)_n + O_2 \longrightarrow CO_2 + H_2O + energy$$

O_2 is a strong oxidizing agent, accepting two hydrogen atoms (which include two electrons) per oxygen atom to form water. The transfer of hydrogen atoms from organic molecules to oxygen releases a great deal of useful energy. In cellular respiration, this energy is released in small, usable quantities by breaking this single redox reaction into a series of less energetic ones.

Oxygen atoms hold electrons so strongly that water does not give up electrons readily. Therefore photosynthesis requires a considerable energy input to make water give up its hydrogen atoms (containing electrons), which in effect are used to reduce carbon dioxide:

$$CO_2 + H_2O + energy \longrightarrow (CH_2O)_n + O_2$$

(It is important not to confuse the *dissociation* of water into H^+ and OH^- with the *splitting* of water into $2\,H$ and $\frac{1}{2}\,O_2$. Because oxygen holds electrons so avidly, it easily accepts the electron left behind when only H^+ [a hydrogen ion] leaves a water molecule.)

In general, energy-rich molecules are highly reduced (hydrogen-rich). Energy-poor molecules are oxidized.

Because both respiration and photosynthesis consist of series of redox reactions, we can trace the flow of energy in these processes by following hydrogen atoms and electrons as they are transferred from one molecule to the next in the series. Indeed, this will be one of our chief activities in Chapters 7 and 8. At some steps in these reaction series, energy to do the cell's work is stored in the form of energy intermediates.

6-E   Energy Intermediates

Cells need to perform many endergonic chemical reactions, which go "uphill" in terms of energy. These reactions are catalyzed by enzymes, but enzymes cannot alter the energy changes that occur during a reaction nor make an endergonic reaction proceed without the necessary added energy. How then do organisms power endergonic reactions? The answer is that energy-requiring endergonic reactions are **coupled** to energy-releasing exergonic ones.

The exergonic reaction of the pair must release more useful energy than the endergonic one requires because of the inevitable losses. However, releasing too much extra energy is wasteful. Also, any unused energy is converted to heat.

Figure 6-7
Oxidation-reduction reactions. Oxygen from the air has accepted electrons from the metal of this statue, forming a discolored surface layer of metal oxides.

■ *Membrane potentials and the chemical energy stored in ATP are energy intermediates in all living things.*

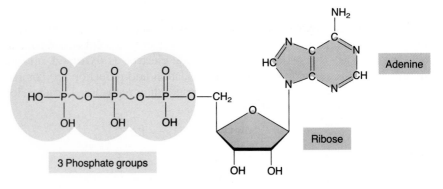

Figure 6-8
Adenosine triphosphate (ATP). Removal of the last one or two phosphate groups, by breaking the bonds shown as red squiggles, releases a great deal of energy.

Figure 6-9
Positions of specialized membranes and H^+ gradients involved in ATP synthesis. Membranes in color contain electron transport systems: the plasma membranes of bacterial cells, and the inner membranes of mitochondria and chloroplasts. The electron transport systems accumulate H^+ on one side of these membranes, in areas shown in yellow. Areas shown in gray, on the other side of these membranes, have low H^+ concentrations. (a, L. Santo, from L. Santo, H. Hohl, and H. Frank, *J. Bacteriol.* 99:824, 1969; c, Herbert W. Israel, Cornell University)

A large burst of leftover heat could damage the enzyme's three-dimensional folding pattern. Reactions that release a lot of energy, such as the oxidation of organic molecules during respiration, are carried out in small steps that release energy gradually. At various steps, some of the energy released is stored as **energy intermediates,** which transfer middle-sized amounts of energy between highly exergonic reactions and endergonic reactions.

The usual energy intermediates in living cells are the electrical energy of a membrane potential and the chemical energy of ATP (adenosine triphosphate). Membrane potentials work much like batteries. For instance, in bacteria the electrical energy from a concentration gradient of H^+ across the plasma membrane powers the motion of flagella. Membrane potentials also provide energy for active transport of some substances. Other active transport systems, such as the sodium-potassium pump, use energy from ATP (Section 4-D). ATP also serves as an energy intermediate when a cell builds proteins from amino acids and in many other energy-requiring processes.

The role of ATP in the cell's energy economy has been compared to that of cash in the human economy. Removing a phosphate group from an ATP molecule releases enough energy to drive many endergonic biochemical reactions. Various enzymes called **ATPases** break down ATP to provide energy for activities such as muscle contraction, active transport, waste excretion, and synthesis of new macromolecules. ATP is also used during photosynthesis as an energy intermediate in the building of carbohydrate molecules, which are the cell's

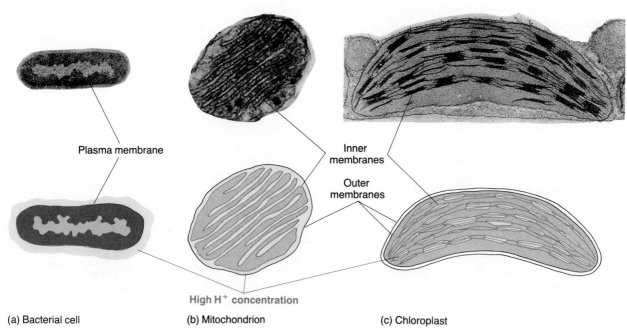

Plasma membrane

Inner membranes

Outer membranes

High H^+ concentration

(a) Bacterial cell   (b) Mitochondrion   (c) Chloroplast savings account of stored energy, broken down as needed to make the ready cash of ATP. The central position of ATP in the economy of life can be illustrated like this:

$$\text{Solar energy} \longrightarrow \text{ATP} \longleftrightarrow \text{Food molecules}$$
$$\downarrow$$
$$\text{Growth, Reproduction, Movement, etc.}$$

Both photosynthesis and respiration, then, can be regarded as means to the same end, the synthesis of ATP.

An ATP molecule consists of the nitrogenous base adenine, a ribose sugar unit, and three phosphate groups. Breaking the bonds shown as squiggles in Figure 6-8, which attach the last two phosphate groups to the molecule, releases a great deal of energy. Energy is usually released from ATP by removing the last phosphate group, yielding ADP (adenosine diphosphate), an inorganic phosphate group (abbreviated as P_i), and about 7 kilocalories of useful energy per mole of ATP:

$$\text{ATP} + \text{H}_2\text{O} \longrightarrow \text{ADP} + P_i + \text{energy (7 kcal/mole)}$$

ADP may be further broken down to AMP (adenosine monophosphate) plus P_i, releasing another 7 kilocalories per mole.

Sometimes instead of releasing a free phosphate group, ATP donates the group to another organic molecule. Attachment of a phosphate group gives this molecule a higher energy content and makes it more reactive, but it has less energy than ATP does.

The number of ATP molecules in a cell is relatively small, and all the energy-requiring reactions of metabolism are continually breaking down ATP. Therefore, energy-yielding reactions (mainly those of cellular respiration) must constantly renew the supply of ATP by joining a phosphate group to ADP, a reaction that reverses the equation above—except that it takes more than 7 kilocalories of energy per mole to drive it in reverse.

Some ATP is made directly, using large amounts of energy released by certain organic molecules when they transfer a phosphate group to ADP. However, most ATP is made using the other energy intermediate, a membrane potential. This electrical energy comes from concentration gradients of hydrogen ions (H^+) across membranes (Figure 6-9):

1. The plasma membranes of bacterial cells (in respiration).
2. The inner membranes of mitochondria (in respiration).
3. The inner membranes of chloroplasts (in photosynthesis).

Each of these membranes contains sets of proteins called **electron transport systems,** which separate hydrogen atoms into electrons and H^+ and carry the electrons away. All the H^+ is left on one side of the membrane, which forms a continuous barrier impermeable to H^+. The result is an H^+ gradient across the membrane, and this gradient provides energy to make ATP. Special channels in the membrane permit H^+ to pass through the membrane, down its concentration gradient. As it does so, **ATP synthetase** enzymes attached to the H^+ channels use the kinetic energy of the moving H^+ to join phosphate and ADP, making ATP. This way of making ATP, using a chemical (H^+) passing through a membrane, is called **chemiosmotic ATP synthesis** (Figure 6-10).

In respiration, the hydrogen atoms entering the electron transport system come from organic molecules in food. Food molecules are oxidized by removal of these hydrogen atoms, and the hydrogens then proceed to the electron transport chain. Here the electrons are passed through a series of redox reactions, and the H^+ gradient is formed and used to make ATP. This is why the energy-releasing redox reactions of respiration are so important: they provide the cell with most of its ATP, which in turn drives most of its energy-requiring metabolic processes.

■ *Most ATP is made by chemiosmotic ATP synthesis, using energy from a gradient of H^+ ions across a membrane.*

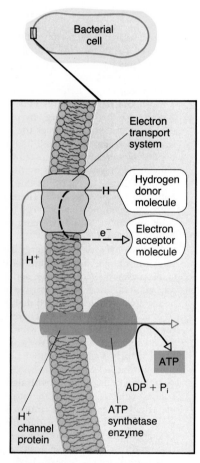

Figure 6-10
Chemiosmotic ATP synthesis. The basic process is shown here using a bacterial cell. The electron transport system in the plasma membrane accepts hydrogen atoms, splits them into H^+ and electrons (e^-), and returns the electrons to the cytoplasm. Meanwhile, H^+ is expelled from the cell, creating a high H^+ concentration outside the plasma membrane. H^+ returns through the membrane via channel proteins associated with ATP synthetase enzymes, which use the energy of the H^+ moving down its concentration gradient to join ADP and P_i, forming ATP.

Nature Invented It First: Solar Power

A JOURNEY INTO LIFE

We use solar energy to run solar-powered calculators and heating systems. Plants have been using solar energy to power much more complicated reactions for millions of years. The amount of solar energy converted into chemical energy by plants each year is about 400 times the amount of energy from fossil fuel and nuclear power used every year by humans. Yet photosynthesis by plants uses less than 0.5% of the visible light that reaches the earth. There is obviously plenty of solar energy left over, and we are slowly learning how to use it directly.

Three main types of energy systems use the inexhaustible fuel of the sun's energy directly: passive, active, and photovoltaic systems. Passive and active solar systems are used mainly to heat buildings. The sun's energy is used to heat stone or brick, which releases its heat slowly, or to heat water, which is then circulated through the building.

Solar cells, or **photovoltaics,** convert solar energy directly into electricity. They are most familiar in small solar cell calculators, but the electricity can be used for any purpose and can be stored in a rechargeable battery until it is needed. A photovoltaic cell consists of thin wafers of silicon with traces of other materials. These emit electrons when sunlight strikes them, and the electrons flow out of the wafer as an electric current.

Generating electricity on a large scale from photovoltaic plants has several disadvantages. Enough cells for even a small power station occupy more land area than a conventional power station, and the net energy efficiency of the cells is low. Photovoltaics also contain small amounts of rare or expensive elements such as gallium and cadmium.

Figure 6-A

Converting solar energy to electricity. Solar collectors like these cover a large area of desert in California. The reflectors focus sunlight onto tubes of a fluid with a high boiling point. The heated fluid then flows to a heat exchanger, where its heat is used to turn water to steam. The steam in turn drives a turbine, which drives an electric generator.

The supply of these elements is limited and their manufacture produces toxic wastes.

The immediate future of solar cells probably lies in a variety of local applications when it is inconvenient or impossible to plug in to the main electricity supply. For instance, solar cells have been used to power systems in space (such as satellites) since 1958. They also supply electricity to some 12,000 homes worldwide, particularly in isolated areas. Solar cells are also useful in many situations where conventional batteries are used at the moment. A solar cell can be used as a battery charger that works merely by exposing it to light. Thus solar cells are common and convenient as battery chargers on sailboats, in recreational vehicles, and in outdoor lighting.

Before photovoltaic electricity can become widespread, the technology for making silicon wafers must be improved to reduce cost. Some analyses show that photovoltaics could produce about one quarter of the world's electricity at reasonable cost by 2050. If this is the case, any country producing photovoltaic cells would have a huge worldwide market to exploit. Since 1981 the United States has cut the budget for federal research into solar energy sharply, while the Japanese have tripled their expenditure on improving photovoltaic technology. It seems rash for the United States to write off solar energy as a future energy source at a time when much of the technology still remains to be explored.

SUMMARY

Living organisms require energy in order to maintain their chemical composition, move, repair damage, grow, and reproduce. Energy cannot be created nor destroyed. However, each time energy is converted from one form to another, some useful energy is degraded into an unusable form. To remain alive, organisms must constantly acquire fresh supplies of energy.

The central energy-processing pathways of life are photosynthesis and respiration. In photosynthesis, the sun's energy is captured and stored in the chemical bonds of food molecules, which can later be broken down during respiration to release the trapped energy:

Both photosynthesis and respiration are essentially oxidation-reduction reactions: many steps in each process involve transferring electrons (or hydrogen atoms, which contain electrons). Many of these redox reaction steps release a great deal of useful energy. The ultimate task of both photosynthesis and respiration is to trap this energy and use it to produce energy intermediates, which can then drive endergonic reactions.

The most common energy intermediate is ATP. Gradients of ions across membranes also serve as energy intermediates for some energy-requiring processes. In fact, most ATP is formed by using the energy of such a gradient to join phosphate groups to ADP.

photosynthesis: $CO_2 + H_2O + energy \longrightarrow (CH_2O)_n + O_2$
respiration: $(CH_2O)_n + O_2 \longrightarrow CO_2 + H_2O + energy$

SELF-QUIZ

1. Does the following equation describe photosynthesis or respiration?

$$CO_2 + H_2O \longrightarrow C_6H_{12}O_6 + O_2$$

2. In the above equation, is carbon oxidized or reduced as a result of the reaction?
3. True or False? The cells of plants contain mitochondria and carry on cellular respiration.
4. ATP is made by joining _____ and _____. This reaction is (endergonic, exergonic). Hence it requires _____ in addition to the reactants.
5. The electron transport system:
   a. makes ATP
   b. contains ATP synthetase
   c. is responsible for separating hydrogen atoms into their components
   d. can work against the second law of thermodynamics
   e. cannot perform redox transfers if the membrane is broken
6. The inner membranes of mitochondria and chloroplasts:
   a. contain channels for movement of H^+
   b. have ATP synthetase enzymes attached
   c. contain molecules of the electron transport system
   d. form closed compartments
   e. all of the above

QUESTIONS FOR DISCUSSION

1. List as many energy-requiring activities carried out by living organisms as you can.
2. Why must photosynthetic plants carry on respiration?
3. Of the light energy reaching the earth from the sun, the earth's plants are believed to convert less than 1% into the form of potential energy stored in the chemical bonds of food molecules. What happens to the rest of the energy?
4. Organisms cannot use heat energy to drive their energy-requiring processes. Does this mean that the heat released by metabolism is of no use to them? Why or why not?

SUGGESTED READINGS

Atkins, P. W. *The Second Law*. Scientific American Books, 1984.

Hinkle, P. C., and R. E. McCarty. "How cells make ATP." *Scientific American,* March 1978. Compares the chemiosmotic mechanism for ATP synthesis in bacteria, mitochondria, and chloroplasts. [Note: to understand this article, you must know that biochemists ignore the outer mitochondrial membrane. When they say, "outside the mitochondrion," they mean "outside the inner membrane."]

Food As Fuel: Cellular Respiration and Fermentation

CHAPTER

7

Flight

Yeast cells

Gymnast

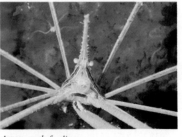

Arrow crab feeding

*L*iving cells constantly carry on energy-requiring activities such as muscle contraction, protein synthesis, active transport, and cell division. The energy to power these processes most often comes from the breakdown of ATP made by the cell. The cell must constantly obtain energy to renew its supply of ATP. It does this by breaking down organic food molecules, usually glucose and other carbohydrates. The energy so released is used to join ADP and inorganic phosphate (P_i) to form ATP. The breakdown of food to release energy for making ATP occurs by two kinds of processes: respiration and fermentation.

Most kinds of cells can carry on **cellular respiration,** the stepwise oxidation of high-energy food molecules to the low-energy molecules carbon dioxide and water. Oxidation is the removal of electrons, and in respiration O_2 accepts the electrons removed from food molecules. The overall equation is:

$$\text{organic molecules} + O_2 \longrightarrow CO_2 + H_2O + \text{energy}$$

The same equation applies to combustion: a fire also uses O_2 in the breakdown of organic molecules—in wood or oil, for example—and produces carbon dioxide and water. A fire also gives off energy, in the form of heat and light. But whereas combustion oxidizes organic molecules all at once, respiration oxidizes food in a series of controlled steps, each releasing a little of the food molecule's energy. This permits cells to capture and store more energy in the form of ATP than they could if the energy were released in one big burst.

Respiration is called an **aerobic** process because it requires molecular oxygen (O_2). Some cells live in **anaerobic** conditions, with little or no O_2 available. Many of these carry out **fermentation**—the breakdown of food molecules using organic molecules instead of inorganic oxygen as the electron acceptors (Figure 7-1). This is less efficient than respiration and yields less ATP.

In this chapter we shall see how cells break down the most commonly used food molecule, glucose, to carbon dioxide and water. Then we shall study two familiar kinds of fermentation. Finally, we shall see how organisms obtain energy from molecules other than glucose—other sugars, and fats and proteins—and how the pathways of respiration connect with other metabolic pathways.

- Cellular respiration breaks down food molecules and releases their stored energy in small steps. Some of this is stored as chemical potential energy in ATP.
- Fermentation makes less ATP than respiration, and breaks food molecules down only part way.
- Other metabolic pathways, such as those for proteins and lipids, feed molecules into the respiratory pathway.

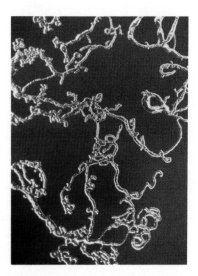

Figure 7-1
Anaerobic organisms. Colonies of tetanus bacteria are strictly anaerobic, obtaining energy by fermenting proteins. (Biophoto Associates)

KEY CONCEPTS

7-A Overview: Breakdown of Glucose under Aerobic Conditions

We begin with a molecule of glucose, a six-carbon sugar. Glucose is broken down to carbon dioxide and water in a long series of energy-releasing reactions. Some of the energy once stored in the glucose molecule's chemical bonds is eventually used to make ATP. Most of the energy transfer occurs as hydrogen atoms (or their electrons) removed from the breakdown products of glucose are passed through a pathway called the electron transport system, in a series of redox reactions (Section 6-D).

Each reaction step is catalyzed by an enzyme. Some of these enzymes require organic molecules called **coenzymes** in addition to their substrates (raw materials). We can think of these coenzymes as carrier or shuttle molecules, because they carry substances from one reaction to another. Many coenzymes

■ *To understand energy flow in the breakdown of food molecules, we must follow the movements of hydrogen atoms and their components— electrons (e^-) and H^+.*

Table 7-1   Coenzymes Used in Respiration

Coenzyme		Function	Made from this vitamin:
Nicotinamide adenine dinucleotide	(NAD^+)	Carry hydrogen atoms to electron transport chain	Niacin
Flavin adenine dinucleotide	(FAD)		Riboflavin
Coenzyme A	(CoA)	Carries acetyl group to citric acid cycle	Pantothenic acid

are made from vitamins in our diets (Table 7-1). Coenzymes last for a long time and can be used over and over again, so they are required only in very small quantities. This is why we need only small amounts of vitamins.

The overall reaction for the complete oxidation of glucose is:

$$C_6H_{12}O_6 + 6\ O_2 \longrightarrow 6\ CO_2 + 6\ H_2O + \text{energy (ATP and heat)}$$

This summarizes a long, complex process, which we can divide into four parts:

1. Glycolysis. Glycolysis breaks down the six-carbon sugar glucose to two three-carbon **pyruvate** molecules. Two important products are made along the way: a small amount of ATP, and hydrogen atoms. The hydrogens are picked up by a coenzyme, **nicotinamide adenine dinucleotide (NAD⁺),** forming NADH.

Glycolysis is an anaerobic process, not actually part of respiration. Respiration consists of three aerobic pathways:

2. Preparation for the citric acid cycle. Each three-carbon pyruvate gives up one carbon atom as CO_2, leaving a two-carbon **acetyl** group. Hydrogens are also released and picked up by coenzyme NAD^+, forming more NADH.

3. Citric acid cycle. The two-carbon acetyl groups left from pyruvate are in effect dismantled to two CO_2 molecules. The cycle's important products are more ATP and a wealth of hydrogen atoms, which are picked up by coenzymes NAD^+ and **FAD (flavin adenine dinucleotide),** forming NADH and $FADH_2$.

4. Electron transport and chemiosmotic ATP synthesis. NADH and $FADH_2$ made in the steps above pass the hydrogens they carry to a system of electron transport molecules, housed in a membrane. Here the hydrogens are separated into H^+ and electrons, and the H^+ is pushed to one side of the membrane, forming an H^+ gradient across the membrane. This gradient serves as an energy source for making ATP from ADP and P_i. Electrons leaving the system join with oxygen, and with still other H^+, to form water.

Inputs	Outputs
Glucose	Pyruvate NADH ATP
Pyruvate	$\begin{cases} CO_2 \\ \text{Acetyl CoA} \\ NADH \end{cases}$
Acetyl CoA	CO_2 ATP NADH $FADH_2$
$\left.\begin{matrix} NADH \\ FADH_2 \end{matrix}\right\} H^+$ gradient $\longrightarrow$ ATP O_2	H_2O

Figure 7-2 outlines the events in the aerobic breakdown of glucose to carbon dioxide and water.

7-B   Glycolysis

The anaerobic process of glycolysis makes a small amount of ATP, and some organisms use glycolysis to make all their ATP when oxygen is unavailable. Many bacteria and fungi can survive indefinitely on the ATP made during glycolysis alone. However, most of these organisms, and all so-called "higher" organisms, also possess the citric acid cycle and electron transport pathways, and use them

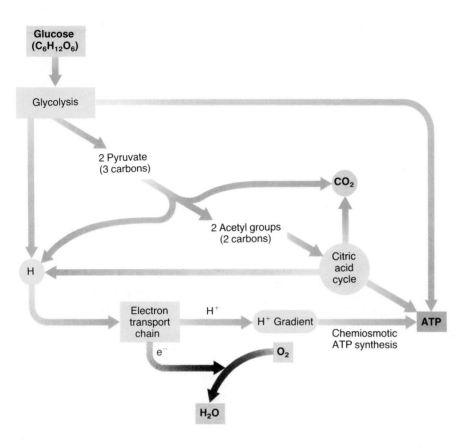

Figure 7-2
Summary of the breakdown of glucose under aerobic conditions. The raw materials are glucose and oxygen (pink boxes). The end products are the low-energy molecules carbon dioxide and water, and molecules of ATP, which store much of the energy released from glucose. Glycolysis and the citric acid cycle make some ATP directly. They also send hydrogen atoms to the electron transport chain. Electron transport sets up a hydrogen ion gradient, which drives the chemiosmotic ATP synthesis that makes most of the ATP derived from glucose.

to make ATP when oxygen is available. In organisms with these respiratory pathways, the main function of glycolysis is to produce three-carbon pyruvate molecules, which can be prepared for the citric acid cycle.

The enzymes of glycolysis are found in the cytoplasm. Interestingly, the first half of the pathway of glycolysis actually *uses* energy—two molecules of ATP. Each ATP donates a phosphate group to the six-carbon glucose, which then breaks into two three-carbon molecules, each with a phosphate group (Figure 7-3). The cell must make this early "investment" of ATP to start a project that will eventually bring in an ATP "profit."

In the second half of glycolysis, two more phosphate groups are added, one to each three-carbon molecule. These phosphates do not come from ATP, but are simply inorganic phosphate (P_i) from the cytoplasm. At the same time, each three-carbon molecule also gives up hydrogen to NAD^+. Each NAD^+ picks up one hydrogen atom and the electron of another, leaving the remaining H^+ in solution. The reduced form of NAD^+ may be written NADH (more completely, $NADH + H^+$). The two NADH molecules from glycolysis carry the first hydrogens we must follow to their energy destination.

We now have two three-carbon molecules, each containing two phosphate groups. These molecules donate their phosphate groups to ADP, forming a total of four ATPs. Two three-carbon molecules of a compound called pyruvate emerge from glycolysis.

Figure 7-3
The important events of glycolysis. In the first series of reactions, glucose receives two phosphate groups from ATP and is broken into two three-carbon molecules, a net energy expenditure of two ATP. The rest of glycolysis recoups this expenditure and releases additional energy for use by the cell. Four ATP are made (for a net gain of two ATP in all of glycolysis), plus two hydrogen-laden coenzymes (NADH), which will yield more energy when the hydrogens are passed to the electron transport chain.

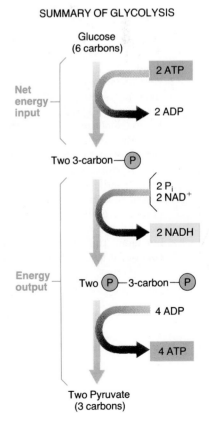

SUMMARY OF GLYCOLYSIS

Glucose
(6 carbons)

Net energy input

2 ATP

2 ADP

Two 3-carbon—(P)

2 P_i
2 NAD^+

2 NADH

Energy output

Two (P)—3-carbon—(P)

4 ADP

4 ATP

Two Pyruvate
(3 carbons)

The Chemical Reactions of Glycolysis

This box presents glycolysis in some detail as an example of the sorts of changes that occur from one step to the next in a metabolic pathway. Follow the steps of glycolysis in the molecular diagrams as we discuss them. The arrows are numbered to correspond with the numbers in the text, and the names beside the arrows are the enzymes that catalyze each step.

1. In the first step of glycolysis, an ATP is actually used up. This is like spending "seed money"; some ATP energy is "invested" to start a series of energy-releasing reactions that will ultimately yield an ATP "profit." A phosphate group from ATP is attached to the sixth carbon atom of glucose, forming glucose-6-phosphate. This reaction activates glucose by transferring some energy to the molecule. In addition, the negatively charged phosphate group traps the glucose molecule inside the cell. It also provides a recognition site that binds to the enzyme which catalyzes the next step of glycolysis. Other carbohydrates, such as glycogen, sucrose, fructose, and galactose, can be converted into glucose-6-phosphate and enter glycolysis at this point.

2. Next the glucose-6-phosphate molecule is rearranged, forming another six-carbon sugar, fructose-6-phosphate. Notice that this leaves the molecule's first carbon poking up from the sugar's ring structure.

3. In a second energy-investing reaction, another ATP donates a phosphate group, which is attached to the newly exposed carbon atom. This produces a molecule of fructose-1,6-bisphosphate.

Figure 7-A
The reactions of glycolysis.

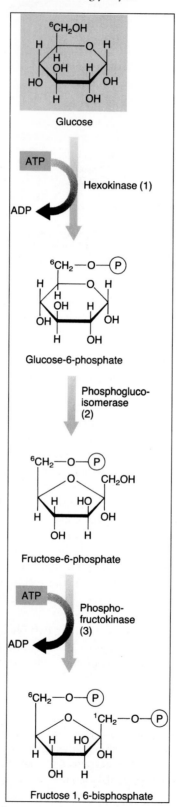

Glucose

ATP → Hexokinase (1) → ADP

Glucose-6-phosphate

Phosphogluco-isomerase (2)

Fructose-6-phosphate

ATP → Phospho-fructokinase (3) → ADP

Fructose 1, 6-bisphosphate

4. Fructose bisphosphate is split into two three-carbon molecules, each with a phosphate group attached to one end. One of these three-carbon molecules, dihydroxyacetone phosphate, is eventually converted into the same form as the other, phosphoglyceraldehyde (PGAL). PGAL is the form used as a substrate by the next enzyme of glycolysis.

   Up to this point, the reaction sequence has *used* energy, in the form of two ATPs. Beginning with the next step, the cell begins to extract its energy profit.

5. Two PGAL molecules are oxidized, and each receives a phosphate group. This requires two inorganic phosphate groups (P_i) that have been floating around in the cytoplasm and two molecules of the coenzyme NAD^+. A phosphate group is added to each PGAL molecule, and two hydrogen atoms are removed, one from PGAL and one from P_i. These two hydrogens reduce NAD^+ to $NADH + H^+$. The product of this reaction is 1,3-diphosphoglycerate. Breaking the bond written as a squiggle ($\sim$) is a highly exergonic reaction that yields enough energy to make ATP from ADP and P_i.

6. In the next step, these newly added phosphate groups are broken off, furnishing the energy to attach them to two ADP molecules. This forms ATP directly—that is, without going through the steps of the electron transport chain and chemiosmotic ATP synthesis.

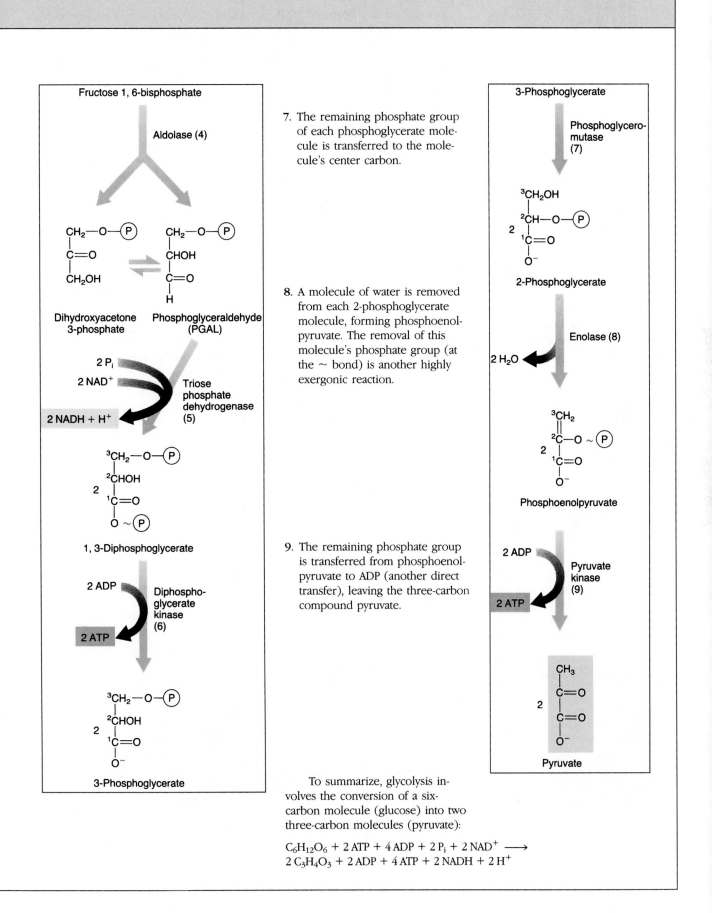

7. The remaining phosphate group of each phosphoglycerate molecule is transferred to the molecule's center carbon.

8. A molecule of water is removed from each 2-phosphoglycerate molecule, forming phosphoenolpyruvate. The removal of this molecule's phosphate group (at the ~ bond) is another highly exergonic reaction.

9. The remaining phosphate group is transferred from phosphoenolpyruvate to ADP (another direct transfer), leaving the three-carbon compound pyruvate.

To summarize, glycolysis involves the conversion of a six-carbon molecule (glucose) into two three-carbon molecules (pyruvate):

$$C_6H_{12}O_6 + 2\,ATP + 4\,ADP + 2\,P_i + 2\,NAD^+ \longrightarrow$$
$$2\,C_3H_4O_3 + 2\,ADP + 4\,ATP + 2\,NADH + 2\,H^+$$

The important products of glycolysis are:

1. **Energy in the form of ATP.** For each glucose molecule broken down, glycolysis produces a net energy gain of two molecules of ATP (four ATPs have been formed but two have been used up).
2. **Energy in the form of electrons and hydrogen carried by NADH.** Under aerobic conditions, the two NADHs from glycolysis can pass their electrons and hydrogen to the electron transport system, where energy can be extracted to produce more ATP.
3. **Pyruvate.** Two pyruvates are formed from each glucose that entered glycolysis. Each three-carbon pyruvate can enter a mitochondrion. Here it will be modified to form a two-carbon acetyl group for the citric acid cycle, where further energy is extracted from these fragments of the original glucose molecule.

7-C   Mitochondria: ATP Factories

■ *A eukaryotic cell's mitochondria carry out respiration and make most of the cell's ATP.*

Glycolysis occurs in the cytoplasm. In eukaryotes, pyruvate and NADH formed during glycolysis enter mitochondria, where respiration occurs. The structure of the mitochondrion plays an important role in carrying out the chemical processes of respiration.

A mitochondrion is separated from the cytoplasm by its outer membrane. Its highly folded inner membrane forms a closed compartment containing a protein-rich solution, the **mitochondrial matrix.** Here lie many enzymes of the citric acid cycle, and the rest are attached to the inner face of the inner membrane. This membrane also contains the electron transport molecules, which create an H^+ gradient by dumping H^+ ions outside the membrane. In addition, the inner membrane contains ATP synthetase enzymes, which make ATP (Figure 7-4). In aerobic bacteria, which lack mitochondria, the plasma membrane carries out these functions.

7-D   Respiration

Figure 7-4
Mitochondrial structure. (a) Electron micrograph of a section of a mitochondrion from a plant cell, and (b) a three-dimensional cutaway drawing of the same mitochondrion. The inner membrane shows the typical deep folds. (In mitochondria of animals the inner membrane usually forms folds like long, thin tubes [see Figure 5-11].) (a, Biophoto Associates)

Respiration is a series of redox reactions, using oxygen as a final electron acceptor, that break down organic molecules and release their energy. Respiration occurs inside the mitochondria of eukaryotes—both plants and animals—and in the cytoplasm of bacteria.

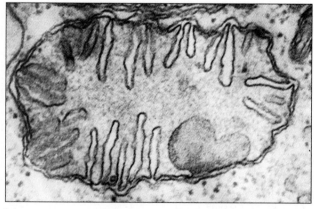

(a)

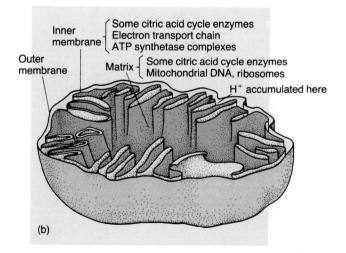

Inner membrane { Some citric acid cycle enzymes
Electron transport chain
ATP synthetase complexes

Outer membrane

Matrix { Some citric acid cycle enzymes
Mitochondrial DNA, ribosomes

H^+ accumulated here

(b)

Preparation of Acetyl Coenzyme A

We left two pyruvate molecules, formed by glycolysis in the cytoplasm, just as they entered a mitochondrion. Here a large complex of enzymes carries out a series of reactions on each pyruvate molecule (Figure 7-5):

1. The pyruvate loses one carbon and two oxygens: a carbon dioxide molecule.
2. The remaining two-carbon acetyl group is attached to a coenzyme A (CoA) molecule, forming **acetyl CoA.**
3. Meanwhile, a molecule of NAD$^+$ is reduced to NADH.

The acetyl group, now attached to CoA, is ready to enter the citric acid cycle.

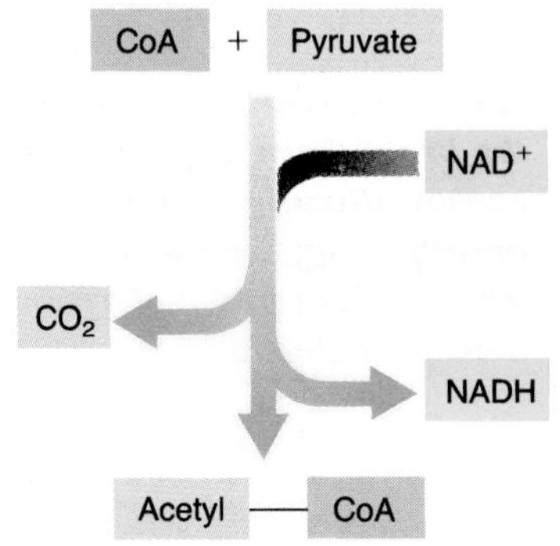

Citric Acid Cycle

The citric acid cycle is also called the **Krebs cycle,** after Sir Hans Krebs, who worked out the cycle in 1937 and received a Nobel Prize for this work in 1953.

Here we summarize the events of the citric acid cycle. Coenzyme A transfers its two-carbon acetyl group to a four-carbon molecule, oxaloacetic acid, forming a six-carbon compound, citric acid. (Coenzyme A goes back to pick up another acetyl group.) The six-carbon molecule gives up two of its carbons, in the form of carbon dioxide. The remaining four-carbon compound is eventually converted into a new molecule of oxaloacetic acid, ready to accept another two-carbon acetyl group from acetyl coenzyme A (Figure 7-6).

Figure 7-5
Preparation of acetyl CoA. Pyruvate loses one of its carbons and two oxygens, which leave as a molecule of carbon dioxide (CO_2). It also gives up hydrogen to NAD$^+$, forming NADH. The two-carbon acetyl group remaining is attached to coenzyme A, forming acetyl CoA, which proceeds to the citric acid cycle.

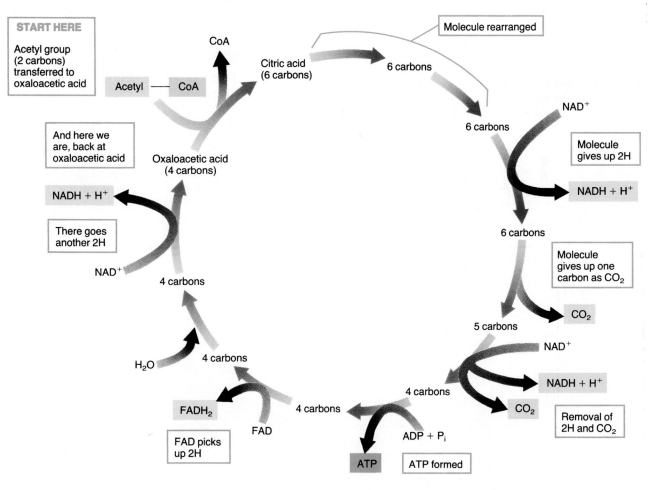

Figure 7-6
The citric acid cycle. Begin at "START HERE," in the top left-hand corner, and follow the cycle clockwise, reading each event in turn (blue boxes).

The other important events of the cycle include:

1. Hydrogen atoms are removed at various stages of the cycle and picked up by coenzymes NAD$^+$ and FAD, forming NADH and FADH$_2$. In terms of energy, this is the most important outcome of the cycle, because these hydrogens later pass to the electron transport system and power the formation of most of the ATP derived from the original glucose molecule.

2. One molecule of ATP is formed directly during each turn of the cycle.

The citric acid cycle may be summarized (the equation's products are set directly under the corresponding reactants so you can compare them easily):

oxaloacetic acid + acetyl   CoA + ADP + P$_i$ + 3 NAD$^+$ + FAD $\longrightarrow$
oxaloacetic acid + 2 CO$_2$ + CoA + ATP   + 3 NADH + 3 H$^+$ + FADH$_2$

Each glucose molecule gives rise to two pyruvates, each of which yields one acetyl group to enter the citric acid cycle. Therefore, it takes two turns of the cycle to break down the remains of one molecule of glucose. At the end of these two cycles, the equivalents of all six carbons from the original glucose have been released as carbon dioxide. For each of the two pyruvates processed, one CO$_2$ is formed when pyruvate is converted to an acetyl group, and the other two are formed in the citric acid cycle. This carbon dioxide leaves the cell and enters the blood. When it reaches the lungs, we breathe it out as a waste product.

The citric acid cycle is also an important metabolic exchange; it can be compared to a traffic circle, where molecules entering from one metabolic pathway can be channelled into another according to the cell's needs (Section 7-F).

Electron Transport and Chemiosmotic ATP Synthesis

Our molecule of glucose has now been completely dismantled. Some of its energy has made ATP during glycolysis and the citric acid cycle. However, most of the energy remains in the electrons carried by coenzymes NADH (from glycolysis, formation of acetyl CoA, and the citric acid cycle) and FADH$_2$ (from the citric acid cycle).

Now these coenzymes are oxidized by passing their electrons (as part of hydrogen atoms) to the **electron transport chain,** a series of carrier proteins in the inner mitochondrial membrane. As the electrons pass down the chain in a series of redox reactions, their energy is gradually released and used to transport H$^+$ out through the membrane. The gradient of H$^+$ so created then pro-

■ *The citric acid cycle completes the breakdown of glucose to carbon dioxide, which is breathed out, and hydrogen, which is sent to the electron transport chain.*

Figure 7-7
Simplified scheme of electron transport and chemiosmotic ATP synthesis in a mitochondrion. At left, NADH + H$^+$ passes two hydrogen atoms (2H$^+$ + 2e$^-$) to the electron transport chain. The electrons make three round trips through the membrane, with 2H$^+$ being dropped off outside the membrane each time the hydrogen atoms reach its outer face. On trips two and three, H$^+$ derived from dissociation of water replaces the original H$^+$ from NADH + H$^+$. At the end of the chain, the spent electrons are accepted by oxygen to form water. H$^+$ from the gradient moves through a protein channel where its energy is used by the enzyme ATP synthetase to join ADP and P$_i$, forming ATP.

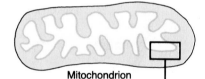

Mitochondrion

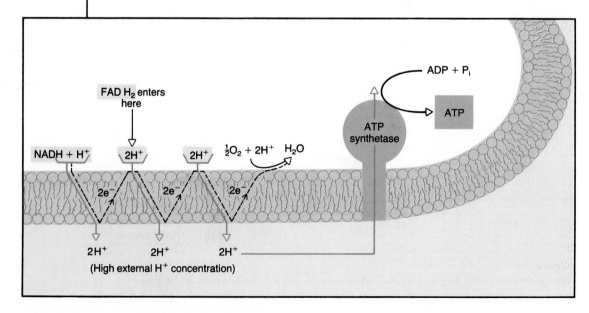

vides energy to join ADP and P_i, forming ATP. Meanwhile, coenzymes NAD^+ and FAD can go back and pick up more hydrogens.

Some of the electron transport molecules carry hydrogen atoms, whereas others carry only electrons. The first molecule in the chain lies at the inner surface of the inner mitochondrial membrane. This molecule accepts two hydrogen atoms from $NADH + H^+$ and transports them through the membrane to the outer surface, where they pass to the next transport molecule. However, this next molecule can carry only electrons, so the H^+ are stranded outside the membrane without a ride back in. Meanwhile, the electrons are carried back through the membrane, and so they cannot recombine with the H^+ (Figure 7-7).

Once back at the membrane's inner face, the electrons are picked up by still another molecule, one that transports entire hydrogen atoms. This molecule gets complete atoms by adding two new H^+ to the two electrons. (These H^+ come from dissociation of H_2O in the mitochondrial matrix into H^+ and OH^-.)

Depending on the original source of hydrogens, the electrons make a total of two or three round trips through the membrane, accompanied by H^+ on the outward journeys only. The significance of all this shuttling of electrons back and forth is that it results in the buildup of a high concentration of H^+ outside the membrane. For each original glucose molecule, a grand total of 64 H^+ is exported to the outside of the inner mitochondrial membrane.

We finally come to the role of oxygen in cellular respiration. Oxygen is breathed into the lungs. Here it enters the blood, which carries it to the body's cells. Oxygen diffuses into a cell and on into a mitochondrion, where it takes part in the last step of respiration.

Electrons arriving at the end of the electron transport chain have given up most of their energy in the redox reactions of the chain. Oxygen takes these spent electrons, plus some H^+ (again, from dissociation of water), forming a new water molecule:

$$\tfrac{1}{2}O_2 + 2\,e^- + 2\,H^+ \longrightarrow H_2O$$

Without oxygen, the last molecule in the electron transport chain would have no way to release electrons. It would be stuck holding electrons, and so it could not accept the next pair of electrons passed down the chain. Its neighbor would have to hold those electrons, and so on: the whole chain would soon be stopped up, and we could not extract further energy from our food by respiration.

The electron transport chain lies in the inner membrane of the mitochondrion. H^+ accumulates outside this membrane, storing energy in the form of an electrical membrane potential. This energy can then be used to make ATP.

In **chemiosmotic ATP synthesis,** H^+ moves inward through the membrane by way of special channels associated with **ATP synthetase** enzymes. These enzymes use the energy of the H^+ moving down its concentration gradient to make ATP from ADP and P_i (see Figure 7-7). Each time ATP is made in this way, the H^+ gradient loses some of its stored energy. The gradient is constantly renewed by the continuous flow of electrons arriving via NADH and $FADH_2$ and the electron transport chain.

Sometimes the lipid layers of the inner mitochondrial membrane contain **uncouplers:** substances that permit H^+ to leak through the membrane without going through the channels where their energy can be used by ATP synthetase. This dissipates the H^+ gradient without using its energy to make ATP. If too much H^+ escapes from the gradient without doing work, a well-nourished cell can literally starve to death. Dinitrophenols, yellow substances once used as food additives to make baked goods look as if they contained more eggs than they really did, are such uncouplers. They were prescribed for a time as a cure for obesity, but were abandoned after several patients ran out of ATP and died during treatment. Another uncoupler is the antibiotic gramicidin.

(a)

(b)

Figure 7-8
Organisms use ATP from respiration for many activities. (a) Fireflies produce flashes of light by reactions using ATP and oxygen. The flashes attract mates, a necessary prelude to reproduction. (b) Fruits, such as these raspberries, carry out rapid respiration as they ripen. This provides ATP for the chemical processes that make the fruit softer, sweeter, and more colorful. (a, Biophoto Associates)

■ *The oxygen we breathe is crucial to life because it is the final electron acceptor for the electron transport chain.*

Uncoupling also occurs in the brown fat cells of hibernating animals. These animals require less ATP than usual. However, they do need heat to stay alive, even though their body temperature is permitted to drop. The mitochondrial membranes in brown fat cells contain uncoupling proteins that allow H^+ to flow back through the membranes without producing ATP. The energy is dissipated as heat instead of being used to make unneeded ATP.

Electron transport and chemiosmotic ATP synthesis can be summarized:

1. Energy is stockpiled in the form of an H^+ gradient across the inner mitochondrial membrane, via a series of oxidation-reduction reactions. The electron transport chain accepts hydrogens from NADH and $FADH_2$ (formed during glycolysis, preparation of acetyl CoA, and the citric acid cycle). The chain separates the hydrogens into H^+ and electrons, and uses the energy released during the redox reactions of electron transport to expel the H^+ across the membrane. The spent electrons emerging at the end of the chain are accepted by oxygen, which also picks up H^+, to form water.
2. The energy of the H^+ gradient is used to join ADP and P_i to form ATP. This happens as H^+ re-crosses the membrane via special protein channels connected to ATP synthetase enzymes.

The Energy Yield of Glucose

How much ATP has the cell obtained from our original glucose molecule? We have seen that glycolysis produces four ATP but uses up two ATP to start the process. So the net gain from glycolysis is two ATP per glucose broken down. The citric acid cycle produces another two ATP, for a total of four ATP made directly per glucose.

The electrons passed from one NADH through the electron transport chain to an oxygen atom provide enough energy to form up to three ATP. Electrons donated by $FADH_2$ provide energy to make up to two ATP. So we can calculate the maximum possible number of ATPs derived from one glucose molecule:

Table 7-2   Maximum ATP Yield from One Molecule of Glucose

Process	Number of reduced hydrogen carriers	Number of ATP per carrier	Number of ATP by chemiosmotic synthesis	Number (net) of ATP made directly
Glycolysis	2 NADH + H^+	3	6 ATP	
				2 ATP
Preparation of acetyl CoA	2 NADH + H^+	3	6 ATP	
Citric acid cycle	6 NADH + H^+	3	18 ATP	
	2 $FADH_2$	2	4 ATP	
				2 ATP
			34 ATP	4 ATP
		GRAND TOTAL:	38 ATP	

7-E   Fermentation

Cellular respiration demands a constant supply of oxygen, to accept electrons from the electron transport chain. If the cell runs out of oxygen, all the transport molecules are soon stuck holding electrons and the chain stops working. The H^+ gradient is quickly used up and then is no longer available to drive ATP

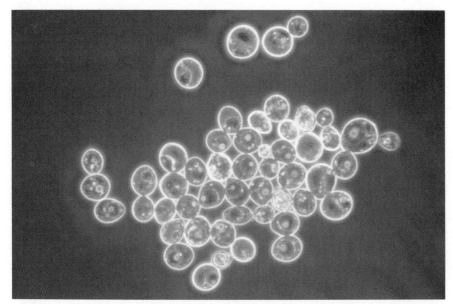

Figure 7-9
Yeasts are unicellular fungi. Those shown here are reproducing rapidly. Small cells bud off from larger ones and may grow so fast that they bud again before they have completely separated from their parents. (Biophoto Associates)

synthesis. The citric acid cycle also backs up and stops. In this situation, many cells can obtain small amounts of ATP by the anaerobic process of **fermentation.** Indeed, some bacteria meet all their energy needs in this way.

We shall examine two kinds of fermentation. Both use the familiar pathway of glycolysis, which makes a net of two ATP in the breakdown of each glucose molecule to two pyruvates (see Figure 7-3).

What happens to glycolysis in the absence of O_2? NAD^+ must be converted to NADH before any ATP can be made (see pp. 114–115). But when electron transport stops for lack of oxygen, NADH cannot pass electrons on to the chain. Cells contain very little NAD^+, and all of it will soon be converted to NADH. Without NAD^+ to go back and accept more hydrogen, glycolysis too will grind to a halt.

In fermentation, pyruvate produced by glycolysis serves as an alternative acceptor of hydrogen from NADH. This frees NAD^+ to go back to glycolysis for more hydrogen, so glycolysis keeps making its small amount of ATP.

Some organisms carry out **alcoholic fermentation,** discovered by Louis Pasteur during his study of the chemistry of wines. In winemaking, juice pressed from grapes is inoculated with yeasts, which are unicellular fungi (Figure 7-9). Yeasts break down the sugars in the juice to pyruvate by glycolysis. Then each pyruvate molecule is dismantled into a molecule of carbon dioxide and a molecule of the two-carbon compound acetaldehyde. Acetaldehyde is next reduced by accepting two hydrogens from $NADH + H^+$, forming the two-carbon alcohol ethanol, also called ethyl alcohol—the active ingredient in alcoholic beverages (Figure 7-10). This transfer of hydrogen frees NAD^+ to go back to glycolysis and pick up more hydrogen, allowing the yeast cell to keep making ATP.

If fermentation continues until the yeast cells have used up all the sugar around them, a dry wine results. Sweet wines still contain sugar because the yeast cells produced enough alcohol to inhibit fermentation before they used up all the sugar. Stoppering a wine bottle before fermentation has finished yields a bubbly liquid because carbon dioxide is still being given off. Such was the young wine that stretched and split the wine skins in the New Testament story. To make a fizzy wine like champagne you use a very strong bottle and cork it up before fermentation has finished. The carbon dioxide dissolves in the wine under pressure and is released as bubbles when the bottle is opened.

Yeasts produce alcohol only when little or no oxygen is present. (Given plenty of oxygen, they use respiration to break sugar down completely to carbon dioxide and water.) When wine is fermenting rapidly, it produces carbon dioxide fast enough to drive off the air above the wine, and this prevents oxygen

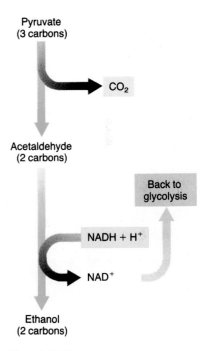

Figure 7-10
Alcoholic fermentation as performed in yeast. Pyruvate produced in glycolysis first gives off a molecule of carbon dioxide, leaving a two-carbon compound, acetaldehyde. This is then reduced by accepting two hydrogens from $NADH + H^+$. The process releases NAD^+, which is needed for glycolysis to continue and produce ATP at a low rate while the cell is deprived of oxygen.

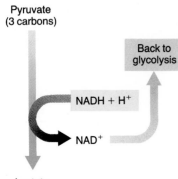

Pyruvate
(3 carbons)

Back to
glycolysis

NADH + H$^+$

NAD$^+$

Lactate
(3 carbons)

Figure 7-11

The conversion of pyruvate to lactate in muscle. When the cell lacks O$_2$, NADH accumulates. Pyruvate produced by glycolysis accepts hydrogens from NADH + H$^+$ and becomes lactate, freeing NAD$^+$ to return to glycolysis and accept more hydrogens.

from dissolving in the wine. But when fermentation slows down, wine must be sealed up immediately, before oxygen can enter.

Alcoholic fermentation by yeasts is also used to produce ethanol for fuels. Currently many researchers are trying to develop yeasts that convert more of the sugars in corn or other plant materials to ethanol.

A second familiar kind of fermentation occurs in muscles during strenuous exercise. The muscle cells make ATP by respiration, using oxygen as fast as the bloodstream can deliver it, but this does not provide enough ATP. So, the muscles make a small additional supply of ATP by fermentation. Pyruvate from glycolysis is reduced by accepting hydrogens from NADH + H$^+$ and so is converted into another three-carbon compound, lactate (Figure 7-11). Again, this frees NAD$^+$ to pick up more hydrogen, allowing glycolysis to proceed.

Lactate produced in the muscles is picked up by the blood. However, the body cannot tolerate a great buildup of lactate, but must eventually oxidize lactate back to pyruvate. After strenuous exercise, we still keep breathing heavily for a time, to repay the body's **oxygen debt.** The liver takes up lactate and converts it back to pyruvate. With oxygen now available, some of the pyruvate is broken down to yield energy via the citric acid cycle and electron transport chain. This provides the energy to convert the remaining pyruvate back to glucose by reverse glycolysis. The glucose re-enters the blood and is carried back to the muscles, where it is converted to glycogen and stored.

Fermentation producing lactate as an end product is also carried out by various bacteria. Some of these are cultured and used to produce fermented food products such as cultured buttermilk and sour cream, yogurt, sauerkraut, pickles, green olives, and some sausages.

Most cells can carry on glycolysis, but not all organisms can carry on respiration. For this reason, and because the atmosphere of the earth probably contained little oxygen when life first evolved, it is generally believed that glycolysis is the more primitive pathway of food breakdown and that aerobic pathways evolved later.

Pasteur and Yeasts

A JOURNEY
INTO LIFE

The word enzyme means "in yeast." For economic reasons, most of the early studies on enzymes and their actions were attempts to understand the alcoholic fermentation by which wine is made. As early as 1785 the Academy of Florence offered a prize for a theory of fermentation that could be applied to keeping wine in better condition while it was transported. However, no real light was

shed on the subject until the French wine industry asked Louis Pasteur to investigate the condition called *"l'amer"* that destroyed large quantities of the best Burgundy every year.

From his experiments, Pasteur concluded that fermentation occurred only when living yeast was present. Justus von Liebig, an influential chemist, thought otherwise and performed many experiments in which he killed yeast cells by boiling them and then tested them to see if they would ferment sugar. They would not, and enzymes, also called "ferments," came to be considered as catalysts that would not function outside a living cell. We now know that enzymes *can* function outside cells. Liebig, in boiling the yeast cells, had not only killed them but also destroyed the tertiary structure (three-dimensional folding pattern) of their enzymes so that they no longer functioned.

Pasteur discovered that *l'amer,* which turned wine sour, was caused

by bacteria. Microscopic examination showed that the wine turned sour when it contained more bacteria than yeast cells. *L'amer* could be prevented by excluding most bacteria. This could be accomplished through greater cleanliness, including sulfur sterilization, which is now standard practice in many winemaking steps. Pasteur also showed why it is important to exclude air during fermentation: wine yeasts produce alcohol under anaerobic conditions, but if oxygen is present, other yeasts and bacteria that convert alcohol into acetic acid (CH$_3$CH$_2$OH $\longrightarrow$ CH$_3$COOH) will turn the wine to vinegar.

Pasteur loved good wines and devoted many years to studies of their fermentation and aging. His book *Etudes sur le Vin,* published in 1866, revolutionized winemaking, giving it a scientific basis for the first time.

7-F Alternative Food Molecules

Presented with a smorgasbord of commonly available food molecules, most cells use glucose to make ATP. However, since all organic molecules contain stored energy, any of them may be broken down to release the energy needed to make ATP.

Many carbohydrates other than glucose are processed by way of glycolysis. Polysaccharides, from food or the body's glycogen stores, can be broken down to glucose. Sugars other than glucose can be converted to glucose or fructose and fed into glycolysis. Thus claims that other "natural" sugars are less fattening than sucrose or glucose are untrue; a cell treats them all alike. However, it is true that sucrose (table sugar) is worse than other sugars for your teeth. Bacteria that cause tooth decay can use sucrose, but not other sugars, to make a "glue" by which the bacteria stick to tooth surfaces.

Fatty acids in the food or stored in body fat do not go through glycolysis, which breaks down only carbohydrates. Instead, they enter respiration at the point where acetyl CoA forms. Fatty acids are broken down into two-carbon acetyl groups, which combine with coenzyme A and enter the citric acid cycle. Here their hydrogens are stripped off and carried to the electron transport chain. Fatty acids are built by the reverse process, linking two-carbon acetyl units, and this is why the vast majority of them contain an even number of carbon atoms.

Protein that is already part of the body is not used for energy except during advanced starvation, after the body's carbohydrate and fat reserves have been depleted. Proteins in food are broken down into amino acids, and any amino acids not needed to build new proteins are **deaminated** by the removal of their amino groups. Depending on its structure, the rest of the molecule is converted into pyruvate, acetyl CoA, or one of the molecules of the citric acid cycle. It then enters the pathway at the appropriate place.

In summary, the citric acid cycle and electron transport chain are a final common pathway for the breakdown of just about any organic molecule to yield energy. Figure 7-12 shows how protein, carbohydrate, and fat pathways are linked and shows that many of these pathways also work in reverse. Many citric acid cycle molecules also take part in various other metabolic pathways. When they can be spared from respiration, these molecules may be diverted from the citric acid cycle into body-building processes. Note especially that pathways lead from both carbohydrates and proteins to fat deposits, by way of acetyl CoA.

■ *Any excess food in the diet, whether carbohydrate, protein, or fat, can end up as stored fat, a "savings deposit" of energy.*

Table 7-3 shows the energy yield per unit of weight for carbohydrates, fats, and proteins. Weight for weight, fats contain more than twice as much energy as carbohydrates or proteins. This is because fats contain a higher hydrogen: oxygen ratio than do carbohydrates or proteins. Most of the energy from food molecules is extracted by passing the electrons from hydrogen atoms along the electron transport chain. So, the higher the proportion of hydrogen, by weight,

Table 7-3 *Energy Yield of Major Food Components*			
Class	*Composition*	*Energy yield*	*Storage*
Carbohydrates	CH_2O	4 kcal/gram	Hydrated: attracts much water
Fats	CHO	9 kcal/gram	Hydrophobic: concentrated fat droplets
Proteins	CHON(S)	~4 kcal/gram	None

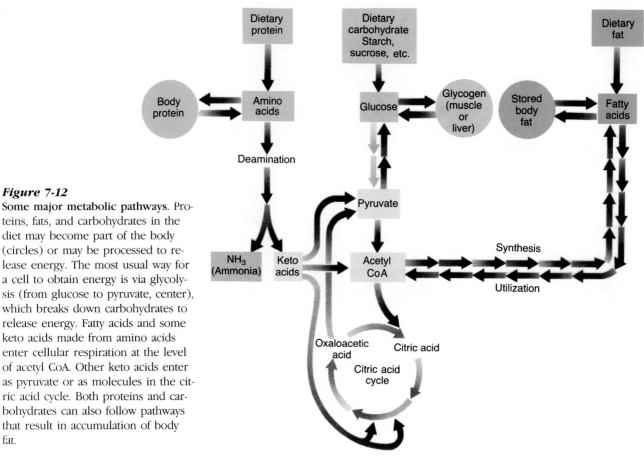

Figure 7-12
Some major metabolic pathways. Proteins, fats, and carbohydrates in the diet may become part of the body (circles) or may be processed to release energy. The most usual way for a cell to obtain energy is via glycolysis (from glucose to pyruvate, center), which breaks down carbohydrates to release energy. Fatty acids and some keto acids made from amino acids enter cellular respiration at the level of acetyl CoA. Other keto acids enter as pyruvate or as molecules in the citric acid cycle. Both proteins and carbohydrates can also follow pathways that result in accumulation of body fat.

in the molecule, the more energy is stored in a given weight of that substance.

Fats are nonpolar, and so they tend to repel water and arrange themselves as concentrated fat droplets. The nonpolar nature and high energy content of fats make it advantageous for animals to store most of their energy reserves as fat (Figure 7-13a). However, a small amount of carbohydrate, in the form of glycogen in the liver and muscles, is kept as a short-term reserve that can be used as needed for a quick energy boost.

Figure 7-13
Food storage. (a) Adipose (fat-storing) cells. Each cell is occupied almost entirely by colorless fat, with the nucleus pushed against the plasma membrane at one side. (b) Starch, stained purple, is stored in numerous amyloplasts in these cells from a buttercup root. (a, Biophoto Associates; b, Ed Reschke)

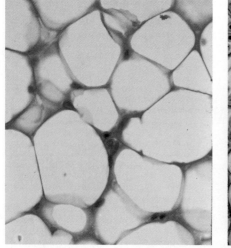
(a)

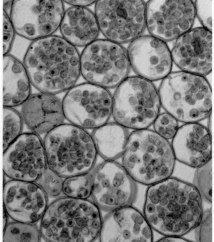
(b)

Plants store most of their energy reserves in the form of carbohydrates, especially starch (Figure 7-13b). The —OH groups of carbohydrates make them polar, and so they attract and hold much water around them. This makes them heavy and bulky to store, but since plants do not have to move around and drag this food supply along, the weight does not inconvenience them as it would an animal. Carbohydrates are also much easier to mobilize into the energy-release pathways than are fats. (This is one reason why it is so hard to lose fat weight.) These properties are probably the reason that plants store their energy reserves in the form of carbohydrate.

SUMMARY

Cellular respiration is the process that extracts useful energy from the energy stored in the chemical bonds of food molecules. This is done in a series of redox reactions, using oxygen as the final electron acceptor. The energy so released is used to make the cell's supply of ATP. ATP, in turn, donates the energy to various energy-requiring processes, such as metabolic reactions, active transport, muscle contraction, or production of new polymers.

During glycolysis, glucose is broken down anaerobically to two molecules of pyruvate, which is processed further in respiration. NAD^+ picks up hydrogens and becomes $NADH + H^+$, which passes two hydrogen atoms to the electron transport chain, where more energy can be extracted. Glycolysis also makes some ATP directly.

Each pyruvate formed during glycolysis loses a carbon dioxide and becomes an acetyl group, which combines with coenzyme A. Coenzyme A transfers the acetyl group to the citric acid cycle. During one turn of the cycle, the equivalents of the acetyl group's two carbons are removed as carbon dioxide, one of the end products of respiration. Some ATP is also produced. NAD^+ and FAD accept hydrogens and become $NADH + H^+$ and $FADH_2$, which carry hydrogens to the electron transport chain.

Most of the ATP derived from respiration is produced by electron transport followed by chemiosmotic ATP synthesis. $NADH + H^+$ and $FADH_2$ (from glycolysis, formation of acetyl CoA, and the citric acid cycle) pass pairs of hydrogen atoms to the electron transport chain. The chain separates hydrogen atoms into H^+ and electrons, and forms a gradient of H^+ that can be used to join ADP and P_i to form ATP. At the end of the chain, the electrons combine with oxygen and H^+ to form water, the other end product of respiration.

Neither glycolysis nor the citric acid cycle requires oxygen directly. However, if a cell has too little of the final electron acceptor, oxygen, most of the NAD^+ in the cell will be tied up as NADH, unable to release its electrons to the electron transport chain. Under such anaerobic conditions, some cells continue to produce ATP during glycolysis by carrying out fermentation. Pyruvate accepts electrons from $NADH + H^+$, forming ethanol or lactate. This releases NAD^+ so that glycolysis can continue. No such mechanism exists for the citric acid cycle, which therefore cannot function under anaerobic conditions.

Many other metabolic pathways feed into glycolysis, the citric acid cycle, and the electron transport chain, enabling cells to use many organic compounds other than glucose as food sources to generate usable energy in the form of ATP.

Table 7-4 *Comparison of Glycolytic Fermentation and Aerobic Breakdown of Glucose*

Feature	Glycolytic fermentation	Aerobic breakdown
Raw materials	Glucose	Glucose, O_2
Final electron acceptor	Pyruvate (from glucose)	O_2
End products	CO_2, ethanol (yeast, etc.) Lactate (muscle, etc.)	CO_2, H_2O
Net energy yield	2 ATP	Up to 38 ATP

SELF-QUIZ

1. NAD$^+$ functions in cell respiration as a(n):
   a. energy intermediate   d. oxidizable substrate
   b. enzyme   e. hydrogen donor
   c. coenzyme
2. Which of the following statements is *not* true?
   a. Most of the ATP in an aerobic cell is formed via the electron transport chain and chemiosmotic synthesis.
   b. In eukaryotes, the chemiosmotic formation of ATP requires that the inner mitochondrial membrane remain intact.
   c. NAD$^+$ is a carrier molecule that travels down the electron transport chain to release ATP.
   d. In eukaryotes, the electron transport chain and the enzymes of the citric acid cycle are located in mitochondria whereas the enzymes of glycolysis are located in the cytoplasm.
   e. The role of oxygen is to act as an acceptor of electrons.
3. The respiratory electron transport chain is found in the _____ membrane in eukaryotes.
4. Give the end products of the following reaction sequences:
   a. glycolysis
   b. the citric acid cycle
   c. yeast fermentation
   d. electron transport chain
   e. lactate fermentation in muscle
5. While a muscle is in the process of reducing an oxygen debt:
   a. lactate is converted into pyruvate
   b. all the NAD$^+$ is in the reduced form
   c. pyruvate is converted into lactate
   d. NADH acts as an oxygen acceptor
6. True or False? Both yeast cells and muscle make two ATP (net) per glucose molecule fermented anaerobically.
7. True or False? Carbohydrate is unnecessary in the human diet since the products of fat and protein breakdown can enter the citric acid cycle to generate energy.

QUESTIONS FOR DISCUSSION

1. The insecticide rotenone inhibits transport in the first part of the electron transport chain. Cyanide inactivates cytochromes, transport molecules near the end of the chain. Explain why these poisons cause death.

2. Why do foods that are rich in fat tend to be expensive (in dollars and cents) compared with carbohydrates?

SUGGESTED READING

Krebs, H. A. "The history of the tricarboxylic acid cycle." *Perspectives in Biology and Medicine* 14:154, 1970. An engaging account of how Sir Hans Krebs worked out the citric acid cycle.

When you have studied this chapter, you should be able to:

1. Name or recognize the necessary raw materials of photosynthesis and the important end products.
2. Name the main photosynthetic pigment and state its function.
3. State which colors of light are most effective in promoting photosynthesis, and explain why.
4. Describe or sketch the structure of a leaf. Show where the chloroplasts are, and explain how the raw materials for photosynthesis arrive there and how the end products leave.
5. Describe or sketch the structure of a chloroplast. Explain where the following are found and the importance of their location to their roles in photosynthesis: chlorophyll, electron transport system, ATP synthetase enzymes, hydrogen ion reservoir, carbon fixation enzymes.
6. Name the raw materials and end products of the energy-capturing reactions and of carbon fixation.
7. State what drives carbon fixation and explain what happens in the C_3 cycle.
8. Summarize the important steps in energy transfer during photosynthesis.
9. List three ecological variants of photosynthesis and explain why each is advantageous to plants growing in certain habitats.

Photosynthesis

CHAPTER

Coconut palm

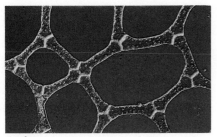

An alga

Aspen leaves

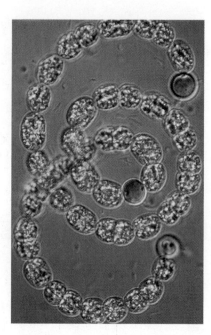

Figure 8-1
A filament of the cyanobacterium *Anabaena*. Cyanobacteria are prokaryotes and hence lack chloroplasts. Their photosynthetic membranes lie in the cytoplasm. The brown cells are not photosynthetic, but convert nitrogen from the air to a form that can be used to make amino acids for proteins. (Biophoto Associates)

KEY CONCEPTS

Figure 8-2
The electromagnetic spectrum. Our eyes can detect light with wavelengths of about 400 to 750 nanometres, and so this range is called visible light. Wavelengths just shorter than visible light are called ultraviolet. Even shorter are x-rays (which overlap with ultraviolet and gamma rays), then gamma rays. Wavelengths on the longer side of the visible spectrum are known as infrared, then microwaves, and radio waves.

A story of a wondrous plant that turns sunlight, air, and water into sugar might seem like a fairy tale, but in fact it's a chapter in a biology textbook. This is the story of **photosynthesis,** the process whereby plants capture the sun's energy and store it in the form of chemical bonds in carbohydrate molecules. The carbohydrates are made by joining carbon dioxide to an organic molecule and then adding hydrogen extracted from water. The green pigment chlorophyll captures light energy to power the process. Photosynthesis consists of a complex series of reactions. The overall reaction can be summarized:

$$CO_2 + H_2O \xrightarrow[\text{chlorophyll}]{\text{light,}} CH_2O + O_2$$

Organisms that carry out photosynthesis according to this equation are the eukaryotic land plants (such as trees and grass) and algae (such as seaweeds) and the prokaryotic cyanobacteria (Figure 8-1). These are not the only photosynthetic organisms. Some bacteria photosynthesize using different pigments and hydrogen sources, and they do not produce oxygen as a by-product. We shall not cover these unusual types of photosynthesis in this chapter.

Photosynthesis occurs in the chloroplasts of eukaryotic plant cells. In prokaryotes, it occurs partly in specialized photosynthetic membranes in the cytoplasm and partly in the cytoplasm itself.

Photosynthesis is vital to life as we know it for two reasons. First, the oxygen in the air today (about 20% of the atmosphere) came from photosynthesis, and photosynthetic plants continue to replenish the oxygen supply. This oxygen is essential for every organism that relies on cellular respiration to break down food and produce ATP. Second, all of our food and the food of almost every other organism comes, directly or indirectly, from photosynthesis. (A few kinds of bacteria obtain their food from other sources [Section 19-B], but these are rare exceptions.) Our dependence on plants for food has stimulated research on how photosynthesis occurs and how we can manipulate plants and their environment to produce more food.

- Photosynthesis makes carbohydrates from carbon dioxide and water, using light energy trapped by chlorophyll.
- The oxygen in the air came from photosynthesis. The vast majority of organisms need this oxygen for respiration.
- All our food, and the food of nearly all other organisms, comes directly or indirectly from photosynthesis.

8-A   Light

To understand how plants convert light energy into chemical energy, we must first consider some properties of light. If sunlight passes through a glass prism, a soap bubble, or a raindrop, it breaks up into a spectrum (a rainbow) of colors.

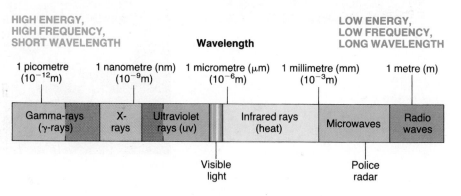

HIGH ENERGY, HIGH FREQUENCY, SHORT WAVELENGTH

Wavelength

LOW ENERGY, LOW FREQUENCY, LONG WAVELENGTH

| 1 picometre $(10^{-12}m)$ | 1 nanometre (nm) $(10^{-9}m)$ | 1 micrometre (μm) $(10^{-6}m)$ | 1 millimetre (mm) $(10^{-3}m)$ | 1 metre (m) |

| Gamma-rays (γ-rays) | X-rays | Ultraviolet rays (uv) | Infrared rays (heat) | Microwaves | Radio waves |

Visible light

Police radar

So visible light, the light that we can see, is actually made up of light of many different colors.

Visible light is only part of the vast range of radiation in the electromagnetic spectrum emitted by the sun (Figure 8-2). Electromagnetic radiation behaves as if it travels in waves, and different parts of the spectrum have different wavelengths. (A wavelength is the distance from the top of one wave to the top of the next.) Electromagnetic radiation ranges from gamma rays and x-rays, with very short wavelengths, through visible light and microwaves, to radio waves with very long wavelengths. Electromagnetic radiation has energy. The shorter the wavelength of the radiation, the higher its energy.

It is not chance that photosynthesis uses visible light rather than any other part of the electromagnetic spectrum. First, visible light is readily available because most of the sunlight reaching the earth's surface is visible light. Second, visible light contains the right amount of energy. Visible light has wavelengths of about 400 to 750 nanometres (a nanometre is a billionth of a metre). Radiation with shorter wavelengths than violet light (about 400 nanometres) contains so much energy that it breaks hydrogen bonds and disrupts the structure of many biological molecules. On the other hand, radiation with wavelengths above about 750 nanometres contains so little energy that it is rapidly absorbed by water before it even gets to a plant's photosynthetic machinery. Low-energy radiation has little effect on the molecules in an organism except to heat them up. Only visible light, with intermediate wavelengths, has enough energy to cause chemical changes without destroying biological molecules.

8-B *Trapping Light Energy: Photosynthetic Pigments*

Before light energy can be used to make food, it must first be absorbed by the plant's photosynthetic pigments. **Pigments** are molecules that appear to be colored because they absorb some wavelengths of light more than others. The photosynthetic pigment **chlorophyll** looks green because it absorbs light of colors other than green, while allowing green light to continue on its way and be reflected or transmitted to our eyes. In plants, chlorophyll occurs in chloroplasts, the organelles of photosynthesis. When you look at a photosynthetic plant cell through a microscope, you can see that only the chloroplasts are green (Figure 8-3).

We can extract pigments from a plant and measure how much light of each wavelength a pigment absorbs. Using these data we can plot the pigment's **absorption spectrum.** The peaks in the graph indicate which wavelengths the pigment absorbs most strongly (Figure 8-4).

In 1883, T. W. Engelmann produced circumstantial evidence that chlorophyll is important in photosynthesis. He used *Spirogyra,* an alga with a long,

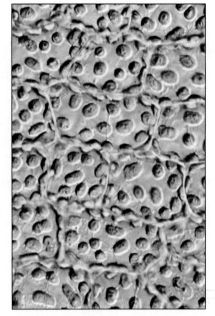

Figure 8-3
Cells from a moss. The green chloroplasts make the entire plant look green. The rest of the cell is colorless (the blue seen here is artificial, and is due to the microscope's optics). (Biophoto Associates)

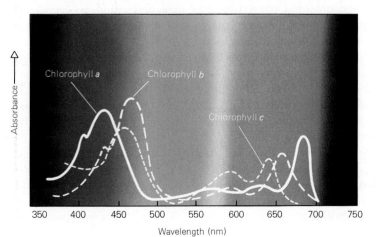

Figure 8-4
Absorption spectra for three chlorophylls. Chlorophyll *a* occurs in all photosynthetic plants, protists, and cyanobacteria. Chlorophyll *b* is found in all land plants and some algae. Chlorophyll *c* occurs in some other algae.

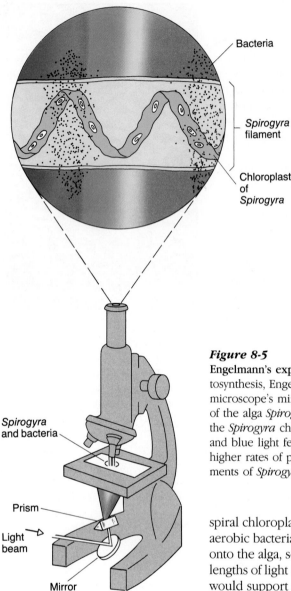

Figure 8-5

Engelmann's experiment. To determine which wavelengths of light best support photosynthesis, Engelmann inserted a prism into the beam of light reflected from the microscope's mirror. The prism dispersed the light into a spectrum, directed at cells of the alga *Spirogyra* on the microscope stage. Oxygen released by photosynthesis in the *Spirogyra* chloroplasts attracted aerobic bacteria, which congregated where red and blue light fell on the alga. Engelmann concluded that red and blue light support higher rates of photosynthesis than do other wavelengths. The photograph shows filaments of *Spirogyra.* (Carolina Biological Supply Company)

spiral chloroplast. He placed the alga in water on a microscope slide and added aerobic bacteria (which require oxygen). Then he passed light through a prism onto the alga, so that different parts of the alga were exposed to different wavelengths of light (Figure 8-5). Engelmann expected that some wavelengths of light would support more photosynthesis than others and that he could detect these wavelengths by the behavior of the bacteria. Since oxygen is given off during photosynthesis, the oxygen-requiring bacteria should cluster around those parts of the alga that gave off most oxygen during photosynthesis. Engelmann observed that the bacteria clustered around the alga where it was exposed to red and blue light, indicating that photosynthesis was occurring fastest at these wavelengths. Since chlorophyll absorbs red and blue light, and so appears green, this result strongly suggested that the green chlorophyll, present in the algal chloroplast, was important in photosynthesis.

There are several types of chlorophyll, with similar molecular structures. A chlorophyll molecule has two main parts. At one end is a complex ring structure with a magnesium ion (Mg^{2+}) bound in the center. This is where light energy is trapped. The rest of the molecule is a long, nonpolar "tail" anchored in the lipid of the photosynthetic membrane. Chlorophyll *a* is the main photosynthetic pigment in all green plants (Figure 8-6).

Sunlight contains wavelengths other than the blue and red ones absorbed by chlorophyll *a*. These wavelengths do not go to waste. Plants also produce various **accessory photosynthetic pigments,** which absorb light energy and transfer it to chlorophyll *a*. Many plants contain chlorophylls *b* or *c*, which absorb different wavelengths in the blue and red parts of the spectrum (see

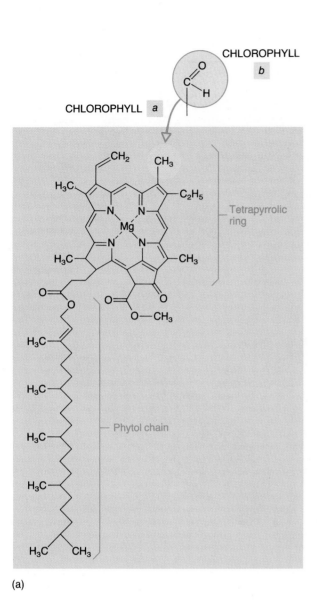

CHLOROPHYLL *a*

CHLOROPHYLL *b*

Tetrapyrrolic ring

Phytol chain

(a)

(b)

Figure 8-6

Chlorophyll. (a) Molecular structure of chlorophyll *a*. Swapping the —CH$_3$ group circled for —CHO would give chlorophyll *b*. The tetrapyrrolic ring is the molecule's "business end," which absorbs light energy. The long, nonpolar phytol chain "tail" lies among the lipids of the photosynthetic membrane. (b) Because chlorophyll absorbs mainly red and blue wavelengths, plant parts containing a lot of chlorophyll appear green. However, in the fall, chlorophyll is destroyed and other pigments, normally masked by chlorophyll, become visible. The yellow seen here comes from carotenoids, which are accessory photosynthetic pigments. Red is from nonphotosynthetic anthocyanins.

Figure 8-4). The **carotenoids** are another important group of accessory pigments found in all green plants. They absorb blue and green wavelengths and so impart yellow or orange hues to the plant.

In many plants, chlorophyll is broken down in the autumn, and its magnesium and nitrogen are moved to storage areas elsewhere in the plant before the leaves drop. This conserves hard-to-obtain elements, which can be used again the following spring. The breakdown of chlorophyll makes the yellow, brown, and orange colors of carotenoids visible in autumn leaves. The red color of some autumn leaves comes from anthocyanin pigments. These are not photosynthetic pigments. Rather, they are produced when sugar is trapped in the leaf as the tree walls off the leaf before it is shed.

8-C   *Tour of a Leaf*

In most familiar plants, the main photosynthetic organs are the leaves. A leaf's broad, flat shape presents a maximum surface area for light absorption at a minimum weight. A layer of tightly packed cells, the **epidermis,** covers the upper and lower leaf surfaces. The epidermal cells secrete a waxy, waterproof **cuticle,** which prevents loss of water vapor from the leaf into the air. However,

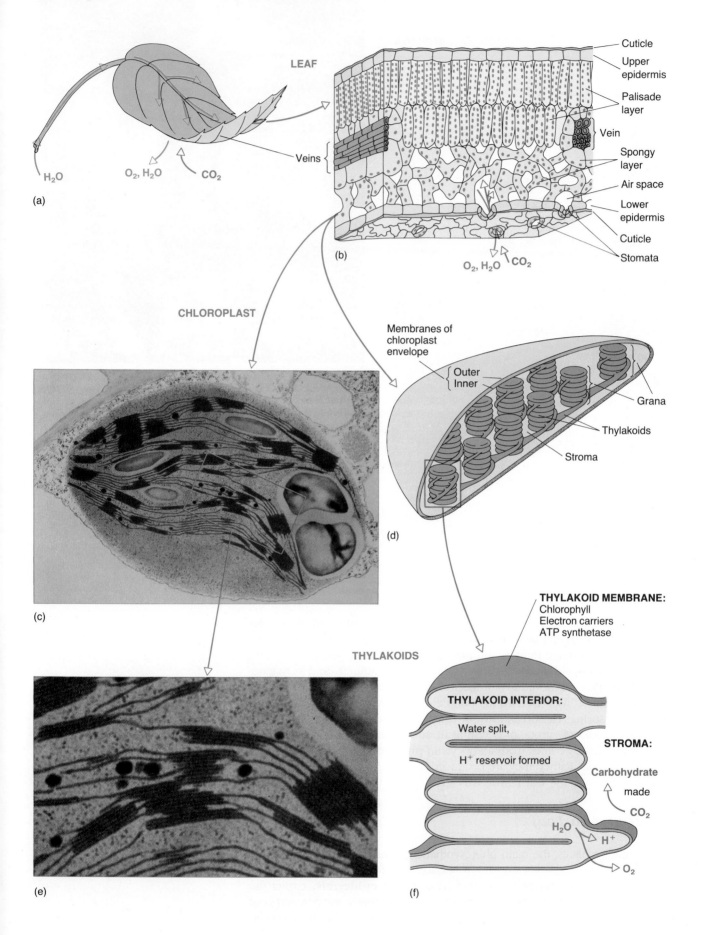

(a)

LEAF

Cuticle
Upper epidermis
Palisade layer
Vein
Spongy layer
Air space
Lower epidermis
Cuticle
Stomata

Veins

H₂O

O₂, H₂O   CO₂

(b)

O₂, H₂O   CO₂

CHLOROPLAST

Membranes of chloroplast envelope
Outer
Inner

Grana

Thylakoids

Stroma

(c)

(d)

THYLAKOIDS

THYLAKOID MEMBRANE:
Chlorophyll
Electron carriers
ATP synthetase

THYLAKOID INTERIOR:

Water split,

H⁺ reservoir formed

STROMA:

Carbohydrate
made
CO₂

H₂O   H⁺

O₂

(e)

(f)

◀ *Figure 8-7*
The location of photosynthesis. (a) In higher plants, photosynthesis occurs in leaves. Water taken in by the roots passes up the stem and through the leaf veins to the cells. (b) Leaf tissues. Photosynthetic cells, which contain chloroplasts (green), are sandwiched between the upper and lower epidermis. Stomata in the lower epidermis admit CO_2 needed for photosynthesis and permit the by-product O_2 to leave. Water vapor also escapes by this route. (c) Electron micrograph of a chloroplast ($\times 4500$), and (d) interpretive drawing of its three-dimensional structure. The photosynthetic thylakoid membranes are green because they contain chlorophyll. (e, f) Closeup of the thylakoid membranes. (c, e, Biophoto Associates)

the cuticle also retards the passage of carbon dioxide, needed by the photosynthetic cells in the leaf's interior. The lower epidermis contains pores, called **stomata** ("mouths"; singular: **stoma**). The stomata permit gas exchange between the atmosphere and the leaf's interior. In most plants, they are open during the day, permitting the uptake of carbon dioxide and release of oxygen, but also inevitably allowing water vapor to escape. Stomata are usually closed at night, and they also close during the day if the plant has lost too much water.

Cells with chloroplasts form two distinct layers sandwiched between the upper and lower epidermis. The **palisade layer** consists of closely packed columnar cells, standing upright just beneath the upper epidermis. Below them lie the loosely arranged cells of the **spongy layer,** with air spaces between them (Figure 8-7). Air enters these spaces via the stomata. Since molecules diffuse faster in gas than in liquid, the air spaces allow for rapid gas exchange between the air and the photosynthetic cells.

The leaf's **veins,** made up of vascular tissue, carry water arriving from the roots and remove newly made carbohydrate to parts of the plant where it is used or stored.

Each of the leaf's photosynthetic cells contains many chloroplasts. A chloroplast is separated from the cytoplasm by its envelope, consisting of two membranes. Inside the chloroplast, a third system of membranes, called **thylakoids,** contains the molecules that trap and use light energy during photosynthesis (Figure 8-7d). Some of the thylakoids occur in stacks, called **grana** ("grains") because a light microscope shows them as little specks. All the thylakoids in one chloroplast appear to be continuous, surrounding a single interior space. This interior space contains a reservoir of hydrogen ions (H^+), built up by electron transport systems housed in the thylakoid membranes. The thylakoids also contain the photosynthetic pigments—chlorophylls and carotenoids. The thylakoid membranes provide a great deal of surface area arranged to intercept as much light as possible.

The thylakoids are surrounded by a protein-rich solution called the **stroma.** The stroma contains the enzymes that make carbohydrate during photosynthesis, as well as the chloroplast's DNA and ribosomes.

8-D   *Overview: What Happens During Photosynthesis?*

The reactions of photosynthesis take place in two main stages: those that capture energy and those that use this energy to make carbohydrate.

Energy Capture

The energy-capturing stage of photosynthesis involves three steps, which all occur in the thylakoid membranes (Figure 8-8):

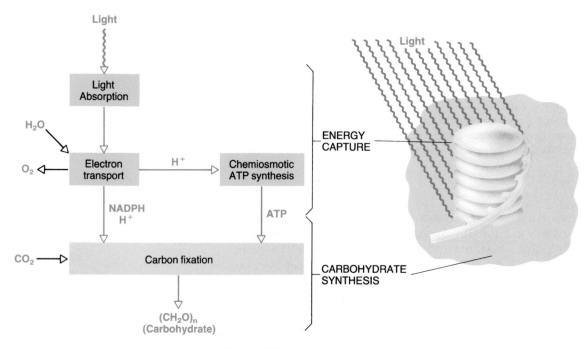

Figure 8-8

A summary of what happens in photosynthesis. The red arrows trace the path of energy from light to carbohydrate molecules. Energy-capturing steps occur in the thylakoid membranes, carbohydrate synthesis in the stroma of chloroplasts.

Inputs	Outputs	
Light energy Pigments	Electrons	**1. Light absorption.** The trapping of light occurs during **photochemical reactions,** chemical reactions powered by light energy. Such reactions are familiar in photography: light striking the film causes chemical reactions that produce an image. The brighter the light, the more reactions occur per unit time. In the photochemical reactions of photosynthesis, light energy boosts electrons in chlorophyll *a* to higher energy levels, so high that the electrons can leave the molecule.
Electrons } $NADP^+$ } H_2O	NADPH { O_2 { H^+ reservoir	**2. Electron transport.** The chloroplast's electron transport molecules accept the high energy electrons from chlorophyll and use some of their energy to build a H^+ reservoir in the thylakoid interior. The electrons finally pass to a hydrogen-carrying coenzyme, $NADP^+$, and some of their energy is used to reduce it to NADPH. ($NADP^+$ is NAD^+ [Chapter 7] with an extra phosphate group.) Chlorophyll's lost electrons are replaced by splitting water molecules (H_2O) and taking electrons from the hydrogen atoms. H^+ is left behind to help fill the reservoir. Oxygen atoms from split water molecules join together, forming O_2.
H^+ reservoir $ADP + P_i$	ATP	**3. Chemiosmotic ATP synthesis.** H^+ trapped in the reservoir in the thylakoid interior space passes through the membrane via channels associated with ATP synthetase enzymes. This flow of H^+ down its concentration gradient powers the synthesis of ATP from ADP and P_i.

Energy capture is now complete. The energy in light has been captured as chemical energy in ATP and NADPH.

Carbon Fixation

Ribulose bisphosphate } CO_2 } ATP NADPH	Sugars $ADP + P_i$ $NADP^+$	During photosynthesis, carbon becomes "fixed" when a gas (carbon dioxide) is incorporated into a solid (carbohydrate). ATP and NADPH, produced during the energy-capturing reactions, provide the energy and hydrogen needed to fix the carbon.

If a plant has already captured light energy as chemical energy in NADPH and ATP, why does it need carbon fixation too? First, ATP and NADPH are short-lived. A foodstore of carbohydrate to be used for respiration whenever energy is needed is much more useful. Second, carbon fixation builds up carbon skeletons from which other organic molecules can be made.

8-E The Energy-Capturing Reactions

The energy-capturing reactions take place in the thylakoid membranes. They involve two sets of light absorption events, which capture the energy used to power two sets of electron transport reactions. One set of electron transport reactions produces NADPH, and the other indirectly produces ATP.

Light Absorption Light absorption occurs in two kinds of **photosystems.** Each photosystem contains hundreds of pigment molecules. Most of these are chlorophyll *a* and *b* **antenna pigments,** which gather light energy and pass it to a "special pair" of chlorophyll *a* molecules bound to protein in the photosystem's **reaction center.** In Photosystem I, the reaction center chlorophyll is known as **P700** (pigment 700, so called because the wavelength of light it absorbs maximally is about 700 nanometres). The chlorophyll *a* in the Photosystem II reaction center is called P680.

Having so many antenna pigments maximizes the photosystems' ability to intercept light energy, which is then funnelled to the reaction center. When light of the proper wavelength strikes an antenna pigment, the light energy is transformed into rapid vibration of the molecule. This vibration energy migrates from molecule to molecule in the photosystem until it reaches the reaction center chlorophyll. Here the energy boosts an electron to a higher energy level. From this energy level, the electron can pass "downhill," in terms of energy, to a neighboring molecule, part of the electron transport system (Figure 8-9).

This electron transfer is a redox reaction, one that is too endergonic to occur when chlorophyll is in its unexcited state. Hence, light energy has been used to power an endergonic redox reaction. As the electron is passed down the transport chain, the remaining redox reactions in the chain are exergonic, releasing the energy derived from light and putting it to use.

Electron Transport In Section 8-D, we saw that electrons lost by chlorophyll *a* end up at NADP$^+$. The chlorophyll's electrons are replaced by splitting water. So the overall path of electron transport in photosynthesis is from water to NADP$^+$, by way of the two photosystems and two electron transport chains:

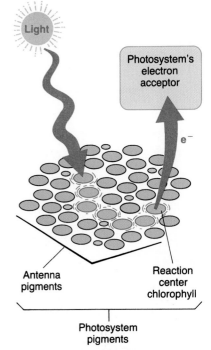

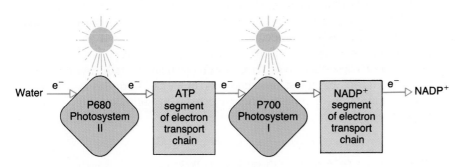

Figure 8-9
Absorption of light energy. The antenna pigments absorb light energy and transform it to vibrational energy, which is passed from one molecule to the next until it reaches a reaction center chlorophyll. Here it excites an electron to a higher energy level. The chlorophyll passes this electron to the photosystem's electron acceptor molecule.

The transfer of electrons from water to NADP$^+$ is a highly endergonic redox reaction. In photosynthesis, it takes two separate boosts from light energy, one from each photosystem, to push electrons all the way up this "energy hill." After each boost, the electrons lose some energy, which is used for the work of photosynthesis.

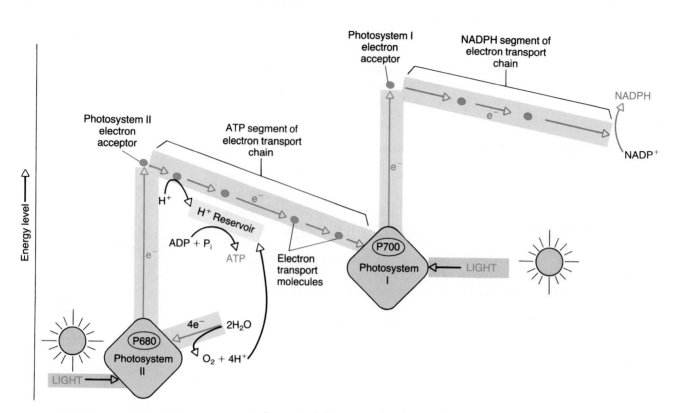

Figure 8-10
The path of energy in the electron transport chain of photosynthesis. Electrons move from water (bottom), through the photosystems and electron transport chain, to NADP⁺ (far right). This transfer is steeply "uphill," in terms of energy, and requires two boosts from light along the way. This results in a zigzag pattern, commonly called the Z scheme (red arrows).

A diagram of the energy level at each step has a zigzag pattern (Figure 8-10). Electrons make the trip in stages, and some are going through the final stage, passing to NADP⁺, while others are just leaving water molecules. Let us follow these reactions, starting with the *final* leg of the electrons' journey, from Photosystem I to NADP⁺.

Light energy striking Photosystem I boosts electrons from P700 into the NADPH electron transport chain. This chain passes the electrons to NADP⁺ in the stroma. Here some of the electrons' energy is used to reduce NADP⁺ to NADPH, as NADP⁺ accepts two electrons from the chain and takes a hydrogen ion (H⁺) from the stroma.

While this has been going on, light energy striking Photosystem II has boosted electrons from P680 to a higher energy level. These electrons pass to the other part of the electron transport chain, which delivers them to P700 in Photosystem I, replacing its lost electrons (see Figure 8-10). Electron transport between the two photosystems releases much of the energy the electrons gained from light. In effect, this energy is used to transport H⁺ from the stroma, through the thylakoid membrane, to the thylakoid interior space. This adds H⁺ to the H⁺ reservoir inside the thylakoid. Since the H⁺ reservoir is used to make ATP, the electron transport molecules between Photosystems II and I may be thought of as an ATP segment of the electron transport chain.

The reactions described so far have moved electrons from the photosystems in the thylakoids out into the stroma, and P680 in Photosystem II is now missing electrons. Photosystem II replaces these electrons by splitting water from the thylakoid interior:

$$2\,H_2O \longrightarrow 4\,e^- + 4\,H^+ + O_2$$

The electrons (e⁻) replace those lost by P680 in Photosystem II.
The H⁺ remain in the thylakoid interior, as part of the H⁺ reservoir.
The O₂ is a by-product of photosynthesis and diffuses out of the cell.

■ *Electrons flow from water in the thylakoid interior, through the photosystems in the thylakoid membrane, to NADPH in the stroma. This flow of electrons is an electric current, powered by solar energy. It provides energy to make NADPH and ATP.*

Chemiosmotic ATP Synthesis The electron transport reactions have built up a high concentration of H⁺ in the thylakoid interior. This H⁺ reservoir powers the chemiosmotic synthesis of ATP. Hydrogen ions pass through special protein

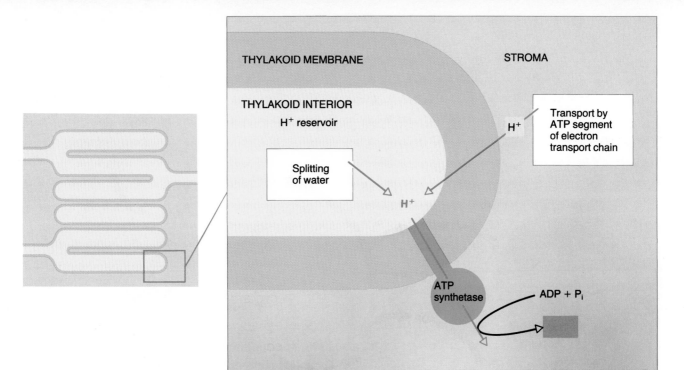

Figure 8-11
Chemiosmotic ATP synthesis in chloroplasts. The H$^+$ reservoir builds up in the thylakoid interior space. H$^+$ then moves out of the reservoir by way of special channels. Here the enzyme ATP synthetase uses the energy of H$^+$ moving down its gradient to join ADP and P$_i$, forming ATP.

channels in the thylakoid membrane. These channels are associated with ATP synthetase enzymes, which use the energy of H$^+$ moving down its gradient to join P$_i$ to ADP, forming ATP (Figure 8-11).

8-F Carbon Fixation

NADPH and ATP, produced by the reactions that capture the energy of light, are used by enzymes in the stroma to fix carbon.

Carbon fixation begins with the attachment of carbon dioxide to a pre-existing organic molecule, **ribulose bisphosphate,** a five-carbon sugar with two phosphate groups. The resulting six-carbon structure is unstable and immediately hydrolyzes to two identical three-carbon molecules of phosphoglycerate (PGA) (Figure 8-12).

These two PGA molecules are next reduced to phosphoglyceraldehyde (PGAL). This takes two steps. First each PGA receives more energy as a second phosphate group is donated by ATP. In the next step, the new phosphate group is removed, releasing this energy, as the molecule is reduced to PGAL by the addition of hydrogen from NADPH.

Many molecules are going through these reactions at the same time in the chloroplast. Some of the three-carbon PGAL molecules resulting from carbon fixation are channelled into a metabolic pathway that joins two three-carbon

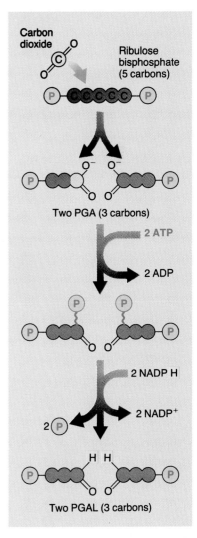

Figure 8-12
Carbon fixation. The six-carbon compound formed by joining carbon dioxide to ribulose bisphosphate is hydrolyzed into two molecules of PGA (phosphoglycerate), shown oriented in opposite directions. Using energy from ATP, and hydrogen from NADPH, PGA is reduced to PGAL (phosphoglyceraldehyde).

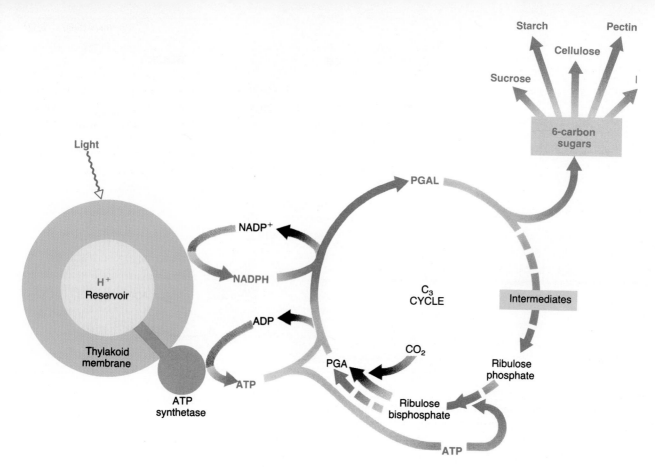

Figure 8-13
The C_3 cycle and its connections with the energy-capturing reactions in the thylakoid membranes. ATP and NADPH, and ADP, P_i, and NADP$^+$, shuttle between the two interdependent sets of reactions (energy-capturing, in the thylakoid membranes; and carbon fixation, in the stroma). Red arrows trace the course of energy from sunlight through temporary energy intermediates to the more permanent energy-storage form of six-carbon sugars and their polymers.

molecules to form six-carbon sugars, such as glucose. Glucose can then be polymerized into starch, an energy-storage compound, or cellulose, which makes up the cell wall. Or, glucose can be processed to make sucrose, which is transported to other parts of the plant. Other three-carbon molecules go into the synthesis of amino acids, by the addition of nitrogen-containing groups. Most of the PGAL molecules produced must be channelled into the formation of more of the five-carbon sugar ribulose, which then goes back to receive more carbon dioxide (Figure 8-13). A long series of reactions reworks five three-carbon molecules to form three molecules of the five-carbon ribulose phosphate. Each of these receives another phosphate group, from ATP formed during energy capture in the thylakoids. It is then ready to accept a CO_2 molecule during carbon fixation.

Because some of the three-carbon end products are converted to new molecules of the five-carbon starting material, the whole process of carbon fixation is actually a cycle. It is called the **C_3 cycle** after its three-carbon products, or the **Calvin cycle** after its discoverer, Melvin Calvin, who received a Nobel Prize for this work in 1961.

The enzyme that attaches carbon dioxide to ribulose bisphosphate is called ribulose bisphosphate carboxylase—rubisco for short. Rubisco makes up about 25% of the protein in chloroplasts. This enzyme is far and away the most abundant protein in green tissue and therefore in the world! It is estimated that there are 20 pounds of this protein for every person on earth.

In summary, the C_3 cycle begins with carbon dioxide and a five-carbon molecule, ribulose bisphosphate, which react to form two three-carbon molecules (PGA). Each PGA molecule uses an ATP and an NADPH as it is converted to PGAL. A third ATP is required to regenerate the starting molecule, ribulose bisphosphate. The overall equation is:

$$RuBP + CO_2 + 2\,NADPH + 3\,ATP \longrightarrow$$
$$RuBP + CH_2O + 2\,NADP^+ + 3\,ADP + 3\,P_i$$

It takes six turns of the Calvin cycle to produce the equivalent of one (six-carbon) glucose molecule, that is, to fix six carbon atoms into organic form.

The ADP, P_i, and $NADP^+$ released by the C_3 cycle are recycled to form more ATP and NADPH. These substances are present in relatively small amounts, and so if either electron transport or the C_3 cycle stops, the other soon stops as well. The stockpile of ATP and NADPH, for instance, will last only a matter of seconds once the light is turned off. After the supply is exhausted, carbon fixation can no longer proceed.

> ■ *In carbon fixation, carbon dioxide is attached to an existing carbohydrate molecule and reduced, using energy from ATP and hydrogen from NADPH, both produced in the energy-capturing reactions.*

8-G What Controls the Rate of Photosynthesis?

Photosynthesis is the sum of many different chemical reactions. Several factors play key roles in some of these reactions and so can affect the rate of the whole process.

Photosynthesis begins with photochemical reactions, which trap light energy and go at increased rates when the light intensity increases (Figure 8-14). However, the remaining reactions of photosynthesis are like the chemical reactions we saw in previous chapters, called **thermochemical reactions** because their rates are increased by heat. Because photosynthesis is made up of both photochemical and thermochemical reactions, both light and temperature influence its rate.

In dim light, the rate of photosynthesis is limited by lack of light. Photochemical reactions that initiate electron transport, and hence chemiosmotic ATP synthesis, go too slowly to produce NADPH and ATP as fast as carbon fixation can use them. As the light becomes brighter, the rate of photosynthesis increases until ATP and NADPH become so plentiful that carbon fixation cannot keep up, and the rate of photosynthesis levels off. This is what happens on bright, cool days. On warmer days, the rate of photosynthesis increases still further. At these higher temperatures, molecules move faster, and the rate of the thermochemical reactions of carbon fixation speeds up.

The rate of a reaction can also be decreased by scarcity of raw materials or excess of products. On bright, warm days, light and temperature are optimum for photosynthesis, but a low concentration of the raw material carbon dioxide often limits its rate. Adding more carbon dioxide to the air in a greenhouse increases the rate of photosynthesis (and so of plant growth) of some greenhouse crops. Above a certain concentration, however, carbon dioxide inhibits photosynthesis.

The other raw material of photosynthesis, water, is so abundant in living tissue that the amount channelled into photosynthesis is negligible. However, a water shortage does limit photosynthesis indirectly: if the plant loses too much water, the stomata of the leaves close. This water-conservation measure also cuts down the rate of CO_2 uptake and hence slows the carbon fixation reactions of photosynthesis.

The carbohydrate end products of photosynthesis are quickly changed into other chemical forms or removed from the chloroplast, and so they do not accumulate and inhibit the reaction. The other end product, oxygen, diffuses out of the cell and leaves the plant. However, when oxygen is being produced rapidly, some of it accumulates at levels high enough to slow photosynthesis, by

Figure 8-14
Catching sunbeams. In bright light, the energy-capturing reactions of photosynthesis proceed rapidly, as evidenced by the vigorous bubbling of the by-product oxygen from these aquatic ferns and duckweeds. (W. Ormerod/Visuals Unlimited)

Experimental Milestones in Photosynthesis

A JOURNEY INTO LIFE

The scientific study of photosynthesis began in 1648, with a simple experiment such as a high school class might do today. The Dutchman Jean-Baptiste van Helmont sought to answer this question: since plants do not eat, how do they grow? To find out, he planted a willow shoot that weighed 2.3 kilograms (kg) in 90.8 kg of dry soil in a large pot. He covered the top of the soil to keep dust in the air from settling on the soil, and watered the plant as necessary with rainwater or distilled water. At the end of five years, the plant weighed 76.8 kg, but the soil had lost only 56 grams. Van Helmont concluded that the tree had gained more than 74 kg from water alone.

Van Helmont had identified one source of plant material, but he was wrong in thinking that the tree's weight gain was due to water alone. He had ignored one other possible source of nourishment to the willow tree—the air.

In 1771, the English scientist and clergyman Joseph Priestley published evidence that animals and plants alter the composition of the air around them in complementary ways. He burned a candle in a covered jar until it went out. If a plant was grown in the jar for several days, a candle would once more burn in the jar. Priestley concluded that "there was something attending vegetation, which restored the air that had been injured by the burning of candles." He found that exactly the same effect occurred if he used a mouse instead of a candle. The mouse and the candle both altered the air inside the jar in some fashion that could be reversed by a plant (Figure 8-A). Today, we know that a photosynthesizing plant gives off oxygen and that a mouse or a burning candle uses up O_2 from the air.

In 1779, Jan Ingenhousz, a Dutch physician at the Austrian court, found that plants need sunlight if they are to produce oxygen. Furthermore, only the green leaves and stems of

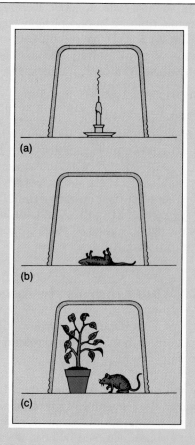

Figure 8-A
The results of Priestley's experiments.

inactivating electron transport molecules and by interfering with the carbon dioxide fixation enzyme.

8-H   Ecological Aspects of Photosynthesis

Plants living in unusual habitats have evolved some interesting variations on photosynthesis.

As we have seen, the stomata permit carbon dioxide to enter the leaves, but at the same time, some of the plant's water inevitably evaporates through the stomata into the air. If the leaves lose water faster than the roots replace it, the plant starts to wilt and the stomata close somewhat, slowing the loss of water but also reducing the entry of carbon dioxide.

The C_3 cycle enzyme rubisco, which adds CO_2 to the five-carbon ribulose, works better at high carbon dioxide levels. However, some plants can carry on photosynthesis rapidly even at low carbon dioxide levels because they can trap and store CO_2. An enzyme attaches available CO_2 to a three-carbon compound, forming a four-carbon molecule that holds the CO_2 temporarily. These plants are called C_4 plants after the four-carbon compound. The C_4 molecules are transported to a central location, where their CO_2 is removed and accumulated at high enough levels for the C_3 cycle to re-fix it into carbohydrate.

The C_4 pathway uses five ATP per molecule of carbon dioxide fixed instead of the three ATP used in C_3 photosynthesis alone, but at high temperatures C_4

plants carried on photosynthesis; fruits used up oxygen instead of producing it. Ingenhousz gave visible proof that photosynthesis results in the production of a gas: when he placed sprigs of willow underwater in bright sunlight, they became covered with gas bubbles.

Ingenhousz also showed that plants, like animals, carry on respiration, using up oxygen and giving off carbon dioxide. However, respiration was obvious only if the plants were kept in darkness; in the light, plants produced more oxygen by photosynthesis than they used in respiration.

In 1782, Jean Senebier, a Swiss clergyman, found that plants use up carbon dioxide when they produce oxygen and suggested that carbon dioxide contributed to the nourishment of the plant: he recognized that air was a source of plant nutrition. In 1804, Nicholas Theodore de Saussure, also a Swiss, found that the amount of carbon dioxide taken up did not nearly account for the increase in dry weight of a growing plant. He decided that the remainder of this weight gain must come from water.

In 1864, the German plant physiologist Julius Sachs placed a photosynthesizing leaf on a microscope stage and watched starch grains growing inside the chloroplasts. He suggested that some of the organic matter produced by photosynthesis was carbohydrate.

By the end of the nineteenth century, the plant could be regarded as a "black box" that took in water and carbon dioxide and, in the presence of light, converted them to carbohydrate and oxygen.

The carbon atoms in the carbohydrate produced during photosynthesis must come from carbon dioxide, and the hydrogen must come from water. However, both carbon dioxide and oxygen gas contain oxygen atoms and, because both are gases, it was long assumed that the oxygen given off during photosynthesis came from carbon dioxide. However, we now know that the oxygen gas released during photosynthesis comes from water. Evidence for this conclusion came from two sources.

In the early 1930s, C. B. van Niel found that purple sulfur bacteria use light to make carbohydrates. They require hydrogen sulfide as a raw material and give off sulfur, which accumulates in the cells as yellow globules. Van Niel, while still a graduate student, speculated that the reactions in sulfur bacteria were analogous to photosynthesis in green plants.

a) General reaction for green plants:

$$CO_2 + 2\,H_2O \xrightarrow{\text{light}} (CH_2O) + H_2O + O_2$$

b) General reaction for sulfur bacteria:

$$CO_2 + 2\,H_2S \xrightarrow{\text{light}} (CH_2O) + H_2O + 2\,S$$

In 1941, Samuel Ruben and Martin Kamen produced evidence that supported van Niel's hypothesis. They used the heavy isotope oxygen 18 to distinguish the oxygen of water from the oxygen of carbon dioxide. When they gave a plant water labelled with oxygen 18 (O) and carbon dioxide containing the more common isotope, oxygen 16 (O), as raw materials for photosynthesis, labelled oxygen was given off:

$$6\,CO_2 + 12\,H_2O \longrightarrow C_6H_{12}O_6 + 6\,O_2 + 6\,H_2O$$

The experiment can also be done the other way around:

$$6\,CO_2 + 12\,H_2O \longrightarrow C_6H_{12}O_6 + O_2 + 6\,H_2O$$

plants can photosynthesize faster than C_3 plants. This sacrifice of efficiency for speed is advantageous because it allows plants to grow and reproduce faster. However, this profligate use of energy is worthwhile only when energy is abundant compared to other resources. C_4 photosynthesis has evolved in plants of warm, sunny, somewhat dry climates such as grasslands, where there is plenty of light energy to drive ATP synthesis but water stress is a frequent problem. In such a situation, the C_4 plant's stomata may close partially, reducing loss of water vapor from the leaves. This also reduces uptake of CO_2 from the air, but the C_4 pathway allows the plant to capture enough carbon dioxide for a high rate of photosynthesis anyway. Not surprisingly, C_4 plants include crabgrass, other important weeds, and some major fast-growing crops such as sugarcane and corn.

Because the rapid growth of C_4 plants depends on a high rate of photosynthesis in bright light, C_4 plants are at a relative disadvantage in cool or shady situations, and C_3 and C_4 plants coexist, with neither having a noticeable edge, in many habitats (Figure 8-15).

Figure 8-15

Photosynthetic rates of C_3 shade plants, C_3 sun plants, and C_4 plants at different light intensities. Shade plants use light more efficiently than sun plants at low light intensities, but reach saturation (their maximum photosynthetic rate) at relatively low light intensity. Saturation of C_3 sun plants occurs at much higher light intensities. Even here, C_4 plants still show increased rates of photosynthesis.

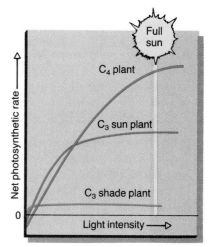

(a)

(b)

Figure 8-16
Water-saving adaptations. (a) Sugar-cane, an important C_4 crop plant. (b) Ice plant, a CAM plant indigenous to South Africa, was introduced to California and now covers countless acres along the California coastline. Note the thick, succulent leaves.

Some plants use a variation of C_4 photosynthesis called **crassulacean acid metabolism (CAM).** CAM occurs mainly in cacti, euphorbs, and other succulents—plants with fleshy, water-storing stems or leaves, adapted to desert conditions (Figure 8-16). The stomata of CAM plants are closed during the hot day. In the cooler night, when water evaporates more slowly, the stomata are opened and carbon dioxide is taken up and fixed into organic acids. During the day, when the stomata are closed, carbon dioxide is removed from the acids, and light energy is used to re-fix it by way of the C_3 pathway. Although CAM plants receive plenty of sun, they must conserve water very strictly, and so they have very low photosynthetic rates. However, their ability to conserve water enables them to survive in habitats too dry for the "fast food" plants that would otherwise crowd these slow growers out.

Even ordinary C_3 plants can be separated into two groups, sun plants and shade plants. While **sun plants,** such as soybeans, cotton, and tomatoes, show increased rates of photosynthesis as light intensity increases, **shade plants** do not. Shade plants, including many ferns, African violets, and philodendrons, simply do not photosynthesize rapidly, even in bright light. On the other hand, shade plants are much more efficient at using very dim light (see Figure 8-15).

SUMMARY

Photosynthesis is the process in which green plants store the energy of sunlight by converting carbon dioxide and water into organic compounds:

$$CO_2 + H_2O \xrightarrow[\text{chlorophyll}]{\text{light,}} CH_2O + O_2$$

These organic compounds are used by plants, and by the animals that eat plants, to build cells and to power other energy-requiring processes. Organic molecules are eventually broken back down to carbon dioxide and water via respiration, and these materials can then be recycled in photosynthesis. Since energy cannot be recycled, however, the sunlight that drives photosynthesis is the ultimate source of energy for nearly all life on earth.

Photosynthesis may be considered in two parts:

1. **Energy capture.** Solar energy is trapped by chlorophyll and other photosynthetic pigments in the thylakoid membranes, initiating a flow of electrons through the membrane's electron transport chain. The overall flow of electrons is from water molecules split inside the thylakoid compartment, through the electron transport systems and photo-

systems in the thylakoid membrane, to $NADP^+$ in the stroma. This flow of electrons reduces $NADP^+$ to NADPH and creates the H^+ reservoir used to make ATP. Oxygen from water is released as a by-product. The NADPH and ATP are released into the stroma of chloroplasts, where they are used to fix carbon dioxide.

2. **Carbon fixation.** During the C_3 cycle, carbon dioxide becomes attached to a five-carbon sugar, ribulose bisphosphate, which then breaks to yield two three-carbon (PGA) molecules. ATP and NADPH are used to reduce these to phosphoglyceraldehyde (PGAL). The resulting ADP and $NADP^+$ are recycled. The three-carbon PGAL molecules may be made into structural or energy-storing molecules, or may be processed, with the use of more ATP, to make more ribulose for carbon fixation.

A photosynthesizing plant captures light energy in two parts of the reaction sequence. In each case, the light energy boosts an electron to a high energy level, and the energy so transferred is channelled into increasingly permanent forms of energy storage:

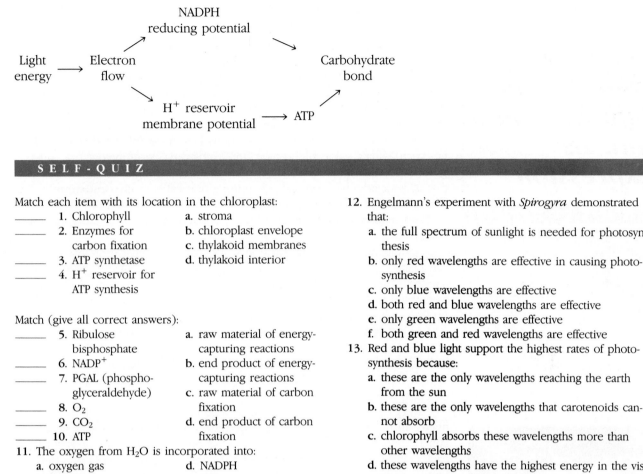

Match each item with its location in the chloroplast:

_____ 1. Chlorophyll
_____ 2. Enzymes for carbon fixation
_____ 3. ATP synthetase
_____ 4. H^+ reservoir for ATP synthesis

a. stroma
b. chloroplast envelope
c. thylakoid membranes
d. thylakoid interior

Match (give all correct answers):

_____ 5. Ribulose bisphosphate
_____ 6. $NADP^+$
_____ 7. PGAL (phospho-glyceraldehyde)
_____ 8. O_2
_____ 9. CO_2
_____ 10. ATP

a. raw material of energy-capturing reactions
b. end product of energy-capturing reactions
c. raw material of carbon fixation
d. end product of carbon fixation

11. The oxygen from H_2O is incorporated into:
    a. oxygen gas
    b. water
    c. carbohydrates
    d. NADPH
    e. ATP

12. Engelmann's experiment with *Spirogyra* demonstrated that:
    a. the full spectrum of sunlight is needed for photosynthesis
    b. only red wavelengths are effective in causing photosynthesis
    c. only blue wavelengths are effective
    d. both red and blue wavelengths are effective
    e. only green wavelengths are effective
    f. both green and red wavelengths are effective

13. Red and blue light support the highest rates of photosynthesis because:
    a. these are the only wavelengths reaching the earth from the sun
    b. these are the only wavelengths that carotenoids cannot absorb
    c. chlorophyll absorbs these wavelengths more than other wavelengths
    d. these wavelengths have the highest energy in the visible spectrum
    e. these wavelengths activate the ATP synthetase enzyme

1. In the early 1930s, C. B. van Niel found that purple sulfur bacteria use light to make carbohydrates. They require hydrogen sulfide as a raw material and give off sulfur:

$$CO_2 + H_2S \xrightarrow{\text{light}} (CH_2O) + H_2O + 2\,S$$

Other kinds of bacteria produce food by processes called chemosynthesis, using energy from inorganic chemical reactions rather than light energy. Why is the kind of photosynthesis outlined in this chapter so much more prevalent among living organisms today than the kind of photosynthesis used by sulfur bacteria or the chemosynthesis used by bacteria?

2. Why is photosynthesis only about 1% efficient in converting the energy in sunlight that strikes a leaf into energy stored in organic molecules? What happens to the rest of the energy?

Govindjee and W. J. Coleman. "How plants make oxygen." *Scientific American,* February 1990.

Wald, G. "Life and light." *Scientific American,* October 1959. A classic article covering how organisms use light, not only in photosynthesis but also in phototropism (the bending of plants toward light), vision, and production of light.

Youvan, D. C., and B. L. Marrs. "Molecular mechanisms of photosynthesis." *Scientific American,* June 1984.

The Unity of Life: Genetic Information and Its Expression

PART 2

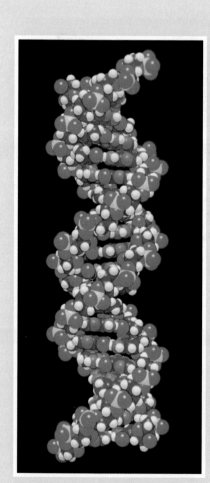

DNA double helix (computer graphic)

*E*ach of us originated from a single cell—an egg, fertilized by a sperm. What directed this fertilized egg to divide into more and more new cells, and the resulting mass of cells to move around, grow, absorb nourishment, and take shape as a unique individual? What makes each of us distinct from other people but gives us all a basic similarity as members of the human species? And what allows discerning relatives hovering over a new arrival to proclaim that it has its father's nose and its mother's shy smile? The answer to all these questions is **genetic information.**

The units of genetic information, units governing inherited characteristics such as hair color, blood type, and embryonic development, are called **genes.** A gene is a length of DNA, part of a much longer DNA molecule that is associated with proteins, forming a unit called a **chromosome.** The nucleus of a eukaryotic cell contains several to many chromosomes. Each chromosome's DNA molecule may contain hundreds or thousands of genes.

Many genes contain information that determines what proteins the cell can make. The order of nucleotide monomers in the DNA of a gene constitutes a "genetic code" that dictates the order in which amino acids are assembled to form a protein. So DNA codes for protein structure, and a protein's structure, in turn, determines its function in carrying out some part of the cell's activities (Section 3-E). This, in turn, has an effect on the organism as a whole, such as what color flowers it has, or whether its heart develops normally. The whole process whereby genes direct the production of proteins, which in turn affect the whole organism, is known as **gene expression.**

Every cell in an organism must be able to produce the proteins it needs. In most cases, it can do this because it contains a complete set of genetic information, identical to that in the original fertilized egg. Before each cell division, the genetic material is copied and distributed precisely into two sets, one for each of the new cells. With a full set of genetic information, each cell then produces some proteins but not others, according to its particular role in the body.

In this part of the book we explore the structure of the genetic material (Chapter 9) and see how it directs the production of proteins (Chapter 10). Our growing understanding of these topics now permits us to manipulate genetic material in many ways, using the modern technology of genetic engineering.

Chapter 11 looks at how chromosomes are passed from a cell to its offspring and from one generation of organisms to another. This behavior of chromosomes explains the patterns of genetic inheritance observed from one generation to the next (Chapter 12). Genes are not the whole story of how an organism becomes what it is, however: the environment in which the genes operate also plays an important role (Chapter 13).

The study of genetic information is a young field of biology but one that has mushroomed into a major area of research. Discovering the structure of DNA, cracking the genetic code for making proteins, and more recently finding ways to make DNA and protein molecules to order, are all feats of the last half of this century. This has been an exciting time, awhirl with fresh insights, new technology, and achievements to rival science fiction. In many cases we have been able to analyze how genetic defects operate at the level of cells and molecules. Eventually we shall be able to change the genetic makeup of individuals, or perhaps even entire species, at will. This new technology brings with it a host of ethical, social, and possibly environmental problems that we must all grapple with now and in the future.

Primula

Sumatran orang-utans

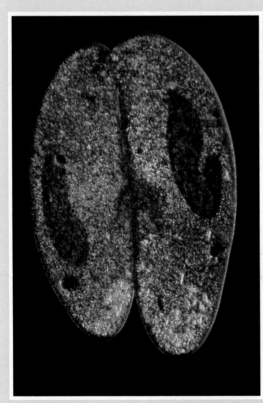

Paramecium *mating*

Growth: red-humped caterpillar

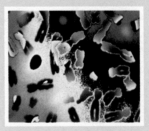

Chromosomes, human male

DNA and Genetic Information

9

Sumatran orang-utans

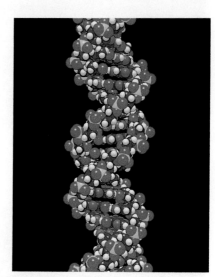

DNA double helix (computer graphic)

Lobelia deckanii, Mt. Kenya

*T*he idea that offspring inherit certain features of their parents is familiar to us all. But did you ever wonder exactly how you got your mother's brown eyes or your father's curly hair? After all, the egg and sperm that gave rise to what is now you did not have eyes or hair. Rather, they contained **genetic information,** which directed the formation of eyes, and later of eye color, and the growth of hair, at the proper time in your development.

During the last century, biologists learned that an individual inherits half its genetic information from each parent. When they studied cells dividing to form eggs and sperm, they noted that each egg or sperm ends up with half of the number of chromosomes from the nucleus of the original cell (Section 11-E). When egg and sperm unite at fertilization, only the sperm nucleus, containing chromosomes, enters the egg. The new individual receives a full set of chromosomes: half from its mother's egg and half from its father's sperm.

This behavior of chromosomes was exactly what biologists expected of structures bearing genetic information. They became convinced that chromosomes do in fact contain the genetic information. By 1940, biochemists knew that chromosomes contain two substances: proteins and DNA (short for deoxyribonucleic acid, Section 3-D). Which carried the genetic information?

Biologists reasoned that organisms must contain a variety of genetic information. They knew that proteins are a diverse and complex group of polymers, made up of 20 different kinds of amino acid monomers, and proteins are very specific. On the other hand, DNA polymers contain only four kinds of nucleotide monomers. It seemed as though chromosomal proteins must be more complex than DNA, and most scientists therefore thought that these proteins carried the genetic information. But this turned out to be wrong: DNA contains the genetic information.

A chromosome's DNA contains hundreds or thousands of units of genetic information, called **genes.** Each gene is a stretch of the chromosomal DNA molecule containing hundreds or thousands of nucleotide monomers. Many genes contain instructions for making proteins. The order of their nucleotides constitutes a **"genetic code"** of directions for the order in which amino acids should be joined to make proteins. Each protein then performs some job, such as making eye pigment or forming part of a hair. So the gene's information is said to be **expressed** in the form of protein, which in turn contributes to the structure or function of the whole organism.

All the cell nuclei in your body contain the same set of genes, but not all genes are expressed in every cell. For example, our eyes, tongues, and internal organs do not normally produce the proteins that would make them sprout hair (although this can happen in some medical disorders). Some genes do not code for proteins but instead regulate which other genes in the cell are expressed.

In this chapter we examine the evidence that the genetic material is indeed DNA, and how the structure of DNA was worked out. This structure is very simple, yet elegant, and it has the remarkable property that it dictates the production of exact copies of itself—copies that are then passed on to future generations of cells.

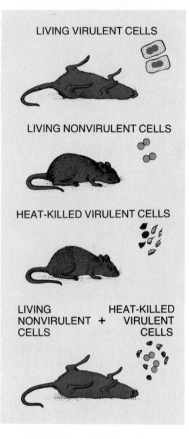

LIVING VIRULENT CELLS

LIVING NONVIRULENT CELLS

HEAT-KILLED VIRULENT CELLS

LIVING NONVIRULENT CELLS + HEAT-KILLED VIRULENT CELLS

Figure 9-1
Griffith's experiments on bacterial transformation. Living virulent bacteria injected into mice cause fatal disease. Living nonvirulent bacteria or heat-killed virulent cells do not cause disease. Mice injected with both living nonvirulent bacteria and heat-killed virulent bacteria develop disease. Something from the heat-killed cells transforms the nonvirulent cells into virulent cells.

9-A   Evidence that DNA Is the Genetic Material

The Riddle of Bacterial Transformation

In 1928 Fred Griffith was studying pneumonia caused by bacteria, a serious disease in the days before antibiotics. He worked with two strains of bacteria, containing different genetic information. One strain's genetic information directed the production of an external capsule that protected the bacteria from attack by an animal's immune system. This strain was **virulent:** when injected into mice, it caused fatal disease. The other strain, which could not produce capsules, was **nonvirulent** and did not kill the mice.

If virulent bacteria that had been killed by heating were injected into mice along with living, nonvirulent bacteria, some of the mice died (Figure 9-1). Even though these mice had received no living virulent bacteria, their corpses contained living virulent bacteria. Griffith concluded that some of the genetic material from the dead virulent bacteria had entered the living nonvirulent bacteria, transforming them to the virulent form. This phenomenon was named **bacterial transformation,** the transfer of genetic information from one bacterium into another.

What factor had transformed the bacteria? Oswald Avery and his colleagues spent a decade on this problem, growing tons of bacteria, which they separated into their chemical components for testing. In 1944, they reported that the transformation studied by Griffith could be produced by DNA from virulent bacteria but not by their proteins or any other substance.

The conclusion that DNA was the genetic material was not immediately accepted. Most biologists at that time did not regard bacteria as "real" organisms. So what if DNA was the genetic material of bacteria? Many felt this information was irrelevant to understanding the genetic material of higher organisms. We now know that DNA is in fact the genetic material of plants, animals, fungi, and protists, as well as of bacteria and some viruses. Indeed, most of the major discoveries in molecular genetics were made using bacteria, which can be grown quickly and easily for experiments. *Escherichia coli,* a common bacterium that inhabits the human intestine, has been particularly important.

Bacteriophages

More evidence that genetic material is DNA was obtained from studies of **bacteriophages** (**phages** for short), viruses that parasitize bacteria. A phage takes over a bacterium's metabolic machinery, causing it to produce new phages and then to burst, releasing the phages to infect more bacteria.

A phage consists of a DNA molecule inside a protein coat. In 1952 Alfred Hershey and Martha Chase tested the hypothesis that phages do not enter bacteria intact; rather, the protein coat attaches to the cell wall and injects its DNA into the bacterium. If this is so, and the bacterium then produces new phages, the DNA must be the phage's genetic material.

Hershey and Chase distinguished between the phages' protein and DNA by using radioactive isotopes of sulfur and phosphorus as labels (*A Journey Into Life,* Chapter 2). Proteins contain sulfur but not phosphorus, whereas DNA contains phosphorus but not sulfur. Hershey and Chase prepared phages that had proteins labelled with radioactive sulfur and DNA labelled with radioactive phosphorus. When these phages infected bacteria, the radioactive phosphorus entered the bacteria, whereas the radioactive sulfur remained outside. The phages' DNA, but not their protein, had entered the bacteria (Figure 9-2). The bacteria produced new phages. This was strong evidence that the phage genetic material consisted of DNA but not protein.

Figure 9-2
Infection of a bacterium by a bacteriophage. The phage's DNA is injected into the bacterium, but its protein coat remains outside. Many new phages are produced inside the bacterium.

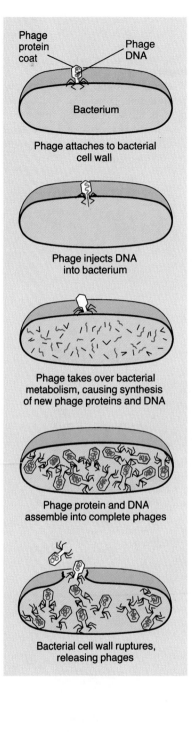

Phage attaches to bacterial cell wall

Phage injects DNA into bacterium

Phage takes over bacterial metabolism, causing synthesis of new phage proteins and DNA

Phage protein and DNA assemble into complete phages

Bacterial cell wall ruptures, releasing phages

The Quantity of DNA in Cells

Circumstantial evidence that DNA is also the genetic material in eukaryotes came from measuring the amount of DNA in different cells. For example, the body (nonreproductive) cells from an animal's liver, kidneys, and other organs contain the same amount of DNA as each other and twice as much as sperm, which are reproductive cells. Since two reproductive cells—sperm and egg—combine to form a new individual, each reproductive cell must have half as much genetic information as a body cell, or else the amount of genetic information would double in each generation. This distribution of DNA is what one would expect of the genetic material.

In addition, DNA shows very little turnover compared to other cell components. Proteins and RNA are constantly being made and destroyed, but DNA, once made, is remarkably stable.

Base Content of DNA

All DNA consists of the same four nucleotides—those containing the nitrogenous bases adenine (A), thymine (T), guanine (G), and cytosine (C). In the late 1940s, Erwin Chargaff and his co-workers discovered that all members of a species have DNA with almost exactly the same nucleotide composition (whereas their proteins' amino acid composition is more variable). Also, the nucleotides occur in different proportions in members of different species (Table 9-1). In addition, the DNA of each species contains equal numbers of adenine and thymine nucleotides, and also equal numbers of guanines and cytosines. This became a major clue to discovering the structure of DNA.

9-B   The Structure of DNA

By the early 1950s, biologists were convinced that DNA indeed carries a cell's genetic information, and many people were trying to work out the structure of the DNA molecule. Any model of DNA structure had to take several experimental findings into account:

Table 9-1   Composition of DNA from Different Organisms

Organism		Percent of nucleotide molecules			
		A	*T*	*G*	*C*
Animals:	Human	30.9	29.4	19.9	19.8
	Chicken	28.8	29.2	20.5	21.5
	Locust	29.3	29.3	20.5	20.7
Plant:	Wheat	27.3	27.1	22.7	22.8
Fungus:	Yeast	31.3	32.9	18.7	17.1
Bacterium:	*Escherichia coli*	24.7	23.6	26.0	25.7
Bacteriophage:	T$_4$	26.0	26.0	24.0	24.0

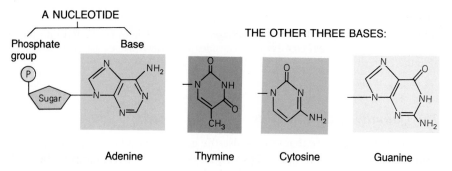

A NUCLEOTIDE

Phosphate group   Base

THE OTHER THREE BASES:

Adenine   Thymine   Cytosine   Guanine

Figure 9-3

Nucleotides found in DNA. One nucleotide unit, containing the base adenine, is shown at the left. In other DNA nucleotides, adenine is replaced by one of three other bases: thymine, cytosine, or guanine.

1. DNA is made up of nucleotides. Each DNA nucleotide has three parts: a five-carbon sugar (deoxyribose), one or more phosphate groups, and one of four possible nitrogen-containing bases: adenine (A), thymine (T), cytosine (C), or guanine (G) (Figure 9-3).
2. When the nucleotides are linked together in a strand of DNA, sugar and phosphate groups alternate to form a sugar-phosphate "backbone" held together by covalent bonds. The bases stick out to one side (Figure 9-4).
3. Chargaff had shown that the number of nucleotides containing adenine (A) equals the number containing thymine (T), and the numbers containing guanine (G) and cytosine (C) are also equal to each other. In the shorthand popular with biologists, A = T, and G = C (Table 9-1).
4. The most direct evidence for the structure of DNA came from x-ray diffraction pictures, made by passing x-rays through crystals of DNA. This produces a pattern of dots that gives information about the molecule's shape. In 1952, Rosalind Franklin produced photographs showing that DNA is twisted into a spiral, or **helix,** with the bases perpendicular to the fiber. These pictures also gave evidence that the sugar-phosphate backbone is on the outside of the helix, with the bases inside. Furthermore, the diameter of the helix showed that it must be composed of more than one such strand.

Two main questions about DNA's structure remained: how many strands are there in the DNA molecule, and how are they put together?

James Watson and Francis Crick fitted all this evidence together. They used a set of scale models of nucleotides to build possible structures until they found one that fitted all the data.

The Watson and Crick model of DNA structure consists of two strands of DNA. (To a biologist, the number two is satisfying because both cells and chromosomes reproduce by the formation of two new entities from the original one.) The two strands are arranged like a ladder, with the ladder's sides being the sugar-phosphate backbones of the two strands and the rungs being the bases (Figure 9-5).

A rung consists of either adenine paired to thymine, or guanine paired to cytosine. The atoms in each pair match up in such a way that hydrogen bonds form between the bases, and these hydrogen bonds hold the two bases in the rung together. Therefore, the bases in these pairs are said to be **complementary.** Each rung has one single- and one double-ring base (see Figure 9-3), and so all the rungs have equal widths. In each rung, either base may be on either backbone strand. The pairing of A and T, and of G and C, explains Chargaff's finding that A = T and G = C in the DNA of any species.

Finally, the whole ladder is twisted to form the spiral detected by Franklin's x-ray photographs. Because the spiral is composed of two strands wound around each other, the DNA molecule is referred to as a double helix (Figure 9-6).

Tremendously excited by this simple yet elegant structure, Watson and Crick published their model. Their paper was only two pages long, but it became a cornerstone of modern molecular genetics. As Watson remarked, "It was too pretty not to be true."

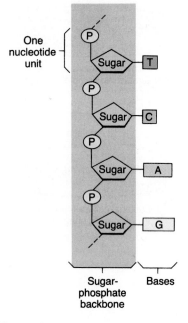

A STRAND OF DNA

One nucleotide unit

P
Sugar   T
P
Sugar   C
P
Sugar   A
P
Sugar   G

Sugar-phosphate backbone   Bases

Figure 9-4

A strand of DNA. Nucleotides are linked together, with the sugar of each nucleotide bonded to the phosphate group (yellow circle) of the nucleotide below, forming part of the sugar-phosphate backbone (gold) that runs the length of a DNA strand. The nitrogen-containing bases of the nucleotides (A, T, C, or G) stick out at the side of the strand.

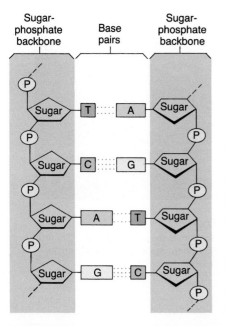

Sugar-phosphate backbone | Base pairs | Sugar-phosphate backbone

Figure 9-5
Two strands of nucleotides in a DNA molecule. Both strands' sugar-phosphate backbones lie on the outside of the molecule, with the nitrogen-containing bases meeting in the middle. Each base is held to a base on the opposite strand by hydrogen bonds (red dots). Adenine on one strand always pairs with thymine on the other, and cytosine always pairs with guanine.

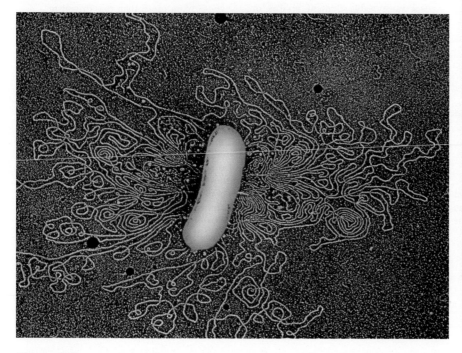

Figure 9-7
Prokaryotic DNA. This electron micrograph shows a broken cell of the bacterium *Escherichia coli,* with its single, circular DNA molecule spilling out in loops around it. If the DNA were stretched out in a straight line, it would be about a thousand times as long as the bacterium itself. (Dr. Gopal Murti/Photo Researchers/Science Photo Library)

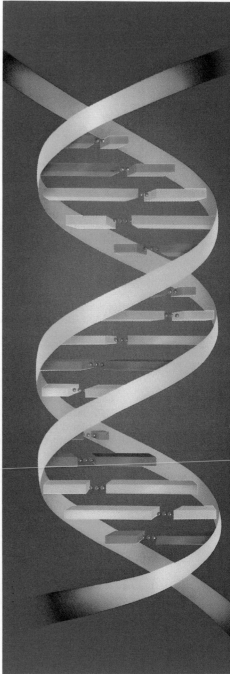

Figure 9-6
The DNA double helix. The sugar-phosphate backbones of the two strands are shown as gold ribbons. Each base sticking out from a backbone is paired with a complementary base on the opposite strand by hydrogen bonds (red dots). Adenine (dark blue) is hydrogen-bonded to thymine (dark green) and guanine (light blue) is paired with cytosine (light green). The base pairs form the "rungs" of a ladder. The whole ladder is twisted into a spiral—a double helix. (Illustration by Lili Robbins)

Both prokaryotes and eukaryotes contain DNA with this double helix structure. A prokaryote's genetic material is one long double helix of DNA with its ends joined to form a circle (Figure 9-7). The circular bacterial DNA is folded many times and occupies a nuclear area about one tenth of the cell's volume.

Eukaryotic DNA is organized into a number of chromosomes, each containing one long DNA double helix (Section 9-F). Eukaryotic cells contain more DNA than prokaryotic cells do.

Genetic Engineering

A JOURNEY INTO LIFE

Although we still have much to learn about genes, recently developed techniques have already given rise to a new technology of molecular genetics. We can isolate a desired gene and grow millions of copies of it. We can analyze these copies to find out the gene's nucleotide sequence. We can also decode this nucleotide sequence to find out the sequence of amino acids in the corresponding protein (Section 10-B). In several cases, we have even transferred functioning genes into cells of bacteria, yeasts, plants, and animals.

We can also make DNA to order, using "gene machines" that can be programmed to produce short strands of DNA in any desired sequence. This tailor-made DNA is a useful tool for studying DNA. It can also be used in protein synthesis experiments. By changing the genetic code so as to eliminate particular amino acids from a protein, we can determine how the amino acids affect the function of the protein as a whole.

These feats are all part of **genetic engineering,** the deliberate manipulation of genetic material. Its applications, today or in the future, include: making safer vaccines by

engineering a weaker version of the disease-causing agent; producing chemicals by harnessing the metabolism of microorganisms; producing enzymes for industry; cleaning up wastes from industry, oil spills, and pesticide accidents by engineering bacteria with enzymes that convert the waste to harmless substances; replacing defective genes in human beings; and creating improved strains of crops and farm animals.

All of these applications rely on our ability to transplant genes into a cell's genome. The new gene may come from another organism, of the same or a different species, or it may contain DNA produced in the laboratory.

Before genes can be transplanted into cells of higher organisms, such as plants and animals, they must first be spliced into a **vector,** a carrier such as a transposable element, a virus, or a **plasmid,** a small circular molecule of DNA that occurs naturally in some bacteria and yeasts (Figure 9-A). A vector must be able to enter a cell. It must then become part of the cell's genome so that it obeys the cell's normal controls over gene expression. It is also vital, of course, that any extra genes the vector may bring into the cell be harmless to the cell. Genes are now routinely transplanted into cells in laboratory culture. However, we still have much to learn about the control of gene expression, that is, how genes control protein synthesis. This is holding up efforts to make transplanted genes express themselves normally (Figure 9-B).

The first practical application of the new genetic technology was the production of useful proteins on a commercial scale using **recombinant**

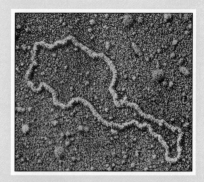

Figure 9-A
A plasmid from a bacterium. (Stanley N. Cohen, Science Source/Photo Researchers)

DNA, DNA produced by combining genes from more than one organism. In 1982, human insulin, produced by bacteria containing transplanted human insulin genes, became the first recombinant-DNA protein to reach the marketplace. Insulin is a hormone needed daily by millions of people with diabetes. Previously all the insulin available was extracted from the pancreas glands of butchered cattle and pigs, a long and expensive process. Although bacteria-grown insulin is still expensive to produce, it is better for some patients, who cannot tolerate the slight differences between human insulin and that of other species.

Another protein produced in this way is human interferon, a protein that interferes with replication of viruses (Section 25-A). So far, it is used mostly in medical research and to treat a few rare cancers.

9-C   DNA Replication

Before a cell divides to form two new cells, its DNA is **replicated,** or duplicated. Each new cell will then receive a copy of the original cell's genetic information. Because the two strands of a DNA molecule have complementary base pairs, the nucleotide sequence of each strand automatically supplies the information needed to produce its partner. For example, if one strand runs A-A-T-G-C-C, then its partner *must* run T-T-A-C-G-G. If the two strands of a DNA molecule are separated, each can be used as a mold, or **template,** to produce a complementary strand. The template and its complement together then form a new DNA double helix, identical to the original molecule. Watson and Crick suggested that this was in fact how DNA replicated.

■ *The structure of DNA contains the information needed to produce exact copies of itself.*

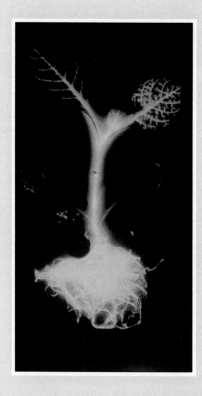

Figure 9-B
A luminous tobacco plant grown to demonstrate gene transplantation in plants. Fireflies produce light when the enzyme luciferase acts on luciferin. The gene for luciferase has been transplanted from a firefly into this tobacco plant. When the plant is "watered" with luciferin, it produces light. This shows that the transplanted gene is behaving normally by causing the tobacco plant to produce protein—the enzyme luciferase. (Courtesy of Dr. Marlene DeLuca, University of California, San Diego. From *Science* 234:856–859, 14 November 1986. © 1986 by The American Association for the Advancement of Science.)

If you own a cat, you may already have purchased genetically engineered vaccine against feline leukemia, one of the first commercially available products of recombinant DNA technology.

How Safe is Genetic Engineering?

The ability to manipulate genes obviously has many potential rewards for human society. On the negative side, the likelihood that we shall one day be able to control the genetic makeup of human beings and other organisms poses ethical problems that are new to our experience and must be faced.

Many people also worry about the possibility of an accident in a genetic engineering laboratory. Suppose a strain of bacteria with a gene for a dangerous toxin were let loose on the world? Most workers feel that the chance of this happening is slight, because safeguards are already in place. The bacteria used in many recombinant DNA experiments are *Escherichia coli,* a species universally found in the human intestine. However, the genetic strains used in the laboratory have been specially developed to be unable to survive outside their test-tube homes. The danger is further reduced by regulation of laboratories doing recombinant DNA research. Government, scientific organizations, and citizens' groups all participate in drawing up the rules that researchers must follow.

Several experiments are already underway in which genetically engineered bacteria have been applied to outdoor test plots. Researchers then watch to see whether they function as designed under field conditions, whether they die out or become established, and whether they stay put or spread beyond the site of application. Genetically engineered crop plants are also being field-tested. There has been much argument about the safety of such experimental releases of genetically engineered organisms into the environment, and they are being closely monitored.

Another concern is whether transgenic crops (those containing recombinant DNA) will be safe to eat. Since the new genes and the proteins they encode contain the same nucleotides and amino acids in all our food, there seems to be little risk from most new genes. However, new crops must be checked to make sure that any new protein they produce does not interact with the plant's normal chemistry to produce toxic substances. In addition, any plant engineered to produce toxins that fend off insects or diseases must be tested to see if its toxin content endangers human consumers. Finally, some plants are being engineered so that they can survive spraying with herbicides, which are used to kill weeds growing in the same field. The use of herbicides on these crops must be regulated to ensure that crops do not reach the table containing dangerous levels of herbicides.

Before replication can occur, the stretch of the DNA double helix about to be copied must be unwound. In addition, the two strands must be separated, much like the two sides of a zipper, by breaking the weak hydrogen bonds that link the paired bases. (For an idea of the complications involved, try unwinding the strands of half a metre of two-ply yarn.) The strands must be held apart, to expose the bases so that new nucleotide partners can hydrogen-bond to them.

The enzyme **DNA polymerase** moves along joining newly arrived nucleotides into a new DNA strand complementary to the template strand (Figure 9-8).

In a prokaryote, replication begins at one point on the DNA circle, the **replication origin.** From this point, replication travels in both directions around the molecule until the whole circle is copied. Linear eukaryotic chromosomes, in contrast, have many replication origins. Replication may start at as many as 1000 places at once and continue until all the DNA has been copied.

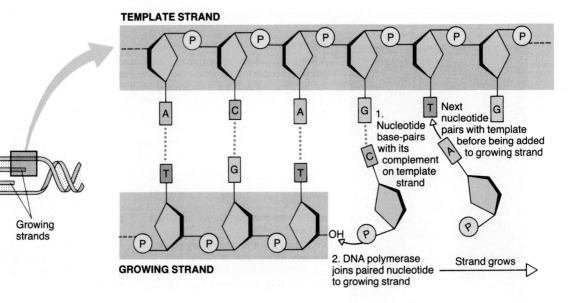

TEMPLATE STRAND

GROWING STRAND

1. Nucleotide base-pairs with its complement on template strand

Next nucleotide pairs with template before being added to growing strand

2. DNA polymerase joins paired nucleotide to growing strand

Strand grows

Template strands

Growing strands

Figure 9-8
Replication of DNA. The area shown in a box on the left is enlarged on the right. The DNA "unzips" along the weak hydrogen bonds joining paired bases on opposite strands. This produces two single strands, each of which serves as a template for a new strand. Complementary nucleotides pair with the bases exposed on the template strands. At 1, a nucleotide containing cytosine forms hydrogen bonds to pair with a complementary nucleotide, containing guanine, on the template strand. At 2, the enzyme DNA polymerase attaches the incoming cytosine nucleotide by its phosphate group to the sugar at the free end of the new, growing strand.

■ *The double helix structure of DNA is vital to organisms' genetic stability.*

■ *Both fidelity of replication and the rare occurrence of mutations are essential features of DNA. The first perpetuates an organism's instructions for leading a particular kind of life in a specific environment, whereas the second gives the potential for innovation.*

DNA replication is extraordinarily accurate. DNA polymerase makes very few errors, and most of those that are made are quickly corrected by enzymes that "proofread" the nucleotides added into the new DNA strand. If a newly added nucleotide is not complementary to the one on the template strand, these enzymes remove the nucleotide and replace it with the correct one.

Thanks to the accuracy of replication and proofreading, a cell's DNA is copied with less than one mistake in a billion nucleotides added to the growing chain.

9-D DNA Repair

Like all other biological polymers, DNA is subject to damage from agents that include the body's own heat and the aqueous environment inside the cell. Any form of damage to DNA can alter its information content and therefore could result in disastrous changes to the cell's proteins.

Although thousands of changes occur in a DNA molecule every day, not more than two or three *stable* changes accumulate in a cell's DNA each year. The vast majority are eliminated by the coordinated effort of a squad of 20 or more different kinds of DNA repair enzymes.

DNA repair depends on the existence of two copies of the genetic information, one in each strand of the double helix. As long as one strand remains undamaged, the repair enzymes can use it as a template to replace a damaged segment on its partner. So most damage is remedied unless both strands are altered beyond recognition at the same time.

9-E Mutations

Mutations are inheritable changes in DNA molecules. They may result from uncorrected errors in replication, from failure to repair damage properly, or from spontaneous rearrangements of segments of DNA (Table 9-2). In any of these cases, the new, "wrong" DNA sequence is copied just as accurately as "right" ones.

The amount of change in the mutated DNA is not necessarily correlated with its effect on the organism. For example, a change of one nucleotide pair for another in a gene coding for a protein may have effects so slight as to be undetectable, or so severe as to cause death. This depends on how the change affects the protein encoded by that gene (Section 10-B).

Many mutations are brought about by **mutagenic agents,** often called **mutagens.** Various kinds of radiation cause mutations (*A Journey Into Life,* Chapter 11). X-rays and radioactive particles may cause breaks in the DNA molecule, and have been implicated as causes of some kinds of cancer. Certain chemicals also alter DNA.

Mutations are inherited when they are copied during DNA replication and passed on to the cell's descendants. Mutation in a body cell may cause changes in the hereditary characteristics of the cell and of body parts made up of that cell's descendants. Mutations in cells destined to form eggs or sperm can be passed on to an organism's offspring. Typical mutation frequencies for a gene in a human egg or sperm range from 1 to 250 per million eggs or sperm, depending on the gene involved.

A small amount of mutation is advantageous, because it produces variation in the genetic material, and this is the raw material of evolution. However, the genetic material of modern organisms has resulted from hundreds of millions of years of natural selection. Most of the changes that occur are more apt to harm than to help the delicate balance of living cells. However, there is always the chance that some new mutation will prove beneficial.

9-F Structure of Eukaryotic Chromosomes

Eukaryotic cells contain many linear chromosomes. For example, human body cells normally contain 46 chromosomes, with a total of about six billion nucleotide pairs divided among them. Each chromosome contains a single DNA molecule extending from one end of the chromosome to the other. The DNA is associated with various proteins, forming a substance called **chromatin.** Chromatin contains roughly equal amounts of DNA and proteins.

A chromosome's long DNA molecule is protected from breaking and tangling by chromosomal proteins called **histones,** which wrap the DNA tidily up into compact chromosomes. Histones are small proteins containing a lot of the alkaline amino acids arginine and lysine (see Figure 3-14). The positively charged R groups of these amino acids bind strongly to the negatively charged phosphate groups of DNA. Histones occur in enormous amounts—there may be 60 million copies of a single type of histone in the chromatin of one cell!

The DNA from the 46 chromosomes of a human cell nucleus has a total length of about 2 metres, an average of 5 centimetres per chromosome. Chromosomes in a dividing cell are ten thousand times shorter than this. Histones and other proteins are responsible for packing these DNA molecules so tightly. The DNA is wound around clusters of histones, forming a string of bead-like particles called **nucleosomes** (Figure 9-9). Most of the nucleosome "beads," in

Table 9-2   Some Types of Mutations

Change of one nucleotide into another
Insertion of one or more nucleotides into DNA sequence
Deletion of one or more nucleotides from DNA sequence
Inversion of part of nucleotide sequence (so that part of the DNA is "backwards")
Breakage of chromosome and loss of fragment
Attachment of part of one chromosome to another
Loss of one or more entire chromosomes
Extra copies of one or more chromosomes

Figure 9-9

Nucleosomes. (a) An electron micrograph of chromatin, showing nucleosomes as dark granules, often described as looking like beads on a string. The thin line between the nucleosomes is DNA. (b) Diagram of two nucleosomes in part of a eukaryotic chromosome. In each nucleosome, the DNA strand is wound around histone proteins. (a, B. Hamkalo)

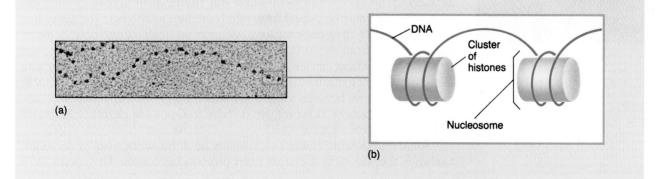

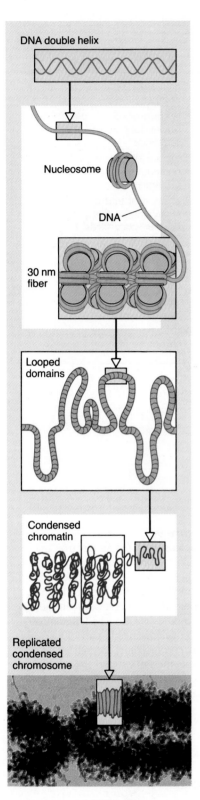

DNA double helix

Nucleosome

DNA

30 nm fiber

Looped domains

Condensed chromatin

Replicated condensed chromosome

Figure 9-10

How DNA is packed in the chromatin of a chromosome. Drawings of the sequence from the stretched-out DNA double helix (top) to the fully condensed chromosome of a dividing cell (bottom). The replicated, condensed chromosome at the bottom is from a light micrograph of a dividing cell. The human chromosomes in Figure 9-12 are also in this fully condensed state. (photo, E. J. DuPraw)

turn, are wound up still more tightly, forming a helical fiber about 30 nanometres (nm) in diameter (Figure 9-10). This fiber is still about 100 times longer than the diameter of a cell nucleus. It is further folded into large loops, called "looped domains," that shorten the chromosome further. Chromatin packed up as tightly as possible is said to be **condensed** (see Figure 9-12).

9-G   Organization of the Genome

A cell's **genome** is the total of all its genes. In 1977, researchers found methods of determining the sequences of nucleotides in DNA (and RNA) molecules. This provided the tools to describe precisely how genes are arranged within a cell's DNA molecules. Many of the findings produced by this new technique, however, were completely unexpected and are still unexplained. It is oddly true that in many ways we now know less about how genes are organized than we thought we did in 1974!

Some parts of the genome are genes that carry instructions for making proteins. Other genes dictate the sequence of nucleotides in RNA molecules found in ribosomes. Still others regulate the activity of some of their fellow genes. In prokaryotes, genes directing RNA and protein synthesis make up most of the genome. However, in many plants and animals, including humans, less than 10% of the genome serves these functions. We simply do not know what most of the rest of the DNA of these organisms does. However, we can guess that some of it probably helps to maintain chromosome structure, and some may signal where genes begin and end, or tell which kinds of cells should make the protein encoded by a gene. (For instance, red blood cells make hemoglobin but other cells do not.)

Jumping Genes

Geneticists once envisioned the genome as a fixed number of genes arranged in specific sequences on the chromosomes. In the 1940s, Barbara McClintock cast doubt on this picture when she discovered transposable genes in maize (corn) plants (Figure 9-11). **Transposable elements,** also called **jumping genes,** are segments of DNA that can move from one part of the genome to another. A gene's position is important because genes often affect their neighbors' activity.

In the 1970s, it became clear that jumping genes are both widespread and important. For instance, bacteria sometimes become resistant to drugs (such as antibiotics) that once killed them. Worse still, from a medical point of view, this drug resistance may be passed from one bacterium to another. The genes that make a bacterium drug-resistant are passengers on small transposable elements that can move from the DNA of one bacterium to that of another. One such transposable element carries a passenger gene that makes any bacterium containing it resistant to the antibiotic ampicillin. No one has yet found transposable elements that move between individual human beings or other vertebrates (animals with backbones). However, we do have transposable elements that move within the genome of a cell.

Some transposable elements apparently lie dormant for many generations, but when they do move they may exert profound effects on the genome. As a

Figure 9-11
Indian corn. The kernels grow in many different colors, which are genetically determined. McClintock discovered that the spotted and streaked patterns seen in many of these kernels result from interaction of transposable elements with genes that govern pigment production. McClintock received a Nobel Prize for her work on transposable elements in 1983.

result of their activities, genes can be moved around, duplicated, lost, split, or merged.

Repetitive DNA

Most genes are present in only one or a few copies in a genome. However, every eukaryotic cell carries many copies of the genes needed to make ribosomal RNA and histones. For example, the human genome contains 400 copies of ribosomal RNA genes and 30 to 40 copies of histone genes. This is adaptive because the cell needs large quantities of these molecules, and their production is speeded by having multiple copies of these genes.

The genome also contains other repeated DNA sequences of unknown function. About 10% of the human genome consists of a huge number of copies of two kinds of transposable elements. One of these, the *Alu* sequence, is about 300 nucleotide pairs long and is repeated a million times in each body cell. Another 10% of the DNA, called **satellite DNA,** consists of millions of copies of very short sequences of nucleotides. Most satellite DNA occurs near the centromere, the region where two replicated chromosomes are held together (Figure 9-12).

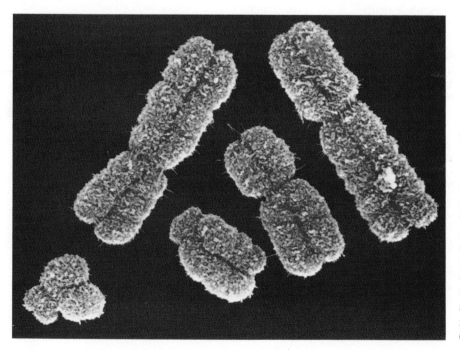

Figure 9-12
Human chromosomes. This scanning electron micrograph shows replicated chromosomes from a cell about to divide. Each chromosome now consists of two identical halves, which will eventually separate along the furrow down the length of the chromosome. Until then, the two halves are held together at an area called the centromere, the "waist" in each chromosome. Each half of the chromosome contains one very long DNA molecule, which is highly coiled. (Biophoto Associates)

SUMMARY

The evidence that DNA is the genetic material in all organisms, from bacteria to oak trees, came from several lines of inquiry:

1. DNA is the substance that transfers genetic information from one cell to another during bacterial transformation.
2. When a phage takes over the genetic machinery of a bacterium, only its genetic material, DNA, enters the cell.
3. In most species, all of the body cells of individuals of the same species contain the same amount of DNA, while reproductive cells contain half this amount of DNA. The cells of members of different species contain different amounts of DNA.
4. The DNA of all members of any species has the same proportions of A and T, and of C and G, which differ from the proportions in other species.

The DNA molecule is a double helix, with two sugar-phosphate backbones forming the sides of a twisted ladder. These strands are connected by crosswise "rungs" consisting of the base-pairs adenine and thymine or guanine and cytosine, with each base hydrogen-bonded to its complement on the opposite strand.

DNA contains the information that dictates its replication. Enzymes separate the two strands of the double helix. Each strand then serves as the template for the formation of a complementary strand of DNA by the enzyme DNA polymerase. Enzymes proofread the newly formed DNA and correct any errors of replication. Damaged DNA is also repaired. These mechanisms ensure that the nucleotide sequence of DNA is very stable and that mutations are rare.

Mutations are inheritable changes in the DNA. X-rays, ultraviolet radiation, and various chemicals are among the mutagens that may cause loss or duplication of parts of the DNA, or changes in the sequence of nucleotides, which are passed on in future replications of the DNA.

Most of a prokaryote's DNA codes for proteins or for ribosomal RNA, but such genes are only a small fraction of the DNA of eukaryotic plants and animals. The genome of eukaryotes also contains much DNA that is not active and whose function (if any) is unknown. Transposable genetic elements can move from one place to another, carrying other genes along.

SELF-QUIZ

1. DNA is believed to be the genetic material because:
   a. all the body cells of an individual seem to have identical amounts and compositions of DNA, while reproductive cells have half the amount of DNA found in body cells.
   b. the proteins are the same from cell to cell in an individual, but the DNA differs; thus the DNA must be the material that makes different tissues different.
   c. DNA is the largest type of macromolecule found in living organisms.
   d. DNA is found in the cell nucleus.
2. A nucleotide consists of:
   a. A, G, T, and C
   b. nitrogenous bases
   c. a sugar, a phosphate group, and a nitrogen-containing base
   d. a sugar-phosphate backbone
3. In a DNA molecule:
   a. nitrogenous bases bond covalently to phosphate groups
   b. sugars bond ionically to nitrogenous bases
   c. sugars bond to nitrogenous bases by hydrogen bonds
   d. nitrogenous bases bond to each other by hydrogen bonds

4. Write the sequence of nucleotide bases that would be found in a strand of DNA complementary to this template strand:
   A-T-C-T-G-T-A-T-G-A

   _____

5. The number of adenine bases in a DNA molecule equals the number of thymine bases because:
   a. Whenever DNA polymerase places a thymine base into a new DNA strand, it always puts an adenine directly after it
   b. A DNA strand consists of alternating adenine and thymine bases
   c. Adenine on one strand hydrogen-bonds to thymine on the other strand
   d. DNA contains equal numbers of each of the four nitrogenous bases
6. A mutation may result from:
   a. addition or loss of one or more nucleotides in a DNA strand
   b. change of one nucleotide to another
   c. loss of an entire chromosome
   d. part of a DNA strand getting turned around "backwards"
   e. all of the above

Tell whether each of the following would be found in the nuclei of your own cells, in the cells of the intestinal bacterium *Escherichia coli,* or both:

___ 7. Circular DNA molecules

___ 8. DNA with adenine base-paired to thymine, and guanine to cytosine

___ 9. DNA in 46 linear molecules

___ 10. DNA wound into nucleosomes

___ 11. Genes carrying instructions for proteins and ribosomal RNA

___ 12. Histones closely bound to DNA

QUESTIONS FOR DISCUSSION

1. Why is the constancy of DNA content from cell to cell in an organism considered to be evidence that DNA is the genetic material? Is it necessary for cells of an organism to contain identical genetic information? Is it possible for an organism to have different genetic information in different cells of the body?

2. What is the biological importance of the fact that the sugar-phosphate backbones of the DNA double helix are held together by covalent bonds, and that the cross-bridges between the two strands are held together by hydrogen bonds?

3. Why is it necessary to limit the amount of x-rays a person is exposed to over a given period of time? Which organs must be especially well shielded from x-ray exposure?

4. Is transplanting genes into people for medical reasons ethically equivalent to transplanting a replacement kidney into someone whose own kidneys have failed?

5. What limits would you place on gene transplant research involving human eggs and embryos? Why?

6. Manipulation of the human genome raises many horrifying possibilities, including populations with many more of one sex, or clones of identical individuals. What ethical guidelines exist to help us determine what we should and should not permit? What additional guidelines would you propose?

7. Growth hormone produced by genetic engineering techniques is now being used to treat children whose bodies do not produce enough. The added hormone enables them to grow normally. How might such substances be abused? What rules do we need to prevent this from happening?

SUGGESTED READINGS

Fedoroff, N. V. "Transposable genetic elements in maize." *Scientific American,* June 1984.

Gilbert, W., and L. Villa-Komaroff. "Useful proteins from recombinant bacteria." *Scientific American,* April 1980. Gives further details of recombinant DNA techniques.

Radman, M., and R. Wagner. "The high fidelity of DNA duplication." *Scientific American,* August 1988.

Watson, J. D. *The Double Helix.* New York: Atheneum, 1968. A personal story of the discovery of DNA structure. Highly readable and human.

Watson, J. D., and F. H. C. Crick. "Molecular structure of nucleic acids. A structure of deoxyribose nucleic acid." *Nature* 171:737, 1953. A classic Nobel prize-winning paper.

Watson, J. D., J. Tooze, and D. T. Kurtz. *Recombinant DNA: A Short Course.* New York: W. H. Freeman, 1983.

RNA and Protein Synthesis

Milk production

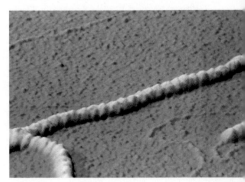

DNA-protein complex

Growth: red-humped caterpillar

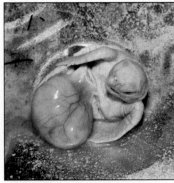

Development: sea turtle embryo

*D*NA contains genetic information, which governs inherited characteristics. But how does DNA produce these inherited features of organisms? In the 1940s, scientists began to suspect that genetic information dictates the structure of proteins. If this is true, then organisms containing different genetic information should have different proteins.

Experiments by George Beadle and Edward Tatum supported this idea. They studied several strains of pink bread mold (*Neurospora*) containing **mutations,** inheritable changes in their genetic material. Beadle and Tatum found that each mutant strain had lost the ability to produce one enzyme. We now know that all proteins, not just enzymes, are made under genetic control.

These experiments were done in the late 1940s, when biologists were coming to realize that DNA is the genetic material (Section 9-A). Scientists then suggested that the order of nucleotides in DNA determines the order of amino acids in proteins.

It also became clear that ribonucleic acid (RNA) plays a role in protein synthesis. Cells that make a lot of protein also make RNA rapidly. In a eukaryotic cell, DNA stays in the nucleus, but RNA occurs both in the nucleus and in the cytoplasm, where ribosomes carry out protein synthesis. It seemed likely that RNA acted as a go-between, carrying genetic information from nuclear DNA into the cytoplasm, where it could be used to make proteins. We now know that this is in fact the case. The transfer of information from DNA to RNA to proteins is one form of **gene expression.**

Important as genes are, it has always been rather difficult to define a gene, partly because our understanding of the term keeps changing. Like "truth," the word "gene" describes a concept that is fuzzy around the edges. For our purposes, we can define a **gene** as a length of DNA that serves as a functional unit. Most of the genes discussed in this chapter are **structural genes,** genes containing information needed to make proteins. Genes with other functions also exist. Some contain the information needed to make molecules of transfer and ribosomal RNA (Section 10-A). Some are regulatory genes, whose function is to control the activity of other genes. Regulatory genes determine whether or not a structural gene will be expressed, that is, be used to direct the production of its protein.

In this chapter we shall examine the structure of RNA and the roles of various types of RNA in protein synthesis. Then we shall see how proteins are assembled and look at how a cell controls which of the many proteins encoded by its genes are actually made.

- Proteins are made on ribosomes using genetic information from DNA.
- The genetic information for protein synthesis is carried from the DNA to the ribosomes by RNA molecules.
- Genes are not always expressed: a cell is programmed to make only the proteins it needs.

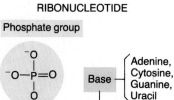
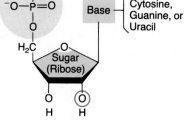

RIBONUCLEOTIDE

Phosphate group

Adenine, Cytosine, Guanine, or Uracil

Base

H_2C

Sugar (Ribose)

Figure 10-1
A ribonucleotide monomer of RNA. A ribonucleotide contains the sugar ribose, which has one more oxygen (red circle) than does the sugar deoxyribose in DNA.

10-A   RNA

Like DNA, RNA comes in long unbranched molecules made up of nucleotide subunits. Each nucleotide is made up of a sugar, a nitrogen-containing base, and a phosphate group (Figure 10-1).

However, RNA differs from DNA in several respects:

1. RNA usually consists of a single strand of nucleotides, while DNA is double-stranded, with two complementary chains of nucleotides.
2. The sugar in RNA is ribose, whereas the sugar in DNA is deoxyribose, with one less oxygen atom than ribose.

3. DNA and RNA differ in the kinds of bases they contain. In DNA we find adenine, guanine, cytosine, and thymine. RNA also contains adenine, guanine, and cytosine; but uracil (which is very similar to thymine) takes the place of thymine.

Three main types of RNA participate in protein synthesis:

1. **Messenger RNA (mRNA)** contains the code for the order of amino acids in a protein. It carries this information from the DNA of a structural gene to ribosomes, where the protein is made.
2. **Transfer RNA (tRNA)** molecules carry amino acids to the mRNA at ribosomes and fit them into the proper place in the growing protein chain.
3. **Ribosomal RNA (rRNA)** is a major component of ribosomes. We do not yet know much about its role in protein synthesis. However, one rRNA molecule recognizes the beginning of the message in mRNA.

Transcription of DNA Into RNA

All RNA is made using information from a DNA template. RNA synthesis is called **transcription** ("written across") because it rewrites the genetic message coded in DNA, into an RNA molecule. In each gene, only one strand of the DNA acts as the template for formation of a complementary strand of RNA. (The template strand may be different in adjacent genes.)

Transcription of DNA into RNA begins when the enzyme **RNA polymerase** binds a specific DNA sequence called a **promoter,** which signals where RNA synthesis should start. The promoter also determines which DNA strand is transcribed. RNA polymerase separates nearby DNA into two strands. It moves along the DNA, using the coding strand as a template and joining the complementary nucleotides (A, C, G, or U) together one by one, to form an RNA strand. The process is similar to the formation of a complementary strand of DNA (see Figure 9-8). When the polymerase reaches a **termination** signal on the DNA, it leaves the DNA, and the newly transcribed RNA strand also detaches. A particular

■ *DNA carries the genetic information for the sequence of nucleotides in each of the three types of RNA. One of these, messenger RNA, carries the code for the sequence of amino acids in a protein.*

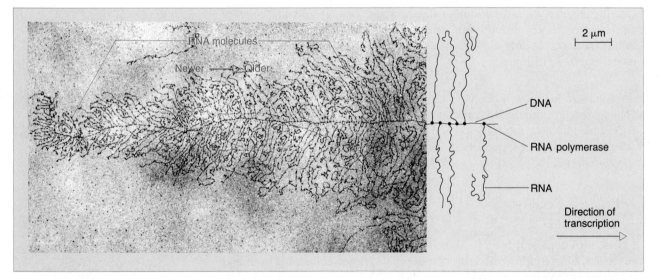

Figure 10-2
Gene transcription. The more or less horizontal line across the middle of this photograph is a strand of DNA. The lines above and below it are many molecules of RNA still attached to the RNA polymerases that are making them. We can tell that polymerase enzymes are moving along the DNA from left to right in the photograph because the longer RNA molecules are on the right. (×15,000; Courtesy of O. L. Miller)

stretch of DNA may be transcribed by several RNA polymerase molecules at once, forming many copies of the complementary RNA (Figure 10-2).

10-B The Genetic Code

Each structural gene carries the code for one **polypeptide,** a long chain of amino acids. A functional **protein** consists of one or more polypeptides. DNA, RNA, and polypeptides are all long, unbranched molecules. It is therefore not surprising that a sequence of nucleotides in DNA is transcribed to a sequence of nucleotides in messenger RNA, which in turn codes for a sequence of amino acids in a polypeptide. The synthesis of protein using this code is called **translation** (because genetic information is "translated" from the nucleotide "language" of DNA and RNA into the amino acid language of proteins).

Biologists speak of the translation of genetic information as if it were the decoding of a secret message. Messenger RNA contains four kinds of nucleotides, and so the genetic "language" must have a four-letter "alphabet." Proteins are made up of 20 kinds of amino acids, so the genetic code must have at least 20 different "words," one for each amino acid. The words cannot be only one nucleotide "letter" long, because in that case there would be only four possible code words, and proteins could contain only four different amino acids. Similarly, the words cannot be just two nucleotides long, because four letters arranged in all possible combinations of two gives only 16 different code words (4^2), still not enough to specify 20 different amino acids (Figure 10-3). The four nucleotides arranged in threes, however, produce 64 possible different code words (4^3), more than enough to produce a unique code word for each amino acid. The smallest theoretical size for a code word in DNA is, therefore, three nucleotides.

Francis Crick and others tested this triplet code theory by adding different numbers of nucleotides into the DNA of a bacteriophage. They reasoned as follows: if the code is a triplet code, then inserting just one or two nucleotides into the middle of a gene will change the entire message after that point into something completely different. For example, if one or two nucleotides containing guanine (G) are added, the DNA message

CAT—CAT—CAT

might become

CA**G**—TCA—TCA—T or CA**G**—**G**TC—ATC—AT

However, inserting three nucleotides into the middle of the gene should merely create a short disruption, after which the message will read like the original version:

CAT—**GGG**—CAT—CAT or CA**G**—**GG**T—CAT—CAT or
C**GG**—**G**AT—CAT—CAT

Experiments bore out this prediction. When a string of three extra nucleotides was added into the DNA, a slightly altered protein was produced. Adding one, two, or four nucleotides changed all subsequent code words, and so the polypeptide made according to these new instructions contained a completely different string of amino acids. This usually made the protein (and hence the bacteriophage) unable to function.

Biochemists worked out the code by making artificial messenger RNAs with known sequences of nucleotides. These were put into test tubes with ribosomes, transfer RNAs, amino acids, and other chemicals needed for protein synthesis, and here they were translated into polypeptides. A long string of uracil (U) nucleotides was translated into a polypeptide containing only the amino acid

THE FOUR CODE LETTERS

A
C
G
U

SIXTEEN DOUBLETS FROM THE FOUR CODE LETTERS

AA	AC	AG	AU
CA	CC	CG	CU
GA	GC	GG	GU
UA	UC	UG	UU

Figure 10-3
Two-letter words. The four different kinds of nucleotides in RNA can be arranged in pairs to form 16 different possible combinations.

Table 10-1 Codons Found in Messenger RNA*

First Base	Second Base				Third Base
	U	**C**	**A**	**G**	
U	UUU ⌉ Phenylalanine UUC ⌋ UUA ⌉ Leucine UUG ⌋	UCU ⌉ UCC ⌉ Serine UCA ⌋ UCG ⌋	UAU ⌉ Tyrosine UAC ⌋ UAA ⌉ STOP UAG ⌋ STOP	UGU ⌉ Cysteine UGC ⌋ UGA STOP UGG Tryptophan	U C A G
C	CUU ⌉ CUC ⌉ Leucine CUA ⌋ CUG ⌋	CCU ⌉ CCC ⌉ Proline CCA ⌋ CCG ⌋	CAU ⌉ Histidine CAC ⌋ CAA ⌉ Glutamine CAG ⌋	CGU ⌉ CGC ⌉ Arginine CGA ⌋ CGG ⌋	U C A G
A	AUU ⌉ AUC ⌉ Isoleucine AUA ⌋ AUG Methionine (START)	ACU ⌉ ACC ⌉ Threonine ACA ⌋ ACG ⌋	AAU ⌉ Asparagine AAC ⌋ AAA ⌉ Lysine AAG ⌋	AGU ⌉ Serine AGC ⌋ AGA ⌉ Arginine AGG ⌋	U C A G
G	GUU ⌉ GUC ⌉ Valine GUA ⌋ GUG ⌋	GCU ⌉ GCC ⌉ Alanine GCA ⌋ GCG ⌋	GAU ⌉ Aspartic acid GAC ⌋ GAA ⌉ Glutamic acid GAG ⌋	GGU ⌉ GGC ⌉ Glycine GGA ⌋ GGG ⌋	U C A G

* To use the table, find the letter of the first base of the codon in the column at the left, and go across this row until you are in the column headed by the letter of the second base. Then find the third base, marked at the far right of the table. The three *STOP* codons signal positions where the ribosome stops reading and terminates the polypeptide chain. The codon AUG initiates synthesis of a polypeptide.

phenylalanine. So, the mRNA "word" UUU must code for phenylalanine. (The DNA code word is the base-pair complement of this, AAA.) Eventually, all the amino acid code words were worked out. For this work, Marshall Nirenberg and H. Gobind Khorana received a Nobel Prize in 1968.

The mRNA code words, or **codons,** appear in Table 10-1. You should note several features of the genetic code from this table:

1. The codons shown are the code words found in messenger RNA. The DNA code triplets are the complements of those shown.
2. Three of the 64 triplets do not code for amino acids: UAA, UAG, and UGA are *Stop* codons, signals for the end of a polypeptide chain.
3. There is more than one codon for most amino acids, and so the code is said to be **degenerate.** This is biologically useful. For one thing, it makes mutations less damaging. If the code were not degenerate, 20 codons would code for amino acids, and 44 would code for nothing (that is, they would act as *Stop* codons). Therefore, most mutations would lead to *Stop* codons, which would stop protein synthesis. Shortening a polypeptide in this way usually leads to inactive proteins, whereas substituting one amino acid for another may be harmless.
4. A codon's third base is often less specific than the first two. For instance, all of the codons for the amino acid proline have CC as the first two bases.

The genetic code contains no punctuation or spaces such as might signal the beginning or end of a codon. In other words, the code must be read from a particular starting point or the whole sequence will be read incorrectly. For instance, the RNA sequence UCUAGAGCUA will produce the amino acid sequence Serine—Arginine—Alanine if read from left to right. If the reading of the RNA sequence starts at the second nucleotide (C), however, instead of at the beginning, it will produce the completely different amino acid sequence Leucine—Glutamic acid—Leucine. It is therefore vital for messenger RNA to have an initiation point, which says in effect "start here." The initiation codon is the sequence AUG, which codes for the amino acid methionine.

Some mutations exchange one nucleotide for another. By examining Table 10-1, you can see that a mutation that changes the third nucleotide in a codon often will not change the amino acid specified by the codon. However, changing the first or second nucleotide is likely to result in placing a different amino acid into that slot, while leaving the rest of the amino acid sequence unchanged.

Mutations that change the genetic material by adding or deleting nucleotides are called **frameshift mutations** because they shift the entire reading frame of the message. This usually changes the sequence of amino acids produced beyond the mutated area, as in the CAT-CAT-CAT example above.

With minor exceptions, the genetic code is universal. The same codons code for the same amino acids in virtually all viruses, bacteria, plants, animals, and fungi. This is compelling evidence that all organisms on earth today evolved from a common ancestor. The major groups of organisms have had separate evolutionary histories for hundreds of millions of years, and so it seems that the code must have been established shortly after life originated, and have continued almost unchanged for the billions of years since. The main exceptions to this rule are mitochondria, organelles that contain their own DNA and ribosomes. Of the 64 triplet codons, up to six may mean something different in mitochondria than they do in the surrounding cytoplasm.

10-C *Overview of Protein Synthesis*

Before we look at the molecules involved in protein synthesis in more detail, let us briefly sum up the whole process. RNA is made on a DNA template. The genetic code for a polypeptide is copied from DNA into messenger RNA (mRNA) molecules. The mRNA carries this code to ribosomes, where protein synthesis takes place (Figure 10-4). The mRNA code specifies the order in which amino acids are joined to form a polypeptide. Messenger RNA cannot recognize an amino acid directly. An adaptor molecule, transfer RNA (tRNA), is needed to bring the two together. Transfer RNA molecules bring the proper amino acids to the ribosome and fit them into their coded positions. As the genetic code in an mRNA molecule is "read" on a ribosome, the amino acids are joined together, one by one.

Figure 10-4
Gene expression. The flow of genetic information from DNA to the amino acid sequence of a polypeptide.

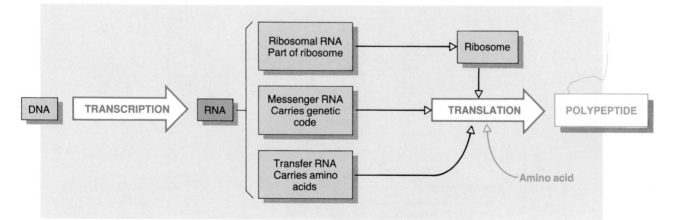

Figure 10-5

Simultaneous transcription and translation. In this electron micrograph of *Escherichia coli* genes, RNA polymerase molecules are transcribing DNA into mRNA. As the front end of each mRNA is formed, a ribosome binds to it and starts to translate it. (The polypeptide molecules cannot be seen.) As the ribosome translates, it moves up the mRNA. As soon as there is room behind it, another ribosome attaches to the mRNA. An mRNA molecule with several ribosomes attached is called a polysome. (Courtesy O. L. Miller. O. L. Miller, Jr., B. A. Hamkalo, and C. A. Thomas, Jr. *Science* 169:392, 1970)

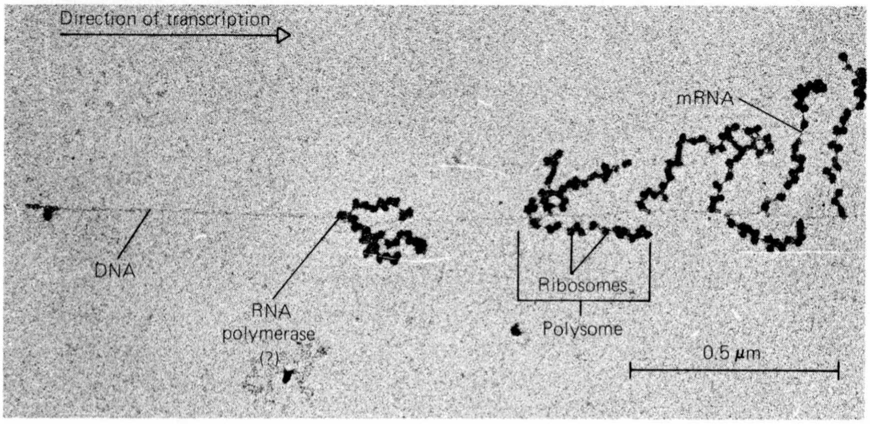

■ *Messenger RNA carries genetic information from genes to ribosomes.*

Figure 10-6

How a mature messenger RNA molecule is produced in a eukaryotic cell.

10-D   Messenger RNA

A prokaryote's DNA lies in the cytoplasm. As messenger RNA is transcribed, ribosomes may attach to the front end of the mRNA and start to translate it into protein while the rest of the molecule is still being made on the DNA (Figure 10-5).

In eukaryotes, on the other hand, transcribed RNA must be processed in the nucleus before it becomes mature mRNA, ready to enter the cytoplasm for protein synthesis. First, a special nucleotide "cap" is attached at the front end. This is the signal that will bind the mRNA to a ribosome. A "tail" of adenine-containing nucleotides is attached to the other end. The tail is thought to extend the lifetime of the RNA somewhat, before it is degraded by ever-present RNA-digesting enzymes.

Many eukaryotic structural genes contain sections called **intervening sequences** or **introns,** which do not code for parts of proteins. Introns are

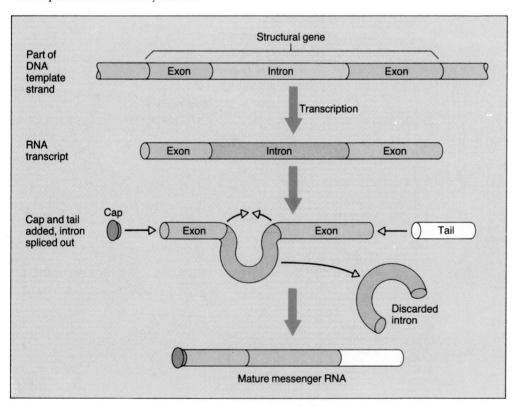

transcribed, but they are then "spliced" out of the RNA molecule, leaving a shorter, mature mRNA that can be translated into protein. Some eukaryotic genes contain no introns, but others have as many as 50 introns. The parts of the gene that are represented in the mature mRNA are called **exons** (because they are e̲xpressed as protein) (Figure 10-6). The function of introns is not well understood. Some think they may play a role in the evolution of proteins, but others regard introns as parasites or as junk DNA.

10-E Ribosomal RNA and Ribosomes

Ribosomes, the sites of protein synthesis, consist of several types of ribosomal RNA and about 70 kinds of polypeptides. In eukaryotes, ribosomes are made in an area of the nucleus called the **nucleolus.** The nucleolus bustles with activity, sometimes churning out several hundred thousand ribosomes per hour.

A functional ribosome consists of two subunits, one large and one smaller (Figure 10-7). The two parts join together only for protein synthesis. The ribosomes of prokaryotes, chloroplasts, and mitochondria are similar to, but smaller than, the ribosomes in eukaryote cytoplasm.

10-F Transfer RNA

Transfer RNA (tRNA) carries amino acids to the ribosomes during protein synthesis. Each kind of amino acid has a specific kind of tRNA molecule to transport it.

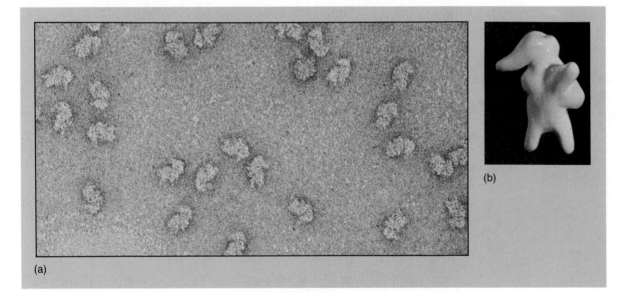

(a)

(b)

Figure 10-7
Working out the structure of ribosomes. (a) Electron micrograph of small ribosomal subunits isolated from human tumor cells grown in cell culture. The subunits are irregular in shape, so they may lie on the specimen stage in various positions. It is difficult to tell exactly where the subunits end. This problem is overcome by measuring photographs of hundreds of subunits and feeding the information into a computer. (b) Computer-generated maps of all sides of the subunit are used to build this clay model of the subunit. Fairly accurate models of both small and large subunits from several other organisms have also been made by this method, revealing that ribosomes from different organisms have somewhat different shapes. (Courtesy Miloslav Boublik)

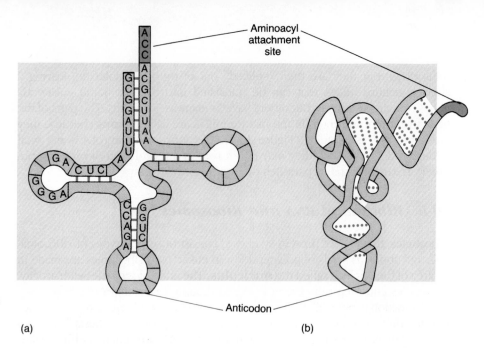

Figure 10-8

Transfer RNA. (a) The structure of a tRNA molecule. (b) The molecule in (a) is twisted into this three-dimensional shape. Hydrogen bonds between base-paired parts of the molecule are shown as red dots. All tRNA molecules have roughly the same shape, but their nucleotide sequences differ somewhat. This particular tRNA carries the amino acid phenylalanine.

(a)

(b)

Aminoacyl attachment site

Anticodon

■ *Each transfer RNA carries a particular type of amino acid and matches up to the correct codon position of an mRNA attached to a ribosome.*

All tRNA molecules have the same general shape. Parts of the molecule fold into loops, held in shape by base-pairing between different areas of the molecule (Figure 10-8). The most important parts of the tRNA molecule are the aminoacyl attachment site and the anticodon.

The **aminoacyl attachment site** is the site where the amino acid is attached to the tRNA molecule, a process called **acylation.** Acylation is carried out by specific enzymes, each of which recognizes the unique shapes of one kind of tRNA and of the corresponding amino acid. Acylation uses energy from ATP. Some of this energy is stored in the bond linking the amino acid to its tRNA. Breaking this bond eventually provides the energy to attach the amino acid to the growing polypeptide chain.

A tRNA molecule's **anticodon** is a row of three nucleotides that base-pairs with the complementary codon of an mRNA molecule attached to a ribosome. This allows the amino acid carried by the tRNA to be placed correctly into the peptide chain.

10-G Protein Synthesis

The basic reaction of protein synthesis is the formation of a peptide bond between two amino acids (see Figure 3-15). This reaction is repeated many times, as each amino acid in turn is added to the growing polypeptide chain. But before protein synthesis can begin, several events must occur.

Initiation

First, mRNA binds to a small ribosomal subunit (Figure 10-9). The first codon (AUG) then base-pairs with the anticodon of a tRNA carrying the amino acid methionine. This methionine eventually becomes the first amino acid in the polypeptide chain. (After the first AUG, any other AUG codons on the mRNA are translated by inserting the amino acid methionine into the chain.) Now a large ribosomal subunit binds to the complex. Initiation is complete and the peptide chain can be made.

Peptide Chain Formation

A ribosome has two sites where codons are translated (called the P and A sites). The events outlined above have placed the mRNA's initiation codon, AUG, at the ribosome's P site (see Figure 10-9). The mRNA codon for the second amino acid

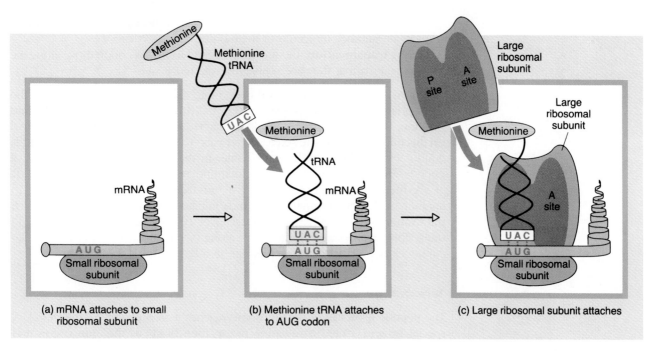

(a) mRNA attaches to small ribosomal subunit

(b) Methionine tRNA attaches to AUG codon

(c) Large ribosomal subunit attaches

Figure 10-9
The initiation of protein synthesis. The colored arrows point to the action at each step.

is lined up at the A site. Three steps then bring in the next amino acid and join the first one to it (Figure 10-10):

1. A tRNA with a complementary anticodon binds to this second mRNA codon, at the A site. The amino acid carried by this second tRNA will become the second amino acid in the peptide chain.
2. An enzyme joins the first two amino acids in the chain to each other by a peptide bond. The first tRNA is now empty and the second is holding both amino acids.

Figure 10-10
The three steps in elongation of a peptide chain.

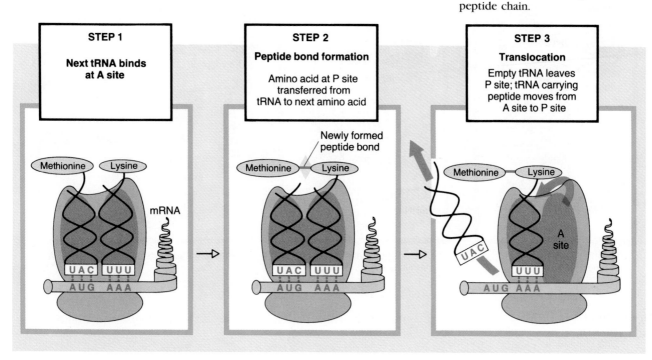

STEP 1

Next tRNA binds at A site

STEP 2

Peptide bond formation

Amino acid at P site transferred from tRNA to next amino acid

STEP 3

Translocation

Empty tRNA leaves P site; tRNA carrying peptide moves from A site to P site

The empty (methionine) tRNA now leaves the ribosome. It eventually picks up another molecule of methionine.

3. During **translocation,** the second tRNA and its mRNA codon move along the ribosome so that they are now at the vacated P site.

Translocation brings the third codon into the ribosome's A site, where the appropriate tRNA attaches by its anticodon, bringing the third amino acid into position. The growing peptide chain is attached to the newly arrived amino acid on this third tRNA, and the sequence repeats. This sequence of three steps adds amino acids to the peptide one by one. In the bacterium *Escherichia coli,* a peptide grows at the rate of about 20 amino acids per second.

As the leading end of the mRNA emerges from the ribosome, it may bind to another small ribosomal subunit, which initiates protein synthesis again. Each mRNA molecule typically has several to over 100 ribosomes attached to it and transcribing its message as they move along (see Figure 10-5). An mRNA molecule does not last long. Its average life in a bacterium is 2 minutes.

Termination

The peptide grows until the ribosome reaches a *Stop* codon on the mRNA. A special protein, called a releasing factor, then binds to the *Stop* codon and causes the mRNA to leave the ribosome (Figure 10-11).

Many newly made proteins must be processed further after they are released from the ribosome. Generally the first methionine, and often some of its neighbors, must be removed. Various chemical groups may be added or changed. Sulfur bridges must often be formed between different parts of the

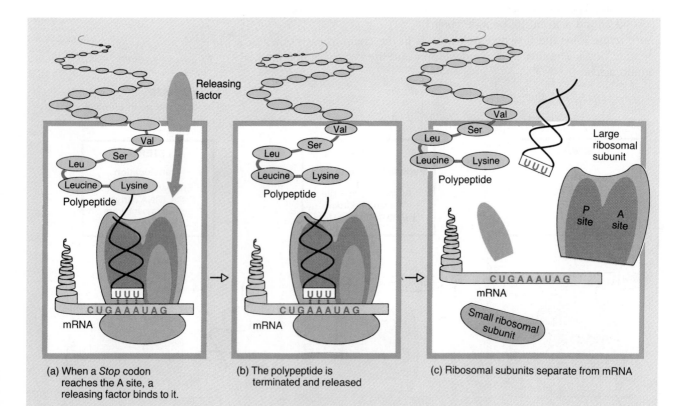

(a) When a *Stop* codon reaches the A site, a releasing factor binds to it.

(b) The polypeptide is terminated and released

(c) Ribosomal subunits separate from mRNA

Figure 10-11
Termination of protein synthesis.

molecule or between two separate polypeptides that make up a larger protein molecule (Section 3-E).

10-H   Control of Protein Synthesis

A cell's DNA contains instructions for making hundreds of proteins, but protein synthesis is controlled so that a cell makes only some of them. A gene may be thought of as being turned on, or expressed, when its DNA is being transcribed into RNA and the RNA translated into protein. Genes that are not being transcribed can be thought of as turned off.

Bacteria switch genes on and off quickly. This permits them to exist in places, such as a rotting leaf or inside an animal's intestine, where the available food changes rapidly. Within minutes, a bacterium can produce the enzymes needed to use a new kind of food molecule that enters the cell. In multicellular eukaryotes, the ability to change protein production quickly is not so important. The surroundings of each cell change only slowly. The most important task is to produce and coordinate many different types of cells.

In nearly all cells, large numbers of "housekeeping genes" remain switched on throughout life. These genes control functions that most cells perform all the time, such as protein synthesis, glycolysis, and food uptake. A much smaller number of specialist genes makes one type of eukaryotic cell different from another, giving a nerve cell its distinctive functions or making a liver cell different from a muscle cell.

Protein synthesis is usually controlled by regulating the transcription of RNA. Here we consider how this occurs in prokaryotes and some additional types of control that have been added during the evolution of eukaryotes.

Control in Prokaryotes

Control of protein synthesis in prokaryotes was first shown in a classic study published in 1961 by François Jacob and Jacques Monod. They worked with *Escherichia coli* bacteria, which can grow in a medium containing only salts and a source of carbon—for example, glucose. The bacteria made enzymes that metabolize glucose and used up most of the glucose. Then Jacob and Monod added a different sugar, lactose, to the medium. The bacteria stopped growing briefly while they started to produce a new set of enzymes that handle lactose. Jacob and Monod studied the control of this switch from producing one set of enzymes to producing another. This work led to a general model of how transcription is turned on and off in bacteria.

The agents that switch genes on and off are **gene regulatory proteins,** proteins that bind to DNA and start or stop transcription. Gene regulatory proteins bind to a specific DNA sequence next to a gene's promoter. Some of these proteins prevent the binding of RNA polymerase to the promoter and so prevent transcription, that is, they keep the gene turned off. Others help RNA polymerase bind and so turn the gene on. Each kind of gene regulatory protein is specific for the promoter site of one to several genes, and so it turns only these genes on or off. (Some even turn one gene on and another off.)

Cells always contain gene regulatory proteins. The presence of other chemicals in the cell determines if and when these proteins control transcription. In Jacob and Monod's experiment, while the bacteria are living on glucose, a gene regulatory protein bound near the lactose genes' promoter prevents transcription. When lactose is added, some of it is converted to another sugar, which binds to this gene regulatory protein and causes it to unbind from the DNA. RNA polymerase can then bind to the promoter and transcribe the genes for enzymes that metabolize lactose. Hence the molecule derived from lactose acts as an

■ *In prokaryotes, transcription is turned on or off by the binding of gene regulatory proteins to sites near the promoter. The proteins' ability to bind to DNA depends, in turn, on the presence of other molecules.*

Proteins as Evolutionary Puzzle Pieces

A JOURNEY INTO LIFE

The discovery of the genetic code and the invention of ways to find the amino acid sequence in a polypeptide opened an exciting new field: the use of proteins as "living biochemical fossils." Traditionally, evolutionary "trees" have been constructed by comparing living and fossil organisms, and by determining the ages of rocks in which the fossils are found. Comparing proteins from various modern organisms can sometimes confirm the fossil evidence and can sometimes shed light on points where the fossil record is inadequate. The sequence of amino acids in a living organism's proteins can be used, just like the structure of its teeth or bones, to trace the organism's probable ancestry.

To construct an evolutionary tree that includes most living organisms, we select a protein found in many different organisms. We then take samples of this protein from many species and determine the amino acid sequence in each sample. A popular subject is cytochrome *c*, an electron transport molecule in cellular respiration. All aerobic organisms make some form of cytochrome *c*. Once the amino acid sequences have been determined, a computer compares the proteins with one another and tabulates their similarities and differences. The more similar the amino acid sequences of the protein in two species, the more closely related the species are likely to be (Figure 10-A).

If the computer is also supplied with the genetic code, it can construct the "missing links" between related species. That is, it can work out the probable structure of the cytochrome *c* in a species that was the common ancestor of the two modern species. For example, suppose one species has the amino acid isoleucine in a particular position in its cytochrome *c*, and another species has proline in that position. The DNA code TAA stands for isoleucine, and GGA stands for proline. If one of these codes was present in the ancestral DNA, then there must have been mutations in two adjacent nucleotides to convert the code found in one species to that found in the other. (That is, the DNA nucleotides TA must have mutated to GG, or vice versa.) If, however, both species came from a common ancestor whose cytochrome *c* had leucine at that position, then each species could have arrived at its present amino acid by mutation of only one nucleotide. The ancestral DNA could have been GAA:

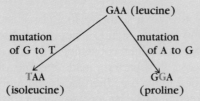

This case is more probable.

Some positions in some proteins are occupied by the same amino acid in every species for which that protein has been analyzed. Is this coincidence? Probably not. It is more likely that these amino acids cannot be replaced without destroying the protein's folding pattern, and hence its function. Variations in some of the other amino acids do not seem to make so much difference, although they may be responsible for the "fine tuning" that adapts the protein to a species' particular environment.

The relationships among vertebrates have been studied by examining the structure of hemoglobin, a protein that is found in all vertebrates. Hemoglobin consists of four polypeptides: two alpha chains and two beta chains. Analysis of the beta chains has shown that humans and chimpanzees have identical beta chains in their adult hemoglobin. Gorilla and human beta chains differ by only one amino acid; pigs differ from humans by about 17 amino acids; and horses and humans differ by 26.

The hemoglobin chains are members of a large family of similar molecules. The alpha and beta chains of hemoglobin are similar to each other and to different types of hemoglobin chains made only in the fetus before birth. They also resemble myoglobin, a muscle protein that stores oxygen.

How did so many different types of molecules with similar structures come into being? Most likely, at some point in the evolutionary past, an ancestral gene for an oxygen-carrying protein was replicated in more than one copy. Such multiple copies of genes may well have undergone **adaptive radiation,** the evolution of divergent features. In this way, one kind of protein could have evolved into several different kinds, each with a slightly different function. An individual that had all of these different, specialized proteins might have had a selective advantage over individuals that had only one, unspecialized protein to do all the jobs. The process of gene duplication, followed by adaptive radiation of the different copies, is believed to play an important role in evolution.

inducer, a signal that turns on the production of proteins needed for lactose metabolism (Figure 10-12).

Control in Eukaryotes

In eukaryotes, once a gene has been switched on, it tends to remain that way for a long time, whether it is a housekeeping gene or a specialist gene. Once a cell has differentiated into, say, a liver cell, its descendants remain liver cells, with the same specialist genes switched on, throughout the many cell divisions that occur during the animal's life.

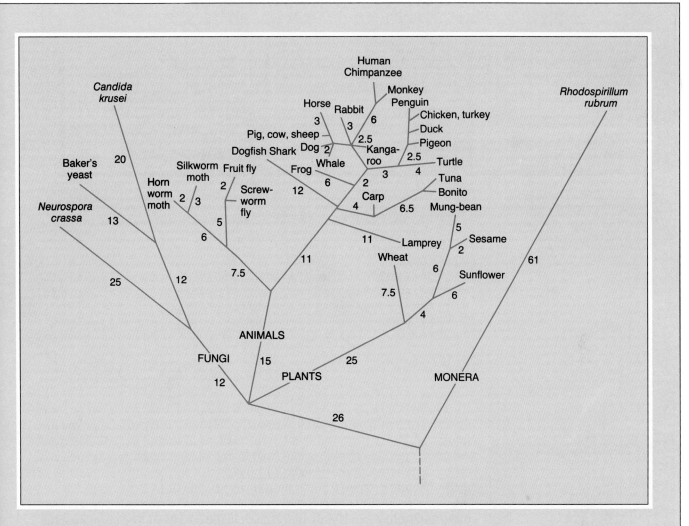

Figure 10-A

An evolutionary tree for cytochrome *c.* The tree was constructed by giving a computer the amino acid sequence of the cytochrome *c* molecule of each organism. Numbers beside the branches indicate the percentage of amino acids changed between forks. Below each branchpoint is the ancestral cytochrome *c* common to all organisms shown above it. (Redrawn from *Atlas of Protein Sequence and Structure,* Vol. 5, 1972)

Eukaryotes produce gene regulatory proteins that work in the same way as those of prokaryotes, but regulation in eukaryotes has some additional complications. As in prokaryotes, gene regulatory proteins bind to DNA sites next to promoters, turning genes on or off. In addition, transcription of many genes is also affected by a DNA sequence some distance from the gene and its promoter. This distant sequence, called an **enhancer,** is the binding site for other, different gene regulatory proteins. Many enhancers can bind more than one gene regulatory protein at a time, and the rate of transcription depends on which regulatory proteins are bound. This permits precise control over the rate of protein synthesis.

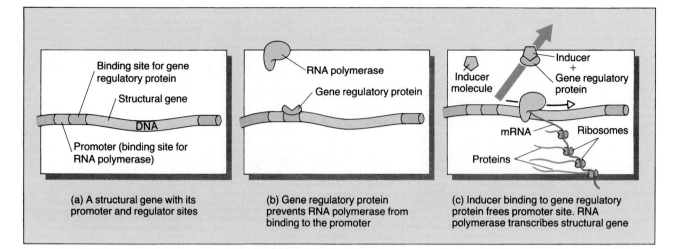

(a) A structural gene with its promoter and regulator sites

(b) Gene regulatory protein prevents RNA polymerase from binding to the promoter

(c) Inducer binding to gene regulatory protein frees promoter site. RNA polymerase transcribes structural gene

Figure 10-12
One of the ways gene regulatory proteins control protein synthesis in prokaryotes.
As in Jacob and Monod's experiments, the gene regulatory protein shown here prevents protein synthesis. It must be removed from the DNA, by binding an inducer molecule, before transcription and translation can occur. (If this were a eukaryotic cell, ribosomes would not be translating the RNA while it was still being transcribed.)

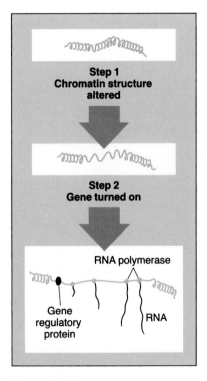

Figure 10-13
The two stages believed to be involved in activation of a eukaryotic gene. In Step 1, the structure of a stretch of chromatin is altered in preparation for transcription. In Step 2, gene regulatory proteins bind to specific sites on the altered chromatin and permit RNA polymerase to start transcribing the DNA.

An enhancer may affect several genes at once, and these genes can be some distance from each other as well as from the enhancer. This permits one **master gene regulatory protein** bound to an enhancer to control the action of many different genes. An example comes from studies of humans that were genetically male but developed into females (a process that involves the production of many different proteins). These individuals turned out to lack a single master regulatory protein, the receptor for the male hormone testosterone. In a normal male, testosterone binds with this receptor in the cytoplasm. The hormone-plus-receptor combination enters the nucleus, where it binds to the appropriate enhancer and controls transcription of the many genes that make an individual male rather than female.

Gene regulatory proteins exert precise control over protein synthesis, but other, more general methods of control seem to determine whether regulatory proteins can reach the DNA in the first place. Control of transcription in eukaryotes seems to be a two-stage process. Before a gene can be expressed, first the structure of the chromosome must be altered, and then gene regulatory proteins can bind to the DNA to turn transcription on and off (Figure 10-13).

Changes in Chromosome Structure Chromosomes are made up of a combination of DNA and protein called **chromatin** (Section 9-F). Tightly condensed (coiled up) chromatin is inactive, probably because it is so tightly coiled that polymerase enzymes cannot move along the DNA. Looser areas of chromatin contain areas of DNA that are being transcribed (Figure 10-14). The histones that hold nucleosomes together do not prevent transcription, which can go on right through the nucleosome. Above the level of nucleosomes, however, it is clear that the way DNA is packed into chromosomes does affect gene activity, but we know little about how this is controlled.

Many structural changes in chromosomes happen on such a small scale that they cannot be studied with a microscope. Other changes are sufficiently large-scale to be visible. Some insect tissues grow by an increase in cell size rather than in number of cells. As these cells grow, their chromosomes replicate up to

Figure 10-14

Gene expression. At any one time, some genes in a cell are active and some are inactive. This is an electron micrograph of one chromosome. The long, thin single lines are lengths of DNA containing inactive genes that are not being transcribed. The short loops and wiggly lines sticking out from the DNA line are a mixture of DNA with RNA molecules that are being transcribed. (From M. M. Lamb and B. Daneholt. *Cell* 17:838, 1979. Copyright © Massachusetts Institute of Technology, published by the MIT Press)

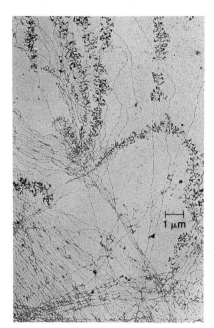

10 times, to produce giant **polytene chromosomes,** which contain hundreds of strands of DNA. Each polytene chromosome has a banded structure. Sometimes the bands appear as "puffs" where the DNA is partly unraveled (Figure 10-15). For instance, in the midge *Chironomus,* four salivary gland cells produce granules while the rest of the cells do not. Only the cells that produce granules have a puff at one end of one chromosome. We now know that puffs represent active genes. They are regions where the structure of a chromosome has changed, permitting RNA to be transcribed.

In bacteria, a little RNA is transcribed from genes that are supposedly turned off. These genes are said to be "leaky." Higher eukaryotes have a mechanism that seems to reduce the leakiness of their genes: **methylation,** the addition of a methyl group (CH_3) to some of the cytosine bases in DNA. Later, as DNA is replicated, a methylating enzyme detects methyl groups on the old DNA strand and attaches methyl groups to its new partner. In this way cells pass down the methylation of genes from one generation to the next, and genes that have been turned off during embryonic development stay turned off. This mechanism would probably not be worthwhile for a bacterium that lives for only a few hours. However, in a vertebrate's much longer life, it saves a lot of energy if cells are prevented from making proteins they do not need.

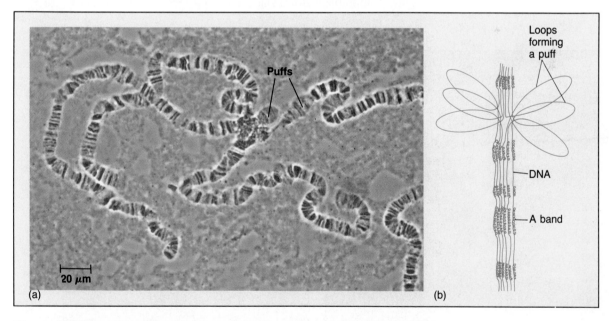

Figure 10-15

Giant polytene chromosomes. (a) Polytene chromosomes from the salivary gland of a fruit fly larva *(Drosophila).* Some bands are expanded into puffs. (b) Drawing of part of a polytene chromosome. The chromosome contains hundreds of strands of DNA. Only a few strands are shown here. The chromosome appears banded, and one band is expanded into a puff, an area where RNA synthesis is going on. (a, P. J. Bryant, University of California, Irvine/BPS)

10-I   *Control of Gene Activity During Development*

Genes are switched on and off throughout life, but the most dramatic examples occur during development of an embryo. Most multicellular organisms start life as single cells, endowed with a particular set of genetic information, from which the walking, talking, or photosynthetic, flowering, adult develops.

As development unfolds, genetically identical cells descended from the original cell **differentiate,** that is, become different from one another, as different genes are switched on and off in different cells. We can view development as a precisely coordinated series of events. Each gene is switched on in its turn, and then something resulting from the expression of that gene, or some environmental influence, switches on the next gene in the program.

Embryonic development is complex, with many different things happening at once, and this makes it difficult to study. Clues as to how differentiation occurs in an embryo have come from the study of less complex systems. We consider one example here.

Metamorphosis

Metamorphosis is a more-or-less abrupt alteration in an animal's anatomy and physiology as it changes from a larva, such as a caterpillar, maggot, or tadpole, into a very different adult (Figure 10-16).

The biological significance of metamorphosis is that it permits the young stage of an animal to have a way of life very different from the adult's. The change from one form to another must be abrupt when the intermediate between the two stages is not adapted to either way of life. In all cases, the adult stage is the one that eventually becomes sexually mature.

The existence of metamorphosis in amphibians (frogs, salamanders, and their relatives) reflects the fact that they are not fully adapted to life on land. As adults, many amphibians live on land and eat insects. As embryos, however, they

(a)

(b)

Figure 10-16
Metamorphosis in a frog. (a) The underside of a tadpole. The hind legs have begun to grow, showing that metamorphosis has begun. (b) Metamorphosis is almost complete in this tiny frog. However, it has still not completely resorbed its tail. (Biophoto Associates)

Table 10-2   *Changes That Occur During Frog Metamorphosis*

Function	Aquatic larva	Terrestrial adult
Locomotion	Tail with fins	Legs, no tail
Obtaining O_2	External gills	Lungs and skin
Transporting O_2	Larval hemoglobin	Hemoglobin has different amino acid sequence and properties
Feeding	Sucking mouth	Big mouth with jaws, sticky tongue
Digestion	Long coiled intestine for digesting algae	Digestive tract short; insect food easily digested
Sensing environment	Small eyes; lateral line organ (row of pressure-sensitive pits along side of body)	Huge eyes, with different visual pigment; ears
Excretion	Ammonia (NH_3)	Urea ($H_2N-CO-NH_2$)

live in water, because the eggs cannot stand exposure to air. Nearly all amphibians lay their eggs in water and have a swimming tadpole larva that feeds on aquatic algae. Table 10-2 lists the main changes that occur as the animal changes from an aquatic plant-eater into a terrestrial insect-eater.

All of these changes result from changes in gene activity during metamorphosis. They are brought about by the hormone **thyroxin,** secreted by the thyroid gland. This hormone must be present for metamorphosis to occur. Removing the thyroid gland prevents a tadpole from going through metamorphosis. Thyroxin acts directly on tissues that change during metamorphosis. If a tadpole's tail is removed and placed in a bath containing thyroxin, the white blood cells in the tail will digest it so that it gets smaller, just as they would in the intact tadpole. In the control experiment, where the bath contains no thyroxin, the tail remains unchanged.

Metamorphosis in insects is also controlled by hormones. In this case, the presence of **juvenile hormone** prevents metamorphosis into the adult animal (Figure 10-17). So in insects, sysynthesis of a hormone must be turned off, rather than on, to produce the change from larva to adult.

Figure 10-17

Metamorphosis in insects. (a) In this bloodsucking bug, the nymphal stages resemble the adult, but they have only small wing pads. (b) At the final molt, the wing pads develop into functional wings and the reproductive organs mature. (c) If the last nymphal stage is treated with juvenile hormone, which is not normally present at this stage, it molts to form a giant nymph, still without functional wings. (d) A 17-year cicada clings to its old nymphal skeleton as it dries and hardens following its final molt. Note that the adult has two pairs of large, transparent wings, whereas the last nymphal stage had only small wing pads, visible near the hind legs of the old skeleton.

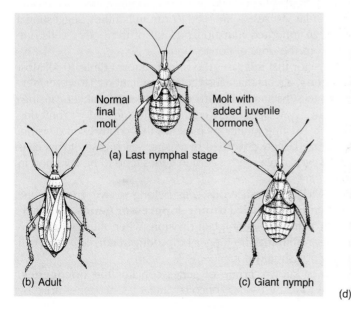

(a) Last nymphal stage — Normal final molt — Molt with added juvenile hormone

(b) Adult   (c) Giant nymph

(d)

10-J Cancer

Most multicellular organisms replace lost or worn-out cells by division of remaining healthy cells. Normally, the division of these cells is under strict control. New cells are produced only when and where they are needed. However, these controls sometimes go awry.

A **tumor** is a clump of cells that grows and divides abnormally. Some tumors are harmless, like the common wart or fibroid cysts of the uterus. Other kinds of tumors may eventually become **malignant,** which means that (1) their cells divide without the normal restraints, and (2) they invade and destroy healthy tissues, often in distant parts of the body. A "cancer" is a malignant tumor.

A cancer starts when a single cell is transformed into a cancer cell by undergoing mutations in some of the genes that control cell division. This cell then divides repeatedly, producing a **clone** of genetically identical cells, which form a tumor. Later, some of these cells undergo further changes and become malignant. Cancer cells may **metastasize,** that is, they detach from their neighbors, travel to other parts of the body, and start new tumors.

How is a normal cell transformed into a cancerous one? The control of cell division may be compared to a high-security system with multiple safeguards. If one part fails, others will still keep the cell from dividing improperly. Hence, it takes more than one change—three to seven, according to some studies—for a cell to escape from the controls on division and become cancerous. A cancer therefore develops in steps, which are not necessarily the same from one case of cancer to another, even for the same kind of cancer.

The first step in development of a cancer is a mutation, which makes a cell abnormal and which is passed on to the cell's offspring. Later, additional changes may occur in some cells of this lineage and also be passed on. Eventually a cell bearing accumulated changes sustains one more that transforms it to a cancer cell. This may take years.

What causes cancers? You can see that many different factors may contribute to a single case of cancer, and some of them may be long gone by the time the cancer appears. The traditional way to probe the causes of cancers is by correlation studies: comparing a group of cancer victims with a matched group of healthy people to see what factors are more common in the background of cancer victims. As long ago as 1775, Percival Potts noted that chimney sweeps suffered from a high incidence of cancer of the scrotum. In the 1960s, cigarette smoking was linked to most cases of lung cancer, and it is now known to increase the risk of several other cancers (Figure 10-18). Some cancers occur only in the presence of particular viruses. Some tend to run in families, suggesting a genetic predisposition: an inherited mutation puts all of the body's cells one step closer to cancer from the time of conception.

A **carcinogen** is a factor that increases the risk of cancer (Table 10-3). Most carcinogens are **mutagens,** agents that cause genetic mutations. However, several cancers are linked to substances that are **carcinogenic** (cancer-causing) without being mutagenic. These apparently contribute to cancer by promoting the expression of already existing mutations in genes that control cell division.

Because mutations are largely governed by chance, each case of cancer is genetically unique. However, mutations of certain genes tend to be common in some kinds of cancer.

■ *A cancer arises from a single cell that has accumulated mutations in the genes governing cell division.*

The mutated genes found in cancerous cells belong to two groups. One group, for lack of a better name, is called **tumor suppressor genes.** Normally, these genes code for proteins that inhibit cell division. When these genes mutate, the corresponding protein may no longer be produced correctly, and cell division may occur when it should not.

The other, and better known, group of genes contains the **oncogenes,** genes with the potential to cause cancer (onco = cancer). Oncogenes arise by

Figure 10-18
Cancer. Normal cells (orange) and cancerous cells (green) in the bronchi, the tubes that carry air into the lungs. The normal cells are covered with cilia, which sweep particles from the air up toward the throat. The cancer cells have lost their cilia but are covered with numerous shorter, finger-like microvilli. (Lennart Nilsson © Boehringer Ingelheim International, GmbH)

mutation of normal cellular genes, called **proto-oncogenes,** which control the production of proteins that stimulate growth and cell division. A proto-oncogene is converted to an oncogene by a mutation that causes the gene to be overexpressed. Often such a mutation does not change the gene itself but produces extra copies of the gene, or brings the gene under the control of a regulatory gene that increases its expression. Overexpression means that too much of the gene's protein is made, and so the cell divides more rapidly than usual and forms a tumor. About 60 proto-oncogenes are known, including genes that code for growth hormones and for dozens of other proteins that control cell division.

Viruses are one of the many factors that can contribute to the development of cancer by causing mutations. They may insert their genetic material into the chromosomes of host cells or move host genes to different chromosomal locations. The first human cancer shown to be caused directly by a virus was a rare form of leukemia, caused by HTLV-1, a member of the retrovirus group, which also includes the AIDS virus.

Table 10-3   Some Known Carcinogens

Carcinogen	Comments
Asbestos dust Chromium compounds Some petroleum products	Workers in these industries have high risk of lung cancer
Tobacco	12–15% of cigarette smokers die of lung cancer. Especially risky when combined with exposure to asbestos or radon
Estrogen	Mammalian hormone. In large amounts, can contribute to uterine and breast cancer
X-rays	Many people have unnecessary medical x-rays
Benzene	Until recently, was a common solvent in labs
Nitrates and nitrites	Converted into carcinogenic nitrosamines in digestive tract. Common as food preservatives and in most green vegetables. Also common water pollutants in agricultural areas, where nitrogen fertilizers run off farmland
Aflatoxins	Produced by fungus *Aspergillus flavus* when growing on food. First found in peanuts. Causes liver cancer
Vinyl chloride	Causes one type of liver cancer

Toward Human Gene Transplants

A JOURNEY INTO LIFE

In Chapter 9 (*A Journey Into Life*) we saw that genes can be transplanted into cells in the laboratory. How does this enable us to produce whole organisms in which every cell carries the new gene, including the germ cells that will give rise to the next generation? To understand this, we need to know more about what happens to genes during embryonic development.

Multicellular organisms develop from a single fertilized egg cell (a **zygote**). It takes hundreds of cell divisions to form all the cells in the adult. Before each division, a cell's DNA is replicated, and a copy passed to each new cell. So, with very few exceptions, each cell of an adult organism contains exactly the same genes as the zygote from which the cells descended. These differentiated body cells can be removed, grown in laboratory culture, and used as hosts for gene transplants. However, differentiated cells such as these do not normally give rise to new individuals.

In the 1950s, Frederick C. Steward cultured carrot root cells in an artificial medium and succeeded in growing whole carrot plants from single suspended cells. Clearly, in this case, a differentiated root cell contained all the genetic information that a carrot needs and had lost no genetic information during its own dif-

ferentiation. The differentiated cell is said to be **totipotent** ("all-powerful"). Many other plant cells appear to have the same ability, which is very useful because it means that one particularly desirable plant (or genetically engineered plant cell) can be cloned to form many genetically identical plants (*A Journey Into Life* "Better Plants Through Engineering," Chapter 36).

No one has yet grown an entire animal from a single differentiated cell, but experiments implying that this might be possible have been done with frogs. In these **nuclear transplantation** experiments, the nucleus is removed from an egg and replaced with the nucleus of a differentiated cell (Figure 10-B). If the nucleus from a cell in a tadpole's intestine is transplanted into a frog egg, the artificial zygote so formed develops normally into a sexually mature adult frog.

Nuclear transplantation puts all the genes of the transplanted nucleus into every cell in the resulting embryo, including the **germ cells,** those that give rise to the eggs or sperm. These transplanted genes will be passed on to future generations.

No one has yet managed nuclear transplantation in a mammal. (A report of nuclear transplantation in mice in 1983 was shown to be fraudulent.) The reason is probably technical. The mammalian egg is tiny and surrounded by numerous membranes. The membranes are broken when a nucleus is injected into the cell, and

Figure 10-B
Nuclear transplantation experiment. A nucleus from a cell in a tadpole's intestine (chosen because it is large enough to be manageable) is injected into an egg and supports the development of a mature frog.

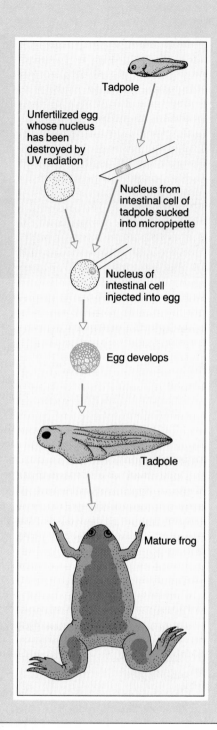

Tadpole

Unfertilized egg whose nucleus has been destroyed by UV radiation

Nucleus from intestinal cell of tadpole sucked into micropipette

Nucleus of intestinal cell injected into egg

Egg develops

Tadpole

Mature frog

After joining the host genome, some viruses enter a **latent** period and are not expressed until activated days or years later. For instance, the virus that causes hepatitis B may cause liver cancer 20 or 30 years after the attack of hepatitis. Papilloma virus infection, which causes harmless warts, is also necessary for development of cancer of the cervix (the neck of the uterus). But most women infected with papilloma viruses will never develop cervical cancer. For one thing, the virus will not cause cancer without additional (unknown) factors.

the cell dies. However, recent experiments have shown that individual genes can be transplanted into zygotes and expressed in the animal that develops from the zygote.

Geneticists believe that we shall one day be able to cure many inherited human diseases by gene transplants. We might be able to transplant a normal gene into a baby or embryo known to have inherited a genetic disease. However, replacing a defective gene is a lot more complicated than replacing a defective kidney, for both technical and ethical reasons. Here we consider some of the experiments on gene transplantation in mammals and some of the problems that remain.

Beta thalassemia is a hereditary disease in which the red blood cells produce defective beta chains (or sometimes no beta chains) for hemoglobin (see Figure 3-19). Victims suffer from acute anemia (lack of hemoglobin) and usually die in infancy. In 1986 experiments, the *human* gene for beta chains was injected into zygotes of mice with thalassemia. Apparently the gene was incorporated into the zygotes' genomes: the zygotes grew into adult mice with normal hemoglobin (except that it contained human beta chains) and passed the transplanted gene on to their offspring when they reproduced.

This remarkable result does not open up the immediate prospect of replacing defective genes in human zygotes. Not all the transplant mice were cured of their thalassemia. Some produced no normal beta chains; others produced too many. Although the transplanted gene was present in all the mice, its expression varied enormously. This is because different host cells spliced the gene into different places in their own chromosomes.

In 1989, workers reported that they had transplanted extra genes for growth hormone into pigs. Growth hormone causes rapid growth in young mammals. Researchers thought that the gene transplant might permit them to develop a variety of pigs that grew faster than usual. It did. The transplant pigs grew rapidly and developed meat containing less fat than normal pigs. But there was a snag: the pigs had many genetic defects, including acute arthritis, which made them unsuitable for use as farm animals. We do not know why this occurred. Perhaps extra growth hormone itself caused the problems, or perhaps the growth hormone genes were inserted into the pig genome in such a location that they interfered with the normal functions of other genes.

Genes are not expressed normally unless they are located where the appropriate regulatory genes can control them. We do not yet know enough to control precisely where a transplanted gene is spliced into a genome or how it will then be expressed. Nuclear transplantation would permit this control indirectly. After a gene had been transplanted in culture, the cultured cells could be tested for gene expression, and only nuclei with appropriate control of protein production would be transplanted into zygotes.

If nuclear transplantation could be made to work in mammals, animals with any desired genome could be produced. The very possibility of such transplants in humans is horrifying. It would mean that we could produce clones of genetically identical people with any genetic makeup we chose. The human race is not, and probably never will be, ready to cope with genetic control over the population. For example, one study showed that if people could choose how many children to have and the sex of each (a simple genetic feature), three quarters of the babies born would be boys and only one quarter girls.

However, there are less contro-versial ways of transplanting genes so that the new genes do not affect every cell and are not passed on in the sperm or eggs. Cells in the bone marrow are prime candidates for such an approach. These cells, which give rise to blood cells and to cells of the immune system, continue to divide throughout life. Consider a person with a genetic defect of the immune system that prevents production of a protein. A sample of bone marrow cells can be removed from the patient. A gene can then be transplanted into the cells, which are injected back into the bone marrow. In theory, the transplant cells would divide and produce a large cell population that would supply enough of the protein to prevent symptoms of the disease. Experiments on genetically engineered bone marrow are now underway in humans.

The problem of gene expression remains. Until we learn more, we cannot ensure that a transplanted gene will produce useful amounts of the necessary protein without causing unacceptable side effects. Where the disease can be prevented only by turning *off* an abnormal gene, the problem is even greater. In that case we would want to transplant a regulatory gene that would turn off the defective gene. That would almost certainly involve inserting the gene into one particular location within the genome.

The likelihood that we shall one day be able to transplant genes from one human to another opens up many possibilities—both good and ill. Scientists believe that organizations such as churches and medical societies should start to consider the moral and ethical implications of genetic engineering techniques such as this one now, before the techniques are ready to be used.

For another, the disease-fighting immune system probably destroys most cancerous cells before they form sizable tumors (Chapter 25).

A potential cancer may be stopped by surgical removal of the tumor while it is still a locally contained mass of cells. Once it metastasizes, it is very difficult to locate and destroy all the cells that may be able to cause tumors throughout the body. Researchers are studying the possibility of using various molecules, or antibodies or cells of the immune system, to search out and destroy cancerous cells selectively. They are also identifying some environmental factors that increase the chances of developing a tumor in the first place and urging changes in workplaces and homes, and in habits, so that people can avoid these factors.

Cancers are the second most common cause of death in the United States, accounting for 20% of all deaths. This figure scares many people but, to put it into perspective, note that an American has almost as great a chance of dying of homicide as of cancer. Before the age of 45, homicide is a much more likely cause of death (although perhaps this is not a particularly comforting thought!). Deaths from cancer have increased in the twentieth century. This is mainly because cancers tend to develop later in life, and people are living longer instead of dying of infectious bacterial diseases in early life as they used to. (Life expectancy in the United States is now nearly 80 years, compared with 45 years in 1900.) There has also been an increase in cancer because carcinogens are becoming more common in our environment. For example, workers in vinyl chloride factories have an increased chance of dying of one form of liver cancer. Cancer of the prostate is much more common in areas with acute air pollution than in other areas.

SUMMARY

DNA carries the genetic information for the order in which amino acids must be joined to produce proteins. This information directs protein synthesis.

RNA is transcribed from the cell's DNA and so has a complementary nucleotide base sequence. The three main types of RNA in a cell are messenger RNA, whose base sequence is translated into the sequence of amino acids in a polypeptide; ribosomal RNA, which makes up part of the structure of ribosomes; and transfer RNA, which carries amino acids to the ribosome for protein synthesis and brings them into their proper position to be joined to the polypeptide chain.

A sequence of three nucleotides in mRNA codes for each amino acid. The genetic code is degenerate in that most amino acids are encoded by more than one codon. The code has no "punctuation" except codons that signal the beginning and end of the polypeptide. Mutations in DNA are also transcribed into RNA and may change the protein produced.

During protein synthesis, the code carried by the sequence of nucleotide bases in messenger RNA is translated into the sequence of amino acids in a polypeptide. The mRNA attaches to a ribosome, and transfer RNAs carrying amino acids attach to the mRNA-ribosome complex by means of base-pairing between the mRNA codons and the tRNA anticodons. Each tRNA in turn donates its amino acid to the growing polypeptide chain. As each successive peptide bond is formed between the growing polypeptide and the newly arrived amino acid, the ribosome moves along the mRNA. This brings the next codon onto the ribosome, where it can bind the anticodon of the tRNA carrying the next amino acid. When a *Stop* codon reaches the ribosome, the completed polypeptide is released and processed into a finished protein.

Protein synthesis is usually controlled by turning transcription on or off. This occurs by means of the binding or unbinding of gene regulatory proteins at specific DNA sites. In prokaryotes, these DNA sites are near the gene's promoter. Other substances, often food molecules, control the ability of regulatory proteins to bind to DNA.

Transcription in eukaryotes is thought to involve two steps:

1. The structure of part of the chromatin changes, by loosening of the DNA coiling.
2. As in prokaryotes, regulatory substances interact with gene regulatory proteins to switch genes on and off. These gene regulatory proteins may control specific genes, by way of their promoters, or groups of genes, by affecting their enhancers.

As eukaryotic cells differentiate during development, gene regulatory proteins turn off genes they will not need, but this control is leaky. Later, these genes may be turned off more securely by being methylated.

Various signals inside and outside an organism cause some of the changes in gene activity that make up differentiation. Amphibian and insect metamorphosis are convenient, non-embryonic systems for studying differentiation.

Cancers arise by the progressive accumulation of mutations in genes controlling cell division. These mutations may be brought about by viruses, radiation, or mutagenic chemical carcinogens. A tumor arises from a single mutated cell that has escaped the controls on division. Some cancerous cells become malignant and metastasize, detaching from their tumor neighbors, invading healthy tissues elsewhere in the body, and starting new tumors there.

1. Using the base-pairing rules, fill in the mRNA sequence that would be transcribed from the following strand of DNA. Then, use Table 10-1 to determine the amino acid sequence that would be translated when the mRNA combines with a ribosome:

   T—A—C—A—A—G—T—A—C—T—T—G—T—T—
   T—C—T—T
   mRNA _____
   amino acids _____

2. Suppose the two guanine (G) nucleotides in Question 1 were changed to cytosine (C) nucleotides. How would this mutation affect the amino acid sequence translated from the mRNA?

3. Suppose the G nucleotides were removed from the DNA in Question 1. How would this mutation affect the amino acid sequence translated from the mRNA?

4. According to current ideas concerning protein synthesis:
   a. transfer RNA molecules specific for particular amino acids are synthesized along a messenger RNA template in the cytoplasm
   b. amino acids line up with their mRNA codons on the ribosome and are then linked together by transfer RNA
   c. enzymes that catalyze protein-synthesizing reactions in the cytoplasm are transcribed from regulatory genes
   d. transfer RNA molecules transport mRNA from the nucleus to the ribosomes
   e. messenger RNA, synthesized on a DNA template in the nucleus, provides information that determines the sequence in which amino acids are linked during translation

5. List three differences between the structures of DNA and RNA.

6. Transfer RNA is synthesized:
   a. on a DNA template
   b. from a messenger RNA template on a ribosome
   c. on ribosomes without a template
   d. in the nucleolus by the interaction of messenger RNA and chromosomal DNA

7. During differentiation, cells with the same DNA:
   a. must develop similarly
   b. divide at equal rates
   c. contain different genes
   d. may transcribe different genes

8. How do food molecules induce prokaryotic cells to make enzymes that metabolize the food? Food molecules:
   a. cause an inducer to bind to the DNA and attract RNA polymerase, which transcribes mRNA coding for the needed enzymes.
   b. cause repressor proteins to leave the DNA, which can then be transcribed to mRNA.
   c. bind to tRNA, which carries them to the ribosome for protein synthesis.
   d. bind to RNA polymerase enzymes and activate them.
   e. change the structure of the DNA so that regulatory substances can bind to it.

QUESTIONS FOR DISCUSSION

1. Why is it important for each type of tRNA to have its own type of enzyme to bind it to an amino acid?
2. Suppose a cell's DNA contained a mutation that changed one of the nucleotides in an anticodon of tRNA. How might this mutation affect protein synthesis?
3. We have seen why the genetic code could not consist of codons with fewer than three nucleotides each. What factors might have selected against codons of more than three nucleotides?
4. Even though some 20% of Americans die of cancer, life expectancy in developed countries is about 80 years. Would the enormous sums spent on cancer research, therefore, be better spent on diseases such as AIDS that kill most of their victims when they are much younger, or diseases such as Alzheimer's that damage the quality of the patients' (and their relatives') lives for much longer?
5. Why are treatments that prevent viruses from reproducing not likely to cure many types of cancer caused by viruses?

SUGGESTED READINGS

Alberts, B., et al. *Molecular Biology of the Cell.* New York: Garland Publishing, Inc., 1989. Chapters 5, 10, and 21 provide clear, modern treatment of the topics of protein synthesis, control of gene expression, and cancer.
Browder, L. W. *Developmental Biology,* 2d ed. Philadelphia: Saunders College Publishing, 1984. An embryology textbook with a good section on differentiation.
Feldman, M., and L. Eisenbach. "What makes a tumor cell metastatic?" *Scientific American,* November 1988.
Halliday, Robin. "A different kind of inheritance." *Scientific American,* June 1989. How the addition of methyl groups to DNA may affect gene expression during development.
Hunter, T. "The proteins of oncogenes." *Scientific American,* August 1984. How proteins encoded by oncogenes make cells containing them cancerous.
Ptashne, M. "How gene activators work." *Scientific American,* January 1989. The action of gene regulatory proteins.
Weinberg, R. A. "Finding the anti-oncogene." *Scientific American,* September 1988. Mutation of a gene that normally suppresses the development of cancer can increase susceptibility to cancer of the eye.

Reproduction of Eukaryotic Cells

CHAPTER

11

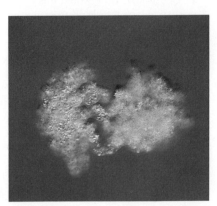

Amoeba dividing

Paramecium *mating*

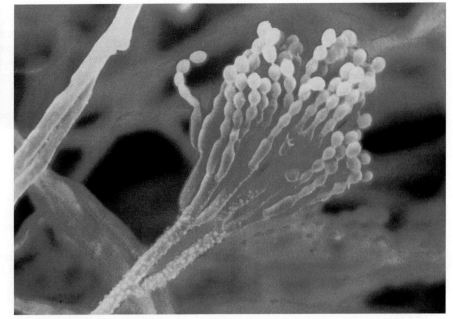

Fungal spores

*L*ife is handed down from one generation of organisms to the next in the form of new cells. A unicellular organism produces more of its kind by dividing in two. A multicellular organism, such as you or the tree outside your window, begins life as a single cell, and repeated cell divisions produce the many cells of the body. Eventually, some cells in the reproductive organs divide to form reproductive cells, which give rise to the next generation.

When a cell divides, it must pass on the genetic information needed to specify the kinds of proteins it can produce. In a eukaryotic cell, the DNA carrying this genetic information is divided among many chromosomes in the nucleus. Before the cell divides, its chromosomes must be replicated and then distributed precisely into two new nuclei, so that each receives a complete set of chromosomes. Each new cell must inherit not only a nucleus containing all the genetic information it will need, but also the cytoplasmic components required to express the genetic information, such as ribosomes to make proteins and mitochondria to supply the necessary energy.

The most complicated part of cell division is nuclear division. Various cell components must interact in such a way that they divide the replicated chromosomes accurately into two complete sets.

The two main types of nuclear division are mitosis and meiosis. **Mitosis** produces two new nuclei with the same number of chromosomes as in the original nucleus. This ensures that each new cell inherits a complete set of the parent cell's genetic information. In unicellular organisms, mitosis produces two genetically identical new individuals. A multicellular organism begins life as a single cell, the fertilized egg, which develops into an embryo and eventually into an adult organism, by repeated mitotic cell divisions. Mitosis occurs all the time in our bodies, as new cells replace old ones, such as worn-out blood cells, or skin cells injured by cuts or burns.

Meiosis is the type of nuclear division that produces new combinations of chromosomes and genes, packaged in nuclei containing only half the number of chromosomes found in the original nucleus. Meiosis is associated with reproduction. In animals, it gives rise to the **gametes,** the sexual reproductive cells: sperm and eggs (Figure 11-1). In plants, meiosis produces **spores,** reproductive cells that are asexual (a = not). Spores divide and produce structures that give rise to the sexual gametes (Chapter 20).

Meiosis is vital to any organism that reproduces sexually. Without meiosis, the gametes would contain as many chromosomes as the other cells of the parent, and the fertilized egg would contain twice as many. Hence the number of chromosomes would double in each generation.

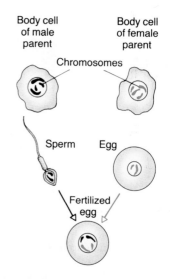

Figure 11-1
Body cells and reproductive cells. As a result of meiosis, sperm and eggs (gametes) contain only half as many chromosomes as the other cells in an animal's body. During fertilization, a sperm fuses with an egg, and the fertilized egg then contains the characteristic number of chromosomes found in normal body cells of the species.

KEY CONCEPTS

- A cell reproduces by dividing into two new cells.
- Before a eukaryotic cell divides, its chromosomes are distributed precisely into two new nuclei.
- Mitosis produces two new nuclei that both contain the same genetic information as the original nucleus had.
- Meiosis produces new nuclei with only half the number of chromosomes found in the original nucleus, and with new genetic combinations.

Before considering the reproduction of eukaryotic cells, we must learn more about how their chromosomes are organized.

11-A *Eukaryotic Chromosomes*

Viewed through a microscope, a cell's chromosomes usually appear as a single, diffuse mass of chromatin. Individual chromosomes can be distinguished only right before and during cell division, when they **condense**—that is, coil up

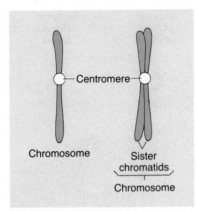

Figure 11-2
Chromosome and chromatids. Replication of a chromosome forms two sister chromatids, which remain joined at the centromere. (Figure 9-12 shows a scanning electron micrograph of chromatids at this stage.)

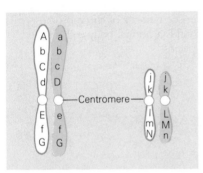

Figure 11-3
Two pairs of homologous chromosomes. The members of a pair bear genes for the same traits at the same locations. For example, the pair on the left bears gene locations A through G. These chromosomes have two different forms of the gene at the A location (*A* and *a,* where *a* might code for a mutant form of the protein dictated by *A*), two of the same form (*b*) at the B location, and so on.

■ *Most plants and animals are diploid, with paired chromosomes. During sexual reproduction, each parent passes on one member of each pair to the offspring.*

tightly into short, thread-like structures (Section 9-F). By this time, the chromosomes have already been replicated as described in Section 9-C. Each consists of two copies, attached at a region called the **centromere.** As long as the two copies remain so attached, they are called **sister chromatids** (Figure 11-2).

Haploid and Diploid Chromosome Numbers

In most higher plants and animals, chromosomes from the body cells can be matched up in pairs. The two chromosomes of a pair are called **homologous chromosomes,** or simply **homologues.** Most homologous chromosomes look alike: they are the same length, their centromeres are in the same position, they show the same pattern of light and dark bands when stained, and they carry genes for the same inherited characteristics, lined up on the chromosome in the same order (Figure 11-3). For example, human chromosome #1 contains the genes for the Rh blood protein and for a starch-digesting enzyme in the saliva. However, the corresponding genes on the two homologues need not be identical. For instance, some chromosomes have a gene for the protein that makes a person Rh-positive, and some have a gene coding for a different version of this protein (Rh-negative), at the Rh location. The number of chromosome pairs varies from one species to another.

Humans have 46 chromosomes in most of their body cells. These can be arranged in homologous pairs according to their length and the position of the centromere. In human males, 22 of these pairs contain look-alike chromosomes, but the twenty-third pair is odd, with two unlike chromosomes, called X and Y (Figure 11-4). In the cells of a human female, both chromosomes in the twenty-third pair are X chromosomes. The X and Y chromosomes are called **sex chromosomes** because they determine their owner's sex (Section 13-F). The other 22 pairs are called **autosomal chromosomes,** or **autosomes.**

The possession of pairs of chromosomes is important in the life history of eukaryotic organisms. At meiosis, the two homologous members of each chromosome pair are separated into different nuclei. As a result, a gamete contains one member of each pair of chromosomes, for a complete set containing exactly half the number of chromosomes. For example, each human egg or sperm contains 23 chromosomes, one from each of the 23 pairs. When an egg and sperm join at fertilization, the new individual receives one member of each pair of chromosomes from its mother, and one member of each pair from its father, for a total of 23 complete pairs of chromosomes.

A cell that contains pairs of homologous chromosomes is said to be **diploid,** that is, having two sets of chromosomes. The diploid number in humans is 46, or 23 pairs. A cell that contains one set of unpaired chromosomes is said to be **haploid,** containing half the diploid chromosome number. An egg or sperm is haploid, and in humans the haploid number of chromosomes is 23. The haploid number of chromosomes in a species is generally designated as **N:** N = 23 in human beings, and **2N** (diploid) = 46. Table 11-1 shows chromosome numbers for several organisms.

In most animals, the fertilized egg and the cells that arise from it by mitotic division contain the diploid number of chromosomes. Some cells in the ovaries or testes eventually undergo meiosis, which produces haploid nuclei. Only the egg and sperm cells have haploid nuclei. Cells that can undergo meiosis are known as **germ cells;** the rest of the body's cells are called body or **somatic** (soma = body) **cells.**

Not all organisms are diploid. Many lower organisms are haploid. Familiar examples include moss plants, many algae and fungi, and male honeybees (drones). Some organisms, especially many plants, are **tetraploid** (4N), with four homologous chromosomes of each type. Other ploidy numbers are also found, but more rarely.

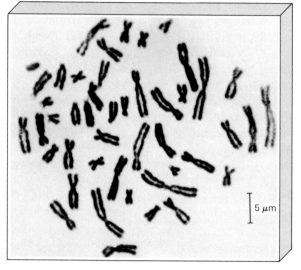

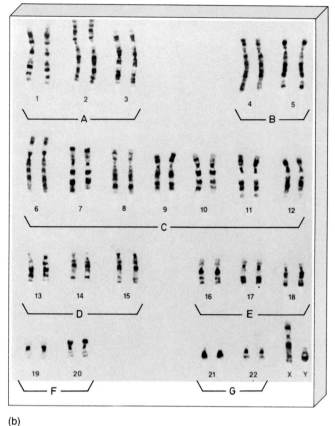

(a)

(b)

Figure 11-4
Genetic material of a normal human male. (a) 46 sets of sister chromatids from a body cell. (b) To determine whether a person has chromosomal abnormalities, a karyotype is made by cutting up a photograph of the chromosomes from a cell and arranging them into standard groups. The chromosomes in this karyotype have been stained by a method that shows patterns of bands. The bands make it easier to match up chromosome pairs by providing information in addition to length and centromere position. (a, Carolina Biological Supply Company; b, Jorge J. Yunis, M.D., Medical Genetics Division, University of Minnesota Medical School)

11-B   The Cell Cycle

Each kind of cell has a typical lifespan, which begins when the cell is formed by division of the parent cell and ends when the cell itself divides or dies. Under good growing conditions, the lifespan of lower eukaryotes varies from about 2 hours for yeast (a unicellular fungus) to a few days for *Amoeba* (a protist). Cells of early animal embryos also divide rapidly, as often as every 15 or 20 minutes. Most dividing cells in an adult have lifespans of about 8 hours to over 100 days, depending on the cell type. Some types of cells cannot divide once they have reached their final differentiated state, and they must eventually die. Examples in our bodies are nerve, skeletal muscle, and red blood cells.

A newly formed cell will usually not divide until it has approximately doubled in size. To do this, it must absorb nutrients and use them to produce more ribosomes, enzymes, cytoskeleton molecules, and membrane material. Mitochondria and chloroplasts reproduce themselves by dividing in two.

Many unicellular organisms grow and divide as fast as they can obtain enough nutrients to do so. This often makes for evolutionary success because the sooner a cell divides, the more descendants it will leave over a period of time. In contrast, multicellular organisms must have strict controls on the number of each type of cell in the body, and the cells control one another's division.

Table 11-1   Haploid (N) and Diploid (2N) Numbers of Chromosomes in Some Organisms

Organism	N	2N
Pea plant	7	14
Corn	10	20
Potato	24	48
Fruit fly	4	8
Chicken	39	78
Cat	19	38
Dog	39	78
Chimpanzee	24	48
Human	23	46

Control of cell division is a complex process that we are only beginning to understand. In animals, cell division seems to require at least two things in addition to nutrients. First, a cell's membrane receptors must bind a critical number of one or more kinds of **growth factors**—highly specific proteins found in the body fluids in minute amounts. Cells compete for molecules of the right kinds of growth factors for their cell type. So, at any one time, only a limited number of cells in any tissue have enough growth factors to begin dividing. The second requirement is attachment to something outside the cell, such as neighboring cells or the membranes or fibers between cells. This requirement may help to ensure that, if a cell becomes detached from its proper place in the body, it cannot divide and form a tumor wherever it comes to rest. Once an attached cell does begin to divide, however, it loses much of its attachment and assumes a rounded shape until division is complete. The two new cells then settle into the parent cell's space, reattach to their surroundings, and resume their typical shape.

Cells that can and do divide have a typical life history, called the **cell cycle,** lasting from the time the cell is formed by division until it divides in its turn (Figure 11-5). The cell cycle has four distinct periods. During the period of mitosis **(M),** the nucleus and cytoplasm divide and form two new cells. The rest of the cycle, known as **interphase,** is divided into the remaining three periods. The period from a new cell's "birth" until it begins to replicate its DNA is called the first gap period, or **G_1.** During this time, the cell is usually growing and carrying on the business of life. The middle period, the **S** (synthesis) period, is the time of DNA synthesis, when the chromosomes are replicated in preparation for the next cell division. The second gap period, **G_2,** lasts from the end of DNA synthesis until the next cell division.

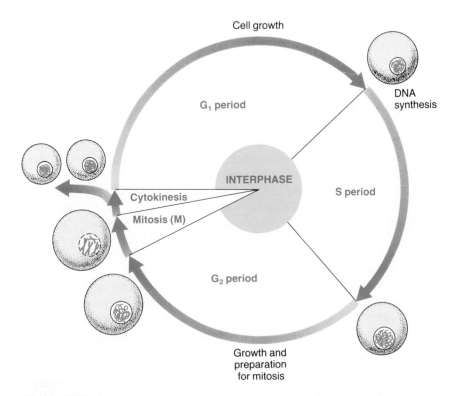

Figure 11-5
A typical cell cycle. The two gap periods (G_1 and G_2) are separated by the synthesis (S) period, during which the chromosomes are replicated. Mitosis, or nuclear division, is followed by cell division (cytokinesis).

Environmental Radiation and Cell Division

A JOURNEY INTO LIFE

Our environment exposes us to many kinds of radiation, some natural, some from modern human inventions. Radiation can be dangerous to living organisms, including ourselves. One of the things it affects is cell division.

From a biological point of view, the most important radiation is **ionizing radiation**—x-rays, gamma-rays, and particles such as neutrons and alpha particles (helium nuclei). These are given off by the decay of radioactive elements in the earth's crust and in space. Ionizing radiation has enough energy to ionize substances it strikes, by knocking off electrons. For instance, ionized water, H_2O^+, is very reactive and can cause peculiar reactions in the cell.

In large enough doses, ionizing radiation causes so much disruption that it can kill a cell. In much smaller doses, the most important effect of radiation is to cause breaks in DNA molecules. If a DNA molecule is damaged too badly for repair enzymes to restore it, it cannot be replicated, and this prevents cell division. On the other hand, the damage may show up as mutations that can be replicated and passed on. Some mutations of this type in somatic cells are believed to cause cancers, which are characterized by unrestrained cell division and invasion of other tissues (Section 10-J).

The ability of ionizing radiation in appropriate doses to block cell division is used to treat cancer. Cells are most sensitive to radiation damage just before mitosis. Because cancer cells divide more often than the normal cells around them, bombarding an organ with radiation kills or blocks division in many more cancerous than normal cells.

Ultraviolet radiation (uv), part of the electromagnetic spectrum (Figure 11-A), is emitted by the sun and by "black light" bulbs. Nucleic acids absorb uv and can be permanently damaged by it. Ultraviolet can kill cells, and in fact it is used to kill bacteria on laboratory equipment that cannot be sterilized by heat or solvents. Less drastically, uv can cause mutations. These mutations may make cells divide more rapidly, as happens in skin cancer, or make them stop dividing. Slowing of cell division is one reason light-colored skin ages so rapidly when exposed to much sunlight (or to sun lamps or tanning booths): damaged cells are not replaced as fast as they otherwise would be. Darker-skinned people are less prone to premature skin aging and skin cancer because the dark pigment, melanin, in their skin absorbs ultraviolet rays and prevents them from penetrating to the DNA of living cells. Ultraviolet radiation does not cause damage to deeper organs of the body because, unlike x-rays and gamma-rays, it is rapidly absorbed by water in living tissue and so does not penetrate beyond the skin.

Both ionizing and uv radiation are part of the natural environment. It has been estimated that over 80% of the ionizing radiation to which the "average" American is exposed is natural, or background, radiation from rocks, cosmic radiation, and the like. About two thirds of this comes from breathing radon gas, and its decay products, in indoor air. This natural radiation averages 260 to 300 millirems per year. For comparison, a person receiving a dose of 350,000 millirems (350 rems) in a month (rather than a year) is thought to have a 50-50 chance of dying from this exposure. (A rem is a unit of ionizing radiation defined by its biological effect.)

"Civilization" exposes us to additional radiation. Most of this comes from medical and dental x-rays. About 1% comes from the fallout from nuclear weapons testing and from nuclear power stations and their waste (even the Chernobyl accident, which killed several dozen people from high short-term doses, increased the long-term radiation exposure of people living more than 30 kilometres from the plant by a negligible amount). About 3% comes from such domestic sources as water, building materials, natural gas, color televisions, and smoke detectors. Most experts think we would be well advised to reduce our exposure to medical x-rays.

It is, however, very difficult to say how much radiation beyond the inevitable background level should be considered a health hazard. First, it is hard to estimate the background level of radiation. Both the kinds and the amounts of radiation a person receives vary with geography, occupation, personal habits, and so forth. Second, the degree of damage resulting from exposure to a particular dose of radiation varies depending on which particular tissues and molecules are struck by the radiation, something we cannot predict.

A third variable, which is very important but poorly understood, is the ability of affected tissue to correct the damage. Cells contain enzymes that can repair either breaks in DNA or mis-pairing caused by radiation damage to particular nucleotides in the molecule. (Interestingly, some enzymes that can repair damage caused by ultraviolet radiation are activated by light, the very agent that causes the damage!) However, this repair is not perfect, and this is why radiation can sometimes cause cancers and other disorders.

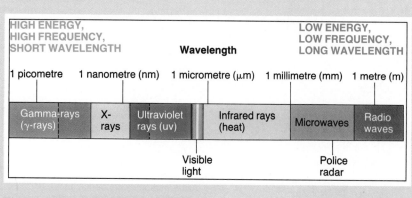

Figure 11-A
The electromagnetic spectrum.

The length of G_1 is the most variable part of the cell cycle. The key point in the cycle occurs late in G_1: some unknown signal molecule(s) is believed to switch the cell into the S period. After this point of no return, the cell is committed to proceed completely through the next mitosis.

During the S period, DNA replication begins at specific points in the DNA and follows a definite pattern until all the DNA is copied. Histone proteins for the new chromosomes are also made during the S period, and the centrioles double (except in higher plants, which lack centrioles) (Section 5-L).

When replication is finished, it is switched off and the cell enters G_2, the time of final preparation for division. During G_2, the cell is believed to make some of the proteins used in mitosis itself. One of these activates another protein, which prepares the cell for mitosis. In particular, it causes the chromosomes, which have been spread out in a loose mass during interphase, to condense very tightly. The cell is now ready for mitosis.

■ *Mitosis ensures that sister chromatids separate precisely into two new nuclei, so that each new cell receives one copy of each chromosome.*

11-C   Mitosis

In mitosis, the replicated, condensed chromosomes are separated into two equal groups. Two new nuclei are formed, each containing a complete set of the genetic information present in the original nucleus. Mitosis is a continuous process, but for convenience it is divided into four phases according to the appearance of the chromosomes as viewed through a light microscope: prophase, metaphase, anaphase, and telophase.

Prophase Looking through a microscope, you can first tell that a cell is about to divide during prophase, when the loose mass of interphase chromatin condenses into distinct chromosomes, visible as sets of sister chromatids (Figure 11-6). This condensation is an impressive process. It is comparable to taking a thin strand some 200 metres long and coiling it into a cylinder about 1 millimetre across and 8 mm long. During prophase, the nucleolus, which is the site of ribosome synthesis, usually disappears because the material in the nucleolus becomes scattered. A complex of proteins, called a **kinetochore,** assembles on each chromatid, in the centromere region (see Figure 11-7). The nuclear mem-

Figure 11-6
Mitosis. The nucleus divides into two new nuclei with identical genetic information. The first cell shown is in interphase, before mitosis begins. The next four cells show the stages of mitosis: prophase, metaphase, anaphase, and telophase. The last cell is undergoing cytokinesis, the division of the cytoplasm that usually follows mitosis.

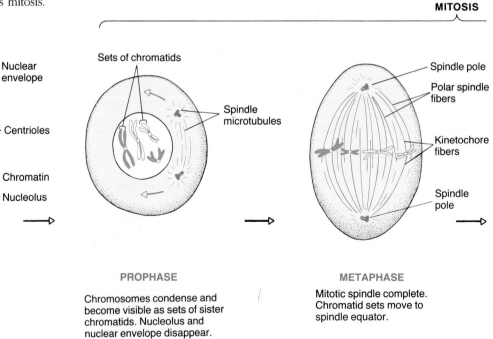

MITOSIS

INTERPHASE
Chromatin spread out in indistinct mass. Nucleus and nucleolus distinct

PROPHASE
Chromosomes condense and become visible as sets of sister chromatids. Nucleolus and nuclear envelope disappear. Spindle microtubules appear.

METAPHASE
Mitotic spindle complete. Chromatid sets move to spindle equator.

brane disappears at the end of prophase. It breaks down into small vesicles and is re-formed from them after mitosis.

The other notable change during prophase is the beginning of a framework of microtubules, the **mitotic spindle,** which will eventually take part in the movement of the chromosomes. As mitosis begins, microtubules of the cytoskeleton break down into their protein subunits. (This loss of the cytoskeleton is why cells become rounded during division.) The microtubule subunits are reassembled to form the spindle. As the spindle microtubules are assembled, they push the ends, or **poles,** of the spindle apart. In animal cells, the spindle poles are occupied by pairs of centrioles, also composed of microtubules (Section 5-L).

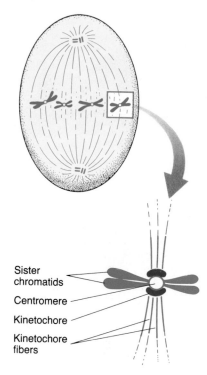

Metaphase During metaphase, the mitotic spindle is completed. The spindle consists of two types of microtubules: polar fibers, which extend from the poles past the equator, and kinetochore fibers, which grow from the poles until their free ends are captured by a kinetochore. The two kinetochores of sister chromatids capture fibers from opposite poles (Figure 11-7).

Each kinetochore is pulled toward its fibers' pole: the farther this is, the stronger the pull. The forces in this tug-of-war balance out midway between the poles, and so all of the chromatids become lined up at the equator of the spindle, the sure sign of a cell in metaphase.

Sister chromatids

Centromere

Kinetochore

Kinetochore fibers

Anaphase Anaphase begins abruptly. All at once, each set of sister chromatids separates, thereby becoming independent chromosomes, which are pulled to opposite poles of the spindle. Each chromosome eventually ends up near one pole of the mitotic spindle, with its sister at the opposite end, so that there is a complete set of chromosomes at each pole, the basis for a new nucleus.

During anaphase, the polar fibers push the poles farther apart, making the cell longer. At the same time, the chromosomes are pulled toward the poles as the kinetochore fibers shorten by losing the subunits nearest the kinetochore. (Exactly how this works without the chromosome's falling off the fiber is not understood.)

Figure 11-7
Attachment of chromosomes to the mitotic spindle at metaphase. A kinetochore forms on each sister chromatid's centromere area. The drawing of the whole cell shows only one spindle fiber attached to each chromatid. However, each kinetochore typically attaches to many spindle fibers radiating from one pole, as seen in the closeup.

Telophase In the last stage of mitosis, telophase, two new nuclei are organized. The chromosomes, now in two groups at the poles of the mitotic spindle,

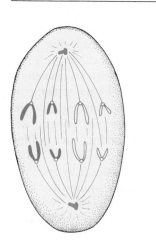

ANAPHASE
Centromeres divide, freeing sister chromatids as individual chromosomes, which then move to opposite poles of the spindle.

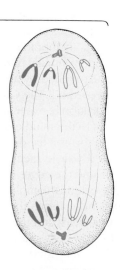

TELOPHASE
Two new nuclei form. Division of the cytoplasm often begins now.

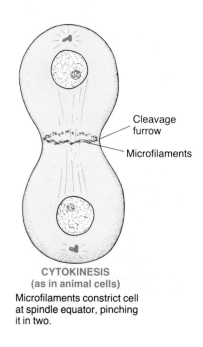

Cleavage furrow

Microfilaments

CYTOKINESIS
(as in animal cells)

Microfilaments constrict cell at spindle equator, pinching it in two.

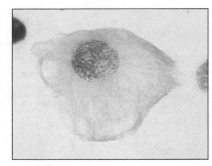

(a) Interphase

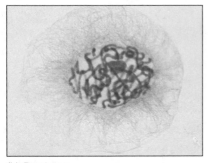

(b) Prophase

(c) Metaphase

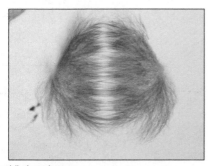

(d) Anaphase

(e) Telophase

Figure 11-8

Dividing cells from a plant, the blood lily. (a) A cell in interphase, with its chromatin dispersed throughout the nucleus. (b) A cell in prophase, with the chromatin condensing into distinctly visible chromosomes. (c) Metaphase. The sets of sister chromatids have lined up at the equator. (d) Anaphase. The sister chromatids have separated into individual chromosomes, which are separating into two groups along the spindle. (e) Telophase. Each group of chromosomes has reached a pole of the spindle and will soon be organized into a new nucleus. (Andrew S. Bajer, University of Oregon)

uncoil into masses of tangled chromatin. A new nuclear envelope forms around each group of chromosomes. Ribosome synthesis resumes, and hence nucleoli become visible in each new nucleus. The nuclei are ready for the normal activities of interphase (Figure 11-8).

Cells of any ploidy (haploid, diploid, tetraploid, etc.) can undergo mitosis.

Colchicine, a chemical derived from the autumn crocus plant, prevents formation of a mitotic spindle and so blocks mitosis and cell division. It is sometimes used in attempts to prevent cancer cells from dividing. In colchicine-treated cells, sister chromatids can still separate from each other, thus doubling the number of chromosomes in the cell. In this way, diploid cells can become tetraploid. Colchicine treatment is used to stop cell division at metaphase so that condensed chromosomes can be collected for analysis (for instance, to make a karyotype like the one in Figure 11-4). Colchicine is also used by plant breeders to make tetraploid plants from diploid ones. Tetraploid plants are often larger and more vigorous than their diploid ancestors. Many cultivated vegetables and flowers are tetraploids that have arisen either naturally or by deliberate treatment with colchicine. Water-processed decaffeinated coffee comes from a diploid species with strongly flavored beans. The milder species used for regular coffee is an artificially produced tetraploid.

11-D Cytokinesis

Mitosis is now complete, but the two nuclei still lie in the same cytoplasm. Division of the cytoplasm is called **cytokinesis:** the original cell forms two new cells, each housing one of the newly formed nuclei (Figure 11-9). In animal cells, cytokinesis begins during early anaphase. A ring of microfilaments, made up of the contractile proteins actin and myosin, forms around the cell's equator, just beneath the plasma membrane. These filaments constrict the cell to form a **cleavage furrow** and eventually pinch the cytoplasm in two (see Figure 11-6).

Plant cells are surrounded by a rigid cell wall, and cytokinesis occurs by a completely different method. The Golgi complexes release vesicles containing material for a new partition between the cells-to-be. The vesicles move along the spindle microtubules to the middle of the cell. Here they fuse together and the material they contain forms a flat disc, the **cell plate,** enclosed by a membrane made up of the fused vesicle membranes. The cell plate and its surrounding membrane grow around the edges by addition of more vesicles (Figure 11-10). Soon the cell plate extends completely across the cell, cutting it in two. The material in the cell plate forms the **middle lamella,** the common partition between the two new cells, and the membrane on either side of the cell plate becomes part of the new cells' plasma membranes. Each new cell builds a new cell wall on its side of the middle lamella.

Cytokinesis divides up not only the cytoplasm, but also the various structures within it, such as ribosomes, Golgi complexes, mitochondria, plastids, and cytoskeleton molecules. Most cells contain many of each kind of structure, distributed throughout the cytoplasm, so that each new cell is bound to receive at least some of every component it needs.

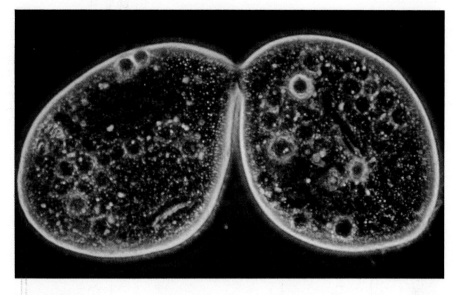

Figure 11-9
Cytokinesis in the protist *Paramecium*. This unicellular organism undergoes cytokinesis as microfilaments constrict the cytoplasm, similar to division of an animal cell. (Biophoto Associates)

11-E   Meiosis

Meiosis is the process of nuclear division in which haploid nuclei are formed from diploid nuclei. DNA synthesis occurs before meiosis as well as before mitosis. Therefore, a nucleus enters meiosis with enough DNA to make four haploid nuclei. Because a nucleus cannot divide into more than two new nuclei

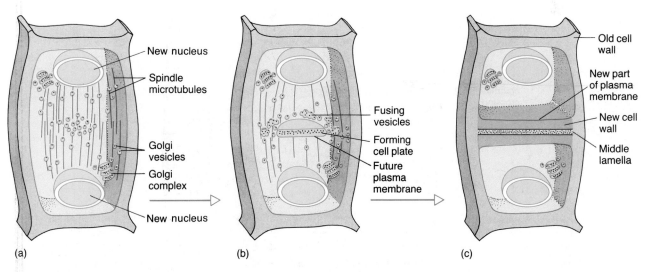

(a)   (b)   (c)

Figure 11-10
Cytokinesis in a plant cell. (a) Vesicles containing material for the cell plate bud off from Golgi complexes. The lingering spindle microtubules serve as tracks for the vesicles to move to the equator. (b) The vesicles fuse, and the material they contain is built into a cell plate, which forms the middle lamella. The fused vesicle membranes on either side of the plate will become the plasma membranes at the ends of the new cells. (c) Each new cell lays down a wall between the middle lamella and the new portion of its plasma membrane.

at any one division, it takes two divisions during meiosis to reduce the DNA content of each nucleus to haploid. We can summarize the movements of one pair of homologous chromosomes through meiosis:

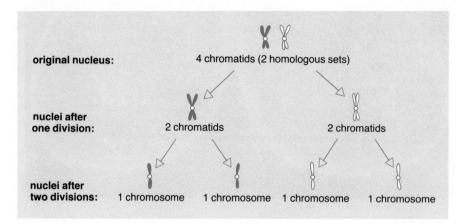

■ *The precise choreography of meiosis I ensures that each new nucleus receives one member of each homologous chromosome pair.*

The two divisions in meiosis are unimaginatively called meiosis I and meiosis II. Like mitosis, both meiotic divisions involve formation of a spindle and movement of chromosomes to the spindle's poles, and so meiosis looks very similar to mitosis (Figure 11-11). The names of the stages are also similar. However, meiosis has some additional features not found in mitosis.

Since meiosis produces haploid nuclei from diploid nuclei, it must provide a way for the cell's homologous chromosome pairs to be parcelled out precisely into two groups, each group containing exactly one member of each homologous pair. The special events of meiosis that allow this precise sorting of the

Figure 11-11
Meiosis. The events of meiosis as it would occur in the formation of male gametes in an animal, or of spores in fungi or plants.

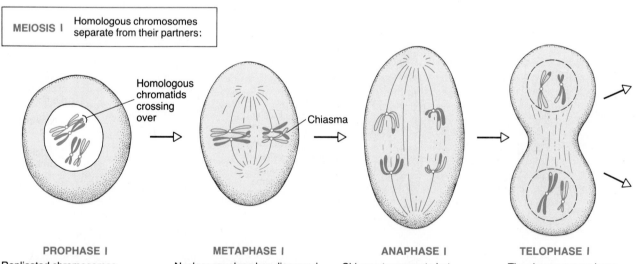

| **MEIOSIS I** | Homologous chromosomes separate from their partners: |

PROPHASE I	**METAPHASE I**	**ANAPHASE I**	**TELOPHASE I**
Replicated chromosomes condense and pair with their homologues to form tetrads. Pairing is necessary for separation of members of each homologous pair in the first meiotic division, so that each resulting nucleus receives one member of each pair. Crossing over occurs during prophase I.	Nuclear envelope has dispersed. Tetrads are held together by chiasmata. Kinetochores of homologous sets of sister chromatids attach to spindle fibers from opposite poles, and tetrads move to the equator.	Chiasmata separate but centromeres do not. Each set of sister chromatids moves toward a pole of the spindle, as its homologue travels toward the opposite pole. Sister chromatids travel as a pair and do not separate until anaphase II.	The chromosomes have formed two groups. In some species, nuclear envelopes reappear and the cytoplasm divides In others, nuclear envelopes remain absent and metaphase II starts immediately.

chromosomes occur during prophase of meiosis I. These events are complex and take a lot of time, making this the longest stage of meiosis. Meiosis frequently takes days to complete instead of the hours or minutes required for mitosis.

During prophase I of meiosis, each chromosome somehow "finds" its homologue among all the other chromosomes in the nucleus, and the two line up next to each other with point-by-point precision, a poorly understood process called **synapsis.** Since the chromosomes have already been replicated, the resulting group consists of four chromatids altogether and is called a **tetrad.** During this tetrad stage, portions of the chromatids are exchanged between the homologous chromosomes, a phenomenon called **crossing over.** This is one source of genetic variation that occurs as a result of sexual reproduction. For a while, the chromatids remain joined at the crossover exchange point, called a **chiasma** ("cross"; plural, **chiasmata**). This holds the homologous pair of chromosomes together, while the centromeres hold the two sister chromatids of each chromosome together; hence the entire tetrad moves as one (Figure 11-11). Research during the 1980s showed that crossing over is a common event. Normally, each tetrad contains at least one chiasma.

In metaphase I, all the tetrads line up at the spindle equator. Each set of sister chromatids has a kinetochore attached to spindle fibers from one pole, and the homologous set, just across the equator, is attached to the opposite pole. This arrangement, much like couples lined up opposite their partners for a barn dance, allows the partners to be separated from one another at anaphase I. As anaphase I begins, the chiasmata come apart, whereas the centromeres remain intact. Hence each set of sister chromatids moves as a unit toward one spindle pole, while the homologous set of chromatids moves to the opposite pole. Each group of chromosomes is then organized into a new nucleus during telophase I.

There is no DNA replication between meiosis I and II.

■ *Meiosis I reduces a diploid nucleus to two haploid ones, each with one complete set of replicated chromosomes.*

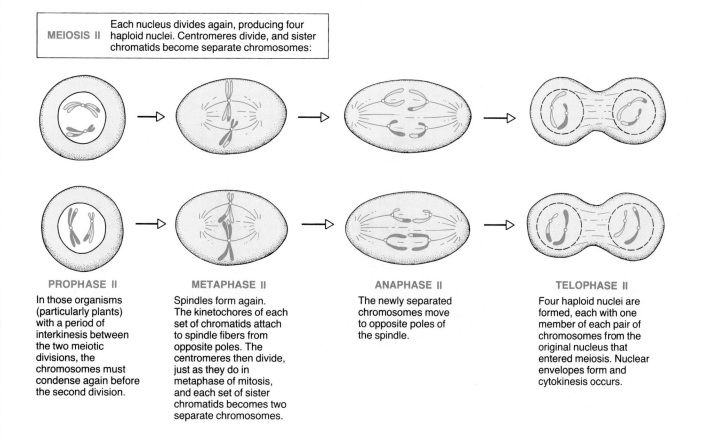

MEIOSIS II | Each nucleus divides again, producing four haploid nuclei. Centromeres divide, and sister chromatids become separate chromosomes:

PROPHASE II

In those organisms (particularly plants) with a period of interkinesis between the two meiotic divisions, the chromosomes must condense again before the second division.

METAPHASE II

Spindles form again. The kinetochores of each set of chromatids attach to spindle fibers from opposite poles. The centromeres then divide, just as they do in metaphase of mitosis, and each set of sister chromatids becomes two separate chromosomes.

ANAPHASE II

The newly separated chromosomes move to opposite poles of the spindle.

TELOPHASE II

Four haploid nuclei are formed, each with one member of each pair of chromosomes from the original nucleus that entered meiosis. Nuclear envelopes form and cytokinesis occurs.

■ *Meiosis II is essentially a mitotic division of a haploid nucleus into two new haploid nuclei.*

During prophase II, a new spindle forms in each of the two new cells. Each set of chromatids now has two separate kinetochores, which attach to spindle fibers from opposite poles. Metaphase II finds the sets of sister chromatids lined up at the spindle equator. At the beginning of anaphase II the centromeres finally divide, releasing the sister chromatids as individual chromosomes. The chromosomes then separate into two groups during anaphase II, and at telophase II they become organized into two haploid nuclei. Since meiosis I produced two nuclei, the division of each one at meiosis II gives a total of four haploid nuclei.

Meiosis is vital in all eukaryotes that reproduce sexually, but it does not always take place at the same stage in the life history. Meiosis is most familiar as part of the process of gamete formation in the life histories of animals.

11-F   Genetic Reassortment

Besides reducing the chromosome number from diploid to haploid, meiosis also shuffles the genetic material, forming new combinations of genes and chromosomes that become the genetic information of the next generation. Meiosis produces this **genetic reassortment** in two ways: by producing new combinations of genes on chromosomes (crossing over) and by producing new assortments of chromosomes.

The production of chromosomes with new combinations of genes occurs during crossing over (Section 11-E). This results from the exchange of segments of DNA between homologous chromosomes, a process called **genetic recombination.** Crossing over occurs while the chromosomes are in tetrads during prophase I: two chromatids, one from each homologue, cross each other and are broken off and joined to the opposite strand. This rearranges genes that were on the same chromosome so that they are on two different chromosomes, and vice versa (Figure 11-12). Since one of the original chromosomes was inherited from each parent, crossing over combines genes from both parents in each of the recombinant chromatids. In humans, crossing over occurs an average of two or three times in each pair of homologous chromosomes during gamete formation. When more than one crossover occurs, these may involve the same two chromatids or different combinations of chromatids (Figure 11-12b).

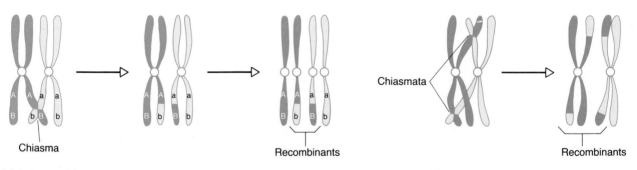

(a) A single crossover

(b) Crossovers of different chromatids

Figure 11-12

Crossing over. (a) Parts of homologous chromatids may cross each other and be broken off and rejoined onto the opposite chromatid. Eventually this produces four chromosomes with different gene combinations from each other (AB, Ab, aB, and ab). (b) If more than one crossover occurs in the same tetrad, different chromatids may be involved in each crossover event. Here, each dark blue chromatid is involved in one crossover. The light blue chromatid on the left takes part in two crossovers, its sister in none. The three chromatids on the left end up as recombinants.

Table 11-2 Differences Between Mitosis and Meiosis

Original nucleus:

AaBb

MITOSIS

MEIOSIS

Occurs in haploid (N) and diploid (2N) cells

Occurs in diploid (2N) cells

Nucleus divides once

Nucleus divides twice

No synapsis

Synapsis, tetrad formation, and crossing over by homologous sets of sister chromatids during prophase I

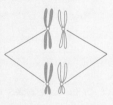

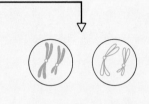

Sets of sister chromatids line up singly at metaphase

Sets of sister chromatids line up in tetrads, paired with their homologues, at metaphase I

Sets of sister chromatids line up singly at metaphase II

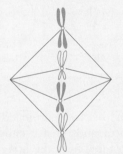

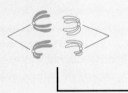

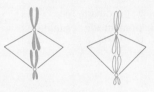

Sister chromatids separate as centromeres divide at anaphase

Homologues separate at anaphase I, but centromeres do not divide at this time

Sister chromatids separate as centromeres divide at anaphase II

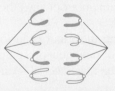

Produces two nuclei with same ploidy (N, 2N, etc.) as original nucleus

Produces four haploid nuclei

New nuclei have same chromosome content as original nucleus and same as one another

New nuclei have half the chromosome content of original nucleus, and are not identical to one another

AaBb   *AaBb*

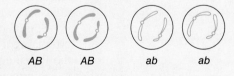

AB   *AB*   *ab*   *ab*

This is one possible outcome of meiosis. A different lineup of tetrads at metaphase I would produce gametes containing chromosome combinations: *Ab*, *Ab*, *aB*, and *aB*.

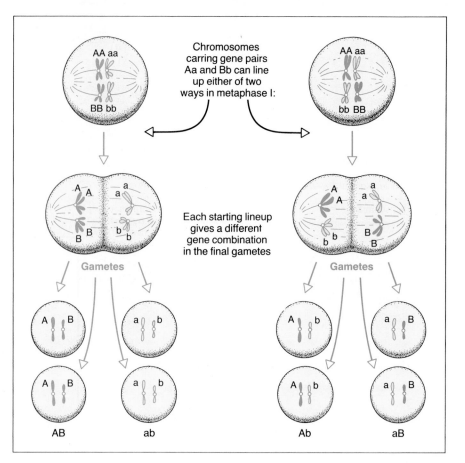

Chromosomes carring gene pairs Aa and Bb can line up either of two ways in metaphase I:

Each starting lineup gives a different gene combination in the final gametes

Figure 11-13
The luck of the lineup. Genetic reassortment is partly due to the various possible arrangements of chromosomes during metaphase I of meiosis. Two chromosome pairs can line up in either of two ways, producing four possible combinations in the resulting gametes.

Another source of genetic reassortment occurs because chromatid tetrads can line up at metaphase I with either set of chromatids nearer either pole. Then, in anaphase I, each set of chromatids is separated from the homologous set and goes into a new nucleus with the members of other homologous pairs that were lined up on the same side of the equator as itself (Figure 11-13). The lining up at metaphase I is random, and so there is an equal chance that any one chromosome will end up in a new cell with either member of any other pair of homologous chromosomes.

Additional genetic reassortment occurs during the random fusion of gametes at fertilization. In this way a steady supply of new genetic combinations arises in sexually reproducing species, furnishing the raw material—variation—for evolution by means of natural selection.

 Meiosis produces genetic reassortment both by swapping genes between chromosomes and by sorting the chromosomes into new combinations.

11-G   Gamete Formation in Animals

The formation of gametes (sperm and eggs) is similar in most animals, although details vary among species. However, the two processes—formation of sperm and of eggs—differ somewhat from each other. Let us begin with the production of sperm, which is in some ways simpler.

Sperm, or spermatozoa, are the male gametes. Since a sperm contains little cytoplasm, it is very small, and in nearly all species the sperm can swim, using the flagellum that forms its tail.

Sperm are produced in the testes by the process of **spermatogenesis** (Figure 11-14a). The male's germ cells, called **spermatogonia,** divide continuously by mitosis. Some of the new cells become **spermatocytes,** the cells that undergo meiosis. Primary spermatocytes go through meiosis I, which produces

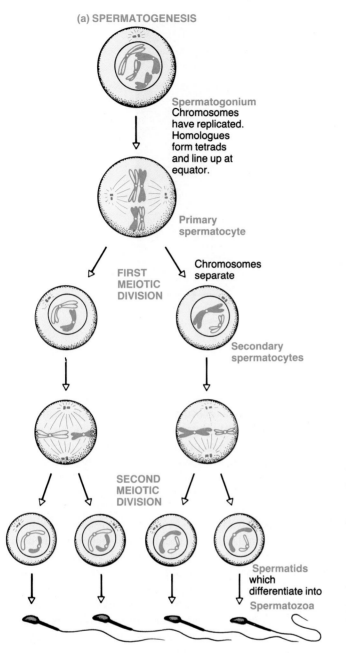

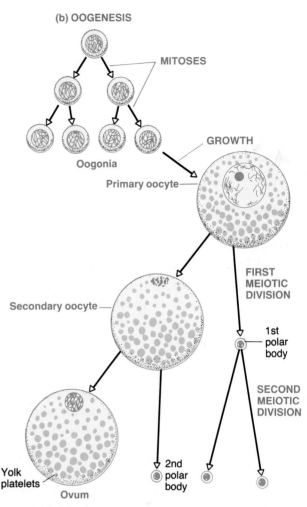

Figure 11-14

Gamete formation in animals. (a) Spermatogenesis. Development of a spermatogonium into four spermatozoa (sperm) through meiosis and differentiation. (b) Oogenesis in a vertebrate. The oogonia divide to form more oogonia by mitosis. Eventually, oogonia grow and differentiate into primary oocytes. The nuclear divisions of meiosis are like those shown in spermatogenesis, but cytokinesis is unequal, forming a large cell and a tiny polar body at each division.

two secondary spermatocytes. During meiosis II, the two secondary spermatocytes divide again and produce a total of four haploid **spermatids.** Although meiosis is now complete, the spermatids must undergo further differentiation into spermatozoa. A mature sperm has a head, which contains the nucleus with its haploid set of chromosomes; a long tail, or flagellum, which propels the sperm through its fluid surroundings; and between these a midpiece containing many mitochondria, which supply the ATP necessary for flagellar motion.

Oogenesis is the formation of female gametes, the eggs or **ova** (singular: **ovum**). Whereas sperm are often the smallest cells in a male animal's body, eggs are the largest cells in a female. This is because the egg cell is the main source of stored food, ribosomes, messenger RNA, and other cytoplasmic components that support the embryo's early development. Oogenesis ensures that the mature egg contains as much as possible of these components. Meiotic

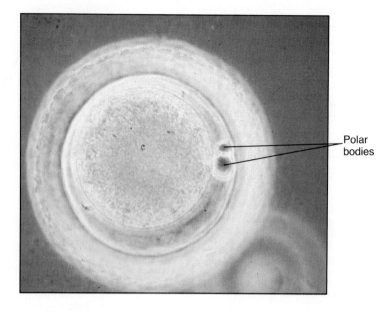

Polar bodies

Figure 11-15
The fertilized egg of a rabbit. Two polar bodies can be seen. (Biophoto Associates)

■ *In the formation of sperm and eggs, the nuclear divisions of meiosis are alike. In sperm production, cytokinesis is equal, forming four cells with equal chances of fertilizing an egg. In egg formation, cytokinesis is grossly unequal, producing only one gamete, provided with the resources to support embryonic development.*

nuclear division is accompanied by unequal cytokinesis, so that the original diploid cell produces only one large ovum and two tiny cells called **polar bodies** (Figure 11-14b).

In the ovary, cells called **oogonia** divide by mitosis for a time. Eventually, they stop dividing and differentiate into primary oocytes. The DNA is replicated, and the primary oocytes enter prophase I, proceeding through the formation of tetrads and crossing over. In humans and other mammals, all this occurs while the female is still an embryo. At this point in prophase I, meiosis is arrested for days or years, depending on the species, and it does not resume until the female reaches sexual maturity. During this time the cell absorbs nutrients from neighboring somatic cells and stockpiles the materials needed for early embryonic development.

In meiosis I of oogenesis, the chromosomes separate in the usual manner, but during cytokinesis the cytoplasm divides unequally. One nucleus is pinched off with a minimum of cytoplasm, forming the first polar body, while the other nucleus is left with most of the cytoplasm, in a cell called the secondary oocyte. This goes through the second meiotic division. Again two haploid nuclei are formed according to the normal events of meiosis, but cytokinesis is extremely unequal, forming a tiny second polar body and an enormous ovum. The polar bodies are really just a means of shedding excess chromosomes from the developing egg, and they soon disintegrate (Figure 11-15).

SUMMARY

Cells are the reproductive units of life. New cells are produced when existing cells divide in two. These divisions are of two kinds: mitotic division, in which a cell (of any ploidy) gives rise to two new cells with chromosome complements identical to that of the parent; and meiotic division, in which a diploid cell divides twice, forming four haploid new cells.

The time from one nuclear division to the next is known as the cell cycle. It can be divided into interphase (G_1, S, and G_2) and mitosis. The initiation of DNA synthesis in the S period of interphase is the

key event committing the cell to undergo mitotic division.

Mitosis is a nuclear division in which precise events ensure that the two new nuclei inherit chromosomes identical to those of the parent nucleus. During prophase of mitosis, the replicated chromosomes, each consisting of two sister chromatids, condense and become visible under the light microscope. The nucleolus and nuclear membrane disperse, and microtubules are assembled to form the mitotic spindle. In metaphase, all of the sets of sister chromatids are lined up at the equator of the spindle. In each set, the two sisters' kinetochores are attached to spindle fibers from opposite poles. During anaphase, each centromere splits into two, releasing the sister chromatids from one another and allowing them to travel to the opposite poles of the spindle. During telophase, the chromosomes at each pole form a nucleus as the nuclear membrane and nucleolus re-form and the chromosomes unravel from their condensed form.

Mitosis is usually accompanied by cytokinesis, the division of the cytoplasm and its components to form two separate cells. In animal cells, a band of microfilaments pinches the cell in two. Cytokinesis in plants involves the assembly of a partition between the two new cells, which then build new end walls on either side.

Meiosis is the series of two nuclear divisions that produces four haploid nuclei from a diploid nucleus. Meiosis halves the number of chromosomes in a cell in such a way that each new nucleus receives one member of each pair of homologous chromosomes. Meiosis also results in genetic reassortment, both by crossing over, in which homologous chromosomes exchange genes, and by forming new chromosome combinations as a result of how the chromosomes line up at metaphase I. Additional genetic variety results from the random combination of gametes at fertilization.

Synapsis and crossing over occur during prophase I of meiosis. Then, the tetrads of homologous sister chromatids line up at the spindle equator in such a way that homologous chromosomes are separated from each other during the first meiotic division. Each of the two resulting nuclei contains one member of each pair of homologous chromosomes. Not until the second division do the centromeres divide, permitting sister chromatids to move into different nuclei.

Gamete formation in animals involves both meiosis and differentiation to form specialized reproductive cells. Each spermatocyte gives rise to four sperm, the male gametes, which are stripped down to the bare necessities: a haploid set of genetic material and the locomotory apparatus to deliver it to the egg. Oogenesis involves unequal cytokinesis, producing only one large egg swollen with material destined to support the early embryo, and tiny polar bodies, which contain little more than the excess chromosomes being shed from the forming egg.

SELF-QUIZ

1. A cell cycle is:
   a. the time from the formation of a cell until its death
   b. the series of events that takes place from the formation of a cell until it divides again
   c. the sequence of events that assures each new cell of a set of chromosomes identical with that of its parent cell (mitosis)
   d. the growth of a cell until it is large enough to divide again
2. For the species depicted in Figure 11-1, what is the value of N? of 2N?
3. A diploid somatic cell:
   a. cannot undergo division again
   b. can undergo mitosis but not meiosis
   c. can undergo mitosis or meiosis
   d. can undergo meiosis but not mitosis
4. A cell in prophase of mitosis can be distinguished from a cell in prophase I of meiosis by:
   a. the presence of only half as many chromosomes in the meiotic cell
   b. the formation of tetrads in the meiotic cell
   c. the presence of twice as many chromosomes in the meiotic cell
5. The function of mitotic cell division in the life history of an organism is:
   a. reproduction of identical individuals if the organism is unicellular
   b. growth of an individual if the organism is multicellular
   c. repair of injured tissue
   d. all of the above

6. Substances that interfere with microtubule function interfere with cell division because:
   a. microtubules must be distributed equally to the new cells
   b. microtubules are involved in the precise separation of the chromosomes, which ensures that a complete set of chromosomes gets into each daughter cell
   c. without microtubules, cytokinesis cannot take place, and a cell with two nuclei is formed
   d. microtubules are essential to the disappearance of the nuclear membrane, and without them the chromosomes have to stay too close together within the nuclear membrane to be able to separate into two new nuclei.
7. Both oogenesis and spermatogenesis involve equal division of the ____. However, unequal division of the ____ occurs during production of ____, whereas in production of ____ this division is equal.
8. The importance of crossing over during meiosis is:
   a. it assures that one member of each homologous pair ends up in each new nucleus
   b. it results in chromosomes containing new combinations of genes
   c. it results in nuclei with too much or too little genetic material
   d. it ensures that the developing egg receives most of the cytoplasm from the oocyte

QUESTIONS FOR DISCUSSION

1. Tetraploid plants are frequently larger and have larger fruits and flowers than their diploid relatives. What might account for the fact that octaploid plants (8N) tend to be tiny and scrawny, and produce few offspring?
2. Why is it necessary for cytokinesis to occur in such a way that each new cell receives some ribosomes, mitochondria, and, in plants, plastids?
3. Since the genetic information is carried equally by egg and sperm, what do you suppose to be the selective advantage of the inequality of size that has evolved between the tiny mobile sperm and the large immobile egg?
4. How would leakage from nuclear waste repositories affect the organisms that come into contact with it? Why is it difficult to design safe nuclear waste disposal facilities?

SUGGESTED READINGS

Mazia, D. "The cell cycle." *Scientific American,* January 1974.

McIntosh, J. R., and K. L. McDonald. "The mitotic spindle." *Scientific American,* October 1989.

Upton, A. C. "The biological effects of low-level ionizing radiation." *Scientific American,* February 1982.

When you have studied this chapter, you should be able to:

1. Define and use these terms:
   parental (P_1), first filial (F_1), and second filial (F_2) generations
   alleles, homozygous, heterozygous, dominant, recessive
   monohybrid cross, dihybrid cross, segregation, independent assortment
   codominance, incomplete dominance
   homologous chromosomes, linkage groups, crossing over
2. Define and compare the terms **phenotype** and **genotype** and their relationship to the terms **dominant** and **recessive**.
3. Use a Punnett square to illustrate a monohybrid or independently assorting dihybrid cross, and work out the genotypic and phenotypic ratios expected from such crosses.
4. Explain what is meant by a test cross, and discuss its significance as a genetic tool. Design a test cross to determine the genotype of an organism with a dominant phenotype.
5. Correlate the pattern of inheritance of genetic characteristics in breeding experiments with the behavior of the chromosomes during meiosis and fertilization.
6. Explain the biological significance of tetrad formation and crossing over during meiosis.
7. In your own words, state the rules of inheritance that were Mendel's most important contribution to genetics.

Mendelian Genetics

CHAPTER

12

Silver leaf monkey and baby

Tulips

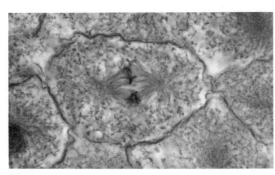

Chromosomes, human male

Mitosis

*P*eople have long understood that plants and animals inherit some characteristics from their parents. Prehistoric people doubtless recognized a child's resemblance to its parents, bred calves from the cows that gave the most milk, and saved the most productive grain for seed. However, the breeding of plants and animals was not put on a scientific basis until the twentieth century. Genetics is the study of patterns of inheritance as hereditary characteristics (also called **characters** or **traits**) are passed from parents to offspring. This information is applied to the practical goal of breeding economically useful varieties of plants and animals.

Genetics is based on the work of Gregor Mendel, a monk (and later the abbot) at the monastery of Brünn, in what is now Czechoslovakia. In 1866, Mendel published a completely new and thoroughly documented model of inheritance. However, the influential scientists then studying inheritance were absorbed in a maze of complex hypotheses, and the few who read Mendel's paper dismissed his model as trivial because it was so simple. Hence Mendel's work received little attention until after his death. It was rediscovered in 1900 almost simultaneously by three different people.

In the meantime, chromosomes had been named and their movements during mitosis and meiosis observed and described (see Sections 11-C and E). In 1902, several scientists realized that the chromosomes moved precisely as would be expected of the structures responsible for the patterns of inheritance Mendel reported. Once this connection between chromosomes and heredity was established, the science of genetics entered a period of productive research.

Mendel was the first person to recognize that genetic traits are inherited as separate particles. He did not actually see these hereditary particles, but he reasoned that they must exist because that would explain the patterns of inheritance shown by genetic traits. He proposed that organisms have a pair of particles for each inherited trait, one from each parent. We now know that the particles of inheritance are segments of chromosomal DNA molecules, and we call them **genes.** Many genes code for proteins, and in this way the microscopic gene manifests itself as some trait of the organism, such as curly hair. Genes are replicated and passed on to new cells as parts of DNA molecules, and this accounts for the observation that offspring inherit genetic traits from their parents. In sexual reproduction, a new individual receives half of its genes from its mother's egg and half from its father's sperm.

- In most familiar organisms, each individual has a pair of genes for each trait, one inherited from each parent.
- Each of the individual's gametes (eggs or sperm), and hence each offspring, receives one of the individual's two genes for each trait.
- The pattern of inheritance of genes from generation to generation reflects the behavior of chromosomes in meiosis and fertilization.

12-A   A Simple Breeding Experiment

Mendel worked with garden peas, which were available in many different varieties. Each variety bred true to type. For example, tall pea plants of one variety always produced tall offspring, and plants of a dwarf variety always produced dwarf offspring. Pea flowers contain both male and female parts, and normally the flower pollinates itself. Hence, each plant is both male and female parent to its seeds. Over the generations, this builds up considerable genetic uniformity. But it is possible to cross-pollinate peas artificially by transferring pollen from the male flower parts of one plant to the female flower parts of another. By crossing (that is, breeding together) plants of two varieties with contrasting traits, such as tall and dwarf varieties, Mendel could trace the inheritance of

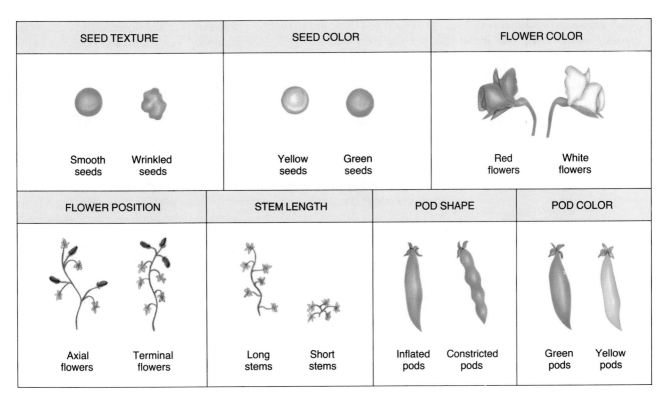

SEED TEXTURE	SEED COLOR	FLOWER COLOR
Smooth seeds / Wrinkled seeds	Yellow seeds / Green seeds	Red flowers / White flowers

FLOWER POSITION	STEM LENGTH	POD SHAPE	POD COLOR
Axial flowers / Terminal flowers	Long stems / Short stems	Inflated pods / Constricted pods	Green pods / Yellow pods

Figure 12-1
Genetic traits. Mendel studied seven different traits of pea plants, each of which appeared in two different forms. In each pair, the dominant form (discussed shortly) is the one on the left.

these traits. In all, Mendel worked with seven traits, each of which occurs in two distinct forms (Figure 12-1).

Mendel began by studying crosses involving only one trait at a time. Let us follow one such experiment, on the inheritance of flower color. In this experiment, Mendel crossed a pure-breeding strain of red-flowered pea plants with a pure-breeding strain that produced white flowers. These plants are referred to as the **parental, or P₁, generation.** A cross between different parental strains, such as these, produces genetically mixed offspring known as **hybrids.** Mendel collected the hybrid seeds and planted them to see what traits this **first filial (F₁) generation** had inherited from the P₁ parents. When the F₁ hybrid plants matured, they all produced red flowers. Mendel allowed these red flowers to self-pollinate, and from them he collected over 900 seeds of the **second filial (F₂) generation.** Most of these F₂ seeds grew into red-flowered plants, but about a quarter of them produced white-flowered plants (Figure 12-2).

Gene Pairs

Mendel saw that these results could be explained if an inherited trait, such as flower color, was governed by two "factors," which we now call genes. A plant received two genes for each of its traits, one from each parent. In turn, each plant passed on one of its two genes at random to each offspring.

Figure 12-2
A cross between pure-breeding red-flowered and pure-breeding white-flowered pea plants (P₁). All the offspring of the first filial (F₁) generation were red-flowered. Self-pollination of these F₁ offspring yielded an F₂ generation of about ¾ red-flowered and ¼ white-flowered plants. (The multiplication sign between the two flowers at the top is the symbol for a mating.)

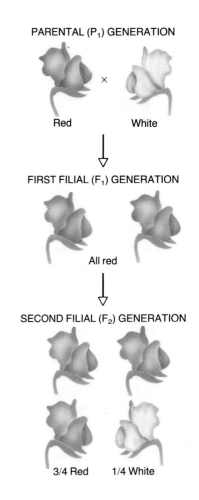

PARENTAL (P₁) GENERATION

Red ✕ White

FIRST FILIAL (F₁) GENERATION

All red

SECOND FILIAL (F₂) GENERATION

3/4 Red   1/4 White

■ *Each organism receives two genes for each trait, one from each parent. During reproduction, each offspring receives, at random, one of these two genes.*

If each red-flowered parent had two genes for red flowers, and each white-flowered parent had two genes for white flowers, then each offspring of the cross between the two received one red-flower gene and one white-flower gene. When these plants reproduced, each egg or pollen grain would receive one of the two genes, so that half the eggs and half the pollen would contain each kind of gene. When the genes from egg and pollen combined, at random, a quarter of the offspring would have two red-flower genes, a quarter would have two white-flower genes, and half would have one red and one white.

This conclusion fits in with what we know about chromosomes and genes (Section 11-A). Diploid eukaryotic cells contain pairs of homologous chromosomes. **Homologous chromosomes** are usually of the same length, their centromeres are in the same position, and they bear genes for the same traits in the same locations. Since chromosomes come in pairs, so do the genes carried by the chromosomes.

A genetic trait can occur in two or more different forms; for example, flower color may be either red or white. Therefore, the genes that govern the trait must come in alternative forms, called **alleles**—the red-flower allele and the white-flower allele. In Mendel's pea plants, one chromosome may have the red-flower allele at the flower-color location, and its homologue can have either the red-flower or the white-flower allele at the same location.

Any one pea plant may have two alleles for red flowers, or two alleles for white, or one of each. An individual with two of the same allele is said to be **homozygous** for that allele. Plants with two alleles for red or two for white are homozygous for flower color. An individual with two different alleles for a trait is said to be **heterozygous;** for example, plants with one red and one white allele are heterozygous for flower color.

In Mendel's crosses, the original P_1 generation came from pure-breeding stock. This means that all the red-flowered plants were homozygous for red flowers, and the white-flowered plants were homozygous for white flowers. Each member of the F_1 generation must have received one allele for red flower color from the red-flowered parent and one allele for white flower color from the white-flowered parent. The F_1 generation was therefore heterozygous with respect to flower color.

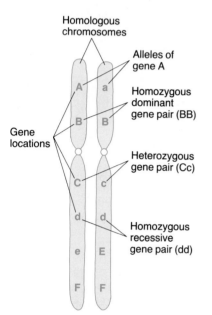

Figure 12-3
The vocabulary of genetics. This diagram shows the relationships among some terms introduced in the text.

(Figure labels:)
Homologous chromosomes
Alleles of gene A
Homozygous dominant gene pair (BB)
Gene locations
Heterozygous gene pair (Cc)
Homozygous recessive gene pair (dd)

Dominant and Recessive Alleles

Mendel found that all of the F_1 plants bore red flowers. What had happened to the white alleles? Since self-crossing of the F_1 plants produced both red- and white-flowered plants, the alleles for white must have been present in the F_1 plants, but masked. Mendel concluded that one allele of a gene may express itself (that is, appear as an observable trait in the organism) and mask the presence of the other allele, when the two occur together in a heterozygote. The allele that expresses itself is called the **dominant** allele, and the masked allele is said to be **recessive.** Homozygous and heterozygous red-flowered plants cannot be told apart just by looking at them. The recessive allele can be detected only in the homozygous condition, when the dominant allele is not present.

Geneticists often use a shorthand, in which genes are designated by letters of the alphabet—capital letters for dominant alleles, and the lower case of the same letter for recessive alleles (Figure 12-3). In the flower-color example, we can use *RR* for the red-flowered parent, *rr* for the white-flowered parent, and *Rr* for the heterozygous F_1 plants.

■ *Many pairs of alleles show a dominant-recessive relationship, with the dominant allele expressing itself and masking the presence of the recessive allele in the heterozygous condition.*

Genotype and Phenotype

Because of dominance, we cannot tell the **genotype,** or genetic makeup, of an individual that shows the dominant trait merely by inspection: pea plants with genotypes *RR* and *Rr* look alike in that both have red flowers. In this case, both

■ *An individual's genotype is fixed at the time of fertilization, but its phenotype results from the interaction of all of its genes with one another and with factors in the environment.*

Figure 12-4
Genotype and phenotype. Individuals of dominant phenotype have more than one possible genotype, but the recessive phenotype indicates a homozygous recessive genotype.

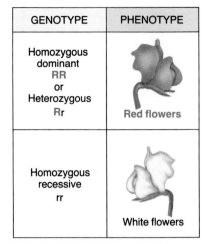

GENOTYPE	PHENOTYPE
Homozygous dominant **RR** or Heterozygous **Rr**	Red flowers
Homozygous recessive **rr**	White flowers

kinds of plants have a red-flowered **phenotype:** the expression of their genes (Figure 12-4). The phenotype can be observed in some way, perhaps visually, as in flower color, or chemically, as in the tests used to find out the blood types of people whose blood looks identical, and so on. An individual with a dominant phenotype may have a genotype that is either **homozygous dominant** (homozygous for the dominant allele) or heterozygous. An individual with a recessive phenotype, however, must have a genotype that is **homozygous recessive** (homozygous for the recessive allele).

Both an organism's genes and its environment can affect its phenotype. For example, a plant may be short because it has "dwarf" genes or because it is so poorly nourished that it cannot grow to the height dictated by its "tall" genes.

Law of Segregation and Meiosis

Mendel recognized that a pea plant's paired genes must separate from each other when the plant reproduces. This is now called Mendel's **law of segregation.** He also saw that the gametes containing the single genes must combine at random to form new gene pairs at fertilization.

Scientists in Mendel's time had not yet discovered the steps of meiosis (Section 11-E). However, we now know that the events of meiosis account for the law of segregation: the members of each pair of homologous chromosomes separate into different nuclei during meiosis, and so the genes carried by these chromosomes also become separated. For example, consider a diploid cell containing a pair of homologous chromosomes, one chromosome carrying the A allele and its homologue carrying the a allele. As the chromosomes proceed through meiosis, A and a are separated into different nuclei, and so they end up in different gametes (Figure 12-5). Each chromosome is replicated before meiosis begins, but the two copies remain attached. During the first division of meiosis, each chromosome lines up with its homologue to form a **tetrad** (a group containing four chromosome copies). Then the homologous chromosomes are separated, one going to each pole of the spindle. Two nuclei are formed, each containing one of the two alleles (A and a). At the second meiotic division, the two copies of each chromosome separate, forming a total of four nuclei.

Monohybrid Cross

Now let us return to Mendel's flower-color cross. Such a genetic cross, in which only one trait of the parents (flower color, in this case) is of interest, is called a **monohybrid cross.** We can now diagram the pattern of inheritance as the genes are passed from one generation to the next (Figure 12-6). As we just saw in our review of meiosis, the red-flowered parent RR forms gametes containing a single R allele, and the white-flowered parent, rr, forms only r gametes. At fertilization, the F_1 offspring inherit a pair of flower-color genes, R from the red-flowered parent and r from the white-flowered one, so that they have the geno-

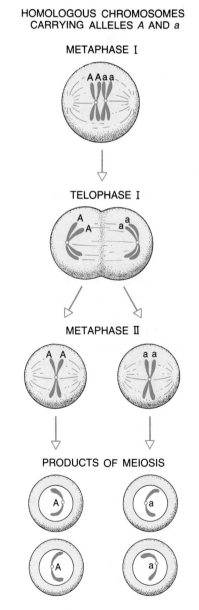

HOMOLOGOUS CHROMOSOMES CARRYING ALLELES A AND a

METAPHASE I

TELOPHASE I

METAPHASE II

PRODUCTS OF MEIOSIS

Figure 12-5
The law of segregation reflects the events of meiosis. Homologous chromosomes are separated so that each member of the pair ends up in a separate gamete, and hence so do the paired alleles (A and a) they carry. Figure 11-11 shows meiosis in more detail.

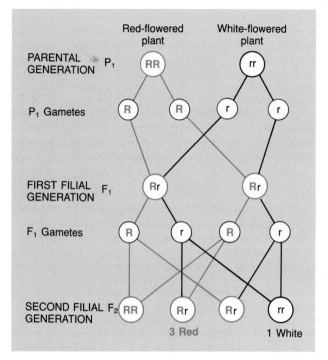

Figure 12-6
A **monohybrid cross**. This diagram shows genotypes of gametes and off-spring when a plant homozygous for red flower color is crossed with one homozygous for white flower color, and the hybrid offspring then self-pollinate. This is the same cross shown in Figure 12-2.

type *Rr*. These F_1 plants are red-flowered because *R,* the dominant allele, expresses itself as red flower color and masks the recessive allele *r*.

When the F_1 generation reproduces, *R* and *r* segregate into separate cells at meiosis: half the gametes carry the allele for red flowers *(R)* and half carry the allele for white *(r)*. An egg has an equal chance of receiving either an *R* or an *r* allele, and it is equally likely to be fertilized by a sperm from a pollen grain containing either *R* or *r*. Therefore, four combinations are possible to produce the F_2 generation (Figure 12-6):

(1) an *R* egg and *R* sperm *(RR)*   (3) an *r* egg and *R* sperm *(rR)*
(2) an *R* egg and *r* sperm *(Rr)*   (4) an *r* egg and *r* sperm *(rr)*

■ *In a monohybrid cross involving alleles with a dominant-recessive relationship, the* F_2 *generation is expected to show a phenotypic ratio of 3 dominant phenotypes: 1 recessive phenotype (or $\frac{3}{4}$:$\frac{1}{4}$). The genotype ratios are 1 homozygous dominant: 2 heterozygous: 1 homozygous recessive.*

Once fertilization is complete, (2) and (3) are indistinguishable. Hence the possible genotypes in the F_2 generation are *RR, Rr,* and *rr*. We expect to find these in a ratio of 1*RR*:2*Rr*:1*rr* (or, equivalently, $\frac{1}{4}RR$:$\frac{1}{2}Rr$:$\frac{1}{4}rr$), since there are two ways to obtain the *Rr* combination and only one way to obtain each of the others. The ratio of phenotypes is three red-flowered plants to one white-flowered plant, since the *RR* and the *Rr* plants all have red flowers. (For convenience, we can write the genotype of a red-flowered plant as *R___*. This notation indicates that at least one dominant allele is present. We use the underline in place of the second gene either to show that we don't know which allele is present, or to show that we are lumping together both *RR* and *Rr* individuals, which are phenotypically indistinguishable.)

12-B   *Predicting the Outcome of a Genetic Cross*

When we know the genotypes of parents used in a genetic cross, we can predict the genotypes of the offspring and their expected ratios. One way to do this is by drawing a **Punnett square** (named after geneticist Reginald Crundall Punnett). To construct a Punnett square, we write the genes present in the gametes of one parent above separate boxes across the top of the square, and those from the other parent down the side (Figure 12-7). We then fill in all the boxes by com-

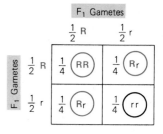

bining the genes shown at the top of the column with those shown at the left of the row. The combinations produced by filling in all of the boxes show the possible genotypes of the F$_2$ individuals and the ratio in which they are expected to occur. In the flower-color cross, the F$_2$ generation includes *RR, Rr,* and *rr* individuals in the ratio of 1:2:1. If we know that one allele is dominant to the other, we can also predict the phenotypes of the F$_2$ generation. In this case, since *R* is dominant to *r,* the F$_2$ generation is expected to have three times as many red-flowered *(R___)* as white-flowered *(rr)* individuals.

We could also calculate this outcome directly from the probabilities of each type of gamete. The gametes produced by each F$_1$ heterozygous parent can be written as $(\frac{1}{2}R + \frac{1}{2}r)$. To find the distribution of genotypes in the next generation, we must multiply the proportions of gametes of each sex together, in the same way we multiply two binomials in an algebra problem:

$$\underbrace{(\tfrac{1}{2}R + \tfrac{1}{2}r)}_{\substack{\text{male}\\\text{gametes}}} \underbrace{(\tfrac{1}{2}R + \tfrac{1}{2}r)}_{\substack{\text{female}\\\text{gametes}}} = \underbrace{\tfrac{1}{4}RR + \tfrac{1}{4}Rr + \tfrac{1}{4}rR + \tfrac{1}{4}rr}_{\substack{\text{offspring}\\\text{genotypes}}}$$

Combining the middle two terms, we come out with $\frac{1}{4}RR + \frac{1}{2}Rr + \frac{1}{4}rr$. This is the same 1:2:1 ratio as the genotypes found using the Punnett square.

12-C Test Cross

If both *RR* and *Rr* plants have the same red-flowered phenotype, how can we find out the genotype of a particular red-flowered plant? The usual method is to cross such a plant with a plant of known genotype and observe the phenotypes of the offspring.

In a **test cross,** an organism of dominant phenotype but unknown genotype is crossed with one that is homozygous recessive for the trait in question. A white-flowered pea plant, in our example, has the homozygous recessive genotype, *rr,* and must pass on an *r* allele to each of its offspring. If the red-flowered plant of unknown genotype were actually heterozygous *(Rr),* we should expect half of the offspring of the test cross to be white-flowered and half red-flowered. On the other hand, if the red-flowered parent were homozygous *(RR),* it could pass only the *R* allele to its offspring and they would all be heterozygous, with red flowers (Figure 12-8). So, if a red-flowered × white-flowered test cross produces any white-flowered offspring, the red-flowered parent must be heterozygous, because white-flowered offspring must have obtained an *r* allele from each parent. However, if all the progeny have red flowers, it is not absolutely

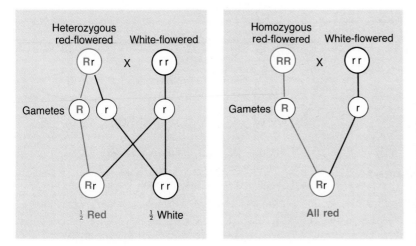

Figure 12-8
Test cross. A red-flowered plant of dominant phenotype but unknown genotype is crossed with a white-flowered (homozygous recessive) plant. By examining the offspring, it should be possible to determine the genotype of the red-flowered parent plant.

certain that the red-flowered parent was homozygous *RR:* it is possible, though unlikely, for a plant to have all red-flowered offspring even if its genotype is *Rr.*

The ratios shown in diagrams of genetic crosses represent *expected* proportions of offspring of different genotypes. The *actual* proportions of genotypes resulting from a particular cross depend on chance events of meiosis and fertilization. But it is true that the more offspring—all red-flowered—we get from such a cross, the more nearly certain we can be that the parent with the dominant phenotype is not heterozygous.

12-D The Dihybrid Cross: Independent Assortment of Genes

In addition to his monohybrid crosses, Mendel performed **dihybrid** crosses of plants with two different pairs of contrasting alleles. In one experiment, Mendel crossed plants homozygous for seeds that were both smooth and yellow with plants homozygous for wrinkled, green seeds. All the F_1 offspring were smooth and yellow, showing that smooth was dominant to wrinkled and yellow was dominant to green.

Self-fertilization of the F_1 plants produced an F_2 generation of seeds with the following phenotypes:

315 smooth yellow	101 wrinkled yellow
108 smooth green	32 wrinkled green

To find the ratio among these F_2 phenotypes, we take the number of offspring in the smallest category—32—and divide it into the number of offspring in each category. Then we round the quotient to the nearest whole number. We find that the phenotypic ratio in the F_2 generation is about $9:3:3:1$. This ratio is typical of a dihybrid cross in which both pairs of alleles show a dominant-recessive relationship.

Mendel explained these data by assuming that genes governing seed color and seed texture move independently during reproduction. In this process of **independent assortment,** each pair of alleles behaved as it would in a monohybrid cross, without reference to the other pair. For example, when we consider only smooth versus wrinkled seeds, we find: $315 + 108 = 423$ smooth, and

Figure 12-9
Independent assortment of members of two gene pairs during gamete formation in dihybrid individuals. (a) The dihybrid (F_1) individuals produce four different types of gametes in equal proportions. Some of the F_1 gametes have the same gene combinations as those received from the P_1 gametes (*SY* and *sy*), but two new combinations are also produced (*Sy* and *sY*). (b) Two different possible tetrad alignments at metaphase I of meiosis result in the production of a total of four types of gametes. Figure 12-10 shows the F_2 generation.

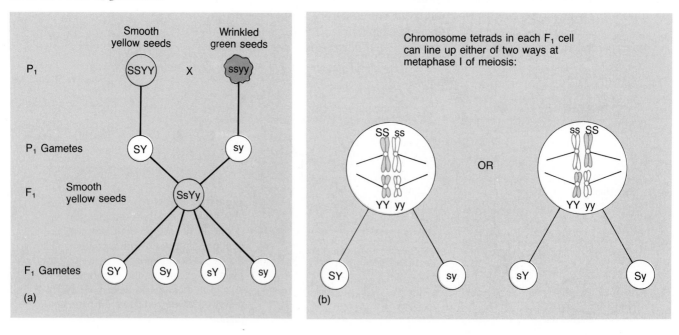

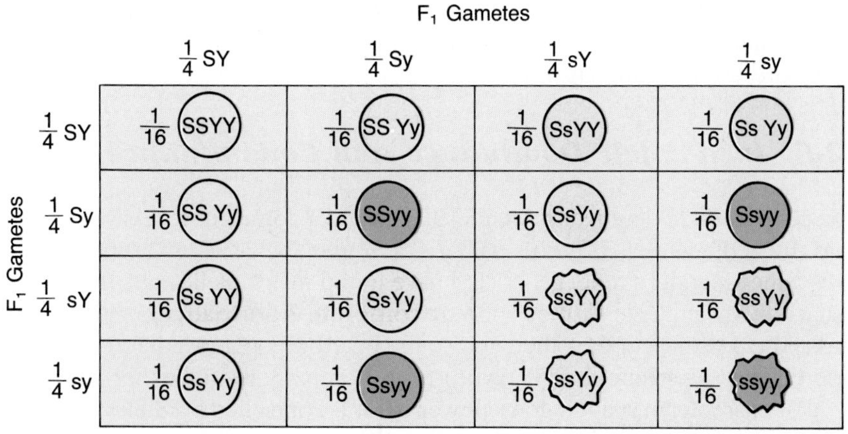

F₁ Gametes

Figure 12-10
Punnett square for the dihybrid cross in Figure 12-9. Gametes produced by the F₁ generation are arranged along the top and side of the square. Combinations in boxes represent possible genotypes (letters) and phenotypes (circles around letters) in the F₂ generation. Phenotypes are denoted by yellow for yellow seeds, green for green seeds, smooth circles for smooth seeds, and irregular circles for wrinkled seeds.

$101 + 32 = 133$ wrinkled. This gives a ratio of $423/133 = 3.18:1$, quite close to the 3:1 ratio of a monohybrid cross, which is what it is. Similarly, the inheritance of yellow versus green seeds behaves like a monohybrid cross. In other words, this dihybrid cross is the product of two separate monohybrid crosses:

(3 smooth + 1 wrinkled) (3 yellow + 1 green) =
9 smooth yellow + 3 smooth green + 3 wrinkled yellow + 1 wrinkled green

In this example, let S = smooth, s = wrinkled, Y = yellow, and y = green. The P₁ plants must have had genotypes *SSYY* (smooth, yellow) and *ssyy* (wrinkled, green). Since each gamete receives just one member of each gene pair, these parents must have produced gametes *SY* and *sy* respectively. All members of the F₁ generation have the genotype *SsYy*, giving a smooth yellow phenotype.

Self-fertilization of the F₁ plants produces a more complex situation. According to Mendel's "law of independent assortment," the members of each gene pair are sorted into gametes independently of the members of the other gene pair. We can see that this must be so from our study of meiosis: when the chromosome tetrads line up for the first division of meiosis (at metaphase I), the four alleles can be arranged in either of two ways (Figure 12-9b). If S and Y line up opposite s and y, the gametes formed are *SY* and *sy*. If S and y line up opposite s and Y, the gametes *Sy* and *sY* form. Either arrangement is equally likely, and so each F₁ plant produces four kinds of gametes—*SY, sy, Sy,* and *sY*—in equal proportions. Each gamete always receives *one member of each pair of genes*.

To find all possible genetic combinations in the F₂ offspring, the F₁ gametes formed by independent assortment can be written along the sides of a Punnett square. Since any female gamete can be fertilized by any male gamete, there is a total of nine possible genotypes, falling into four phenotypes, in the F₂ generation (Figure 12-10 and Table 12-1).

■ *The 9:3:3:1 dihybrid ratio is the product of two separate 3:1 monohybrid ratios, one for each gene pair.*

■ *The members of a gene pair move into gametes independently of members of other gene pairs. This is Mendel's law of independent assortment.*

Table 12-1   Phenotypes and Genotypes, and Their Ratios, Resulting from a Self-Cross of SsYy Individuals

Phenotype	Ratio	Genotype		
smooth, yellow	$\frac{9}{16}$	$\frac{1}{16}SSYY : \frac{2}{16}SSYy : \frac{2}{16}SsYY : \frac{4}{16}SsYy$	**or**	$S\_\_Y\_\_$ *
smooth, green	$\frac{3}{16}$	$\frac{1}{16}SSyy : \frac{2}{16}Ssyy$	**or**	$S\_\_yy$
wrinkled, yellow	$\frac{3}{16}$	$\frac{2}{16}ssYy : \frac{1}{16}ssYY$	**or**	$ssY\_\_$
wrinkled, green	$\frac{1}{16}$	$\frac{1}{16}ssyy$	**must be**	$ssyy$

* The notation __ indicates that either allele may be present; e.g., $S\_\_yy$ represents either *SSyy* or *Ssyy*, both of which will produce a smooth green phenotype.

(Independent assortment is not the invariable rule. It applies to genes carried on different chromosomes. In Section 12-F, we shall see what happens if the genes are on the same chromosome.)

12-E   *Incomplete Dominance and Codominance*

■ *In both incomplete dominance and codominance, heterozygotes have a different phenotype from homozygotes for either allele.*

The pairs of alleles studied by Mendel all showed a dominant-recessive relationship. Indeed, Mendel chose his pairs because they behaved as distinct alternatives. Since Mendel's time, geneticists have found many allelic pairs that do not behave this way. Instead, they show **incomplete dominance:** neither allele masks the presence of the other, and so the heterozygote has a different phenotype (as well as a different genotype) from homozygotes for either allele.

For example, in snapdragons, flower color is controlled by alleles that show incomplete dominance. Plants with red or white flowers are homozygous. When a red-flowered plant is crossed with a white-flowered plant, the F_1 plants all have pink flowers, and they are all heterozygous for red and white flower color. Half their gametes will contain the allele for red flowers, and half will contain the allele for white flowers. The F_2 phenotype and genotype ratios will both be 1:2:1, just the same as the F_2 *genotype* ratios for any other monohybrid cross (Figure 12-11). Since the heterozygote produces pink flowers, the expected phenotype ratio for the F_2 plants is:

1 red-flowered : 2 pink-flowered : 1 white-flowered

The term **codominance** describes a similar situation, one in which both alleles are expressed in the heterozygote's phenotype. A case in point is sickle cell anemia: heterozygous peoples' red blood cells contain both normal and sickle hemoglobin (an oxygen-carrying protein, Section 13-B). These people

Figure 12-11
Incomplete dominance. In snapdragon plants, the red- and white-flower alleles show incomplete dominance: heterozygotes have pink flowers.

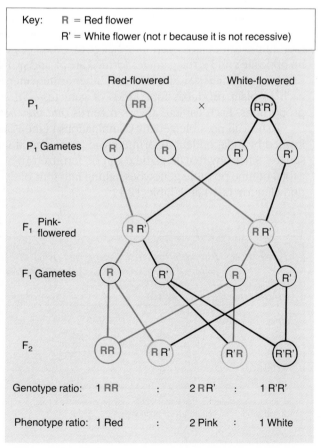

Key: R = Red flower
       R' = White flower (not r because it is not recessive)

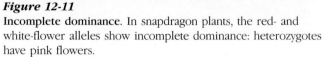

enjoy the benefits of normal oxygen transport, thanks to their normal hemoglobin, and their sickle hemoglobin gives them resistance to malaria.

12-F Linkage Groups

In individuals heterozygous for two pairs of genes (for example, *SsYy*), Mendel had found that four types of gametes—*SY, Sy, sY,* and *sy*—occurred with equal frequency. This led to the law of independent assortment. However, later researchers found many pairs of genes that did not assort independently: offspring with two of the combinations showed up in higher proportions than expected, and those with the other two combinations were much rarer than expected.

How can we explain this? Looking back at Figure 12-9b, we can see that the alleles *S* and *s* assort independently from *Y* and *y* because the S and Y gene pairs are on different pairs of homologous chromosomes. But there are many genes

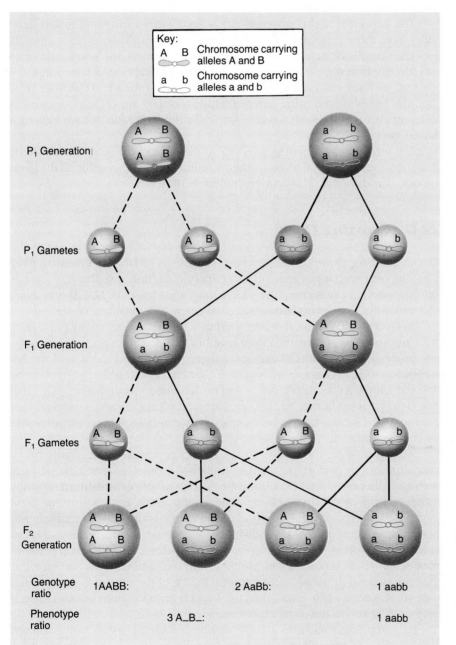

Figure 12-12
Dihybrid cross involving linkage. Since the A and B gene locations are on the same chromosome, the alleles on each chromosome move as a unit rather than assorting independently (compare with the S and Y pairs in Figure 12-9b). Hence the genotype and phenotype ratios in the F_2 generation are the same as for a monohybrid cross rather than for an unlinked dihybrid cross.

on each chromosome (see Figure 12-3). What happens if a cross involves two gene pairs carried on the same pair of chromosomes?

The chromosome moves as a unit during meiosis. Hence, we would expect genes on the same chromosome to stay together throughout the process and end up in the same haploid nucleus, rather than assorting independently, as genes on different chromosome pairs do. In other words, they will act as though they are linked together. Cases of **linkage,** in which genes are inherited as a pair or group, are common because every chromosome carries many genes.

Let us consider an example of genetic linkage. In a cross between parents *AABB* and *aabb* in which A and B (and a and b) are linked, the only possible gamete from the homozygous dominant parent is *AB* because A and B are linked. Similarly, the other parent can form only gametes containing *ab*. The F$_1$ individuals are heterozygous *AaBb,* but the gametes they form must be like those of the P$_1$ individuals because A is still linked to B and a to b; these pairs will not assort independently. Instead, A and B are carried into one gamete by the chromosome that bears them both, while a and b are carried into the opposite gamete by the homologous chromosome (Figure 12-12).

The genotypic and phenotypic ratios from a cross involving linked genes differ from those expected if the genes were not linked. This deviation from the expected results is the clue showing that two gene pairs are linked. Hence we can identify linked genes by studying the ratios of offspring obtained in the F$_2$ generation. If the ratio is the Mendelian ratio 9:3:3:1 for a two-character cross, then the genes are assorting independently, and they are probably located on different chromosomes. However, if we find a different ratio, we are looking at linked genes.

The term **linkage group** refers to all the genes with inheritance patterns that show they are linked to each other, so a linkage group is really all the genes on one chromosome. This may be hundreds of genes.

12-G   Crossing Over

During meiosis, chromosomes exchange segments of DNA by **crossing over.** Crossing over rearranges genes that were previously linked so that they are now on opposite chromosomes, and vice versa (see Figure 11-12). The resulting chromosomes are **recombinants,** bearing new combinations of alleles.

As far as we know, a crossover is equally likely to occur at any point along the chromosomes. Hence, the closer together two genes are on a chromosome, the fewer possible points of crossover there are between them, and the less frequently such a crossover will occur. If they are close together, the two genes will stay on the same chromosome and be inherited together more often than not. Genes that are farther and farther apart on the same chromosome, however, are more and more likely to be swapped between homologous chromosomes by crossing over.

We can estimate the relative distances between genes on the same pair of chromosomes by performing a large number of crosses, counting the offspring with each phenotype, and calculating the percentage of **recombinant** offspring, those showing new, "crossed-over" gene combinations. For instance, we could perform a test cross of *AaBb* dihybrid offspring from our previous example with homozygous recessive *(aabb)* individuals. We would then determine the percentage of offspring with phenotypes showing that they had received recombinant *Ab* or *aB* chromosomes in gametes from the heterozygous parent, rather than the original *AB* or *ab* chromosomes. If 10% of the offspring have crossed-over chromosomes *(Ab + aB)*, then the A and B gene locations are said to be 10 map units apart on the same chromosome.

By compiling data from a large number of such crosses, using different combinations of gene pairs, geneticists have been able to construct **chromo-**

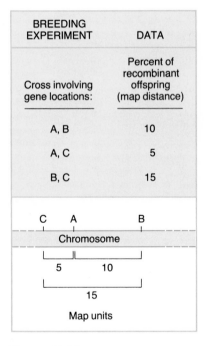

■ *All the genes on one chromosome end up in the same gamete, unless crossing-over occurs during meiosis.*

Figure 12-13
How positions of genes on chromosomes are mapped. Experimental crosses are performed and the percentage of recombinant offspring from each cross is calculated. Each 1% of crossing over is arbitrarily set equal to one map unit of distance on the chromosome. By performing many crosses to obtain many map distances, experimenters can deduce the order of genes on the chromosome and their relative distances from each other.

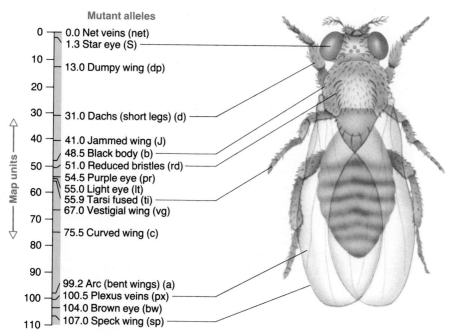

Mutant alleles

Map units	
0	0.0 Net veins (net)
	1.3 Star eye (S)
10	
	13.0 Dumpy wing (dp)
20	
30	31.0 Dachs (short legs) (d)
40	41.0 Jammed wing (J)
	48.5 Black body (b)
50	51.0 Reduced bristles (rd)
	54.5 Purple eye (pr)
60	55.0 Light eye (lt)
	55.9 Tarsi fused (ti)
	67.0 Vestigial wing (vg)
70	
	75.5 Curved wing (c)
80	
90	
	99.2 Arc (bent wings) (a)
100	100.5 Plexus veins (px)
	104.0 Brown eye (bw)
110	107.0 Speck wing (sp)

Figure 12-14

A map of some of the genes on chromosome 2 of the fruit fly *Drosophila*. The map shows mutant alleles that have been identified at the gene sites shown, along with their distance in map units from one end of the chromosome. Note that two of the mutations, star eye and jammed wing, are dominant, but that most are recessive. Also note that the genes' locations are not arranged in any particular anatomical order, although for the sake of neatness we have drawn lines to the affected body parts so that they don't crisscross.

some maps for several organisms (Figure 12-13). That is, they have determined which genes are together on which chromosomes, in what linear order, and approximately how far apart. We now know a great deal about the chromosome maps of such organisms as the fruit fly *Drosophila* (Figure 12-14), laboratory mice, corn, and the pink bread mold *Neurospora,* all popular subjects for genetic experiments. Somewhat different techniques have also allowed geneticists to prepare genetic maps for some viruses and for the circular DNA of some bacteria.

Ethics forbids setting up controlled crosses of humans. Therefore few features of the human chromosome map were known until recently. However, it is now being filled in rapidly, using genetic engineering techniques to produce millions of copies of DNA segments, and then analyzing these copies to determine their nucleotide sequences.

In 1986, the United States committed itself to the formidable undertaking of mapping the entire human genome. This is expected to take 15 years and cost $3 billion. To print the nucleotide sequence for the entire human genome will require the equivalent of 200 Manhattan telephone books. A large number of researchers are working on this project, using high-technology equipment to sequence the DNA, as well as powerful computers to handle the resulting data.

By 1987, 1400 genes had been mapped as to which chromosome they occupy in the human genome, and many had been localized to specific parts of their chromosome. New genes were being discovered at the rate of one every three days. One finding that surprised some researchers was that genes with related functions are often located near each other on a chromosome. With the broad outlines of a human chromosome map established, research will fill in more detail between major landmarks already known.

Secrets of Mendel's Success

A JOURNEY INTO LIFE

It is one of the ironies of science history that between 1860 and 1900 many biologists (including Charles Darwin) struggled in vain with the problem of how characters are inherited, which Mendel had solved by 1866. Mendel's 1866 paper put forth a simple yet clear theory that has withstood the test of time: each character is determined by a pair of "factors" (genes), one from each parent, which segregate in the gametes and combine randomly at fertilization. Why did Mendel, working alone, discover this, where people in the midst of the scientific community failed? And why did it take biologists 34 years to realize that he had solved the problem?

In Mendel's day, plant breeders had begun to study the inheritance of variation in plants such as melons and peas, but most professional biologists were interested in evolution and the differences between species. Many of them had no grasp of scientific method. Botanist Carl von Gärtner performed thousands of crosses and amassed mountains of results from which he drew not a single conclusion. Another influential botanist, Carl Nägeli, wrote dozens of papers and books, speculating wildly without

Figure 12-A
Gregor Mendel, 1822–1884.
(V. Orel, Mendelianum of the Moravian Museum)

ever performing an experiment! Mendel was entirely different. He was a meticulous worker with a sure grasp of the scientific method.

One of the secrets of Mendel's success was that he had an extensive background in physics as well as biology. He was fascinated with numbers and kept records of weather, sunspots, and other phenomena throughout his life. Franz Unger, Mendel's botany professor at the University of Vienna, believed that new species evolved from varieties within existing species. Mendel set out to discover what a variety was—for example, what made a tall variety of pea plants different from a short variety.

Because Mendel was interested in species and evolution, he studied

populations of plants, as well as individuals. This was important because it revealed patterns that would not have shown up if he had only studied individual plants. This aspect of his work was partly a result of his having studied with Unger, but the remainder of his success was due to genius, if genius is, as Thomas Edison contended, "one per cent inspiration and ninety-nine per cent perspiration." It is obvious from the way Mendel went to work that, after a few years of growing peas, he developed a hypothesis and proceeded to test it.

Mendel knew from the start that it was very important to choose the right subject for his experiments. He decided that he needed a plant in which (a) reproduction could be readily controlled; (b) there were contrasting varieties that bred true; and (c) the offspring of crosses between different varieties were just as fertile as their parents.

The reproduction of peas is easy to control because of their flower structure. Most familiar flowers have male parts (**stamens**) and female parts (**pistils**) exposed to the air, and the pistils can receive pollen blown or rubbed off neighboring plants or carried by insects. However, in peas and their relatives, a modified petal, the **keel,** completely surrounds the reproductive parts, separating them from the outside world. Pollen from other flowers cannot enter, and each flower normally pollinates itself (Figure 12-B).

Mendel could permit self-pollination or he could artificially cross-pollinate, taking pollen from flowers of one variety of peas and placing it on the pistils of flowers of another variety. In order to do this, Mendel had to open up one flower and pluck off a stamen. Then he had to open

SUMMARY

The experiments of Gregor Mendel were the foundation of the modern science of genetics—the study of the patterns of inheritance of genetic traits. Mendel succeeded in discovering the rules of inheritance largely because of his shrewd choice of an experimental organism, the garden pea plant; his painstaking breeding of large numbers (hundreds) of plants; and

his use of mathematics to analyze his results. We can summarize Mendel's conclusions in modern terms in this way:

1. Genetically based traits are determined by discrete units, called genes, which are passed from parent to offspring during reproduction.

another flower and dust some of the first flower's pollen onto the second flower's pistil. (To be certain that the pistil was not fertilized by pollen from its own flower, Mendel had already opened these flowers earlier and amputated the stamens before they produced pollen.) In order to obtain large numbers of offspring, Mendel hand-pollinated hundreds of flowers. This tedious process ensured that Mendel knew the parentage of every seed he collected. Equally tedious and important, he kept meticulous records of these crosses and their outcomes.

Mendel had studied mathematics and probability theory. Hence he realized the importance of obtaining a large number of offspring in his experiments to minimize the effects of "sampling error," which may result from looking at too small a number of cases. He also began by studying just one genetic trait at a time, and he followed each trait through many generations. In this way he was able to discern the patterns of dominance and recessiveness, the paired nature of genes, and gene segregation. When he came to study two traits at a time, Mendel's mathematical ability quickly helped him to grasp the fact that he was working essentially with two independent one-character crosses simultaneously, and to formulate the law of independent assortment.

Perhaps Mendel's most important contribution, however, was his recognition of genes as discrete particles. Virtually all biologists before 1900 had the mistaken notion that the inheritable characters of parents blended in their offspring. They also thought that each character was determined by numerous identical particles (genes) in each cell. If this were so, no consistent ratios would ever

be found in crosses and it is hard to see how any theory of genetics could ever be developed.

Why did these mistakes persist even after Mendel's work was published? Mendel's modesty did not help. He made little effort to publicize his work and once referred to his seven years' labor, involving more than 30,000 plants, as "one isolated experiment"! It also seems that even nineteenth century biologists did not keep up with their reading. Although Mendel published very little, his most important paper went to 115 libraries.

A major reason Mendel was ignored was the arrogance of a professional biologist toward an amateur. Mendel sent his paper to Nägeli with a cheerful and enthusiastic letter. Nägeli either did not understand Mendel's theory or, more likely, rejected it because it conflicted with his own theory of blending inheritance. Instead of supporting Mendel, Nägeli suggested that Mendel should repeat his pea plant experiments using hawkweeds, which, we now know, do not always follow the rules of sexual reproduction. Hawkweeds sometimes produce seeds without benefit of pollen, in which case the offspring have no male parent. It would obviously be extremely confusing to try to sort out patterns of heredity if you think you know which plants are parents but really don't! Unfortunately, Mendel took Nägeli's advice, to his great confusion. Nägeli's influential book on evolution and inheritance, published in 1884, did not mention Mendel or his work.

Mendel became abbot of the monastery where he spent most of his life, and died before his work was recognized and became the foundation of modern genetics.

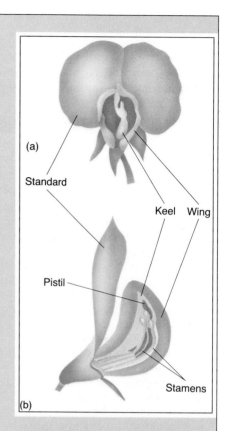

Figure 12-B

Flower of the garden pea. (a) External view; the petals are modified into shapes which suggest their names: standards, wings, and keel. (b) Cutaway view of flower from one side, showing the position of the reproductive parts (stamens and pistil). Since these parts are enclosed by the keel, the pea flower normally self-pollinates.

2. A plant or animal contains pairs of genes that determine its genetic characteristics.

3. During meiosis, the two members of each gene pair separate from one another and pass into different cells (law of segregation).

4. At fertilization, each offspring receives a pair of genes for each characteristic, one member of each pair from the gamete of each parent.

5. Genes for a trait may occur in different allelic forms, and one allele of a gene (dominant) may mask the presence of another allele (recessive) with which it is paired in a heterozygous individual.

6. The genes from each parent remain distinct in the offspring and may reappear in the phenotype of later generations even if they are masked by the phenomenon of dominance in some individuals in intervening generations.

7. During meiosis, the genes of one pair assort independently of genes of other pairs, so long as they

are located on different chromosomes (law of independent assortment).

The behavior of genetically determined traits in breeding experiments is paralleled by the behavior of the chromosomes during meiosis. This parallelism provides part of the evidence that genes are carried on chromosomes. Genes located on the same chromosome are linked and are inherited together except when they are separated by crossing over during meiosis.

SELF-QUIZ

The following problems will test your understanding of the ideas in this chapter.

1. In humans, the ability to taste phenylthiourea (PTU) is dominant. "Tasters" *(TT)* or *(Tt)* perceive an extremely bitter taste from very dilute solutions of PTU, while "non-tasters" *(tt)* experience no sensation even at much higher concentrations.
   a. What are the genotypes of Mr. and Mrs. Gagglebud, who can taste PTU, and who have three children, one of whom is a non-taster?
   What offspring phenotypes would be expected from the following crosses, and in what ratios?
   b. heterozygote × heterozygote
   c. homozygous taster × heterozygote
   d. heterozygote × non-taster
2. Two *Drosophila* (fruit flies) with normal wings are crossed. Among 123 progeny, 88 have normal wings and 35 have "dumpy" wings.
   a. What inheritance pattern is shown by the normal and dumpy alleles?
   b. What were the genotypes of the two parents?
3. If a dumpy-winged female (from Question 2) is crossed with her father, how many normal-winged flies will be expected among 80 offspring?
4. A number of plant species have a recessive allele for albinism. Homozygous albino (white) individuals are unable to make chlorophyll. If a tobacco plant heterozygous for albinism is allowed to self-pollinate and 500 of its seeds germinate:
   a. how many of these offspring will be expected to have the same genotype as the parent plant?
   b. how many seedlings will be expected to be white?
5. Sniffles, a male mouse with a colored coat, was mated with Esmeralda, an alluring albino. The resulting litter of six young all had colored fur. The next time around, Esmeralda was mated with Whiskers, who was the same color as Sniffles. Some of Esmeralda's next litter were white.
   a. What are the probable genotypes of Sniffles, Whiskers, and Esmeralda?
   b. If a male of the first litter were mated with a colored female of the second litter, what phenotypic ratio might be expected among the offspring?
   c. What would the expected results be if a male from the first litter were mated with an albino female from the second litter?
6. A kennel owner has a magnificent Irish setter, which he wants to hire out for stud. He knows that one of its ancestors was Erin-go-bragh, who carried a recessive allele for atrophy of the retina. In its homozygous state, this gene produces blindness. Before he can charge a stud fee, he must check to make sure his dog does not carry this allele. How can he go about this?

7. For a long time, human eye color was thought to be controlled by a single gene: brown *(B)* dominant over blue *(b)*. Using this assumption,
   a. can brown-eyed parents have a blue-eyed child?
   b. can blue-eyed parents have a brown-eyed child?
   It is now known that at least two other pairs of genes can affect eye color, making it possible for blue-eyed parents to have a brown-eyed child, although this is rare.
8. In cats, the allele for black fur *(B)* is dominant to the allele for brown *(b)* and the allele for short hair *(S)* is dominant to the allele for long hair *(s)*. Make a Punnett square for each of the following crosses:
   a. *BbSs × Bbss*
   b. *BBSs × Bbss*
   c. *BbSs × bbss*
   d. What proportion of the offspring from the cross shown in part b would be expected to be black with short hair?
9. In tomato plants, the gene for purple stems *(A)* is dominant to its allele for green stems *(a)* and the gene for red fruit *(R)* is dominant to its allele for yellow fruit *(r)*. If two tomato plants heterozygous for both traits are crossed, state what proportion of the offspring are expected to have:
   a. purple stems and yellow fruits
   b. green stems and red fruits
   c. purple stems and red fruits
10. If 640 seeds resulting from the cross in Question 9 are collected and planted, determine how many are expected to grow into plants with:
    a. red fruit
    b. green stems
    c. both green stems and yellow fruits
11. If one of the parents from Question 9 is crossed with a green-stemmed plant heterozygous for red fruits, what proportion of the offspring would you expect to have:
    a. purple stems and yellow fruits?
    b. green stems and yellow fruits?
    c. green stems and red fruits?
12. A peony plant with straight stamens and red petals was crossed with another plant having straight stamens and streaky petals. The seeds were collected and germinated, and the following offspring were obtained:
    62 straight stamens, red petals
    59 straight stamens, streaky petals
    18 incurved stamens, red petals
    22 incurved stamens, streaky petals

a. Which allele in each pair (straight vs. incurved stamens, red vs. streaky petals) is dominant?

b. What were the genotypes of the parental plants?

c. What further crosses would you make in order to get a definite answer for part a?

13. In a plant heterozygous for two pairs of genes *(AaBb)*, state the chance that a pollen grain it produces will carry:

a. an *A* allele

b. an *a* allele and a *b* allele

c. an *a* allele and a *B* allele

14. Pooh had a colony of tiggers whose stripes went across the body. His American pen-pal, Yogi, sent him a tigger whose stripes ran lengthwise. When Pooh crossed it with one of his own animals, he obtained plaid tiggers. Interbreeding among the plaid tiggers produced litters of a majority of plaid members, but some crosswise- and lengthwise-striped animals were also produced. Diagram the crosses made by Pooh, showing the genotypes of the tiggers that account for the coat patterns observed.

15. In cattle, the gene for straight coat *(S)* is dominant to its allele for curly coat *(s)*. The gene pairs for red *(RR)* or white *(R'R')* coat color show codominance; heterozygotes have a roan coat *(RR')* (red lightened by intermixed white hairs).

a. If a curly red cow is mated to a homozygous straight white bull, what will the genotype and phenotype of the calf be?

b. If the calf is mated to a roan animal with curly hair, what are the possible offspring phenotypes?

16. A farmer has three groups of cows: white ones in the clover patch, red ones in the alfalfa field, and roan in the cornfield. He has a roan bull, Ferdinand, who services the cows in all three fields. (Refer to Question 15 for more information.)

a. What color calves should he expect in each field, and in what proportions?

b. Ferdinand dies from a bee sting and the farmer decides to make his herd of cows exclusively roan coat in memory of his beloved bull. He sells all the red and white cows, and vows to sell any red or white calves born later. What color bull should he buy to replace Ferdinand, if he wants to sell as many calves as possible?

17. The allele for pea comb *(P)* in chickens is dominant to the allele for single comb *(p)*, but the alleles for black *(B)* and white *(B')* feather color show codominance, *BB'* individuals having "blue" feathers. If birds heterozygous for both pairs of genes are mated, determine what proportion of the offspring are expected to be:

a. single-combed

b. blue-feathered

c. white-feathered

d. white-feathered and pea-combed

e. blue-feathered and single-combed

18. A female *Drosophila* heterozygous for the recessive genes sable body and miniature wing was mated with a sable-bodied, miniature-winged male, and the following progeny were obtained:

> 249 sable body, normal wings
> 20 normal body, normal wings
> 15 sable body, miniature wings
> 216 normal body, miniature wings

From these results, would you conclude that the two gene pairs involved are linked or unlinked?

If you decided they are linked, which statement below describes the linkages found in the female parent?

a. The genes for sable body and miniature wings were on one chromosome, and the genes for the normal forms of these traits were on its homologue. Some crossovers occurred during meiosis.

b. The genes for sable body and normal wings were on one chromosome, and the genes for normal body and miniature wings were on its homologue. Some crossovers occurred.

19. In *Drosophila,* the gene for red eyes is dominant to the gene for purple eyes and the gene for long wings is dominant to the gene for dumpy wings. A female fly heterozygous for both traits is crossed with a male which has purple eyes and dumpy wings. The F_1 are:

> 109 red eyes, long wings
> 114 red eyes, dumpy wings
> 122 purple eyes, long wings
> 116 purple eyes, dumpy wings

From these results, would you conclude that the two gene pairs involved are linked or unlinked?

If you decided they are linked, which statement below describes the linkages found in the female parent?

a. The genes for red eyes and long wings were on one chromosome, and the genes for purple eyes and dumpy wings were on its homologue. Some crossovers occurred during meiosis.

b. The genes for red eyes and dumpy wings were on one chromosome, and the genes for purple eyes and long wings were on its homologue. Some crossovers occurred.

QUESTIONS FOR DISCUSSION

1. In performing a cross to determine the genotype of an organism having a dominant phenotype *(A__)*, why is it preferable to mate it with a homozygous recessive individual rather than a known heterozygote?

2. Figure 12-9 shows that two pairs of chromosomes can line up two different ways during meiosis. Therefore, an individual heterozygous for two gene pairs on different pairs of chromosomes *(AaBb)* can form four different kinds of gametes. Consider an individual heterozygous for three gene pairs *(AaBbCc)* on three different chromosome pairs. How many ways can the chromosomes line up at metaphase I, and how many different kinds of gametes will be formed? How many different kinds of gametes are possible from a human being, with 23 chromosome pairs?

3. Mendel worked with two pairs of genes that were on the same chromosome but gave results not much different from the 9:3:3:1 ratio expected of unlinked genes. How frequently must such genes cross over in order for their linkage to go undetected in experiments like Mendel's? How could you tell that these genes were really linked after all?

4. Evaluate the saying "alike as two peas in a pod" in light of your study of this chapter.

5. Many scientists question the value of committing so much money and talent to the human genome mapping project: are we doing it simply because it is now technically feasible? They feel we would do better working as in the past: focusing on specific genetic defects, pinpointing the responsible genes, and studying these genes and their neighbors to discover ways to alleviate the suffering of people born with the defect. How do you feel our resources should be allocated?

SUGGESTED READINGS

Mendel, G. J. *Experiments in Plant Hybridisation.* Edinburgh, Scotland: Oliver and Boyd, 1965. An English translation of Mendel's original paper, with comments and a biography of Mendel by others.

Miller, J. A. "Mendel's peas: a matter of genius or of guile?" *Science News* 125:108, February 18, 1984. Discusses the accusation by modern statisticians that Mendel fudged his data.

Strickberger, M. W. *Genetics,* 3d ed. New York: Macmillan, 1984.

White, R., and J.-M. Lalouel. "Chromosome mapping with DNA markers." *Scientific American,* February 1988.

When you have studied this chapter, you should be able to:

1. Explain how mutations may affect the protein encoded by a gene, and how this is related to phenotypic expression of mutant alleles.

2. Given data from an appropriate breeding experiment, recognize the 1:2:1 and 2:1 ratios characteristic of lethal alleles, and demonstrate knowledge of the inheritance patterns expected from parents carrying lethal alleles by working out crosses correctly.

3. State the possible genotypes of people with blood types A, B, AB, and O, and use your knowledge of these genotypes to solve problems.

4. Explain what is meant by the term **multiple alleles**, and how this differs from **polygenic characters**; give or recognize examples of each.

5. State the pattern of sex determination (sex chromosome complement of each sex) and inheritance of sex-linked genes for mammals and for birds, and use this information in working out sex-linkage problems.

6. Explain the difference between sex-linked and sex-influenced characteristics, and give examples of each.

7. List at least five factors that may affect the expression of a particular gene in an organism.

8. Describe the inheritance pattern found in the human genetic disorders hemophilia, red-green colorblindness, sickle cell anemia, Tay-Sachs disease, and phenylketonuria.

9. Describe nondisjunction and translocation of chromosomes.

Inheritance Patterns and Gene Expression

CHAPTER

13

Primula

Ocelot

Male mallard

*M*any abnormalities of humans and other organisms are the phenotypic expression of mutant alleles of genes. Much research has been devoted to finding out what these mutations are, how the mutant alleles are inherited and expressed, and how their expression is affected by factors in the organism's environment.

Genes are lengths of DNA that act as units of hereditary information. Many genes code for the sequences of amino acids in polypeptides and proteins (Chapter 10). Proteins must fold up properly to work, and their folding depends on having the correct amino acids in the correct order (Section 3-E). Many mutations change the DNA of a gene in such a way that it ends up coding for a different sequence of amino acids. As a result, a different protein (or perhaps no protein) is made. This in turn may alter the organism's metabolism or structure, and this is how a mutated DNA sequence results in a difference in the observed phenotype.

This chapter considers how some genetic differences between individuals produce different phenotypes and looks at some of the factors that control phenotypic expression of genes. It also introduces some new genotypic and phenotypic ratios that provide clues to the nature of particular genes.

- The effect of a mutation of a gene depends on how much it changes the protein encoded by the gene and on how important the protein is to life.
- A gene's expression depends on the other genes present in the genome and on factors in the organism's external environment.

13-A   Phenotypic Expression of Mutations

A **mutation** is a rare, random, and inheritable change in a cell's genetic material (Section 9-E). The mutation becomes part of the genotype of the cell and of all its descendants, and this may result in an abnormal phenotype. Mutations in somatic (body) cells may cause damage, including cancer, to the parts of the body that arise from the mutated cells. Mutations in germ (reproductive) cells may cause no noticeable abnormality in the individual in which they occur, but they will be passed on to its offspring and may be expressed in the offspring's phenotype.

Whatever the change wrought by a mutation, it is much more likely to harm than to help the delicate evolutionary engineering of the protein encoded by the mutated gene. Why is this? Even though mutations are rare, evolution has been going on so long that a particular mutation is likely to have occurred many times before. If the mutation were advantageous, it is likely that it would have been preserved through the process of natural selection and passed on to future generations. Therefore it would now be a common allele in the population.

Most mutations produce recessive alleles (see Figure 12-14). Why are these mutations recessive? Often this is a difficult question to answer on a molecular level, but in some cases the reason is clear. For example, one study analyzed seven different mutations from men with **hemophilia,** a genetic disorder in which blood fails to clot normally (Section 13-G). Three mutations had resulted in premature *Stop* codons, causing production of an incomplete version of a protein needed for clotting; three others had deleted thousands of nucleotides from the gene. The seventh had substituted one amino acid for another, resulting in a milder form of hemophilia.

An allele that codes for no protein or for an inactive protein will be at least partly recessive to an allele that codes for a functional protein molecule. In a heterozygote, the normal allele directs the production of the normal protein, while the recessive allele contributes no functional protein. If enough normal protein is produced, the individual's phenotype appears normal, and the normal

■ *Most mutations produce unfavorable, recessive alleles.*

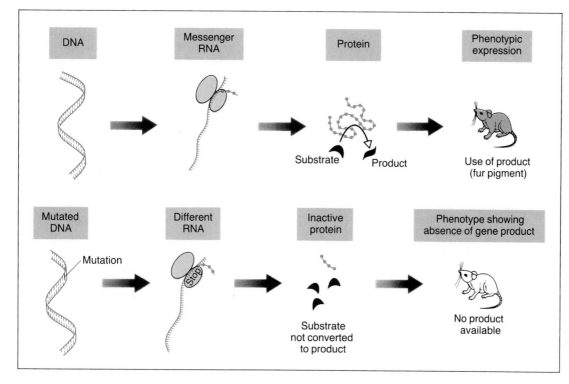

Figure 13-1
How some mutations may result in recessive alleles. The DNA in this example codes
for an enzyme needed to make a pigment in the fur. The mutant makes an incomplete enzyme, and so the animal does not produce the pigment. An albino animal
results.

allele is dominant. In an individual homozygous for the inactive allele, no protein is made. Therefore, the trait is not expressed, and the recessive phenotype
is the absence of the normal trait (Figure 13-1).

If one copy of the normal allele does not ensure synthesis of enough protein to produce the normal phenotype, a heterozygote might be noticeably different from either homozygote. Hence the normal and mutant alleles would
show incomplete dominance.

In codominance, both alleles code for functional proteins, but these may
have different properties. As a result, the phenotype of the heterozygote may be
distinct from that of either homozygote.

In the past, it has been very difficult to distinguish an abnormal protein from
the thousands of normal proteins in a cell (especially if we do not know the
protein's function, as is often the case). But genetic engineering is changing that.
Instead of laboriously trying to isolate a particular protein, we can now determine the protein's amino acid sequence indirectly. First, the gene for the disease
must be located, isolated, and cloned to produce multiple copies of its DNA,
which are then transcribed and translated to produce enough of the abnormal
protein to analyze. The protein's structure and properties may provide clues to
the most effective treatment of the disease.

In 1986, researchers used this method to detect the protein responsible for
chronic granulomatous disease (CGD), whose victims suffer from chronic bacterial infections. Eventually, researchers hope to cure CGD by transplanting the
gene for the normal protein into CGD sufferers. Many other genes have now
been detected by similar techniques and analyzed to deduce the nature of the
proteins they encode. These include one gene responsible for inherited Alzheimer's disease, and genes for Duchenne muscular dystrophy and cystic fibrosis.

13-B Lethal Alleles

Suppose a mutation destroys the genetic code for a protein essential to life. An organism that fails to produce an active form of that protein will die prematurely, and the allele that fails to encode a functional protein is called a **lethal** allele. Dominant lethal alleles are possible, but most are rapidly eliminated because they cause death before the individual carrying them reproduces. (Exceptions are those not expressed until later in life, as in the case of Huntington's disease in humans.) However, recessive lethal alleles are eliminated by selection only when they occur in homozygotes. They usually occur heterozygously, masked by functional dominant alleles that permit the individual to survive and reproduce. Hence nothing prevents recessive lethal alleles from spreading to future generations and becoming quite common. It has been calculated that the average human is heterozygous for perhaps three to five lethal recessive alleles. This is higher than the figure for many other organisms, and is part of the reason that marriages between close relatives produce a higher proportion of offspring with lethal inherited traits in humans than in most other species.

If just one copy of a normal allele does not produce enough of its protein for normal body functioning, the heterozygote has a different phenotype from either homozygote. An example in humans is the lethal allele that causes shortening of the middle bone in the fingers (brachydactyly) in heterozygotes. This makes the fingers appear to have only two bones instead of three. In homozygotes, this allele results in abnormal development of the skeleton. Homozygous babies lack fingers and show other skeletal defects that cause death in infancy.

In a marriage between two brachydactylic people, one out of every four children would be expected to be homozygous for the lethal allele and die during infancy; half would be expected to be heterozygous and show brachydactyly; and one fourth would be expected to be normal (Figure 13-2). This 1:2:1 ratio among offspring is typical of lethal alleles when the normal allele does not mask the mutant one completely.

Some lethal alleles are mutations of genes that code for proteins so essential that without them the embryo does not develop normally. In pregnancies with more than one offspring, embryos that die early may be resorbed back into the uterus, and a 2:1 ratio may be observed when the remaining offspring are born: two thirds heterozygotes to one third homozygous normal offspring (Figure 13-3). In mice, for example, the short-tail allele *(T')* causes early embryonic death in the homozygote. The embryo is then resorbed. If such embryos are taken from the uterus early in pregnancy, before they can be resorbed, they are seen to have no backbone and none of the tissue that later forms the muscles,

Figure 13-2

A lethal allele in humans. The normal allele, *B*, is incompletely dominant to the allele for brachydactyly, *B'*. Note the characteristic genotype and phenotype ratios for crosses of (heterozygous × heterozygous) and (normal × heterozygous).

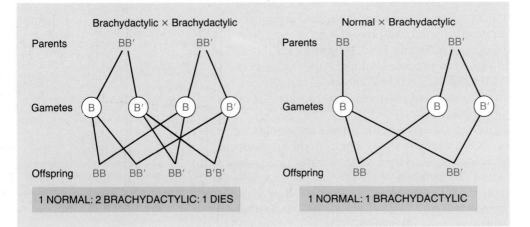

Brachydactylic × Brachydactylic	Normal × Brachydactylic
Parents BB' BB'	Parents BB BB'
Gametes B B' B B'	Gametes B B B'
Offspring BB BB' BB' B'B'	Offspring BB BB'
1 NORMAL: 2 BRACHYDACTYLIC: 1 DIES	1 NORMAL: 1 BRACHYDACTYLIC

kidneys, and many other important organs. Heterozygotes *(TT')* have shorter tails than normal mice *(TT)*.

A similar lethal allele exists in cats. A Manx cat is heterozygous for this allele, and its backbone is so short that the cat has no tail. The last vertebrae of the back and the last part of the digestive tract may be abnormal, and in this case the cat may have problems that prevent it from living out a full nine lives.

Sickle Cell Anemia A famous human allele that is frequently lethal in the homozygous condition is the one responsible for sickle cell anemia. The gene involved codes for the beta polypeptide chain of hemoglobin—the oxygen-carrying protein found in red blood cells and responsible for their red color (see Figure 3-19). The sickle allele results from a change in just one nucleotide pair. This results in the substitution of the amino acid valine for glutamic acid as the sixth amino acid in the hemoglobin beta chain (see Table 10-1 and Figure 3-14).

This seemingly small change has drastic consequences. When red blood cells containing sickle hemoglobin are exposed to low oxygen levels, the hemoglobin molecules aggregate and form fibers. These fibers distort the cells into odd shapes, such as sickles (Figure 13-4). Sickled cells become stuck in the smaller blood vessels and impede circulation to the areas supplied by these vessels. The sickled cells also break down easily, leaving the victim with fewer red blood cells than normal, a condition known as anemia. Poor circulation and anemia deprive the tissues of needed oxygen, producing symptoms such as tiredness, headaches, muscle cramps, poor growth, and eventually perhaps failure of organs such as the heart and kidneys.

The sickle allele shows codominance with the normal allele: heterozygous people produce both normal and sickle beta chains. Their red blood cells sickle only when the oxygen level is extremely low—for instance, at very high altitudes. Without special blood tests, heterozygotes may not know that they carry the sickle allele. People homozygous for the sickle allele are more severely affected because all of their beta chains are abnormal.

An individual heterozygous for a genetic condition is referred to as a **carrier,** whereas an individual homozygous for the condition is called an **affected individual.** People heterozygous for the sickle allele are sometimes referred to as "having sickle cell trait." This terminology is unfortunate, since it suggests that the carrier is less fit than the normal homozygote, which is not usually the case.

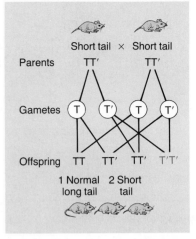

Figure 13-3
A cross involving a lethal allele that shows a 2:1 ratio among the off-spring. The short-tail allele *(T')* in mice causes death and resorption of homozygous recessive embryos early in development (red). These never appear among offspring born to short-tailed parents. One third of the progeny are normal long-tailed, and two thirds are short-tailed.

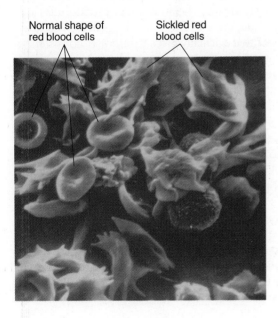

Figure 13-4
Blood cells from a person with sickle cell anemia. When the blood is low in oxygen, the red blood cells assume long, thin, jagged "sickle" shapes. Here, some red blood cells still have their normal shape: round, flattened disks with pushed-in middles. Normal blood cells are about 7.5 micrometres across. (David M. Phillips/Visuals Unlimited)

About half of the people homozygous for the sickle allele die by the age of 20. Furthermore, women in this group have fewer babies than do heterozygous or homozygous normal women. We might expect natural selection to keep such a lethal allele at a very low frequency in the population, as many people homozygous for the sickle allele die without producing offspring (Section 15-A). Yet in large areas of tropical Africa, 20 to 40% of the people are heterozygous for the allele. This strongly suggests that heterozygotes have some selective advantage so great that it offsets the disadvantage of recessive homozygotes. In 1953 it was noted that these people live in precisely the areas with the highest rates of death from a virulent form of malaria, a disease caused by a parasite of red blood cells.

Having at least one copy of the sickle allele lowers a person's chances of developing malaria. Red blood cells containing sickle hemoglobin sickle more readily when they are infected with malaria parasites. When a cell sickles, the parasites inside it die. The body's defenses may then be able to destroy the remaining parasites before a full-blown case of malaria develops. In malaria-infested regions, therefore, it is advantageous to be heterozygous for the sickle allele, which protects against a common deadly disease, even though the sickle allele is usually lethal in the homozygous state.

The same explanation may account for the high frequency of thalassemias, a group of genetic disorders in which too little hemoglobin is produced, in districts of Italy, Greece, and other areas with high incidences of malaria.

Tay-Sachs Disease   Tay-Sachs disease, a metabolic disorder resulting in deterioration of the brain and death by about the age of four, is also the result of a lethal recessive allele. A homozygous recessive child lacks an enzyme that metabolizes a certain lipid in the brain's nerve cells. Without this enzyme, the lipid accumulates and destroys the cells' ability to function. So far, this condition is incurable. One in 30 people of East European Jewish extraction is a carrier (heterozygous) for this disorder. However, about one third of the Tay-Sachs cases in the United States are among non-Jewish people.

Cystic Fibrosis   The most common lethal allele in the Caucasian population of the United States is the one responsible for cystic fibrosis. About one in 20 Caucasians is a carrier, and one in 2000 babies is affected by this disorder. In 1986, researchers reported the cause of this disease: a defect in cell-membrane channel proteins that permit chloride ions to pass through the membrane. Normally, chloride ions leave cells in the lung surface via these channels, and water follows by osmosis. In cystic fibrosis victims, the movement of chloride (and hence of water) is impeded. As a result, the mucus on the lung surface is unusually thick, because it lacks the water that normally gives it a thinner consistency. Victims suffer from improper lung function and recurrent respiratory infections, and 50% die of these causes by age 21. In addition, thick mucus in the digestive tract may interfere with digestion and absorption of food, and victims may look poorly nourished despite a voracious appetite.

In 1989, cystic fibrosis genes and their normal counterparts were isolated and analyzed. About 70% of alleles from cystic fibrosis patients have a mutation in which the code for one amino acid is deleted. This amino acid is thought to play an important role in control of the channel's activity. Various other defects occur in the remaining cystic fibrosis alleles.

Huntington's Disease   Huntington's disease is unusual because it is a dominant lethal allele. Hence, a person with only one allele for this condition may develop the disease. Its symptoms include involuntary twitching, degeneration of part of the brain, depression, and irritability. The disease progresses slowly for 10 to 20 years, finally causing death. The cause is not certain, but it may be a variant enzyme in the brain which produces too much of its product; this in turn kills certain brain cells. The first symptoms usually do not appear until 35 to 45 years of age. By that time, most victims have children, who in turn have a 50:50 chance of having also inherited the allele.

13-C   *Inborn Errors of Metabolism*

Many genes code for proteins that are enzymes in the body's metabolism. Mutation of such a gene may result in an inherited genetic abnormality known as an **inborn error of metabolism.** We have seen that the metabolic disorders of Tay-Sachs and Huntington's diseases are lethal, but others are less severe, and some do little or no apparent harm to affected individuals.

Phenylketonuria (PKU) and **albinism** are two hereditary disorders that result from defects of enzymes in the same metabolic pathway (Figure 13-5). Since this pathway is not vital, the responsible alleles are not lethal.

Victims of PKU are homozygous recessives who lack the enzyme that normally converts the amino acid phenylalanine to another amino acid, tyrosine. Phenylalanine is converted instead to phenylpyruvic acid, which builds up to toxic levels in the blood and eventually is excreted in the urine, giving it a characteristic odor. However, the kidneys do not excrete phenylpyruvic acid fast enough to prevent its damaging various organs, especially the brain, and victims of PKU become mentally retarded if left untreated. PKU can now be controlled by a special diet low in phenylalanine during childhood. This prevents most brain damage, but victims may still have some learning disabilities. When brain development is complete, the PKU victim can adopt a normal diet. Many states now require that newborns receive a blood test for PKU and several other metabolic disorders.

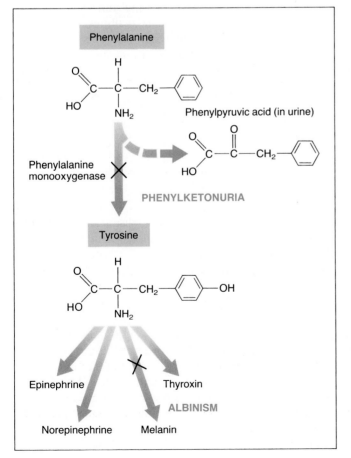

Figure 13-5
Inborn errors of metabolism. This diagram shows the metabolic pathway that converts the amino acid phenylalanine to tyrosine, which in turn can be converted to several other substances. The solid arrows are enzyme-catalyzed steps in the pathway. "Metabolic blocks" (black X's) result from the absence of the corresponding enzymes. Two metabolic blocks are shown here—those responsible for phenylketonuria and albinism.

Figure 13-6
Albinism, a homozygous recessive condition. Here an albino peacock spreads his tail in a mating display while a normally colored male strolls past. (D. J. Cross/BPS)

If a woman homozygous for PKU becomes pregnant, although her child may or may not actually have the genotype for PKU, it has a greatly increased chance of mental retardation or microcephaly (small head). This is probably due to high phenylalanine levels in the mother's blood. Some such women have returned to a low-phenylalanine diet during pregnancy, but it is not yet clear whether this eliminates the risks to the fetus. Whether or not the children of a PKU-affected woman are also affected, they will all be PKU carriers, since each must inherit one copy of the recessive allele from her.

Albinism is a condition characterized by absence of melanin, the dark pigment that makes eyes, hair, and skin brown or black. True albinos have white hair (or feathers, as in Figure 13-6) and very light skin and eyes. They lack functional enzymes to convert the amino acid tyrosine to melanin.

You may wonder whether victims of PKU are also albino, since they cannot make the tyrosine that is eventually converted to melanin. The answer is no, because tyrosine can be obtained in the diet as well as from conversion of phenylalanine. However, people homozygous for PKU usually have light coloring. A person could, of course, be both albino and PKU-affected if he or she were homozygous for both pairs of alleles.

13-D   Multiple Alleles

Up to this point, we have considered only genes with two distinct alleles. However, a gene contains hundreds of nucleotides. If mutations occur in different parts of the gene in different individuals, the population as a whole will contain a number of different alleles, known as **multiple alleles.** Each allele may produce a different phenotype. Any one individual can contain no more than two of the different alleles of a gene, one on each chromosome of the homologous pair carrying that gene.

■ *A population may contain multiple alleles of a gene. Any one individual has at most two different alleles (one pair) of each gene.*

A familiar case of multiple alleles is that of the human ABO blood groups, with three main alleles, I^A, I^B, and i. I^A and I^B code for two different enzymes, each of which attaches a different sugar to a protein on the surface of red blood cells. Both enzymes are produced in a person having both the I^A and I^B alleles. That is, I^A and I^B are codominant. The i allele does not code for an enzyme; it is recessive to both I^A and I^B. Table 13-1 shows the possible genotypes and phenotypes in this blood group system.

Blood types must be matched when a person receives a blood transfusion. The ABO and rhesus (Rh positive or negative) blood groups are the best known and the most medically important, but more than 20 different human chromosome locations are known to carry genes coding for various blood proteins.

In the past, blood groups were sometimes used to decide questions of parentage, such as in paternity lawsuits or in cases of suspected mix-ups of babies in a hospital nursery. Only a few drops of blood are needed to determine the blood types of the child and its supposed parents. This genetic evidence reveals whether a particular person or couple could have had a child of a particular blood type. Such evidence can never be used to determine that a particular person definitely is the father or mother of a particular child, but it will often rule out a particular person as the child's parent. For instance, a man with blood type AB has the genotype $I^A I^B$, and so he could not have been the father of a baby with blood type O. A baby with blood type O must have the genotype *ii*, and its father must therefore have had at least one *i* allele to pass on (see Table 13-1). If the baby's blood type is A or B, the same man could have been the father. However, there are many other men in the world with blood types such that they could have fathered the baby, and so the baby's parentage can never be established conclusively on ABO blood group evidence.

Modern paternity testing uses proteins coded by another series of multiple alleles: the MHC (short for major histocompatibility complex) antigens, proteins found on the surfaces of most human cells (Section 25-D). There are six different gene pairs, each with several to many different possible alleles, coding for these proteins. This results in such a large number of possible genotype combinations that everyone (except identical twins) has an essentially unique "chemical fingerprint" on his or her cells' plasma membranes. Because of their great variety, these proteins can usually be used to settle questions of paternity when ABO blood tests are inconclusive. For convenience, white blood cells are used for these tests. This technique was first used in 1980 to determine the paternity of identical twins born to a rape victim. The twins' MHC antigens matched those of the woman and her husband and were distinctly different from those of the accused rapist. The MHC technique is very accurate, and as a result many more fathers are now being forced to accept financial responsibility for their offspring.

Table 13-1 Human ABO Blood Groups	
Blood Group (Phenotype)	**Genotype**
A	$I^A I^A$ or $I^A i$
B	$I^B I^B$ or $I^B i$
AB	$I^A I^B$
O	*ii*

13-E Polygenic Characters

In the case of multiple alleles, a particular location on a chromosome can be occupied by any one of several different alleles of a single gene. In contrast, the term **polygenic character** describes the case of a single phenotypic trait governed by more than one pair of genes. These genes may occupy two or more different locations on the same homologous chromosome pair or on nonhomologous chromosomes. Familiar examples in humans include height, intelligence, body build, and hair and skin color, all determined by the interactions of many genes.

Human skin color is a polygenic trait. There is some debate whether three or four different gene pairs are involved. Very dark-skinned people have alleles coding for production of melanin at all their skin-color-gene locations, whereas in light-skinned people these locations are occupied by alleles that don't code for melanin production. For example, if there are three gene pairs involved, a very dark-skinned person would have six alleles for melanin *(AABBCC)*, whereas a very light-skinned person would have none *(aabbcc)*. The alleles are thought to be additive: the more alleles for melanin production a person has, the more melanin is produced, and the darker the skin.

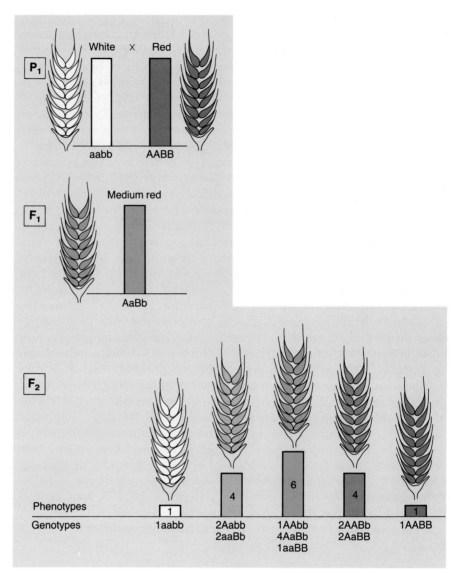

Figure 13-7

A cross involving a polygenic trait.
The two parental (P₁) strains of wheat
differ in two gene pairs for kernel
color. One strain has four alleles for
uncolored kernels *(aabb),* and the
other strain has four alleles for red
color in the kernels *(AABB).* F₁ plants
have two red-color alleles *(AaBb),* and
so their color is midway between
those of the two parental strains. F₂
plants have kernels of five different
colors (white and four shades of red),
depending on whether they have
zero, one, two, three, or four red-
color alleles. In this example, both P₁
strains have the alleles *cc* in the third
gene pair governing kernel color. A
cross of *AABBCC × aabbcc* would
give an F₂ with seven different kernel
colors.

A similar example is known in wheat. Three different gene pairs govern the
color of kernels, from white to deep red. Figure 13-7 shows a cross of two strains
of wheat that differ in only two of these gene pairs: *AABB × aabb* (in both
strains, the third pair is assumed to be *cc,* coding for no pigment).

Polygenic characters are difficult to study because it is hard to disentangle
the effects of the many interacting alleles that influence a single phenotypic
character. Environment may further muddy the waters; for instance, in our example, whether a person has light or dark skin, its color can become lighter or
darker depending on exposure to the sun. Traits such as human height, human
intelligence, and the size of an ear of corn are strongly influenced by environmental factors, such as nutrition, as well as by the many genes that determine the
possible range of variation in the phenotype.

Polygenic characters are so common that early observers were misled into
believing in blending inheritance. Polygenic traits frequently appear to blend
because the phenotypes of various individuals show a wide range, and the offspring of parents at the extreme ends of the range generally come out about
midway on the scale.

13-F Sex Determination

In many species, the most obvious difference in phenotype between individuals is their sex. It is also one of the most far-reaching differences, for the sex hormones influence the phenotypic expression of many other genes, and hence sex affects many organs that are not directly involved in sexual reproduction. In most familiar animals, sex is genetically determined, and about half the individuals are male and half female.

The simplest cross to produce half males and half females (a 1:1 ratio in the offspring) is one between a homozygote and a heterozygote. In humans and most other mammals, and in birds, this is what happens: one sex is heterozygous and one homozygous for a pair of chromosomes, called the **sex chromosomes** (Figure 13-8). In most mammals, males are heterozygous for the sex chromosomes, having one X and one Y chromosome (see Figure 13-10). Females have two X chromosomes. Most birds and some reptiles are the other way around: females are heterozygous, with one Z and one W chromosome, while males are homozygous ZZ. Many insects are like mammals in having XX females and XY males, but some have a reversed system much like that of birds. Higher plants often have reproductive organs of both sexes in the same individual, but those with separate sexes tend to have XX females and XY males.

During the formation of gametes in a man (or in XY males of other species), the X and Y chromosomes segregate and end up in different sperm. All eggs produced by an XX female contain one of her two X chromosomes. At fertilization, an X-bearing egg can combine with an X-bearing sperm (forming a female) or with a Y-bearing sperm (forming a male).

Apparently all individuals have the genes needed to develop into a member of either sex. For example, embryonic birds and mammals have been induced to grow into members of the genetically "wrong" sex by hormone treatment. Under normal circumstances, hormones from the ovary or testis maintain the correct sex of each individual.

How do the sex chromosomes influence sex? The answer is not the same for all organisms. In humans and most other mammals, rudimentary "indifferent

> ■ *In most species, one sex is heterozygous and the other homozygous for one pair of chromosomes, the sex chromosomes. The offspring's sex is determined by which sex chromosome it receives from the heterozygous parent.*

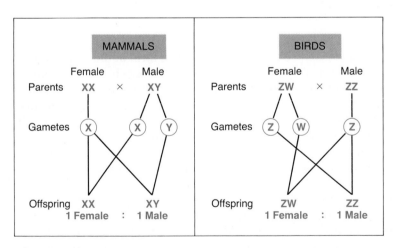

Figure 13-8
Systems of sex determination in mammals and birds. In mammals, females have a pair of X chromosomes; males have one X and one Y chromosome. In birds, males have a pair of like chromosomes, called Z to emphasize that the male/female homozygote/heterozygote system is reversed from the situation in mammals. Female birds have a single Z chromosome and a W chromosome.

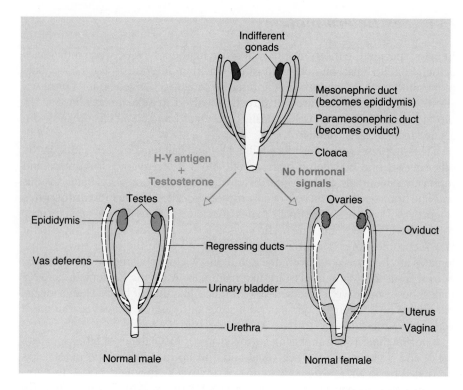

Figure 13-9

Sex differentiation in human embryos. Early embryos have indifferent gonads. Near them lie two sets of undifferentiated ducts (mesonephric and paramesonephric ducts). H-Y antigen causes the indifferent gonads to develop into testes. The testes produce the male hormone testosterone, which stimulates the mesonephric ducts to differentiate into the male reproductive tract. The end of the duct near the testis becomes the epididymis, and the rest becomes the vas deferens. Without this chemical stimulation, the gonads develop into ovaries and the paramesonephric ducts into the oviducts of the female tract. In both sexes, the unused pair of ducts regresses.

gonads," which can become either testes or ovaries, form in the early embryo (Figure 13-9). Which way they develop depends on the sex chromosomes present. The mammalian Y chromosome carries two genes that make the embryo develop as a male. One gene codes for testis-determining factor, a gene regulatory protein. The other codes for a protein called H-Y antigen, which binds to cell-surface receptors in the indifferent gonads and causes the gonads to differentiate into testes. The testes, in turn, produce the hormone testosterone, which must be present to induce differentiation of the male reproductive tract.

In human embryos, the testes begin differentiating in the sixth week of development. If this does not occur, the gonads differentiate into ovaries in the following week. In this case, the rest of the female reproductive tract develops automatically, without hormonal signals from the ovaries. If a mammalian embryo's ovaries or testes are removed before the reproductive tract differentiates, the embryo develops a female tract.

So, in mammals, the Y chromosome carries at least two genes that give the embryo its first "push" toward becoming a male. Having a Y chromosome leads to maleness, while embryos with only X chromosomes become phenotypically female. This is true even in the rare cases of people born with more or fewer

Table 13-2 *Phenotypes for Various Sex Chromosome Complements in Humans*

Sex Chromosomes	Phenotype*
XX	Normal female
XY	Normal male
XXX	Female; fertile or sterile; usually normal
X (Turner's syndrome)	Female; short, with webbed neck; sterile, ovaries rudimentary or absent
XXY (Klinefelter's syndrome)	Male; sterile; eunuchoid; possible mental retardation
XXXY	Male
XYY	Male; tall, acne-prone; impaired fertility; possible mild mental retardation

*Defects in various genes involved in hormone production can alter the phenotype normally exhibited by a particular sex chromosome combination.

than two sex chromosomes (Table 13-2). The genes on the Y chromosome that trigger differentiation of the testes seem to be the only genes involved that are on the sex chromosomes. Many other genes come into play later, and all of these seem to be located on the **autosomes,** the non-sex chromosomes.

At the beginning of embryonic development, both X chromosomes in a female mammal are active. Later, one of the two X chromosomes in each cell (except those cells that develop into the ovaries and eggs) is inactivated. Which X chromosome is inactivated in a particular cell is determined at random, but the same one is inactive in all that cell's descendants.

As a result of this inactivation, each cell of a female mammal has only one active X chromosome. The cells of male mammals also have one active X chromosome each, since they contain one X and one Y chromosome.

Various unusual methods of sex determination occur in some organisms. In honeybees and some of their relatives, females are diploid but males are haploid, developing from unfertilized eggs.

In some animals, sex is determined by environmental factors. In the American alligator, snapping turtle, and some other reptiles, sex is determined by the environmental temperature during a particular stage of development. (This discovery led to speculation about the mysterious extinction of the dinosaurs and other ancient reptile groups. If these reptiles had a similar mode of sex determination, a prolonged change in climate could have resulted in production of offspring of only one sex. Eventually this would have doomed any species whose members could reproduce only sexually.)

Environmental determination of sex may prove useful to animals that cannot move far to find a mate. For instance, in the marine worm *Bonellia,* the developing larva swimming in the sea belongs to neither sex. Eventually it drifts to the bottom and becomes an adult. If it settles down alone, it develops into a relatively large female, but if it lands near an existing female, it is attracted to her, and she produces a chemical that causes the larva to develop into a microscopic male. The male migrates into the female's excretory organ and lives there as a parasite.

The snail-like slipper shell lives in stacks of individuals. Young individuals are males, which turn into females as they grow larger and older. Chemicals appear to influence sex determination in this situation as well. If a stack consists entirely of males, some of them turn into females. In contrast, some fish, such as the saddleback wrasse and bluestreak cleaner, may start out as females, but when there are no larger males around, the largest females become males. All of these systems guarantee that when two or three are gathered together, some will be male and some female.

13-G   *Sex Linkage*

Like other chromosomes, the sex chromosomes carry genes. We know of four genes on the human Y chromosome: one for testis-determining factor, one for H-Y antigen, one that affects size of the teeth, and one for a cell-surface protein. The X chromosome carries various genes, including a partner for the Y chromosome's cell-surface protein gene.

In mammals, part of the X chromosome is homologous with part of the Y chromosome. This permits the X and Y chromosomes to find each other at the beginning of meiosis and segregate into different sperm cells (Section 11-E). Because genes located in these homologous areas of the sex chromosomes are paired, they behave like the autosomal genes we have studied before, and only detailed chromosome mapping will reveal that they are on the sex chromosomes.

Mammalian X chromosomes also have large **nonhomologous portions,** which contain genes that have no mates on the Y chromosome (Figure 13-10). Genes located in these nonhomologous areas of the X chromosome are said to be **sex-linked.**

In male mammals, any recessive allele on a nonhomologous part of the X chromosome will be expressed in the phenotype, since there is no gene on the Y chromosome that could mask it. Therefore, it is possible for a single recessive allele to express itself in the male. A female must have two copies of such a recessive allele before it shows in her phenotype. Recessive sex-linked phenotypes, therefore, are more common in male mammals than in females. Since many recessive alleles are harmful, this is one of the reasons that more male than female mammals of any age die.

Red-green colorblindness, hemophilia, and the Duchenne type of muscular dystrophy are well-known recessive sex-linked traits in humans. Let us consider a cross between a woman homozygous for normal color vision and a colorblind man. All the children of this marriage will have normal color vision, since all receive an X chromosome bearing a normal allele from their mother (Figure 13-11). Imagine that the daughters in this family marry men with normal color vision (like their brothers), while the sons marry women who are carriers of the gene for colorblindness (like their sisters). In the next (F_2) generation, we would find the ratio of three offspring with normal vision to one colorblind, as expected from a monohybrid cross. Our results have this added twist, however: all the colorblind children are male! Girls can also be colorblind, but only if they inherit an X chromosome with an allele for colorblindness from their fathers as well as from their mothers. Therefore, colorblind girls are much rarer than colorblind boys.

■ *The hallmark of a sex-linked recessive trait is its more frequent appearance in the sex having heterozygous sex chromosomes.*

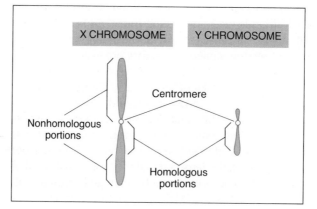

Figure 13-10

Human sex chromosomes. The X and Y chromosomes have homologous portions, containing the same gene pairs. The X chromosome also has two known nonhomologous portions, with no corresponding gene partners on the Y chromosome. The Y chromosome appears to have nonhomologous parts too, but their location is unknown.

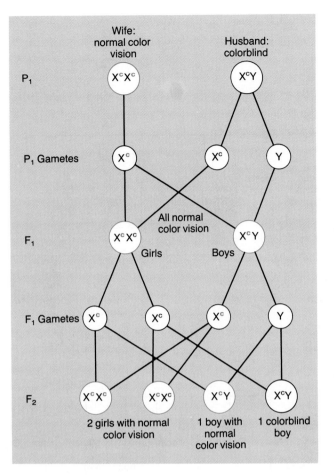

Figure 13-11
Inheritance of red-green colorblind-ness, controlled by genes on the X chromosome. In a marriage of a woman homozygous for normal color vision and a colorblind man, all the children have normal color vision. If the children marry people with the same genotypes as their siblings, half of the male offspring are expected to be colorblind, inheriting the maternal X chromosomes bearing the color-blindness allele. Girls that inherit this maternal X chromosome receive an X chromosome with a normal allele from their fathers, and so they are not colorblind.

Certain forms of hemophilia are also caused by a recessive allele on the X chromosome. A person with hemophilia lacks a protein needed to make blood clot, and so may bleed to death after even a slight cut (Section 13-A). If a woman has one recessive hemophilia allele, she usually also has a dominant, normal allele on her other X chromosome, and so she does not have hemophilia. A man with the allele, on the other hand, has no second X chromosome bearing a normal allele to code for the vital clotting protein. Hence his hemophilia allele will be expressed. In the past, men with hemophilia seldom lived to become fathers, and therefore few hemophiliac females are known. However, hemo-philia can now be controlled (but not cured) by regular injections of clotting factor extracted from normal blood. Genetically engineered clotting factor will also be available soon. As a result, more men with hemophilia now live to grow up and reproduce. More hemophiliac females can be expected, as some men with hemophilia may marry women heterozygous for the allele.

Queen Victoria of England was the world's most famous hemophilia carrier. Her hemophiliac son, Leopold, Duke of Albany, and her two carrier daughters, Princesses Alice and Beatrice, spread the gene to the royal houses of Russia, Prussia, and Spain. For a time hemophilia was called the "royal disease," but fortunately none of Queen Victoria's modern descendants appears to have in-herited the allele (Figure 13-12).

Another familiar example of sex linkage occurs in one of the many gene pairs that affect coat color in cats. The sex-linked orange allele, located on the X chromosome, diverts molecules into a metabolic pathway that makes them into orange pigment instead of black. A male cat with the orange allele on his X

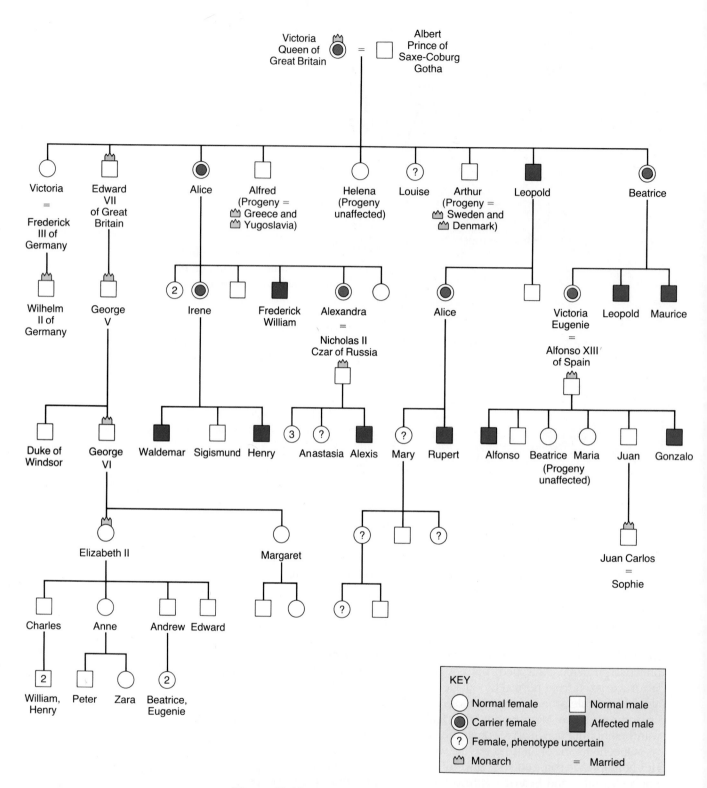

Figure 13-12
Pedigree of hemophilia among the descendants of Queen Victoria. Queen Victoria was the first identified carrier of this allele in her family. She passed it on to at least three of her children. Only some of her descendants are shown, including the royal family of Great Britain, which has not inherited the allele, and those lineages that did contain it. Notice the standard symbols used to denote affected and normal males and females in pedigree diagrams (key). Numbers inside symbols indicate more than one offspring showing the phenotype.

Figure 13-13
A tortoiseshell cat. Patches of orange and black fur in this heterozygous female resulted from the random in-activation of one X chromosome in each cell during embryonic develop-ment.

chromosome, or a female with orange on both X chromosomes, is some shade of orange. The non-orange allele allows black pigment to form, and the non-orange male or homozygous non-orange female is some shade of black (includ-ing brown or gray, depending on its other genes). A cat with an orange allele on one X chromosome and a non-orange allele on the other has a mottled pattern of orange and black spots called tortoiseshell (Figure 13-13). Such a cat is almost always female, since only females normally have two X chromosomes (see Table 13-2 for males with extra X chromosomes).

The tortoiseshell pattern results from X chromosome inactivation (Section 13-F) early in the cat's embryonic development. If the chromosome with the orange allele remains active, the cell's descendants form a patch of orange fur, and if the chromosome with the non-orange allele is active, the patch is black. Since the X chromosomes are inactivated randomly, tortoiseshell cats vary in the amount and pattern of the colors in their coats.

13-H   Sex-Influenced Genes

The main role of sex hormones is influencing the reproductive system and related organs, but these hormones also affect many other characters. Genes that are expressed to a greater or lesser degree as a result of the level of sex hor-mones are called **sex-influenced genes.** These genes are usually (but not necessarily) located on the autosomes, and so there is no difference in genotype between the two sexes. However, males and females with the same genotype may differ greatly in phenotype because the expression of the genes depends on the levels of sex hormones, and hence on the individual's sex. For example, a bull may have genes for high milk production, but he will not produce milk because he has only low levels of female hormones. However, these genes would make him a useful sire for a dairy herd. Similarly, males and females both have the genetic potential to produce the organs characteristic of the opposite sex, but they develop organs typical of their own sex because they have more of the appropriate hormones. (Both males and females do have the hormones characteristic of the opposite sex, but at lower levels.)

In humans, the allele for male pattern baldness is autosomal, and its expres-sion is influenced by the presence of male hormones (Figure 13-14). A man will become bald if he has only one allele for baldness. In other words, the allele acts

Figure 13-14
Some factors that affect gene expres-sion. Male sex hormones promote the expression of the allele for baldness in men. The child's blond hair is turning darker as she grows older, a common age-related inherited trait.

Genetic Counseling and Fetal Testing

Genetic abnormalities can bring much pain and suffering to the victims and to their families. Parents of victims may have feelings of guilt that lead to alcoholism, drug addiction, or divorce. The time, energy, and money needed to care for afflicted children may also deprive the family's other children of a normal home life.

Genetic counseling can help couples to determine their chances of having children afflicted by a particular defect, an event that is more likely if the couple or their relatives have already had such a child. Blood tests can now determine whether or not prospective parents are carriers for such traits as Tay-Sachs disease, cystic fibrosis, or sickle cell anemia, or are destined to develop Huntington's disease.

Some couples, faced with the knowledge that each child has a 25% chance of being homozygous recessive for a condition that will bring years of suffering or incapacity followed by an early death, choose not to have families. Others begin pregnancy and have the fetus tested to determine whether or not it is affected; if it is, they may choose abortion and hope that a later pregnancy will have happier results. In fact, the availability of such testing has increased the birth rate among at-risk couples.

How is a fetus tested for genetic defects? In the technique of **amniocentesis,** the tip of a syringe is inserted through the mother's abdominal wall and uterus into the sac of fluid (amniotic sac) surrounding the fetus (Figure 13-A). Cells that have sloughed from the fetus's skin or respiratory passages into the fluid are sucked into the syringe. These cells can be examined for chromosomal abnormalities, such as those resulting from translocation or nondisjunction of chromosomes during meiosis (Section 13-J). The cells can also be cultured in the laboratory, and in about two weeks will produce enough cells for geneticists to examine for disorders such as PKU, sickle cell anemia, cystic fibrosis, and Huntington's disease. A drawback of this technique is that it cannot be performed until the sixteenth week of pregnancy, when the amniotic sac is large enough for a doctor to extract the required fluid and its cells without accidentally damaging the fetus with the needle. By the time the cells have been grown in laboratory culture and analyzed, it may be too late for a legal abortion to be performed.

A newer technique, not yet widespread, can be performed during the eighth to tenth weeks of pregnancy, when abortion is safer for the woman. In **chorionic villus sampling (CVS),** cells are removed from chorionic villi, part of the fetus's life support system attached to the wall of the uterus. These cells are sucked into a catheter inserted through the cervix (the neck of the uterus) (Figure 13-A). Unlike amniocentesis, this supplies a mass of many rapidly dividing fetal cells, and so a diagnosis of chromosomal or genetic defects can be made quickly, within a few days. However, chorionic villus sampling cannot detect a common class of birth defects resulting from failure of the neural tube to close (Section 27-G). These defects can be found by amniocentesis.

Recently, some human fetuses have been treated before birth for deficiency of the B vitamin biotin, an inborn error of metabolism. Other fetuses have had surgery to correct blockage of the urethra, the tube that empties the urinary bladder, or diaphragmatic hernia, a hole in the diaphragm that permits abdominal organs to protrude into the space where the lungs should be developing (see Figure 23-8). We can expect that more genetic disorders and other conditions will be detected and treated before birth in the future.

as a dominant in men because the male sex hormones somehow stimulate expression of the baldness allele. However, in women, the allele acts as a recessive, so that a female must have two alleles for baldness before she loses her hair.

Gout is another trait whose expression is influenced by sex. In gout, painful deposits of uric acid salts build up in the tissues, especially in the joints of the big toes. The allele for gout is expressed much more in the presence of male than of female sex hormones. In Victorian literature, gout figured largely as a reason for the temper tantrums of irascible old men. Avoiding red wine and rich and spicy foods was supposed to alleviate the condition, but this treatment tried the tempers of its victims still further. Gout can now be treated, but it is still a painful nuisance to many sufferers.

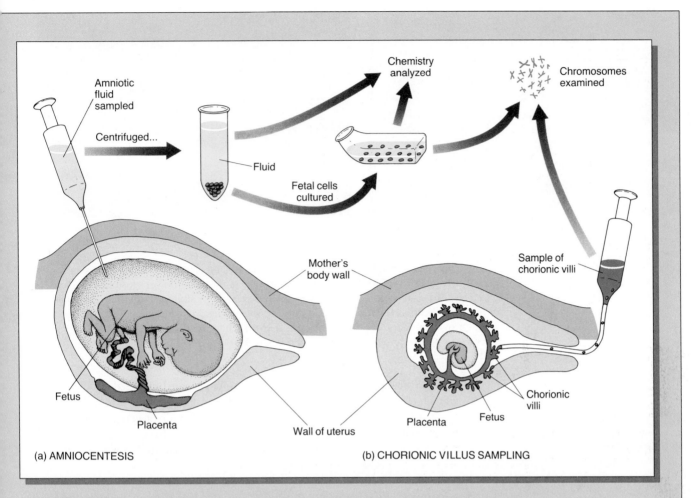

Figure 13-A

Diagnosing genetic defects in a fetus. (a) Amniocentesis. A needle is inserted through the mother's abdomen into the amniotic sac, and fluid containing fetal cells is withdrawn into a syringe. Fetal cells are cultured and examined for chromosomal abnormalities. The fluid and cells can also be tested chemically for metabolic defects. Amniocentesis may also be done late in pregnancy to see whether the lungs are producing the chemicals that will enable them to breathe air. If so, the fetus is ready to be delivered by cesarian section, in cases where early delivery is deemed advisable. (b) Chorionic villus sampling. By about three weeks after fertilization, the fetus's chorionic membrane has developed branched villi, some of which will form part of the placenta. At 8 to 10 weeks there are enough fetal cells here that some can be sucked up to provide a sample for genetic analysis.

13-I Some Factors That Affect Gene Expression

All the genes possessed by an individual determine its genetic potential: what might be. What actually happens is another matter. The expression of a gene is influenced by the other genes present, either by way of proteins encoded by these other genes, or more indirectly. The influence of enzyme-produced sex hormones on gene expression is one example of such an indirect effect.

Hormone production varies with age, and so age may play a part in gene expression. Consider the many changes accompanying puberty, such as voice change and growth of the testes in males; breast enlargement and the characteristic pattern of body fat deposition in females; and growth of hair in the armpits and pubic area in both sexes.

(a)

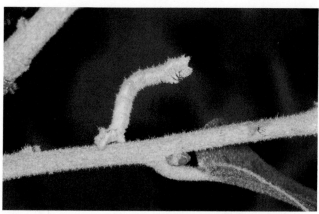

(b)

Figure 13-15
Phenotypes dependent on diet. (a) This oak-feeding caterpillar, fed a diet of oak flowers, has developed a resemblance to one of the male flower clusters, complete with dots that look like the flowers' pollen sacs. (b) A sibling of the first caterpillar, fed oak leaves, resembles the oak tree's twigs. In nature, caterpillars that hatch in the spring look like the flowers they eat; those hatched after the flowers wither eat leaves and become twig mimics. (Erick Greene)

Many traits are controlled mainly by one gene pair but are also influenced by the products of other genes, called **modifier genes.** It was long believed that eye color in humans was controlled by a single pair of genes, with brown eyes dominant to blue. It is now known that there are also at least two pairs of modifier genes involved, and it is possible, though extremely uncommon, for blue-eyed parents to have brown-eyed children.

The external environment also plays an important role, both in embryonic development and in later life. In the last few decades there have been several documented cases in which drugs taken by pregnant women have caused improper development of the fetus, or cancers later in the baby's life.

Many other external factors also affect gene expression. For instance, a good diet is necessary if a person is to reach the height made possible by his or her genes. In many countries, young adults tower above their parents or grandparents as a result of improved nutrition. Farmers and gardeners know that proper nourishment is just as important for plants. One species of caterpillar even develops a phenotype that mimics the appearance of the food it eats, thereby blending in with it (Figure 13-15).

Light also influences gene expression. The development of plants is especially sensitive to light. A human being becomes darker (or redder!) when exposed to bright sunlight for a time.

Temperature affects the expression of some genes. Himalayan rabbits and Siamese cats are normally light-colored, with black feet, ears, nose, and tail. What makes the fur different colors on different parts of the body? We now know that the darker color is due to the activities of an enzyme which is unstable at higher temperatures. The extremities of these animals are cool enough for the enzyme to function and produce dark fur, but the body itself is warm enough to inactivate the enzyme (Figure 13-16).

Figure 13-16
Effect of skin temperature on expression of coat color genes in the Himalayan rabbit. Black fur grows on parts of the body with skin temperature below 33°C. If fur is shaved from a warmer part of the body, and an ice pack applied while the fur grows back, the new fur is also black.

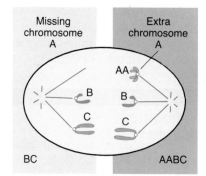

(a) NONDISJUNCTION

Figure 13-17
Nondisjunction and translocation. The cells are in the second division of meiosis (see Figure 11-11). (a) In chromosome A, the sister chromatids fail to separate, but remain attached and move to the same pole, eventually ending up in the same gamete. (b) One copy of chromosome A becomes attached to chromosome B, which moves into a gamete along with a separate copy of chromosome A. In both cases, the gametes formed on the left will lack chromosome A, whereas those on the right will contain two copies of chromosome A.

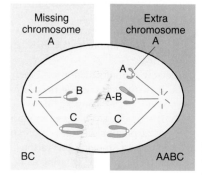

(b) TRANSLOCATION

13-J Nondisjunction and Translocation

Occasionally, chromosomes behave abnormally during meiosis. **Nondisjunction** is the failure of chromosomes to separate properly. **Translocation** is the attachment of all or part of a chromosome to a nonhomologous chromosome. Either of these abnormal events produces gametes with one chromosome too many or too few (Figure 13-17). Most of the resulting fetuses die, but sometimes an individual with an extra chromosome in each cell survives. Such a person is usually mentally retarded if the extra chromosome is an autosome.

The most common cause of mental retardation in the United States is Down's syndrome, which results from having an extra copy of all or part of chromosome 21 (see Figure 11-4). The symptoms of Down's syndrome include a small brain and mental retardation, a "mongoloid" eye fold, impaired immunity to disease, short stature, and flaccid muscles. Defects in the heart and the lenses of the eyes are also common. It has long been known that older women have an increased risk of bearing children with Down's syndrome. Recent studies show that older men are also more likely to father such infants, regardless of the mother's age. Translocations of chromosome 21, or variants of chromosome 21 with a tendency to nondisjunction, run in some families and produce Down's syndrome children regardless of the parents' ages. About 1 baby in 700 live births has Down's syndrome.

Failure of the sex chromosomes to segregate properly during meiosis results in such abnormal sets of sex chromosomes as XXY (Klinefelter's syndrome), XYY, XXX, or a single X chromosome (Turner's syndrome). Some of these conditions produce sterility or mental retardation (see Table 13-2).

SUMMARY

Genes express themselves by coding for the sequences of amino acids in polypeptides or proteins. The severity of a mutation depends on how much it affects the protein encoded by the gene and on how important that protein is in maintaining life. Some mutations result in lethal alleles, which cause premature death. Most familiar lethal alleles are recessive and cause death only in the homozygous condition. Heterozygotes (carriers) survive and may pass the allele to future generations.

Changes in less vital proteins may cause metabolic disorders (inborn errors of metabolism), such as albinism and phenylketonuria.

Several different alleles of a gene may exist in a population as a result of different mutations in different individuals. Such multiple alleles are found in the human ABO blood group and in cell surface proteins. Polygenic characters are determined by the interaction of several different gene pairs. These polygenic characters show a wide range of phenotypes.

In most familiar organisms, sex is determined by one sex being homozygous and the other heterozygous for an entire pair of chromosomes, the sex chromosomes. In humans and most other mammals, females have the sex chromosome combination XX and males are XY, whereas in birds females are ZW and males ZZ. Traits carried on nonhomologous portions of the sex chromosomes are said to be sex-linked. Sex-influenced characters are carried on the autosomes (usually), but depend on the balance of sex hormones for their expression, and hence are more common in one sex than the other.

An individual's phenotype depends on what mix of genes it has, how these genes are influenced by the products of other genes (enzymes, enzyme products such as hormones or pigments, or non-enzyme pro- teins), and what factors it encounters in its external environment. External factors influencing gene expres- sion include nutrition, light, and temperature.

SELF-QUIZ

1. In the homozygous condition, a recessive lethal allele in cattle produces "amputated" calves with malformations of the limbs, skull, and internal organs. These calves die soon after birth.
   a. What proportion of the normal offspring from a cross of two heterozygotes would be expected to be carri- ers for this trait?
   b. How could a farmer eliminate this trait from his herd if some "amputated" calves have been born to his cows?
2. Review the information on brachydactyly, Section 13-B.
   a. If two brachydactylic people marry, what are their chances of having a child with normal fingers?
   b. If a brachydactylic person marries a normal person, what phenotypic ratios are expected in their offspring?
3. A geneticist studying the various gene pairs that govern coat color in mice is trying to develop true-breeding strains of each possible coat color. He carries out several generations of matings among mice with yellow coats and always obtains some offspring with other colors of coats.
   a. What does this indicate about the genotype of yellow mice?
   b. The geneticist tallies up his results over several gener- ations and finds that he has obtained a total of 184 yellow mice and 95 of other colors. What does this suggest about the nature of the yellow allele?
   c. Why did the geneticist never obtain a homozygous yellow mouse?
   d. How could he prove what became of the homozygous yellow offspring?
4. Below is a pedigree of ABO blood groups for several generations of humans. Circles represent females,

squares males. Marriages are shown by horizontal lines directly connecting two people, and children are con- nected to their parents by a vertical line down from the marriage line. For example, (b) and (c) are married to each other, and (d) is one of their two sons. Give the possible genotype(s) for each individual marked with a letter. (Hint: start at the bottom of the diagram and use what you know about children's genotypes to determine those of their parents.)

5. Ms. Smith and Ms. Jones gave birth to baby boys (named John and Tom, respectively) on the same day in a large city hospital. After Ms. Smith took her baby home, she began to suspect that it was Ms. Jones's baby, and that the hospital had somehow mixed the infants up. Blood tests revealed that Mr. Smith had blood type O, MN, and Rh^+; Ms. Smith had blood type B, N, Rh^+; and John Smith had blood type B, M, Rh^-. Mr. Jones had blood type A, M, Rh^+; Ms. Jones had blood type AB, MN, Rh^+; and Tom Jones had blood type O, MN, Rh^+. The Rh^+ al- lele is dominant to the Rh^- allele; the M and N alleles are codominant. Had a mixup occurred?
6. In rabbits, normal coat color (C) is dominant to chin- chilla (c^{cb}), which is dominant to Himalayan (c^h), which is dominant to albino (c). What offspring are expected from the following crosses, and in what ratios?
   a. $Cc^h \times c^{cb}c^h$ b. $c^{cb}c \times c^hc$ c. $c^{cb}c \times c^{cb}c$
7. In chickens a sex-linked dominant allele causes a feather pattern known as "barred." If a barred hen is mated with a nonbarred rooster, what will be the feather pat- tern and sex of the offspring?
8. What are the expected genotypic ratios among the chil- dren of a normal woman whose father was a hemophil- iac, and whose husband is normal?
9. Under what circumstances is it possible for both a father and his son to be hemophiliacs?
10. Red-green colorblindness in humans is a sex-linked re- cessive trait.
    a. In a large family in which all the daughters have nor- mal vision and all the sons are colorblind, what are the probable genotypes of the parents?
    b. If a normal-sighted woman whose father was color- blind marries a colorblind man, what is the probabil- ity that their son will be colorblind?
    c. What is the probability that the couple in part b will have a colorblind daughter?
11. If a species of mammal has some members which carry a sex-linked lethal trait that causes early death and re- sorption of the embryo, what sex ratio would be ex- pected among the offspring of a female carrier and a normal male?
12. It is often said that men inherit baldness from their ma- ternal grandfathers via their mothers. In light of what you have learned about this trait, is this a valid state- ment? Explain.

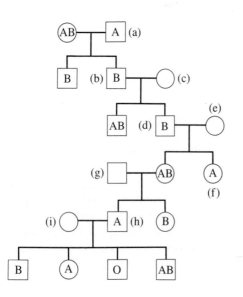

QUESTIONS FOR DISCUSSION

1. One problem with genetic counseling is that people who learn they are carriers for genetic diseases such as hemophilia, Tay-Sachs disease, sickle cell anemia, or phenylketonuria may consider this a terrible stigma. Men have been known to deny paternity of their children and divorce their wives for infidelity when told that the child had inherited a harmful recessive gene from each parent. What kinds of arguments and counseling would you use, if you were a genetics counselor, in an attempt to induce a healthier, more productive response to such a discovery?

2. People who are carriers for sickle cell anemia face the risk that their red blood cells may sickle when they are in environments with low oxygen levels. Hence these people are probably exposed to greater than usual risks if they become divers, jet pilots, or mountaineers. Otherwise they have no physical handicaps; nevertheless, they have frequently been denied access to various professions as a result of ignorance and prejudice against genetic disorders. Since this is the case, a proposed nationwide screening for sickle cell carriers might well do more social harm than good. Is it better for carriers to remain in ignorance of genetic conditions for which there is no cure at the moment? If not, why not?

3. Until the advent of modern technology, hemophiliac men usually died before they reached reproductive age. Nowadays they can be provided with "clotting factor," a blood extract that permits them to lead normal lives and live to have children. The treatment costs about $6,000 to $10,000 a year per person, and there are about 20,000 hemophiliacs living in the United States. Can society or should society insist that such men be sterilized, so that they cannot perpetuate their disease, if taxpayers have to pay the bill for their medication?

4. The genes for both normal clotting factor (missing in hemophiliacs) and normal chloride channels (defective in cystic fibrosis patients) have been isolated and cloned. Hence both proteins can now be produced in the laboratory. Injections of the normal clotting protein will be an effective treatment for hemophilia. Why will it not be possible to treat cystic fibrosis with laboratory-produced chloride channel proteins?

5. Table 13-2 shows that individuals with a single X chromosome are known to occur, but not individuals with only a Y chromosome. Why do you think this is?

6. Every so often the Ann Landers column has a letter from a mother whose husband or in-laws have been chiding her for having daughters instead of sons. Is this censure justified? Why?

7. Name some factors besides those mentioned in the chapter that may influence gene expression.

8. Propose an explanation of the genetic basis for the phenotypes shown in each of these photographs:

(a) The lion's mane, and the black tips of its hairs

(b) Variations of color and pattern in the leaves of one plant

SUGGESTED READINGS

Friedmann, T. "Prenatal diagnosis of genetic disease." *Scientific American,* November 1971. A thoughtful article explaining the technology of amniocentesis and pointing out the ethical problems it presents.

Greene, E. "A diet-induced developmental polymorphism in a caterpillar." *Science* 243:643, 1989. Easily understood experiments show that what this caterpillar looks like depends on what it eats.

Hartl, D. L. *Human Genetics.* New York: Harper and Row, 1983.

Lawn, R. M., and G. A. Vehar. "The molecular genetics of hemophilia." *Scientific American,* March 1986.

Murray, J. D. "How the leopard gets its spots." *Scientific American,* March 1988. A speculative article presenting a model of how coat patterns might develop in spotted and striped animals.

Nathans, J. "The genes for color vision." *Scientific American,* February 1989. How we see color, how this ability evolved, and how mistakes in crossing over during meiosis result in colorblindness.

Patterson, D. "The causes of Down syndrome." *Scientific American,* August 1987.

Sayers, Dorothy L. *Have His Carcass.* London: Harcourt, Brace Jovanovich, 1932. A mystery novel about a human genetic trait.

Evolution and the Diversity of Life

PART

3

Our study of cells and genetics has shown that the information each organism needs to survive, develop, and reproduce is packaged in its genetic material and passed from each organism to its offspring. This is the background we need to understand the central idea that unifies all of biology: groups of organisms can evolve, changing genetically from generation to generation as a result of influences in the world they live in. Without the idea of evolution, biology would be little more than a list of organisms and how they work—as it was 150 years ago. In the context of evolution, everything else makes sense. Why do we enjoy a high-fat diet when it is bad for our health? Because humans are descended from animals that ate mostly fruit, a diet short of lipids and lipid-soluble vitamins. We have inherited the genes that endowed them with a craving for fat, which drove them to seek out the fat they needed. Why do some people suffer from sickle cell anemia? Because their ancestors came from populations that survived only because some individuals contained the gene for sickle cell anemia, which protected them from dying of malaria. All this, and much more, starts to make sense when we understand evolution.

From considering how evolution occurs in Chapters 14 through 16, we go on to explore what evolution has produced: the beginnings of life on earth (Chapter 17) and the millions of species of organisms that have descended from the first cells to arise in the primordial ooze. We classify organisms by similarities of structure, function, chemistry, and behavior that show their evolutionary relationships to each other. A lifetime (much less a single biology course) is too short to study the diverse adaptations of even a single group of organisms, and so we can offer only a glimpse of this fascinating field. Hence our survey of the diversity of life (Chapters 18 through 21) will concentrate on the major groups of organisms and on the important evolutionary trends that resulted in the organisms that share the earth with us today.

Peacock

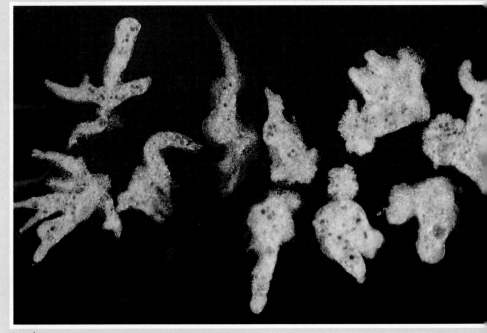

Amoebas

244

Cupid Terrace, Mammoth Hot Springs

King vulture

Frilled lizard

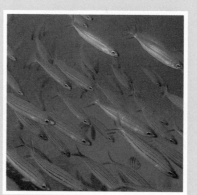

Boga

245

Evolution and Natural Selection

CHAPTER 14

Allosaurus

Chimpanzee

King vulture

Fossil fish

*I*n the last few chapters we have seen how genes are passed from grandparents to parents to offspring. In the next few chapters we consider these same events, but on a much larger scale. We shall no longer focus on pairs of genes passed through a few generations of a single family. Instead we consider all the genes passed on over thousands of generations within a whole population of a particular species of organisms.

We shall see that the proportions of various genes in a population may change over this broad sweep of time. Sometimes this causes changes in the population so slight that they pass unnoticed, but at other times, genetic changes result in organisms that differ markedly from their ancestors. For instance, the octopus shown in Figure 14-1 is believed to have descended from a snail-like ancestor with a shell. By a series of genetic changes over many generations, octopus ancestors lost the shell and developed long, manipulative arms. Such changes in the overall genetic makeup of a population over time are called **evolution.** When members of a population have evolved sufficient differences from their ancestors, we may consider them new species.

Notice that the unit that evolves—that changes over time—is the population, not its individual members. An individual cannot evolve. Each individual's genetic makeup is fixed at the time of fertilization, when egg and sperm unite to begin the new individual. The individual's role in evolution is to pass on some of its own share of the population's genes to its offspring and so contribute genes to the next generation.

The accumulation of genetic changes over many generations may result in dramatic changes in the population over time. This gives a more general definition of evolution: **evolution** is the origin of organisms by descent and modification from previously existing forms of life.

One mechanism that produces evolution is **natural selection,** the differential reproduction of genotypes from one generation to the next. By "differential reproduction," we mean that individuals with some genotypes produce more offspring than those with others. This results in evolution because it changes the proportions of different genetic alleles in the population (Section 12-A). Therefore it causes the population to change genetically over time.

To take an example of evolution by means of natural selection, an animal's coloring is largely determined by its genes. An animal that blends into its natural background is less likely to be noticed by predators, and therefore is more likely to survive and produce offspring, than is a similar animal of a conspicuous color. Because more animals with good camouflage breed and pass on the genes that dictate this coloring, a larger proportion of individuals in the next generation of the population will be camouflaged. The genetic makeup of the population has changed somewhat from one generation to the next, and that is evolution.

Figure 14-1
An octopus. This soft-bodied animal has long arms with suckers, which it uses to catch its food. It has evolved numerous differences from its ancestors, which had shells but no arms. (Matthew Gilligan)

KEY CONCEPTS

- The theory of evolution states that organisms arise by descent and modification from previously existing organisms.
- Evolution is a change in proportions of one or more alleles in a population from one generation to the next.
- Natural selection, the differential reproduction of genotypes, is the most important cause of evolution.
- Populations of organisms evolve adaptations to their environments.

14-A History of the Theory of Evolution

For thousands of years, most people believed that each separate species of organism had been specially created. From time to time philosophers proposed that the living world changed over time, but until the mid-seventeenth century this idea gained little ground in the Western world. From about 1750 on, however, many people became convinced that species changed over the ages.

Figure 14-2
Why does the giraffe have a long neck? Lamarck suggested that its ancestors stretched their necks to browse on the leaves of trees, and that this increase in length was passed on to succeeding generations.

Lamarckism

In 1809, the French biologist Jean-Baptiste de Lamarck elaborated the common beliefs of the day into his own proposal. Lamarck is an example of an early evolutionist whose idea about how the process occurred was not convincing. He is remembered more for his incorrect explanation than for his belief in evolution.

Lamarck suggested that organisms could acquire traits that made them better adapted to their environments, and could pass such traits on to their offspring. Hence a population changed from one generation to the next. This is the theory of evolution by the inheritance of acquired characters. His most famous example was the long neck of the giraffe (Figure 14-2). Lamarck suggested that giraffes had evolved their long necks because they strained to reach leaves growing above their heads as they ate, thereby stretching their necks; this added length was passed on to their offspring. This idea dovetailed nicely with pre-1900 beliefs, which held that different parts of the body contributed to eggs and sperm by sending minute particles through the bloodstream to a collection point in the reproductive organs.

Darwin did not refute Lamarck's idea when he proposed natural selection as the mechanism of evolution. He did not understand that individuals inherit discrete genes, and he thought that the inheritance of acquired characters might have a minor role in evolution. This view was discarded when Mendel's work on genetics was rediscovered and expanded (Chapter 12).

This is not to say that nothing an organism does in its lifetime can affect its offspring's genotype. Taking drugs that destroy chromosomes, or being exposed to high levels of radioactivity, may alter the genes passed on to the offspring. However, it is clear that, with a few possible exceptions, nothing an organism does will make its offspring inherit the same characteristic that it has acquired.

Darwin and Wallace

The theory of evolution by natural selection was put forward in a joint presentation of the views of Charles Darwin and Alfred Russel Wallace before the Linnaean Society of London in 1858. As we have seen, Darwin and Wallace were not the first to suggest that evolution occurred. Their names are linked with evolution because they proposed natural selection as the mechanism that brought it about. We are more likely to believe in a process when people give a convincing explanation of *how* it happens than if they merely assert that it *does* happen. Darwin's explanation of natural selection eventually convinced the world that evolution occurred.

We know little about Wallace's early life, but it seems likely that Darwin as a young man believed in special creation. Such a belief formed part of the faith of most Christian denominations of his day, and Darwin at one time began training for the clergy. Years of observation and reading, however, presented Darwin with a more compelling explanation of the origin of species.

Darwin and Wallace came to the same conclusion about evolution as a result of very similar experiences. First, both Wallace and Darwin were influenced by reading the works of geologist Charles Lyell and economist Thomas Malthus. Lyell wrote that the world was an ancient arena in which rock formations slowly appeared, changed, and disappeared. He recognized that competition between species leads to a "struggle for existence" and even discussed the extinction of species caused by human activities. All the information needed to formulate the theory of evolution was present in Lyell's work. Malthus similarly argued that there is competition between organisms. He wrote that every human population must eventually outgrow its food supply and then be reduced by disease, starvation, or war.

(a)

(b)

Figure 14-3

The Galapagos Islands. (a) Darwin's observations of the islands' peculiar plants and animals became an important part of his thinking about evolution and how it might have been brought about. (b) A small-billed ground finch, found only in the Galapagos. In this hot, dry, isolated cluster of islands, Darwin noted a wealth of species found nowhere else on earth.

Second, both Wallace and Darwin observed plant and animal life in several parts of the world. Wallace traveled in South America, and later in the islands of Indonesia. It was here, in 1854, that the idea of natural selection came to him as he lay in bed with a fever. In the 1830s, Darwin obtained a position on *H.M.S. Beagle,* a British naval ship embarking on a five-year mapping and collecting expedition. This trip took Darwin to South America and the nearby Galapagos Islands (Figure 14-3), where he collected much of the evidence he later used to support the theory of evolution by natural selection.

In 1845, Darwin published *The Voyage of the Beagle,* an account of his travels. He showed that he already held the clue to how evolutionary change was brought about, a mechanism that he was not to publish until more than a decade later. He wrote, "some check is constantly preventing the too rapid increase of every organized being left in a state of nature. The supply of food, on average, remains constant; yet the tendency in every animal to increase by propagation is geometrical."

Upon his return to England in 1837, Darwin settled down to a lifetime of writing and thought. By the next year, he had formulated the theory of evolution by means of natural selection, but he pondered it and accumulated supporting evidence before presenting it publicly. Twenty years later, in 1858, he received a manuscript from Wallace, describing natural selection. Wallace had written his paper in three days. Darwin passed Wallace's paper to Lyell and to the botanist Joseph Dalton Hooker, who persuaded Darwin to let them present a version of his theory and Wallace's paper at a scientific meeting in 1858. Darwin then worked feverishly to finish his book, *The Origin of Species by Means of Natural Selection,* which was published in 1859. In it, he marshalled an impressive array of evidence to support his theory, the result of a quarter of a century of observation and inquiry. The book sparked immense controversy, a fitting tribute to the most original and important biology book ever written. Although evolution was accepted in Darwin's day, not until the twentieth century did most biologists fully accept the idea that evolution occurs by means of natural selection.

14-B   The Evidence for Evolution

Several different lines of evidence convinced Darwin and Wallace, and many of their contemporaries, that modern organisms have arisen by evolution from more ancient forms of life.

The Evidence from Artificial Selection

■ *Breeders select which animals and plants shall reproduce, and so change the proportions of various genes in populations of organisms.*

Darwin illustrated selection with examples drawn from the selective breeding of domestic plants and animals. These organisms do not usually breed randomly. Breeders and gardeners save seed only from the largest, prettiest flowers and the tastiest melons. Dairy farmers mate the cows that produce the most milk with bulls whose mothers were good milk producers. Modern hybrid corn is very different from its inbred parents (Figure 14-4). Breeders and farmers exert **artificial selection** on domesticated animals and plants by determining which members of the population shall reproduce and which shall not. The striking changes they produce over relatively few generations are powerful proof that organisms can evolve.

However, this evolution results from deliberate manipulation by breeders with definite ends in view. It is more difficult to show that a similar process accounts for changes in natural populations. A weakness of Darwin's evidence for evolution was that he never provided a convincing demonstration that selection actually occurs in nature. His detractors pointed out that nature has no mind, no goal nor purpose. How could a haphazard series of accidents result in organisms that appeared as though they were designed specifically for the place they hold in nature? The examples of selection in wild populations described in Section 14-C were not worked out until a century later.

The Evidence from the Fossil Record

Figure 14-4
Artificial selection. (a) A cow in the Swiss Alps, bred to produce large quantities of milk. (b) Trials of hybrids between different varieties of corn.

Usually, when an organism dies, scavengers and decay organisms rapidly destroy it. Occasionally, however, a body may come to rest in an acid bog, or be buried under a layer of mud that cuts off oxygen, conditions that prevent decay

(a)

(b)

Figure 14-5
Fossil-bearing rocks. The badlands near Drumheller, in Alberta, Canada, are the site of an ancient shallow sea. Layers of sediment collected on the sea floor, forming the layers seen here. Before they could harden into rock, the river eroded them extensively. The black layers are seams of coal, the fossil remains of ancient plants. Many dinosaur skeletons have also been found in this area. The farm at the top of the photograph gives an idea of the scale.

and may permit the body to be preserved. A **fossil** is any preserved evidence of life long past: a body or body part, an impression of the surface of the body such as a footprint; organic molecules such as oil, which are chemical remains of organisms; or even a coprolite (preserved excrement).

Fossil-hunting was a popular recreation in nineteenth-century England. Drawings in Victorian magazines portray ladies in long skirts, and gentlemen in jackets and ties, scrambling over rocks with geological hammers in their hands and fanatical gleams in their eyes. Most important fossils of the time were found by amateurs such as these.

These finds captured the popular imagination, and newspapers printed articles and letters arguing about the religious and scientific implications of fossils. Some people suggested that God had fashioned the fossils and scattered them in the rocks to delight fossil-hunters. However, geologists were beginning to produce a very different explanation. The growing collections of fossils in North America and Europe provided strong evidence that organisms had changed over the centuries.

First, nearly all of the fossils ever found are of species that are now extinct. Second, some species are older than others. Many fossils occur in formations made up of several layers of rocks, and geologists realized that the bottom layer had usually been formed first and contained older fossils than the overlying layers (Figure 14-5). Assigning relative dates to fossils in this way made it clear that some groups of species were older than others. (In this century, it became possible to calculate the actual age of rock formations by using isotopes [Chapter 2, *A Journey Into Life*].)

The abundance of members of different groups also changed with the age of the rock. It appears that some groups of organisms originated much earlier than others, and that some of the more recent groups have largely replaced more ancient ones over the course of time (see this book's endpaper).

In a few cases, fossils allow us to reconstruct the family tree of a particular group. For instance, we can trace the origin of mammals from reptiles in great detail in the fossil record. We can also follow the changes from dinosaur-like reptiles to ancient birds with feathers, but also with teeth (Figure 14-6), and eventually to modern birds with feathers and no teeth.

The classic case of fossil genealogy is the story of horse evolution, published by Othniel C. Marsh in 1879. Marsh described older and older fossils linking

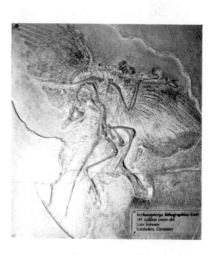

Figure 14-6
A fossil of *Archaeopteryx*. This extinct animal from the Jurassic period is one of the first known birds. Note the impressions left by the feathers, characteristic of birds, in the area of the forelimbs, and the vertebrae of the long, unbird-like tail. Also notice the somewhat twisted position.

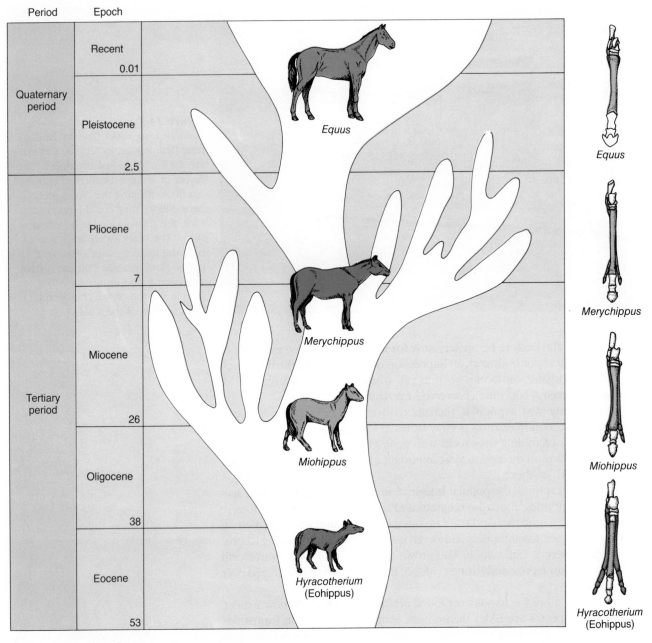

Figure 14-7
Evolution of the modern horse. The ancestral *Hyracotherium* gave rise to many species. Most became extinct, but some have survived to the present day. The legs of the four species drawn here show progressive development of the central toe and loss of the side toes. They are also progressively larger, although some evolutionary branches not shown produced species smaller than their ancestors. The numbers given under each epoch are dates (million years ago).

■ *The fossil record shows that different organisms have lived at different times during the earth's history, and that some kinds of organisms have changed over geological time.*

modern horses with the tiny dog-sized *Hyracotherium,* the "dawn horse" found in Eocene rocks. Through the fossil record he traced the major changes in the teeth, legs, and feet of ancestral horses (Figure 14-7). Many similar examples have now been described in which the ancestry of modern species can be traced through successive rock layers, with the youngest rocks containing those fossils most like the modern forms.

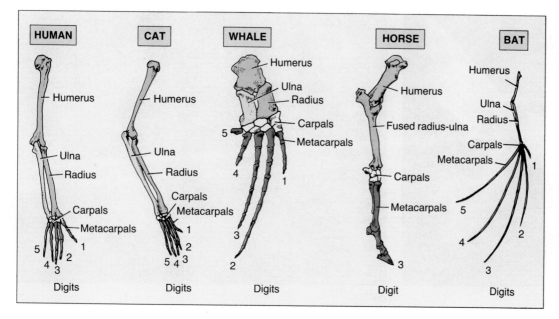

Figure 14-8

The bones of the forelimbs of a human, cat, whale, horse, and bat. All show the same basic pattern. In different species, the various bones have evolved different proportions as the limbs became adapted to different functions. The digits (fingers) are numbered 1 through 5. Note that the horse has lost all but the third digit, which is greatly enlarged.

The Evidence from Comparative Anatomy

Even without fossil evidence that different organisms have lived at different times in the past, we might suspect that organisms had evolved by comparing the structures of species alive today. Not surprisingly, similar kinds of organisms have very similar structures. For example, the skeletons, teeth, and muscles of different members of the cat family are very similar, and the same is true of different species of bats or of whales. However, a comparison of a cat's bones with those of a bat or a whale reveals that these three groups of animals all have skeletons composed of quite similar groups of bones, despite their adaptations to very different ways of life. The forelimb bones of cats, bats, and whales are arranged in the same pattern: a bat's wing, a cat's front leg, and a whale's flipper all contain bones identifiable as humerus, radius, ulna, and so on (Figure 14-8). Indeed all mammals, birds, reptiles, and adult amphibians have forelimbs with this same basic framework, although the limbs may perform very different functions in animals as different as a pigeon, a penguin, a turtle, or a human. Furthermore, all of these forelimb bones originate from the same part of the embryo. Such structures, with the same origin, but occurring in different species, are said to be **homologous** to each other. Homologous organs may perform the same or different functions.

The converse of homologous organs are **analogous** organs, which have similar functions but are constructed differently and appear unrelated. The wings of birds and of insects, for instance, may both be used for flying, but they are completely different structurally (Figure 14-9).

Darwin saw that homologous and analogous organs posed problems to a creationist viewpoint. It made no sense that several different types of wings should be invented. Even the homologous wings of birds and bats differ somewhat in structure. Surely one design must be superior to the other. Why create

(a)

(b)

Figure 14-9

Analogous structures. The wings of an elephant hawk moth (a) and a blue tit (b) are both broad, flattened, lightweight structures used for flight. However, they have very different developmental and evolutionary origins. (Biophoto Associates)

several different sorts of wings? Similarly, why did so many animals contain apparently inefficient homologous structures? Why do whales have heavy bones, like those of terrestrial mammals, in their flippers, instead of the lighter, folding fins of a fish, apparently so much better designed for propulsion in water?

Eventually Darwin came to realize that evolution made sense of all these paradoxes. Organisms were not created from a clean slate. They arose from ancestors with characteristics already determined by their own evolutionary histories. Whales have bony flippers because they evolved from land mammals with bony forelimbs. Insects have no bones to support their wings because they evolved from animals with external skeletons of chitin and without bones. The very imperfection of adaptations, the feeling that so many of them could have been designed better, became, to Darwin, the most convincing evidence that evolution has occurred.

Anatomy provides a further argument for evolution in the form of **vestigial** structures: organs useless to their present owners but homologous with structures that serve important functions in other species. The most familiar example is the human appendix, a worm-like blind sac near the junction of the small and large intestines (see Figure 22-2). The appendix is homologous to the caecum, a large, blind chamber in which leaves and grasses are digested in many other mammals. Another example is the minute pelvic and hind limb bones in the skeletons of whales and of boa constrictors, even though these animals have no true hind limbs. All these vestigial organs are the evolutionary remnants of organs that were larger, and useful, in their owners' ancestors.

The Evidence from Comparative Biochemistry

In more modern times, the arguments for evolution based on comparative anatomy have been paralleled by evidence from other fields. Studies of the similarities and differences in the structures of homologous proteins and genetic material show that organisms known to be related are very similar in these respects. For instance, humans and chimpanzees, long thought to be closely related, have proteins that are 99% alike. DNA and proteins are much less similar when they come from organisms that are only distantly related (see *A Journey Into Life,* "Proteins as Evolutionary Puzzle Pieces," Chapter 10).

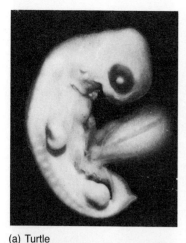
(a) Turtle

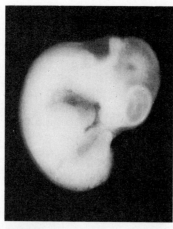
(b) Mouse

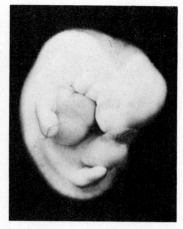
(c) Human

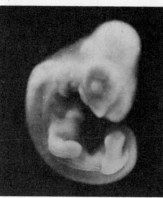
(d) Pig

(e) Chick

Figure 14-10
The similarities of different vertebrate embryos. Structures that are homologous in all vertebrates are obvious in these early embryos. Each has a tail that projects behind the anus and traces of gill slits in the neck region. (Roberts Rugh)

The Evidence from Embryology

A similar line of evidence for evolution comes from studying embryonic development, especially of animals. In many instances, the embryo contains structures that will not be found in the adult. For example, the early embryos of reptiles, birds, and mammals, including humans, develop a row of vestigial gill slits just behind the head (Figure 14-10). This suggests that these groups of animals descended from the fishes, in which gill slits persist and function throughout life. Similarly, the embryos of baleen (whalebone) whales and of birds develop tooth buds, even though the adult animals are toothless. Sometimes human babies are born with short tails, or with several nipples in two rows down the front of the body, characteristics that are common in other mammals but not in human adults.

Embryology also reveals that certain apparently "new" features of higher vertebrates developed, not from scratch, but from the remodeling of ancestral structures. For instance, some of the embryo's gill arches, which in fish develop into structures supporting the gills, become in mammals parts of the lower jaw, the ear bones, and parts of the air passages of the respiratory system.

> ■ *The anatomy and embryology of animals reveal many peculiarities that are most easily explained by assuming that animals have evolved from other kinds of animals.*

The Evidence from Biogeography

Biogeography is the study of the distribution of organisms across the face of the globe. Both Darwin and Wallace, in their travels, noticed that the present-day distribution of organisms made no sense seen from a creationist point of view,

(a)

(b)

Figure 14-11

Species endemic to the Galapagos. (a) A male land iguana. Darwin wrote, "we could not for some time find a spot free from their burrows on which to pitch our single tent . . . they are ugly animals, of a yellowish orange beneath, and of a brownish red colour above. These lizards, when cooked, yield a white meat, which is liked by those whose stomachs soar above all prejudices." (b) The flower of the endemic *Opuntia* cactus. The pads of this cactus are a favorite food of the land iguana. Each pair of iguanas may live under its own cactus. The cactus plant itself is the size of a small tree (see Figure 14-3a).

■ *Darwin and Wallace became convinced that only an evolutionary origin for species could reasonably explain the distribution of modern plants and animals, and biologists since have agreed with them.*

but could be explained by evolution. Why did the Galapagos, a group of small islands off the west coast of South America, contain more different species of finches than the entire South American continent? Another puzzle was the distribution of mammals. Why were marsupial (pouched) mammals found only in Australia and South America? (Opossums are marsupials, but have colonized North America from South America.) Why did Australia contain none of the placental mammals found throughout the rest of the world?

Both men became convinced that these, and dozens of other puzzles, could be explained as the result of the evolutionary histories of these modern organisms—including where their ancestors lived. From an ancestral group living in a particular place, descendant populations could spread, or radiate, into other areas. In doing so, they would encounter new environmental conditions that would bring about the evolution of new adaptations. Such an evolutionary process, giving rise to new species adapted to new habitats and ways of life, is called **adaptive radiation.** In Australia the adaptive radiation of marsupials gave rise to a variety of species that closely resemble equivalent placental mammals elsewhere. Australian marsupials include the rabbit-like bandicoot, the woodchuck-like wombat, the Tasmanian "wolf," and the flying squirrel-like flying phalanger, as well as unique forms such as koalas and kangaroos. However, the spread of organisms may be limited by geographical boundaries. For instance, marsupials in Australia could not move to other continents because oceans barred the way.

Darwin was profoundly struck by the flora and fauna of islands, particularly the Galapagos Islands about 1000 kilometres west of South America. What caught his attention was the remarkable numbers of **endemic** species, species found nowhere else, even on other apparently similar islands nearby (Figure 14-11).

That the tiny, relatively barren Galapagos Islands (and other isolated islands visited by the *Beagle*) housed such large numbers of endemic species seemed a profligate waste of effort by the Creator. Darwin saw that the existence of so many endemic organisms could be explained by assuming that members of certain mainland species had colonized the islands, and there evolved into new forms of life. Some of the original settlers had even given rise to several new species through adaptive radiation, as different groups of their descendants be-

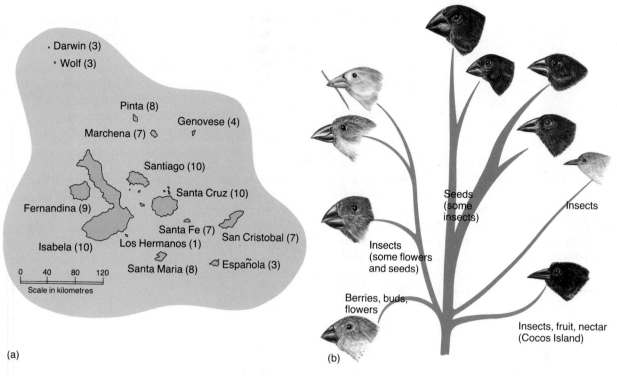

(a)

(b)

Figure 14-12
Darwin's finches. (a) Map of the Galapagos Islands showing the number of finch species found on each island. (b) Probable family tree of some of the finches that have evolved in the Galapagos Islands and on Cocos Island, 800 kilometres away. The birds probably evolved from a common ancestor. They are very similar, differing mainly in the adaptations of their bills for eating different types of food.

came adapted to different habitats on the same or different islands. Such was the case with the 13 species of Darwin's Galapagos finches: the ancestral finches must have come from the South American mainland, and their descendants underwent adaptive radiation in the Galapagos, becoming adapted to living in different habitat zones and to eating different sorts of food (Figure 14-12). A fourteenth species evolved on Cocos Island, about 800 kilometres away. Local conditions on islands differ from those on the mainland, so natural selection on an island inevitably produces different adaptations.

14-C   Evolution by Means of Natural Selection

All of this evidence for the occurrence of evolution is quite convincing, but the feature of Darwin and Wallace's theory that convinces most people is the idea that natural selection produces evolution. We shall see in Chapter 15 that factors other than natural selection can also produce evolution, but natural selection is by far the most important reason that evolution occurs.

Natural selection is a simple idea. What it amounts to is that some genotypes are reproduced more frequently than others. What is not so simple is grasping how natural selection affects populations and brings about evolution. This is one area of biology where thinking about a subject will teach you more than reading about it. (The Questions for Discussion at the end of this chapter list some examples to think about.)

We start with the logical argument that natural selection occurs and brings about evolution. This may be summarized:

1. Individuals of a species vary.
2. Some variations are genetically determined.
3. More individuals are produced than live to grow up and reproduce.
4. Individuals with some genetic variations are more likely to survive and reproduce than those with others.

Conclusion: From the preceding four premises, it follows that those hereditary traits that make their owners more likely to grow up and reproduce will become more common in a population from one generation to the next.

To take an example, if part of a child's genetic variation is an inherited mutation that causes a severe liver disease, the child has much less chance of living to grow up and reproduce than somebody without this mutation. And only by reproducing does an individual pass on its inherited characteristics. If an organism does not reproduce, it plays no direct role in the evolution of future generations.

Inherited characteristics that improve an organism's chance of living and reproducing will be more common in the next generation than those that decrease its chance of reproducing. Various combinations of genes will be naturally selected for or against, from one generation to the next, depending on how they affect survival and reproduction. For natural selection to cause a change in a population from one generation to the next (that is, to cause evolution), it is not necessary that all genes affect survival and reproduction. The same result occurs if there are just some genes that make an individual more likely to grow up and reproduce.

■ *Logically, evolution is bound to occur by means of natural selection if individuals vary, if some variations are inherited, and if more individuals are born than live to grow up and reproduce.*

The Peppered Moth

A classic example of natural selection in the wild, documented by observation and experiment, is the case of the peppered moth, which lives in all parts of England. In nineteenth-century England, many people collected moths and butterflies, and collectors avidly sought rare specimens of the peppered moth that were a dark, almost black color rather than the usual pale, mottled gray. We now know that each moth's genes determine whether it is the normal gray form or the black form—called the **melanic** form after the black pigment melanin. By looking at collections made from about 1850 to 1950, biologists found that melanic moths became more and more common during that time, and gray ones scarcer, particularly near industrial cities. This change in a population of organisms over time is, in itself, evolution.

Moths fly, feed, and mate at night. During the day they rest on tree trunks or other surfaces, protected from predators by camouflage. Biologists proposed that before industrial pollution, the typical gray form of the peppered moth had been well camouflaged against tree trunks covered with pale, plant-like lichens. In polluted areas, however, where industrial smoke had killed the lichens and blackened the tree trunks, the gray form stood out in contrast to its background (Figure 14-13). Here, many more gray than melanic moths would be found and eaten by predators. The most likely predators were birds, which hunt by sight,

Figure 14-13
Different forms of the peppered moth. (a) A melanic moth is highly visible against the pale lichens of a tree trunk in an unpolluted area, whereas the normal gray form is well camouflaged. (b) Moths on a blackened tree trunk covered with soot, which has killed all the lichens. Against this background, the gray moth is much more visible than the melanic one. (Michael Tweedie/Photo Researchers)

(a)

(b)

and against whom camouflage, or lack of it, would be important. The evolution of darker populations of an animal in the presence of industrial pollution is known as **industrial melanism.**

In the 1950s, Bernard Kettlewell decided to use these moths to study natural selection experimentally. He raised large numbers of both black and gray forms of the moth in the laboratory, marked them, and released them in two places: one an unpolluted rural area where the black form was more visible, the other a polluted industrial area where the gray form was easier to see against the blackened tree trunks. Kettlewell then recaptured as many of the marked moths as he could. The percentage of melanic moths recovered was twice that of gray moths in the industrial area, but only half that of gray moths in the unpolluted countryside. This agreed with the prediction that the gray moths were more likely to survive (and so to be recaptured) in the country, and melanic moths were more likely to survive near the town.

This experiment was done with a human "predator" (the person catching the moths), but humans are not normally much of a threat to survival of the peppered moth. Does the differential camouflage work against the moths' real predators? To find out, Kettlewell hid in a blind and watched moths he had placed on tree trunks. On one occasion, he watched equal numbers of gray and black moths in an unpolluted area. Birds caught 164 of the melanic and only 26 gray moths.

In a polluted area, a larger proportion of melanic than of gray moths will live long enough to reproduce. Since the color of the moths is inherited, the next generation will contain proportionally more melanic moths. In other words, the frequency of the gene for black color increases in the population with time—and that is evolution.

The selective pressure that brings about this evolution is clear: in polluted areas birds kill a higher percentage of moths with the gene for gray color than of moths with the gene for black color. Natural selection over many generations has produced populations of the peppered moth that are well adapted to survive in their environments, populations whose characteristics change as the environment changes.

On the basis of this evidence, we would predict that if pollution were reduced, melanic moths would become rarer and gray forms more common in industrial areas. In fact, the Clean Air Act of 1952 reduced air pollution in England, and collections of peppered moths from industrial Manchester in the next 20 years revealed a dramatic decrease in the ratio of melanic to gray individuals in the moth population. The ability to predict events in this way is the most impressive evidence that can be produced for a scientific theory.

■ *The first well-documented case of natural selection causing evolution in a wild population was selection by birds for camouflaged peppered moths.*

14-D   Genetic Contribution to Future Generations

The phrase "survival of the fittest," often used in discussion of evolution, suggests that natural selection selects mainly for survival. It does not. It selects for the contribution of genes to future generations. Survival is important, in that an individual that dies young will not reproduce, but even reproduction is no guarantee of evolutionary success.

Consider Table 14-1, which shows how many young starlings survived for three months after hatching. The female starlings that seemed to be reproducing most efficiently—those laying nine or ten eggs in one brood—could actually be doomed to evolutionary failure and strongly selected against because hardly any of their young survived. Females laying four or five eggs per brood had a higher number of offspring surviving for at least three months after they hatched.

Young birds from the larger broods weigh less than those from the smaller broods, presumably because the parents could provide adequate food for no

Table 14-1	Survival in Swiss Starlings in Relation to Number of Eggs Laid*	
Brood size (number of eggs in nest)	**Number of young marked**	**Recoveries per 100 birds marked†**
1	65	0
2	328	1.8
3	1278	2.0
4	3956	2.1
5	6175	2.1
6	3156	1.7
7	651	1.5
8	120	0.8
9, 10	28	0

* The number of eggs laid during one nesting period is genetically regulated and, like other genetic variations, is acted upon by natural selection. David Lack marked all the nestlings in all the nests he could find, and then recaptured them months later when they had left the nest.

† The only recoveries scored are those for birds over three months old when they were recaptured.

Source: Lack, D. *Ecology* 2, 1948

more than five or six nestlings. A shortage of food was probably a major cause of death of young from larger broods. Table 14-1 also shows that the most common brood sizes produce the nestlings with the lowest mortality rates, as we would predict from the action of natural selection. It seems reasonable to suppose that in years when there is more (or less) food available to the birds than in the year studied in this example, selection would favor birds with broods larger (or smaller) than the average. This accounts for the fact that the population contains birds producing broods larger and smaller than the average brood: these genes persist because in some years they are favored.

Plainly, the reproductive success of a starling is not fully told by the story of one brood. Selection optimizes reproductive success over a lifetime, and the adaptations that produce this success are many.

14-E Adaptations

Selection ensures that the members of future generations are the descendants of reproductively successful members of the present generation and, therefore, well adapted to their environment. "Environment" in this evolutionary context encompasses all of the factors that can affect whether or not an organism lives to reproduce.

For example, let us consider an acorn. Whether it successfully resists the selective pressure of its environment depends on the speed and normality of its germination and development, whether bacteria or fungi infect it as a seed or seedling and destroy it at this stage, whether as a seedling it has enough stored food for rapid growth, whether it escapes being eaten, whether the soil in which it grows can support a large plant, and whether the young tree avoids death by disease, trampling, or browsing. The genome of a successful oak will contain genes that adapt it to withstand all these selective pressures (Figure 14-14).

Resistance to Pesticides and Antibiotics

Several dramatic examples of natural selection in action today, and the adaptations it produces, are provided by the evolution of resistance to pesticides and antibiotics.

Figure 14-14
An oak tree. This individual has withstood all selective pressures for many years. (Matthew Gilligan)

A scale insect feeds on citrus trees in California. In the early 1900s, growers sprayed the trees with cyanide gas, and this killed the scale. But in 1914 some of the insects survived the spraying. The cyanide did not kill them because they possessed a single gene, newly apparent in the population, that permitted them to break cyanide down into harmless compounds. As spraying continued, more insects with the new gene than without it survived to reproduce, and they passed on the gene to their offspring. The frequency of the new gene in the population increased until the whole population was resistant to the spray. Because scale insects, like many other insects, have more than one generation a year, they evolve quickly. Resistance to pesticides is a very expensive problem for agriculture. To combat the evolution of resistance, growers are encouraged to spray pesticides on their crops only when necessary and to use different chemicals in different months and years.

Precisely the same thing happens with antibiotics used to kill bacteria that cause human disease. When a bacterial population meets a particular drug, bacteria susceptible to that drug are killed (Figure 14-15). Sometimes a population contains one or a few individuals with mutations that confer resistance to the drug; they will survive, and they multiply rapidly once competing bacteria have died. In addition, many genes conferring resistance to antibiotics are now known to be carried in plasmids (*A Journey Into Life,* Chapter 9), which can be duplicated and passed to other members of the population that previously lacked genetic resistance. Soon these genes become widespread. Since antibiotics and disease-causing bacteria frequently meet in hospitals, it is not surprising that some hospitals harbor drug-resistant bacteria. In many countries, women are now encouraged to give birth at home whenever possible, because mother and infant are safer from bacterial infection at home than in the hospital.

Most countries have outlawed the use of antibiotics in cattle feed. Cattle fatten faster if fed antibiotics, but they also become breeding grounds for antibiotic-resistant bacteria. Antibiotics are still added to cattle feed in the United States, and drug-resistant bacteria in cattle are becoming increasingly common.

Disease-causing organisms don't have things all their own way. In 1915, nearly all the oysters in the Malpeque Bay of Prince Edward Island, Canada, were killed by a disease. However, a few oysters survived, and began to re-establish the population. Fifteen years later, the disease-causing organism was still present, but most of the oysters now had genetic resistance to it; only one oyster in 1000 was susceptible. By 1938, the oyster harvest was higher than it had been before the disease struck, and when the disease appeared elsewhere, oysters from Malpeque Bay were sent to contribute their genetic resistance to the newly afflicted populations.

These examples of the evolution of adaptations illustrate merely a few of the less obvious selective pressures that are always acting on all organisms and the adaptations that have evolved in response to them, adaptations that appear ingenious but are really the result of natural selection among randomly produced variations of genes.

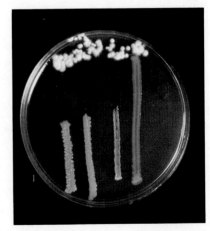

Figure 14-15
Resistance to antibiotics in bacteria. The fuzzy dots across the top of this dish are clumps of the fungus *Penicillium,* which produces the antibiotic penicillin. The penicillin spreads out through the dish. The four lines are rows of different varieties of bacteria. Three of the bacterial varieties have been killed as the penicillin reached them; the fourth (far right) is penicillin-resistant, and it continues to grow. (Biophoto Associates)

Coevolution of Plants and Herbivores

A JOURNEY INTO LIFE

Coevolution is the evolutionary change that occurs in two or more different species when they act as selective pressures on one another. This process provides some fascinating examples of evolution in action.

Acacia trees grow in tropical areas throughout the world, and many species have spines that protect them against herbivorous (plant-eating) mammals. Some Central American acacias have a more unusual protection: a mutually beneficial relationship with the ants that live on them. Ant acacias have several structures that benefit the ants. These include hollow thorns in which the ants live, glands on the leaves that produce sugary nectar, and Beltian bodies, swollen, nutrient-rich leaf tips, which the ants cut off and feed to the larvae (Figure 14-A).

It is advantageous for the acacia tree to host an ant colony because the ants reduce the damage done to the tree by other herbivores. The ants react to anything that touches the tree. They remove dust, fungal spores, pollen grains, and spider webs. They destroy the seedlings of other plants that sprout under their tree, and sting other insects or mammals that try to eat the tree (Figure 14-B). When its ants are removed, the tree usually dies after a few months. Fungi invade it, and it grows more slowly and becomes choked with vines.

A few species of insects have evolved defenses that permit them to survive on an acacia tree guarded by ants. Some seem immune to ant stings and ignore the ants. Others can pick up the ants and throw them off the tree. Still others have hard cuticles that an ant's sting cannot penetrate.

The coevolution of ant acacias and their insect populations probably went something like this: ants invaded an acacia and fed on the leaf parts and nectar of the tree. The ants also removed other insects, allowing the acacia to grow faster, and making it able to shade out other plants and to produce more offspring. Acacias that were more attractive to ants reproduced more rapidly. The availability of food and shelter, in turn, exerted selection on the ants to protect the tree with increasing efficiency. Of the insects that fed on the tree before the ants arrived, most species were expelled, but a few evolved defenses against ant attacks.

The leaves of acacia species that are not defended by ants contain cyanides and other chemicals toxic to herbivores. These seem to have been lost in the ant acacias, presumably because in the presence of ants the tree need no longer expend the energy to produce toxic chemicals. There is selection for the tree to use the energy to produce ant food or more offspring, instead.

We can find examples of coevolution much closer to home. Cooked cabbage, broccoli, or mustard greens give off the distinctive odor of mustard oils—the group of toxic chemicals characteristic of the crucifer plant family. In the intact plant, mustard oils are usually bonded to sugar molecules. This makes them much less toxic and allows them to be stored in the plant without damaging the tissues they are defending. When a cell is damaged—as it is when an insect bites into it—an enzyme cleaves off the sugar molecule. This is analogous to pulling the pin on a hand grenade, and the mustard oil is released.

How toxic are mustard oils? They are plainly not very poisonous to humans or to the cabbage white butterfly caterpillars that sometimes wipe out entire plantings of crucifers in the home garden. The main difficulty in answering this question is

Figure 14-A
Ant and plant. An acacia ant feeds on the yellowish-brown Beltian bodies on an acacia plant. A nectar gland is also visible on the stem below the ant. (Paul Feeny)

that insects usually do not eat anything except their normal food plants. You cannot take a caterpillar from an oak tree and plunk it on a cabbage leaf to see if it will be poisoned, because the caterpillar will not eat the cabbage. In an experiment to get around this problem, black swallowtail butterfly larvae, which normally feed on plants of the carrot family, were raised on some rather special carrot leaves. These leaves were cultured in solutions containing various concentrations of mustard oils. The larvae were therefore feeding on their usual carrot diet, plus compounds from a family of plants that the larvae do not normally eat.

At mustard oil concentrations that occur naturally in crucifer plants, the larvae lost a lot of fluid in their feces and soon died. Clearly, then, mustard oils can be an effective defense against insects that do not normally attack the plants that contain them. These compounds are also toxic to various fungi and bacteria. The selective pressure for mustard oils in crucifers seems to be the protection they give the plants against several natural enemies.

However, insects that are not poisoned by mustard oils can eat crucifers, and they have evolved the ability to detect mustard oils and thereby find the plants. For example, some flea beetles can home in on the odor of their crucifer food plants in a field containing many other crops. Another insect that moves toward the scent of mustard oil is a wasp that does not eat crucifers but is a parasite of an aphid that does. Moving toward mustard oils permits the wasp to find the aphid. Experiments have shown that the aphid escapes the wasp when it is not on a crucifer, for instance when it lives on a beet plant. When it is on a crucifer, or on a beet plant smeared with mustard oil, it is attacked by the parasitic wasp.

These and other experiments have shown that the same compound in a plant can act as both an attractant and a repellent. Thus mustard oils are feeding stimulants to flea beetles, and are egg-laying stimulants to the cabbage white butterfly, who lays her eggs on crucifers. Both are attracted by the mustard oils that crucifers produce. On the other hand, mustard oils repel those herbivores that cannot eat crucifers efficiently.

Other families of plants have their own, different kinds of defensive chemicals and retinues of insect species adapted to cope with them. Examples are carrot and parsley (umbellifer family); onion, leek, and garlic (lily family); and the mint family.

These examples show that plants and the animals that eat them influence each other's evolution. Furthermore, the selective pressure exerted by one species may result in a variety of different adaptations in other species.

(a)

(b)

Figure 14-B
Ants defending the plants they live on. (a) The ground around this young ant-defended acacia in Costa Rica is kept bare by the ants, thus protecting the tree from competing plants and from fires in the dry season. (b) This armyworm caterpillar, placed on an ant acacia, was stung to death in minutes by the resident ants. Also seen here are the hollow swollen thorns in which acacia ants make their nests. (Paul Feeny)

SUMMARY

The theory of evolution states that new types of organisms arise by descent and modification from pre-existing forms. Members of any species of organism differ from one another, and some of their differences are inherited. Natural selection is the differential reproduction of genetically different individuals. It leads to evolution, a change in the proportions of genes in a population from one generation to the next.

The theory of evolution by natural selection was put forward in 1858 by Darwin and Wallace. Their thinking was stimulated by the writings of Lyell and Malthus and by observations they made during their own travels.

Biologists realized that artificial selection by farmers and breeders had produced rapid evolution in domesticated plants and animals. They also recognized the compelling logical arguments that evolution had occurred from observations of the fossil record and biogeography, which provided information on the distribution of organisms through time and space. Comparative anatomy and embryology provided evidence that various structures in ancestral organisms had been modified in their descendants and had become adapted to different functions, or had even been lost when a new way of life rendered them unnecessary. The evolution of wild populations by means of natural selection was not convincingly shown until the twentieth century, when predation by birds was shown to be the selective pressure that led to the evolution of dark color in populations of the peppered moth in polluted areas of England.

The anatomical, behavioral, and physiological traits that survive natural selection may be thought of as adaptations that fit an organism to its particular environment. Adaptations are many and various. The only consistent result of selection is that it maximizes the genetic contribution of a "successful" individual to future generations.

SELF-QUIZ

1. In light of the definition of evolution, which of the following is *not* capable of evolving?
   a. a population of deer
   b. the color of a population of moths
   c. your biology teacher
   d. a population of chickadees
   e. the millions of bacteria in your large intestine
2. Which of the following did Kettlewell conclude from his studies on industrial melanism in moths?
   a. a black moth lays more eggs than a gray moth in industrial areas
   b. black moths are more resistant to pollution than are gray moths
   c. pollution caused some moths to become darker than others
   d. black moths are more likely to survive in polluted areas than are gray moths
   e. birds prefer the taste of black moths over gray moths
3. Which bird is most evolutionarily successful?
   a. lays 9 eggs, 8 hatch and 2 reproduce
   b. lays 2 eggs, 2 hatch and 2 reproduce
   c. lays 5 eggs, 5 hatch and 3 reproduce
   d. lays 9 eggs, 9 hatch and 2 reproduce
   e. lays 7 eggs, 5 hatch and 4 reproduce
4. Suppose that you have a pack of 50 assorted dogs. You select the largest male and the largest female, mate them, and sterilize the other dogs. Assuming that food supplies remain adequate, you should expect that, in the next generation:

   a. the young dogs will be, on the average, larger than their two parents
   b. the young dogs will be, on the average, larger than the older members of the pack
   c. the young dogs will be the same average size as the older dogs
   d. all of the young dogs will be larger than the older dogs
5. Explain how Darwin would have accounted for the evolution of the long necks of giraffes.
6. Penicillin and other antibiotics were introduced in the 1940s and were effective in combatting infections caused by *Staphylococcus* bacteria. In 1958, however, there were several outbreaks of *Staphylococcus* infection. People with the infections did not respond to treatment with any antibiotic, and many people died. The most likely explanation for this situation is:
   a. the bacteria reproduced in hosts that were not contaminated by antibiotics
   b. bacteria from other animals (such as deer, birds, and cats) migrated into human hosts
   c. the bacteria exposed to nonlethal doses of antibiotics quickly learned to avoid them
   d. each generation of bacteria acquired the ability to use the antibiotics as nutrients
   e. bacteria containing a gene for antibiotic resistance survived and multiplied and these were the forms causing the lethal infections

QUESTIONS FOR DISCUSSION

For Questions 1 to 5, consider Table 14-1.

1. From what brood size do the greatest number of young survive?
2. Is this also the most frequent brood size? (Assume that the experimenter marked every bird that could be found.)
3. What do you suppose is the disadvantage to a starling of laying a smaller than average clutch of eggs?
4. Suppose the environment changed so that only half as much food was available to the starlings. Would you expect a gradual change in the most frequent brood size? How would this change be brought about?
5. Which female starlings will leave more young per head in the population and hence make the greatest contribution to the genes of the next generation?
6. Are all causes of death natural selection? For example, when organisms die in an earthquake, have they been selected against?
7. The embryologist Charles H. Waddington treated fly larvae with heat shock. As a result of this treatment, some of the adult flies showed the abnormal condition "crossveinless" (some of their wing veins were missing). After many generations of this treatment, he let a generation of flies develop without heat treatment and many of them were also crossveinless. Does this experiment provide convincing proof of Lamarckism? If not, what other explanation can you suggest, and what experiments would you perform to test your suggestion?
8. Is human evolution subject to the same pressures as the evolution of other species? Why or why not?
9. Is there any time in its life history when an organism is not subject to selective pressure? Are gametes subject to selective pressure? Are eggs? Embryos? Is there selective pressure on young animals that are fed and protected by their parents?
10. Some insects lay eggs on more than one species of larval food plant. There is some evidence that a female is more likely to lay her eggs on the plant species on which she grew as a larva than on any other kind of plant. Is this an example of Lamarckian inheritance? Why?
11. Scientists are beginning to breed crop plants to have built-in chemical defenses against insect pests. In your opinion, how well will this work?
12. What is the adaptive advantage to a plant of a contact irritant (such as the oil on poison ivy leaves that makes a rash on the skin of passing animals)?

SUGGESTED READINGS

Bishop, J. A., and L. M. Cook. "Moths, melanism, and clean air." *Scientific American,* January 1975. The peppered moth story and how the moth can be used to monitor air pollution.

Cook, L. M., G. S. Mani, and M. E. Varley. "Postindustrial melanism in the peppered moth." *Science* 231:611, 1986. A follow-up to the peppered moth story.

Darwin, C. *The Origin of Species by Means of Natural Selection.* New York: The Modern Library, Random House, Inc., 1982. A reprint of the 1859 first edition in one volume together with the sequel, *The Descent of Man.*

Mayr, E. "Darwin and natural selection." *American Scientist* 65:321, 1977. An eminent geneticist's discussion of the logical argument for evolution by natural selection.

Nelkin, D. *The Creation Controversy: Science or Scripture in the Schools.* New York: W. W. Norton, 1982. The controversy between creationism and evolution. Nelkin examines the tactics by which creationists attempt to impose their views on educational systems. She contends that the battle is political (rather than scientific or religious). Judge Overton's decision in the Arkansas case is discussed and reprinted.

Scientific American, September 1978 issue, *Evolution.*

Stebbins, G. L., and F. J. Ayala. "The evolution of Darwinism." *Scientific American,* July 1985. Two eminent geneticists discuss how views of evolution have changed since Darwin's day.

Stone, I. *Origin.* New York: Plume/New American Library, 1981. A very readable historical novel based on the life of Charles Darwin.

Population Genetics and Speciation

Green turtle

Daffodils

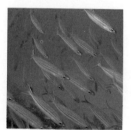

Boga

*I*n Chapter 14 we saw that populations (not individual organisms) are the units that evolve. This is because evolution involves changes in the mix of genes present in entire populations of organisms. The difference between a population of ancient dinosaurs and a population of their descendants, modern birds, lies in the different genes of members of the two populations.

A **population** consists of all the members of a species that occupy a particular area at the same time—for example, the perch population of a lake, the dandelion population of a hillside, or the penguin population of an island (Figure 15-1). The members of a population are much more likely to breed with one another than with members of other populations of the same species. Therefore, populations form breeding groups, and genes tend to stay within the same population for generation after generation.

All the genes in all the members of a population are collectively called the population's **gene pool.** According to one definition, evolution is the change in the frequency of genes in a population's gene pool from one generation to the next. So, if we can discover how a population's gene pool changes with time, we shall understand how evolution occurs.

Different populations of the same species sometimes become isolated from one another. When this happens, one of the most important processes in evolution may occur: the formation of a new species, whose members, by definition, do not exchange genes with members of other species.

15-A   *The Hardy-Weinberg Law*

One way to see how a population evolves is to construct a model of a population that does not change genetically from one generation to the next, and then see how a real population differs from this model. The **Hardy-Weinberg Law** pro-

Figure 15-1
A population. These Adelie penguins live on an island in the Antarctic. (Robert W. Hernandez, Science Source/Photo Researchers)

The Hardy-Weinberg Law

The Hardy-Weinberg Law states that the frequencies of alleles *A* and *a* will remain the same from generation to generation in a population if the conditions listed in Section 15-A are met. To satisfy ourselves that this is true:

Let p = the frequency of allele *A* in the population (the proportion of all alleles that are *A*).

Let q = the frequency of allele *a*.

Since all chromosome locations for this gene in all members of the population must be occupied by either the *A* allele or the *a* allele, $p + q = 1$. Thus if *a* occurs at 20% of the chromosome locations (q = frequency of a = 0.2), the other 80% must be occupied by *A* (p = frequency of *A* = 0.8), and the two frequencies together equal one ($0.2 + 0.8 = 1$).

Now let us see what happens during reproduction. The frequencies of *A* and *a* alleles in the gametes produced by this population are the same as the frequencies of the alleles in the population. *AA* homozygotes, of either sex, produce only *A* gametes; *aa* homozygotes produce only *a* gametes; *Aa* heterozygotes produce equal numbers of both types of gamete. What are the chances that an *A* sperm will fertilize an *A* egg? Since the frequency of *A* gametes is *p,* the frequency at which *A* sperm fertilize *A* eggs is $p \times p = p^2$. Similarly, the frequency of $a \times a$ fertilizations = q^2. The frequency of fertilizations between *A* sperm and *a* eggs = $(p \times q)$, and the frequency of fertilization of *a* sperm and *A* eggs = $(q \times p)$. Adding all this up, we get the expected frequencies of the three genotypes in the next generation:

$$
\begin{array}{ll}
AA & p^2 \\
Aa & 2pq \\
aa & q^2
\end{array}
$$

Because the frequencies of the three genotypes must add up to 1, $p^2 + 2pq + q^2 = 1$.

We could have done our calculations using algebra:

Allele frequencies	$p + q = 1$
Squaring both sides of the equation gives	$(p + q)^2 = (1)^2$
Genotype frequencies	$p^2 + 2pq + q^2 = 1$

Why do we square $p + q$ to find the genotype frequencies? Since p is the proportion of *A* alleles in the population, it is also the proportion of *A* gametes produced by each sex. Likewise, q is the proportion of *a* gametes. By arranging these on a Punnett square, we obtain the genotype frequencies:

	pA	qa
pA	p^2 *AA*	pq *Aa*
qa	pq *Aa*	q^2 *aa*

To see how this might work in practice, we will work through a simple example. Suppose we have a population in which the frequency of allele *A* = 0.6 = p, and the frequency of allele *a* = 0.4 = q. Note that $p + q = 0.6 + 0.4 = 1$. The frequencies of the three genotypes in the next generation can therefore be calculated as follows:

Frequency of *AA* = p^2 =	0.6×0.6	= 0.36
Frequency of *Aa* = $2pq$ =	$2 \times 0.6 \times 0.4$	= 0.48
Frequency of *aa* = q^2 =	0.4×0.4	= 0.16
Total: $p^2 + 2pq + q^2$		= 1.00

Now let us check the Hardy-Weinberg prediction that the frequencies of alleles *A* and *a* are still p and q, respectively. The frequency of *A* is p^2 (from the *AA* homozygotes) + $\frac{1}{2}(2pq)$ (from the heterozygotes; half of their alleles are *A* and half are *a*). So the frequency of *A* is now:

$$p^2 + \tfrac{1}{2}(2pq) = p^2 + pq$$

vides such a model. It shows that random mating in a large population of diploid organisms that is not undergoing mutation or natural selection cannot produce evolution. (The Hardy-Weinberg Law is described in more detail in the box above.)

Consider a population whose gene pool contains two alternate alleles, *A* and *a,* either of which can occupy one particular location on a chromosome. Every member of the population has one of three possible genotypes: *AA, Aa,* or *aa.* The Hardy-Weinberg Law shows that the next generation will contain the two alleles in the same frequencies (their proportions relative to each other in the population's gene pool). This will remain true through all successive generations, if the population meets *all of the following conditions:*

1. **No net mutation.** The alleles in question must not mutate. (If they do, they become other alleles, and so their frequencies automatically change.)
2. **No selection pressure.** There must be no natural selection with respect to the alleles in question (no genotype has a reproductive advantage over the others).

Since $q = 1 - p$, from our earlier definition, we can substitute $(1 - p)$ for q in this expression, and we get:

$$p^2 + p(1 - p) = p^2 + p - p^2 = p$$

So the frequency of A is still p.

A similar calculation shows that the frequency of the a allele in this new generation must still be q. In other words, as predicted by the Hardy-Weinberg Law, the frequencies of A and a have not changed. Hence both the allele and genotype frequencies will stay the same in this population, generation after generation, as long as the population meets Hardy-Weinberg conditions.

The Hardy-Weinberg Law has an interesting implication for lethal alleles (alleles that kill individuals who bear them). A lethal *dominant* allele expressed before reproductive age is removed from the population every time it arises by mutation, since it causes the individual's death. The allele will arise only by mutation and will therefore be extremely rare. On the other hand, a lethal *recessive* allele is expressed only in homozygous recessive individuals, who will always die. But the Hardy-Weinberg equation shows that many more heterozygous ($2pq$) than homozygous recessive (q^2) individuals will occur in each generation. However rare the homozygous recessives become, there will still be many heterozygous carriers in the population. This means that a harmful recessive allele, even one so unfavorable as to be lethal, will hardly ever be completely eliminated from a large population (Figure 15-A).

The frequency of genotype aa is q^2, and so the frequency of the a allele (q) in the population is the square root of the proportion of homozygous recessive individuals. The number of heterozygous carriers of allele a is $2pq$. (We find p by subtracting the value of q from 1.0, and then multiply to find $2pq$.)

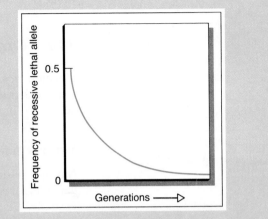

Figure 15-A
The frequency of a recessive lethal allele in a population. Even if we start with a population in which every individual is heterozygous for the allele (that is, the allele has a frequency of 0.5), the frequency will drop rapidly in succeeding generations. This is because many copies of the allele will be eliminated by selection, as all homozygous recessive individuals die without having a chance to reproduce and pass the allele to future generations. However, heterozygous carriers of the allele will persist in the population indefinitely. They become so rare that matings between two carriers are extremely infrequent, and therefore few homozygous recessive individuals will be produced.

For example, if 1% of the population is homozygous recessive for a certain trait, the frequency of $aa = 0.01$, and q is the square root of this, or 0.1. Therefore, p is $1 - 0.1 = 0.9$, and the proportion of heterozygote carriers of allele a is $2pq = 2 \times 0.9 \times 0.1 = 0.18$.

3. **No mating preferences.** The population must reproduce sexually, and mating must be random with respect to genotype (so that, for instance, an AA female does not prefer aa to AA or Aa males when she mates).
4. **Isolation.** There must be no exchange of genes (gene flow) between the population and any other population.
5. **Large size.** The population must be very large, because the law is based on statistical probabilities. Random sampling errors are more likely to occur in small populations.

If all these conditions are fulfilled, A and a will remain in the population indefinitely at the same frequencies and there will be no evolution.

The Hardy-Weinberg Law points out the fact that sexual reproduction, with its reshuffling of genes, is not by itself enough to cause evolution. Evolution is a change in allele frequencies from one generation to the next, and under the conditions of the Hardy-Weinberg Law there is no change.

The model has a useful application. Since it gives the conditions under which evolution will not occur, it implies that if these conditions are not met,

■ *According to the Hardy-Weinberg law, evolution will not occur in a large, genetically isolated population with no mutation, no selection pressures, and no mating preference.*

evolution is likely to occur. In other words, evolution is likely to occur if natural selection occurs, if mating is not random, if there is gene flow between populations, or if the population is small.

15-B   Causes of Evolution

Mutation violates the Hardy-Weinberg conditions because it changes one gene into another and therefore alters the frequency of a gene in the next generation. Let us see how the other factors produce evolution.

Natural Selection

Natural selection is the nonrandom differential survival and reproduction of genotypes from one generation to the next. Lethal recessive alleles provide one example. Whenever a lethal recessive allele occurs homozygously, the affected individual dies before it can reproduce, and therefore those particular copies of the allele do not have a chance to be passed on to the next generation. This violates condition 2 for the Hardy-Weinberg Law. Homozygous dominant individuals and those heterozygous for a lethal allele have a reproductive advantage over homozygous recessives, which are said to be **selected against.** Hence the frequency of the lethal recessive allele decreases over the generations (see Figure 15-A on page 269).

Individuals in a population show a range of phenotypes. When an agent of natural selection, such as predation or competition, is at work, some of these phenotypes are more likely to survive and reproduce than others. Usually the surviving traits are at least partly under genetic control. Hence the genes responsible for these phenotypes will be more common in the next generation. These genes are said to have been **selected for,** which means the same as "not selected against." In the simplest case of a single pair of alleles, if possession of one allele confers even a slight reproductive advantage, its frequency in the population will increase from one generation to the next, at the expense of the less favorable alternate allele. The more favorable allele is said to have greater **fitness** than the less favorable allele. In this way, selection can change the gene frequencies in a population from one generation to the next, causing evolution. Natural selection is by far the most important and potent cause of evolution, and its action has been emphasized throughout this book.

Two common forms of natural selection are stabilizing selection and directional selection. **Stabilizing selection** exists when average phenotypes have a selective advantage over extremes in either direction. It is exceedingly common in nature. For instance, several studies have shown that babies of average birth weight (between 7 and 8 pounds) have a much higher chance of surviving to the age of 5 than babies with weights significantly above or below the average.

Directional selection occurs when the phenotypes at one extreme have a selective advantage over those at the other. Figure 15-2 shows the actions of these kinds of selection, using a population with a normal (bell-shaped) distribution of phenotypes from a polygenic trait—a trait controlled by many genes, such as human height or the weight of seeds.

Let us consider an example. If, in a population of seeds, seeds of average size have a better chance of germinating and of growing than seeds that are unusually large or small, and if seed size is inherited, stabilizing selection is acting and the next generation will contain a lower proportion of unusually large or small seeds. On the other hand, if birds tend to eat large seeds and ignore small ones, they will exert directional selection in favor of small seeds.

Disruptive selection takes place when the extremes of a range of phenotypes are favored relative to intermediate phenotypes (Figure 15-2). It might

■ *The most important cause of evolution is natural selection, the differential reproduction of genotypes.*

DISTRIBUTION OF WEIGHTS IN THE POPULATION BEFORE SELECTION

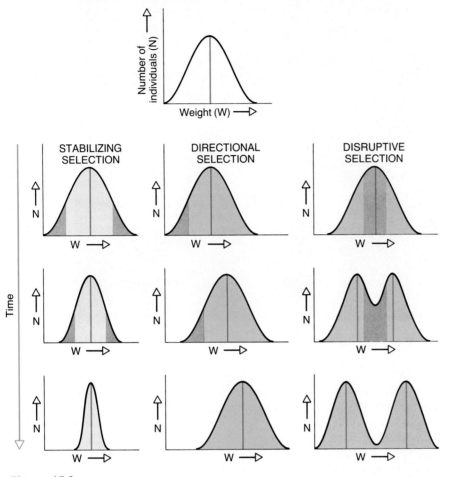

Figure 15-2

Effect of different types of selection on a polygenic character. The population has a normal distribution of individuals of different weights (shown by the bell-shaped curves). The red lines show the median weight (the value that falls between the lowest and the highest 50% of measurements.) The individuals eliminated by selection are shown in gray. Stabilizing selection eliminates very heavy and very light individuals, leading to a population with less weight variation. Directional selection is shown eliminating lightweight individuals, leading to a population with a higher median weight. Disruptive selection eliminates individuals of median weight, producing two populations with different median weights.

happen to our seeds, for example, if a particular kind of beetle specialized in feeding only on seeds of intermediate size, ignoring the very small and very large seeds.

Mating Preferences

Mating that is not random with respect to genotype can also bring about evolutionary change. If females consistently choose to mate with males with certain genetic traits, they exert selection in favor of the alleles for those traits. Such nonrandom mating can have bizarre results. For instance, over the centuries peahens have preferred to mate with peacocks that produce brilliant displays

Figure 15-3
Courtship display of a peacock. The peacock's tail is the product of mate selection. The tail reduces the male's chance of surviving because it is heavy and awkward, and has taken much protein and energy to produce. But females prefer to mate with males who display fabulous tails during courtship, and so the genes for producing this remarkable decoration survive. (Biophoto Associates)

Figure 15-4
A cloud of yellow pollen blows away from a fir tree. Gene flow between populations of plants may occur when some pollen grains are blown away and carry their genes from one population to the next. However, most pollen grains will fertilize plants within their own population.

■ *Gene flow between popula-tions may change gene fre-quencies and, therefore, may result in evolution.*

with their tails (Figure 15-3). This has selected for ever larger and more colorful tails. Such mating preference is really only a form of natural selection because it gives one genotype a reproductive advantage over another. However, the agent of selection is different from the kind of selective pressure we usually think of. This points up what is probably a universal situation: a population's gene pool represents a balance between opposing selective forces. In this case, female choice favors males with large gaudy tails, whereas predation tends to eliminate such males. The colorful tail makes the male more conspicuous to predators, and the tail's size doubtless hampers his attempts to escape. The outcome is stabilizing selection held in balance by two opposing selective agents, female preference and predation.

Gene Flow

The gene pools of most populations of the same species exchange genes, result-ing in **gene flow** between the populations. Animals may leave one area and contribute their genes to the gene pool of a neighboring population, or a high wind may disperse plant seeds or pollen far beyond the bounds of the local population (Figure 15-4). Gene flow between populations is generally second only to selection as a cause of evolution in local populations. Gene flow between populations tends to increase their similarity. Natural selection has the opposite effect: it tends to make every population uniquely specialized for its particular habitat. One possible outcome of these two conflicting forces is a gradient of variation from one population to the next (Figure 15-5). The closer together two populations are, the more genetically similar they are likely to be.

Genetic Drift

Evolution can occur simply by chance: random events may bring death or par-enthood to some individuals regardless of their genetic makeup. The resulting random change in the gene pool is called **genetic drift** (Figure 15-6).

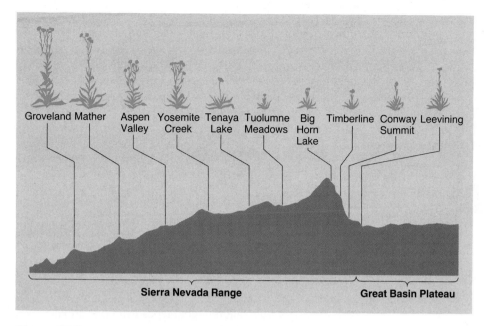

Figure 15-5
Genetic variation between populations. These yarrow plants *(Achillea)* were collected from different populations in California and Nevada and then grown under uniform conditions to reveal genetic differences between them. The plants show gradients of differences from their neighbors.

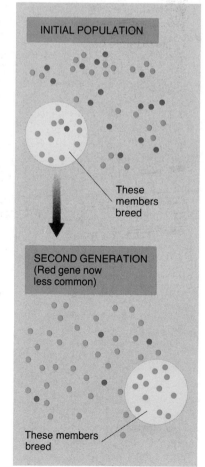

Genetic drift is much more important in a small population than in a large one. Consider a population of five individuals, in which only two breed. The chances are quite good that any particular allele is represented in only one member of the population. If this individual does not breed, the allele will not be present in the next generation. If the individual does breed, the frequency of the allele may increase in the next generation. In either case, since a change in gene frequency from one generation to the next has taken place, evolution has occurred. By contrast, if the population is large—say we multiply each of these numbers by 1000—then the 2000 who breed are likely to include about 400 who contain the allele. This is the same proportion as in the general population, and so evolution by genetic drift is much less likely in a large population than in a small one.

An interesting case of genetic drift is the **founder effect,** in which the ancestors of a new population do not contain all the genes present in the old one. For instance, when a few individuals leave a large population and colonize a new area, the chances are good that the founders of the new population do not carry a representative sample of all the genes in the old population. The gene pools of the old and new populations will be different. The founder effect is thought to play an important role in the evolution of island species (Figure

Figure 15-6
A model of genetic drift. The dots represent individuals. Red individuals carry a rare allele. Suppose that in any one year only 25% of the population breeds—those in the yellow circle. By chance, only one individual carrying the red allele breeds this year, and so the allele is much rarer in the next generation. The breeding population in this second generation is quite likely not to contain any individuals with the rare allele, in which case this allele will disappear from the population.

Figure 15-7
Island species. The genes carried by founders of island populations are not likely to be a representative sample of the genes in the mainland population. This lava heron is a member of a species found only in the Galapagos Islands.

15-7). It may also come into play when a population is reduced to very few individuals, who then become the ancestors of a later, larger population (see *A Journey Into Life* on cheetahs, this chapter).

A new population's environment will inevitably be somewhat different from the one its founders left, so the new population will experience different selective pressures, and therefore evolve in a new direction. In practice, it is usually impossible to tell how much of the genetic difference between the old and new populations results from the founder effect and how much results from different selective pressures in the two environments. The founder effect will have a great influence on a population of plants that populate an island from a single seed, or animals such as domestic hamsters, most of which have descended from one original pregnant wild female. (Recently, new hamsters have been introduced from the wild.)

15-C What Promotes and Maintains Variability in Populations?

Evolution can occur only in a population that contains genetic variation, in which at least some members have different genotypes from one another. Why do the members of a population not become more and more alike genetically,

■ *Evolution can occur by the chance events of genetic drift. This is most significant in a small population.*

Table 15-1 Factors that Increase and Decrease Genetic Variation in a Natural Population

	Factor	Effect
Increasing variation	Mutation	Introduces variation
	Sexual reproduction	Genetic reassortment occurs at gamete formation and at fertilization
	Polymorphism, disruptive natural selection, and heterozygote superiority	Retain more than one genetic form of a character in the population
	Immigration and outbreeding	May introduce new genes or gene combinations
	Increased population size	Occurs when selective pressures are relaxed; hence more variants survive in the breeding population
	Geographic variation	Adaptation to several different habitats increases variation
Decreasing variation	Natural selection (both stabilizing and directional)	Limits number of genotypes passed on to the gene pool of the next generation.
	Inbreeding	Reduces number of heterozygotes
	Emigration	May remove genotypes from gene pool
	Decreased population size	Usually due to increased selection so there is less variation in breeding population. (Also, loss of variation by genetic drift is more likely in small populations.)

as less fit alleles are eliminated by selection? The answer must be that factors promoting genetic variation are constantly at work in natural populations (Table 15-1). Mutation is one obvious source of genetic variation. But mutation is such a rare event that a small population may survive for many generations without mutations occurring. There must be additional sources of variation.

Genetic polymorphism (poly = many; morph = form) is the occurrence in a population of two or more genetic variants in such proportions that the rarest of them cannot be due to mutation alone. Polymorphism may occur for a time while one new, favorable allele (from mutation or immigration) steadily replaces other alleles in a population. More commonly, polymorphism persists indefinitely. This can result from the opposing effects of selection and gene flow. Or it can result when different genotypes are favored at different times because of changes in the environment.

Sex is probably the most widespread polymorphism. Most individuals are either male or female, and both forms are almost always present in the population at the same time. The human ABO blood group is a polymorphism involving three main alleles (see Table 13-1). The proportions of each blood type vary widely in different human populations. For instance, the frequency of the B allele is highest in central Eurasia—India, Mongolia, and western Siberia—and generally decreases with increasing distance from these areas (Figure 15-8).

To explain the distribution of ABO blood groups, we might suggest gene flow as populations mix, a gradient of selective pressures, or perhaps a combination of these factors. There is some evidence that people of blood type A are more susceptible to smallpox. This fits in with the finding that in India, where smallpox was long endemic, only about 27% of the people contain the A allele, compared with 46% in England and 48% in Germany. Smallpox was officially

■ *Populations do not become genetically homogeneous as they evolve adaptations to specific environments. Genetic variation is maintained.*

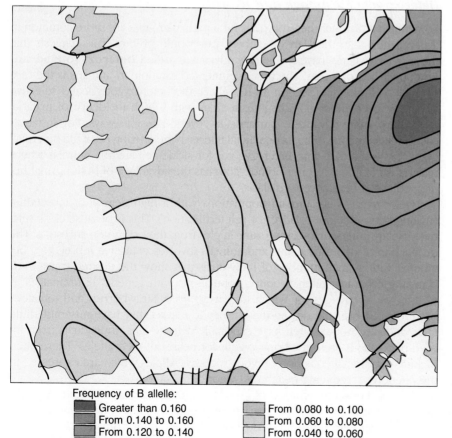

Frequency of B allele:

■ Greater than 0.160	▨ From 0.080 to 0.100
▨ From 0.140 to 0.160	▨ From 0.060 to 0.080
▨ From 0.120 to 0.140	□ From 0.040 to 0.060
▨ From 0.100 to 0.120	

Figure 15-8
Frequency of the B allele of the human ABO blood group system in various European populations. In general, the frequency of the allele decreases progressively from east to west. Populations in central Asia and India, to the east of the area shown, have especially high frequencies of this allele. (After Schreiber, D. E., and R. Matessi)

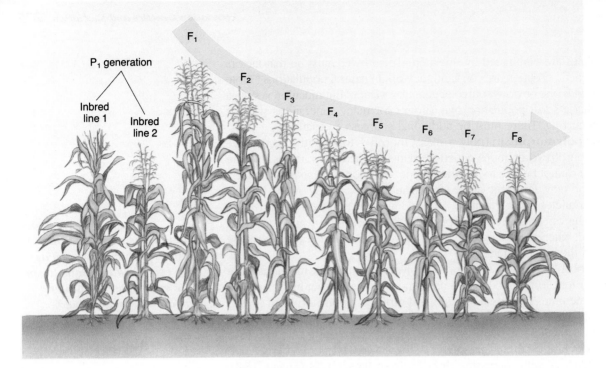

Figure 15-9
Hybrid corn (the F_1 generation) is superior to the different inbred parental lines crossed to produce it. Each parental strain has become homozygous for many gene pairs. The F_1 hybrids are heterozygous for many more gene pairs than their parents and enjoy many genetic advantages over either parental line. Inbreeding of the F_1 and succeeding generations produces many individuals homozygous for various traits. This results in a gradual decrease in the desirable genetic qualities of the average plant. (After D. F. Jones. From Goodenough, U.: *Genetics,* 3d ed. Philadelphia: Saunders College Publishing, 1984)

declared extinct in the late 1970s. It will be interesting to observe any changes in the frequency of blood groups over the next few generations in India, as might be expected if smallpox was one of the selective pressures determining the frequency of blood groups.

Another cause of genetic polymorphism in a population is the phenomenon known as heterozygote advantage.

Heterozygote Advantage and Hybrids

Sometimes individuals heterozygous for a particular gene are more common in a population than the Hardy-Weinberg Law would predict. This suggests that these heterozygotes have a selective advantage, called **heterozygote advantage,** over homozygotes. The best documented example in humans is the case of sickle cell anemia (Section 13-B), where the heterozygote is at a selective advantage over either homozygote in areas with a high incidence of malaria.

Heterozygote advantage can arise in several possible ways. For instance, each allele may contribute its beneficial effects to the phenotype of the heterozygote. This is the case with heterozygotes for sickle cell anemia in areas where malaria is common. Neither homozygote has the advantage of both normal hemoglobin and resistance to malaria.

Heterozygotes are rare in populations with a high degree of inbreeding (mating between close relatives or self-fertilization). This is because close relatives usually inherit many of the same alleles from their common ancestors. The production of domestic plants and animals involves extensive inbreeding, beginning with a small number of individuals that show the desired traits. Such breeding programs often produce populations with genetic disadvantages as well as the advantages for which they were bred. Strawberries and tomatoes bred to resist bruising during the journey to market may have miserably little taste, and race horses often have especially delicate legs. Individuals from inbred populations are often homozygous for nearly all their alleles. These often include disadvantageous homozygous recessive alleles which are expressed in the phenotype, reducing health and vigor.

Matings between members of two different inbred strains produce **hybrid** offspring, which may be superior to their parents in many ways. Mongrels tend to be healthier than "pure-bred" dogs, which often suffer from genetic ailments. Hybrid corn is valued for its reliable uniformity as well as for the specific qualities of the parental lines (Figure 15-9).

Genes and Environment: Double Trouble for Cheetahs

A JOURNEY INTO LIFE

The rapid growth of the human population today is causing many problems in our environment, including drastic declines in the wild populations of most other large mammals. In attempts to save these species from extinction, zoos and game parks have established programs to breed many of them in captivity. But efforts to breed cheetahs have fared poorly (Figure 15-B). Seeking the reason for this, researchers discovered that cheetahs face an additional threat from within: a high degree of genetic uniformity. In fact, cheetahs have less genetic variability than most laboratory mice or other deliberately inbred livestock.

This suggested to researchers that modern cheetahs are descended from a very small population: their lack of genetic variation is a result of the founder effect. The fossil record shows that there were once several species of cheetahs, and our present-day species was distributed worldwide. About 10,000 to 12,000 years ago (late Pleistocene), many species of mammals became extinct. It is thought that cheetahs narrowly escaped the same fate. Very few animals, and a correspondingly limited gene pool, survived. These survivors repopulated Africa, and the Near East from Arabia to India. Then, heavy hunting about a century ago exterminated the cheetah from much of this range, leaving populations only in parts of Africa south of the Sahara Desert.

The lack of genetic variation in cheetahs has had three serious consequences:

1. **Low fertility.** The males produce sperm of poor quantity and quality. The semen of cheetahs has a sperm concentration only 10% of that found in other members of the cat family, such as housecats, and 71% of cheetah sperm cells are abnormal (compared to 29% in housecats). Sperm quantity and quality are believed to be under genetic control because they are adversely affected by inbreeding in many species.

2. **High death rate among cubs.** In the wild, cheetah cubs suffer a 70% death rate. Even in captive breeding programs, where predators (such as lions and leopards) and starvation are no threat, 30% or more die by six months of age. This suggests that deleterious recessive genes are taking an unusually high toll among cheetahs.

3. **Susceptibility to disease.** Most populations of animals have a high degree of variability among the genes involved in fighting diseases. When a disease strikes a population, it kills those individuals whose genes do not enable them to fight that particular disease effectively. Other individuals, with a different mix of disease-fighting genes, survive. The genetic uniformity of cheetahs means that a very high proportion of the population is susceptible to any disease that their genes do not equip them to fight. An outbreak of feline infectious peritonitis in an Oregon wildlife park killed nearly half the cheetahs. Ten lions in the same compound did not even fall ill, and this disease is seldom more than 10% fatal when it strikes colonies of housecats.

With only 12 of 200 cheetahs in U.S. zoos actively reproducing, and with the danger of disease epidemics among relatively dense captive populations, it is clear that human efforts to sustain captive cheetah populations face many challenges.

Wild cheetah populations face the same problems from the lack of genetic variability. They also face the dangers of predation on cubs, starvation, drought, and parasite infestations in nature. A further threat is loss of natural habitat. When humans take over wild land for ranching, cheetahs and other predators are killed to prevent them killing livestock. In 1988, researchers convinced farmers in Namibia to trap cheetahs alive and turn them over to captive breeding programs. By studying the cheetah's diet and behavior, providing veterinary care, keeping careful pedigree (family tree) records, and promoting outbreeding, researchers hope to be able to increase the genetic diversity of the cheetah population and hence its chances of survival.

Figure 15-B
A cheetah.

■ *Heterozygous, hybrid individuals are often more fit than homozygous, inbred individuals.*

What is the genetic basis of this **hybrid vigor?** Since hybrids are heterozygous for many gene pairs where their inbred parents were homozygous, their superiority might result from heterozygote advantage in many gene pairs. Perhaps more often, hybrid superiority results simply from the fact that hybrids have a high percentage of gene pairs with at least one dominant, advantageous allele. Each dominant allele will mask disadvantageous effects of its recessive partner.

The pitfalls of inbreeding make it clear that inbreeding is often selected against, and why adaptations that promote outbreeding are common. For example, many plants have adaptations promoting cross-pollination rather than self-pollination. Most human societies have taboos against incest. In many animals, such as monkeys and lions, young males leave the social group in which they were born and, with luck, eventually join another group, where they breed with the resident females. In chimpanzees and some rodents, it is the females who emigrate and join other groups.

15-D What Is a Species?

The different kinds of organisms in the world can be divided into species. Traditionally, a **species** has been defined as a group of organisms capable of breeding with one another to produce fertile offspring and unable to breed with members of other species. Thus, even though many species live side by side, they are reproductively isolated from one another. The consequence of reproductive isolation is genetic isolation: populations that do not interbreed cannot exchange genes. This means that the gene pools of different species are isolated from each other, whereas members of the same species share a common gene pool.

■ *The members of a species share a common gene pool, which is different from the gene pools of other species.*

A more modern definition states that a **species** consists of one or more populations that share a common gene pool. This definition emphasizes the genetic continuity of related individuals and populations. The members of a species that reproduces sexually usually share a common gene pool because there is gene exchange among them due to interbreeding. Members of other species (including species that reproduce only asexually) share a common gene pool because they are descended from a common ancestor.

As long as gene flow occurs between two populations, they belong to the same species even if their members cannot breed together (Figure 15-10). For instance, stretching across Europe is a string of populations of the European cherry fruit fly, with interbreeding and gene flow between the populations. However, when flies from eastern and western Europe are brought together to breed, the resulting eggs fail to develop. Nevertheless, these populations are members of the same species because there is, at least in theory, some exchange of genes between them, and so they share a common gene pool.

In practice, it is seldom possible to test directly whether two populations share the same gene pool. However, the range of phenotypes a species' gene pool can produce is limited, and therefore members of the same species usually look alike. The eighteenth-century naturalist Carolus Linnaeus developed the first working definition of a species, based on **morphological** (structural, anatomical) differences. If two organisms were sufficiently different, they were considered to belong to different species. In practice, this is how most organisms are classified today.

Morphological characters are convenient to work with and they allow for clear communication. People who discover a new species can write a precise description of it. They can also select **type specimens** of the species, individuals that are preserved in a museum for future reference. The morphological characters that distinguish species are also the natural criteria to use in con-

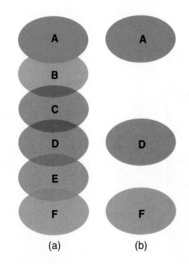

Figure 15-10
How a series of populations may form distinct species. (a) Gene flow occurs between neighboring populations, so all populations are considered one species as long as continuous, interbreeding populations link them. (b) If intermediate populations B, C, and E become extinct, then populations A, D, and F might become separate species.

I

II

KEY TO THE COMMON CLASSES OF ARTHROPODA (ADULT SPECIMENS)

To use the key, begin at the first pair of statements and decide which one pertains to the specimen you are trying to identify. Each member of the pair indicates either a class or the number of the next pair of statements to consult. Continue until you arrive at a statement that gives the class of the specimen.

1a.	Three pairs of jointed legs on the thorax (part of the body between the head and the abdomen)	**Class Insecta**
b.	Other	2
2a.	Four pairs of jointed legs	**Class Arachnida**
b.	Other	3
3a.	Two pairs of antennae	**Class Crustacea**
b.	One pair of antennae	4
4a.	Two pairs of jointed legs per body segment	**Class Diplopoda**
b.	One pair of jointed legs per body segment	**Class Chilopoda**

III

IV

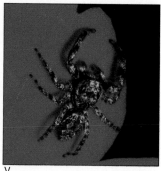

V

structing a dichotomous key (Figure 15-11), one of the most important tools of a field biologist.

Today, other characteristics, such as DNA sequences, can also be used to describe species. We can store genomes, in the form of eggs, seeds, or tissues, at low temperatures. In theory, we could deposit gene samples from all the millions of species of organisms alive today in immense gene banks. The specimens could be used as type specimens for the concept of a species based on a shared gene pool.

Gene banks of frozen material are generally used for other purposes. Strains of microorganisms and tissues are stored for use as standardized "guinea pigs" in research. Seeds of crop plants and their wild relatives are collected for use in plant breeding. The Center for Plant Conservation is attempting to preserve the roughly 700 U.S. plant species that are in critical danger of becoming extinct within the next ten years. Seeds are collected from as many populations as possible (to maximize genetic diversity) and then frozen. The seeds can later be used to reintroduce the plants to nature preserves, the nursery trade, and botanical gardens.

We tend to talk as if organisms can be neatly divided into species. In reality, some populations are only partially isolated from their neighbors, genetically or sexually.

Figure 15-11

A dichotomous key, so called because each step gives two alternative choices. To see how it works, follow the instructions given at the top. The pictures are all of animals from the large group known as arthropods, which have external skeletons made of chitin, and jointed appendages. Arthropods can be divided into a number of classes. Use this key to identify the classes of the specimens in the photographs around the key. For the answers, see the Chapter 15 Self-Quiz answers at the back of the book.

15-E Barriers to Breeding Between Species

When two different populations live in the same place, they are said to be **sympatric** (sym = same; patria = homeland). If they live in different places, they are **allopatric** (allo = other). In nature, sympatric populations of different species do not interbreed because each species has its own, unique reproduc-

Table 15-2 Barriers that Prevent Interbreeding Between Members of Different Species

Barrier	Effect
Prezygotic isolation	Prevents mating (and therefore fertilization to form a zygote) between species
Habitat differences	Individuals of two species never meet
Different breeding times	Members of two species not in breeding condition at the same time (see Section 16-C)
Mechanical barriers	Shape of genitalia (copulatory structures) prevents fertilization by members of other species (common in insects)
Behavioral specificity	Mating cannot occur without species-specific behavior (see Section 16-C)
Postzygotic isolation	Prevents successful reproduction after fertilization (and therefore zygote formation)
Hybrid inviability	Hybrid offspring dies before reaching sexual maturity
Hybrid sterility	Hybrid offspring survives but is sterile
Hybrid breakdown	Hybrid offspring fertile but many of its offspring are not

tive mechanism, and the mechanisms of different species are not compatible. Reproduction involves many adaptations of anatomy, physiology, and behavior. Members of the two sexes must come into breeding condition at the same time, usually as a result of hormonal changes within their bodies. Then they must often produce the appropriate steps in the courtship and copulation behavior of the species before the male and female gametes come together. Fertilization involves biochemical recognition between sperm and egg or between pollen and female flower parts. Finally, on a molecular level, the genetic information and cytoplasmic messengers of sperm and egg must be compatible if embryonic development is to proceed normally and the resulting offspring survive.

The types of reproductive barriers that usually prevent members of different species from breeding with each other are listed in Table 15-2.

15-F Speciation

Speciation is the formation of one or more new species from an existing species.

Allopatric Speciation

Allopatric speciation occurs when a population becomes geographically separated from the rest of its species and then changes so much that it becomes a new species. This can occur when a small number of individuals colonize a new area. For instance, in the uplands of Hawaii, there is a species of goose found nowhere else. This goose closely resembles the Canada goose of North America and almost certainly evolved from Canada geese that migrated to Hawaii (Figure 15-12).

Suppose a few seeds or a few birds are blown onto an island where they grow and breed. The most likely fate of this new population is that it will become extinct because its members are poorly adapted to their new island home. But sometimes a few individuals will survive and found a new, isolated population.

■ *Most new species form when a small population is cut off from the rest of the species and evolves adaptations to its new habitat in isolation.*

(a)

(b)

Figure 15-12
Parent and derived species. (a) The Hawaiian goose (nene), Hawaii's
state bird, is adapted to life on rugged upland lava flows far from
water. This species is thought to be derived from a small population of
(b) the Canada goose of North America. (a, M. J. Rauzon/VIREO; b,
R. Villani/VIREO)

The new population will tend to become genetically distinct from the rest of
the species for two main reasons. First, the founder effect (Section 15-B) will
often apply, and the new population will have a unique gene pool from the start.
Second, the new population will be subject to a new set of selection pressures
(Section 15-B) and so it will evolve adaptations to its new home. Its reproduction is one of the many features that may be changed as a result of this selection.

Eventually, the differences between the original population of the species
and the new population on the island may become so great that the two can no
longer interbreed. At this stage, it may be impossible to tell, just by looking at
them, whether the two populations are different species or not. The test of
speciation comes if the two populations ever come together again. If they do not
then interbreed, the two populations are considered "good" species.

It may occur to you that new species might also form when one large,
widespread species becomes split into two, for instance by a landslide, by construction of a highway, or when continents drift apart (see *A Journey Into Life* on
page 286). Apparently, however, this does not usually occur. For instance, North
American and European sycamore trees have been separated for at least 30
million years, but they have not formed separate species. The two are very
similar, and they can interbreed to produce normal, fertile offspring. It seems
that species containing large numbers of individuals and covering wide areas of
a continent may remain essentially unaltered for millions of years. This is because they are subject to gene flow and stabilizing selection. The genetic
changes that lead to the formation of new species are much more likely to occur
under the influence of directional selection acting on a small population of a
species adapting to a new habitat. (Directional selection acting on *all* the members of a species changes that species, but it does not produce a new species.)

Pleistocene Glaciations A dramatic example of directional selection producing speciation in small isolated populations comes from the effects of the

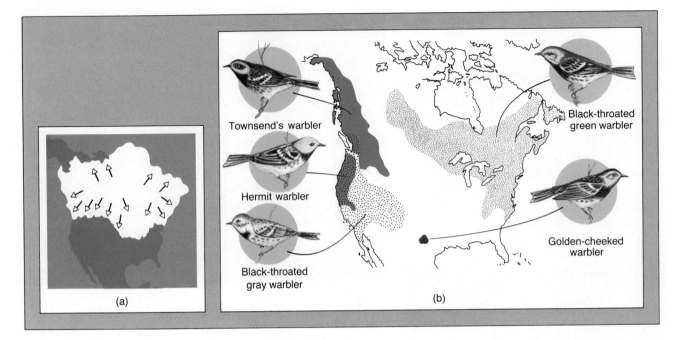

(a)

(b)

Figure 15-13

Speciation during the Pleistocene glaciations. (a) A map showing the extent of the Wisconsin ice sheet, last of the four ice sheets that pushed south into the United States during the four Pleistocene glaciations. (b) Breeding areas of members of the black-throated green warbler group today. Note that there is very little overlap between breeding ranges. These five species probably evolved from a single ancestral species. (From Lancaster, *The Living Bird,* Vol. 3, 1964. The Cornell Laboratory of Ornithology)

great glaciations of the Pleistocene Epoch. Over the past million years, and ending only a few thousand years ago, four major glaciations (Ice Ages) covered most of Canada, the northernmost United States, and the western mountains (as well as Northern Europe) with ice, often thousands of feet deep.

Each time the ice sheets spread south from the Arctic, they pushed many species of animals and plants into new habitats and cut them off into isolated populations. During the glaciations, separated populations of a single species evolved adaptations to their new habitats. The glaciers then retreated, permitting some of these populations to come back into contact. In some cases, the populations had evolved into two or more "good" species from one species present before the glaciation.

The wood warblers of North America underwent impressive allopatric speciation in this way. Two dozen or more species can be found hunting insects in the foliage of even a single forest in parts of North America. The ancestral black-throated green warbler was probably distributed across the continent before the Pleistocene. At each glacial advance, its range was pushed down to the southeast, pinching off a population in the west that evolved into a new species (Figure 15-13). When the glacier retreated, the black-throated green warblers again spread across most of the continent, and small populations were again isolated by the next glacial advance, and so on.

Sympatric Speciation

Sympatric speciation is the formation of new species within a single population without isolation. The example of sympatric speciation usually cited is that of polyploidy. **Polyploidy** is multiplication of the normal chromosome number. This can happen when chromosomes fail to segregate at meiosis, producing diploid instead of haploid gametes, or when a cell in a plant stem replicates its chromosomes without undergoing mitosis. The resulting polyploid cell may later give rise to a stem bearing polyploid flowers, which in turn form diploid gametes. (Some geneticists do not think polyploids should be considered separate species from their diploid relatives. They contain *more* genes than their diploid parents, but not *different* genes, at least at first.)

Polyploid plants are often able to tolerate some adverse environmental condition, such as cold or drought, better than their normal parents. Indeed, the number of polyploid species increases with increasing latitude (that is, with colder climate). Probably more than a third of all plant species have arisen by polyploidy. Nearly all domesticated varieties of plants with larger fruits and flowers than those of their wild ancestors are polyploid. Polyploidy is rare among organisms that must be fertilized by another individual in order to reproduce. In such organisms, a polyploid individual would usually be an evolutionary dead end for lack of a genetically compatible mate. Therefore, polyploidy is most common in self-fertilizing plants and in plants that reproduce only vegetatively, such as bananas (Figure 15-14). It also occurs in **parthenogenetic** animals, animals that reproduce by means of unfertilized eggs. Some parthenogenetic animals are polyploid: various beetles, moths, shrimp, goldfish, and lizards fall into this category.

In theory, sympatric speciation without polyploidy can occur in animals. If a population is polymorphic such that two varieties are advantageous but hybrids between them are selected against, the two varieties may evolve into separate species. Aubrey Manning demonstrated this by using artificial selection in a laboratory population of fruit flies. However, there is no known example of this situation in nature.

Unhappily for those with tidy minds, it will probably never be possible to produce convincing evidence that any animal species actually has arisen sympatrically. This is because sympatry is so hard to demonstrate. Many pairs of populations appear to live together but close examination shows that they are effectively isolated. For instance, cichlid fish inhabit the Great Lakes of the African Rift Valley. There are about 126 species of cichlid in Lake Tanganyika and more than 200 species in Lake Malawi. The species are distinguished mainly by their feeding adaptations. Some feed in deep and some in shallow water, some eat algae, some molluscs, some plankton, and some other cichlids. Despite the fact that all are swimming in the same lake, these fish species are isolated from each other by specific habitat within the lake and by reproduction. To demonstrate sympatric speciation, one would have to show that two species belonged to one population before speciation occurred, and this would involve showing gene flow or reproduction between them—for which one is not likely to find evidence.

Figure 15-14
A red banana flower and many green bananas. Most cultivated bananas are triploid (3N). With this uneven number of chromosome sets, they cannot undergo the delicate process of meiosis (Section 11-E). The nutritious fruits develop without fertilization, and hence without seeds, and the plants multiply vegetatively (asexually).

Selection Against Hybrids

When two previously separated populations of related organisms come together, they may interbreed. This situation has various possible outcomes. First, if the two are already "good" species, they will not interbreed at all (or may interbreed in some areas to produce populations of hybrids). Second, they may interbreed freely so that they merge into one big population and all genetic distinction between them disappears. A third possibility might theoretically exist but probably occurs very seldom: members of the two populations may interbreed at first but later form separate species. This might occur if the hybrid offspring between the two populations are at a selective disadvantage compared with members of either parent population. In this case, selection would favor those individuals that mate with members of their own population over those that produce hybrids.

15-G   How Quickly Do New Species Form?

Biologists sometimes divide evolution into microevolution and macroevolution. **Microevolution** is evolution within populations and the formation of new species—the kinds of evolution we have discussed so far in this chapter. **Macroevolution** is the evolution of new groups of species, such as flowering plants or dinosaurs. The main way of investigating macroevolution is to study the fossil

Figure 15-15

Look-alikes. This fossil of a fern that died about 300 million years ago *(left)* is strikingly similar to one species of living fern *(right)*. (Robert J. Lynch)

record of life on earth. Examining fossils reveals patterns in macroevolution which we might not predict from what we know of microevolution. For instance, when we measure the rate at which new species appear in the fossil record, we find that evolutionary changes may be fast or slow, or anywhere in between.

Many fossil species existed for millions of years with very little change and without giving rise to new species. Then, in a "short" period of time (which in geological terms may mean thousands of years), related but different species appeared. Sometimes these new species replaced the older one, and sometimes the old and new species coexisted at least for a while. The evolution of the horse is sometimes shown as a gradual process, with body size steadily increasing and the number of toes becoming reduced from three to one (see Figure 14-7). Closer study shows that speciation occurred many times during the evolution of horses, leading both to species with smaller bodies and to species with larger bodies. Since larger-bodied species survived longer than smaller-bodied species and gave rise to more new species, what we see in the fossil record is the *appearance* of a gradual increase in body size.

In many cases, fossils millions of years old are strikingly similar to species alive today (Figure 15-15). For example, half of the fossil seashells from seven million years ago apparently belonged to species still alive today. The sycamores discussed in Section 15-F are another example. At other times, many new species have evolved in a relatively short time (one to a few thousand years), as in the cases of the warblers (see Figure 15-13).

This "stop and go" pattern of speciation is called **punctuated equilibrium**. We can see how it probably comes about. A widespread species exists unchanged for a long time under the influence of stabilizing selection. Eventually, one of its small, isolated populations evolves into a particularly successful new species. This new species spreads rapidly, either driving the parent species to extinction or existing with it.

This model helps to explain a riddle that puzzled Darwin and many other people: the sudden appearance of various major groups of organisms in the fossil record. Any new group of organisms evolves from a previously existing one, but often we do not find fossils of gradually changing intermediate forms spanning the time until the new group is clearly distinct from its ancestors. Instead, some groups, such as the flowering plants, seemed to appear almost full-fledged in the fossil record, with little evidence to show how they originated or which organisms were their ancestors.

Major groups such as this usually have evolutionary novelties that distinguish them from their ancestors. It is noteworthy that these often include reproductive adaptations, such as flowers in plants, and waterproof eggs in reptiles. These new adaptations proved so successful that the species in which they evolved underwent wide adaptive radiation, spreading across the globe and spawning hundreds of new populations, many of which also formed new species. This swift burst of speciation appears as rapid change in the group's fossil record. When organisms with the new adaptations had spread into most of the habitats they were equipped to exploit, speciation became less frequent. This is reflected in a slower rate of change in the group's fossil record. Since such evolutionary novelties usually arise in small, local populations, the chances of the intermediate forms being preserved as fossils, and of our finding them if they were, are very small.

SUMMARY

Natural selection acts on individuals, but only populations can evolve. This is because the population is the smallest unit with a gene pool in which the frequency of alleles can change—that is, evolve.

The Hardy-Weinberg Law shows that the proportions of different alleles and genotypes in a population will remain the same as long as:

1. there is no net mutation;
2. there is no selection for or against the traits being considered;
3. mating is random with respect to genotype;
4. there is no gene flow to or from other populations; and
5. the population is large.

Under these conditions, evolution will not occur. If any of these conditions is not met, evolution is likely to occur.

Populations of the same species in different geographical areas tend to differ somewhat in their genetic makeup. Adaptations to local areas increase this difference; gene flow between adjacent populations decreases it.

Random changes in the gene pool as a result of genetic drift may be an important cause of evolution in small populations that are not subject to strong selective pressures.

Genetic variation in a population can be increased by mutation, by polymorphism maintained by hetero-zygote advantage or other factors, by gene flow between populations, and by sexual reproduction. The most important factor that decreases variation in a gene pool is natural selection, which adapts a population to local conditions. Genetic drift may also decrease genetic variation by randomly eliminating some alleles from a small population.

A species is an actually or potentially interbreeding group of organisms with a common gene pool and similar morphology, physiology, and behavior.

A population of an existing species may evolve into a new species. Allopatric speciation occurs when a population is separated from the rest of a species, cutting off gene flow between the two. This isolated population evolves under the influence of local directional selection, and may become so different from the parent population that it is considered a new species. Whether it is actually a "good" species or not will be determined only when the population again becomes sympatric with members of the original species.

The formation of polyploid races is usually described as sympatric speciation.

Most species probably form in small, isolated local populations. Such a population may diverge very rapidly from the parent population. If it does not become extinct, it may form a new, rapidly evolving species.

SELF-QUIZ

1. The Hardy-Weinberg Law allows us to predict that:
   a. sexual reproduction is necessary for evolution
   b. sexual reproduction may be a cause of evolution
   c. sexual reproduction plays no role in evolution
   d. sexual reproduction will cause evolution if individuals prefer mates with one genotype over those with other genotypes

2. In certain parts of Africa, people with one normal hemoglobin and one lethal sickle hemoglobin allele are more likely to survive than homozygotes for either allele. These populations are *not* experiencing:
   a. natural selection
   b. heterozygote advantage
   c. polymorphism
   d. genetic drift
   e. violations of the Hardy-Weinberg Law

3. Suppose that mosquito control measures completely eliminate the threat of malaria from an area where it was once prevalent. If you followed human allele frequencies for the next 30 generations, what changes would you expect to see in the frequencies of the normal and sickle alleles, once selection against the homozygous normal individuals is removed?

4. Selection will not eliminate a lethal recessive allele from a large population of diploid organisms because:
   a. there will always be some heterozygote carriers for the allele
   b. heterozygotes are at a selective advantage
   c. the allele will have some good effects
   d. the rate of mutation producing new copies of the lethal allele is higher in a larger population

(Self-Quiz continues on page 288)

Continental Drift

A JOURNEY INTO LIFE

Geological changes that produce new barriers between populations tend to promote the formation of new species. The most dramatic geological change of all has been the gradual drifting apart of the continents. This movement of land masses separated different populations of many plant and animal species. The separate populations then evolved independently and in many cases became distinct species.

Biologists believed in continental drift before geologists accepted the idea. This was because biologists found it easier to explain the distribution of species by assuming that the continents moved than by assuming that the animals and plants did. Lungfishes, for instance, live only in fresh water and are not noted for their ability to travel. Yet the three species of lungfishes in the world today, while clearly related, live in South America, South Africa, and Australia, respectively. Biologists never really believed that ancestral lungfishes traveled thousands of miles through oceans or over dry land to reach the places where they are found today. Flightless birds—the ostrich in Africa, the rhea in South America, and the emu in New Zealand—have closely related species of flightless insect parasites in their feathers. How did this strange situation arise?

The distribution of some ancient groups of organisms is even more surprising. Fossils of *Glossopteris,* a genus of Permian plants (see endpaper), occur in South America, Africa, India, and Australia (Figure 15-C). It does not seem likely that seeds of these plants were dispersed over vast ocean distances. To account for these and many other distributions, theories involving land bridges between the continents developed. These theories were replaced by the belief that the continents themselves had moved.

During the 1960s the earth sciences underwent a revolution with the general acceptance of the theory of **plate tectonics.** The part of plate tectonic theory that concerns us here is the thesis that the outermost layer of the earth is made up of about 15 rigid plates, which are pushed past, over, and under one another. The plates move because they are "floating" on the mantle of molten rock inside the earth.

Geological evidence strongly supports the evidence from fossil plants and animals, suggesting that in Permian and Triassic times, 200 to 250 million years ago, the southern continents (South America, Africa, Antarctica, and Australia), along with India, were united as one large mass called Gondwanaland (Figure 15-D).

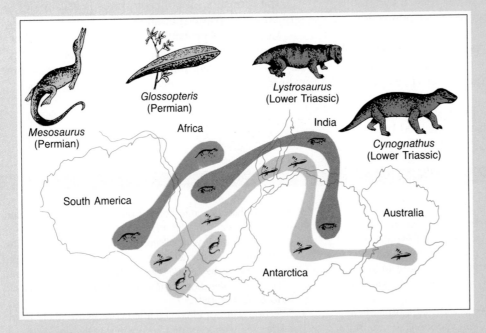

Figure 15-C
The distributions of four fossil species found on more than one of the southern continents. This distribution suggested to biologists that the southern continents were once joined. (You can find the Permian and Triassic periods on the endpaper of this book.) (From Colbert, E. H. *Wandering Lands and Animals.* © 1973 by Edwin H. Colbert. Reprinted by permission of E. P. Dutton)

North America, Europe, and Asia were meanwhile united as a northern land mass called Laurasia. Moreover, these two supercontinents were themselves united for a time to form a single world land mass, Pangaea. The rest of the earth was covered by ocean.

The theory of **continental drift** is now generally accepted. It proposes that the continents reached their present positions by moving apart after the Triassic Period. Gondwanaland and Laurasia separated from each other during the Jurassic Period. Before Gondwanaland began to split up into the present-day southern continents and India, the dinosaurs and coniferous trees had become supreme, ancestral mammals were well established, and flowering plants had already evolved.

After the continents separated, each group of organisms evolved in a different way. Marsupial (pouched) mammals speciated in Australia and South America, while placental mammals dominated the other land masses. India drifted north and collided with Asia. The force of the collision pushed up the Himalaya Mountains. At about the same time, Laurasia was splitting apart. By this time, many mammals had evolved in Laurasia. Thus it is not surprising that North American, Asian, and European mammals are related to each other more closely than any of them are related to the mammals of South America or Australia.

The Central American land bridge was pushed up from the sea 3 to 5 million years ago. This provided a route for animal migrations between North and South America. Only a few South American animals, such as the opossum (a marsupial) and the armadillo, succeeded in colonizing North America. By contrast, many placental mammals invaded South America, causing the extinction of several of its native marsupials.

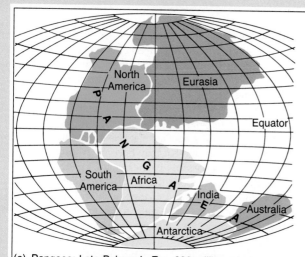

(a) Pangaea: Late Paleozoic Era, 230 million years ago.

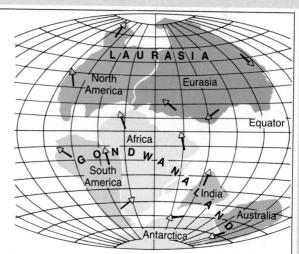

(b) Laurasia and Gondwanaland: Mesozoic Era, 180 million years ago.

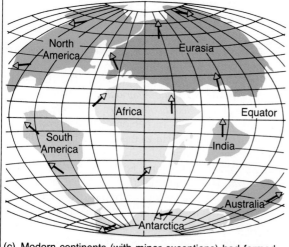

(c) Modern continents (with minor exceptions) had formed by the end of the Mesozoic, 110 million years ago.

Figure 15-D
Continental drift. Movement of the continents (indicated by arrows) produced the modern continents from a single land mass—Pangaea—that existed in the Paleozoic Era (from about 600 million to about 200 million years ago).

5. Genetic drift is most important in which situation?
   a. gene flow
   b. polymorphism
   c. small populations
   d. breeding of flying insects
   e. relaxation of selective pressures

6. Genetic drift is more likely to cause evolution in a small population because:
   a. mating is nonrandom in small populations
   b. random events are more apt to happen to small populations
   c. there is no natural selection in small populations
   d. deviations from statistical averages are more likely to be seen in small populations than in large ones

7. In each of the following situations, tell whether genetic variability in a population would increase, decrease, or remain the same.
   ____ a. increased mutation rate
   ____ b. decreased natural selection
   ____ c. increased variability in the environment
   ____ d. sexual reproduction

8. A new species of organism has evolved when:
   a. the climate of a population's range has changed greatly
   b. variation has been introduced into a population by mutation
   c. members of a population can no longer interbreed in the wild with individuals who are genetically their third cousins
   d. selection pressure has produced a group of hybrid individuals

9. All of the following conditions would result in a change in the frequency of a specific allele in a population *except:*
   a. selection against the homozygous recessive phenotype
   b. selection against the dominant genotype
   c. genetic drift
   d. random mating
   e. mutation of the dominant allele to the recessive allele

QUESTIONS FOR DISCUSSION

1. What, in biological terms, is a "race" of people or any other animal?

2. Evolutionist Theodosius Dobzhansky wrote that a totally uniform physical environment could support only one species of organism. Do you agree? Why or why not?

3. What is the effect on the human gene pool of the discovery of expensive medical treatments that permit individuals with previously lethal phenotypes to live? Should society permit research to find ways to treat more such conditions?

4. Huntington's disease is a genetic condition caused by a dominant gene. The brain deteriorates, and the victim loses control over both mind and muscles. A period of insanity accompanied by jerky movements of the face and limbs is finally followed by death. (Interestingly, at least seven women accused of witchcraft in New England during the 1600s were related to families now known to contain Huntington's disease.) However, the condition usually does not set in until the victim's 30s or 40s, and by then he or she may have produced children, half of whom will also receive the allele and are therefore doomed to suffering the disease. Can selection operate against such a late-acting genetic trait?

5. Why are mutations almost always deleterious?

6. What characteristics would you expect to see in the genetic makeup of a species that has the ability to colonize a variety of habitats that may become available?

7. Suppose an impassable barrier cuts off gene flow between two populations that inhabit areas with somewhat different climates. What change in the two populations' gene pools would you expect to observe as time passes?

8. Suppose a population has frequency of allele $A = 90\%$, and $a = 10\%$. If the population meets the Hardy-Weinberg conditions, what are the frequencies of genotypes AA, Aa, and aa?

SUGGESTED READINGS

Ayala, F. J., and J. A. Kiger. *Modern Genetics,* 2d ed. Menlo Park, CA: Benjamin/Cummings, 1984. Several up-to-date chapters on population genetics.

Colbert, E. H. *Wandering Lands and Animals.* New York: E. P. Dutton, 1973. A readable introduction to continental drift.

Mlot, C. "Blueprint for conserving plant diversity." *BioScience* 39 (6): 364, 1989. The efforts of the Center for Plant Conservation to save seeds of U.S. plant species that are on the verge of extinction.

Nance, R. D., T. R. Worsley, and J. B. Moody. "The supercontinent cycle." *Scientific American,* July 1988. A discussion of the processes that cause continental drift and how they influence climate and evolution of organisms.

O'Brien, S. J., D. E. Wildt, and M. Bush. "The cheetah in genetic peril." *Scientific American,* May 1986.

Stanley, S. M. *The New Evolutionary Timetable.* New York: Basic Books, 1981. Layperson's guide to evolutionary thought from Darwin to the present by a leading punctuational evolutionist.

Stone, C. P., and D. B. Stone, Eds. *Conservation Biology in Hawaii.* Honolulu: University of Hawaii Press, 1988. A resource on the biology of Hawaii, highlighting the ignorance of visitors and residents about these remarkable islands, which are rapidly being destroyed, and stressing the need for research and education.

Evolution and Reproduction

CHAPTER
16

Peacock

Male gorilla playing with baby

Male stag beetle

Honey bee queen and workers

An organism's genes may enable it to obtain food, avoid being eaten, and cope with the climate, but ultimately there is only one measure of its evolutionary success: the proportion of its genes present in future generations. Without genes that make it reproduce, all the adaptations resulting from its other genes are useless, in evolutionary terms. The reproductive adaptations of organisms are many and varied. We shall consider some of them in this chapter as examples of the remarkable effects of natural selection.

Successful reproduction automatically selects for perpetuation of the genes that brought it about—and the various modes of reproduction in the living world are indeed wonderful.

KEY CONCEPTS

• An organism can be viewed as a vehicle by which genes are passed from one generation to the next.
• The reason sexual reproduction is so widespread is probably that it produces more genetic variation in organisms than does asexual reproduction.
• The ecology of a species plays a major role in determining the type of sexual system it develops.
• A species' sexual system determines how members of the species are related to each other and thus the types of behavior toward other individuals that can evolve.

16-A   Is Sex Necessary?

At first glance it seems that sex *is* necessary. Most animals, and many higher plants, rely exclusively on sexual reproduction to perpetuate their genes. On the other hand, sexual reproduction is much less common among protists, fungi, and lower plants, which usually reproduce in other ways. Perhaps in these organisms sex is not necessary but is valuable under some circumstances.

Asexual reproduction is any means of multiplying that does not involve both eggs and sperm. Many unicellular organisms reproduce asexually simply by dividing into two identical, smaller cells. Many plants can reproduce vegetatively (Section 36-I). Asexual reproduction in some lower animals occurs by the budding off of smaller individuals, which eventually detach from the parent (Figure 16-1). **Parthenogenetic** organisms develop from unfertilized eggs (see Figure 16-2).

When an organism reproduces asexually, each offspring represents a considerable investment of the parent's energy. However, the energy is efficiently used in that nearly all of it goes into the growth of a new individual. Furthermore, the relatively large size of the offspring produced by many asexual methods gives them a good chance of surviving to reproductive age.

By contrast, sexual reproduction wastes a lot of energy. If sperm, pollen, or eggs are released into the water or air to find each other by chance, millions of them fail to find mates or fall prey to other organisms, and the energy spent to make them is wasted. With internal fertilization, sperm are introduced directly into the female's body near the eggs. Fewer gametes are lost, but the organism must invest energy in other ways. For example, plants may produce flowers and nectar, thereby attracting animals that carry pollen to the female parts of other flowers. Animals must spend a lot of time and energy finding and courting mates. Overall, sexual reproduction is apt to waste more energy than asexual reproduction does.

As a result, many asexually reproducing organisms can produce more surviving offspring per season than can similar organisms that reproduce sexually. Why, then, do so many living things use so much energy and so many wasted cells reproducing sexually? Sexual reproduction must have tremendous adaptive value, or it could not have become so common.

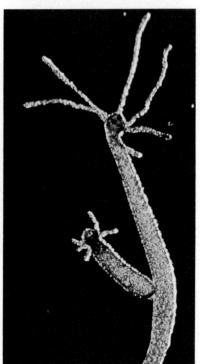

Figure 16-1
Asexual budding. *Hydra* is a commonly studied invertebrate animal. When the bud is large enough, it breaks off and starts an independent life. (Biophoto Associates)

The main biological difference between sexual and asexual reproduction is that sexual reproduction involves genetic recombination and reassortment during meiosis and fertilization, and so it produces more genetic variation in a population (Section 11-F). In most organisms that reproduce asexually, the offspring result from mitosis. Hence the parent passes on all its own genes to all its offspring, and therefore they are all alike genetically—that is, the original organism plus its descendants are said to form a **clone** (Figure 16-2). The only genetic variation that can arise in such a population comes from mutations. (An exception is the asexual production of male hymenopterans [Section 16-E], which receive half of their mothers' genes as a result of meiosis.)

Some organisms can reproduce both with and without sex. Many protists, algae, and small invertebrates (such as the aphids in Figure 16-2) reproduce asexually during the summer and then reproduce sexually when the temperature drops and days become shorter in the fall. These organisms switch from asexual to sexual reproduction because of changing environmental conditions.

Under this system, an organism that is well adapted to its environment can produce numerous, equally well-adapted copies of itself while conditions remain constant during the summer. When conditions then change in the fall, the organism reproduces sexually, creating many genetically different offspring, some of which will probably survive. Often, sexual reproduction gives rise to an egg or cyst in a weatherproof covering, which protects the enclosed individual until good growing conditions return.

Many kinds of bacteria and other "lower" organisms have existed virtually unchanged for more than 500 million years. Such organisms can tolerate a wide range of conditions. It is hard to imagine an environmental change so catastrophic and widespread that it would threaten them with extinction. Most higher plants and animals, on the other hand, can live only in the few places that supply their particular needs. A specialized species with only asexual reproduction (the common dandelion is a good example) may do very well for a while, but is ultimately doomed to extinction much more surely than is a species that reproduces sexually. A change in the environment that kills dandelions will kill all dandelions in the area because they are all very similar genetically (Figure 16-3).

The genetic variability produced by sexual reproduction is important to the formation of new species. If a group of organisms does not give rise to new species, it may become extinct. We think of the ruling reptiles (including dino-

Figure 16-2
A clone. The aphids on this branch are sap-sucking insects that produce many generations in a summer by parthenogenesis: unfertilized eggs develop in the mother's body and are born as miniature versions of her. These offspring soon produce their own daughters, genetically identical to their sisters, cousins, and aunts, who are all descended from the same ancestor. Eventually, overcrowding or other cues may cause a female to lay eggs, which develop into winged males and females that disperse to new locations.

Figure 16-3
Dandelions. These widespread and successful plants reproduce asexually and are therefore nearly identical genetically. Dandelion pollen is sterile. The eggs develop without fertilization. Therefore, each seed carries the same genes as the parent plant.

■ *Asexual reproduction is found among many common, widespread species that tolerate a broad range of environmental conditions and reproduce rapidly. More specialized species are in danger of dying out, leaving no descendants, if their members do not have the genetic diversity that sexual reproduction provides.*

saurs) as a large group of higher animals that became extinct. But today the ruling reptiles' descendants, the birds, are alive and well and living all over the world. Without the genetic variation produced by sexual reproduction, the reptiles could not have left these very different, successful descendants.

There may be occasions when asexual reproduction would be selectively more favorable even in higher organisms. But living things are trapped by their evolutionary history. Sexual reproduction may be so entrenched that it is hard to get rid of.

16-B   Evolution of Sexual Reproduction

Sexual reproduction, involving meiosis and fertilization at some stage in the life history, occurs only in eukaryotes. The first sexual reproduction probably resembled that seen even today in organisms like the filamentous alga *Ulothrix,* which can also reproduce asexually. This alga consists of a string of haploid cells (having unpaired chromosomes). During the growing season, *Ulothrix* reproduces asexually by means of small flagellated cells called **zoospores.** The zoospores are released through a pore in the parent cell. They swim to suitable locations, where they settle and divide, starting new filaments (Figure 16-4).

Sexual reproduction in *Ulothrix* probably evolved as a modification of this program. Some cells of the filament divide to produce flagellated gametes, cells smaller than zoospores. All the *Ulothrix* filaments in an area release their gametes at the same time, and they swim around in the water. When two gametes collide, they stick together and fuse to become one. This is a sexual event, and it forms a diploid zygote (having pairs of homologous chromosomes). The *Ulothrix* zygote eventually undergoes meiosis to produce haploid zoospores, which can develop into new filaments. This is a very simple form of sexual reproduction in that all the gametes look alike. Fertilization probably originated as an accidental fusion of two undersized spores which, as a result, gained the selective advantage of larger size by pooling their resources.

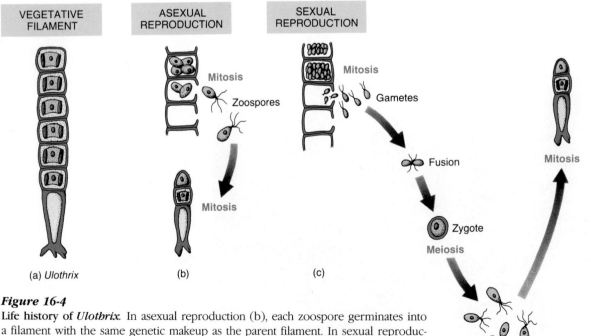

VEGETATIVE FILAMENT

ASEXUAL REPRODUCTION

SEXUAL REPRODUCTION

Mitosis

Zoospores

Mitosis

Mitosis

Gametes

Fusion

Mitosis

Zygote

Meiosis

Zoospores

(a) *Ulothrix*   (b)   (c)

Figure 16-4
Life history of *Ulothrix.* In asexual reproduction (b), each zoospore germinates into a filament with the same genetic makeup as the parent filament. In sexual reproduction (c), gametes fuse to form a diploid zygote, which undergoes meiosis to form four zoospores of new genetic types.

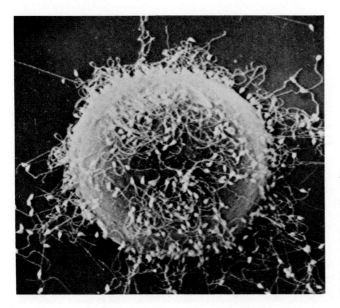

Figure 16-5
The relative sizes of egg and sperm.
This sea urchin egg contains many
thousand times more material than
any of the sperm swarming around it.
Sea urchins release eggs and sperm
into the sea at the same time, and
fertilization occurs outside the body.
(Biophoto Associates)

In the sexual reproduction of most plants and all animals, the egg produced
by the female is a very large cell that stays in one place. The male sperm is a
small cell that travels to the egg. The egg contains not only stored food but also
messenger RNA carrying the information necessary to direct the early stages of
embryonic development. The major advantage of this system is that the embryo
can be larger and better developed before it has to provide its own food. Higher
animals could probably never have evolved without an egg containing stored
food and genetic information for embryonic development. It is hard to imagine
even a worm developing if the tiny embryo had to form a mouth and feed itself
when it contained only two or four cells.

Thus, an egg that contains food and information, and is therefore too large
to be motile, is of enormous selective advantage. But a nonmotile egg is no use
without a sperm that can find and fertilize it (Figure 16-5).

> ■ *Sperm and egg vitally comple-*
> *ment each other and have*
> *evolved together.*

16-C   Evolution of Mechanisms that Ensure Fertilization

When an organism reproduces asexually, it is independent of other individuals.
In sexual reproduction, however, male and female must cooperate. One mechanism that coordinates male and female activity is the existence of breeding
seasons, with members of a population all coming into breeding condition in
response to some environmental cue, such as temperature or daylength. Another is mating behavior, including chemical secretions by some members of
the population that stimulate others to come into breeding condition.

An adaptation found among organisms that are fixed in place, or at least
cannot move very far to find mates, is the existence of reproductive organs of
both sexes in the same individual. This **hermaphrodism** is common in plants,
and it also occurs in animals such as snails, leeches, and earthworms (Figure
16-6). In addition to ensuring that any other member of the species in the
vicinity is a potential mate, hermaphrodism also gives each organism a chance to
be both a mother and a father! Hermaphroditic animals and plants may be able
to self-fertilize, but there is often selective pressure to cross with other individuals. This reduces the chance of inbreeding, which would eliminate the genetic
variation produced by sexual reproduction (Section 15-C).

Figure 16-6
Mutual copulation between two earthworms. Earthworms are hermaphroditic, and they exchange sperm. Hence each worm uses its partner's sperm to fertilize its own eggs. (R. K. Burnard, Ohio State University/BPS)

16-D   Evolutionary Roles of Male and Female

The female is, in a sense, a limiting resource for the sexual reproduction of a species. At the simplest level, because eggs are bigger than sperm, it takes more energy to produce an egg than to produce the sperm that fertilizes it. A male's evolutionary success is usually limited, not by the number of sperm he can produce, but by his ability to deliver his sperm to as many eggs as possible. The female's evolutionary success is limited by the number of her eggs that survive to become part of the breeding population.

The fact that a female's parental investment in each future offspring is usually greater than the male's has a fascinating consequence: the selective pressures acting on a female may conflict with those acting on a male of the same species. While it may be advantageous for a male animal to copulate with as many females as possible in order to raise his chances of fathering surviving offspring, it is apt to be advantageous for a female to be much more choosy. She produces fewer eggs and so has fewer second chances to reproduce successfully if her first mate is genetically unfit.

Under this selective pressure, it is not surprising that females of all species of animals studied show discrimination in their choice of mates. The female who discriminates, and copulates only with genetically fit males, will be at a selective advantage. On the other hand, it may be to a male's advantage to appear genetically fit even when he is not, because females may then be deceived into mating with him.

Sexual Differences

A female's reproductive success is not usually limited by her inability to find mates. It is more likely to be limited by her inability to rear her young. A male who demonstrates that he can contribute to raising offspring will be attractive to females. Among birds, the male of choice is often the holder and defender of a territory that provides food and shelter needed by the female and her young. Males may also compete for control over valued resources other than territories, because possession makes them irresistibly appealing to females. For instance, a man with wealth and status attracts some women no matter how unattractive he is.

■ *A female's reproductive success is usually limited by the number of offspring she can raise to maturity. A male's is more often limited by the number of eggs his sperm can fertilize.*

Figure 16-7
Sexual differences. As in many species, this male red-capped robin, from Australia, is more colorful than his mate. (Biophoto Associates, N.H.P.A.)

The different sexual roles of male and female may lead to their having different appearances, a phenomenon called **sexual dimorphism.** For example, female birds are more likely than males to have drab colors (Figure 16-7). Because females are vulnerable as they sit on their eggs, it is advantageous for them to be inconspicuous. This camouflage works. When the male is the more conspicuous sex, mortality is invariably higher among males than among females. When defending a territory, a male may flaunt vivid coloration or unusual, exaggerated postures, making him more visible not only to other males who might think of invading but also to females, who may notice what a nice territory he has.

An interesting variation on males' use of color to advertise their valuable property to females is found among some bowerbirds and weaverbirds. The male African village weaverbird, for instance, is dull colored, but he builds a colorful nest and jumps up and down beside it saying, in effect, not "look at me" but "look at the gorgeous nest I have built for you." If no females are attracted, or if the color of the nest starts to fade, he will tear it to pieces and build another one.

Another type of sexual dimorphism is the possession of weapons by the male but not the female. Features such as large antlers or horns in many hoofed animals, long tusks in boars, and the enormous size of male seals give a male an edge in combat against other males for mates or breeding territories (Figure 16-8). Hence there is selective pressure for males, but not for females, to possess these traits.

Mating Systems

Each species of animals has a characteristic courtship behavior. One of its functions is to ensure that prospective mates recognize each other as members of the same species, so that a female does not copulate with a member of the wrong species. In **polygamous** species, those in which each animal may mate with more than one other, males frequently have a vast sex drive and little discrimination. They will court almost anything vaguely appropriate, and females must recognize and pick out the right male. The male's appearance, physique, and courtship behavior assist in this. As a corollary, in **monogamous** animals, which mate with only one other individual, the sex drive in both sexes

Figure 16-8
A male destined for death or glory. The horns of this male impala are used in fights with other males for the right to mate, and to identify him to females. Horns take a lot of energy to produce and are heavy to carry.

Figure 16-9
Sexually monomorphic, monogamous animals: Fischer's lovebirds. (William Dilger)

is about equal, courtship is mutual and equally selective, and the sexes are often indistinguishable in behavior and appearance (Figure 16-9).

Polygamy may be divided into polyandry, in which one female mates with more than one male, and the much more common polygyny, in which one male mates with many females.

■ *Polygyny gives dominant males greater reproductive potential than does monogamy, but it can evolve only when the female can raise the young alone.*

Polygyny **Polygyny,** in which one male mates with many females, may evolve where a female gets a better share of some limited resource for her offspring by joining a mated pair, or a male and his harem, than by mating with an unmated male. The resource she gains may be nothing but better genes for her offspring. For instance, male swallowtail butterflies fight for and defend territories at the tops of hills. This is purely a mating area, and any male that can defend it will mate many times. Females fly up the hill, fighting off males who attempt to mate with them on the way. If a female makes it to the top, she will mate with the dominant male there. As a result, her sons get good fighting genes and have a good chance of fathering the next generation of butterflies.

In a polygynous social system, one or a few males live in a group with a number of females. The dominant male is usually the only one that mates with the females when they are at their most fertile, and he must defend his harem not only against predators but also against other males who try to depose him as head of the harem (Figure 16-10). Defending his dominant position requires a lot of energy and constant vigilance. In many species the dominant male is displaced by a rival several times in a year.

Dominance is worth fighting for because dominant males have an enormous reproductive advantage. For instance, male elephant seals guard harems of females as they haul out of the ocean onto rocky coasts to bear their pups and then to mate. In one study, 4% of the males were responsible for 88% of copulations observed in such a polygynous group.

All other factors being equal, polygyny is a more favorable system than monogamy for the male of a species (or at least for the few males who succeed in reproducing). However, polygyny is possible only where the female does not need the male's full-time help in bringing up the young. Its presence or absence, therefore, usually depends on the advantages or disadvantages to the female.

Figure 16-10
The rivals. Red deer stags fight for leadership of a herd of females, a prerequisite for reproduction. The energy he puts into fighting is part of a stag's parental investment in his off-spring. (Biophoto Associates)

Polyandry **Polyandry,** the mating of one female with more than one male, is much less common than polygyny, but it has evolved at least five times in different groups of birds. Consider the case of the jacana, a long-legged bird that runs around on the waterlily pads covering some lakes in Central America. These birds defend small territories in which females lay eggs and then abandon them to their mates to incubate and raise. A female may mate with, and lay eggs for, several males in one breeding season.

There has been much debate as to how polyandry evolves. The most widely accepted theory is that it evolves from a monogamous situation, common in birds, in which both parents share equally in nest-building, incubation, and parental care. If it became advantageous for only one parent to be at the nest at any one time, pure chance might determine which leaves and which stays.

Polyandry also evolves in environments where eggs and young are frequently destroyed. This is the case with some birds that breed at the edges of rivers and streams where flooding often destroys the nest. The female's energy is more usefully devoted to laying replacement eggs than to incubating existing clutches. The male incubates the eggs and the female can devote most of her energy to producing a new clutch of eggs. These will be fertilized and incubated by the original male if the original clutch has been destroyed, or by another male if it has not. In this case, the female also benefits from having her eggs in several different nests, because this increases the chance that at least one clutch will survive. Females compete for males, which are generally scarce because many of them are busy incubating eggs at any one time.

Monogamy When the combined energy of both parents is needed to raise the young, there will be selection for males with **monogamous** behavior (Figure 16-11). For example, monogamy is the most common form of human sexual

Figure 16-11
Monogamy. Like most other birds, swans mate for life, and both parents care for their young. Many birds are more completely monogamous than humans.

297

system (although polygamy arises quite frequently). We can infer that monogamy predominates because the human infant is so demanding to raise. Humans mature more slowly than most other animals, and it seems likely that, throughout much of human history, both parents have had to do their share if they were to raise offspring to sexual maturity with consistent success. It is in the man's interest as much as the woman's that his offspring reach maturity, and so the man, too, will be better off monogamous unless he can provide for the children of more than one wife.

■ *Human monogamy reflects the energy it takes to raise children and the selective advantage of teaching cultural traditions and passing on material goods to the offspring.*

The slow maturation of humans is related to the large size of the human brain and the long time needed to educate it. The large body of culture in any human society is a major factor in the survival and reproduction of its members. However, this culture is not transmitted genetically but must be passed down by example and by language. A particular aspect of culture, unique to humans, probably reinforced the tendency toward monogamy in early humans: the possession of material goods, which are of vast importance to human evolutionary success. It would be strongly advantageous to bequeath things like clothes, a cave or house, or land to your genetic children, and the majority of men throughout history have not amassed enough to split up among the children of more than one wife.

Ecology and Mating Systems

■ *A species' sexual system reflects its ecology, since the location of resources also determines where the organisms spend most of their time and how they behave.*

An important factor in determining the mating system of a species is its **ecology,** the relationship of the organism to its environment. Factors such as the distribution of food, water, nesting sites, and shelter in the environment affect the distribution and social behavior of individuals.

Where food is scarce and found in small, isolated pockets, individuals tend to be solitary and come together only for a short time, in the breeding season. (Bears, badgers, and moose behave like this.) Couples may come together only to mate, or they may form short-lived pairs or colonies while they raise the young.

Alternatively, some animals may live as monogamous pairs defending a territory with widely scattered resources. In this way they can help each other hunt or watch for predators, and they need not spend time and energy searching for a mate when the time comes. Most wild members of the dog family fall into this group, as do various other mammals.

On the other hand, most monkeys and apes can live in troops because their diet is mainly plentiful plant food. The all-important (in evolutionary terms) females and young can eat well and still enjoy the protection of living in groups. As we would expect from this, in situations where food is occasionally inadequate, groups containing only one male are the rule and there is considerable competition among males to enter a troop, since this is the only way they can breed.

An interesting illustration of this point came from a study of baboon troops in Africa (Figure 16-12). Most troops live on the open plain. One troop, however, lived in a woodland area that contained food and water all year. This troop treated their woodlot as a territory, and both females and males defended its boundaries. (This is unusual. Baboon troops are usually protected almost exclusively by males.) Individual members came and went from year to year, but the size of the troop remained the same, and there were fewer males than in any other troop studied. This is almost certainly because the territory provided food for a limited number of individuals, so that nonreproductive males were regularly driven out of the troop.

In contrast, troops on the open plain travel from one feeding ground to another. They are not limited by food and water, but they are exposed to more predators. Here, females must stay closer to their young to protect them. Extra

Figure 16-12
Members of a baboon troop. The troop is a social group that feeds and travels together.

males are welcome in the troop because they supply added protection from predators. The troop can afford to let them eat some of the food, and the extra males also mate occasionally.

16-E Selfishness and Altruism

We have seen that a species' ecology and evolutionary history are often reflected in its mating system. Mating systems have evolutionary effects of their own. For one thing, the mating system has a profound effect upon how closely members of a species are related to one another. In a polygynous herd of horses where only one male mates, all of one year's offspring will be more closely related than they would be in a monogamous group where the year's offspring have different fathers. The degree to which members of a species are related can have some unusual evolutionary results. One of these is selection for **altruistic behavior,** behavior that favors the reproductive success not of the altruistic individual, but of another member of the species. (As far as we know, other animals do not consciously think of their behavior as altruistic, but they may act in ways that we would describe by this term.)

We can understand altruism by thinking of selection as acting, not on individuals, but on genes. An allele for altruism may spread through a population at the expense of a particular individual that carries it. For instance, an allele that favors altruistic behavior toward close relatives enhances its own survival in the population because there is a good chance that closely related individuals also contain the allele. To take an extreme example, if an individual dies to save ten close relatives, one copy of the "kin-altruism" allele is lost but ten or more other copies are saved.

This kind of indirect selection, for genes present in related individuals, is known as **kin selection.** Kin selection is the basis of parental care (which is a dramatic example of altruistic behavior). Suckling her young costs a female mammal energy and benefits her nothing directly, but she is enhancing the survival of her own genes, which are carried by her offspring. The more closely individuals are related, the more likely altruism is to evolve between them because the more likely it is that they share some of the same genes. The relatedness between a parent and its offspring, for instance, is $\frac{1}{2}$, since half the parent's genetic material is inherited by its offspring. This figure, describing how closely two individuals are related, is called an **index of relatedness.**

Indices of relatedness can be calculated for other relatives. In a monogamous species, the chance that a brother or sister also has a particular gene of yours is $\frac{1}{2}$ (50%). The chance that a grandparent, uncle, aunt, nephew, or niece also has a particular allele of yours is $\frac{1}{4}$. The index of relatedness with a first cousin or great grandchild is $\frac{1}{8}$, with a second cousin $\frac{1}{32}$, and with an identical twin 1. Third cousins are much less special: the index of relatedness is only $\frac{1}{128}$, probably not much greater than the chances of sharing the allele with any other individual in the whole population.

We would therefore expect the degree of altruistic behavior to tail off toward relatives who are less close. An allele that prompted an individual to lay down his or her life for a relative would have to save more than two brothers, sisters, or children, more than four uncles and aunts, or more than eight first cousins, and so on, to be selectively advantageous. Otherwise it would not survive, on the average, in enough other bodies to compensate for its loss in the altruistic individual.

Altruistic suicide is not common (indeed, if it were, it would be heavily selected against). What counts more often is the statistical risk of death or loss. Even a third cousin may be worth saving if the risk to the altruist is small. The most common acts of altruism, such as taking a turn at sentry duty, cost the altruist relatively little but contribute markedly to the welfare of related individuals. Alleles that promote the appropriate level of altruism toward other individuals are statistically likely to survive in the population (Figure 16-13).

Research shows that organisms as diverse as honeybees, tadpoles, rodents, and monkeys can recognize their kin, even if they have never met them before. Furthermore, they can distinguish different degrees of kinship. In many cases the clue to kinship is smell, but in visually oriented animals such as monkeys the animals are thought to recognize kin by appearance.

In most animals, altruism between brothers and sisters is not so marked as between parents and their offspring. The reason may be partly that the parent/

Figure 16-13
Conditions for the evolution of altruism, selfishness, and spite. The family is represented by an individual and her sister. The sister in purple is the actor. On average, her sister shares half her genes (purple half of body). The hatchet indicates behavior that harms another. The plate of food symbolizes any valuable resource. (Adapted by permission from *Sociobiology: The New Synthesis,* by E. O. Wilson, Harvard University Press, 1975)

ACT	EVOLUTIONARY CONSEQUENCE
ALTRUISM	The altruist reduces her own fitness but increases that of her sister so much that the genes they share are more common in the next generation.
SELFISHNESS	The selfish individual reduces her sister's fitness but increases her own to an extent that more than compensates.
SPITE	Our heroine (with hatchet) lowers the fitness of an unrelated (unshaded) competitor while damaging her own position: however, the reduction of competition benefits her sister to an extent that more than compensates.

child relationship is highly predictable. There is little probability that the relationship is not really what it seems to be. In most species, a mother can be more sure that her offspring are her own than can a father. A father is vulnerable to deception (unbeknownst to him, his mate may have copulated with another male, who is the true father of her offspring). Therefore fathers may be expected to put less effort into caring for young than mothers do. Similarly, maternal grandmothers are more sure of their grandchildren than are paternal grandmothers. A grandmother can be sure of her daughter's children, but her son might have been deceived into rearing young that do not contain his (and therefore her) genes.

Reproduction is costly. Females in particular invest considerable energy, nutrients, and time in offspring. What determines how much, on the average, the evolutionarily successful parent spends on each offspring?

Clearly, parents must not spread the energy they invest in their young too thinly over many offspring, because then few or none of them may receive enough to survive. Equally clearly, devoting too much investment to too few offspring will be selected against because more prolific parents will leave more progeny in the next generation. As an example, it takes more energy to raise the youngest of a group of offspring than to finish raising an elder brother or sister. If food is scarce, it will clearly pay a parent to concentrate on saving the elder offspring first. On the other hand, if food is ample, it is better to feed the smallest offspring first, because the others are better able to survive a short period without attention. Offspring will presumably be pushed out to care for themselves when the parental investment involved in further attention to them would be more advantageously spent on raising new offspring.

Parental behavior is altruistic. It was a prerequisite to the evolution of animals such as birds and mammals, which produce young that do not survive unless their parents invest a lot of energy in caring for, feeding, and protecting them. There is also strong selection for such behavior, because individuals who do not help raise young do not contribute genes to future generations.

However, some biologists question whether other altruistic behaviors truly exist. Many cases of apparent altruism, when examined more closely, turn out to benefit the "altruistic" individual at a later time. And many feel that humans operate primarily through selfish motives, performing helpful acts either because they count on future gain for themselves when the recipient returns the favor, or because of a selfish desire to feel good about having helped someone else or to avoid unpleasant feelings of guilt for not helping. These views can be reconciled. We would expect altruistic genes to operate by making the actor feel virtuous for helping, or guilty for not helping.

The Social Insects

Biologists have long been intrigued by the specialized division of labor in an insect colony and by the degree of cooperation between individuals. Most members of a colony are sterile workers, who will never leave any offspring. Instead, the workers devote their lives to raising the offspring of other individuals. This is altruism on a grand scale.

Colonies of wasps, ants, termites, and social bees, especially honeybees, are impressive societies. Food is shared communally, and information is exchanged in elaborate ways, including a great range of chemical signals and the "dances" by which worker honeybees indicate the direction and distance of a food source to their nestmates. Workers are fearless in the defense of a colony; many sacrifice their lives on its behalf.

A colony of social insects is a huge family, sometimes numbering several million, all descended from the same mother—the reproductive female, or **queen.** Among the ants, bees, and wasps (all in the insect order Hymenoptera),

■ *A species' mating system determines the degree of relatedness among individuals. This, in turn, determines the extent of selection for altruistic behavior.*

Social Insects and their Environmental Role

A JOURNEY INTO LIFE

Social insects, such as ants, termites, wasps, and social bees, play important roles in human economies. Termites destroy wooden structures, leaf-cutter ants are the scourge of South American agriculture, and honeybees produce honey and beeswax. But the most important environmental role of social insects is that they are vital to the reproduction of thousands of species of plants. In northern agriculture, bees are the most important species. They pollinate tomatoes, cucumbers, fruit trees, and dozens of other crops. In tropical areas, the 9000 species of ants are more important. Some ants are pollinators, and others eat and collect seeds, distributing plants to new areas.

In the United States, crops pollinated by bees are worth about $10 billion annually. Commercial beekeepers transport beehives from one crop to another as each flowers. But beekeepers are now alarmed by the spread of Africanized bees, which are aggressive toward people approaching or handling their hives and are apt to attack *en masse,* chasing what they perceive as intruders for hundreds of metres. A few hundred people have been stung to death; hence the often-heard nickname "killer bees."

What is the source of this problem? The honeybees in the Americas were originally imported from Europe, where they had long been domesticated. However, these bees are not well suited to tropical climates. In an effort to boost honey production, 46 African honeybee queens were imported to southern Brazil in 1956 for crossbreeding with better tempered European bees. But 26 of these African queens escaped and started taking over hives of the more docile European bees, in a population explosion that swept throughout South and Central America at a rate of about 320 kilometres (200 miles) per year. By spring of 1990 they were 240 km (150 miles) south of Brownsville, Texas, where agricultural officials mounted a watch for the first swarms to arrive.

The United States has not awaited this invasion passively. Teams sent to Mexico have killed more than 13,000 African bee swarms, and geneticists have analyzed bees to determine the extent of hybridization between African and the much more numerous European bees. The results are confusing. Some researchers find little evidence for interbreeding between African and European honeybee strains. Others find data suggesting that African genes are spreading both by African drones mating with European queens and by African female lineages migrating into new territory. (Geneticists can tell the difference because mitochondria are inherited only through the female lines: sperm contribute genes, but not mitochondria, to each offspring. The offspring's mitochondria are all derived from the mother's egg. So, when a bee contains African mitochondrial DNA—as do nearly 100% of bees caught in some parts of Mexico—we know that it is descended from an unbroken African female line.)

Some evidence suggests that African and European strains do hybridize, but later generations shift back toward more African genes, at least in tropical habitats. African genes appear less well adapted to temperate climates. African traits are expected to establish a persistent presence in the southern states, but not in the north, with a hybrid zone between the two like one now found in Argentina. Beekeepers will try to swamp the African genes by mass releases of European drones, by periodic replacement of hive queens with known European queens, and by destruction of African swarms. The bad news is that these programs will increase agricultural costs. The good news is that the number of human deaths from Africanized bee stings is dwindling, although it is not clear whether this is because European genes are diluting their aggressive tendencies or because people are learning to avoid close encounters with bees.

workers are infertile females (Figure 16-14). Termite workers also include infertile males. How do the workers gain an advantage great enough for such altruistic, even Kamikaze, individuals to have evolved?

In ants, bees, and wasps, the answer lies in a curious aspect of sex determination. The single mature queen in the colony receives enough sperm during her mating flight to last for the rest of her life. She uses these sperm to fertilize eggs as she lays them, producing female offspring, most of which develop as workers (a few become new queens). But not all of the queen's eggs are fertilized. The unfertilized ones develop as males, which thus have no father. All they have is a single (haploid) set of chromosomes derived from their mother. Thus a male's sperm cells all contain the same haploid set of chromosomes, and his degree of relatedness to his mother is 100%. His mother, on the other hand, is diploid and her son receives only 50% of her genes so, strange as it may seem, her relatedness to him is only 50%.

A colony contains workers who are full sisters, sharing the same (haploid) father (whose sperm are genetically identical), so any allele a worker obtained

(a)

(b)

from her father is also present in each of her sisters. However, there is only a 50% chance that any allele a worker obtained from her mother will also be present in any one sister. In other words, the relatedness between sisters averages $\frac{3}{4}$ (not $\frac{1}{2}$ as in normal sexual animals).

This remarkable situation means that a worker bee is related more closely to her sisters than to her mother! She can therefore increase her evolutionary success more by farming her mother as a sister-making machine than she can by reproducing. More accurately, an allele that promotes making sisters will replicate more rapidly than will an allele for making offspring. This selects for sterility of workers, which has evolved independently at least 11 times in the Hymenoptera and once elsewhere (the termites), an extraordinary example of the complex effects that can be produced by a particular sexual system.

Figure 16-14
Altruistic behavior in honeybees. (a) A worker foraging for nectar to contribute to the hive's communal honeycombs. (b) A queen (center) surrounded by workers—the daughters who raise her offspring, their sisters and brothers. (a, Carolina Biological Supply Company; b, Biophoto Associates)

SUMMARY

All species have a certain amount of genetic variation among their members as a result of mutation. The amount of such variation is slight, and members of species that reproduce asexually are, therefore, genetically very similar to one another. Members of sexual species are much more variable because genes are reshuffled during meiosis and form new combinations at fertilization. Asexual reproduction uses energy more efficiently than sexual reproduction and is common in species that occur in many places in the world, produce many offspring in a short time, and are adaptable to changing conditions. For more localized and specialized species, the energy wasted in sexual reproduction is worthwhile because at least some of the genetically different individuals so produced can usually survive and evolve in changed conditions.

Sexual reproduction probably originated as the accidental fusion of two asexual reproductive cells. The advantage of having one gamete stuffed with a food supply for the new individual selected for the evolution of large immobile eggs and small sperm that travel to the egg. Adaptations such as specific mating seasons and hermaphrodism help to overcome the difficulties encountered when two individuals must act in concert to achieve reproduction.

The sexual system of a species is determined by the amount of energy each sex puts into producing and rearing offspring and by ecological factors such as the distribution of food and the prevalence of predators. Because it takes more energy to produce eggs than sperm, females are usually more selective than males in their choice of mate. Sexual dimorphism may arise when the selective pressures on the two sexes conflict, or when males and females have different roles in reproduction. Sexual dimorphism tends to be greater in polygamous species. In monogamous species, the members of both sexes must choose their mates with care, and the roles and behavior of the two sexes are more similar.

A species' sexual system determines how closely members of a population are likely to be related to each other. An allele that makes an individual perform altruistic behaviors detrimental to itself can be selected for when the behavior enhances the survival of other copies of the allele in the individual's relatives, allowing that allele to spread through the population. Altruistic behavior is most likely to arise in species in which closely related individuals spend much time together. Parental behavior is the best example.

SELF-QUIZ

1. An advantage of sexual reproduction over asexual reproduction is that it:
   a. increases the mutation rate
   b. increases genetic variability in a population
   c. produces larger offspring
   d. reduces the offsprings' risk of death during development
   e. gives organisms something to do on Saturday night

2. Some organisms reproduce asexually when environmental conditions are (favorable, unfavorable) and sexually when conditions are (favorable, unfavorable) for growth.

3. The main advantage of having a large egg and a small sperm is that it:
   a. has separate male and female sexes
   b. provides for the nourishment of the growing embryo
   c. assures cross-fertilization
   d. involves two immobile gametes

4. It is generally true that males increase their chances of evolutionary success by trying to mate with as many females as possible. One possible exception to this generalization might occur when:
   a. there are many more females than males
   b. there are about equal numbers of males and females
   c. the father's care is required to raise the offspring
   d. there are many predators
   e. the male holds a territory against other males.

5. When food is distributed in such a way that an animal must spend a large part of its day traveling from one place to another to find enough to eat, what type of mating system would you expect it to have?
   a. monogamy
   b. polyandry
   c. polygamy
   d. polygyny

6. A childless human male is most likely to enhance his evolutionary success by altruistic behavior toward the children of:
   a. his sister
   b. his brother
   c. his mother
   d. his grandmother
   e. his niece

7. Monogamy is most likely to be found among birds and mammals whose young are:
   a. born or hatched helpless and in need of much parental care
   b. born or hatched precocial (able to care for themselves, like a horse)
   c. nourished on milk
   d. part of a litter or clutch of more than 50
   e. carried by the female from conception to birth

QUESTIONS FOR DISCUSSION

1. List as many possible selective advantages as you can think of for courtship rituals.

2. Do humans have courtship rituals? Is courtship in humans mutual or is it carried out predominantly by one sex? Why?

3. Female walruses must bear their young on land, but suitable stretches of beach are scarce. Similarly, seaside nest sites for gulls are scarce. Both have crowded breeding grounds. Walruses eat mussels, clams, and so forth, whereas gulls will eat almost anything—including the egg or chick next door. Explain what mating system you would expect each of these animals to show.

4. What is the advantage of hermaphrodism over separation of the sexes? What selective pressure might have led to evolution of species with separate sexes? Why are there never more than two different sexes in a species?

5. Some people do not reciprocate a favor when it is their turn. What might be some possible outcomes if such a nonaltruistic allele were present in a population together with an altruistic allele?

6. Explain sibling rivalry for parental favors and parental impartiality toward offspring in terms of degree of relatedness.

7. Do you think that altruistic behavior in humans is genetically controlled?

8. Researchers Trivers and Hare weighed the fertile males and females produced in colonies of 20 species of ants and found that the investment in females was three times the investment in males (by weight). Can you explain how this situation may have been selected for?

9. How might the invention of birth-control methods and labor-saving household appliances affect the monogamous mating system of humans?

10. Martin Daly and Margo Wilson analyzed statistics on homicides within human families in the United States, Canada, and elsewhere. Can you explain the genetic or evolutionary theory that might account for the following examples of their findings:
    a. In nonindustrial societies, infants are most often killed by parents for the following reasons: (1) doubt that the child is the parent's own; (2) conviction that the child is weak and unlikely to produce offspring as an adult; (3) external pressures such as food scarcity and burdensome demands from older siblings that reduce the baby's chance of surviving.
    b. In industrial societies, parents are more likely to kill infants than to kill older children.
    c. Mothers are more likely than fathers to kill infants.
    d. Disproportionate numbers of child killings are by stepparents.

SUGGESTED READINGS

Borgia, G. "Sexual selection in bowerbirds." *Scientific American,* June 1986. A study of female choice of mates in a species that attracts females by building elaborate nests.

Cole, C. J. "Unisexual lizards." *Scientific American,* January 1984. An interesting study of parthenogenetic vertebrates.

Dawkins, R. *The Selfish Gene.* New York: Oxford University Press, 1976. Natural selection from the gene's point of view; good discussion of altruism.

Grandjean, E. C. "Love among the icebergs." *The Living Bird Quarterly,* Summer, 1986. A discussion of the ecological and evolutionary basis of polyandry.

Wilson, E. O. *Sociobiology: The New Synthesis.* Cambridge, MA: Harvard University Press, 1975. After years of work on social insects, Wilson produced this large book on social behavior and its genetic basis. The book caused a furor because it treats human behavior as just another example of the social behavior of animals. Fascinating, but not very easy reading.

Origin of Life

CHAPTER

17

Cupid Terrace, Mammoth Hot Springs

Amargosa River, Death Valley

Chambered nautilus

*I*f we could trace the ancestry of living organisms, we would find a long line of cells stretching back billions of years. Each cell came from division of a previously existing cell . . . but where did the first cell come from?

Some people say the first organisms came to earth in spaceships or meteorites, but this only moves the question of how life began to a more distant arena, beyond our reach to study. Most scientists think that life on earth started here. If we assume this is true, we can search our earth for evidence of how and when life arose (Figure 17-1).

Geologists who study ancient rocks tell us that the early earth was very different from what we see now—even if we were to subtract all the living things. In fact, conditions on the early earth would have killed the vast majority of modern organisms almost instantly. The atmosphere was toxic, the heat intense. Yet this poisonous inferno gave rise to life: chance chemical events, over a period of hundreds of millions of years, built the simple compounds of the primordial earth into organic monomers, then polymers, then aggregates of many polymers. By a few billion years ago, some of these aggregates had become cells. This scenario was first proposed in 1924 by a Russian, Alexander Oparin. In 1929 Oparin's ideas, not yet translated from the Russian, were echoed by J. B. S. Haldane in England.

Research supports these predictions. Scientists have simulated the prebiotic ("before life") world, using their best guesses about the temperature and the available chemicals and energy sources. Amazingly, the nonliving systems formed in the laboratory show many features typical of life.

What properties would we seek in a chemical aggregate on its way to becoming a cell? In earlier chapters, we saw the basic features of present-day cells. The forerunners of true cells should have shown at least the rudiments of these:

1. A large number and variety of organic molecules, many of them joined into long polymer chains.
2. A lipid-protein membrane that separates the cell from its environment, selectively controls the exchange of materials between the two, and maintains a difference in the concentration of molecules and ions between them.
3. Proteins that help to exchange substances with the environment, form cell structures, and most importantly, catalyze the cell's metabolic reactions.
4. Nucleic acids that contain precise instructions for making proteins.

In this chapter, we consider what the prebiotic earth was like and outline how nonliving chemicals may have become organized into living cells. Many interesting experiments shed light on aspects of the problem and show how various features of life could have evolved. This chapter omits a great deal of evidence that requires advanced knowledge of chemistry and of the metabolism of obscure types of bacteria. Hence the evidence given here looks much scantier than it is. But it is true that the subject matter of this chapter, perhaps more than any other in the book, is highly speculative. Future work will fill more gaps in the picture, but we shall never really know whether or not it is a good likeness of the events that actually took place.

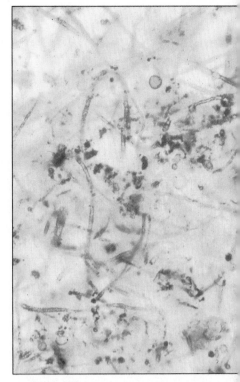

Figure 17-1
Fossils of early cells. These fossil prokaryotes, found by microscopic examination of thin sections of rock, lived 2 billion years ago. They came from the Gunflint Iron Formation, Ontario. (Andrew H. Knoll. Barghoorn and Tyler, *Science* 147:563, 1965)

- Chemical and physical conditions on the early earth differed from those here today.
- Random chemical reactions built up supplies of organic chemicals, which assembled into large aggregates. Some of these eventually came to have the organization, function, and reproduction characteristic of life.
- Living organisms themselves are responsible for many of the differences between early and present-day environmental conditions on earth.

17-A   Conditions for the Origin of Life

Under what conditions can life arise? Scientists believe there are four basic requirements: the right chemicals, including water, various inorganic ions, and organic molecules; an energy source; absence of oxygen gas (O_2); and eons of time. Of the necessary chemicals, water is abundant on earth, and the inorganic ions occur in rocks, volcanic gases, and the atmosphere. How were organic molecules produced from these simple chemicals without the enzymes of living organisms, which catalyze most of these reactions today? Before we answer this question, let us look at the last two conditions.

Time   It may take millions of years for a given quantity of some chemical to undergo a reaction that an enzyme could catalyze in a second or two. In the prebiotic era, inorganic chemicals reacted to form organic molecules without the help of enzymes—that is, extremely slowly. Once made, the simple organic chemicals had to come together in larger, more complex structures. The chances of this happening are minuscule.

Given enough time, however, even very improbable events are almost bound to occur. For example, if the probability that an event will occur in a year is one in a thousand, the probability that it will not occur is 0.999; the probability that it will not happen in two years is $(0.999)^2$; in three years, $(0.999)^3$, and so on. Table 17-1 shows that there is a very small probability that the event will not happen at least once in 8128 years. Conversely, there is a very high probability (0.9997) that it *will* happen *at least once*—and once may have been enough for the origin of life on earth. The events required for the origin of life were much less probable than one in a thousand, but they had plenty of time to happen. The earth formed about 4.6 billion years ago, and signs of life appear in rocks laid down about 1.1 billion years later (that is, about 3.5 billion years ago). So, unlikely as living systems are, they had so much time to evolve that their origin was probably inevitable!

Table 17-1	Probability that an Event Will Not Happen

Given this probability:	in 1 year	0.999
Then:	in 2 years	0.998
	in 3 years	0.997
	in 4 years	0.996
	in 1024 years	0.359
	in 2048 years	0.129
	in 4096 years	0.017
	in 8128 years	0.0003

■ *The origin of life required a supply of simple inorganic chemicals in the absence of molecular oxygen; an energy source to promote their reaction; and eons of time. All these conditions existed on the early earth.*

Absence of Molecular Oxygen   The origin of life required the absence of oxygen because O_2 is a powerful oxidizing agent. Oxidation would have broken down organic molecules, or at least made them useless to a prebiotic system, at a relatively rapid rate. Organic molecules exposed to O_2 on the early earth would not have lasted long enough to form more complex structures. This is one reason why we do not find new organisms arising from organic matter on earth today. (Another is that free organic molecules are usually absorbed and used as food by bacteria and fungi even before oxygen can damage them.) Fortunately for us, the earth's atmosphere was originally devoid of oxygen.

17-B   The Prebiotic Earth

The earth formed by solidification and accretion of matter from space about 4.6 billion years ago. Chaos prevailed for the next 0.2 to 0.3 billion years, as bombardment by meteorites added material to the forming planet. About 4.3 billion years ago, conditions began to stabilize.

At this time the early earth had a **reducing atmosphere,** so called because the reducing agent hydrogen was the most common element. Scientists believe that this reducing atmosphere consisted of H_2 (molecular hydrogen) and hydrogen compounds of other elements, such as H_2O (water vapor), NH_3 (ammonia), and CH_4 (methane). This view is based on several lines of evidence. Hydrogen is enormously abundant in the solar system, especially in the gases that formed the sun and planets. Even today the atmospheres of Jupiter and Saturn are mostly H_2, water vapor, and ammonia. The composition of gases now emerging from volcanoes, as well as chemical calculations, also support the idea of a primitive reducing atmosphere.

However, H_2 is so light that it soon escaped the earth's gravity and went off into space. Sunlight, much brighter on earth than on the outer planets, decomposed the ammonia into H_2 (which also escaped) and nitrogen gas (N_2). Similarly, methane was replaced by CO_2. By the time life was evolving, the atmosphere was only mildly reducing. The most abundant gas was probably water vapor, which later condensed to form the oceans. Next were carbon dioxide and carbon monoxide, which later became locked in rocks in the form of carbonates, where they are today. Also present were nitrogen (which makes up about 80% of the air today), smaller amounts of other gases, and negligible O_2. Sunlight striking water or carbon dioxide released small amounts of O_2, but this oxygen did not last long: it soon reacted with iron or sulfur compounds. The oxidizing atmosphere we breathe today is about one fifth O_2, produced almost exclusively by the photosynthesis of green plants.

17-C Production of Organic Monomers

In 1953 Stanley Miller, then a graduate student, built a small-scale model of conditions on the early earth, including an "ocean" and a primitive reducing "atmosphere" (Figure 17-2). Electrodes in the "atmosphere" chamber gave off electric sparks, representing lightning, a possible source of energy to drive chemical reactions on the early earth. After a week, the "ocean" contained many small organic compounds, including amino acids, carried from the "atmosphere" by the condensation of "rain."

Miller's amino acids generated much excitement. Amino acids are the building blocks of proteins, and biochemists recognized the enormous variety of proteins and their overwhelming importance in the activities of living cells. Indeed, proteins were regarded as *the* class of substances necessary for life.

Many people have done variations of Miller's experiment, using different proportions of starting gases, and sometimes including hydrogen sulfide (H_2S), carbon dioxide, or inorganic ions. Energy sources used, besides electric sparks, include heat, bright sunlight, ultraviolet light, and radioactivity, all possible

Figure 17-2

Miller's apparatus for simulating prebiotic conditions. The "atmosphere" was a mixture of hydrogen gas (H_2), methane (CH_4), and ammonia (NH_3) in a glass chamber the size of a soccer ball. Sparks from electrodes represented lightning, a source of energy. The "ocean" was heated using a gas-jet "sun." The resulting water vapor could provide oxygen to the reaction and also carry organic molecules back to the sea in "rainfall." (Photo courtesy of Dr. Stanley Miller)

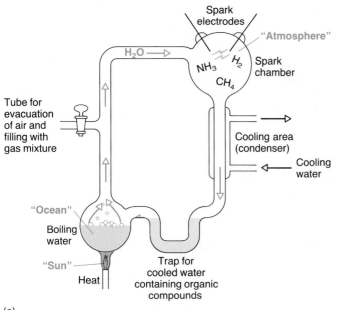

(a)

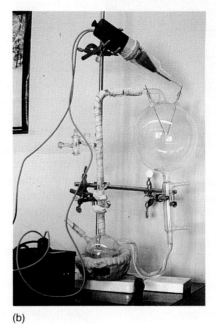

(b)

Table 17-2	Important Precursors of Organic Compounds Found Beyond the Earth	
Molecule	**Name**
NH_3	Ammonia
$HC\equiv N$	Hydrogen cyanide
$N\equiv C-C\equiv N$	Cyanogen
CH_4	Methane
H_3C-CH_3	Ethane
$H_2C=CH_2$	Ethylene
$HC\equiv CH$	Acetylene
CO	Carbon monoxide
H_2CO	Formaldehyde
CH_3CH_2OH	Ethanol

sources of energy on the prebiotic earth. As long as no O_2 is present, these simulations yield a variety of organic products, including a host of compounds not formed in Miller's mixture.

Astronomy and geology also provide evidence that organic monomers can form without the agency of living organisms. Several important small molecules that are raw materials or intermediates in the synthesis of organic monomers are found in stars, dust clouds, space, and the atmospheres of other planets (Table 17-2). The European Space Agency's Giotto spacecraft, which flew past Comet Halley, even detected polymers of one of these small molecules (formaldehyde). Meteorites, chunks of material that fall from space, contain a wide variety of more complex organic monomers, including amino acids, alcohols, sugars, the nitrogenous bases found in nucleic acids, and lipid-like molecules capable of forming films similar to membranes. Small amounts of six common amino acids were also found in material brought back from the moon. Even now, organic compounds are formed on earth **abiotically** (without life: neither within living cells nor under their influence). Hot metallic carbides in volcanic gases and lava form hydrocarbons when they react with water, but this occurs at a very low rate.

All these lines of evidence support the contention that organic compounds could have formed on the prebiotic earth by the action of available forms of energy. Without oxygen to destroy them or organisms to absorb them, these compounds would have accumulated. Haldane suggested that eventually the sea had the composition of a "hot, dilute soup."

Nowadays, many scientists favor a different view: organic molecules formed not a soup, but a sludge, attracted to the complex, charged surfaces of clay or other minerals in the pores of underwater rocks (Figure 17-3). These minerals could serve both as templates holding organic molecules in place, and as catalysts of chemical reactions. Indeed, the chemical industry today uses mineral catalysts to produce many organic compounds, and many enzymes also use mineral ions as coenzymes.

17-D Formation of Polymers

Sidney Fox found that heating a dry mixture of amino acids produced **proteinoids,** protein-like polymers of about 100 amino acids each. The heat (about 60°C) quickly evaporated the water released as the amino acids linked together. This prevented the proteinoids from hydrolyzing back into amino acids. Such events might have happened on the early earth if tide pools trapped seawater containing amino acids, which would have polymerized as the water evaporated on hot, sunny days.

Clay and other minerals also enhance the formation of polymers. Many kinds of monomers adhere to the surfaces of mineral particles. This increases their local concentration. Polymers form readily when the minerals are dried and then warmed.

Some proteinoids exhibit enzyme-like properties. For example, they can catalyze some chemical reactions, and their catalytic properties can be destroyed by overheating and by chemicals that inhibit enzymes. The conclusion must be that molecules similar to enzymes, which are so vital to life, could have been produced on earth before living organisms existed.

17-E Formation of Aggregates

Natural or artificial polymers placed in water may aggregate and form larger structures (Figure 17-4). These aggregates show some features of living cells. Any lipids present form membrane-like coatings around the aggregate's exte-

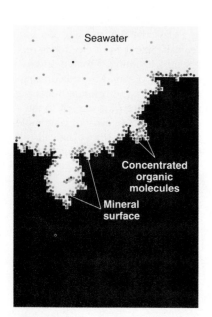

Figure 17-3
Concentration of organic molecules from a dilute solution. Clay and other minerals have irregular, electrically charged surfaces. Dissolved ions or polar molecules become concentrated at the interface between the minerals and the surrounding solution. Minerals not only hold other molecules, but also act as catalysts for chemical reactions between them.

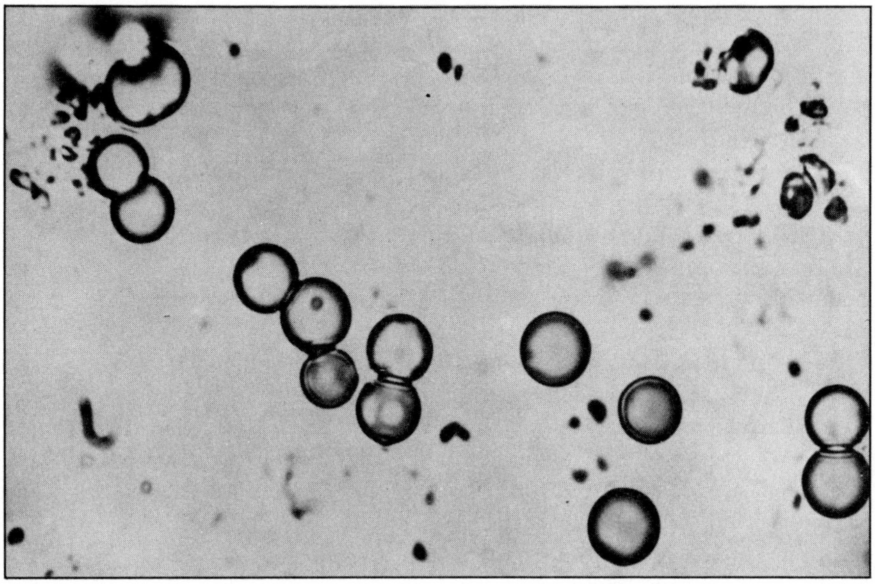

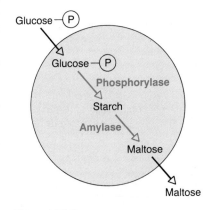

Figure 17-4
Proteinoid microspheres. These microspheres formed when water was added to proteinoids. Each is 1 to 2 micrometres in diameter. The microspheres have a double layer of proteins as a boundary between the interior and the external environment. This boundary is selectively permeable and admits polynucleotides (in theory, the precursors of nucleic acids) very readily. Like living cells, microspheres have internal structure: watery areas, lipid-like areas, and boundary-layer areas provide sites for distinct chemical activities. (Sidney Fox)

rior. Aggregates are selectively permeable and accumulate some kinds of monomers, but not others, from the surrounding medium. They also catalyze certain reactions in their interiors, and they grow, eventually breaking up into smaller aggregates.

These aggregates illustrate the saying, "the whole is greater than the sum of its parts." An aggregate has both structures (such as the "membrane") and functions (such as the ability to "choose" molecules, collect them, and grow) not found in its component molecules.

Aggregates are extremely limited in the function of their "membranes" and "metabolism," and they lack a reliable means of reproducing any advantageous molecules they might contain. Nevertheless, they do show that organic polymers can associate and interact in discrete units set apart from their surroundings by distinct boundaries. Such units are prerequisites for the evolution of life, because selection must have entities to "choose" among, and it seems likely that these entities must have been combinations of interacting molecules with potentially life-like properties, rather than individual polymers.

17-F Beginnings of Metabolism

Aggregates studied in the laboratory have a simple metabolism of one or a few reactions—a far cry from a living cell with its thousands of reactions (Figure 17-5). Oparin wrote: "The path followed by nature from the original systems of protobionts [pre-living aggregates] to the most primitive bacteria . . . was not in the least shorter or simpler than the path from the amoeba to man."[1] Along the way, the protobiont must have added hundreds of chemical reactions to its metabolism; feedback mechanisms evolved to regulate the various metabolic pathways; each metabolic enzyme became more efficient; and mechanisms for protein synthesis and replication of genetic material evolved, allowing the protobiont to make many copies of its metabolic enzymes and other proteins.

Protobionts arrived at the threshold of life by **chemical selection,** a process similar to natural selection but acting on nonliving systems. At first, selection probably favored mere longevity: aggregates with the most stable com-

Figure 17-5
A rudimentary metabolism. This two-step "metabolic pathway" was observed by Oparin in an aggregate (gold circle) made from natural polymers, the enzymes phosphorylase and amylase. The aggregate absorbed glucose phosphate from the surrounding solution. Enzymes taken up by the aggregate polymerized the glucose phosphate into starch and then digested the starch to maltose. The appearance of maltose in the aggregate's surroundings showed that this simple "metabolic pathway" had been completed.

[1] This statement displays poetic license and should not be taken to imply that an amoeba was a direct ancestor of humans.

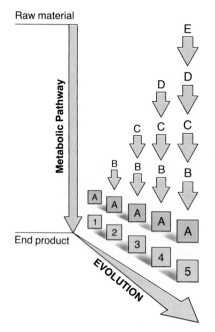

Raw material

Metabolic Pathway

End product

EVOLUTION

Figure 17-6
Evolution of metabolism. This diagram shows five stages in the evolution of a metabolic pathway that makes end product A, which was originally obtained from a pool of abiotically formed molecules.

binations of chemicals lasted longer before disintegrating. As more and more of these aggregates accumulated, it was a real asset for them to have catalysts for chemical reactions that made them more stable.

At first, the raw materials for these reactions were probably abundant compared to the few aggregates that needed them. However, as these successful, stable aggregates grew and fragmented, and as new ones formed spontaneously, more and more similar systems competed for fewer and fewer raw materials. Now selection favored the aggregates that were most efficient at competing for the now-scarce material, and for those that could convert a second, abundant material to the first, now-scarce one. For example, suppose that the crucial molecule A was in short supply because of intense competition among aggregates taking it up faster than abiotic forces were producing it. Any aggregate containing a catalyst that converted abundant molecule B to the now-scarce A would have an advantage—until there were so many systems using the reaction B $\longrightarrow$ A that B also became scarce. Now a system that evolved a catalyst for the reaction C $\longrightarrow$ B as well as B $\longrightarrow$ A would survive, while B-users would disappear, and so on. Hence metabolic pathways became ever longer. In addition, they evolved "backwards," from the useful end product to less directly useful raw materials (Figure 17-6).

Origin of Energy Metabolism

One of the first requirements of a metabolic system is to trap and use energy to drive various reactions. All organisms living today have hydrogen ion (H^+) pumps housed in membranes. Perhaps similar pumps provided energy that primitive aggregates or cells used to transport materials through their membranes. Such pumps might have been driven by light, an abundant energy source used by simple H^+ pumps in some kinds of bacteria today. The modern electron transport systems of respiration and photosynthesis are more elaborate H^+ pumps.

An older, more conventional view holds that energy originally came from ATP in the primordial soup. At first, ATP was a relatively common and available energy source. As ATP ran low, systems that could use their H^+ membrane potentials, or energy released by chemical reactions, to make ATP from ADP and P_i would have had a selective advantage.

The need for an ATP-regenerating system probably selected for anaerobic pathways of fermentation, such as glycolysis, the fermentation of sugars (Section 7-B). Glycolysis is undoubtedly extremely ancient, judging by its presence in nearly all modern organisms. The citric acid cycle and electron transport chain of aerobic organisms provide highly efficient ways to extract further energy from the products of fermentation. This aerobic respiration arrived relatively late in the evolution of metabolism.

17-G   The Beginnings of Biological Information

Self-organized aggregates of polymers show some similarities to modern cells, but they cannot be considered "living" because they lack genetic information. In modern organisms, **genetic information** allows cells to make proteins to precise specifications and to pass on copies of this information to their offspring. A cell's DNA contains this biological information. DNA directs the formation of RNA, which in turn directs the linking of amino acids into proteins (Chapter 10). Biologists diagram this information flow as:

$$\text{DNA} \longrightarrow \text{RNA} \longrightarrow \text{protein}$$

In this system, the three kinds of substances show division of labor. DNA is very stable, and so it can serve as a file copy of the information needed to make

RNAs in as many copies as needed. It also spends some time being replicated before the cell divides. RNA takes part in protein synthesis. Proteins, in turn, perform the actual work of the cell's metabolism, exchanging substances with the environment, catalyzing metabolic reactions, and even assembling new proteins and nucleic acids.

When we ask how such a complex system could have evolved, we find a problem. In the synthesis of nucleic acids and proteins, the nucleotide and amino acid monomers must be joined to each other by enzymes. However, these enzymes must first be made according to instructions provided by already-existing nucleic acids! Which came first, the enzymes or the nucleic acids?

In the 1980s, a possible answer was suggested by a surprising discovery: not only can RNA carry genetic information, but it can also act as a catalyst. (This work won a 1989 Nobel Prize for Thomas R. Cech and Sidney Altman.) Experiments show that RNA polymers can form without enzymes. In addition, short RNA strands complementary to existing RNA polymers form readily. RNA can catalyze RNA splicing and the joining of short RNA chains into longer ones. This and other evidence suggest that the first genetic information was RNA, and that it could replicate without aid from proteins.

The "RNA world" model of early life proposes that the first cells used RNA as their genetic information and had a complex metabolism using RNA catalysts. Gradually, segments of catalytic RNA molecules were replaced by more efficient protein strands. Today, RNA catalyzes only a few kinds of reactions; the vast majority are catalyzed by proteins.

There are two problems with this scenario. First, the monomers needed to make RNA polymers do not form readily under abiotic conditions. The other problem is that apparently RNA can catalyze only reactions involving RNA and its nucleotide monomers. It seems that RNA cannot catalyze the wide range of reactions catalyzed by modern protein enzymes.

Suppose, however, that a self-replicating RNA found itself in an aggregate with proteinoids that catalyzed reactions useful to the aggregate or to the RNA's reproduction. It is quite likely that the RNA and proteinoids would interact with each other to some extent. As time went on, chemical selection would preserve, not favorable proteinoids nor favorable RNAs alone, but rather those aggregates in which the two became more and more precisely associated.

Experiments have shown possible steps in this association. It now appears that the genetic code did not arise randomly, but is based on slight chemical differences that make certain amino acids and nucleotides more likely to associate with each other. If these associations helped amino acids or nucleotides join together, then their order in the new polymers was not random but directed by their companions.

Even a rudimentary system of protein synthesis would have helped a prebiotic aggregate greatly because it could produce at least rough copies of proteins that catalyzed favorable reactions. At first, replication of nucleic acids was probably quite inaccurate. Frequent changes in the order of nucleotides produced new amino acid sequences in proteins, some of them better catalysts than self-ordered proteinoids. So proteins and RNA must have evolved together.

DNA was added to the informational store later than RNA. DNA is more stable than RNA, and it is easier to copy faithfully. RNA eventually became the "go-between" carrying instructions from DNA to proteins today.

Reproduction and metabolism must have evolved together, although we have discussed them separately. Reproduction relies on metabolism to provide the energy and raw materials needed to replicate the genetic information and to produce many copies of proteins. Metabolism, in turn, relies on genetic information to direct the production of all the enzyme catalysts needed to make these raw materials. Once a reliable means of protein synthesis and gene replication evolved, the story of the origin of life was complete.

■ *The proteins of metabolism and protein synthesis, and the nucleic acids carrying the codes for these proteins, must have evolved together, in increasingly precise association.*

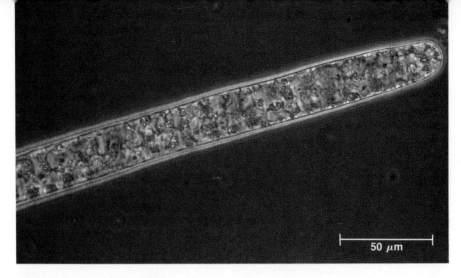

Figure 17-7

Figure 17-7
A chemosynthetic bacterium. The relatively large cells grow in long filaments. The cells get energy, not from light, but by oxidizing H_2S to sulfur, which builds up as yellow particles visible in the cell. Some species of these bacteria are important in the culture of rice because they remove toxic H_2S from the water of the rice paddies. (P. W. Johnson and J. McN. Sieburth, University of Rhode Island/ BPS)

■ *The evolution of autotrophy vastly increased the amount of life the earth could support.*

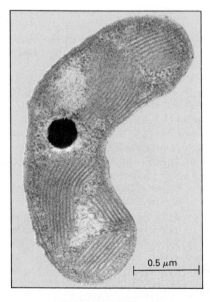

0.5 μm

Figure 17-8
A purple sulfur bacterium. In this lengthwise section, the dark circle is a storage granule. The many sets of parallel lines are internal photosynthetic membranes, which contain the photosynthetic pigment molecules used to trap light energy. (Compare these with the photosynthetic membranes of the chloroplast in Figure 8-7.) (S. C. Holt, University of Massachusetts/BPS)

17-H Heterotrophs and Autotrophs

The advent of the first true organisms marked the end of the era of chemical selection and the beginning of the era of natural selection. Competition grew more intense, and primitive cells evolved faster ways to get energy and convert it into offspring. For a long time all organisms were anaerobic, and they were all **heterotrophs,** feeding on organic molecules made outside their own bodies. At first, they absorbed food from the surrounding soup (or sludge). But as this food became scarce, some cells began to devour their neighbors to obtain nutrients. Others evolved ways to acquire energy from other sources. They became **autotrophs,** organisms that make their own food from inorganic molecules.

The evolution of autotrophy was tremendously important to the evolution of life. It freed the living world from dependence on the slow production of food by abiotic means. An organized metabolic assembly line made food much faster, and in great abundance. Autotrophic organisms could therefore grow and reproduce more rapidly. Inevitably, some of them fell victim to their heterotrophic neighbors, whose population also grew as they exploited this new food supply.

The photosynthesis of green plants is a very advanced form of autotrophy (Chapter 8). Other, less successful evolutionary "experiments" in do-it-yourself food production are still found among some modern bacteria. **Chemosynthetic** bacteria use energy released by various inorganic chemical reactions to make CO_2 into organic compounds, even in the absence of light (Figure 17-7).

Various kinds of bacteria also show a range of ability to trap light energy. Studying these bacteria allows us to reconstruct the steps in the evolution of photosynthesis. At first, light energy was probably used to create an H^+ membrane potential, which supplied energy to make ATP and transport substances through the membrane. This met some energy needs, but the cell still had to get food from its surroundings. The next step was using light energy to obtain hydrogen atoms to reduce CO_2 as it was fixed—the hallmark of true photosynthesis. Early photosynthetic bacteria used easy hydrogen donors: their own organic waste products, or inorganic hydrogen gas or hydrogen sulfide (H_2S) (Figure 17-8). The use of water as a hydrogen donor came quite late because it is hard to extract hydrogen atoms from water. However, water is much more abundant than other hydrogen donors, and so organisms that can use water as a hydrogen source have an enormous selective advantage.

17-I Respiration

Photosynthesis that uses water as a hydrogen donor also releases O_2. Indeed, virtually all of the O_2 in the air today has come from water-splitting photosynthesis. When early organisms first evolved this type of photosynthesis, they pro-

Time (billions of years ago)	Event
4.6	Earth originates
4.3	Conditions on earth stabilize
3.8	Ocean mineral content similar to today's; atmosphere like today's but without O_2; carbon as CO_2 (Isua formation rocks, Iceland)
3.5	Earliest known stromatolites (mats of prokaryotic cells) (Australia)
3.5–3.3	Probable cyanobacteria with oxygen-generating photosynthesis
2.8	Oxygen-generating photosynthesis (for sure)
2.5–2.0	Deposits of banded iron, oxidized by O_2 from photosynthesis
2.0	Great diversity of bacteria (Gunflint chert, Ontario)
	Atmospheric oxygen reaches 1%
	Respiration appears
1.45	Eukaryotic cells; sexual reproduction allows more rapid evolution
0.7	Soft-bodied animals (jellyfish, worms)
0.6	Hard animal skeletons (Cambrian Period begins)

Table 17-3 Some Important Milestones in the Origin of Life

Subsequent events are outlined in the book's endpaper, "Geologic Time Scale."

duced small amounts of O_2. At first, the oxygen reacted with iron-rich minerals dissolved in the surrounding shallow seas, and the resulting oxidized minerals settled to the bottom in layers that later formed rock. Once these minerals were exhausted, oxygen began to accumulate in the atmosphere (Table 17-3). This produced a grave environmental crisis: O_2 destroys flavins, an important group of coenzymes, and without their flavins organisms could not survive. The threat posed by oxygen was probably the selective pressure that brought about the next important evolutionary advance, respiration.

Aerobic respiration probably began with cells that combined their wastes (such as the pyruvate resulting from glycolysis) with O_2, thereby disposing of two toxic substances at once. As time went on, more and more steps were inserted into this process. The modern citric acid cycle and electron transport chain release and capture more of the energy from the waste products of glycolysis. Glycolysis yields two ATPs per glucose molecule, whereas respiration yields more than ten times as much. This tremendous energy bonus from respiration made it possible for organisms to grow and reproduce much faster. It also allowed them to "experiment" with new enzymes and new structures that used a lot of energy but made them superior competitors, able to outstrip their anaerobic neighbors. Today anaerobic organisms are restricted to habitats lacking enough oxygen to support aerobic forms of life.

17-J Origin of Eukaryotes

The next big jump in evolution was the rise of eukaryotic cells, containing membrane-bounded organelles (Figure 17-9). Mitochondria and chloroplasts probably originated as bacteria that came to live within other cells (Section 19-E).

A major advance shown by eukaryotes is sexual reproduction, which involves meiosis and fertilization. This gives rise to genetic variation by reshuffling existing genes, in addition to the variation produced by mutation to form new genes. Genetic variability is the raw material for natural selection, and so for

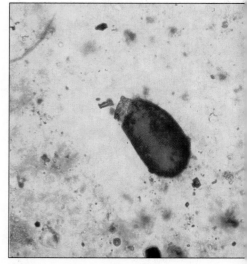

Figure 17-9
Earliest known heterotrophic protist. This organism, which lived 800 million years ago, was found in rock from the Backlundtoppen Formation, Spitsbergen (an island several hundred miles north of Norway). (Andrew H. Knoll. Knoll and Calder, *Palaeontology* 26:467, 1983)

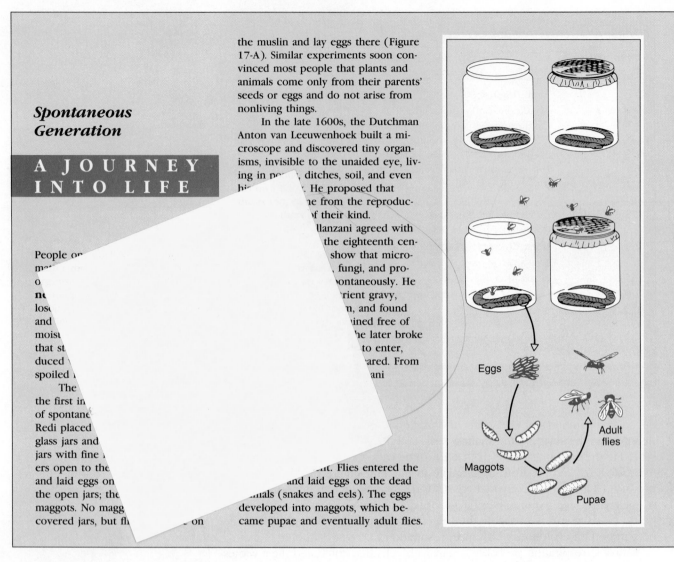

Spontaneous Generation

A JOURNEY INTO LIFE

People o
m
o
n
los
and
moist
that st
duced
spoiled
The
the first in
of spontane
Redi placed
glass jars and
jars with fine
ers open to the
and laid eggs on
the open jars; the
maggots. No magg
covered jars, but fl ... on

the muslin and lay eggs there (Figure 17-A). Similar experiments soon convinced most people that plants and animals come only from their parents' seeds or eggs and do not arise from nonliving things.

In the late 1600s, the Dutchman Anton van Leeuwenhoek built a microscope and discovered tiny organisms, invisible to the unaided eye, living in po... , ditches, soil, and even hi... . He proposed that ...ne from the reproduc- ...f their kind. ...llanzani agreed with ... the eighteenth cen- ... show that micro- ... , fungi, and pro- ...ontaneously. He ...rient gravy, ...n, and found ...ined free of ...he later broke ... to enter, ...ared. From ...ani

...nt. Flies entered the ...and laid eggs on the dead ...mals (snakes and eels). The eggs developed into maggots, which became pupae and eventually adult flies.

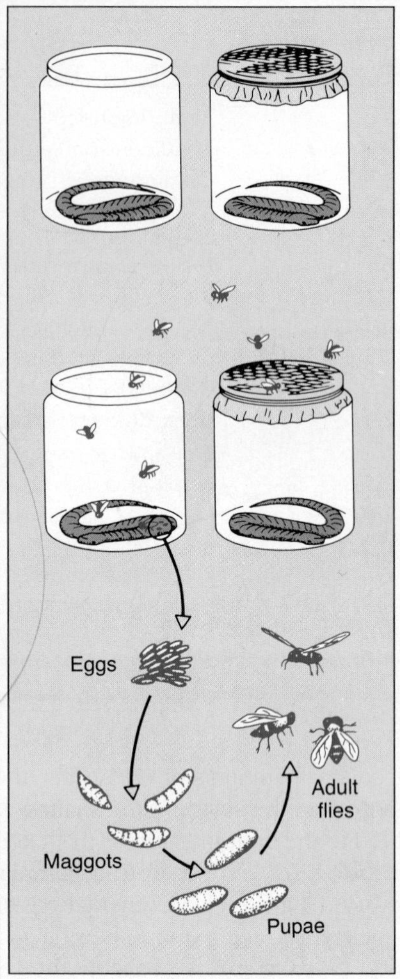

evolution. The fossil record indicates that eukaryotes evolved much faster than prokaryotes, diversifying into organisms with a variety of sizes, shapes, and lifestyles. Although eukaryotes have dominated the earth for at least 700 million years, prokaryotes have also survived and flourished. Many of them appear to have changed very little in more than a billion years.

17-K Early Fossils

Until 1954, the oldest known fossils came from the Cambrian Period of geological time, which began about 600 million years ago. Most major groups of organisms had evolved by this time. Their origins were veiled in mystery because most Precambrian rocks have been either deeply buried, extensively eroded, or altered by heat and pressure great enough to destroy fossils.

By using microscopes to examine thin sections of rock, scientists have found many microfossils of unicellular organisms even in Precambrian deposits (see Figures 17-1 and 17-9). The oldest definite fossils are stromatolites (mats of bacteria), about 3.5 billion years old, similar to those shown in Figure 17-10. Fossils believed to be photosynthetic cyanobacteria, possibly able to carry out claimed that microorganisms could not arise spontaneously. His opponents claimed that heating the flasks had destroyed "vital molecules" that float around in the air until they enter matter, giving it life. Since Spallanzani's sealed, heated flasks sometimes came out teeming with microorganisms, the issue remained in doubt. (We now know that some bacteria make resting spores that survive heating and grow afterward, accounting for Spallanzani's erratic results.)

In the nineteenth century, Louis Pasteur in France and John Tyndall in England showed that air contains bacteria, and that if air is purified before entering a flask of sterilized broth, no bacteria will appear in the broth. Pasteur drew the necks of his glass flasks out into S-shaped curves. Air could enter a flask freely, but it did not travel fast enough to carry bacteria along with it. Any cells in the air were trapped at the bottom of the curve (Figure 17-B). Tyndall sterilized air entering his flasks by passing it through a flame or through absorbent cotton. These treatments, too, removed bacteria from the air and kept the broth clear. By 1880, all but a few diehards agreed that living organisms, of whatever size, come from reproduction of previously existing organisms.

Pasteur is often credited with disproving the theory of spontaneous generation. But he himself once said that his fruitless 17-year search for spontaneous generation did not prove it was impossible. What Pasteur showed was that life did not arise in his flasks under the conditions he used (sterilized broth, clean air) in the time he waited. He did not show that life could *never* arise from nonliving matter under *any* set of conditions.

Indeed, modern scientists believe that life did arise from nonliving matter, but under conditions very different from those of today, and requiring vast eons of time. Many scientists view the origin of life as an inevitable stage in the evolution of matter, one that has probably happened many times in suitable parts of the universe.

Figure 17-B

Pasteur's swan-necked flasks. Air entered the flask's open tip freely, but not fast enough to carry bacteria through the curved neck along with it. Bacteria were trapped at the bottom of the curve, while air continued on into the flask. Only if the neck was broken off could bacteria enter the flask and putrify the nutrient broth.

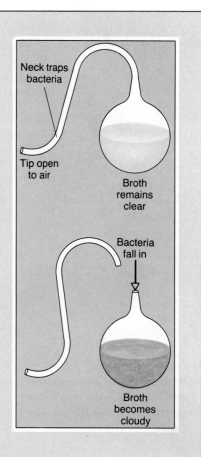

Neck traps bacteria

Tip open to air

Broth remains clear

Bacteria fall in

Broth becomes cloudy

---

oxygen-generating photosynthesis, also date from up to 3.5 billion years ago. A 1-billion-year-old rock formation contains prokaryotes (and possibly eukaryotes) of about 30 different species. Some of the presumed cyanobacteria in this assemblage look exactly like forms alive today!

Further microscopic and chemical analysis of ancient rocks will give us a better picture of the early history of life on earth.

Figure 17-10

Fossil stromatolites. This photo shows a section of rock from the Draken Formation, Spitsbergen, as it appears to the naked eye. The rock formed from mats of autotrophic prokaryotes that grew in successive layers, 800 million years ago. The section shown here is a slice down through these layers. Each layer contained a vast population of microscopic prokaryotes. Similar prokaryote communities are still found today in unusual habitats too harsh to support eukaryotes, which would break up the mats by burrowing in them or by eating the cells. (Andrew H. Knoll)

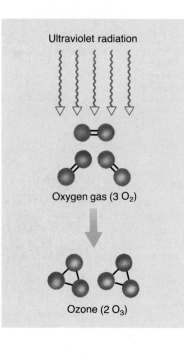

Ultraviolet radiation

Oxygen gas (3 O_2)

Ozone (2 O_3)

Figure 17-11
Formation of ozone. Ozone (O_3) forms when oxygen (O_2) molecules are exposed to ultraviolet radiation. Each oxygen atom forms two covalent bonds.

17-L *Organism and Environment*

Organisms often appear to be at the mercy of their environment: they must succeed within the conditions that exist in the environment or perish. In fact, however, organisms alter their environments in many ways.

What changes have organisms made on the face of the earth? Early heterotrophs consumed all the food made by abiotic processes. Autotrophs have removed much of the atmosphere's carbon dioxide, and the first water-splitting autotrophs added constant, low levels of O_2. This O_2, in turn, acted as a selective force favoring the evolution of respiration.

The oxygen also formed the ozone layer of the atmosphere. Ozone (O_3) forms when ultraviolet light from the sun hits O_2 (Figure 17-11). The ozone layer acts as a filter, preventing much of the ultraviolet light from reaching the earth's surface. Ultraviolet light (uv for short) is very destructive to DNA and proteins. The first organisms lived in water, which absorbs the energy of uv and so shields its inhabitants. Before the ozone layer formed, uv was probably one of several factors that kept early organisms from moving out of water onto land.

On land, vast resources of sunlight and minerals were available to plants that could make the move, and both plant and animal pioneers of terrestrial life found little competition. The trees and grasses that now clothe much of the land continue to add oxygen to the atmosphere. They also change the patterns of water flow from the land to the seas and speed the formation of soil from rock.

■ *Organisms and environment have molded each other during the history of life on our planet.*

SUMMARY

We can't go back to the early earth to see exactly what it was like and how life began, but we can gather evidence and make some intelligent guesses.

The conditions when life began were very different from those on earth today. Evidence from geology and astronomy suggests that the primitive earth probably had a mildly reducing atmosphere, composed of the gases in today's atmosphere except for oxygen. Such an atmosphere would have been a place where organic compounds could form. Gradually, these molecules could have polymerized, come together in aggregates of polymers, and evolved systems of metabolism, information transfer, and reproduction, eventually becoming living organisms.

Five questions about the origin of life have been addressed:

1. How did enzymes arise without previous enzymes to make them?
2. How did cells come into existence without cells to make them?
3. How did membranes originate?
4. How did informational macromolecules arise before the complex genetic code?
5. How did reproduction begin?

Some important events during the early history of life were the evolution of photosynthesis and respiration, the acquisition of intracellular organelles, and the beginning of sexual reproduction. Organisms have changed their environment from an earth of barren water and rock under a mildly reducing atmosphere to one of teeming oceans and verdant landscapes in an oxidizing atmosphere. Each environmental change caused by organisms exerted selective pressures to adapt to the new environment, which in turn changed the environment even more. Thus living organisms and their environment have shaped each other during the evolution of life on earth.

SELF-QUIZ

1. For each gas listed below, tell whether it was *more* or *less* abundant in the mildly reducing atmosphere of the early earth than it is today:

   ___ oxygen

   ___ carbon dioxide

   ___ water vapor

2. In Stanley Miller's classic experiment:

   a. nucleic acids were formed

   b. ultraviolet radiation was used

   c. oxygen was one of the starting ingredients

   d. water was strictly excluded from the system

   e. amino acids were formed

3. Number the following structures and processes in the order in which they are believed to have evolved:

   ___ respiration

   ___ polymers

   ___ water-splitting photosynthesis

   ___ organic monomers

   ___ acquisition of intracellular organelles

   ___ fermentation

4. RNA rather than DNA is believed to have been the first form of genetic information because:

   a. RNA is more stable than DNA

   b. RNA is single-stranded, whereas DNA has two strands

   c. RNA catalyzes some steps in its own synthesis and replication

   d. RNA mutates less than DNA and is easier to copy without mistakes

   e. DNA is complementary to RNA

5. The earliest autotrophs were important to the evolution of life on earth because:

   a. they blocked harmful ultraviolet rays from the sun

   b. they provided a self-renewing food supply

   c. they rid the environment of toxic substances

   d. they provided oxygen for respiration

   e. all of the above

6. List two changes in the environment that resulted from evolution of water-splitting photosynthesis. What effects did these have on the later evolution of living organisms?

7. Respiration was important to early life on earth because:

   a. it rid the environment of toxic ozone

   b. it provided much more energy than photosynthesis

   c. it provided much more energy than fermentation

   d. it permitted experimentation with the genetic code

   e. all of the above

QUESTIONS FOR DISCUSSION

1. Suppose that a scientist claimed to have produced life from nonliving materials under laboratory conditions. What criteria must such an "organism" meet before *you* agreed that it was truly living?

2. Some organic compounds formed in simulations such as Miller's are not found in living organisms. Can you explain this?

3. Simulations of RNA replication in the absence of enzymes show mutation rates several orders of magnitude higher than the rates for DNA in modern cells. How might such high mutation rates have affected prebiotic evolution? Why are mutation rates so much lower today?

4. What is the evidence that respiration evolved before the eukaryotic condition?

SUGGESTED READINGS

Cairns-Smith, A. G. "The first organisms." *Scientific American,* June 1985. A view of the origin of life very different from the one in this chapter.

Cech, T. R. "RNA as an enzyme." *Scientific American,* November 1986.

Cloud, P. "The biosphere." *Scientific American,* September 1983. How geological and fossil evidence provides information about early life on earth.

Dickerson, R. E. "Chemical evolution and the origin of life." *Scientific American,* September 1978.

Eigen, M., W. Gardiner, P. Schuster, and R. Winkler-Oswatitsch. "The origin of genetic information." *Scientific American,* April 1981.

Gurin, J. "In the beginning." *Science 80,* July-August, 1980. A lively account of recent finds of ancient fossil microorganisms.

Schopf, J. W. "The evolution of the earliest cells." *Scientific American,* September 1978. An authority on early microfossils explains the evidence for current beliefs about the early history of life.

Waldrop, M. "Did life really start out in an RNA world?" *Science 246*:1248, 1989. Arguments against the model of early life based on RNA without protein-like molecules.

Classification of Organisms and the Problem of Viruses

CHAPTER

18

Amoebas

Kangaroo

Horn shark

*T*he earth today holds an estimated 5 million different species of organisms. So far, scientists have officially described only about 1.5 million of them. We are presently destroying natural habitats so rapidly—by clearing, draining, filling, damming, and polluting—that many of the remaining species will probably be extinct before they can be discovered and catalogued. We are now losing an estimated four species of plants and animals per day!

The first comprehensive system for classifying organisms was invented by the Swedish botanist Carolus Linnaeus (Figure 18-1). Linnaeus based his classification on morphology (structure, anatomy). For example, two species of trees with similar leaves and flowers fell close together in his scheme. After Darwin's theory of evolution gained acceptance, biologists decided that classifying organisms according to their evolutionary relationships would be more natural. Fortunately, the morphological features used by Linnaeus usually reflect evolutionary relationships. Hence biologists have been able to retain many of the names and groupings of Linnaeus's thorough and meticulous system.

Our system of naming and classifying organisms is inadequate to the task. The sheer number of species strains its capacity. More importantly, classification attempts to force the natural world of life into artificial "bins," and the boundaries of these bins can never be defined to the satisfaction of all. However, until better systems of classification come along, we must make do with what we have. Scientists need official names and definitions for organisms in order to know when they are talking about the same species. We also need a classification scheme to make sense of the vast diversity of organisms all around us.

In this chapter we consider how living things are classified, and how biologists try to decide where in the classification scheme particular organisms belong. We then introduce the five-kingdom system of classification that we use in this book. Finally, we look at a peculiar group, the viruses, which show some features of life and yet are not living.

- Classification of organisms is based on evolutionary relationships.
- Every species of organism known to biologists has an official two-word Latin name designating its genus and species.
- Viruses are tiny particles made of a nucleic acid genome enclosed in a protein coat. They have orderly structures, and they can evolve and become adapted to their environment. They show none of the other features of life, but rely on the cells they invade and parasitize to provide the energy and materials needed for their reproduction.

Figure 18-1
Carolus Linnaeus (1707–1778). Linnaeus, who described thousands of species of plants and animals, described himself thus: "Brown-eyed, nimble, hasty, did everything promptly." Linnaeus classified plants by their sexual parts with group names such as *Polyandria,* meaning "twenty or more males with one female." This emphasis on sex shocked some of his contemporaries. The Bishop of Carlisle wrote: "To tell you that nothing could exceed the gross prurience of Linnaeus's mind is perfectly needless," and Goethe worried about the embarrassment chaste young people might suffer when reading botany textbooks. (Biophoto Associates, National Portrait Gallery, London)

Figure 18-2
Not so horrible? This grizzly bear cub shows the dish-faced profile, white-tipped ("grizzled") fur, long curved claws, and hump above the shoulders characteristic of members of the species *Ursus horribilis.* (Mark Newman/ Tom Stack & Associates)

18-A Binomial Nomenclature

The basic unit for classifying organisms is the species. In the case of organisms that reproduce sexually (and many do not), a **species** is a group of organisms that share a common gene pool and do not interbreed with members of other species under natural conditions (Chapter 15).

Linnaeus began the practice of giving every species its own, unique Latin **binomial,** a two-word name. The first word in this binomial is the **genus** (plural, **genera;** adjective, **generic**). A genus contains one species or a group of very similar species. The second word in the binomial denotes the species itself. For example, the binomial for the grizzly bear is *Ursus horribilis* (horrible bear) (Figure 18-2). Another species in the genus *Ursus* is *Ursus arctos,* the Alaskan brown bear.

Note that the genus is always capitalized, the species usually not. Both are italicized (or underlined).

18-B Taxonomy

Taxonomy is the branch of biology concerned with the classification of organisms. Linnaeus arranged organisms into a hierarchy of ever larger and more inclusive categories, a system borrowed from the highly disciplined Swedish military of his day. The most inclusive categories are the **kingdoms.** The other main categories, in descending order, are **phylum, class, order, family, genus,** and **species.** (You can remember this sequence by memorizing the sentence "**K**ing **P**hil **c**ame **o**ver **f**or **G**ene's **s**pecial.") Botanists use divisions, instead of phyla, as categories in the plant kingdom.

A **taxon** (plural, **taxa**) is a group of organisms defined by the classification scheme, such as a particular species or class. For example, Ursidae (a family) is a taxon including the species in the genus *Ursus* as well as those in other genera of bears, such as the polar bear, *Thalarctos maritimus,* and the American black bear, *Euarctos americanus.* Some taxa contain only one group at the next lower level; for example, many families contain only one genus and one species.

Table 18-1 gives the classification for human beings, and here you can see the seven important levels of the hierarchical classification.

Table 18-1 Classification of Homo sapiens

Category	Taxon	Characteristics
Kingdom	Animalia	Heterotrophic, multicellular organisms lacking cell walls, and possessing a motile stage in the life history
Phylum	Chordata	Animals with a dorsal, hollow nerve tube, a notochord, and pharyngeal gill slits at some stage in life
Class	Mammalia	Chordates with only one bone in each side of the lower jaw, hair or fur, young nourished by milk from the mother's mammary glands
Order	Primates	Originally arboreal (tree-living) mammals with flattened fingers and nails, vision the most important sense, poor sense of smell
Family	Hominidae	Primates with bipedal locomotion, flat faces, binocular color vision
Genus	*Homo**	Hominid with large brain, speech, long childhood
Species	*Homo sapiens*†	High forehead, body hair reduced, prominent chin

* *Homo* Latin: man.

† *Sapiens* Latin: knowing, wise.

Linnaeus named and classified all of the plants and animals known to him in his massive books *Systema Naturae* and *Species Plantarum*. He believed that a species could be described by listing the morphological characteristics of a "perfect" member of the species. This led to the practice of selecting and preserving a typical individual of each newly described species, which became the official **type specimen** of the species. Today, authors of new species preserve several specimens showing a typical range of the species's characteristics. Type specimens then become the definition of the species. If later workers need to know whether they are really working on the same species of daisies that the original author described, they compare their daisies with the type specimen(s).

Nowadays, some microorganisms are preserved alive, as clones of individuals with identical genomes, by freezing them in liquid nitrogen ($-196°C$). Researchers can then order a culture of a particular species, and even of particular genetic strains within a species, for their experiments.

Thousands of Linnaeus's species and taxon names are still in use. However, an understanding of evolution has led modern biologists to classify organisms by relationships rather than by structure. An organism's evolutionary history is its **phylogeny.** Phylogenetic taxonomists try to classify organisms by their phylogenetic relationships.

The two methods can be distinguished by their goals. Linnaeus's classification was an artificial system, designed to be helpful in organizing and retrieving information about organisms. The goal of the phylogenetic system is to produce a classification that is both easy to use and informative about evolutionary history. The phylogenetic method amounts to drawing an evolutionary family tree for an organism. In many cases, this leads to the same result as an artificial classification, since organisms that have evolved from a common ancestor are likely to be more similar than those that have not.

Phylogenetic taxonomy has some inherent problems. Even if we can work out the true evolutionary history of a group of organisms, no taxonomist can classify them in a way that everyone will accept. Phylogeny depicts the natural and continuous process of evolution, but the taxonomist must carve this continuum into artificial and separate taxa, for human convenience. There are bound to be disagreements about how, where, and why boxes should be drawn around parts of the evolutionary tree. Are all species of the dog family close enough relatives to be in the same genus? If not, how many genera should we use? How alike must different members of a family be? There are no simple answers to such questions. As a result, there are many different systems of classification in use today.

18-C *Interpreting the Characteristics of Organisms*

Every organism has a variety of different features, often giving conflicting information about its phylogenetic position. Biologists need ways to interpret the significance of different features, giving some more weight than others, as they reach a decision. Suppose your biology teacher asked you to classify a flea, a frog, and a kangaroo in a phylogenetic scheme. We hope you would ignore the fact that all three have hind legs strongly developed for jumping, and give more emphasis to features like the general plan of the skeleton, the type of body covering, and the mode of reproduction. By this reasoning you might arrive at the conclusions made by biologists: fleas belong among the insects, close to the flies; frogs among the amphibians, with salamanders and newts; and kangaroos among the marsupial mammals, with the koala and opossum.

As a species evolves, some characters change faster than others. Every organism shows some **ancestral** characters, which have persisted essentially un-

(a)

(b)

Figure 18-3

Ancestral and derived characters.
(a) This newt, a pond-dwelling amphibian, has a tail, an ancestral characteristic, but only four fingers on the forelimbs, a derived trait. (b) In contrast, the orang-utan and other apes have five fingers, an ancestral trait, but have lost their tails, a derived feature. (a, Biophoto Associates; b, Tom McHugh, Science Source/Photo Researchers)

changed from its remote ancestors, and some **derived** characters, which have evolved more recently. For instance, let us compare apes and newts with their common ancestors, the amphibians of the Carboniferous Period. We find that newts have a tail, an ancestral feature, but they have a derived number (four) of fingers on the forelimb, whereas apes lack a tail, a derived trait, but have the ancestral number (five) of fingers (Figure 18-3).

Traits that have changed little during evolutionary history are called **conservative** characters. These are usually features that cannot change very much if the organism is to survive. They are useful for defining the various taxa. The shape, size, and number of teeth is the conservative character used to define many mammalian orders. For instance, beavers, kangaroo rats, and some fossil species all have similar incisors (front teeth) and are therefore all classified as rodents (order Rodentia).

To identify ancestral and derived characters, we must decide which ones share a common evolutionary origin. Characters found in two species and derived from a common ancestor are said to be **homologous.** Homologous characters have the same genetic basis in the two species, but not necessarily the same appearance or function. A classic example is the forelimbs of vertebrates. A human arm, a horse's foreleg, a whale's flipper, and a bird's wing all look different and do different jobs. However, they share a similar underlying bone structure, and the fossil record shows that they all have a common evolutionary origin from the forelimbs of ancient amphibians. The limbs are homologous, their basic bone structure is ancestral, and their detailed structures are derived. For phylogenetic purposes we conclude that all animals with this type of forelimb are related. However, it is not always easy to disentangle the derived, ancestral, and homologous characters of organisms.

One difficulty is sorting out cases of **convergent evolution,** the development of similar adaptations by organisms of different ancestries, in response to similar environmental pressures. Striking examples occur among plants. Many desert-dwelling members of the New World family Cactaceae (cactuses) resemble desert-adapted members of the Old World family Euphorbiaceae (see Figure 1-13). In both groups, the advantage of being able to conserve water in desert habitats has led to the evolution of thick, water-storing stems and of spiny leaves that deter animals from using the stems as the source of their own water. Among animals, hares and rabbits (mammalian order Lagomorpha) were once placed with the similar-looking rodents (order Rodentia), but the fossil record shows that the two groups had separate origins and have converged. Recent evidence also confirms that the Old World and New World vultures evolved their distinctive features independently (Figure 18-4).

(a)

(b)

Figure 18-4
Convergent evolution. (a) Old World vultures have strong, hooked bills and small naked heads adapted to poking into animal carcasses for food. These vultures are closely related to eagles. Also present at the feast are two taller storks, members of the group most closely related to New World vultures (b). The New World black vultures in (b) have features so similar to those of Old World species that they have traditionally been classified with them, near eagles and hawks, instead of with their stork relatives.

To classify a newly discovered organism, many kinds of evidence in addition to appearance must be considered, such as:

1. Life history. Among invertebrates, the genetic program for development tends to be conservative. Hence similarities between larvae often give better clues to relationships than do resemblances (or non-resemblances) between adults.
2. Biochemical studies. Comparing the pigments, proteins, or nucleic acids of different species may provide useful information.
3. Behavior.
4. Geographic distribution.
5. The fossil record.

Even with all this information, taxonomists still face the old problem of how much weight to give characters that point to different conclusions about how the organism should be classified.

Monophyletic or Polyphyletic?

Many taxonomists consider that the most useful taxon is a **clade,** containing a common ancestral species and all the species descended from it. Such a taxon is **monophyletic,** meaning that it represents one evolutionary line. In practice, many taxa in use today are **polyphyletic,** made up of several evolutionary lines but not including their common ancestor (if there is one). One example is the mammals. The class Mammalia contains all vertebrates with only one bone in each side of the lower jaw. By this definition, mammals probably arose at least three, and possibly five, times from different groups of reptiles in ancient times. Mammals also differ from reptiles in the way the jaw is hinged to the skull, and in the way the limbs attach to the rest of the skeleton. The differences arose during the evolution of fast-moving, dog-like carnivorous mammals from reptiles that had less efficient locomotion and weak jaws liable to be broken by

■ *In theory, taxonomy reflects phylogeny. In practice, many taxa are based mainly on morphology, for convenience, and are in fact polyphyletic.*

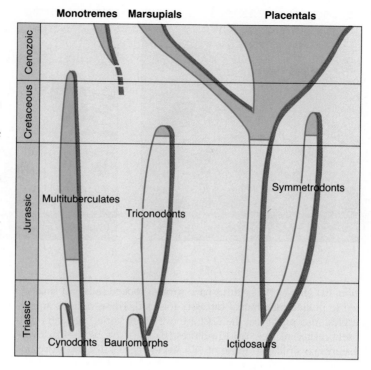

Figure 18-5
Grades and clades in one view of the origin of mammals. Names of the three groups of modern mammals are in bold type at the top of the figure. Animals that have reached the mammal grade of organization are in areas shaded in green. Three clades (groups with their common ancestors) are shown. (The origin of monotremes is much debated.) Monotremes are those mammals that lay eggs (e.g., the duck-billed platypus). Marsupials are those mammals with pouches where the young complete their development (e.g., opossum, kangaroo, koala, Australian wolf). Placentals are the majority of mammals, and the most familiar.

struggling prey. Several different groups of reptiles evolved the same method of strengthening the jaw and of moving faster, all of them giving rise to descendants classified as mammals. Mammals, then, are not members of a monophyletic clade but of a polyphyletic **grade** of organization attained more than once by related, but separate, evolutionary lines (Figure 18-5).

18-D   The Five Kingdoms

The most inclusive taxa are the kingdoms. Linnaeus's system of classification had two kingdoms, the plants (Plantae) and animals (Animalia). This seemed reasonable in his day, since the familiar land plants and animals were clearly very different. Plants did not move around; they did not eat, but seemed to need only water in order to grow. Animals were **motile;** that is, they could move from place to place. Animals had to eat plants, or each other, in order to stay alive. On a microscopic level, plants could be seen to have cell walls, which animal cells lacked. Fungi seemed to be aberrant plants, since they had cell walls and root-like structures but lacked the green pigments of the other "plants."

Today, however, it is apparent that many forms of life do not fit neatly into either the plant or the animal camp. Some organisms, such as *Euglena,* seem to fit both descriptions. *Euglena* (see Figure 19-15) has a rather stiff covering, not as thick as a plant cell wall but certainly giving more protection than a plasma membrane. *Euglena* also has chloroplasts and carries on photosynthesis. However, it also has animal-like features: it has a flagellum that it uses to swim, and it can engulf other organisms and digest them as food. Bacteria present another taxonomic problem, since they have cell walls but may also have flagella, used to move around. Most cannot make their own food, but some can carry on photosynthesis. These and other organisms give evidence that the division into plant and animal kingdoms is artificial, with a confusing zone between the two.

Modern attempts to revise biological classification at the kingdom level have been many and varied. For consistency, this book uses a scheme based on the five-kingdom system popularized by ecologist Robert Whittaker.

The kingdoms are separated according to two main criteria: degree of cellular complexity and mode of nutrition. The kingdom Monera contains the organisms with prokaryotic cells. Eukaryotes belong to the other four kingdoms. The kingdom Protista is composed of unicellular (one-celled) eukaryotic organisms. Eukaryotes with multicellular (many-celled) bodies are placed into one of the other three kingdoms—kingdom Plantae, kingdom Animalia, and kingdom Fungi—which are separated from one another largely on the basis of nutrition. Most of the members of the kingdom Plantae are photosynthetic; they make their own food. Most members of the kingdom Animalia are **ingestive;** they engulf or swallow their food and digest it internally. Members of the kingdom Fungi are **absorptive;** they absorb organic molecules from outside their bodies directly through their exterior plasma membranes. We shall consider each kingdom in turn.

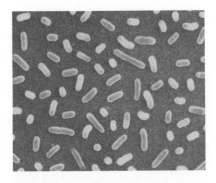

Figure 18-6
Monera. Small, prokaryotic cells of *Escherichia coli,* a bacterium that inhabits the human colon. This species is also used for laboratory research, including genetic engineering. (David M. Phillips/Visuals Unlimited)

Kingdom Monera

All prokaryotic organisms—the bacteria and cyanobacteria—are placed in the kingdom Monera (Figure 18-6). Prokaryotes lack nuclear membranes and mitochondria, chloroplasts, and other membrane-bounded organelles found in eukaryotic cells. Their DNA consists of one circular double helix, and they divide and reproduce without the nuclear divisions of meiosis or mitosis found in eukaryotes. Prokaryotes also differ from eukaryotes biochemically in such things as the materials and structure of their cell walls, the size and composition of their ribosomes, and some of their metabolic pathways.

Kingdom Protista

The Protista are the one-celled eukaryotic organisms (Figure 18-7). As eukaryotes, they have nuclear membranes and linear chromosomes (composed of DNA and proteins) that can go through mitosis and, in most forms, meiosis. The members of the three higher kingdoms undoubtedly evolved from protist ancestors. Modern protists show ways of life that closely parallel those of higher eukaryotes, some being photosynthetic, some ingestive, and some absorptive, some motile and some nonmotile, some with cell walls and some without. All combinations of these features occur in various members of this group.

Figure 18-7
Protista. A diatom, a photosynthetic protist with an intricately patterned cell wall. (Carolina Biological Supply Company)

Kingdom Fungi

The fungi are often classified as plants because they are nonmotile, and because they have external cell walls. However, fungi cannot make their own food; they must absorb food from a living or nonliving organic source. In many cases, they secrete digestive enzymes that digest food outside their bodies before they can absorb it. As nutrients around their bodies become depleted, they must grow outward into fresh food sources. Whittaker felt that fungi should be separated from plants because they are completely absorptive, whereas most plants absorb only the simple raw materials they need to make their own food. In addition, fungi and green plants differ in cell wall composition, body plan, and reproduction (Figure 18-8).

Figure 18-8
Fungi. The reproductive structure of the aptly named *Phallus impudicus.* (Biophoto Associates)

Mimicry and Convergent Evolution

A JOURNEY INTO LIFE

As an example of the evolutionary twists that taxonomists must unravel in classifying organisms, consider the two insects in Figure 18-A. They look quite similar, but close inspection shows that the resemblance is only superficial. One is a fly (Order Diptera). The other is a honeybee, a member of a completely different order of insects (Hymenoptera) that also includes wasps and ants. The fly is a **mimic** of the bee. Like other flies, it has only one pair of wings,

whereas hymenopterans have two pairs. Most flies also lack the distinct "waist" of bees and wasps, but flies that mimic hymenopterans look as if they have waists, either because their bodies really do have constrictions or because they are colored to look that way. A mimic fly's wings may also be so large that they look like two pairs. These flies are often found hovering over flowers, with rapidly beating wings, just where we would expect to find bees or wasps. The resemblance is so close that you yourself have probably reacted by moving carefully away from a harmless fly wearing black and yellow stripes.

What is the reason for this strange evolutionary convergence? Many animals with effective defenses, such as hymenopterans, skunks, and poisonous fish and frogs, are strikingly colored. This **aposematic coloration** warns potential predators to stay away and not attack them. Aposematic coloration protects best against predators that can learn. Jane and Lincoln Brower showed that a toad stung once by a bumblebee will

not attack anything looking remotely like a bumblebee for a long time afterwards. Similarly, dogs will seldom tangle with a skunk or a toad more than once.

During the course of evolution, members of several different well-protected species have come to resemble one another, a phenomenon called **Mullerian mimicry.** For instance, many hymenopterans—honeybees, bumblebees, wasps, hornets, and yellowjackets—have black and yellow stripes, as do some noxious beetles (Figure 18-B). Mullerian mimicry is advantageous to all these noxious species because it reduces the numbers of each that are killed while predators learn to avoid their aposematic coloration. If each predator had to learn to avoid any insect with yellow stripes or pink dots or red wings, because all species of noxious insects looked different, many more insects would be killed during the predator's learning process, and aposematic coloration would be much less valuable.

(a) Bee

(b) Fly

(c)

(d)

Figure 18-A
Two bees or not two bees? Resemblance of a mimic fly (b and d) to a bee (a and c) is good enough to fool some of the people some of the time. Careful examination, though, shows differences between the two. The fly has large eyes and stubby, club-shaped antennae. The bee's eyes are smaller and her antennae more slender. A bee also has two pairs of wings to a fly's one pair, but this may be hard to see.

Figure 18-B
Mullerian mimicry. Blister beetles, noxious insects that have a color scheme of black and yellow stripes, similar to that of many stinging hymenopterans. (Biophoto Associates, N.H.P.A.)

How can we decide if two similar-looking animals are examples of Batesian mimicry, and which species is the model and which the mimic? One way would be to perform experiments like the Browers', but other clues may often be used. First, a model usually looks and behaves like closely related species, whereas a mimic may be enormously different from its relatives. Thus if an insect looks and acts like a wasp, but on close examination turns out to be a fly, it is a fair guess that it is a mimic.

Atypical behavior is another good clue to mimicry. There are black-and-yellow-striped flies that, if you touch them, will rapidly curve their abdomens around and stab your hand. In fact they have no sting, so the gesture is meaningless, but the fly looks so much like a wasp about to sting that you are likely to shake it off your hand without harming it. Most moths fly by night and rest by day, but moths that mimic butterflies have reversed this behavior pattern and fly by day with the butterflies they mimic.

The convergent evolution of Batesian mimics so that they resemble their models is just one of many things that can obscure the relationships (or lack of them) between organisms and keep taxonomists on their toes. As a further complication, an assemblage such as the hymenopteran Mullerian mimics probably do not illustrate convergent evolution. Rather, the various black and yellow striped species have all inherited similar color patterns from a common ancestor, so that in their case black and yellow stripes are a homologous feature!

If a predator can learn to avoid eating all insects with black and yellow stripes, then there might be enormous advantage to a tasty, unprotected species if it too had black and yellow stripes. **Batesian mimicry** is the resemblance of an unprotected mimic species to a **model,** an aposematically colored, protected species. This is the type of mimicry shown by our fly mimic and its hymenopteran model.

The Browers showed that Batesian mimicry does indeed protect its owners from insect-eating birds and toads. Many of their experiments were done with the monarch butterfly, which contains a chemical that makes predators ill. An unrelated butterfly, the viceroy, mimics the monarch (Figure 18-C). The Browers found that a Florida scrub jay that was raised in captivity, and that had never seen either butterfly before, would eat the viceroy quite cheerfully. After one or two mouthfuls of a monarch, however, the bird would not even peck at a monarch, and after it had learned to avoid the monarch, it would not eat any more viceroys either.

Figure 18-C
Batesian mimicry. The monarch butterfly (larger) is protected from birds by the cardiac glycosides it contains. The viceroy (smaller) is edible, but birds avoid it because they mistake it for the monarch which it resembles. (Biophoto Associates, N.H.P.A.)

Kingdom Plantae

All members of the plant kingdom are eukaryotes with cell walls containing cellulose. Most contain chlorophyll and carry on photosynthesis inside chloroplasts, although a few species have lost their chlorophyll and obtain all of their nutrients by absorption. Most plants are immobile and remain in one place throughout life (Figure 18-9). They depend on water and air to bring nutrients to them, and they also grow out and intercept more sunlight and nutrients. The plant kingdom includes the multicellular algae as well as all the familiar multicellular land plants—the mosses, ferns, grasses, shrubs, and trees.

Kingdom Animalia

Animals are multicellular, eukaryotic, heterotrophic organisms that obtain food mainly by ingestion (Figure 18-10). Most animals can move, and this permits them to acquire food from their environment by *going* for it—in contrast to plants and fungi, which must either *wait* for it or *grow* for it! All but the simplest animals produce gametes (eggs and sperm) in multicellular organs, and the fertilized eggs develop into multicellular embryos.

Difficulties With the Five-Kingdom System

We have seen that all classification schemes pose problems, and the five-kingdom system is no exception. There is clearly a distinction between prokaryotes and eukaryotes. However, the four eukaryotic kingdoms are unsatisfactory for two main reasons. First, modern taxonomy should aim to classify organisms so that the members of a group are related more closely to each other than to members of other groups. The kingdom Protista violates this rule by containing organisms that are related more closely to some plants than to any other protists. Second, many different protistan lines gave rise to multicellular groups, and so none of the three higher kingdoms contains the single ancestor (if there is one) of all its members. For instance, some biologists think that different groups of fungi arose from different protistan groups rather than from a common fungal ancestor. In fact, multicellular lines originated from protists at least 17 different times, and it seems likely that the eukaryotic condition itself originated independently more than once. Breaking up the world of life into truly monophyletic groups would give us more kingdoms than we wish to bother with for most purposes. Despite its inadequacies, the five-kingdom system is widely used today, not because it is natural but because it is convenient.

Table 18-2 Differences Between Viruses and Cells

Character	Viruses	Cells
Structure	Virus particle: nucleic acid core inside protein capsid	Cell containing nucleic acids, lipid-protein membrane, ribosomes, cytoplasm, etc.
Nucleic acids	DNA or RNA, but not both	Both DNA and RNA
Enzymes	One or a few; e.g., lysozyme (digests bacterial cell wall), polymerase (replicates viral genome)	Many enzymes; diverse functions
Metabolism	None; relies on host cell metabolism for monomers, protein synthesis machinery, and some enzymes of nucleic acid synthesis	Makes own ribosomes and enzymes needed for synthesis of proteins, nucleic acids, etc.
Reproduction	Nucleic acid genome and capsid proteins produced separately, then assembled into virus particle	Division into two similar cells following growth

18-E The Problem of Viruses

Viruses are tiny particles composed largely of nucleic acid and protein, but lacking many of the features of living cells. They occupy a strange limbo somewhere between the living and nonliving worlds. Viruses are like living organisms in possessing genetic material, composed of nucleic acids and capable of mutation and recombination. Viruses can therefore evolve and adapt to their changing environments. On the other hand, viruses are not made up of cells, and they have no ribosomes, nor the metabolic machinery for protein synthesis and energy generation.

Lacking these components, viruses are invariably parasites. They can reproduce only inside living cells, and even here their reproduction is unique. Cells reproduce by growing and eventually dividing into two new cells. By contrast, viruses are disassembled into their separate components: nucleic acid genomes and protein coats. The virus takes over the host cell and causes its metabolic machinery to produce dozens to hundreds of new viral genomes, and thousands of protein subunits to make new viral coats. Then these components are assembled into new viruses, the same size as the original one: unlike cells, viruses do not grow (Table 18-2).

Another bizarre feature of viruses is that many of them can be crystallized, a common enough property of minerals and even of fairly complex organic molecules, but certainly not of living cells. Furthermore, crystallized viruses, when wetted and exposed to living host cells, soon establish infections and get back to the business of causing the cell to make more viruses.

Because of all these odd characteristics, viruses are not considered real living organisms and do not belong in the five kingdoms. Nevertheless, since viruses are active only inside living cells, and indeed may have devastating effects on their hosts, the study of viruses is clearly the province of biology.

■ *Viruses are tiny parasitic particles that can be produced only inside living cells. Viruses do not eat, grow, metabolize, make proteins, or reproduce by themselves, but they do evolve.*

18-F Virus Structure

Viruses are very small, from about 15 to a few hundred nanometres in diameter. Only the largest surpass the smallest bacteria in size. The "basic" virus consists of a nucleic acid molecule and a surrounding protein coat, the **capsid.** In addition, some viruses have a membranous outer envelope of glycoprotein and lipid

■ *A virus consists of a DNA or RNA molecule inside a protein capsid, which is sometimes enclosed in a membranous envelope.*

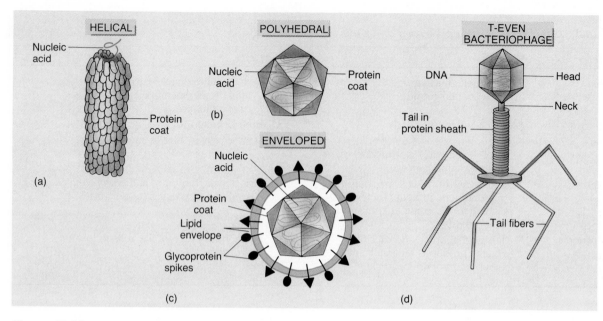

Figure 18-11

Virus structure. (a) The nucleic acid of a helical virus is wound inside a coat of repeating protein subunits arranged in a helical pattern. (b) A polyhedral virus has a protein coat in the shape of an icosahedron, a geometric solid having 20 faces. (c) Enveloped viruses have a membranous outer envelope around a helical or (as here) polyhedral protein coat. (d) T-even bacteriophages, which attack *Escherichia coli,* have more complex protein coats. The head encloses the DNA genome. The tail fibers attach to the bacterial cell wall and contract to inject the DNA into the host cell.

covering the capsid. Most of these enveloped viruses infect animals. Viruses may also contain one or a few enzymes—those needed for invasion of a cell and for replication of the viral nucleic acid.

The capsid protects the genetic material in its passage from one host cell to another. Viral capsid or envelope proteins also bind specifically to receptor molecules on the host cell's surface, the first step in invading a cell. The capsid is made up of a number of protein subunits, and their organization determines the virus's shape: most are either helical or polyhedral, or a combination of the two. A more complex structure occurs in some **bacteriophages**—viruses that infect bacteria, often called just **phages** (Figures 18-11 and 18-12).

The viral genome consists of a molecule of DNA or RNA, but not both, which may be either single- or double-stranded. Some RNA viral genomes consist of more than one molecule. The largest viral genomes contain hundreds of genes, whereas the smallest have only a handful. In something as small as a virus, space is extremely limited. There is room only for genes that code for basic necessities, such as viral coat proteins, virus-specific nucleic acid polymerase enzymes, enzymes needed to take over the host cell, and regulatory genes for rapid production of new viral components. In enveloped viruses, the envelope proteins are encoded by the viral genome, whereas the lipids are appropriated from the host cell's plasma membrane.

Viruses are classified according to whether they have DNA or RNA genomes, by capsid shape and size, and by presence or absence of an envelope.

18-G   Viral Reproductive Cycles

Three main types of reproductive cycles have been found among viruses.

A **lytic cycle** occurs when a virus invades a cell, destroys the cell's DNA, takes over its metabolic machinery, and causes the cell to make as many as several thousand new virus particles (Figure 18-13). Viral enzymes then cause the cell to break, or **lyse.** The released viruses disperse to infect new host cells. A cell invaded by a lytic virus is almost invariably killed by it within a very short time. This is typical of many phages and also of viruses that cause colds and poliomyelitis ("polio").

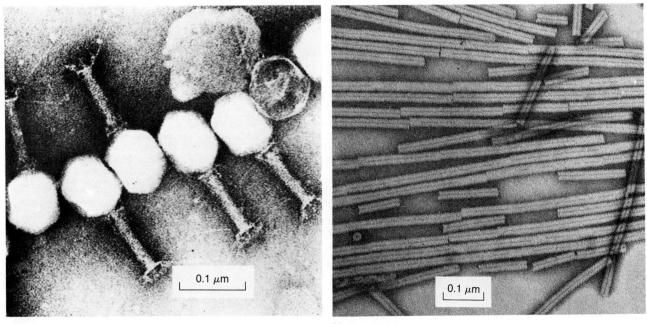

(a) (b)

Figure 18-12
Electron micrographs of virus particles. (a) Bacteriophage T4, which infects *Escherichia coli.* (b) Tobacco mosaic virus, the first virus shown to have RNA as its genetic material. (a, Carolina Biological Supply Company; b, Biophoto Associates)

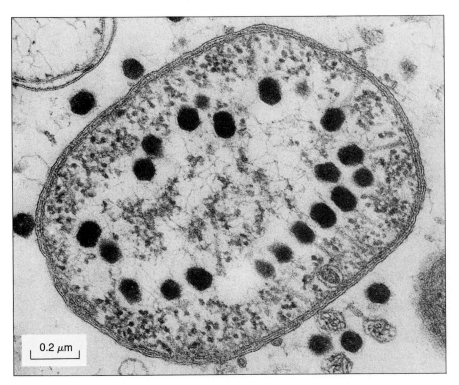

Figure 18-13
A cell of the bacterium *Escherichia coli* infected with a bacteriophage. In this electron micrograph, the black hexagons inside the cell are new phages produced after the phage genetic material was injected into the bacterium. Note the empty phage coats attached to the cell wall. The thin fibrous material in the cell is DNA. (L. D. Simon, *Virology* 38:287, 1969)

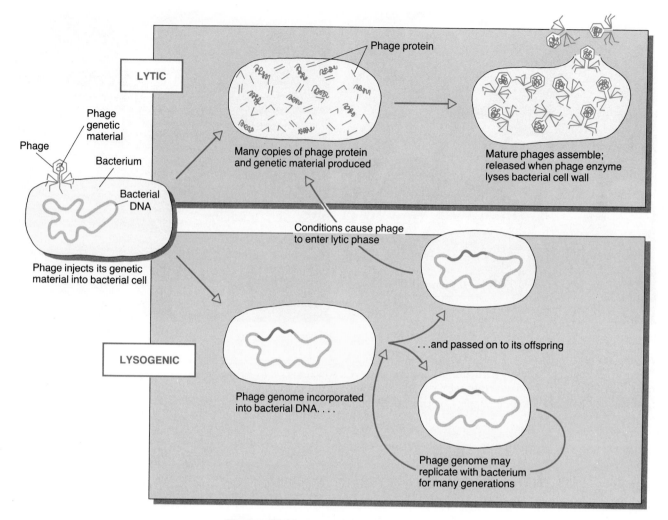

Figure 18-14

Reproduction of a bacteriophage. In a lytic cycle, the phage takes over the host cell and destroys it. In a lysogenic cycle, the host may survive for many generations with the phage genetic material incorporated into the host genome, until some condition triggers the phage to become lytic.

A **lysogenic cycle** is typical of some phages, called temperate phages. These may either go through a lytic cycle and destroy the host cell, or may instead enter a dormant phase in which their DNA is joined to the host's (Figure 18-14). Here the viral DNA is replicated with the host's in each cell generation. Certain external stimuli can cause a lysogenic cell's phage DNA to enter the lytic cycle, releasing intact phages. Herpes viruses in animals, such as those that cause cold sores in humans, show comparable cycles.

Some lysogenic bacterial cells are of importance to human health. The bacteria that cause diphtheria, botulism, and scarlet fever produce the toxins responsible for these diseases only when they contain particular phages, which carry the genes encoding the toxins.

Phages that are released when a lysogenic cell finally lyses may carry along part of the bacterial DNA, which is later inserted into the DNA of a new host bacterium. This process, called **transduction** ("leading across"), produces genetic recombination in the new host. Viruses that can carry DNA from one cell to another are used to study bacterial genetics.

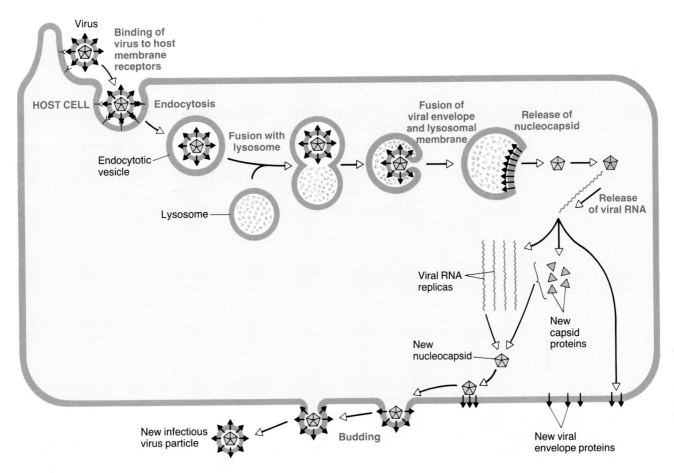

Figure 18-15
Continuous production of viruses. An animal cell becomes infected by a virus (top left) and produces many new viral genomes and protein coats, which combine in the cytoplasm to form nucleocapsids. These move to the plasma membrane, where they associate with new viral envelope proteins, also produced by the host under the direction of viral RNA. The finished virus particle buds off from the plasma membrane (bottom).

In the lytic and lysogenic cycles, viruses are assembled inside the host cell and released when the cell lyses. A few phages, and many animal viruses, are produced and released continuously by budding from intact host cells. New copies of the viral genome and capsid combine in the cytoplasm and then move to the host's plasma membrane. Here they attach to viral envelope proteins that the host has made and inserted into its own membrane. The host's plasma membrane then bulges out around the forming virus particle, until at last the virus, surrounded by its new membranous envelope, buds off from the host cell (Figure 18-15). Enveloped animal viruses that bud from their host cells in this way include influenza, measles, mumps, and rabies viruses.

Viruses with genomes of RNA instead of DNA have novel mechanisms of replication. If the RNA is single-stranded, it usually serves as the template for a complementary RNA strand. This second strand then acts as a template for new copies of the viral genome identical to the original RNA strand.

Other viruses with single-stranded RNA genomes code for an unusual enzyme called **reverse transcriptase.** Contrary to the usual flow of genetic information, this enzyme uses the viral RNA as a template to make a complementary DNA strand! Next, the DNA acts as the template to make double-stranded DNA,

■ *Some viruses destroy their host cells shortly after invading them. Others "go underground" in the host's DNA for long periods before they are reproduced and destroy the host. Still others cause living host cells to produce new viruses continuously.*

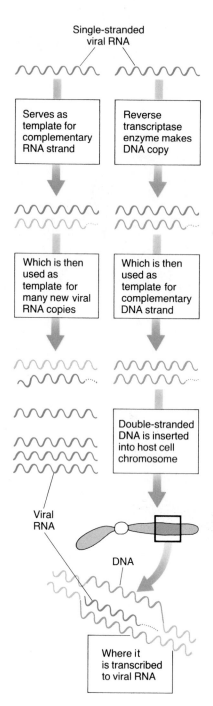

Single-stranded
viral RNA

| Serves as template for complementary RNA strand | Reverse transcriptase enzyme makes DNA copy |

| Which is then used as template for many new viral RNA copies | Which is then used as template for complementary DNA strand |

| | Double-stranded DNA is inserted into host cell chromosome |

Viral RNA

DNA

| | Where it is transcribed to viral RNA |

Figure 18-16

Replication of genomes of RNA viruses. (left) RNA ⟶ RNA ⟶ RNA. (right) RNA ⟶ DNA ⟶ DNA ⟶ RNA.

which is inserted into the host cell genome. Here it is transcribed into many new viral RNA molecules (Figure 18-16).

Viruses that produce reverse transcriptase are called **retroviruses.** Since their genomes spend some time joined to the host DNA, new retrovirus particles may also incorporate host genes and carry them to new host cells. More than 20 different vertebrate **oncogenes,** mutated genes from a former host cell that can cause cancer in new host cells, have been identified in various retroviruses (Section 10-J).

The retrovirus that causes AIDS (acquired immune deficiency syndrome) can go through reproductive cycles resembling all three types we saw earlier. The AIDS virus enters the bloodstream and its capsid binds specifically to receptors on a T helper cell (part of the immune system; Chapter 25). Inside the cell, the virus loses its protein coat, releasing its RNA genome and reverse transcriptase enzyme into the cytoplasm. The enzyme makes a DNA molecule complementary to the RNA genome. This DNA enters the nucleus and inserts itself into a cell chromosome. After this, one of three things may happen. First, the virus may quickly direct the cell to produce a flood of new AIDS viruses, which are released so rapidly that the cell is destroyed. Or, the virus may instead enter a latent period lasting up to ten years before it is activated and causes the cell to churn out a destructive horde of viruses. Third, the virus may cause a persistent infection, with production of new viruses slow enough so that few host cells are killed. AIDS occurs when so many T helper cells have been destroyed that the body's immune system can no longer fight off diseases.

18-H *Viral Diseases of Plants and Animals*

Before a virus can invade a cell, its outer proteins must bind to protein receptors on the host cell surface. Because of this specificity, each kind of virus can attack only particular kinds of host cells. Polio viruses can infect only humans and a few other primates (monkeys, apes, etc.); the common cold virus attacks the cells lining the human respiratory tract.

Virus diseases of plants include tobacco mosaic and necrosis, alfalfa mosaic, and wound tumor. Plant viruses may be spread by wind or insects. It is usually impossible to cure diseased plants. Instead, farmers try to prevent the spread of viruses, and breeders develop virus-resistant strains of important crop plants.

Viral diseases of animals include rabies, chickenpox, polio, colds, influenza, warts, AIDS, and some forms of cancer (Table 18-3). Various herpes viruses that cause eukaryotic cells to become lysogenic may cause some human cancers. Different herpes viruses cause cold sores, venereal infections, and mononucleosis. Even after the symptoms of the infection have disappeared, the virus apparently remains in a lysogenic form, and can enter the lytic cycle when the person is ill or stressed, causing a fresh outbreak of cold sores or genital sores.

Most viruses cause disease by disturbing the cell's metabolism and eventually destroying it. The virus is reproduced in the dying cell (Figure 18-17). The new viruses then invade neighboring cells. The symptoms of viral infections, such as fever and swollen lymph nodes, are caused not by the viruses but by the activity of the body's immune system, which destroys most viruses before they can cause serious damage (Chapter 25).

Because viruses rely so heavily on host cell machinery for their reproduction, it is extremely difficult to find drugs that will destroy viruses without dam-

Table 18-3 *Major Groups of Viruses That Infect Animals*

Nucleic acid	Group	Some diseases caused
DNA	Poxviruses	Smallpox, cowpox, myxomatosis in rabbits, diseases in fowl
	Herpesviruses	Human oral and genital infections, Epstein-Barr infections, tumors
	Adenoviruses	Human respiratory and intestinal infections, conjunctivitis, sore throat, tumors
	Papovaviruses	Human warts; cancers in other animals
RNA	Paramyxoviruses	Human rubeola, mumps; canine distemper; Newcastle disease of chickens
	Myxoviruses	Influenza of humans, other animals
	Retroviruses	Rous sarcoma of chickens; mouse mammary tumor; feline leukemia; AIDS
	Rhabdoviruses	Rabies, various infections
	Reoviruses	Vomiting and diarrhea in children; Colorado tick fever
	Togaviruses	Human rubella, yellow fever, dengue, equine encephalitis, etc.
	Picornaviruses	Intestinal infections (enteroviruses), poliomyelitis, common cold (rhinoviruses)

aging the host equally. The antiviral drug acyclovir, used externally to treat herpes simplex infections and warts, works by interfering more with the virus's DNA polymerase than with the host's. Because it is so difficult to treat viral diseases, our best defense against them is still prevention: good hygiene, vaccination, and quarantine of infectious cases. Thanks to such practices, the age-old scourge of smallpox became extinct in the late 1970s, and new vaccines have reduced the number of cases of the viral diseases polio, measles, and rubella (German measles).

■ *Viral diseases are best fought by preventive methods such as cleanliness and vaccination.*

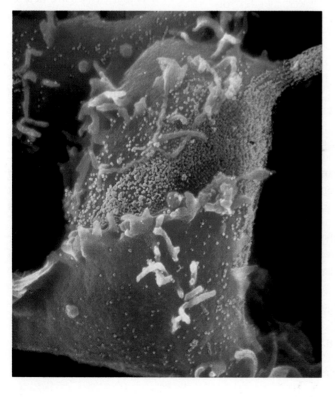

Figure 18-17
A human cell rupturing as it releases newly formed viruses (blue) into the surrounding fluid. (Photo by Lennart Nilsson, copyright Boehringer Ingelheim International, GmbH).

18-I   Viroids

In recent years, infectious, disease-causing particles even smaller than viruses have been isolated from plants. A **viroid** consists of a short, single strand of RNA, which folds into a shape with many double-stranded areas maintained by base-pairing (Figure 18-18). Viroids have no capsids—not surprisingly, since they contain too few nucleotides to code for the proteins to make such a coat. Indeed, infected host cells do not appear to contain any proteins coded by the viroid genome. How, then, does a viroid cause disease? We don't yet know, but there are two main hypotheses: that it somehow interferes with the regulation (turning on and off) of host cell genes, or that it interferes with intron splicing (Section 10-D). In fact, it has been proposed that viroids are escaped introns, because the two contain some nucleotide sequences in common.

Many wild and cultivated plants harbor viroids without showing symptoms of disease. When these viroids invade susceptible crops growing nearby, they can wreak havoc. Viroids have killed more than 10 million coconut trees in the Philippines, and they decimated the United States chrysanthemum trade in the 1950s. Viroids also cause spindle tuber of potatoes, exocortis of citrus trees, and several other diseases of important crop plants.

18-J   Viruses and Evolution

The peculiarities of viruses raise the question: what is their evolutionary origin? Several answers have been proposed. The first was that viruses are evolutionary relics, descended from ancestors that never evolved into true cells. When biologists realized that viruses depend totally on the very complex protein-synthesis and energy-generating machinery of living cells, most discarded this idea. Second, viruses may be reduced cells, which became parasites inside other cells and eventually jettisoned most of their own cell components and genes. These things were, after all, readily available in their host cells. Some fairly large viruses, containing dozens of genes and surrounded by a lipid-protein membrane, might be viewed as stripped-down cells. A third view is that viruses are neither retarded pre-cells nor regressed cells, but renegade genes that must return "home" to be replicated. This view is supported by the fact that the genetic similarity seems to be much closer between virus and host than between one virus and another. Or, this similarity could be explained by assuming that viruses have captured host cell genes during their evolution. In this view, viruses may have started as **plasmids,** small independently replicating nucleic acid molecules found today in some bacteria, yeast, and mammalian cells (see Figure 9-A).

Whatever the case, viruses clearly play an important role in the evolution of cellular organisms. First, they exert selective pressure. Second, many viral genomes are inserted into the host DNA and later released from it, often carrying part of the host's genome along into new host cells of the same or different species. Third, studies show that viral genes have become permanent parts of most species' genomes.

Viruses themselves often evolve very rapidly. Those that cause colds or influenza mutate often, producing offspring with novel genes. Hence some of the viruses can always find some hosts that have not yet built up immunity to the proteins produced by their particular genes.

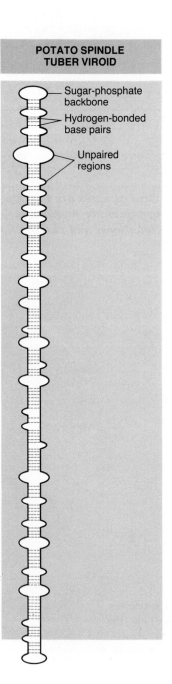

POTATO SPINDLE TUBER VIROID

Sugar-phosphate backbone

Hydrogen-bonded base pairs

Unpaired regions

Figure 18-18
The RNA of the viroid that causes spindle tuber of potatoes. The molecule is 359 nucleotides long and is shown folded up into a structure held together by hydrogen bonds between nucleotide bases (red lines).

SUMMARY

Taxonomy is the branch of biology concerned with relationships among organisms and with their classification. The basic unit of classification is the species; each species is given a unique Latin binomial, denoting its genus and species.

Species are grouped into progressively more inclusive taxa. The main levels in the taxonomic hierarchy, from most to least inclusive, are: kingdom, phylum, class, order, family, genus, and species. A taxon in each higher level contains one or more taxa of the next lower level. Taxa are not units intrinsic in nature, waiting to be discovered by humans, but instead are artificial inventions to help us think about living organisms in an orderly manner. Biologists often disagree about how the rules of taxonomy should be applied and where the lines should be drawn to define taxa.

In theory, living things are classified by phylogenetic relationships, but these are often difficult to disentangle, and the sheer number of existing species precludes drawing up a phylogenetic tree that encompasses all known organisms. In practice, therefore, living things are usually classified by morphology. Other features, such as physiology, biochemistry, behavior, geographic distribution, and fossil evidence, are also used.

This book uses the taxonomic system that divides organisms into five kingdoms: Monera, the prokaryotes; Protista, the unicellular eukaryotes; Fungi, the fungi; Plantae, the plants; and Animalia, the animals. This classification is based largely on the mode of nutrition and cellular organization of organisms. Several evolutionary lines of organisms are almost certainly grouped in each kingdom.

Viruses do not share all the features of cellular organisms and do not belong in the five kingdoms of living organisms. Viruses resemble cells in that they contain nucleic acid and protein molecules, and some are surrounded by membranous envelopes of lipid and protein. Unlike cells, viruses lack the metabolic machinery to make proteins and to generate energy, and many viruses can be crystallized. Cells reproduce by dividing in two, but hundreds of viruses may be produced in a host cell after infection by a single virus particle.

A virus's capsid or envelope proteins bind to the surface of a host cell by means of specific protein-receptor interactions. The viral genome, consisting of either DNA or RNA, but not both, enters the cell interior. In a lytic cycle, the virus immediately takes over the host cell's metabolic machinery, causing it to produce new viruses, which are then freed by lysis of the host cell. In a lysogenic cycle, the viral genome becomes incorporated into the host's DNA and is replicated and passed along to the cell's progeny. Eventually the viral genome is released from that of the host, possibly taking along some of the host's genetic material, and many new viruses are produced and released as in the lytic cycle. A third kind of cycle involves the gradual production and release of viruses, leaving the host cell intact, at least for a while.

Viral disruption of host cells is responsible for many diseases, including some tumors, in plants and animals. Some plant diseases are caused by viroids, which are short, naked strands of RNA that do not code for protein. The mechanism by which viroids cause disease is still unknown.

SELF-QUIZ

1. Modern classification is based on:
   a. taxonomy
   b. phylogeny
   c. morphology
   d. fossils
   e. autotrophy

2. All of the following make it difficult to construct acceptable classification schemes *except:*
   a. deciding where to impose artificial cutoffs in the midst of a naturally continuous series of organisms
   b. convergent evolution
   c. differences in rates of evolution for different characters
   d. persistence of conservative characters
   e. the large numbers of species that must be accommodated

3. The Latin binomial for the common dog is properly written:
   a. canis familiaris
   b. Canis Familiaris
   c. Canis familiaris
   d. *Canis familiaris*
   e. *canis familiaris*

4. In which of the following lists are the levels of the taxonomic hierarchy *not* arranged in correct descending order?
   a. phylum, order, family
   b. class, family, genus
   c. class, order, family
   d. family, class, order
   e. order, family, genus

5. Characters of two different organisms that have evolved from the same structure in an ancestral form but now have very different appearance and function can properly be termed (give all correct answers):
   a. derived
   b. homologous
   c. polyphyletic
   d. conservative
   e. convergent

6. You are given a microscope slide on which is mounted some biological material. On examining it, you observe that there are numerous individual cells containing chloroplasts and swimming around rapidly. This material belongs in the kingdom ___.
7. One reason that viruses are considered nonliving is that:
   a. they lack replicable nucleic acids
   b. their nucleic acids do not code for proteins
   c. they cannot make their own food molecules
   d. they cannot carry out their own reproduction
   e. they do not undergo mutation and so do not become adapted to changes in their environment
8. A virus always contains:
   a. DNA, RNA, and proteins
   b. nucleic acids and lipids
   c. proteins and nucleic acid
   d. DNA and proteins
   e. nucleic acids, proteins, lipids, and carbohydrates
9. Examine each statement below, and tell whether it is true of lytic viruses, lysogenic viruses, or both.
   ___ a. Many virus particles may be released from each host cell.
   ___ b. Part of the host's DNA may be carried to a new host cell by the virus.
   ___ c. The viral DNA may be incorporated into the host's genome.
   ___ d. Viral proteins are made on host ribosomes.
10. True or False. The genomes of some RNA viruses are replicated by the formation of RNA templates, whereas the genomes of others are replicated by the formation of DNA templates. In both cases, the viral RNA genome is then transcribed from the template.

QUESTIONS FOR DISCUSSION

1. Although modern taxonomy tries to classify organisms on the basis of their phylogenetic relationships, in practice most organisms are classified according to their morphology. Why?
2. Review the characteristics of life listed in Section 1-C. What features do viruses share with cellular organisms? Which do they lack?
3. Why is it difficult to produce convincing evidence that viruses cause cancer in humans?
4. One theory of the origin of viruses is that they began as stray pieces of host cell DNA or RNA, much like the "naked" viroids discussed in Section 18-I. Outline how such an escaped fragment of the cellular genome could have evolved into one of the complex modern viruses.

SUGGESTED READINGS

Diener, T. O. "Viroids." *Scientific American,* January 1981.

Hirsch, M. S., and J. C. Kaplan. "Antiviral therapy." *Scientific American,* April 1987. How drugs such as acyclovir (used against herpes infections) and AZT (used against AIDS) work.

Margulis, L., and K. V. Schwartz. *Five Kingdoms: An Illustrated Guide to the Phyla of Life on Earth,* 2d ed. San Francisco: W. H. Freeman, 1988.

Sibley, C. G., and J. E. Ahlquist. "Reconstructing bird phylogeny by comparing DNA's." *Scientific American,* February 1986. Direct studies of genes have revealed that some birds long thought to be close relatives have instead developed close resemblances by convergent evolution.

Simons, K., H. Garoff, and A. Helenius. "How an animal virus gets into and out of its host cell." *Scientific American,* February 1982.

Weiss, R. "The viral advantage." *Science News* 136:200, September 23, 1989. A chilling account of recent outbreaks of viral diseases.

Bacteria, Protists, and Fungi

CHAPTER

19

1. Use the following terms correctly:
   microorganism
   plasmid, **spore** (of bacterium)
   aerobe, obligate anaerobe, facultative anaerobe
   autotroph, heterotroph
   chemosynthesis
   saprobe, decomposer, pathogen
   symbiosis, parasitism, commensalism, mutualism, symbiont
   hypha, mycelium
   spore (of fungus), **sporangium**, fruiting body

2. Give the criteria used to place organisms in the kingdoms Monera, Protista, and Fungi.

3. List and describe the four sources of genetic change known in prokaryotes; discuss the contributions of these sources and of reproductive rate to the evolutionary success of prokaryotes.

4. Describe the distinguishing features of cyanobacteria.

5. Summarize the ecological significance of prokaryotes, including their roles in nitrogen fixation, nitrification, and decomposition.

6. State what the microbiota of an animal is, and list three reasons why it is important to the animal.

7. Give evidence for and against the theory that mitochondria and plastids arose from free-living prokaryotes.

8. Describe the three main types of locomotion found among protists.

9. Explain what phytoplankton are and state their ecological importance.

10. Describe the major characteristics of fungi, and tell how fungi differ from plants and animals.

11. Describe the asexual reproduction of a fungus such as black bread mold *(Rhizopus)*.

12. State what (a) lichens and (b) mycorrhizae are, and explain their ecological roles.

13. List or recognize ways in which bacteria and fungi are economically beneficial and ways in which they are economically harmful.

14. Give some examples of the use of bacteria and fungi in food production.

Kingdom Monera: The Bacteria
19-A Reproduction and Evolution in Prokaryotes
19-B Bacterial Metabolism and Ways of Life
19-C Classification of Prokaryotes
19-D Symbiotic Bacteria
19-E Origin of Mitochondria and Plastids

Origin of Eukaryotes and the Kingdom Protista
19-F Kingdom Protista: Unicellular Algae and Protozoa
19-G Photosynthetic Protists
19-H Protozoa (Heterotrophic Protists)
19-I On To Multicellularity!
19-J Slime Molds: Protists or Fungi?

Kingdom Fungi
19-K Classification of Fungi
19-L Symbiotic Relationships of Fungi
19-M Bacteria and Fungi: Friends and Foes

A Journey Into Life: Prokaryotes and Pollution

A Journey Into Life: Sexually Transmitted Diseases

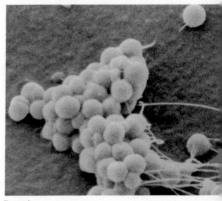

Bacteria

Fungus

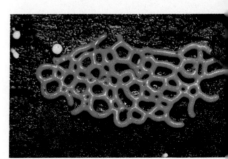

Pretzel slime mold

*T*he kingdoms Monera, Protista, and Fungi contain mostly microscopic organisms **(microorganisms).** Despite their small size, many of them have specialized cell structures, metabolism, or ways of life not found among larger organisms. Studying them, we also find clues to two important evolutionary advances, the origin of eukaryotic cells and of multicellularity.

These points are of biological interest, but there is an even more compelling reason to study microorganisms: their importance in the economy of nature and of human society. Autotrophic bacteria and protists are the basic food for many other organisms, particularly in aquatic habitats, where they are eaten by some small animals and heterotrophic protists. Other heterotrophic microorganisms live by recycling: they are **saprobes,** organisms that break down dead organic matter and thereby release nutrients that are reused by other organisms. Without the activities of these decomposing bacteria and fungi, plants (and animals that eat plants) could not exist. Nor could they survive without nitrogen-fixing bacteria, which provide nitrogen in a form plants can use to make amino acids for their proteins.

Although many microorganisms are independent, others cannot live without forming close associations with members of other species. These **symbiotic relationships** may be mutually beneficial, or one member may be a parasite that damages its larger host. Some of these parasites cause devastating diseases of humans and of our domestic animals and crops.

KEY CONCEPTS

- Microorganisms include free-living autotrophs (bacteria, cyanobacteria, and unicellular algae), saprobes (bacteria and fungi), and predators (protozoa). Many members of all three kingdoms live in symbiotic relationships with other species, as parasites or mutualistic symbionts.
- All three kingdoms contain many disease-causing members, some of which have changed the course of human history and caused untold loss of life, labor, and property.
- Bacteria and fungi play important ecological roles as decomposers. Some bacteria fix nitrogen and so support all plant life on earth.

Kingdom Monera: The Bacteria

The kingdom Monera contains the bacteria. Fossil bacteria dating from 3.5 billion years ago are the earliest signs of life on earth. Today, bacteria still outnumber other organisms, thriving in almost every conceivable habitat. But because of their tiny size, they were the last major group of organisms to be discovered—some large ones were first seen in 1676. In the 1860s and 1870s, Louis Pasteur and Robert Koch discovered the role of bacteria in causing food spoilage and many diseases. To most people, bacteria mean disease and decay—banes of human health and wealth. However, we are now coming to appreciate the broader roles of bacteria in the environment. Bacteria change and recycle mineral nutrients, help clean up pollution, and combat many organisms harmful to humans.

The prokaryotic cells of bacteria are generally much smaller, and much simpler in structure, than eukaryotic cells. Their definitive feature is the absence of a nuclear envelope separating the genetic material from the cytoplasm (Section 5-A). Bacterial cells come in three general shapes: rods, spheres, and spirals (Figure 19-1). In some species, new cells stick together after the parent cell divides, forming characteristic filaments (strings) or clusters. These cells are usually independent, but in some species of prokaryotes the cells cooperate, with a limited degree of division of labor.

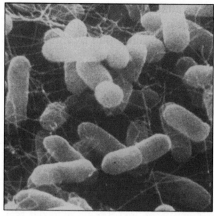

(a)

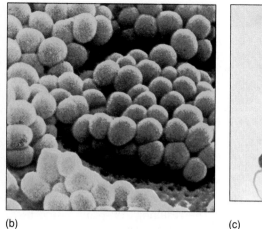

(b)

(c)

Figure 19-1

Cell shapes of bacteria. (a) Short rods: *Escherichia coli,* an inhabitant of the human intestine and a favorite research subject. (b) Spheres: cells of *Staphylococcus aureus* live on human skin, where they stick together in clumps. (c) *Spirillum volutans,* a large bacterium (about 40 micrometres long but very thin). It has a flagellum at each end, used in locomotion. (a, David M. Phillips/Visuals Unlimited; b, CNRI/Science Photo Library/Photo Researchers; c, Ed Reschke)

19-A   Reproduction and Evolution in Prokaryotes

Prokaryotes usually reproduce by **binary fission,** division into two genetically identical new cells (Figure 19-2). In the ideal environment of a laboratory culture, some bacteria divide every 20 to 30 minutes. However, such good growing conditions rarely occur in nature.

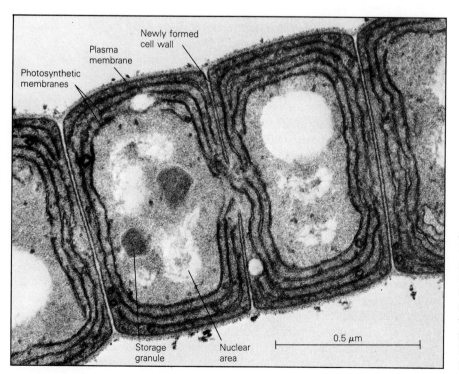

Figure 19-2

Binary fission. This electron micrograph shows part of a filament of the cyanobacterium *Oscillatoria*. The cell in the center is dividing into two new cells of equal size as the new cell wall forms at the edges and moves toward the center. Notice the several layers of photosynthetic membranes inside each cell. (Biophoto Associates)

Bacteria do not reproduce sexually, as eukaryotes do, by fusion of two gametes with roughly equal amounts of genetic material. In prokaryotes, genetic information from two individuals can be combined into one by three means:

1. **Transduction.** In the most common means of genetic recombination, a bacteriophage (virus) carries bacterial DNA from one bacterium to another, attached to the viral genome (Section 18-G).
2. **Transformation.** DNA from a broken cell is taken up by a living one.
3. **Conjugation.** DNA passes from one cell to another by way of a protein strand called a **pilus** (see Figure 5-2c). The amount of DNA transferred varies from part of a gene to an entire genome.

Once inside the recipient cell, the donor's DNA lines up with the homologous part of the resident DNA. Enzymes then replace the old DNA with the new. The resulting "offspring"—the recipient cell—contains genetic material from two parents. However, the parents gave unequal, and variable, amounts of DNA, in contrast to the nearly equal contributions in eukaryotic sex.

Besides their genome DNA, some bacteria contain **plasmids,** smaller, separate circles of DNA bearing additional genes. Some plasmids carry genes for resistance to **antibiotics,** substances that destroy microorganisms. Plasmids are replicated and passed on to the offspring at cell division. They can also be passed to other bacteria by any of the three means of gene transfer.

New gene combinations are the raw material of natural selection. The transfer of genes from cell to cell must play a role in prokaryote evolution, but mutations are probably more important. Mutations occur rarely—a given gene mutates once in every 10 thousand to 10 billion individuals. How can this provide enough variability to explain, say, the widespread evolution of resistance to antibiotics in the last few decades?

■ *A huge population, growing at a high rate, can produce a great deal of genetic variation among individual prokaryotes.*

The secret of prokaryote evolution lies in rapid reproduction: under ideal conditions, a single bacterium can produce millions of offspring in a week. With so many cells in a population, even rare mutations can produce so many genetic variants that genetic recombination is almost superfluous.

19-B   *Bacterial Metabolism and Ways of Life*

Bacteria are the most numerous organisms on earth; a handful of soil may contain billions of them. They occur in habitats as remote as the sea floor, icebergs, and hot springs, or as close as your mouth (Figure 19-3).

This far-flung distribution is possible because many bacteria have metabolic abilities not found in any eukaryote. Some carry on fermentation, others respiration (Chapter 7). **Aerobes** require O_2 for respiration; **obligate anaerobes** are killed by O_2; **facultative anaerobes** can take it or leave it. Some bacteria use nitrate (NO_3^-), sulfate (SO_4^{2-}), or iron (Fe^{3+}) instead of O_2 in their respiration. Most of these forms of **anaerobic respiration** yield more energy than fermentation but less than aerobic respiration (see Table 7-4).

Because of their rigid cell walls, bacteria cannot eat large food particles. Rather, they absorb small molecules that filter through their walls. Many bacteria can make any organic molecule they need, starting with only one kind of organic molecule, such as glucose or fatty acids. Others require a more complex diet.

When conditions are unfavorable for growth, some bacteria form thick-walled, resting **spores** that are resistant to heat and drying. Most bacterial spores are not means of reproduction, because no increase in cell number occurs. Rather, spore formation permits cells to survive adverse conditions and to disperse to new locations, where a new vegetative cell may germinate from the spore.

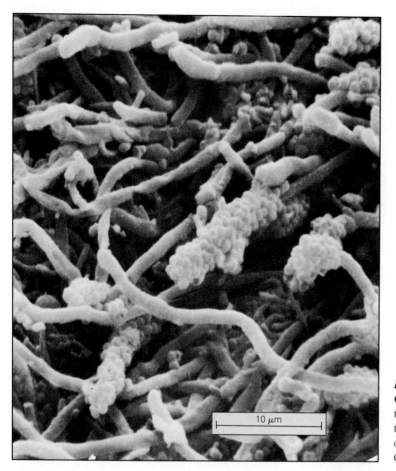

10 µm

Figure 19-3
Oral bacteria. Plaque from a human
tooth that has not been brushed for
three days, showing a dense growth
of bacteria. (Z. Skobe, Forsyth Dental
Center/BPS)

Bacteria undoubtedly owe their biological success to their varied metabolic
abilities, coupled with small size, rapid reproductive rate, and the ability to form
resistant spores. These features permit bacteria to live in many habitats that are
here today and gone tomorrow. A raindrop on a leaf evaporates in a few hours,
but by then a bacterium has divided several times and its descendants have
formed spores that can blow away to new homes.

As we discuss the metabolism of bacteria, notice how the various species
interact: one microorganism's wastes are another's food and drink.

Autotrophic Bacteria

Autotrophs can make their own food from simple inorganic compounds, such
as carbon dioxide, mineral ions, and water. Purple and green bacteria use light
energy to drive primitive forms of photosynthesis, which differ from that of
green plants. First, bacterial photosynthesis does not obtain hydrogen by splitting water. Hence it does not produce O_2. Various photosynthetic bacteria obtain
hydrogen from hydrogen gas, hydrogen sulfide, fatty acids, or alcohols. Second,
the chlorophylls of purple and green bacteria absorb, not visible light, but light
in the near-infrared range (see Figure 8-2). Hence they can photosynthesize in
what we would consider darkness. Third, many photosynthetic bacteria are
anaerobic.

Cyanobacteria use the kind of photosynthesis found in eukaryotes. They get
hydrogen by splitting water and release O_2 as a by-product (Chapter 8).

Chemosynthesis is a form of autotrophy found only in certain bacteria.
They make food using energy, not from light, but from inorganic chemical reac-

(a)

(b)

Figure 19-4
Animals dependent on autotrophic bacteria. (a) Near a thermal vent in the sea floor, chemosynthetic bacteria use hydrogen sulfide as an energy source to make their own food. These bacteria, in turn, support the animal life seen here. Some of the bacteria live inside the huge white worms with red tips seen here, in a mutually beneficial symbiotic relationship. (b) Thousands of lesser flamingoes at Lake Nakuru, Kenya, feast on filaments of the cyanobacterium *Spirulina,* one of the few photosynthetic organisms that can survive in the lake's alkaline water. (a, John B. Corliss)

tions, involving the oxidation of ammonia, nitrites, sulfides, or iron. Chemosynthetic bacteria are the sole food producers of some areas near vents in the deep sea floor that spew forth warm sulfide-laden water (Figure 19-4).

Nitrifying bacteria are chemosynthetic bacteria that oxidize ammonia (NH_3) or ammonium (NH_4^+) to nitrites (NO_2^-), and nitrites to nitrates (NO_3^-). These bacteria play a crucial role in the nitrogen cycle (Section 38-E).

■ *All life on earth depends on bacteria to provide nitrogen for proteins by nitrogen fixation and nitrification.*

Nitrogen-Fixing Bacteria

Nitrogen fixation is a complex set of reactions that reduce nitrogen gas (N_2) from the air to ammonia (NH_3). The only organisms that can fix nitrogen are some bacteria and cyanobacteria. All other life on earth is directly or indirectly dependent on their activities. Green plants need nitrogen, from ammonium or nitrates, to produce amino acids for their proteins (some of which later become animals' proteins). However, most of the earth's nitrogen exists as N_2 in the air, which plants cannot use. The nitrogen-fixing prokaryotes link the vast N_2 supply of the air to the rest of the living world. They provide ammonia, which nitrifying bacteria (discussed above) convert to nitrites and then to nitrates, which can be used by plants.

Nitrogen-fixing bacteria have great importance to agriculture. Some live in nodules on the roots of **legumes** (for example, peas, beans, clover, alfalfa, vetch) (Figure 19-5). These bacteria use sugars produced by the legume's photosynthesis and supply the plant with ammonium. Growing legumes as part of crop rotation adds nitrogen compounds to the soil and so improves its fertility. Genetic engineers are working to make other crops able to house such nitrogen-fixing bacteria.

Figure 19-5
Root nodules. These swellings on the roots of a pea plant house nitrogen-fixing *Rhizobium* bacteria. (Hugh Spencer/Photo Researchers)

Prokaryotes and Pollution

A JOURNEY INTO LIFE

In natural environments, decomposers recycle the wastes produced by other organisms. However, a large population of humans can produce so much waste that it disrupts this system.

For example, anaerobic, photosynthetic sulfur bacteria thrive in mud at the bottom of stagnant water. This anaerobic mud usually contains hydrogen sulfide (H_2S, which smells like rotten eggs) produced by sulfide bacteria living on dead plants. Normally, sulfur bacteria use the H_2S (instead of water) for photosynthesis, as fast as sulfide bacteria make it. However, if sewage is dumped into the water, the sulfide bacteria grow rapidly on nutrients in the sewage. Soon there are so many sulfide bacteria that they make the water cloudy and cut off light from the sulfur bacteria, which then grow poorly and leave much H_2S unused. The H_2S rises to the surface, producing the characteristic stench of a sewage-polluted lake.

Population explosions (**"blooms"**) of some cyanobacteria are another unpleasant result of water pollution. These organisms thrive on phosphates and nitrates in untreated sewage. Because these cyanobacteria produce toxins and also encase themselves in slimy inedible sheaths, fish cannot control the blooms by eating them. In fact, some fish are killed by cyanobacterial toxins. As cyanobacteria die, aerobic bacteria consume them and use up all the oxygen in the water, and so any remaining fish suffocate.

Laws passed during the last few decades have required local governments to install improved sewage facilities, which remove much of the nutrients from waste water. This has improved conditions in many lakes and rivers.

Bacteria that thrive on organic pollutants become our allies when humans produce environmental disasters such as oil spills. Oil that humans do not retrieve during cleanup operations is eventually eaten by bacteria adapted to that type of food. For example, the remains of the 68 million gallon spill off the coast of France in 1978 disappeared within three years.

After the 11 million gallon *Exxon Valdez* spill in Alaska's Prince William Sound in 1989 (Figure 19-A), fertilizer containing nitrogen and phosphorus was sprayed onto the oil-coated shore to balance the diets of bacteria able to live on the hydrocarbons in oil. It was hoped that this would enable bacteria to multiply and consume the oil before it spread farther, threatened more fish hatcheries, and killed even more animals such as puffins, ducks, loons, sea otters, seals, and whales. However, this fertilizer is toxic and it killed some shore organisms that survived the oil itself. The ultimate effect of these added nutrients on populations of other organisms living on the shore and in nearby ocean environments is still unknown.

Another experiment was tried during the *Mega Borg* spill and fire in the Gulf of Mexico in 1990. The oil slick was sprayed with extra bacteria of a type specialized to eat oil, in hopes of clearing the mess up faster. The best way (if any) to manage our bacterial allies in such situations is not yet clear.

Figure 19-A
Oil spill. The *Exxon Valdez,* surrounded by miles of spilled oil. The floating booms, used to contain the spill at first, can be seen as a diagonal line in this photograph. (World Wide Photo)

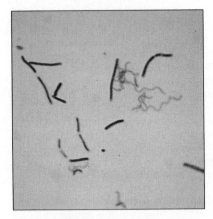

Figure 19-6
Gram-stained bacteria. The Gram stain, invented by H. C. Gram, is used to help identify bacteria. Because of differences in their cell walls, the Gram stain procedure turns Gram-negative cells pink and Gram-positive cells purple. The difference between Gram-positive and Gram-negative bacteria is medically important: Gram-positive forms are more susceptible to most antibiotics and to lysozyme, an enzyme in our body fluids that destroys their cell walls. (Carolina Biological Supply Company)

Heterotrophic Bacteria

Most bacteria are **heterotrophs,** using food made by other organisms. **Saprobes** live on dead organic material, such as dead animals or plants, feces, leaves, or bark. They secrete hydrolytic (digestive) enzymes into the food around them and absorb the resulting small organic molecules and inorganic ions. Saprobic bacteria and fungi are the most important **decomposer** organisms. When a dead leaf drifts to the forest floor or an animal dies, fungal and bacterial spores floating in the air have already settled on it. The spores quickly begin to grow and break down the dead organism, recycling it in the form of carbon dioxide, water, and minerals that autotrophs can use to produce more food. Many heterotrophic bacteria are **parasites,** obtaining food directly from the bodies of living organisms.

19-C   Classification of Prokaryotes

Biologists classify most organisms by their structure. However, many kinds of bacteria look so much alike that they cannot be told apart by structure alone. Because bacteria reproduce asexually by dividing in two, it is also impossible to define bacterial species by their ability to interbreed.

Bacteria have long been classified by both structural and metabolic features: cell size, shape (rods, spheres, or spirals), and arrangement in filaments or clumps; number and position of flagella; ability to use oxygen or to ferment various organic compounds; type of photosynthesis; reaction to Gram staining (Figure 19-6), and so on. This gives us a useful way to identify unknown bacteria, a necessary first step in practical situations such as diagnosing a disease or finding the cause of food spoilage. However, such characters do not always reflect evolutionary relationships.

Recently, researchers have drawn up an evolutionary tree for prokaryotes, based on nucleotide sequences in their nucleic acids.

Archaeobacteria

Archaeobacteria are so different from other prokaryotes that some biologists think they belong in a sixth kingdom by themselves. In structure they resemble other bacteria, but in some details of protein synthesis they are more like eukaryotes. Still other features are like nothing else on earth today, including their membrane lipids, cell walls, coenzymes, and transfer RNAs. Such basic biochemical features tend to be highly conservative (Section 18-C). Therefore, many biologists think that archaeobacteria diverged from other organisms very early in evolution, after the origin of the genetic code but while the machinery of metabolism and protein synthesis was still evolving.

There are three main types of archaeobacteria. One group makes methane (CH_4) in anaerobic habitats such as the bottoms of bogs, lakes, and sewage treatment ponds, and the digestive tracts of animals, especially cows and other ruminants (Section 22-D). The other two groups are aerobic. Salt-loving bacteria inhabit extra-salty environments such as salt evaporation ponds, Great Salt Lake, and the Dead Sea. The third group lives in hot, acidic springs, at temperatures of 80 to over 90°C and pH of less than 2!

Eubacteria

The vast majority of prokaryotes falls into a second main evolutionary group, the **eubacteria** ("true bacteria"). Eubacteria include the photosynthetic and chemosynthetic autotrophs. However, most of them are harmless saprobes, some are normal residents of humans or other eukaryotes, and still others cause

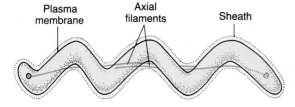

Figure 19-7
Drawing of a spirochete. Each cell has at least two axial filaments, and sometimes hundreds. The filaments are attached at either end of the cell and lie between layers of the cell wall.

diseases such as plague, tetanus, botulism, gangrene, diphtheria, tuberculosis, and leprosy. Leprosy is the most widespread human bacterial disease in the world today, with over 20 million cases. The leprosy bacterium divides only once every 12 days, and so the disease takes a long time to develop. A few groups of eubacteria deserve special mention.

Actinomycetes produce branching, multicellular filaments that resemble fungi. They are the source of many valuable antibiotics.

Mycoplasmas have two notable features: they have lost their cell walls, and they are easily the smallest living cells, only 0.1 to 0.25 micrometre in diameter. These features permit them to pass through filters that trap other bacteria. The lack of cell walls also makes them resistant to penicillin (Section 5-A). Mycoplasmas live as parasites in plant or animal cells. They cause one kind of human pneumonia, and many diseases of other animals.

Spirochetes are spiral bacteria with distinctive flagella, called axial filaments, between layers of the cell wall (Figure 19-7). Parasitic spirochetes include those that cause syphilis (see *A Journey Into Life* at the end of this chapter), yaws (a tropical skin disease), and Lyme disease.

Rickettsiae are tiny, parasitic bacteria that usually live inside other cells. They are carried by ticks or insects, which transmit them to mammals by bites. Rickettsial diseases include typhus, one of the all-time great killers of humans, transmitted by lice, and Rocky Mountain spotted fever, carried by ticks.

Most **cyanobacteria (blue-green bacteria)** are photosynthetic. Like eukaryotic plants, they get hydrogen by splitting water and give off O_2. Most are blue-green because they contain green chlorophyll *a* plus blue accessory photosynthetic pigments called phycobilins. Cyanobacteria may form filamentous or clustered colonies (Figure 19-8). Some colonies show division of labor: besides photosynthetic cells, there may be spore-producing cells, or cells specialized for attachment to a rock or other substrate or for nitrogen fixation.

Prochlorophytes are a group of photosynthetic bacteria first discovered in the late 1970s. Like cyanobacteria, they carry on water-splitting, oxygen-producing photosynthesis. However, they differ in containing two chlorophylls—*a* and

Figure 19-8
Cyanobacteria. (a) Filaments of *Nostoc,* a nitrogen-fixing form useful for enriching the soil in rice paddies. Numbers of these filaments often form mucilaginous masses on the soil surface. (b) *Merismopedia,* a colonial form. (a, Sinclair Stammers/Science Photo Library/Photo Researchers; b, J. R. Waaland, Univ. of Washington/BPS)

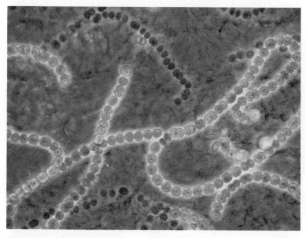

(a)

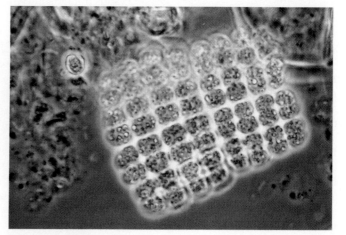

(b)

b—which also occur in most chloroplasts. For this reason, it seems likely that some member(s) of the group gave rise to the chloroplasts in most members of the plant kingdom (Section 19-E). They are also important ecologically: one tiny species of prochlorophyte, discovered in 1988, is one of the two most numerous photosynthetic organisms in the world's oceans (the other is a cyanobacterium). Ten gallons of water contain about 5 billion of these cells—the same as the earth's current human population!

19-D Symbiotic Bacteria

We have seen that many bacteria exploit the resources of the nonliving environment. The living environment—other organisms—also contains a wealth of resources. Many members of all five kingdoms have evolved ways to tap their living neighbors' resources by forming close associations with them. Such an intimate relationship between members of different species is called **symbiosis** ("together living"). It is often divided into three categories:

1. **Parasitism.** One species, the parasite, benefits at the expense of the other, the host species, as in bacterial diseases.
2. **Commensalism.** (Com = together; mensa = table) One species benefits, and the other is not harmed (for example, some small fish live attached to the bodies of sharks and feed on scraps that "fall from the table" of these predators).
3. **Mutualism.** Both species benefit from the association. When one species is much smaller than the other—such as a bacterium or protist associated with a plant or animal—it is called a **symbiont** of the larger host species.

Because of their small size, bacteria can form especially intimate symbioses with a variety of other organisms. Hence the symbiotic relationships of bacteria make a good introduction to a topic that will be a recurrent theme in our study of all five kingdoms.

■ *Many prokaryotes live as symbionts with eukaryotes, sometimes even inside eukaryotic cells.*

Many symbiotic bacteria are part of the **microbiota** of an animal: organisms that normally live on or within it. They may be saprobes, living on dead skin cells or digested food in the intestines, or parasites or symbionts absorbing food from living tissue. An animal's normal microbiota may benefit the host by producing enzymes that can digest cellulose, by making vitamins, by producing antibiotics, or by competing with disease-causing bacteria (all organisms grow better without competition for food and space).

On the other hand, members of an animal's microbiota may cause disease if they settle down in the wrong places. *Escherichia coli* (from the intestine) causes cystitis if it gets into the urinary bladder. *Staphylococcus aureus* (from the skin)

Figure 19-9
A symbiotic relationship. This deep sea lanterneye fish emits light thanks to bioluminescent bacterial symbionts. The bacteria are sheltered in tube-like light organs under the fish's skin, and some of them find new homes in the fish's offspring when it reproduces. The pattern of light organs on male and female fish is different, and so researchers believe the fish recognize each other by their light patterns.

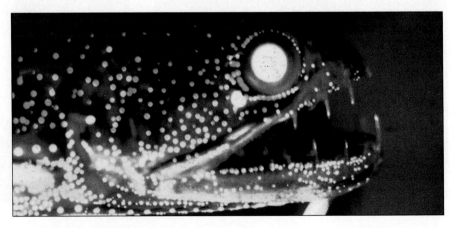

can cause serious infections if it gets into a wound; this is one reason why surgeons wear gowns, masks, and gloves in the operating room.

Other symbiotic bacteria include those that live in the root nodules of legumes and in bioluminescent fish (Figure 19-9). Perhaps the most fascinating symbiosis is that between the anaerobic protist *Mixotricha paradoxa* and four different kinds of bacteria—one that acts much like mitochondria, breaking down the protist's fermentation products, and others that serve in locomotion (Figure 19-10). The whole assemblage lives in the gut of a termite: *M. paradoxa* itself is a symbiont, digesting the termite's meals of wood! The termite cannot produce wood-digesting enzymes, and without its symbionts it would starve.

In many symbioses, the partners become genetically interdependent. For example, root nodules of legumes contain leghemoglobin, a pigment that binds O_2 and keeps it from interfering with nitrogen-fixing enzymes. The heme part of leghemoglobin is produced by the nodule bacteria, the globin protein by the plant. Often the smaller member of a symbiosis loses genes for proteins similar to those made by its partner, and some of its genes move into the partner's genome. As a result, many symbiotic organisms are hard to grow without their partners because they rely on their partners for vital molecules.

19-E   *Origin of Mitochondria and Plastids*

Many biologists think that plastids and mitochondria evolved from prokaryotes living inside larger host cells, the ancestors of modern eukaryotes. This **endosymbiont theory** (endon = within) is supported by the fact that some prokaryotes live inside eukaryotic cells today (Figure 19-11). These bacteria are enclosed in a vacuole made up of host-cell membrane—comparable to the outer membranes of chloroplasts and mitochondria. Furthermore, plastids and mitochondria are about the size of bacteria; like bacteria, they contain circular DNA without histone proteins; and they reproduce by binary fission. They also make some of their own proteins. Their ribosomes are similar to prokaryote ribosomes in both size and function. Like many symbiotic bacteria, these organelles depend on host nuclear genes to supply some of their proteins.

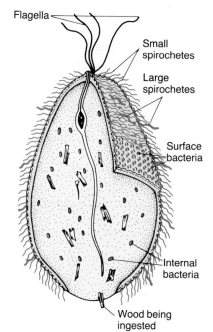

Figure 19-10

A symbiont's symbionts. The protist *Mixotricha paradoxa,* a symbiont in the gut of Australian termites, and its symbiotic bacteria. The cell's four flagella do not move it, but serve as rudders. The cell moves by the activities of up to half a million small spirochetes on its outer surface. Larger spirochetes are interspersed among these. The small spirochetes are anchored to the cell by bacteria that live in little depressions on the surface. Still other bacteria, inside the cell, are thought to play a metabolic role, breaking down the end products of the cell's fermentation.

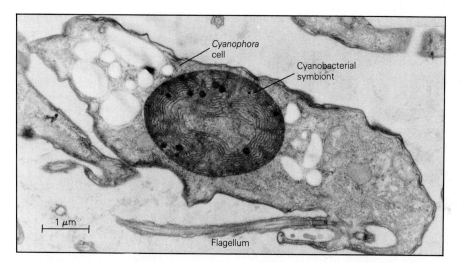

Figure 19-11

The protist *Cyanophora* with a symbiotic cyanobacterium in its cytoplasm. The cyanobacterium produces food for itself and its host; *Cyanophora* uses its flagellum to move to areas where light is available for the photosynthesis of its symbiont. Notice the many layers of photosynthetic membranes in the cyanobacterial cell. (Biophoto Associates)

■ *Many biologists think that plastids, and probably mitochondria, evolved from intracellular symbiotic bacteria.*

The theory's critics argue that mitochondrial and prokaryotic protein synthesis differ somewhat. The mitochondrial genetic code differs slightly from the usual system, and mitochondrial genes sometimes contain introns, whereas prokaryotic genes do not (Section 10-D). The membranes of some bacteria contain molecules needed for respiration, and attachment sites for DNA. Mitochondria could have begun when fragments of such membranes broke off and formed separate sacs in the cytoplasm, enclosing attached DNA (perhaps a plasmid), ribosomes, and respiratory molecules. Cytoplasmic and organelle DNA and ribosomes then evolved at different rates.

Chloroplasts are more similar to bacteria than mitochondria are. This leads some biologists to believe that chloroplasts began as symbionts, but mitochondria did not. The theory's champions feel, on the contrary, that this reflects a history in which mitochondria became symbionts long before chloroplasts. They point out, first, that all organisms with plastids also have mitochondria, but not vice versa. Second, we would expect mitochondria to be less like bacteria if they had been symbionts for a longer time than chloroplasts.

The endosymbiont theory contends that mitochondria originated from symbiotic aerobic bacteria. Much later, photosynthetic symbionts moved into some mitochondria-containing cells. What was the original host cell? The best guess is a bacterium similar to a present-day acid hot springs archaeobacterium that has proteins similar to histones and actin, which are characteristic of eukaryotes. Because it lacks a cell wall, such a bacterium could easily have engulfed other bacteria. This theory can account for the apparently rapid appearance of eukaryotic cells in the fossil record, and for the lack of organisms intermediate between prokaryotes and eukaryotes.

Origin of Eukaryotes and the Kingdom Protista

The evolution of the first eukaryotic cells, about 1.45 billion years ago, was a tremendous advance. Specialized organelles, surrounded by membranes, kept different cell activities separate in distinct areas, where they did not interfere with each other. They also permitted a dramatic increase in cell size. As we have just seen, mitochondria and chloroplasts may have evolved as symbionts inside eukaryotic cells. The membranes of the nuclear envelope, endoplasmic reticulum, and Golgi complexes probably evolved from invaginations of the plasma membrane (Figure 19-12).

The hallmark of eukaryotic cells is the nuclear envelope around the genetic material, separating it from the cytoplasm. Eukaryotic cells have a number of linear chromosomes, containing much more genetic information than a prokaryote's one DNA molecule. The evolution of mitosis and meiosis guaranteed an orderly sorting of chromosomes into new nuclei during cell division. Most eukaryotes can engage in true sexual reproduction, with its varied range of new genetic combinations. This in turn permitted the tremendous adaptive radiation of eukaryotes.

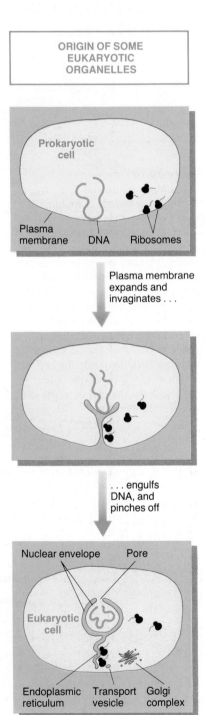

Figure 19-12

How some eukaryotic organelles may have evolved. The plasma membrane of a prokaryote, with sites for DNA and ribosome attachment, expands into the cytoplasm. Eventually it surrounds the DNA, forming the double membrane layers of the nuclear envelope. The outer layer is continuous with the endoplasmic reticulum. Golgi complexes and transport vesicles form from sacs pinched off from parts of this membrane. The interiors of these organelles are in some ways part of the "extracellular" medium (blue).

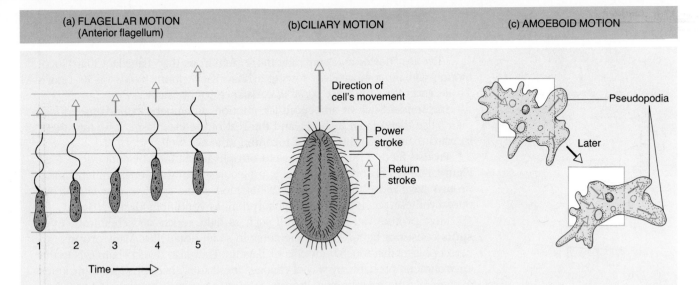

Direction of cell's movement

Power stroke

Return stroke

Pseudopodia

Later

1 2 3 4 5

Time ———▷

Figure 19-13

Modes of locomotion found among protists. (a) Flagellar motion. Numbers 1 through 5 show the cell's position with respect to the reference lines as its anterior flagellum undulates, pulling the cell forward through the water (arrows). (b) Ciliary motion. Cilia are held straight on the power stroke; as they push backward, the cell moves forward. By bending on the return stroke, the cilia avoid pushing the cell backward. The cell's many cilia move in a coordinated sequence. (c) Amoeboid motion. Streaming of the cytoplasm pushes out temporary pseudopodia, and the rest of the cell follows. In the second drawing, the cell has moved with respect to the reference square.

19-F   Kingdom Protista: Unicellular Algae and Protozoa

Most unicellular eukaryotes are classified in the kingdom Protista. However, some are so closely related to particular plants or fungi that they are placed in those kingdoms, with their multicellular relatives. Conversely, although most protists live as solitary cells, many form **colonies** of similar, but largely independent, cells.

Protists have radiated into many different habitats and ways of life. They live wherever there is water: in the sea, in fresh water, in moist soil, or in the bodies of animals.

Photosynthetic protists, and simple multicellular plants, are often called **algae.** Other protists, called **protozoa,** ingest their food and lack chloroplasts and cell walls. Before protists were separated into a kingdom by themselves, the autotrophs were considered plants (unicellular algae), and the heterotrophs were considered animals (protozoa).

Physiology of Protists

The Protista are so heterogeneous that few generalizations can be made about their physiology (how they work). Some have a cellulose cell wall, a shell, or other covering that provides protection and support and may also keep the cell from taking in too much water by osmosis (Section 4-C). Many protists without cell walls have **contractile vacuoles** that collect excess water and expel it from the cell (see Figure 19-21). Nearly all protists can carry on respiration, but many can also live indefinitely using fermentation when oxygen is not available.

Under adverse conditions, many protists can enclose themselves in thick walls, forming **cysts,** resistant to desiccation (drying out) and temperature changes. Here the cell can survive until favorable conditions return, or until it is carried to a new home.

Some protists are **sessile** ("sitting"), living attached to objects such as rocks or plants. Others float passively in their watery homes. However, many can move by means of cilia, flagella, or **pseudopodia** (flowing extensions of the cytoplasm).

In flagellar locomotion, wave-like bending motions pass from one end of the flagellum to the other. This pulls or pushes the cell through the water, depending on whether the flagellum is at the anterior or posterior (front or rear) of the cell (Figure 19-13).

Cilia are shorter and generally more numerous than flagella. Ciliary locomotion can be compared to rowing a boat. Each cilium bends on its return stroke and offers the least possible resistance to the water.

In pseudopodial or amoeboid locomotion, an amoeba extends pseudopodia ("false feet") from its body, and the rest of the cytoplasm flows into these forward extrusions. This type of locomotion is not well understood.

Protists feed in various ways. Heterotrophs ingest food by endocytosis (see Figure 4-14) or absorb small organic molecules from their environment. Many protists have more than one type of nutrition: some species can switch from photosynthesis to endocytosis to absorption, as conditions dictate.

Most protists can sense stimuli such as light, touch, and chemicals. **Eyespots** consist of light-sensitive pigments in small organelles. Many protists can detect objects that touch their cilia or flagella. They also detect chemicals in the environment, probably by way of changes these substances produce in proteins in their plasma membranes. Protists respond to stimuli by moving toward or away from them as appropriate for their particular way of life. For example, a protist with flagella and chloroplasts swims toward light, which it needs for photosynthesis. A heterotroph that eats decaying organic matter might avoid light and so move toward a food supply at the bottom of the pond.

Protists usually reproduce asexually by dividing in two after a mitotic division of the nucleus. Most can also undergo some kind of sexual reproduction.

19-G Photosynthetic Protists

Many cyanobacteria (Monera) and algae (Protista) are members of the **phytoplankton** (phyton = plant; plankton = wanderer), photosynthetic organisms floating near the surfaces of oceans, lakes, and ponds. Phytoplankton are ecologically important as food for many aquatic animals. They also produce an estimated 30 to 50% or more of the oxygen in the atmosphere.

Table 19-1 The Kingdom Protista	
Unicellular Algae (Mainly Autotrophs)	
Phylum Pyrrophyta* (~2000 species)	Dinoflagellates. Pectin and cellulose cell walls; mostly marine; two flagella; chlorophylls *a* and *c* and carotenoids; food stored as starch
Phylum Euglenophyta* (~800 species)	*Euglena* and its relatives. No cell wall; flexible pellicle of protein; mostly freshwater; two anterior flagella; chlorophylls *a* and *b* and carotenoids
Phylum Chrysophyta (~10,000 species)	Diatoms and their allies. Cellulose and pectin cell walls, containing silica in diatoms; marine or freshwater; diatoms lack flagella, allied forms may have one or more flagella; chlorophylls *a* and *c* and carotenoids; food stored as oils
Protozoans (Mainly Heterotrophs)	
Phylum Zoomastigina*	Flagellates without chloroplasts. No cell wall; freshwater, symbiotic, or parasitic; one or more flagella; e.g., *Trypanosoma*
Phylum Sarcodina* (~40,000 species)	Amoebas, foraminiferans, heliozoans, and radiolarians. No cell wall but some secrete shells containing silica, calcium carbonate, etc.; marine and freshwater; pseudopodia
Phylum Apicomplexa (~3900 species)	Parasitic protists. Cell with distinctive, complex arrangement of several components (apical complex) at one end; locomotion by bending or gliding, flagella in male gametes; complex life history; e.g., *Toxoplasma, Plasmodium*
Phylum Ciliophora (~8000 species)	Ciliates. Two types of nuclei; cilia for locomotion and food collection; no cell wall; freshwater and marine, some parasites; e.g., *Paramecium, Didinium*

* All groups marked * are sometimes combined into the Phylum Sarcomastigophora: protists moving by means of flagella or pseudopodia.

Phytoplankton need light, but because water absorbs much light, they can survive only if they remain near the surface. Some use flagella to swim upward. Others have projections that keep them afloat by expanding their surface area and hence providing resistance to sinking. Still others store their extra food as oil, which buoys them up near the surface.

When oil-storing phytoplankton die and sink to the bottom, they may be covered by sediment and water that exert enormous pressure. Such phytoplankton, of hundreds of millions of years ago, made the oil we use as fossil fuel today.

Photosynthetic protists are placed in three phyla on the basis of conservative characters such as the type of cell wall, flagella, photosynthetic pigments, and the form in which food is stored (Table 19-1).

Phylum Pyrrophyta (Dinoflagellates)

Phylum Pyrrophyta ("fire plants") takes its name from some species that are bioluminescent, able to give off light (like fireflies). Members of this phylum have two flagella and are commonly called dinoflagellates (dinos = whirlpool) (Figure 19-14). Most are **marine** (ocean-dwelling). Some forms lose their flagella and live as symbionts of corals, sea anemones, and clams, supplying food made by photosynthesis in exchange for protection. However, most are free-living and make up one of the two main groups of eukaryotes in the marine phytoplankton. Ancient planktonic dinoflagellates formed large oil deposits.

Some members of the genus *Gonyaulax* produce a nerve poison deadly to vertebrates. During population explosions of these forms, shellfish that have eaten large quantities of these cells become unfit for human consumption. Since *Gonyaulax* contains a red pigment, such population explosions may tinge the water pink, a fitting warning signal called a "red tide."

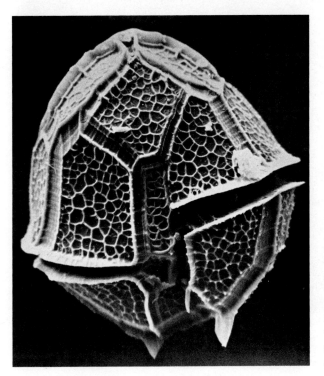

Figure 19-14
A dinoflagellate. This scanning electron micrograph, at more than 2000× magnification, shows the many "plates" of the cell wall. One of the two flagella lies in the groove around the cell's middle. (Biophoto Associates)

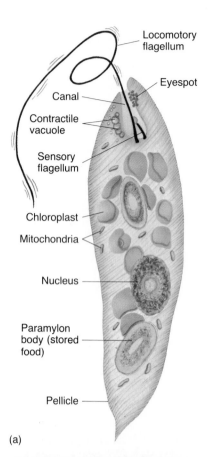

(a)

(b)

Figure 19-15
Euglena (Euglenophyta). The eyespot (a cluster of pigment) and the sensory flagellum are thought to play a role in sensing and responding to light. The cell moves toward light, using its locomotory flagellum. This positions the chloroplasts favorably for photosynthesis, to make the cell's food. The contractile vacuole expels excess water taken in by osmosis. Other views of *Euglena* appear in Figure 5-A. (b, Biophoto Associates)

Phylum Euglenophyta

The Euglenophyta are named after the common genus *Euglena* (Figure 19-15). Most members of this group live in fresh water, especially water polluted by excess nutrients. There are two flagella. These protists lack cell walls but may have an elastic, transparent **pellicle,** made of protein, just beneath the plasma membrane.

Euglenophytes are usually photosynthetic. However, if *Euglena* is raised in the dark, it loses its green color and becomes a heterotroph, ingesting food.

The Euglenophyta are considered to be the protists most resembling the ancestors of both plants and animals. This is partly because most other protists are highly specialized for their own ways of life, and partly because euglenophytes have flagella. Flagella are also found in the sperm of animals and of most lower plants.

Phylum Chrysophyta (Diatoms and Golden Algae)

Chrysophytes include the diatoms and the "golden algae" (chrysos = golden). They live as single cells or simple colonies in the sea, in fresh water, and in wet spots on land. Diatoms and pyrrophytes are the two main groups of eukaryotic

Figure 19-16
Diatoms (Chrysophyta). The intricately patterned walls of diatoms were once used to test the resolving power of microscopes. Victorian microscopists painstakingly arranged diatoms to form designs on microscope slides; this photo shows part of one such arrangement. (Biophoto Associates)

marine phytoplankton. In unpolluted fresh water, diatoms (with green algae, Chapter 20) are the most important group of phytoplankton. Chrysophytes store much of their surplus food as oil and were important in the formation of petroleum deposits.

Diatoms have a rigid cell wall impregnated with silica (SiO_2), a compound also found in glass and sand. Often the cell wall has an intricate pattern of pits or ridges (Figure 19-16). These walls are very resistant to decay, and accumulate on the ocean floor in enormous numbers. Such deposits may later be raised above sea level by geological activity. The resulting "diatomaceous earth" is used as a fine abrasive in silver polish and toothpaste and as the packing in air and water filters.

19-H   Protozoa (Heterotrophic Protists)

Phylum Zoomastigina (Zooflagellates)

The phylum Zoomastigina (zoon = animal; mastix = whip) contains heterotrophic flagellates. These include free-living organisms, symbionts, and parasites. The parasites often have complex life histories with two host species. Termites and wood roaches are totally dependent on their flagellate symbionts to digest their meals of wood.

A major threat to human welfare are parasitic trypanosomes, which live in the blood of vertebrates, mainly mammals (Figure 19-17). Human trypanosome diseases include sleeping sickness (Africa) and Chagas' disease (Latin America). Sleeping sickness is transmitted by bites of tsetse flies, hosts for part of the trypanosome life history. In Latin America, Chagas' disease affects about 12 million people. It too is transmitted by bloodsucking insects, which hide in houses or other buildings, crawling out at night to feed on people, dogs, cats, or guinea pigs (which are raised for food).

Trypanosomes are especially difficult for the body to conquer because they keep changing the cell-surface proteins (antigens) that the immune system must recognize in order to kill the parasites. As quickly as the host develops a defense against one antigen, the parasite switches to another. The immune system must start all over to combat what is essentially a new disease.

Leishmaniasis is caused by parasitic flagellates transmitted by sand flies. This disease afflicts millions of people in Africa and Asia. Symptoms range from skin sores, to loss of soft tissues of the face, to fatal disease.

Vaccines against these flagellate diseases will take years to develop. Meanwhile, the best available countermeasure is to control the insect carriers necessary for the parasites to complete their complex life histories.

Closer to home for most of our readers is *Giardia,* carried to lakes and reservoirs by expanding populations of beavers and muskrats. To avoid the diarrhea, cramps, fatigue, and weight loss of giardiasis, campers in back country should boil all water before drinking it.

Phylum Sarcodina

The Sarcodina move and engulf their prey with pseudopodia. *Amoeba proteus,* a freshwater species, has been thoroughly studied. It is easy to raise on a diet of bacteria and is large enough to be a good subject for experiments on amoeboid

■ *Trypanosome infections and leishmaniasis are two of the "big six" diseases on the worldwide health list. (The others are leprosy [caused by bacteria], malaria [protists], schistosomiasis [flatworms], and filariasis [roundworms]).*

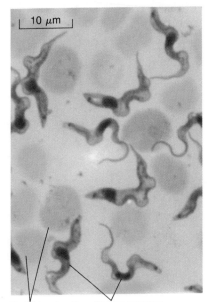

Figure 19-17
Phylum Zoomastigina: trypanosomes, parasites of vertebrate blood. Here, members of the species that causes African sleeping sickness swarm among red blood cells. (Ed Reschke)

Figure 19-18
A foraminiferan. This member of the phylum Sarcodina extrudes long, thin pseudopodia out through its shell. (Manfred Kage/Peter Arnold, Inc.)

movement, endocytosis, and nucleus-cytoplasm interactions. However, this amoeba is not typical of the phylum. Most sarcodines have shells, made of calcium carbonate (lime), silica, or bits of sand, through which they extrude long, thin pseudopodia to catch prey (Figure 19-18). Some forms are colonial.

The limy shells of one group, the **foraminiferans,** sink to the bottom of the sea when the organism dies. Millions of years' worth of this debris has formed chalk rocks or limestone, such as the famous white cliffs of Dover in England. Foraminiferan fossils are also common in deposits where oil has accumulated from ancient phytoplankton. As an oil well is drilled, the bit passes through layers of foraminiferans that lived at different times. By identifying the species in a particular layer, a geologist can estimate the age of the rock and decide where oil deposits are likely to be found.

Phylum Apicomplexa

All apicomplexans are parasites, usually with the complicated life histories typical of this way of life. Some need two different host species. Most move by body flexion or gliding; some have flagellated gametes.

Toxoplasma may well be the most common human parasite. Half of all adults in the United States have probably been infected at some time. This protozoan invades body cells and usually causes mild symptoms (enlarged lymph nodes) that pass unnoticed. However, it can cause serious disease in a fetus or newborn, leading to blindness, mental retardation, or death. *Toxoplasma* is usually spread in the feces of cats or in undercooked meat from infected pigs.

Human malaria, caused by four species of the genus *Plasmodium,* illustrates the complex life histories typical of parasites. Malaria parasites require two different hosts: humans, and female mosquitoes of the genus *Anopheles,* which transmit the disease from one person to another (Figure 19-19).

Malaria was attacked vigorously earlier in this century by draining swamps where mosquitoes breed, by spraying houses with DDT to kill mosquitoes, and by treating victims with drugs to destroy the parasites. The disease was eradicated from much of its former range, including much of the United States, and drastically lowered in most other areas. But just when victory seemed near, some parasites evolved resistance to the drugs, and mosquitoes to DDT, bringing a resurgence of malaria to many areas of Asia, Latin America, and Africa. About 3.5 million people are infected and 2 million die each year, mostly children. Most victims live in developing nations and neither they nor their governments can afford new insecticides, nor drainage programs to combat mosquitoes. Current work on a vaccine against sporozoites may take years to finish. Furthermore, sporozoites are vulnerable only on the short trip from the bite site to the liver, and they continuously shed their surface proteins and make replacements. The immune system ends up attacking the empty coats rather than live parasites.

Phylum Ciliophora (Ciliates)

Ciliates are complex, heterotrophic protists with many cilia (Figure 19-20). The body wall contains a pellicle and often numerous **trichocysts,** barbed or poisoned thread-like organelles that can be discharged to the outside. Trichocysts serve for anchorage, defense, or capture of prey.

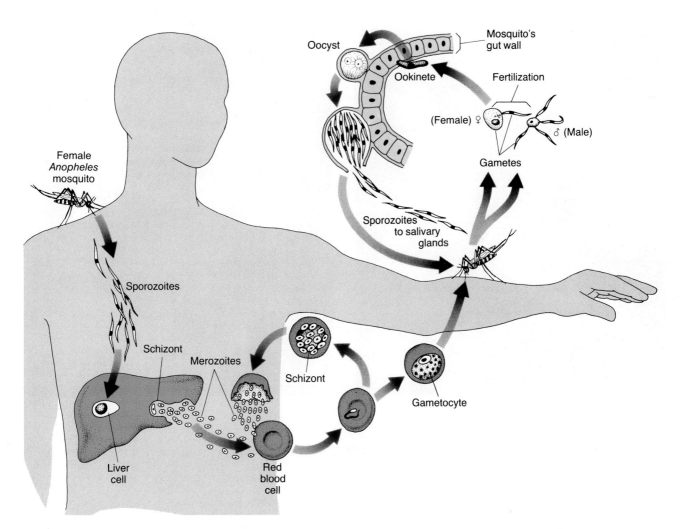

Figure 19-19
Life history of *Plasmodium vivax,* **which causes human malaria.** Parasites in the sporozoite stage enter the bloodstream with saliva from a female *Anopheles* mosquito. Within a few hours, these parasites invade cells in the liver. For a week or two, they grow, divide, and develop into forms called schizonts, each of which eventually ruptures and releases 20,000 merozoites. Each merozoite enters a red blood cell and repeats the sequence of development into schizonts and merozoites, with infection of even more red cells. Repeated cycles of rupture and invasion occur every two or three days, producing the symptoms of malaria: periodic bouts of shaking, chills, fever, and sweating. Some of the parasites in the red blood cells develop into sexual gametocytes, which may be taken up by a mosquito when she bites a malaria victim. Fertilization occurs in the mosquito's gut, and the zygote develops into a stage that encysts in the gut wall. After about 12 days, each cyst releases many sporozoites, which migrate to the salivary glands, ready to be injected into a new human host at the mosquito's next meal.

Most ciliates prey on bacteria, small animals, or fellow protists; some eat organic particles from the water, some are symbionts, and a few are parasites. Cilia around the mouth sweep food into a gullet. Here food enters a vacuole, which fuses with a lysosome full of digestive enzymes. The products of digestion are absorbed into the cytoplasm. Contractile vacuoles discharge excess water at specific sites on the cell surface.

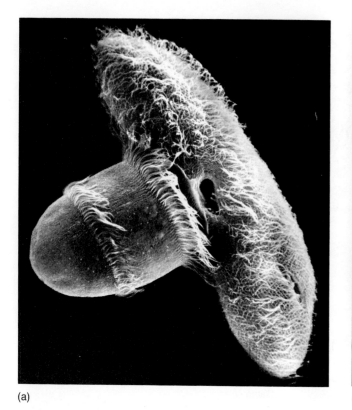

(a)

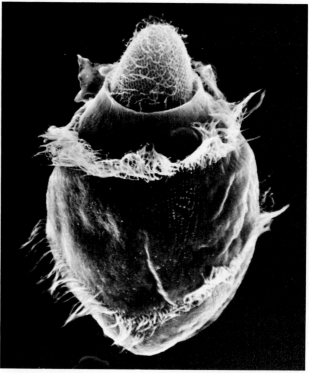

(b)

Figure 19-20
Ciliate eats ciliate. (a) A *Didinium* (at left) attacks a *Paramecium* that looks much too big for it. The cavity halfway along the *Paramecium* is its gullet. (b) The *Didinium* has stretched enormously and engulfed all but the tip of the *Paramecium*. (Biophoto Associates, courtesy of Drs. G. Antipa and E. Small)

Figure 19-21
***Paramecium*, a ciliate.** The body is completely covered with cilia, used in locomotion and also to sweep food particles into the gullet. The cytoplasm at the end of the gullet engulfs food into vacuoles by phagocytosis. Food is digested by enzymes in the food vacuoles as they move around in the cytoplasm. The two star-shaped contractile vacuoles collect water taken up by osmosis and expel it from the cell. Trichocysts are defensive, barbed threads that the cell discharges at its surface when it is disturbed. (b, Biophoto Associates)

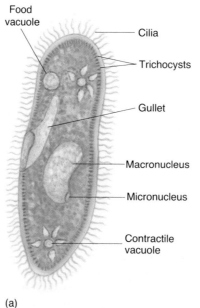

Food vacuole

Cilia

Trichocysts

Gullet

Macronucleus

Micronucleus

Contractile vacuole

(a)

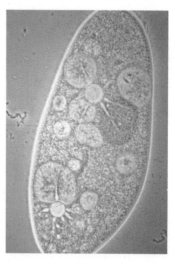

(b)

Figure 19-22
Surface-to-volume ratios. Surface area increases with the square of the linear dimension of the object, whereas volume increases with the cube of the linear dimension. If the objects are of the same shape, the larger one will have less surface area per unit volume. That is, the surface area-to-volume ratio (SA/V) decreases as an object increases in size while retaining the same shape.

Cube side length	Surface area (SA)	Volume (V)	SA/V
1 cm	6 cm²	1 cm³	6:1
2 cm	24 cm²	8 cm³	3:1
3 cm	54 cm²	27 cm³	2:1

Surface area = 1 cm × 1 cm × 6 = 6 cm²
Volume = 1 cm × 1 cm × 1 cm = 1 cm³

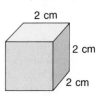

Surface area = 2 cm × 2 cm × 6 = 24 cm²
Volume = 2 cm × 2 cm × 2 cm = 8 cm³

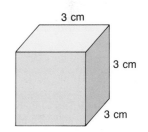

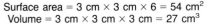

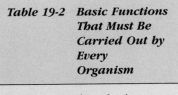

Surface area = 3 cm × 3 cm × 6 = 54 cm²
Volume = 3 cm × 3 cm × 3 cm = 27 cm³

Ciliates have more than one nucleus: each cell has one or more small **micronuclei,** each containing one copy of the genome, and a large **macronucleus,** containing up to 500 times more DNA. Apparently the micronucleus controls sexual reproduction and heredity, while the macronucleus controls growth, metabolism, and asexual reproduction. *Paramecium* is the most familiar genus of freshwater ciliates (Figure 19-21).

19-I   On To Multicellularity!

Once upon a time, all organisms were unicellular. Wherever such life was abundant, there must have been strong selective pressure for increased size. A large organism could eat more of its neighbors and be eaten by fewer. However, a single cell cannot just become larger and larger. Eventually the center of the cell is too far from the outside to obtain the substances it needs from its environment fast enough. As cell size increases, the ratio of cell surface area to cytoplasmic volume decreases (Figure 19-22). Because food, oxygen, and wastes are exchanged through the surface, the decrease in the surface-to-volume ratio also decreases the rate at which supplies reach each unit volume of cytoplasm. Thus the size of single cells is limited. Body size must be increased by an increase in cell number.

Some protists are colonial, with many independent cells joined into a larger unit. Each cell is small and exchanges substances with its environment efficiently, but the cells are still more or less identical (see Figure 20-4c). True multicellularity implies division of labor among the cells of an organism.

Every cell and every organism must carry out certain basic life functions (Table 19-2). Usually each cell can carry out its own basic functions, but in multicellular organisms each cell is also specialized to help carry out one of these functions for the entire body. The story of the kingdoms Fungi, Plantae, and Animalia can be viewed as the evolution of increasingly specialized groups of cells to form tissues and organs. The first sign of division of labor is usually the specialization of some cells for reproduction, protection, or anchorage, while others acquire energy by feeding or photosynthesis.

The simplest multicellular organisms are small, with no cell very far from the watery environment that provides food and removes wastes. As organisms became larger, the inner cells were farther from the environment. Organisms that evolved ways to transport substances between the environment and these cells were able to grow larger, but those without such systems had to remain small. Finally, a large-bodied organism must coordinate its various parts into a working system. Coordination exists in even the smallest of single cells, but it is most impressive in the complex systems of sense organs, nerves, hormones, and muscles found among the higher animals.

We tend to view increases in size and complexity as "progress" over yesteryear's smallness and simplicity. However, unicellular organisms do have some advantages over larger creatures. A single cell can live in a tiny space and needs only a little food before it is ready to reproduce. This allows unicellular organisms to exploit many habitats that larger forms could not. The ubiquity and diversity of bacteria and protists today attest to their evolutionary success.

Table 19-2   Basic Functions That Must Be Carried Out by Every Organism

Feeding or making food

Gas exchange

Waste removal

Internal transport of food, gas, etc.

Sensing environmental stimuli

Dispersal (locomotion, scattering seeds or larvae)

Support and protection

Coordination of all functions (nerves, hormones, etc.)

Reproduction

(a)

(b)

Figure 19-23
Acellular slime molds. (a) A plasmodium, the feeding stage of an acellular slime mold. (b) Developing fruiting bodies of another species. (b, Biophoto Associates)

19-J   *Slime Molds: Protists or Fungi?*

Slime molds are a classic problem in taxonomy. They spend part of their lives in a mobile, amoeba-like state, engulfing organic matter and bacteria, much as protozoa do. Hence many people would assign slime molds to the kingdom Protista. On the other hand, slime molds also produce fungus-like reproductive structures: some cells form a stalk, and others become spores with cellulose walls. This cooperation and division of labor show a primitive degree of multicellularity.

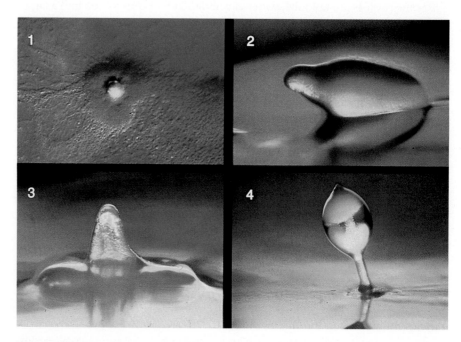

Figure 19-24
Reproduction in a cellular slime mold. (1) Individual amoeba-like cells congregate into a single mass, visible as a spot in the center of the frame. (2) The slug-like mass of coalesced cells crawls about until it (3) settles in one spot and (4) forms a reproductive structure, a stalk topped by a rounded sporangium (spore-producing structure). (Carolina Biological Supply Company)

The feeding stage of an **acellular slime mold** (phylum Myxomycota) is a mass of cytoplasm containing many nuclei but not separated into individual cells (Figure 19-23).

In **cellular slime molds** (phylum Acrasiomycota), the feeding stage is a uninucleate amoeba. When these amoebas run out of food, some of them secrete a chemical that attracts others, and they congregate and form a reproductive structure (Figure 19-24).

Kingdom Fungi

The kingdom Fungi contains eukaryotic, mostly multicellular organisms that absorb food molecules from their surroundings. All fungi are heterotrophs, living as saprobes, parasites, or partners in a mutualistic symbiosis. They obtain food as bacteria do: they secrete digestive enzymes, which hydrolyze the organic matter around them into small organic molecules and minerals that the fungus can absorb. As decomposers, the fungi are vitally important to plants and animals. Fungi are also important in the human economy. Some fungi cause tremendous losses of food and crops every year, but fungi are also used to produce foods and medicines.

Fungi almost always have cell walls, and they reproduce and disperse to new habitats by means of various types of spores.

Body Plan

The absorptive lifestyle of fungi is intimately linked with two important characteristics: production of spores and mycelial growth. A **spore** is a tiny, usually haploid, cell that disperses the fungus to new habitats, usually by floating through the air. The production of many tiny spores increases the chance that at least a few will fall onto a suitable food source. When this happens, the spore germinates, starts absorbing food, and grows into a thread-like **hypha.** The hypha grows rapidly and branches to form a tangled mass called a **mycelium** (Figure 19-25).

The mycelium, with its high surface-to-volume ratio, is well suited to absorbing food. A hypha grows from the tips and releases chemicals that cause other hyphae to grow away from it. As a result, the fungus spreads out through its food source. Parasitic fungi absorb nutrients from the host's body fluids, and parasites of plants may produce specialized hyphae called **haustoria** that penetrate the plant's cell walls and lie against the plasma membranes, where they can absorb food.

Some fungi are **coenocytic,** with many nuclei lying in the same cytoplasm, not divided into separate cells. Others are divided by **septa** into compartments containing one or more nuclei.

Fungi have rigid cell walls. Most groups have walls containing the polysaccharide chitin, also found in the external skeletons of insects and their kin. Some fungi have cellulose cell walls, like those of plant cells.

Reproduction

Fungi may spread vegetatively, by growth or fragmentation of the mycelium. Spores may be formed asexually or as a result of sexual processes. Spores are often produced on aerial structures that hold them away from the food source, up where they can catch an air current for a ride to a new home.

The parts of a fungus we normally see are reproductive structures. If you use a microscope to examine the green, white, pink, or black fuzz on moldy

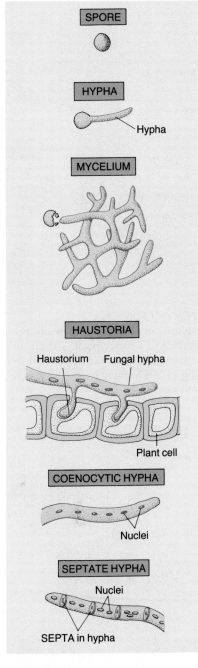

Figure 19-25
Some basic features of fungal anatomy.

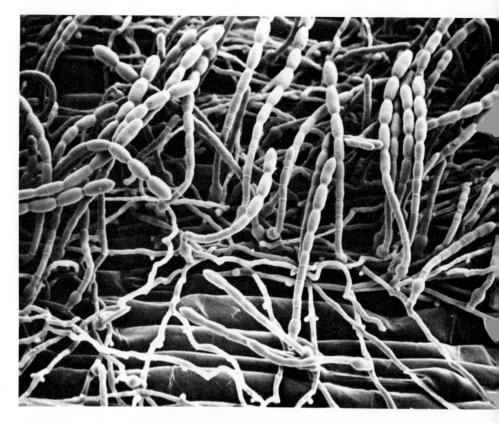

Figure 19-26
Mycelia of powdery mildew on the surface of a barley plant. The tips of the vertical, aerial hyphae are pinching in and forming spores. (Biophoto Associates)

food or diseased plants, you will see that it consists of masses of hyphal threads tipped with **sporangia:** structures bearing strings or spherical clusters of rounded, thick-walled spores (Figure 19-26). Similarly, the above-ground parts of mushrooms and cup fungi are **fruiting bodies**—large, complex reproductive structures composed of many hyphae. Fruiting bodies disperse spores produced by sexual processes; the vegetative mycelia grow hidden in the food source.

19-K Classification of Fungi

Fungal taxonomy follows the rules for plants, which use the term **division** instead of phylum in the taxonomic hierarchy. The names of fungal divisions end in -mycota (mykes = fungus), but common names of these groups end in -mycetes, harking back to an older classification scheme. Fungi are assigned to divisions on the basis of the characteristic form of microscopic sexual reproductive structures—one of the few consistently conservative features in each division (Table 19-3).

Division Oomycota

Oomycota means "egg fungi," referring to the appearance of the sexual reproductive structures (as do all the names of the divisions of fungi). The oomycetes are familiar as the white fuzz on diseased aquarium fish or on organic matter sitting in water. But despite the group's common name of "water molds," many oomycetes live on land, including some damaging plant parasites, such as late blight of potatoes and downy mildews. Some of these can disperse by means of airborne spores. However, the distinctive features of oomycetes are asexual, unwalled, flagellated spores, which can swim through water or raindrops to new food sources.

Table 19-3   The Kingdom Fungi and Its Divisions

Kingdom Fungi	All divisions contain saprobes and parasites with cell walls and (usually) hyphae. Unless otherwise indicated, fungi are terrestrial and reproduce both sexually and asexually by means of spores
Division Oomycota (~475 species) 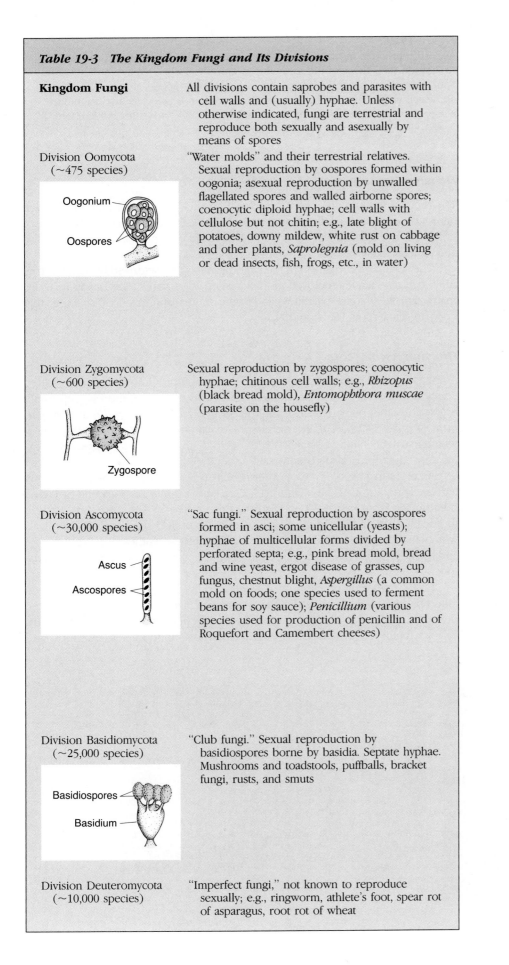	"Water molds" and their terrestrial relatives. Sexual reproduction by oospores formed within oogonia; asexual reproduction by unwalled flagellated spores and walled airborne spores; coenocytic diploid hyphae; cell walls with cellulose but not chitin; e.g., late blight of potatoes, downy mildew, white rust on cabbage and other plants, *Saprolegnia* (mold on living or dead insects, fish, frogs, etc., in water)
Division Zygomycota (~600 species)	Sexual reproduction by zygospores; coenocytic hyphae; chitinous cell walls; e.g., *Rhizopus* (black bread mold), *Entomophthora muscae* (parasite on the housefly)
Division Ascomycota (~30,000 species)	"Sac fungi." Sexual reproduction by ascospores formed in asci; some unicellular (yeasts); hyphae of multicellular forms divided by perforated septa; e.g., pink bread mold, bread and wine yeast, ergot disease of grasses, cup fungus, chestnut blight, *Aspergillus* (a common mold on foods; one species used to ferment beans for soy sauce); *Penicillium* (various species used for production of penicillin and of Roquefort and Camembert cheeses)
Division Basidiomycota (~25,000 species)	"Club fungi." Sexual reproduction by basidiospores borne by basidia. Septate hyphae. Mushrooms and toadstools, puffballs, bracket fungi, rusts, and smuts
Division Deuteromycota (~10,000 species)	"Imperfect fungi," not known to reproduce sexually; e.g., ringworm, athlete's foot, spear rot of asparagus, root rot of wheat

Diagram labels: Oogonium, Oospores; Zygospore; Ascus, Ascospores; Basidiospores, Basidium

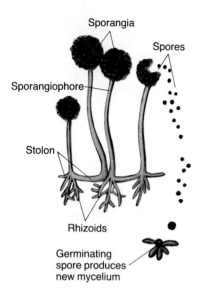

Sporangia

Spores

Sporangiophore

Stolon

Rhizoids

Germinating
spore produces
new mycelium

Figure 19-27

Asexual reproduction in *Rhizopus* (black bread mold). The vegetative hyphae are stolons, on the surface of the food source, and rhizoids, penetrating into the food. Sporangiophores are aerial hyphae bearing sporangia. Spores are shed into the air, which wafts them to new habitats. The mycelium is colorless; the common name of this organism comes from the masses of black spores it produces. The organism usually reproduces by this asexual process.

Division Zygomycota

The remaining fungi are mostly terrestrial, with chitinous walls, haploid nuclei, haploid spores, and no flagellated cells. Most of these fungi commonly reproduce by forming asexual spores atop aerial hyphae (hyphae that stick out in the air, such as those of bread mold) (Figure 19-27).

The sexual reproduction of the Zygomycota is characterized by formation of diploid zygote nuclei enclosed in thick-walled zygospores. This group includes many saprobes, some parasites, and some mutualistic mycorrhizal fungi that grow in plant roots (Section 19-L).

Division Ascomycota

Among the more than 30,000 species of Ascomycota are unicellular forms, called yeasts, and a great variety of multicellular forms, including morels, cup fungi, ergots and other plant parasites, and the common molds *Aspergillus* and *Penicillium* (Figure 19-28). The characteristic sexual structure is an **ascus** ("sac"), which usually contains a stack of eight haploid spores, called ascospores. Asexual reproduction is also common.

An ergot in a head of rye or other cereals is an ascomycete containing many extremely toxic substances (Figure 19-29a). Humans may be poisoned by eating bread made from infected rye. Ergotism, also called St. Anthony's Fire, caused the death of thousands in medieval Europe. Ergot also supplied the chemicals from which lysergic acid diethylamide (LSD) was first synthesized. Ergot toxins, and the notorious poisons of some mushrooms, protect these fungi from predators.

In 1927, Alexander Fleming noticed that his bacterial cultures had been killed by the ascomycete *Penicillium notatum* (Figure 19-29b). This led to the discovery of the world's most widely used and effective antibiotic, **penicillin.** Various fungi are also used to produce vitamins, amino acids, enzymes, and

Figure 19-28

The ascomycete *Penicillium*. (a) Asexual spores called conidia impart the green color seen here. While some hyphae were producing these spores, others grew out into new areas of the food source. The white fuzz consists of immature reproductive hyphae that have not yet formed spores. Vegetative parts of the mycelium are buried in the orange, turning it to mush. (b) Hyphae and conidia of *Penicillium* as seen through the light microscope. (b, Biophoto Associates)

(a) Orange infected with *Penicillium*

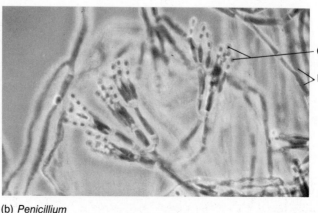

Conidia

Hyphae

(b) *Penicillium*

(a)

(b)

Figure 19-29
Ascomycetes with chemical defenses. (a) An ergot in a head of rye. (b) *Penicillium* (white fuzzy mycelium) growing in the center of a Petri dish containing many bacterial colonies (red spots). *Penicillium* produces penicillin, which kills bacteria. Diffusion of penicillin from the fungal mycelium through the agar in the dish has killed the nearest colonies of bacteria. (a, Carolina Biological Supply Company; b, Biophoto Associates)

other organic compounds on a commercial scale. Even the substances produced by ergots are valuable drugs. Although they are deadly in large amounts, small doses are used to induce labor, control bleeding, and treat migraine headache, high blood pressure, and varicose veins.

Division Basidiomycota

The Basidiomycota include rusts, smuts, mushrooms, puffballs, bracket fungi, and coral fungi (Figure 19-30). The sexual reproductive structure is a **basidium** ("club") bearing four haploid basidiospores.

Mushrooms are the most familiar basidiomycetes. In these fungi, a well-fed mycelium forms an underground mass of hyphae with a bulbous base, a stalk, and a knob-like cap. Some morning after a heavy rain we awake to find that the hyphae of the stalk have swelled with moisture and elongated, lifting the cap

Figure 19-30
Basidiomycete fruiting bodies. (a) Bracket fungi on the surface of wood which is being decomposed by the vegetative mycelium growing inside it. (b) A coral fungus, the reproductive structure of a mycelium growing on dead plant matter in the soil.

(a)

(b)

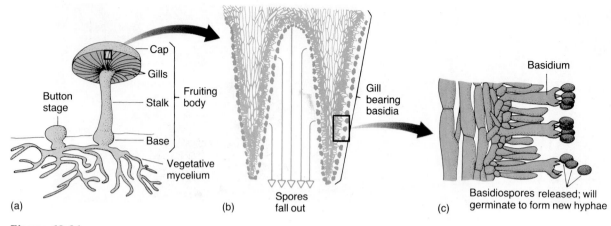

(a)   (b)   (c)

Figure 19-31
Reproduction of a mushroom. (a) A mushroom is the fruiting body of a basidiomycete. Both the fruiting body and the vegetative, feeding mycelium are made up of masses of hyphae. (b) Cross section of two gills, showing basidia with basidiospores all over the outer surfaces. (c) Closeup of basidia and basidiospores. (Adapted from Wilson, C. L., *et al. Botany,* 5th ed. Holt, Rinehart and Winston, 1971.)

above ground. The cap opens like a belated umbrella, and numerous basidia along the edges of the gills or pores beneath the cap shed their spores (Figure 19-31).

Division Deuteromycota

The division Deuteromycota contains a rummage-sale collection of fungi that cannot be assigned to any other group because their sexual reproduction (if they have any) has never been observed. Fungi that cause ringworm, athlete's foot, and other skin infections belong to this group. Other deuteromycetes cause important diseases of crops, including strawberry leaf blight and bitter rot of grapes.

19-L   Symbiotic Relationships of Fungi

Many fungi are harmful parasites, but some form mutualistic associations, in which both partners benefit.

Mycorrhizae

Many fungi grow associated with plant roots in a symbiosis called a **mycorrhiza** ("fungus root") (Figure 19-32). The fungal hyphae branch out into the soil and use their superior absorptive ability to take up mineral nutrients. By using radioactive elements as tracers, researchers found that the fungi pass on some of these minerals to the plant roots and receive food from the plant. Most land plants form mycorrhizal associations, usually with zygomycete or basidiomycete partners. When pine trees were introduced into new areas, such as Puerto Rico and Australia, they grew very poorly until supplied with soil from pine forests, containing the appropriate mycorrhizal fungi; after this, they grew rapidly.

Some fungi help protect their plant partners from acid rain. Acid promotes two changes in the soil unfavorable to plants: leaching (washing away) of required nutrients, and increased solubility of toxic minerals such as zinc, copper, aluminum, and manganese. The right mycorrhizal fungus can absorb nutrients

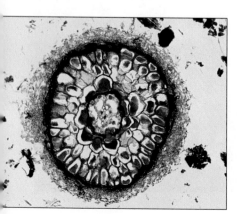

Figure 19-32
A mycorrhiza. Fungal hyphae, stained aqua, surround a cross section of a plant root. The fungus absorbs soil minerals and passes them to the plant, receiving organic substances made by the plant in return. (Carolina Biological Supply Company)

(a) (b) (c)

Figure 19-33
Lichens come in three main body forms. (a) Crustose forms resemble thick blotches of paint. (b) Foliose forms resemble leaves, flat with curling edges. (c) "British soldiers" is a colorful example of the fruticose (shrubby, branched) form. (c, Biophoto Associates, N.H.P.A.)

from depleted soil water and pass them to the plant, and can also protect the plant from toxic substances in the soil.

Lichens

Lichens look like plants (Figure 19-33). However, a lichen is really an intimate symbiosis between two kinds of organisms: ascomycete or basidiomycete fungi, and photosynthetic cells—either cyanobacteria or green algae (Sections 19-C and 20-C). The fungus obtains organic compounds from the photosynthetic partner, but it is unclear what it provides in return. Possibly it absorbs water and minerals.

Lichens have slow metabolism and growth, but they are extremely resistant to drought and cold. They are the most important autotrophs in the low-growing vegetation of the **tundra,** found in the Arctic and also at high altitudes on mountains. Lichens absorb minerals from the air and so can grow without soil, on stones or rocky ocean islands where nothing else survives. However, lichens are quickly killed by air pollution, and the state of the lichens in an area can be used as an indication of its air quality.

People have found other odd uses for lichens. Archaeologists have estimated the age of the mysterious stone heads of Easter Island by measuring patches of lichens growing on them. Lichens also produce pigments traditionally used to dye Harris tweed.

19-M Bacteria and Fungi: Friends and Foes

Diseases

A **pathogen** is an organism able to produce disease. Pathogens are specific for particular hosts because infections begin with the specific binding of certain pathogen and host surface molecules, much like the binding between an enzyme and its substrate (Section 3-F). Some pathogenic bacteria destroy the host's cells, but most cause disease by producing **toxins** (poisonous substances). Often the toxin alone can cause symptoms typical of a bacterial disease.

Most fungi require more oxygen than bacteria do. Hence, human fungal infections, such as ringworm, athlete's foot, and vaginal infections, are nearly all restricted to body surfaces exposed to air (Figure 19-34). Fungal diseases of plants are much more numerous. Fungi can easily enter plants through injured areas or natural pores in leaf surfaces. Air spaces within leaves permit the fungus to obtain all the oxygen it needs and still send haustoria into the plant's cells for food.

The Irish potato famine was a major disaster caused by a fungal disease, late blight of potatoes. The potato, native to South America, was brought to Europe in the sixteenth century. Potatoes require little labor and produce high yields of one of the most nutritious plant foods. By the nineteenth century, they were almost the only crop grown in Ireland. However, **monocultures** (plantings of only one crop) are especially susceptible to the rapid spread of disease. In 1845 and 1846, late blight destroyed virtually the whole Irish potato crop, leading to a devastating famine. From 1845 to 1851, a million people died in Ireland and a million and a half emigrated, mainly to the United States and Canada. The European potato shortage stimulated agriculture in North America, where grain for export has been grown in large quantities ever since.

Control of Disease   Preventing the spread of pathogenic bacteria depends on knowing how they disperse. Diseases like diphtheria, scarlet fever, whooping cough, and tuberculosis are caused by airborne bacteria, usually released in cough or sneeze droplets. This is why victims are quarantined. The most important preventive measures against bacterial disease are hygiene and sanitation. In the last century, maternal deaths following childbirth were reduced about tenfold when doctors and midwives learned to wash their hands and their instruments between patients. Also, Joseph Lister developed aseptic surgical techniques. Even today, keeping food and water reasonably free of bacteria saves many more lives than antibiotics do.

Heating to 60°C for 30 minutes destroys protein toxins and kills most bacteria. For this reason pasteurization (heating) has proved both easy and effective

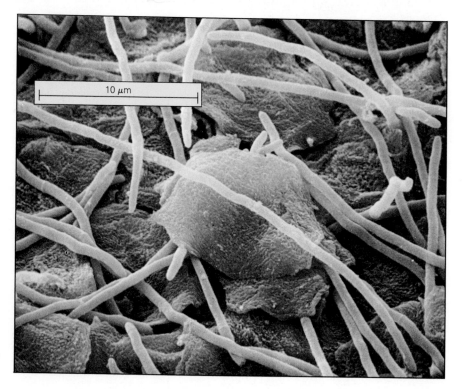

Figure 19-34
Fungal infection of human skin. Slender, tubular hyphae grow over and under flat human epidermal cells. (Biophoto Associates)

in protecting against botulism in canned food, brucellosis and tuberculosis in milk, and dysentery in drinking water contaminated with human feces. In addition, antibiotics have saved thousands of lives since about 1940. During all wars before World War II, more lives were lost to disease (from unsanitary camps and infected wounds) than to enemy action.

Once a fungus establishes itself in a plant or an animal, or in wood, paper, or leather, it is virtually impossible to eradicate. Measures that prevent spores from germinating, such as keeping things dry and using fungicides, are the most effective defenses against fungi.

Bacteria, Fungi, and Food

Bacterial and fungal spores are everywhere, and some of them inevitably land on our food. Milk spoils when bacteria ferment the milk sugar lactose and produce lactic acid, which coagulates the milk proteins. Pasteurization retards spoilage by killing many of these bacteria. On the other hand, bacteria are necessary to make dairy products such as cottage cheese and yogurt, and other foods such as sauerkraut, pickles, and vinegar. Fermented foods like these keep longer, and are often more nutritious, flavorful, and digestible than their raw materials.

Brewers' yeast, a unicellular ascomycete fungus, ferments sugars to make alcohol in the production of wine and beer. A different strain of yeast is used in bread-making. Growing under aerobic conditions, its respiration gives off carbon dioxide, which forms bubbles in the dough and makes the bread rise.

English Stilton and French Roquefort, Brie, and Camembert cheeses all get their flavors from specific ascomycete fungi. In the Orient, soy sauce is traditionally made by fermenting boiled soybeans and wheat with another ascomycete for about a year. This produces a flavorful sauce rich in vitamins and amino acids, a valuable addition to a low-protein diet of rice.

Growing edible mushrooms (basidiomycetes) is a million-dollar industry in many parts of the world. Morels (Figure 19-35) and truffles are ascomycetes. In France, truffle hounds and pigs are trained to hunt the underground truffles by smell. In 1990, the most prized truffles sold for $720 per pound.

"Food poisoning" comes from toxins produced by bacteria growing in food. The most serious (usually fatal) form of food poisoning is botulism. Since botu-

Figure 19-35
A morel, an edible fruiting body of an ascomycete. It releases spores produced by sexual reproductive processes. (Biophoto Associates, N.H.P.A.)

Sexually Transmitted Diseases

A JOURNEY INTO LIFE

Sexually transmitted diseases (STDs) are transmitted from one person to another primarily by contact between the genital organs during sexual activity. A dozen or more such diseases occur in humans.

Historically, the most famous sexually transmitted disease was syphilis, caused by a spirochete bacterium. In some people the disease causes damage to the nervous and circulatory systems. Nervous degeneration caused by syphilis is believed to have been responsible for the strange behavior of England's King Henry VIII, the insanity of the composer Smetana, and the blindness of the composer Frederick Delius.

Syphilis is not very common because, even if it is not treated, it is not very contagious. Like herpes infections, syphilis may become **latent,** producing no symptoms. Syphilis cannot be transmitted to another person when it is in its latent stages, which is most of the time. Furthermore, the chance of catching syphilis from a single sexual encounter with someone who has an active case is only about 1 in 40. However, new cases of syphilis increased 23% in early 1987, reversing a previous downward trend.

Gonorrhea, one of the most widespread STDs, is somewhat more contagious than syphilis. It too is caused by a bacterium (Figure 19-B). In men the disease is readily detected because it produces a discharge of pus from the penis and a burning sensation during urination. Most infected men therefore seek treatment. The disease is widespread because it produces few or no symptoms in women, where the cervix is the area usually infected. In both sexes gonorrhea can cause sterility. Gonorrhea has usually been treated with penicil-

lin. Predictably, penicillin-resistant gonorrhea bacteria have evolved, as only those that happened to have mutations conferring resistance to penicillin survived and reproduced. There are now some strains of the disease that are incurable because they are resistant to all known antibiotics.

Infections caused by bacteria of the genus *Chlamydia* are also widespread, although many cases involve no symptoms, and others are mistaken for gonorrhea. However, *Chlamydia* is not susceptible to the penicillin used to treat gonorrhea, and so *Chlamydia* infections cannot be cured until properly diagnosed. As with gonorrhea, infections can cause scarring of the reproductive tissues and hence sterility.

Many people today are concerned about infectious genital blisters caused by Type 2 herpes simplex virus, very similar to the virus that causes cold sores and fever blisters on the lips. The blisters rupture, leaving painful ulcers. These heal, and the disease then becomes dormant. The virus is still there, however, and the

disease may recur at any time. Herpes is infectious only when it is active. People with herpes can therefore easily avoid transmitting the disease by avoiding sexual contact during the relatively short periods when they have herpes blisters. The greatest danger posed by herpes lies in the high mortality rate (about one third) among those newborn babies who catch the disease from their mothers during childbirth. (Most women with genital herpes, however, do not communicate the disease to their infants.)

Acquired immune deficiency syndrome (AIDS) was added to the list of STDs in the 1970s. This particularly terrifying STD is discussed in *A Journey Into Life,* Chapter 25.

Various other sexually transmitted diseases are caused by viruses, bacteria, yeasts, and protists. It is extremely difficult to stop the spread of these diseases because victims often pass a disease on to others before they learn that they are themselves infected. The growing resistance to antibiotics among pathogens also makes treatment increasingly difficult.

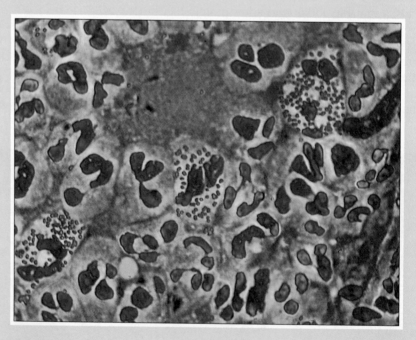

Figure 19-B
Human cells infected with the bacteria that cause gonorrhea. Here, the bacteria appear as many tiny blue dots in the cytoplasm of three cells (lower left, center, and upper right). (Carolina Biological Supply Company, courtesy of Dr. Cecil Fox, Centers for Disease Control)

lism bacteria are obligate anaerobes, they grow in canned goods but not in fresh and frozen foods. Most cases of botulism are due to inadequate heating of home-canned goods. The most effective safeguard is to can only acid foods like fruit and pickles, since acid kills the bacteria.

Salmonella bacteria in pork, poultry, or eggs cause diarrhea shortly after the contaminated food is eaten. The culprits are living bacteria, not toxins, and so this is called a food infection, not food poisoning.

SUMMARY

The kingdom Monera contains the bacteria. Mutation and rapid reproduction, coupled with small size and metabolic diversity, account for the evolutionary success of bacteria.

Three billion years ago, bacteria were the earth's only inhabitants, living in communities of many interdependent species. Some were autotrophs, making their own food; photosynthetic forms used solar energy, whereas chemosynthetic bacteria obtained energy from chemical reactions. Other bacteria exploited their neighbors for food. Dead cells or cell wastes provided raw materials for other bacteria, and so nutrients recycled in the bacterial community.

Today, bacteria are still the most numerous and ubiquitous organisms on earth, and they still play all of these roles. However, eukaryotes (which may themselves be the descendants of cooperative prokaryote symbioses) have become the dominant photosynthetic and heterotrophic organisms. Nitrogen-fixing bacteria and cyanobacteria support eukaryotic life by converting N_2 into a form that plants can use to make amino acids, and saprobic bacteria play the vital ecological role of decomposers. Many bacteria are free-living, while others form parasitic or mutualistic symbiotic relationships with other bacteria or with eukaryotes. Some bacteria are pathogens, producing toxins that cause disease, and some are part of the normal microbiota of animals. Eukaryotic chloroplasts, and possibly mitochondria, may have evolved from intracellular symbiotic bacteria.

Eukaryotic cells appeared about 1.5 billion years ago and evolved into a wide array of unicellular heterotrophs and autotrophs in the kingdom Protista. Autotrophic protists live either free in the phytoplankton, or attached to moist surfaces, or as symbionts of larger organisms. The clues to their phylogeny include such conservative features as photosynthetic pigments, form of food storage, cell wall structure, and type of flagella (if any).

Heterotrophic protists may be free-living predators on other protists, bacteria, and small multicellular organisms, or may be symbionts or parasites of animals. Three major phyla are characterized by their type of locomotion: Zoomastigina move by means of flagella, Sarcodina by pseudopodia, and Ciliophora by cilia. The fourth, phylum Apicomplexa, contains parasites with complex life histories. Some parasitic protists have an enormous impact on human health and on the human economy, in the form of medical expenses and lost labor capacity.

Small organisms, such as protists and bacteria, are well adapted to exploiting habitats and ways of life that provide only limited food and space. Tiny organisms can reproduce rapidly, and many of them can form spores or cysts that are resistant to adverse conditions such as drying or cold.

Selective pressure for large size probably favored the evolution of some ancient protists into multicellular fungi, plants, and animals. Here, division of labor among cells is added to division of labor among each cell's organelles.

Slime molds live like amoeboid protozoa but reproduce like fungi, forming multicellular fruiting bodies with walled spores.

Fungi are eukaryotic saprobes and parasites that obtain food by absorption and usually have cell walls. They reproduce by sexual and asexual spores. On the basis of sexual reproductive structures, fungi are classified into four main divisions: Oomycota, Zygomycota, Ascomycota, and Basidiomycota. A fifth division, Deuteromycota, contains fungi not known to reproduce sexually.

Lichens—symbioses between fungi and photosynthetic cells—are important producers of food and soil in cold or barren areas. Other fungi form mycorrhizal associations with the roots of higher plants, to which they supply minerals.

Fungi and bacteria share the vital ecological role of decomposing dead matter and recycling nutrients to green plants. Microorganisms cause great economic harm by destroying crops, food, and material possessions, and by infecting humans and livestock. On the positive side of the ledger, some fungi produce tasty fruiting bodies, and other fungi and bacteria are used to make fermented foods, drugs, antibiotics, and various organic chemicals.

SELF-QUIZ

1. An organism should be placed in the kingdom Monera if:
   a. it consists of a single cell
   b. it has a cell wall
   c. it forms spores
   d. it lacks a nuclear membrane separating its genetic material from the cytoplasm
   e. it causes diseases
2. Newly started rice paddies produce poor crops until they have established a flourishing population of cyanobacteria. This is probably because:
   a. the rice needs nitrogen fixed by the cyanobacteria
   b. the rice cannot compete with weeds, which are poisoned by toxins produced by the cyanobacteria
   c. the cyanobacteria use up surplus nutrients from sewage in the rice paddies
   d. the cyanobacteria provide plasmids carrying genes that increase fertility
   e. cyanobacteria form a protective coating on the rice plants
3. True or False. Many bacteria protect their hosts from pathogens by producing antibiotics.
4. True or False. All food containing bacteria is unsafe for consumption and should be thrown out.
5. Bacteria are used in the production of:
   a. wine
   b. vinegar
   c. gelatin
   d. pasteurized milk
   e. marshmallows
6. An organism should be placed in the kingdom Protista if it is ____, ____, and is not clearly related to members of the kingdom Fungi, Plantae, or Animalia.
7. State whether each description below applies to locomotion by means of cilia, flagella, or pseudopodia:
   ____ a. cytoplasm flows out into temporary extrusions from the cell body
   ____ b. oar-like beating propels the cell through the water
   ____ c. wave-like undulations pull or push the cell through the water

8. Commercial mushrooms are grown in soil enriched with old straw (stems of dead plants). These mushrooms are:
   a. autotrophic
   b. parasitic
   c. saprobic
   d. chemosynthetic
9. The organism below that would have hyphae is:
   a. a slime mold
   b. black bread mold (*Rhizopus*)
   c. yeast
   d. a trypanosome
10. The taxonomy of the fungi is based on:
    a. life history
    b. sexual reproductive structures
    c. mode of nutrition
    d. complexity of vegetative structures
11. The bracket fungi found on trees are:
    a. fruiting bodies of mycelia growing hidden in the tree trunk
    b. mycelia absorbing nutrients from the exposed surface of the wood
    c. sporangia
    d. lichens
12. In the mycorrhizal association between a pine tree and a fungus, the fungus:
    a. eventually depletes the tree's mineral supply
    b. secretes toxic materials that inhibit the growth of nearby trees
    c. absorbs nutrients from the soil
    d. converts nitrogen into a form the tree can use
13. Fungi are probably most widely spread by:
    a. airborne spores
    b. being eaten by animals and later deposited in their feces
    c. spores or bits of hyphae stuck to insects
    d. fragmentation of vegetative mycelia
    e. water currents

QUESTIONS FOR DISCUSSION

1. Why are fossil prokaryotes more difficult to find and study than fossils of other organisms?
2. What is the adaptive value to an organism of producing a toxin? What is the disadvantage?
3. Many prokaryotes can make all the molecules they need when provided with a nutrient medium containing a supply of inorganic salts and one type of small organic molecule (for example, a monosaccharide or fatty acid) as an organic carbon source. How can prokaryotes be so self-sufficient if their DNA is so small compared to even one of the many chromosomes found in eukaryotic cells?
4. Explain how a bloom of photosynthetic cyanobacteria can lead to shortage of oxygen in a lake or pond.

5. If mitochondria and plastids evolved from symbiotic bacteria, these have lost their cell walls and some of their genes, while other genes have moved into the host cell's nucleus. (The genes that remain in mitochondria code for polypeptide chains that are very insoluble in water.) What adaptive advantages would have selected for these changes?
6. Can you think of selective pressures other than the one discussed in this chapter that might have selected for evolution of multicellular organisms? What other advantages do multicellular organisms have?
7. How do the protists and fungi you studied in this chapter carry out each of the functions listed in Table 19-2?
8. How is the structure of fungi related to their way of life?

9. When plants are moved from their original habitat into a new one, a disease that was a minor nuisance in the home country may become a major disaster. This was the case with the late blight of potatoes, downy mildew on grapes, and white pine blister rust. What possible reasons can you think of for this?
10. Why do you think monocultures of an agricultural crop are more susceptible to disease than mixed plantings?
11. Fresh water heavily polluted with industrial wastes contains few fungi, compared to unpolluted waters. How might this affect the life in a lake or stream?
12. Is it valid to conclude that *Penicillium* secretions aid the fungus by reducing competition from bacteria in nature because they do so in the laboratory? What experiments could you do to investigate this question?

SUGGESTED READINGS

Baker, D. *"Giardia!" National Wildlife,* August-September 1985.

Carson, R. L. *The Sea Around Us.* New York: Oxford University Press, 1951. Contains a delightful discussion of the importance of algae in the sea.

Curtis, H. *The Marvellous Animals.* Garden City, NY: Natural History Press, 1968. A delightfully written introduction to the kingdom Protista.

Habicht, G. S., G. Beck, and J. L. Benach. "Lyme disease." *Scientific American,* July 1987.

Hardy, A. *The Open Sea.* Part I, The World of Plankton. Boston: Houghton Mifflin, 1965. Anecdotal and wonderfully illustrated.

Margulis, L. *Symbiosis in Cell Evolution.* San Francisco: W. H. Freeman, 1981. The case for the theory of symbiotic origin of organelles.

Shapiro, J. A. "Bacteria as multicellular organisms." *Scientific American,* June 1988.

Woese, C. "Archaebacteria." *Scientific American,* June 1981.

The Plant Kingdom

CHAPTER

20

Pollen and seed cones of pine

Micro-orchids

Kelp (brown alga)

*T*he plant kingdom contains the multicellular, photosynthetic eukaryotes (and their very close unicellular relatives). In all plants, the cells are surrounded by cell walls containing cellulose. Plant cells also contain plastids. Most plants have chloroplasts in at least some of their cells, but a few kinds have lost the ability to carry on photosynthesis. This way of defining the plant kingdom was used in Whittaker's proposal for a five-kingdom system. Some people now feel that only land plants (embryophytes) belong in the plant kingdom. They classify the multicellular algae as protists (using a different definition of that kingdom from the one used in this book).

What kinds of selective pressures do plants meet? First and foremost, plants must obtain light for photosynthesis. The multicellular bodies of many plants have specialized cells that hold the photosynthetic parts in positions where they receive a reliable supply of sunlight. Second, most plants cannot move about. They are too large to swim with flagella, and their stiff cell walls preclude the evolution of muscle tissue. This makes it doubly important for a plant to grow where it can obtain enough sunlight, and many plants have adaptations that increase their offsprings' chances of starting life in favorable habitats. Since they cannot move, plants also cannot come together to mate, and so they must have other ways to bring their gametes together for sexual reproduction. However, many plants do not face this problem because they have only asexual, or vegetative, reproduction.

During the course of evolution, plants have become able to exploit just about every habitat with enough light, moisture, and minerals to support the growth of photosynthetic organisms.

- Most members of the plant kingdom are multicellular, photosynthetic, and immobile, and all plants have cellulose cell walls.
- Because most plants cannot move around, obtaining light for photosynthesis and bringing gametes together for sexual reproduction have been dominant themes in their evolution.
- Plants originated in the sea, and some forms later colonized land. The most successful land plants have vascular tissues that transport food and water through the plant and support the plant body in the air.
- The most successful vascular plants have adaptations freeing their reproduction from dependence on water.

The Multicellular Algae

The groups of algae with at least some multicellular members are placed in three divisions: Rhodophyta (red algae), Phaeophyta (brown algae), and Chlorophyta (green algae) (Table 20-1). (Divisions in the plant kingdom are equivalent to phyla in the animal kingdom.) The members of these three divisions are believed to have evolved from different groups of protistan ancestors.

Two of the three divisions, the red and brown algae, are mostly marine, with only a few freshwater forms. Both groups include species that have attained some size and complexity. The third division, the green algae, is well represented in both marine and freshwater environments, and also in moist areas on land. Multicellular algae provide food for animals such as snails and sea urchins.

Multicellular algae vary a lot in structure, and so they are classified by conservative biochemical characteristics such as kinds of photosynthetic pigments and food storage molecules. Like cyanobacteria and photosynthetic protists, all multicellular algae contain chlorophyll *a*, but the other forms of chlorophyll (*b, c*) and the accessory photosynthetic pigments (which gather other wavelengths of light energy) are used to divide the algae into their main groups.

Table 20-1 Divisions of Multicellular Algae

Division	Common name	Characteristics
Rhodophyta (~4000 species)	Red algae	Chlorophyll *a*, phycobilin accessory pigments, store starch-like polymer; no flagellated cells; unicellular or multicellular; most marine, some freshwater, a few terrestrial; e.g., Irish moss, nori
Phaeophyta (~1500 species)	Brown algae	Chlorophylls *a* and *c*, fucoxanthin accessory pigment; store carbohydrates and lipids; all multicellular, some very large; almost all marine; e.g., *Fucus*, giant kelps
Chlorophyta (~7000 species)	Green algae	Chlorophylls *a* and *b*, store starch; unicellular to multicellular; mostly freshwater, many marine, some terrestrial; e.g., *Spirogyra, Ulothrix, Ulva*.

20-A Division Rhodophyta: Red Algae

The division Rhodophyta contains single-celled forms as well as plants that grow as filaments, branching structures, and broad flat plates or ruffles (Figure 20-1). Some of the more complex red algae grow up to a metre long, but most are small and delicate. All live attached to some surface: a rock, coral reef, animal shell, or even a larger alga. No flagellated cells occur among red algae.

The chloroplasts of red algae show strong evidence of descent from cyanobacteria (Section 19-C). Both have chlorophyll *a* as the only chlorophyll, and both contain similar phycobilin accessory pigments. The arrangement of photosynthetic membranes is also similar.

Red algae are prominent residents of tropical coral reefs, where they play an important part in reef-building. The construction of reefs used to be attributed solely to coral animals, which secrete calcium carbonate (lime) tubes around themselves. However, many algae, especially red algae, become encrusted with calcium carbonate which they extract from the sea. This adds material to the reef.

Some red algae are eaten by people in the Orient and in coastal areas of Europe (Figure 20-2). Carrageenan, a substance used in puddings, candies, and

Figure 20-1
A red alga. *Plumaria* grows on rocks in the intertidal zone. (Biophoto Associates)

Figure 20-2
Nori *(Porphyra)*, a red alga. These nets at a Japanese seaweed farm are positioned to expose the alga at low tide, as in its natural intertidal habitat. The dried algae are a common food in the orient. (Biophoto Associates)

ice cream, comes from the red alga called Irish moss. Agar, also from a red alga, is used as a base for nutrient media in the laboratory culture of microorganisms.

20-B   Division Phaeophyta: Brown Algae

Members of the division Phaeophyta are all multicellular, ranging from microscopic filaments to Pacific kelps that may be 70 metres long—the biggest and most complex algae. The carotenoid pigment fucoxanthin imparts the typical brown to olive drab hue of most phaeophytes. They also contain chlorophylls *a* and *c*. These pigments also occur in members of the protistan phylum Chrysophyta (see Table 19-1). Both groups have similar food storage molecules as well, and so brown algae are thought to have evolved from chrysophyte ancestors.

Brown algae are especially noticeable in cool, shallow waters along the sea coast in temperate and subpolar areas. The larger members of the group are interesting because of their tissue differentiation. Members of the genus *Fucus* are good examples. A foot or more long, they grow on rocks in the intertidal zone, attached by a specialized **holdfast** (Figure 20-3). A stalk-like **stipe** connects the holdfast to the flat, photosynthetic **blades.** When the tide is in, the blades float near the surface of the water, close to the light, buoyed up by gas-filled **air bladders.** Numerous swellings at the ends of the blades contain chambers where reproductive cells develop. The plant has a gelatinous covering, which reduces evaporation and keeps it from losing too much water when it is exposed to the air at low tide.

Some kelps grow much larger than *Fucus,* and can live anchored in deeper water and still have parts that float at the surface, near the sunlight. Their holdfasts keep them from being swept out to sea.

The brown algae include many forms of economic importance, used as fertilizer and as food for livestock or humans. Some kelps are processed to extract a component of their cell walls called alginate. This substance is an ingredient in about half the ice cream produced in the United States; it imparts smoothness and helps prevent formation of ice crystals. Alginate is also used in various drugs and cosmetics.

20-C   Division Chlorophyta: Green Algae

Members of the division Chlorophyta (green algae) show great diversity of form and live in a variety of habitats. Most are quite small and relatively simple compared with the larger brown algae. Many chlorophytes are single-celled; others are simple or branched filaments of cells, with or without holdfasts (Figure 20-4). Most Chlorophyta inhabit fresh water, both still and running, and some live on moist rocks, soil, and tree trunks on land.

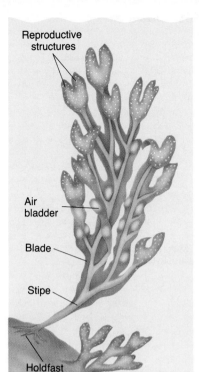

Figure 20-3
The brown alga *Fucus.* When the tide is in, the plant is buoyed upright in the water by its air bladders. During low tide, it lies limp on the rocky shore.

Figure 20-4
Members of the division Chlorophyta. (a) A desmid, a single-celled form with two mirror-image halves. (b) A calcium-secreting marine form. (c) *Volvox,* a colonial form. The colony rotates in the sunlight by the beating of flagella on the individual cells. Daughter colonies form inside the colony and are released by rupture of the parent colony's single layer of cells. (a, Biophoto Associates; b, Steven Webster; c, Carolina Biological Supply Company)

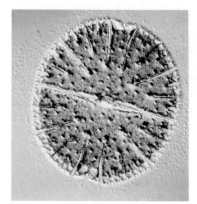

(a)

(b)

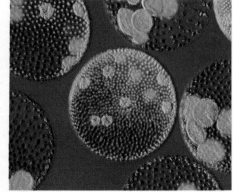

(c)

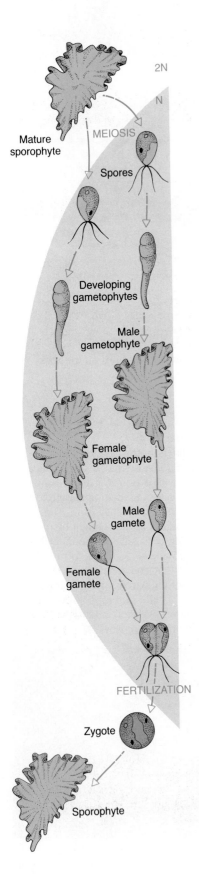

The Chlorophyta appear to belong to the same evolutionary line as the land plants. The chloroplasts of both groups contain chlorophylls *a* and *b*. Both also have many carotenoids and store their food as starch.

The 7000 species of green algae include several lines that clearly show evolutionary trends from unicellular to multicellular forms. Another trend is in sexual reproduction, from forms in which two similar gametes unite at fertilization, to those with large, immobile eggs and small, motile sperm.

Some unicellular, flagellated green algae look much like the flagellated sperm or spores of larger species. Not all green algae have flagellated stages, however. Those without flagella include the single-celled desmids (Figure 20-4a) and filamentous forms such as *Spirogyra* (see Figure 8-5).

20-D   Life Histories

Life histories show some important trends in plant evolution. In the familiar human life history, the body's cells are diploid, and the only haploid cells are the gametes (egg and sperm) produced by meiosis. The same situation occurs in the brown alga *Fucus* and in many other algae. A completely opposite type of life history is also found in some algae, for example, in the green alga *Ulothrix* (see Figure 16-4). The cells of a *Ulothrix* filament are haploid, and the zygote is the only diploid cell in the life history. Most plants have life histories intermediate between these two types: a multicellular haploid stage alternates with a multicellular diploid stage, a situation known as **alternation of generations.**

In these plants, as in *Ulothrix,* meiosis does not give rise to gametes. Rather, it produces **spores,** haploid cells that divide by mitosis to produce multicellular haploid plants. Eventually, the haploid plants produce gametes. Since the plants are already haploid, however, the gametes are produced by mitosis rather than by meiosis. The haploid gametes then undergo fertilization and produce diploid zygotes. The zygotes divide by mitosis and grow into multicellular diploid plants. Some of the resulting diploid cells undergo meiosis to produce haploid spores.

Figure 20-5 shows a life history with both a multicellular diploid form and a multicellular haploid form, in the green alga *Ulva* (sea lettuce), which grows near the tide line. The multicellular diploid body is called a **sporophyte** because it produces spores. Each haploid spore grows into a multicellular haploid plant, called a **gametophyte** because it produces gametes. Fusion of two haploid gametes produces a diploid zygote, which grows into a multicellular diploid sporophyte:

$$\text{sporophyte} \overset{\text{(meiosis)}}{\longrightarrow} \text{spore} \rightarrow \text{gametophyte} \rightarrow \text{gamete} \overset{\text{(fertilization)}}{\longrightarrow} \text{zygote} \rightarrow \text{sporophyte}$$

This is basically the life history found in all higher plants.

Figure 20-5
Alternation of generations in the life history of *Ulva.* In this green alga, the multicellular diploid plant (sporophyte) and the multicellular haploid plant (gametophyte) look alike. In many other species, sporophyte and gametophyte are distinctly different. In *Ulva,* the female gamete is slightly larger than the male gamete, but not too large to move by means of its flagella, and certainly much smaller than the nonmotile egg of higher organisms. (photo, J. M. Kingsbury)

Land Plants

Both land plants and green algae have the same photosynthetic pigments, chlorophylls *a* and *b* and beta-carotene, and store most of their food reserves as starch. In addition, most groups of green algae have flagellated sperm, as do the so-called "lower" land plants (those that do not produce seeds).

Land plants probably evolved from green algae growing at the edges of oceans or lakes. Such plants face constant danger from extreme conditions. When the water recedes, they may lose too much water to the air and die. On the other hand, a high tide or a heavy rain may sweep them into an unsuitable habitat, or bury them under mud and silt. Such conditions select for several adaptations: structures that anchor the plants in one place, a large size that keeps them from being buried completely, and a surface coating that reduces water loss. Plants that evolved such adaptations could survive better not only at the water's edge but also higher up on the shore, on dry land.

Land offers many advantages to plants that can function out of water (Table 20-2). More light is available on land because water itself absorbs much of the sun's energy. Carbon dioxide for photosynthesis is also present in much higher concentrations in air than in water. In addition, the first land plants had to compete for these resources only against their fellow colonists, and predation was probably limited to a few animals venturing out of their watery homes to feed at night.

The main disadvantage of moving out of water is that water becomes hard to obtain and is easily lost by evaporation from the plant's surface. Also, air is much less dense than water, and so it gives virtually no support to the plant body. Finally, water is no longer available for reproduction. Many algae require water for their flagellated sperm to swim to eggs, and for the young plants to disperse to new locations. On land, plants must carry out these functions without a constant supply of water, either by taking advantage of rain or dew to carry the reproductive cells or by providing these cells with a waterproof coating before they travel through the air.

On land, the resources a plant needs are segregated: water and minerals lie below the surface of the soil, while light and air are above it. The division of labor in the bodies of land plants reflects this division of resources. Underground structures serve as anchors and absorb water and minerals for the entire plant. The photosynthetic structures above ground produce enough food for all the plant's cells (Table 20-3).

With this separation of resources, and of the plant parts specialized for obtaining them, comes a new problem: moving substances between the two areas. Most land plants have **vascular tissue,** which transports substances

Table 20-2 Comparison of Water and Land as Habitats for Plants

	Water	Land
Water	Close to each cell	Under land surface; evaporates quickly above surface
Minerals	Close to each cell	On or under land surface
Gases	Dissolved at low concentrations	Plentiful in the air
Support	Provides buoyancy, support	Much less support for parts in air
Light	Cuts out some wavelengths, and lowers intensity	More light available
Temperature	Changes slow, narrow range	Changes more rapid, wider extremes
Reproduction	Motile gametes swim	Water seldom available for swimming gametes
Dispersal	Water carries offspring to new locations	Water seldom available to carry offspring to new locations

Table 20-3 *Adaptations of Land Plants to Terrestrial Environments*	
Problem	**Adaptation**
1. Obtaining water and mineral nutrients when they no longer surround the entire plant	Roots
2. Transporting water within the plant	Xylem
3. Transporting food from sites of manufacture to sites of use	Phloem
4. Preventing evaporation from surfaces exposed to air	Cuticle
5. Obtaining gases for photosynthesis and respiration	Stomata
6. Obtaining sunlight for photosynthesis	Leaves
7. Supporting body in medium lacking buoyancy	Xylem
8. Coordinating growth and response to environment	Hormones
9. Getting gametes together without reliable supply of water for sperm	Pollen
10. Dispersing new individuals to suitable locations	Airborne spores; seeds

within the plant, especially between the food-making parts above ground and the water-absorbing parts below. There are two types of vascular tissue. **Phloem** conducts organic materials, mainly food, from sites of manufacture to sites of use or storage. **Xylem** transports mainly water and minerals from the roots to the stems and leaves. Xylem also provides support, so that the plant body stands up in the air, where the leaves can be deployed effectively for photosynthesis. A plant with xylem may be compared to a building whose plumbing pipes double as supporting columns.

Above ground level, the plant's surface is covered by a waxy **cuticle,** which is nearly impermeable to water and reduces evaporation of precious water from the plant into the air. However, the cuticle is also impermeable to carbon dioxide and oxygen, which a plant must exchange with the air. The leaves and stems of vascular plants can exchange gases through tiny pores called **stomata** ("mouths"; singular: **stoma**). Stomata are surrounded by pairs of guard cells, which can regulate the size of the opening (Figure 20-6). Some water is inevitably lost through the stomata, but much less than if evaporation proceeded freely from the entire above-ground surface of the plant.

Land plants also have a variety of **hormones,** chemicals that coordinate the activities of the plant and its response to environmental cues. Hormones make

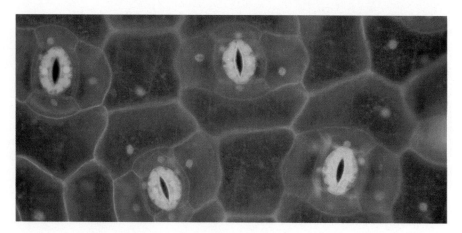

Figure 20-6
Cells from the epidermis on the underside of a leaf. Each stoma is a pore (black) between two guard cells (yellow), which resemble lips. The orange lines are cell walls of other cells in the epidermal layer that covers the underside of the leaf.
(P. Dayanandan/Photo Researchers)

Table 20-4 Summary of the Embryophytes

Group	Common name	Characteristics
Nonvascular plants		
Division Bryophyta (~23,000 species)	Mosses, liverworts	Small, nonvascular, anchored by rhizoids; gametophyte dominant, sporophyte dependent on it; flagellated sperm require water to reach egg; mostly moist habitats
Vascular plants		Small to huge; vascular tissue in sporophyte, most with roots, stems, and leaves; sporophyte dominant, independent, usually photosynthetic; wide range of habitats
Lower vascular plants:		Vascular plants without seeds; mostly small to medium size; gametophyte free-living, independent of sporophyte; flagellated, swimming sperm; dispersal by airborne spores
Division Lycophyta (~1200 species)	Club mosses, ground pine	Small, but some tree-like fossil forms; scale-like leaves with single vein
Division Sphenophyta (~25 species)	Horsetails	Most short, some tree-like fossil and tropical forms; jointed stems with vertical ribs, tiny single-veined leaves around joints
Division Pterophyta (~12,000 species)	Ferns	Most short, some tropical tree-like forms; large, many-veined leaves, often with divided shapes (fronds)
Gymnosperms:		Medium to huge; woody xylem; sexual reproduction by air-borne pollen; dispersal by seeds
Division Cycadophyta (~100 species)	Cycads	Palm-like shrubs and small trees; tropical and subtropical; pollen and seeds borne in cones
Division Gnetophyta (~70 species)	*Gnetum, Ephedra, Welwitschia*	Desert plants with similarities in structure of vascular tissue
Division Coniferophyta (~550 species)	Conifers	Shrubs and trees with needle- or scale-like leaves; most produce pollen and seeds in cones
Division Ginkgophyta (1 species)	*Ginkgo*	Tree with broad, fan-shaped leaves, smooth naked seeds, sexes separate
Angiosperms:		
Division Anthophyta (~275,000 species)	Flowering plants	Tiny to huge; efficient vascular tissue, most with broad leaves; flowers containing sexual reproductive parts, pollen carried by wind or animals; double fertilization; dispersal by seeds, which develop inside fruits

roots grow down into the soil, and stems turn up toward the light, and they initiate reproduction and dormancy (Chapter 35).

One division of land plants, the Bryophyta, contains nonvascular plants (plants without vascular tissue), and the other divisions contain vascular plants (Table 20-4). Bryophytes and vascular plants, botanists speculate, evolved in different directions from green algal ancestors.

Reproduction on Land

All land plants have multicellular reproductive structures. Here developing reproductive cells—sperm, eggs, and spores—are protected, at least for a time, by a surrounding jacket of sterile cells. Land plants are sometimes called **embryophytes** because the zygote remains within the parent plant as it develops into an embryo.

We shall encounter a few important trends in the evolution of the life histories of land plants:

1. The earliest land plants probably had life histories with gametophytes and sporophytes similar in size and appearance and living independently of one another, as in *Ulva*. In the bryophytes—the mosses and their relatives—the haploid gametophytes became the dominant generation, and the diploid sporophytes became reduced to smaller forms growing on the gametophytes.

Figure 20-7
Male delivery in land plants. (a) The eggs of lower land plants are fertilized by flagellated, swimming sperm, such as this fern sperm, which has many flagella. (b) In higher land plants, such as pines, pollen is carried through the air. Note the pale "wings" at each side of the pollen grain. Once arrived near the female gametophyte, the pollen grain grows a pollen tube, through which the sperm nucleus migrates to reach the egg. (Carolina Biological Supply Company)

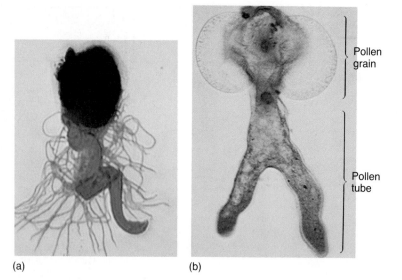

(a)   (b)

The vascular plants show a trend in the opposite direction: from forms in which the small gametophyte and much larger sporophyte are independent (as in ferns), to the condition in "higher" (seed-bearing) plants where the sporophyte is dominant and the greatly reduced gametophyte grows within the sporophyte and is dependent on it.

2. In the lower land plants, spores are the main means of dispersal; in higher plants, there is a new structure, the seed, which is well adapted for dispersing a new individual to a new location and establishing it there.

3. Lower land plants produce flagellated sperm that must swim to the eggs in a liquid medium; in seed-bearing plants, pollen grains are carried to female reproductive structures by the wind or by animals (Figure 20-7).

20-E Division Bryophyta: Mosses and Liverworts

The most familiar members of the division Bryophyta are the mosses. The haploid gametophyte (the familiar green moss plant) is the dominant generation (Figure 20-8). Many bryophytes have conducting cells that carry out slow transport of water and food in both gametophyte and sporophyte. However, they do not have true vascular tissue. Because of this, and because their swimming sperm require water to reach the eggs, bryophytes have remained small and

Figure 20-8
Bryophytes. (a) Green gametophytes of a moss, with sporophytes (red stalks with pale capsules) growing on them. Spores shed from the sporophyte capsules are carried away in the breeze. (b) The liverwort *Marchantia* growing in a bed of moss. Note the flat, branching body and the little gemmae cups, asexual reproductive structures containing balls of cells that may be splashed out and grow into new plants.

(a)

(b)

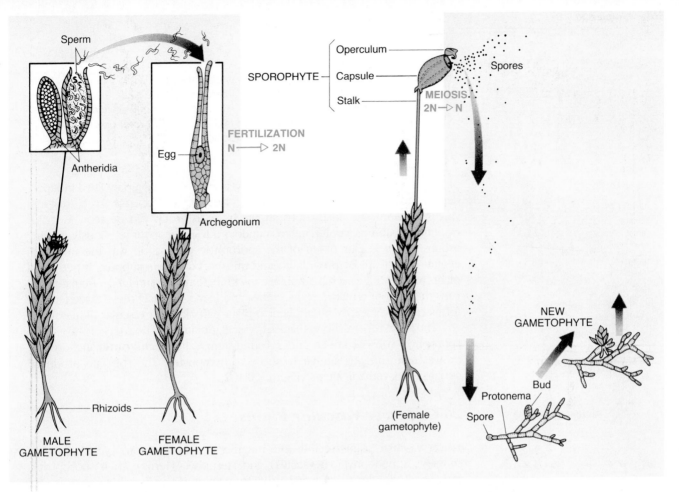

Figure 20-9

Life history of a moss. Stages shown in the blue area are haploid (N); those in the white area are diploid (2N). Sperm formed in the antheridia of the male use their flagella to swim to eggs, formed in organs called archegonia at the top of female gametophytes. Fertilization forms a diploid zygote, which develops into a sporophyte, consisting of a stalk bearing a capsule. Cells inside the capsule undergo meiosis and form haploid spores, which are shed into the air when the operculum covering the tip of the capsule opens. A spore germinates to form a thread-like protonema, which produces buds of new gametophytes.

generally confined to moist habitats. Some bryophytes, though, survive surprisingly hot or dry conditions.

Bryophytes have root-like organs called **rhizoids,** which anchor the plant. The green, leaf-like structures where photosynthesis occurs are only one or two cells thick, and very susceptible to drying out. Moss gametophytes grow close together, holding each other up and also creating a network of tiny spaces that hold water like a sponge. The leaf-like parts absorb most of the plant's water from these spaces. Water in the spaces also aids reproduction.

Bryophytes need water to reproduce because their sperm are flagellated. Sperm shed from male organs swim through a film of moisture to the top of a plant where a female organ, containing an egg, has developed. A sperm fertilizes an egg, forming a diploid zygote. The zygote, still at the top of the gametophyte, grows into a sporophyte, a long stalk with a capsule on top. Cells inside the capsule undergo meiosis, producing haploid spores. When the spores are mature, the capsule opens and flings them into the air. If a spore lands in a suitable place, it germinates, forming a filament that grows along the surface of the ground. Buds arising from the filament develop into a clump of new gametophytes (Figure 20-9).

The liverworts are the other major group of bryophytes. Most liverworts look much like flattened mosses, but some have flat, ribbon-like gametophytes with rhizoids on the underside (see Figure 20-8). Liverworts have sexual reproduction much like that of mosses.

■ *Bryophytes, lacking vascular tissue and dependent on water for swimming sperm to reach the eggs, live in moist areas on land.*

20-F Vascular Plants

The vascular plants include the most advanced and complex forms of plant life. For convenience, we can arrange the divisions of vascular plants into three groups: lower vascular plants, gymnosperms, and angiosperms (flowering

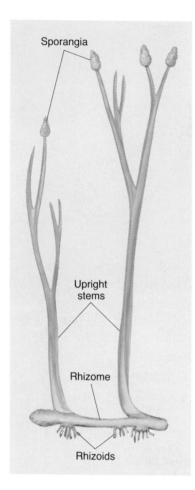

Figure 20-10

The sporophyte of *Rhynia*. This primitive vascular plant lived over 350 million years ago (Lower Devonian Period). The plant had an underground stem, the rhizome, with rhizoids similar to those of bryophytes. Upright photosynthetic stems bore sporangia, which produced spores. There were two species of *Rhynia*, one standing about knee-high, and the other about half that height.

plants) (see Table 20-4). Although the divisions in each group evolved independently, their members are similar in life history, reproductive structures, and vascular tissue. The adaptations shown by each group can be seen as major evolutionary jumps, conferring tremendous advantages over previously existing groups. The vascular tissue of the sporophyte generation was the first such evolutionary leap. By providing rapid transport and firm support, it permitted plants to grow taller, a big advantage in competition for sunlight. Vascular tissue runs throughout a plant's roots, stems, and leaves—in fact these three familiar terms cannot properly be applied to plant parts lacking vascular tissue.

The fossil record shows that early vascular plants had no roots or leaves. The sporophyte consisted of an underground stem, called a **rhizome,** anchored by nonvascular rhizoids similar to those of bryophytes. The rhizome produced aerial photosynthetic stems (Figure 20-10).

20-G *Lower Vascular Plants*

Lower vascular plants include the members of the divisions Lycophyta (club mosses), Sphenophyta (horsetails), and Pterophyta (ferns). Their sporophytes have small, slender roots, which grow from an underground rhizome or from a trailing above-ground stem. Aerial stems, with leaves, grow from the rhizome.

Lycopods and horsetails have small leaves, each containing a single **vein,** a strand of vascular tissue that brings water to the leaf from the roots and carries away excess food made by the leaf. Ferns and seed-bearing plants have large leaves with many veins, often in highly branched, intricate patterns. The sturdy veins support the more delicate photosynthetic tissue.

Division Lycophyta: Club Mosses and Ground Pines

The sporophytes of club mosses and ground pines have aerial stems covered with many small, scale-like leaves (Figure 20-11). **Sporangia** (singular: **sporangium**), the spore-producing structures, appear on the upper surfaces of leaves.

Lycopods reached their heyday as prominent members of the coal forests in the Carboniferous Period (see the geological timetable on this book's endpaper). Many lycopods of that time grew to the size of large trees, up to 30 metres tall. When the climate later became drier, these enormous lycopods died out, probably because their vascular systems could not cope with the demands of such large plants for water in the new climate. Many of the lycopods that survived this period are still with us today. All are less than half a metre high, but they are quite common and easy to find in fields and forests once you know what to look for.

Figure 20-11

Club moss. The green photosynthetic stems, covered with tiny scale-like leaves, sprawl over the ground. Leaves bearing sporangia are arranged in golden clusters held aloft on stems like branched candelabra.

Division Sphenophyta: Horsetails or Scouring Rushes

The members of the division Sphenophyta are nicknamed "horsetails," because some of them resemble horses' tails, or "scouring rushes," because people used them to scrub dirty pots and pans before the invention of steel wool. Horsetails were ideal for this because some of their cells are impregnated with silica, a very abrasive material.

(a)   (b)   (c)

Figure 20-12
Horsetails. (a) *Equisetum arvense,* with its thin green branches, is often mistaken for a clump of pine seedlings. (b) Fertile shoots of *E. arvense* are nonphotosynthetic. They grow in early spring, before the vegetative shoots; this gives the spores more free air space for their travels. Note the jointed stems ringed by tiny dark leaves. (c) Closeup of the tip of a fertile shoot. Sporangia are borne on the undersides of the numerous brown and white "umbrellas." The vertical ribs of the stem are clearly visible. (b, Biophoto Associates, N.H.P.A.)

Like lycopods, horsetails have an underground rhizome that produces tiny roots and aerial stems. The stems are mostly hollow and are easily recognized by their jointed appearance and the fine vertical ribs between joints. A ring of leaves, so small that they are often overlooked, grows around the stem at each joint. Some species have slender branches that look like the needles of young pine trees. Depending on the species, sporangia grow in clusters atop green vegetative stems or nongreen reproductive stems (Figure 20-12).

Like lycopods, horsetails flourished in the Carboniferous forests, with some members attaining heights of about 15 metres. However, only one genus, *Equisetum,* with about 25 species, has survived to represent the group today.

Division Pterophyta: Ferns

The ferns are the first plants we have met that have large, many-veined leaves. Fern leaves are called **fronds,** and they usually arise directly from a rhizome, which also has many small roots. Ferns in temperate areas seldom grow more than a metre high, but tropical tree ferns may be quite tall (Figure 20-13). Although they were abundant in the Carboniferous coal forests, ferns have probably never been a dominant part of the vegetation.

The life history of ferns is similar to that of lycopods and horsetails, and can be taken to represent all three groups. In ferns, sporangia usually form on the undersides of the green vegetative fronds (where worried owners sometimes mistake them for a disease), or on separate nongreen fertile fronds. Cells within the sporangia undergo meiosis, forming haploid spores, which are shed into the air. A spore grows into a small, green photosynthetic gametophyte, anchored to

■ *The sporophytes of lycopods, horsetails, and ferns have roots that take up water from the soil and vascular tissue that transports it to leaves and stems up in the sunlight. However, these plants must still live in somewhat moist areas because their sexual reproduction still depends on water for flagellated sperm to swim in.*

(a)

(b)

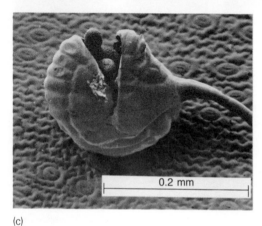

0.2 mm

(c)

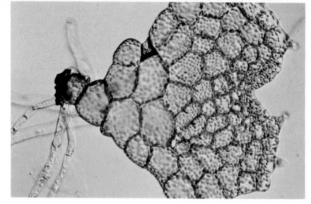

(d)

Figure 20-13

Ferns. (a) A tropical tree fern. (b) Sporangia clustered on the back of a fern frond. (c) Closeup of one sporangium, which has split open to release its spores. (d) A fern gametophyte. Note the photosynthetic cells with their green chloroplasts, and the colorless rhizoids. (b, c, d, Biophoto Associates, N.H.P.A.)

the soil by rhizoids (Figure 20-13d). The gametophytes produce sperm and eggs. Fern sperm have flagella, and must have moisture on the surface of the soil to swim through. The zygote remains within the (female) gametophyte while it develops into a sporophyte embryo. The young sporophyte pushes a leaf up into the air and a root down into the soil, and it soon establishes itself as an independent plant (Figure 20-14).

20-H   Gymnosperms

The redwoods of California's coastal mountains, Canadian pines and hemlocks, cypress standing knee-deep in Southern swamps, *Ginkgo* trees on city streets, wiry *Ephedra* in the western deserts, and palm-like cycads of the tropics are all members of the divisions lumped together under the term gymnosperm ("naked seed"). What is probably the oldest tree in the world, a 4900-year-old bristlecone pine in the mountains of eastern Nevada, is a gymnosperm. So are the tallest tree, a coast redwood *(Sequoia sempervirens)* over 100 metres high, and the tree with the greatest bulk, a giant sequoia *(Sequoiadendron giganteum)* nicknamed "General Sherman," which is over 80 metres tall, 20 metres around at its base, and 3500 to 4000 years old (Figure 20-15).

In contrast to lycopods, horsetails, and ferns, gymnosperms are fully adapted to life on land and can live in dry places. One of their evolutionary achievements is woody tissue, made up of xylem. New wood is added each year

■ *Gymnosperm advances include stronger and more efficient vascular tissue, windborne pollen in place of swimming sperm, and dispersal by seeds rather than by spores.*

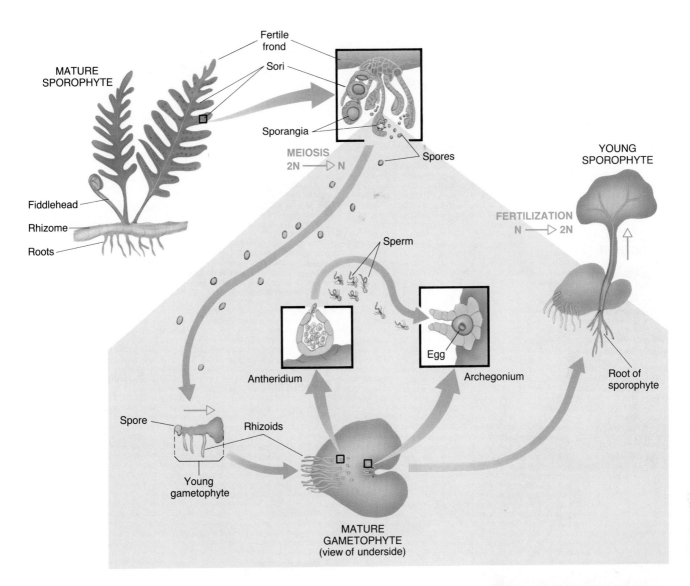

Fertile
frond

Sori

MATURE
SPOROPHYTE

Sporangia

MEIOSIS
2N ——▷ N

Spores

YOUNG
SPOROPHYTE

FERTILIZATION
N ——▷ 2N

Fiddlehead

Rhizome

Roots

Sperm

Antheridium

Egg

Archegonium

Root of
sporophyte

Spore

Rhizoids

Young
gametophyte

MATURE
GAMETOPHYTE
(view of underside)

Figure 20-14
Life history of a fern. The large sporophyte and tiny gametophyte (only a few millimetres across) are independent of each other. The sporophyte bears sporangia in
clusters called sori. Meiosis of cells inside the sporangia forms spores, which are
shed into the air and then grow into tiny gametophytes. Male and female organs (antheridia and archegonia) develop on the underside of the gametophyte. The male
gamete is a sperm with many flagella; it swims through environmental moisture to an
archegonium and fertilizes the egg. The zygote develops into a new sporophyte.

so that the plant grows in diameter as well as in height. Wood strengthens the
stem and allows the plant to grow tall and compete for sunlight. And, since
xylem is the water-transporting tissue, it also delivers water to the leaves efficiently. As a further adaptation to dry habitats, the leaves produce a thick cuticle.
　Gymnosperm reproduction also has two new evolutionary advances that
free it from requiring liquid water: wind-borne pollen, and seeds. The repro-

Figure 20-15
Gymnosperm giants. Mature giant sequoias are the largest of living plants, dwarfing
the 6-foot man standing beside a young specimen (perhaps 50 years old) in the foreground.

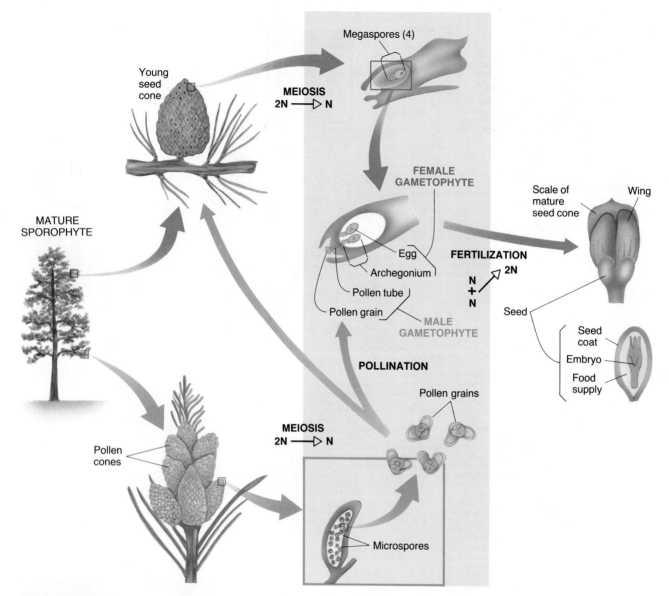

Figure 20-16

Life history of pine. The mature sporophyte (pine tree) produces megaspores in the seed cones, microspores in the pollen cones. Microspores develop into pollen grains (male gametophytes), which are carried by the wind to seed cones. Here, each pollen grain grows a pollen tube, which releases sperm (see Figure 20-7). Megaspores develop into female gametophytes within the seed cone. After fertilization, the zygote grows into the embryonic sporophyte, surrounded by a food supply and a protective seed coat. The winged, wind-blown seed is shed when the seed cone scales open and serves as the dispersal unit in the life history.

duction of a pine tree may serve as an example (Figure 20-16). The tree forms two kinds of cones, in which meiosis produces two kinds of spores (small **microspores** in the pollen cones and larger **megaspores** in the seed cones). Megaspores remain in the seed cones, and develop into female gametophytes, each containing two to several eggs. Microspores develop into immature male gametophytes, otherwise known as **pollen grains.**

Pollination occurs when wind carries the pollen to a seed cone on another tree. A seed cone's shape creates eddies that cause pollen to settle out of wind currents. Each species produces cones and pollen of a distinct size and shape, such that cones tend to filter more pollen of their own than of other species out of the air. As a result of these adaptations, pollen of the same species is deposited at the bases of the cone scales, near the entrance to the female gametophyte.

The pollen grain now completes its development, growing a slender pollen tube toward the female gametophyte. As the tube approaches its destination, one of its haploid nuclei divides by mitosis and forms two sperm nuclei. **Fertilization** occurs when a sperm nucleus leaves the pollen tube and fuses with an egg nucleus. The resulting zygote and the surrounding female structures now develop into a **seed,** with three main parts. Innermost is the multicellular diploid **embryo** of the new sporophyte generation, which develops by mitosis from

(a)

(b)

Figure 20-17
Cycads. Cycads look like palm trees (a), but they have cones (b).

the zygote. Around the embryo is a **food supply,** which will provide energy for the embryo to grow until it is large enough to make its own food. Around the outside is the **seed coat,** which develops from the parent sporophyte. The seed is an extraordinary evolutionary invention that permits terrestrial plants to distribute their offspring widely without relying on water.

Division Cycadophyta

Cycads are gymnosperms of tropical and semitropical regions. With their large, palm-like leaves they might easily be mistaken for palm trees, except that they have cones (Figure 20-17). Some human populations eat cycad stems and seeds, but some of these must be prepared specially to eliminate toxins. The living cycads are mere remnants of a group that was once much more abundant.

Division Gnetophyta

The division Gnetophyta contains the closest living relatives of flowering plants. The living members of this group are alike only in certain details of their vascular tissue. *Gnetum,* a woody vine with leaves that look like those of a flowering plant, lives in tropical deserts and mountains. The members of the genus *Ephedra* are small, wiry shrubs with jointed stems and scale-like leaves (Figure 20-18). They are desert dwellers in Europe, Asia, and the Americas. Native Amer-

Figure 20-18
Ephedra viridis, also known as Mormon or Mexican tea, growing on the southern rim of the Grand Canyon. The leaves of this desert-adapted gymnosperm are much reduced; the green stems carry on photosynthesis. (Paul Feeny)

(a)

Figure 20-19
Ginkgo. (a) A male specimen. (b) A branch of a female tree, showing a seed and the distinctive fan-shaped leaves.

(b)

icans reportedly used these shrubs for flour and tea, and for a medicine that constricts blood vessels, which was useful for respiratory problems.

Welwitschia is a bizarre resident of southern African deserts. It has an extremely long main root, a woody trunk up to a metre wide but only a few centimetres high, and two enormous strap-like leaves that grow from the trunk throughout the plant's life of perhaps centuries.

Division Coniferophyta

The most familiar gymnosperms are conifers—pines, firs, cedars, yews, hemlocks, junipers, larches, spruces, redwoods, and cypresses. Most have cone-like reproductive structures and needle-like or scale-like leaves with little surface area and thick cuticle. These drought-resistant leaves appeared during the Permian, and were probably a major reason for the conifers' success during those times of global aridity. Many conifers do well in poor or shallow soil, where nutrients are scarce, and in areas subject to regular cold or dry spells. This is probably one reason conifers are so common today.

Division Ginkgophyta

Ginkgo biloba (Figure 20-19) is the sole surviving species of a group that flourished during the Mesozoic Era. With its broad, fan-shaped leaves and round, smooth seeds, it bears little resemblance to the conifers but has similar vascular tissue. Wild *Ginkgo* trees may still exist in remote mountainous parts of China, but for centuries this species was known only from cultivated specimens in oriental gardens and temples. Today it is very popular for street plantings in many American and European cities, since it is remarkably resistant to urban smog and to insect pests. Those in the know prefer to plant only male trees, because the females produce foul-smelling seeds.

20-I Division Anthophyta: Angiosperms or Flowering Plants

Flowering plants include an estimated 275,000 species, six times the number of species of all other plants! They are impressively varied, ranging from duckweeds to dogwoods, onions to oak trees. They range in size from tiny *Wolffia,*

(a)

(b)

(c)

(d)

(e)

(f)

Figure 20-20
The variety of flowering plants. (a) The prickly pear cactus receives little water in its arid home. (b) At the opposite extreme, water lilies live surrounded by water. (c) This rain forest in the Andes contains tall trees, shrubs with vividly colored flowers, and epiphytes—plants that grow on other plants—in this case many bromeliads. (d) Wheat is a member of a major family, the grasses. (e) Mistletoe makes its own food by photosynthesis but parasitizes the branch of a host tree for its water and minerals. (f) This snow plant, from the Sierra Nevada, cannot photosynthesize, but obtains food from dead matter by way of a mycorrhizal fungus. (e, Carolina Biological Supply Company)

A Tree That Changed History

A JOURNEY INTO LIFE

Figure 20-A
Cinchona.

The discovery of a plant that controlled malaria had dramatic effects on human history. Malaria is one of the most devastating human diseases (Section 19-H). Carried from one person to another by blood-sucking female mosquitoes, it was endemic in southern Europe, Asia, and northern Africa during the Middle Ages. European travellers took malaria all over the world. Thousands died of it during the settlement of the United States. In nineteenth-century India, it killed about 2% of the population each year and made another 2% too ill to work. Even in this century, 10 million cases were reported during an epidemic in the Soviet Union in the 1920s, and each year 2 million people worldwide die from it.

The treatment for malaria came from Peru, where malaria was introduced by European settlers. In 1638, the Spanish Countess of Cinchon recovered from a bout of malaria after taking an extract of the bark of the tree now called *Cinchona* (Figure 20-A). Native Peruvians had recommended this to Jesuit missionaries as a cure for fever. The bark was soon imported to Europe, but many Protestants, such as English leader Oliver Cromwell, distrusted the Catholic Jesuits, refused their drug, and died of malaria. In 1852, the indefatigable Louis Pasteur discovered that quinine, the active principle in the bark, was a mixture of substances similar to strychnine and morphine. Not surprisingly, quinine is somewhat toxic; it can cause temporary ringing in the ears when taken, and large quantities can cause permanent hearing loss.

We have seen (Section 13-B) that some west Africans (and some southern European populations) had already evolved sickle cell hemoglobin, which affords some protection from malaria. Genetic resistance to malaria was a major reason west Africans were prized as slaves in the southern United States and the Caribbean, because quinine was expensive. During the nineteenth century, the cost of quinine for the British Army occupying India was over $320 million each year (in 1990 dollars).

Tonic water is quinine dissolved in carbonated, sweetened water. Few who enjoy "gin and tonic" today realize that this drink was invented to make daily doses of bitter quinine palatable to people living in malarial areas. Quinine permitted northern Europeans, who had no natural defenses against the disease, to establish vast empires in tropical areas better defended by malaria than by any army.

Quinine also permitted Europeans to move about 20 million Indian and Chinese laborers to tropical areas where they would have died without quinine. The great European empires of the nineteenth century in Africa, Madagascar, Malaya, and Ceylon were based on plantations worked by the cheap labor of these immigrants, the ancestors of large modern populations in these areas.

Huge new industries were based on these population movements: sugar in the Indian Ocean and the Caribbean, tin and rubber in Malaysia, and tea in India and Ceylon were all made possible by quinine. The drug also probably permitted the United States to win the Second World War in the Pacific against Japan. During that war, 25 million Allied troops travelled to areas where malaria was epidemic, including most of the Pacific, the Mediterranean, and northern Africa, areas where they could not have survived without quinine.

A German scientist, Paul Ehrlich, first treated malaria with an artificial substitute in 1881. His efforts to make quinine led to a wealth of artificial fertilizers, pharmaceuticals, and plastics which, in the 1920s, became the basis for the modern chemical industry.

However, most quinine is still extracted from the bark of *Cinchona* grown in plantations. The natural product is cheaper and pleasanter tasting than the synthetic variety and remains one of the world's important drugs. If one South American tree affected world history in all these ways, no wonder scientists are sure that in destroying unexplored tropical forest, we are depriving ourselves of products whose value we cannot even imagine.

Table 20-5   Economic Importance of Land Plants

Bryophytes	Sphagnum moss: Used as fuel (peat) in Ireland, Scotland, etc.; used as mulch and planting medium in gardening and nursery industries
Lycopods	Formerly used as Christmas greens, the ground pine is now rare and protected in most areas
	Waterproof spores once used to dust pills so that they would not stick together in humid weather; highly inflammable, they were also used in fireworks
Ferns	Foliage used in florist industry; plants sold for house and garden
Gymnosperms	Lumber: Douglas fir, hemlock, spruce, various pine species, cedar, redwood
	Turpentine: Distilled from pine trees
	Pulp for paper: Various conifers
	Christmas trees: Spruces, pines, firs, eastern red cedar
	Landscape plants: Spruces, junipers, yews, cedars, cypress, hemlock, pines, cycads, *Ginkgo*
	Gin: Sometimes flavored by redistilling spirits with juniper "berries"
Flowering plants	Food: Fruits, berries, seeds, nuts, grains, stalks, leaves, roots, tubers; extracted juices, syrups, fats
	Clothing: Cotton, linen
	Lumber: Oak, maple, ash, birch, poplar, walnut, cherry, pecan
	Fuel: Wood, charcoal
	Landscaping: Grass, oak, maple, magnolia, birch, and many other trees; flowering shrubs; annual and perennial herbaceous flowers
	Beverages: Coffee; tea; fermentation of many angiosperm species to make beer, wine, and liquors
	Drugs and medicines: Tobacco; aspirin (originally derived from willow bark); morphine, opium; marijuana; atropine (from *Belladonna* plant); digitalis (from foxglove); various tonics, from sassafras, dandelion, coltsfoot, etc.; quinine (from *Cinchona* bark) used to treat malaria

about 1 millimetre long, to towering *Eucalyptus* trees that vie with redwoods for botanical height records. Representatives of the Anthophyta grow in deserts, on mountain tops, and in polar regions, salt marshes, lakes, and streams. Their flowers borrow every hue of the rainbow.

Flowering plants are crucial to the existence and economy of human beings (Table 20-5). We may occasionally snack on pine seeds or spruce beer or sauté a batch of fern fiddleheads (the unrolling fronds) as a novel spring vegetable. However, almost all of our plant food comes from flowering plants, as does most of the food for domestic animals. Flowering plants also provide housing, clothing, dyes, medicines, and spices.

The fossil record provides little evidence about the origin of the Anthophyta. The oldest identified fossil flower dates from the early Cretaceous Period, 110 million years ago, although fossil flower pollen several million years older is known. Flowering plants radiated rapidly into a variety of different forms during the late Cretaceous, around 70 million years ago. The dinosaurs, most of the cycads, and some conifers became extinct at this time, while mammals and flowering plants started to dominate the living world.

Why were flowering plants so successful? One theory holds that they were superior competitors: thanks to more efficient vascular tissue, they could get more water to the leaves faster, and so their leaves could be larger and carry out more photosynthesis. This allowed them to grow and reproduce more quickly.

At least part of the reason for the success of flowering plants comes from their coevolution with animals. When earlier land plants were evolving, terrestrial animals were few, but when flowering plants appeared, land animals were already well established and diversified. Many flowering plants depend on animals, especially insects, for pollination (Figure 20-21). They may also rely on animals to disperse their seeds. However, many flowering plants rely on the wind for one or both of these functions.

■ *Compared to earlier land plants, flowering plants have more efficient vascular tissue, faster growth, and more protection for delicate reproductive and dispersal stages of the life history.*

(a)

(b)

Figure 20-21

Angiosperm flowers. (a) The colorful flowers of a hardhead thistle attract animal pollinators. (b) The flowers of a grass are wind-pollinated. Pollen is released from the dangling yellow anthers into the breeze. The small, feathery white structures are female flower parts, which must receive pollen from passing air currents. (Biophoto Associates, N.H.P.A.)

The most distinctive feature of angiosperms is the **flower.** Its parts, including the male and female reproductive structures, are modified leaves of the sporophyte generation. The pollen grains, as in gymnosperms, are immature male gametophytes, transported to female flower parts by animals or wind. The female gametophytes, enclosed inside the female flower parts, are even smaller and simpler than those of gymnosperms.

The term angiosperm means "hidden seed," a reference to the fact that the seeds develop inside **fruits,** which in turn grow from female flower parts. Fruits are not necessarily juicy and delicious. The pods of peas and milkweed are fruits, and so are pumpkins and peanut shells.

A third reproductive feature found among flowering plants is **double fertilization.** The pollen tube releases two sperm nuclei. One of them fertilizes the egg, forming a zygote which develops into an embryo in the seed. The other sperm nucleus fuses with a nucleus near the egg, in the female gametophyte. The resulting nucleus then divides and develops into **endosperm** tissue, which provides the seed's food supply. Recent evidence shows that double fertilization, but not endosperm formation, also occurs in the gymnosperm genus *Ephedra* (Section 20-H).

We shall study reproduction of flowering plants in detail in Chapter 36.

SUMMARY

The three groups of multicellular algae have not evolved the adaptations that enable other plants to live on land, and so they are restricted to living in water (or very moist places on land). Because they require light for photosynthesis, algae live only in shallow water where enough light is available. Holdfasts or adhesive secretions attach most forms to rocks or other substrates, and air bladders in some forms allow the photosynthetic parts of the plant to float near the surface where there is adequate light.

Alternation of diploid and haploid generations, with similar or different body structure, is characteristic of plants.

Land offers plants advantages that are not found in water: more sunlight and more carbon dioxide. However, land plants must have adaptations that allow them to cope with the problems of a terrestrial existence: lack of water, evaporation of the water that is available, and lack of support. In the vascular plants, vascular tissue transports water taken in by the roots, supports the stems and leaves in the air, and transports food from the photosynthetic parts to the roots. A waxy cuticle retards evaporation from the leaves and stems, and stomata allow gases to enter the leaves with minimal water loss. Leaves expose large surface areas for photosynthesis, and hormones coordinate the activities of different parts of the plant with one another and with cues from the environment. In the most advanced land plants, pollen grains and pollen tubes permit sexual reproduction without the need for water in which sperm can swim, and seeds supply food and protection to young individuals, increasing their chances of survival.

The adaptations of plants to life on land show several evolutionary trends:

1. In bryophytes, the gametophyte became dominant and the sporophyte dependent on it. In vascular plants, the (vascular) sporophyte became the dominant stage in the life history. During the evolution of vascular plants, the size of the sporophyte progressively increased, while that of the gametophyte decreased.
2. The increase in size of the vascular plant sporophyte was accompanied by an increase in strength and efficiency of its vascular tissue.
3. As tissues of vascular sporophytes became more and more specialized, new organs were added progressively:

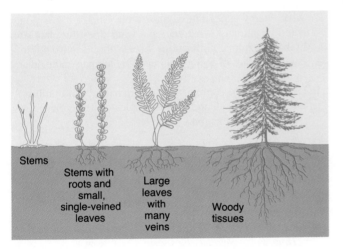

Stems

Stems with roots and small, single-veined leaves

Large leaves with many veins

Woody tissues

4. The male gametophyte evolved into a waterproof pollen grain, which travelled to the female gametophyte before releasing the sperm. Sperm no longer faced a long, dangerous swim to the egg.
5. The female gametophyte of all land plants retains the egg and protects the zygote as it develops into an embryo. In seed-bearing plants, the female gametophyte itself is retained on or in the sporophyte parent, which protects and nourishes it. The sporophyte also contributes food and protective coatings to the seed—a new dispersal structure containing the embryo of the next sporophyte generation.
6. A spore is a single haploid reproductive cell that is the dispersal stage of bryophytes, lycopods, horsetails, and ferns. In gymnosperms and angiosperms, megaspores give rise to female gametophytes, and microspores give rise to pollen grains (male gametophytes). Instead of being shed, the megaspores remain in the sporophyte, which protects the female gametophyte and plays a major role in the production of the seed. Seeds replaced spores as the dispersal stage in the life history.
7. As plant structure and reproduction became more and more independent of moisture in the environment, plants underwent adaptive radiation and spread into more and more different habitats.

SELF-QUIZ

For each adaptation of algae on the left, pick its function from the list on the right. The functions may be used one or several times or not at all.

Adaptation
___ 1. Accessory photosynthetic pigment
___ 2. Gelatinous secretion
___ 3. Holdfast
___ 4. Air bladder

Function
a. maintains a favorable location
b. obtains more nutrients
c. protects from drying out
d. obtains more energy

5. A haploid reproductive cell that may divide and give rise to a new plant is called a(n):
   a. spore
   b. gamete
   c. zygote
   d. embryo

6. Which of the following stages in the life history of a plant are diploid?
   a. gamete
   b. zygote
   c. gametophyte
   d. sporophyte
   e. spore

7. Beginning with the zygote stage, arrange the terms in Question 6 in the order found in a plant's life history.

8. In the evolution of land plants, sporophytes became dominant over gametophytes due to what adaptation?
   a. airborne pollen
   b. seeds
   c. vascular tissue
   d. diploidy
   e. stomata

9. A seed encloses an individual of the next sporophyte generation in the form of a(n):
   a. spore
   b. gamete
   c. zygote
   d. embryo
   e. pollen grain

10. An adaptation of land plants that reduces evaporation of water from the body surface into the air is:
    a. roots
    b. rhizoids
    c. stomata
    d. cuticle
    e. wood

11. Mosses are adapted to live in a land environment in that:
    a. they have means of vegetative reproduction
    b. they have alternation of generations
    c. they have dependent sporophytes
    d. they hold water near their bodies
    e. they are no more than a few inches tall

12. An advance shown by gymnosperms that was not found in previous groups of plants was:
    a. growth exceeding a few feet high
    b. dominance of the sporophyte over the gametophyte generation
    c. protection of the gametophyte within the sporophyte body
    d. presence of large leaves
    e. production of reproductive structures at tips of the plant

13. Where would you find the gametophyte of
    a. a moss?
    b. a fern?
    c. a pine tree?
    d. a lily?

QUESTIONS FOR DISCUSSION

1. Nonmotile plants must be able to get gametes together for fertilization. How does *Ulva* (Figure 20-5) accomplish this?

2. Red algae have no flagellated cells; how do you think their gametes might get together for sexual reproduction?

3. What modifications in the structure and reproduction of aquatic algae would be necessary for them to live permanently on land?

4. Algae lack roots; how do they obtain nutrients?

5. Parental care of the young was an adaptation that played a major part in the evolutionary success of birds and mammals. What parallels can you find in the plant kingdom?

6. Most conifers in temperate climates keep their leaves through the winter. Most angiosperm trees in the same environment drop their leaves each fall and produce a new set each spring. What are the advantages and disadvantages of being evergreen? Of dropping and regrowing the leaves?

7. What energy expenditures must a plant make if it is pollinated by animals? What energy savings does the plant gain by having animal pollination? What is the adaptive advantage to a plant of being pollinated by animals rather than by wind?

8. List all possible combinations of animal or wind pollination and dispersal, and give examples of flowering plants with each combination.

SUGGESTED READINGS

Hobhouse, H. *Seeds of Change: Five Plants that Transformed Mankind.* New York: Harper and Row, 1986. The story of quinine—and potatoes, tea, sugar cane, and cotton.

Jensen, W. A., and F. B. Salisbury. *Botany: An Ecological Approach,* 2d ed. Belmont, CA: Wadsworth, 1984. A fresh approach, enjoyable reading.

Niklas, K. J. "Aerodynamics of wind pollination." *Scientific American,* July 1987. Despite its daunting title, this article is more biology than physics.

Wilkins, M. B. *Plantwatching: How Plants Remember, Tell Time, Form Relationships, and More.* New York: Facts on File, 1988. A survey of various plant groups and how plants live and work.

The Animal Kingdom

CHAPTER

21

Frilled lizard

Eclectus parrot, female

Nudibranch

Figure 21-1
A marine invertebrate. This large colonial animal swims by contractions of its many bells. It traps prey, often quite large fish, with stinging structures on its trailing tentacles. Its transparent body is invisible to predators and prey. (Phylum Cnidaria). (J. M. King, courtesy of Alice Alldredge)

*T*he animal kingdom contains a wide variety of organisms. All are multicellular heterotrophs, and all produce **gametes** (eggs and sperm) in multicellular structures, the **gonads** (ovaries and testes). After fertilization, the zygote develops into an embryo inside the egg or inside its mother's body, until it is mature enough to be hatched or born. Most animals ingest their food—eat now, digest later—rather than digesting food externally and then absorbing it, as fungi do. Animals are also **motile,** able to move around in at least one stage of the life history. Even those that spend most of their lives in one place can move parts of the body and so obtain food.

The first animals arose during the Precambrian Era. Many new animal phyla appeared at the start of the Cambrian Period, about 570 million years ago. These included most phyla alive today, and many long since extinct.

Depending on who is counting, the animal kingdom contains up to 33 living phyla, each with a distinct body plan. Here we cover only the nine major ones. One subphylum of the phylum Chordata contains all the **vertebrates,** animals with backbones. All other animals are **invertebrates,** animals without backbones.

More than half of the invertebrate species other than insects live in the sea, and these introduce us to the marine world in which life evolved (Figure 21-1). Competition for food, or predation pressure, eventually drove many animals out of the sea. Some moved from bays up into rivers, and evolved adaptations to cope with the hypotonic environment of fresh water (Sections 4-C and 26-B). Other animals, living on the shore, evolved the ability to survive on ever higher and drier sites and became **terrestrial** (land-dwelling).

The earliest animals took small food items directly into their cells. Later, some animals had a space inside the body able to hold larger food, such as entire animals. At first this was a sac with only one opening. Eventually there evolved a tube with two openings—mouth and anus—encased in an outer tube, the body proper. Internal digestion frees the outside of the body to deal with the environment. Here we find sense organs and protective structures. Between the inner and outer tubes, most animals have a middle layer containing muscles used in locomotion, feeding, and defense against predators; structures for internal functions such as circulation and excretion; and reproductive organs.

The most active and resourceful animals are predators on other animals. Capturing prey requires keen senses to find food and speedy response and coordination of nerves and muscles to catch it. Under such evolutionary pressures, predatory animals have made many important evolutionary advances. Their descendants have often then evolved adaptations of the new body plan to many other ways of life.

In this chapter we look at some common selective pressures and evolutionary trends in the animal kingdom. We then examine the general body plans and ways of life in members of nine major animal phyla. Later chapters cover the evolution of some animal structures and functions more specifically.

KEY CONCEPTS

- Life evolved in the sea. The few animal groups that successfully colonized fresh water and land evolved many adaptations of structure, function, and reproduction to these difficult environments.
- Most animals are made up of three layers: an outer protective and sensory tube, an inner digestive tube, and between them the organs of motion, reproduction, excretion, and transport.
- Active carnivores made many of the major evolutionary advances among animals, but most groups then radiated to produce members with other lifestyles such as herbivory, sessile filter feeding, or parasitism.

Figure 21-2
Dinner in the Namib desert. A scorpion kills a grasshopper by injecting it with poison from a gland at the tip of its abdomen. (Phylum Arthropoda. Scorpion, Class Arachnida; grasshopper, Class Insecta) (Biophoto Associates, N.H.P.A.)

21-A Animal and Environment

The sea is a more stable and hospitable environment for life than either fresh water or land. For instance, the salt concentration of the sea is very similar to that of a cell. Most marine invertebrates therefore have no osmotic problems of water gain or loss. Although light and temperature vary, largely with depth, the temperature in any one area of the sea changes slowly, within fairly narrow limits. Furthermore, the small algae, protozoa, and animals in the **plankton** (floating organisms) provide a constantly renewed source of food.

In contrast, fresh water usually contains lower concentrations of nutrients and so supports less life. It is also hypotonic to living cells: a freshwater animal must constantly expend energy to retain its salts and expel the excess water that enters its body by osmosis.

Land is an even more difficult environment because water is often in short supply, making death by dehydration a constant danger. Relatively few animals have invaded fresh water successfully, and only two groups, the terrestrial arthropods (spiders, insects, and their kin) and the higher vertebrates (reptiles, birds, and mammals), have really solved the problems of life on land. Many members of these groups move about freely even in the dehydrating conditions of warm, sunny days (Figure 21-2). The terrestrial representatives of other animal groups must retreat to cool, moist places on such days, emerging only at night.

Despite its advantages, life in the sea poses certain difficulties. Photosynthesis requires light, but water absorbs light; thus most plant life—the main food for marine animals—lives only near the surface, which is constantly tossing about. Fish, which are vertebrates, swim well enough to maintain their positions despite the waves, but few invertebrates are powerful enough swimmers to do this. Invertebrates have evolved two sorts of adaptations that permit them to cope with the problem of being swept away by the sea. One is to remain small enough to float around close to their food supply (Figure 21-3). The alternative is for an animal to burrow in the sea floor, to cling to rocks or other stable objects using structures such as claws or suckers, or to be **sessile** ("sitting"), anchored in one place.

Most sessile animals are **filter feeders.** They use cilia or muscles to set up water currents past (or through) their bodies and filter out or seize any food

■ *Sea water is isotonic with the body fluids of marine invertebrates. Fresh water is hypotonic, and animals need osmotic adaptations to colonize it. Life on land requires adaptations against loss of water by evaporation.*

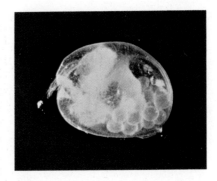

Figure 21-3
An ostracod. This tiny planktonic crustacean is about 1 millimetre long. The pink spheres are eggs. (Phylum Arthropoda) (Bruce Robison)

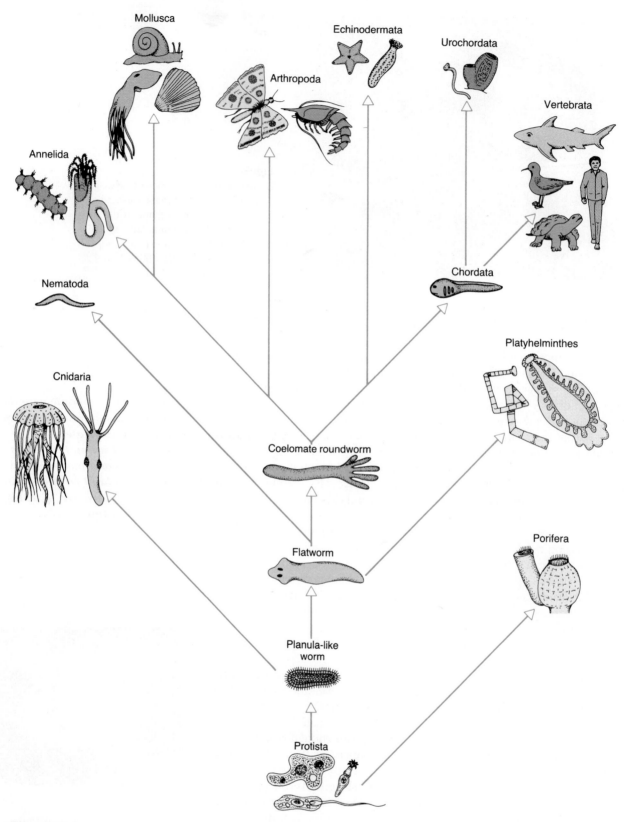

Figure 21-4

Evolution of major animal phyla. This evolutionary tree represents one widely held theory.

Table 21-1 The Major Invertebrate Phyla

Phylum	Common names	Characteristics
Porifera (~10,000 species)	Sponges	Simple, sessile animals with specialized cells but no tissues or organs; no mouth or digestive cavity; collar cells used in filter feeding; most marine
Cnidaria (~9000 species)	Jellyfish, sea anemones, corals, *Hydra*	Free-swimming medusa or sessile polyp; two tissue layers; mouth surrounded by tentacles bearing nematocysts; gastrovascular cavity; no anus; predatory; radially symmetrical; solitary or colonial; most marine
Platyhelminthes (~12,700 species)	Flatworms: turbellarians, flukes, tapeworms	Free-living or parasitic; bilaterally symmetrical; three body layers, with organs; gut with mouth but no anus; most marine or freshwater
Nematoda (~10,000 species)	Roundworms	Free-living or parasitic; gut with mouth and anus; fluid-filled body cavity; marine, freshwater, and damp soil habitats
Annelida (~6200 species)	Segmented worms: polychaetes, earthworms, leeches	Segmented body with coelom and circulatory system (except leeches); setae (bristles) in some; gut with mouth and anus; gas exchange through skin or gills; marine, freshwater, or damp soil
Mollusca (~87,000 species)	Snails, slugs, nudibranchs, clams, oysters, scallops, mussels, squids, octopuses	Segmentation reduced; body covered by a mantle, which may secrete a shell; head and muscular foot usually present; gas exchange by gills or lining of mantle cavity; circulatory system with heart; marine, freshwater, or terrestrial
Arthropoda (~800,000 species)	Spiders, mites, crustaceans, insects, millipedes, centipedes	Segmented, with jointed exoskeleton containing chitin; jointed appendages; gas exchange by body surface, gills, or tracheae; marine, freshwater, or terrestrial
Echinodermata (~6000 species)	Sea stars, brittle stars, sea urchins, sea cucumbers	Spiny calcareous skeletons; tube feet; pentaradial symmetry; all marine
Chordata (~1300 invertebrate species)	Tunicates, amphioxus	Notochord; hollow dorsal nerve tube; pharyngeal gill slits; post-anal tail at some stage; ventral heart; marine

carried by the current. Other sessile animals hang out a net of tentacles or mucus and trap passing food. A sessile organism needs protection from mobile predators. It may have active protection such as stingers, or passive protection such as a coating of toxic mucus or a thick shell.

Sessile animals cannot move around to find a mate. Instead, most of them have behavioral adaptations in which all members of a species release their sperm and eggs at the same time in response to environmental cues, such as a particular water temperature and a full moon. In most sessile or slow-moving marine invertebrates, the embryo develops into a tiny, motile animal that can disperse the organism to new habitats. This **larva** is a stage in the life history that differs from the adult in its structure, habitat, and diet. The disadvantage of larvae is their high mortality rate. Predators take many, and many others do not end up in a suitable habitat. These selective pressures have favored the production of vast numbers of offspring in many invertebrates. The danger of being carried into unfavorable habitats by currents is especially great in fresh water. Most freshwater invertebrates no longer have a larval stage, but hatch from the egg as miniature adults.

Many sessile animals can also reproduce asexually by dividing, usually into two parts, or by budding. This permits an individual that has found a good spot to populate the area with a clone of individuals containing its own genes.

21-B Animal Structure

As we study the major animal phyla, we shall see some evolutionary trends in body structure. Let us preview some of these.

Body Symmetry

Some lower animals have body plans with **radial symmetry,** that is, the animal can be divided into mirror-image halves by several different planes passing through its long axis (Figure 21-5). Radial symmetry permits an animal to detect the approach of food or danger from any side. However, animals with well-developed muscular systems can move and obtain food more efficiently if they are **bilaterally symmetrical,** with only one plane that divides the body into two mirror-image halves. Bilateral symmetry allows an animal to have a more streamlined shape and to concentrate the power of its muscles and appendages into producing motion in one direction.

Along with the trend toward bilateral symmetry in animals goes the evolution of **cephalization**—the development of a head, with a concentration of sensory and nervous tissue that monitors the area the animal is entering.

Body Layers

The structure of most animals is basically a tube within a tube. The inner tube is the digestive tract. The cells lining this tube are specialized to digest food, absorb it, and push it through the body. The outer tube consists of tissues specialized for dealing with the outside world. Here we find protective structures, such as skin, shells, and horns, as well as sense organs and the nervous system, which tell an animal what is going on around it. Between these inner and outer body layers are packed the other organs of the body: the muscles, blood vessels, reproductive organs, and so on. These deal with internal functions and have no direct contact with the outside world or the gut. In all except a few primitive invertebrate phyla, this basic body structure develops in the early embryo from three layers of cells: the **endoderm,** which will form the lining of the gut; the **ectoderm,** which will form the body's outer layer; and the **mesoderm,** which forms between the other two layers (Figure 21-6). (Endon = within; ektos = outside; mesos = middle; derma = skin)

The Coelom

The **coelom** is a fluid-filled body cavity that develops as a space in the mesoderm of higher animals (Figure 21-6). The origin of the coelom was one of the most important steps in animal evolution. First, the coelom separates the muscles of the gut from the muscles of the body wall. As a result, the gut can move independently of the body wall, and movement of food down the gut does not depend on locomotory movements. Second, the fluid that fills the coelom may act as a simple circulatory system, transporting waste, food, and gases around the body. More importantly, the coelom provides space where a true circulatory system with blood vessels can develop. A constant blood flow would be impossible if the heart were squashed by other organs every time the animal moved a muscle.

RADIAL SYMMETRY

BILATERAL SYMMETRY

Figure 21-5
Radial and bilateral symmetry. In radial symmetry, a cut made in any of several planes would cut the animal into mirror-image halves. A bilaterally symmetrical animal can be cut through only one plane of symmetry to yield mirror-image halves.

Figure 21-6
Body plans of members of some animal phyla, in cross section. (a) Cnidaria (Section 21-D) have a two-layered body, with a mesoglea ("middle glue") between the endoderm and ectoderm layers. All higher animals have three full layers (b). (c) Annelids and all higher groups have three layers, with a coelom that forms as a space in the mesodermal layer. Mesenteries suspend various organs in the coelom.

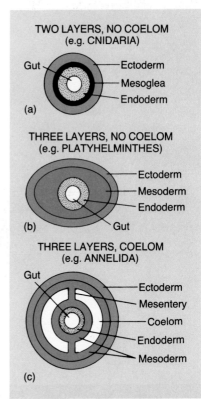

With these principles of animal structure and ways of life in mind, let us look at the major phyla and see what they show about animals and their evolution.

Invertebrates

21-C Phylum Porifera: Sponges

Sponges are simple, sessile animals living singly or in colonies (Figure 21-7). Their bodies do not have the kinds of layers described in the previous section. They have some cell specialization, but no tissues, and their cells are remarkably

(a) Sponges

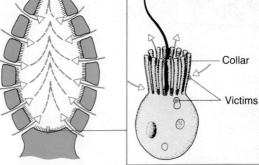

(b) Body plan

(c) Collar cell (choanocyte)

Figure 21-7
Sponges. (a) These golden tubes are sponges, growing attached to coral on the sea floor. (b) The sponge body plan. The black fringe in the body cavity represents the flagella of collar cells, whose beating draws the feeding current through the body. Water enters through numerous tiny openings on the sides and leaves through the opening at the top (arrows). (c) A collar cell, a type of cell characteristic of sponges. Lashing of the flagellum draws a current of water through the collar, a ring of microvilli that trap food particles. Food is ingested by endocytosis and digested intracellularly. (a, Alice Alldredge)

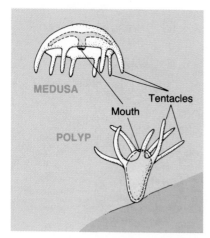

Figure 21-8
The two body forms found in cnidarians. The free-floating medusa is an inverted version of the sessile polyp. Both have tentacles surrounding the mouth, which leads into a gastrovascular cavity.

independent of each other. For instance, a living sponge can be pushed through fine silk, which breaks it up into individual cells and cell debris. If these cells are left to stand in sea water, they will form a functional sponge again!

Sponges range from a few millimetres to the size of a barrel. The body is held in shape by a skeleton of fibrous protein (as in natural sponges sometimes used for bathing) or of little spikes (spicules) of calcium carbonate or silica. The sharp, brittle spicules also serve as a defense against predators; in addition, the sponge may produce toxic chemicals. Such defenses are vital for sessile animals. The simplest sponge body form is like a vase with porous sides. Water enters through these pores and exits through the large opening at the top (Figure 21-7b).

The distinctive feature of sponges is a type of cell called a **collar cell** or **choanocyte,** with a flagellum that beats and propels the water current through the body (Figure 21-7c). Food particles swept into the body with this current are strained out by the collar cells, which pass food on to amoeba-like cells that

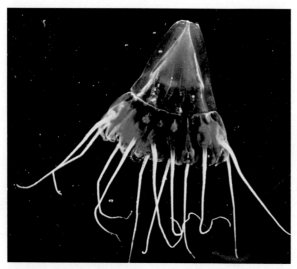

(a)

(b)

Figure 21-9
Phylum Cnidaria. (a) A jellyfish, showing the medusa body form. (b) A sea anemone, a large solitary polyp. The tentacles trap food and pass it to the mouth in the center. (c) The tiny polyps of a coral, part of a vast colony sharing a common skeleton, made up of calcium carbonate secreted by the polyps. (a, Bruce Robison; b, Biophoto Associates; c, Steven Webster)

(c)

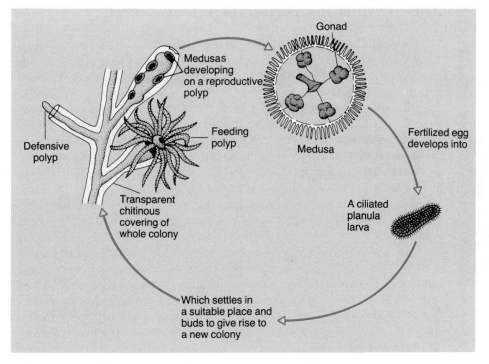

Figure 21-10
Life history of a colonial cnidarian. Different individuals in the colony are specialized for different functions. Medusas bud off from reproductive polyps and swim away, develop gonads, and then release gametes.

digest it and distribute it to other cells. The water current provides the cells with oxygen and removes wastes. Most sponges are marine.

21-D Phylum Cnidaria: Jellyfish, Corals, Sea Anemones

Members of the phylum Cnidaria have soft bodies with a basic radial symmetry. The inner and outer layers are well developed, but the middle layer is not. The two body layers contain several cell types, but there are no full-fledged organs. The gut is not a complete tube, with mouth and anus, but rather a blind sac, with one opening serving both to ingest food and to expel indigestible remains. It is called a **gastrovascular** cavity (gastro = stomach; vascular = vessel) because it also serves as a circulatory system, distributing food around the body.

Cnidarians come in two basic, and essentially similar, body forms: **polyp** and **medusa** (Figure 21-8). In both forms, a ring of tentacles surrounds the mouth, which leads into the gastrovascular cavity. Polyps are sessile, and most medusas are free-swimming, moving by weak contractions of their bell-shaped bodies. Jellyfish are medusas. *Hydra,* a species often studied in biology class, is a polyp, as are sea anemones and corals (Figure 21-9). Some cnidarians have both polyp and medusa stages in their life history (Figure 21-10), but others have lost one stage or the other. Many are colonial, consisting of numerous polyps and/or medusas attached to one another (Figures 21-1 and 21-9c).

Most cnidarians are **carnivores,** animals that eat other animals. They do not chase their prey but sit (or float) waiting to trap victims. The tentacles bear special offensive and defensive structures called **nematocysts,** unique to cnidarians (Figure 21-11). The nematocyst may twist about bristles on the body of the prey, entangling it, or may secrete a sticky or paralytic substance. Nematocyst toxins can cause a nasty sting, occasionally fatal to swimmers.

In most cnidarians, the tentacles pull prey trapped by the nematocysts and stuff it into the mouth. Having nematocysts and a mouth and gastrovascular cavity permits cnidarians to eat much larger prey than a protozoan or sponge can manage.

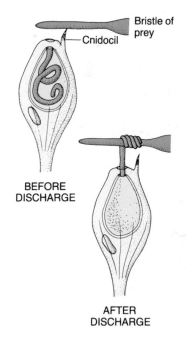

Figure 21-11
Nematocysts. These specialized structures are characteristic of Cnidaria. The cnidocil acts as a trigger. Touching the cnidocil causes the nematocyst to eject its thread, which may entangle the prey (as shown) or secrete a sticky or poisonous substance that immobilizes the prey.

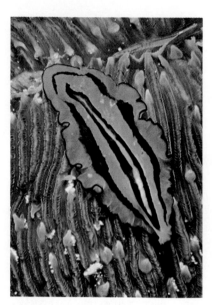

(a)

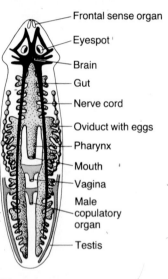

- Frontal sense organ
- Eyespot
- Brain
- Gut
- Nerve cord
- Oviduct with eggs
- Pharynx
- Mouth
- Vagina
- Male copulatory organ
- Testis

(b)

Figure 21-12

Flatworms. (a) A turbellarian crawling across coral. (b) The internal anatomy of a free-living platyhelminth worm. Notice that the mouth is not at the animal's front end, but in the middle of the body. Both male and female reproductive organs are present. (a, Biophoto Associates, N.H.P.A.)

■ *Flatworms and roundworms have bilateral symmetry, the beginnings of cephalization, and three body layers with well-developed tissues and organs. Both phyla include many important parasites of humans.*

21-E   *Phylum Platyhelminthes: Flatworms*

The flatworms are the most primitive group of animals with bilateral symmetry, the beginnings of cephalization, and the development of true organs, in three full body layers (Section 21-B). The middle layer contains the reproductive organs, an excretory system, and distinct layers of muscles. However, the digestive tract still has only one opening (Figure 21-12b). There is no coelom, circulatory system, nor blood to transport food around the body. Hence food is distributed by the digestive system, which branches throughout the body. Gas exchange and waste excretion occur over the entire surface of the extremely flat body.

Turbellarians are free-living flatworms with cilia over the entire body surface (Figure 21-12). They feed on smaller organisms or dead organic matter.

The other two groups of flatworms are entirely parasitic. **Flukes** live as external or internal parasites, usually attached to the host by suckers. As is typical for parasites, reproductive organs occupy most of the body, reproductive rates are high, and the life history is often complex, involving more than one host (see *A Journey Into Life:* Parasitism). Blood flukes of the genus *Schistosoma* cause the devastating disease schistosomiasis (or bilharzia), which affects some 200 million people in over 70 tropical nations. These flukes live in freshwater snails for part of the life history. They cycle between their human and snail hosts wherever people drink and wash in water that also contains wastes from infected people.

Tapeworms are highly specialized parasites found in the intestines of probably every species of vertebrate. A tapeworm has no head, mouth, or digestive system. It absorbs digested food from the host's gut through its body surface, and its body contains little except reproductive organs. The secondary hosts for the most common tapeworms that infect humans are cattle, pigs, and fish. Tapeworm eggs or larvae enter these hosts with the food. The larvae burrow through the host's gut wall, enter the blood, and travel to the muscles, where they lodge as immature worms. They infect humans that later eat this flesh unless they are killed by thorough cooking.

21-F   *Phylum Nematoda: Roundworms*

Most nematodes, or roundworms, are so small that they attract little notice. However, they can be found by the million in every environment. Any handful of good soil contains thousands of tiny white or transparent roundworms (Figure 21-13).

Nematodes have advanced over the flatworms in possessing a complete digestive tract, with an anus as well as a mouth, but they still lack a circulatory system. A tough nonliving cuticle covers the body.

About 50 species of roundworms parasitize humans. One causes trichinosis, often contracted by eating undercooked, infected pork, although an outbreak in Canada was traced to eating undercooked bear meat. Among the more harmful species are the intestinal roundworm (*Ascaris lumbricoides),* the guinea worm, and the filaria worm, which lodges in the lymph nodes, and in extreme cases causes elephantiasis (enormous swelling of the leg). Hookworms enter the skin of people walking barefoot on soil contaminated with feces from infected humans. These worms caused the anemia, weakness, and fatigue—formerly attributed to laziness—of thousands of poor laborers in the southern United States.

21-G   *Phylum Annelida: Segmented Worms*

The advances shown by annelid worms include a coelom and a circulatory system with closed blood vessels, some of them enlarged as pumping "hearts" (Figure 21-14). Their bodies are **segmented,** divided into repeated sections by

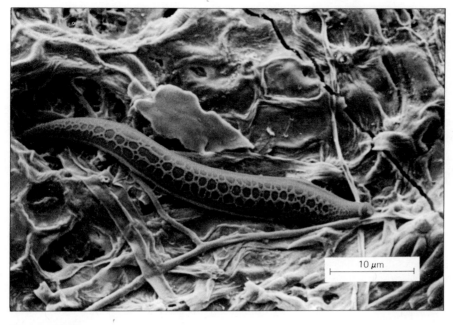

Figure 21-13
A roundworm. This scanning electron micrograph shows a soil nematode that inhabits leaf litter. (Biophoto Associates)

partitions called **septa.** Annelids exchange gases with their environment through the body surface. The area available for this function may be increased by **gills,** extensions of the body surface.

The largest group of annelids, the **polychaetes** ("many bristles"), are mostly marine. Polychaetes burrow in the sea floor or paddle lightly over reefs. Some are tube-dwelling carnivores or filter feeders (Figure 21-15).

Earthworms and their kin make up a second group of annelids. Most earthworms burrow in mud or soil. They ingest soil as they burrow, digest any organic matter, and void the rest as worm casts, which can often be found in a garden. Earthworms play a crucial role in loosening, aerating, and mixing soil.

Leeches, the third annelid group, live mainly in fresh water. Some prey on other invertebrates or scavenge dead matter, but the most famous are external parasites of aquatic vertebrates. Only a few species have the rasping "teeth" needed to break through the tough skin of a mammal. In the last century, doctors often used medicinal leeches to bleed patients as a treatment for all manner of diseases. Indeed, "leech" was used as a nickname for doctors, and doctors kept aquaria with leeches that could be popped into containers in their medical

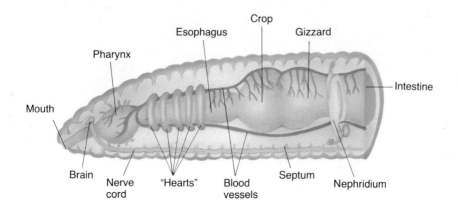

Figure 21-14
Annelid anatomy. This drawing of the front end of an earthworm shows the main features of the digestive, circulatory, and nervous systems. The body is divided into segments, each containing its own set of blood vessels and nerves and a pair of nephridia (excretory structures).

(a)

(b)

Figure 21-15
Polychaete annelids. (a) This polychaete, wriggling across a coral, clearly shows the segmented body and numerous tufts of white setae (bristles) on each segment. (b) A polychaete that lives in a tube. Its feathery, spiral arms filter food out of the water. (Steven Webster)

bags when they went out on housecalls. Leeches are still sometimes used to prevent bruising and blood pooling after surgery (Figure 21-16).

21-H Phylum Mollusca

Molluscs include chitons, snails, octopuses, and "shellfish"—clams, oysters, limpets, scallops, and their kin. The molluscan body consists basically of a muscular **head-foot,** with the body on top loosely covered by a **mantle,** a flattened piece of tissue, which in many species secretes a shell containing calcium (Figure 21-17).

The **gastropods** include marine, freshwater, and terrestrial snails and slugs. Most gastropods have a well-defined head, usually with eyes and tentacles, and an elongated, flattened foot on which they creep around. Most have gills inside

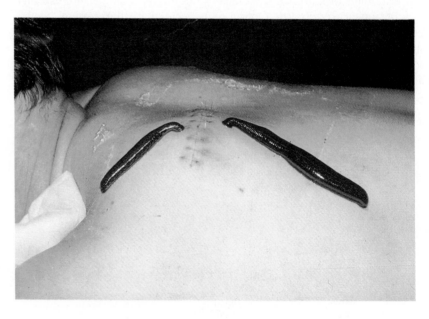

Figure 21-16
Medicinal leeches. By injecting an anticoagulant and sucking up blood, leeches keep an incision free from clots and accumulated blood. (St. Bartholomew's Hospital/Science Photo Library/Photo Researchers, Inc.)

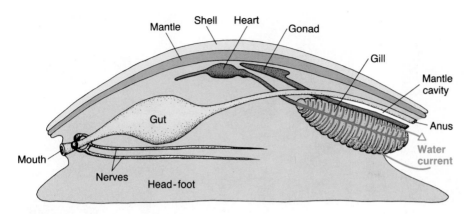

Figure 21-17
Hypothetical ancestral mollusc. This diagram shows the general anatomy of members of the phylum Mollusca. The muscles of the lower body form a continuous mass from which both the head and foot are formed.

the mantle cavity. However, in some forms the mantle cavity acts as a lung, permitting the animal to obtain oxygen from air instead of from water. Some of the world's loveliest animals are **nudibranchs**—commonly known by the unlovely name of sea slugs. They have lost the mantle, shell, and gills, leaving a naked body that is often brilliantly colored (Figure 21-18).

The **bivalves** are a large group of marine and freshwater molluscs with the body flattened between the two valves (halves) of a hinged shell (Figure 21-19). The edge of the mantle is drawn out to form two **siphons,** one for water entering the mantle cavity and one for water leaving. Most bivalves are filter feeders. Cilia draw a water current across the gills, where food particles are strained out and trapped in strands of mucus. The gill cilia then move these strands to the mouth.

The marine, carnivorous **cephalopods** (octopuses, squids, and nautiluses) are among the most advanced of all invertebrates. Behavioral studies have

Figure 21-18
Gastropods. (a) A terrestrial snail. Note the head-foot and the coiled shell. (b) A nudibranch, a marine gastropod without a shell. This form can swim by undulating its body. (Biophoto Associates, N.H.P.A.)

(a)

(b)

Figure 21-19
A marine bivalve. Note the edges of the mantle visible within the shell, and the long foot protruding out toward the left. (Biophoto Associates)

Figure 21-20
Three octopuses on a rocky ledge. Their flexible bodies and the suckers on their eight tentacles permit these animals to move about in rocky crevices where they wait to ambush their prey. (Biophoto Associates)

shown that octopuses are quite intelligent and can solve many problems. Cephalopods appear to be quite closely related to gastropods, but the body has been rearranged. The mouth is now in the middle of the foot, whose edges are drawn out to form tentacles lined with rows of suckers (Figure 21-20). Most cephalopods swim by jerky jet propulsion. They take water into the mantle cavity and force it out through a siphon, which the animal can point in various directions to determine which way it will move. However, octopuses more often crawl using their tentacles.

21-I   Phylum Arthropoda

The phylum **Arthropoda** ("jointed foot") contains more than three times as many species as all other animal phyla combined (Table 21-2). Some groups of arthropods have solved the problems of life on land more completely than any other group except higher vertebrates. Their major step forward was the improvement of the cuticle, which also covers the body in many other invertebrates. In arthropods, the cuticle is composed of layers of protein complexed with the strong, flexible polysaccharide **chitin.** It forms an **exoskeleton** (external skeleton) consisting of a set of plates, which cover not only the body but also the jointed appendages. Hinges between adjacent plates permit movement, while the tough plates protect against attack and injury (Figures 21-21 and 21-22). In many species, wax in the outer layer also prevents water loss.

Arthropod survival often depends upon the efficiency of their waterproofing. Since Roman times, people in Africa and Europe have mixed fine dust with stored grain to keep it free of insects. The dust abrades the soft cuticle between segments, destroying the insect's waterproofing so that it dries up and dies.

■ *The chitinous exoskeleton of arthropods, with its many jointed appendages modified for a variety of jobs, proved so versatile that the arthropods have undergone impressive adaptive radiation, with more species and individuals than any other animal phylum.*

Table 21-2   The Phylum Arthropoda and Its Major Classes

Phylum Arthropoda	Segmented animals with jointed exoskeletons containing chitin; jointed appendages; respiration through body surfaces or by gills or tracheae; marine, freshwater, and terrestrial
Class Arachnida (~57,000 species)	Body with 1 or 2 main parts; 6 pairs of appendages (chelicerae, pedipalps, 4 pairs of walking legs); most terrestrial; e.g., spiders, scorpions, ticks, mites
Class Crustacea (~25,000 species)	Body of 2 or 3 parts; antennae, chewing mouthparts, 3 or more pairs of legs; most marine; e.g., shrimp, krill, lobsters, crabs, barnacles, ostracods, copepods
Class Insecta (~700,000 species)	Body divided into head, thorax, and abdomen; antennae; mouthparts modified for chewing, sucking, or lapping; adults with 3 pairs of legs and usually 2 pairs of wings; breathing by tracheae; most terrestrial; e.g., beetles, flies, butterflies, ants, termites, dragonflies, aphids
Class Diplopoda (~7000 species)	Body with distinct head bearing antennae and chewing mouthparts; most segments of body grouped in pairs covered by a single skeletal plate, each apparent segment bearing 2 pairs of walking legs; breathing by tracheal system; terrestrial, eating dead or living plants. The millipedes (Figure 21-21)
Class Chilopoda (~2000 species)	Body with distinct head bearing large antennae and chewing mouthparts; appendages of first body segment modified as poison claws; remaining segments bearing a pair of walking legs each; terrestrial in damp areas including houses; predaceous on insects. The centipedes (Figure 21-22)

One result of possessing an exoskeleton that cannot expand is that arthropods must **molt,** or shed their cuticles, in order to grow. In many arthropods, successive molts take the animal through a series of larval stages.

The primitive arthropod body plan has one pair of jointed appendages per body segment. The appendages became variously modified during the course of evolution into specialized antennae, mouthparts, walking legs, claws, or swimming paddles. Segmentation of the body also became modified. In some arthropods, the segments became grouped to form distinct parts of the body, such as the **head, thorax,** and **abdomen** in insects.

Figure 21-21
Made for walking. This millipede ("thousand feet") is a member of the arthropod class Diplopoda (see Table 21-2). The rounded head bears a pair of antennae. The remaining appendages visible here are the "thousand" legs. The legs move in coordinated waves, which progress along the body from one pair to the next. Despite the abundance of legs, millipedes travel slowly, and when disturbed they curl up and secrete noxious repellent chemicals. (Biophoto Associates)

Figure 21-22
Long and low. This centipede ("hundred feet") is a member of the arthropod class Chilopoda (see Table 21-2). The antennae are larger than in millipedes; the legs are fewer but move faster. A disturbed centipede usually darts away before a predator has a chance to pounce. (Biophoto Associates)

Figure 21-23
Class Arachnida. This wolf spider shows the six pairs of appendages characteristic of its class. Note also the simple eyes: two large ones, with four smaller ones in a row beneath them, all facing front; and on each side of the head, another small one around the corner from the large one. (P. J. Bryant, Univ. of California, Irvine/BPS)

Class Arachnida

The **arachnids** include the spiders, ticks, mites, and scorpions (Figure 21-23). Arachnids have six pairs of jointed appendages. The first pair are adapted for feeding, and often have associated poison glands that inject a substance that anesthetizes or kills the prey. The second pair act as sense organs to detect touch and chemicals and also help hold food. The other four pairs are walking legs.

All spiders have spinnerets with which they spin silk, and which may be used to make webs that trap prey and the cocoons that protect their eggs. All spiders are carnivorous, usually preying on insects.

Some ticks and mites transmit diseases, including Rocky Mountain spotted fever, Lyme disease, and Asian scrub typhus. Mites themselves cause the itching of mange and scabies. Some mites are also serious plant pests (Figure 21-24).

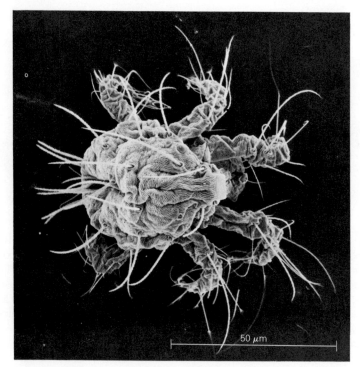

Figure 21-24
A red spider mite. This scanning electron micrograph shows a miniature member of the arthropod class Arachnida. (Biophoto Associates)

(a)

(b)

Figure 21-25
Class Crustacea. (a) A lobster, a crustacean with a cuticle reinforced with calcium carbonate. (b) A barnacle. These odd crustaceans attach to a solid surface by their heads and secrete calcareous shells around themselves. They use their legs to shovel food into their mouths. (a, Biophoto Associates; b, Steven Webster)

The vast majority of mites live around us unnoticed because of their tiny size—a species that reaches a millimetre or two is enormous as mites go.

Class Crustacea: Lobsters, Crabs, Wood Lice, and Their Relatives

The class **Crustacea** includes lobsters, shrimp, crabs, crayfish, wood lice, pill bugs, water fleas, and barnacles (Figure 21-25). In general, crustaceans are aquatic arthropods with forked appendages: two pairs of antennae, three pairs of feeding appendages formed for food handling and chewing, and several pairs of legs. Typically, as we know from our caution in the vicinity of crabs, anterior appendages are modified to grasp food and convey it to the mouth. Planktonic crustaceans are the main food of many vertebrates and include the shrimp-like krill upon which the biggest whales feed.

Class Insecta

Insects are, without doubt, the most successful terrestrial invertebrates and the only major competitors against humans for dominance on land. It has been estimated that the insects on earth weigh 12 times as much as the humans, and that there are 300 million insects for every person alive. Nearly a million insect species have been described, more than the number of all other animal species put together. Insects, bats, and birds are the only living animal groups with members that can fly.

Insects range in size from tiny beetles only 0.1 millimetre long to tropical moths with a wingspan of 30 centimetres. The insect body is usually divided into

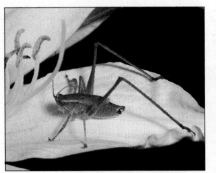
(a)

(b)

(c)

Figure 21-26

Insect development. (a) This immature katydid is a smaller, wingless, version of the adult. Its wing buds will develop into full-size wings when it undergoes its last molt to adulthood. This kind of development is typical of many insects. (b) In insects with complete metamorphosis, the eggs hatch into larvae, in this case a black swallowtail butterfly caterpillar, which looks nothing like the adult. (c) The caterpillar eventually becomes a pupa and develops into an adult. Here, an adult black swallowtail butterfly waits for its wings to dry and harden after it emerges from its tan pupal case. In a few minutes, it will be able to fly.

head, thorax, and abdomen. The thorax of most adult insects bears three pairs of walking legs and two pairs of wings (flies and mosquitoes have only one pair, and fleas, lice, and silverfish have none). The head bears one pair of antennae, specialized mouthparts, and, usually, compound eyes (Figure 21-26).

Insects show several adaptations to life on land, the first being the waxy, waterproofed cuticle, which also supports the body in the air. Insects have an internal system of **tracheae,** air-filled tubes that branch throughout the body and carry oxygen to the cells. Specialized excretory tubules produce nearly solid wastes, minimizing the loss of body water. Since insects are basically terrestrial, they have internal fertilization. The eggs are laid with a waterproof covering, which protects them from dehydration. Females lay their eggs where the young will find food when they hatch. The young undergo a series of molts, and the higher insects pass through larval stages (such as caterpillars or maggots) that are completely different from the adult in appearance and way of life (Figure 21-26).

Why are most insects so small? The mechanics of an exoskeleton, and the fact that flight requires less energy with a lighter body, must impose some theoretical upper limit on the size of a flying insect. Nevertheless, modern insects do not reach this limit, and some extinct insects were much larger. The answer appears to be that small size permits insects to occupy habitats where vertebrates, the other main group of land animals, cannot compete with them.

Several species of insects have been domesticated, including honeybees, which produce honey and wax, and lac insects, source of the main ingredient in shellac. Silk, from the cocoon of the silkworm pupa, has been a major product of China for centuries. Cochineal insects are the source of a bright red dye that the Aztecs and Incas, and their Spanish conquerors, prized nearly as much as gold and silver. The star-spangled banner and the red coats of the invading British Army were both colored with this dye.

Crime investigators have also found uses for insects. A thirteenth-century Chinese law enforcement manual recounted how a murderer was caught when his sickle (the murder weapon) was the only one in the village to attract flies sensitive to the odor of decaying flesh. In New Zealand, the bodies of Asian insects in a load of confiscated marijuana provided the crucial evidence to con-

Insects in Our Environment

A JOURNEY INTO LIFE

Insects perform many roles vital to human life. Without bees and other insects, for instance, many flowering plants would never be pollinated—a prerequisite to producing crops such as apples, citrus fruits, berries, and cucumbers. Many beetles, ants, and flies are important decomposers, breaking down the dead bodies of plants and animals.

Nevertheless, people have devoted more time to killing insects than to praising them. It is perhaps unduly gloomy to conclude that we are losing the battle against insects, who will one day inherit the earth, but those who believe this have good reason for their opinion.

Insects attack human beings directly with bites and stings. Much more important, blood-sucking insects transmit many diseases. Malaria, river blindness, and sleeping sickness carried by insects blind and kill millions of people a year. Insects probably do more damage indirectly, however, by transmitting plant diseases, such as Dutch elm disease and many viral diseases of crop plants and by eating crops and killing trees (Figure 21-A). In the United States alone, during 1975 the gypsy moth, tussock moth, southern pine beetle, and spruce budworm destroyed enough forest trees to build nearly a million houses.

Insects destroy more than 10% of all crops grown in the United States, but the damage is even worse in the tropics, where hot weather throughout the year permits insects to grow and reproduce faster. In Kenya, officials estimate that insects destroy 75% of the nation's crops. A locust swarm in Africa may be 30 metres deep along a front 1500 metres long, and will consume every fragment of plant material in its path, leaving hundreds of square kilometres of country devastated.

Pesticides have not solved the insect problem. This is partly because pesticides act as selective pressures for the evolution of resistant strains of insects, which evolve too fast for expensive pesticide research to keep up. The list of pesticide-resistant insects nearly doubled between 1970 and 1980. Workers now direct much of their effort to using a combination of chemical and biological methods to control damaging outbreaks of insects. Biological controls include raising, sterilizing, and releasing large numbers of males (used for species in which females will mate only once), using sex attractant chemicals (pheromones) to attract males to traps instead of to females, breeding pest-resistant plants, and introducing specific predators and parasites of pest insects. Yet despite the fact that human beings have waged war on insects since the two have existed together, human efforts have apparently not succeeded in exterminating even a single species of unwanted insects.

(a)

(b)

Figure 21-A

Insect devastation. (a) The voracious caterpillar of the gypsy moth, a species imported into Massachusetts in the nineteenth century in hopes of starting a silk industry. Escapees were the ancestors of populations now found from Maine to Florida and coast to coast. During periodic population explosions, hordes of caterpillars strip leaves from millions of acres of forest. (b) A gypsy moth outbreak in the late 1980s killed the trees shown in the foreground, as well as the background forest, in Shenandoah National Park, Virginia.

vict drug dealers on charges of importation (a more serious offense than mere possession). Police have also found that the age of maggots, and the other kinds of insects present, can sometimes provide a remarkably accurate estimate of how long a corpse has been dead.

21-J Phylum Echinodermata

Sea stars, brittle stars, sea cucumbers, sea lilies, sea urchins, and sand dollars belong to the phylum **Echinodermata.** From looking at the adult animals, we would hardly guess that vertebrates are more closely related to echinoderms than to annelids, molluscs, and arthropods. Yet studies of embryonic development suggest that such is indeed the case.

The name Echinodermata, meaning "spiny-skinned," refers to the spines and plates of calcium carbonate that form a skeleton just under the skin in all members of the phylum. Another characteristic feature is suction-cup tube feet, used for locomotion, for gas exchange, and, in predatory forms, for feeding. Most adult echinoderms also have **pentaradial symmetry:** the body is divided into five parts around a central area where the mouth lies (Figure 21-27). All echinoderms are marine. Most are bottom-dwelling and able to move about slowly.

Most sea stars are carnivorous, using their tube feet to grip their prey. Those that are predators on bivalves can exert enough suction to pry open a very narrow slit between the valves. The animal then everts its stomach, which squeezes into the shell and digests the prey. The "crown of thorns" sea star feeds on cnidarian polyps and is notorious for the damage it does to such coral reefs as the Great Barrier Reef of Australia.

Sea urchins and sand dollars live mouth-down on the bottom of the sea, protected from intrusion by brittle calcareous (calcium carbonate) spines, which easily penetrate soft flesh and are very hard to remove (Figure 21-28a). The calcareous skeletal plates are fused into an envelope around the animal, pierced by holes for the mouth, anus, and tube feet. The sausage-shaped sea cucumbers also lack arms (Figure 21-28b). The mouth, at one end of the long body, is sometimes surrounded by modified tube feet called tentacles, used for filter feeding.

Figure 21-27

Echinoderm characteristics. (a) Sharp spines and flexible tube feet, with suction-cup tips. (b) A brittle star, showing the pentaradial symmetry characteristic of adult echinoderms. Brittle stars look much like sea stars, but their arms are sharply marked off from the central disk. They move by wriggling their arms rather than using their tube feet. Most eat organic detritus from the sea floor. (a, Charles Seaborn; b, Biophoto Associates)

(a)

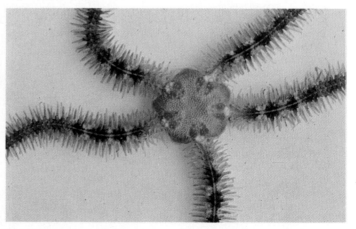

(b)

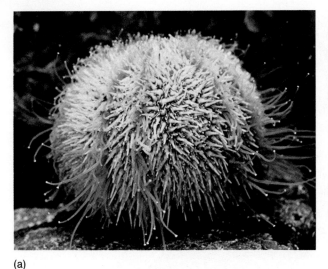

(a)

(b)

Figure 21-28
Echinoderms. (a) A sea urchin of the rocky shore. The gonads of this species are prized by gourmets. Long tube feet can be seen extended from the lower half of the body. (b) A sea cucumber. (Biophoto Associates)

21-K *Phylum Chordata*

The phylum **Chordata** contains some invertebrates and all the vertebrates. Although we have a remarkable fossil record of the evolution of vertebrates into a wide range of forms, details of the origin of the earliest chordates are lost, probably forever, owing to the poor fossil record from the Precambrian and Cambrian geological periods. The only way to reconstruct early vertebrate history is to study the invertebrate groups—especially the echinoderms and invertebrate chordates—most closely related to the vertebrates.

All chordates share several important features (Figure 21-29):

1. At some stage in the life history, all chordates have a stiff, rod-like **notochord,** which serves as an internal skeleton. In the embryonic development of vertebrates, the notochord is surrounded or replaced by a column of vertebrae that form the backbone.
2. At some time in their lives, all chordates have **pharyngeal gill slits** leading from the **pharynx,** the throat cavity behind the mouth, to the exterior.
3. The nerve cord forms a hollow tube running from head to tail on the dorsal side of the body, in contrast to most invertebrates, whose main nerve cord is solid and ventral (Figure 21-30 explains terms describing anatomical positions).

These three features define the chordates. But, in addition, most chordates have more or less segmented bodies, an endoskeleton (internal skeleton) and, at some stage of life, a tail that extends behind the anus.

These features comprise a particularly successful set of adaptations. The internal notochord, working with segmented blocks of muscle (see Figure

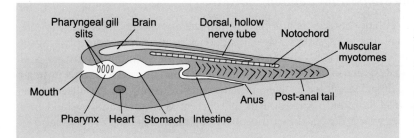

Figure 21-29
Chordate characteristics. This diagram of a generalized, primitive chordate shows the notochord, with the dorsal, hollow nerve tube above it; pharyngeal gill slits; segmentally arranged blocks of muscle (myotomes); and tail extending behind the anus.

Figure 21-30
Anatomical terms describing the relative positions of parts of an animal's body.

Anterior: of or toward the front
Posterior: of or toward the rear
Dorsal: of or toward the upper surface
Ventral: of or toward the lower surface
Caudal: toward the tail
Lateral: of the side
Proximal: of the part of an appendage nearer to the point of attachment to the body
Distal: of the part of an appendage farther from the body

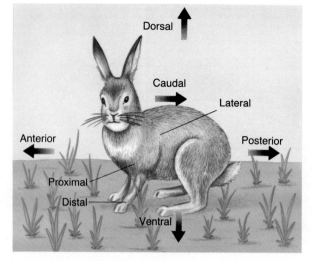

■ *Muscles acting on a rod along the animal's back made early chordates good swimmers.*

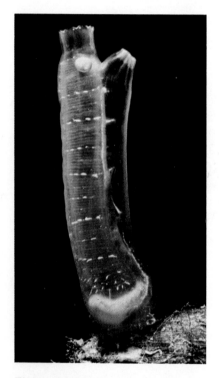

Figure 21-31
A **solitary sea squirt**. The striped "basket" filling the body is pharyngeal gill slits arranged to filter food out of the water. The pink object near the base is the stomach. Water enters the body through the large opening at the top and leaves via the transparent tube at the right. (Biophoto Associates)

21-29), allowed early chordates to swim quickly and efficiently by side-to-side wiggles of the body. As they swam forward, they took in food and water through the mouth and let the extra water escape through the gill slits. On its way out, the water gave up oxygen to the blood passing through the gills. Sense organs in the head detected where the animal was going and found food, and the brain and nervous system became well developed.

Chordate Subphylum Urochordata: Tunicates

Urochordates, the sea squirts and their relatives, are a group of entirely marine invertebrates. Most sea squirts are sessile filter feeders. Many live in colonies, which may share a common mouth. The tadpole-like larval stage could almost have posed for our drawing of a generalized chordate (see Figure 21-29). Upon hatching, it swims to the surface with efficient fish-like wriggles, using the action of its muscles against its notochord. It drifts a short distance in the plankton, turns, and swims down to search for a suitable rock or dock piling. Here it attaches by adhesive projections on the tip of its nose and metamorphoses into an adult, losing its notochord and tail, while the gill slits expand tremendously (Figure 21-31).

Since this larva is the most primitive known chordate, it looks as if the characters typical of chordates evolved, not as adaptations of an adult to its way of life, but in a larva. Most biologists think that vertebrates evolved from a tadpole-like creature that failed to metamorphose but became sexually mature while still a larva. This idea is supported by the facts that a number of living animals develop some degree of sexual maturity while still larvae, and that groups other than vertebrates also seem to have originated in this way.

Chordate Subphylum Cephalochordata

The only cephalochordate is amphioxus, the lancelet. Adults look like tunicate tadpoles with the gill system vastly expanded to form an enormous pharyngeal gill basket (Figure 21-32). This leaves little room for swimming muscles, and these animals swim poorly. An amphioxus lives buried in sand with only its head end protruding. Here it feeds, like a urochordate, by using cilia to pull a current of water into its mouth. Any food in the current is filtered and trapped in a mucous net which passes down the pharynx into the gut, where the food is digested.

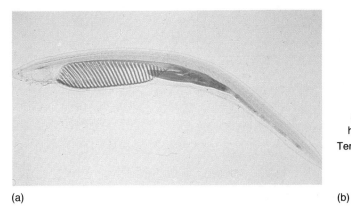

(a)

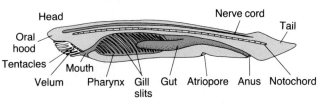

(b)

Vertebrates

Members of the subphylum Vertebrata differ from other chordates in having a backbone, the vertebral column, which replaces the notochord to a greater or lesser extent (Table 21-3). In addition to the basic chordate features listed before, all vertebrates have some sort of a liver, endocrine organs that secrete hormones, kidneys, a ventral heart, closed blood vessels, and some degree of segmentation. Cephalization is pronounced, with sense organs and nerves concentrated at the front end of the body so that vertebrates have very obvious heads. Early vertebrate fossils lacked movable jaws and probably fed like amphioxus.

The evolution of a backbone in vertebrates permitted rapid, efficient locomotion. Primitive vertebrates were fish that swam as sharks do today, by throwing the body into S-shaped curves, with the segmented muscles pulling against the vertebral column and pushing the paddle-like tail against the water. An animal that can move rapidly can be carnivorous, feeding on other animals, which are more nutritious than plants, weight for weight. Many vertebrate evolu-

Figure 21-32
Amphioxus. (a) Photograph of an amphioxus with the gill slits and the gut stained red. (b) Diagram showing the chordate characters of notochord, dorsal nerve cord, and pharyngeal gill slits. The tentacles waft food and water into the mouth. Excess water is pushed out through the gill slits into an atrium (not shown) and leaves the body via the atriopore. (a, Biophoto Associates)

■ *Vertebrates almost certainly evolved from a tadpole-like invertebrate larva.*

Table 21-3 The Classes of the Subphylum Vertebrata

Class Agnatha (~45 species)	Jawless fishes; gill openings separate; skeleton cartilaginous; notochord persists throughout life; marine and freshwater. Lampreys and hagfishes
Class Chondrichthyes (~275 species)	Cartilaginous fishes; cartilaginous skeletons; jaws; notochord replaced by vertebrae in the adult; gill openings separate; paired pectoral and pelvic fins; tail fin usually asymmetrical; most marine. Sharks, skates, and rays
Class Osteichthyes (~25,000 species)	Bony fishes; bony skeletons and jaws; gill openings all covered by a single operculum; paired pectoral and pelvic fins; tail fin usually symmetrical; many have a swimbladder; marine and freshwater; e.g., herring, salmon, sturgeon, eels, sea horse, electric eel
Class Amphibia (~2500 species)	Tetrapods that lay eggs without an amnion or shell; respiration via lungs and skin; scales absent; most freshwater or terrestrial. Salamanders, newts, frogs, and toads
Class Reptilia (~6000 species)	Tetrapods with amniotic eggs and scaly skin. Snakes and lizards, turtles, and crocodilians
Class Aves (~8600 species)	Birds. Tetrapods with feathers; oviparous, laying amniotic eggs; high body temperature; bipedal, most species have more than one mode of locomotion; forelimbs usually modified to form wings; e.g., sparrows, penguins, ostriches
Class Mammalia (~4400 species)	Tetrapods with young nourished by milk from mammary glands of females; most viviparous; high body temperature; body usually covered with hair; only one bone in each side of lower jaw, teeth differentiated and specialized. The monotremes (echidnas and platypus), marsupials (e.g., opossum, kangaroo), and placental mammals (e.g., humans, bats, whales, rodents, dogs, cattle, elephants)

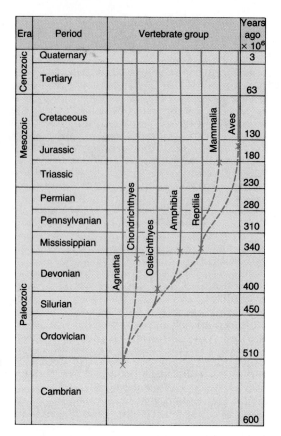

Figure 21-33
Fossil record and probable family tree of the classes of vertebrates. The crosses indicate when members of each group first appear in the fossil record. Note that birds are the class that originated most recently.

tionary advances were associated with a carnivorous way of life, starting with the adaptive radiation of the fishes (Figure 21-33).

21-L   Three Classes of Fishes

Class Agnatha: Jawless Fishes

The most primitive vertebrates were **ostracoderms,** agnathan fishes covered with heavy bony plates, known only as fossils. Their modern relatives are the jawless **cyclostomes** ("round mouths"), the lampreys and hagfishes. Adult lampreys and hagfishes are long, cylindrical creatures without paired fins. The adult lamprey is a semiparasite, and hagfishes are scavengers.

Adult lampreys have a sucking mouth and a rasping tongue covered with teeth, which are used to break the skin and suck the blood of bony fish (Figure 21-34). They do not usually kill their prey. The larva, called an **ammocoete,** lives for up to seven years as an amphioxus-like filter feeder buried in the mud of a stream. Its mouth and pharynx greatly resemble those of an amphioxus, except for one significant difference: in an amphioxus the water current is propelled by cilia on the gills, but in the ammocoete the gills have muscles that pump the feeding current, at a much faster rate. Hence an ammocoete takes in food faster, and it can grow to a greater size on filtered food.

Evolution of Fishes

Agnathans gave rise to two other large and successful groups of fishes: the **Chondrichthyes,** the sharks and rays, and the **Osteichthyes,** or bony fishes, such as salmon and haddock. These two groups made two major evolutionary

(a)

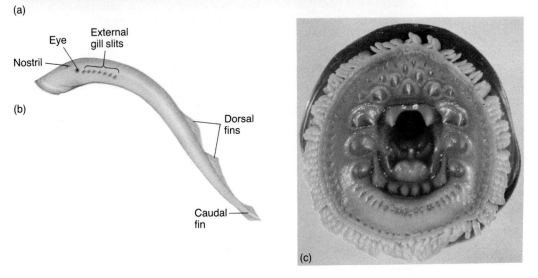

(b)

(c)

Figure 21-34
Lampreys. (a) Three on one: lampreys attached to a carp (bony fish) by their round, sucking mouths. Their attack will weaken the fish, but usually not kill it. (b) Drawing to show the external gill slits and unpaired fins. (c) The round, suction-cup mouth contains many rasping teeth. (a, Tom Stack/Tom Stack & Associates; c, courtesy of Dr. Kiyoko Uehara)

advances over their agnathan ancestors, increasing their efficiency as fast-moving carnivores. First, part of the gill skeleton moved forward and evolved into **jaws,** permitting the fish to bite and chew its food instead of sucking or filtering it. Second, both groups have two pairs of **lateral fins: pectoral fins** at the front and **pelvic fins** at the back, as well as other, unpaired fins (Figure 21-35). Paired fins allowed these fishes to balance and maneuver in new ways. Paired fins also played an important role in evolution: they eventually gave rise to the paired forelimbs and hindlimbs of terrestrial vertebrates.

Class Chondrichthyes: Cartilaginous Fishes (Sharks and Rays)

Sharks, dogfish, rays, and skates have skeletons composed entirely of cartilage, rather than bone, and almost all are marine. Sharks swim in primitive vertebrate fashion, by sinuous waves of the body, using their segmental muscles and jointed backbones.

The sharks are notorious carnivores and scavengers. One killed in the Adriatic Sea had in its stomach two raincoats, part of a horse, an automobile license

SHARK

Gill slit — Dorsal fins

Lateral line

Tail fin

Pectoral fin — Stomach — Spiral valve — Pelvic fin — Anal fin

Liver — Intestine

BONY FISH

Operculum — Dorsal fins

Lateral line

Tail fin

Pectoral fin — Stomach — Intestine — Anal fin

Pelvic fin — Swimbladder

Figure 21-35

Comparison of a shark and a bony fish. Externally, the shark has a row of separate gill slits, while the gills of the bony fish are covered by a common operculum with a single opening at its rear edge. Both fish have dorsal, anal, and tail fins, and paired pectoral and pelvic fins. Note the characteristic asymmetry of the shark's tail fin. Compared to the shark, the pelvic fins of the bony fish are placed far forward. Internally, most of the body consists of swimming muscles, and the body cavity is relatively small. The shark's large, fat-filled liver and the bony fish's swimbladder both provide buoyancy. A bony fish has a long, thin intestine, whose walls provide a large surface area for absorption of digested food into the body. The shark's intestine is shorter but its internal absorptive surface is increased by a winding spiral valve.

plate, and a length of rope. The stomach of a small dogfish shark, such as you might dissect in the laboratory, is more likely to contain crustaceans and bony fish. The two largest sharks (like the largest whales) are not predaceous but gentle filter feeders. The whale shark lives mainly on plankton, filtering more than a million litres of water an hour.

The flattened skates and rays, which live on the sea floor, feed mostly on invertebrates. Their pectoral fins are greatly enlarged and are used for locomotion (Figure 21-36b). Some rays have poison spines on the back or tail, which they use to defend themselves. The electric ray repels intruders with an organ that can produce quite a powerful electric shock.

Class Osteichthyes: Bony Fishes

The number of species of Chondrichthyes has declined since the Permian Period, but the bony fishes are still expanding and diversifying, thanks to their versatile anatomy and physiology. Most modern bony fishes are members of one very successful group, the **teleosts.**

Figure 21-36

Class Chondrichthyes. (a) A white tip reef shark. Note the separate gill slits, asymmetrical tail, and triangular dorsal fin. (b) An eagle ray. Note the row of gill slits on the ventral surface of the body and the greatly enlarged, wing-like pectoral fins. (Biophoto Associates, N.H.P.A.)

(a)

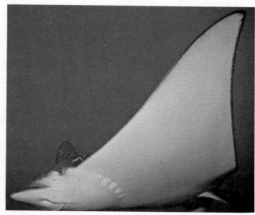

(b)

(a)

(b)

Figure 21-37
Class Osteichthyes. (a) Note the placement of the pelvic fins far forward on the body compared to those of the shark in Figure 21-36. (b) Puffer fish are protected not only by their spines, which stick out when the fish puffs up, but also by tetrodotoxin, a nerve poison so potent that less than a microgram will kill a human being. The poison is secreted into the jelly surrounding the eggs, and so is also found in the female's ovaries. Puffer fish are considered a delicacy in Japan. Chefs must be specially licensed to prepare the fish so that no tetrodotoxin is served to diners; nevertheless, fatal accidents sometimes occur. (Biophoto Associates, N.H.P.A.)

The diversity among teleosts is almost as amazing as that of the insects. Among them are filter feeders like herrings, parrot-fish that crunch up coral, insect-eaters like trout, and predaceous carnivores like barracuda and blennies. Teleosts come in all shapes, sizes, and colors. Boxfish are nearly spherical, moray eels are snake-like, and a stonefish looks like a lump of rock.

The impressive adaptive radiation of teleosts depends largely on a few evolutionary innovations. One of the most important was the **swimbladder,** a gas-filled sac formed as an outgrowth of the pharynx. By altering the gas pressure in the bladder, a fish can alter its buoyancy so that it floats at any depth in the water without exerting its muscles. In some bony fish, the swimbladder is used, like a lung, to breathe.

In most teleosts, the tail provides much of the push during swimming; the paired fins provide fine control. The pelvic fins are usually farther forward and higher on the body than those of a shark (Figure 21-37). You can always tell a bony fish from a shark because the gills of a bony fish do not open separately to the exterior, as do those of a cartilaginous fish. Instead, the bony fish's gills are all covered by a common operculum. Water for respiration moves in through the mouth and out through the gills, pumped by muscles in the head and at the base of the operculum. The gills of bony fishes also have osmotic adaptations that help to control the body's water content (Section 26-B). Bony fishes occur in both fresh and salt water, and forms such as trout, salmon, and eels can move from one to the other.

Many teleosts have sharp, protective spines on their dorsal fins. In some species, these spines are connected to poison glands and can inject poisons powerful enough to kill human beings or large fish.

Many deep-sea fishes are luminescent. Light flashes are probably used to signal the opposite sex and to startle attackers. In the deep-sea angler fish, the luminous tip of a fin is used as a lure to attract prey. In addition, many fish can change color; tiny muscles in the skin alter the sizes of different chromato-phores (color cells).

When the surrounding water is low in oxygen, some fish come to the sur-face and gulp air into their swimbladders. Other fish have lungs or other ar-rangements for getting oxygen from air. It is not difficult to imagine that some air-breathing fish, using their pectoral and pelvic fins to move about on land, were the ancestors of the first terrestrial vertebrates. In fact, some modern fish do leave the water and crawl about on land.

21-M The Move to Land: Tetrapods

Many selective pressures probably contributed to the evolution of vertebrates that could live, at least part-time, on land. In the Devonian Period, the seas teemed with carnivorous fish, and any fish that could remove itself or its eggs onto land would lower its mortality rate impressively. Plant life on land was well established, and terrestrial insects were evolving and multiplying rapidly. A ver-tebrate that ate plants or insects, and that could survive on land, would have had little competition for food. In addition, air contains more oxygen than water does. However, any fish that survives on land for even a short time must have adaptations to existing surrounded by air.

To support the body on land, a fish needs sturdy bones, especially in its pectoral and pelvic fins and in its backbone, along with strong muscles to move these bones. Furthermore, gills cannot be used to breathe on land. The surface tension of water makes the feathery gill filaments stick together when a fish comes out of water into the air. A respiratory surface that keeps its shape in air is necessary.

The body surface must be waterproofed to reduce dehydration, but this is not enough. The respiratory surface must be kept moist because gases can cross plasma membranes only in solution. Internal respiratory surfaces—lungs and swimbladders—lose less water by evaporation than do gills.

The fossil record shows that there evolved in the Carboniferous Period, about 300 million years ago, a number of animals that looked much like modern lungfish. These creatures had fish-shaped bodies, short stubby legs, and no gills. They were the first land vertebrates, ancestors of modern amphibians and rep-tiles.

Modern land vertebrates—amphibians, reptiles, birds, and mammals—are called tetrapods ("four feet"). Only reptiles, birds, and mammals are fully adapted to life on land. Most amphibians are still largely dependent on water.

Class Amphibia: Frogs, Salamanders, Toads, Newts

The amphibians are still tied to water because their eggs dry out easily and because most still have a fully aquatic larval stage. Most amphibians must return to water to reproduce (Figure 21-38).

The two largest groups of living amphibians are the **urodeles** (newts, sala-manders, hell-bender, mud puppy) and the **anurans** (frogs and toads). The urodeles are more generalized and show the transition from fish to tetrapod more clearly. Their limbs contain small bones and muscles, like those found at the base of the pectoral and pelvic fins of lungfish. In addition, they have pecto-ral and pelvic limb girdles (shoulder and hip girdles), which eventually evolved until they formed a strut between the backbone and limbs in all higher verte-brates. Adult anurans have very specialized skeletons, with shortened back-bones, loss of the tail, limb girdles firmly attached to the backbone, and leg bones and muscles developed for jumping (Figure 21-39).

Amphibians have a soft glandular skin, which is used for gas exchange in most species, despite the fact that most amphibians also have small lungs. (The

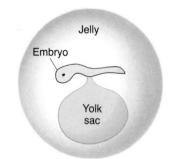

Figure 21-38

The egg of an amphibian (or of a teleost). The embryo obtains its food from the yolk sac and is protected and buoyed up by the jelly coat. The egg lacks the membranes that protect the embryos of terrestrial vertebrates (see Figure 21-40).

(a)

(b)

Figure 21-39
Amphibians. (a) A fire salamander, a urodele, looks very similar to a lizard, which is a reptile. The salamander's skin is moist, but a lizard's is covered with dry scales. (b) A Central American tree frog, an anuran. Note the long hind legs adapted for jumping and the absence of a tail, both typical features of adult anurans. (a, Biophoto Associates, N.H.P.A.; b, Mark Rausher)

aquatic larvae have gills.) Amphibians were also the first vertebrates with true tongues. In most frogs and toads, the tongue is long and sticky and can be shot out rapidly to catch flies.

Biologists are concerned because populations of most amphibians in North America (and many elsewhere) have declined dramatically since about 1975, and no one knows why. Since so many species are affected, some environmental change must be to blame. The increase in acid rain and in periods of drought, human destruction of habitat, stocking of lakes with fish that eat tadpoles, and accumulation of toxic chemicals are all known or suspected threats to one species or another of vanishing amphibians.

Class Reptilia: Lizards, Snakes, Turtles, Crocodiles

The Mesozoic Era is sometimes known as the "age of reptiles" because reptiles of all shapes and sizes are the main animals found in marine, freshwater, and terrestrial fossil beds laid down during this time. With the insects, reptiles dominated animal life on land for about 200 million years and are still very much with us today.

Reptiles are better adapted to life on land than are amphibians. Their main advantage is an egg that can be laid on land because it is protected from dehydration. Reptiles, birds, and a few mammals lay **amniotic eggs** (Figure 21-40).

Figure 21-40
The amniotic egg. (a) The embryo lies in a fluid-filled sac formed by a membrane, the amnion. The yolk sac provides the embryo's food, and the allantois holds its wastes. The chorion surrounds all these, and outside of all is the waterproof egg shell. (b) A snake hatching. Unlike the hard, brittle shells of birds' eggs, this egg's shell is leathery. (Biophoto Associates)

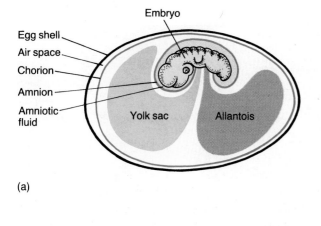

Embryo
Egg shell
Air space
Chorion
Amnion
Amniotic fluid
Yolk sac
Allantois

(a)

(b)

The developing embryo is surrounded by a membrane, the **amnion,** enclosing amniotic fluid, which protects the embryo from dehydration and from being jolted around. Two membranous sacs are attached to the embryo. The **yolk sac** contains yolk, the embryo's food. The **allantois** stores the embryo's nitrogenous waste until hatching and is found only in reptiles, birds, and mammals. Blood vessels grow out from the embryo through the membranes of the yolk sac and allantois until they come close to the surface of the egg, where they take in oxygen from the environment and release carbon dioxide. The embryo, amnion, yolk sac, and allantois are all surrounded by a membrane called the **chorion,** which controls the overall permeability of the egg. The egg is permeable to gases, but relatively impermeable to water. Around the chorion is the outer egg shell. Because the egg is laid in a leathery shell or the young develop within the mother's body, reptiles have internal fertilization, and the male has a penis (or even two).

The first reptiles were carnivores that looked rather like small dogs. Their limbs were stronger and were tucked further under their bodies than those of amphibians, enabling them to move faster. Their jaws were more firmly attached to the skull, permitting them to subdue and eat larger prey, and their skins were waterproofed and scaly, minimizing water loss. Without a moist skin, reptiles had to breathe entirely with their lungs.

The adaptive radiation of the early reptiles is a fascinating story. There were reptiles that swam, walked, and flew—some the size of small airplanes. The dog-like ancestors of the mammals became fast-running carnivorous quadrupeds (animals walking on four legs). Many members of the group that gave rise to the birds had a tendency to bipedalism (walking on two legs) and had reduced forelimbs. Some reptiles were insect-eaters, some carnivores, and some placid herbivores of enormous size.

There is still some mystery about the extinction of many of the reptiles. Large numbers of them, including all the dinosaurs and all the flying reptiles, disappeared from the fossil record during a short time at the end of the Cretaceous Period. Some people think these extinctions resulted from competition with early mammals. Others speculate that they were caused by climate changes, perhaps resulting from the collision of a large meteorite with the earth. This would have sent huge dust clouds into the atmosphere, cutting off sunlight from the earth's surface, which, in turn, would have both killed many plants and cooled the global climate. It is quite likely that more than one factor contributed to these extinctions.

It is incorrect to call reptiles "cold-blooded." Most maintain a body temperature considerably higher than their surroundings, but they are **ectothermic,** taking most of their heat from the environment. Many reptiles must lie in the sun before they warm up enough to be active. Birds and mammals, on the other

Figure 21-41

Lizards and snakes. (a) This lizard shows the dry scales and clawed toes characteristic of reptiles. (b) A snake skeleton. The many vertebrae, with their attached ribs free at the distal end, contribute to the flexibility of the snake's body. The loosely hinged jaw can open wide enough to let the snake swallow prey bigger around than the snake itself. (Biophoto Associates, N.H.P.A.)

(a)

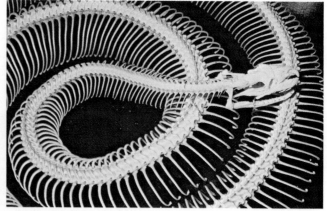

(b)

(a)

(b)

Figure 21-42
More reptiles. (a) An alligator. The many teeth are of various sizes but all the same conical shape. (b) A green sea turtle. This species is almost extinct, killed on its nesting beaches to make turtle soup, oil for cosmetics, and leather for belts and handbags.

hand, are **endothermic,** generating most of their heat by their metabolism. Thus they can be active at any time, which gives them an enormous advantage over reptiles.

Of the vast array of prehistoric reptiles, only three groups have members alive today: lizards and snakes; turtles and tortoises; and the crocodile clan.

The lizards and snakes are the modern reptiles most like their prehistoric ancestors. Lizards are easy to mistake for salamanders, which are typical amphibians. Snakes have highly specialized anatomy. Most species move by using their muscles to throw the body into curves. Scales on the ventral surface, or the curves of the body itself, provide traction. The group includes expert swimmers, burrowers, and tree climbers. The backbone is greatly elongated, and most of the vertebrae bear long, flexible ribs that hold the body in shape (Figure 21-41).

A snake's tongue flicks in and out, carrying chemicals from the air or ground to sense organs located in the roof of the mouth. Although snakes' ears have no external openings, they are well developed, responding mainly to vibrations of the ground detected through the lower jaw. Pit vipers and some boas also have heat-detecting organs on the head, which allow them to strike warm-blooded prey accurately on dark nights or in deep burrows.

Crocodiles and their kin are the closest living relatives of the extinct ruling reptiles and of their descendants, the birds. Although crocodiles are quadrupeds, they plainly belong to this line of evolution toward bipedalism because their hind limbs are longer than their forelimbs. Crocodiles, alligators, and gavials all spend much of their time in water, and have a special arrangement of their nostrils that permits them to breathe while the rest of the body is submerged (Figure 21-42). All are carnivorous.

The turtles, terrapins, and tortoises are one of the most ancient reptilian groups, specialized by the development of a protective bony or leathery shell. Most are herbivorous. Various species are adapted to life on land, in fresh water, and in the sea. Sea turtles are famous for their annual migrations to the beaches where they lay their eggs, and all species are endangered, many of them on the verge of extinction (Figure 21-42).

Birds and mammals both originated from different, early groups of reptiles.

Class Aves: The Birds

Birds can be simply defined as the only organisms with feathers. In addition, all birds are oviparous, meaning the females lay eggs. (In contrast, many female fishes and reptiles retain their eggs in their bodies until the embryo is developed enough to survive on its own.)

Figure 21-43
Archaeopteryx, **the first known bird**. This drawing is based on fossil skeletons and feather imprints (see Figure 14-6). *Archaeopteryx* was the size of a pigeon. Note the heavy tail, the teeth, and the claws on the digits in the wing. These features are not present in modern birds (except that in one modern species the young have claws on some wing digits).

The earliest known bird, *Archaeopteryx,* lived 150 million years ago, in the Jurassic Period. *Archaeopteryx* was a bird because it had feathers, but in other ways it looked much more like one of the small, bipedal dinosaurs from which the birds presumably originated. *Archaeopteryx* had a long tail, with separate vertebrae (modern birds have a greatly reduced tail skeleton), teeth (birds have a beak), and small wings with claws on the ends of the toes (the forelimb skeleton had not completed the change from legs into wings) (Figure 21-43).

How did bird flight begin? One theory holds that birds used their forelimbs at first to stabilize themselves when jumping from branch to branch, and later as parachutes (as in flying lizards and flying squirrels). Another theory holds that birds were bipedal insectivores, running along the ground catching insects and waving their arms to jump higher after escaping prey. A third theory merges the first two.

■ *Feathers permit birds to fly while retaining other means of locomotion. They also contribute to endothermy, which has permitted birds to colonize most parts of the world.*

Birds owe much of their success to feathers, a remarkable evolutionary innovation. For their weight, feathers are among the strongest known materials, and they also provide flexibility, excellent insulation, and an admirable covering for a flying surface (Figure 21-44). In the other flying vertebrates, the bats, the web of skin that forms the wing stretches from forelimb to hindlimb. Birds use only the forelimbs for flight. The legs are free to be used for running or swimming, and nearly all birds have two different types of locomotion (such as flying and swimming, running and flying). The colorful feathers of many birds are used in courtship displays and for other social signals.

The strongly social nature of many birds and their complex behavior patterns result from two main factors. First, a bird's brain is quite large and complex, as it must be to control the intricate muscular movements of flying. Second, birds are endothermic, and they use a lot of energy to generate heat for themselves and for their eggs, which must also be kept warm. Therefore birds must feed more often than most reptiles, and they must take time off from babysitting to do so. As a result, parent birds usually collaborate in building nests, incubating eggs, and feeding the young (Figure 21-45).

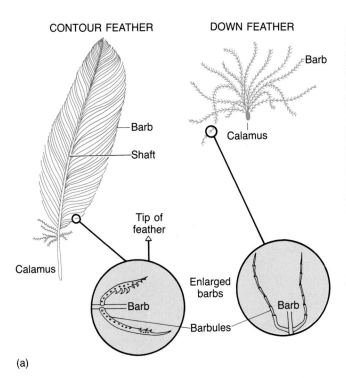

CONTOUR FEATHER

DOWN FEATHER

(a)

(b)

Figure 21-44

Feathers. (a) Contour feathers provide a smooth, streamlined surface for flight. When the bird preens, it draws the barbs through its bill from base to tip. As a result, the hooked barbules on one barb hook onto the ridged barbules on the next, linking the barbs into a smooth but flexible surface. Down feathers are the last word in insulation. Their long, unhooked barbules provide a mass of spaces where air is trapped and warmed by the body. They are the first feathers to develop on young birds, such as the masked booby chick (b), and they underlie the contour feathers on parts of the body of most adult birds. Contour feathers have a bit of down at the base; thus they supply insulation as well as a smooth body surface.

Figure 21-45

Birds. (a) Family life is important among the birds. These blue tits share the task of feeding their nestlings. (b) Most birds can fly, as demonstrated by this white pelican. (a, Biophoto Associates, N.H.P.A.)

(a)

(b)

Bird anatomy is conservative, without the great range of structural modifications found in other vertebrate classes. This is undoubtedly because most birds fly, and flight is structurally demanding. Bird flight ranges from flapping flight, such as that of sparrows, robins, chickadees, and so forth, to the soaring flight of hawks, vultures, and albatrosses. Soaring birds ride the thermal currents in the air, like a glider (though more efficiently), and flap their wings infrequently.

Birds that live on land usually eat seeds and fruit (parrots, fowl, grosbeaks), insects and their larvae (thrushes, swifts, woodpeckers), smaller vertebrates (owls, eagles, hawks), or carrion (vultures and crows). Many birds find most of their food in water, either by wading (sandpipers and herons) or by swimming and diving, with feet modified as paddles (gulls, pelicans, ducks, geese, and cormorants). Penguins, the most highly specialized water birds, cannot fly because their wings as well as their feet are modified as paddles. The other flightless birds are typified by ostriches, which rely on their running speed to escape predators.

Class Mammalia

Mammals originated from early reptiles some 200 million years ago, before the first birds. Early mammals were about the size of small mice, with teeth adapted to eat mainly insects. Their large eye sockets suggest that they were nocturnal (active at night). Well on the way to becoming endothermic, they could be active during the cooler night, avoiding competition with reptiles.

A burst of mammalian evolution followed the extinction of many reptile species over a span of several million years at the end of the Cretaceous Period (see Figure 18-5). Many forms became larger, and mammals came to exploit many of the resources formerly monopolized by reptiles. Two features—fast quadrupedal locomotion and new adaptations to carnivory—are the secrets of mammalian success.

Most mammals are **viviparous** ("alive-bearing")—the young develop in the mother's uterus, nourished and supplied with oxygen by her blood, which flows through vessels close to those of the embryo. Viviparity permits the mammalian mother to remain mobile while incubating embryos that must be warm to survive. All female mammals nourish their young with milk produced in mammary glands.

Mammals, like birds, are endothermic. The body is insulated by hair or fur and by a layer of fat beneath the skin. Endothermy gives a carnivore an enormous advantage, making it ready for action at all times. It also permits birds and mammals to live in extreme temperatures that other land vertebrates cannot survive: penguins and polar bears inhabit polar areas, and camels and vultures are among the few animals active at noon in the desert.

An additional reason for mammalian success is the **integument,** consisting of the skin and associated structures. The integument helps control body temperature. The hair on the skin, and the fat under it, provide insulation, and the sweat glands permit cooling. In many species, individuals communicate via chemical signals produced by glands in the skin. Other integumentary structures include claws, nails, hoofs, horns, and antlers.

Another mammalian advance was the evolution of specialized teeth. Whereas reptiles and fish have teeth all roughly the same size and more or less conical, early mammals evolved different kinds of teeth: chisel-like incisors for cutting, pointed canines for gripping and tearing, and grindstone-like molars for crushing and breaking. During vertebrate evolution, the number of separate bones that make up the lower jaw has been steadily reduced until mammals have just one bone on each side (the **mandible**). These alterations in the jaw and teeth permit a carnivore to grip a struggling victim more firmly, with less chance of damaging the jaw (Figure 21-46).

■ *The successful basic mammalian plan is of a fast-moving, endothermic carnivore, protecting its developing young within the female's body.*

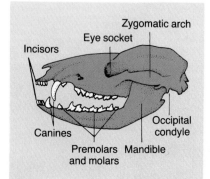

Figure 21-46
Specializations of mammalian teeth. The jaws contain teeth of different sizes and shapes, specialized to perform different functions. The zygomatic arch protects the eye from being jogged by the jaw muscles. The backbone attaches to the skull at the occipital condyle.

(a)

(b)

Figure 21-47
Marsupials. (a) The koala, a native of Australia, lives in trees and feeds only on certain species of eucalyptus. (b) These young opossums are completing their development attached to a teat in the marsupium, or pouch, of their mother. (b, Carolina Biological Supply Company)

Although the earliest mammalian design is that of a fast, quadrupedal carnivore, many modern mammals are neither fast, quadrupedal, nor carnivorous. The success of this group led to extensive adaptive radiation, during which herbivory, bipedalism, flight, and many other adaptations evolved.

Monotremes The monotremes are one of the three groups of modern mammals. The only three species now living are the duck-billed platypus and two kinds of spiny anteaters (echidnas). Unlike other mammals, monotremes lay eggs, which are very similar to reptilian eggs. Females suckle the hatched young with milk from mammary glands in typical mammalian fashion.

Marsupials Marsupials are mammals whose young are born at an early stage of development and finish developing in a pouch, or marsupium. They live only in Australia and America, although fossil evidence shows that they once inhabited Europe. The most familiar American marsupial is the opossum, which has colonized North America from Central America. Other marsupials live in Central and South America, but the greatest variety, including kangaroos and the koala, inhabit Australia (Figure 21-47).

Placental Mammals Most modern mammals are placental. The egg has hardly any yolk, and the first organ to form during embryonic development is a **placenta,** formed from tissues of both mother and embryo. Two membranes found in reptile eggs, the chorion and allantois, evolved in mammals into the embryonic part of the placenta. This grows into the wall of the mother's uterus, and its blood vessels carry food and gases between the embryo and the mother's blood.

Hedgehogs, shrews, and moles are members of the order **Insectivora,** small active mammals that feed on insects and other invertebrates (Figure 21-48). Because their small bodies lose heat rapidly, the insectivores must eat almost constantly to provide metabolic fuel. Insectivores evolved early in mam-

Figure 21-48
Order Insectivora. The eyes of this short-tailed shrew are so small that they cannot be seen, and there are no external ears, adaptations to a life spent in tunnels. The shrew preys on insects and other invertebrates, and its saliva is poisonous.

malian history and have changed very little since the Cretaceous Period. Some early insectivores are believed to have given rise to the bats and primates.

Bats are the only mammals with true flapping flight (Figure 21-49). Their wings consist of a web of skin stretched over very long thin bones of the fore-limb digits and attached also to the hind limbs. Bat flight is slower but more maneuverable than that of most birds, enabling most species of bats to live by catching insects. Most bats are nocturnal. They navigate and catch food in the dark using an echolocation system, in which they detect echoes of their own voices that have bounced off other objects. These habits reduce competition between bats and birds, most of which fly by day, using vision as the primary sense. Surprisingly, the 850 species of bats account for 22% of the species of mammals. Adaptive radiation among bats has produced forms that catch fish or frogs, blood-feeding vampires, pollen and nectar feeders, and large diurnal fruit eaters with good vision but lacking echolocation.

Primates include lemurs, monkeys, apes, and humans. They retain many features of early mammals, with additional adaptations to living in trees (Section 21-N).

Figure 21-49
Flying bats. The wings are made up of skin stretched over the elongated fingers and extending to the hind limbs. (Dr. Merlin Tuttle/Milwaukee Public Museum/Photo Researchers)

Figure 21-51
A cetacean. This Atlantic bottle-nosed dolphin, a toothed whale, shows many adaptations to a totally aquatic existence: the blowhole (nostril) in the top of the head, the streamlined, fish-shaped body, the dorsal fin, and limbs modified as flippers. (Paul Feeny)

Figure 21-50
Order Rodentia. A mouse. (Biophoto Associates)

Rodents—mice, squirrels, and their relatives, with gnawing teeth—are the most widespread and generally successful modern mammals apart from humans (Figure 21-50). Most rodents have remained small and reproduce very rapidly.

Rabbits, hares, and **pikas** belong to another group with rodent-like features.

Cetaceans are mammals highly adapted to a permanent life in the sea. Toothed whales, including porpoises, sperm whales, and killer whales, feed mainly on fish and large invertebrates (Figure 21-51). Baleen whales are filter feeders. They engulf huge mouthfuls of plankton and water and use the sieve of whalebone or baleen plates lining the jaws to trap planktonic crustaceans, while their huge tongues push the water out. Surprisingly, these are the largest whales, and include the blue whale, the largest animal that has ever lived. Whales are intelligent, sociable animals, able to communicate with one another by sound. Echolocation helps them identify food and other objects in the water.

The most specialized mammalian hunters belong to the order **Carnivora:** the cats, dogs, skunks, bears, and so on (Figure 21-52). Their behavior is complicated and involves the ability to learn much of their hunting skill.

The **ungulates,** mammals that walk on the tips of their toes, are the ultimate vertebrate herbivores. One group includes the elephants, and two others contain the hoofed mammals, divided according to whether they walk on an even number of toes (deer, cattle, and so forth) or an odd number (horses,

Figure 21-52
Order Carnivora. Lionesses sharing a kill with their young. (Biophoto Associates)

Figure 21-53
Ungulates. (a) A mother elephant pauses to graze while her calf nurses and other members of the herd forage nearby. (b) Odd-toed ungulates: zebras form herds, where some individuals keep a lookout for predators while others graze. (c) This male elk, an even-toed ungulate, displays an impressive pair of antlers.

rhinoceroses, and tapirs) (Figure 21-53). Because their teeth are flattened and used to crush and grind tough plant material, many ungulates are not well equipped to fight a potential predator. Most rely on running fast to escape their enemies, and they also tend to feed in herds, where every animal watches out for danger.

The even-toed ungulates owe much of their evolutionary success to a digestive system in which bacteria break down plant cellulose (Section 22-D). In addition, most are keen of sense and fleet of foot. Their adaptive radiation is impressive: they have given rise to such diverse forms as pigs, hippopotamuses, and camels. Cattle, sheep, and deer often have horns or antlers, which probably evolved as defensive weapons, although escape is usually the preferred course of action.

21-N   *Human Evolution*

Human beings belong to the species *Homo sapiens,* in the mammalian order Primates. Primates can be divided into the prosimians and the anthropoids. Prosimians ("before monkeys") include tree shrews, tarsiers, lemurs, and lorises. Anthropoids are the monkeys, apes, and humans.

The most notable characteristics of primates are adaptations to **arboreal** (tree-dwelling) life, although humans and some other primates have become ground-dwelling. A major feature is the enormous expansion of the brain, especially the frontal lobes of the cerebral hemispheres, the seat of intelligent behavior and an important area in the control of muscular dexterity (Figure 21-54). This is important to an arboreal animal, which must climb and leap among the branches of trees.

In addition, birds and monkeys that must jump and land on branches usually have excellent vision. In most primates, both eyes face forward and therefore see the same thing. The superimposed images from the two eyes provide **stereoscopic** (three-dimensional) vision. (It is difficult to judge the distance to an object with one eye closed.)

Primates have five digits (fingers or toes) on each limb. One digit (our thumbs) can touch or nearly touch the other four. This enables the feet to grasp tree limbs, food, and other objects. The digits end in sensitive tips, often covered by flattened nails rather than the curved claws of most other mammals (Figure 21-55). The rest of the primate skeleton is much like that of early ancestral mammals. In contrast, many other modern mammals, such as deer with their antlers and hoofs, whales with their flippers, or bats with their wings, have highly specialized skeletons.

Our Primate Relatives

Early in primate history, a mouse-like prosimian started living in trees (Table 21-4). The most advanced prosimian is the spectral tarsier, an arboreal, nocturnal creature with huge eyes, stereoscopic vision, and nails instead of claws (Fig-

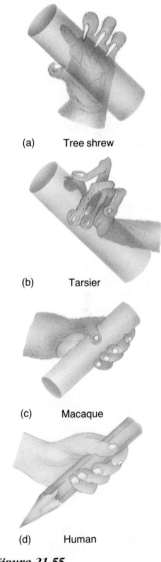

(a)   Tree shrew

(b)   Tarsier

(c)   Macaque

(d)   Human

Figure 21-55
Hands of primates. This series of drawings shows the evolutionary trend from relatively immobile digits with claws to the human hand, with a thumb that can touch the other four fingers and with fingertips protected by nails.

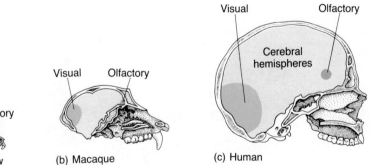

(a) Tree shrew   (b) Macaque   (c) Human

Figure 21-54
Evolution of the primate brain. The cerebral hemispheres (yellow) and the visual center of the brain expanded during primate evolution. By comparison, the olfactory area, which governs the sense of smell, remained small. Tree shrews are arboreal prosimians. Macaques are ground-dwelling monkeys.

Table 21-4   *Classification of the Order Primates*

Suborder Prosimii ("before monkeys")
  Tree shrews, lemurs, lorises, bush baby, tarsier
Suborder Anthropoidea: Monkeys, apes, humans
  Superfamily Ceboidea: American monkeys, including marmosets, capuchin
  Superfamily Cercopithecoidea: Eurasian and African monkeys, including
    macaques, baboons
  Superfamily Hominoidea
    Family Pongidae. The anthropoid apes:
      Gibbons, orang-utans, gorillas, chimpanzees
    Family Hominidae:
      Australopithecus (extinct prehumans)
      Homo habilis, H. erectus, H. sapiens

■ ***Primate features important to human evolution include an unspecialized skeleton, stereoscopic vision, and mobile fingers.***

Figure 21-56
Primates. (a) A spectral tarsier, a small, nocturnal, tree-dwelling prosimian from southeast Asia. (b) An African olive baboon sitting upright so that its hands are free to handle a steinbok (a small antelope) it has caught. Although most primates are chiefly vegetarian, many consume small amounts of meat, which provides them with B vitamins. (a, Gary Milburn/Tom Stack & Associates; b, Timothy W. Ransom/BPS)

ure 21-56a). In addition, the upper lip is free of the gums. This feature is a large part of the reason higher primates have such mobile and expressive faces. We use facial expressions for communication. This reflects the fact that vision is our main sense, rather than smell, which is used in communication by most other mammals. During primate evolution, the snout (nose) became progressively reduced. This is probably an adaptation that gives the forward-looking eyes a clear view of the world.

Monkeys, apes, and humans are the anthropoid primates. An obvious feature of anthropoids is their upright posture. Even quadrupedal monkeys such as baboons sit upright for long periods, freeing their hands to manipulate food, handle their young, and perform other tasks (Figure 21-56b). Some arboreal monkeys also spend long periods in a vertical position as they swing through the trees by their arms, a type of locomotion called **brachiation.**

Monkeys are similar to most other mammals in their basically quadrupedal locomotion, in the proportions of their limbs, and in the compression of their rib cages from side to side. In contrast, the arms of apes are long compared with their hind limbs (adapted to brachiation?) and their rib cages are flattened from front to back. Apes' tails are reduced to the few fused vertebrae of the coccyx.

(a)

(b)

The loss of the tail makes sitting upright more comfortable. The brains of apes are also relatively larger than those of monkeys.

There are only four genera of modern apes: gibbon, orang-utan, gorilla, and chimpanzee. All live in Africa and Asia, and their structure and behavior bridge the gap between monkeys and humans.

There are more similarities than differences between apes and hominids (members of the human family). Nevertheless, the four genera of modern apes share some features not found in hominids. For instance, these apes have arms and backbones more adapted to brachiation, powerful canine teeth, large incisors, and grasping feet, features not shared by humans. Gorillas and chimpanzees spend a lot of time on the ground, walking on their hind feet and the knuckles of their hands. This allows them to use their fingers to carry things.

It is clear that chimpanzees are our nearest living relatives (Figure 21-57). Not only is there a strong physical resemblance, but human and chimpanzee proteins and genes are about 99% similar. We are not as closely related to gorillas or orang-utans.

Human Origins

Few areas of research have produced as much argument and confusion as the search for the fossils of our ancestors. This is partly because the fossil evidence of human ancestry is fragmentary. The most spectacular find seldom consists of more than part of a jawbone and a few teeth. Much of the muddle, however, can only be ascribed to human vanity. Researchers inevitably hope to discover vital clues to human ancestry. As a result, many fossils now identified as monkeys (and even modern humans!) were at first hailed as "the missing link" between apes and humans.

Since we are closely related to the African apes, the search for fossils of the presumed common ancestor of apes and humans has centered in Africa. Studies of fossil pollen grains show that many forested areas in Africa became open grassland some 15 million years ago during the Miocene Epoch (see the book's endpaper). Between 10 and 4 million years ago, human ancestors apparently moved out of the forests and roamed the open savanna in bands, like modern baboons. The ancestors of the apes remained in the forest. In Tanzania, Mary Leakey found footprints made 3.75 million years ago by primates that were bipedal (walking on two feet) (Figure 21-58). Although the selective pressure that led to ground-dwelling bipedalism is hotly debated, the result was to free the hands for carrying food and infants, throwing stones, and other activities. This permitted the beginning of technology, the human use of tools and machines.

The first definite hominid fossils are almost 4 million years old (from the Pliocene). The first specimen, found in 1924, was named *Australopithecus africanus*. In 1979, a related group of fossils was discovered in the Ethiopian desert, including the most nearly complete skeleton of an australopithecine found so far. This specimen, nicknamed "Lucy," was a full-grown female that lived about 3.6 million years ago. She stood upright and was about a metre tall, although males of her species were much larger. There is some debate as to whether these australopithecines were completely ground-dwelling or spent some of their time in trees because, compared with humans, they had long arms and short legs.

A battle rages among the experts over how many species of *Australopithecus* existed and which of them, if any, gave rise to the genus *Homo*. Australopithecines had brains little larger than those of modern apes. They were ape-like in having large, heavy jaws, showing that they ate a lot of tough plant food. There is no convincing evidence that they used tools, but we would not expect to find the remains of the earliest tools. By analogy with the tools used by chimpanzees

Figure 21-57
A chimpanzee. This is a member of the modern-day species *(Pan troglodytes)* most closely related to humans. (Biophoto Associates)

Figure 21-58
Fossil evidence of bipedalism. This 3.75-million-year-old deposit of volcanic ash in Tanzania contains the fossilized footprints of three bipedal primates: an adult and young walking side by side, and another adult walking in the footprints of the first. There is some debate about which species of early hominid made these tracks. (Photo by Peter Jones)

Parasitism

A JOURNEY INTO LIFE

Many phyla of animals include parasites, from microscopic body lice to tapeworms several metres long. **Parasites** are organisms that extract their food from living **hosts.** Some are external parasites (**ectoparasites**) that attach to the outside of the host's body. These include leeches (annelids) and ticks, lice, and fleas (arthropods) (Figure 21-B). Others are **endoparasites** that live inside the host. Flukes and tapeworms (flatworms) and many roundworms belong in this group (Table 21-A).

Finding food is often difficult for parasites because appropriate hosts may be few and far between. Many tapeworms and flukes compensate by producing hundreds of thousands to millions of eggs, the large numbers ensuring that at least a few find a host; most starve to death. Pinworms have a hand-to-mouth method of transmission: these small, short-lived worms infect mainly young children, who scratch the worms' eggs from the anal area and transfer them to the mouth on the fingers, allowing the next generation of worms to reach the digestive system of the same host. Female fleas have an interesting adaptation: they are most attracted to pregnant females of host mammal species. The flea feeds on the pregnant female's blood and lays eggs, which hatch into larvae that live on scraps of skin or dung in the burrow of the mother-to-be. When the young mammals leave their mother, the next generation of fleas hops aboard, provided with new hosts and a means of dispersal.

Parasite life histories are often complex and involve more than one host. Some immature parasites change the behavior of their host, making it more apt to be eaten by a second host in which the parasites can be-

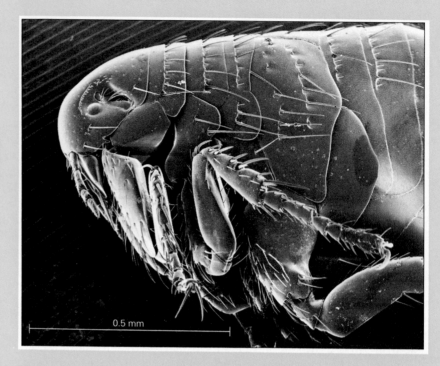

Figure 21-B

An ectoparasite. This scanning electron micrograph shows the front end of a human flea, an insect that lives as an ectoparasite during adulthood. The body is very flat, so that the flea looks very thin when viewed from above. This is an adaptation to running quickly between hairs on the host's skin. (Biophoto Associates)

come adults. For example, worms that form cysts in muscles may make the host slower and less likely to escape a predator. Larvae of the canine tapeworm invade the nervous system of hosts such as sheep, causing them to totter around in circles and become separated from the herd. They are easily picked off by wolves, which then provide the tapeworm with its adult home.

Intestinal parasites are surrounded by digested food, and other parasites feed on nutritious blood. There is not much for the parasite's digestion to do, and most parasites have reduced digestive systems. The energy freed by this savings is devoted to expansion of the reproductive system. In fact, tapeworms have no digestive system whatever: they absorb all their food through the body wall, and their bodies are little more than egg factories (Figure 21-C).

From the parasite's point of view, the ideal host-parasite relation-

ship is one in which the host remains alive at least long enough to permit the parasite to complete its development and reproduce. Thus there is often strong selection for parasites not to kill their hosts. The worst outbreaks of diseases occur when parasites—be they viruses, bacteria, protists, fungi, or animals—first come into contact with a particular population of hosts. History is full of examples of parasites, such as the plague or syphilis bacteria, that killed huge proportions of new-found host populations, but this also, of course, killed most of the parasites. Such a first encounter selects for those hosts with defense mechanisms against the parasite, and for those parasites that are less virulent, until, after a while, the parasite will do less damage to its new-found host population than it did initially.

A slightly different way of life is found among some insects known as **parasitoids** ("parasite-like"). The

Table 21-A   *Some Parasitic Worms Common in Humans*

Name	Symptoms	Means of infection
Platyhelminthes		
Chinese liver fluke	None in mild cases; destruction of liver, bile stones, and clogging of liver ducts in severe cases	Eating raw fish
Blood fluke (schistosomes)	Enlargement of liver and spleen Urinary disorders Bloated abdomen, wasted arms and legs	Drinking or wading barefoot in water containing infected person's urine. Infects about 200 million people in 70 nations. Not found where there are modern sewage disposal systems
Swimmer's itch	Itching after exposure of skin to infested water	Burrowing of fluke larvae of species that cannot successfully infect humans
Bladder worm (immature stage of a worm that lives in dogs when adult)	Cysts up to the size of an orange; symptoms depend on part of body invaded	Infected dogs licking people's hands or faces or contaminating drinking water
Pork, beef, and fish tapeworms	Immature worms: cysts Adult worms may cause diarrhea, loss of weight, perforation of intestine	Eating undercooked meat containing worm cysts
Nematoda		
Pinworm	Anal itching	Females lay eggs around anal opening; hands may transfer eggs to mouth, maintaining infection in same person. Physical contact may also transfer to other people
Hookworm	Anemia, lethargy	Young worms burrow through skin (bare feet) from moist soil and grass contaminated by feces of infected humans

female lays an egg in a host (usually the larva, pupa, adult, or egg of another insect species), and the egg hatches and uses the host as a food source in its own development. Just as the young parasitoid reaches adulthood, it kills its host and cuts its way out of the host's body. So, although the developing parasitoid does feed on a living host, killing the host is a programmed part of its life rather than an accident. Females of some species of parasitoids can tell whether a host is already parasitized, and lay eggs only on uninhabited hosts. This ensures that their own offspring will have enough food to develop. Many kinds of parasitoids attack only one or a few closely related species of hosts. Parasitoids of pest insects are sometimes raised and released as part of pest management programs.

Figure 21-C

An endoparasite. This photograph of one end of a tapeworm shows that there is no head, but the crown of hooks and suckers attaches to the wall of the host's digestive tract. The long string of short, wide sections contains little but reproductive organs (not visible here). Ripe sections detach from the other end of the worm and are shed in the host's feces. (Biophoto Associates)

Figure 21-59
Human ancestors at home. An artist's impression of the life of *Homo habilis.* (Carolina Biological Supply Company)

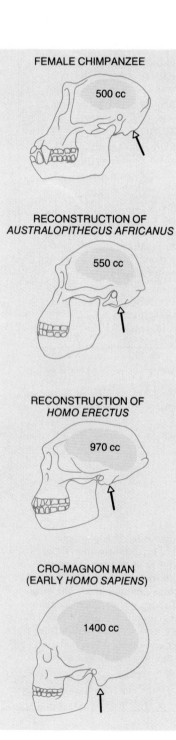

FEMALE CHIMPANZEE

500 cc

RECONSTRUCTION OF
AUSTRALOPITHECUS AFRICANUS

550 cc

RECONSTRUCTION OF
HOMO ERECTUS

970 cc

CRO-MAGNON MAN
(EARLY *HOMO SAPIENS*)

1400 cc

today, the first tools of hominids must have been "found" materials—rocks, bones, large thorns, and lengths of vine—which either would not have been preserved or, if they were, might not be recognizable as tools.

About 2 million years ago, at least one species of australopithecine lived at the same time as the first definite species of *Homo (Homo habilis),* which had a larger brain than an australopithecine. Piles of animal bones found with *H. habilis* skeletons show that meat had been added to plants as a regular part of the diet by this time, but whether it came from hunting or scavenging is not clear. These hominids also used crude stone tools (Figure 21-59). We cannot tell whether language had evolved by this time. We can only speculate that the advantages of cooperation in hunting and group defense may have selected for the development of language for communication.

Homo habilis lasted only a few hundred thousand years. By about 1.75 million years ago it had been replaced by a similar species, *Homo erectus. H. erectus* was a fully bipedal, tool-using hominid with an omnivorous diet (eating both plants and animals). The skull was thick, with heavy jaws and teeth, and a low forehead, although some individuals had brains almost as large as those of modern humans (Figure 21-60). Some *H. erectus* bones were found in caves, indicating that this species used at least temporary home bases. Besides animal bones and quite advanced stone tools, some of the caves contain heaps of charcoal and charred bones, showing that fire had been domesticated and brought indoors by this time. Presumably this habit originated in the use of natural fires (started by lightning or spontaneous combustion) to keep warm, cook food, or split stones to make tools.

Colonization of Colder Areas

Tools and fire, the beginnings of technology and culture, contributed to the success of *Homo erectus.* The species spread widely, and emigrants from Africa colonized other, colder areas in Europe and Asia. Behavioral adaptations or

Figure 21-60
Changes in proportions of the skull from great ape to human. Note the increase in size of the brain case (blue), change in angle and position of the attachment between the neck and the skull (arrows) as hominids became more bipedal and upright, and relative reduction in size of the teeth and jaws as hominids changed from a purely herbivorous to an omnivorous diet.

technological expertise are necessary if human beings are to survive winters as cold as those of Central Europe and China. It seems that the prehuman brain of *H. erectus* could produce social and technological solutions—such as fire, clothing, stored food, and communal living in caves—to the problems of surviving cold winters.

Homo sapiens probably evolved from *H. erectus* a few hundred thousand years ago. The Neanderthals, a species with human-sized brains but heavier skulls and teeth, appeared about 100,000 years ago. Neanderthals were more similar to *H. erectus* than to *H. sapiens,* but nothing is known of the origin of this short-lived group. Remains of completely modern *H. sapiens* go back about 41,000 years.

■ *Human changes from ape-like ancestors include bipedalism, an even larger brain, omnivorous diet, tool use, and language.*

■ *Human ancestors invented technological solutions to the problems of living in cold climates.*

SUMMARY

The lower invertebrates evolved in the sea, where most of them still live. These animals are either planktonic, or bottom-dwelling with planktonic larvae. Some have adapted to life in fresh water, in damp terrestrial habitats, or in the watery interior of a host's body.

Porifera (sponges) and Cnidaria (jellyfish, sea anemones, and corals) display the least specialization and cooperation among cells. Sponges are sessile filter feeders, and cnidarians are sessile or slow-moving predators with stinging, radially arranged tentacles. In the Platyhelminthes and Nematoda the main organ systems found in most animals are present, except for a circulatory system and a skeleton. These flatworms and roundworms are free-living or parasitic, with bilateral symmetry and some cephalization.

A coelom is a fluid-filled body cavity within the mesoderm. The coelom permitted the evolution of more efficient modes of digestion and circulation. The most important phyla of animals with a coelom are:

1. Annelida, segmented worms including polychaetes, earthworms, and leeches.
2. Mollusca, largely unsegmented animals with a muscular foot, and a mantle covering the body and usually secreting a calcareous shell: gastropods (snails and slugs), bivalves (shellfish), and cephalopods (nautiluses, squids, and octopuses). Most molluscs are marine, but many are freshwater or terrestrial.
3. Arthropoda, the most successful animal phylum, with segmented bodies, chitinous exoskeletons, and a varied array of jointed appendages. The crustaceans are mainly marine, and the arachnids and insects are mainly adapted to terrestrial life.
4. Echinodermata, sluggish marine animals with a spiny calcareous skeleton, tube feet, and usually pentaradial symmetry. Echinoderm embryology shows that they are close relatives of the chordates.
5. Chordata, with a notochord, a hollow dorsal nerve cord, and pharyngeal gill slits sometime in life. Living invertebrate chordates include tunicates and amphioxus. Most chordates living today are vertebrates, with the notochord surrounded or replaced by a vertebral column of cartilage or bone.

Vertebrates probably evolved from animals resembling the tadpole-like larva of tunicates, using their gills for filter feeding and their segmental muscles, attached to the notochord, to throw the body into curves as they swam.

The agnathan fishes, the first vertebrates, gave rise to the bony and cartilaginous fishes. These groups showed two major advances over the agnathans: (1) jaws that could snap and bite their food; (2) paired pectoral and pelvic fins that provided balance while the body or tail muscles still gave the main thrust for swimming.

Amphibians, the first terrestrial vertebrates, evolved from bony fishes with air-breathing lungs and sturdy paired fins able to support the body on land. With their thin, moist skin still used as a major gas exchange surface, and eggs that must be laid in water, amphibians have remained in moist habitats.

With the evolution of well-developed air-breathing lungs, waterproof integuments, and the amniotic egg, the adaptive radiation of land vertebrates began. Reptiles gave rise to mammals, and later to birds. With the extinction of many reptiles toward the end of the Cretaceous Period, birds and mammals inherited the earth and have radiated widely ever since.

Birds' feathers provide lightweight insulation as well as strong flight surfaces. Both birds and mammals are endothermic, using rapid metabolism to generate body heat, which is retained by a layer of fat under the skin and an outer layer of feathers or hair. Parental care is well developed in both groups.

Consult this chapter's tables for more detailed summaries of each group.

Human beings, members of the species *Homo sapiens,* are primates with arboreal ancestors. Arboreal life produced adaptations such as forward-facing eyes with stereoscopic color vision, bipedal tendencies, and precise control over the digits. These features are seen today in anthropoid apes. Human ancestors abandoned life in the trees for bipedal locomotion on the open plain. From this point, human evolution revolved around increasingly sophisticated use of the enlarging brain and of the hands to collect food, to make tools, to develop language, and finally to make fires and clothes, which permitted hominids to spread away from their original home in Africa.

SELF-QUIZ

1. An animal that must move a great deal will experience selective pressures favoring (bilateral, radial) symmetry.
2. The chief function of the larval stages of marine invertebrates is ____.
3. The evolutionary importance of a coelom is that:
    a. it permitted animals to have a circulatory system and other internal organs that move
    b. it permitted animals to move onto land with an internal storage place for extra body fluid
    c. it provided the possibility of evolving a hard, protective exoskeleton
    d. it allowed organisms to have excretory systems
    e. it paved the way for evolution of locomotory appendages
4. Filter feeding is *not* found in:
    a. adult sea squirts (tunicates)   d. larval agnathans
    b. adult amphioxus   e. sponges
    c. adult agnathans
5. You would be most likely to find an adult tunicate:
    a. in a mountain stream
    b. in a large river such as the Mississippi
    c. preying on clams
    d. in a seacoast town, attached to the piling of a dock
6. Which of the following chordate characteristics contributes *least* to its efficiency of locomotion?
    a. myotomes   d. post-anal tail
    b. pharyngeal gill slits   e. streamlined body shape
    c. notochord
7. Which of the following is *not* a vertebrate?
    a. an amphioxus   d. a kangaroo
    b. a lamprey   e. a duck-billed platypus
    c. a shark
8. What is the most characteristic feature of Aves, found in no other class of living vertebrates?

9. Which pair of animals is most closely related?
    a. sea anemone and sea urchin
    b. bat and elephant
    c. duck-billed platypus and Canada goose
    d. leech and tapeworm
    e. barnacle and tick
10. List three problems of terrestrial life that previously aquatic vertebrates had to overcome before they could invade the land.
11. The first major vertebrate class to be totally independent of bodies of water during reproduction was the ____.
12. A vertebrate that has a body covering of scales, glandless skin, no limbs, and internal fertilization, and that suns itself to increase its body temperature in the morning belongs to the class ____.
13. List at least two differences in body structure and at least two differences in reproduction between reptiles and amphibians.

    Matching: For the features listed below, pick out the group or groups that possess(es) them from the list at the right.
    ____ 14. Mantle and calcareous exoskeleton   a. Annelida
    ____ 15. Collar cells   b. Arthropoda
    ____ 16. Jointed chitinous exoskeleton   c. Chordata
    ____ 17. Only two cell layers developed   d. Cnidaria
    ____ 18. No anus   e. Echinodermata
    ____ 19. Tube feet   f. Mollusca
    ____ 20. Segmented worms   g. Nematoda
        h. Platyhelminthes
        i. Porifera
21. Explain how each of the following contributed to the further development of *Homo:* "taming" of fire, invention of tools, maintenance of a home base.

QUESTIONS FOR DISCUSSION

1. Why is filter feeding such a common way of life among invertebrates?
2. Many invertebrates reproduce by parthenogenesis, budding, or other asexual means. Why is this so common?
3. Cephalization is pronounced in the vertebrates but not in the tunicates, amphioxus, or echinoderms. What differences in selective pressures may have caused this difference in degree of cephalization?
4. We often talk as if *Homo sapiens* were the most highly

evolved and specialized mammal. In what ways is this true, and in what ways is it not true?
5. Many mammals have adapted to a life permanently at sea. Why haven't they gone back to using gills for respiration?
6. What are the advantages of social and parental behavior that have made it profitable for organisms to spend some of their energy in these activities? What are the drawbacks?
7. Why do so few birds live as grazers on leaves and grass?

SUGGESTED READINGS

Buchsbaum, R. *Animals Without Backbones,* 2d ed. Chicago: University of Chicago Press, 1976. A classic elementary textbook on invertebrates.

Ehrlich, P. R., D. S. Dobkin, and D. Wheye. *The Birder's Handbook: A Field Guide to the Natural History of North American Birds.* New York: Simon and Schuster, 1988. A field guide embellished with essays on all aspects of bird biology and conservation.

Evans, H. E. *Life on a Little-Known Planet.* New York: E. P. Dutton, 1978. An entertaining and enlightening account of our insect neighbors.

Goreau, T. F., N. I. Goreau, and T. J. Goreau. "Corals and coral reefs." *Scientific American,* August 1979.

McFarland, W. N., F. H. Pough, T. J. Cade, and J. B. Heiser. *Vertebrate Life,* 3d ed. New York: Macmillan Publishing, 1989. A readable general text.

Moore, J. "Parasites that change the behavior of their host." *Scientific American,* May 1984.

O'Toole, C., ed. *The Encyclopedia of Insects.* Facts on File Publications, 1986. Includes all terrestrial arthropods.

Pilbeam, D. "The descent of hominoids and hominids." *Scientific American,* March 1984. A clear account of how thinking on the subject changed during the preceding five years, pointing out what we know and don't know about our human "roots."

Rukang, W., and L. Shenglong. "Peking man." *Scientific American,* June 1983. Traces the evolution of a *Homo erectus* population that inhabited a large cave for over 200,000 years.

Weiss, R. "Incrimination by insect." *Science News* 134:90–91 (August 6, 1988). How insects help police analyze evidence of crimes.

Wellnhofer, P. *"Archaeopteryx." Scientific American,* May 1990. What the few known fossils of this organism tell us about the evolution of bird flight.

Animal Biology

PART

4

Animals are multicellular heterotrophs that obtain food by hunting, searching, filtering, or ambushing—in short, by performing some action. As a result, they have evolved increasingly sophisticated means of finding food and of moving, or building traps, so that they can capture it. Our survey of the animal kingdom (Chapter 21) reveals evolutionary trends toward larger size, greater complexity, faster, more efficient locomotion, and colonization of new habitats—the land and even the air. Along with these trends go changes in **physiology,** the workings of the body's organs.

In a protist or bacterium, all bodily functions are carried out by a single cell. But as fungi, plants, and animals evolved, these functions became increasingly divided up among different cells, tissues, and organs. This trend is carried to its farthest extreme in the animal kingdom, which contains organisms with the most specialized organs and organ systems. In this part of the book, we consider animal organ systems according to the physiological functions they perform.

Animals ingest their food from the outside world and convert it into the molecules that their bodies' cells need (Chapter 22); obtain oxygen, which the cells need to release energy from their food (Chapter 23); transport food, oxygen, and other substances to their cells (Chapter 24); and collect and eliminate cell wastes and excess salts and water (Chapter 26). Animals also have specialized immune systems, found in no other organisms, that help protect them from disease (Chapter 25).

Animals must also detect what is going on inside and outside the body and respond to these events. Detecting events in the outside world is necessary if an animal is to find food, water, or a mate, and avoid being eaten. Changes within the body also have to be detected and coordinated so that different organ systems cooperate to produce a smoothly functioning individual. All animals have nervous systems (Chapter 28) that process information detected by their sense organs (Chapter 29) and direct the appropriate responses, carried out by muscles and skeletons (Chapter 30) and by hormone systems (Chapter 31).

All of these organ systems play a role in **homeostasis,** keeping the environment of the body's cells in a range suitable for life. Within the body, the levels of organic molecules, salts, gases, pH, and temperature fluctuate. When any of these factors approaches either the upper or the lower limits of the body's tolerance, mechanisms intervene to return it to a more favorable level. This process is called **negative feedback.** Negative feedback mechanisms coordinate all aspects of animal physiology and so maintain homeostasis.

The reproductive system (Chapter 27) is unique. It contributes only marginally to the body's welfare (by producing hormones), but all the adaptations of the other systems exist because, during the course of evolutionary history, they have contributed to the reproductive system's ability to produce new individuals. This, of course, is the bottom line of biology.

Some invertebrates have organ systems that are very simple, making it easy to understand their vital features. So in each chapter, we look at the evolution of a body function, working our way from these simple systems to the sophisticated systems found in mammals, usually using the human body as an example. We shall find many similarities in the physiology of all animals, from sea anemones to humans, because we are all faced with accomplishing similar tasks to keep our bodies going.

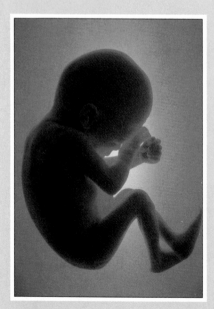

Five-month-old fetus

446

Adelie penguins, Antarctica

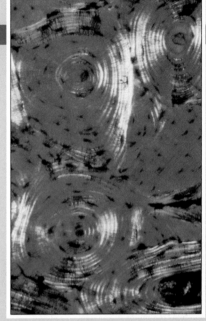

Human bone, Haversion canal

Blue and red sea cucumber

447

Animal Nutrition and Digestion

CHAPTER

22

Blue and red sea cucumber

Grasshopper

Jaguar

Feeding spoonbill

Animals are heterotrophs. They cannot make their own food from inorganic substances but must ingest organic molecules from the environment. Animals can be broadly divided into **herbivores,** which eat plants; **carnivores,** which eat animals; and **omnivores,** which eat both. Animals with each mode of nutrition have digestive systems suited to handling and digesting the type of food they eat.

Feeding is a necessary evil. The longer an animal spends eating, the less time it has for more advantageous activities, such as reproduction and raising its offspring. Thus, there is strong selective pressure for an animal to feed as rapidly as possible.

Digestion is the mechanical and chemical breakdown of food into small organic molecules, which can be absorbed from the gut. In this chapter we shall consider what food animals need, how animals obtain food, and the role of the digestive system and liver in processing food so that it can be used by the body.

KEY CONCEPTS

- Animals need macronutrients (proteins, carbohydrates, and fats) and micronutrients (vitamins and minerals) in their diets.
- Animal digestive systems have evolved into disassembly lines where food is broken down first mechanically, and then chemically by digestive enzymes, after which it is absorbed and distributed throughout the body.

22-A Nutrients

The nutrients that any animal must ingest as food may be divided, for convenience, into **macronutrients**—nutrients needed in large quantities—and **micronutrients**—nutrients required only in small amounts.

Macronutrients

The macronutrients are proteins, fats, and carbohydrates. All three can serve as energy sources because they can be broken down into molecules that are respired to produce ATP (Chapter 7). The amount of energy available from a given amount of a macronutrient is commonly measured as the number of Calories of heat it yields when fully oxidized (Table 22-1). All three classes of macronutrients also provide carbon atoms used by the body to form organic polymers. In addition, proteins supply amino acids for building the body's own proteins.

Macronutrients can be stored until the body needs them for energy. Carbohydrates are stored as the polysaccharide glycogen, in muscle and liver, and fats are stored as fat. Proteins cannot be stored. Excess protein is broken down to amino acids, which are deaminated—that is, their amino ($—NH_2$) groups are removed. The rest of the molecule is then processed like fat or carbohydrate.

Problems With Macronutrients The most common dietary problem for people in industrialized countries is obesity: if more calories are ingested in the food than are used up, the excess is stored as fat. Overeating and lack of physical activity are the usual causes of obesity. Once obesity sets in, it may alter the body's metabolism in such a way as to perpetuate itself in a vicious circle of overeating and inactivity. Women have a greater tendency than men to store fat under the skin. This is biologically valuable, if currently unfashionable, because it provides food reserves that can carry not just the woman but also her unborn child or nursing infant through times of shortage.

Probably the most common dietary problem for most animals is protein deficiency. This usually occurs, not because the diet contains too little total protein, but because it does not contain enough of the essential amino acids. All animals can convert some amino acids into others, but the **essential amino**

■ *Macronutrients supply energy, carbon atoms, essential amino acids, and essential fatty acids.*

Table 22-1	The Caloric Values of Macronutrients	
Macronutrient	**Calories per gram***
Protein	~4.0
Fat	9.0
Carbohydrates	4.0

* A calorie is the amount of heat needed to raise the temperature of 1 gram of water by 1°C. The "Calories" in food are actually kilocalories, often indicated by a capital C.

1 kilocalorie (Kcal) = 1 Calorie = 1000 calories.

acids must be supplied in the diet because an animal cannot make them from other amino acids.

One of the best-known protein deficiency diseases is kwashiorkor, often found in African populations where the diet consists primarily of cornmeal. Such a diet contains very little of the essential amino acids tryptophan and lysine. Victims of kwashiorkor (particularly growing children, who need much protein) show symptoms such as failure to grow normally, lethargy, and edema (swelling due to excessive retention of fluid) in parts of the body.

Fat deficiencies can also cause nutritional diseases. Certain fatty acids are essential in the diet. They are constituents of cell membranes and of some hormones. People who live on diets of fish, rice, or fruit, which are all very low in fat, develop a craving for fats and treat them as a delicacy.

Nowadays many people are interested in the relationship between diet and health. It has been observed that immigrants tend to die of the causes of death common in their adopted countries. People in Japan have a high incidence of death from stomach cancer, but people of Japanese origin living in the United States are more likely to die of breast cancer and heart attacks. Researchers asked if the change to an American diet could account for the shift in the cause of death among these immigrants. The answer to this question is still not clear.

We are deluged by advertising for foods such as high-fiber cereals, polyunsaturated fats, and cholesterol-free foods. Claims that eating these foods improves health are unsubstantiated, and indeed illegal in the United States (since they have not been approved by the Food and Drug Administration). The National Academy of Sciences was asked to review all the studies on diet and health and to produce recommendations for improving the American diet. The Academy reported that there was not enough evidence to draw conclusions about precise relationships between diet and health. However, there was considerable evidence that the typical Western diet contained too much fat for perfect adult health. (Children require more fat than adults.) The Academy conclusion was simple: Americans might improve their health if they ate more fresh fruits and vegetables.

Micronutrients

Micronutrients are the substances an organism must have in its diet in small quantities because it cannot make them for itself or because it cannot make them as fast as it needs them. Micronutrients can be divided into **vitamins,** which are organic compounds, and **minerals,** which are inorganic. Various deficiency diseases result from shortage of certain vitamins and minerals in the diet.

Vitamins The vitamin needs of various animals differ. For instance, many other animals can make ascorbic acid from other molecules, and so they do not need to eat vitamin C as humans do. The vitamins humans need are generally divided into two categories: water-soluble and fat-soluble.

Most water-soluble vitamins (Table 22-2) are coenzymes needed in metabolism. They are easily excreted by the kidneys. The fat-soluble vitamins (Table 22-3) have various poorly understood functions. Fat-soluble vitamins that are not needed immediately may be stored in fatty tissue. Because these vitamins are not soluble in water, the body's enzymes must process them before they can be excreted by the kidneys. As a result, some of them can accumulate to toxic levels if consumed in amounts larger than the body can handle. Some diet plans recommend dangerous doses of certain fat-soluble vitamins. (In 1978, for instance, two people on high-vitamin diets died of vitamin A poisoning.) The key to

Table 22-2 Functions and Deficiency Symptoms of Water-Soluble Vitamins

Vitamin	Metabolic role	Deficiency symptoms
Thiamine (B_1)	Coenzyme in carbohydrate metabolism	Beriberi, loss of appetite, fatigue
Riboflavin (B_2)	Part of FAD, coenzyme in respiration and protein metabolism	Inflammation and breakdown of skin
Niacin	Part of NAD and NADP, coenzymes in energy metabolism	Pellagra, fatigue
Pyridoxine (B_6)	Coenzyme in amino acid metabolism	Anemia, nerve problems
Pantothenic acid	Part of coenzyme A, in carbohydrate and fat metabolism	Similar to other B vitamins
Biotin	Coenzyme in addition of carboxyl groups	Rare; tiny amounts required
Folic acid	Coenzyme in formation of nucleotides and hemoglobin	Some types of anemia
Cobalamin (B_{12})	Coenzyme in formation of proteins and nucleic acids	Pernicious anemia
Ascorbic acid (C)	Helps build intercellular cement for bones, cartilage, skin	Scurvy, anemia, slow wound healing

avoiding such dangers is common sense and moderation. Almost anything can be poisonous if consumed in excessive amounts.

Vitamin-rich foods include fresh fruits and vegetables and whole or enriched grain products, as well as fortified milk, liver, and fish oil.

Minerals We need some minerals in relatively large amounts. Sodium and potassium, for instance, are vital to the working of every nerve and muscle in the

Table 22-3 Functions and Deficiency Symptoms of Fat-Soluble Vitamins*

Vitamin	Physiological role	Deficiency symptoms
A (retinol)	Part of visual pigments in eye	Night blindness, drying of mucous membranes
D (calciferol)	Absorption of calcium and phosphorus, building bones	Rickets in children
E (tocopherol)	Protects blood cells, vitamin A, etc., from oxidation	Lysis of red blood cells, anemia
K (menadione)	Needed in synthesis of prothrombin for blood clotting	Hemorrhage in newborns who lack gut bacteria that produce vitamin K

*Vitamins A, D, and K are toxic in large amounts.

Diet and Cardiovascular Disease

A JOURNEY INTO LIFE

Cardiovascular diseases, those that affect the heart and blood vessels, cause about half the deaths in North America. These diseases include **hypertension** (high blood pressure) and atherosclerosis. In **atherosclerosis,** deposits of lipids develop on artery walls, reducing the diameter of the blood vessel and making its walls less elastic (Figure 22-A). Cardiovascular diseases cause death in many ways. They increase the risk of strokes and heart attacks (including "coronaries," blockage of the arteries that supply blood to the heart muscles).

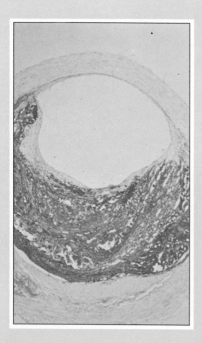

Figure 22-A
An artery (gray) three-quarters blocked by an atherosclerotic plaque (red). (Biophoto Associates)

Hypertension can be controlled by losing weight or, if that fails, by drugs. Doctors often tell hypertension patients to eat less sodium. However, studies published in 1990 show that the amount of sodium chloride (table salt) in the diet does not affect blood pressure. No matter how much salt the diet contains, the sodium content of the blood is controlled within narrow limits by hormones that regulate how much sodium the body excretes (Section 26-F).

Susceptibility to heart attacks is correlated with blood levels of cholesterol-carriers called HDLs and LDLs. Cholesterol is a small lipid that the body needs to produce new membranes. Most cholesterol is carried by low-density lipoproteins (LDLs). Some is carried by high-density lipoproteins (HDLs). Studies of men of many nationalities and races have shown that the risk of heart attack is greater the higher the LDL concentration and the lower the HDL concentration in the blood. People with high levels of HDL or low levels of LDL because of their genetic makeup apparently never die of atherosclerosis.

The factors correlated with high HDL levels are those long known to be associated with a low risk of heart attack—being slim, exercising, not smoking, consuming little alcohol, and being female. (Women's HDL levels average about 10% higher than men's, probably because of the female hormone estrogen, which raises HDL levels).

You will notice that these studies talk about correlations, not about cause and effect. It is difficult to show cause and effect for these diseases. This is because cardiovascular diseases take a long time to develop, and few experiments can be performed on human subjects. Even an obvious correlation may not prove much. For instance, runners have a low incidence of cardiovascular disease. Is this because people with high HDL levels are predisposed to take up running, or does running raise HDL levels? Runners are also more likely to consume alcohol and less likely to smoke than other members of the population. Is this cause, or effect, or neither?

Since 1965, deaths from heart attacks in North America have decreased by 34%. Researchers do not know why this has happened. It would seem that some change in lifestyle must be responsible. Some have suggested that people now eat less fat, exercise more, and smoke less than in the 1960s. Most of these changes are known to reduce the risk of heart attack, but their effects are too minor to account for a 34% decrease in deaths from heart attacks.

The most effective way to reduce the risk of a heart attack is for people with a couch potato lifestyle to get more exercise. Even mild exercise, such as gardening or walking regularly, has been shown to reduce the death rate from cardiovascular disease significantly. However, it is hard to believe that people are much more active now than they were in 1965, since the average American is now more overweight.

Among doctors, lowering blood cholesterol levels is the fashionable approach to preventing heart attacks, but it is not very effective. Lowering blood cholesterol does reduce the risk of heart attack slightly, but it does not lead to longer life. In 1990, researchers reviewed studies on more than 27,000 people who had lowered their blood cholesterol by diet or drugs. These people suffered slightly fewer heart attacks than members of a control group with uncontrolled blood cholesterol levels, but they did not live any longer. (They died, on average, slightly sooner.)

It is generally true that reducing deaths from heart attacks does not increase life expectancy (the average number of years that a newborn baby can be expected to live). Life expectancy in developed countries is now about 80 years. If infant deaths, homicides, and accidents were eliminated, it would be even higher. People would die of old age—the general failure of their organ systems—at between 50 and 120 years of age. Interestingly, even if cardiovascular disease and cancer were cured, life expectancy would not increase by more than a few years.

Table 22-4	Physiological Roles of the Important Minerals
Mineral	**Major physiological roles**
Sodium (Na)	Main extracellular positive ion; active transport, osmotic and acid-base balance, nerve and muscle activity
Potassium (K)	Major intracellular positive ion; acid-base balance, nerve and muscle activity
Calcium (Ca)	Component of bones and teeth; membrane permeability, blood clotting, nerve and muscle activity
Phosphorus (P)	Bone formation; part of DNA, RNA, ATP, etc.; energy metabolism
Magnesium (Mg)	Bones and teeth; carbohydrate and protein metabolism
Chlorine (Cl)	Major extracellular negative ion; osmotic and acid-base balance; stomach acid
Sulfur (S)	Protein structure; detoxification reactions

Table 22-5	Physiological Roles of Trace Minerals
Mineral	**Some physiological roles**
Iron (Fe)	Component of heme group in hemoglobin, cytochromes
Copper (Cu)	Needed to make hemoglobin and bone; part of cytochromes
Iodine (I)	Component of thyroid hormone
Manganese (Mn)	Needed in urea formation, protein metabolism, glycolysis, citric acid cycle
Cobalt (Co)	Component of vitamin B_{12}; required for red blood cell formation
Zinc (Zn)	Component of many enzymes; senses of smell and taste
Molybdenum (Mo)	Component of some enzymes
Fluorine (F)	Prevents tooth decay
Selenium (Se)	Needed in fat metabolism
Chromium (Cr)	Needed in glucose metabolism
Boron	Needed for calcium metabolism and use of copper

body (Table 22-4). Large quantities of these minerals (particularly sodium) are excreted in the urine every day. Sodium excretion is also a vital part of sweating, which is necessary to the regulation of body temperature in some mammals. Calcium is required for muscular activity and, with phosphorus, is needed in large amounts for bone formation.

Other minerals are known as **trace minerals** (Table 22-5). Some of them are needed in tiny amounts for the activities of enzymes in various metabolic pathways. The functions of other trace minerals are poorly understood.

Foods rich in minerals include meats, milk and cheese, grains, nuts, legumes (members of the pea, bean, and peanut family), and spinach.

22-B Digestive Systems

Digestion is the hydrolysis of food macromolecules into monomers, particularly amino acids, monosaccharides, glycerol, and fatty acids. Food can be digested either **intracellularly,** that is, in vacuoles within cells, or **extracellularly,** in a digestive cavity in the animal's body.

Sponges have mainly intracellular digestion. They are limited to food items small enough to be taken into a cell, and this often means that they must feed continuously.

Extracellular digestion becomes more specialized and efficient as we progress up the evolutionary scale, although intracellular digestion remains an important final stage in digestion. Cnidarians, such as *Hydra* and sea anemones, pull surprisingly large prey into the **gastrovascular cavity,** which doubles as a transport system (see Figure 21-8). Cells lining the cavity secrete digestive enzymes into the cavity, and digestion begins. As small particles separate from the food, they are picked up by cells lining the cavity, and digestion is completed intracellularly.

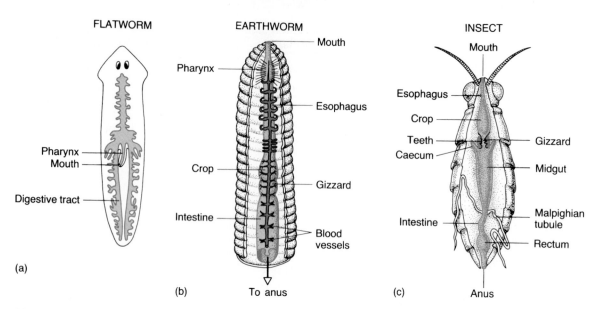

FLATWORM

Pharynx
Mouth

Digestive tract

(a)

EARTHWORM

Mouth
Pharynx
Esophagus
Crop
Gizzard
Intestine
Blood vessels

(b)
To anus

INSECT

Mouth
Esophagus
Crop
Teeth
Caecum
Gizzard
Midgut
Malpighian tubule
Intestine
Rectum

(c)
Anus

Figure 22-1

Invertebrate digestive tracts. (a) A freshwater flatworm (platyhelminth) has a mouth but no anus. The branched gut also acts as a circulatory system, distributing food to all parts of the body. (b) Front end of the gut of an earthworm. The muscular pharynx sucks food into the mouth. Food passes down the esophagus to the crop, a storage chamber that passes small amounts of food to the grinding surfaces in the gizzard. Most of the worm's digestive tract consists of a long intestine, which extends all the way to the anus at the posterior tip of the body. (c) An insect's digestive tract. This has many of the same kinds of structures as an earthworm's. Several caeca may be found just before the midgut. (Malpighian tubules are excretory structures. They discharge wastes into the gut.)

Free-living flatworms have a muscular **pharynx** that allows them to suck up food. They also have a highly branched intestine, with a large surface area for secretion of digestive enzymes and absorption of food (Figure 22-1a). Digestion is completed intracellularly. Both cnidarians and flatworms must discharge undigested remains the same way they came in, through the mouth, because the digestive cavity has only one opening.

In all higher animals, the **digestive tract** is basically a tube with a mouth for ingestion and an anus for **egestion.** Such a tubular digestive system allows food processing to be organized as a one-way flow. Additional food can be taken in while previously eaten food is being digested, and different parts of the tube specialize in performing different functions. The food is broken down step by step, and the nutrients released are absorbed farther down the tube.

Somewhere near the front end, most animals have the tools necessary to break food into smaller parts—teeth in mammals, fish, and sharks; a bill in turtles and birds; a muscular **gizzard** containing stones or grinding edges in earthworms, insects, most birds, and crocodiles (Figure 22-1b, c). A thin-walled **crop** often stores food and releases it in small amounts to the gizzard, where it is ground up.

The intestine is the next part of the digestive tract. The anterior part of it may be specialized as a **stomach** to store food and secrete digestive enzymes, while following areas are increasingly specialized to absorb food molecules. The digestive tracts of many animals have outpocketings called **caeca** (singular: **caecum**), blind sacs that hold food destined for a longer stay in the intestine. Many molluscs and arthropods have large digestive glands, where food is digested intracellularly.

22-C Human Digestion

The Human Digestive Tract

The digestive tract of humans may also be called the **gut, alimentary canal,** or **gastrointestinal tract** (Figure 22-2). Basically, it has five functions: (1) food intake; (2) food storage and transport; (3) mechanical breakdown and digestion of food; (4) absorption of nutrients; (5) formation and evacuation of feces.

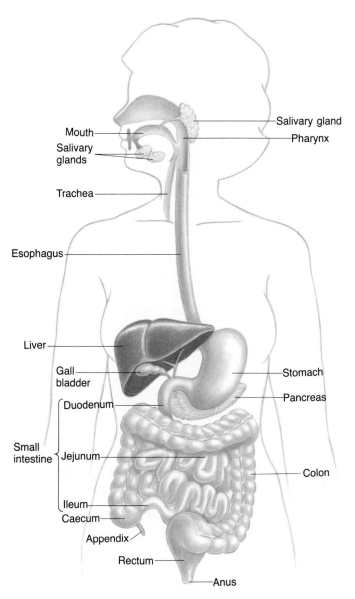

Figure 22-2
The human digestive tract and associated structures. The salivary glands, liver, gall bladder, and pancreas play important roles in processing food.

Food enters the human gut by manipulations of the mouth and associated structures such as the lips, tongue, teeth, and jaw muscles. Once a bite of food has been swallowed, the rest of its journey is completed automatically. The muscular walls of the alimentary canal contract in waves, starting from the end closest to the mouth. Movement of this type is called **peristalsis.**

Besides taking in food, the mouth begins the process of mechanical breakdown, using teeth of various shapes and sizes (Figure 22-3). The **incisors,** chisel-like teeth in the front of the mouth, cut bite-size pieces of food from a larger portion. The tongue pushes the food back to the **molars,** the millstones of the mouth, which mash the food into small particles. Meanwhile, **saliva** is released into the mouth. Saliva moistens the food, sticks it together into a bolus (ball) for swallowing, and provides lubrication so that the food will not scratch the delicate mucous membranes of the mouth and digestive tract. Saliva also contains a starch-digesting enzyme (amylase).

As a bolus of food is swallowed, it passes through the pharynx, gliding over the epiglottis, a sort of trap door that prevents food from entering the trachea (windpipe), where it could block off the air supply to the lungs. Food then drops

Figure 22-3
A human skull, showing the teeth, which are specialized for various roles in the mechanical breakdown of food. Starting from the center front and working back, there are two chisel-shaped incisors, a bluntly pointed canine, two premolars, and two or three molars. (The third molars, "wisdom teeth," never grow in some people.) (Biophoto Associates)

(a)

(b)

(c)

Figure 22-4

Signs of peristalsis in an ostrich. The ostrich takes a mouthful of grass (a), which is formed into a bolus in the mouth. The bolus is swallowed into the esophagus, where it is pushed down to the stomach by peristalsis (b, c).

into the **esophagus,** a long tube with thin muscular walls. A wave of peristalsis passes down the walls of the esophagus, pushing the lump of food down to the stomach (Figure 22-4).

The **stomach** serves several functions. It is the widest part of the digestive tract, in keeping with its function as a holding chamber that stores food and releases it into the intestine in small servings. Having a stomach enables us to feed at intervals and do other things between meals. The stomach's muscular walls churn the contents around. Glands in the stomach wall release acid and digestive enzymes that start to digest any protein in the food (Figure 22-5).

From the stomach, food passes to the **duodenum,** the first part of the **small intestine.** A circular muscle called a **sphincter** closes the stomach off from the intestine except when it relaxes briefly to permit a small amount of food to squirt through, propelled by the stomach's muscular contractions.

In the intestine, more digestive enzymes are added to the food. Some are secreted by the lining of the intestine itself, and some by the **pancreas,** which empties its digestive enzymes into the duodenum by way of a duct. Bile from the gall bladder also moves down a duct and enters the intestine at the same place. Digestion continues in the small intestine, and nutrients are absorbed through its lining. The remainder of the food passes into the **colon,** or **large intestine,** where millions of bacteria live and work. The large intestine absorbs water, minerals, and vitamin K (produced by intestinal bacteria) and pushes the remaining fecal matter into the **rectum,** where it is held until it is voided. Defecation, or expulsion of the feces from the body, depends on the contraction of the walls of the rectum and relaxation of the anal sphincter, a circular muscle at the very end of the digestive tract.

Human Digestive Enzymes

Digestive enzymes hydrolyze food, breaking it into smaller and smaller molecules by adding water to the bonds between monomers. The enzymes are produced by two types of glands: those with ducts and those without. The glands with ducts are the pancreas and **salivary glands.** In addition, ductless, enzyme-secreting tubular glands line the stomach and small intestine. The entire gut is also lined by millions of mucous glands producing mucus, which lubricates the food and sticks it together (Figure 22-6).

Each day the three pairs of salivary glands discharge more than one litre of saliva into their ducts, which lead to the mouth. Saliva is mainly mucus, but it also contains a digestive enzyme that hydrolyzes starch. The importance of mucus becomes clear if the salivary glands fail: without saliva, it is extremely difficult to swallow solid food, even with lots of water to wash it down.

When the stomach contains food, it secretes a fluid containing hydrochloric acid (HCl) with the remarkably low pH of less than 2.0. Other cells in the stomach lining secrete **pepsin,** a name covering a number of enzymes that hydrolyze peptide bonds in proteins. These enzymes work only at a very low pH. One of the most important actions of pepsin is to digest the protein collagen, a major constituent of fibrous tissue in meat, which is soluble at the low pH found in the stomach.

It is remarkable that the stomach does not digest itself, because the hydrochloric acid and pepsin it secretes can digest most flesh. The stomach protects itself by secreting acid- and enzyme-proof mucus, which coats the stomach wall. Even so, a stomach cell's life is short. The stomach surface loses about half a million cells a minute, and all its cells are replaced every three days. Ulcers form when the stomach does not secrete enough mucus to protect itself.

Most of the digestive enzymes are produced in the pancreas and empty into the small intestine close to the stomach. Because pancreatic enzymes work best at a pH of 7 to 8, the first role of the pancreas is to secrete a concentrated solution of sodium bicarbonate, which neutralizes the hydrochloric acid entering the intestine from the stomach.

Pancreatic juice also contains enzymes that work on all three major food types: proteins, carbohydrates, and fats. Protein-digesting enzymes are made by the pancreas in inactive forms that do not digest the pancreas. The inactive enzymes become active after they reach the intestine. For instance, **trypsin** is released when another enzyme cleaves off part of the inactive molecule **trypsinogen.** Trypsin itself activates other pancreatic enzymes.

Still other digestive enzymes are produced by cells lining the small intestine. Some of these enzymes are built into the plasma membrane facing the intestinal lumen (the hollow inside a tube). Others stay in the cell interior, where they apparently digest small molecules absorbed from the lumen in transit to the bloodstream. The intestinal enzymes digest mainly small molecules, many of them products of the attack by stomach or pancreatic enzymes on larger food molecules of all three major food types.

Food entering the small intestine is mixed not only with sodium bicarbonate and digestive enzymes, but also with bile. **Bile** is a solution of salts, bilirubin (made when hemoglobin from red blood cells is broken down in the liver), cholesterol, and fatty acids. Bile is made in the liver, stored in the **gall bladder,** and enters the small intestine by way of a duct. (Gall stones sometimes develop when large amounts of water are absorbed from the bile, leaving solid cholesterol in the gall bladder; other gall stones are made up mainly of bilirubin. Gall stones may block the bile duct from the liver to the intestine, causing great pain.) Bile has two functions in the intestine. First, it acts as a detergent, breaking fat into small globules that can be attacked by digestive enzymes. Second, and more important, bile salts aid in the absorption of lipids from the intestine. Removal of the gall bladder sometimes causes difficulty with lipid absorption.

Digestion is regulated by parts of the nervous system and by hormones—chemical messengers that affect only specific target cells. The hormonal interactions are complicated, but one example will illustrate how this works: if the stomach contains peptides or if its wall is distended by food after a meal, cells in the stomach lining secrete the hormone **gastrin.** Gastrin stimulates other stomach cells to secrete acid and pepsin. These digest the peptides and permit the stomach to get smaller, thereby removing the stimuli (peptides and stretching) that cause gastrin to be released. Hence, gastrin secretion stops.

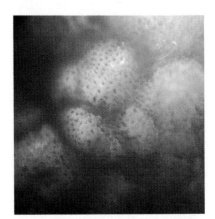

Figure 22-5
Surface view of the interior of the human stomach. The small openings are gastric pits, leading into narrow, dead-end tunnels lined by gland cells that secrete mucus, acid, or digestive enzymes. (Photo by Lennart Nilsson, © Boehringer Ingelheim International GmbH)

Lumen of colon

Muscle

Figure 22-6
Section of the colon. Mucus-producing cells are stained blue. Between them lie cells that absorb water and minerals from the contents of the colon. The band of muscle in the wall of the colon contracts and relaxes to push food through the colon by peristalsis. (Biophoto Associates)

Absorption of Food

In a sense, the lumen of the digestive tract can be considered as a continuation of the external environment: its contents still lie outside the body's cells. Only after a period of digestion do small molecules released from the food pass through plasma membranes into living cells. So we can say that the food we eat does not enter our bodies until some time after we have eaten it.

Anything we swallow reaches the stomach quickly. However, the stomach absorbs only lipid-soluble substances such as alcohol and a few drugs. Nearly everything we digest is absorbed into the body in the small intestine.

The small intestine, about 3 metres long, is highly adapted for absorption. Its lining is thrown into circular folds, and the surface of the lining forms finger-like extensions called **villi.** The surface of each villus, in turn, is covered with a carpet of tiny **microvilli** formed by extensions of the plasma membranes of cells lining the lumen (Figure 22-7). These folds on folds on folds enormously increase the surface area available to take up food molecules.

In this large surface is an array of transport molecules, which absorb food molecules from the intestinal lumen selectively. The absorptive layer of a villus is only one cell thick. Inside it lies a network of blood vessels waiting to pick up absorbed molecules and carry them throughout the body.

Many molecules are moved out of the lumen by active transport or facilitated diffusion. Glucose and amino acids are taken up by active transport carriers powered by the high concentration of sodium ions from pancreatic juice and bile (*A Journey Into Life,* Chapter 4). The intestinal cells can also absorb peptides and digest them intracellularly before passing them on to the bloodstream.

■ *Eating gets food into our mouths but not into our bodies. It takes the digestive tract some hours to reduce a square meal to the nutrient soup that the body's cells can absorb, distribute, and use.*

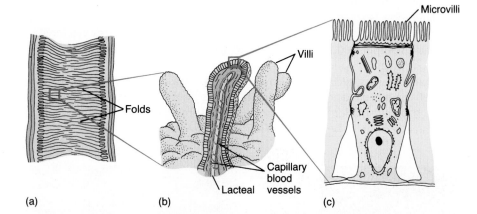

(a)   (b)   (c)

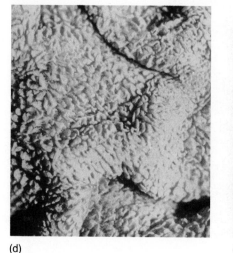
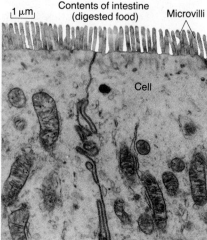

(d)   (e)

Figure 22-7
The large surface area of the small intestine. (a) The lining of the intestine is highly folded. (b) Finger-like villi line the intestine. The villi contain blood capillaries and lymphatic vessels (lacteals), both of which transport food absorbed from the intestinal lumen. (c) The plasma membranes of cells covering the villi are folded into microvilli, which further increase the absorptive surface area facing the lumen. (d) Surface view of villi in the small intestine of a mouse. (e) Microvilli on cells from the small intestine of a rat. (Biophoto Associates)

Lipid-soluble molecules enter cells of the intestinal lining by diffusing through the plasma membrane. Here they are converted mainly into fats and are then released by exocytosis from the far end of the cell as tiny fat droplets, called **chylomicrons.** Chylomicrons are coated by a layer of protein, which makes them water-soluble and easily transported in the blood. Small fatty acids are absorbed directly into the bloodstream near the intestine without being converted into fats.

In addition to all these food molecules, about 10 litres of fluid must be absorbed from the digestive tract into the blood every day. Of this, about 1.5 litres consists of fluid we have drunk, and about 8.5 litres consists of the fluid secreted into the lumen as digestive enzymes and mucus. Most of these 10 litres are absorbed in the small intestine.

Water is absorbed indirectly as a result of sodium transport. Sodium is actively transported out of the intestinal cells at the end farthest from the lumen. Negatively charged chloride ions follow, attracted by the positively charged sodium ions. The accumulation of sodium and chloride ions lowers the osmotic potential outside the intestine, and water moves out of the lumen by osmosis.

The main substances absorbed in the large intestine are sodium, small amounts of other ions, vitamin K, and water. This absorption ensures that only about 100 millilitres of water and small amounts of inorganic ions are lost in the feces every day. The feces are about three-fourths water and one-fourth solid matter. Of the solid matter, about 30% is made up of bacteria (normal residents of the intestine), 15% is inorganic matter, 3% is protein, 20% is fat, and 30% is undigested roughage.

22-D *Feeding and Digestion in Herbivores*

Most of the aquatic plant life on earth consists of tiny floating plants. It is therefore not surprising that most aquatic herbivores are filter feeders, straining these minute plants out of large volumes of water (Figure 22-8). The filtering system usually traps the plants in mucus, which is then moved to the gut by cilia.

Animals that eat terrestrial plants must cope with the tough cell walls that help support plants in air. These cell walls are largely cellulose, but animals do

(a)

(b)

(c)

Figure 22-8
Feeding methods of herbivores. (a) A sabellid, a marine annelid worm, feeds by filtering food out of the water with its feathery tentacles. (b) A leafhopper uses its tubular mouthparts to suck fluid from plants. (c) The caterpillar of a black swallowtail butterfly chews the leaves of carrots, celery, parsley, and related plants. (a, Biophoto Associates; c, Paul Feeny)

not produce enzymes that digest cellulose. Hence, breaking cell walls and digesting cellulose are major problems for terrestrial herbivores.

Most herbivorous insects have mouthparts adapted to breaking or piercing cell walls so that they can feed on the cytoplasm. One of the reasons herbivorous insects such as locusts and grasshoppers are so destructive of crops is that they digest only a small fraction of the food they eat. Termites use their food more efficiently, thanks to symbionts living in their guts (see Figure 19-10). Because these symbionts secrete cellulose-digesting enzymes, termites can live on wood, which contains little except tough cell walls—a food source most insects cannot use.

Herbivorous mammals have greatly enlarged molars that grind plant cell walls (see Figure 22-9). But their most effective adaptation is a collection of symbiotic microorganisms in their guts. These include cellulose-fermenting bacteria, which can use polysaccharides as food under anaerobic conditions. Since the food must be exposed to the bacteria for fairly long periods of time, it must pass through the gut slowly. Consequently, herbivores tend to have longer intestines than do omnivores or carnivores of similar sizes. (Some bacterial fermentation of food probably occurs in the intestines of all terrestrial vertebrates, including omnivores such as pigs, rats, and humans.)

Many vertebrate herbivores house their gut microorganisms in a caecum, a sac set off to one side at the junction of the small and large intestines, and used as a fermentation chamber. (In humans, the caecum is a blind sac at one end of the large intestine. The narrow appendix was probably once a functioning caecum, which has become reduced in size as we evolved toward a less herbivorous diet.)

Digestion aided by microorganisms has reached its greatest complexity in the **ruminants.** In these mammals, the end of the esophagus and beginning of the stomach form a large sac, the **rumen,** containing microorganisms in an alkaline fluid. Most ruminants are even-toed ungulates, including sheep, cattle, and deer. Ruminant digestion has also evolved in unrelated animals, including some marsupials, colobus monkeys, sloths, and even in the hoatzin, an unusual bird.

There are many advantages to using symbionts for digestion. Some microorganisms can make amino acids from urea and ammonia. Animals do not have enzymes that can do this. Hence microorganisms are valuable when the diet is low in protein. In addition, symbionts produce many vitamins, especially B vitamins, which the host can use. Herbivores such as baboons, which do not have ruminant digestion, have to eat meat occasionally (grasshoppers, snakes, baby monkeys) to replenish their B vitamins. A ruminant needs few dietary vitamins except vitamin A (which can be made from a molecule common in plants) and vitamin D (which is less common).

Figure 22-9
Tooth form and function. Skulls of a mammalian carnivore and a herbivore show teeth specialized for different functions. The carnivore's incisors and canines are pointed and blade-like, specialized for cutting and pulling. The saw-like molars and premolars are used to slice and tear as the jaws move mainly up and down, like scissors; they have limited ability to chew. In the herbivore, in contrast, the incisors are specialized to clip vegetation. The molars and premolars are broad and flattened, and the jaws move sideways to grind leaves and grasses.

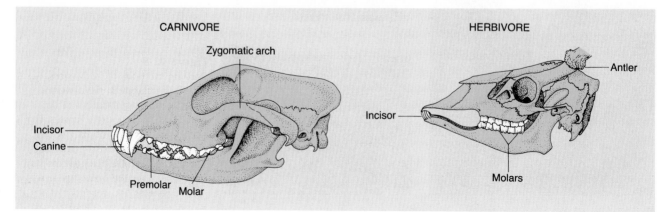

CARNIVORE

Zygomatic arch

Incisor
Canine
Premolar   Molar

HERBIVORE

Antler

Incisor

Molars

22-E Adaptations of Mammalian Carnivores

Herbivores eat food that is hard to digest. Carnivores, on the other hand, encounter their main nutritional problems in catching their food in the first place. Since the food of carnivores consists of other animals, its chemical composition is very similar to that of their own bodies, with little of the waste that results from eating plants with thick cell walls and a high water content.

Carnivorous mammals have teeth adapted to killing their prey and shredding it into bite-size pieces (Figure 22-9). The canine teeth are often elongated into fangs that can inflict extensive damage. The muscles and bones of the skull are powerful, enabling these animals to subdue their meals without damaging their jaws. The molars are modified so that they resemble the blades of short saws, adapted to shredding meat into chunks that can be swallowed. Extensive chewing is not necessary, since there are no thick cell walls to break. The strong stomach acid and powerful protein-digesting enzymes make short work of the food, and the intestine is short compared with that of herbivores and omnivores of the same size.

■ *Herbivores eat food (plants) that is often abundant but hard for digestive systems to deal with. In contrast, the animals eaten by carnivores are harder to capture, but more nutritious and easier to digest and use.*

22-F Feeding in Birds

A bird's jaws are composed of a beak (or bill) made up of bone and keratin, a protein also found in hair and fingernails. The beak is modified according to the feeding habits of the species (Figure 22-10). Modern birds have no teeth, and so birds that eat hard food grind it in a muscular gizzard, the hindmost part of the stomach. The gizzard usually contains stones that the bird picks up and swallows. This arrangement moves weight from the head (where teeth would have been) closer to the center of gravity, an adaptation that makes flying more efficient.

A special organ found particularly in seed-eating birds is the crop, a storage sac between the esophagus and the stomach. Birds use a lot of energy in flying and maintaining a high body temperature. Storing food in the crop ensures an almost continuous supply of food to the stomach, and so the bird does not have to spend all its time eating. Birds generally use their food more efficiently than do mammals. A three-week-old stork converts about 33% of the weight of its diet

Figure 22-10
Bird bills. (a) The thick, stout bill of a cardinal is suited to crushing seeds. (b) A pelican's bill is huge, with a soft, fleshy bag on its undersurface. The bill is used to scoop up water and fish when the bird dives. When it surfaces, the pelican points its bill down to drain out the water and then tilts it up to swallow the fish. (a, Kirtley-Perkins/Visuals Unlimited; b, Biophoto Associates, N.H.P.A.)

(a)

(b)

of fish into stork. This compares with a food use efficiency of about 10% in a young mammal.

22-G Functions of the Mammalian Liver

■ *The liver is a major regulatory center, controlling the organic components in the body's blood and other fluids.*

The liver's main function is to control the level of many substances in the blood. Digested food molecules absorbed into the bloodstream from the intestine pass directly to the liver by way of the **hepatic portal vein.** Before these molecules pass to the rest of the body, the liver may change their concentration and even their chemical structure. The liver detoxifies some otherwise poisonous substances. For instance, it produces enzymes that metabolize various drugs. It also stores food molecules, converts them biochemically, and releases them back into the bloodstream as they are needed. For instance, the liver removes glucose from the blood under the influence of the hormone insulin and stores it as glycogen. When the level of glucose in the blood falls, the hormone glucagon causes the liver to break down glycogen and release glucose into the blood.

Liver cells make many of the blood proteins. In addition, they convert nitrogenous wastes into urea, which can be excreted by the kidneys. Together with the kidneys, the liver is vital in regulating what the blood contains when it reaches all the other organs of the body. Because the liver is the body's major organ for making all these biochemical adjustments, severe liver damage or loss of the liver is rapidly fatal.

22-H Stored Food and Its Uses

The body's carbohydrate stores (glycogen in the liver and muscles) would supply its energy needs for only about 12 hours if they were used alone. However, a human being of normal weight can usually survive without food for about six weeks, using fat reserves for energy (Figure 22-11). A hibernating animal, with a lowered body temperature, can survive for months on the fat reserves that it built up by eating extra food during the autumn.

Because a given weight of fat provides about twice as many calories as the same weight of carbohydrate or protein, energy is stored most compactly in the form of fat. Fat is stored in the fat cells of **adipose tissue,** a storage tissue found under the skin and elsewhere in the body. Fat is constantly exchanged between the bloodstream and adipose tissue: every molecule of fat in adipose tissue is replaced about every three weeks.

22-I Regulation of Feeding

Feeding is governed by two kinds of controls: long-term and short-term. Long-term regulation ensures that enough stored food is maintained in the body. It can be altered by the action of hormones that ensure, for instance, that an animal

Figure 22-11
The fate of stored food in a starving human being whose initial body weight was 15% fat. (A human being of average weight takes weeks to die from starvation, although death from thirst occurs in a few days.)

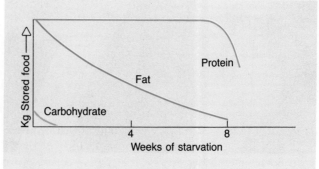

Figure 22-12
A sheep feeding. Sheep are widely used in studies of feeding behavior and physiology in herbivores.

builds up its fat reserves before its hungry young are born or hatched, and before hibernation. Short-term regulation ensures that an animal eats regularly on a day-to-day basis, so that food passes through the gut more or less continuously.

The control of feeding is poorly understood. The brain contains various centers that, when stimulated, start or stop feeding and control the selection of food. There are various hypotheses, but no real answers, as to what stimulates these brain centers in a normal animal.

Why an animal eats *what* it does is even more complicated. For carnivores, things are reasonably simple because carnivores eat food that is nutritious and nontoxic. Carnivores rapidly learn to avoid food that tastes unpleasant, as you realize if you have ever seen a cat attack a toad.

Herbivores face a bigger problem because many plants contain toxic chemicals, and even more (including some of our common foods) are toxic when eaten in large amounts. Furthermore, few herbivores can eat just one kind of plant because no single plant species contains the complete mix of nutrients that animals need. A sheep presented with a field of grasses, most of which are new to it, adds only one new species to its diet at a time and eats very little of that (Figure 22-12). Presumably, this gives the long-term learning system time to "tell" the sheep whether the new plant is a good thing or not. Using this system, a herbivore can work up to a varied and nutritious diet.

Small, slow-moving herbivores, such as caterpillars, may eat only one plant in their lives—the one where they hatch from the egg. Special arrangements make up for the nutrient deficiencies of the food plant. For example, the caterpillar of a tiger swallowtail butterfly cannot obtain all the essential amino acids and sodium it requires from its diet of cherry tree leaves. The caterpillar hatches with extra supplies of these nutrients, deposited in the egg by its mother, who obtained them as an adult by drinking nectar from flowers.

Specific Appetites

One of the least-understood aspects of feeding is the development of an appetite for a specific substance. Often an animal with a long-term deficit of fat in its diet will develop a craving that can be satisfied only by fat. Children with calcium deficiencies have been known to eat the plaster off walls (which is mainly $CaCO_3$). Laborers in the tropics frequently develop salt deficiency because of the volume of sweat they produce. They drink salt water or salty beer and find it delicious, although it tastes repulsive to anyone who is not short of sodium.

Plainly, specific appetites are valuable because they lead animals to seek and eat food containing the nutrients they need. Somehow, the deficiency of a particular nutrient changes the reaction of the sense organs or the brain to potential food, so that food containing that nutrient tastes or looks much more appetizing to an animal that needs that substance, but we have no idea how this happens.

SUMMARY

Because animals are heterotrophs, their diet must contain all of the organic and inorganic substances they need for metabolism, growth, and energy production. Animals obtain fats, carbohydrates, proteins, vitamins, and minerals in their food.

The function of digestion is to break food down into molecules that can be absorbed from the digestive tract into the body. In the vertebrates, digestive enzymes are made in the salivary glands, pancreas, and lining of the stomach and small intestine. Many animals, particularly herbivores, harbor symbiotic microorganisms that digest food in the alimentary canal.

Digested food is absorbed into the body by diffusion, facilitated diffusion, and active transport across the enormous surface area of the small intestine.

The liver plays a major role in controlling the fate of newly absorbed food molecules. It stores excess glucose as glycogen, produces many blood proteins, and converts nitrogenous and other wastes into a form that can be excreted by the kidneys. Excess carbohydrate or protein is converted into fats and stored in adipose tissue.

Feeding is regulated by long-term and short-term control mechanisms that are not well understood. These controls ensure that the alimentary canal is efficiently occupied most of the time and that the animal maintains its body reserve of fat without spending unnecessary time feeding. All animals have complicated regulatory systems that control what, as well as how much, they eat.

SELF-QUIZ

Matching: For each numbered phrase below, choose the letter of the correct class of nutrient on the right. More than one letter may be correct; choose all that apply.

____ 1. Inorganic nutrients
____ 2. Macronutrient that cannot be stored in the body
____ 3. May be the source of energy for the body's metabolism
____ 4. Source of material for cell membranes
____ 5. Coenzymes for metabolic enzymes
____ 6. Digested by enzyme in saliva
____ 7. Absorbed in large intestine

a. protein
b. carbohydrate
c. fat
d. water-soluble vitamins
e. fat-soluble vitamins
f. minerals

8. The main advantage of having a digestive tract with a mouth and anus is:
   a. it permits different parts of the gut to become specialized to perform different parts of the digestive process in turn
   b. it permits an animal without teeth to have a means of grinding its food
   c. it permits animals to eat a great deal at once and digest it while doing something else
   d. it permits animals to eat larger organisms as food
   e. it permits animals to eat food in larger chunks

9. In humans, digestion of food is completed in the:
   a. small intestine
   b. mouth
   c. large intestine
   d. stomach
   e. rectum

10. In humans, protein digestion is carried out by enzymes secreted by the:
    a. stomach, pancreas, and salivary glands
    b. liver, salivary glands, pancreas, and small intestine
    c. salivary glands, stomach, pancreas, and small intestine
    d. liver, stomach, pancreas, and small intestine
    e. stomach, small intestine, and pancreas

11. A portion of the stomach that has evolved extremely thickened muscular walls and is quite efficient at grinding hard food is called a(n):
    a. rumen
    b. gizzard
    c. crop
    d. pancreas
    e. caecum

12. Which of the following is probably *not* an action of symbiotic microorganisms of the gut?
    a. use of the host's food for their own nutrition
    b. extracellular digestion
    c. respiration
    d. breakdown of substrates that the host cannot digest
    e. manufacture of vitamins needed by the host animal

13. Which of the following is *not* a function of the mammalian liver?
    a. secretion of digestive enzymes for export to the gut
    b. regulation of blood glucose and amino acid content
    c. production of the nitrogenous waste urea
    d. production of plasma proteins for the blood
    e. detoxification of poisonous substances

QUESTIONS FOR DISCUSSION

1. Herbivores can seldom survive by eating only one species of plant (for example, corn is low in the amino acids tryptophan and lysine; many plants contain too little sodium). This is probably no evolutionary accident. What's in it for the plant?
2. Why does an alcoholic drink affect the body so quickly?
3. Why does it take longer to become hungry after a protein-rich meal than after a meal that is mostly carbohydrate?
4. Some kinds of stress can upset an animal's normal nutrient balance. For example, infection increases the rate at which vitamin C is used. How might the organism compensate for this disturbance?

SUGGESTED READINGS

Brown, M. S., and J. L. Goldstein. "How LDL receptors influence cholesterol and atherosclerosis." *Scientific American,* November 1984.

Cohen, L. A. "Diet and cancer." *Scientific American,* November 1987. Discussion of the National Research Council recommendation that a U.S. diet containing less fat and refined sugar and more fiber would lower the risk of cancer. Discusses what we can learn about our dietary needs from the diets of our Stone Age ancestors.

The Surgeon General's Report on Nutrition and Health, 1988. Washington, DC: Science News Books, 1988. A summary of recent research on the relationship between diet and health.

Uvnas-Moberg, K. "The gastrointestinal tract in growth and reproduction." *Scientific American,* July 1989. A treatment of the digestive tract as the largest hormone-secreting organ in the body. Discusses the role it plays in adjusting digestion and diet to pregnancy and fetal and infant growth.

Gas Exchange in Animals

CHAPTER

23

Jellyfish

Alligator

Angelfish

Versicolored emerald hummingbird

From a climber struggling to the top of Mount Everest with an oxygen cylinder, to a shark gliding through the depths of the ocean, every animal is continually exchanging gases with its environment. Why is this constant interchange necessary? As we saw in Chapter 7, living cells obtain energy to drive their activities from the oxidation of food molecules, usually by cellular respiration. Respiration requires oxygen, which must be obtained from the environment, and produces carbon dioxide, which the body must expel. Small animals can obtain oxygen and give off carbon dioxide directly through plasma membranes on the outer surfaces of their bodies. These gases do not have far to go to reach any cell in the body. The evolution of larger animals has been possible, however, only because they have evolved specialized respiratory and circulatory systems. The respiratory system provides a special, large surface area to take up oxygen from the environment and eliminate carbon dioxide from the body. The circulatory system distributes oxygen to the body's cells and collects their carbon dioxide for the return trip to the respiratory surface. This chapter looks at the behavior of gases and describes some of the arrangements for gas exchange found in animals.

- Gas exchange between cells and their environment occurs by diffusion. In large or active animals, gas exchange is speeded up by ventilation of a respiratory surface, facilitated diffusion, and transport by respiratory pigments in a circulatory system.
- The main respiratory organs in animals are the body surface, gills, tracheae, and lungs.

23-A Supplying Oxygen

Most of the earth's oxygen is in the air, but some is dissolved in water. Either water or air may serve as an animal's **respiratory medium,** the immediate source of oxygen.

The respiratory medium gives up oxygen to the body at the body's **respiratory surface,** which may be the general body surface or a specialized area (Figure 23-1). Most of an animal's cells lie some distance from the respiratory surface and obtain their oxygen from the **extracellular fluid,** the fluid that bathes every cell. They rely on the blood to bring oxygen from the respiratory surface to the extracellular fluid and to remove carbon dioxide by the reverse route (Figure 23-2).

In many animals, the process of **ventilation** moves the respiratory medium and brings a fresh supply of oxygen past the respiratory surface. Ventilation is necessary because gas exchange depends on diffusion. The greater the concen-

Figure 23-1
A nudibranch (a marine mollusc). The finger-like projections increase the surface area of the body available to obtain oxygen from the surrounding seawater. The animal's blood then carries the oxygen from the body surface to all the cells and picks up carbon dioxide for the return trip. (Charles Seaborn)

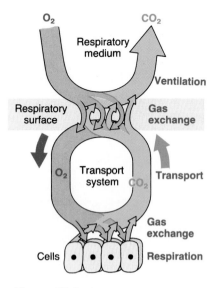

Figure 23-2
Summary of gas exchange. Gas is exchanged between the respiratory medium and body fluids at the respiratory surface (top), and between the body fluids and living cells (bottom). Oxygen enters the body fluids from the respiratory medium, whereas carbon dioxide enters the system as it is produced by body cells. CO_2 is picked up from the cells by the body fluids and leaves the body at the respiratory surface.

■ *Air contains more oxygen than water does and also weighs less. However, air-breathing animals face the danger of dehydration.*

Table 23-1	Oxygen Content of Some Respiratory Media
Medium	**Oxygen content (millilitres per litre)**
Sea water at 5°C	6.4
Fresh water at 5°C	9.0
Fresh water at 25°C	5.8
Air	209.5

tration gradient of gases across the respiratory surface, the faster gases will diffuse between the respiratory medium and the blood. Near the respiratory surface oxygen is quickly removed from the respiratory medium, and carbon dioxide rapidly builds up. This reduces the concentration gradient of both gases and slows their diffusion across the respiratory surface. Ventilation creates a current that brings a fresh supply of air or water, which renews the oxygen and removes the carbon dioxide at the respiratory surface, permitting gas exchange to occur as rapidly as possible.

Oxygen moves from the environment into the blood partly by diffusion and partly by **facilitated diffusion,** using a carrier protein (Section 4-D). The carrier is a cytochrome, P450, which speeds up diffusion, allowing the blood to pick up oxygen faster.

P450 also oxidizes toxic substances, rendering them less harmful to the body. (Although the liver is the main organ of detoxification, the lungs also contribute to this activity.) The ready supply of oxygen at the lung surface makes it an ideal site for oxidizing foreign substances.

23-B Problems of Gas Exchange

What problems are involved in moving oxygen from the external medium to the extracellular fluid surrounding a cell? First, the amount of oxygen available from the environment varies. Compared with air, water contains little oxygen, and the warmer or saltier the water, the less oxygen it can hold (Table 23-1). Oxygen also diffuses half a million times faster in air than in water. On the other hand, there are difficulties with breathing air. Gas molecules must cross plasma membranes in solution, and so respiratory surfaces must always be moist. Air-breathing animals lose a lot of their precious water by evaporation from the respiratory surface into the air.

How fast an animal can absorb oxygen depends on the amount of oxygen available and on the area of respiratory surface exposed to the respiratory medium. All but the smallest animals have adaptations that increase the area of the respiratory surface. Since carbon dioxide is much more soluble in water than is oxygen, any respiratory surface large enough to supply an animal with oxygen is well able to dispose of carbon dioxide quickly enough.

A final consideration in gas exchange is how much oxygen an animal needs. This depends on its size (that is, how many cells must be supplied with oxygen) and on its activity. An active animal uses energy faster than a sluggish one, and so must obtain oxygen faster. A human being uses oxygen 15 to 20 times faster during exercise than at rest. So the respiratory and circulatory systems must be able to respond to the increased oxygen demand when an animal is active.

An animal's **metabolic rate** is the rate at which it releases energy from its food to drive its metabolism. Since most energy is released by cellular respiration using oxygen, metabolic rate is usually measured as the volume of oxygen used per unit of body weight per unit of time. Thus:

$$\text{oxygen requirement} = \text{metabolic rate} \times \text{body weight}$$

Homeothermic (warm-blooded) animals maintain constant body temperatures that are usually higher than those of the surroundings. They have high metabolic rates because they continually oxidize food. This produces heat that replaces what is lost across the body surface. Since surface-to-volume ratio increases with decreasing size (see Figure 19-22), small homeothermic animals lose relatively more heat than do large ones and so have higher metabolic rates (Figure 23-3). (However, a large animal still requires more *total* oxygen per unit time than does a small one.)

Poikilotherms, animals that do not maintain a constant body temperature, may show a drastic change in oxygen consumption with changing environmen-

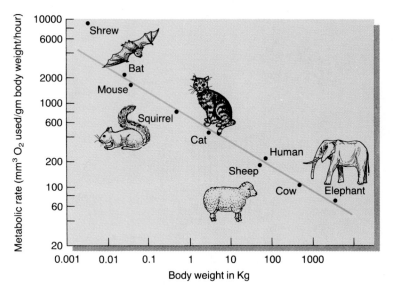

Figure 23-3

Size and metabolic rate. The metabolic rates of homeothermic animals are inversely proportional to their size. Notice that both axes in this graph have logarithmic scales. Both body weight and metabolic rate range over several orders of magnitude for the mammals indicated.

tal temperature (Figure 23-4). The higher the temperature, the faster the chemical reactions of metabolism occur, the higher the animal's metabolic rate, and the more oxygen it needs. An animal that obtains its oxygen supply from water has a dual problem as the temperature rises. It needs more oxygen because of the rise in its metabolic rate, but less oxygen is dissolved in water at higher temperatures (see Table 23-1). A few active fish, such as trout, can obtain enough oxygen only in very cold water. Thus the discharge of waste heat (for instance, from a power plant) into a lake or stream may ruin the trout fishing.

23-C Respiratory Surfaces and Ventilation

Four main types of respiratory surface are used by animals: the body surface, gills, lungs, and tracheae (Figure 23-5).

The Body Surface

Some animals, such as earthworms, obtain all the oxygen they need across the general body surface. To do this, the animal must be fairly small, so that it has a high surface-to-volume ratio. The body surface must also be kept moist. Third,

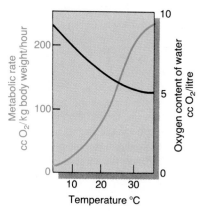

Figure 23-4

How temperature affects the metabolic rates of poikilotherms. Metabolic rates vary with the temperature of the environment (blue line). For an aquatic animal, higher temperatures decrease the amount of oxygen available, because less oxygen dissolves in warm than in cold water (black line).

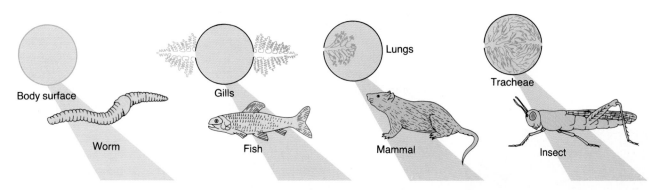

Figure 23-5

The four main types of respiratory surfaces. Animals such as sponges, cnidarians, and many worms use the general body surface as the respiratory surface. The skin is also used for gas exchange in amphibians. Gills are outgrowths with large surface areas. They may be exposed directly to the water or covered by other body parts, as in many molluscs, arthropods, and fish. The lungs of many vertebrates and some molluscs, and the tracheae of insects, are internal respiratory surfaces. Lungs provide localized gas exchange, whereas tracheae branch throughout the body, bringing air close to all cells.

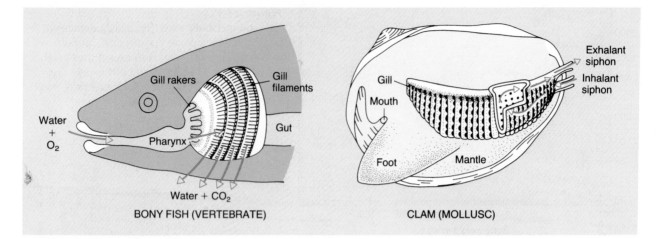

BONY FISH (VERTEBRATE)   CLAM (MOLLUSC)

Figure 23-6
Gill-breathing animals. A one-way current of water crosses the respiratory surface. In the fish, water enters through the mouth and passes out across the gills. In the clam, water enters through the inhalant siphon, passes through pores in the gills, and leaves through the exhalant siphon.

the animal must have a fairly low metabolic rate. Fourth, the thin moist surface must be protected from injury; it is often covered with a slimy mucus that makes it too slippery to be damaged by sharp objects.

The body surface is also used to supplement gas exchange by some vertebrates with lungs or gills, especially amphibians (frogs, toads, salamanders, and newts). Many amphibians obtain 25% or more of their oxygen across their thin moist skin. Indeed, members of one group of salamanders have no lungs and carry out all of their gas exchange through the skin.

Gills

Animals such as fish and many aquatic arthropods, molluscs, and amphibians carry out gas exchange through the thin membranes of **gills,** feathery tissue outgrowths that are exposed to the water.

Water is heavy for the amount of oxygen it contains. It takes considerable energy to push a constant current of water across the gill membranes. It would take even more energy to stop the water, reverse its direction, and pass it back out of the gill area the same way it came in. Instead, in most aquatic animals, water enters the gills through one opening and exits by another in a continuous one-way flow.

The gills of a bony fish are located behind the head (Figure 23-6). Water enters through the mouth, passes across the gills, and leaves via the opening behind the **operculum,** which covers the gills.

In animals such as many bivalve molluscs and the lower chordates (amphioxus, tunicates [Section 21-K]), ventilation of the gills is linked to feeding; food is filtered out of water drawn into the gill area, and gas exchange occurs at the same time. Some animals use respiratory water currents for locomotion by squeezing water forcibly out of the gill area. Squid, for instance, can eject water from the siphon with considerable force, creating a jet-propulsion stream that moves them rapidly forward or backward, depending on which way the siphon is aimed (Figure 23-7).

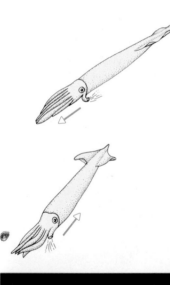

Figure 23-7
The squid can move forward or backward (blue arrows) depending on which way the siphon is aimed as it ejects water from the mantle cavity surrounding its gills. The photograph shows a young squid swimming. (Langdon Quetin)

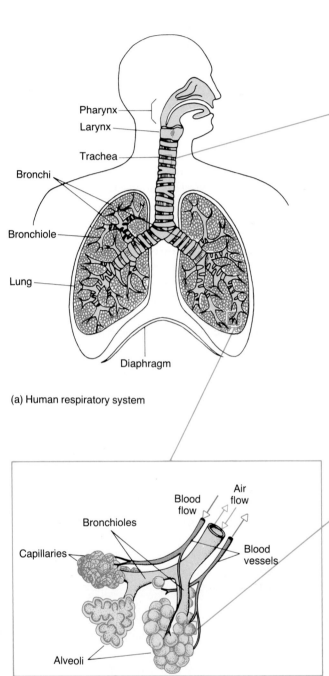

(a) Human respiratory system

Labels on diagram (a): Pharynx, Larynx, Trachea, Bronchi, Bronchiole, Lung, Diaphragm

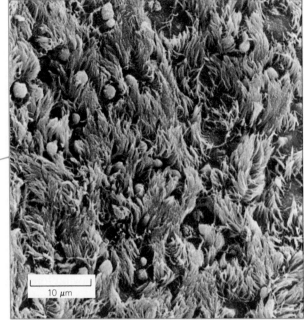

10 μm

(b) Ciliated lining of trachea

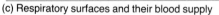

Labels on diagram (c): Blood flow, Air flow, Bronchioles, Capillaries, Blood vessels, Alveoli

(c) Respiratory surfaces and their blood supply

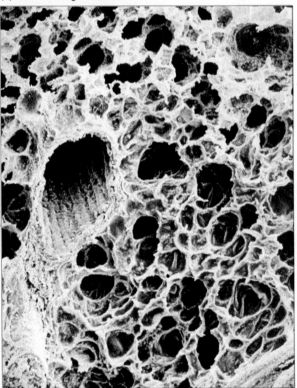

(d) Bronchiole and alveoli

Figure 23-8

The human respiratory system. (a) The anatomy of the respiratory system. Air travels from the nose or mouth to the lungs by way of the air passages: the trachea, bronchi, and bronchioles (lavender). The actual respiratory surfaces are the walls of the numerous alveoli in the lungs (red). (b) A scanning electron micrograph of the inside of the trachea. The globular structures are glands that secrete mucus. Most of the surface is covered with cilia, which move the mucus and any foreign particles it traps up into the throat. (c) The bronchioles end in tiny sac-like alveoli throughout the lungs. Each alveolus is surrounded by capillaries of the blood system. Gas exchange takes place across the moist surfaces of the alveoli, between air inside the alveoli and blood flowing through the capillaries. (d) Scanning electron micrograph of a section of lung tissue, showing part of a bronchiole surrounded by spongy alveoli. Note the vast alveolar surface area. (b, d, Biophoto Associates)

Lungs

Because air is less dense than water, it takes less energy to move it to and from the respiratory surface. In addition, air contains about 30 times as much oxygen as the same volume of cold water, and oxygen molecules diffuse 500,000 times faster in air than in water. Thus air-breathing animals can ventilate their lungs by a **tidal** (in-and-out) flow rather than by the one-way stream usual with gills. This makes it unnecessary to have two respiratory openings on the surface of the body. On the debit side, breathing air presents the problem of losing water by evaporation from the respiratory surface.

As an example of a respiratory system with lungs, let us examine our own (Figure 23-8). Air enters the body through the nose or mouth. It passes through the **pharynx,** a common passageway for both air and food, and enters the **trachea** (windpipe) by way of the **larynx,** also known as the voice box or Adam's apple. The walls of the trachea contain rings of cartilage which hold the tube open. On the inner surface of the trachea, cilia keep the air passages clear by moving foreign particles up into the pharynx, where they can be swallowed. The lower end of the trachea divides into two **bronchi,** which divide into finer and finer tubes, the **bronchioles.** The smallest bronchioles end in a myriad of tiny sacs, the **alveoli,** whose thin walls are the actual respiratory surfaces. A vast network of capillaries surrounds the alveoli. Blood in these capillaries picks up oxygen for transport to the rest of the body and gives up carbon dioxide.

Much air stays in the alveoli even when we breathe out as much as possible. This air keeps the walls of the alveoli from sticking together and collapsing.

How does an air-breathing vertebrate ventilate its lungs? There are two main ways, positive pressure breathing and negative pressure breathing. A frog breathes by a **positive pressure** mechanism (Figure 23-9). At the beginning of a ventilation cycle, the frog opens its nostrils and lowers the floor of its mouth. Enlargement of the mouth cavity creates a partial vacuum there, and air enters through the nostrils. (This part of the frog's breathing cycle actually operates on the negative pressure principle, described in the next paragraph.) The frog then closes its nostrils and raises the floor of its mouth, pushing the air into the lungs. After holding the air in its lungs, the frog pushes the air back out by opening its nostrils and contracting its abdominal muscles.

Mammals have **negative pressure breathing** (Figure 23-10). They have a respiratory muscle, the **diaphragm,** not found in other vertebrates. The diaphragm extends across the bottom of the chest cavity, beneath the lungs, closing off the chest cavity from the abdominal cavity below. During inhalation, the muscles between the ribs contract and lift the ribs outward. At the same time, the diaphragm contracts and so moves lower. These movements increase the volume of the chest cavity and so decrease the pressure within it. Air then rushes into the nose or mouth, down the trachea, and into the lungs. During exhalation, relaxation of the rib muscles and diaphragm decreases the volume of the chest cavity, increasing the pressure inside it and forcing air back out of the lungs. (You can get an idea of how negative pressure breathing works by closing your mouth and holding your nose while you expand your rib cage and lower your diaphragm. You will feel the partial vacuum created. Then, remove your fingers from your nose and you can hear and feel the air rushing in as the pressure equalizes.)

Negative pressure breathing allows an animal to eat and breathe at the same time. If it were necessary to push air from the mouth into the lungs, any food in the mouth might be pushed into the trachea and cause an obstruction. Negative

■ *Animals that obtain oxygen from water use gills or the general body surface as the respiratory surface. Air-breathing animals use lungs, a tracheal system, or the moist body surface.*

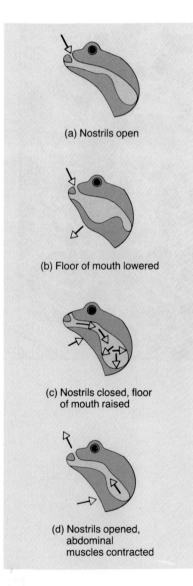

(a) Nostrils open

(b) Floor of mouth lowered

(c) Nostrils closed, floor of mouth raised

(d) Nostrils opened, abdominal muscles contracted

Figure 23-9
Positive pressure breathing in a frog. Movements of the mouth force air into the lungs.

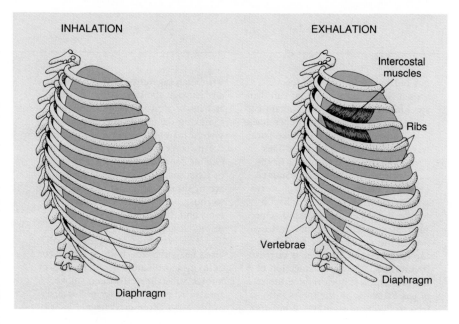

INHALATION   EXHALATION

Intercostal muscles

Ribs

Vertebrae

Diaphragm

Diaphragm

Figure 23-10
Negative pressure breathing. During inhalation, the ribs are lifted up and out and the diaphragm is lowered. This increases the size of the chest cavity and decreases the pressure within it. Air rushes in through the nose, mouth, or both. Exhalation occurs when the breathing muscles relax, forcing air back out; it is a passive process.

pressure breathing creates a more gentle stream of air, which is less apt to pull food along into the air passages.

Tracheal Systems

Air-breathing vertebrates have lungs. The other major group of land animals, the terrestrial arthropods (insects, centipedes, millipedes, and some spiders), breathe air by means of **tracheae,** air tubes that extend throughout the body. The tracheal system does not rely on the circulation to transport gases. Instead, the tracheae start at **spiracles,** tiny openings at the body surface, and branch into all parts of the body, ending close to every cell (Figure 23-11). Chemicals given off by cells that are short of oxygen induce tracheae to grow branches into that area.

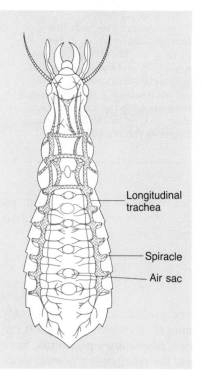

Longitudinal trachea

Spiracle

Air sac

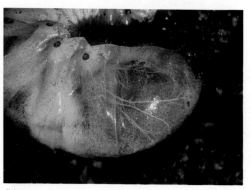

(b)

Figure 23-11
Tracheal systems. (a) The tracheal system of an insect. (b) The abdomen of a beetle larva. Branching white tracheae look like string when seen through the transparent cuticle. The red-brown circles are thickened areas around the openings of spiracles.

473

Air Pollution and Smoking

A JOURNEY INTO LIFE

In Athens, Greece, six times as many people as usual die on days when the air is heavily polluted. The Hungarian government attributes one in 17 deaths to outdoor air pollution and the American Lung Association blames 120,000 U.S. deaths each year on this cause. Most urban air pollution comes from motor vehicles and industry. But the air inside a building is nearly always more polluted than the air outside it. Sources of indoor air pollution include fungi and dust in air conditioning and heating ducts, glue in chipboard and other building materials, plastic in furniture and carpets, tar, particles, and smoke from open fires, stoves, cooking ranges, and furnaces. All these pollutants can damage our bodies.

Most of the research on health effects of air pollution has focused on smoking tobacco and marijuana because smoking introduces pollutants directly into the lungs at high concentrations.

Since smoke is drawn into the lungs by way of the respiratory air passages, the respiratory system comes into closest contact with smoke and is most directly affected by the chemicals it contains. Most of the particulate matter in smoke set-tles out in the lungs. **Tars** form a brown, sticky substance that can damage lung tissue. They also contain carcinogenic (cancer-causing) chemicals.

Nicotine in tobacco smoke paralyzes the cilia in the bronchi. This permits debris to accumulate in the air passages. Hence smokers suffer excessive damage to the respiratory system from the effects of tobacco smoke and other air pollutants. One such effect is **emphysema**, destruction of the gas exchange surface of the lungs' alveoli. Another is **chronic bronchitis**, a low-level infection of the bronchioles. Both reduce the rate of gas exchange in the lungs and can lead to disabling shortage of breath. Bronchitis and other respiratory infections result partly from failure of paralyzed cilia to sweep away disease-causing bacteria, and partly from smoke's inhibition of bacteria-fighting immune cells that reside in the lungs.

All smoke also contains carbon monoxide, produced by incomplete burning. This gas occurs in smokers' blood at 4 to 15 times the level found in nonsmokers (depending on how much the person smokes). Carbon monoxide combines with hemoglobin and reduces the amount of oxygen the blood can carry. Hemoglobin binds carbon monoxide very tightly, and so the gas lingers in the blood, robbing the body of oxygen, for as long as 6 hours after the cigarette is finished.

The effects of carbon monoxide can be seen from smoking just one cigarette. Smoking one cigarette provides enough nicotine to cause blood vessels to constrict (become narrower). This cuts down the flow of blood and oxygen to the body. It also speeds up the heart rate and increases blood pressure. A steady smoking habit renews these effects with each cigarette. This can eventually damage the circulatory system and lead to cardiovascular disease. (Smoking damage to the circulatory system may have other consequences. For instance, smokers are usually not eligible for procedures such as kidney or liver transplants. Transplant organs are in short supply and must be given to those with healthy circulatory systems and, therefore, a greater chance of recovering from the operation.)

Another effect of breathing polluted air is that the body is forced to detoxify inordinately large amounts of foreign substances. Tobacco smoke, anesthetic gases, or other pollutants in the lungs force cytochrome P450 (Section 23-A) to use up a lot of oxygen detoxifying these substances, instead of transferring this oxygen to the blood.

Pregnant women are especially affected by pollutants in the air. Not only do pollutants reduce the amount of oxygen transported from the lungs into the mother's bloodstream, but they also restrict the oxygen supply to the fetus. If the P450 in the placenta is tied up detoxifying substances from the mother's blood, the fetus loses a large percentage of its oxygen supply. This may account for the fact that the rate of miscarriage and birth defects is 30% higher for nurses who work with anesthetics than for women in other occupations, and that women who smoke tend to give birth to small babies.

Fetuses are not the only nonsmokers affected by smoking. The smoke from an "idling" cigarette contains more tar, nicotine, and cadmium (a toxic metal) than does the smoke inhaled by a smoker. People inhaling cigarette smoke in the air around them are affected by the same kinds of changes as the smoker, but to a

Large or active insects ventilate the tracheal system by pumping air in and out with their abdominal muscles. The smallest insects need only open their spiracles and let diffusion bring oxygen to their cells.

23-D Respiratory Pigments

Blood is largely salt water, which cannot carry much dissolved oxygen. Many animals have adaptations that increase the amount of oxygen their blood can carry. The most common adaptation is to possess **respiratory pigments**, large protein molecules that bind and release oxygen. The respiratory pigment hemo-

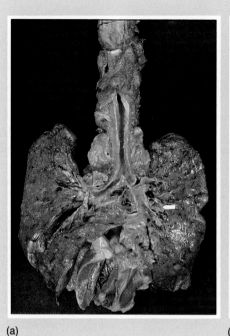

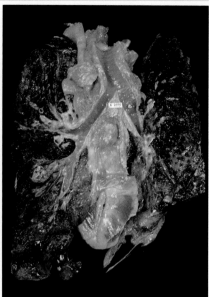

(a)

(b)

Figure 23-A
Effects of smoking on human lungs. (a) Normal human lungs. (b) Lungs of a cigarette smoker. (Martin M. Rotker/ Taurus Photos Inc.)

lesser degree. Children exposed to such "passive smoking" in the home have twice as many respiratory infections as children of nonsmokers.

In 1919, doctors in a Boston hospital hastily called medical students to an autopsy to see a type of cancer so rare that they might never witness it again: lung cancer. Today, lung cancer is one of the most common types of cancer. There appear to be many carcinogens in tobacco smoke. They cause several types of lung cancer, most of them highly malignant and rapidly fatal. Cells in the lungs of smokers show a number of chromosomal abnormalities, and cells in the nearby larynx and esophagus, and even more distant organs such as the kidneys and liver, also show DNA damage. In addition, the breakdown

products of the radioactive gas radon become attached to particles in smoke and lodge with them in the lung. Because smoking exacerbates the effect of radon in this way, about 85% of the cancers attributed to radon occur in smokers. Smokers are similarly much more likely to die of lung cancer caused by inhaling asbestos fibers than are nonsmokers. Studies have implicated smoking in various cancers other than lung cancer. Benzene in cigarette smoke is blamed for about 550 cases of adult leukemia every year.

In summary, smoking is a major cause of emphysema, chronic bronchitis, cancer, and heart disease. Smokers can expect to live about ten years less than nonsmokers. About one in every 15 smokers can expect

to die of lung cancer (some 40,000 people each year), and several times that number will die prematurely of cardiovascular disease. In fact, smoking tobacco causes greater health damage than any other drug, legal or illegal.

With all these disadvantages, why do so many people smoke? For reasons that are not fully understood, smoking is highly addictive. However, after learning of the adverse affects of smoking, over 50 million people have managed to stop. In people who quit, the physiological damage done by smoking is gradually reversed. Ten years later, a person's chance of suffering adverse health effects from a previous smoking habit are hardly any higher than those of a lifelong nonsmoker.

globin in the red blood cells of a mammal carries 98% of the oxygen in the blood.

Hemoglobin (Hb) is the general name for a group of oxygen-carrying molecules that all contain a heme group (Figure 23-12). The iron atom in its center is the actual binding site for oxygen. When oxygen is bound to the iron atom, the pigment is said to be **oxygenated** (not oxidized) and is called **oxyhemoglobin (HbO).** Oxyhemoglobin appears bright red, and deoxygenated hemoglobin is a darker, purplish red. The hemoglobin of vertebrates consists of four polypeptide chains, each with a heme group attached (see Figure 3-19).

Hemoglobins are found in almost all vertebrates, and in various invertebrates, including earthworms and some insects. Other invertebrates, such as

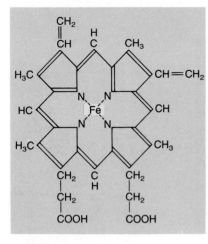

Figure 23-12
The heme group of a hemoglobin molecule. Oxygen attaches to the iron (Fe) atom (yellow box). The structure around the iron atom is attached to the hemoglobin's protein chains.

many arthropods and molluscs, have other kinds of respiratory pigments: copper-containing hemocyanins.

As blood moves through the lungs, its hemoglobin picks up oxygen until it is nearly saturated. That is, almost all the iron atoms in all the heme groups have bound oxygen molecules. The degree of saturation depends on the **partial pressure of oxygen:** the portion of the total air pressure attributable to oxygen. The higher that oxygen pressure, the more oxygen binds to hemoglobin in the blood, up to the saturation point.

The oxygen pressure is lower in the rest of the body than in the lungs because the body's cells continually use up oxygen in respiration. So blood moving through the tissues encounters lower oxygen pressures, which cause hemoglobin molecules to give up some of their oxygen—normally about 37%. During exercise, when the cells use more oxygen than usual, the oxygen pressure drops lower and the hemoglobin releases almost 20% more of its oxygen.

■ *Hemoglobin's oxygen satura-tion level depends mainly on the oxygen pressure in the surrounding fluid and, to a lesser extent, on pH.*

Oxyhemoglobin gives up oxygen more readily in an acid environment, such as active tissues. An exercising muscle releases carbon dioxide from respiration, which reacts with water to form carbonic acid. It also releases lactic acid from fermentation (Section 7-E). These acids cause oxyhemoglobin to give up 10% more of its oxygen. Hence, oxyhemoglobin releases more oxygen to the tissues that need it most.

An animal's hemoglobin shows adaptations to its way of life. For instance, the hemoglobin of a small mammal gives up more oxygen at a given oxygen pressure than that of a large mammal. This is an adaptation to the higher meta-bolic rate and correspondingly greater oxygen demands of small mammals.

An animal may have hemoglobin with different properties at different times in its life. Before birth, human fetuses produce several kinds of hemoglobin that are not made by the body of an adult. After birth, fetal hemoglobin is gradually replaced by adult hemoglobin. All of a fetus's oxygen comes from its mother's bloodstream. If fetal and adult hemoglobins had the same affinity for oxygen, the fetus could not pick up very much of the oxygen released by the mother's blood. But a fetus's hemoglobin has a higher affinity for oxygen than the mother's, permitting it to pick up oxygen at oxygen pressures low enough to cause the mother's hemoglobin to release oxygen.

23-E Carbon Dioxide Transport

Carbon dioxide produced during respiration must be carried by the blood to the lungs, where it is excreted. Some of this carbon dioxide travels combined with hemoglobin and other proteins in the blood.

■ *Carbon dioxide is not simply a waste product. It provides bicarbonate ions, the most important buffer in the blood.*

Most of the carbon dioxide in the blood is carried as dissolved bicarbonate ions. The enzyme **carbonic anhydrase,** in red blood cells, forms carbonic acid from water and carbon dioxide. Carbonic acid dissociates into hydrogen and bicarbonate ions:

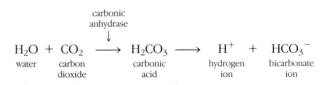

$$\underset{\text{water}}{H_2O} + \underset{\substack{\text{carbon} \\ \text{dioxide}}}{CO_2} \xrightarrow{\substack{\text{carbonic} \\ \text{anhydrase}}} \underset{\substack{\text{carbonic} \\ \text{acid}}}{H_2CO_3} \longrightarrow \underset{\substack{\text{hydrogen} \\ \text{ion}}}{H^+} + \underset{\substack{\text{bicarbonate} \\ \text{ion}}}{HCO_3^-}$$

If all of these hydrogen ions remained in the bloodstream, the blood would be very acidic indeed. But most of the hydrogen ions are neutralized by combining with hemoglobin, which is negatively charged, to form **acid hemoglobin.** This reduces the acidity of the blood:

$$\underset{\substack{\text{hydrogen}\\\text{ion}}}{H^+} + \underset{\substack{\text{bicarbonate}\\\text{ion}}}{HCO_3^-} + \underset{\substack{\text{hemoglobin}\\\text{ion}}}{Hb^-} \longrightarrow HCO_3^- + \underset{\substack{\text{acid}\\\text{hemoglobin}}}{HHb}$$

If the blood becomes too basic, acid hemoglobin dissociates, releasing hydrogen ions.

The blood gives up only about 10% of its carbon dioxide as it passes through the lungs. The other 90% is retained, mostly in the form of bicarbonate ions, which act as important blood **buffers,** substances that keep the pH from fluctuating (Section 2-G). Therefore, although carbon dioxide is a waste product, its presence is essential in regulating the pH of the blood.

23-F Regulation of Ventilation

The partial pressures of oxygen and carbon dioxide in our blood must be continually adjusted so that they stay within so-called physiological limits. This adjustment is made by altering the depth and rate of breathing. As blood passes through the capillaries of the lungs, the gases in the blood come into equilibrium with those in the alveoli. The faster and deeper we breathe, the greater the percentage of oxygen in the alveolar air, and the lower the percentage of carbon dioxide, because the air in the lungs is replaced more completely and more frequently. With slow or shallow breathing the reverse is true. The breathing rate is controlled by the carbon dioxide and oxygen content of the blood and cerebrospinal fluid (the fluid around the brain and spinal cord). Receptor organs monitor the pressures of these two gases in the body fluids. The brain responds to this information by sending nerve signals to the diaphragm and rib muscles, speeding or slowing the rate of breathing (Figure 23-13).

Perhaps surprisingly, the level of carbon dioxide has more effect on breathing than does the level of oxygen. If the carbon dioxide content of the blood drops below a certain critical level, breathing is inhibited. By purposely hyperventilating, that is, taking several deep breaths in swift succession, you can hold your breath longer. Swimmers often do this so that they can swim underwater for a longer time. At each breath, however, the carbon dioxide content of the blood goes down, and if it goes too far, you lose consciousness. This is an important mechanism that prevents you from lowering the carbon dioxide pressure of the blood to a dangerous level.

Similarly, when you hold your breath, the carbon dioxide level in the blood rises, and if it goes above a certain level, you will lose consciousness. Once this happens, the breathing reflexes take control again, and cause you to inhale.

Breathing during sleep is governed by somewhat different mechanisms, which sometimes permit breathing to lapse. Periods of apnea (cessation of breathing) occur several times each night, causing the person to wake up briefly. The mechanisms that govern breathing while the person is awake then take over and cause breathing to resume. The frequency of apnea is increased by consumption of alcohol, antihistamines, and tranquilizers.

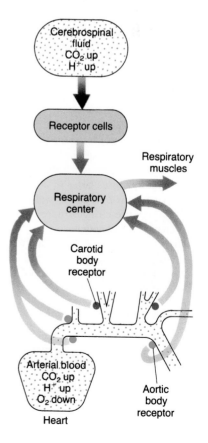

Figure 23-13

Action of CO_2 receptor cells. An increase in the percentage of CO_2 and H^+ in the blood and cerebrospinal fluid is detected by receptors in the nervous system (top) and blood vessels (aortic and carotid body receptors). These receptors send the information, via nerves (red arrows), to the respiratory center in the medulla of the brain. The respiratory center sends impulses that stimulate the respiratory muscles, increasing the breathing rate.

SUMMARY

All animals whose cells carry out respiration must obtain oxygen from their environments and expel carbon dioxide. Obtaining enough oxygen is the more difficult task because oxygen is much less water-soluble than carbon dioxide. The total amount of oxygen an animal needs depends largely upon its total number of cells and on its metabolic rate. The metabolic rate increases with activity and, in a warm-blooded animal, with increasing surface-to-volume ratio and therefore decreasing size.

Gas exchange with the environment can take place only across a moist surface. Small or inactive animals can obtain all their oxygen by diffusion across the general body surface. A high surface-to-volume ratio is found in animals whose bodies are small, flattened, or covered with projections. Larger or more active animals usually have part of the body specialized as an expanded respiratory surface, and they transport gases to and from this surface by way of a circulatory sys-

tem. An exception is the tracheal system of insects, in which the respiratory surface itself branches throughout the body.

The four main types of respiratory organs are the body surface, gills, lungs, and tracheal systems.

In animals with lungs, the gas content of the blood is regulated by controlling the depth and rate of breathing and, therefore, the gas composition in the alveoli.

Many animals have respiratory pigments in their blood that vastly increase its oxygen-carrying capacity. The oxygen saturation of the respiratory pigment hemoglobin depends on the oxygen pressures of the nearby fluid and air.

Carbon dioxide is transported mainly in the form of bicarbonate ion, an important buffer that helps keep the pH of the body fluids at a constant level. Normally, carbon dioxide levels in the body determine the breathing rate.

SELF-QUIZ

1. Both hyperventilation and holding the breath can cause unconsciousness. Under *normal circumstances*, why does this occur?
   a. change in carbon dioxide levels in the blood
   b. change in oxygen levels in the blood
   c. loss of hemoglobin from red blood cells
   d. distress of the lungs
   e. excess release of oxygen from hemoglobin
2. Most oxygen in the blood is carried in the form of ___; most carbon dioxide in the form of ___.
   a. carbonic acid
   b. bicarbonate ion
   c. oxyhemoglobin
   d. oxygen pressure
   e. hemoglobin ion
3. A disadvantage of using air as a respiratory medium is:
   a. it carries less oxygen than water does
   b. it increases the risk of drying out
   c. oxygen diffuses faster in air than in water
   d. air contains nitrogen as well as oxygen
   e. air pressure changes more than water pressure with changes in temperature
4. The metabolic rate of a poikilothermic animal increases:
   a. with increasing environmental temperature
   b. with decreasing environmental temperature
   c. with increase in size
   d. with decrease in muscular activity
   e. with increase in age

5. The main difference between the insect tracheal system and most other types of respiratory systems is:
   a. tracheal systems do not rely on the blood to transport oxygen to the tissues
   b. insects do not ventilate their tracheal systems
   c. insects do not dispose of carbon dioxide via their tracheal systems
   d. insects exchange both carbon dioxide and oxygen via their tracheal systems
   e. oxygen need not be in solution to cross the membranes in the tracheal systems of insects
6. The main factor that determines the saturation of hemoglobin with oxygen is:
   a. oxygen concentration in the blood
   b. carbon dioxide concentration in the blood
   c. pH of the blood
   d. hemoglobin concentration in the blood
   e. breathing rate
7. All of the following affect the amount of oxygen the blood delivers to the tissues *except:*
   a. the concentration of red blood cells in the blood
   b. the pH of the blood
   c. the concentration of CO_2 in the blood
   d. the concentration of white blood cells in the blood
   e. the oxygen pressure in the air

QUESTIONS FOR DISCUSSION

1. Ice fish are a family of bony fishes that inhabit Antarctic waters. These fish have no hemoglobin in their blood. How do you think they are able to survive without this respiratory pigment that all other adult vertebrates possess? What characteristics would you expect a member of this family to show?

2. Many adult amphibians use gills for breathing. There are many reptiles, birds, and mammals that spend almost all their time in water. Why do you think it is that none of them has evolved so that it retains the embryonic gills and uses them for gas exchange in the adult stage?

3. The evolution of coverings for the gills necessitated arrangements for openings to and from the gill area as well as muscles to draw a current of water across the gills. Yet very few animals have gills without some sort of covering. What is the advantage of a gill covering that has selected for evolution of the covering plus all these accessory arrangements?

4. Do you consider that the respiratory surfaces of your lungs are exposed to the environment? Why or why not?

5. A long-term smoking habit can destroy the cilia in the lining of the air passages. How might this affect health?

SUGGESTED READINGS

Avery, M. E., N-S. Wang, and H. W. Taeusch, Jr. "The lung of the newborn infant." *Scientific American,* April 1973. Describes the physiological changes in the lung just before birth and efforts to speed these changes in premature infants.

Comroe, J. H., Jr. "The lung." *Scientific American,* February 1966. Anatomy and physiology of the human lung and respiratory tract, describing physiological measurement techniques and applications.

Transport and Temperature Regulation

CHAPTER

24

Mountain lion

Jerusalem cricket

Adelie penguins, Antarctica

*T*he earliest organisms lived in the sea, which provided them with oxygen, carried away carbon dioxide and other wastes, and surrounded them with a relatively constant environment. The **extracellular fluid,** the fluid surrounding every cell of a higher animal, is like a tiny captive sea. Cells obtain food and oxygen from this fluid and discharge their wastes into it. The body's transport system supplies the extracellular fluid with fresh food and oxygen and removes wastes from it, ensuring that the fluid remains an environment where cells can flourish.

Even the smallest animal must transport substances within its body. Oxygen must be moved into the body, and then to the fluid around each cell. Food molecules must move from the site of digestion to the extracellular fluid, and waste products must be removed from it and expelled. Various fluid systems, called **vascular systems,** transport substances in most members of the animal kingdom. A **circulatory system** is a vascular system in which the transport fluid moves in a particular direction, usually because it is propelled by a muscular, pumping heart.

Having an efficient transport system permits the animal's cells to obtain and use food and oxygen more rapidly. This permits the animal to be larger and more active than it could otherwise. Besides transporting gases, food, and wastes, the circulatory system carries hormones, as well as molecules and cells that help protect the body from disease (Chapter 25). It also distributes heat generated by metabolism or absorbed from the environment.

- Vascular systems transport substances and heat from one part of the body to another and between the animal's external environment and the extracellular fluid, which surrounds every cell.
- Animals with faster, more direct transport systems can metabolize faster and lead more active lives than those with less efficient transport.

24-A Transport in Invertebrates

Cnidaria

Cnidarians are slow-moving creatures with low metabolic rates. Most of their oxygen enters through the body's thin outer layer of cells, and soluble wastes diffuse out across the general body surface. Food is transported by the **gastrovascular cavity,** which does double duty as a digestive (gastro) and transport (vascular) system.

In *Hydra* the gastrovascular cavity extends into each tentacle from the center of the body (Figure 24-1). In larger jellyfish, the cavity extends into a system

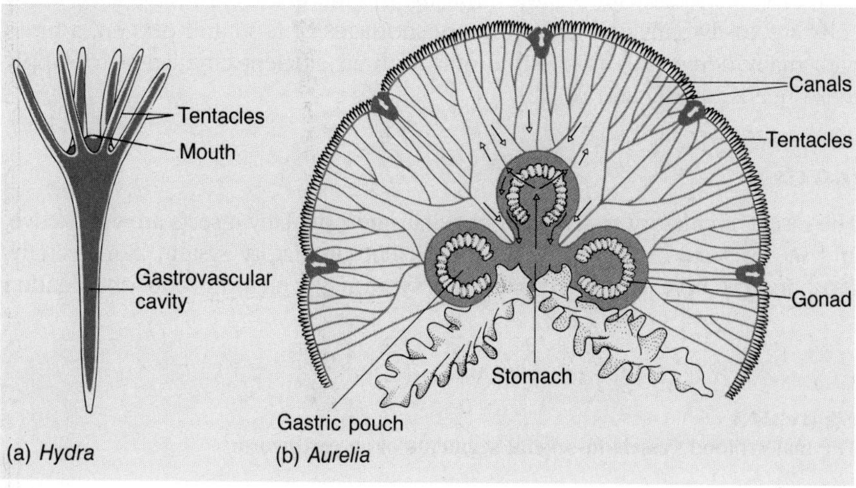

■ *Many lower invertebrates have no distinct circulatory system. They transport substances in systems that also have other functions, such as digestion or excretion.*

Stomach

Gastric pouch

(a) *Hydra*   (b) *Aurelia*

Figure 24-1
Gastrovascular cavities (blue) of two cnidarians. Arrows show the movement of digested food in *Aurelia.* In this jellyfish, a branching system of canals carries digested food from the gastrovascular cavity to the tentacles and other outlying parts of the body.

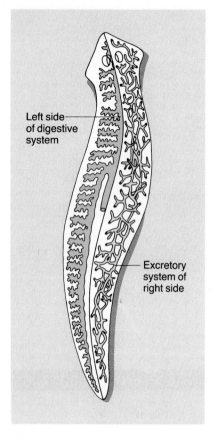

Left side of digestive system

Excretory system of right side

Figure 24-2

Transport in a planarian. The digestive system (blue) and the excretory system (yellow) branch throughout the body and perform their separate functions. Only half of each system is shown.

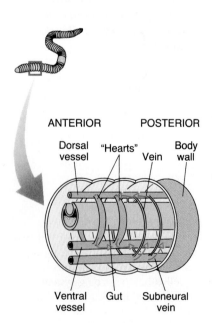

ANTERIOR   POSTERIOR

Dorsal vessel   "Hearts"   Vein   Body wall

Ventral vessel   Gut   Subneural vein

of canals that branch throughout the body. The fluid in the system is pushed in defined pathways, propelled by cilia. A cnidarian's metabolic rate increases when it moves or feeds, but the contraction of its muscle fibers during these activities also speeds the flow of fluid through the canals, automatically speeding delivery of food as the demand increases.

Planaria

Free-living flatworms have higher metabolic rates than cnidarians. However, because they have flattened bodies, the general body surface still provides enough surface area for gas exchange. Flatworms have no separate circulatory system or blood. Food is distributed by the digestive cavity, which branches throughout the body, providing a large surface area from which food can be absorbed. The excretory system of a planarian also acts as its own transport system. It too branches throughout the body and collects waste substances that must be expelled (Figure 24-2).

Annelida

An annelid has a **coelom,** a fluid-filled body cavity in which organs can move independently of one another. The coelom provides room for blood vessels and space for a heart to move as it pumps. The beating of an earthworm's dorsal blood vessel and five pairs of muscular "hearts" (enlarged blood vessels) moves the blood even while the animal is at rest. Muscles are more powerful than cilia, and so the circulation is faster in an earthworm's blood vessels than in the ciliary canals of cnidarians.

An earthworm has a simple **closed circulatory system,** in which the blood never leaves the vessels (Figure 24-3). The walls of the blood vessels are thin, and substances can diffuse easily between the blood and the extracellular fluid bathing the cells. The blood of earthworms contains a type of hemoglobin, which allows the blood to transport more oxygen.

A circulatory system that moves the blood steadily at all times permits division of labor among organs and tissues. In platyhelminths, which have no circulatory system, the gut and excretory organs branch throughout the body, serving as their own circulatory systems. In an annelid, by comparison, the gut is a simple unbranched tube, specialized only for digestion. The separate circulatory system carries the digested food to all of the body cells.

The localization of function in specific organs is one of the major reasons for the efficiency of higher animals. An outstanding example is the clumping of nerve cells to form a brain. Information is processed more rapidly and the animal responds faster and more precisely than it could if messages had to travel around a nerve network scattered throughout the body. However, since nerve cells are easily damaged by temporary shortages of food and oxygen, a brain with many nerve cells is possible only with an efficient circulation to supply these needs rapidly and reliably.

Insects

The insect circulatory system is of peculiar interest. Many insects are very active, and so we would expect to find an efficient circulatory system. Surprisingly, most insects have **open circulatory systems,** with only one open-ended

Figure 24-3

The major blood vessels in several segments of an earthworm.

Figure 24-4

Dorsal blood vessel of a cicada. This cicada has just molted, shedding its old exoskeleton. Until it hardens, the new exoskeleton is transparent, revealing the organs beneath. The insect's dorsal blood vessel can be seen in the thorax and abdomen. The posterior part of the vessel is the heart, which pumps blood forward toward the head.

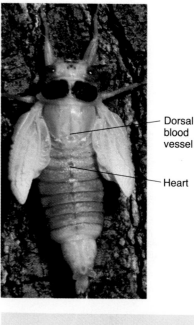

Dorsal blood vessel

Heart

blood vessel. This vessel releases blood into the body cavity, where it bathes the tissues directly. An insect's blood is moved by contractions of a long, thin-walled heart, which is really just the posterior end of the blood vessel (Figure 24-4), and by movements of the body muscles.

How can such an open circulatory system work efficiently enough to supply an insect's needs? The secret is that the circulatory system does not transport oxygen. Oxygen reaches cells by way of the **tracheal system,** a series of air-filled tubes that branch throughout the body. Other substances can travel more slowly via the blood.

24-B   Circulation in the Vertebrates

All vertebrates have closed circulatory systems. Exchange of substances between the blood and the extracellular fluid occurs only across the thin walls of the **capillaries,** the narrowest blood vessels. Contractions of a strong, muscular heart exert the pressure needed to force the blood through the capillaries. Blood travels from the heart to the capillaries through large vessels called **arteries** and returns from the capillaries to the heart through **veins.**

Fishes

In fishes, the heart consists of a series of four chambers, which collect the blood and then pump it out (Figure 24-5). Blood leaves the heart via a short, muscular artery, the **ventral aorta,** and travels to the gills. Here the aorta branches into smaller vessels, and eventually capillaries, where the blood picks up oxygen. From the gill capillaries, blood flows into the **dorsal aorta,** whose branches distribute blood to the capillaries of all the body organs (Figure 24-6a). Blood returns to the heart through the veins.

This type of circulation, in which blood passes through the heart only once in a complete circuit around the body, is called a **single circulation.** Such a system has the advantage that all of the blood going to the body has already been oxygenated in the gills. A disadvantage is that the narrow gill capillaries slow down the blood flow, so that blood leaves the gills at a much lower pressure than when it entered. No matter how hard the heart pumps, the blood in a fish's dorsal aorta is travelling at a relatively low pressure, since it has had to pass through the gill capillaries. This slows the rate of oxygen delivery to the cells and limits the metabolic rate that fish can attain.

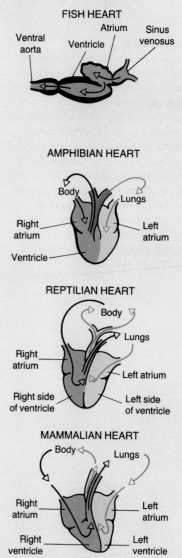

FISH HEART

Ventral aorta — Ventricle — Atrium — Sinus venosus

AMPHIBIAN HEART

Body — Lungs — Right atrium — Left atrium — Ventricle

REPTILIAN HEART

Body — Lungs — Right atrium — Left atrium — Right side of ventricle — Left side of ventricle

MAMMALIAN HEART

Body — Lungs — Right atrium — Left atrium — Right ventricle — Left ventricle

Figure 24-5

Blood flow through the hearts of various vertebrates. Deoxygenated blood is colored blue, oxygenated blood red. The heart of a fish has one atrium and one ventricle. The amphibian heart has two atria but only one ventricle. In most reptiles, the ventricle is partly divided, and oxygenated and deoxygenated blood are largely kept separate. The heart of mammals and birds has two atria and two completely separate ventricles.

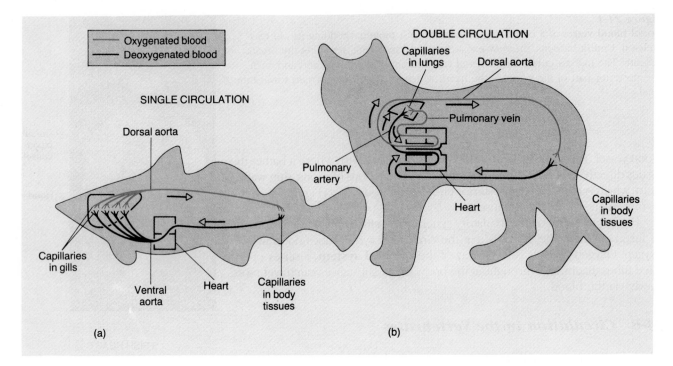

Figure 24-6

Single and double circulation. (a)
The single circulation of a fish. Blood
passes through the heart once during
each circuit of the body. Pressure in
the dorsal aorta is low since the narrow gill capillaries offer considerable
resistance to the blood as it passes
through, picking up oxygen. (b) In
the double circulation of birds and
mammals, blood must pass through
the heart twice before it returns to
the same point. Blood returning to
the heart from the lungs is pumped
at high pressure into the systemic
(body) circulation.

Amphibians and Reptiles

In higher vertebrates, the problem of low blood pressure in the body capillaries
is overcome by a **double circulation,** which passes blood through the heart
twice in each complete circuit around the body. Blood is first pumped from the
heart to the lungs, where it is oxygenated. Blood then returns to the heart and is
pumped through it a second time. This raises the blood pressure again before
the blood goes out to the rest of the body. The hearts of birds and mammals are
divided into two sides, right and left. Each side has an **atrium** (receiving chamber) and a **ventricle** (pumping chamber). The right side of the heart receives
deoxygenated blood from the body and sends it to the lungs. Oxygenated blood
returns to the left side of the heart, which then pumps it to the body (Figures
24-5 and 24-6).

The hearts of amphibians and reptiles (except crocodilians) are not fully
divided into two halves (see Figure 24-5). In amphibians, some blood that has
just come from the lungs returns to the lungs instead of passing out to the rest of
the body. This is not so inefficient as it looks. Most amphibians absorb more
oxygen through the skin than through their lungs or gills. Blood returning to the
heart from the skin often contains more oxygen than that returning from the
lungs, and there would be little advantage in keeping the blood from these two
sources separate.

In reptiles the circulatory system is more advanced. The blood obtains oxygen only in the lungs, and is immediately returned to the heart. There are two
atria, but in most reptiles the ventricle is only partially divided. However, the
valves in the ventricle work in such a way that there is little mixing of blood from
the two sides. Thus, the heart is functionally, if not structurally, divided. In
reptiles, blood travels from the heart to the body through paired dorsal aortae,
one on each side of the body.

Mammals and Birds

Blood leaving the heart for the body by way of two aortae, as it does in reptiles,
loses pressure faster than it would if it were travelling in one large vessel. It is
not surprising that mammals and birds, with their very high metabolic rates,

have only one aorta: the other has disappeared. Birds have retained the right, and mammals the left, of the reptilian paired aortae.

Both birds and mammals have double circulations with the ventricles completely separated. This has two important effects. First, keeping oxygenated and deoxygenated blood separate in the heart ensures that blood reaching the body organs from the aorta contains as much oxygen as possible. Second, to animals with a high metabolic rate, it is important for the blood in the aorta to be under considerable pressure. The blood loses pressure as it passes through the capillaries of the lungs. Returning it to the heart after it passes through the lungs permits the heart to raise the pressure again before the blood goes out to the rest of the body. Higher blood pressure means faster circulation. Oxygen and food reach the tissues faster, and waste is removed more rapidly.

■ *In the double circulation of higher vertebrates, the blood passes through the heart twice in one circuit of the body. As a result, blood pressure is high in every artery and so substances are transported rapidly.*

24-C The Mammalian Circulation

Blood Vessels

Arteries, capillaries, and veins are the pipes through which blood travels to the tissues. **Arteries** are vessels that carry blood away from the heart. Their walls are muscular and highly elastic. The arteries branch and rebranch into capillaries.

Capillaries are so narrow that blood cells must pass through them in single file. The capillary walls are only one cell thick, in keeping with their role as the sites where substances pass between the blood and the extracellular fluid. Shortly, the capillaries rejoin to form larger vessels which finally combine to form **veins,** blood vessels leading back to the heart. The walls of veins contain connective tissue and muscle, as do those of arteries, but veins are much less elastic, and tend to have larger internal diameters (Figure 24-7).

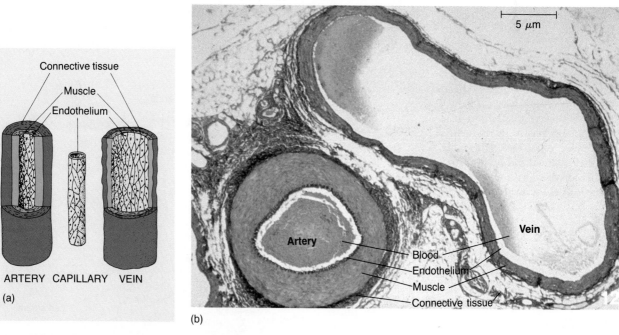

(a)

(b)

Figure 24-7
Blood vessels. (a) The layers in the walls of blood vessels. The walls of capillaries consist of a single layer of epithelium, called the endothelium (endo = within). In veins and arteries the endothelium is surrounded by muscle and connective tissue. The walls of veins are thinner and flabbier than those of comparable arteries. (b) A thin-walled vein and a thick-walled artery. (b, Biophoto Associates)

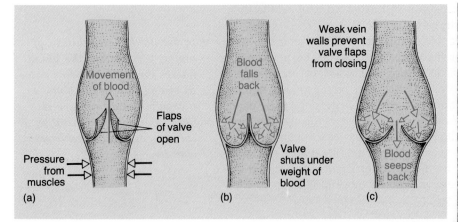

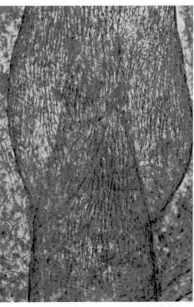

Figure 24-8
Valves in the circulation. (a) When blood flows up from a leg toward the heart, the valve opens and allows it to pass. (b) If blood moves in the reverse direction, it fills the cup-like flaps of the valve and presses the edges together, preventing backflow. (c) The walls of a varicose vein are weak and allow blood to collect and distend them so that the edges of the valve flaps cannot meet. Blood may then return through the valve, and circulation is impaired. (d) A valve in a lymph vessel (Section 24-E). (d, Carolina Biological Supply Company)

Another important difference between veins and arteries is that veins in the lower body contain **valves,** flaps of tissue that help to keep blood flowing in one direction. These valves open under the pressure of blood going toward the heart, and close when the blood begins to go backward in response to the pull of gravity (Figure 24-8).

When the walls of a vein are weakened, blood may collect in the vein and distend it so much that the valve flaps cannot meet. Since the valve cannot now prevent blood from flowing backwards, pools of blood collect in the weakened vein. Such **varicose veins** can be very painful if the weakness is in a large vein. **Hemorrhoids** are varicose veins in the walls of the rectum. These veins have been damaged by pressure from conditions such as constipation or pregnancy.

Figure 24-9
The mammalian heart. (The heart's owner is facing us, so that the heart's right side is on our left.) (a) Basic structure of the heart. The two atria receive blood from veins and empty it into their respective ventricles. The ventricles contract and push blood into the arteries. (b) A more realistic view, showing the major blood vessels.

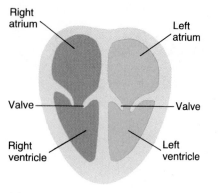

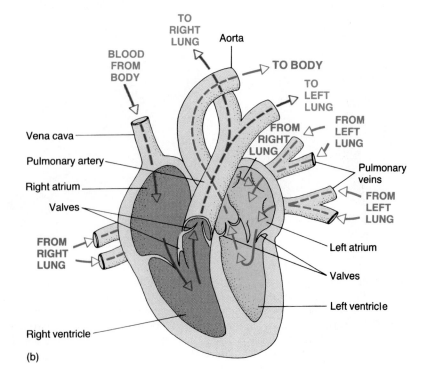

The heart too has valves that direct the flow of blood in a one-way path. Valves between the atria and ventricles prevent backflow of the blood into the atria when the ventricles contract, and valves between the ventricles and arteries prevent blood from falling back into the heart when the ventricles relax after pumping the blood out (see Figure 24-9).

The Circuit of Blood in the Body

Closed blood vessels, with strategically placed valves preventing backflow, ensure that the blood of a vertebrate flows in only one direction and in definite channels. Blood returns to the heart from the body via two large veins, the **venae cavae;** it then flows through the right atrium, and continues on into the right ventricle (Figure 24-9). Contraction of the right atrium sends more blood in to "top off" the ventricle. When the right ventricle contracts, it pushes the blood through a valve into the **pulmonary artery,** whose branches carry it to the lungs. In the lungs, the blood flows through capillaries surrounding the air-filled alveoli. Blood picks up oxygen and loses carbon dioxide across the thin walls of the alveoli and lung capillaries. The freshly oxygenated blood then flows through the **pulmonary veins,** back to the heart. This time, the blood enters the left atrium and passes through the valve into the left ventricle. When the left ventricle contracts, the oxygenated blood passes through a valve into the **aorta,** the main artery to the body. The wall of the left ventricle is much thicker and more muscular than the wall of the right ventricle; it must push the blood throughout the body, not just on the short journey to the lungs.

The aorta gives rise to many branch arteries, which take blood throughout the body (Figures 24-10 and 24-11). Blood leaving capillaries in the body enters veins, all of which empty into the venae cavae before they join the right atrium of the heart.

■ *The path of blood in the mammalian circulation can be summarized:*
right atrium →
right ventricle →
pulmonary artery →
lung capillaries →
pulmonary vein →
left atrium →
left ventricle →
aorta →
body arteries →
body capillaries →
veins →
venae cavae →
right atrium

The Heart Cycle

As we have seen, the heart of a bird or mammal is really two pumps joined side by side. Each pump consists of a thin-walled atrium, which receives blood from veins and pumps it into the adjoining ventricle, and a thick-walled ventricle, which pumps blood into arteries. The right side of the heart receives blood from the body and pumps it to the lungs; the left side receives blood returning from the lungs and pumps it to the rest of the body.

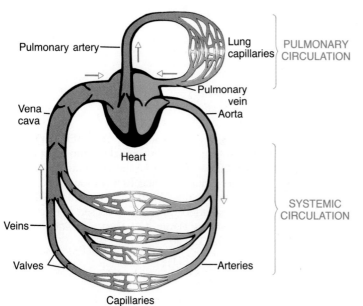

Figure 24-10
Outline of the circulation in a mammal. The heart and its blood vessels have been rearranged somewhat for clarity. Deoxygenated blood is colored blue, oxygenated blood red. Follow the direction of the arrows through a complete circuit, noting that the blood passes through the heart twice before it returns to the starting point. The systemic circulation supplies all parts of the body except the lungs.

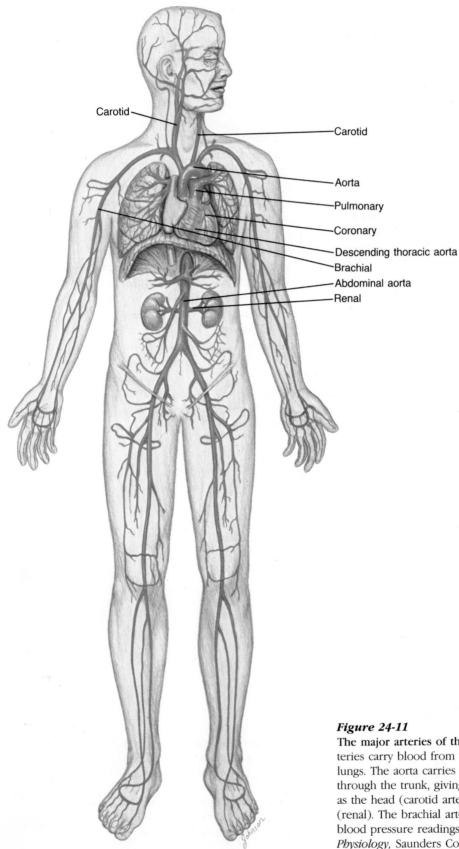

Carotid

Carotid

Aorta

Pulmonary

Coronary

Descending thoracic aorta

Brachial

Abdominal aorta

Renal

Figure 24-11

The major arteries of the human body. The pulmonary arteries carry blood from the right ventricle of the heart to the lungs. The aorta carries blood from the left ventricle down through the trunk, giving off branches to various areas, such as the head (carotid arteries), heart (coronary), and kidneys (renal). The brachial artery, in the arm, is used to take blood pressure readings. (From A. C. Guyton, *Anatomy and Physiology,* Saunders College Publishing, 1985)

The heart beats continuously throughout an animal's life. Each heartbeat is initiated by a "pacemaker," a small mass of tissue called the **sinoatrial node,** located at the entrance to the right atrium.

When the ventricles contract, they exert considerable pressure on the blood. This point in the heartbeat is called **systole,** and the pressure of the blood during ventricular contraction is known as the **systolic pressure.** Following contraction, the heart relaxes and blood rushes in from the venae cavae and pulmonary veins, partially filling the ventricles. This part of the heartbeat cycle is known as **diastole,** and the blood pressure at this time is called the **diastolic pressure.**

The brachial artery in the arm, just above the elbow, is usually used for measuring blood pressure. Blood pressure is expressed as the ratio of the systolic pressure over the diastolic pressure (both measured in millimetres of mercury). For instance, 120/80 is considered the "average" blood pressure. Systolic pressure indicates the force with which the left ventricle pushes blood. Diastolic pressure indicates the resistance of the blood vessels; it is useful in diagnosing hardening of the arteries or strain on their walls.

Blood Pressure and Circulation

Blood pressure is determined mainly by the rate and force of the heartbeat, the volume of blood pumped at each stroke, and the resistance of the blood vessels to the flow of blood. The greater any of these, the higher the blood pressure.

As the ventricles push blood into the arteries, the artery muscles relax and the elastic walls expand to accommodate the blood. As the blood passes, the muscular artery walls contract and exert pressure on it, helping the heart to push the blood along. After the blood leaves the main arteries, the walls of the other blood vessels offer only frictional resistance to its flow. Muscles in the walls of a blood vessel may contract or relax, changing the vessel's diameter, and this changes the blood pressure by making it harder or easier for the blood to pass through the vessel. The narrower the vessel, the greater the resistance, and so the speed of blood flow drops almost to nothing in the capillaries. The veins are flabby and do not help to push the blood back to the heart. In addition, blood in the veins below the heart must be returned against the pull of gravity. Thus, blood tends to collect in the veins.

Blood eventually does return to the heart, propelled mainly by the muscles of the body. When muscles contract, they squeeze against the outsides of veins, forcing blood to move along inside. The valves in the veins permit blood to flow only toward the heart. When the muscles again relax, the valves keep the blood from falling back. Since muscular contraction is needed to push blood through the veins, it is more tiring to stand still than to walk for an equal period. Standing allows blood to collect in the veins of the feet and legs. The feet swell with stranded blood, and the body temporarily loses the use of blood that should be distributing oxygen and nutrients to other tissues. Studies have shown that students who jiggle their feet are more alert, and perform better on long exams, than their peers who sit still.

Several different short-term mechanisms raise and lower the pressure in the arteries to maintain homeostasis. When the blood pressure rises above 60 millimetres of mercury, it stimulates **baroreceptors,** pressure receptors in the walls of arteries in the head and chest. Acting by way of the nervous system, the baroreceptors trigger (a) **vasodilatation,** or widening, of the small arteries, (b) decreased heart rate, and (c) vasodilatation of the veins (which decreases the return of blood to the heart). All of these lower the blood pressure. When the blood pressure falls rapidly, for instance as a result of heavy bleeding, several hormones constrict the blood vessels and usually restore the blood pressure to normal within a few minutes.

One of these hormones, vasopressin, is also involved in the long term regulation of blood pressure. Vasopressin decreases the volume of water excreted by the kidneys, retaining more water in the blood and raising the blood pressure (see Section 26-F). Many people suffer from **essential hypertension,** long-term high blood pressure of unknown origin. Hypertension increases the likelihood of a stroke or heart attack. The usual treatment is **diuretic drugs,** drugs that increase the amount of water excreted by the kidneys. A diet very low in sodium may have the same effect. Drugs or diet both counteract the effect of vasopressin, increasing the volume of water excreted by the kidneys. The kidneys of people with hypertension require higher than normal blood pressure to function properly. The causes of this kidney abnormality are not known, but the condition appears to be hereditary and to involve damage to the blood vessels that supply the kidneys.

The Circulatory System's Adjustment to Exercise

The circulatory system adjusts in various ways to changes in physiological conditions. These adjustments are usually controlled by **negative feedback,** a mechanism whereby a change in some condition, such as blood pH, stimulates activity that brings the condition back to its normal range. Negative feedback systems ensure that the composition of the extracellular fluid remains almost constant. We shall consider, as an example, some of the circulatory system's responses to vigorous exercise.

As exercise begins, the nervous system sends impulses to the **adrenal glands,** near the kidneys, causing them to release the hormone **epinephrine** (also called **adrenalin**) into the bloodstream. Epinephrine causes blood vessels in the skin and abdominal organs to constrict, decreasing the blood supply to these organs and sending blood that is normally "stored" in these areas into more active circulation. This in effect increases the volume of blood available. Epinephrine also causes local vasodilatation of the small arteries and capillaries

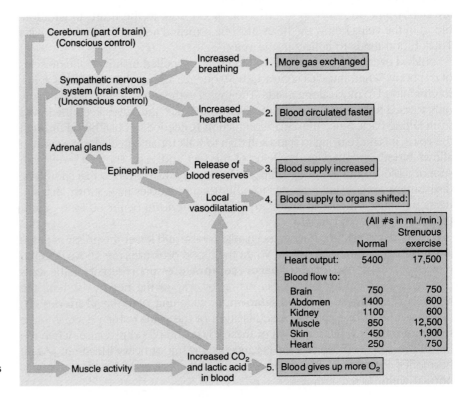

Figure 24-12
Some of the human body's responses to exercise.

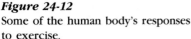

in the muscles and heart, increasing the blood supply to these organs (Figure 24-12). This trade-off of blood supplies helps to maintain the blood pressure. There is not enough blood to fill the whole circulatory system in the dilated state. Epinephrine also stimulates faster breathing and heartbeat rate, speeding both delivery of extra oxygen to the muscles and removal of wastes.

Exercising muscles produce more carbon dioxide and lactic acid than usual. These substances make the blood more acidic as it passes through the muscles, and an increase in acidity does three things: it makes the blood give up more of its oxygen in the muscles, it increases the dilatation of the blood vessels in the muscles, and it also stimulates the nervous system to increase the secretion of epinephrine, and the breathing and heartbeat rates, still further. Intense muscular activity also generates a great deal of heat. When the hypothalamus (part of the brain) becomes too warm, it sends nerve impulses that cause dilatation of blood vessels in the skin. The resulting increase in blood flow to the body surface allows the extra heat to be given off to the environment.

These are only a few of the interactions involved in the body's adjustment to exercise, but they illustrate the complexity of the physiological mechanisms that adjust the body's vital functions to changes in its activity.

Diseases of the Circulatory System

Cardiovascular diseases are diseases of the heart and blood vessels (cardio = heart). More than half of all cardiovascular deaths are caused by heart attacks. A heart attack occurs when the blood supply to part of the muscle that makes up the heart fails. With their blood supply cut, the cardiac muscle cells stop contracting and may die. A heart attack may occur when one of the heart's arteries is obstructed by a blood clot. It can also be caused by atherosclerosis, a condition in which the artery walls are thickened, and the passageway for blood narrowed by the growth of cells and deposits of lipids and other materials (see *A Journey Into Life,* Chapter 22). Even if the patient recovers from the heart attack, part of the heart muscle may have been killed and the heart permanently weakened.

A stroke occurs when the blood supply to some part of the brain is damaged. As with a heart attack, this is apt to kill the cells deprived of the blood's life-giving oxygen. A stroke may result from a blood clot in one of the brain's blood vessels, or from the rupture of a weak blood vessel. Its severity depends on what part, and how much, of the brain is damaged.

24-D Blood

The familiar red fluid called **blood** is really a tissue made up of a liquid containing several types of cells (Table 24-1). About half the volume of blood is made up of a fluid called **plasma,** and the other half is blood cells. The plasma contains various salts and a great variety of plasma proteins. **Serum** is plasma from which the proteins involved in clotting have been removed.

Blood cells can be divided into three main groups: the **white cells, red cells,** and **platelets** (Figure 24-13). Each microlitre (millionth of a litre) of blood contains 4.7 to 9.7 thousand white blood cells. Most of the many types of white blood cells help protect the body from disease (Chapter 25).

Red blood cells are by far the most numerous cells in the blood (3.6 to 5.5 million per microlitre). Their main function is oxygen transport. Mature mammalian red cells have no nuclei and contain mostly hemoglobin, a protein that binds oxygen. Red cells are produced from nucleated, dividing cells in the bone marrow. Red blood cells usually survive for about four months in the bloodstream. Then they break up, and white blood cells destroy their remains by phagocytosis. If the number of red blood cells falls, the resulting oxygen short-

Table 24-1	Main Components of the Blood
Water	45–54% vv*
Salts	
Sodium	2400 mg/l
Potassium	80 mg/l
Calcium	80 mg/l
Magnesium	26 mg/l
Chloride	2600 mg/l
Bicarbonate	1500 mg/l
Plasma Proteins	7–9% wv†
Blood Cells	40–50% wv
White cells	
Red cells	
Platelets	
Substances Transported by Blood	
Sugars	
Amino acids	
Fatty acids, glycerol	
Hormones	
Nitrogenous wastes	
Carbon dioxide	
Oxygen	

* vv means volume per volume; e.g., 12 ml per 100 ml is 12% vv.

† wv means weight per volume; e.g., 13 g per 100 ml is 13% wv.

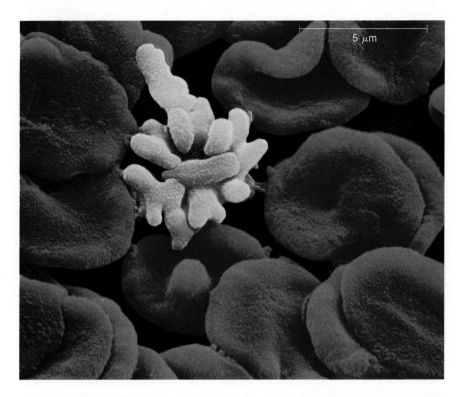

Figure 24-13
Red and white blood cells. False color scanning electron micrograph. (Biophoto Associates)

age causes kidney cells to secrete the hormone **erythropoietin** into the blood. This hormone stimulates the bone marrow to increase red blood cell production. These new cells boost the blood's oxygen-transporting capacity. As the blood's oxygen level returns to normal, erythropoietin production stops and red blood cell production returns to normal.

Anemia is a condition in which the blood contains fewer red blood cells or less hemoglobin than usual as a result of unusually slow production or fast destruction of red cells or hemoglobin. Anemia is a symptom that may be caused by a variety of diseases.

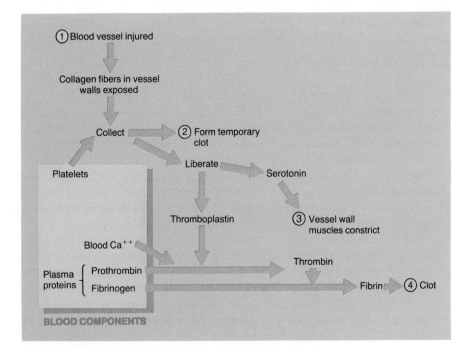

Figure 24-14
A simplified diagram of some of the reactions involved in the clotting of blood.

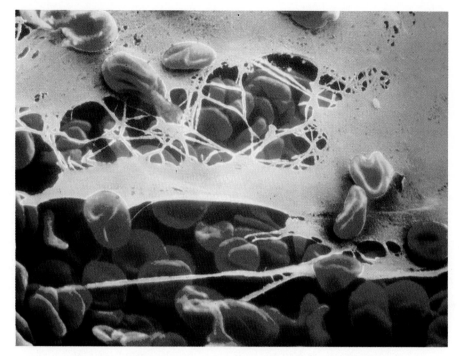

Figure 24-15
Blood clotting. False color scanning electron micrograph of part of a blood clot, showing red blood cells caught in a network of fibrin. (Photo by Lennart Nilsson, © Boehringer Ingelheim International GmbH)

The platelets are important in blood clotting. Platelets are not really cells, but are formed by the pinching off of parts of the cytoplasm of large cells in the bone marrow.

Blood Clotting

Clotting begins when the wall of a blood vessel is broken or damaged. The injured cells release substances that attract blood platelets. When the platelets come into contact with fibers of the structural protein collagen exposed by the injury, they disintegrate and form a temporary plug for the injured vessel. The platelets also release two substances. The first is **serotonin,** which causes the muscles in the blood vessel wall to contract and constrict the vessel, reducing blood loss. Platelets also release the enzyme **thromboplastin,** which changes one of the plasma proteins, **prothrombin,** into **thrombin.** Thrombin is also an enzyme; it changes another plasma protein, **fibrinogen,** into **fibrin.** Strands of fibrin form a meshwork around the disintegrated platelets. Still another plasma protein converts the loose fibrin meshwork into a tough, hard, permanent plug or clot, which seals off the injured part of the blood vessel from the exterior (Figures 24-14 and 24-15).

24-E   The Lymphatic System

In many ways, the body's capillary beds are the most important parts of the circulatory system, for it is here that the exchange of substances between blood, extracellular fluid, and cells takes place. Most substances, such as glucose and oxygen, leave the blood and enter the extracellular fluid by diffusing down the concentration gradient between the two. Wastes and carbon dioxide enter the blood in the same manner. In addition, water and larger molecules, such as hormones and small proteins, enter and leave the blood either by moving through spaces between the cells of the capillary walls or by passing through the cell itself by way of endocytotic vesicles.

Water leaves the capillaries under the pressure generated as blood is forced through a tube of small diameter. Toward the end of a capillary, so much water has been lost that the proteins left behind are quite concentrated and most (but not all) of the water returns to the capillary by osmosis.

The remaining fluid is collected and drained away through the **lymphatics,** thin-walled vessels with valves that ensure one-way flow. The lymphatics eventually join to form the **thoracic duct** and the **right lymph duct,** which empty into veins near the heart (see Figure 25-4). Often these are the only lymph vessels large enough to be visible. The lymphatics perform several vital functions:

1. They drain excess water from the extracellular fluid back into the circulatory system.
2. They temporarily store fluids taken into the body. Some of the fluid absorbed from the digestive tract finds its way into the lymphatic system, which releases it gradually so that the kidneys do not have to perform sudden surges of urine excretion.
3. They carry large molecules, such as proteins and hormones, from the cells where they are produced to the bloodstream. Such molecules are too large to cross capillary walls and so cannot reach the bloodstream directly.
4. Some food molecules, especially fats, move into the lymph rather than into the blood when they are absorbed from the small intestine. The lymphatics form the main route by which such molecules reach the blood.
5. Lymph nodes occur in several areas of the body. These nodes are an important part of the body's defense against disease (Chapter 25).

24-F Temperature Regulation

Heat spreads rapidly through any volume of water. Since cytoplasm, extracellular fluid, and blood are mostly water, heat spreads rapidly through an animal's body, and the circulation of the blood enhances this spread of heat. In fact, heat is one of the important things transported by blood.

The temperature of a living cell determines the rate of its metabolic processes. An organism can grow faster and respond to the environment more rapidly if its cells are kept warm, and the ability of some animals to maintain a

Figure 24-16
Heat gain and loss. (a) Heat exchange between an animal lying in the sun and its environment. Numbers represent temperatures in °C. (b) This panting dog lying in the sun cools its body by the evaporation of water from the surfaces of the nasal passages and mouth.

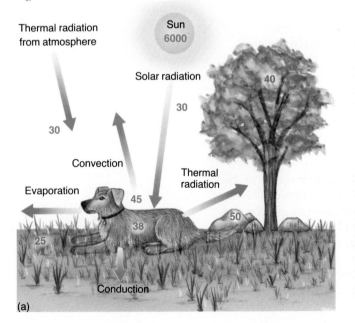

constant, relatively high body temperature is believed to be a major reason for their evolutionary success. We usually think of all animals except mammals and birds as "cold-blooded," but in fact, many lower vertebrates and invertebrates also have adaptations that permit them to regulate their body temperatures. This is called **thermoregulation**—controlling the temperature of the body.

The vast majority of reptiles, amphibians, fishes, and invertebrates can tolerate body temperatures in the range of about 0 to 35°C. However, a rapid change of 30°C will inactivate most enzymes and kill the animal. When the temperature changes slowly, many of these animals can undergo **acclimation**—adjustment to a change—and remain active. They produce new enzymes that function at the new temperature. In a number of fishes and aquatic invertebrates, this changeover from one set of enzymes to another occurs regularly as the water temperature changes with the seasons.

All animals produce **metabolic heat** from the chemical reactions in their bodies. Most land animals regulate their body temperatures by controlling the rate at which this heat is produced and lost (Figure 24-16). **Ectotherms** use mainly behavioral adaptations for thermoregulation. For instance, they move into areas where the environmental temperature is appropriate. **Endotherms,** mainly birds and mammals, depend more on physiological adaptations and insulation to control body temperature.*

The distinction between ectothermy and endothermy is seldom clear-cut: most animals fall somewhere between the two. Thus, desert rodents, which are endotherms, escape the midday heat by burrowing, a behavioral adaptation. Moths, which are primarily ectotherms, fly poorly unless their wing muscles are at 35°C or more, and they warm the muscles by physiological means, contracting them (fluttering their wings) before they take to the air. Behavioral and physiological thermoregulation, then, are found in both ectotherms and endotherms.

■ *Animals regulate their body temperatures by behavioral and physiological means. Ectotherms use more behavioral mechanisms, endotherms more physiological ones.*

Behavioral Thermoregulation

A dog lying in the shade on a hot day and bees shivering in their hive on a cold one both exemplify behavior patterns that control body temperature. The dog avoids being further warmed by the radiant heat of the sun. The bees produce metabolic heat by muscle movement. The bees also close the entrance to the hive and huddle together, slowing the loss of their metabolic heat.

Behavior that decreases the loss of metabolic heat to the environment can help terrestrial animals keep their bodies warm, but this is not practical for aquatic invertebrates and most gill-breathing vertebrates. These animals get oxygen from a current of water, which has an extremely high capacity to absorb heat—in this case, from the animal's body. The animal loses heat almost as fast as it is produced. Only a few large fish, such as tunas, can raise their body temperature above that of the surrounding water. These fish have countercurrent heat exchangers that work on the same principle as those of endotherms, discussed later.

The vast majority of land ectotherms move into areas that have favorable temperatures. Many earthworms, reptiles, and arthropods retire to burrows during the hottest or coldest parts of the day. This is advantageous because soil changes temperature more slowly than air does. Millipedes, houseflies, and crabs escape the heat of the sun by avoiding light.

Most land animals, rather than avoiding heat, need to capture as much as possible from their environments. This is probably one reason many more species of ectotherms live in the tropics than in cooler climates.

*****Poikilothermic** and **homeothermic** (Section 23-B) do *not* mean the same as ectothermic and endothermic. Poikilothermic ("cold-blooded") and homeothermic ("warm-blooded") describe the temperature of the body. Ectothermic and endothermic describe where the body heat comes from.

(a)

(b)

Figure 24-17

Thermoregulation. (a) A sea lion and her pup warm up by sleeping in the sun on warm rocks. A thick layer of fat under the skin, and the body covering of fur, help to retain the body's own metabolic heat. (b) A lizard raises its body temperature by sunbathing. Unlike the sea lions, it raises its body temperature almost entirely with environmental heat.

Sunbathing is the most common behavioral adaptation to capturing environmental heat (Figure 24-17). Even though the air may be very cold, an animal can often absorb much radiant heat by sitting directly in the sun's rays. Many ectotherms are also fairly dark in color, and so they absorb heat instead of reflecting it.

Physiological Thermoregulation

Birds and mammals maintain high body temperatures, from 35 to 42°C depending on the species. They do this by producing a great deal of metabolic heat and regulating the rate at which this heat is lost to the environment. When the environment is too cold, for example, the brain induces shivering, muscular movement that produces metabolic heat. When the environment is too warm, the animal becomes less active, and it may rest in the shade or in a burrow.

Layers of fat under the skin, plus a body covering of fur or feathers, insulate the interior of the endotherm's body from the environment and decrease heat exchange. Such insulation is not found in ectotherms. Cold-climate endotherms, such as whales, seals, and polar bears, tend to have thicker layers of insulation than their relatives in more temperate areas. At the other extreme, a camel living in a hot desert is protected against gaining too much heat from the sun by the thick layer of fur and fat insulating its back.

Several physiological mechanisms enable an endotherm to decrease its heat loss when its body temperature falls. Because heat is lost across the body surface, blood coming to the skin from the interior of the body is usually warmer than the skin. When more blood passes through the skin, more heat is lost from the body. When the body is too cool, nervous and hormonal signals constrict the surface blood vessels and decrease the blood flow to the skin, reducing heat loss.

Extremities, such as the ears, nose, and legs, are often slim and streamlined, with little insulating fat. Endotherms stand to lose considerable heat as the blood flows through these areas. To take extreme examples of this problem, consider how much heat must be lost through the legs of a bird fishing in an icy mountain stream, or through the paws of an arctic fox trotting across a snowfield. Endotherms cope with this difficulty by means of **countercurrent heat exchang-**

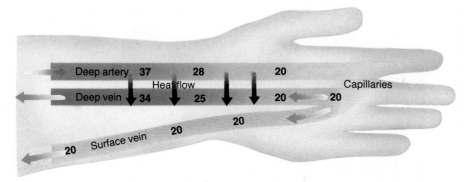

Figure 24-18
Countercurrent heat exchanger between an artery and a vein deep in the human arm. Numbers give the temperature (in °C) of blood in various parts of the blood vessels. Heat is lost in the capillaries of the hand. As blood runs through the adjacent deep vein and artery, it exchanges heat so that the arterial blood is cooled by the time it reaches the hand, and venous blood is warmed as it returns to the body. When heat conservation is unnecessary, the heat exchanger can be bypassed, and most of the blood is returned to the body via a surface vein where its temperature does not change. The nervous system controls the degree of constriction of the two veins and so controls heat loss.

ers, arrangements in which the blood vessels entering the ear or leg run next to those that are leaving. This allows blood travelling in the vessels to exchange heat with blood running in the opposite direction: blood on its way to the limb gives up some heat to blood returning to the body. By the time outgoing blood reaches the limb, it has been cooled and it has little heat left to lose to the environment. Its heat has been given up to blood returning to the body, and this blood has been warmed almost to body core temperature (Figure 24-18).

When they need to cool themselves, most endotherms radiate heat by increasing the supply of warm blood to the skin and by rearranging their fur or feathers so that more heat escapes. Most also employ evaporative cooling as a second line of heat control: as water evaporates from the skin or lungs, the heat needed to vaporize the water is removed from the body. Human beings and some other mammals have sweat glands in the skin for evaporative cooling. Many carnivores, ungulates, and primates pant to increase evaporation from the lungs and nasal passages. The metabolic heat generated by a flying bird is tremendous, and air on its way to and from the lungs passes through numerous air sacs, which provide extra surfaces for evaporative cooling.

Such physiological and behavioral mechanisms enable birds and mammals to maintain a high body temperature. These endotherms can move about at environmental temperatures so low or so high that other animals cannot be active. The coldest land areas of the world are populated only by mammals and birds, such as seals, polar bears, penguins, and eider ducks.

Torpor and Hibernation

There is no advantage in being warm and active when neither food nor mates are available, and many animals, both ectotherms and endotherms, maintain high body temperatures only at certain times of the day or year. At other times, they conserve energy by allowing their body temperatures to fall.

A period of daily **torpor** (inactivity) is common in most terrestrial invertebrates, amphibians, and reptiles. A torpid animal's body temperature varies with the temperature of its surroundings. Daily torpor is also common in small mammals. The loss of heat to the cooler environment is more acute for a smaller mammal because it has a larger ratio of surface area, which loses heat, to body volume, which produces the heat. The smaller the animal, the higher its metabolic rate must be to maintain a given temperature. A small mammal such as a shrew must eat more than its weight in food each day, and more than 80% of its food calories go just to maintain its body temperature. A shrew or a bat is never more than a few hours away from starving to death. A period of torpor, when its body temperature falls to the level of the environmental temperature, relieves the constant drain on the animal's energy resources.

Some vertebrates hibernate in winter or **estivate** in summer. Metabolic rate and body temperature decrease, and the animal saves energy by avoiding activity at a time of year when food or water is scarce.

Adaptations of Diving Mammals

Figure 24-A
An Atlantic bottlenose dolphin (also known as a porpoise). The blowhole through which the animal breathes is visible on the top of its head. Thousands of these animals are killed every year when they become trapped in fishing nets and drown.

Many mammals have become adapted to diving, for food or protection. These animals face a variety of problems. First, they must still breathe air, but their trips to the surface may be infrequent (Table 24-A). Second, as an animal dives and then resurfaces, it encounters changes in pressure of about one atmosphere (760 millimetres of mercury) for every 10 metres of depth. At high external pressures the gas in the lungs is compressed and forced into solution in the blood. Then, when the animal returns to the surface, pressure is reduced and the dissolved gas, mainly nitrogen, comes out of solution and forms bubbles that may block blood vessels (caisson disease or "the bends"). Third, water has a high heat capacity, and it absorbs heat faster than air, threatening a warm-blooded animal with death from **hypothermia:** chilling of the body. The adaptations of diving animals that permit them to overcome these problems are exaggerations of features found in other animals.

Some diving mammals show **bradycardia,** or slowing of the heart rate, when they dive; for example, a seal's heart rate drops from 150 to 10 beats per minute. Bradycardia is seen in other animals, including humans, when they submerge, and fish show it when removed from water; the advantage of bradycardia is probably that it saves the body energy and oxygen. In addition, blood vessels constrict and reduce the circulation to the kidneys, gut, and so forth. This conserves oxygen for use by the brain, which must not be allowed to experience oxygen deficiency. Furthermore, although circulation to the head is maintained, the respiratory center in the nervous system of a diving animal tolerates relatively high levels of carbon dioxide, and thus it does not stimulate the animal to breathe while submerged. Other body organs carry out fermentation rather than respiration of their food, and most of the carbon dioxide and lactic acid they produce is retained in the tissues; when the animal resurfaces, a surge of metabolic wastes enters the bloodstream. In seals, this anaerobic metabolism occurs only during occasional long dives. During more common, short dives, the muscles continue to use oxygen.

Many diving mammals can store extra oxygen for use during a dive. They have more red blood cells, which permit the blood to carry more oxygen. Their muscles also contain extra quantities of **myoglobin,** an oxygen-binding pigment related to hemoglobin. Seals carry most of their oxygen store in the blood, whereas whales store more oxygen in myoglobin.

It is perhaps surprising that the lungs do not carry more oxygen than usual during a dive; extra-large lungs would permit an animal to carry down more oxygen. In fact, seals exhale as they submerge, and whales have much less lung volume per unit of weight than do other mammals. These are adaptations that prevent caisson disease; the less air in the lungs, the less nitrogen will dissolve in the blood during a dive (air is 79% nitrogen). Furthermore, as the animal dives, compression of its lungs forces much of the air into the air passages, whose walls are impermeable to gases. This air cannot dissolve in the blood. However, some nitrogen inevitably does enter the blood. The tissues of dolphins have been found to tolerate levels of dissolved nitrogen that would be dangerous to a human diver (Figure 24-A). An important behavioral adaptation to diving is slow resurfacing, so that the nitrogen dissolved in the blood comes out of solution gradually and returns to the lungs instead of blocking the blood vessels.

Diving mammals reduce heat loss by reducing the flow of blood to the skin during a dive. Their bodies are often shaped with a low surface-to-volume ratio, and the blood vessels to their appendages are arranged in such a way that they can act as countercurrent heat exchangers (see Figure 24-18).

On its return to the surface, a diving mammal must spend several minutes breathing to expel carbon dioxide and replenish its oxygen supply before it can dive again.

Table 24-A	Duration and Depth of Diving for Some Mammals	
Animal	**Duration (minutes)**	**Depth (metres)**
Beaver	15	shallow
Muskrat	12	shallow
Walrus	10	80
Gray seal	20	100
Bottle-nosed whale	120	unknown
Blue whale	49	100
Most people	1	shallow
Trained skin divers	2.5	20

SUMMARY

Most animals have transport systems that move substances within the body. In some primitive animals, transport of food is carried out by the gastrovascular cavity. Animals with coeloms have true circulatory systems. An open circulatory system has few blood vessels, and blood bathes the cells directly. The closed circulatory systems of many invertebrates and of all vertebrates have blood vessels through which blood is pumped by the heart.

The circulatory systems of vertebrates show an evolutionary trend from single to double circulation. The double circulation of birds and mammals provides complete separation of oxygenated and deoxygenated blood and raises the blood pressure in the body's capillaries.

The quick and orderly flow of blood is accomplished by a muscular pump, the heart, and a set of pipes, the blood vessels. Blood flows through the circuit from the region of high pressure, the contracting ventricles of the heart, through the vessels at progressively lower pressure, until it returns to the heart. Valves in veins and in the heart prevent backflow. Blood pressure and blood supply in various parts of the body may be regulated by dilatation and contraction of small arteries and capillaries, and by changes in heartbeat rate and volume of blood pumped by the heart at each stroke.

The circulatory system responds to changes in the body's activities so that the body's new needs are met. During exercise, the amount of blood flowing and the rate of flow are increased, and more blood is diverted to the active muscles.

Blood is a liquid tissue consisting of a watery fluid that carries salts, proteins, and blood cells. White blood cells defend the body against disease. Red blood cells contain hemoglobin, a protein that transports oxygen by combining with O_2 in the lungs and releasing it in the capillaries of the body tissues. The blood platelets are important components in the clotting mechanism of the blood. Clotting helps to plug the vessel walls after injury, preventing loss of vital fluids or entry of disease organisms.

The lymphatic system consists of vessels that collect extracellular fluid, proteins, and digested fats, and empty them into the venous system.

Living cells can function only at certain temperatures. Animals cope with adverse temperature changes either by tolerating them or by making behavioral or physiological responses that help to maintain their bodies within a suitable temperature range. Ectotherms have more behavioral than physiological adaptations for thermoregulation. They may move into hotter or colder areas, or they may sunbathe, burrow, or huddle together to maintain a favorable body temperature.

Thermoregulation by physiological mechanisms is most apparent in endotherms. They produce large quantities of metabolic heat and regulate its escape into the environment by such means as insulation, alteration of the blood supply to the skin, regulation of the temperature of blood reaching the extremities, and evaporation of water from body surfaces.

Torpor, hibernation, and estivation are temporary reductions in metabolic rate that allow many animals to reduce the energy they would have to expend on thermoregulation under extreme conditions.

SELF-QUIZ

1. Place an "X" in the boxes to indicate which transport systems exhibit the feature mentioned:

	cnidarian	earthworm	insect	fish	mammal
food transport					
oxygen transport					
high-pressure fluid picks up food					
muscular circulatory pump(s)					

2. Select *two* advantages of a double circulation over a single circulation:
   a. In the double circulation, all the blood going to the tissues is oxygenated, whereas in the single circulation it is not.
   b. In the double circulation, the blood can transport twice as many types of substances.
   c. In the double circulation, the blood is at higher pressure when it enters the body tissues.
   d. In a double circulation, the blood travels around the body faster.
   e. In a double circulation, there are twice as many blood vessels servicing the body tissues.

3. The greatest amount of oxygen will be lost from the blood while it is travelling through:
   a. the capillaries around the alveoli
   b. the left atrium of the heart
   c. the arteries
   d. the capillaries in the body
   e. the veins

4. If you were asked to dissect an animal so as to reveal a valve, all of the following places would be good to try *except:*
   a. the opening between the right atrium and the right ventricle
   b. the fork where the pulmonary artery splits and one branch goes to each lung
   c. the base of the aorta where it leaves the left ventricle
   d. a vein in the leg
   e. a lymph vessel that empties into the thoracic duct

5. Which of the following organs will receive a decreased flow of blood during strenuous exercise?
   a. brain   d. heart
   b. skin   e. lungs
   c. liver

6. When a person exercises hard, all of the following occur *except:*
   a. blood glucose decreases
   b. ADP increases
   c. glycogen increases
   d. lactic acid increases
   e. CO_2 increases

7. Which sentence below states the main difference between an ectotherm and an endotherm?
   a. An ectotherm obtains much of its body heat from the environment, whereas an endotherm retains its own internally generated heat.
   b. An ectotherm's body temperature is always lower than that of an endotherm.
   c. An ectotherm always has the same temperature as the water or air around it, whereas an endotherm always has the same body temperature.
   d. An ectotherm picks up heat from its environment, whereas an endotherm loses heat to its environment.
   e. Ectotherms live in water, whereas endotherms live on land.

8. State whether each statement below is characteristic of endotherms, ectotherms, or both.
   ____ a. Regulation of body heat is mainly by moving to locations with favorable temperatures.
   ____ b. Heat is generated by the body's metabolism.
   ____ c. An insulating body covering reduces heat exchange with the environment.

QUESTIONS FOR DISCUSSION

1. What forces, besides contraction of the heart, may move fluids in the bodies of animals?
2. What restrictions in size and activity are imposed on animals that possess an open circulation combined with a tracheal system?
3. Can you think of any reasons why cephalopods (squid, octopus, and so forth) are the only molluscs with closed circulatory systems, and why other molluscs (snails, clams, and so forth) manage with open systems?
4. Birds and mammals have four-chambered hearts and maintain high body temperatures. In what way might these two characteristics be linked?
5. Arteries usually lie deep in the body, whereas veins lie near the surface. What is the advantage of this arrangement?
6. Why does blood flow to the skin increase during strenuous exercise?
7. Misinformed people often define arteries as blood vessels that contain oxygenated blood, and veins as vessels that contain deoxygenated blood. What is wrong with these definitions?
8. Why do you suppose the body core of an alligator basking in the sun warms up faster when the animal is moving than when it lies still?

SUGGESTED READINGS

Cessins, A. R., and K. Bowler. *Temperature Biology of Animals.* London: Chapman and Hall Publ., 1987. This book contains an excellent summary of all aspects of how animals are influenced by, and adapt to, the temperatures of their environments.

Kanwisher, J. W., and S. H. Ridgway. "The physiological ecology of whales and porpoises." *Scientific American,* June 1983.

Ramsay, J. A. Chapter 2: Circulation. *Physiological Approach to the Lower Animals,* 2d ed. New York: Cambridge University Press, 1968. A brief, but charming, comparative invertebrate physiology text.

Schmidt-Nielsen, K. "Countercurrent systems in animals." *Scientific American,* May 1981.

Wood, J. E. "The venous system." *Scientific American,* January 1968.

Zapol, W. M. "Diving adaptations of the Weddell seal." *Scientific American,* June 1987.

Defenses Against Disease

CHAPTER

25

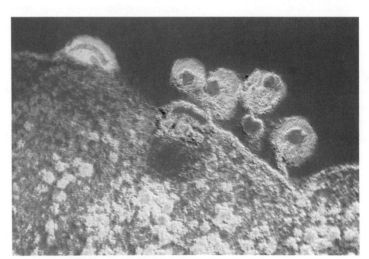

T4 lymphocyte releasing AIDS viruses

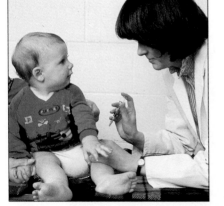

Vaccination

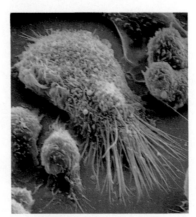

Macrophage SEM

501

A living body is an ideal incubation chamber, providing food, shelter, and just the right combination of water, minerals, and temperature for cells. It is no wonder that many small organisms have adapted to life inside the bodies of their larger neighbors. Some, like symbiotic bacteria in digestive tracts, cause their hosts no trouble or are beneficial. Other organisms, called **pathogens,** cause disease.

Probably all organisms have defenses against pathogens, but here we consider only the defenses of animals. Some defenses are **nonspecific,** that is, they protect the animal from many different diseases. For instance, nearly all animals contain **phagocytes,** wandering cells that engulf and destroy any pathogens and debris they encounter. Other defenses are **specific,** working against particular pathogens. These include the reactions of the **immune systems** of vertebrates, whose cells can mount an attack tailored specifically to each different disease-causing invader. When the disease has been overcome, cells of the immune system "remember" how to fight that disease more effectively in the future.

Occasionally, the immune system is damaged or fails to perform its normal functions. Two main types of disease may result: either the immune system does not protect the body against pathogens, or its cells damage parts of its own body.

KEY CONCEPTS

- Animals defend themselves against disease by nonspecific mechanisms and by immune reactions, which recognize specific foreign substances in the body, destroy them, and remember them.
- Failures of the immune system often lead to devastating diseases, either because the immune system fails to defend the body against the attack of foreign organisms or because it attacks its own body.

25-A   Nonspecific Defenses

All animals have protective measures that resist invasion by other organisms, or that kill those foreign organisms that do manage to enter the body.

External Defenses

The body surfaces in contact with the outside world are the skin, and the mucous membranes lining the eyelids, nose, digestive tract, vagina, and urethra. These surfaces are barriers that prevent most pathogens from entering the body.

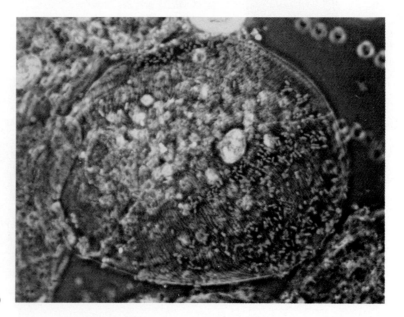

Figure 25-1
Our resident bacteria. An epithelial cell from the mucous membranes of the mouth with its population of bacteria (red rods). (Biophoto Associates)

In addition, millions of bacteria live on these surfaces (Figure 25-1). These resident bacteria produce substances that protect their homes—our skin and mucous membranes—from foreign organisms, many of which might cause disease.

Skin is the human body's largest organ and its main protection against infection (Figure 25-2). The **epidermal cells** of the skin's surface are constantly dying and sloughing off. We notice this when a sunburned nose peels, but it also goes on all over the body in undamaged skin. The lost cells are replaced by division of cells that lie under them. In this way the body soon repairs minor scrapes and cuts and maintains a constantly renewed barrier against infection.

The skin also produces chemical defenses: oil and wax from its sebaceous glands, and sweat from its sweat glands. These secretions contain lactic acid and fatty acids, which make the pH acidic enough to kill or slow the growth of many fungi and bacteria.

Mucous membranes cover body surfaces that must be kept moist, such as the linings of the nose, mouth, and vagina. Mucous membranes secrete **mucus,** a fluid that traps microorganisms and that contains bactericidal (bacteria-killing) enzymes such as **lysozyme,** found in tears, nasal mucus, and saliva. (Lysozyme was discovered by Alexander Fleming when he noticed that bacterial cultures died after he sneezed on them.) Douching the vagina frequently is unhealthy because it removes the bactericidal mucus.

Pathogens in the respiratory tract are likely to be trapped in mucus and swept into the pharynx by cilia. They are then usually swallowed and enter the stomach, where they encounter protein-digesting enzymes and an extremely acid environment. If they survive this and reach the large intestine, they are attacked by gut microorganisms, which secrete antibiotics that kill many fungi and bacteria.

Figure 25-2
A section through human skin. The cells in the outer layer of the epidermis constantly die and slough off. They are replaced by division of cells in the basal layer. Chemical defenses of the skin include acids in the oily secretion of sebaceous glands and in sweat from the sweat glands. These acids combat fungi and bacteria. Evaporation of sweat cools the body. The hair erector muscles can raise and lower the hairs, altering the air circulation next to the skin and helping to control the body temperature. (Lili Robbins)

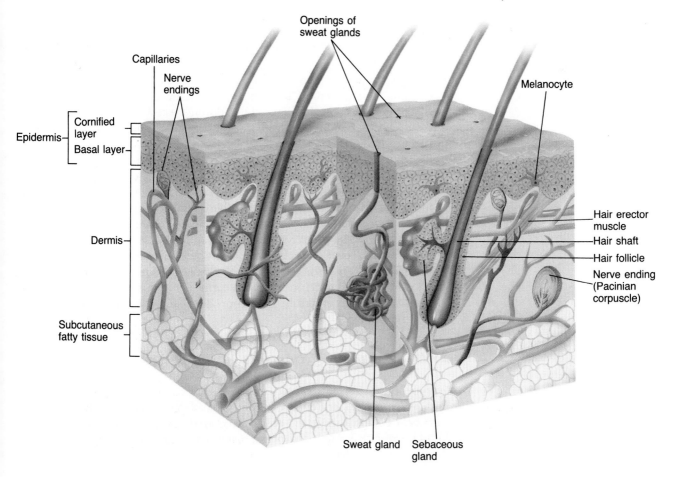

Despite these defenses, some pathogens do enter the body, usually by crossing the mucous membranes, which are thinner and more vulnerable than the dry, oily skin.

Internal Defenses

Once past the barrier of the skin or mucous membranes, pathogens face several nonspecific physiological reactions, capable of destroying a variety of organisms. In many situations, these nonspecific reactions work together with specific immune responses (Section 25-B).

1. **Inflammation.** A swollen red area occurs where the skin or a mucous membrane has been wounded, and pathogens have entered the body. A pimple or a boil is the result of inflammation from such a local infection. Inflammation helps to fight infection and heal the wound. The wound attracts **mast cells,** which are found in most organs and which contain a variety of hormone-like substances (Section 31-E). The mast cells release **histamine,** an amino acid derivative that causes heat, redness, and swelling. It increases the blood supply to the area and attracts large numbers of phagocytes, which engulf bacteria and dying body cells. Eventually they die and may accumulate as **pus.**

   The **complement reactions** complement (that is, round out) the pathogen-destroying effects of inflammation. **Complement** consists of about 20 kinds of proteins, produced mainly by the liver, which circulate in the body fluids. When they encounter a microorganism, these proteins undergo a series of reactions that may break down the microorganism's plasma membrane or attract phagocytes which engulf the pathogen.

2. **Fever.** An infected area often feels warm to the touch. Heat is one of the body's ways of fighting pathogens. Normally, the brain keeps the human body at about 37°C (98.5°F). However, when the body is infected by pathogens, some white blood cells respond by releasing hormones that act as **pyrogens** ("fire-producers"). If enough pyrogens reach the brain, the body's thermostat is reset to a higher temperature, allowing the body temperature to rise, which we call a **fever.**

   Very high fevers are dangerous and must be reduced. But a study of children with influenza ("flu") showed that a few degrees of fever helps the body fight infection. In this study, children treated with aspirin to reduce their fever were compared with those whose temperature was permitted to rise naturally. The children whose fever was kept down were ill longer, and had more serious symptoms, than those with the natural fever.

   Cells metabolize faster at higher temperatures, so fever increases the rate at which the immune cells fight infection. In addition, many bacteria require more iron in order to reproduce at these temperatures. Pyrogens not only raise the body temperature but also reduce the concentration of iron in the blood, slowing bacterial reproduction even further.

3. **Interferons.** Fever increases the production of virus-fighting proteins called **interferons.** When some cells in the body are invaded by viruses, they produce interferons, which help to protect healthy neighboring cells from viruses. Interferons stimulate cells to produce substances that interfere with viral replication.

25-B   Overview of Immune Responses

Few people suffer twice from diseases such as measles, chickenpox, and mumps. The body's first encounter with the pathogens that cause these diseases somehow equips it to get rid of the same kinds of pathogens when they next invade. A response that is bigger and better the second time around tells us that this reaction is produced by the immune system.

If you have had measles, you are immune to further attacks of measles but not to rubella (German measles) or mumps: the immune response to measles is **specific** for the measles virus. The body's attack on the measles virus, but not on its own cells, shows that the body can distinguish the measles virus as **foreign,** that is, not a normal part of the body. The immune response to a bout of measles confers **immunity,** protection against measles viruses encountered later in life. This shows that immunity involves some sort of **memory:** the body "remembers" that it has previously encountered this type of virus.

The role of the immune system is to recognize and destroy foreign antigens that invade the body. An **antigen** is any substance that can stimulate the body to mount an immune response against it. The most common antigens are substances from another organism, such as toxins produced by bacteria or the protein coats of viruses. Many substances are antigens for one person but not for another: the glycoproteins on your liver cells are not antigens to your own body, but they would act as antigens if injected into another person.

An immune reaction is specific because it involves the binding of antigen by receptors produced by white blood cells called **lymphocytes.** Each lymphocyte produces only one kind of antigen-binding receptor, which can bind only one kind of antigen. Different lymphocytes produce different receptors. So among all the lymphocytes in the body, there are receptors that can bind almost any antigen (Figure 25-3).

The binding of antigen to receptors on its surface may stimulate a lymphocyte to reproduce, forming a clone of cells all making molecules that bind that antigen. These cells can fight the pathogen that formed the antigen, which is also multiplying in the body and causing disease. The clone of lymphocytes also accounts for the body's memory of a pathogen. If the pathogen infects the body again, the many cells of the memory clone can mop up the pathogen before it has a chance to reproduce and damage the body.

It is essential that the immune system should not attack the body's own cells. The body is said to have **tolerance** for its own antigens. Tolerance develops

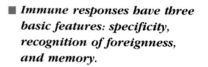

■ *Immune responses have three basic features: specificity, recognition of foreignness, and memory.*

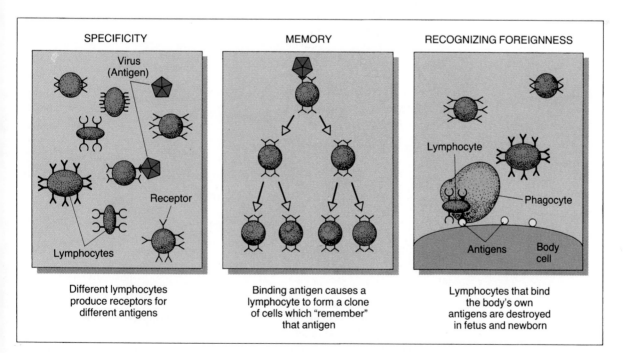

Figure 25-3
The three main features of an immune response. The response is specific to a particular antigen, permits the body to remember an antigen, and distinguishes foreign antigen from normal parts of the body.

during embryonic life. As a fetus or newborn, an animal destroys developing lymphocytes if they produce receptors that bind the body's own antigens.

25-C The Immune System

The immune system is not a set of organs like the digestive or respiratory system. It consists of many sites in various parts of the body, where the cells involved in immune reactions are produced or housed or work. **Bone marrow,** in the middle of many bones, produces nearly all of the blood cells (Section 24-D). Some of the white blood cells produced here are part of the immune system. Once formed, these cells move out to live and work throughout the body, travelling by way of the blood and lymph vessels.

Many immune responses take place in lymph nodes (sometimes called lymph glands), in the spleen, and in the main places where pathogens invade the body—the linings of the respiratory, digestive, genital, and urinary tracts.

In Chapter 24 we saw that fluid filters out of the blood in the body's capillary beds and joins the extracellular fluid, or lymph, that surrounds all cells. The lymph drains slowly into thin-walled lymphatic vessels, which drain back into the blood via the thoracic duct near the heart (Figure 25-4). At various places, lymph travelling toward the heart passes through **lymph nodes,** which contain a mesh lined with white blood cells. The lymph nodes filter pathogens out of the

> ■ *An immune response to an antigen has three parts:*
> 1. *recognizing it*
> 2. *destroying it*
> 3. *remembering it*

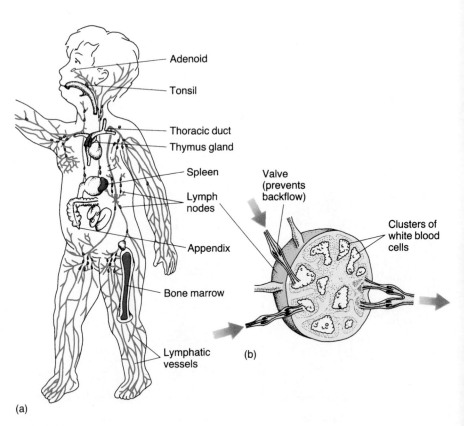

Figure 25-4
The human immune system. (a) The lymphatic system. Lymphatic vessels are drawn in red. Lymph nodes and other organs of the immune system are shown in purple. The groups of lymph nodes in the groin, armpits, and neck often feel tender and swollen during an illness. (b) A lymph node. Lymph enters the node through several lymphatic vessels. In the node, the lymph filters through a network of spaces containing large clusters of white blood cells until it reaches the one or two lymphatic vessels that carry it away from the node. (Thom Smith)

lymph for attack by the cells of the immune system. Tonsils and adenoids are lymph nodes in the throat and nose, respectively. Lymph nodes also occur in the armpits, neck, and groin. The spleen acts as a similar filter for the blood.

Cells of the Immune System

Two major groups of white blood cells take part in immune responses:

1. **Phagocytes,** the cells that engulf and destroy pathogens. These are of two main types: **neutrophils,** found in the blood (Figure 25-5), and larger **macrophages,** which can leave the blood and enter the tissues and body cavities. Macrophages are big enough to take up large microorganisms, including protists.
2. **Lymphocytes,** smaller white blood cells responsible for the immune system's recognition and memory of foreign antigens. Lymphocytes circulate throughout the body, from the bloodstream through the lymph and back into the blood, with extended stays in the lymph nodes and spleen.

There are two main types of lymphocytes. Both originate in the bone marrow. Later, **T lymphocytes** (also called **T cells**) move to the thymus gland, a lymph gland at the base of the neck under the breastbone (see Figure 25-4). Here they undergo part of their development. **B lymphocytes** (or **B cells**) undergo similar processes in the bone marrow.

There are three main steps in the immune system's response to an invader: recognizing the invader, attacking it, and remembering it. These steps are the subjects of the next three sections.

25-D Recognizing Foreign Antigens: Receptors and Antibodies

Foreign antigens are recognized by the body when they bind to specific receptor proteins produced by lymphocytes. The first of these to be discovered were **antibodies** produced by B cells. B cells produce some antibodies that remain attached to the B cell surface and some that are released into the blood or lymph. T cells, in contrast, produce only **T cell receptors,** which remain attached to the T cell surface.

The body produces millions of different antibody molecules, but all have the same basic Y-shaped structure (Figure 25-6). Each is made up of four peptide chains: two identical heavy chains and two identical light chains, all joined by

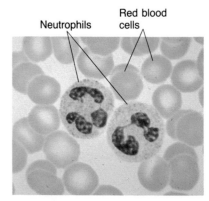

Figure 25-5
Human phagocytes. This photograph of a blood smear shows neutrophils, with their characteristic large, variably shaped nuclei. (Biophoto Associates)

Figure 25-6
Antibodies. (a) An antibody molecule bound to two molecules of antigen. Each antibody molecule consists of two identical light peptide chains and two identical heavy chains joined by disulfide bonds (yellow). Part of each chain has a variable amino acid structure (orange) and part is constant (blue). (b) Enlarged view of the antibody's antigen-binding site. The ribbon represents the polypeptide chain. (c) How the forked nature of an antibody permits clumping. Each antibody can bind to two antigen molecules (green), which may be either loose in the body fluids or exposed on the surface of a particle such as a bacterium or foreign blood cell (gray), as in the example shown here.

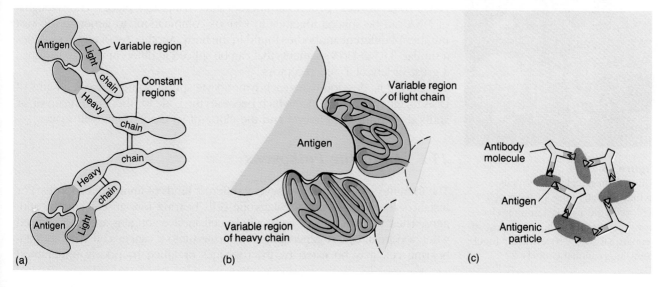

(a)   (b)   (c)

Table 25-1	The Antibody Groups
Group	**Major characteristics**
A	Found in mucous secretions and gut; defends external body surfaces
M	First antibody produced during immune response; sticks bacteria together and immobilizes them; stimulates complement reactions and macrophages
G	Main antibody in blood; combats microorganisms and their toxins; stimulates complement reactions and phagocytes; can cross the placenta
E	Responsible for symptoms of allergy; effective against parasitic worm infections
D	Rare; found on the surfaces of lymphocytes; function unknown

■ *An antibody's variable region determines which antigen(s) it will bind. Its constant region determines its general role in the body.*

disulfide bonds. Each of the four chains consists of a so-called constant region at one end, and a variable region at the other. The variable regions form binding sites so specific that each kind of antibody can bind only one (or a few closely related) antigens.

Antibodies are classified into five main groups—A, M, G, E, and D—according to which constant regions occur in their heavy chains. An antibody's group determines its general biological function (Table 25-1). For instance, group G antibodies combine with bacteria and viruses in the blood. The group G constant region attracts macrophages that engulf and destroy an invading bacterium. It also enables a G antibody to cross the placenta, so that antibodies produced by the mother's body can reach the fetus and protect it from disease.

The Genetics of Antibodies and T Cell Receptors

Each lymphocyte produces only one type of antibody or receptor. However, the mammalian body can recognize millions of different antigens because it produces millions of different antibodies. This poses an interesting genetic problem. Antibodies are proteins, and a traditional rule of thumb in genetics is "one gene, one polypeptide"—that is, there should be one gene for each different kind of polypeptide produced. But a mammal produces more different antibodies than it has genes. How can this be? The answer lies in the fact that antibody genes come in "kits" of several different interchangeable pieces. These lengths of DNA can be spliced together in various combinations to generate the vast number of different antibodies found in the body. Another set of genes provides a similar "kit" of DNA segments that can be spliced in different ways to produce a great variety of T cell receptors.

During its development, each lymphocyte splices some of these lengths of DNA together in a sequence which becomes the code for the T cell receptor or antibody that the lymphocyte and the clone of its descendants will make.

25-E Attacking Pathogens

The immune system produces several different kinds of immune responses. For instance, the body destroys eukaryotic cells bearing foreign cell-surface antigens. These include cancer cells, whose cell-surface antigens are often altered when a normal cell is genetically transformed into a cancer cell. The antigen-bearing cell may be eaten by macrophages or killed by poorly understood

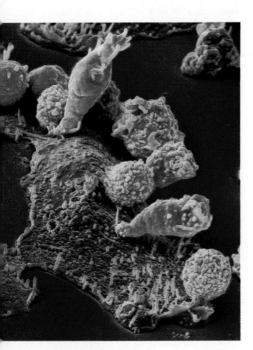

Figure 25-7
Killing cancer cells. Killer cells (white) are attacking a cancer cell (colored gold in this scanning electron micrograph). (Photo courtesy of Lennart Nilsson, © Boehringer Ingelheim International GmbH)

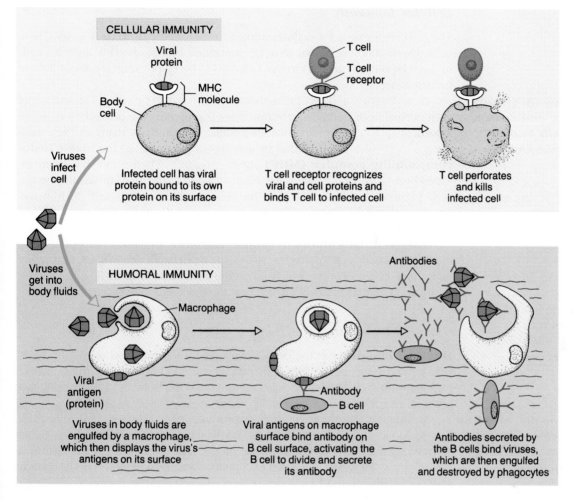

Figure 25-8
Cellular and humoral immunity. The body fights viruses (or other pathogens, such as bacteria, fungi, protists, or foreign cells) in these two main ways. If the virus manages to invade a body cell, it will usually provoke cellular immunity: attack by T cells (top). Viruses travelling in the body fluids provoke humoral immunity: attack by soluble antibodies produced by B cells (an event that usually occurs after a phagocyte such as a macrophage has engulfed some of the viruses).

killer cells (lymphocyte-like immune cells), which are apparently specialized to destroy abnormal body cells (Figure 25-7).

The best-understood immune responses are those produced by T and B cells—cellular and humoral immunity. Both are involved to some extent when the body is fighting most types of infection. We can illustrate the difference between the two by considering what may happen when a virus invades the body (Figure 25-8):

1. **Cellular immunity.** If the virus invades a body cell, that cell can be recognized, attacked, and destroyed by T cells. Destroying the cell prevents the virus from replicating.
2. **Humoral immunity.** If the virus has not yet invaded a body cell, it may be bound by antibody molecules that have been secreted into the body fluids by B cells. The virus-antibody complex will then be engulfed and destroyed by a phagocyte.

Cellular Immunity

The cells responsible for cellular immunity are called **cytotoxic T cells.** These cells destroy body cells that have been invaded by viruses. This includes cells invaded by cancer-causing viruses, believed to be responsible for about 20% of human cancers.

How does a T cell recognize that a cell is infected by virus? The surface of each normal body cell bears glycoproteins that identify the cell as belonging to a particular tissue in a particular individual. In mammals, many of these cell-surface molecules are specified by a group of genes called the **major histo-compatibility complex (MHC)** (histo = tissue). When a cell has been invaded by a virus, viral proteins are often attached to the MHC molecules on the cell's plasma membrane. This gives the cell a double identity: "self" MHC markers show that it is a member of the body, but at the same time viral antigen indicates that it is foreign.

In the **cellular immune response,** a cytotoxic T cell recognizes this combination as a body cell infected by virus. This T cell bears one of the body's many different T cell receptors, which binds the combined virus-MHC marker. The cytotoxic T cell may then kill the cell by punching holes in its plasma membrane.

■ In cellular immunity, cytotoxic T cells recognize and destroy infected body cells bearing combined self and foreign surface markers.

Humoral Immunity

Humoral immunity is the body's main defense against pathogenic bacteria, viruses, and fungi in the blood or other body fluids (humor = fluid). These pathogens are bound by antibodies secreted into the body fluids by B cells, and are then engulfed and destroyed by phagocytes.

Throughout life, the bone marrow produces large numbers of B cells. Each B cell produces one type of antibody and displays copies of this antibody on its surface. Many B cells circulate in the blood, and most of them die within a month or two. However, if a B cell encounters an antigen that binds to its surface antibody, it is activated in a series of steps. As an example, consider what happens when a bacterium gets into the bloodstream:

■ In humoral immunity, macrophages and T helper cells activate B lymphocytes to secrete antibody specific for an invading antigen.

1. One or more of the millions of different antibodies in the blood binds to antigen on the surface of the bacterium.

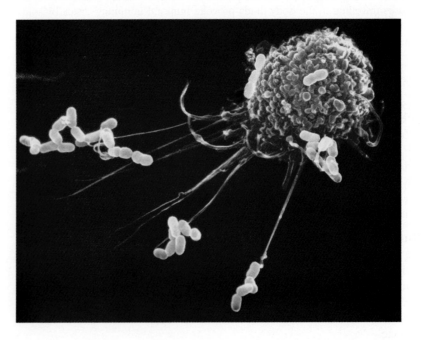

Figure 25-9
Phagocytosis. A macrophage (colored red) engulfing bacteria (colored green). (Photo courtesy of Lennart Nilsson, © Boehringer Ingelheim International GmbH)

Figure 25-10
B cell activation and clone formation. Some members of the clone become long-lived memory cells. Others become plasma cells and secrete antibody, which binds antigens on bacterial surfaces, forming complexes that are engulfed by phagocytes.

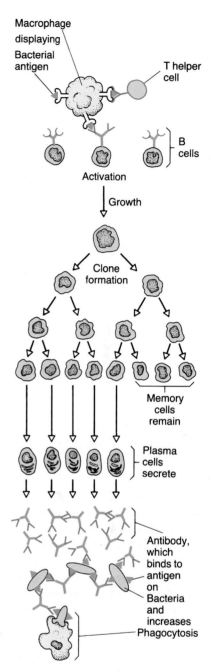

2. A phagocytic macrophage ingests the bacterium coated with antibody (Figure 25-9). The macrophage then displays the bacterium's antigen by pushing it out onto its own plasma membrane. In its travels through the body, the macrophage reaches a lymph node, where many B cells spend a lot of their time.
3. Antibody on the surface of one of the B cells binds the bacterial antigen exposed on the macrophage surface. This holds the macrophage and the B cell together so that the B cell can be activated to divide.
4. The macrophage now secretes **interleukin,** a protein essential to B cell activation.
5. T lymphocytes known as "helpers" are also needed for activation. T helper cells bearing the appropriate T cell receptors bind to the bacterial antigen on the macrophage. Once bound, these T helper cells produce **B-cell growth factor,** another protein necessary for B cell activation.
6. Each activated B cell divides to form a clone (Figure 25-10).
7. Some members of the B cell clone set to work churning out antibody specific to the bacterial antigen and secreting it into the body fluids. These cells, called **plasma cells,** can be recognized by their large areas of rough endoplasmic reticulum, which produces antibody for secretion from the cell. Within a few days of the bacterial infection, a great deal of antibody to the bacterium enters the blood. The bacteria will be bound by the antibody and engulfed and destroyed by macrophages.

   Other members of the dividing population of B cells will produce **memory cells** that are highly specific for the invading bacterium.

25-F   *Immunological Memory: Primary and Secondary Responses*

A cellular or humoral response to the body's first encounter with a foreign antigen is called a **primary immune response.** During a primary response, the antigen will eventually disappear from the blood, bound by antibody and destroyed by cytotoxic T cells or macrophages. The plasma cells that secreted antibody will also die. However, the memory cells produced during the primary response remain in the body for life. If the same antigen enters the body again, the memory cells permit the immune system to mount a **secondary immune response,** faster and more extensive than the primary response (Figure 25-11). The secondary immune response quickly eliminates the antigen again.

   Memory cells from the humoral response display a sample of their antibody on their surfaces. If the antigen invades the body again, the memory cells with matching surface antibodies bind the antigen. This stimulates them to divide rapidly and produce a clone of many antibody-secreting plasma B cells.

   The body must build up a memory clone for each antigen it encounters before it can produce secondary responses to most microorganisms. This is why babies and children starting school have so many colds and infections: they must encounter many antigens, and build up many clones of memory cells, before they are immune to as many diseases as the average adult.

■ *Memory cells from B cell clones formed during the primary immune response remain in the body, ready to produce antibody in an overwhelming secondary response if the same antigen enters the body again.*

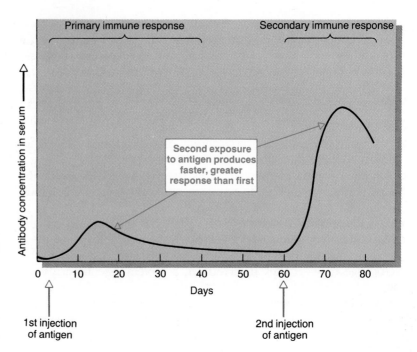

Figure 25-11
Primary and secondary immune responses. The black curve shows the amount of antibody to a specific antigen found in the blood of a rabbit at different times. Red arrows indicate the times at which a first and then a second injection of the antigen were given. After the second injection, the rabbit produces the specific antibody more rapidly and in greater amounts.

 Vaccination causes a primary immune response against a harmless form of a disease-causing antigen, so that the body will produce a powerful secondary response if the live pathogen that makes the antigen ever invades the body.

25-G   Medical Aspects of Immune Responses

Vaccination

Vaccination against a specific disease works by inducing the immune system to mount a primary immune response and to produce memory cells, ready to trigger a secondary response at the body's first real battle against the disease antigen. The practice of vaccination, however, began long before people understood how it works. Arabic and Chinese manuscripts more than a thousand years old refer to vaccination against smallpox. Lady Mary Wortley Montagu, wife of the British ambassador to Turkey, introduced this ancient custom into England in 1718. She had her children vaccinated by rubbing part of the scab from a healed smallpox sore into a small wound in the skin. This introduced a few live smallpox viruses into the body, stimulating a primary immune response and thereby conferring immunity to smallpox in later life. The snag was that vaccination with even a small amount of live virus sometimes caused a case of smallpox, which could be fatal.

Edward Jenner, an English physician, found a way around this problem in 1796. Jenner noticed that dairy workers who had caught the relatively mild disease cowpox from cows seemed to be immune to smallpox. He found that rubbing pus from cowpox sores into scratches in the skin prevented people from coming down with smallpox later. In this case, the antigens of smallpox and cowpox are so similar that the same antibodies work against both of them. Almost a century later, Louis Pasteur found a safer way to prepare vaccines. He discovered how to disable microorganisms by heat and other treatments. This damaged the microorganisms so that they could no longer cause disease but left their antigens intact and able to stimulate a primary immune response.

Nowadays, we have vaccines for a number of bacterial and viral diseases. The first successful immunizations against cancer (in cats and chickens) are based on the fact that cells infected with cancer-causing viruses usually display two new antigens on their surfaces, one of them specific to the virus. Unfortunately, attempts to produce antibodies to cancers not caused by viruses have had very little success. Several important infectious diseases also remain without effective vaccines, including malaria, trypanosome infections, and Acquired Immune Deficiency Syndrome (AIDS; see *A Journey Into Life,* this chapter). All these diseases are caused by pathogens whose cell surface antigens change frequently. As a result, no single antibody is effective against very many cases.

Smallpox was not only the first disease to be prevented by vaccination but also the first disease to be officially declared wiped out by human efforts. The last known outbreaks of smallpox occurred in India and Africa in the late 1970s. International vaccination programs had greatly reduced the number of smallpox cases. The final conquest came after health officials adopted a different strategy: searching out pockets of infection (people were rewarded for each case they reported), quarantining the victims, and vaccinating their friends and relations.

Passive Immunity

An animal is said to be passively immune when it contains antibodies that were not produced in its own body. A newborn baby is passively immune, temporarily protected from disease by antibodies that reached it from the mother's blood before birth. These antibodies are used up during the first few months of life, until the baby's immune system is sufficiently mature to take over.

Passive immunity can also be used medically. Some antigens are so virulent that the body's own primary immune response has little chance of preventing serious damage or death. If by some mischance such an antigen enters the body, the victim can sometimes be protected temporarily by injections of antibodies. Antibodies against potent antigens such as tetanus toxin or snake venom may be prepared by injecting the antigen into a horse and later collecting samples of the horse's blood, which now contains antibodies to that antigen. A more modern system is to harvest the antibodies from laboratory-grown clones of antibody-producing cells.

However they are acquired, the antibodies involved in passive immunity eventually disappear from the recipient's body, and the immunity is lost.

Tissue and Organ Transplantation

The rejection of transplanted organs is caused by cellular immune responses. Immunologists study this process using skin grafts, which are easy to work with and do not harm the recipient. If skin is transplanted from one mouse to another, it looks healthy for several days. However, the graft is eventually invaded by T cells, and within a day or two it sloughs off. A second graft from the same donor to any part of the same host is rejected faster than the first one.

Organs can be transplanted from one animal to another without being rejected only if the two animals have compatible MHC antigens (Section 25-E). Identical twins have identical MHC antigens, but even close relatives such as nonidentical siblings, or parent and child, often do not have antigens similar enough to permit a successful transplant. So when someone needs a skin or organ transplant, the first step is to find a donor with antigens that match those of the recipient as closely as possible.

Rejection of transplanted tissue is normally prevented by **immunosuppressant drugs,** drugs that suppress the body's immune responses. Such drugs are always used after heart, liver, or kidney transplant operations. Most immunosuppressant drugs work by preventing lymphocyte cell division.

25-H Malfunctions of the Immune System

The immune system is vital in protecting the body from disease. As we know in the case of AIDS, when something goes wrong with the immune system, the consequences are often fatal.

AIDS

Acquired immune deficiency syndrome (AIDS) is a deadly disease that was first identified in 1981. AIDS is caused by human immunodeficiency virus (HIV or AIDS virus), which probably passed from a monkey host into human populations in Africa during the 1960s.

Most people infected with the AIDS virus have no immediate symptoms. They go about their lives, passing the virus to others, sometimes before the virus can be detected by blood tests. AIDS develops years later. How quickly this happens seems to depend upon the general state of the immune system. The young and healthy may take 10 years to develop the disease. Older people, those with other infections, and newborn babies may develop the disease within a year. As far as we know, AIDS is always fatal, usually several years after symptoms of the disease first appear.

The AIDS virus is transmitted almost exclusively via semen and blood. The virus can pass from one person to another during anal or vaginal intercourse. Blood-to-blood transmission may occur when the same hypodermic needle is used to inject drugs or vaccines into more than one person. It can also occur when someone receives a transfusion of infected blood or blood products, or even when someone accidentally passes blood to another person through a lesion, such as a cut, in the skin. Similarly, blood vessels break during childbirth, allowing blood from an infected mother to reach her baby. Blood donors are in no danger in developed countries, because a new needle is used for each donor. Since mid-1985, blood for transfusions has

been screened for AIDS virus antibodies in most countries and is largely safe (although an individual may not develop viral antibodies for as much as 3 years after infection, which means that some virus particles escape detection and may be present in transfused blood).

The AIDS virus is not spread by casual contact, hugging, changing diapers, using the same toilet seats, or even sharing toothbrushes. Doctors, nurses, friends, and family members who live with AIDS patients almost never catch the disease. The virus has been found in urine, tears, saliva, breast milk, and vaginal secretions, but it seems not to be transmitted by these fluids unless it gets into a cut. The virus apparently will not cross intact mucous membranes (in the mouth, vagina, or rectum) or skin. However, there are often gaps in mucous membranes which have been stretched, or which have been injured by ulcers or by fungus, herpes, or gonorrhea infections. These gaps in the membrane form a route by which the virus can pass from one person to another.

In Western nations, infection from a man was the first method of transmission recognized, particularly between male homosexuals. AIDS has spread to heterosexual men and women, and to babies, mainly by way of intravenous drug users who share needles. In one study, 56% of female prostitutes in New Jersey were found to carry the AIDS virus. All of them had been infected by intravenous drug use, sex with bisexual men, or sex with male drug users. Infected women can pass the virus to their sexual partners, especially during menstruation. As many women as men will undoubtedly be infected eventually.

In Africa, the virus is often transmitted by sexual promiscuity, the vaccination of many people with the same needle, and transfusion of unscreened blood into malaria patients. Africa also has a high incidence of sexually transmitted diseases, which produce lesions in genital areas. For all these reasons, as many African women as men are already infected with AIDS.

The AIDS virus is a retrovirus (a virus with an RNA genome). It attacks cells in the brain, which is why many AIDS patients experience brain damage and insanity before they die. However, the usual cause of death from AIDS in adults is that the virus kills lymphocytes (Figure 25-A). AIDS patients have normal numbers of macrophages and B lymphocytes, but the number of T helper lymphocytes is drastically reduced. With a disabled immune system, the AIDS patient dies from pathogens that a normal immune system can control. The patient may die of diarrhea, rare forms of cancer, pneumonia, or tuberculosis. (The speed with which AIDS patients die of these infections shows what a good job the immune system usually does of defending us from the pathogens that always surround us.)

Drugs to combat AIDS are slowly being developed. Antibiotics, which help the immune system combat bacterial infections, have no effect on viruses. The drug zidovudine (also called azidothymidine [AZT]) resembles the nucleotide thymidine and prolongs the life of many AIDS patients. It also postpones the disease in people infected with the virus but not yet showing symptoms of AIDS. The drug kills replicating viruses, probably by being incorporated into their genetic material. This increases the number of lymphocytes that survive the virus and restores some immune functions. However, some patients' AIDS viruses have already mutated to AZT-resistant forms. Other drugs with effects similar to AZT are now being tested. None of them is a cure for the disease.

We are left with three defenses against the AIDS virus, or any other virus: vaccination, immune responses, and prevention. It is not easy to produce a vaccine to the AIDS virus because the virus evolves rapidly and changes its antigens frequently. A vaccine is a sample of the antigen that makes up part of the virus's surface coat. It is no use vaccinating someone with a particular antigen if they will later be exposed to a different one.

The AIDS virus does not completely disable the immune system.

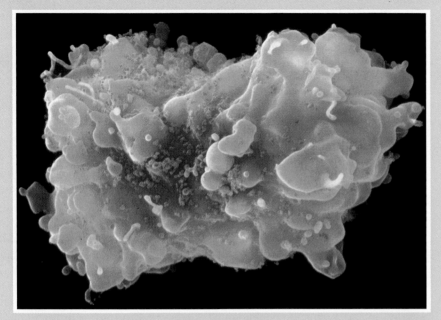

Figure 25-A
AIDS infection. AIDS viruses (colored purple) on a T helper cell (colored orange). (Photo courtesy of Lennart Nilsson, © Boehringer Ingelheim International GmbH)

People infected with the virus produce antibodies to the virus, and some people infected with the virus apparently never develop AIDS. Some of these people, presumably, already have immune systems that defend them against the virus. This is natural selection in action, and we can see that even if a cure for AIDS is never found, a human population resistant to the virus will eventually evolve.

However, this is no consolation to those now infected by the virus or to health officials overwhelmed by the scope of the problem. In parts of West Africa, one quarter of the population is infected, and millions of children have been orphaned by AIDS. The U.S. Public Health Service estimates that about 2 million people in the United States are already infected with the virus. Even if AIDS spreads less rapidly than it did from 1984 to 1988, more than 400,000 Americans will have been diagnosed with AIDS by 1995. AIDS is a staggering public burden. AIDS patients require frequent hospitalization and treatment for their many symptoms, and their average hospital stay is twice as long as that of other patients. The psychological strain on those caring for so many patients with no hope of recovery is another inestimable burden.

Most cases of AIDS result from sex with an infected partner or sharing hypodermic needles. No one is completely safe from AIDS if, during the last ten years, they have had sex with anyone who has had sex with a third person. Chastity, lifelong monogamy, and a clean needle for each injection are the only ways to prevent the spread of AIDS. Even these practices would not stop AIDS for many years (because of the risk from blood transfusions and the number of individuals already infected who do not know it).

The most effective way to combat AIDS is education. This was started by homosexual organizations in the United States and is now recommended for the entire population by scientists and the U.S. Surgeon General. Education aims to (1) explain how the virus is transmitted, (2) promote chastity and monogamy, (3) encourage the use of condoms, and (4) point out the dangers of sharing needles. People who use condoms properly, throughout sexual intercourse, are almost completely safe from transmitting the virus and from being infected. It also seems that people infected by the AIDS virus but who do not have AIDS are less likely

to develop AIDS if they are not exposed to the virus again. Condoms reduce the chance of a second exposure. Another measure, common in Europe, that drastically slows the spread of the disease, is to make disposable hypodermic needles readily available so that drug users are less likely to share needles.

The AIDS epidemic is bound to get worse, because AIDS education has not had much effect on the behavior of people at risk of contracting AIDS. A 1989 study of sexually active U.S. college students showed that fewer than half were monogamous and that condom use (by about 40% of the study group) had increased only slightly in the previous five years.

Although we do not know exactly how fast AIDS is spreading through the population, it is difficult to exaggerate the danger of this disease. Unless clean needles are made available to drug addicts, and unless millions of people change their sexual habits almost immediately, we may well face the modern equivalent of the medieval plague—which killed a quarter of the population of Europe.

For instance, if the thymus gland is abnormal, T lymphocytes fail to develop. Without T helper cells, B lymphocytes cannot form clones. A baby born without T lymphocytes fails to produce B cell clones to combat invading organisms, and so it usually is killed by the first pathogens it encounters. A few such babies have been saved by keeping them in sterile environments and by transplanting bone marrow cells and thymus tissue into them, which may permit them to make antibodies.

Autoimmune Diseases

Autoimmunity is a dangerous condition in which the self-recognition system breaks down and the body develops antibodies to some of its own antigens. In some cases this happens because the body is stimulated to produce antibody in response to foreign antigen very similar to one of the body's own antigens. In this case, the antibody may destroy the body's similar protein as well as the foreign antigen.

For example, when the body fights infections by streptococcal bacteria, it forms antibodies that may break down the body's own proteins, frequently damaging or destroying the heart valves. This is why "strep throat" is a serious disease that should be treated quickly with antibiotics. A number of other devastating (although fairly rare) diseases are also thought to be caused by autoimmunity, including insulin-dependent diabetes, rheumatoid arthritis, multiple sclerosis, pernicious anemia, Addison's disease, myasthenia gravis, and ulcerative colitis. Some of these diseases occur almost exclusively in people with specific MHC antigens—presumably ones that are easily mistaken for foreign by the body's immune cells. Thus, people with the genes for these MHC antigens have an inherited risk of developing the corresponding autoimmune disease.

Allergies and Anaphylaxis

About 10% of the human population suffers from allergies: inappropriate immune responses to harmless substances encountered in the environment, or in food or medicine—for example, milk, chocolate, pollen, penicillin, cat saliva, or mites in house dust. Generally the first exposure to the allergy-producing antigen (an **allergen**) produces no symptoms, but it **sensitizes** the body by evoking a primary immune response. The next encounter with the allergen evokes a secondary immune response known as an allergic, or hypersensitivity, response.

■ *Allergic reactions are secondary immune responses to substances that have provoked inappropriate primary responses.*

Allergic reactions are due to group E antibodies. In most people, the first time an allergen enters the body, T cells recognize it as harmless and prevent B cells from responding to it. In a person who will produce an allergic reaction, however, something goes wrong. The B cells produce group E antibodies, which bind to surface receptors on mast cells (Section 25-A). When the same allergen next enters the body and reaches such a bound mast cell, the antigen binds to the E antibody and the mast cell self-destructs, releasing histamine, which causes a hypersensitivity reaction (Figure 25-12). Histamine makes blood vessels dilate and increases the permeability of capillary walls, so that fluid escapes and swells the tissues.

The normal role of E antibodies is not to cause allergy but to protect the body from infection by platyhelminth parasites (tapeworms and flukes). Although flatworm infestations are not a major threat to human health in developed countries today, this is largely due to improved hygiene in the last two hundred years. Such parasites are still important in many areas.

Anaphylaxis is a severe secondary allergic reaction. The first time an antigen such as egg albumin is injected into a guinea pig, it has no obvious effect. If the injection is repeated three weeks later, the sensitized animal produces the

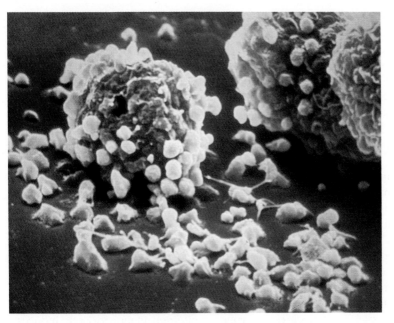

Figure 25-12
An allergic reaction. A mast cell contains many granules of histamine. During an allergic reaction, the cell releases these granules rapidly, and they cause the symptoms of allergy. (Photo courtesy of Lennart Nilsson, © Boehringer Ingelheim International GmbH)

symptoms of general anaphylaxis: the muscles of the bronchiole walls contract, constricting the air passages to the lungs, and the capillaries dilate. The animal will probably die unless injected with the hormone epinephrine. Similar **anaphylactic shock** sometimes occurs in human beings who are allergic to such things as penicillin or insect stings.

We do not know why some people have allergies while others do not, but allergies tend to run in families, suggesting that they have a genetic basis. Studies also suggest that breast-fed infants are less prone to develop some kinds of allergies later in life, compared to infants fed on baby formula. Those who suffer from allergies may derive some consolation from the thought that they are apparently less likely to develop tumors than are other members of the population.

Allergies are sometimes treated by attempting to induce tolerance to allergens (Section 25-B). Small amounts of the allergen are repeatedly injected, in the hope that the body will cease to react to it. This treatment does not always work. Part of the explanation probably lies in the fact that the immune response and the genes that control the immune system vary enormously among individuals.

SUMMARY

All animals have nonspecific defense mechanisms that protect them from a variety of diseases. Examples of nonspecific mechanisms include the skin and mucous membranes, the inflammatory reaction, fever, and interferons. Vertebrates are also protected by immune responses, which are defenses against specific diseases, each tailored to the particular pathogen it destroys.

Immune responses are characterized by specificity, by the ability to recognize antigens either as part of the body or as foreign, and by formation of a "memory" that the body has encountered a particular foreign antigen before.

The bone marrow produces the white blood cells that interact to produce immune responses. These cells travel in the blood and lymph to all parts of the body but are particularly concentrated in the lymph nodes. During an immune response, these cells recognize a foreign antigen, destroy it and the cells bearing it, and remember it. The body does not normally produce antibodies to its own molecules. The immune system comes to tolerate the body's own molecules during embryonic development.

The vast diversity of antibodies and T cell receptors in the body is produced by recombination of relatively few genes, each of which specifies part of an antibody or receptor molecule. The variable regions of these molecules bind foreign antigens with great specificity.

Killer cells kill abnormal body cells such as cancer cells. Cellular immune responses by cytotoxic T cells destroy body cells whose cell surface markers include both major histocompatibility complex (MHC) molecules and viral antigens, indicating a viral infection.

Humoral immunity is the main defense against invading pathogens in body fluids. B cells specific to an invading antigen are activated by macrophages and T helper cells to divide to form a clone containing plasma cells and memory cells. Plasma cells from the B cell clone make and release antibody molecules. The foreign antigen is destroyed by phagocytes which engulf the antigen when it is bound by this antibody.

The immune system's first exposure to an antigen causes a primary immune response, in which lympho-

cytes specific to that antigen are stimulated to form a clone of cells capable of combatting that antigen. After the invasion has been defeated, memory cells remain, so that a secondary immune response to the same antigen is greater and more rapid than the first, often destroying the antigen before it can cause illness.

Vaccination stimulates a primary immune response to a pathogenic antigen so that the body responds with an effective secondary response if it later encounters the pathogen itself.

Tissue and organ transplants are usually rejected because the host's T cells recognize the donor's antigens as foreign and destroy the transplanted organ. Matching MHC antigens of donor and host, and drugs that suppress immune responses, are needed to prevent transplanted organs from being rejected.

Autoimmune diseases, which destroy body tissues, occur when the body develops antibodies to some of its own antigens. Allergic reactions or anaphylactic shock occur when an allergen inappropriately induces mast cells bound to group E antibodies to release histamine.

SELF-QUIZ

Match the following structures with the function each performs:

_____ 1. A foreign macromolecule that may endanger the body
_____ 2. Site of a filter that removes invaders from the body
_____ 3. Long-lived cell that helps the body respond quickly to previously encountered antigens
_____ 4. Macromolecule that binds foreign molecules in the bloodstream

a. antibody
b. antigen
c. B lymphocyte
d. lymph node
e. memory cell
f. thymus gland

5. The skin protects the body against disease by:
a. repairing breaks in its surface
b. secreting acid
c. forming a barrier between the body and the external environment
d. all of the above

6. The introduction of a bacterial antigen into the body triggers a response specifically against that antigen by:
a. causing antibody molecules to assume a shape that permits them to bind the antibody
b. causing mutations in cells so that they produce antibodies to the antigen that caused the mutation
c. causing cells with the proper antibody to disintegrate and release the antibody
d. stimulating reproduction of cells that make the antibody to that antigen

7. Skin can be grafted from one identical twin to another, time after time, without being rejected because:

a. antibodies in the blood do not react to antigens on the other twin's cells
b. the twins have the same MHC genes and antigens, and so the twins' cells do not stimulate cellular immune responses in each other
c. the twins have been exposed to each others' MHC antigens as fetuses and, as a result, do not mount immune responses against each other
d. macrophages cannot destroy the twin's cells because the cells are not bound to antibody
e. B cells do not react to the cell surface antigens of an individual with the same MHC genes

8. Which of the following is *not* a role of lymphocytes?
a. producing antigens
b. producing B-cell growth factor
c. forming immunological memory
d. destroying body cells that are infected by viruses
e. producing plasma cells

9. Vaccination protects the body against catching a disease because:
a. it provides antibodies made by another animal
b. it makes the disease organism histocompatible with the body
c. it produces an enlarged clone of memory cells against that disease
d. it builds up an immunological tolerance for the disease antigen
e. it releases large amounts of nonspecific defensive secretions

QUESTIONS FOR DISCUSSION

1. What might be the selective advantage of a baby's being born before its immune system has matured?
2. Studies have shown that poliomyelitis, mononucleosis, and Hodgkin's disease are more common among children and young adults who have few or no siblings, few play-mates, uncrowded homes, and well-educated, well-to-do parents. How might these factors affect their immune systems' ability to fend off the viruses that are probably responsible for these diseases?
3. Breast cancer nearly always develops in women past childbearing age. Does this mean that natural selection cannot increase the immune response to tumors of the breast?
4. The blood of people who use large amounts of opiates (heroin, opium, codeine) contains fewer T cells than the blood of nonusers. What characteristics would you expect users to show as a result?
5. If a mouse (#1) that has rejected a skin graft from mouse #2 is then given a new graft from mouse #3, would you expect the rejection of this new graft to follow the same pattern as would be seen with a second graft from mouse #2? Explain.

SUGGESTED READINGS

Ada, G. L., and G. Nossal. "The clonal-selection theory." *Scientific American,* August 1987. The history of experiments showing how the body responds to antigen by producing specific antibodies.

Buisseret, P. "Allergy." *Scientific American,* August 1982. Why allergies appear to be malfunctions of the immune system.

Cohen, I. R. "The self, the world and autoimmunity." *Scientific American,* August 1988. A discussion of autoimmune diseases and the possibility that the immune system can be induced to control them.

Marrack, P., and J. Kappler. "The T cell and its receptor." *Scientific American,* February 1986.

Scientific American, October 1988 issue, "What Science Knows About AIDS." An entire issue devoted to AIDS, including articles on the prospects for a vaccine, origin of the virus, epidemiology of the disease, international issues, treating AIDS, social and public health aspects of the disease.

Young, J. D.-E., and Z. A. Cohn. "How killer cells kill." *Scientific American,* January 1988.

Excretion

CHAPTER

26

Soft coral

Banded gila monster

Zebras drinking

*A*n animal's body fluids must be maintained as a medium in which its cells can live. This requires regulation of temperature, pH, and the amounts and proportions of salts and water in the body fluids. It is relatively easy for animals that live in the sea to maintain **homeostasis** ("same-standing") of their body fluids because the sea has a relatively stable salt composition and pH, and its temperature changes little and slowly (Figure 26-1).

A great deal of evidence suggests that life began in the sea. First, most invertebrates and primitive plants are marine. Second, the body fluids of most invertebrates resemble sea water in their salt concentration. When organisms invaded fresh water and land, however, body fluid homeostasis demanded new adaptations. For instance, early vertebrates invaded fresh water, which contains few salts. They evolved the ability to live with body fluids containing a low salt concentration. Today, the body fluids of all vertebrates have a salt composition similar to that of sea water but only about one-third as concentrated. The early freshwater fish evolved organs that permitted them to excrete excess water that entered the body by osmosis—the kidneys, organs found only in vertebrates.

All the body's fluid compartments are connected. Blood and extracellular fluid (which surrounds all cells) exchange substances as the blood passes through the capillaries of the circulatory system. Blood is also the source of fluid and solutes in the lymph and in the **cerebrospinal fluid,** which bathes the brain and spinal cord. Both of these fluids drain back into the veins and return to the heart. Because all the body's fluid compartments connect with one another, an animal can regulate the composition of all its body fluids by controlling the content of any one of them. In the vertebrate body, the liver and kidneys are the most important organs that monitor and adjust the composition of the blood and thus keep the composition of all the body fluids constant. The liver regulates the blood's content of food molecules (Section 22-G). The kidneys dispose of nitrogenous and other wastes and regulate salts and water.

Homeostasis might be easier if an organism were a self-contained system, but every organism must constantly take in substances from its environment, use them in the chemical reactions of metabolism, and discharge the resulting wastes back into the environment (Table 26-1). With this constant flow of substances, the task of maintaining the constant composition of the fluid surrounding the cells is formidable. It is further complicated by the fact that wastes and substances needed by the body are mingled in the body fluids. The excretory system must remove waste, especially toxic nitrogenous wastes from the breakdown of proteins, while maintaining the body fluids' pH, water, and salt balance.

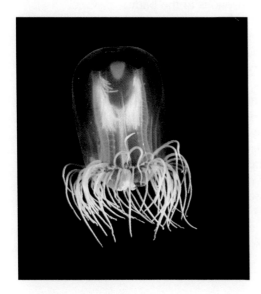

Figure 26-1
A jellyfish. Homeostasis of the body fluids is no problem for this marine cnidarian because its body fluids are very similar to the sea water that surrounds it. Gases, waste molecules, water, and salts diffuse readily between its body fluids and the sea water. (Langdon Quetin)

Table 26-1   Exchange of Substances Between the Human Body and Its Environment

Substance	Daily intake	Daily output
Water	1 litre in fluids	1 litre in urine
	1 litre in foods	0.75 litre in perspiration
	0.35 litre from oxidation of food	0.5 litre in expired air
		0.1 litre in feces
Solid food	2 kilograms	0.15 kilogram fecal weight
Oxygen and carbon dioxide	12,450 litres (about a 2.3-metre cube)	12,450 litres

- Living cells require an almost constant chemical composition in their immediate environment, the extracellular fluid.
- An animal's body fluids are in contact with one another. Regulating the content of any one body fluid (usually the blood) therefore maintains homeostasis in all of them.
- An animal's excretory organs must rid its body of nitrogenous wastes while keeping salts and water in proper balance.

26-A   Substances Excreted

The chief wastes produced by the body are carbon dioxide and water from oxidation of organic molecules, and nitrogenous wastes from breakdown of proteins (Table 26-2). Carbon dioxide is excreted across the body's respiratory surfaces. Excretory organs such as kidneys have the two major functions of removing nitrogenous wastes and regulating the body's salt and water content. In addition, excretory organs control the excretion of substances like spices, drugs, and hormones. Onions, garlic, and some other spices have volatile components that leave the body through the lungs, whereas the rest leaves through the kidneys. Penicillin and other drugs are removed primarily via the kidneys. The kidneys, liver, and lungs also carry out **detoxification,** converting toxic substances into forms that are not poisonous to the body.

Although the kidneys control the salt concentration and composition of the blood, some salts are also excreted through the skin in sweat, and some leave with the feces. When a lot of water and salt exits through the skin as sweat, the kidneys form less urine. A person working in the desert may lose more than 1 litre of sweat per hour, with a loss of 10 to 30 grams of sodium chloride per day. This heavy loss of salt causes no immediate difficulty because water is also lost, and so the salt concentration of the body fluids is maintained. However, drinking water after such heavy sweating dilutes the extracellular fluid, a condition sometimes called "electrolyte imbalance." This may lead to muscle cramps—an example of the importance of maintaining the composition of the body fluids.

Undigested food from the gut does not appear on our list of substances excreted by the body. Food that passes down the digestive tract and out the anus is not excreted but **egested**—that is, it travels through and is expelled from the body without ever passing through a plasma membrane to become part of the body. The term "excretion" applies only to substances that must cross plasma membranes to leave the body.

Table 26-2   Important Substances Excreted by the Human Body

Substance excreted	Excreting organ(s)
Nitrogenous wastes	Kidneys
	Skin (small amount in sweat)
Water	Kidneys
	Skin
	Lungs
Salts	Kidneys
	Skin (in sweat)
Carbon dioxide	Lungs

Nitrogenous Wastes

Nitrogenous wastes are a by-product of the breakdown of proteins. First, proteins are hydrolyzed to amino acids. This occurs in the digestive tract as food proteins are digested and in the body cells, where proteins are constantly made and destroyed. Fats and carbohydrates are stored for future use, but the body cannot store proteins or amino acids. Amino acids that the body cannot use immediately are **deaminated;** that is, their amino ($-NH_2$) groups are removed. The remaining organic acids can be used for energy or converted into carbohydrate or fat and stored. Each $-NH_2$ group removed picks up another hydrogen atom and becomes NH_3, ammonia. Since ammonia is the first metabolic breakdown product of amino acids, it can be produced with very little energy.

Ammonia is toxic except in a very dilute solution. Many aquatic animals dissolve their ammonia in large quantities of water from their environment and excrete this dilute solution.

Land-dwelling animals are often short of water. Mammals and adult amphibians conserve water by converting ammonia to urea and excreting the urea (Figure 26-2). The additional metabolic steps cost energy but are still worthwhile because they save water. Since urea is much less toxic than ammonia, it can accumulate at higher levels without damaging the tissues and can be excreted in a more concentrated form, using less water.

In mammals, most urea is formed by the liver, which takes excess amino acids out of the blood, deaminates them, and incorporates them into urea molecules. The brain and kidneys also form urea, but in lesser amounts.

Other land animals, notably reptiles, birds, and insects, incorporate their ammonia into **uric acid** (Table 26-3). This takes about 15 enzymatic steps and requires a lot of energy. However, this energy investment pays off in terms of water conservation because uric acid can be excreted in almost solid form.

Since the kidneys can handle nitrogenous waste only in solution, birds and reptiles pass a dilute solution of uric acid from the kidneys into the **cloaca** ("sewer"), a common reservoir at the end of the urinary, digestive, and reproductive tracts. Here most of the water is resorbed, and uric acid crystals precipitate and mix with the feces; the two are then voided together. Insects have a similar arrangement.

Most mammals do not excrete uric acid, but small amounts do appear in human urine. Our bodies produce uric acid by the breakdown of adenine and guanine (nucleotide bases), not of proteins. Persons with certain metabolic dis-

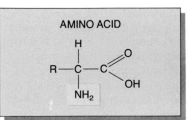

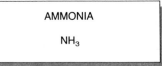

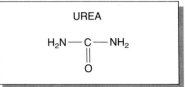

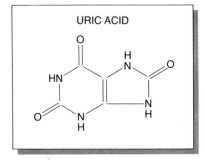

Figure 26-2
Nitrogenous waste. An amino acid and the three main forms of nitrogenous waste that animals produce from amino groups removed from excess amino acids. Urea and uric acid molecules contain nitrogen and hydrogen atoms derived from more than one amino group (yellow).

Table 26-3 *Advantages and Disadvantages of Nitrogenous Wastes in Relation to Habitat*

Waste	Advantages	Disadvantages	Habitat	Excreted by
Ammonia	Production requires little energy	Toxic in concentrated solution. Must be excreted in lots of water.	Water	Marine and freshwater invertebrates, bony fish, amphibian larvae
Urea	Less toxic than ammonia. Less water needed to excrete it.	Requires some energy to produce	Land	Adult amphibians, mammals
			Sea	Chondrichthyes
Uric acid	Can be excreted with very little water	Requires a lot of energy to produce	Land	Reptiles, birds, insects

Figure 26-3
An exception to the rule. Although this sea turtle is a reptile, it excretes up to half its nitrogenous wastes as ammonia rather than the uric acid typical of land-dwelling reptiles. The turtle's blood has only about one-third the salt concentration of sea water. Excess salt is excreted in a concentrated salt solution from glands near the eyes. This looks like tears running down the face of a female turtle while she is ashore laying her eggs. The white coral formations below the turtle are the skeletons of tiny cnidarian polyps, marine animals that excrete ammonia through the general body surface.

■ *The production and excretion of nitrogenous wastes involves a trade-off between energy expenditure and water conservation. Excreting ammonia takes little energy but uses a lot of water, whereas excreting uric acid costs lots of energy but uses very little water. Urea excretion uses intermediate amounts of energy and water.*

orders produce more uric acid than usual, and this causes the disease known as gout. In gout, uric acid crystals accumulate in some of the joints of the body, especially in the toes, causing great pain.

Lower animals with access to a lot of water excrete most of their nitrogenous wastes as ammonia. Some groups of higher animals evolved the metabolism and excretory organs to produce and excrete urea or uric acid, and many members of these groups were able to move into drier habitats. However, animals do not always excrete the form of nitrogenous waste predicted by their taxonomy; habitat plays an important part, too. For example, most reptiles excrete uric acid, but aquatic turtles also excrete a good deal of ammonia and urea (Figure 26-3).

Why do reptiles and birds excrete uric acid, whereas mammals turn their nitrogenous wastes into urea? The production of urea or uric acid is linked to the mode of reproduction. Birds and reptiles lay amniotic eggs, with membranes and shells that enclose all the water the embryo will have until it hatches (see Figure 21-40). The embryo produces uric acid, which sits as a solid mass, leaving the water in the egg free for other uses. A mammalian embryo, by contrast, has access to its mother's fluids via the placenta, and so it can rely on her system to dispose of its wastes. Any water available to the mother can also be used by the embryo. Mammals appear to have lost the metabolic pathway by which birds and reptiles produce uric acid.

26-B   Osmoregulation in Different Environments

The removal of nitrogenous wastes from the body fluids is inextricably tied to **osmoregulation**—regulation of the fluids' osmotic properties, that is, the balance between salts and water. The form of nitrogenous waste excreted depends on the availability of water, which in turn depends on the water balance between the animal's body fluids and its environment.

The body fluids of most marine invertebrates are **isotonic** with the sea water in which they live (that is, no net gain or loss of water occurs between the body fluids and sea water across the membranes separating them). Freshwater invertebrates and most vertebrates have body fluids about one-third as concentrated as those of marine invertebrates. Hence their fluids are **hypotonic** to sea water, but **hypertonic** to fresh water. These animals tend to lose water from their body fluids by osmosis if they are placed in sea water, and to gain water from fresh water (Section 4-C).

Marine fish and other marine vertebrates must prevent osmotic water loss to their hypertonic environment and uptake of too many salts by diffusion. Fresh-

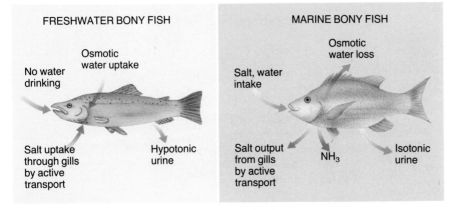

Figure 26-4
Osmotic adaptations of bony fish living in fresh water and in salt water. (What changes must a fish such as a salmon undergo as it returns from the sea to spawn in the stream where it hatched?)

water organisms have just the opposite problems: they must prevent loss of salts by diffusion and uptake of water by osmosis. They do this in part by excreting large volumes of dilute urine, but they must also conserve salts while ridding their bodies of nitrogenous wastes.

How do vertebrates live with these osmotic problems? Freshwater fish are covered with mucus, which retards passage of water and salts through the body surface. Freshwater bony fish do not drink water, but they must pass water over the gills to get oxygen, and water inevitably enters through the permeable gill membranes. These fish eliminate water by producing copious dilute urine, but they lose salts both via the urine and via diffusion from the gills. This is counteracted by active transport of salts into the body by special cells in the gills (Figure 26-4). Freshwater fish also take in salts as part of their food.

In a sense, although a marine bony fish is surrounded by water, it actually lives in a physiological desert, because it tends to lose water to its hypertonic environment. How does the fish conserve water and survive in this desert? The marine fish loses water both through the gills and in its urine. It takes in salt and water as part of its food. It achieves osmotic balance by excreting much of the salt by active transport through its gills (Figure 26-4).

The Chondrichthyes (sharks, skates, and rays) have evolved an interesting and unusual method of coping with the marine environment. Like most other vertebrates, they have body fluids with a salt concentration about one-third that of sea water, but they also produce and retain large quantities of urea (Figure 26-5). Their tissues have become adapted to levels of urea that would kill most other organisms. The combination of salts plus urea makes their body fluids slightly hypertonic to sea water: these fish actually absorb some water from the sea through their gills by osmosis and use it for excretion.

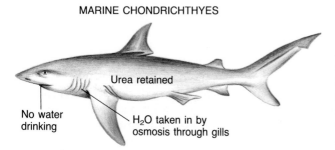

Figure 26-5
Osmotic adaptations in marine **Chondrichthyes**. Retention of urea makes the body fluids hypertonic to sea water. Hence water enters the body by osmosis through the gill membranes.

Figure 26-6
Osmotic adaptations of marine birds. A salt gland secretes a highly concentrated solution of the salt taken in with sea water. The kidneys produce a concentrated urine. These features permit the bird to conserve water for use in the body.

Seafaring birds drink sea water, and get rid of their excess salt by way of a **salt gland** in the head. This excretes a very concentrated salt solution, which drips out of the nostrils. Birds also excrete uric acid, conserving as much water as possible (Figure 26-6).

Marine mammals, such as whales and porpoises, take in sea water along with their food. Their kidneys can produce a urine several times as concentrated as sea water (Figure 26-7). This is especially important for carnivorous marine mammals, because their high-protein diet yields much urea to excrete.

Some land vertebrates can also produce highly concentrated urine. Laboratory rats can live indefinitely when all they are given to drink is sea water. Sea water is too concentrated to support human life. Although the human kidney can produce urine slightly more concentrated than sea water, this is not enough to offset other water losses through the lungs and skin. Furthermore, the high content of magnesium and sulfate in sea water may cause diarrhea and increase water loss with the feces. For every swallow of sea water, even more of the precious body water must be used to excrete the salts taken in. Thus humans lost at sea are indeed surrounded by "water, water everywhere, nor any drop to drink."

If our hypothetical mariners had read this far in the chapter, they might recall that bony fish have body fluids only one-third as concentrated as sea water, and think that eating fish would be easier on the kidneys. This is of little help, however, because fish is high in protein that will force the body to produce a lot of urea, again requiring more water for urine production. However, it is possible to improve the osmotic situation by drinking the dilute body fluids squeezed from bony fish. In addition, shipwrecked sailors can eat algae, which contain more carbohydrate and less protein than fish. It usually takes only a few days for a human to die of thirst without drinking at all, but people have survived at sea with no fresh water for more than two months by eating a low-protein diet and drinking any available hypotonic fluids. (Human urine may be one of these.)

26-C How Excretory Organs Work

Any excretory system does three things:

1. It collects fluids from somewhere inside the body, usually from the blood or from spaces between organs.
2. It modifies the fluid by resorbing substances the body needs to keep, and by transporting waste substances into the excretory product.
3. It provides a way to expel the excretory product from the body.

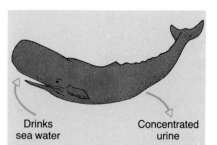

Figure 26-7
Marine mammals drink sea water but excrete the salts in a very concentrated urine, conserving the water.

Figure 26-8

Excretory system of a planarian (phylum Platyhelminthes). The beating of cilia in flame cells collects body fluids and passes them down a system of excretory ducts to pores on the surface of the body. The system serves primarily in osmoregulation, ridding the body of excess water.

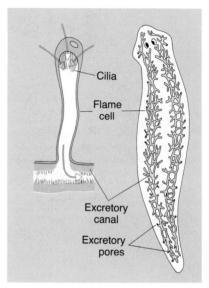

During excretion, an organism expends metabolic energy. First, it uses energy to break down proteins and in many cases to form urea or uric acid. Second, energy is used in active transport to modify fluids collected from the body into final excretory products. Although human kidneys make up less than 0.5% of the body weight, they use 7.2% of the oxygen consumed by the body. Pumping blood from the heart to the kidneys takes another 2.7%, so that about 10% of the human body's energy is spent just moving blood to the kidneys and cleansing it.

26-D Excretory Organs of Invertebrates

Corals and other members of the marine phylum Cnidaria live in an isotonic environment and lose most of their wastes by diffusion. Freshwater flatworms (phylum Platyhelminthes), on the other hand, have organized, multicellular excretory systems, which expel excess water (Figure 26-8). Body fluids are collected into **flame cells** by the beating of cilia. The fluid then passes through a series of tubules to an excretory pore at the body surface.

The functional unit of excretion in an earthworm is the **nephridium** (Figure 26-9). The nephrostome, an open, ciliated funnel, draws fluid from the coelom into the long, thin tubule. As fluid flows through the nephridium, substances that the body needs are reclaimed and passed into surrounding capillar-

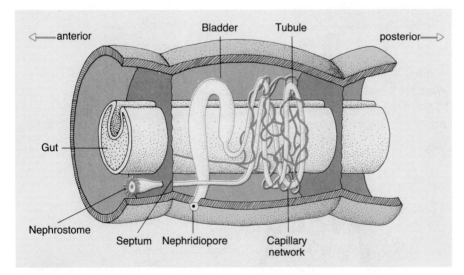

Figure 26-9

An earthworm nephridium. The nephrostome is a funnel-shaped opening that collects fluid. The rest of the nephridium lies in the next segment back. Capillaries of the circulatory system intertwined with the tubule resorb needed substances before the fluid reaches the enlarged bladder, from which the urine is discharged through the nephridiopore. Each segment, except a few in the anterior end, contains a pair of nephridia.

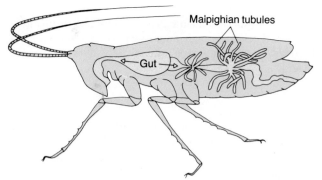

Figure 26-10
Malpighian tubules of an insect. These tubules discharge uric acid into the gut, whence it is voided with the feces.

ies of the circulatory system. Fluid expelled at the body surface contains water, nitrogenous wastes, and any salts that have not been resorbed.

Like the earthworm, most other invertebrates have nephridia, but insects have evolved a completely different system: long, slender **Malpighian tubules,** attached at one or both ends to the gut (Figure 26-10). Nitrogenous wastes from the body fluid are converted into uric acid, which is then moved down the Malpighian tubule into the gut. Cells in the rectum resorb water from these wastes before they are eliminated as fairly dry fecal pellets.

26-E The Kidney

Every vertebrate has a pair of kidneys. A kidney's functional units are **nephrons,** which closely resemble the nephridia of earthworms except that nephrons collect fluid filtered from the blood. A human kidney contains more than a million nephrons, each intertwined with capillaries of the circulatory system.

In the human excretory system, the **renal artery** carries blood from the aorta into the kidney. Here fluid, salts, and wastes are removed from the blood to form urine. Purified blood leaves the kidney via the **renal vein,** whereas

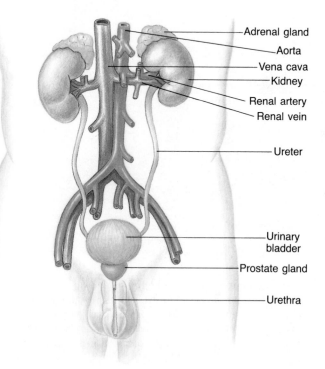

Adrenal gland
Aorta
Vena cava
Kidney
Renal artery
Renal vein

Ureter

Urinary bladder
Prostate gland
Urethra

Figure 26-11
The human urinary system and associated blood vessels. The kidneys lie in the small of the back, against the spinal column.

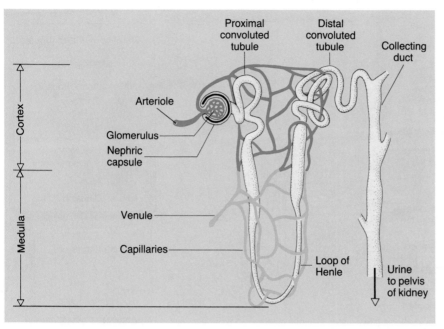

Figure 26-12

A nephron and its associated blood supply. A human has about 2 million nephrons. Blood pressure pushes fluid out of the blood in the glomerulus and into the nephric capsule. The filtrate so formed travels through the nephron tubule while substances are exchanged between the filtrate, the extracellular fluid surrounding the tubule, and the blood in the nearby capillaries. The fluid that reaches the collecting duct is urine, which flows through the pelvis of the kidney and down the ureter to be stored in the urinary bladder.

urine leaves via a **ureter** and is stored in the **urinary bladder.** Eventually the urine is expelled from the body via the **urethra** (Figure 26-11).

The human body contains about 5.6 litres of blood; 1.2 to 1.3 litres pass through the kidneys each minute, for a daily total of about 1600 litres. This is nearly one quarter of all the blood pumped by the heart, and so a very high proportion of all the blood is passing through the kidneys at any time. Each day, about 180 litres of fluid filter out of the blood and enter the nephrons. Most of this fluid is resorbed, leaving a daily urine output of about 1 litre. Table 26-4 shows the approximate composition of urine. This varies a lot depending upon what has been taken into the body and what needs to be excreted to restore the body fluids to within normal range.

Functions of the Nephron

Let us follow the process of urine formation in a nephron (Figure 26-12). The nephron's cup-shaped **nephric capsule** surrounds a knot of blood capillaries called a **glomerulus** (Figure 26-13). Since the blood is under pressure and the walls of the capillaries are permeable, much of the fluid from the blood filters

Table 26-4	Concentration of Various Substances in Human Urine
Substance	**Concentration**
Sodium (Na⁺)	128 mmol/L*
Potassium (K⁺)	60 mmol/L*
Calcium (Ca²⁺)	5 mmol/L*
Magnesium (Mg²⁺)	15 mmol/L*
Chloride (Cl⁻)	134 mmol/L*
Bicarbonate (HCO₃⁻)	14 mmol/L*
Phosphate	50 mmol/L*
Sulfate (SO₄²⁻)	33 mmol/L*
Glucose	0 mg/100 ml†
Urea	1820 mg/100 ml†
Uric acid	42 mg/100 ml†
Creatinine	196 mg/100 ml†

*mmol/L = millimoles per litre (see Section 2-C)

† mg/100 ml = milligrams per 100 millilitres

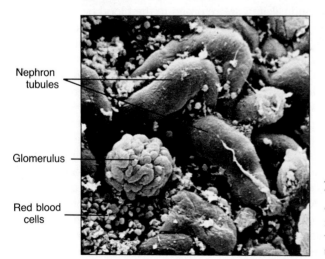

Nephron tubules

Glomerulus

Red blood cells

Figure 26-13

The kidney cortex. The red blood cells visible in this scanning electron micrograph have leaked from blood vessels broken during preparation of the specimen. (Biophoto Associates)

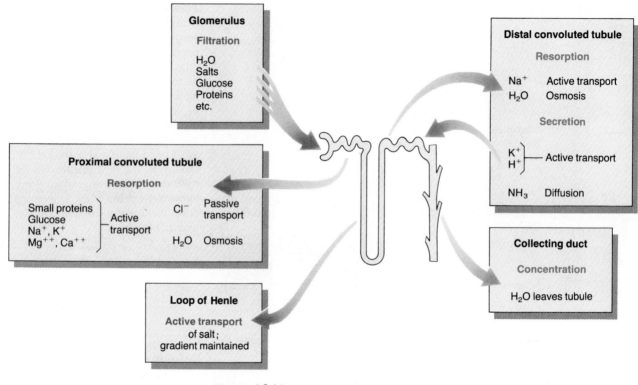

Figure 26-14
Summary of the activities in different parts of the nephron.

■ *The intimate association of capillaries and nephrons in the kidneys permits substances to pass between blood in the capillaries and the forming urine in the nephrons. In this way, the kidneys control the composition of the blood.*

into the capsule. Left behind are large proteins and whole cells, which are too large to pass through the filter, along with the rest of the blood fluid (plasma).

From the capsule, the **filtrate** (filtered fluid) passes into the nephron tubule. Meanwhile, the remainder of the blood follows along in capillaries outside the nephron. The cells of the nephron tubule modify both fluids, **resorbing** some substances from the filtrate and returning them to the blood, and **secreting** others from the blood into the urine-to-be. The nephron tubule has four main parts: the **proximal convoluted tubule,** the U-shaped **loop of Henle,** the **distal convoluted tubule,** and finally the **collecting duct.**

In the proximal convoluted tubule, a considerable amount of resorption takes place (Figure 26-14). Small proteins, glucose, and ions such as sodium, potassium, magnesium, and calcium are returned to the blood by active transport. Negatively charged chloride ions follow passively after the positively charged ions. Water follows these solutes passively by osmosis. About 75% of the salt and water in the filtrate returns to the blood from the proximal convoluted tubule. Normally, all of the glucose also returns here. However, if glucose in the filtrate exceeds the **kidney threshold level,** some glucose will remain and appear in the urine. This happens when the active transport carriers for glucose, working at top speed, still cannot pick up all the glucose and move it back to the blood as quickly as the filtrate passes through the proximal tubule.

In many mammals and birds the loops of Henle are quite long, and their activities (discussed later) permit the production of relatively concentrated urine. As the fluid passes through the distal convoluted tubule, more sodium is resorbed. This is the chief area of secretion into the tubule, chiefly of potassium and hydrogen ions by active transport, and of ammonia by diffusion. Secretion of hydrogen ions adjusts the pH of the blood. Various drugs, such as penicillin, are also secreted into the urine here. The collecting duct, along with the loop of Henle, plays a vital role in water balance.

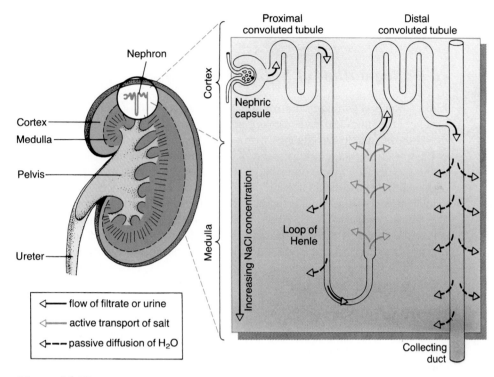

Figure 26-15
How urine is concentrated. The loops of Henle and collecting ducts lie in the medulla of the kidney. Here the loops of Henle maintain a high sodium chloride gradient (colored shading) in the extracellular fluid by actively transporting salt ions into the extracellular fluid. As the urine passes down the collecting duct, it loses water by osmosis to the extracellular fluid.

Concentration of the Urine The actual concentration of urine takes place in the collecting ducts, but the process depends on the activities of the loops of Henle. To understand how this works, it is necessary to realize that the loops of Henle and collecting ducts lie in the **medulla** of the kidney (Figure 26-15). The other parts of the nephron lie outside the medulla in the kidney's outer region, the **cortex.** The extracellular fluid in the medulla contains an osmotic gradient, with solutes steadily more concentrated in the direction away from the cortex. The gradient contains two kinds of solutes: salt (sodium and chloride ions) and urea.

The salt gradient is created by the loops of Henle, which actively transport salt ions out of the filtrate. Active transport starts when the filtrate reaches the thick section of the ascending part of the loop. As the filtrate moves up, less salt is available to be transported out, and so the gradient has a higher salt concentration near the bottom of the loop (colored shading in Figure 26-15). This part of the loop of Henle is relatively impermeable to water, and so water cannot follow the ions out here.

From the loop of Henle, urine passes through the distal convoluted tubule to the collecting duct, where the actual concentration of the urine takes place. The walls of the collecting duct are permeable to water. As the urine passes down the collecting duct, it re-enters the medulla of the kidney. Here water moves from the urine into the extracellular fluid by osmosis in response to the high concentration of solutes in the medulla. From here it returns to the blood. The last part of the collecting duct is also permeable to urea, which by now is quite concentrated in the urine. Some urea diffuses out into the medulla, down its own gradient, adding to the solute concentration there. Urine leaving the

■ *The role of the loop of Henle is to set up a high concentration of sodium chloride in the medulla of the kidney.*

collecting duct passes through the pelvis of the kidney and down the ureter to the urinary bladder, where it is stored.

26-F Regulation of Kidney Function

Kidney function is under the minute-by-minute control of many systems. For instance, the filtration rate remains nearly constant because certain cells in the kidneys can detect changes in blood pressure and adjust the contraction of the muscles in the blood vessel walls, maintaining the proper blood pressure in the glomeruli. On the other hand, the rate of urine formation and the urine composition change dramatically, depending on circumstances. For instance, drinking a lot of water dilutes the blood but raises the filtration rate only slightly. The major change is a drop in the rate of water resorption, so that the excess water is excreted from the body. Urine composition and the rate of urine formation are largely regulated by the hormones vasopressin, aldosterone, and angiotensin, and the enzyme renin.

Vasopressin (ADH)

Vasopressin (also known as **antidiuretic hormone, ADH**) is a hormone released from the posterior pituitary gland in the brain (Section 31-D). Its presence increases resorption of water from the urine. Loss of water from the body stimulates vasopressin secretion and so slows the loss of water via the urine. The body may detect a decrease in its water content in one of two ways: either as a reduction in blood volume (for example, caused by severe bleeding) or as an increase in the concentration of the blood due to loss of water (for example, from sweating).

Vasopressin increases the permeability of the collecting ducts to water. In the absence of vasopressin, the walls of the ducts are nearly impermeable to water, and very little water is resorbed from the urine.

Insufficient vasopressin production results in diabetes insipidus, a disease characterized by thirst and production of large amounts of dilute urine. It is much less common than diabetes mellitus, characterized by glucose in the urine.

Aldosterone, Renin, and Angiotensin

Sodium accounts for about 90% of all the positively charged ions in the body fluids. Since water moves passively (by osmosis) following the movement of salt ions, the volume of water in the extracellular fluid depends on the amount of salt. Thus the amount of sodium in the body is the chief factor determining the volume and concentration of the blood and extracellular fluid.

The body's sodium content (and hence the volume of body water) depends upon the balance between intake in the diet and loss in the urine. (Sodium is also lost in feces and sweat, but the body has less control over these.) Vertebrates have nervous and hormonal means of regulating how much sodium they eat and how much leaves the body.

Angiotensin and aldosterone are hormones that together control how much sodium is resorbed from the filtrate in the nephron. **Aldosterone** is one of several steroid hormones secreted by the cortex of the **adrenal gland,** which is attached to the kidney (see Figure 26-11). Aldosterone promotes resorption of sodium by the distal tubule. The rate of aldosterone secretion is determined by the blood's salt content. A slight decrease in the sodium content of blood causes the adrenal cortex to increase its aldosterone secretion. This leads to resorption of more sodium from the filtrate, and hence to a decrease in the urine's sodium content.

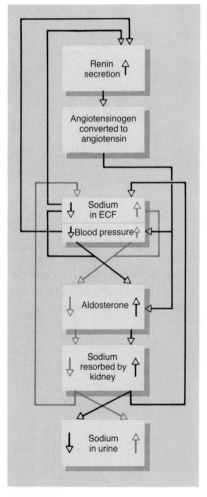

Figure 26-16
Interaction among aldosterone, renin, and angiotensin. Black arrows indicate events that tend to return low sodium levels to normal, and red arrows indicate adjustments that lower the sodium levels when there is too much sodium.

Adaptations of Mammals to Sodium-Deficient Environments

A JOURNEY INTO LIFE

Figure 26-A
Impalas obtain sodium by eating soil near a water hole in Kenya. Minerals are concentrated in the soil when water evaporates around the edges of the water hole, leaving the minerals behind.

Many arid and mountainous areas contain little sodium. Although no animal can survive with much less than the usual amount of sodium in its body, most mammals have adaptations that permit them to survive in such sodium-deficient areas. Some of these adaptations increase the dietary intake of sodium, whereas others reduce loss of sodium from the body.

Herbivores, including butterflies, reindeer, elephants, gorillas, elk, moose, sheep, and kangaroos, have been seen eating soil, drinking sea water, or consuming other improbable things whose only redeeming value appears to be their high sodium content (Figure 26-A). Plainly an appetite for salt when the body is deficient in sodium is a useful adaptation, and it is common in mammals.

A number of physiological differences are found between kangaroos, sheep, foxes, cattle, and rabbits in sodium-deficient mountainous areas of Australia and members of the same species in coastal areas, where there is plenty of sodium (from the sea). Whenever possible, animals in mountainous areas select food plants containing sodium, and in addition hold sodium loss from the body to a minimum. For instance, the urine of mountain animals contains virtually no sodium, whereas kangaroos and wombats on the coast excrete a lot of sodium in their urine. This difference is undoubtedly due to the higher levels of renin, angiotensin, and aldosterone found in the mountain-dwelling animals. During the day rabbits produce soft feces, which they eat to extract further sodium (and vitamins). The fecal pellets egested by highland rabbits after the second digestion contain practically no sodium.

Sodium in the feces may be reduced by resorption from the gut. Sodium-deficient mammals with ruminant digestion (Section 22-D) also reduce the sodium loss in their feces by changing the composition of their saliva. Ruminants such as cattle and sheep produce many litres of saliva a day. This usually contains a high concentration of sodium bicarbonate, which produces an alkaline environment in the stomach for the microbial symbionts that ferment the animal's food. Much of this sodium bicarbonate is absorbed back into the body as the feces pass down the intestine, but some is lost with the feces. Aldosterone decreases the amount of sodium in the saliva and causes potassium to be secreted in its place. Although the need to secrete large quantities of sodium in the saliva every day seems to impose a sodium-supply problem on ruminants, the effect may, in fact, be the other way around. Sheep survive temporary sodium shortages better than do nonruminants such as humans or foxes. This is probably because the ruminant stomach contains a large store of sodium, which can be steadily replaced by potassium and used for more vital functions in the body during times of sodium shortage.

While aldosterone secretion can be controlled directly by the blood's sodium concentration, it is also controlled indirectly by renin and angiotensin. The kidneys secrete the enzyme **renin** in response to a decrease in blood pressure or in sodium or potassium in the blood. Renin converts the protein angiotensinogen in the plasma into the hormone angiotensin. **Angiotensin** causes constriction of blood vessels and an increase in aldosterone secretion. Both of these actions increase the blood pressure, either directly or by raising its sodium concentration. An increase in blood sodium concentration decreases loss of water in the urine and therefore maintains the volume, and hence the pressure, of the blood (Figure 26-16).

SUMMARY

Living cells require a relatively constant chemical environment. Cells are continuously altering their environment by taking things from it and converting them into new substances—some useful, some wastes that must be removed from the cell and from its immediate environment. Animals have evolved various specialized mechanisms that maintain a constant internal chemical environment in the face of this continuous flow of materials between the external environment, the extracellular fluid, and the interiors of cells.

All of an animal's body fluids are in contact with one another. By cleaning any of these fluids, the excretory system can keep the fluid environment of the body's cells constant. All excretory systems work by collecting fluid (from the extracellular fluid or blood plasma), retrieving the substances needed by the body, and expelling the remaining fluid with the waste it contains.

An animal's excretory system must rid the body fluids of toxic nitrogenous wastes by using some of the body's own water. However, many animals live in habitats where it is hard for them to obtain or keep water or salts. These animals must have adaptations for disposing of nitrogenous wastes while at the same time keeping the proper osmotic balance of salts and water in their body fluids. Most of the energy they spend on excretion goes, not to collection and disposal of wastes, but to retaining the substances they

need. Gills, salt glands, or kidneys use energy for active transport of salts into or out of the body as necessary. Animals that must conserve water generally also use energy to convert the nitrogenous waste ammonia into urea or uric acid, which can be excreted using less water.

The function of the vertebrate kidney is to maintain the composition of all of the body's fluids within the narrow limits necessary for the body's cells to function. The basic unit of the kidney is the nephron, a long tube closely associated with capillaries of the circulatory system. Blood plasma is filtered, under pressure, into one end of this tube. As the filtrate passes through the nephron, substances needed by the body are resorbed through the cells of the nephron tubule into the extracellular fluid, and then into the capillaries, either by diffusion or by active transport. Substances that the body does not need are secreted from the blood into the filtrate. After being changed in these ways during its passage through the tubule, the filtrate is collected as urine.

Homeostatic mechanisms under hormonal control regulate the amount and composition of the urine produced. The hormones vasopressin and aldosterone help the body conserve water and sodium, respectively, by reclaiming more of these substances from the urine.

SELF-QUIZ

1. The pack rat, a rodent, often goes for long periods without drinking, eats leaves of juicy plants, and moves about in the open only in the evening and at night. From these habits, you can guess that it lives in:
   a. desert areas
   b. the Arctic tundra
   c. the woodlands of the eastern United States
2. The main nitrogenous waste substance excreted by the pack rat will probably be:
   a. ammonia b. urea c. uric acid
3. An advantage of excreting nitrogenous wastes in the form of uric acid is that:
   a. uric acid can be excreted in almost solid form
   b. the formation of uric acid requires a great deal of energy
   c. uric acid is the first metabolic breakdown product of amino acids
   d. uric acid may be excreted through the lungs
   e. uric acid is highly toxic, so it is important for the animal to get rid of it
4. The main excretory structure in houseflies is the:
   a. Malpighian tubule d. loop of Henle
   b. flame cell e. nephridium
   c. nephron

5. Salmon have gills that are more permeable to water than to salts. Salmon hatch in freshwater streams, and then migrate to the ocean. Once they reach the ocean, you would expect the rate of uptake of water into their bodies through the gills to:
   a. increase
   b. decrease
   c. remain the same
6. Urine leaves the kidney via:
   a. the renal vein   d. the ureter
   b. the urethra   e. the collecting duct
   c. the bladder

In questions 7 through 11, match each structure on the left with its function from the list at the right.

_____ 7. Loop of Henle
_____ 8. Renal artery
_____ 9. Proximal convoluted tubule
_____ 10. Glomerulus
_____ 11. Distal convoluted tubule

a. carries blood into the kidney
b. area where a considerable amount of resorption takes place
c. main area of secretion
d. filtration of blood
e. plays a role in concentration of urine

12. Filtration into the kidney tubule is accomplished by means of:
    a. active transport
    b. blood pressure
    c. an osmotic gradient
    d. secretion
    e. diffusion
13. Severe dehydration causes an increased concentration of solutes in the blood. This causes a(n) (increase/ decrease) in the amount of urine produced. This change in urine production is caused primarily by:
    a. an increase in the amount of water filtered out of the blood
    b. a decrease in the amount of water filtered out of the blood
    c. an increase in the amount of water resorbed
    d. a decrease in the amount of water resorbed

QUESTIONS FOR DISCUSSION

1. Why does an increase in aldosterone secretion increase the volume of extracellular fluid and increase blood pressure?
2. If we divide animals into marine and freshwater invertebrates, marine and freshwater bony fish, cartilaginous fish (class Chondrichthyes), amphibians, and terrestrial vertebrates, which animals are in the following osmotic situations?
    a. approximately in osmotic equilibrium with their environment
    b. must have adaptations to guard against dehydration (water loss)
    c. must have adaptations to prevent gain of excess water from the environment
    d. in danger of taking up too many salts from the environment
    e. must have adaptations to prevent loss of salts to the environment
3. Many fish lack a urinary bladder, but urinary bladders are found in amphibians and all higher vertebrates. What is the advantage of having a urinary bladder?
4. Certain reptiles have kidneys with no glomeruli in the nephric capsules. How would this affect urine formation? What type of habitat are such animals adapted to?

SUGGESTED READINGS

Baldwin, E. *An Introduction to Comparative Biochemistry,* 4th ed. New York: Cambridge University Press, 1964. A short and provocative book with emphasis on osmoregulatory problems and their solutions by animals.

Guyton, A. C. *Physiology of the Human Body,* 6th ed. Philadelphia: W. B. Saunders, 1984. A brief textbook covering all aspects of human physiology.

Schmidt-Nielsen, K. "Salt glands." *Scientific American,* January 1959. An account of the structure and function of salt glands and their contribution to osmoregulation, particularly in sea birds.

Smith, H. W. *From Fish to Philosopher.* Garden City, NY: Doubleday, 1961. Evolutionary history of vertebrate internal homeostasis delightfully written by an eminent renal physiologist.

Sexual Reproduction and Embryonic Development

CHAPTER

27

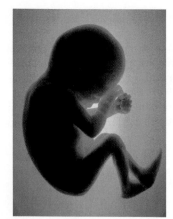

Five-month-old fetus

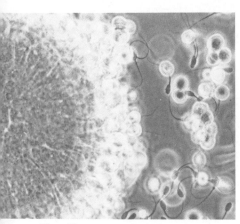

Egg with sperm

Leatherback turtle hatching

Egg arrangement

Animals have evolved many adaptations for obtaining food, exchanging gases, regulating body temperature, and retaining body water and salts. These adaptations allow some animals to survive in a variety of challenging environments. But the only true measure of evolutionary success is reproduction. Many animals can reproduce asexually (Chapter 16), but sexual reproduction is much more common.

The crucial event in sexual reproduction is **fertilization,** the union of two haploid reproductive cells, the **gametes,** to form a diploid **zygote.** In animals, the gametes are always different from each other. The female gamete is a large, nonmotile **egg.** The male gamete is a **sperm,** which actively moves toward the egg.

After fertilization, the zygote undergoes a period of embryonic development before it is born or hatched (Figure 27-1). Embryonic development involves cell division, cell movement, and differentiation into various cell types, as the genetic information in the zygote expresses itself and forms the young animal.

In most animal species, the two gametes that unite at fertilization come from different parents. The parents' anatomy, physiology, and behavior are coordinated so that their gametes are produced at the same time and brought together in the same place. Hormones often play an important role in this coordination. Many animals also engage in **courtship behavior** before they release gametes, increasing the chance of fertilization.

- Sexual reproduction involves (1) getting sperm and egg together for fertilization, and (2) ensuring that the fertilized egg has a suitable environment where it can develop until the young animal is ready to survive on its own. In animals, fertilization and development may be external or internal, or fertilization may be internal, followed by external development.
- Human fertilization and embryonic development are both internal. The male reproductive tract produces sperm and the penis inserts them into the female tract. The female tract produces eggs and provides the sites of fertilization and embryonic development.
- Hormones coordinate production of gametes in both sexes. In females, hormones maintain pregnancy and, after childbirth, stimulate milk production.
- Understanding how pregnancy occurs can help people avoid unwanted pregnancy or enhance the chances of beginning a desired pregnancy.
- Embryonic development involves division of the fertilized egg into many new cells, movement of these cells into new positions, and differentiation of cells to form the tissues and organs of the body.

27-A Reproductive Patterns

Animal reproduction falls into three main patterns, depending on whether fertilization and embryonic development occur within or outside the body of the (female) parent:

1. External fertilization and development.
2. Internal fertilization followed by external development.
3. Internal fertilization and development.

In the first pattern, found in many aquatic animals, eggs and sperm are released from the parents' bodies into the surrounding water, where they must find each other. For this to succeed, male and female must release their gametes at the same time and place.

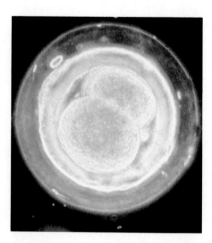

Figure 27-1
Sexual reproduction. The fertilized egg of a mouse completing the first cell division as it begins embryonic development. (Biophoto Associates)

(a)

(b)

Figure 27-2
External and internal fertilization. (a)
The yellow material draped around
these brown tube sponges consists of
masses of eggs, exuded from the
body to await fertilization by sperm
shed into the surrounding water. (b)
A male "ladybug" (actually a beetle)
introduces sperm into the body of a
female, a prelude to internal fertilization. The female later lays eggs with
waterproof shells, which develop externally without further parental attention. (a, S. K. Webster, Monterey Bay
Aquarium/BPS; b, Peter J. Bryant, University of California at Irvine/BPS)

In animals with internal fertilization, sperm are passed from the male to the
female. Male squids and spiders use some of their many appendages to pass a
packet of sperm to the female. In other invertebrates, sharks, reptiles, birds, and
mammals, sperm is transferred by **copulation,** direct passage of sperm from
inside the male's body to inside the female's body (Figure 27-2). In most animals, this transfer involves an intromittent ("into-sending") organ such as a
penis.

Internal fertilization has several advantages. The female reproductive tract
provides a confined, protected space where sperm and egg can get together
without danger of being eaten or washed away. If development is internal, fertilization must also be internal. But even if the egg develops externally, internal
fertilization gives the sperm easy access to the egg before the final preparations
for laying. As the fertilized egg then passes down the female reproductive tract
to the exterior, it can be surrounded with secretions, membranes, or shells that
will protect the developing embryo.

Internal development gives an animal embryo even more advantages. The
mother's body provides exactly the right chemical conditions and, in mammals,
warmth as well. Since the mother carries the embryo everywhere she goes, it is
not vulnerable to the predators that plunder the egg masses or nests of animals
that develop externally. In most species other than mammals, internal development simply means keeping the egg inside the body instead of laying it. As with
animals that develop externally, the food stored in the egg must support the
entire embryonic development. In mammals, however, the embryo shares whatever food its mother takes in during her pregnancy. Such a continuously renewed food supply may permit the embryo to reach a larger size before birth.

Human Reproduction

27-B   Human Reproductive Organs

We can illustrate reproduction in animals with internal fertilization by considering human reproduction. Human reproductive tracts are very similar to those of
other vertebrates, except that only mammals have a uterus, the part of the female
tract where embryos develop, and most male vertebrates except mammals and
reptiles lack a penis.

The human reproductive organs can be roughly divided into the internal
organs and the external, or accessory, sex organs. In addition, each sex has
secondary sexual characteristics, such as breasts in the human female or a

higher metabolic rate in the human male, which are not part of the actual reproductive apparatus.

Female Reproductive Organs

Figure 27-3 shows the external sex organs of a woman, collectively known as the **vulva.** The **labia majora** and **labia minora** (labia = lips) cover and protect the urinary and genital openings. Note that the openings of the urethra, from the urinary tract, and of the vagina, from the reproductive tract, are separate. Human females are born with a membrane, the **hymen,** covering the vaginal orifice. A slit in the hymen allows the menstrual flow to pass. The hymen presumably performs a protective function, but it is essentially absent in many women. During embryonic development, the **clitoris** arises from the same structure that develops into the penis in a male embryo. Like its male counterpart, the clitoris produces a pleasurable sensation when it is stimulated by touch and, indeed, this seems to be its only function.

The internal female reproductive organs consist of the ovaries and oviducts (fallopian tubes), uterus, and vagina (Figure 27-4).

The two **ovaries** are the female **gonads,** or gamete-producing organs. The ovaries produce eggs. An ovary contains many follicles, each consisting of an immature egg surrounded by nutritive follicle cells (Figure 27-5). At **ovulation,** a mature follicle ruptures and the egg pops out into the coelom. This release of a ripe egg stimulates the oviduct's finger-like endings to move and grasp the egg. The beating of cilia lining the oviduct draws the egg into the tube and on toward the uterus. Fertilization usually occurs about halfway down the oviduct. An unfertilized egg disintegrates as it passes through the uterus about 72 hours after ovulation.

The **uterus** is a highly elastic organ whose main function is to hold a developing embryo and expel it during childbirth. The external opening of the uterus is the **cervix,** made up largely of the biggest, most powerful circular muscle in the body. The strength of the cervical muscle is necessary to hold about 15 pounds of fetus and fluid in the uterus against the pull of gravity during pregnancy. The cervix protrudes into the upper end of the vagina.

The **vagina** is the receptacle for the penis during copulation, or sexual intercourse, and the pathway to the exterior for the baby during childbirth. In keeping with these functions, it has extremely elastic walls.

■ *A ripe egg's path through the female reproductive tract is:*

ovary
↓
oviduct
↓
uterus
↓
vagina

If the egg is fertilized in the oviduct, it develops in the uterus and the fetus leaves via the vagina at birth.

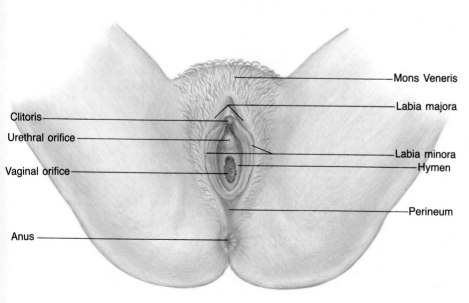

Clitoris
Urethral orifice
Vaginal orifice
Anus

Mons Veneris
Labia majora
Labia minora
Hymen
Perineum

Figure 27-3
Vulva of a woman. Note the three separate openings characteristic of females of higher mammals—the urethral orifice from the urinary tract, vaginal orifice from the genital tract, and anus from the digestive tract. The clitoris is anterior to the urethral orifice. The mons Veneris ("mountain of Venus") is a fatty pad over the pubic bones.

Figure 27-4
The internal reproductive organs of a woman. (a) The position of the reproductive tract. (b) Closeup of the reproductive organs, with one side cut open to show the pathway taken by the egg from the ovary to the vagina. (c) The relationship of the reproductive tract to other lower abdominal structures.

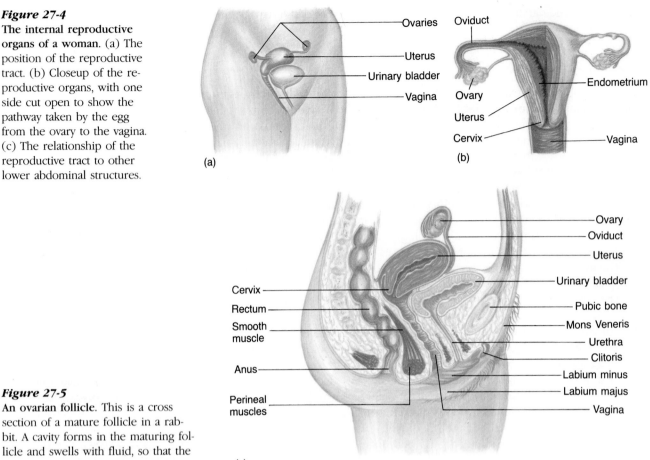

(a)

(b)

Figure 27-5
An ovarian follicle. This is a cross section of a mature follicle in a rabbit. A cavity forms in the maturing follicle and swells with fluid, so that the mature follicle balloons out on the surface of the ovary and eventually ruptures, expelling the egg into the abdominal cavity. (Biophoto Associates)

(c)

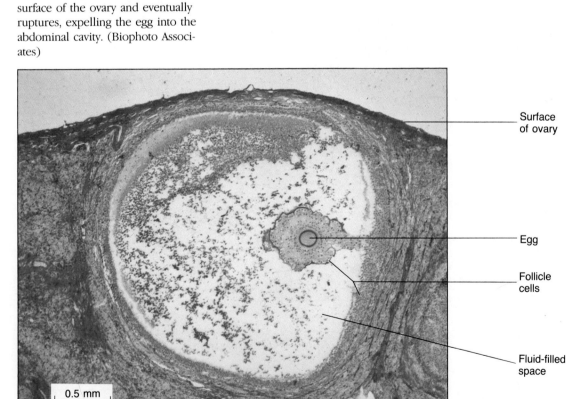

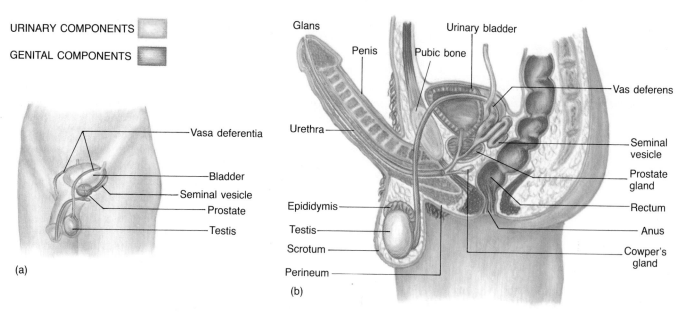

URINARY COMPONENTS

GENITAL COMPONENTS

(a)

Glans

Penis

Urethra

Epididymis

Testis

Scrotum

Perineum

(b)

Pubic bone

Urinary bladder

Vas deferens

Seminal vesicle

Prostate gland

Rectum

Anus

Cowper's gland

Vasa deferentia

Bladder

Seminal vesicle

Prostate

Testis

Figure 27-6

Male reproductive tract. (a) Position of the reproductive organs in a human male. Sperm are formed in the testes and travel through the vasa deferentia to the urethra, leaving the body through the opening of the urethra at the tip of the penis. (b) Cross section through the pelvic region of a man during erection. Notice that the urethra and vasa deferentia pass through the prostate gland. Enlargement of the prostate can constrict this passageway and interfere with urination.

Male Reproductive Organs

The gonads of a human male are the **testes** (singular: **testis**). They produce sperm from the time of sexual maturity, at puberty, until death. The testes of most mammals lie in the **scrotum,** a sac outside the body cavity, which provides the cool temperature required for sperm production.

Sperm develop from cells lining the **seminiferous tubules** of the testes (Section 11-G). Muscular action in the walls of these tubules carries sperm to the **epididymis.** Here they are stored and complete their maturation. During sexual stimulation, the sperm move through the **vas deferens** (plural: **vasa deferentia**) by contraction of its walls. The sperm then move into the urethra, where they are joined by secretions from the **seminal vesicles** and the **prostate** and **Cowper's glands.** These secretions activate the sperm and provide a fluid medium for them to swim in. The sperm and their attendant secretions, collectively called **semen,** leave the penis via the urethra (Figure 27-6). The **penis** is an external sex organ that introduces semen into the vagina during sexual intercourse. Urine from the bladder also leaves the male's body via the urethra, but the two fluids cannot pass through the urethra at the same time.

27-C Physiology of Sexual Intercourse

During copulation (often called sexual intercourse in humans), the male's penis introduces sperm into the female's vagina. Before the penis can enter, it must become at least partly erect under the influence of sexual stimulation. Sexual stimulation can be brought about by any of the senses, usually most effectively by touch. The external genitals, and especially the glans of the penis, are the most sexually sensitive areas.

The penis consists of three cylinders of spongy tissue that extend into the body (Figure 27-7). Within these cylinders are many hollow bodies, which are usually empty and collapsed. During sexual stimulation, about ten times the usual volume of blood is carried from the arteries into the hollow bodies. The blood filling the spaces presses against the outsides of the veins, blocking much of the flow of blood leaving the penis. The penis thus enlarges and becomes rigid.

Sexual arousal of a human female can also result from many different stimuli. The clitoris is the organ most sensitive to touch. The vulva contains erectile

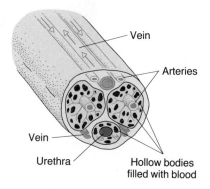

Vein

Arteries

Vein

Urethra

Hollow bodies filled with blood

Figure 27-7

Cross section of a human penis. During erection, the hollow bodies fill with blood, partly blocking the veins so that little blood can leave the penis. The engorged penis is large and stiff.

■ *The path of sperm through the male reproductive tract is:*

testis

↓

epididymis

↓

vas deferens

↓

urethra

tissue like that in the male's penis. In a sexually aroused woman, the vulva becomes swollen because of an increased blood supply, and the walls of the vagina exude fluid. Glands just inside the vaginal opening secrete mucus, which acts as a lubricant for entrance of the penis.

When the erect penis has been inserted into the vagina, stimulation by the movement of the genitals against each other may result in **orgasm.** Orgasm is characterized by an increase in heart rate and blood pressure, engorging of various tissues with blood, and faster and deeper breathing, finally resulting in an explosive burst of involuntary muscular contractions. In men, orgasm is accompanied by **ejaculation,** the forceful ejection of semen, propelled by peristaltic waves of contraction of the muscles in the sperm ducts. In women, orgasm is characterized by rhythmic spasms of the muscles surrounding the vagina.

Fertilization can occur without either sex experiencing orgasm. Often a small amount of semen is released before ejaculation, and this may contain enough sperm to fertilize an egg.

27-D Hormones and Reproduction

Hormones play a major role in the reproduction of nearly all animals. Hormones control gamete production. They also control the superficial characteristics that differentiate the sexes, and they are necessary for sexual behavior. In addition to producing the hormones of their own sex, members of both sexes produce small amounts of the hormones characteristic of the opposite sex.

Male Hormones

Hormones with masculinizing effects are called **androgens.** The most important is **testosterone,** a steroid hormone secreted primarily by the testes.

A boy's testes remain dormant until the onset of puberty, at the age of 10 to 14. From then on, the pituitary gland, located beneath the brain, releases a gonadotropic ("gonad-feeding") hormone into the circulation. This is **luteinizing hormone (LH).** Another gonadotropic pituitary hormone, **follicle-stimulating hormone (FSH),** also helps to regulate sperm production. (We shall see that these two hormones play important roles in females as well, and their names come from their effects on the ovaries.)

Luteinizing hormone promotes secretion of testosterone by the testes. Testosterone, in turn, causes a reduction in the pituitary's secretion of LH. As a result, the level of testosterone in the blood is constantly kept within narrow limits.

Testosterone, with the help of FSH, promotes sperm formation. It also stimulates development of the male secondary sexual characteristics, features that are not directly related to reproduction but are characteristic of a sexually mature male. The secondary sexual characteristics of human males include deepening of the voice, development of male musculature, and growth of hair on the face and other parts of the body.

Female Hormones

Hormones control women's secondary sexual characteristics and menstrual cycles. As in the male, the pituitary hormones FSH and LH are secreted at puberty, triggering the development of the secondary sexual characteristics and the onset of menstrual cycles. The hormonal progress of the menstrual cycle determines a woman's fertility.

Table 27-1 Hormones Involved in Human Reproduction

Source	Hormone	Role
Pituitary	Follicle-stimulating hormone (FSH)	Production of gametes in both sexes
	Luteinizing hormone (LH)	Secretion of sex hormones by gonads in both sexes
		Ovulation in females
	Oxytocin	Contractions of uterus during childbirth
		Letdown of milk from mammary glands to nipple
	Prolactin	Growth of mammary glands and milk production (lactation)
		Inhibition of FSH and LH
Testes	Testosterone	Sperm formation, with FSH
		Development of male secondary sexual characteristics
Ovaries	Estrogen	Development of female secondary sexual characteristics
		Preparation of uterine lining for implantation of embryo
	Progesterone	Preparation of uterine lining for implantation of embryo
		Enlargement of breasts for lactation
Placenta	Human chorionic gonadotropin (HCG)	Maintenance of pregnancy

The secondary sexual characteristics of human females include breasts and the rounded contours imparted by a thick, widespread layer of subcutaneous ("under skin") fat. This fat provides food during periods of starvation, and insulation against cold, giving women (and the unborn babies or nursing infants that depend on the mother's body for food) a survival edge over men. Subcutaneous fat may be unfashionable, but it is nevertheless vital for fertility. Women with insufficient fat do not ovulate and cannot become pregnant.

The Menstrual Cycle   At puberty (usually 10 to 14 years of age), the pituitary gland starts a series of hormonal cycles that periodically render a woman fertile (capable of becoming pregnant) until the cycles cease at the menopause, some 30 to 40 years later. These hormonal changes and the effects they produce are called menstrual cycles.

Human menstrual cycles are notoriously variable, but the "model" cycle lasts 28 days. The days are numbered from the first day of blood flow in the menstrual period (Figure 27-8). At the beginning of the cycle, the pituitary secretes increasing amounts of follicle-stimulating hormone (FSH). FSH causes an ovarian follicle to mature and produce the steroid hormone **estrogen.** A surge of estrogen from the follicle stimulates a dramatic increase in secretion of luteinizing hormone (LH) by the pituitary. This LH, together with FSH, brings about the final maturation of the follicle, culminating in ovulation, when the follicle ruptures and releases a mature egg. Ovulation occurs on about the fourteenth day of the cycle.

Still under the influence of LH, the cells of the ruptured follicle grow and form a **corpus luteum,** which secretes more estrogen and yet another steroid hormone, **progesterone.** Progesterone and estrogen prepare the endometrium, the lining of the uterus, to receive a fertilized egg. The levels of progesterone and estrogen at this time inhibit secretion of LH and FSH from the pituitary. The drop in LH and FSH deprives the corpus luteum of the hormones that stimulate it. It begins to degenerate, and its hormone secretion falls so low that the surface of the endometrium dies and sloughs off in a menstrual period. Without the corpus luteum hormones to inhibit it, the pituitary increases secretion of FSH once more, and the cycle repeats. This sequence is interrupted if fertilization and pregnancy occur.

■ *The menstrual cycle results from the interaction of pituitary and ovarian hormones. Pituitary FSH and LH act on the ovary, promoting maturation of an egg follicle, ovulation, and formation of the corpus luteum. The ovarian hormones estrogen and progesterone prepare the uterus for pregnancy. They also inhibit secretion of the pituitary hormones. Hence if pregnancy does not occur, the cycle begins anew.*

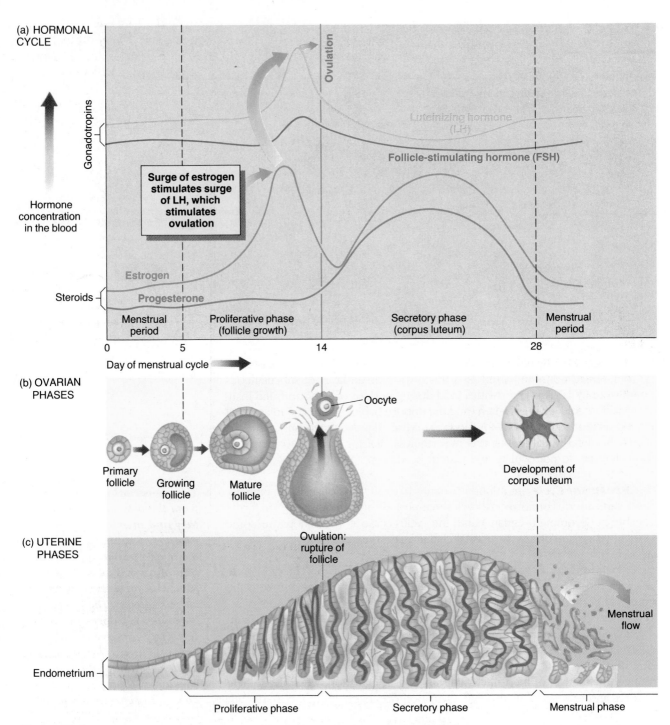

(a) HORMONAL CYCLE

Gonadotropins

Luteinizing hormone (LH)

Follicle-stimulating hormone (FSH)

Surge of estrogen stimulates surge of LH, which stimulates ovulation

Ovulation

Hormone concentration in the blood

Estrogen

Steroids

Progesterone

| Menstrual period | Proliferative phase (follicle growth) | Secretory phase (corpus luteum) | Menstrual period |

0   5   14   28

Day of menstrual cycle

(b) OVARIAN PHASES

Oocyte

Primary follicle — Growing follicle — Mature follicle

Ovulation: rupture of follicle

Development of corpus luteum

(c) UTERINE PHASES

Endometrium

Menstrual flow

Proliferative phase   Secretory phase   Menstrual phase

Figure 27-8

Events of the menstrual cycle. (a) Levels of sex hormones in the bloodstream during the phases of one menstrual cycle in which pregnancy does not occur. (b) Development of an ovarian follicle during the cycle. A fluid-filled space develops as the follicle grows. At ovulation the follicle ruptures, releasing the egg (oocyte). The remaining follicle walls then develop into a corpus luteum ("yellow body"), which secretes progesterone for a time and then degenerates. (c) Phases of development of the endometrium, the lining of the uterus. The endometrium thickens during the proliferative phase, developing long narrow glands and a rich blood supply. In the secretory phase, after ovulation, the endometrium continues to thicken and the glands secrete nutritive material in preparation to receive an embryo. When no embryo implants, the new outer layers of the endometrium disintegrate and the blood vessels rupture, producing the menstrual flow.

Figure 27-9
A lactating sow and her litter.
(Biophoto Associates)

Hormones of Pregnancy At the beginning of pregnancy, the developing embryo and the lining of the uterus jointly form a special organ, the **placenta,** which provides for nourishment of the embryo. It also secretes **human chorionic gonadotropin (HCG).** Increased levels of HCG, as of LH, maintain the corpus luteum and therefore production of progesterone, which inhibits the pituitary's secretion of FSH and LH. This prevents the next ovulation, and thus prevents formation of any new embryo while the first one is developing. If the embryo is abnormal, or dies, secretion of HCG by the placenta stops, permitting the body to **abort** or **miscarry** the pregnancy as the uterine lining degenerates and sloughs, along with the embryo. It is estimated that as many as three out of five human embryos that implant are abnormal and abort naturally in this manner. If a hormone from the mother's body maintained pregnancy, there would be no way to discard dead or abnormal embryos.

Likewise, a hormone produced by the developing fetus determines when birth occurs. The average period of **gestation,** or pregnancy, is 270 days in humans, but this varies a lot. It is important that the baby be born when it is mature, and not when the mother feels like it. The fetus signals that it is mature by secreting hormones from its adrenal glands. These fetal hormones diffuse across the placenta and build up in the mother's bloodstream until they cause secretion of the hormone **oxytocin** by the mother's pituitary. Oxytocin stimulates the muscles of the uterus to contract and cause birth. Later, it also causes milk to be let down from the mammary glands to the nipples in response to the baby's suckling.

The placenta is expelled within an hour after the baby is born. Many mammalian mothers eat their placentas, thereby re-using the nutrients and avoiding the risk of attracting predators. In addition, hormones eaten in the placenta trigger maternal behavior. Whether or not women would make better mothers if they ate their placentas has not, as far as we know, been tested!

Expulsion of the placenta during childbirth causes the mother's pituitary to secrete **prolactin,** a hormone that initiates **lactation** (milk production) by the mammary glands (Figure 27-9). Prolactin also inhibits the release of LH from the pituitary and counters the effects of FSH and LH on the ovarian follicles. Hence in nursing mothers the menstrual cycles are suppressed. In some cultures breast feeding is an important means of birth control, although it is not as reliable as most methods of artificial birth control. Breast feeding is also the best way to ensure that the baby receives adequate nutrition and avoids intestinal infections.

27-E   Birth Control

Human beings and other (particularly social) animals regulate their reproduction, so that the number of offspring bears some relation to the parents' ability to raise them. A bird may abandon her eggs if her neighbors of the same species have many young already hatched. Severely malnourished women are infertile.

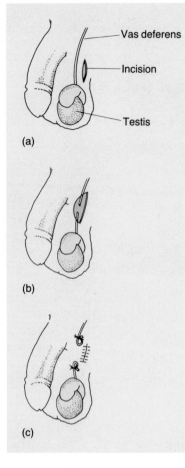

- Vas deferens
- Incision
- Testis

(a)

(b)

(c)

Figure 27-10
Vasectomy. In this surgical procedure, the vas deferens is severed so that sperm can no longer enter the semen.

Rabbit and mouse embryos die when their mother is stressed, as by overcrowding or the presence of a strange male.

Strictly speaking, **contraception** refers to birth control methods that prevent fertilization. The only methods of contraception that are 100% reliable are abstention from sexual intercourse and sterilization.

The condom is the contraceptive device most commonly used by men. A condom is rolled onto the erect penis shortly before intercourse. It catches the semen so that sperm do not enter the female reproductive tract. The condom is also the only form of birth control that can reduce or prevent the spread of AIDS and other sexually transmitted (venereal) diseases from one partner to the other (Chapter 19, *A Journey Into Life,* "Sexually Transmitted Diseases").

Birth control pills (oral contraceptives) are very reliable contraceptives if they are properly used (Table 27-2). "The pill" contains synthetic estrogen and progesterone at a dose high enough to prevent ovulation. Pregnancy is avoided because there is no egg to be fertilized. The pill does not reliably prevent ovulation and pregnancy until it has been taken regularly for at least 2 weeks. Women who have intercourse during this period, or who experience condom failure, can often protect themselves from pregnancy by taking a number of oral contraceptive pills within 72 hours of unprotected intercourse.

The medical risks of using the pill are largely the same as those of pregnancy. The pill increases the risk of such things as high blood pressure and excessive blood clotting, mainly among women over 35 who smoke. Conversely, oral contraceptives reduce their users' chances of getting certain kinds of cancer.

Another method of contraception is the use of a rubber diaphragm, which is smeared with a spermicidal (sperm-killing) jelly or cream each time it is used. The woman then inserts it into the vagina so that it covers the cervix. The diaphragm blocks the entrance to the uterus so that sperm cannot reach the egg. The cervical cap is similar.

The much less reliable rhythm method consists of avoiding intercourse during the woman's "fertile period," the part of the menstrual cycle when there is an egg present to be fertilized. The difficulty is in deciding when ovulation will occur, because the menstrual cycle is so variable.

Intrauterine devices (IUDs) are small plastic or metal objects inserted into the uterus by a doctor. IUDs are not true contraceptives. They somehow act on

*Table 27-2 Contraceptive Use in the United States**		
Method	*Estimated % use*	*% Accidental pregnancy in one year of use†*
Male sterilization	15	0.15
Female sterilization	19	0.4
Oral contraceptive pill	32	3
Condom	17	12
Diaphragm + spermicide	5	18
"Rhythm" (periodic abstinence)	4	20
IUD	3	6
Contraceptive sponge	3	18
Vaginal foams, jellies	2	21

* Data from *Developing New Contraceptives: Obstacles and Opportunities.* Washington, DC: National Academy Press, 1990.

† About 89% of women using no contraceptive become pregnant within one year.

the uterine lining so that the already developing embryo cannot implant. IUDs are as effective as contraceptive pills in preventing pregnancy. On the other hand, a disquieting number of IUDs become deeply imbedded in the wall of the uterus, causing dangerous abdominal infections and requiring surgery to remove the device. Such infections may result in infertility.

European women are preventing pregnancy for five years at a time by having a contraceptive implanted under the skin of the upper arm. West Germany, Mexico, and China make injectable one-month contraceptives that are marketed in 40 countries. The French have an abortion pill, RU 486. In contrast, birth control in the United States has improved little since 1960, largely because of our legal liability system which makes developing and marketing new birth control devices financially risky.

Vasectomy and Tubal Ligation

Sterilization is any more or less permanent change that prevents an animal from reproducing sexually. Sterilization of either sex is the fastest-growing method of contraception in the world. The most common sterilization operation for men is **vasectomy,** that is, severing and tying off the vasa deferentia. This is a simple operation, usually performed under local anesthesia (Figure 27-10). Afterwards, sperm are still produced but are resorbed into the body, and the fluid ejaculated contains only the secretions of various glands.

Sterilization of a woman usually involves **tubal ligation,** the cutting and tying off of the oviducts. This operation can also now be performed under local anesthesia. After tubal ligation the ovary continues to function as it did before, but sperm cannot reach the eggs, and thus fertilization cannot occur.

Abortion

Induced abortion is one of the oldest human birth control methods. Abortion was largely accepted, and fairly common, in the United States and Europe until the early nineteenth century. Before this time, a woman had up to a one-third chance of dying in childbirth, whereas abortions were less risky for her, so that abortions often saved women's lives. After midwives and doctors found that washing their hands and clothes improved their patients' survival rates, childbirth became less dangerous to a woman than a nineteenth-century abortion, and doctors started to oppose abortion. Most religious and ethical opposition did not develop until some time later. The situation has changed since World War II, now that the danger of abortion to a woman is again less than the risk of a completed pregnancy, and abortion is, once again, legal in most countries.

Development

27-F Fertilization and Implantation

Sperm released into the vagina during ejaculation swim through the cervix and uterus into the oviduct, where fertilization usually occurs. (The description of human pregnancy in this section applies generally to other mammals as well.)

As a sperm penetrates the egg cell membrane, a rapid electrical reaction, followed by a slower chemical change, runs through the membrane. After this, the egg cannot be penetrated by another sperm (Figure 27-11).

As a fertilized egg moves slowly toward the uterus, it is already dividing rapidly into many cells. At this stage, the embryo is called a **morula.** About a week after fertilization, the morula starts to **implant** in the endometrium of the

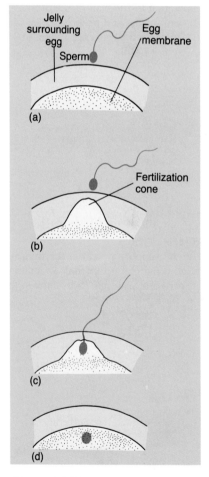

Figure 27-11
A sperm entering a frog's egg. (a) The sperm sticks to the jelly coat that surrounds the egg. (b) The egg forms a fertilization cone, which (c) engulfs the sperm nucleus and then (d) retracts, as microfilaments in the egg cytoplasm shorten.

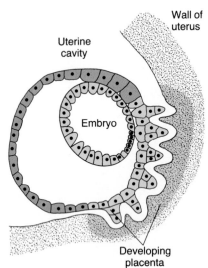

Uterine
cavity

Wall of
uterus

Embryo

Developing
placenta

Figure 27-12
**A developing mammalian embryo
implanting in the wall of the uterus.**
The embryonic and maternal cells
that form the placenta are shown in
green and purple.

uterus, establishing an intimate link between the mother and embryo that lasts throughout the pregnancy.

During evolution, selection has favored the birth of only one human baby at a time. Babies borne singly tend to be larger and healthier, with a better chance of survival, than those with womb-mates. However, multiple births result from a small percentage of pregnancies. Identical twins are produced when the mass of cells that will develop into an embryo separates into two groups during the first week of development. Each group of cells develops into a separate embryo. Since these cells contain identical genes, the resulting embryos are genetically identical, and so must be of the same sex. Identical twins are also called monozygotic twins because they originate from a single zygote. Multiple births may also result when more than one egg is produced and fertilized. Each develops into a separate embryo, and they are no more genetically alike than any other siblings.

When the morula reaches the wall of the uterus, it consists of the cells of the future embryo, surrounded by a sphere of cells also derived from the fertilized egg (Figure 27-12). An interesting feature of the development of all mammals is that some of these outer cells become support systems for the embryo and never become part of the adult. Some of these outer cells invade the wall of the uterus, anchoring the embryo and beginning the development of the most important organ of pregnancy, the placenta. Other cells in this outer sphere give rise to the three **extra-embryonic membranes,** the amnion, chorion, and allantois, some of which also form part of the placenta. In addition to these embryonic tissues, the placenta also contains part of the mother's endometrium.

At first the placenta is tiny, but by the time of birth it will develop into an organ about the size of a stack of four dinner plates. The membranes joining the embryo to the placenta develop into a cord, the **umbilicus.** The arteries and vein within this cord carry blood from the fetus to the placenta and back. In the placenta, blood capillaries of the mother and fetus lie close together. Here the fetal blood picks up food and oxygen, and gives up wastes. Other substances present in the mother's blood may also enter the fetal circulation, and may retard development or, if they interfere with differentiation, cause birth defects. Many drugs are known to harm developing fetuses, and doctors urge pregnant

Figure 27-13
The life support system of the human embryo. (a) Arrangement of the extra-embryonic membranes to form the umbilical cord and placenta. (b) A fetus in the uterus shortly before birth.

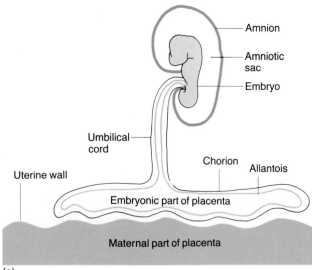

(a)

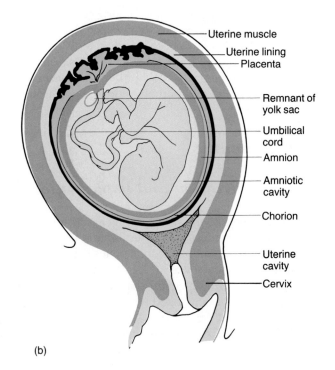

(b)

women to avoid smoking, drinking alcohol, using cocaine or other drugs, and taking medication.

The three extra-embryonic membranes, the amnion, chorion, and allantois, are important in the development of all terrestrial vertebrates—reptiles, birds, and mammals (Figure 27-13). The **amnion** is a fluid-filled sac around the embryo that cushions it against bumps. This is the sac that bursts when the "waters break" during childbirth. Since the amniotic fluid contains some cells sloughed from the embryo, drawing fluid from the sac into a hypodermic needle (**amniocentesis**) permits a geneticist to check for chromosomal defects in the embryo (*A Journey Into Life,* Chapter 13).

The **allantois** is a sac connected to the embryo's gut. In reptiles and birds it stores the embryo's waste products until the egg hatches. In humans it becomes riddled with blood vessels and makes up most of the embryonic part of the placenta. The third membrane, the **chorion,** which surrounds the fetus outside the amnion, also makes up a part of the placenta. The chorion secretes the hormone human chorionic gonadotropin, which maintains pregnancy and can be detected in pregnancy tests.

27-G Stages of Embryonic Development

The complex events of embryonic development are usually divided into four main stages, involving rather different cellular events: cleavage, gastrulation, neurulation, and organogenesis. These stages are found in all vertebrate embryos and have been studied most thoroughly in frogs and chickens. *A Journey Into Life,* "Human Development," describes the timetable for these and other events in human embryos.

Cleavage

After fertilization, the zygote divides to form 2, 4, 8, and then 16 cells, and so on. This period of cell division is known as **cleavage,** and the cleaving embryo as a morula.

The main functions of cleavage are to produce a large number of cells by rapid division and to segregate different substances in the cytoplasm into different cells. These substances determine how the various cells develop later. Most cells grow in size before they divide, but no cell growth occurs during cleavage: the zygote divides into ever smaller cells. Eventually, a fluid-filled cavity called the **blastocoel** appears in the center of the embryo in most animals. After the blastocoel has formed, the embryo is known as a **blastula.**

The pattern of cleavage can usually be predicted from the amount of yolk in the egg. **Yolk** is a store of nutrients used by the developing embryo. Its yellowish color comes from an iron-storage protein. In eggs with little yolk, such as those of mammals and of many invertebrates, cleavage produces many cells of roughly the same size. Frog eggs contain more yolk than mammalian eggs, and the **cleavage furrow,** formed as the cells divide, passes more slowly through the yolk at the bottom of the egg than through the clearer area above (Figure 27-14). As a result, the top of the morula gets ahead of the bottom, and many small cells have been formed in the top half of the embryo while there are still

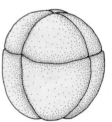

(a) 4-cell stage

(b) 8-cell stage

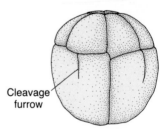

Cleavage furrow

(c) 4th cleavage

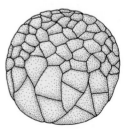

(d) Blastula

Figure 27-14
Cleavage in a frog egg. The egg contains quite a lot of yolk and so cell division occurs more rapidly in the top half of the morula than in the yolky bottom half. (d) Because of the slower cleavage in the bottom half of the morula, there are many more cells in the top than in the bottom half of the embryo by the time the blastula forms.

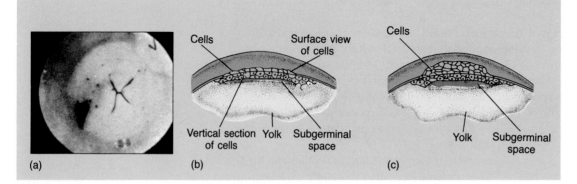

Figure 27-15
Cleavage in a bird embryo. (a) Light micrograph of the surface of a chicken egg after two cleavage divisions. The cleavage furrows move so slowly through the yolk that no complete cells have yet formed. (b) Side view of the top of the egg during later cleavage. (The subgerminal space is a fluid-filled cavity that separates the developing embryo from the yolk.) (c) Chick blastula in a side view of the egg. Note that even at this stage the embryo occupies only a small area at the top of the egg. (a, M. W. Olsen, *Journal of Morphology* 70:533, 1942)

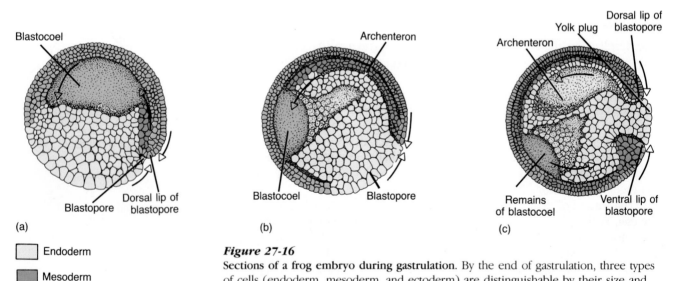

Endoderm

Mesoderm

Ectoderm

Neurectoderm

Figure 27-16
Sections of a frog embryo during gastrulation. By the end of gastrulation, three types of cells (endoderm, mesoderm, and ectoderm) are distinguishable by their size and position. (A fourth cell type, neurectoderm, is also labelled here. These are the ecto-derm cells that will form the nervous system.) (a) The blastopore forms where small cells from the top of the embryo adjoin larger ones in the bottom half. The larger, yolky cells at the bottom of the embryo (yellow) move through the blastopore into the embryo, where they will become endoderm cells, lining the digestive tract. (b) The archenteron forms within the endoderm and eventually becomes the cavity in-side the digestive tract. Smaller cells at the top of the embryo roll over the dorsal lip of the blastopore into the embryo. The first cells to enter will become the meso-derm (orange). The small cells remaining at the top (blue and green) spread out until they cover the surface of the embryo. These become ectoderm cells (blue), which will form the skin and other external structures, and neurectoderm (green), which will form the nervous system during neurulation.

only a few larger ones below. Bird eggs contain so much yolk that the cleavage furrow cannot pass through it at all, and the egg cleaves only in a small area on top of the yolk (Figure 27-15).

Gastrulation

After formation of a blastula, the next stage in embryonic development is **gastrulation,** during which the cells rearrange themselves into distinct layers. Although cell division continues during gastrulation, the most obvious event is cell movement. An opening called the **blastopore** forms at the side of the blastula. Cells from the surface of the blastula around the blastopore move through the blastopore into the hollow interior (Figure 27-16a).

During gastrulation, the single-layered, hollow blastula becomes a **gastrula.** The cells eventually arrange themselves into three layers called the **germ layers.** (Germ in this context means something small that will give rise to more elaborate structures later.) The gastrula contains a new central cavity, the **gastrocoel** (or **archenteron**), which will eventually form the lumen of the digestive tract (Figures 27-16b, c, 27-17).

The outermost germ layer is the **ectoderm.** This will form both the **neurectoderm,** which develops into the nervous system, and the **epidermis,** which gives rise to skin, hair, nails, sweat glands, and other structures on the outside of the body. The innermost layer of cells is the **endoderm,** next to the gastrocoel. From the endoderm will come the lining of the gut, the digestive glands, and other internal structures. Between the endoderm and the ectoderm lie the cells of the third germ layer, the **mesoderm,** which will form the skeleton, muscles, gonads, kidneys, and allied structures in the adult. At a later stage, yet another cavity forms in the embryo, this time in the middle of the mesoderm. This is the **coelom,** the body cavity in which internal organs are suspended.

Neurulation

After the three germ layers have formed, the neural tube and head begin to develop in the process of **neurulation.** The part of the ectoderm destined to form the nervous tissue first forms two parallel folds that rise up and finally join

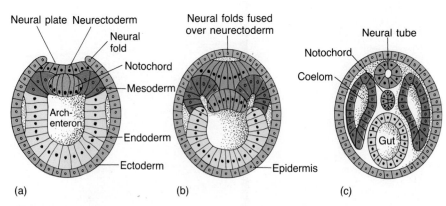

(a)   (b)   (c)

Figure 27-17
Fate of the germ layers in a mammal. (a) At the end of gastrulation, ectoderm surrounds the embryo, with neurectoderm beginning to sink below the ectoderm. The notochord forms from the mesoderm. (b) Ectoderm has grown up to cover the neurectoderm. Pouches that will become the coelom are forming in the mesoderm. (c) Neurectoderm has rolled up to form the neural tube, the gut has formed as a tube surrounded by endoderm, and the coelom has formed within the mesoderm on each side of the body.

to form the **neural tube,** from which all the nervous system develops (Figures 27-17, 27-18).

Organogenesis

The final stage of development is **organogenesis,** organ formation. The nervous, digestive, and circulatory systems begin to form. The **notochord** forms from mesoderm. The notochord is a rod that runs from the embryo's head to its

Test Tube Babies and Surrogate Mothers

After fertilization but before implantation, the embryo lives free from attachment to the mother. Medicine and animal husbandry take advantage of this brief period to intervene. Eggs or developing morulas can be removed from the ovaries or oviducts, treated in the laboratory, and then returned to the uterus without damage, if they are provided with a suitable fluid environment during their stay in the outside world. The biggest problem with these techniques has been finding a suitable chemical solution. Mammalian eggs and early embryos, in particular, are easily damaged. (Sperm survive better because they are protected by the semen—their special fluid environment.)

"Test tube" babies do not develop in test tubes, but merely undergo fertilization and a few rounds of cell division outside a woman's body. The scientific term for this is *in vitro* ("in glass") fertilization. If a woman has blocked oviducts or if her vagina produces spermicidal secretions, or if her husband's sperm count is low, fertilization may not occur naturally in her body. However, a doctor may be able to remove eggs from her body, add the husband's sperm in a laboratory dish, check that development has begun, and then return the morulas to the uterus. This technique succeeds less than 20% of the time. To increase the chances of success, doctors usually treat the pro-

spective mother with fertility drugs. These usually cause several ovarian follicles to mature at the same time, so that more than one egg is released. Hence this technique results in a high proportion of multiple births.

Once the embryo is outside the female's body, it can just as easily be inserted in the uterus of a different female, provided her hormones are in the proper phase of the reproductive cycle for implantation to occur. To date, human babies and young of many other species have been born to such surrogate mothers. This permits a woman who has a normal uterus but damaged ovaries to bear a child (which of course will have the genes of the woman who donated the egg).

However, the greatest use for embryo transplants is not for humans but for other mammals. With it, a breeder can obtain many embryos from a genetically superior cow or mare, for instance, and use inferior stock as surrogate mothers, increasing the rate of production of desirable young. It also turns out that female mammals can successfully gestate embryos of closely related species. This allows zoo keepers to produce more young of endangered species, using surrogate mothers of species that are common and easily bred in captivity (Figure 27-A).

Figure 27-A

A newborn bongo and its surrogate mother, an eland. As a young embryo, the bongo was transplanted into the eland's uterus, where it implanted and developed. Bongos are a rare and elusive species inhabiting dense forests in Africa. The larger and more common elands, members of the same genus, inhabit open areas. (Cincinnati Zoo)

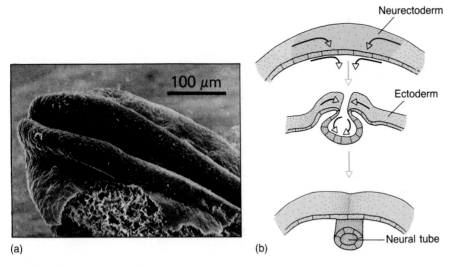

(a)

(b)

Figure 27-18

Neurulation. (a) Scanning electron micrograph of part of the back of a chick embryo during neurulation. The lip-like ridges that will fuse to cover the neural tube are moving toward each other at the head end of the embryo (left). (b) Sections through an embryo during neurulation to show how the neurectoderm (green) folds in to form the neural tube, which is covered by ectoderm (blue). The lip-like ridges in (a) can be seen in the middle diagram. (a, Kathryn Tosney)

tail. Later in development it becomes encased in cartilage and bone and is converted into the backbone. Muscle blocks (somites) appear in a regular row along the backbone, and then in other parts of the body. The heart, shaped like a lumpy tube, starts to pulsate (Figure 27-19). After the beginning of organogenesis, the embryo grows in size and the nervous, circulatory, and respiratory systems mature.

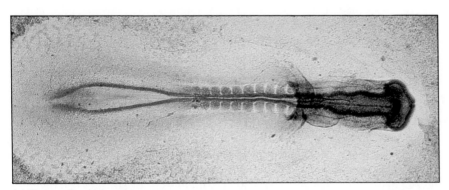

(a) Ten-somite chick embryo

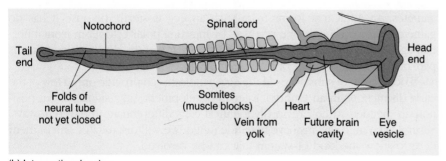

(b) Interpretive drawing

Figure 27-19

Organogenesis in a chick embryo. These external views show the embryo after 30 hours of development. (a) Light micrograph of the embryo. (b) Interpretive drawing. The most visible features are the rudimentary nervous system (brain and spinal cord), which runs from the head to the tail, and the ten somites, muscle blocks of the body wall. Behind the somites, the neural tube is still open, with the notochord visible below it. The heart forms from a tube that develops thicker and thicker walls. It brings nourishment from the yolk below it, through the large veins seen here, to the embryo. The heart, therefore, develops very early compared with most other organs; it is already beating at this stage. (Biophoto Associates)

Vertebrate embryos hatch or are born at various stages of maturity. A mouse, which will live in a nest with its mother for some time after birth, is born blind, naked, and almost helpless. A horse, which will run with the herd from the day it is born, can stand and walk soon after birth and looks much like a miniature adult.

To summarize, the stages of embryonic development are:

1. **Cleavage,** division of the zygote into many smaller cells of the morula, ending with a hollow blastula.
2. **Gastrulation,** movement of cells of the blastula to form a three-layered gastrula.
3. **Neurulation,** formation of the neural tube.
4. **Organogenesis,** formation of organs and systems.

27-H   Determination and Induction

With this brief description of the embryonic stages, we can discuss some of the cellular and molecular mechanisms involved in development.

One of the most important changes during development is **differentiation,** whereby the similar-looking cells of a cleaving embryo give rise to all the different cells in the body of the fetus. Differentiation results from the expression of different genes in different cells. As a result, each cell contains only the particular proteins encoded by its active genes, and these proteins determine how the cell develops.

The path by which a cell will develop is sometimes fixed, or **determined,** in development long before the cell shows any outward sign of differentiation as a particular type of cell. Determination can be demonstrated by transplantation experiments. For instance, a block of cells that would normally become brain cells can be transplanted in an early amphibian embryo to another area from which cells that would normally form skin have been removed. The transplanted cells form skin. They differentiate in accordance with their new, rather than their original, position in the embryo. We can conclude that these cells had not been determined as brain cells before their transplantation. However, if the same experiment is made three days later in development, the transplanted cells develop as brain, completely inappropriate to their new position. We conclude that they had been determined as brain cells at some point during the three-day period, although they had not yet differentiated into brain cells.

Determination is heritable. When a cell determined as a liver cell divides, its offspring are all committed to differentiating as liver cells.

In many species, gradients of chemicals in the egg cytoplasm are the first thing that determines differences between cells in different positions. When the zygote divides into two, four, then eight cells, these cells contain different amounts of any substance that is unequally distributed in the egg cytoplasm. For instance, in *Drosophila* (a fruit fly), a region of cytoplasm at one pole (end) of the egg determines that cells that end up with this cytoplasm after cell division will eventually give rise to the gametes. Shining ultraviolet light on that pole of the egg destroys something in the cytoplasm that is necessary to determine gametes. An adult that forms from such an egg is sterile because it has no gametes. The damage can be reversed by injecting polar cytoplasm from a normal egg into an irradiated egg. This makes it clear that the determining factor is part of the cytoplasm, not of a nucleus.

Localization of substances in the egg cytoplasm can provide only a few of the early distinctions in an embryo. Later in development, new signals about position are needed when previously identical cells differentiate in different ways. Several such mechanisms have been identified. We will discuss just one of them here to show the kind of system that can be involved.

Embryonic Induction

Embryonic induction is an interaction between cells, in which some cells alter the fate of others. For example, the lens of the eye is formed from the ectoderm of the head as a result of contact with part of the brain, which grows out to form a bulge called the optic vesicle (Figure 27-20). When the optic vesicle touches part of the ectoderm of the head, it induces that bit of the ectoderm to form the lens. If some barrier, such as a piece of cellophane, is placed between the growing optic vesicle and the ectoderm, the lens never develops. If optic vesicles are transplanted from their normal site in the head to elsewhere in the body, they induce lens tissue to form in any ectoderm that they touch. They cannot, however, induce tissue other than ectoderm to form a lens. Hence the fate of a tissue is determined by the activity of its own genes as influenced by its environment: genetic activity determines that a cell has become ectoderm rather than mesoderm or endoderm, but unless it touches the optic vesicle, ectoderm will not develop into a lens. The whole process is called induction because contact with the optic vesicle *induces* ectoderm to form lens instead of skin or whatever else it would otherwise have formed.

The general plan for the body is specified very early in development. New levels of detail are added progressively as the cluster of embryonic cells that will form the body part grows to a size at which further information on a cell's position can be supplied by stimuli such as hormones, adhesion to other cells, or inductive interactions. Determination occurs when genes that control the activity of groups of protein-coding genes are switched on or off more or less permanently, so that the state of the control gene is passed to the cell's offspring.

27-I Birth

The process in which uterine contractions expel the baby and the placenta is called **labor** and can be divided into three main stages. The first stage, **dilation,** usually lasts from 2 to 20 hours and ends when the cervix of the uterus is fully open or dilated. The second stage, **expulsion,** lasts from about 2 to 100 minutes. It begins with **full crowning,** the appearance of the baby's head in the cervix, and continues while the baby is pushed, head first, down through the vagina into the outside world, where it draws its first breath.

The third, or **placental,** stage begins when the baby has been born. The uterus continues to contract while the umbilical cord is clamped, and some 5 to 45 minutes after the baby is born the uterus expels the placenta. The umbilical cord can now be severed, and the baby's independent existence begins.

27-J Maturation, Aging, and Death

Animals continue to change in various ways after leaving the protection of the egg or of the mother's body at hatching or birth. Some animals undergo metamorphosis, a change in body form, and most animals take some time to become sexually mature. Maturation involves the same processes of genetic differentiation, growth, and change of form as those that occurred in the embryo. Aging appears to be different.

Aging is the sum of changes that accumulate with time and make an organism more likely to die. It begins even before an individual officially completes development and sexual maturity. Slower healing, for instance, is one of the signs of aging, and even a human teenager's broken bones heal less rapidly than those of a young child.

Aging and death are genetically programmed. A human dies of old age at about 80 to 100 years of age, and some insects at a few days or weeks. Aging

■ *An embryo's body develops from a general layout to more specific details as the number of cells increases and various sets of genes are turned on and off in sequence.*

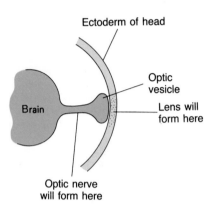

Figure 27-20
Induction. Contact between the optic vesicle and ectoderm induces formation of the lens of the eye from ectodermal tissue.

Human Development

A JOURNEY INTO LIFE

Human embryonic development is similar to that of other mammals with long gestation periods. However, medical names, instead of the usual terms from embryology, are sometimes applied to some of the developmental stages.

After fertilization, the cleaving morula travels down the oviduct into the uterus. Once a cavity has developed inside it, the embryo becomes a blastula (sometimes known as a blastocyst in humans). Some of the cells of the blastula will develop into the actual embryo. Others form the embryonic part of the placenta and the extra-embryonic membranes that protect the embryo.

Part of the extra-embryonic membranes, known as the embryonic trophoblast, grows into the endometrium of the uterine wall. The trophoblast has remarkable qualities. Because the embryo is genetically different from its mother, we would expect the mother's immune system to recognize the trophoblast as foreign and reject it as if it were an organ transplanted from another individual. Obviously, this does not happen. The trophoblast is **immunologically privileged,** one of the few tissues that the body does not recognize and reject as foreign. It invades the uterine wall almost as if it were a cancerous growth. Occasionally something goes wrong with this remarkable system, and the trophoblast forms one of the most malignant and invasive of cancers.

By about three weeks after fertilization, a quarter of the wall of the uterus is occupied by the placenta—spongy endometrium full of blood vessels penetrated by chorionic villi, finger-like projections of the embryonic trophoblast (Table 27-A). The capillaries of mother and embryo do not join in the placenta, but they lie so close to each other that they can exchange gases, nutrients, and wastes

by diffusion. The embryo, now undergoing neurulation, is attached to the placenta by the elongating umbilicus. At this time, the embryo is about 2 millimetres long.

Cleavage, gastrulation, and neurulation all take place during the first trimester (first three months) of pregnancy. Organogenesis begins immediately after neurulation, and the 4-week embryo has a simple heart, and buds where limbs will develop. It also has a tail and gill slits, vestiges of its early vertebrate ancestors which will disappear later in development (Figure 27-B). By 6 weeks, a simple eye is visible and the tail is still present (Figure 27-C). After the second month, the embryo is recognizable as a primate because it has features, such as a neck and large head, that distinguish it from other vertebrate embryos (Figure 27-D). From this stage on, the embryo is sometimes referred to as a fetus.

The second trimester is occupied mainly by further organ development. During this period, limb and face muscles move, and the heart becomes large enough so that it can be heard through a stethoscope placed on the mother's abdomen. Bone continues to replace the carti-

---

Table 27-A   Outline of Human Development

Medical name	Days after fertilization (approximate; varies a lot)	What's happening?
First Trimester (Embryo)	0–8	Cleavage
	6	Implantation begins
	21	Neurulation
	24	Nervous system, gut, and blood vessels start to develop
	28–35	Embryo most susceptible to damage by rubella, thalidomide, most drugs, etc.
(Fetus)	45	Testes differentiating in male
	70	Arms and legs move
	75	Primary oocytes enter first meiotic division in female
Second Trimester	90	All major organ systems formed but tiny and growing in size
	140	Heart can be heard with a stethoscope
Third Trimester	180	Temperature regulation, central nervous system, and lungs complete development; fetus most susceptible to damage by alcohol
	270	Birth (parturition)

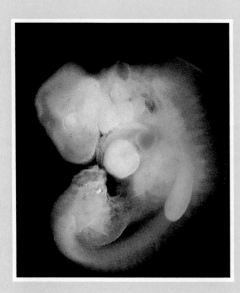

Figure 27-B

A human embryo 29 days after fertilization. The embryo is about 7 millimetres (less than half an inch) long. The tail, pharyngeal gill slits, and limb buds can be seen. The segmented myotomes that form along the backs of all vertebrates curve around the right side of the embryo. The red object in the middle is the developing heart. (Lennart Nilsson, *A Child is Born.* New York: Dell Publishing Co., Inc., 1977)

Figure 27-D

A 6- to 7-week human embryo, about 2 centimetres long. The embryo is now recognizable as a primate, with a neck and large head. The red object in the center of the embryo is the developing liver. Chorionic villi can be seen around the edges of the embryonic part of the yellow and red placenta at the bottom. (Lennart Nilsson, *A Child is Born.* New York: Dell Publishing Co., Inc., 1977)

lage that forms the early skeleton, and the body is covered with hair (the lanugo).

At the beginning of the third trimester (6 months), the fetus is still tiny. It will normally more than triple in size before birth. By 7 months, the fetus looks from the outside like a tiny normal baby, but it is not. The lungs and temperature regulation systems are not functional. More important, the central nervous system is not fully developed. Nearly all babies born at this age will be severely handicapped if they survive. By the end of the third trimester (9 months), the lungs and nervous system will develop. Then the fetus will usually be born, with almost a 99% chance of surviving to live a normal life. At birth, the full-term baby weighs about 3000 grams (7 pounds) and is about 52 centimetres (20 inches) long.

Figure 27-C

A 5- to 6-week human embryo, about 1 centimetre long. The eyes and fingers are beginning to develop. Notice the cavity where the brain will be. The sac on the right overlapping the head is the yolk sac, which nourishes the early embryo of all vertebrates. The sac gets smaller as the yolk is used up and the placenta takes over the task of nourishing the embryo. (Guigoz, Petit Format/Photo Researchers, Inc.)

assures death at this characteristic age if disease or predators do not kill the individual first.

Although we can describe many processes that are characteristic of aging, it is still unclear which of these are cause and which effect. For instance, with age, the immune system becomes less efficient at defending the body against disease. Is this why disease becomes more common with age, or vice versa?

As people grow older, their bodies cope less effectively with stress and disease. Because the ability to survive changing conditions depends largely on the immune, nervous, and hormonal systems, researchers have concentrated on these systems, using the hypothesis that degeneration here results in the loss of adaptability seen in the rest of the body: slower healing, hardening connective tissue, brittle bones, and so forth.

One example of this comes from studies of people with diabetes, who age faster than normal. Diabetics have more glucose in their blood than other people, and one symptom of aging is the formation of glucose links between structural proteins such as collagen in the skin, tendons, ligaments, and the walls of blood vessels. The glucose linkages make these tissues rigid and inelastic. The immune system's macrophages destroy cells with glucose-linked proteins on their surfaces. However, it appears that macrophages in older individuals have fewer receptors for these proteins and are less efficient at removing the damaged cells from the body.

Despite a long list of hypotheses, searches for a single cause of aging have all ended in failure. It seems likely that the "aging genes" do not control one single system but instead control many minor degenerations. Because the body's systems all interact with one another, minor deficiencies anywhere in the body can accumulate to produce aging of the body as a whole and of its individual systems.

SUMMARY

Almost all animals can reproduce sexually, as a large, nonmotile egg is fertilized by a small, motile sperm. This usually involves the coordination of the anatomy, physiology, and behavior of two parents.

In a human female, an egg is ovulated from a follicle in the ovary and travels through the oviduct to the uterus. If it is fertilized during this journey, it starts to divide and will implant in the endometrium lining the uterus.

The male's testes, which produce sperm, lie in the scrotum. After they leave the testes, sperm pass in turn through the epididymis, vasa deferentia, and urethra on their journey to the exterior. Glandular secretions are added to sperm, forming semen. The penis is composed of spongy tissue which becomes engorged with blood during sexual stimulation. It can then be inserted into the vagina. Sexual stimulation may eventually result in orgasm by both sexes and in ejaculation of semen by the male.

Sexual maturation during puberty, maturation of gametes, and the female menstrual cycle are controlled by hormones. Hormones also control pregnancy, birth, and lactation.

Humans commonly control their reproduction by techniques that either prevent fertilization or prevent a developing embryo from completing its growth in the uterus. Abstinence and sterilization are the only fully effective forms of contraception. Contraceptive pills work almost as well, as do condoms, and diaphragms when they are properly used.

Embryonic development involves several kinds of processes: cell division; differentiation into various types of cells due to activation of different sets of genes; growth of the embryo; and changes in the shape of the body as a whole and of the organs that form within it.

The stages of development are:

1. Cleavage, when the embryo divides rapidly into ever-smaller cells to form a hollow blastula.
2. Gastrulation, when the cells rearrange themselves into layers. The three germ layers of the gastrula are the ectoderm, which will develop into the skin and other external structures as well as the nervous system; the mesoderm, which forms internal organs such as the kidneys, heart, and blood vessels; and

the endoderm, which forms the digestive tract and organs associated with it. A split in the mesoderm will eventually form the coelom.

3. Neurulation, when the neurectoderm rolls up to form the neural tube, from which the brain and the rest of the nervous system will develop.
4. Organogenesis, which produces all the organs of the body by a complicated series of cell interactions.

The way a cell will develop is often determined some time before the cell actually differentiates into a specialized cell. A cell's determination involves a relatively permanent genetic change, which is inherited by all its descendants. Determination is caused partly by various substances in the egg cytoplasm, which be-

come encased within different cells of the morula during cleavage, and partly by interactions with other cells.

Part of the morula develops into the new individual, and part forms the amnion, chorion, and allantois, the membranes surrounding the embryo. In humans, the chorion and allantois become the fetal part of the placenta. The placenta anchors the fetus and secretes the hormones that permit pregnancy to continue. Nutrients, waste, gases, and hormones pass between the fetal and maternal blood vessels in the placenta.

Aging is an important part of development after birth, but we understand very little about it. It is possible that aging has many causes whose interactions are complex and difficult to disentangle.

SELF-QUIZ

Associate the reproductive organs on the right with the descriptions on the left:

____ 1. Tube for conducting sperm
____ 2. Receptacle for penis
____ 3. Production of seminal secretions
____ 4. Conducts eggs
____ 5. Holds baby in uterus
____ 6. Produces sperm
____ 7. Prepares nutritive lining for embryo

a. cervix
b. Cowper's gland
c. ovary
d. oviduct
e. prostate gland
f. testis
g. urethra
h. uterus
i. vagina
j. vas deferens

8. From the list of reproductive organs and passages above, construct a (correct) route for the passage of sperm from the site of production to the site of fertilization:

_____

For each of the following birth control methods, choose the correct means of interference with reproduction:

____ 9. Diaphragm and jelly
____ 10. Vasectomy
____ 11. "The pill"
____ 12. IUD
____ 13. Tubal ligation
____ 14. Induced abortion
____ 15. Condom

a. prevents fertilization of egg
b. prevents embryo implantation
c. prevents completion of embryonic development of implanted embryo
d. prevents ovulation
e. prevents sperm formation
f. prevents release of sperm into seminal fluid

From the list of hormones on the right, choose the ones with the functions listed on the left:

____ 16. Maintains pregnancy
____ 17. Ovulation occurs in response to a surge of ____, which in turn is stimulated by a surge of ____.
____ 18. Primary hormone in sperm production
____ 19. Induces labor
____ 20. Produced by the placenta

a. estrogen
b. follicle-stimulating hormone
c. human chorionic gonadotropin
d. luteinizing hormone
e. oxytocin
f. progesterone
g. testosterone

Match the correct stage(s) of development with each of the following characteristics:

C = Cleavage   N = Neurulation
G = Gastrulation   O = Organogenesis

____ 21. Pattern depends on amount and distribution of yolk
____ 22. Embryonic nerve tube first forms
____ 23. Results in formation of skeleton and muscles from mesoderm
____ 24. Rapid cell division with no increase in size
____ 25. Produces gastrocoel

QUESTIONS FOR DISCUSSION

1. Does vasectomy affect male potency (ability to engage in sexual intercourse)?
2. Does abortion affect a woman's subsequent fertility?
3. Can a woman become pregnant the first time she has sexual intercourse?
4. Why do you think the venereal disease gonorrhea is now epidemic in the United States and Western Europe?
5. A bird's egg has a hard, dry shell. How do sperm manage to fertilize these eggs?

6. Most countries have rules that extraordinary attempts should not be made to save babies born before 8 months of gestation because nearly all those born before $7\frac{1}{2}$ months are severely handicapped, both mentally and physically. They become a severe burden to the state and to their families. The American medical profession devotes staggering amounts of time and money to saving babies born before 8 months of gestation. Why do you think this is? Do you think this is a wise use of our medical resources?

SUGGESTED READINGS

Avery, M. E., N-S. Wang, and H. W. Taeusch, Jr. "The lung of the newborn infant." *Scientific American,* April 1973. Describes the physiological changes in the lung just before birth and efforts to speed these changes in premature infants.

Browder, L. *Developmental Biology,* 2d ed. Philadelphia: Saunders College Publishing, 1984. A straightforward, well-illustrated text.

Cerami, A., H. Vlassara, and M. Brownler. "Glucose and aging." *Scientific American,* May 1987. Explains how the nonenzymatic linkage of glucose to structural proteins, and possibly DNA, might account for many symptoms of aging.

Djerassi, C. *The Politics of Contraception.* New York: W. W. Norton, 1980. Why are modern contraceptives relatively primitive compared with our other technology? The author argues that this results from the control of contraceptive development by drug companies and governments.

Frisch, R. E. "Fatness and fertility." *Scientific American,* March 1988.

Lein, A. *The Cycling Female.* San Diego: University of California Press, 1979. A readable book on menstrual cycles, their variations, and how they affect women.

Short, R. V. "Breast feeding." *Scientific American,* April 1984.

Wassarman, P. M. "Fertilization in mammals." *Scientific American,* December 1988.

When you have studied this chapter, you should be able to:

1. Use these terms correctly:
   ganglion, cephalization
   central nervous system
   peripheral nervous system
   autonomic nervous system
   nerve, spinal nerve, cranial nerve
2. Describe the basic structure and function of neurons.
3. Explain the role of the sodium-potassium pump in maintaining the neuron's membrane potential.
4. Draw a graph of the potential changes that occur during an action potential, and relate them to the flow of sodium and potassium ions across the axon's membrane.
5. Describe the myelin sheath, and state its effect on impulse conduction.
6. Draw a model synapse with its principal components, explain the function of each component, and explain how information is transmitted to the postsynaptic cell.
7. Explain the difference between an excitatory and an inhibitory synapse in terms of effect on the postsynaptic membrane.
8. Describe how the brain of a fish and the brain of a human being differ in the relative sizes and functions of these structures: medulla, cerebellum, thalamus, hypothalamus, and cerebral hemispheres.
9. Describe the reticular formation in vertebrates, and explain the relationship of the reticular activating system to arousal.
10. Draw a labelled diagram of a simple reflex arc, and describe how it works.
11. State the contrasting roles of the parasympathetic and sympathetic nervous systems.

Nervous Systems

CHAPTER

28

Polar bear

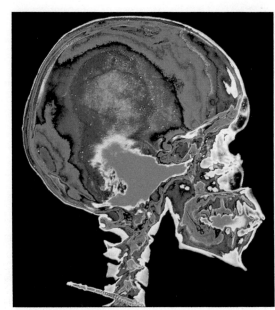

X-ray human head, density slicing

Learning

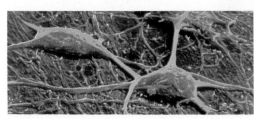

Neurons

561

*D*igestion, respiration, circulation, excretion, and reproduction may be studied separately, but in a living animal they must all work together. An animal's activities are coordinated mainly by its nervous and hormonal systems. We have already seen several examples of this type of coordination, such as control of breathing and of body fluid volume.

The remaining chapters on animal biology deal with the nervous system and with the sense organs, muscles, and glands that work with it in coordinating the animal's activities. The body's **receptors,** in the sense organs, detect **stimuli**— changes in the body's internal or external environment, such as blood chemistry, sound, or light. This information is converted into electrical impulses, the form in which information is handled by the nervous system. The nerve cells that carry these electrical impulses are called **neurons.**

In an animal's nervous system, neurons are arranged to carry messages in pathways (Figure 28-1). **Sensory neurons** receive information from receptors in the sense organs and transmit it to **interneurons,** neurons that relay messages from one neuron to another. Often a nervous pathway involves many interneurons, perhaps receiving and processing information from various parts of the body. Eventually, the message passes to **motor neurons,** which send out signals to the body's **effectors,** the organs that carry out a response. The most common effectors are glands, which secrete hormones, digestive enzymes, and so on, and muscles, which contract and move parts of the body. The responses made by effectors do two main things: they maintain internal homeostasis, or they make suitable behavioral responses to the external environment.

KEY CONCEPTS

- The nervous system works with the hormonal system to coordinate the physiology and behavior of the various organs and systems in the body.
- The progress of information through any nervous system occurs in three main steps:

1. Sensory activities, the collection of information from outside and inside the body
2. Central processing of all this information in the nervous system
3. Responses appropriate to the results of the first two steps.

Figure 28-1
The flow of information into, through, and out of the nervous system.

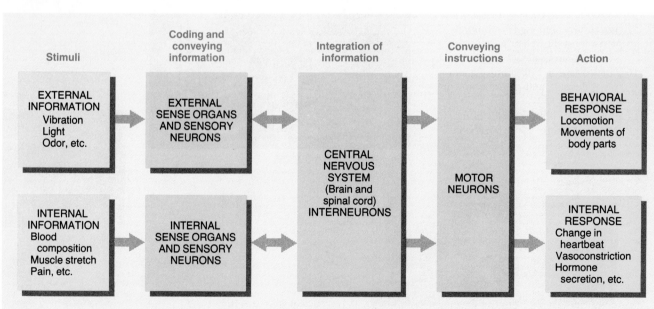

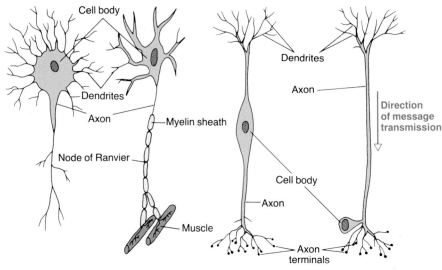

Figure 28-2
Structure of different types of neurons. The arrow shows the direction in which nerve impulses are transmitted in all the cells shown here. (a) A multipolar neuron has a more or less central cell body, many dendrites, and a branched axon. (b) A motor neuron to a vertebrate muscle. This one has an axon surrounded by myelin (see Figures 28-9 and 28-10). (c) A bipolar neuron has two main branches, on opposite sides of the cell body. (d) A monopolar sensory neuron, such as might carry impulses from the skin to the central nervous system.

Neurons

Neurons are the cells that transmit messages in the nervous system. Every neuron has a **cell body** containing the nucleus and most of the cell's organelles. A neuron also has long, thin extensions of two types (Figures 28-2 and 28-3). The neuron's many, branching **dendrites** are extensions that receive information from other cells or from the external environment. A typical neuron also has one **axon,** the extension that carries information to other cells. The end of the axon divides up into many terminals.

A neuron's axon terminals make connections to other neurons (or, if it is a motor neuron, to muscle or gland cells). Each neuron may make connections with many others, and so a single neuron may receive information from, or send information to, many parts of the body. In the vertebrate brain, the dendrites of a single cell receive thousands of connections from the axon terminals of other neurons.

A neuron may carry messages over a considerable distance. For example, the axons of some neurons extend from the spinal cord to the toe of an elephant. The longest cells in any animal are some of its neurons.

In addition to neurons, the nervous system contains **glial cells,** which support the neurons, convey food to them, and perform other service functions. In fact, the human nervous system contains about ten times as many glial cells as neurons. Glial cells account for up to half the brain's volume.

28-A   Overview: How Neurons Work

A neuron's distinctive feature is its **electrical excitability,** that is, its ability to generate and transmit a rapidly moving electrical impulse. It can do this because ions are distributed unequally on the two sides of its membrane. These ions are poised to rush through the membrane, down their concentration gradients,

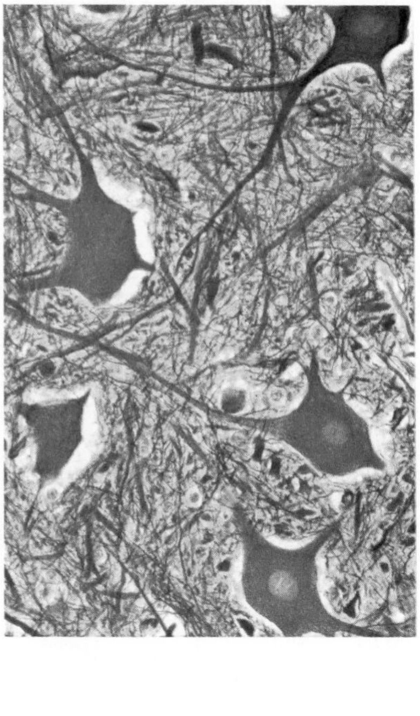

Figure 28-3
Neurons. This section, from the gray matter of the spinal cord, shows several cell bodies and some of their extensions—dendrites and axons. The cells have been stained and photographed using a light microscope. (Ed Reschke)

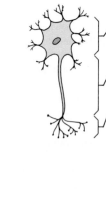

when the membrane's permeability changes and allows them to pass. The result is an electrical **nerve impulse,** the form in which information travels in the nervous system. Transmitting information in this way takes three steps:

1. The neuron's dendrites and cell body receive and "add up" incoming information.
2. If the neuron receives enough input, its axon fires a nerve impulse, a self-propagating electrical signal that travels all the way to the axon terminals.
3. The axon terminals release chemicals that carry the signal to the next cell(s) in the nerve pathway. These chemicals are the incoming information for the next neuron(s), which go through these three steps in their turn.

Nerve impulses can also be generated by applying electrical impulses to neurons, a common experimental technique.

28-B *Electrical Properties of Neurons*

The Resting Potential

Like most other cells, a neuron has an asymmetric distribution of ions across its plasma membrane. This is mainly due to active transport by the **sodium-potassium pump,** which uses energy to pump sodium ions (Na^+) out of the cell and potassium ions (K^+) in (Section 4-D). By making these two stockpiles of ions—sodium outside and potassium inside—the pump has stored electrical potential energy. This potential energy can be converted into useful energy—an electric current carrying information—if the ions are permitted to flow back through the membrane, down their concentration gradients (Section 6-A).

In a resting neuron the membrane is largely impermeable to these ions. However, it does leak slightly, and potassium escapes faster than sodium enters (Figure 28-4). In this way a net positive charge builds up outside the membrane, until it repels any further exodus of positively charged potassium ions.

At this balance point, called the membrane's **resting potential,** the inside of the neuron is about 70 millivolts more negative than the outside. Because of the difference in electrical charges on the two sides of the membrane, the membrane or cell is said to be **polarized.**

Receiving Information: Local Potentials

When a stimulus impinges on a neuron, the balance of the resting potential is temporarily upset. The membrane's permeability changes in a small area as "gates" in some of the membrane's ion permeability channels open briefly. The gates are thought to be channel proteins that work by changing shape. Opening more sodium or potassium gates permits more of the ion in question to cross the membrane. But soon these gates close, and those for the other ion open briefly, permitting that ion to move in the opposite direction and restore the resting potential.

Meanwhile, though, the flow of ions has changed the membrane potential in a local area, and this change in potential causes nearby gates to open. Here, in turn, the membrane potential changes as ions flow across. So a brief change in membrane permeability at one point causes a rapidly moving disturbance in the membrane's resting potential. This disturbance is called a **local potential,** and its size is proportional to the intensity of the stimulus.

Some local potentials **depolarize** the membrane (decrease the difference in electrical charge across it). They do this by opening sodium gates. Positive sodium ions enter and make the membrane potential temporarily less negative (Figure 28-5). Stimuli that depolarize the membrane are **excitatory,** making the neuron more likely to fire a nerve impulse.

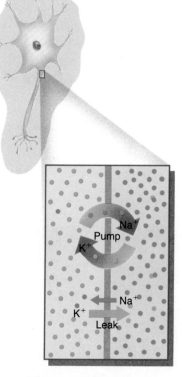

Figure 28-4
Source of the resting potential. Active transport by the sodium-potassium pump keeps sodium (Na^+, green) more concentrated outside the cell and potassium (K^+, blue) more concentrated inside, despite the passive leakage of these ions in the opposite direction (straight arrows).

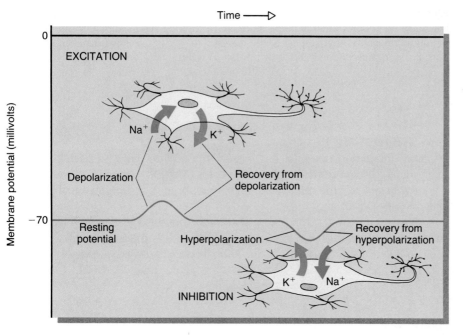

Time ⟶

0

EXCITATION

Na⁺

K⁺

Depolarization

Recovery from depolarization

−70

Resting potential

Hyperpolarization

Recovery from hyperpolarization

K⁺   Na⁺

INHIBITION

Membrane potential (millivolts)

Figure 28-5

Local potentials. The red graph line shows membrane potential changes during excitatory (depolarizing) local potentials and inhibitory (hyperpolarizing) ones. In a local potential, the extent of change in membrane potential from its resting level is proportional to the intensity of the stimulation received. The cell diagrams show the flow of ions— sodium (Na^+) and potassium (K^+)—through the membrane as the resting potential is disturbed, and flow of the other ion in the opposite direction as the resting potential is restored.

Some local responses are **inhibitory,** making the neuron less likely to generate a nerve impulse. Inhibitory responses open more potassium and chloride (Cl^-) gates, permitting the escape of more potassium and the entry of more chloride, which is normally more concentrated outside than inside the cell. Movement of these ions down their gradients causes the membrane potential either to become more negative **(hyperpolarization)** or to remain near the resting potential despite excitatory input.

Local changes in membrane potential spread along the membrane from the point of stimulation, much as ripples spread when you throw a stone into a pond. However, the cytoplasm does offer some resistance to ion flow. Hence a local potential dissipates within a short distance—unless it is reinforced by other local potentials.

Transmitting Information: Action Potentials

Excitatory local potentials in a neuron's dendrites and cell body may add together and cause the axon to fire a nerve impulse. If inhibitory local potentials are occurring, it takes more excitatory input to overcome them. Each axon has a **threshold,** a level of depolarization that must be exceeded before it will fire. Once this threshold is reached, the resulting depolarization is **self-propagating.** That is, it causes an above-threshold depolarization in nearby parts of the membrane, and so on; a uniform wave of depolarization sweeps down the axon to the terminals.

This self-propagating wave of depolarization down an axon is called an **action potential.** All stimuli above the threshold level cause identical action potentials. Since a neuron fires an action potential maximally or not at all, an action potential is an "all-or-none" response.

What happens in an action potential? A threshold depolarization opens sodium gates, and Na^+ ions rush into the more negative axon. This depolarizes the membrane further, and opens even more sodium gates. So the inrush of sodium is a self-reinforcing process (Figure 28-6). This quickly reverses the local membrane potential, with the inside now positive relative to the outside. As sodium ions enter the axon, they spread sideways to nearby, resting areas of the membrane. Here they cause more sodium gates to open and so depolarize these

■ *A neuron processes and transmits information in the form of spurts of ions moving through the membrane, briefly disturbing its resting potential.*

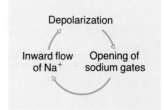

Depolarization

Inward flow of Na^+

Opening of sodium gates

Figure 28-6

Sodium flow. The circular interaction between permeability (opening or closing of ion gates) and potential during propagation of a nerve impulse.

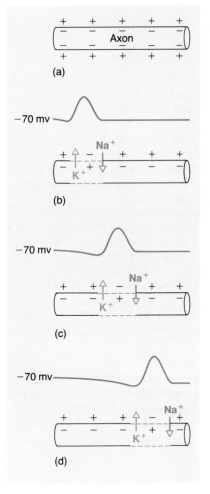

(a)

(b)

(c)

(d)

−70 mv

Na⁺
K⁺

−70 mv

Na⁺
K⁺

−70 mv

Na⁺
K⁺

Figure 28-7
Spread of an action potential along an axon. (a) Resting state, with the inside of the membrane negative with respect to the outside. (b, c, d) Successive stages in conduction of the impulse along the axon from left to right.

areas too. As the cycle continues, a wave of depolarization sweeps along the axon (Figure 28-7).

After the action potential has passed, the sodium gates in each patch of membrane close. Meanwhile, potassium gates open. Now more potassium ions than usual can leave the cell, restoring the membrane's resting potential (Figure 28-8).

All axons transmit action potentials rapidly, but this speed can be increased in two ways. First, axons of larger diameter have less internal electrical resistance and so transmit impulses faster. Animals have axons of various diameters.

Figure 28-8
Recording an action potential. Stimulating electrodes deliver an electric shock to the axon of a neuron, generating an action potential which is detected as it sweeps past the recording electrodes. One recording electrode is placed inside the membrane and the other outside. These electrodes are attached to an oscilloscope, which records the difference in electrical charges at the two electrodes, and thus measures the electrical potential difference across the membrane (red curve in the graph). Superimposed on the graph are the flows of sodium and potassium.

In the resting neuron, the difference between the two recording electrodes is −70 millivolts. (The minus sign means that the inside is more negative than the outside.) The rising phase of the action potential begins at t_1. This signals depolarization of the membrane as sodium rushes in. The potential difference across the membrane declines to zero, and reverses briefly as the inside becomes positive with respect to the outside (at t_2). Now potassium rushes out, and the inside of the cell once more becomes negative with respect to the outside.

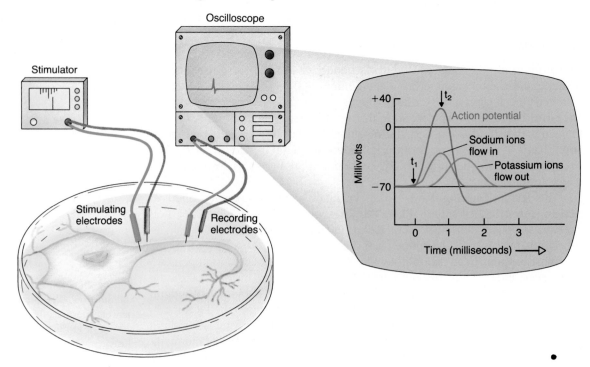

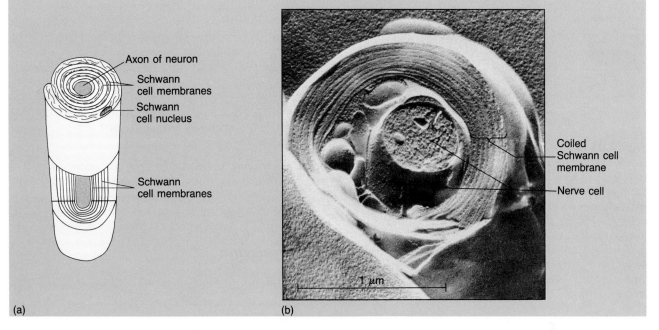

Figure 28-9
The myelin sheath. (a) Diagram of a myelin sheath formed around an axon by a
Schwann cell, a type of glial cell. The plasma membrane of the Schwann cell forms
several layers of wrapping around the axon. The myelin sheath of the entire axon
consists of several sections, each formed from one Schwann cell (see Figure 28-2b).
(b) An electron micrograph of a myelin sheath. (b, Biophoto Associates)

The thickest are the "giant" axons involved in escape reactions in invertebrates.
When you swat at a cockroach and miss, you are foiled by such a giant fiber,
which runs from the cerci (the roach's twin "tailpipes") to the brain, and in-
forms the brain of the air currents that disturb the cerci on your downstroke.

The second adaptation, found mainly in vertebrates, is a **myelin sheath**
around the axon. Each segment of the sheath consists of several layers of lipid-
rich membrane, produced by a glial cell and wrapped around the axon (Figure
28-9). The naked axon is exposed between these cells in very short gaps called
nodes of Ranvier (see Figure 28-2b).

In axons without myelin sheaths, the speed of impulse conduction is limited
by how fast the membrane's ion gates can open and close, and by how quickly
the ions flow through them and then spread sideways to open nearby gates. The
fatty layers of myelin act like insulation around an electric wire. The tight sheath
prevents ions from flowing through the membrane. The electrical impulse is
forced to move down the interior of the axon from one node of Ranvier to the
next, passing through the ion-rich fluid as a much faster electric current. (In this
situation, the ion-rich axon is analogous to a pipe tightly packed with marbles: if
you force a marble [ion] in one end, a marble at the far end pops out almost
instantly.) When the current reaches a node of Ranvier, it opens sodium gates
and generates another action potential (Figure 28-10).

The result is an extremely fast form of conduction termed **saltatory** ("jumping") **conduction.** The action potential at each node regenerates the original
signal so that it reaches the next node strong enough to start an action potential
there. Although some unmyelinated giant axons in squids conduct action potentials at speeds of nearly 20 metres per second, myelinated mammalian axons of
much smaller diameter may transmit impulses at up to 100 metres per second.

■ *Local potentials may add together and initiate an action
potential, which travels down
the axon. Large diameter or a
myelin sheath increases the
axon's speed of conduction.*

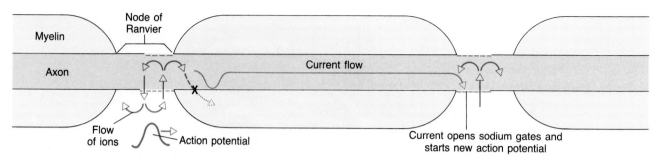

Figure 28-10

Conduction along a myelinated neuron. Red and blue arrows represent the flow of current. The action potential travels across the node of Ranvier only as fast as the ions can cross the membrane and change its potential (red arrows). In parts of the axon surrounded by myelin, the current cannot cross the membrane and thus spreads down the axon's interior to the next node. This conduction is more rapid because it is not limited by the speed of ion flow (blue arrow). When the current reaches the next node of Ranvier, it opens sodium gates. Inflow of sodium ions lowers the membrane potential, starting another action potential.

■ **Messages travel along a neuron's length in the form of electrical impulses in its membrane, and between neurons usually in the form of chemical transmitters.**

Bundles, or tracts, of myelinated, fast-conducting axons occur in various parts of the nervous system. These areas, called **white matter,** owe their color to the abundant lipid in their myelin sheaths. **Gray matter** consists of unmyelinated nervous tissue.

28-C Synaptic Transmission

When an action potential reaches the axon terminals, its message passes to other cells by way of **synapses,** junction areas between neurons. Most synapses are chemical synapses, where the signal is carried between cells in the form of chemical messengers called **neurotransmitters.** These chemicals occur in tiny membranous sacs, or **vesicles,** in the axon terminals (Figure 28-11). The plasma membrane of the axon terminal is called the **presynaptic membrane.** It lies close to the **postsynaptic membrane,** part of another neuron, or part of a muscle or gland cell. Between the two membranes lies a very narrow space, the **synaptic cleft.**

As an action potential reaches the end of an axon, it causes some of the neurotransmitter vesicles to discharge their contents into the synaptic cleft. Transmitter molecules cross the cleft and bind to receptor molecules in the postsynaptic membrane. Many postsynaptic receptors have gated ion channels, which open when neurotransmitters bind to the receptor. This allows ions to flow through the channel, causing a local potential in the postsynaptic membrane.

Synapses may be excitatory or inhibitory, depending on the postsynaptic receptors. Receptors at excitatory synapses have sodium channels, which let sodium ions enter and depolarize the postsynaptic cell. Often the cell must receive excitatory signals from more than one presynaptic cell before it will fire an action potential.

When inhibitory synapses are active, it takes more depolarization than usual to reach the threshold for an action potential. Examples of inhibitory synapses occur in sense organs, where they suppress some incoming information and so sharpen the ability to pick out important features of stimuli.

A neuron can transmit an action potential in both directions from its starting point. However, information passes through the nervous system in only one direction because of the layout of chemical synapses. Only the presynaptic membrane can release neurotransmitters, and only the postsynaptic membrane has receptors for them. Synaptic transmission is much slower than conduction along the axon. So, in general, the more chemical synapses in a neural pathway, the more slowly information passes along that pathway.

In an **electrical synapse,** the membranes of the two cells are connected by gap junctions (see Figure 4-17). Electric current flows directly from one cell to the other in the form of moving ions in the cytoplasm. In contrast to chemical synapses, electrical synapses work rapidly and in both directions.

If electrical synapses are so efficient, why do animals have chemical synapses at all? The reason seems to be that chemical synapses permit more com-

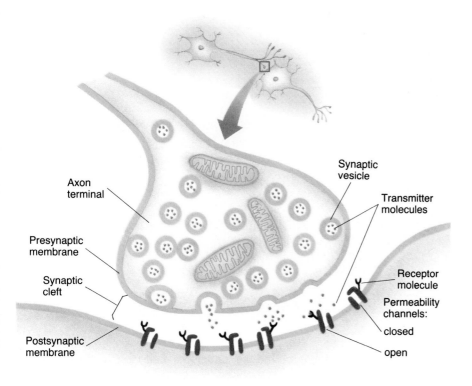

Axon terminal

Synaptic vesicle

Transmitter molecules

Presynaptic membrane

Synaptic cleft

Receptor molecule

Permeability channels:

closed

Postsynaptic membrane

open

Figure 28-11

Transmission across a synapse. This diagram shows one axon terminal and the post-synaptic membrane (part of a dendrite or cell body of another neuron or, perhaps, part of a muscle cell membrane). When an action potential arrives, synaptic vesicles in the axon terminal fuse with the presynaptic membrane and discharge the neurotransmitter they contain into the synaptic cleft. Transmitter molecules cross the synaptic cleft and attach to receptor molecules in the postsynaptic membrane. Binding of the transmitter to the receptor opens permeability channels and permits an increased flow of certain ions across the membrane.

plexity and variety in information processing. Electrical synapses can only be excitatory, but chemical synapses can be either excitatory or inhibitory. By activating different combinations of the excitatory and inhibitory input coming to a particular neuron, it is possible to block, enhance, shorten, or prolong the neuron's response to a signal. This makes it possible for the nervous system as a whole to adjust the intensity of the animal's response, or to make a different response, depending on circumstances.

28-D Neurotransmitters

About 30 different neurotransmitters have been identified. Each occurs in specific sets of neurons, and each neuron typically makes and releases one kind of transmitter. Many transmitters can be either excitatory or inhibitory, depending on which receptors occur at the synapse where they are released. Other transmitters appear to occur only at excitatory or at inhibitory synapses. (Presumably only receptors with excitatory or inhibitory effects exist for these transmitters.)

Some important transmitters are:

1. **Acetylcholine,** the best-known transmitter. It occurs at many synapses in the brain and in other parts of the nervous system, and also at **neuromuscular junctions,** synapses between neurons and skeletal muscles (muscles that

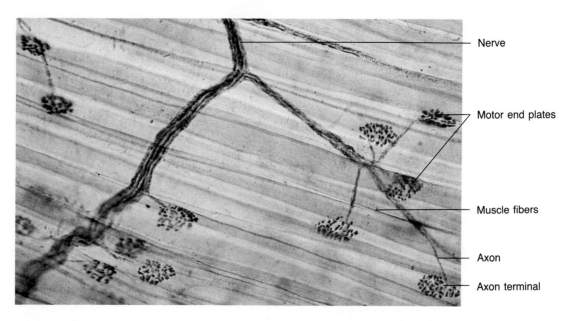

Nerve

Motor end plates

Muscle fibers

Axon

Axon terminal

Figure 28-12

Neuromuscular junctions. These sites are used to study the action of the transmitter acetylcholine. A nerve is composed of the axons of many neurons. Near the muscle, the nerve divides into individual axons. Each axon may divide into several branches, each of which ends in a cluster of axon terminals called a motor end plate. (Figure 28-11 shows a single terminal.) (Biophoto Associates)

move parts of the skeleton and are under our conscious control) (Figure 28-12).

2. **Norepinephrine,** important in the body's response to stress (Section 28-I).

3. **Dopamine,** a molecule similar to norepinephrine. Dopamine is the transmitter for a small group of neurons concerned only with muscular activity. Parkinson's disease, with its bursts of uncontrollable muscular movement, may sometimes be caused by lack of this transmitter; it is often treated with L-dopa, the substrate of the enzyme that makes dopamine. Dopamine also occurs in neurons in some other areas of the brain.

4. **Serotonin,** produced by a group of cells in the medulla, part of the brain just above the spinal cord (see Figure 28-15). Serotonin seems to be concerned with functions such as sensory perception, regulation of body temperature, sleep, consciousness, and emotions. Deficits of serotonin are linked to depression and anxiety; excess can cause nausea.

5. **Neuropeptides,** the largest group of neurotransmitters, each containing 2 to 39 amino acids. Some neuropeptides are identical to local chemical messengers in other body areas (Section 31-E) or to hormones released from the pituitary gland, hypothalamus (both shown in Figure 28-15), or gut. Their roles in the nervous system are poorly understood.

The larger **endorphins** and smaller **enkephalins** have been called the brain's "natural morphine" because the brain's receptors for them also bind morphine, presumably with the same effect on the neurons involved. This explains why the brain is so sensitive to morphine, opium, and related drugs. Enkephalins occur in neurons that process information relating to emotion, mood, and pain. They apparently act by suppressing the response to other neurotransmitters. Endorphins are responsible for the "runner's high" experienced by people who run or perform other intense excercise to the point of producing physical pain.

6. **GABA** (short for gamma-aminobutyric acid), a transmitter at inhibitory synapses throughout the brain. Studies of GABA suggest that as much as 90% of the brain may be devoted to inhibition.

After transmitter molecules have acted on the postsynaptic membrane, they must be removed or destroyed. Otherwise, their action would continue indefinitely, and all useful information would be lost. Norepinephrine is resorbed by the presynaptic membrane and reused. Acetylcholine is broken down by the

Drugs and Neurons

A JOURNEY INTO LIFE

Many drugs affect the nervous system, in most cases by regulating the activity at synapses. However, some have more general effects. For example, caffeine (in coffee, tea, cola, and chocolate) increases the metabolic rate of neurons, thereby increasing alertness and thought. Alcohol is a depressant, acting on all neurons and affecting many functions such as orientation, alertness, coordination, judgment, memory, and mood.

Opium, from the seedpod of a poppy, has been used as a drug since ancient Greek times, not only because it is the most effective pain-killer known, but also because it induces a state of euphoria. Addiction to opiates (opium and related compounds) has been a social problem in the United States ever since the Civil War, when they were used as pain-killers. The search for a nonaddictive opiate has been intense, but all

known opium derivatives—including morphine, Demerol, methadone, codeine, and heroin—eventually produce addiction in many people who take them. They are therefore most useful as pain-killers for the terminally ill, in whom relief from pain outweighs possible addiction.

Opiates bind to postsynaptic receptors in the brain and block the binding of neurotransmitters. This prevents the transmission of nerve impulses along a tract of nerves by which the body normally "tells" the brain that it is in pain. (Pain is a useful biological reaction, for when the brain is informed that some part of the body is in pain, perhaps from a cut or burn, it causes the body to move away from the source of the damage.) Opiates also depress the immune system, especially cells that fight cancer and virus infections.

Marijuana interferes with short-term memory by binding to receptors in the part of the brain involved in the formation of memories. It also binds to receptors in the cerebral cortex (see Figure 28-15), where it impairs thought and reasoning, heightens sensory perception, changes the perception of time, and produces mild euphoria, relaxation, and relief from anxiety. Marijuana also lowers the levels of sex hormones, interferes with the menstrual cycle, and suppresses the immune system.

Many other drugs also act on synapses in the brain. For example,

barbiturates depress the activity of inhibitory synapses, permitting more excitatory activity, which produces mild euphoria. LSD and mescaline are thought to produce hallucinations by binding with receptors for the neurotransmitter serotonin, a molecule they resemble. Cocaine and amphetamines (Benzadrine, Dexedrine, and "speed") mimic or enhance the effects of dopamine and norepinephrine. Cocaine prevents the normal removal of the transmitter dopamine from synapses, so that the synapses continue to fire long after they should have stopped. At low doses it stimulates behavior and elevates mood, but overdoses are fatal.

Addiction to drugs has two components, one physiological, the other psychological. Addicts who stop using drugs, or gradually decrease their dosage, go through a period of withdrawal with unpleasant physical reactions such as shaking, vomiting, hallucinations, pounding heart, and pain. But after their bodies have adjusted to working without the drug and returned to normal, many people still have a psychological craving for the drug that never goes away.

Drug-taking by humans has parallels among other animals. Animals as different as birds, rabbits, deer, and elephants have been observed to seek out and eat such things as fermented fruit (containing alcohol) or intoxicating mushrooms.

enzyme **acetylcholinesterase.** Many insecticides inhibit this enzyme, permitting acetylcholine in the synapses to keep on stimulating the postsynaptic membranes. The nervous system soon goes wild, firing one nerve impulse after another. These impulses in turn cause contraction of the muscles in uncontrollable spasms and, eventually, death. (Such substances must be used with care. They are also toxic to other organisms, such as people and pets, that use acetylcholine as a transmitter.)

Variations in levels of neurotransmitters can affect people's moods and abilities. One form of depression results from having too many receptors for acetylcholine, so that postsynaptic neurons overreact to it. Depression can also result from deficiencies in norepinephrine, dopamine, or serotonin, whereas excesses of these transmitters may result in manic behavior. In fact, anything that affects postsynaptic receptors, or the enzymes that make or destroy transmitters, can lead to malfunction of the relevant part of the nervous system. This is the basis of many mental illnesses and is also how some drugs affect the nervous system. By understanding how the various transmitters and receptors work, we can use medication to treat some mental illnesses or to treat the pain of physical illness.

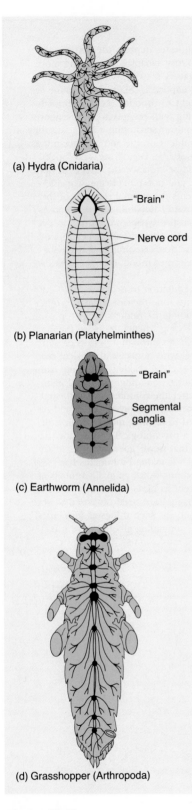

(a) Hydra (Cnidaria)

"Brain"

Nerve cord

(b) Planarian (Platyhelminthes)

"Brain"

Segmental ganglia

(c) Earthworm (Annelida)

(d) Grasshopper (Arthropoda)

Figure 28-13
Nervous systems of some invertebrates. Whereas Cnidaria (a) have a diffuse nerve net, the nervous tissue of higher invertebrates (b, c, d) is arranged in well-defined systems.

28-E   *Organization of Neurons Into Nervous Systems*

In an animal's nervous system, neurons are "wired" to each other at synapses, forming nervous pathways. Some synapses are excitatory, others inhibitory. Any one neuron may synapse with many others, and may make excitatory synapses with some and inhibitory synapses with others. The number, arrangement, and synaptic connections of neurons determine how the animal responds to stimuli and the kinds of behavior it can perform.

During the course of animal evolution, nervous systems became increasingly complex, mainly because animals became larger and more mobile. Both trends call for more neurons. A tiny, sluggish, parasitic roundworm may have as few as 160 neurons, and a correspondingly small range of behaviors. An octopus, with precise control over its eight tentacles and considerable ability to learn new behavior patterns, has more than a billion. A human has about 100 billion neurons.

Large numbers of neurons are not enough; their organization is also crucial. The simple nerve nets of cnidarians have neurons scattered throughout the body (Figure 28-13). Though primitive, this arrangement serves the needs of a radially symmetrical animal, whose food or enemies may approach from any direction. A cnidarian's reaction to most stimuli is generalized because each neuron transmits impulses to all of its neighbors. The animal's reaction also depends on the strength of the stimulus; only part of the body responds to a weak stimulus, the entire animal to a strong one. Surprisingly, even this primitive nerve net is capable of simple forms of learning.

The platyhelminths (flatworms) show the beginnings of some important trends. The first is consolidation: instead of the diffuse nerve net of cnidarians, some neurons became arranged together into nerve cords running the length of the body, with cross-connections between them. This permits faster, more direct processing of information collected from outlying areas, and quick signalling of actions to be taken by effectors. A second advance is specialization: cells took on distinct roles as sensory or motor neurons or as interneurons. The patterns of synapses became more precise, so that incoming information is passed to specific neurons. A third important trend in flatworms is **cephalization,** formation of a head. Having a particular front end correlates with the evolution of bilateral symmetry. A head bears the major sense organs, such as the eyes and ears, which detect what is happening in the outside world; the animal's leading end can sample the new environment for food or safety as it moves. The closer the "decision-making" neurons are to these sense organs, the faster the animal can react. Consequently, nervous tissue became more concentrated in the head, with neurons forming clusters called **ganglia** (singular: **ganglion**). The ganglia in the head are generally called a **brain,** the body's main nervous control center.

In annelid worms, arthropods, and other higher invertebrates, ganglia also occur along the nerve cord and govern particular parts of the body. The nerve cord itself is now single, running down the body's ventral midline.

Some invertebrate nervous systems have relatively few neurons, with a correspondingly limited repertoire of possible behaviors. In a few species of leeches, slugs, and crayfish, we now have maps showing exactly which neurons do what. In a leech, for instance, one particular pair of motor neurons everts the penis, and four neurons in each ganglion sense "pain." Researchers use these animals to examine basic principles of nervous system organization and its effects on behavior. But even these "simple" nervous systems are challenging problems to study, and we are nowhere near ready to tackle such a cell-by-cell analysis of the vastly more complex nervous systems of higher animals. Instead, we approach the nervous systems of vertebrates mostly by determining the functions of groups of cells in various areas.

The Vertebrate Nervous System

The vertebrate nervous system can be divided into the central and peripheral nervous systems. The **central nervous system** consists of the brain and spinal cord. The vast majority of all neuron cell bodies, and in most cases the dendrites and axons too, lie in the central nervous system. Most of these neurons are interneurons, which relay information from one neuron to another. In the process, sensory information is **integrated**—compared, changed, added up, or suppressed—assuring that the response made will be appropriate for existing conditions. (For example, your response to a plate of pastries depends on how full your stomach is, as well as how delicious the pastries look and smell.) The central nervous system is also responsible for **association**—channelling sensory input into appropriate motor pathways.

In the human nervous system, some of these activities go on automatically, without our being aware of them. Others are carried out by the "higher centers," parts of the nervous system at the level of consciousness, where we are aware of the existence of sensory stimuli and of our own thoughts and emotions, and may deliberately choose one course of action over another. We have no way of knowing whether other animals have distinctly conscious nervous functions.

The brain is protected by the skull, the spinal cord by the vertebral column. Inside these bony coverings, the brain and spinal cord are covered by three layers of membranes, the **meninges.** Within the meninges, the **cerebrospinal fluid** bathes the central nervous system and cushions it from jarring. Disorders of the central nervous system can sometimes be diagnosed by doing a "spinal tap" to withdraw a sample of the fluid around the spinal cord. Inflammation of the meninges (meningitis) is often a serious condition because the meninges are so close to vital nervous tissues.

The **peripheral nervous system** connects the central nervous system with the rest of the body, including sense organs that detect stimuli in the outside world. It consists of **nerves,** bundles of axons covered with protective sheaths of connective tissue, and some **peripheral ganglia,** clusters of neuron cell bodies lying outside the central nervous system.

The peripheral nervous system has two parts, which play different roles. The **somatic nervous system** contains both motor and sensory neurons. It controls the activity of the skeletal muscles, those under "conscious" control, such as the muscles we use to smile, run, sing, or draw. The **autonomic nervous system** contains only motor neurons. It controls muscles and glands that usually operate without our being aware of them. These effectors control things like blood pressure and the movement of food in the gut.

■ *The somatic system permits us to react consciously to the outside world, while the autonomic system copes mainly with homeostatic mechanisms within the body.*

VERTEBRATE NERVOUS SYSTEM

Central nervous system	Peripheral nervous system	Divisions of peripheral nervous system:
Brain	Nerves	Somatic (voluntary)
Spinal cord	Ganglia	Autonomic (involuntary homeostatic control)

28-F The Vertebrate Brain

The central nervous system grows from a hollow tube formed during early embryonic development. The walls of the front part of the tube enlarge in a series of uneven bulges, with three main parts, unimaginatively called the forebrain, midbrain, and hindbrain, followed by the long, straight spinal cord.

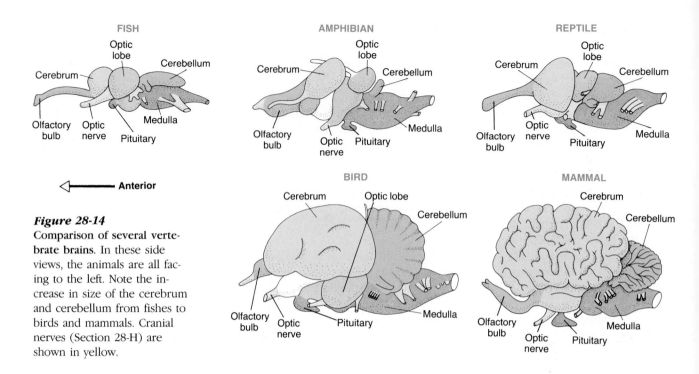

Figure 28-14
Comparison of several vertebrate brains. In these side views, the animals are all facing to the left. Note the increase in size of the cerebrum and cerebellum from fishes to birds and mammals. Cranial nerves (Section 28-H) are shown in yellow.

All vertebrates, from fish to mammals, have brains with the same basic structures (Figure 28-14). In the course of evolution, some parts of the brain changed very little, while others virtually exploded overnight (on an evolutionary time scale). Some areas of the brain retained their primitive functions, whereas others took on new functions as vertebrates evolved (Table 28-1).

Table 28-1   Functions of Major Parts of the Vertebrate Brain

Derivation	Name	Function
Hindbrain	Medulla	Passage of messages between brain and spinal cord; control of visceral reflexes
	Cerebellum	Coordination of equilibrium and movement
Midbrain	Tectum (in lower vertebrates)	Association of sensory and motor pathways
	Anterior colliculi (mammals)	Reflexes of iris and eyelid
	Posterior colliculi (mammals)	Receive sensory information from ear
Forebrain		
Diencephalon	Thalamus	Relays olfactory messages to midbrain (in fishes)
		Area of sensory integration (in higher vertebrates)
	Hypothalamus	Controls emotional states and drives (pleasure, pain, thirst, sex, rage)
	Posterior pituitary	Releases hormones (Section 31-D)
Telencephalon	Olfactory bulb	Receives olfactory information (most important telencephalon area in fishes)
	Corpus striatum	Complex behavior patterns (in birds)
	Cerebral hemispheres (cerebral cortex)	Well developed only in mammals. Sensory and motor association, visual and auditory processing, seat of "intelligence," and, in humans, of ability to use language, both written and spoken

Hindbrain

The most obvious part of the hindbrain of a fish is the **medulla,** the enlargement where the spinal cord enters the brain (Figure 28-14). Through it pass many neurons carrying messages to or from parts of the brain that integrate sensory information. The medulla also contains many neurons that receive sensory input and send out motor signals for automatic, or **reflex,** functions such as breathing, swallowing, vomiting, constriction of blood vessels, and regulation of heartbeat rate. The medulla has changed little during evolution.

The **cerebellum** is an outgrowth of the medulla. During evolution it has grown noticeably and taken on much of the central control of balance and movement.

Midbrain

During evolution, the midbrain has changed more in function than in size or structure. In fish and amphibians it is the principal area for association of sensory input with suitable motor output. In these lower vertebrates, a major part of the midbrain is the **optic tectum,** which receives signals from the **optic nerves,** carrying visual information from the eyes.

In mammals, the analysis of vision has moved out of the midbrain and into part of the forebrain. The midbrain of mammals controls reflexes of the iris of the eye and the eyelids, and analyzes and relays information coming in from the ear via the auditory nerve.

Forebrain

The forebrain has changed a great deal during vertebrate evolution. It has two major parts, the diencephalon and the telencephalon. Lying just in front of the midbrain, the **diencephalon** contains the thalamus, the hypothalamus, and the posterior lobe of the pituitary gland (Figure 28-15). In fishes, the **thalamus** relays information to the midbrain from the **olfactory** (sense of smell) organs. However, in other vertebrates it is one of the centers that integrate all sensory

Figure 28-15
The human brain. (a) The brain has been sliced through the middle, showing the major structures mentioned in the text. The diencephalon is shown in blue. (The hippocampus lies at the level indicated here, but between the central plane of the brain, shown here, and the external surfaces shown in Figures 28-14 and 28-16.) (b) Photograph of the same view. (b, Biophoto Associates)

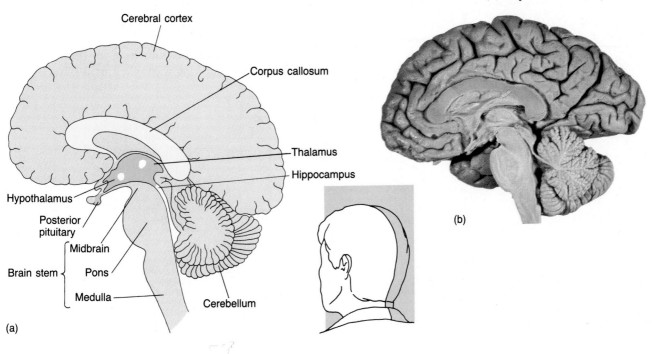

information. Immediately below the thalamus lies the **hypothalamus,** a vitally important area where the nervous and hormonal systems interact (Section 31-D). This area controls homeostatic functions: regulation of body temperature, growth, sexual drive and maturity, hunger, thirst, and salt and water balance.

While the diencephalon slowly expanded to handle increased sensory input, the **telencephalon,** the front part of the forebrain, grew astoundingly in both size and complexity. In fish and amphibians it handles mostly olfactory information, which plays a major role in the lives of these aquatic animals. In birds, however, the most important part of the brain is the **corpus striatum,** which is responsible for their complex behavior patterns. The corpus striatum of birds occupies much of the inside of the **cerebrum,** which is divided by a fissure into the right and left **cerebral hemispheres.**

In mammals, there is a progressive increase in the size and importance of the **cerebral cortex,** the outer layer of the cerebral hemispheres. The original, deeper layers of the hemispheres, the **hippocampus** and other **limbic structures,** regulate emotional state and probably short-term memory. Above these areas, the cerebral cortex lies like the cap of a wrinkled mushroom over the rest of the brain. The gray matter of the cerebrum is composed of thick layers of unmyelinated cells.

How do we know which parts of the brain do what? Using laboratory animals or human patients, researchers can stimulate different areas of the brain with electricity, to make the neurons fire, and observe what happens. Also, they can observe what functions are lost when certain parts of the brain are deliberately or accidentally destroyed. A newer method is to determine which parts of the brain accumulate an experimental chemical during a particular mental activity.

It turns out that many functions involve cells in several areas of the brain. However, certain areas tend to be "in charge" of certain functions. For example, the cerebral cortex has primary sensory areas and primary motor areas (Figure 28-16). Other areas of the cerebral cortex are involved in perception of visual or auditory stimuli and, in humans, in use of symbols and language.

In primitive mammals, each area of the cerebral cortex has specific sensory or motor functions, and this is true to some extent in more advanced mammals. However, in mammals such as primates (monkeys, apes, humans), large areas of the cerebral hemispheres have no known specific function. Most researchers believe that these areas are important to versatility of behavior, abstract thinking, and personality.

Interspersed with the gray matter of the cerebrum are many tracts of white matter, bundles of myelinated axons connecting with other areas of the brain.

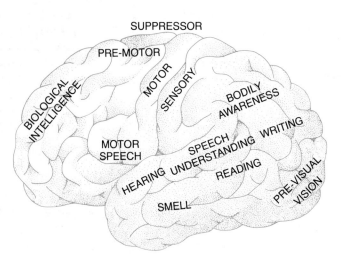

Figure 28-16
Functions of the human cerebral cortex. The functions assigned to several areas are written on this view of the brain's surface, as seen from the left side.

The right and left cerebral hemispheres are linked by way of a large tract of myelinated axons, the **corpus callosum** (see Figure 28-15). Its function is to tell the right half of the brain what the left half is doing (and vice versa).

Roger W. Sperry and his co-workers showed that the two cerebral hemispheres can operate as two different brains; for this work, Sperry shared in a 1981 Nobel Prize. To understand it, you must know that each half of the brain controls structures on the opposite side of the body (the nerve tracts cross over at lower nervous centers). Sperry severed the corpus callosum in experimental animals, and also worked with people who had undergone similar surgery to treat epilepsy. When tasks are presented to the left eye and hand, which are controlled by the right side of the cerebral cortex, the person may show no sign of recognizing them even though the right eye and hand, and left cerebral cortex, had previously mastered the task.

Furthermore, the two sides of the cerebral cortex do not have exactly the same functions. For instance, a person may be able to perform a task set for the right half of the brain correctly, but not be able to speak or write about it, whereas the left half of the brain can both learn to perform the task and generate a verbal explanation of what it is doing. In general, the left side has a strong tendency to assume control of language, whether written, spoken, or the sign language of the deaf. However, attempts to divide functions neatly between the two halves of the brain are too simplistic. To produce normal functioning in most activities, both sides must work together, each contributing expertise in certain aspects of the activity.

Reticular Formation

One interesting subsystem of the brain is the **reticular formation** (Figure 28-17), extending from the medulla into the midbrain. Its cells receive input from all types of sensory neurons entering the brain. Its output goes back down the spinal cord, where it amplifies or reduces incoming sensory signals (much like a volume control on a radio). This adjusts the nervous system's sensitivity to stimuli.

The **reticular activating system** extends from the reticular formation to the cerebral cortex. It acts as a filter to determine which sensory information reaches the level of consciousness in the cortex. This system receives stimuli from all kinds of sense organs, but does not keep the different senses separate: it simply notifies the cortex, where it may induce alertness and attention. We have all at some time awakened suddenly from sleep, not knowing what woke us, but certain that something did. Whatever the stimulus was, it made its way into the reticular activating system, which flashed an "all points" arousal message throughout the cerebral cortex.

Sensory Analysis

In mammals, the various sensory systems relay information to the cerebral cortex for integration and use in making decisions. The processing of sensory information has been particularly well studied in the visual system. In 1981 David Hubel and Torsten Wiesel shared a Nobel Prize with Sperry for their work on the visual system.

The organization of neurons in the visual system shows two general features. First, information from each area in the retina of the eye ends up in a corresponding location in the visual cortex, the posterior part of the cerebral cortex (see Figure 28-16). Second, information reaches the visual cortex by passing through several consecutive layers of neurons. These intervening cells do not simply pass messages through unaltered. Rather, the arrangement of synapses from one layer of cells to the next is such that information is processed

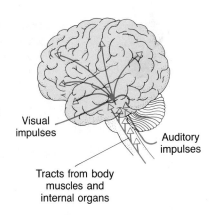

Figure 28-17
The reticular formation. Many sensory pathways provide input to the reticular formation, which in turn communicates with association areas in the cortex. Reticular formation cells are involved in wakefulness and arousal, but because many different senses feed into the system, it is often hard to tell what sort of stimulus caused arousal.

Visual impulses

Auditory impulses

Tracts from body muscles and internal organs

■ *The reticular formation and reticular activating system are responsible for general levels of lethargy or liveliness.*

■ *Information from the eye passes through a hierarchy of neuron layers with synapses arranged to pull out important features of the image automatically for the attention of the visual cortex.*

in increasingly complex ways at successive layers. Cells in the first layer detect (that is, fire most rapidly in response to) small spots of light or dark in the area of the retina that they monitor. Higher order cells detect larger circles or ovals of light, still higher cells edges or corners of light of a certain size and orientation, then moving edges, and so on. Just how the brain interprets the shape, motion, and color we see is still a mystery.

An animal's sensory system may be "hard-wired" to pick out specific features from the masses of stimuli in its environment. For instance, a frog's visual system is specialized to pick out small, dark moving objects—which in the frog's world tend to be juicy flies—and large darkenings of the visual field—danger! These "feature detectors" save time and energy by permitting the frog to react quickly to events significant to itself, while ignoring other aspects of its visual environment.

Motor Pathways

In the cerebral cortex, information about sensory stimuli is passed to association neurons, which connect with the motor pathways that will eventually cause appropriate action by effectors.

The motor area of the cerebral cortex sends out commands for voluntary movements. These signals do not directly stimulate muscle contraction or gland secretion. As in sensory pathways, the command from the cortex may cross anywhere from three to thousands of synapses, allowing the message to be adjusted on its outward path.

The cerebellum plays an important role in coordinating movements, especially rapid movements such as running or typing. It receives sensory information about where each part of the body is, makes moment-by-moment predictions of where each part will be next, and sends out instructions for fine adjustments so that actions are carried out smoothly. People with damaged cerebellums move jerkily, and their motions invariably overshoot the intended final position. For instance, a hand may reach too far to pick up a desired object, and then the next movement, intended to compensate, will bring the hand too far back.

28-G The Spinal Cord

The spinal cord extends from the base of the hindbrain to the end of the vertebral column (Figure 28-18). It is a relay system carrying information between the brain and the peripheral nervous system. It is also the seat of the many **spinal reflexes** that allow the body to make quick responses.

Reflex Arcs

Reflexes save time and energy by performing routine actions without involving higher centers of the nervous system.

Although we have all learned to perform many complex activities, our responses to certain stimuli are simple, unvarying, and quick. For instance, it is hard to hold your lower leg still when the doctor hits you below the knee with a little rubber mallet. The knee jerk, and the rapid withdrawal of a hand from a hot stove, are controlled by **reflex arcs,** pathways of a few neurons each, under little control from higher centers. Reflex arcs contain sensory and motor neurons, and usually one or more interneurons between them (Figure 28-19). A reflex pathway saves time because it has a small number of synapses. The fewer synapses between receptor and effector, the sooner the effector responds to a stimulus. In addition, the message need not make the lengthy trip to areas of consciousness in the brain and back before an appropriate motor response

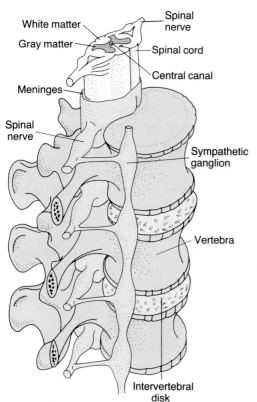

White matter
Gray matter
Meninges
Spinal nerve
Spinal nerve
Spinal cord
Central canal
Sympathetic ganglion
Vertebra
Intervertebral disk

Figure 28-18

The human spinal cord. The cord is surrounded and protected by the meninges and by the vertebral column. The paired spinal nerves protrude through spaces between the vertebrae. The butterfly-shaped area of gray matter is composed of cell bodies. White matter is bundles of myelinated axons.

begins. By the time you actually feel the mallet below the knee, your foot has already begun to swing out.

The organs that regulate physiological homeostasis are controlled by more or less simple reflexes of the autonomic nervous system. Heart rate, breathing, dilation of the pupils, and digestion are all controlled in this way. More complicated activities, such as posture control, locomotion, sexual behavior, and defensive responses, which are governed mainly by the somatic nervous system, also involve many reflex arcs.

A reflex may be quite complex, using many neurons and moving many muscles. For example, if you raise your arm to protect your face, muscles in your back must also adjust their contraction or you would lose your balance.

The spinal cord plays an important role in the integration of reflex behavior. For instance, a "spinal" animal, one whose brain has been destroyed or removed, still shows reflexes. If a piece of acid-soaked paper is touched to the back of a spinal frog, one leg will come up and kick it away; the behavior is repeated no matter how many times the paper is placed on the skin. This response, involving the coordinated action of many muscles, clearly demonstrates one of the chief characteristics of a reflex: unvarying repetition. A frog with an intact brain might make the response two or three times, but eventually the higher centers would intervene and the frog would do something else—perhaps hop away.

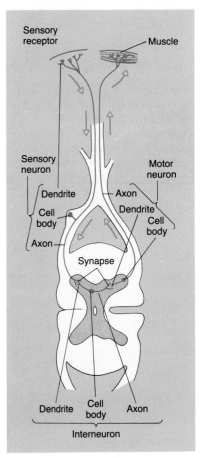

Sensory receptor
Muscle
Sensory neuron
Motor neuron
Dendrite
Axon
Cell body
Dendrite
Axon
Cell body
Synapse
Dendrite
Cell body
Axon
Interneuron

Figure 28-19

A simple reflex arc. The arc shown here contains two synapses and three neurons: sensory neuron, interneuron, and motor neuron. Sensory neurons such as the one shown here are some of the few neurons whose cell bodies lie outside the central nervous system. Ganglia just outside the spinal cord house the cell bodies of its sensory neurons.

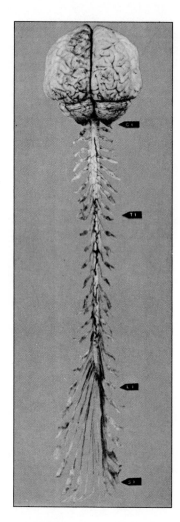

Figure 28-20
The human nervous system. Note the roots of the spinal nerves still attached to the spinal cord. (Dissection by Dr. M. C. E. Hutchinson, Department of Anatomy, Guy's Hospital Medical School, London, England. From Williams and Warwick (eds.): *Gray's Anatomy.*)

28-H Cranial and Spinal Nerves

The vertebrate peripheral nervous system consists of paired nerves that branch from the central nervous system. In reptiles, birds, and mammals, 12 pairs of **cranial nerves** connect the brain with various structures, mostly in the head and neck; fish and amphibians have only the first 10 pairs. The thickest cranial nerves are the olfactory, optic, and auditory, which carry only sensory information coming to the brain from the major sense organs—the nose, the retinas of the eyes, and the ear. The other cranial nerves carry both motor and sensory information to and from the tongue, muscles of the eye and face, and so on (see Figure 28-14).

Human beings have 31 pairs of **spinal nerves,** which branch out from the spinal cord between adjacent pairs of vertebrae (Figures 28-18 and 28-20). Each spinal nerve serves the skin, muscles, and internal organs in its particular segment of the body and also overlaps adjoining segments. So, if a spinal nerve is cut, the corresponding part of the body does not completely lose sensation and the ability to move.

28-I The Autonomic Nervous System

The autonomic nervous system governs most of the body's homeostasis: it regulates the heartbeat and controls contraction of the muscles in the walls of the blood vessels and the digestive, urinary, and reproductive tracts. Autonomic nerves also stimulate glands to secrete mucus, tears, and digestive enzymes.

The autonomic system has two divisions, sympathetic and parasympathetic, with different functions. The **sympathetic system** dominates in time of stress. It initiates the **"fight or flight" reaction:** increases in blood pressure, heartbeat rate, breathing, and blood flow to the muscles, and decreases in the flow of blood to the digestive organs and kidneys. These changes increase the supply of oxygen to the muscles at a time when they may be called upon to use a lot of energy. In contrast, the **parasympathetic system** acts as a counterbalance. It conserves energy by slowing the heartbeat and breathing rates and promotes digestion and elimination.

Another difference between the sympathetic and parasympathetic systems is in the chemical transmitters released at the synapse with the effector: norepinephrine in the sympathetic system, acetylcholine in the parasympathetic system.

Although the autonomic nervous system can carry out its tasks automatically, it is by no means completely independent of the animal's voluntary control. For example, it is possible to decide to stop breathing for a short time. Humans and animals can also be trained to change their heart rates, blood pressures, and digestive reflexes voluntarily. However, any voluntary control that endangers life quickly disturbs homeostasis of the brain tissue, resulting in unconsciousness. Then the autonomic system takes over again and restores normal functions.

28-J Learning and Memory

Learning and memory are complex brain functions. **Learning** may be defined as changes in the nervous system (and its responses) as a result of experience, and **memory** as the retention of these changes over time. The basis of learning is the linking together of two (or more) different pieces of information, perceived at more or less the same time, into a pattern, which is stored as a memory. For example, a person's face, figure, voice, name, and so on become linked to form the pattern that is the memory of that person. It seems likely that this requires information from the eyes, ears, nose, and other sense organs to come together in some part of the nervous system. Once the memory is formed, any one of these stimuli can evoke the rest of the pattern, as when hearing a friend's voice on tape permits us to recall the speaker's name and face. What kinds of changes occur in the nervous system during learning, and how do they become more or less permanent features of the system? Although we still know very little about this, answers are beginning to emerge.

For years scientists destroyed various parts of the brains of trained animals to find out where memory is located. They concluded that memory is nowhere and everywhere. Recent experiments give clues to this paradox. Memories are formed as information passes along neural circuits that pass through many areas of the forebrain. Destroying any of these areas or cutting the connections between them impairs the ability to form new memories but leaves existing ones intact. Existing memories are probably stored as patterns of altered synapses in sensory analysis areas of the cerebral cortex, where the memory-forming circuits began. Cells in these areas can trigger recall of the memory pattern when they again receive a stimulus that is part of the memory.

These findings fit in with what is known about memory from observations of people afflicted with amnesia (loss of memory) as a result of disease or damage to certain areas of the brain. For instance, "memory" has many components, such as verbal, spatial, and emotional, and these seem to be stored in different areas of the brain. Patients with partial loss of memory may recognize faces but not names, or vice versa. They may be unable to identify a person as someone they know, yet feel the appropriate emotion depending on whether or not they liked the person before the memory loss.

As young children we learn many habits and skills, such as walking and talking ("motor" memories), but we do not remember the actual process of acquiring them. So, it seems that our ability to acquire habits develops before our ability to store "memories."

There are three kinds of memory: immediate, short-term, and long-term. When we take lecture notes, we use the first, remembering what the speaker said just long enough to write it down. Short-term memory lasts for minutes to hours, and is used for things like remembering to do an errand or "cramming" for a test. Information to be stored for any length of time must be transferred to long-term memory, where it may remain, much of it in subconscious form, for life.

Experimenters have started to unravel the molecular events of memory by training sea snails to respond to stimuli and tracing the resulting changes in their neurons. It now appears that memory depends on activating postsynaptic receptors, which in turn begin a cascade of changes in enzyme activity in the neuron. In the formation of short-term memories, enzymes operate on the ion gates of neurotransmitter receptors, changing their opening or closing properties. This in turn changes the membrane's responsiveness to the arrival of future messages. In the bigger picture, such a change in individual neurons might result in the animal's being more (or less) likely to perform a particular activity in response to a given stimulus.

Long-term memory involves new RNA and protein synthesis as well as changes in the structure of synapses and in the organization of their proteins. The synaptic changes occur in particular branches of a neuron, not in the entire cell. This makes those synapses more efficient at responding to specific patterns of stimuli on later occasions.

Thanks to our ability to learn and remember, human thought and behavior are constantly adaptable, although young brains are more adaptable than old.

(a)

(b)

Figure 28-21

Sleep. Many animals sleep, although some relax less than we do. (a) A koala must remain still and grasp a tree while it naps, or it would fall to the ground. (b) Flamingoes remain standing, although they tuck their heads under their wings.

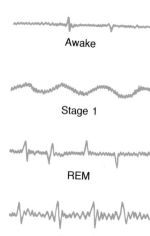

Figure 28-22

Recordings of the brain's electrical activity. These lines are parts of electroencephalograms taken from someone awake and in three different stages of sleep. Each recording covers 20 seconds. The electrical activity is recorded by electrodes attached to the skin.

28-K   Sleep

Sleep is another mysterious nervous function. Most mammals and birds sleep, although many of them keep more reflexes active than humans do and remain upright (Figure 28-21). Reptiles, amphibians, and fishes also show periods when they are very unresponsive to stimuli. Sleep is undoubtedly related to circadian (daily) rhythms, which all animals display (Section 31-G). Indeed, the daily cycle of body temperature is synchronized with sleep, and most people wake spontaneously when their body temperature begins to rise.

With all the progress of biology, we still have no idea why we must sleep each night, nor why sleeping has such a profound effect on our temper, efficiency, alertness, ability to learn, and emotional stability. Nor do we know why some individuals need more sleep than others. Guinea pigs and humans with brain damage have lived for years without sleep, so sleep is obviously not necessary to survival. Since a sleeping animal is highly vulnerable to predators, sleep must have some powerful counteracting selective advantage.

All we can do now is describe the changes that occur during sleep. Heart rate and blood pressure drop, and breathing becomes more shallow. Body temperature drops slightly, but the temperature of the big toes rises about 5°C. More important, changes in the brain's electrical activity (Figure 28-22) can be detected by way of electrodes taped to the head. The cessation of a pattern called "alpha waves" indicates the onset of sleep. The patterns of brain activity recorded from a sleeping person show four different stages of sleep. The most striking of these is **REM** or **paradoxical sleep,** which occurs every 80 to 120 minutes.

REM is an acronym for "rapid eye movements." During REM sleep the reticular activating system (Section 28-F) arouses the brain, and brain activity resembles that of someone who is awake. Paradoxically, though, motor activity is strongly inhibited; the muscles are more relaxed, and the sleeper harder to awaken, than at any other stage. People awakened from REM sleep nearly always recall dreams, although dreaming can also occur during the other stages of sleep. REM sleep seems to be the sleep stage most crucial to our psychological well-being; people deprived of REM sleep become extremely tired, and they compensate for loss of REM sleep by increasing periods of REM sleep on subsequent nights. Alcohol, barbiturates, and morphine all decrease REM sleep.

The concentrations of various neurotransmitters in the brain are different in waking and in different kinds of sleep. Norepinephrine, dopamine, and acetylcholine occur at higher levels during wakefulness, whereas serotonin levels rise in the brain during sleep. Cause and effect are difficult to distinguish here: does

Alzheimer's Disease

A JOURNEY INTO LIFE

Alzheimer's disease is expected to provide considerable insight into the basis of human memory. This disease, characterized by progressive memory loss, afflicts up to 10% of Americans over age 65. Aside from its medical importance, Alzheimer's is of particular interest because it appears to affect only memory and reasoning ability. Other brain functions, and the rest of the body, are normal until late stages of the disease.

We do not yet know how Alzheimer's disease causes loss of memory, but the changes in the brain have been described. An Alzheimer patient's brain (and blood vessels) contain abnormal deposits of glycoproteins called amyloid proteins, which are commonly found in smaller amounts in the brains of elderly people. In Alzheimer patients, some of the amyloid deposits are found within neuritic plaques, clusters of degenerating axons of neurons that produce the transmitter acetylcholine. Also present are tangles of fibers. These changes occur particularly in the hip-pocampus (below the cerebral cortex), the cerebral cortex, and the brain stem (see Figure 28-15). The brain also atrophies (shrinks). The same changes are also found in the brains of individuals with Down's syndrome after about age 35.

After years of work, researchers think that these features are merely symptoms resulting from some as yet unknown underlying cause, possibly not even originating in the brain. There are at least two forms of Alzheimer's disease, and one is known to be genetic. People who develop the disease at a young age (under 50) usually have this form of the disease, thought to be inherited as a dominant gene. Their relatives are more likely to develop Alzheimer's than are members of the general population. In 1987, researchers found that this gene for inherited Alzheimer's is carried on chromosome 21, the same chromosome found in an extra copy in people with Down's syndrome (Section 13-J). Other researchers found that the gene for amyloid protein is in the same area of chromosome 21. However, it is not yet clear whether the Alzheimer's and amyloid genes are one and the same or merely near neighbors.

After years of work, researchers have found no support for the hypothesis that Alzheimer's is caused by an infectious agent such as a slow virus. The toxic element aluminum is found in higher than usual levels in Alzheimer's brains, but it is not clear whether this is a cause of the disease or an effect; perhaps the disease changes the blood vessel walls and permits more aluminum than usual to seep into the brain.

Efforts to replace the missing acetylcholine have not helped most Alzheimer's patients, and so the loss of acetylcholine-releasing neurons is now thought to be a side effect, not a cause of the disease. Other studies have looked at nervous pathways using other neurotransmitters, at mutations in mitochondria, and at the activity of the immune system, which gets involved in the disease at some stage in a manner not yet understood.

Some memory loss is a normal part of aging; that seen in Alzheimer's is more pronounced. Many other diseases produce similar symptoms, and patients must be carefully tested to rule out other disorders, many of which can be treated. No good treatment for Alzheimer's has yet been found, although drugs may be given for secondary symptoms such as depression and sleeplessness. The disease can be diagnosed for sure only by autopsy.

Alzheimer's disease is becoming more common because people in developed nations are living longer than they used to. We can expect research on the disease to provide important insights into memory in the future. So far, this research has tended to confirm what we already know: memory is a diffuse function that involves the healthy functioning of neurons in many different parts of the brain.

Some information used in this box was provided by the Alzheimer's Association, Chicago, Illinois.

an increase in serotonin put one to sleep, or does sleep increase the activity of neurons that produce serotonin?

It seems likely that sleep permits the restoration of biochemical functions depleted by the day's activity and also permits the processing and reorganization of information already present in the nervous system, but how and why these things happen will probably take a long time to discover.

SUMMARY

Information passes along a neuron in the form of fast-moving electrical changes across the plasma membrane. To transmit its messages, the neuron uses the stored energy of its resting membrane potential, the result of a high external concentration of sodium ions (Na^+) and a high internal concentration of potassium ions (K^+). A stimulus applied to the neuron changes its membrane's resting potential briefly by opening the gates of ion channels through the membrane. This allows certain ions to cross the membrane and depolarize or hyperpolarize that part of the membrane. Such small local changes in the membrane potential in the dendrites or cell body may add up until they exceed the threshold needed for the axon to produce an action potential. An axon transmits action potentials faster if it has a large diameter or if it is electrically insulated by a myelin sheath. When an action potential reaches a synapse, its message is usually transmitted as a chemical that crosses the synaptic cleft and disturbs the electrical balance of the next cell(s) in line.

During the evolution of the nervous system, progressive cephalization resulted in the formation of a brain and major sense organs at the front end of the body. A vast increase in the number of neurons permitted better control of the many muscles in a large body and more flexibility of response to stimuli in a complex and changing environment.

The nervous system of a vertebrate consists of the brain and spinal cord, which together compose its central nervous system, and the peripheral nervous system in the rest of the body.

The vertebrate brain has three major parts: the forebrain, midbrain, and hindbrain, whose functions are summarized in Table 28-1. During evolution, the vertebrate brain increased in size and complexity. Some parts of the brain retained their primitive functions, while others took on new functions as body structure and behavior became more complex, and as intelligence increased. The brain's main role is to "make decisions." Using information coded as patterns of action potentials coming from the external or internal environment via the sense organs, the brain produces a set of directions coded as another set of action potentials that cause the effector organs to respond. Information passes through various levels of organization in both sensory and motor areas as the brain analyzes and integrates sensory input and determines and executes appropriate responses.

The spinal cord is primarily a relay station connecting the brain with the peripheral nervous system, although there are spinal reflexes in which sensory and motor components interact through the spinal cord without input from the brain. These reflexes make quick, local responses to potentially harmful stimuli without waiting for analysis by higher levels in the brain.

The peripheral nervous system is divided into the somatic nervous system, which largely serves the muscles under conscious control, and the autonomic nervous system, which carries motor impulses to the muscles and glands of the internal organs, under little conscious control.

Memory is the way the brain stores patterns of information that may be useful later. It takes the form of temporary or permanent changes in some of the synapses in neuron circuits that pass through many parts of the brain.

Sleep is a mysterious but apparently necessary function of the nervous system.

SELF-QUIZ

1. Impulses leave a neuron via the:
   a. dendrites
   b. nucleus
   c. myelin sheath
   d. axon
   e. cell body
2. The myelin sheath around the axons of some vertebrate neurons:
   a. is rich in lipids because it is formed by many layers of membranes
   b. is a secretory product of glial cells
   c. is produced inside the axon and extruded out through the membrane
   d. is continuous all along the length of the axon
   e. secretes neurotransmitter substances for release at the axon terminals
3. Write a short sentence describing the function and importance of each of the following components of a synapse:
   a. neurotransmitter substance
   b. neurotransmitter vesicle
   c. receptor molecules
   d. enzymes that destroy neurotransmitters
4. In an inhibitory synapse:

a. information travels from the postsynaptic to the presynaptic cell
b. the neurotransmitter used is different from the neurotransmitter used in excitatory synapses
c. there are no receptor molecules on the postsynaptic membrane
d. the postsynaptic receptor molecules may cause hyperpolarization of the membrane rather than depolarization when they bind neurotransmitters
e. the stimulus is transmitted electrically, not chemically
5. Regulatory control of deep body temperature, osmoregulation, thirst, and hunger occurs in the:
   a. anterior colliculi
   b. hypothalamus
   c. thalamus
   d. cerebellum
   e. tectum
6. In looking at the evolution of the brain from fishes to humans, the greatest increases in size are seen in the:
   a. medulla and cerebellum
   b. cerebellum and tectum
   c. tectum and cerebral hemispheres
   d. cerebral hemispheres and cerebellum
   e. medulla and thalamus

7. The reticular formation:
    I. acts as a filter for sensory input
    II. regulates mechanisms of arousal
    III. affects spinal cord activity via a feedback system
    a. I only
    b. II only
    c. I and II
    d. I, II, and III

8. Label the parts of the reflex arc indicated by letters A through E.

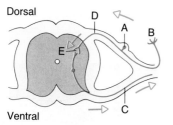

9. The parasympathetic nervous system:
    a. uses norepinephrine as the neurotransmitter at the synapse with the effector
    b. is part of the central nervous system, located in the spinal cord
    c. tends to lower blood pressure, slow the heartbeat rate and breathing, and promote digestion
    d. predominates over the sympathetic system in times of stress
    e. is composed of sensory neurons running parallel to the motor neurons of the sympathetic system

10. The major transmitter between neuron and effector in the sympathetic system is ____, while in the parasympathetic system it is ____.

QUESTIONS FOR DISCUSSION

1. After an action potential arrives at the presynaptic membrane, what factors determine whether the postsynaptic cell will fire an action potential?

2. Why don't all the neurons in an animal's nervous system have either giant or myelinated axons? What is the selective advantage of having some neurons with axons that conduct impulses more slowly?

3. Tetrodotoxin is a chemical produced by puffer fish that interferes with the opening of sodium gates involved in transmission of an action potential. It has no effect on chemical synapses. Why is tetrodotoxin a powerful poison?

4. Multiple sclerosis (MS) is characterized by patchy destruction of myelin. What symptoms would you expect this to produce?

5. What possible disadvantages are there in the evolutionary trend toward cephalization of the nervous system? What advantages?

6. During embryonic development of mammals, the nervous system forms all the neurons an animal will ever have. After birth, neurons never divide to form new neurons. However, synapses do form and decay. What is the adaptive advantage of this state of affairs? What are some drawbacks?

7. The human visual cortex uses increasing amounts of energy under the following series of conditions: eyes closed; looking at diffuse white light; looking at a checkerboard pattern; looking out the window at a park. Explain these findings using the information on visual processing in Section 28-F.

8. What is the advantage of having the ability to form habits before one can lay down memories? What are the disadvantages of this state of affairs?

SUGGESTED READINGS

Adrian, R. H. "The nerve impulse." *Carolina Biology Reader.* Burlington, NC: Carolina Biological Supply Company, 1980.

Axelrod, J. "Neurotransmitters." *Scientific American,* June 1974. Covers neurotransmitters and aspects of drugs affecting the nervous system.

Franklin, J. *Molecules of the Mind.* New York: Atheneum, 1987. A thought-provoking look at the neurochemical revolution, mental illness, and society's attitudes to both.

Julien, R. M. *A Primer of Drug Action,* 5th ed. New York, W. H. Freeman, 1988.

Lent, C. M., and M. H. Dickinson. "The neurobiology of feeding in leeches." *Scientific American,* June 1988. An interesting and readable case study of the nervous system's role in a vital activity. Outlines experiments that detected the nervous pathway responsible for feeding.

Mishkin, M., and T. Appenzeller. "The anatomy of memory." *Scientific American,* June 1987.

Scientific American. "The Brain." September 1979 issue. A collection of articles about neurons and brain structure and function.

Sense Organs

CHAPTER

29

Fly eye

Platypus

Blue shark

Bullfrog

*S*ense organs permit an animal to detect changes in its own body and objects and events in the world around it. Information collected by sense organs is passed to the nervous system, which determines and initiates an appropriate response.

A popular expression holds that we have five senses. In fact, we have more than a dozen different types of sense organs, which monitor conditions both outside and inside our bodies. Internal sense organs detect and report changes in conditions such as body temperature, osmotic relationships, and pH. This provides information used to maintain homeostasis. Sense organs that report sights, sounds, and chemicals in the outside world are used in feeding, finding mates, avoiding enemies, and making other adaptive responses to the environment.

Our studies of other animals' sense organs are inevitably biased by our own senses, which give us only one of many possible perceptions of the world. Humans are creatures of vision and, to a lesser extent, of hearing. It is hard for us to empathize with, say, an earthworm, which perceives neither line nor color.

Many animals have sense organs far more sensitive than ours. Some even respond to stimuli that we cannot detect at all. Bats flying about in total darkness can avoid obstacles, and platypuses can catch prey in murky water at night with their eyes, ears, and nostrils closed. Until recently, people were baffled by these powers, which humans could not duplicate without sophisticated electronic equipment. Now we know that bats emit squeaks in the ultrasonic range, too high-pitched for us to hear. They detect the faint echoes of these sounds with hearing so acute that they can skirt telephone wires and catch flying moths on the darkest nights. The platypus's duck-like bill can locate freshwater shrimp and insect larvae by detecting the weak electric currents they generate. We cannot imagine what it would be like to have such senses. In terms of perception, we live in a different world from that of the worm burrowing in the garden or the bat roosting under the eaves. Nevertheless, there are many similarities in the types of sense organs found among different animals and in how these organs work.

- Sense organs gather information about changes in the animal's body and in its external environment and pass it to the nervous system for evaluation.
- The particular collection of sense organs in each animal species gives its members a unique perception of their bodies and of their environment.

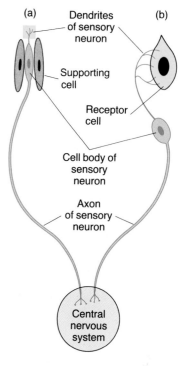

Figure 29-1
Receptors and sensory neurons. Sensory neurons convey information from sense organs to the central nervous system. The receptors for stimuli (yellow) are either (a) the dendrites of the sensory neuron itself, or (b) a non-nervous receptor cell that synapses with a sensory neuron's dendrites.

KEY CONCEPTS

29-A   Sense Organs and Their Functions

The key elements of sense organs are **receptor cells.** In some sense organs, the receptor consists merely of the dendrites of the sensory neuron. In others, the receptor is a separate, non-nervous cell close to the sensory neuron's dendrites (Figure 29-1). Either way, receptor cells respond to stimuli by producing electrical activity in the nervous system. A **stimulus** is some form of energy; various receptors respond to energy in the form of pressure, light, electric current, chemical changes, osmotic potential, and heat (Figure 29-2).

Each sense organ is particularly sensitive to one type of stimulus. The eye, for example, responds to light but not to sound waves or chemicals in the air, which also strike the eye. Sense organs also function well over a very wide range

Figure 29-2
Sensing the environment. The sense organs of most animals are concentrated in the head. The head of a lion bears organs that sense mechanical stimuli (ears and whiskers); light (eyes); heat (tongue); and chemicals (nose and tongue).

Figure 29-3
An optical illusion. Each stripe in this series is of uniform darkness. However, the human visual system is set up in such a way that it accentuates edges. Hence it misinterprets these stripes, perceiving each one as having a brighter edge where it meets its darker neighbor.

■ *Receptors absorb stimulus energy and convert it to electrical energy, the form "understood" by the nervous system.*

of stimulus levels. Many sense organs are also incredibly sensitive to stimuli. The receptor cells called rods in our eyes can detect a single photon (the smallest quantity of light), and a male gypsy moth can detect a single molecule of the female's sexual attractant chemical that comes into contact with his antenna!

Most sense organs are geared to alert the nervous system to *changes* in stimuli. Our eyes send particularly strong signals to the brain when they detect movement, changes in light intensity (such as a sudden shadow or flash of light), and edges between light and dark areas (Figure 29-3). This is adaptive. Detecting movements of predators or prey, and outlines of objects in the surroundings, is more important to an animal than detecting a constant stimulus.

An animal responds to a stimulus in a three-step process:

1. **Receptor cells** are changed by the energy of a stimulus in such a way that the stimulus energy is converted into electrical activity, which can travel in the nervous system. These events are covered in this chapter.
2. The **nervous system** processes the electrical impulses it receives from all of the sense organs (Chapter 28). Eventually the resulting signals pass to appropriate effectors.
3. **Effectors** produce suitable responses. Effectors are usually muscles or glands. Muscles contract, or glands secrete chemicals, as a result of the information passed to the nervous system by the sense organs (Chapters 30 and 31).

"What about when I just sit and listen to music?" you may ask. "The music goes in (step 1) and the nervous system processes it (step 2), but no response occurs—I just enjoy." Not so; in fact, tiny muscles in your ears react to noises and adjust the ears' sensitivity, depending on the loudness of the music. In addition, depending on the type of music, your heart may beat faster or slower, and your other muscles may become more tense or relaxed. Even stopping the natural urge to tap your feet to the beat is a response!

Receptors may be classified by the form of stimulus energy they detect:

Mechanoreceptors detect mechanical energy in the form of movement, pressure, or tension (pull or stretch).
Photoreceptors detect light energy.
Thermoreceptors detect heat.
Chemoreceptors detect chemicals.
Electroreceptors detect electrical energy (Table 29-1).

Table 29-1 *Classification of Some Receptors by Stimuli*

General name	Examples	Effective stimulus
Mechanoreceptors	Meissner's corpuscles	Touch on skin
	Proprioceptors	Position of parts of body
	Hair cells of vertebrates	
	Lateral line organs in fish	Pressure waves and currents in water
	Utriculus and sacculus	Gravity; linear acceleration
	Semicircular canals	Angular acceleration
	Cochlea	Airborne sound waves
Photoreceptors	Ommatidia of arthropods	Wavelength of light
	Rods and cones of vertebrate retina	Wavelength of light
Thermoreceptors	Pit organs of pit vipers; nerve endings in the tongue and skin of mammals	Increasing and decreasing infrared radiation
	Krause's end bulbs	Cold on skin
Chemoreceptors	Taste buds and olfactory organs of vertebrates	Unknown features of the chemistry of specific molecules in air or water
	Chemoreceptors of invertebrates	Same
Electroreceptors	Organs in the skin of some fish; bill of platypus	Electric currents in surrounding water

Many animals, such as pigeons, dolphins, and honeybees, can also respond to the earth's magnetic field.

A receptor's most important role is to convert the energy of a stimulus into electrical energy, the only form of energy that can be transmitted by the nervous system. The receptor produces a local electrical potential, called a **receptor potential,** with a magnitude proportional to the intensity of the stimulus. If the receptor potential reaches the sensory neuron's threshold, it will cause the neuron's axon to fire action potentials (also called nerve impulses, Section 28-B).

29-B *Coding and Interpretation of Messages*

A nerve impulse is an all-or-none phenomenon; it either occurs or it doesn't. However, the stimulus that initiates a nerve impulse may be large or small, and animals do detect differences in intensity of stimuli. How can a neuron with only two modes—either "on" (transmitting an action potential) or "off" (not firing)—send information about the magnitude of stimuli?

The nervous system encodes stimulus strength in three ways (Figure 29-4):

1. **Frequency of action potentials.** A weak stimulus initiates only a few action potentials per second, a strong stimulus many. The frequency has an upper limit because a neuron cannot fire while it is depolarized from a previous action potential. However, stimuli greater than threshold intensity will cause the neuron to fire again before it has completely recovered from firing a previous action potential.

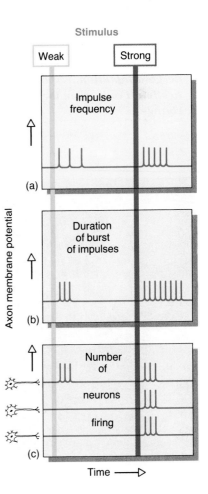

Figure 29-4
Coding of stimulus intensity by neurons that produce only all-or-nothing responses. A weak stimulus is given at the time indicated by the light blue line, a strong stimulus at the dark blue line. A strong stimulus will cause a neuron to (a) fire impulses more rapidly or (b) fire a longer burst of impulses. (Impulses are shown as spikes on the graphs.) Another possibility (c) is that more cells will fire.

■ *Sensory neurons encode stim-*
ulus intensity by:
1. frequency of nerve im-
pulses;
2. duration of a burst of im-
pulses;
3. number of neurons firing.

2. **Duration of a burst of action potentials.** A weak stimulus may give rise to a short burst of pulses in the neuron, a strong stimulus to a longer burst.

3. **Number and kinds of neurons firing.** The threshold needed to initiate a nerve impulse varies from one neuron to another. Thus, a weak stimulus will cause only a few neurons to fire, whereas a stronger stimulus will fire all of these neurons, plus others with higher thresholds.

The nervous system receives all of its information in the form of trains of action potentials. How does the brain know whether an incoming train of action potentials indicates a light shining, a sound, pressure on the skin, or an odor? Such qualitative information is built into the system in the synaptic connections ("wiring") of the nervous system: a neuron "knows" that it is responding to light when it gets a message from a neuron that receives its information from light receptor cells. Neurons on the visual pathways may send impulses for other reasons, for instance, when something hits you on the head or when you press your fingertips gently against the corners of your eyes. In these cases you "see stars" because the brain interprets these pressures as light, even though the eye received no light.

The synaptic connections of neurons also convey other information. For example, a particular part of the brain receives impulses only from neurons in a certain part of the eye. This "tells" the brain the angle at which the light entered the eye. There are also particular neurons that send signals only when the stimulus impinging on the eye has a certain size or shape, or is moving in a certain way (Section 28-F). For example, some cells in the visual cortex of the brain signal only when the eye sees a line with a specific orientation. Thus the brain "knows" that the eye has seen that particular pattern.

Thus, a relatively simple event, the passage of an all-or-none action potential, can be used to transmit information about the intensity and type of stimulus by using the timing of impulses and the wiring of neurons as additional sources of information.

■ *Receptors and neurons un-*
dergo adaptation, responding
to stimuli only when they
change.

If a constant stimulus is applied to a receptor for any length of time, the sensory neuron shows a diminishing response as time passes. This **sensory adaptation** allows an animal to ignore an unchanging stimulus. For example, when you dress in the morning you notice the contact between your clothing and skin. This sensation rapidly disappears, and you are not distracted by the feel of your clothes all day. However, if a fly crawls down your neck, your skin's touch receptors respond immediately to this new stimulus.

29-C Mechanoreceptors

Animals have many kinds of mechanoreceptors, which respond to many kinds of movement, including changes in pressure. The stretching of the receptor membrane causes the receptor potential.

Invertebrates have receptors at the bases of tactile hairs, which are sensitive to touch (Figure 29-5), and receptors that detect the pull of gravity. Some insects also have "ears," which detect sounds (pressure waves in the air).

Vertebrates have many different types of mechanoreceptors. Human touch receptors are concentrated on the tongue, lips, face, and fingertips, reflecting the importance of the head and of the manipulative fingers. Inside the body, we find pressure receptors that constantly monitor blood pressure in the large arteries. **Proprioceptors** are mechanoreceptors that continually monitor the position and movements of parts of the body. Vertebrates have three main types

Figure 29-5
Hair receptors on the back of a caterpillar. (Biophoto Associates)

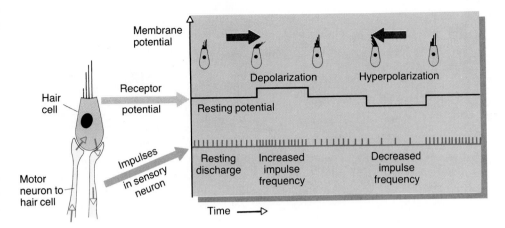

Figure 29-6

Response of a vertebrate hair cell. When the cell is not stimulated, it produces a resting discharge of impulses (action potentials) in the sensory neuron (vertical marks on the red graph line). When its tuft of microvilli is bent one way (black arrow pointing right), the cell produces a depolarizing receptor potential, which increases the impulse frequency in the sensory neuron. When the tuft is bent the other way, the cell becomes hyperpolarized. This decreases the frequency of firing in the sensory neuron. The frequency of impulses in the sensory neuron tells the nervous system which way, and how much, the tuft is bent. The nervous system can change the hair cell's sensitivity to stimulation by sending signals via the motor neuron (blue arrows).

of proprioceptors—in the joints, tendons, and muscles—which detect movement and degree of stretch.

Hair Cells of Vertebrates

Vertebrates have remarkable mechanoreceptors called **hair cells,** each bearing a tuft of large microvilli of graduated lengths (Figure 29-6). Bending this tuft appears to open gated ion channels in the receptor cell membrane and allow ions to flow through the channels. This rapidly produces the electrical activity of a receptor potential in the hair cell, and then in the dendrites of a sensory neuron near the base of the hair cell. Let us look at a few examples of organs that contain hair cells.

Lateral Line Organs The **lateral line organs** of fish and larval amphibians contain hair cell pressure receptors. The lateral line consists of a row of water-filled canals or tunnels extending along the animal's side and onto its head (Figure 29-7). In the canals are hair cells, with their microvilli tufts embedded in a gelatinous substance to form clumps called **cupulae** (singular: **cupula**). When the water in a lateral line canal moves, it pushes a cupula and bends the tuft, generating a signal in the sensory neurons. The lateral line organ is a remarkably sensitive system by which the animal can detect such things as water currents, the movements of other animals in the water, and pressure waves bouncing off stationary objects nearby.

Labyrinth of the Ear The front end of the lateral line organ evolved into the **labyrinth** found in the inner ears of all vertebrates (see Figure 29-9). Hair cells in different parts of the labyrinth detect sounds, the direction of gravity, and

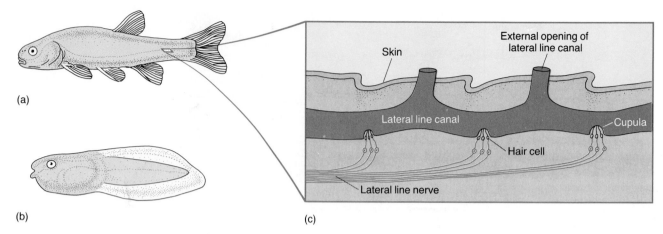

(a)

(b)

(c)

Figure 29-7
The lateral line system. (a, b) Blue
dots show the distribution of openings into the lateral line canal.
(c) Longitudinal section through the
lateral line organ. The lateral line
canal connects with the water outside
the fish via openings in the skin. The
actual pressure receptors are hair
cells with their tufts embedded in
cupulae lining the canal. Dendrites
from sensory neurons synapse with
the hair cells and detect their signals.

acceleration of the head. (**Acceleration** is a change in the speed or direction of
motion.)

Two chambers, the **sacculus** and **utriculus,** contain hair cells that detect
both gravity and linear acceleration (changes in speed when the body is moving
in a straight line). Each chamber contains crystals of calcium carbonate, which
shift in response to gravity and stimulate the tufts of the hair cells.

The three **semicircular canals** detect angular acceleration (rotation, or
turning). They are the sense organs of equilibrium. Since the canals lie in three
different planes, they can detect acceleration of the head in any direction. Each
canal is a hollow ring filled with fluid. At one point in the ring is a cupula, which
operates like a swinging door across the canal. Embedded in each cupula are the
microvilli tufts of hair cell mechanoreceptors.

The semicircular canals can detect changes only in the speed or direction of
rotation of the head; they do not react if the head is rotating at a constant speed
or moving in a straight line. Travelling in a car at constant speed in one direction
gives no sense of movement; only if the car turns do the semicircular canals
relay a sensation of motion. This is due to the inertia of the fluid in the canals.
When the head is stationary, or is moving at a constant speed or in a straight line,
the fluid does not move with respect to the canals. When the head changes
direction, however, the fluid tends to keep going in its original direction, and so
it pushes the cupula into a new position (Figure 29-8).

Hearing The **cochlea** of the inner ear contains another set of mechanoreceptor hair cells, which detect pressure waves in the fluid inside the cochlea. In
terrestrial vertebrates, accessory structures in the outer and middle ear transform the stimulus of sound waves in air to pressure waves in the cochlear fluid.
In the human ear, for example, air-borne vibrations strike the **tympanic membrane,** or eardrum. Vibration of the tympanic membrane moves three small
bones that span the cavity of the middle ear. The third bone presses against the

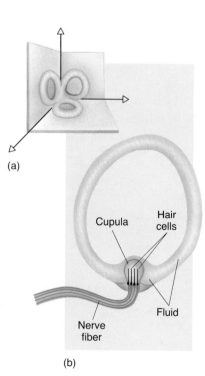

(a)

(b)

Figure 29-8
Semicircular canals in vertebrates. (a) Semicircular canals are arranged in three
planes at right angles to each other. Angular acceleration of the head is interpreted
by combining the information received about changes in rotation speed or direction
detected in each plane. (b) Each semicircular canal contains a cupula, located in an
expanded chamber. Hair cell tufts embedded in the gelatinous material of the cupula
are bent by movement of the fluid. Bending the tufts one way increases the rate of
action potentials in the nerve fiber; bending the opposite way decreases the rate of
action potentials.

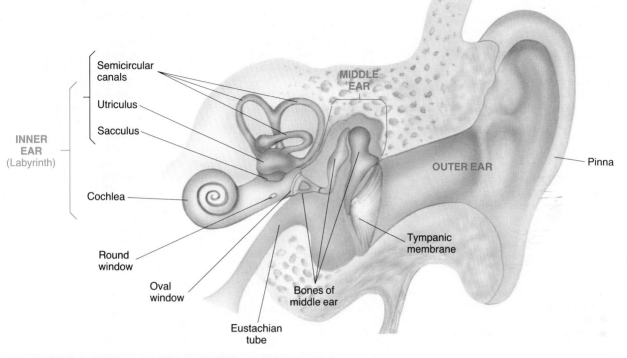

Figure 29-9
The human ear. Sound waves enter the outer ear, causing vibration of the tympanic membrane, which transmits them to the three bones in the air-filled middle ear. This, in turn, causes vibrations of the oval window, and hence of the fluid in the cochlea in the inner ear. The inner ear also contains the semicircular canals, utriculus, and sacculus, where acceleration and gravity are detected.

oval window of the cochlea. Vibration of the oval window in turn moves the fluid inside the cochlea (Figure 29-9).

The entire cochlea is shaped like a snail shell, a coiled tube of gradually changing diameter. This shape permits the nervous system to tell the pitch of a sound—how high or low it is. Each pitch produces a maximum vibration at one point in the cochlea, with a particular width, and the hair cells at that point are stimulated most intensely. The brain "knows" the pitch of the sound because it knows the location of the sensory neurons that are firing in response to signals from these hair cells. The volume (loudness) of the sound is signalled by the frequency of action potentials in the neurons. The ear is incredibly sensitive: the faintest sounds we can hear move the tips of the hair cells only about the diameter of a helium atom!

The hair cells in the cochlea are delicate and easily damaged by loud noises, such as banging, the roar of motors, and even loud music, be it rock or classical. As we age, we progressively lose hair cells, and therefore hearing, especially at higher pitches. Much of this hearing loss can be attributed to noise (see *A Journey Into Life,* "Sounds Make Silence," found in this chapter).

29-D   *Photoreceptors*

Almost all animals have **photoreceptors,** which absorb light energy by using **pigments,** colored molecules that undergo chemical changes when they are struck by light. In flatworms and annelids, photoreceptors are merely **eyespots,**

Calling noise a nuisance is like calling smog an inconvenience. Noise must be considered a hazard to the health of people everywhere. (William H. Stewart, former U.S. Surgeon General)

One hazard of environmental noise is hearing loss. A baby is born with up to 20,000 hair cells in each cochlea—not very many when you consider that this is the total lifetime supply; damaged or killed hair cells are never replaced. Loud noise damages the hair cells' microvilli by disrupting their internal skeletal framework, so that they flop like wet noodles. Noise also causes the microvilli surfaces to develop blister-like vesicles, which rupture as the noise continues. Once the damage starts, it takes *less* sound pressure to do an equal amount of additional damage.

Hearing loss caused by noise is insidious because it is painless and so gradual that it is not noticed. At first, sounds must simply be louder (or closer) in order to be heard. Since the first hair cells to die detect high pitches, above the range of the human voice, victims can still hear conversations. Then the higher-pitched speech sounds fade out, and these people can no longer distinguish between similar words (this often gives rise to complaints that speakers are mumbling). Hearing loss is so widespread that it is regarded as a natural part of aging. However, much of the blame should be placed

on noises of modern civilization. The elders of a Sudanese tribe living in a quiet Stone Age culture hear better at 80 than most Americans do at 30.

Tests of over 4000 students entering college in the early 1980s showed that more than one in three already had measurable loss of hearing for high-pitched sounds. These were mostly not serious yet, but today's young people can expect to suffer greater hearing loss by the time they reach their 50s and 60s than we see in people now in that age group.

Noise is the most widespread occupational hazard. The federal government regulates noise in the workplace, and employers have found it cheaper to provide workers with ear protectors than to pay damage awards from lawsuits for work-related hearing loss (Figure 29-A). However, many workers do not realize the damage noise can do and do not use the equipment properly. Everyone is exposed to dangerous levels of noise off

the job. The worst offenders are gunfire (from military duty, target shooting, or hunting), vehicles (aircraft, subways, heavy city traffic, snowmobiles, and speedboats), and loud music. The noise level in front of a speaker at a rock band concert is equivalent to that of a thunderclap, and efficient stereo headphones, turned to maximum volume, can deliver sound louder than a jet taking off at the other end of a football field.

Hearing loss is not the only biological damage done by noise. It also causes symptoms of stress—high blood pressure, increases in the heartbeat and breathing rates and in muscle tension, and changes in the digestive tract that can lead to ulcers. The brain can learn to tune out noise, so that it no longer reaches the level of conscious awareness, but the body cannot adapt in this way. Prolonged exposure to loud noise keeps the body in a constant state of stress.

Figure 29-A
Ear protection. Employers are required to supply employees in noisy jobs, such as airport baggage handlers, with equipment to protect their hearing. (Karen Roeder)

which detect light as opposed to darkness. **Eyes** detect more detail. In an eye, light from the visual field projects an image onto light-sensitive receptor cells. True eyes occur only among vertebrates, arthropods, and molluscs.

In the eyes of vertebrates, the curved **cornea** focuses light entering the eye (Figure 29-10). The focus is adjusted by the **lens,** which is somewhat elastic and changes shape when it is pulled by muscles around its edges. The **iris** is a diaphragm that controls the size of its central opening, the **pupil,** and so regulates the amount of light entering the eye. The **retina** is a delicate layer contain-

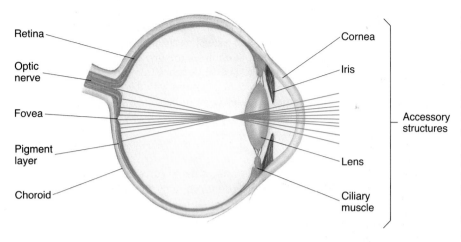

Figure 29-10
The human eye. The eye has been sectioned through its equator and the top half removed. The retina contains the receptor cells and their associated neurons, which detect light (blue lines) focused by the cornea and lens. Neurons in the eye send messages to the brain via the optic nerve. The dark pigment layer absorbs light that has passed through the retina, thereby preventing it from reflecting back to the receptors (this would blur the image). The choroid layer contains blood vessels. The ciliary muscle adjusts the curvature of the lens.

ing the photoreceptors and their associated neurons. Visual information passes to the brain by way of the **optic nerve.**

The actual photoreceptors of the retina are specialized receptor cells called **rods** and **cones** (Figure 29-11). Each receptor synapses with the dendrites of a sensory neuron.

Rods can detect very dim light, but produce poorly defined images with little discrimination among colors. In bright light, we use the cones and their associated neurons, which give good resolution for detail. The cone system is also much better at interpreting color. Cones are especially densely packed in an area of the retina called the **fovea;** we tend to move our eyes so that the image of the object we want to see most clearly falls on the fovea. The fovea contains only cones, and most of the other cones of the retina are nearby. Rods are most concentrated in a ring about 20° from the fovea. This distribution explains why you can see faint stars at night out of the corner of your eye, but not when you move your eyes to look straight at them. These dim stars seem to disappear because the cones in the fovea cannot detect them.

The light-absorbing molecule in animals' eyes is **retinal.** The raw material for retinal is vitamin A, which in turn is formed by breaking down beta-carotene,

Figure 29-11
Rods and cones in the vertebrate retina. Pigment-filled rods and cones (left) are the actual receptor cells. They synapse with bipolar neurons, which synapse with ganglion cells, neurons whose cell bodies lie in the retina and whose axons form the beginning of the optic nerve. Note that light entering through the lens (blue arrows, far right) must pass through several layers of cells before it reaches the receptors.

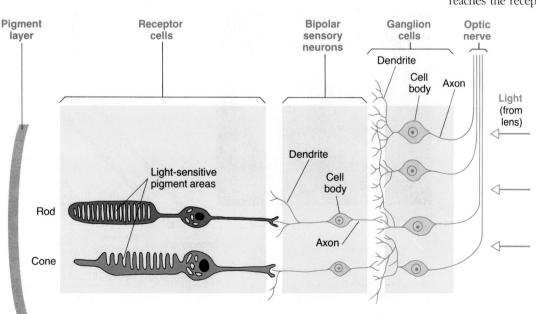

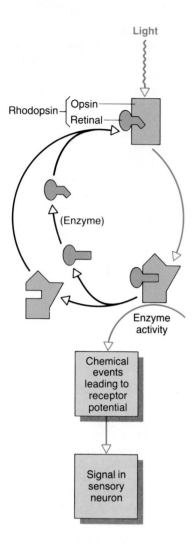

Light

Rhodopsin { Opsin / Retinal }

(Enzyme)

Enzyme activity

Chemical events leading to receptor potential

Signal in sensory neuron

Figure 29-12

The visual cycle. Red arrows show the path by which the stimulus of light is detected in the nervous system. Light causes retinal to change shape, converting rhodopsin to an enzymatically active form. It catalyzes the first reaction in a series that leads to a signal in the sensory neuron. Retinal detaches from opsin and an enzyme converts it back to the bent shape, which recombines with the protein opsin to form the visual pigment rhodopsin, ready to detect more light.

a common plant carotenoid. Animals cannot make carotenoids but must obtain these crucial precursors of visual pigments directly or indirectly from plants.

A visual pigment consists of a retinal molecule nestled inside a protein. Different pigments have slightly different proteins, and this makes the pigments sensitive to different wavelengths of light. The pigments in our eyes absorb wavelengths of about 400 to 750 nanometres, which are therefore called visible light.

Retinal can exist in two molecular forms. In rods, one of these combines with the protein **opsin** to form the visual pigment **rhodopsin.** When light strikes rhodopsin, it causes retinal to change to its other form. This change in shape shifts rhodopsin into an enzymatically active form, which starts a chain of chemical events leading to impulses in the sensory neuron that synapses with the rod. Retinal detaches from opsin, and is converted back to the first form. It can then recombine with opsin to form rhodopsin again. This completes the sequence of reactions called the **visual cycle** (Figure 29-12).

Figure 29-13

Compound eyes. (a) A wasp's compound eyes allow it to view most of its surroundings at once and to detect slight movements as new patterns of ommatidia become excited. The head also bears many hair-like pressure receptors, which respond to air movement or touch. (b) Ommatidia are the units of compound eyes. Each ommatidium consists of a cornea, a lens, and a light-sensitive rhabdom surrounded by retinal cells, which transmit the sensory stimulus. Pigment cells around each ommatidium prevent light from passing into surrounding ommatidia. Only light rays that are parallel to the rhabdom will excite the retinal cells. (a, Biophoto Associates)

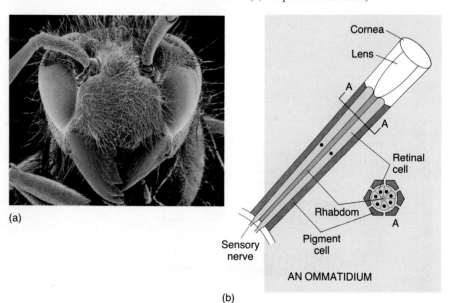

(a)

Cornea

Lens

A

A

Retinal cell

Rhabdom

A

Sensory nerve

Pigment cell

AN OMMATIDIUM

(b)

Cones have a similar system but require more light energy in order to produce a response. Because a series of chemical reactions occurs between absorption of light and production of a receptor potential, rods and cones respond slowly to stimuli compared with hair cells in the ear.

All vertebrates seem to have at least a sprinkling of cone cells in their retinas, but many of them do not respond to different colors (that is, they are in effect colorblind). An animal with color vision has two or more types of cone cells, each producing a different pigment. Humans and other primates (monkeys, apes) have three types. The different pigments absorb light of different wavelengths—red, green, and blue—most strongly. Therefore, each kind of cone is most sensitive to a particular range of wavelengths. The visual system detects color by comparing the levels of excitation in cones containing different pigments. Various degrees of colorblindness result if one or more cone pigments is missing.

The compound eyes of many arthropods may be composed of as many as 20,000 units called **ommatidia** (Figure 29-13). Each ommatidium has a lens-like structure that focuses incoming light rays onto a receptor. Opaque walls around the receptor cut out light coming in from the side. Hence each receptor detects only a narrow beam of light parallel (or nearly so) to its long axis. Such an eye appears to present a rather crude mosaic picture of the outside world. Its main advantages are that it permits an animal to see things very close to its eyes and to detect rapid movement with a sensitivity about five times that of the human eye.

Although vision is the sense humans rely on most, other animals can see things we can't. The corneas and lenses of our eyes, which focus the light, also filter out ultraviolet light so that it does not reach the retinal receptors. Animals such as honeybees, which do not have ultraviolet filters in their eyes, can perceive and react to ultraviolet light (Figure 29-14). It is not clear whether any vertebrates can see ultraviolet light.

The light in the sky is polarized in different directions relative to the sun and to the observer. Squids, octopuses, amphibians, and sea turtles are among the many animals that can detect polarized light. Some animals use this ability for determining compass direction and so for navigation. This built-in equivalent of polarizing sunglasses is also very useful in eliminating the glare of sunlight shining through water or reflecting from its surface.

Figure 29-14
Different eyes see different things. The human eye sees a marsh marigold with no markings (left). An insect sees big patches on the same flowers (right, taken with ultraviolet-sensitive film) because its eyes react to ultraviolet light energy, whereas ours do not. (Biophoto Associates)

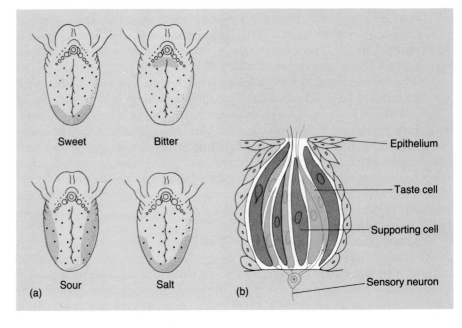

Figure 29-15
Chemoreceptors involved in the sense of taste in humans. (a) The human tongue, showing the distribution of taste buds sensitive to sweet (front), bitter (back), sour (sides), and salt (front and sides). (b) A taste bud in the tongue.

29-E Thermoreceptors

Thermoreceptors detect heat by responding to infrared radiation, which has longer wavelengths than visible light. Blood-sucking insects and ticks, which seek out warm-blooded animals, have well-developed heat detectors. But the most thoroughly studied thermoreceptors are the pits in the faces of pit vipers and the labial (lip) organs of some boas. These snakes can detect the body heat generated by a small mammal at distances of up to about 1 metre.

Mammals have thermoreceptors scattered over the surface of the body, particularly on the tongue. Thermoreceptors in the hypothalamus of the brain detect internal body temperature. Information from internal and external thermoreceptors is integrated in the hypothalamus to produce appropriate behavior such as shivering, sweating, or panting.

29-F Chemoreceptors

Chemoreceptors detect various chemicals. Our bodies have many internal receptors for chemicals such as nutrients, oxygen, carbon dioxide, hormones, and neurotransmitters (Chapters 23, 28, and 31). Our external chemoreceptors participate in the senses of smell **(olfaction)** and taste **(gustation).** The number of possible tastes is believed to be only four: sweet, sour, bitter, and salt. The "taste" of more complicated flavors is in fact a combination of olfactory and gustatory sensations. The importance of smell in "tasting" is evident during a bad cold; when your nose is stopped up, everything tastes like cardboard. The "hot" sensation of foods such as chili peppers is detected by pain receptors, not chemoreceptors.

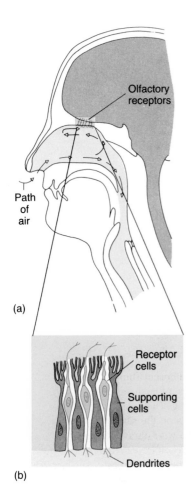

Figure 29-16
Olfaction. (a) Olfactory receptors lie in the top of the nasal cavity in a region of spongy tissue. To reach this area, air must travel through a winding passageway (arrows). (b) Chemoreceptors in the nose are free dendrites of neurons in the olfactory epithelium.

The vertebrate tongue is covered with numerous bumps, or **papillae.** In the grooves around the sides of a papilla lie the clusters of cells known as **taste buds.** A taste bud consists of sensory hair cells and supporting nonsensory cells (Figure 29-15). A single neuron sends branches of its dendrites to all of the hair cells in several taste buds.

A vertebrate's olfactory nerve endings lie in the upper reaches of the nasal cavity. Since this area is not on the main pathway to the lungs, air must travel a circuitous route to reach them. Sniffing helps the sense of smell by moving air more rapidly up into the olfactory area. Here olfactory receptors are quite densely packed; a dog has up to 40 million such endings per square centimetre. The dendrites of the olfactory sensory neurons extend into the air, and their ends have fringes where the initial chemoreception takes place (Figure 29-16).

Molecules in the air bind to specialized receptor macromolecules in the olfactory neuron's membrane, making the membrane temporarily more permeable to sodium ions and so generating a receptor potential. Taste works by a similar mechanism in the membranes of taste cells. Artificial sweeteners produce a sensation of sweetness because their molecular shape permits them to bind to taste cell receptors for sugars. The taste of salt is an exception; salt is not detected by receptors, but by the flow of sodium ions through sodium channels in taste cell membranes. Since these channels are specific for sodium, efforts to find a salty-tasting substitute for people on low-sodium diets have failed.

Some snakes and lizards have chemoreceptors in the **organ of Jacobson** in the roof of the mouth. The animal's tongue flicks out, gathers chemicals from the environment, and transfers them to the organ (Figure 29-17).

Among invertebrates, insects have the best-studied chemical senses. Insect chemoreceptors are found on the mouthparts, legs, and antennae (Figure 29-18).

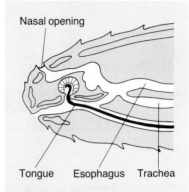

Figure 29-17
Chemoreception in the organ of Jacobson. A lizard or snake shoots out its tongue to pick up chemicals in the environment. The tongue is then drawn back into the mouth and inserted into the organ of Jacobson (red), which is lined with chemoreceptor cells.

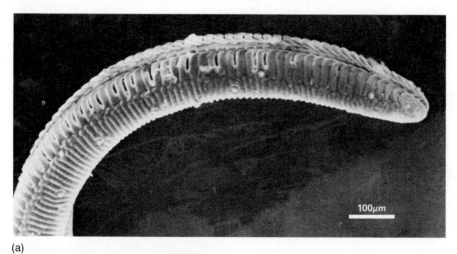

(a)

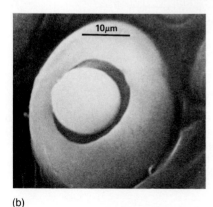

(b)

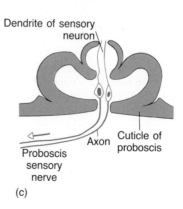

(c)

Figure 29-18
Chemoreceptors on a butterfly's proboscis. (a) Scanning electron micrograph of the tip of the proboscis of a black swallowtail butterfly. The button-like structures are chemoreceptors. (b) Closeup of one chemoreceptor. (c) A schematic cross section of the chemoreceptor. The hair-like dendrites of sensory neurons are the actual receptors.

Moths and Bats

Sense organs are vital in the perennial battle between predators and their prey. One of the most thoroughly studied systems of predator-prey sense organs is found in nocturnal bats and in some of the moths they eat.

It is a warm spring night. As a bat awakes, the receptors in her empty stomach stimulate her to go in search of food. As she flies, she utters four or five short chirps per second, far too high-pitched for us to hear. The bat then listens for the echo of each chirp. Since the echoes are very faint, it seems strange that the bat is not deafened to them by the chirp she made a split second earlier. But every time she chirps, a structure moves in her middle ear, and the three bones that conduct sound from the tympanic membrane to the inner ear slide aside so that they no longer transmit sound to the cochlea. When the chirp is finished, the three bones slide back into place, and the bat can hear the echo of her chirp bouncing off nearby objects. The bat can tell both the direction of the echo (and so of the object) and the time elapsed between uttering the chirp and hearing its echo (which indicates the distance to the object). Hence she can use the echoes of her chirps as a sonar system to avoid obstacles and to detect other moving objects, an ability called **echolocation.** Eventually the bat hears echoes which she identifies as coming from a fat moth.

The moth is looking for a mate, using the sensitive chemoreceptors on his antennae to detect her scent. He happens to be a member of a family (Noctuidae) that has very simple ears, located in the rear of the thorax, just in front of the abdomen. His ears have only two receptor cells apiece, but the cells are sensitive to the high-pitched sound of a bat. One receptor is sensitive to faint sounds; the other responds only to loud sounds. In effect, the first receptor

detects distant sounds, the second nearby sounds.

As the bat approaches, her chirps stimulate the moth's "long-distance" auditory cells more and more strongly. If the bat approaches from the moth's left, his left-hand long-distance cell receives a stronger stimulus. The moth's defensive reaction is to turn as he flies, until the stimulus intensity is the same in both ears (Figure 29-B). When this is the case he will be flying either directly away from or directly toward the bat. He can tell whether the bat is ahead of him or behind him because the sound coming from certain directions is deflected as his wings pass over the ears during flight. The informa-

tion about differences in sound intensity, combined with information about wing position, gives the moth an accurate indication of where the bat is. If the bat is still far away, the moth may escape by turning and flying away.

However, when the bat is very close, the "loud" cells in the moth's ear begin to fire. This signals the moth's brain that immediate evasive action is necessary. He may execute a loop, or may close his wings in a crash dive. Sometimes these tactics succeed, sometimes not. The battle of the sense organs continues night after night, with survival as the stake, a strong selective pressure for sensitive and efficient sense organs.

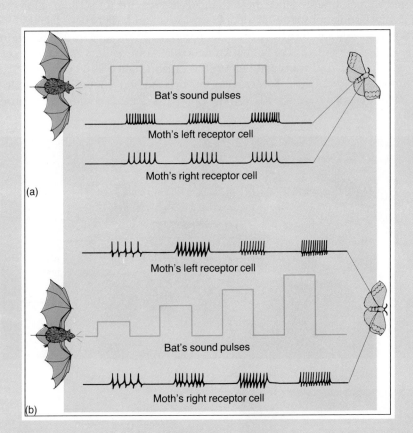

Figure 29-B
Interactions between a bat and the moth it pursues. The nerve impulses recorded from the moth's right and left receptor cells are shown with the pulses of sound produced by the bat. (a) When a bat approaches from the left, the moth's left receptor receives more intense stimulation and fires more rapidly than the right receptor does. (b) As the bat approaches the moth directly from the rear, the stimulus intensity is the same in both ears, and both receptors fire at equal rates. Also, the sounds emitted by the bat become louder (from the moth's point of view), and so the rate of firing increases in both receptors.

29-G Electroreceptors

In a number of bony and cartilaginous fishes, parts of the lateral line organ have become modified to detect electric currents in the surrounding water. Every living organism generates weak electrical fields, and the ability to detect these fields may permit a fish to capture prey or to avoid predators. Experiments have shown that a dogfish shark can use its electroreceptors to detect a worm that the shark can neither see nor smell. Such an ability is especially valuable in turbulent, deep, or murky water, where vision and olfaction are of little use. Some fish generate electrical fields and then use their electroreceptors to detect how surrounding objects distort the field; this allows the fish to navigate in the muddy rivers where they live (see *A Journey Into Life,* "Electric Fish," Chapter 30).

Among mammals, the platypus uses receptors in its duck-like bill to detect weak electric currents from the small aquatic animals that are its prey.

SUMMARY

An animal must be able to detect changes inside its body and in the world around it in order to maintain homeostasis and to produce suitable behavior. Sense organs are collections of cells specialized to react to particular forms of energy by producing electrical impulses in the nervous system. Sense organs may detect mechanical stimuli, light, heat, chemical changes, electric current, or the earth's magnetic field.

The actual receptors in sense organs may be the dendrites of sensory neurons or specialized non-nervous cells. The intensity of a stimulus is coded largely in terms of the frequency of action potentials in the sensory neurons, the total length of a burst of action potentials, and the number of neurons firing. The type of stimulus received is coded by the specific wiring of the nervous system. If a stimulus remains constant, the sense organs and their neural connections tend to undergo adaptation, reducing their response until the stimulus changes.

Mechanoreceptors detect mechanical distortions such as pressure, touch, sound, muscle stretch, movement of joints, and blood pressure. Hair cells are mechanoreceptors found in lateral line organs and the inner ear. Movement of the tuft of microvilli causes a change in the hair cell's membrane potential, which is transmitted to dendrites of a sensory neuron that synapses with the receptor.

Photoreceptors in eyespots, eyes, and ommatidia detect light by means of pigments whose molecular structure is changed by light of particular wavelengths. This photochemical reaction leads to a change in the electrical potential of the receptor cell membrane and in an adjacent sensory neuron. Thermoreceptors are sensory neurons that respond directly to heat, radiation in the infrared region.

Chemoreceptors respond to changes in the concentration of various internal or external chemicals. The detected chemical combines with membrane receptors and changes the membrane's permeability, and hence its electrical potential.

Electroreceptors are found in some bony and cartilaginous fishes; they detect electric currents generated by organisms.

SELF-QUIZ

Match the type of receptor on the right with the sense organs listed on the left:

_____ 1. Nose   a. chemoreceptors

_____ 2. Retina   b. photoreceptors

_____ 3. Pit organ   c. electroreceptors

_____ 4. Proprioceptor   d. mechanoreceptors

_____ 5. Semicircular canals   e. thermoreceptors

6. A sensory neuron transmits patterns of action potentials that indicate (choose two):
   a. the type of stimulus received
   b. the intensity of stimulus received
   c. the duration of the stimulus
   d. the source of the stimulus
   e. the significance of the stimulus

7. A receptor that detects the position of parts of the body is called:
   a. a proprioceptor
   b. a hair cell
   c. a mechanoreceptor
   d. a rod
   e. a cupula

8. Draw the basic structure of a vertebrate eye, and label the structures that modify the light before it reaches the photoreceptors, and the site of transmission of the stim-

ulus to the central nervous system. Also name the two main types of receptors and indicate their distribution.

9. When a sensory neuron ceases to respond to a stimulus that is still present, it is said to have undergone ___.

Choose one word from each pair in parentheses:

10. The visual cycle involving rhodopsin occurs in photoreceptors called (cones/rods). These receptors and their associated neurons are primarily used for vision in (bright/dim) light, and they are (good/poor) discriminators of color.

QUESTIONS FOR DISCUSSION

1. Imagine that you were a dog, and therefore that smell was your dominant sense. How would your perception of the world differ?

2. As tilt of the head increases, how do the nerve impulses convey the sensation of increased tilt to the brain?

3. Why do we say that sense organs detect *changes* in an animal's internal or external environment rather than just detecting the state of the environment?

4. In humans, the ability to taste phenylthiourea (PTU) is inherited as a dominant trait. Can the theory of chemoreception presented in this chapter account for this genetic fact?

5. Why do some people like certain foods that others do not?

6. Why does vitamin A deficiency result in "night blindness"?

7. What visual problems would be experienced by people whose eyes lacked rods? Whose eyes lacked cones?

8. Thermoreceptors of mammals are particularly concentrated on the tongue. These receptors keep human beings from burning their mouths with hot food, but cooking is a recent (in evolutionary terms) invention, practiced only by humans. What is the advantage to a wild mammal of having so many thermoreceptors on the tongue?

SUGGESTED READINGS

Bekesy, G. von. "The ear." *Scientific American,* August 1957. Covers normal structure and function and how age, damage, and disease affect hearing.

Horridge, G. A. "The compound eye of insects." *Scientific American,* July 1977.

Hudspeth, A. J. "The hair cells of the inner ear." *Scientific American,* January 1983.

Koretz, J. F., and G. H. Handelman. "How the human eye focuses." *Scientific American,* July 1988.

Lissman, H. W. "Electric location by fishes." *Scientific American,* March 1963.

Renouf, D. "Sensory function in the harbor seal." *Scientific American,* April 1989. Seals face problems because they must be able to use their senses both on land and under water. Interesting adaptations permit this transition.

Roeder, K. D. "Moths and ultrasound." *Scientific American,* April 1965. Explains the anatomy of the moth's acoustic sensory cells and how these cells gather information to help a flying moth evade hungry bats.

Roeder, K. D. *Nerve Cells and Insect Behavior.* Cambridge, MA: Harvard University Press, 1967. Includes the story of how moths escape bats, by the scientist who performed much of the work.

Rushton, W. A. H. "Visual pigments and color blindness." *Scientific American,* March 1975.

Stryer, L. "The molecules of visual excitation." *Scientific American,* July 1987. The visual pigments and how they work to produce chemical reactions that lead to a receptor potential.

Muscles and Skeletons

30

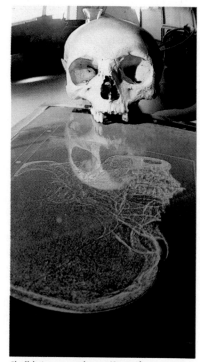

Skull being x-rayed onto Xerox sheet

Runners

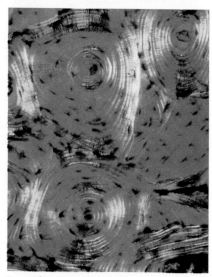

Human bone, Haversian canal

An animal detects things with its sense organs, processes the information in its nervous system, and reacts with its **effectors:** organs that do things in response to signals from the nervous system. Muscles and glands are the most widespread effectors, but electric organs and light-emitting organs also respond to information gathered by sense organs. Glands are considered in Chapter 31. In this chapter, we concentrate on muscles.

Muscles react to signals from motor neurons by contracting. Muscle contractions may move part of the body, or even the whole body during locomotion. Muscles also play important roles in maintaining the body's homeostasis, by circulating blood, pushing food down the digestive tube, and so on. However, muscle contraction does not always result in movement. Sometimes it maintains the status quo, as when a scallop contracts the muscle that keeps its shell closed and a would-be predator out. In all these cases, muscle contraction makes adaptive responses to conditions in the animal's internal or external environment.

The force of muscle contraction cannot do useful work without something to pull or push against: the scallop's shell, the bones of the human body, blood in the heart, or food in the stomach, for instance. An animal's skeleton provides places for its muscles to attach, and a system of levers and pivots for them to pull against, permitting them to perform useful work in moving body parts or in locomotion. The muscular and skeletal systems, working together, lead an animal into temptation—in the form of food or mates—or deliver it from danger—such as predators or too much sun. Even the muscles of the internal organs, which are not attached to the skeleton, have more or less firm attachments to other structures, which serve the same sort of "skeletal" functions.

The rigidity of a skeleton also provides support for the body and protects delicate parts such as the brain and other soft organs.

- An animal's effectors—most commonly muscles and glands—"do something" to maintain homeostasis or make suitable responses to events in the environment.
- Whether executing a motion or holding something steady, muscles and skeletons work together: the muscle provides a force, and the skeleton provides an object for the force to work against.
- In addition to their role in movement, skeletons provide support and protection to other organs.

30-A Muscle Tissue

Muscle tissue has two distinguishing properties: contractility and electrical excitability. Excitability is also a characteristic of neurons (Chapter 28). In both muscle and nervous tissue, electrical excitability is due to the energy stored in an electrical potential difference across the membrane. The excitable muscle membrane depolarizes in response to a chemical transmitter released at a **neuromuscular junction,** the synapse between a motor neuron and a muscle. This excitation in the membrane initiates contraction in the muscle (Figure 30-1). Muscle cells are packed with contractile proteins, which interact in such a way as to cause the cell to grow shorter.

Vertebrates have three types of muscle (Table 30-1 and Figure 30-2):

Smooth muscle lines the walls of many of the internal organs.
Cardiac muscle makes up the heart.
Skeletal, or **striated muscle** is responsible for locomotion and change of position. It is generally the only type of muscle under the animal's voluntary control.

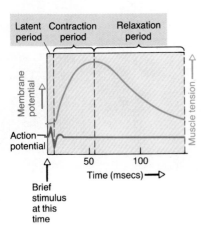

Figure 30-1
Recordings from a skeletal muscle fiber. Electrical activity is shown in red, mechanical activity in blue. When a skeletal muscle is electrically stimulated, it responds by firing an action potential (electrical activity) and by contracting with a rapid twitch (mechanical activity).

Table 30-1   Distribution and Functions of Muscular Tissues and Organs in the Human Body

Function	Structures	Muscle type
Circulation	Heart	Cardiac
	Walls of arteries, veins	Smooth
Excretion	Walls of renal pelvis, ureter, bladder	Smooth
	Internal sphincter between bladder and urethra	Smooth
	External sphincter near exit from body	Striated
Digestion	Tongue, muscles of jaw and pharynx	Striated
	Walls of esophagus, stomach, intestines	Smooth
	Internal sphincter of anus	Smooth
	External sphincter of anus	Striated
Ventilation (Breathing)	Diaphragm, intercostal muscles	Striated
Ejaculation	Walls of genital ducts	Smooth
	Skeletal muscles at base of penis	Striated
Parturition (childbirth)	Uterine wall, cervix	Smooth
	Abdominal muscles, diaphragm	Striated
Heat production	Skeletal muscles (exercise, shivering)	Striated
Maintenance of posture Change of position Locomotion	Skeletal muscles	Striated

Smooth Muscle

Smooth muscle is made up of sheets of individual muscle cells. It occurs in many internal organs, including the walls of the arteries, veins, digestive tract, urinary bladder, and reproductive organs. Smooth muscle usually exerts pressure on the contents of these organs: it constricts blood vessels, moves food

Figure 30-2
Types of muscle. (a) Smooth muscle consists of individual cells. The cell nuclei are the most obvious features in the photograph. (b) Cardiac muscle also consists of individual cells with highly visible nuclei. The contractile proteins are arranged in regular arrays that give the cells a faintly striped appearance. (c) Skeletal (striated) muscle. During development, individual cells fuse, forming long, slender muscle fibers, each with many nuclei lying just within the membrane. Parts of three fibers can be seen, with two dark oblong nuclei visible in the center, in the area between two fibers. The rest of the fibers are filled with regularly arranged contractile proteins (see Figure 30-6) that give the entire fiber its striped ("striated") appearance. (Biophoto Associates)

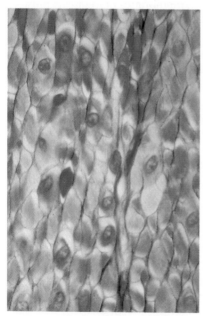

(a)

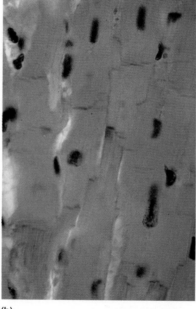

(b)

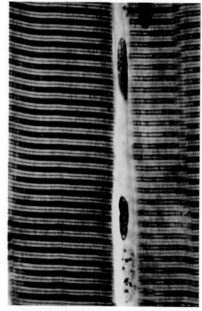

(c)

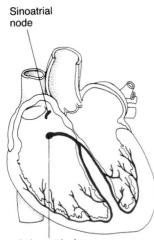

Sinoatrial
node

Atrioventricular
node

Figure 30-3
The electrical conduction system of the human heart. The sinoatrial node, located on the right atrium of the heart, initiates each beat (contraction) of the heart. The impulse from the sinoatrial node spreads throughout the walls of the atria and to the atrioventricular node, located near the atria on the partition between the two ventricles. From the atrioventricular node, the impulse spreads rapidly through the ventricular walls, triggering simultaneous contraction throughout the ventricles.

along the gut, and expels urine from the bladder, semen from the seminal vesicles, or a baby from the uterus.

Some smooth muscles contract spontaneously, but others must first be stimulated by nerves or hormones. Most smooth muscle receives signals from motor neurons in the sympathetic part of the autonomic nervous system, which use the neurotransmitter norepinephrine (Section 28-I). These muscles also contract when a very similar chemical, the hormone epinephrine, arrives from the adrenal glands via the bloodstream. The parasympathetic system also sends signals to some smooth muscles, using acetylcholine as a transmitter.

Smooth muscle produces a gradual contraction of variable force. A smooth muscle cell's electrical activity is not an all-or-none event but a graded response, and how much the cell contracts depends on how much its membrane is depolarized. Smooth muscle contraction often lasts for a long time without fatigue.

Cardiac Muscle

Cardiac muscle occurs only in the vertebrate heart. Here, the individual muscle cells are arranged in long columns of fibers.

Cardiac muscle illustrates the similarity between nerve and muscle tissue. Some cells in the heart are so specialized to conduct electrical impulses that they have lost their contractile proteins and behave much more like neurons than like muscle cells. These cells are part of the mechanism that controls the heartbeat.

Each heartbeat is initiated by spontaneous electrical activity of the heart's pacemaker, the **sinoatrial node** in the wall of the right atrium (Figure 30-3). The pacemaker cells fire in their rhythmic pattern because of a special feature: they are unusually leaky to sodium ions and so they rapidly become depolarized and reach the threshold to fire an action potential. This resets them, and the process repeats itself.

An impulse generated at the sinoatrial node spreads to all parts of the atrium, and then to the **atrioventricular node,** on the partition between the two ventricles. From here the impulse spreads rapidly through the ventricular walls, triggering simultaneous contraction throughout the ventricles (Figure 30-4).

The electrical impulse travels by way of the membranes of the heart muscle cells themselves, and it passes between these cells electrically, not by way of chemical transmitters. The cells are specialized so that their electrical transmission is particularly fast. They are joined together by structures in their plasma membranes called **gap junctions** (Section 4-G). These tiny tunnels permit ions, and therefore electric current, to pass from one cell to its neighbor rapidly, with very little electrical resistance.

The main nerve to the heart is the tenth cranial nerve, the **vagus.** It contains branches of both parts of the autonomic nervous system. Nerve impulses from the sympathetic system speed up the heart, and those from the parasympathetic system slow the heart rate. However, if all of the nerves from the nervous system to the heart are removed, a vertebrate can survive in an apparently normal

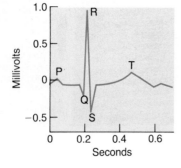

Figure 30-4
An electrocardiogram (EKG). The waves of electrical excitation that spread through the heart and cause it to beat are also conducted throughout the body in the body fluids. These currents can be monitored by electrodes attached to the skin. An EKG is used to determine whether the heart's electrical activity is normal. The P wave is produced by depolarization of the atria; Q, R, and S by depolarization of the ventricles. T is a result of ventricular repolarization.

condition, and its heartbeat alters with the changing demands of exercise just as it does in an intact animal. The rate and force of heartbeats are governed partly by hormones that reach the heart through the bloodstream, and partly by reflexes in a system of nerves that lie completely within the heart and work even though they are not in contact with the rest of the nervous system.

Cardiac muscle must clearly be highly resistant to fatigue if it is to beat throughout an animal's lifetime. Not surprisingly, cardiac muscle has abundant mitochondria (producing ATP which supplies energy for muscle contraction), and its comparatively large blood supply guarantees adequate oxygen.

Skeletal Muscle

The muscles that attach to, and move, the skeleton of a vertebrate are called skeletal, or striated (striped), or voluntary muscles. Most of the research on vertebrate muscle has been done on this readily available tissue.

A skeletal muscle is made up of many **muscle fibers,** each running the entire length of the muscle. Each skeletal muscle fiber arises from the fusion of many embryonic muscle cells. Their plasma membranes coalesce to form a continuous membrane, the **sarcolemma,** around the whole muscle fiber. The many nuclei lie along the outside of the fiber, just under the sarcolemma (Figure 30-5). The axon terminals of a single motor neuron may form neuromuscular junctions with several to over a hundred muscle fibers.

A muscle fiber consists of a bundle of **myofibrils.** Each myofibril is made up of units called **sarcomeres** arranged end to end along its length. A sarcomere is the part of the myofibril between two adjacent **Z lines,** protein-

Figure 30-5
Anatomy of a vertebrate skeletal muscle. This diagram of the deltoid muscle of the shoulder shows serial enlargements of the muscle components, from the intact muscle to the pattern of contractile protein filaments that gives the muscle its striped appearance. The muscle is made up of bundles of muscle fibers. In each fiber, many myofibrils and nuclei are enclosed within a common sarcolemma. Each myofibril contains numerous sarcomeres (the area between one Z line and the next), arranged end to end in single file.

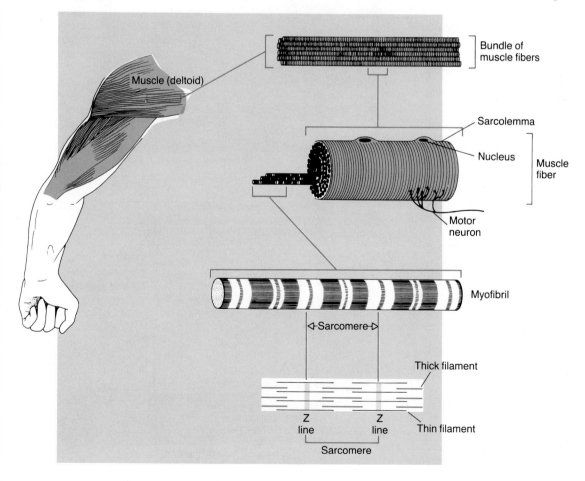

Muscle (deltoid)

Bundle of muscle fibers

Sarcolemma

Nucleus

Muscle fiber

Motor neuron

Myofibril

◁–Sarcomere–▷

Thick filament

Z line

Z line

Thin filament

Sarcomere

containing structures extending across the myofibril. Each sarcomere contains a well-developed cytoskeleton, consisting of a precise arrangement of two kinds of protein filaments. Attached to each side of a Z line are **thin filaments,** which extend less than halfway to the center of the sarcomere. Their free ends overlap **thick filaments,** which are centered in the sarcomere. Viewed with a microscope, the Z lines and the areas where thick and thin filaments overlap appear dark. The sarcomeres of neighboring myofibrils are lined up side by side, so that their light and dark areas produce a visible, striped pattern extending across the muscle fiber (see Figure 30-6).

30-B   Muscle Contraction

■ *In muscle contraction, the heads of myosin molecules in thick filaments attach to actin molecules in thin filaments and push them with a ratchet-like motion. The movement is much like that of a jack used to raise an automobile when you change a flat tire.*

The contraction of a skeletal muscle fiber results from the movement of the protein filaments of the cytoskeleton in the sarcomeres. Each sarcomere's protein filaments slide along each other and mesh together more closely. In this **sliding filament** mechanism of muscle contraction, the filaments stay the same length, but the free ends of the thin filaments move closer to the center of the sarcomere. Since the other ends of the thin filaments are attached to the Z lines, this sliding moves the Z lines of each sarcomere closer together. This shortens the sarcomeres, and therefore the whole muscle (Figure 30-6). Contraction does

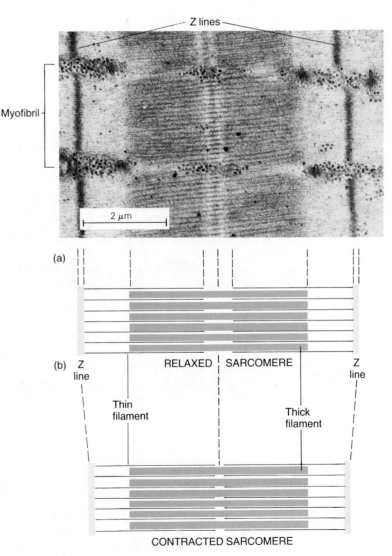

Figure 30-6
The sliding filament model of muscle contraction. (a) An electron micrograph of parts of three relaxed skeletal myofibrils, showing one complete sarcomere in each fibril. Sarcomeres are the contractile units of the fibrils. (b) Diagram of the filaments of contractile proteins responsible for the striped appearance of the myofibril. The thin filaments are attached to the Z lines. The thick filaments lie between the thin filaments. (c) The sarcomere contracted. During contraction, the thick and thin filaments slide past each other, reducing the distance between adjacent Z lines. The filaments themselves do not shorten: the myofibril contracts because all its sarcomeres contract. (a, Biophoto Associates)

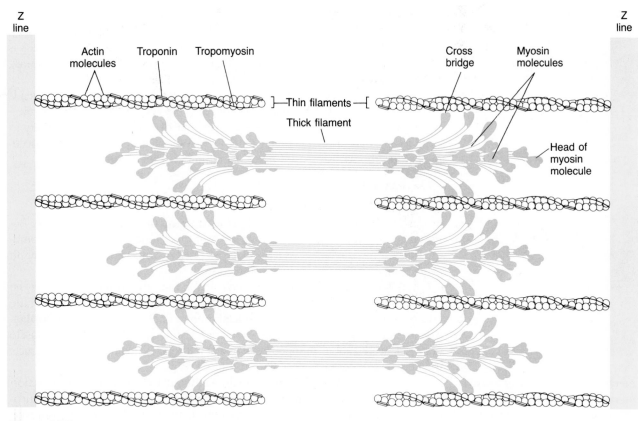

Figure 30-7
Molecular structure of thick and thin filaments. The filaments interact chemically
when the heads of myosin molecules that make up the thick filaments attach to actin
molecules in the thin filaments, forming cross bridges between the filaments. The
heads then swivel, causing the muscle to contract (see Figure 30-8).

not make a muscle smaller. It gets shorter but thicker so that its volume remains
the same.

Each thin filament is made up of many molecules of the globular protein
actin, with smaller amounts of the proteins **troponin** and **tropomyosin** (Figure 30-7). A thick filament contains hundreds of molecules of the protein **myosin,** each shaped like a golf club with a double head.

How do these filaments slide past each other? The heads of the myosin
molecules in the thick filaments form cross-bridges to the actin in the thin
filaments. The cross-bridges swivel, pushing the thin filaments toward the center
of the sarcomere. The cross-bridges then detach, reattach farther along the thin
filaments, and push them still further toward the center of the sarcomere (Figure 30-8). As the myosin heads complete one swiveling cycle, the sarcomere is
shortened by about 1% of its original length. Since there are many cross-bridges
and they do not all move at the same instant, the thin filament cannot slip back
while individual myosin heads are detached.

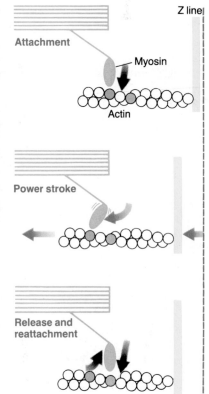

Figure 30-8
How filaments slide during muscle contraction. The head of a myosin molecule in a
thick filament attaches to an actin molecule in a thin filament, swivels, detaches, and
forms a new cross-bridge to another actin. Note that the Z line (yellow) has moved
closer to the end of the thick filament (blue).

Muscle contraction is powered by ATP, produced mainly in the muscle's many mitochondria. ATP binds to the heads of myosin molecules and is broken down to provide energy for the swiveling of the cross-bridges. This process is very efficient, wasting only 30 to 50% of the energy from ATP as heat. In contrast, car engines waste 80 to 90% of the energy available from gasoline.

Curiously, muscles need ATP to relax as well as to contract. Myosin heads cannot detach from the thin filaments until they have bound new ATP molecules. When an animal dies, its muscles soon run out of ATP and lose the ability to contract or relax. They become rigidly locked in whatever position they occupied when the ATP was used up, a phenomenon called **rigor mortis.**

Control of Contraction

A muscle usually contains ATP, and yet it contracts only some of the time. Between contractions, the protein tropomyosin blocks the actin molecules' binding sites for myosin heads. As a result, cross-bridges cannot form and the myosin heads cannot swivel and cause contraction.

A muscle fiber contracts only when calcium ions are present to unblock the actin binding sites. Calcium ions (Ca^{2+}) are stored inside the membranes of the endoplasmic reticulum, which in skeletal muscle fibers is called the **sarcoplasmic reticulum.** Calcium leaves the sarcoplasmic reticulum when a muscle fiber is activated and is pumped back in when it relaxes. A muscle fiber contracts if its cytoplasm contains enough Ca^{2+} and ATP.

■ *Electrical excitation of muscle membranes leads to the mechanical events of contraction.*

A skeletal muscle fiber is normally activated by the arrival of an action potential in the axon of a motor neuron. The axon terminals release the neurotransmitter acetylcholine, which crosses the neuromuscular junction, binds to receptors on the sarcolemma, and causes the sarcolemma to depolarize. An electrical impulse spreads through the sarcolemma, which has extensions that form an internal membrane system (the T tubules) in intimate contact with the sarcoplasmic reticulum (Figure 30-9). The electrical signal causes opening of calcium channels in the sarcoplasmic reticulum, and this releases calcium ions into the cytoplasm. Here they bind to troponin. This causes tropomyosin to move, exposing the actin binding sites for myosin. Myosin heads attach to the actin and swivel, so that the muscle contracts.

Calcium is continually pumped back into the reservoirs of the sarcoplasmic reticulum by active transport. When most of the calcium ions have left the troponin, tropomyosin moves out and blocks the binding sites on the actin again, and the fiber relaxes.

Tetanization and Fatigue

A single nerve impulse arriving at the neuromuscular junction of a mammalian skeletal muscle fiber causes an all-or-nothing muscle twitch, a swift contraction followed immediately by relaxation. The electrical events of depolarization and return to normal in the sarcolemma take a short time compared to the mechanical events of contraction and relaxation (see Figure 30-1).

The smooth, sustained muscle contractions that allow you to carry a cup of coffee across the room without mishap result when the nervous system sends continuous trains of nerve impulses through the motor neurons. Now muscle fibers receive new electrical stimulations before they have relaxed from the previous impulse, and the individual muscle twitches fuse into a smooth, steady contraction, a state known as **tetanization** (Figure 30-10). The muscle contractions also undergo **summation,** producing more force than a single twitch would generate by itself; how this happens is not understood.

A skeletal muscle that is stimulated to contract repeatedly for a long period will eventually become unable to respond to further stimulations, and it gradu-

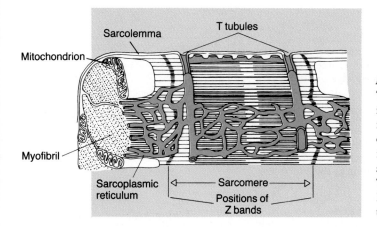

Figure 30-9
T tubules and sarcoplasmic reticulum in part of a muscle. The sarcoplasmic reticulum is a reservoir containing calcium ions. When the sarcolemma is depolarized, the depolarization spreads down the membranes of the T tubules to the sarcoplasmic reticulum, which releases calcium ions into the myofibril.

ally returns to its resting length. This phenomenon, known as **fatigue,** occurs only in striated muscle. Complete fatigue seldom occurs in an intact organism, but it is easily induced in a muscle that has been removed from the body.

Graded Response of an Intact Muscle

It takes more force to carry a whole tray of dishes across the room than to carry just one cup of coffee. How is the strength of muscle contraction controlled? Different animals have different mechanisms for producing this graded response.

Most invertebrate skeletal muscle fibers produce graded contractions, which increase as more impulses arrive via the motor neuron. Similar muscle fibers, called "slow" fibers, are also common in lower vertebrates—fishes, birds, and reptiles. However, most mammalian skeletal muscle fibers are of the "fast," or "twitch," variety, which respond to stimulation in an all-or-nothing manner, as described above. These fibers contract faster than "slow" fibers and so permit rapid movement. The flight muscles of some insects contract even more rapidly than mammalian "fast" muscles.

How can "fast," all-or-nothing fibers produce a graded muscle contraction? A skeletal muscle contains hundreds of **motor units,** each made up of a motor neuron and the muscle fibers it stimulates. The force exerted by the whole muscle can be altered by controlling the number of motor units in action at any one time. Also, firing different motor units in turn permits sustained contractions. For example, in the posture muscles, which work for long periods of sitting or standing, several different fibers are contracting at any one time. These then relax while others contract. As a result, the entire muscle remains partially contracted, but no one fiber stays contracted until it has exhausted its supply of

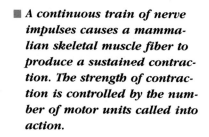

■ *A continuous train of nerve impulses causes a mammalian skeletal muscle fiber to produce a sustained contraction. The strength of contraction is controlled by the number of motor units called into action.*

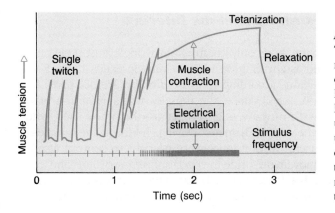

Figure 30-10
Tetanic contraction. The muscle's motor neuron is stimulated electrically at increasing frequency (red line). When the rate of stimulation is low, the muscle contracts in individual twitches (blue curve). As the stimulus frequency increases, the muscle does not have time to relax fully between stimuli. Summation occurs, finally resulting in a smooth, sustained, tetanized contraction.

Electric Fish

A JOURNEY INTO LIFE

The usual function of muscles is to move parts of the body. In some fishes, however, skeletal muscles have been modified for another function: producing electrical discharges. Electric organs have evolved independently in some rays and skates (Chondrichthyes) and in several different families of the Osteichthyes (bony fishes).

In the electric ray *Torpedo,* the electric organ consists of a stack of flattened muscle fibers, called **electrocytes,** that have lost their contractile proteins and consist of little more than large neuromuscular junctions. The organ works much like a battery, with the electrocytes coupled in series. The motor neurons all fire at the same instant, and the depolarizations of the electrocytes sum to produce considerable voltages—up to 50 volts in *Torpedo.* Firing all its electric organs together, an electric eel can deliver a shock of more than 600 volts, but 350 volts is more usual. Strongly electric fishes—electric eels, catfish, and rays—use their electrical discharges to stun prey or to discourage predators. Why they don't electrocute themselves is still a mystery.

A number of different fish can detect electric currents in the water around them. Their electroreceptors are cells that probably evolved from mechanoreceptors in the lateral line organ (see Figure 29-7). Cartilaginous fish have receptors called **ampullae of Lorenzini** embedded in a gelatinous substance at the bottoms of pits in the lateral line system. The function of these organs has been debated for many years because they are sensitive to small changes in pressure, osmotic potential, and temperature. They are even more sensitive to weak electric fields, such as those generated by every living organism. Dogfish sharks can use this sensory ability to find prey, such as worms that they cannot see or smell, from a distance of many centimetres.

Members of some families of freshwater bony fish have electric organs that produce weak electrical discharges. Behavioral experiments showed that these fish use their electric organs for orientation. The electrical discharges last only a few milliseconds and are repeated regularly several times a second. Each discharge creates an electric field around the animal. The fish detects the shape of the field through its electroreceptors (Figure 30-A). Any object in the electric field with an

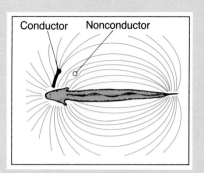

Figure 30-A
The electric fish *Gymnarchus*. The fish generates an electric field with its tail and detects the field with electroreceptors, which are most concentrated in the head. The field is distorted by nearby objects.

electrical conductivity different from that of the water around it distorts the field, and so is detected by the fish. The system is phenomenally sensitive. By rewarding a fish with food, H. W. Lissman showed that it could distinguish between glass and metal rods 2 millimetres in diameter and detect currents of about 3×10^{-15} ampere.

In the turbulent muddy waters where they usually live, sight and smell are of little use to these weakly electric fish. Working together, their electric organs and electroreceptors provide them with a navigation system with which they can move around freely and find their food.

ATP. This system has the disadvantage of requiring many more motor neurons than does a "slow" muscle system, since many different motor units are necessary.

30-C   How Muscles and Skeletons Interact

Muscles can move parts of the body only because they work against skeletons. A **skeleton** may be defined as anything on which a muscle exerts force. Some structures that act as skeletons according to this definition are not part of an animal's "official" skeleton. For instance, the muscles that surround a blood vessel work against the walls of the vessel, and against the blood itself, when they contract and reduce the vessel's diameter. During childbirth, the muscles of the uterus could do no useful work if there were no baby to contract against. Many muscles that cause movement inside the body shorten very little, and exert little force. Their attachment to nearby cells gives them enough leverage to operate. On the other hand, the muscles used in locomotion may contract force-

(a)

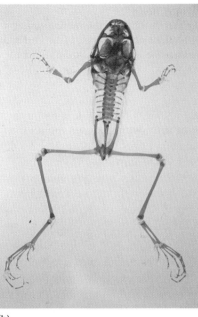

(b)

Figure 30-11
Skeletons. (a) The relatively thin but tough exoskeleton ("outside skeleton") of an insect or other arthropod provides a lot of strength for its weight and also supports and protects soft body parts. However, the arthropod exoskeleton cannot grow. Hence the animal, such as this 17-year cicada, must molt its exoskeleton several times during its life. (b) The endoskeleton ("internal skeleton") of a frog consists of bone (stained red) and cartilage (blue). The bones grow as the animal grows, and bone replaces most of the cartilage at the ends of the bones. (b, L. L. Sadler)

fully and dramatically, and they usually exert their force on what we generally think of as a skeleton: a hydrostatic skeleton, such as the fluid-filled cavities of a cnidarian or annelid; an exoskeleton, such as the cuticle of an insect or the shell of a snail; or an endoskeleton, such as the bones of a vertebrate (Figure 30-11).

Antagonistic Muscles

All movements are controlled by the action of two sets of **antagonistic muscles**—muscles with opposite effects. In systems lacking hard skeletons, such as soft-bodied invertebrates and the internal organs of vertebrates, these antagonistic muscles are arranged in circular and longitudinal sheets. Longitudinal muscles run lengthwise along the body (or organ) and make it shorter and wider when they contract. Contraction of circular muscles, which run around the body (or organ), makes it longer and narrower. The antagonistic action of longitudinal and circular muscles against the contents of the alimentary canal moves food down the gut in most animals. Similar forces exerted against the fluid-filled coelom of an earthworm permit the worm to move (Figure 30-12).

Muscles attached to hard skeletons, such as the exoskeletons of arthropods or molluscs or the endoskeletons of vertebrates, may be arranged in bundles rather than sheets. A skeletal muscle is attached to the skeleton either directly or by way of a **tendon.** In some cases—for instance, with the muscles that move our fingers—the tendons may be almost as long as the muscle.

Antagonistic muscles run parallel to each other across a joint in the skeleton. The muscle that causes the joint to stretch out is called an **extensor,** and its

Start　　　　　　**Finish**

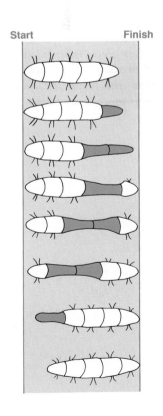

Figure 30-12
An earthworm crawling, using its hydrostatic skeleton. Each segment acts as a closed sac of fluid surrounded by two sets of muscles, which can contract alternately so as to shorten or lengthen the segment. When the segment is short and wide, its bristles are extended, pushing against the substrate and keeping the worm from slipping backward. Each segment is extended in turn (color). At the end of the sequence, the worm has moved forward.

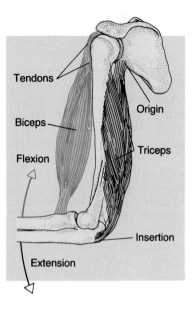

Figure 30-13
Skeletal muscles in the human upper arm. Each muscle is attached to the skeleton by tendons at its insertion (the end attached to the bone that moves) and origin (the end attached to the less mobile bone).

antagonist, which causes the joint to close up, is called a **flexor.** The biceps is the main flexor across the human elbow joint, and the triceps is its antagonistic extensor (Figure 30-13).

Locomotor muscles are so powerful that if a flexor and its antagonistic extensor were both to contract strongly at the same time, they could easily break a bone. This is prevented by **reciprocal inhibition,** a process involving a reflex arc from each muscle to its antagonist. Proprioceptors in the muscle detect how far it is contracted (Section 29-C). Their sensory neurons send this information to interneurons in the spinal cord, which in turn inhibit the firing of motor neurons to the antagonistic muscle. Thus any stimulus that causes a muscle to contract also inhibits the contraction of its antagonist, and the joint moves smoothly.

This is the simplest reflex involved in locomotion. Much more complex reflexes come into play when we walk. For instance, reflexes from the opposite limb ensure that one leg moves after another. Such reflexes are even more complicated in animals that have many legs or that use their tails for balance.

30-D *The Vertebrate Skeleton*

The vertebrate skeleton has two major portions. The **axial** (axis) skeleton consists of the skull, backbone, ribs, and tail. The **appendicular** (limb) skeleton includes the bones in the limbs and in the girdles that attach the limbs to the backbone. The **pectoral girdle** includes the collarbones (clavicles) and shoulder blades (scapulas), and the **pelvic girdle** includes the large, fused hip bones (ilium, ischium, and pubis) (Figure 30-14).

■ *The skeleton supports the body, protects internal organs, and provides attachment sites, pivots, and levers for the muscles to move body parts.*

The bones of the skeleton anchor various muscles and permit them to move the body, but they serve many other functions as well. First, they provide a framework that supports the body against the pull of gravity, and land vertebrates must have skeletons more substantial than those of vertebrates that live supported by the buoyancy of water. Parts of the skeleton also protect delicate internal organs. The skull protects the brain and major sense organs, the backbone surrounds and protects the spinal cord, and the pectoral girdle and ribs protect the heart and lungs.

Joints in the Vertebrate Skeleton

Joints between bones in the skeleton can be of many different types and degrees of rigidity. At one extreme are **sutures,** wiggly lines in the skull where bones meet at interlocking projections that hold them very tightly together. However, most bones are joined to each other by **ligaments.** An extremely flexible joint in females is the one between the two bones in the front of the hip girdle, which are held loosely together by a ligament that stretches considerably and allows the bones to separate during childbirth.

The joints of the limbs must permit smooth movement in various directions. The ends of the bones in these joints are covered by smooth, elastic cartilage, and the whole joint is surrounded by a sac filled with **synovial fluid,** which lubricates the joint (Figure 30-15).

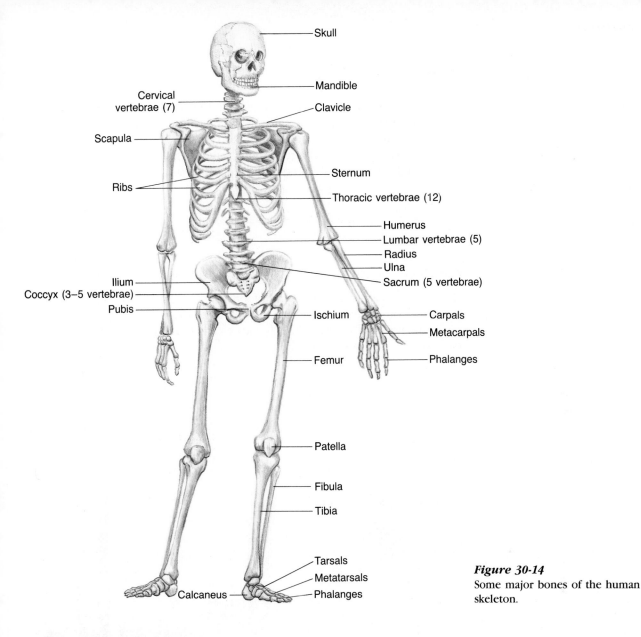

Figure 30-14
Some major bones of the human skeleton.

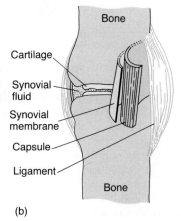

(a) (b)

Figure 30-15
Structure of a joint. (a) This photograph of the finger joint of a human child shows large, pale areas of cartilage at the ends of the bones. In an adult, the cartilages would be smaller, as they are in the drawing in (b), but would still be present to cushion the ends of the bones. The space between the bones is filled with synovial fluid, kept in place by the surrounding membrane. A tough, fibrous capsule surrounds the membrane. Ligaments bind the bones together and limit the movement between them. (a, Biophoto Associates)

30-E   Connective Tissue

Bones, cartilage, tendons, and ligaments are all made up of various types of **connective tissue,** composed mainly of intercellular ("between cells") substances secreted by scattered cells. The intercellular substance contains fibers of a tough, elastic protein, **collagen** (see Figure 3-20). Collagen is the most common protein in mammals, accounting for about one fourth of all the body's protein (it is also the main ingredient of gelatin). Bone is hard and brittle because it has mineral deposits in addition to these fibers. Cartilage has a great deal of a firm intercellular jelly surrounding the protein fibers and cells, while tendons and ligaments are mostly composed of fibers, with very few cells.

Cartilage

Unlike bone, cartilage receives no blood supply; its cells rely on diffusion of nutrients from capillaries in nearby tissues. Cartilage makes up the entire skeleton in members of the class Chondrichthyes (sharks, skates, and rays) and in all early vertebrate embryos. During embryonic development of vertebrates with bony skeletons, minerals are deposited in most of this embryonic cartilage, a blood supply and other typical features of bone appear, and the cartilage is

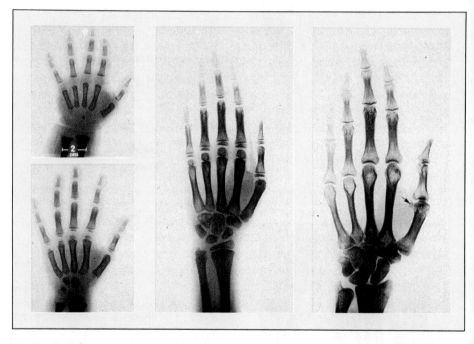

Figure 30-16
Growth of a human hand. This series of x-ray photographs shows the hand at 2 years (top left), 2 years 9 months (bottom left), 12 years (center), and 16 years (right). Note that at 2 years of age, there appear to be large gaps between the bones of a finger. These gaps actually contain the cartilaginous ends of the bones and the finger joints, but cartilage shows up only faintly in an x-ray. In the hand of a 12-year-old (center), the cartilage on either side of the joints has been almost completely replaced by bone, leaving only a thin cartilage cap over the end of the bone at the joint. In the fully developed hand (right), the ends of the bones are enlarged into knobs. The arrow points to a small spur of bone on the thumb. Many clinical conditions, from injury to arthritis, can cause such abnormal deposition of minerals in the body. When such deposits occur in joints, they may interfere with movement. (Biophoto Associates)

replaced by bone (Figure 30-16). Cartilage persists in the adult only in areas where flexibility is necessary. These areas include:

1. The ends of the bones where they form synovial joints (see Figure 30-15).
2. The discs between the vertebrae in the backbone (see Figure 28-18).
3. The ends of the ribs where they join the breastbone.
4. The rings that keep the walls of the trachea (windpipe) from collapsing.
5. The larynx ("voice box") in the throat, at the front end of the trachea.
6. The external ear.
7. The eustachian tube, which connects the throat to the middle ear.
8. The tip of the nose.

Bone

Although the mineral matter of the dried human skeleton weighs only about 5 kilograms, the skeleton of a living human being is much heavier because bone is a living tissue that contains cells, blood vessels, nerves, fluid, and fat deposits.

Bone varies in structure depending upon its position and function in the body. At one extreme is **spongy bone,** composed of an irregular network of mineralized bars, and at the other is **compact bone,** composed of tubular units called Haversian systems (Figure 30-17).

The hard part of bone is made up of organic matter and inorganic salts. The organic matter is mainly collagen fibers, which give bone most of its ability to withstand pull (tension). The inorganic salts are mainly calcium phosphate and smaller amounts of other ions. These give bone its considerable ability to withstand compression and side-slippage.

Distributed throughout a bone are the cells that lay down the collagen and mineral deposits as the bone grows. These cells last into adult life and are responsible for repair and replacement of broken bone and for the formation of **calluses,** which develop at points of pressure on a bone. People who perform one action many times, such as squatting instead of sitting, throwing a baseball, or carrying a baby on the hip, develop particular alterations to their bones. Anthropologists studying ancient humans can look for such bone-structure markers as clues to how these people lived.

Bones grow throughout life as the body increases in size and strength, and as they grow they are also remodeled. New material is added to the outer surfaces and the ends, while old material is destroyed to enlarge the internal cavities for the bone marrow. The differentiation of bone is controlled by a number of local hormones, produced in growing animals and also in response to physical stress, fractures, and other stimuli that indicate the need for more bone growth. **Bone growth protein,** for instance, is a hormone that occurs in the intercellular material of bone. When released, by a fracture or similar trauma, it induces undifferentiated cells to differentiate into bone-forming cells. The amount of bone growth protein declines with age, contributing to osteoporosis (see *A Journey Into Life,* "Osteoporosis").

Nonskeletal Functions of Bones In the centers of many bones are cavities filled with blood vessels and **marrow,** a soft tissue with a number of different functions. Some marrow is primarily a fat depot; some produces red blood cells, platelets, and white blood cells.

Bone also plays a role in regulating the level of calcium ions in the blood. The calcium phosphate in bone is in equilibrium with that in the surrounding extracellular fluid. Thus, if the calcium level in the extracellular fluid rises, some of it is deposited in bone; if the calcium concentration in the extracellular fluid falls, calcium from the bones dissolves into the fluid. The chemical equilibrium between solution and deposition determines the general level of calcium in the extracellular fluid and in the blood.

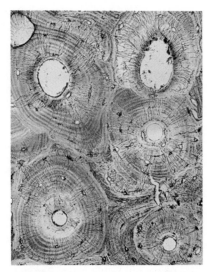

Figure 30-17
A cross section of compact bone. Each set of concentric rings is a Haversian system, surrounding a central channel, the Haversian canal. In the living bone, the Haversian canals contain blood vessels and nerves. The many tan rings around each Haversian canal are mineral deposits, laid down by living cells that inhabit the small, purple, spider-shaped spaces. (Biophoto Associates)

■ *Bones serve as the body's calcium repositories and contain nurseries for the production of blood cells.*

Osteoporosis

A JOURNEY INTO LIFE

In the United States, more than 600,000 bone fractures a year are attributed to brittle bones. Researchers hope to reduce this expensive toll by studying the formation of bone in the embryo and the healing of bones in later life.

Part of the reason osteoporosis is so common, is that people live much longer nowadays, and degeneration of bone is a serious problem in many older people. **Osteoporosis** is a condition in which bones have lost so much inorganic mineral matter that they become light, brittle, and easily broken (Figure 30-B). The disorder has at least two distinct forms. Post-menopausal osteoporosis, involving loss of bone in the spine and forearms, afflicts women 50 to 65 years old. An important contributing factor is the great reduction in the hormone estrogen at the time of menopause. This may be exacerbated by smoking, which reduces the body's estrogen level. Senile osteoporosis occurs in people of both sexes aged 75 and older, and involves loss of bone in the hips and legs. The most effective ways to prevent or treat osteoporosis appear to be exercise, which stimulates addition of material to bone, and increasing calcium in the diet, especially from organic sources such as dairy products. It is more effective to drink milk than to take calcium pills because very little of the calcium from pills is taken up by the body.

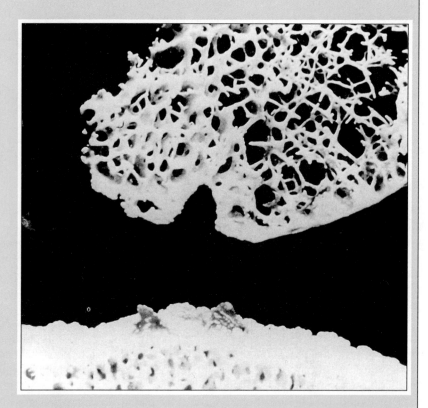

Figure 30-B
Osteoporosis. Bone weakened by osteoporosis (top). Normal bone (bottom). (Courtesy of Wyeth-Ayerst Laboratories, Philadelphia, PA.)

Fine tuning of the calcium level in the body fluids is under the control of hormones. The parathyroid glands, which lie behind the thyroid gland in the neck (see Figure 31-1), secrete **parathyroid hormone.** This stimulates the cells in bone to take additional calcium out of the bone and release it into the blood. Parathyroid hormone also causes the kidneys to reclaim more calcium from the forming urine. Third, it stimulates cells in the kidneys to convert vitamin D to an active form, which promotes absorption of calcium from food in the intestine. A drop in the blood calcium level stimulates the parathyroid glands to produce this hormone; calcium is then released from bone, resorbed from the urine, and absorbed from food to make up the deficit.

A rise in the blood calcium level is counteracted by the hormone **calcitonin,** secreted by the thyroid gland. Calcitonin inhibits the removal of calcium from the bone deposits. This permits the opposite process, deposition of new mineral material in the bone, to bring the blood's calcium level back down to normal.

SUMMARY

Most movement in an animal's body is due to the action of its muscles. A muscle works by shortening so that it pulls against the skeleton or against adjacent tissues.

Muscle contraction has been most extensively studied in vertebrate skeletal muscle. In the presence of ATP and calcium ions, filaments of the protein myosin form cross-bridges to filaments of the protein actin; the swiveling of the cross-bridges moves the filaments past each other, reducing the distance between the Z lines of the sarcomere. ATP supplies the energy needed for contraction.

Contraction of a muscle is stimulated by depolarization of the membranes of muscle cells or muscle fibers. This happens when a neurotransmitter is released from the axon of a motor neuron or, in smooth or cardiac muscle, when the hormone norepinephrine is present. In mammalian skeletal muscle, the arrangement of motor neurons and muscle fibers into motor units allows for graded muscle contractions. Smooth, sustained contractions are brought about by trains of closely spaced impulses from motor neurons.

The vertebrate skeleton supports the body, protects internal organs, and permits the muscles to move parts of the animal's body. Pairs of antagonistic muscles move bones back and forth at joints. Reciprocal inhibition prevents the two members of a pair of antagonistic muscles from contracting strongly at the same time.

The vertebrate skeleton is made up of relatively flexible cartilage and harder bone. Cavities in the bone are filled with marrow, which stores fat and forms blood cells. Tiny spaces in the bony material itself house single cells that deposit or release calcium. Bones store or release calcium to maintain equilibrium with the body fluids. Hormones maintain a fine tuning of the blood/bone calcium balance.

SELF-QUIZ

For each phrase in questions 1 through 5, give the type(s) of muscle that show the characteristic.

_____ 1. Multinucleate fibers

_____ 2. Receives nerves from autonomic nervous system

_____ 3. Can contract without nervous stimulation

_____ 4. Typically found in sheets rather than in bundles

_____ 5. Unicellular organization

a. Cardiac
b. Skeletal
c. Smooth

6. Indicate which of the following items would be found at each of the places indicated on the following diagram:

actin	troponin
calcium	Z line
mitochondria	
myosin	
tropomyosin	

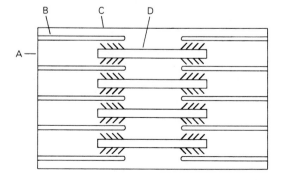

7. During the time that a muscle is in tetanic contraction, what are the motor neurons to the muscle doing?

8. Why doesn't a muscle relax between the arrival of nerve impulses when it is in tetanic contraction?

9. The extent of contraction of a mammalian skeletal muscle is controlled by:
   a. interaction of excitatory and inhibitory nervous input to individual fibers
   b. contraction of some sarcomeres in a fiber but not others
   c. contraction of some entire muscle fibers while others remain relaxed
   d. reciprocal inhibition
   e. all of the above

10. Which of the following is *not* a role of bones?
   a. maintenance of simple chemical equilibrium between dissolved and deposited calcium without intervention of living cells
   b. removal of calcium deposits carried out by living bone cells in response to a hormone
   c. production of red blood cells
   d. production of white blood cells
   e. production of hormones to regulate calcium levels in body fluids
   f. storage of fat
   g. providing attachment sites for muscles
   h. protection of internal organs

QUESTIONS FOR DISCUSSION

1. If you read detective stories, you know that it takes a variable length of time for rigor mortis to set in after death. What are some reasons for this variation?
2. Rigor mortis lasts for several hours and then disappears. Why does it eventually go away?
3. What is the advantage of having cartilaginous tissue in each area of the human body where it is characteristically found? (See Section 30-E.)
4. What are some advantages of having muscles attached to bones via tendons rather than directly?
5. Comment on the biological validity of the sayings "I can feel it in my bones" and "dry as a bone."
6. The disease rickets is characterized by bending of the bones due to lack of calcium deposits to keep them stiffened into the proper shape. Why is it advantageous for the bones to give up these calcium deposits even though doing so results in permanent skeletal deformity?
7. Would you expect levels of parathyroid hormone or of calcitonin to be elevated in a victim of rickets? In a pregnant woman? In her fetus?
8. Pumiliotoxin B, produced by dart-poison frogs, promotes the release of Ca^{2+} from muscle storage areas and inhibits its return. What symptoms would you expect to see in an animal dying after being shot with a dart tipped with this poison?

SUGGESTED READINGS

Caplan, A. I. "Cartilage." *Scientific American,* October 1984.

Lissman, H. W. "Electric location by fishes." *Scientific American,* March 1963.

McLean, F. C. "Bone." *Scientific American,* February 1955. A brief, clear description of bone structure, growth, and function.

McMahon, T. A. *Muscles, Reflexes, and Locomotion.* Princeton, NJ: Princeton University Press, 1984. Chapters on the mechanics and neural control of locomotion.

Schiefelbein, S., et al. *The Incredible Machine.* Washington, DC: National Geographic Society, 1986. Fascinating account of human biology and life.

Vogel, S. *Life's Devices: The Physical World of Animals and Plants.* Princeton, NJ: Princeton University Press, 1989. An entertaining book on how organisms adapt to the physical forces of the world they live in.

Animal Hormones and Chemical Regulation

CHAPTER

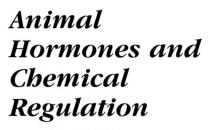

31

Canada goose

Impala bucks fighting

A young man's fancy . . .

Mother dingo nursing

W herever there is division of labor, there must also be an exchange of information to coordinate the work of different units. On a construction site, dozens of people may perform different jobs. They must know what their co-workers are doing and interact appropriately with one another to build the structure properly. In any multicellular organism, the jobs to be done are divided among many different types of cells. Like the different workers on a construction site, these cells must exchange information and work together if the organism is to survive. The nervous system and the body's chemical messengers work together in coordinating the activities of all the different cells in the body.

We can define a **chemical messenger** as a substance produced by one cell that affects other cells. **Hormones** are chemical messengers (such as estrogen and insulin) that travel in the blood and affect cells some distance from those that produced them. Hormones are produced by cells that are usually organized into **endocrine glands** (such as the ovary and parts of the pancreas), glands whose secretions move to their sites of action via the blood rather than through ducts (Figure 31-1).

Chemical messengers carry out several kinds of regulation. First, many contribute to the body's physiological homeostasis by acting as part of a negative feedback control system. In this way they raise or lower the body fluids' level of some chemical—including other hormones. Second, some chemical messengers are involved in repair, growth, and differentiation of the body, particularly during embryonic development. Third, hormones play a role in adaptive responses to events outside the body. For instance, hormones ensure that an animal will be in reproductive condition when environmental conditions favor

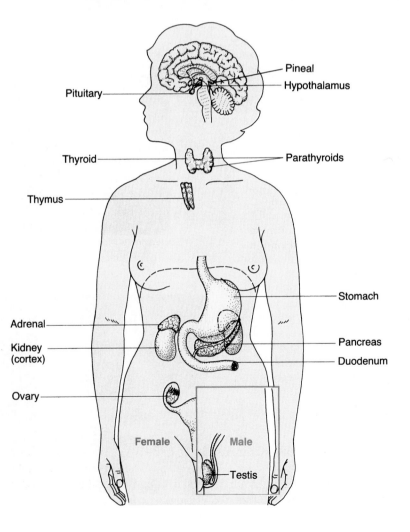

Figure 31-1
Endocrine glands. The locations of the major endocrine glands and other organs with endocrine functions in humans.

the survival of the young. The most obvious interactions between the nervous and hormonal systems occur in this type of control. An animal can detect environmental events only through its sense organs, and the nervous system must convey this information to the glands that produce hormonal changes. In the vertebrates, the hypothalamus in the brain (part of the nervous system) communicates with the pituitary gland (part of the endocrine system), which in turn regulates most of the other endocrine glands.

A wide variety of chemical messengers occurs throughout the body, but each cell responds only to some of the messages. This is because different kinds of cells have receptor molecules that recognize and bind particular kinds of messenger molecules, initiating changes in the cells' activities.

- Neurons and chemical messengers coordinate activities within an animal's body by affecting cells in various ways.
- Chemical messengers are involved in maintaining homeostasis within the body, in growth and differentiation, and in the body's response to outside stimuli.
- The body maintains homeostasis of various chemical components by negative feedback control, using hormones as messengers to raise or lower the levels of other chemicals, including other hormones.
- The need to secrete a hormone is determined, in many cases, by the nervous system, which stimulates secretion by some glands.
- Only those cells with receptors for a hormone can respond to that hormone.

31-A   Chemical Messengers

We now know that animals have dozens of chemical messengers, including:

1. **Hormones** (produced by neurons and by endocrine cells in endocrine organs or in other tissues such as the stomach and kidneys). In Chapter 28 we saw that neurotransmitters, such as acetylcholine and norepinephrine, carry nerve impulses across synapses. Some neurons, however, respond to stimulation by releasing neurohormones (such as oxytocin, which induces labor during childbirth) from their axons into the bloodstream.
2. **Local chemical messengers.** Some chemical messengers have only local effects, within a few millimetres of the cell that produced them, because they are rapidly removed from the extracellular fluid. These local messengers include histamine, which participates in inflammatory and immune reactions (Section 25-A); growth factors, which stimulate growth by particular tissues; prostaglandins, lipids with a variety of effects; and short-range neuroregulators released into a local area by neurons.
3. **Pheromones.** These are chemical signals produced by one animal that affect the behavior of other individuals of the same species.

31-B   Hormones

How do we know that a hormone exists and what it does? Most hormones have been discovered by removing the relevant endocrine gland and observing how this affects the animal. For instance, an amphibian tadpole without a thyroid gland never metamorphoses into a frog; it is reasonable to hypothesize that a hormone produced by the thyroid causes metamorphosis. This can be tested by transplanting a thyroid gland into a tadpole that lacks one. Since this operation is followed by metamorphosis, the theory is strengthened.

Table 31-1   *Some Vertebrate Pituitary Hormones, Their Sources, and Main Actions*

Hormone	Source	Effect	See Section
Releasing factors (various)	Hypothalamus	Release of hormones from anterior pituitary	31-D
Oxytocin	Hypothalamus via posterior pituitary	Uterine contractions in mammals Letdown of milk to nipple in mammals	27-D
Vasopressin (=ADH, antidiuretic hormone)	Hypothalamus via posterior pituitary	Water resorption by nephron tubules Increase in permeability of skin to water in amphibians	26-F
Adrenocorticotropic hormone (ACTH)	Anterior pituitary	Secretion of corticosteroids by cortex of adrenal gland	31-D
Thyrotropin (TSH, thyroid-stimulating hormone)	Anterior pituitary	Secretion of hormones by thyroid	31-D
Follicle-stimulating hormone (FSH)	Anterior pituitary	Production of gametes in both sexes	27-D
Luteinizing hormone (LH)	Anterior pituitary	Secretion of sex hormones by gonads in both sexes Ovulation in females	27-D
Prolactin (=LTH, luteotropic hormone)	Anterior pituitary	Mammary gland growth and lactation in mammals Maintenance of corpus luteum in mammals Migration to water in amphibians Reproductive functions in birds	27-D
Somatotropin (growth hormone)	Anterior pituitary	Body growth in reptiles and mammals Increased blood sugar in mammals	31-D

*All pituitary hormones are proteinaceous.

More convincing evidence for the hormone can then be produced by breaking down thyroid tissue into various chemical fractions and showing that some of these, but not others, produce metamorphosis after they are injected into a tadpole. Finally, it may be possible to isolate the pure hormone from the thyroid gland.

Such experiments are much harder when the endocrine gland involved, such as the vertebrate pituitary gland, produces many hormones (see Table 31-1). In such situations, the most common approach is to take blood from an animal that is thought to contain the hormone and transfuse it into an animal that does not. For instance, the vertebrate pituitary was long suspected of producing vasopressin, a water-conserving hormone, if the animal became dehydrated. To test this, blood from a dehydrated dog was transfused into a dog that had an adequate water supply. Urine production in the second animal decreased; it returned to normal only when the transfusion was stopped. This suggested that the blood of the first dog did in fact contain a water-conserving hormone. Showing that the hormone comes from the pituitary is more difficult. However, such an assumption would be strengthened if chemical analysis showed that the hormone is found only in the pituitary and in the bloodstream.

Some of the most important hormones secreted by vertebrates are listed in Tables 31-1 and 31-2.

Feedback Control of Secretion

Many of the hormones listed in Tables 31-1 and 31-2 control the composition of the body fluids or the rate of metabolism. For these homeostatic mechanisms to function properly, the hormones must be secreted into the bloodstream only

Table 31-2   *Some Vertebrate Hormones, Their Sources and Main Actions* *

Hormone	Source	Effect	See Section
Thyroxin	Thyroid	Stimulation of growth and metabolism Metamorphosis in amphibians	10-I
Calcitonin	Thyroid	Decrease in blood calcium by suppressing its resorption from the bones	30-E
Parathyroid hormone	Parathyroids	Increase in blood calcium by release of calcium stored in bones	31-B, 30-E
Insulin	Pancreas	Decrease in blood sugar	22-G
Glucagon	Pancreas	Increase in blood sugar	22-G
Gastrin	Stomach	Secretion of HCl by stomach	22-C
Epinephrine (adrenalin)	Adrenal medulla	Dilatation of blood vessels Increase in blood pressure Increase in blood sugar	24-C, 31-D
Norepinephrine (noradrenalin)	Adrenal medulla	Same as epinephrine Also serves as a neurotransmitter	
Cortisol, etc.	Adrenal cortex	Metabolism of carbohydrate, protein, fat	
Aldosterone	Adrenal cortex	Na^+ and K^+ retention by kidney Sex drive	26-F
Chorionic gonadotropin	Placenta	Maintenance of all body functions necessary for pregnancy	27-D
Progesterone	Corpus luteum of ovary	Maintenance of uterine endometrium in mammals Enlargement of breasts during pregnancy in mammals	27-D 27-D
Estrogen	Ovary	Initiation and maintenance of sexual maturity and behavior in female mammals	27-D
Testosterone	Testis	Initiation and maintenance of sexual maturity and behavior in male mammals	27-D

* Unless a particular class of vertebrates is specified, the action applies to members of all classes as far as we know.

when they are needed. Their secretion is under **negative feedback control,** whereby a process automatically limits itself.

A thermostat controls a familiar negative feedback system. The thermostat responds to a drop in temperature by turning the furnace on. When the thermostat detects that the temperature has risen to the set level, it turns the furnace off. As a biological example, a drop in the level of calcium in the blood causes secretion of **parathyroid hormone** by the parathyroid glands. Parathyroid hormone stimulates release of calcium from the bones, decreases excretion of calcium by the kidneys, and increases absorption of calcium from the intestine. These actions usually increase the calcium concentration of the blood so that it is back to normal within a few hours. The rise in calcium, in turn, tends to decrease the secretion of parathyroid hormone (Figure 31-2a). This negative feedback loop is one of the control systems ensuring that the level of calcium in the blood remains constant.

A hormonal feedback loop may involve two or more hormones, instead of a hormone and some other substance (for example, calcium). This is how hor-

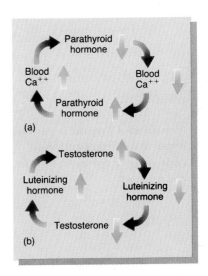

Figure 31-2

Two examples of negative feedback control: (a) between a hormone and calcium ions (Ca^{2+}); (b) between two hormones. In both cases, each of the two substances serves to regulate the level of the other in the body fluids. Blue arrows indicate increases or decreases in levels of the substances in the body fluids.

■ *Secretion of hormones involved in homeostasis is controlled by negative feedback.*

mones secreted by the pituitary gland control the release of other hormones. For instance, a male vertebrate must have the right levels of both **luteinizing hormone (LH)** and **testosterone** in the blood for the testes to produce sperm and function properly. LH, secreted by the pituitary, stimulates testosterone secretion by the testes, but testosterone inhibits the secretion of LH. So, if the level of testosterone in the bloodstream rises, it inhibits the pituitary's secretion of LH, and less testosterone is produced until the testosterone level falls low enough that the inhibition of LH is turned off. LH secretion then rises again, stimulating the secretion of more testosterone. Thus, a rise or fall in the level of either hormone is automatically corrected by way of the other (Figure 31-2b).

31-C How Chemical Messengers Affect Cells

Endocrine cells release hormones into the extracellular fluid. From there, the hormones diffuse into the bloodstream, which carries them throughout the body. However, hormones are specific, influencing only certain cells—called their **target cells.** Only the target cells of each chemical messenger have receptor molecules to bind that messenger. For instance, although the hormone **insulin** travels throughout the body, only certain types of cells, such as muscle and liver, have insulin receptors and respond to insulin by taking up glucose. Kidney and brain cells, in contrast, can get all the glucose they need from the extracellular fluid and do not have the receptors to respond to the presence of insulin.

■ *A hormone circulating throughout the body affects only its target cells because only these cells have receptors for the hormone.*

A hormone delivers its "message" to the target cell by changing the shape of the receptor that binds it. Other molecules in the cell are already set up in such a way that the receptor's new shape initiates certain changes in the cell, such as changes in permeability, enzyme activity, or gene transcription. All of these changes involve large numbers of ions or molecules, compared to the number of messenger molecules the cell received. In effect, the altered receptor **amplifies** the signal of a tiny amount of hormone into a much larger response by the cell.

The reaction of the target tissue to a hormone differs at different times. For example, injection of the hormone prolactin into a sexually immature female will not cause lactation. Presumably, her mammary tissue has not yet produced receptors for the hormone.

We can divide hormones and local messengers into two groups, which differ in the way they affect target cells: (1) lipid-soluble steroid and thyroid hormones, and (2) water-soluble hormones.

Steroid and Thyroid Hormones

Steroid hormones are small lipids made from cholesterol (see Figure 3-7). The even smaller thyroid hormones are made from iodine and amino acids. (We need iodine in our diets, from seafood or iodized salt, in order to make thyroid hormones.) Both kinds of hormones control many developmental and physiological processes in animals.

■ *Steroid and thyroid hormones enter their target cells and change gene transcription in the nucleus.*

Both steroid and thyroid hormones are lipid-soluble. They diffuse through the plasma membrane and bind to receptors inside the target cell (Figure 31-3a). This binding activates the receptor so that it can bind with DNA. The hormone-receptor complex enters the nucleus, where it binds with genes known as enhancers (Section 10-H). The effect of a small number of hormone molecules is amplified by turning on or off the cell's production of many RNA and protein molecules. Amplification of the hormonal message may be increased still more if the product of transcription is a gene regulatory protein that goes on to activate other genes (Section 10-H).

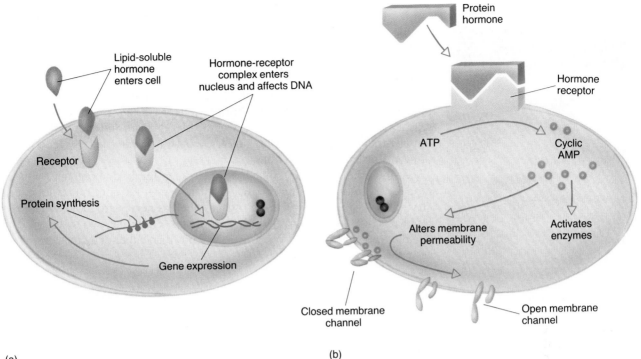

(a)

(b)

Figure 31-3

Two ways hormones affect cells.
(a) Lipid hormones can cross the plasma membrane. Bound to their receptors, they usually cause the cell's DNA to transcribe messenger RNA for production of new proteins. (b) Proteinaceous hormones bind to receptors on the cell surface. This binding causes ATP inside the cell to be converted into cyclic AMP. The cyclic AMP may have many effects, such as altering the cell's metabolism and its permeability.

Water-Soluble Hormones

Most water-soluble hormones and local messengers are proteins (including amino acid derivatives and glycoproteins). They cannot cross the target cell's plasma membrane but bind to receptors on the outside of the cell. This may alter the cell's permeability or cause changes in enzyme activity within the cell. For instance, when the hormone epinephrine binds to its receptors on muscle and liver cells, it activates the enzymes that break down the cell's store of glycogen into glucose.

Most cell-surface receptors that bind these hormones act by way of small signalling molecules within the cell known as **second messengers.** Two of the most important second messengers are calcium ions (Ca^{2+}) and the nucleotide cyclic adenosine monophosphate (cyclic AMP) (Figure 31-3b).

Cyclic AMP is made by a roundabout process. First, the hormone binds to its receptor. This activates another membrane protein, which in turn activates a membrane enzyme to convert ATP into cyclic AMP. Cyclic AMP may cause many changes within the cell: it activates enzymes, alters membrane permeability, and even initiates protein synthesis, depending on the type of cell.

Calcium is another widespread second messenger, often acting in concert with cyclic AMP. For instance, binding of the hormone angiotensin to cells in the adrenal cortex triggers aldosterone secretion. The first effect of angiotensin is to increase the cells' permeability to calcium ions, which flow into the cells from the extracellular fluid. Calcium then alters the activities of various enzymes, including the one that produces cyclic AMP! The triggering of muscle contraction is another example of calcium as a second messenger (Section 30-B).

■ *Tiny amounts of hormones may produce many effects in target cells by using cyclic AMP and calcium ions as second messengers inside these cells.*

31-D   Hormonal and Nervous Control

It usually takes longer for a hormone to act on its target cells than for the nervous system to activate an effector. This is because it takes longer for a hormone to reach its target tissue, and because the hormone causes slower responses in the target tissue.

Figure 31-4
The letdown response. Letdown of milk from the mammary gland is a comparatively rapid hormonal response to suckling by the young.

By having both nervous and endocrine control systems, the body is equipped to cope with a variety of occasions. The nervous system enables an animal to escape from an enemy in a fraction of a second, and some hormones also act relatively swiftly. For example, suckling by young mammals causes milk to let down (Figure 31-4). Suckling of the nipple sends sensory signals to the hypothalamus in the mother's brain. Cells in the hypothalamus extend into the pituitary, where they release the hormone oxytocin into the blood. Oxytocin reaches the mammary glands by way of the bloodstream and stimulates smooth muscles there to contract, pushing milk toward the nipple. This whole sequence of events takes less than a minute.

At the other extreme, other hormones keep the body pregnant for many months. Hormones are more suitable than nerves for controlling long-term changes involving many different organs, whereas nerves are more suitable for rapid reactions involving relatively few organs. Often the two systems acting together can control a situation more efficiently than either of them acting alone, as in the next example.

Fight or Flight

A hot, red face, perspiring hands, and a rapidly beating heart commonly precede a stage appearance or an important exam. These symptoms are part of the "fight or flight" reaction, which prepares the body to meet stress or danger. We shall mention only a few aspects of this reaction here (Figure 31-5).

When a vertebrate senses danger or stress, the central nervous system stimulates the adrenal medulla (inner part of the adrenal glands) to release the hormones epinephrine and norepinephrine into the bloodstream. In addition, much of the sympathetic nervous system is activated, releasing more norepinephrine. Epinephrine and norepinephrine cause the heart to beat faster and to increase the volume of blood pumped per stroke. This raises the blood pressure and circulates the blood more rapidly. In addition, extra glucose is released into the bloodstream by the liver, raising the blood sugar level. Blood vessels in the muscles dilate, increasing the muscles' blood supply and preparing them for action by supplying them with extra oxygen and glucose. At the same time, constriction of the vessels supplying the kidneys and digestive tract reduces their supply of blood.

Nervous control evokes these reactions very rapidly in time of danger. Hormones provide a backup that can maintain the response for a long period. This explains why the state of "nervous energy" persists even after the performance is over or the exam paper handed in.

Figure 31-5
A few of the changes involved in the "fight or flight" reaction, mediated by both the nervous system and hormones.

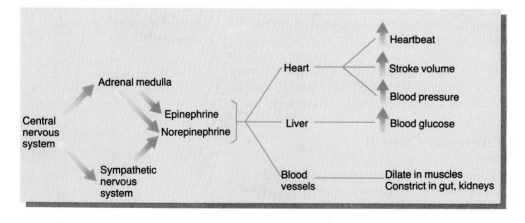

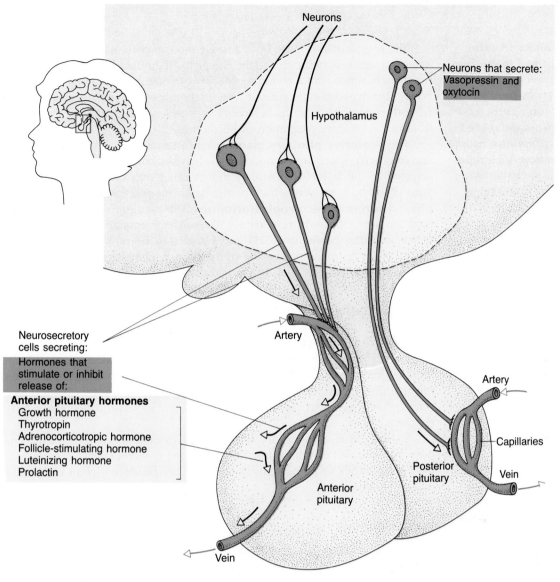

Neurons

Hypothalamus

Neurons that secrete:
Vasopressin and oxytocin

Neurosecretory cells secreting:

Hormones that stimulate or inhibit release of:

Anterior pituitary hormones
Growth hormone
Thyrotropin
Adrenocorticotropic hormone
Follicle-stimulating hormone
Luteinizing hormone
Prolactin

Artery

Artery

Capillaries

Posterior pituitary

Vein

Anterior pituitary

Vein

The Hypothalamus–Pituitary Connection

Most of the endocrine glands in the vertebrate body are controlled, directly or indirectly, by the brain. The nervous and endocrine systems interact by way of the connections between the hypothalamus in the brain and the pituitary gland. The **hypothalamus** is a small area of the forebrain lying in front of the midbrain. It receives input from all parts of the brain. Stimulation of various cells in the hypothalamus elicits sensations and behaviors such as sex drive, pleasure, rage, fear, satiation, hunger, and thirst.

The hypothalamus produces some hormones, but it does not release them directly into the bloodstream. Rather, these hormones travel down neurons from the hypothalamus to the posterior lobe of the pituitary gland, where they are released into the blood. The hypothalamus also produces hormones called **releasing factors,** which control the release of other hormones from the anterior lobe of the pituitary. Secretory neurons in the hypothalamus discharge the releasing factors into capillaries, which carry them to the anterior pituitary (Figure 31-6). Here each releasing factor causes its target cells to release hormones into the bloodstream.

Figure 31-6
The interrelations of hypothalamus and pituitary in the vertebrate brain. The left-hand part of the diagram shows neurons that release hormones into the blood as it passes through capillaries in the stalk of the pituitary gland. This blood then travels to the anterior pituitary, where it passes through a second set of capillaries. Here the hypothalamic hormones enter the extracellular fluid and stimulate release of corresponding pituitary hormones into the bloodstream. At the right are other neurons whose cell bodies lie in the hypothalamus, while their axons extend into the posterior pituitary, where they release hormones into the bloodstream. Most of the hormones shown here are described in Table 31-1.

■ *The nervous and endocrine systems are linked by the connections between the hypothalamus and the pituitary gland. The hypothalamus receives information from the rest of the nervous system and stimulates the secretion of circulating hormones by the pituitary.*

The hypothalamus is the body's single most important control center. Messages from sensory neurons, and the chemistry of the surrounding fluid, provide the hypothalamus with a continuous flow of information about the state of the body. The hypothalamus reacts to these stimuli by producing activity in the nervous system, by initiating behaviors such as feeding or mating, and by controlling the pituitary's secretion of hormones.

The **posterior pituitary gland** releases hormones such as **vasopressin** and **oxytocin,** which are produced in the hypothalamus. Many of the hormones secreted by the **anterior pituitary** induce other endocrine glands in various parts of the body to secrete their particular hormones. For instance, **thyrotropin** and **adrenocorticotropic hormone** cause hormone secretion by the thyroid and adrenal glands, respectively. The anterior pituitary also produces follicle-stimulating hormone (FSH) and luteinizing hormone (LH), which are necessary for hormone production by the gonads in all vertebrates, and growth hormone, which is essential for normal growth of young vertebrates.

31-E Local Chemical Messengers

Hormones travel in the bloodstream and so they reach, even though they do not affect, most of the body's cells. By contrast, many other chemical messengers are secreted into the extracellular fluid and absorbed by neighboring cells so rapidly that they never reach the bloodstream. Many examples of these local messengers have been discovered since the 1960s, and more probably remain to be described.

Some local messengers are secreted only by cells specialized for the purpose. In wounded tissue, for instance, histamine is secreted almost entirely by mast cells. Other local messengers are probably produced by all cells under certain conditions.

Prostaglandins

Prostaglandins are a group of about 20 molecules made from fatty acids. They occur in most vertebrate tissues and have many different actions as local chemical messengers.

Some prostaglandins stimulate smooth muscle to contract, and others cause it to relax. Some cause constriction of capillaries, and others cause dilatation. Prostaglandins are involved in many different aspects of reproduction, and in menstrual cramps, allergic reactions to food, and the inflammatory response to infection. Aspirin inhibits prostaglandin synthesis, and this is thought to be why aspirin inhibits inflammation, lowers fever, and reduces pain.

Some prostaglandins have medical uses: to induce labor in childbirth, to promote healing of stomach and duodenal ulcers, to relieve asthma, and to synchronize the reproductive cycles of livestock for breeding.

Unlike most other chemical messengers, prostaglandins are not stored. They are produced in membranes, and are continuously released to the exterior of the cell. They are also continuously destroyed in the extracellular fluid. When a cell is activated by changes in its environment, however, it may increase the rate of prostaglandin synthesis and release enough prostaglandin so that the messenger influences both the cell that produced it and its neighbors.

Neurotransmitters as Local Messengers

Many neurotransmitters are deactivated as soon as they have crossed the synapse to the postsynaptic cell, and so they never escape from the synapse (Section 28-C). If all signals in the nervous system reached only specific postsynaptic cells in this way, very few neurotransmitters would be needed because there would be no chance of confusion. But in fact more than 30 different neurotransmitters have already been identified in the vertebrate brain alone. Biologists are beginning to realize that many neurotransmitters diffuse out of synapses and so come into contact with many different cells, some of which have receptors for them. Neurotransmitters that behave in this way really act as local hormones and may be distinguished by the special name of **neuroregulators.**

Sleep is at least partly controlled by a neuroregulator. If cerebrospinal fluid is extracted from a sleeping animal and injected into one that is wide awake, the second animal promptly goes to sleep; the fluid appears to contain a "sleep neuroregulator." Sleep appears to require the suppression of a large group of specific nerve cells in a particular part of the brain. A neuroregulator can cause this suppression by affecting all the neurons that bear receptors for the sleep neuroregulator. The neurons need not be completely "turned off" by the neuroregulator: they may remain responsive to other neurotransmitters.

■ *Local chemical messengers act near the cells that secrete them. They include prosta-glandins, neuroregulators se-creted by some neurons, and some growth factors.*

Growth Factors

In many tissues, cell division is controlled by conventional circulating hormones, such as the anterior pituitary hormone **somatotropin** (also called **growth hormone**). Infants who produce too little somatotropin become dwarfs.

Other growth factors do not circulate in the blood but act as local messengers. For example, **bone growth protein** (Section 30-E) and **nerve growth factor** are produced locally during development and are necessary for bones and nerves to develop normally.

31-F Hormones and Animal Life

Animals respond to information about the outside world that is sent to the nervous system by way of the sense organs. Although most reactions, such as eating or running away, are carried out by muscle contractions, some are mediated by hormones. Most of the responses to the environment involving hormones are slow changes, such as occur when an animal comes into breeding condition in the spring.

Environmental Control of Reproduction

There is selective pressure for animals to reproduce under conditions that favor survival of their offspring. For most animals this means birth or hatching in the spring, when warm weather and plentiful food offer the best possible conditions. Breeding often involves dramatic changes in an animal's anatomy, physiology, and behavior.

Animals come into breeding condition in response to environmental conditions such as daylength, temperature, rainfall, food, and so on. Animals detect the external stimuli that induce breeding by way of their sensory receptors. For instance, light may be detected by the eyes or by the **pineal eye,** the "third eye" that lies in the middle of the top of the head in many lower vertebrates. The appropriate stimulus causes hormone production by way of the hypothalamus. The hypothalamus does two things: first, it stimulates the pituitary to release hormones that cause the gonads to grow and produce sex hormones. Second, the sex hormones feed back to the hypothalamus, which then initiates reproductive behavior by sending out the appropriate signals in the nervous system.

31-G   Biological Rhythms

Reproductive cycles are examples of rhythmic or cyclical events in an animal's life. A number of other physiological and behavioral cycles also exist.

Circadian Rhythms

All eukaryotes, even unicellular protists, show daily cycles in many physiological processes. For instance, in most vertebrates the metabolic rate, body temperature, blood sugar level, urine composition, general level of nervous activity, and many other functions alter regularly in a 24-hour cycle (Figure 31-7).

Because 24 hours is the period of the earth's rotation and of one light-dark cycle, it might seem obvious that daily cycles would be controlled by the onset of light or of dark. Is this the case, or is the rhythm **endogenous** ("built into" the animal)? To answer this question, investigators isolate an animal from cues that might tell it the time of day, and see whether the daily rhythm persists. When the animal is kept in darkness with constant temperature and humidity, it still maintains these rhythms, but the cycles are no longer exactly 24 hours. Because the rhythms repeat approximately, but not exactly, every 24 hours in the absence of external cues, they are called **circadian rhythms** (circa = about; dies = day). These endogenous rhythms must be related to an "internal clock."

In a normal environment, light resets the clock every day to produce a 24-hour cycle. The eyes detect light, which stimulates events in the nervous system that adjust the metabolism of certain cells in the hypothalamus of the brain. When the eyes are exposed to light at night, changes in DNA transcription can be detected in these hypothalamic cells, suggesting that protein synthesis may be involved in resetting the clock.

It is possible to reset the circadian clock artificially. For instance, if a vertebrate is kept in a controlled environment where the light is switched on only during the night, night becomes day and vice versa for the animal, and its daily rhythm soon becomes reset exactly 12 hours later. Many zoos do this to nocturnal animals (those that are usually active only at night) so that they become active during the day when visitors want to see them. These animals are exposed to bright light at night, and to a dim red light during the day. Their internal clocks are set so that the animals act as though it is night during what is really daytime.

Figure 31-7
Circadian rhythms. This graph shows circadian rhythms of oral temperature (black line) and reaction time (red) in humans. (Reaction time is the time it takes to react to a stimulus when you are told to react as fast as possible, for instance, to press a button when you see a red light.)

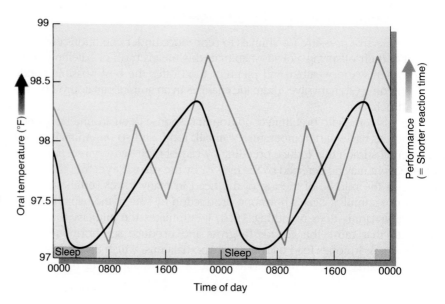

Why did circadian rhythms evolve in the first place? Night and day always follow each other on a 24-hour cycle, and so there is a perfectly good stimulus available to trigger daily cycles. Why should an organism also have an internal cycle of its own? The answer is probably that an animal's internal rhythm permits it to anticipate regular daily events before environmental cues appear. This ability is valuable in many ways. It permits a bat to start hunting at the time of day when its insect prey will also be active, without wasting the energy it would take to check on the light level or insect activity in the area every few hours. Similarly, it permits a reptile to start raising its body temperature before dawn so that it is already prepared to be active at daybreak.

Annual Rhythms

In addition to circadian rhythms, some animals have yearly cycles (Figure 31-8). Many mammals continue to hibernate at roughly the right time even if they are deprived of environmental cues that could tell them the time of year. Deer kept under constant conditions continue to grow, and later to shed, their antlers at the same time every year. Several species of birds start migratory behavior at the same time of year even if they are deprived of environmental cues. In this last case, the adaptive advantage of the annual rhythm is probably that it permits a bird to return north for the breeding season even though it receives few cues of seasonal change in the relatively constant tropical environment where it winters.

Biological Clocks

The existence of circadian and annual rhythms raises the question, "How does it work?" Organisms plainly have internal "clocks" of some sort that keep track of time and can act as "pacemakers" to run these rhythms independently of environmental stimuli.

The vertebrate body appears to contain many different clocks (including at least two in the hypothalamus), each with its own endogenous rhythm. Normally, environmental cues keep all the body's clocks set to the 24-hour daily cycle but we are still far from understanding how all this works.

Circadian rhythms may be remarkably persistent. For example, a reptile continues to show daily fluctuations in body temperature even when it has been kept in the dark at a constant temperature for months. Another intriguing feature of circadian rhythms is their ability to adjust to changes in temperature. Since the rates of biochemical processes usually vary with temperature, we would expect that at lower temperatures the rhythm would be slower. When the tem-

Figure 31-8
Annual rhythms. Each year, a male moose produces antlers and later sheds them, growing another set the next year.

Our Daily Spread

A JOURNEY INTO LIFE

An American will tell you that the normal adult human body temperature is 98.6°F, or 37°C. European fever thermometers, on the other hand, give the normal temperature as 98.4°F. In fact, normal body temperature is anywhere from 36 to 37.6°C (96.8 to 99.5°F). If you take your temperature first thing in the morning, it is likely to be at the low end of this range, and if you keep taking it every hour or two, it will probably show the pattern in Figure 31-7, reaching a high point late in the evening.

Many other physiological factors vary during the day, including blood pressure, heartbeat rate, excretion of various ions in the urine, secretion of different hormones, alertness and reaction time, and the ability to detect faint sounds. Each factor has its own predictable daily pattern, which may differ from the patterns of other factors (Figure 31-A). For example, one hormone may peak during the middle of the night, and another first thing in the morning, whereas body temperature peaks shortly before bedtime.

Are these regular daily patterns in human body functions triggered by external cues, or are they endogenous? To distinguish between these two possibilities, experimenters have put people into isolation in caves, underground apartments, or windowless rooms for periods of several weeks. The subjects have no clocks or watches, no sunlight, no radio or television—nothing to tell them what time of day it is. They eat, sleep, read, or exercise when they want to. Under these conditions, people drift from their almost exactly 24-hour pattern of normal daily life into a new rhythm that depends on the person, frequently about 25 hours.

Ignorance of circadian rhythms can be hazardous to our health. For instance, blood pressure tends to be low in the morning and to rise during the day. People who always visit the doctor in the morning may have blood pressure readings that fall in the normal range, but their blood pressure may rise dangerously in the afternoons and evenings. These people will not receive the treatment needed to control their blood pressure.

Our circadian rhythms probably make themselves felt most acutely when we try to reset our internal clocks. People who have travelled by airplane across several time zones experience "jet lag," and people who work rotating shifts, such as airline flight crews, air traffic controllers, police, and military and hospital personnel, have similar problems. The trouble is that some rhythms reset to the "new time" more quickly than others, and so the body goes through a period when its rhythms are out of synchronization with one another. This can have serious consequences. The famous accident at the Three Mile Island nuclear power station would probably have been quite minor if the people on duty had been more alert. Their errors in handling the emergency may well have arisen from the shift rotation schedule followed at the plant. In one study, more than half of rotating-shift workers reported falling asleep on the job, including truck drivers and nuclear power plant operators. Some people have also developed ulcers because of the continuous stress of having desynchronized body rhythms.

The study of circadian rhythms has produced information that can help in the scheduling of shift rotations for the 20–30 million people in the United States who work at such jobs. For example, it is easier to reset the internal clock to a later time than to an earlier time. This explains why we adjust more easily to travel from east to west than in the opposite direction. It also suggests that rotating shifts should be scheduled so that people work the day, evening, and night shifts in that order, rather than switching to nights and then evenings after the day shift. Furthermore, since it takes up to a week for a person to adjust to the new schedule, workers should be left on each shift for as long as possible. Workers at a mineral mine in Utah were switched from spending a week on each shift and then rotating to an earlier one, to a new schedule, on which they worked three weeks at each shift and then rotated to a later one. They reported increased job satisfaction and better health on the new schedule, and no one fell asleep during work hours any more. The employer gained, too, in less employee turnover and in increased output.

Related research shows that people tend to wake up when their body temperature begins to rise, regardless

perature of the environment is lowered, the animal's rhythm does slow down, but within a few days the rhythm adjusts to the new temperature and resumes its natural frequency. How this temperature compensation is made is an interesting puzzle.

31-H   Pheromones

A **pheromone** is a chemical that travels outside the body, carrying information to other members of the same species (Figure 31-9). The first pheromones described were sex attractants from insects. In many species of moths, beetles,

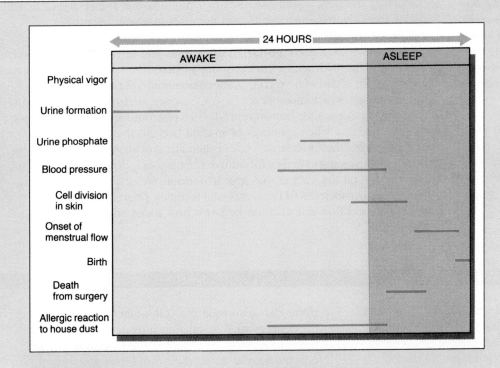

24 HOURS

| AWAKE | ASLEEP |

Physical vigor

Urine formation

Urine phosphate

Blood pressure

Cell division in skin

Onset of menstrual flow

Birth

Death from surgery

Allergic reaction to house dust

Figure 31-A

Human circadian rhythms. Activities of the human body are not uniform throughout the day: different factors or events tend to be higher or more frequent at different times of the day in most people. The 24-hour time span shown begins, at the left, at the time when people get up in the morning (this varies from person to person, so no clock times are given). The red bars show the periods during the day when the high value of each function occurs. For example, urine formation is highest soon after rising, and cell division in the skin is highest just after going to sleep. Likewise, babies are not born uniformly throughout the day; there is an excess of births in the hours just before the mother would normally get up.

of how long they have slept or how long they were awake before going to sleep. This explains why people whose travel or work schedules force them to go to sleep at unusual times may wake up exhausted: their internal alarm, which is linked to deep body temperature, wakes them up when the temperature begins to rise though they have not had enough rest to recover from previous waking periods.

Researchers are experimenting with ways to reset the human biological clock using meals, bright lights, or exercise at specific times of the day. This may eventually help people adjust to work shifts and jet lag more quickly and with less discomfort and danger.

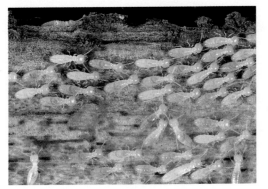

Figure 31-9

Pheromones. Pheromones organize the activity of individuals in insect societies, such as this termite colony. (Biophoto Associates, Holt Studios)

■ *Pheromones function like hormones except that they act on other individuals of the same species, instead of acting on different cells in the body of the animal that secretes them.*

cockroaches, and flies, the female releases a chemical that attracts the male. He finds his mate by flying or crawling up the odor gradient toward her.

Many vertebrates, and particularly mammals, use pheromones in urine or feces, or from special scent glands, to mark trails and territories. When a dog urinates on a fire hydrant, he is depositing a pheromone that tells other dogs that the hydrant is part of his territory. Pheromones also permit members of one sex to distinguish which members of the opposite sex are in breeding condition.

The best evidence for human reproductive pheromones comes from anecdotal evidence that when numbers of women live together, in dormitories or similar situations, their menstrual cycles eventually synchronize. Other than this, there is no convincing evidence for human pheromones, despite the large number of reports on the subject that appear periodically.

In the insect societies of bees, ants, and termites, pheromones organize not just the reproduction but also the behavior and social structure of a colony (Section 32-K).

SUMMARY

The nervous and hormonal systems carry messages that travel between an animal's cells and coordinate their activities. Neurons affect specific cells briefly. Hormones may act on a wide range of cells over a long time.

Animals have three types of chemical messengers: hormones, produced by endocrine cells or by secretory neurons, and carried all over the body by the blood; local hormones such as histamine, prostaglandins, growth factors, and neuroregulators, which usually act near the cells that produce them; and pheromones, which carry information between different individuals of the same species.

Hormones are involved in homeostatic mechanisms within the body, in growth and differentiation of tissues, and in many of an animal's responses to its environment, such as reproductive cycles.

The effects of each hormone or other chemical messenger are limited to those specific target cells with receptors for the hormone. Lipid and thyroid hormones attach to receptors in the target cell's cytoplasm and enter the nucleus, where they affect gene expression. Water-soluble hormones do not enter the cell but instead cause their cell-surface receptors to

make changes inside the cell, often by way of second messengers such as calcium ions or cyclic AMP. These hormones usually change the cell's permeability or enzyme activity. In all these ways, tiny amounts of hormones can drastically affect the target cell's activity.

Hormone secretion is usually controlled by negative feedback in response to some disturbance in the body or to the level of another hormone in the blood.

Hormones and the nervous system often interact with one another to control both long- and short-term aspects of an animal's response to a stimulus. The interaction between the two systems occurs mainly in the hypothalamus. The hypothalamus receives nervous signals from the sense organs by way of other areas in the brain, and also detects changes in blood chemistry. It initiates appropriate responses by way of the nervous system and by way of the pituitary gland, which releases hormones that carry messages to many of the body's other endocrine glands.

Animals have internal biological clocks that tell them the time of day. The 24-hour clock is reset every day by stimuli that include light. The clock interacts with the body's nervous and hormonal mechanisms.

SELF-QUIZ

1. Which of the following techniques would be *least* likely to help in determining the function of a hormone produced by an endocrine gland?
   a. removal of the gland and subsequent analysis of what functions are lost
   b. transplantation of the gland into an animal that lacked the gland

   c. transfusions of blood from an animal lacking the gland into an animal that has the gland and observation of its effects
   d. observing effects of gland extract on various tissues grown in culture
   e. observing the condition of animals with a tumor that causes the gland to be overactive

2. All of the following commonly serve as signals that stimulate hormone secretion *except:*
   a. conditions outside the body
   b. rising levels of another hormone
   c. rising levels of the hormone in question
   d. falling levels of the hormone in question
   e. falling levels of another hormone
3. Hormones are known to cause all the following changes in target cells *except:*
   a. changes in genetic makeup
   b. changes in permeability
   c. changes in metabolic rate
   d. increase in cyclic AMP concentration
   e. synthesis of different messenger RNA and proteins
4. An advantage to having the endocrine system as well as the nervous system involved in the "fight or flight" response is:
   a. the endocrine system responds faster
   b. the endocrine response usually lasts longer
   c. the endocrine system is tuned more precisely to the degree of need
   d. the endocrine system affects only the target organs whose response is needed to meet the emergency
   e. response by the endocrine system frees the nervous system to think of a way out of the situation instead of simply maintaining the body in an alert state
5. a. Thyroid-stimulating hormone (TSH) is released from the anterior pituitary in response to low levels of the hormone thyroxin in the blood. Higher levels of thyroxin reduce the release of TSH. This regulation of levels of TSH and thyroxin is called ____.
   b. Receptor proteins for TSH exist only on cells in the thyroid gland. These thyroid cells are thus designated the ____ of TSH.
6. Using the secretion of thyroxin as an example, diagram the feedback control system for hormone production (see Question 5).

QUESTIONS FOR DISCUSSION

1. Thyroxin levels are generally at about 100 units. A patient has only 80 units in the bloodstream. Normal therapy for this situation is to inject the extra 20 units. Why might this therapy not be effective?

2. Why does the blood of a male mammal contain a much higher concentration of luteinizing (and follicle-stimulating) hormone after he has been neutered than it did before the operation? (Hint: see Figure 31-2)

SUGGESTED READINGS

Berridge, M. J. "The molecular basis of communication within the cell." *Scientific American,* October 1985. Second-messenger pathways in cells.

Coleman, R. M. *Wide Awake at 3:00 A.M.: By Choice or by Chance?* New York: W. H. Freeman, 1986. An entertaining account of research on human biological clocks and recommendations for coping with time shifts.

Crapo, L. *Hormones: Messengers of Life.* San Francisco: W. H. Freeman, 1985.

Uvnas-Moberg, K. "The gastrointestinal tract in growth and reproduction." *Scientific American,* July 1989. The GI tract treated as the largest endocrine organ in the body and the role it plays in adjusting metabolism to pregnancy and infant growth.

Behavior

CHAPTER

32

Mountain gorilla

Learning

Tiger

Polar bear males get acquainted

W e are prone to conclude that a dog is "ashamed" when it puts its tail between its legs and sneaks into a corner after a spanking, and "happy" when it wags its tail. This type of thinking is called **anthropomorphism,** ascribing human emotions to animals. At the opposite extreme, we may say that a bird sings from instinct, because it is incapable of behaving intelligently. The view of human behavior as intelligent and that of other animals as instinctive is reflected in the tendency to ascribe actions of which we are ashamed to "animal instincts." Neither of these approaches to animal behavior gives much useful insight into why animals behave as they do. Recent research has attempted to study animal behavior with as little bias as possible from our human prejudices (Figure 32-1).

In many ways, natural selection acts more directly on behavior than on anything else. Dozens of different behavior patterns may distinguish the individual that reproduces from the one that does not, and all the adaptations of an animal's anatomy and physiology are useless if the animal does not feed itself, escape predators, and find a mate.

An animal's genes determine the range of characteristics it can develop, but just which traits develop depends upon interactions between the genes and the environment. In this chapter we shall consider how behaviorists may try to disentangle the genetic and environmental influences on an animal's behavior and the kinds of selective pressures that have produced the varied behavioral repertoires of different animals.

- An animal's genes determine the range of behavior patterns it can develop in response to environmental stimuli.
- Behaviors that must be produced perfectly the first time they are performed are usually innate. Learning produces behavior that must be flexible to meet local or changeable conditions.
- Many behavior patterns become programmed into the nervous system.
- An animal responds to a particular stimulus only when it is in an appropriate physiological state.
- Some insect and some vertebrate species have evolved societies made up of related individuals, which communicate and cooperate with each other.

Figure 32-1
Why do animals behave as they do? Are these porpoises leaping for joy, to escape predators, or what? (John D. Cunningham/VU)

KEY CONCEPTS

32-A Immediate and Ultimate Causes

A frog is sitting in the grass when a fly buzzes past. Zip! the frog's tongue flicks out and pulls the fly into the frog's mouth. How and why does the frog do this? The question "how" can be answered by describing the frog's sensory, nervous, and muscular systems. The stimulus of a moving fly before the eyes sends impulses along sensory neurons to the central nervous system, which in turn activates and directs the tongue muscles used to catch the fly.

The question of "why" the frog catches the fly is different because it can be answered on two levels. The immediate reason is that the behavior pattern results from a nervous reflex. Seeing a fly activates a reflex that results in the frog's striking at the fly. However, there is also a longer term answer to the question "why?" That particular behavior pattern exists because it has been selected for during the course of evolution.

Three main selective pressures have brought about the behavior patterns we see today:

1. Ultimately, an animal's behavior patterns will be selected for as they contribute to its reproductive success. It is occasionally possible to see how (why) a behavior pattern has evolved by showing that not performing the behavior is selected against. For instance, many birds remove the empty egg shell from the nest after the young has hatched (Figure 32-2). Niko Tinbergen showed

Figure 32-2
Adaptive behavior. A black-headed gull removing an egg shell.

Figure 32-3
Reacting to a stimulus. A lacewing takes off when prodded by the stick at the bottom of the picture. Many behavior patterns are immediate responses to environmental stimuli. (Biophoto Associates)

■ *Genes influence behavior by affecting the structure or function of sense organs, nerves, muscles, hormones, and other physiological systems.*

■ *Environment influences behavior by providing stimuli that may evoke a behavioral response immediately or may have longer term effects on the development of the behavior.*

that adult black-headed gulls that did not remove the shells lost more chicks to predators; the white inside of the egg shell allowed certain predators to discover the otherwise camouflaged nest and chicks.

2. Behavior patterns must allow an animal to solve immediate problems. Hungry animals must feed, and hunted animals must escape predators, if they are to survive and reproduce.

3. Sights, sounds, and other environmental stimuli continually bombard all animals. Behavioral adaptations permit an animal to detect stimuli that are important to survival or reproduction, and then to carry out behavior patterns appropriate to those stimuli (Figure 32-3). Means for discriminating between stimuli and for ensuring that an animal completes a behavior pattern are crucial parts of any animal's behavioral makeup.

Let us begin our attempt to disentangle immediate and ultimate causes of behavior by considering some demonstrations that genes and the environment do indeed affect behavior.

32-B Genes and Environment

Most genes influence behavior because they control an animal's anatomy and physiology, and therefore determine how it can and does respond to its environment. In laboratory populations of the fruit fly *Drosophila,* single genes have been identified that cause a fly to do such things as court members of the same sex, copulate for much longer than normal, or follow a 19-hour rhythm of activity instead of the usual 24-hour cycle. These behaviors result from mutations that cause some abnormality in the sense organs or in the nervous or muscular systems.

Environment affects behavior in two main ways. In the short run, animals perform many behavior patterns only when they are induced to do so by environmental stimuli. In the long run, the environment influences gene expression in the development of many behavior patterns.

One reason behavior is difficult to study is that, in animals with complex nervous and muscular systems, genes and environment continue to interact and alter an animal's behavior throughout its life.

32-C Development of Behavior

An animal's behavior, like its anatomy and physiology, forms during its development, through the interaction between its genes and its environment.

There are often critical periods when a particular environmental influence must be present if a particular behavior pattern is to appear. An example occurs during **imprinting** of young animals. Goslings (young geese) and ducklings learn to follow their parents, and to respond to their parents' signals, during a critical period after they hatch. Konrad Lorenz found that young birds would follow him as if he were their mother if they saw him, rather than their mother, during the critical period. Many animals learn what their future mates will look like by a similar process of sexual imprinting during a critical period (Figure 32-4). (Lorenz had a tame jackdaw [Figure 32-5] that unfortunately became sexually imprinted on him before he understood how the process worked. It caused Lorenz great inconvenience by stuffing regurgitated worms into his ear during its "courtship feeding.")

The interactions between genetics and environment in the development of behavior have been studied intensively in the case of bird song. Peter Marler and Masakazu Konishi showed that male white-crowned sparrows reared in isolation

Figure 32-4
Imprinting. Sexual partners are often determined by imprinting. The large bird is a male Lady Amhurst pheasant that was reared by a bantam hen. He is inappropriately displaying his sexual allure to an unappreciative bantam. (Biophoto Associates, N.H.P.A.)

sing only the inherited song of their species, whereas in the wild, the sparrows learn the dialect of their own local population by listening to adult birds sing. To complicate matters, a bird must hear the dialect during a critical period when it is about three months old if it is to produce the dialect when it first begins to sing, at the age of one year. And even if it has heard the dialect during the critical period, it will never sing this dialect correctly if it is deafened before it has also sung the dialect. Once a bird has sung the full dialect song, however, deafening

Figure 32-5
A jackdaw. (Biophoto Associates, N.H.P.A.)

(a)

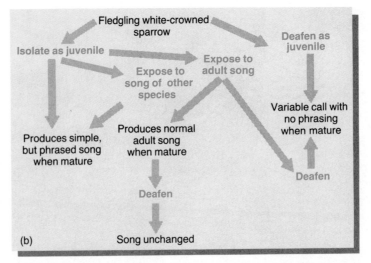

(b)

Figure 32-6
Development of behavior. Summary of Marler and Konishi's findings on song development in white-crowned sparrows. (Photograph: Cornell Laboratory of Ornithology)

■ *Many behavior patterns develop from interactions of the developing nervous system with environmental factors normally encountered during development.*

■ *In general, innate behavior develops where the stimulus is always the same, speed of reaction is important, and the cost of an initial mistake is high. Learning plays a large part in responses to local or changeable conditions.*

has no effect on its further performance (Figure 32-6). This example shows that many factors may be involved in the development of a normal adult behavior pattern.

One further conclusion from this work is that an animal inherits a tendency to learn some behavior patterns but not others. White-crowned sparrows learn the dialects of their own species, but exposing them to the songs of other (even closely related) species during the critical period does not make them learn the songs of these other species. Birds treated in this way end up with a song like that of the completely isolated male.

The environmental stimulus necessary to the normal development of a behavior pattern may be very precise, or it may be rather general. Rats or mice that have been picked up and returned to the nest once or twice a week in their youth mature more quickly, in behavioral terms, than do those never handled. This is a rather unspecific stimulus to the maturation of behavior.

32-D   *Instinct Versus Learning*

The idea that a behavior pattern is either instinctive or learned (or, more often, a combination of the two) is common among behaviorists and in everyday life. **Instinctive**, or **innate** (inborn), behavior is genetically programmed into the nervous system and is difficult to alter. Learned behavior is acquired or disappears as a result of experience.

On the one hand, experimental psychologists have demonstrated that most animals are capable of learning many things. On the other hand, field behaviorists have shown that members of the same species tend to show identical behavior patterns in the wild, suggesting that much of behavior is instinctive. "Instinct" is a difficult concept, however, because it is hard to define except by negatives: instinctive behavior develops without the animal's having to learn it. Such negative definitions are notoriously difficult to use. Furthermore, the only possible experiment to determine whether or not a behavior pattern is instinctive is to deprive the developing animal of as many environmental stimuli as possible and see if the behavior pattern still appears. Even if the pattern does appear under such circumstances, it may still not be instinctive. The experimenter might merely have failed to remove the stimuli that permit the animal to learn the behavior.

Adaptive Value of Learned and Innate Behaviors

One way to make sense of the diversity of behavior is to consider the selective advantages of the two extremes: innate and learned behavior.

As an example, kittiwakes are sea birds that nest on narrow ledges. The chicks keep still from the moment they hatch, whereas related herring gull chicks move around. This innate behavior (or nonbehavior) of kittiwakes is clearly adaptive because a false step means death to a kittiwake chick. There is no room for learning. Innate behavior also saves energy. There is clearly a selective advantage to the animal that does not have to waste energy learning responses that are sure to be required frequently and without variation.

With all of these advantages of innate behavior, why are so many behavior patterns learned? In particular, why are vital behavior patterns, such as recognizing a mate or learning to fly, so often partly learned? Learning gives an animal the flexibility to adapt to a changing environment by acquiring new behavior patterns as they become appropriate, or by responding in new ways to old stimuli. For animals such as vertebrates, which have relatively long lifespans and so experience changing environmental conditions, this flexibility often means the difference between life and death.

Learning is also necessary whenever a stimulus differs for individual members of a species. For instance, every mobile animal with a home base must learn to find that home. No amount of genetic programming will permit a crab to find its own burrow among all the holes in a sandy beach (Figure 32-7). In addition, many social animals live in environments where the relationships between individuals change constantly, and such relationships must almost always be learned.

A species' way of life also determines whether its members evolve learned or innate behaviors. Consider a solitary wasp, which hatches alone, develops as a larva and matures without interacting with other members of her species. The behavior by which she finds a male, mates, builds a nest, and lays her eggs must be largely innate in order for her to perform each action perfectly the first, and perhaps the only, time in her life. On the other hand, a social animal such as a cat can learn much of its behavior from observing other members of its group. It would, however, be an enormous oversimplification to say that the behavior of an insect is innate, and the behavior of a mammal is learned. In any group of animals above the annelids, both types of behavior are vitally important. Even the solitary wasp learns to search for food, to find her way back to her nest, and many other behavior patterns during her short life (Figure 32-8). Similarly, mammals have many innate behavior patterns.

32-E The Neural Basis of Behavior

At a physiological level, a behavior pattern is the action of an animal's effectors (muscles, glands, and so on) in response to a stimulus detected by its receptors. Between receptor and effector lies the nervous system, which determines what information travels from one to the other. In many ways, the nervous system is still the "black box" of behavior. The stimulus that goes in and the behavior that comes out can often be defined, but precisely what goes on inside the nervous system is, in most cases, a mystery. However, the characteristics of a behavior pattern must reflect the organization of the nerve cells that control it.

Many workers have looked for simple behavior patterns that they hope will reveal the essential features of more complex activities. We are now finding out how certain individual neurons function in locomotion and escape reactions of invertebrates such as cockroaches, crayfish, the sea slug *Aplysia,* and leeches, but the picture is still far from complete.

Figure 32-7
An invertebrate that learns. This ghost crab learns to find its home and also learns feeding spots on the beach. It will dig down into a loggerhead turtle nest several feet below the surface of the sand, and return, night after night, to eat the eggs.

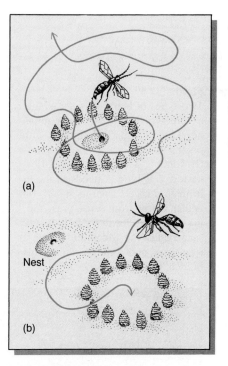

Figure 32-8
Learning by a solitary insect. A digger wasp finds her nest by visual landmarks. (a) The wasp makes an orientation flight over a nest entrance that the investigator has surrounded by pine cones. (b) She returns to the center of the ring of cones, which the investigator has moved in her absence.

Reflexes and more complex behavior patterns share a number of properties that result from the way neurons operate. For instance, both show **latency,** a time delay between stimulus and response. Latency is due to the time necessary for reception by the sense organ, conduction through the nervous system, and excitation of the effector.

Stereotyped Behavior

A striking feature of the behavior of any animal is its repertoire of **stereotyped behaviors**—acts, involving the use of many muscles in a precisely timed sequence, that are always performed in an essentially identical pattern. Reflexes are the simplest examples, but more complicated activities such as locomotion, sound production, breathing, and feeding also fall into this category. For instance, a cockroach or cricket will leap forward in a standard escape reaction when receptors on its abdomen are stimulated by a puff of air. This is an example of an innate stereotyped behavior, sometimes known as a **fixed action pattern.** Other stereotyped behaviors are learned.

Studies of invertebrate neurophysiology suggest that stereotyped behaviors differ from other behavior in two ways: they are controlled by very few neurons in the central nervous system, and they can occur without feedback from the sense organs, although such feedback is available and is often used. For example, the fixed action patterns by which a crayfish flexes its abdomen when it swims are induced by the activity of single, identifiable neurons in the central nervous system. At least some fixed action patterns are "hard-wired" into the nervous system. They are always produced in identical fashion because only one or a few control cells trigger all of the motor neurons for the entire behavior pattern.

By contrast, consider what happens when you pick up a glass of water. This behavior, which is not stereotyped, is a complicated series of interactions between the sense organs and muscles in the arm. The sense organs signal how much the water is slopping about and how far and how fast the glass is rising. These sensory messages reach thousands of neurons in the central nervous system, and hundreds of motor neurons respond, controlling the muscles in the arm so that the glass rises steadily and the water does not spill. Hundreds of central neurons and continual feedback from the sense organs are necessary for this behavior pattern.

An example of a learned stereotyped behavior is a rat's pressing a lever for food. Each rat presses the lever with a characteristic gesture. One uses a fist, another one finger, and each uses its own gesture time after time. Similarly, how

Figure 32-9
Two types of behavior. (a) Walking is a stereotyped behavior. These giraffes need little feedback from their sense organs to stroll across the savanna. (b) Landing is *not* a stereotyped behavior. This blue tit needs all the information it gets from its superb binocular vision, from sense organs in the wing muscles that detect air pressure against the wings, and so on, to stop flying and come to rest on this branch. (Biophoto Associates)

(a)

(b)

you hold your pen, walk, play the piano, or ride a bicycle is a learned stereotyped behavior that is very conservative and unique to you (Figure 32-9).

The selective advantage of behavior patterns programmed into the nervous system is probably that they reduce the number of neurons used in a relatively complex task that must be performed perfectly and often. Some stereotyped behaviors, such as escape movements, must be performed perfectly to work at all. Others save energy because they ensure that the muscular movements of, say, writing or feeding need not be worked out with sensory feedback every time they are performed. All animals seem to be able to "write" programs for behavior patterns into the nervous system during embryonic development and, in many cases, in later life.

Sign Stimuli

What sorts of stimuli trigger behavior patterns? In the 1940s, Niko Tinbergen was studying male three-spined sticklebacks which sported the red belly characteristic of these fish in breeding condition. Every time a red truck drove past a nearby window, all of the fish made frantic attempts to swim through the glass of their tanks toward the truck, as if they would attack it. A male stickleback in reproductive condition will also attack other breeding males.

What stimulus provoked this attack behavior? Tinbergen presented various models to the sticklebacks to find out. When he showed wooden models of sticklebacks to males in reproductive condition, they attacked crude models with an eye and a red belly in preference to life-like models without the red belly (Figure 32-10). The red belly of the male stickleback in reproductive condition is thus the sign stimulus that triggers the fixed action pattern of attack by another breeding male. A **sign stimulus** (also known as a **releaser**) is that portion of the total stimulus which releases a particular behavior pattern.

An interesting extrapolation of the theory of how sign stimuli work is that it is possible to produce a model called a **supernormal stimulus,** which provokes a behavior pattern more effectively than does the normal stimulus. For instance, herring gull chicks peck at the stimulus provided by a red spot on the parent's bill. (This induces the parent to regurgitate fish to feed the chick.) When models of the stimulus were tested, it was found that a bar with big red and white stripes provoked more pecks from young chicks than did a realistic model of the bill (Figure 32-11). The bar was a supernormal stimulus for the fixed action pattern.

Drive and Motivation

A particular stimulus may evoke different responses in the same animal at different times. For example, an animal that sees food will eat if it is hungry, but may ignore the food if it has just eaten. Something inside the animal, which we may call **motivation** or **drive,** is different at these two times. Since different behaviors are appropriate at different times even when the stimulus is the same,

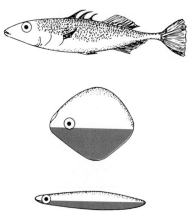

Figure 32-10
The sign stimulus for attack by a male stickleback. Niko Tinbergen found that male sticklebacks in breeding condition do not attack the life-like model lacking a red belly, but they will attack either of the crude models with a red undersurface. The eye is also necessary for the model to act as a stimulus. The presence of an eye is often necessary for an animal to identify an object as another animal.

■ *Many behavior patterns are triggered by specific stimuli, provided the animal is also in a state of nervous or hormonal motivation for that response.*

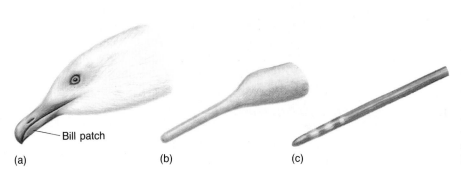

Bill patch

(a)   (b)   (c)

Figure 32-11
Models used in testing for sign stimuli. Young herring gull chicks peck at a colored patch on the parent's bill (see Figure 32-18b). This induces the parent to feed the chick. (a) A life-like (though flat) model of the parent's head releases fewer pecks by a newly hatched chick than (b) a model in which the bill is longer and thinner than normal. This model is less effective than (c), a model that is long and thin and emphasizes the contrast between bill color and bill patch.

variations in motivation help to ensure that an animal's behavior changes to fulfill its short-term needs.

In the vertebrate brain, the hypothalamus appears to control motivation. Attack, escape, and sexual behavior can be evoked by electrical stimulation of parts of the hypothalamus. The hypothalamus seldom acts alone to determine motivation. Its activities can be modified by input from other parts of the brain and by hormone levels. For instance, only when they are in breeding condition, with high levels of the hormone testosterone, do male sticklebacks attack other fish with red bellies.

32-F   Learning

Learning produces adaptive changes in an individual's behavior as a result of its experiences. It occurs in so many different ways that we have to classify them somehow, although there is no evidence that the categories used here bear any relationship to the largely unknown physiological basis of learning.

Habituation is the loss of old responses. Animals may learn not to respond to stimuli that occur often and are unimportant to them. Young animals often show alarm behavior at a variety of stimuli, most of which they rapidly learn to ignore.

Conditioned reflexes are behavior patterns evoked by a previously neutral stimulus which an animal has learned to associate with the stimulus that normally elicits the reflex. The Russian physiologist Ivan Pavlov showed that there is a reflex that causes hungry dogs to secrete saliva when they see food. Pavlov rang a bell when he showed food to the dogs, and after several trials the dogs would salivate when the bell was rung even though he stopped showing them food. The dogs had learned to respond to the new stimulus, the bell, to which they had not previously responded. Pavlov called the bell the **conditioned stimulus.**

Trial and error learning is what its name implies. An animal's spontaneous movements may by chance produce a reward, and the animal learns by trial and error to repeat that behavior pattern. The reward may often be the "pleasure" of performing an action more accurately than before. Trial and error is probably the most appropriate category for the learning of new motor skills (Figure 32-12). Young mammals and birds perfect their prey-catching movements, and humans learn to play the piano, by a trial and error form of practice.

All of these types of learning are varieties of **associative learning. Reinforcement** (reward or punishment) is a central feature of associative learning. Another characteristic of associative learning is that it improves with repetition.

Latent learning occurs without any obvious reward or punishment. It is learning that produces no obvious behavior at the time it occurs. This often happens during exploratory behavior. A recently fed animal may give no sign that it has noticed a new food source until it later returns to feed there.

Insight learning is a form of reasoning that draws on the results of past experiences to arrive at the solution of a novel problem. The classic example of insight in animals came from the work of Wolfgang Kohler on chimpanzees. Presented with a bunch of bananas too high to reach, they would pile up boxes to make a stand from which they could reach the bananas (Figure 32-13). Reasoning of this sort has been shown in many mammals and in some birds, although it is often difficult to distinguish from other forms of learning.

32-G   Territorial Behavior

Now that we have seen something of the evolutionary origins and physical basis of behavior, we go on to consider particular behavior patterns that illustrate these ideas.

Figure 32-12
Trial and error learning. Animals, especially juveniles, learn many things by trial and error.

Figure 32-13
Insight learning. In Kohler's experiment, a chimpanzee was left in a room with a number of boxes and a bunch of bananas hanging from the ceiling. After a period (perhaps of thought) the chimpanzee piled the boxes on top of one another, climbed on them, and reached the bananas.

Figure 32-14
Territorial behavior. Two families of Brent geese threaten each other across the boundary between their territories. (Biophoto Associates, N.H.P.A.)

Many animals defend **territories,** areas where they have a monopoly on resources such as food or nesting sites. Holding a territory has a clear evolutionary advantage, because the resources contribute to the successful production of young. A territory holder attacks and drives away other members of the same species (Figure 32-14). It is to an animal's advantage to defend the territory with a minimum of attack behavior, since every attack carries the risk that the attacker will be injured or spotted by predators. Animals have evolved several features that minimize damage during territorial confrontations. For instance, fighting is infrequent because there are "rules" about who wins encounters between two individuals.

Consider a male thrush defending a territory before the female arrives in the spring. The male is most aggressive near the center of his territory. As he moves toward the boundary, his attacks on a trespassing neighbor become less violent, until he reaches a point at which he is as likely to retreat as to attack when he sees another male thrush. This point marks the boundary of his territory. When two neighbors meet at the boundary between their territories, they both act as if they have conflicting retreat and attack motivations. These tendencies are manifested as conflict behaviors.

Conflict behavior usually contains elements of the two conflicting tendencies (in this case movements toward retreat and toward attack), as well as containing movements apparently unrelated to the issue at hand, such as pecking at the ground (Figure 32-15) or preening (maintenance of the feathers by oiling and smoothing them with the bill). In many species, patterns of conflict behavior appear to have evolved into ritualized **threat displays** that are directed toward

Figure 32-15
Conflict behavior. A gull involved in a territorial clash violently pulls up a clump of grass. The bird acts as if it were caught in a conflict between tendencies to attack and to flee. Instead of doing either, it engages in apparently irrelevant "displacement activity"—pulling up grass. A more placid form of grass-pulling is part of its nest-building behavior.

Figure 32-16
An emperor goose at his nest threatens an intruder. (Biophoto Associates, N.H.P.A.)

intruders (Figure 32-16). Threat is obviously more advantageous than actual fighting in that it does not injure the animal. In the case of a mutual threat display between two animals (which is effectively a ritualized fight), an experienced observer can predict which animal will win by deciding which animal incorporates more attack movements in its display. The loser will eventually move away from the winner.

32-H   Conflict and Courtship

■ *Territorial and courtship behavior appear to have evolved from conflicting behavioral tendencies.*

Most animals, even those that live in social groups, maintain a minimum **individual distance** from one another (Figure 32-17). For example, swallows sitting on a telephone wire are always a certain minimum distance apart—determined by the reach of the neighbor's bill. The invasion of individual distance is perceived as a threat, and the invading animal is usually attacked. The conflicting tendencies to attack and to permit another animal to come close enough to mate are often evident in **courtship behavior,** the behavior patterns

Figure 32-17
Individual distance. These mallard ducks maintain a minimum distance from each other as they rest on a log.

648

Oblique—long call intimidates other males and attracts unmated females.

She pecks him and flies off; he adopts a submissive posture and then tries again.

An unmated female lands and they greet each other with appeasement gestures.

This time with more success: both birds face away and copulation and nest building can now proceed without further hostility.

(b)

He turns his head away with bill down in another appeasement gesture. But she does not reciprocate; her head still points forward and her wings are raised in an aggressive posture.

(a)

Figure 32-18
Courtship in the black-headed gull. (a) The sequence of courtship behavior. Follow the arrows around from the top. (b) A gull uttering the long call by which it claims a territory and, if it lacks a mate, attracts unmated females to begin courtship. (Notice the red bill patch, the sign stimulus which prompts gull chicks to beg for food.)

that precede mating in most animals. The courtship displays of many species seem to have evolved from such conflict behavior.

In a well-studied example of courtship behavior, the male black-headed gull attracts a female to his territory. She alights near him, and both gulls adopt a series of postures that resemble, but are slightly different from, the characteristic threat display of the species. If neither bird attacks the other, both display appeasement gestures, which imply that the hostility between them has lessened. Eventually the female flies off, but she may return many times, and each bird will display fewer threatening and more appeasement gestures with each visit. Eventually the greeting ceremony ceases entirely, and the male feeds the female. After this, copulation can occur and a permanent pair bond forms (Figure 32-18).

In their initial encounters, the birds are displaying behavior that reveals three conflicting tendencies—to attack, to flee, and to stay together. The behavior patterns that result from the conflict have evolved into an elaborate courtship ritual.

32-I Migration and Homing

Many animals have remarkable navigational ability, sometimes travelling over hundreds of miles of land and sea. Migrating animals can do many things that we cannot do ourselves. A Manx shearwater (a sea bird), which had never been more than 10 miles from home, was removed from her nest on an island off the coast of Wales, flown to Boston, and released. She was back on her nest before the letter announcing her release reached the observers in Wales. To perform an equivalent feat, such as sailing from Boston to Wales, a human being would have to spend hours learning to use navigation instruments to cross the ocean, and would still need a map to find the nest on the other side. Birds, monarch butterflies, fish, and salamanders all perform equivalent journeys without mechanical aids and with little or no learning (Figure 32-19).

Many animals, like horses, dogs, cats, and humans, orient themselves by landmarks that they learn and recognize visually. Many animals can also find their way around by using their chemical receptors. Dogs can follow long and complicated scent trails in unfamiliar territory. Moths find mates and ants find their nests by following odor gradients. A dramatic case is that of the salmon, which hatches in a freshwater stream and matures hundreds of miles away in the ocean. Years later, when the time comes to spawn, each salmon "smells" its way back to the very stream in which it hatched.

Many animals can move in a specific compass direction. Bees can tell direction from the sun, and birds can orient themselves in a specific compass direction in the same way. Since the sun appears to move from east to west during the day, an animal using the sun as a reference to maintain a constant compass direction must also know the time of day. It must also be able to compensate for the sun's apparent movement throughout the daylight hours.

Figure 32-19
Migration. (a) A flock of several different species of terns migrating up the coast of South Africa. (b) Monarch butterflies on Cape Cod resting during their autumn migration. (a, Biophoto Associates, N.H.P.A.)

(a)   (b)

Animals have internal clocks that control their circadian rhythms (Section 31-G). By exposing migrating animals to artificial daylight that began and ended 6 hours (one fourth of a day) later than natural daylight, it proved possible to "clock-shift," or reset, their internal clocks. The animals then interpreted the sun's position incorrectly and oriented themselves in a compass direction that was 90°(one fourth of a circle) clockwise away from the correct direction for migration. These experiments confirmed that by combining information from the direction of the sun and their internal clocks, birds can point themselves in a particular compass direction. Birds that migrate at night can also use star patterns to find a compass direction.

The ability to fly on a constant compass setting using the sun or stars, however, does not explain the ability of the Manx shearwater to cross the Atlantic or of a homing pigeon to return to its loft after it has been released in unknown country. In order to get from A to B using a compass you have to know whether A is north, south, east, or west of B. This is called a map sense because it means that you must know the relative positions of A and B on a (hypothetical) map. Many animals obviously have a map sense, which tells them where on earth they are.

Studies with homing pigeons have shown two other cues that some individuals use to navigate, and that could provide them with a map sense: odors, and the earth's magnetic field. Bits of magnetite, a magnetic iron ore, have been found in the heads of pigeons, and in dolphins, whales, monarch butterflies, tuna, mackerel, and some bacteria. These particles are presumed to be involved in the ability to sense a magnetic field.

Studies of how pigeons navigate are complicated because pigeons have an innate ability to learn to find their way home using any one of several cues. Apparently local conditions near a pigeon's home loft determine which cues it tends to pay attention to. Because learning plays a part in homing, experienced birds reach home faster than naive ones. Inexperienced birds use a simple hierarchy of navigational cues. If the sun is shining, they use a sun compass, so that if they have been clock-shifted, they fly off in the wrong direction. Experienced birds are less likely to be fooled by a clock shift. They appear to cross-check the information from their sun compass with other cues, such as odor or the magnetic field, which give conflicting information if the birds have been clock-shifted. An experienced bird solves this dilemma by going to sleep in the nearest tree until the effect of the clock-shift wears off. It then flies straight home.

Much has been learned about animal navigation and orientation in the last two decades, but we are still far from understanding this remarkable collection of behavior patterns.

32-J *Adaptive Decision-Making*

We do not know whether other animals consciously consider their actions. Nevertheless, all behavior involves deciding among many alternatives: to migrate this winter or stay home, to continue feeding on this lawn or move on to the next, to fight for a territory or wait until one falls vacant. The first studies on how animals make these decisions involved the behavior of animals searching for food.

Consider a chickadee eating insects in a patch of forest (Figure 32-20). When will it leave that patch and fly to the next one? If it left every patch as soon as it arrived, it would spend all its time moving and find little food. But if it stayed in one patch all day, it would use up the food there and also take in little food that day. There is selection for the bird to perform some sort of cost-benefit analysis. The costs of moving include the energy used to find and fly to the next patch and

Figure 32-20
A black-capped chickadee. This species was used in the first studies of "cost-benefit analysis." Here, an adult returns to its nest hole after a successful foraging trip. (Uve Hublitz/ Cornell Laboratory of Ornithology)

the risk that the next patch will contain little food. The benefit of moving is the chance of finding more food than in the patch the bird has already picked over. The result of this cost-benefit analysis will vary with the environment. The bird will improve its rate of feeding if it spends longer in patches containing many insects than in patches containing few. So selection will favor flexible decision-making that changes depending on the circumstances.

Computer models predict that a chickadee will find food most rapidly if it leaves every patch (no matter how much food it contains) as soon as its rate of feeding begins to slow down. When the bird enters a patch, it will first find all the obvious insects. Later, it will start to spend longer and longer looking for each additional insect. It should leave at this point. Studies show that chickadees in a forest and parasitoids searching for a host (Chapter 21, *A Journey Into Life,* "Parasitism") behave very much as this model predicts. The chickadee's behavior can be described by saying that it remembers how long it takes to find an insect. When that time has increased for each of a few insects, it moves to a new food patch.

Behaviorists are beginning to apply cost-benefit analysis to more complicated decisions. For instance, animals tend to perform altruistic behavior (behavior that benefits others, Section 16-E) when the cost is slight and the potential benefit to their kin is high. Territorial behavior may vary with the probable costs and benefits of attempting to raise a family.

Most tawny owl pairs that reproduce successfully defend territories year round. One study over many years examined a habitat containing 25 territories, each with its pair of owls. In a typical year, eight of the pairs did not reproduce at all, another nine abandoned the eggs they had laid, and two more pairs allowed their chicks to starve to death. The remaining six pairs raised 18 young among them. It appears that at each stage of reproduction (copulating, incubating eggs, feeding chicks) the adults in effect perform a cost-benefit analysis and decide whether or not to invest the necessary energy in the next stage. Environmental cues determine which decision will be made. Tawny owls feed mainly on mice, and the number of mice in the territory at the time appears to be the main factor in each decision. This is adaptive because if there are not enough mice to raise the young, any mice invested in the attempt will be wasted. The adults must eat to keep themselves alive whether they raise the young or not. If they eat surplus mice themselves they improve their chances of surviving to future years, when mice may be plentiful enough to raise young to maturity.

■ *The processes by which animals make decisions produce remarkably flexible behavior, which changes with changing circumstances so that it enhances reproductive success.*

32-K   Social Behavior (Sociobiology)

Some animals have little contact with members of their own kind, but in many species, individuals do interact to some degree, either for a short time or throughout life. **Social behavior** is the cooperative interaction of members of the same species.

It is often difficult to tell whether particular groups of animals are truly social. For instance, gray herons spend much of their lives alone or with their mates. Occasionally, however, these birds spend days, and even weeks, together feeding. Is this feeding group social? Do the birds communicate with each other or help each other with the fishing? It turns out that the herons are all in the same place only because each of them has found food there. They do not interact any more than human diners at separate tables in a restaurant do. Thus it is not safe to assume that a group of animals constitutes a society. A group of animals is truly social only when there is considerable cooperation and communication among the individuals (Figure 32-21). Examples of extremely social animals are honeybees, humans, and wolves, which form cooperative, long-lived societies upon which the individual's very life depends.

Figure 32-21
Social animals. A lion cub (right) greets its mother. Young lions learn much of their behavior from their elders, and hone their motor skills during trial and error "play." The nucleus of a pride is several related females. Males born in other prides, working in groups of two to five brothers or cousins, become members of the pride until ousted by another coalition of males. Members of the pride cooperate in most activities, notably in hunting.

Communication

All animals that live in societies communicate with other individuals. Human beings, for instance, use sound and hearing (when we speak, clap, or laugh), and visual stimuli and vision (advertising posters, "body language," dressing up, shaking a fist) among our means of communication. Birds, like humans, have highly developed vision. They communicate largely by movement and color (also by sound). Communication by sound can be highly elaborate even if it does not involve language, which is the most important aspect of our own sound communication.

Because our sense of smell is poor, we pay little attention to the chemical communication so common in other animals. Many mammals, such as dogs, mark their territories, determine another animal's mood, find their mates and food, and, for all we know, communicate in many other ways, by scent. A **pheromone** is a chemical that affects the behavior of another member of the species (Section 31-H).

Honeybee Societies

Many insects are more or less social, but honeybee (and ant) societies are the most elaborate and widely studied. The unit of social organization is a family of related individuals. A society of honeybees typically consists of a reproductive female (the queen) and her daughters (and sometimes her sons).

A honeybee hive may contain 80,000 individuals, each with its own job. The queen lays eggs, drones (males) produce sperm needed to fertilize the eggs of new queens, and workers (sterile females) tend larvae, clean the hive, and forage for food (Figure 32-22). The tasks a bee performs, the number of queens produced, and the founding of a new hive are all organized by pheromones.

Figure 32-22
The life and times of honey-bees. (a) A worker "fanning." She stands on her forelegs and uses her wings to disperse pheromones secreted by the scent gland on her abdomen. (b) Workers tending larvae (white objects). (c) A swarm of honeybees on a tree. (Biophoto Associates, N.H.P.A.)

(a)

(c)

(b)

A honeybee queen takes one mating flight, during which she mates with several males. She stores the sperm and uses them to fertilize the thousands of eggs she lays during her life of seven years or more. Her eggs hatch into larvae, cared for by the workers. The diet fed to a larva determines whether it develops into a queen or a worker. A new worker usually first serves as a nurse, preparing cells for eggs and feeding the larvae after they hatch. After about two weeks the worker becomes a house-bee, cleaning, secreting wax for the honeycomb, and guarding the hive. After this she forages outside the hive for the remaining five or six weeks of her life.

As we would expect from such a complex society, honeybees communicate extensively. Karl von Frisch found that foraging bees returning from a successful trip "dance" on the honeycomb, recruiting other bees to harvest a new good food source, and indicating where to find it. Pheromones permit bees to identify their own hive and serve as alarm signals. A pheromone produced by the queen prevents the workers from producing any more queens and ensures that all the female larvae are fed so that they develop as workers. This continues until the hive is overcrowded, when the queen stops producing that particular phero-mone, and the workers start to raise new queens.

Eventually the old queen may leave, for only one queen can survive in a hive. As she leaves, the queen secretes a swarming pheromone, which attracts many of the workers and keeps them with her. The swarm lands somewhere and may remain several days while scouts search for a new site. The scouts return to "dance" a description of the location of a possible new nest. The intensity of her dance conveys the scout's impression of the site's merits. Other workers go to inspect the sites, and finally a consensus emerges when all the scouts are dancing for one site. The swarm then flies to the new site and settles in. This method of making a decision impresses us by its resemblance to the way we like to think humans act.

Vertebrate Societies

Like insect societies, most vertebrate societies consist of genetically related individuals. Unlike insect societies, however, all members of the vertebrate society are fertile, and competition to reproduce is important in determining the social system. A typical vertebrate society consists of genetically related females and their offspring plus unrelated males who join the society for shorter or longer periods. An interesting exception is chimpanzees: here a territory belongs to a group of related males, and females born in other families move into the area (see *A Journey Into Life,* this chapter).

Most vertebrates learn a much greater proportion of their behavioral repertoires than do insects. How does this difference affect the social behavior of the two groups? Members of vertebrate societies can identify each member of the group individually, whereas insects probably cannot. There is usually a dominance hierarchy, or "pecking order," which ensures that dominant individuals have first choice of desirable but limited things such as food, shelter, or mates. An individual's role in the society is largely determined by its position in the hierarchy, and so ability has more effect on an individual's role in a vertebrate society than in an insect society (Figure 32-23). Individuals may fall in rank as the result of age or disability. In one baboon troop, the top male changed five times in two years. Position in the hierarchy is not determined solely by an individual's size or fighting ability, however. In many species (probably most primates), having a mother of high status gives one an initial boost up the social ladder.

Threat displays (Section 32-G) and related behavior patterns are important in maintaining dominance hierarchies. A dominant individual displaces a subordinate by threat behavior. A subordinate responds with appeasement gestures, which inhibit other animals from attacking. Appeasement gestures show submis-

Figure 32-23
Primate society. Nearly all primates, like these long-tailed macaques, live in social groups with dominance hierarchies. Grooming behavior, possible only between social animals, removes parasites and contributes to the health of all members of the group. (Biophoto Associates, N.H.P.A.)

Chimpanzee Societies

A JOURNEY INTO LIFE

The social structure of chimpanzees differs from that of most other social animals. This is of interest not only in itself, but also because chimpanzees are the nearest living relatives of humans: we share about 99% of the same genes. Chimpanzee behavior has some interesting parallels with, and differences from, human behavior.

Chimpanzees live in groups with 50 or more members who occupy a territory of 10 to 30 square kilometres of tropical rain forest in Africa. A chimpanzee spends more than three fourths of its feeding time eating fruits, preferring figs, which are rich in protein. Chimpanzees also eat various other plant parts and insects, and hunt mammals, including monkeys (which also reduces competition for fruit!).

In one study, the territory of a group contained 100 different species of trees, but only ten of these species bore edible fruit. Tropical trees produce fruit at unpredictable intervals, and the crop is ripe only for a short time. Hence chimpanzees spend much of their time travelling in search of trees with fruit. They must

also fend off fruit-eating birds and monkeys, which are smaller and more agile in the treetops. Chimpanzees move efficiently through the forest: they walk on the ground and then climb the fruiting trees, rather than moving through the branches of the forest canopy as monkeys do.

Like other social animals, chimpanzees tend to spend time in large groups. However, few trees bear enough fruit to feed many chimpanzees at once. Hence when chimpanzees travel in search of food, they go alone or in groups of 3 or 4 animals. By scattering in different directions, each little group manages to find enough food while minimizing the energy spent travelling. If a group finds a tree with more than enough fruit to feed its members, the males give loud cries, which soon attract other chimpanzees to feed on the tree too.

Chimpanzees spend their "resting" time in larger groups. Much of this time is spent grooming: searching another animal's fur and picking lice off the skin. This closeness is thought to promote social bonds, but it also keeps the animals healthy by removing parasites that might carry disease. A grooming animal mostly picks lice off parts of the body that the animal being groomed cannot see or reach easily itself—the head, neck, and back.

In most social animals, males are expelled from the group as they approach maturity. The group's nucleus is the females and their daughters, while males come and go, fighting one another for the chance to join the female group and hence to reproduce. In chimpanzees, however, this situation is reversed.

Male chimpanzees remain in the community of their birth. They become its defenders and the fathers of the group's new offspring. Over the generations, this means that the males in a chimpanzee community are all closely related to each other, and that the territory is in effect held and passed down through the male line.

Females, on the other hand, join new groups on reaching maturity. Hence the females are unrelated to their mates in the new community. The females may or may not be related to each other.

As a result of this social structure, it is to a male chimpanzee's evolutionary advantage to promote the welfare not only of himself, but also of other individuals. The group's other adult males are his close relatives and share many of his genes, and the group's young are his own offspring or the offspring of his near relatives. By helping any of these individuals, who share many of the same genes as his own, the male contributes to the welfare of many copies of his own genes. Hence, a male's calling when he finds an abundant food source, which attracts other troop members, is selected for.

An adult female, on the other hand, can best ensure her own evolutionary success by looking out for herself and her own offspring. Indeed, motherhood takes up a lot of her time and energy: nine months for each pregnancy (starting at about 15 years of age), and four or five years of nursing, protecting, teaching, and carrying the infant on her food-finding expeditions, until it becomes independent and she can begin a new pregnancy (Figure 32-A).

■ *In animal societies, individuals interact and so enhance reproductive success of the genes they share.*

sion, often by turning away weapons (teeth or beak) or by presenting the vulnerable throat or belly to the dominant animal.

The evolutionary advantage of a social hierarchy is probably that it reduces the harmful effects (such as injury from fights) of the inevitable competition between related individuals living in the same area. The society also ensures that at times when resources (such as food) are in short supply, some individuals will get all the food they need to survive instead of the whole group becoming half-starved and likely to die, as happens with honeybees. When members of a group are related, such behavior will be selected for because individuals carry many of the same genes, and an individual that starves while a relative lives to reproduce is actually contributing to the survival of many of his or her own genes in future generations.

The social structure of chimpanzees probably explains why males tend to spend more time with other males, females with other females and their offspring. It also accounts for the fact that male chimpanzees do not compete intensely for opportunities to copulate with females in estrus and so become fathers of the next generation. Males often ignore copulations occurring only a few metres away, and an estrous female may copulate with all the males in a party within a short time. However, a dominant male sometimes takes an estrous female "on safari" apart from the group, thereby excluding other males from mating during her receptive period.

Although chimpanzees have ten times more friendly than aggressive interactions with members of their own community, they are hostile to members of other groups. Groups of females have been observed to drive away "foreign" females. However, male aggression is much more frequent and more brutal. A community's males patrol the boundaries of their territory and keep members of other groups out. Males encountering a foreign female and her young have been observed to kill the infant. This behavior is selectively advantageous to the male on two counts: first, it eliminates competition from unrelated genes. Second, having lost her infant, a female quickly comes into estrus, and so the murderer or his relatives may be able to mate with her and father her next offspring if she remains in their territory. Hence natural selection favors perpetuation of genes that make a male more likely to kill young born into other groups, and of genes that make

a female more likely to stay well within the boundaries of the territory protected by the males of her community. Males of one community have even been known to kill the males of a smaller group and take

over their territory—a form of warfare.

These observations of chimpanzee behavior give much food for thought about our own behavior and its genetic consequences.

Figure 32-A
A chimpanzee and her young. (Visuals Unlimited)

SUMMARY

The genes that an animal inherits determine the range of behavior patterns it can develop. In addition, most behavior patterns, innate or learned, develop normally only if the animal is exposed to the appropriate environmental conditions.

The immediate reason that an animal behaves in a particular way is that it has been exposed to environmental stimuli that induce the behavior pattern while the animal was in the appropriate physiological state. Ultimately, behavior patterns that must be produced

perfectly at the first exposure to the stimulus are usually innate. Learning requires time and energy and is reserved for behavior that must be flexible in meeting local or changing conditions. Many behavior patterns, both innate and learned, become programmed into the nervous system. These stereotyped behaviors may be triggered by sign stimuli and controlled by a small number of neurons with minimal sensory feedback.

Animals are always exposed to a variety of stimuli, which may or may not evoke a response, depending on factors such as the animal's physiological state. Conflict behavior, frequently seen in courtship and territorial displays, is one possible outcome of mutually exclusive behavioral tendencies.

Animals make adaptive decisions about which behaviors to perform, using cues that help them maximize the benefit and minimize the cost of the behavior.

Most animals seldom cooperate with other members of their own species, but true societies have evolved in some species of insects and of vertebrates. A society usually consists of genetically related individuals, whose cooperation enhances the survival of genes they share. Communication between individuals is most highly developed in social animals. Vertebrate societies are characterized by hierarchies that determine an individual's access to limited resources. The society provides an individual with protection, and its members cooperate in various aspects of their lives.

SELF-QUIZ

From the list of types of behavior patterns below, choose the one exemplified by each of the following situations.

_____ 1. A male cardinal attacks any other male cardinal that tries to come into your back yard.

_____ 2. A puppy rolls on its back when a strange adult dog growls at it.

_____ 3. A cat meeting a strange (and not overly large or fierce) dog arches its back, fluffs its fur, and hisses.

_____ 4. A student in a typing class makes fewer errors on the tenth homework assignment than on the first.

_____ 5. Your signature is the same every time you write it.

_____ 6. Newly hatched ducklings follow a windup toy as if it were their mother.

_____ 7. One chicken pecks any other member of the flock that gets too close, but no other chicken pecks it back.

_____ 8. A skilled musician can play a tune she or he has never heard before, after someone hums a few bars.

a. appeasement
b. dominance
c. imprinting
d. insight
e. conditioned reflex
f. stereotyped behavior
g. territoriality
h. threat
i. latent learning
j. trial and error

9. Courtship behavior is said to show conflict because:
   a. the two mates fight frequently
   b. the mates cannot immediately agree on a nest site
   c. the mates are sexually attracted to each other but do not normally permit another animal to get as close as copulation requires
   d. the mates must choose each other from a large number of members of the opposite sex
   e. the mates are in competition for food in a territory of limited size

10. The part of the nervous system important in the control of drives for such behavior patterns as feeding, drinking, and sexual behavior is the:
    a. pituitary
    b. sympathetic system
    c. parasympathetic system
    d. reticular activating system
    e. hypothalamus

11. Which of the following is *not* true of stereotyped behaviors?
    a. They may be triggered by sign stimuli.
    b. They are initiated by one or a few neurons.
    c. They exhibit latency.
    d. They can be learned or innate.
    e. None of the above.

QUESTIONS FOR DISCUSSION

1. Despite everything you have read in this chapter, you probably still think that most of an insect's behavior is innate, and that most of a human's or chimpanzee's is learned. Can you justify this position?

2. What are some of the possible selective advantages to defending a territory, an activity that consumes time and energy and increases the risk of injury?

3. William Dilger studied the nest-building behavior of parakeets of the genus *Agapornis*. He crossed members of a species that carries nest-building material in its beak with members of a species that carries its nesting material tucked under its tail feathers and observed the behavior of the hybrid offspring. These offspring showed hybrid behavior and usually dropped the material whether they carried it in their beaks or their feathers. One particular bird tried to build a nest 48 times and failed. On the 49th try it was successful. What does this tell you about whether nest-building behavior in this genus is innate or learned?

SUGGESTED READINGS

Dilger, W. C. "The behavior of lovebirds." *Scientific American,* December 1962. The mixed up lovebirds referred to in Discussion Question 3.

Ghiglieri, M. P. "The social ecology of chimpanzees." *Scientific American,* June 1985. A fascinating study of our nearest relatives.

Gould, J. L., and P. Marler. "Learning by instinct." *Scientific American,* January 1987. An excellent account of how the innate tendencies of different animals to learn different kinds of information is adaptive for the species' way of life.

Gwinner, E. "Internal rhythms in bird migration." *Scientific American,* April 1986. A biological clock is involved in migration and navigation.

Krebs, J. R., and R. H. McCleery. *Behavioural Ecology: An Evolutionary Approach.* 2d ed. Oxford: Blackwell Scientific Publications, 1984. Contains an analysis of predicted and observed decision-making during foraging behavior.

Lorenz, K. Z. *King Solomon's Ring.* London: Methuen, 1942. Delightfully written autobiographical account of life with animals.

Southern, H. N. "The natural control of a population of tawny owls (*Strix aluco*)." *Journal of Zoology* 162:197, 1970. The study of factors that determine tawny owl breeding behavior.

Tinbergen, N. *The Animal in its World.* Cambridge, MA: Harvard University Press, 1972. Selections from Tinbergen's work, including experiments and general papers. A very readable description of the types of experiments you might perform if you became a behaviorist.

Tinbergen, N. *Curious Naturalists.* Garden City, NY: Doubleday & Company, Inc., 1969. Autobiographical description of Tinbergen's life as an animal behaviorist.

Plant Biology

Harebell

We now turn to the most complex and successful group of plants, the flowering plants, to see how they carry on the business of living. Plants do the same basic things as all other organisms—acquire energy, exchange gases, grow, develop, respond to the environment, reproduce, and so on. However, the plant body plan, and its component organs, differs markedly from those of the animals we just studied. Plants have no digestive, excretory, nervous, muscular, or endocrine systems. Their arrangements for energy and nutrient procurement, gas exchange, reproduction, and hormone production are not centralized but scattered over many local areas. Even the transport system distributes substances within a plant's body without the cilia or muscles that propel fluids in an animal. Flowering plants have only three main types of organs—roots, stems, and leaves. At times they also produce reproductive organs, the flowers, parts of which develop into fruits and seeds.

Plants are autotrophs: they make their own food by trapping the sun's energy during photosynthesis (Chapter 8). A plant's structure, physiology, and way of life all result from the demands of autotrophy.

Foremost among these demands is a large surface area spread to intercept the energy of sunlight. Most plants produce this surface not in one vast sheet, but in an array of modules, the leaves. Leaves are broad and flat, a shape that provides a lot of surface area for very little weight. A plant's leaves are deployed in a pattern that intercepts as much sunlight as possible, with little overlap and shading of each other.

Besides light, leaves also absorb carbon dioxide, a gas in short supply in the air. Leaves admit air through minute pores in their surfaces, but at the same time they lose enormous amounts of water. So the second requirement of plants is for water, which they absorb through the large surface area of another set of modules, the roots.

So far, so good, but a plant is not alone: it is surrounded by other organisms. Neighboring plants also produce leaves, which may shade those on our first plant and so reduce the amount of energy it captures. Animals eat some of the leaves, and fungi may invade and damage others. Therefore, the plant must grow throughout its life, producing new sections of stem, with new leaves. This replaces lost or damaged leaves and keeps up with the competition for light. To keep pace with the loss of water from the increased number of leaves, the plant's growth must also add new roots, spreading through as much of the soil volume as possible and competing for water and mineral nutrients. As we saw in the case of animals, a plant must also maintain homeostasis of its cells' environment and coordinate the activities of its parts. However, in plants the emphasis of coordination is on the balanced growth of parts, so that the root system below ground and the shoot system above support each other, in both structure and function.

In Chapter 33 we shall study the growth and development of the plant's vegetative modules—roots, stems, and leaves—and their structure when they reach maturity. Chapter 34 examines how plants absorb water and minerals from the soil and how the transport system moves water from the roots to the leaves, and food from leaves to other parts of the plant. Then, in Chapter 35, we shall see how the production and growth of various plant parts are coordinated and how the plant responds to its environment. Finally, we shall look at the process of reproduction and the development and dispersal of new individuals (Chapter 36).

Composite flower

Beach primrose

Field of corn poppies

Rainbow cactus

Plant Structure and Growth

CHAPTER

33

Rainbow cactus

Bark

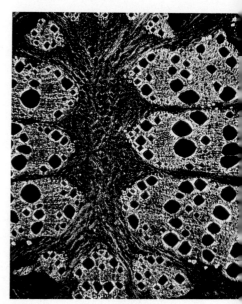

Lilac stem cross section

Redwood sorrel

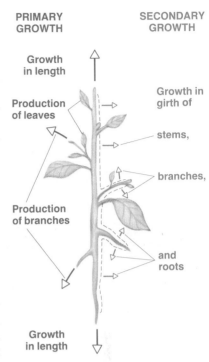

*M*ost vascular plants are autotrophic: they make their own food, using the energy of sunlight and the materials of carbon dioxide from the air, and water and minerals from the soil (Chapter 8). The structure of plants is adapted to their autotrophic way of life. A typical vascular plant has a **root system** that absorbs water and minerals from the soil and a **shoot system** made up of one or more stems with leaves. Stems hold the leaves up where they can intercept sunlight, and the leaves' vast surface area allows them to capture a great deal of light energy, which they store in food molecules.

A vascular plant grows larger and adds new parts throughout its life. Most of the plant body consists of mature, differentiated cells, with specialized functions. Normally, these cells will not divide again. But the plant can grow indefinitely because it has tissues called **meristems,** with cells that retain the ability to divide and produce new cells. Some of these cells differentiate and become new parts of the plant. Others remain meristematic.

Plants vary a lot in size and shape. A plant's structure depends on which parts grow, and how much; on where branches arise, and how much they grow. Ultimately, all of this depends on the activities of the meristems, which lay down the cells for each part.

The growth of plants can be divided into two aspects, resulting from the activities of different meristems (Figure 33-1). **Primary growth** is growth in length of the shoots and roots, production of new root and shoot branches, and production of leaves. The meristems responsible for primary growth occur at the ends of the growing stems and roots. The cells laid down during primary growth make up the primary plant body. **Secondary growth** is growth in girth, or thickness, of the stems and roots produced by primary growth. The meristems that produce secondary growth lie a short distance under the surface of stems and roots. The tissues contributed by secondary growth strengthen the plant and provide the support necessary for new primary growth, both in height and in the spread of the branches. Although all vascular plants exhibit primary growth, secondary growth varies a lot: some species have none while others produce extensive secondary growth.

A plant's meristematic tissues produce cells that grow and differentiate to form three main kinds of mature tissues (Figure 33-2).

Figure 33-1
Primary and secondary growth. Primary growth is mainly growth in length and production of new branches as a result of division and enlargement of cells in meristems at the tips of the plant (red arrows). Primary growth produces the plant's primary tissues, which comprise the primary plant body shown here. Secondary growth produces growth in girth (blue arrows and dashes).

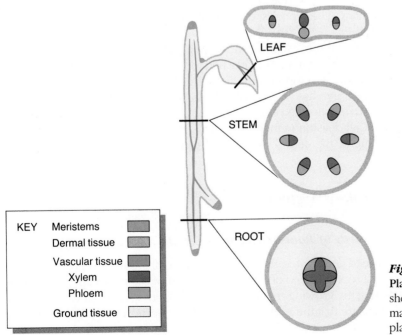

Figure 33-2
Plant tissues. This schematic diagram shows the distribution of the three main kinds of tissues in the primary plant body.

1. **Dermal** ("skin") **tissue** covers and protects the outside of the plant. This layer contains close-fitting cells that often form structures specialized for defense or for obtaining materials from the environment.
2. **Vascular tissue** transports substances over long distances within the plant. There are two types of vascular tissue: **xylem** conducts water and minerals absorbed by the roots up to the stems and leaves, and **phloem** conducts food from sites of production to sites of use or storage. Conducting cells in the xylem and phloem are long and narrow, like sections of a pipe.
3. **Ground tissue** makes up the bulk of the primary plant body, filling the space between the dermal and vascular tissues. Ground tissue consists mainly of a type of cell called **parenchyma,** rather rounded in cross section, with relatively thin cell walls and with many spaces between cells. Parenchyma cells often contain many plastids.

KEY CONCEPTS

- A plant grows as its meristems lay down new cells, which then grow, differentiate, and mature as specialized cells in new plant tissues.
- The vascular plant body consists of roots, stems, and leaves. Each of these organs contains three kinds of tissue:
  1. Dermal tissue, which forms an outer, protective layer.
  2. Vascular tissue, which transports water and food in the plant body.
  3. Ground tissue, which fills in between.

33-A   The Bean Seed

We begin our study of plant structure by following the growth of a bean seed into a young plant, to see how its meristems produce its primary structure. A bean seed is a reproductive package containing a plant embryo in an arrested stage of development, along with its food supply (Figure 33-3). The "skin" of the bean is the protective **seed coat.** Within this coat is the embryonic bean plant. If we peel the seed coat off, we can see the two "halves" of the bean, which are really the embryo's two large "seed leaves," or **cotyledons.** The presence of two seed leaves shows that beans belong to the group of flowering plants called **dicotyledons** (**dicots** for short). A bean's cotyledons are large because they have absorbed the food supply that will nourish the growing bean seedling. In the seeds of many other species, the food is stored separately, outside the embryo itself (see Figure 33-20a).

If we separate the cotyledons, we can see that they are part of the embryo, attached to its tiny main axis. Also attached to this axis are the two **first foliage leaves,** which will develop into the first "leafy" leaves.

The bean embryo has two meristems, one at each end of its axis. Since each of these meristems is located at a tip, or apex, of the plant, they are called **apical meristems.** Cell division in these meristems, followed by growth and differentiation of some of the cells formed, results in growth of the plant's root and shoot systems.

33-B   The Root System

Primary Growth of Roots: Growth in Length

As the bean seed germinates, the first part of the embryo to start growing is the seedling's first root. The end of the root is covered by the **root cap,** a thimble of cells protecting the growing root tip. Cells in the root cap detect the direction of gravity and ensure that the root grows downward (Section 35-C). The root cap's outer cells are scraped off as the root pushes its way down among the rough soil particles.

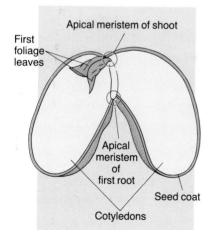

Figure 33-3
A bean seed, opened to reveal the embryo. Food is stored in the bean embryo's seed leaves, the cotyledons. The apical meristem of the shoot is hidden between the first foliage leaves.

The root cap cradles the apical meristem, where cells divide and produce new cells for both the root cap and the root itself (Figure 33-4). Cells destined to become part of the actual root become arranged into definite strands. Above this area, in the **zone of elongation,** the cells are dividing more slowly and are growing longer, pushing the root cap and apical meristem through the soil. The lengthening of these cells is the actual force of growth in the root: the root tip is literally pushed through the soil. Above the zone of elongation, cells in the **zone of maturation** have reached full size and are developing into specialized cells that will perform particular functions. Meanwhile, the apical meristem continues to divide and produce new cells, which follow the same sequence of division, growth, and development.

Primary Structure of Roots

When mature, the cells laid down by the activity of the apical meristem form the root's **primary tissues.** The outermost layer consists of dermal tissue called the **epidermis.** Its cells fit together tightly and form a protective covering, one cell thick, around the root. Some of the epidermal cells in the zone of maturation become **root hairs** by forming extensions that grow out among the soil particles (Figure 33-4). Root hairs anchor the plant in the soil and increase the surface area for absorption of water and minerals.

The bulk of the root's primary tissue consists of a thick layer of ground tissue, the **cortex,** just inside the epidermis (Figure 33-5). The cortex is made up of many parenchyma cells, which may contain **amyloplasts,** plastids that store starch. There are many spaces between the cells of the cortex.

The **endodermis** ("inner skin") is a single layer of cells at the inner edge of the cortex. The endodermis helps control the movement of substances between the root cortex and the root's interior (Section 34-C). Just within the endodermis is another single layer of cells, the **pericycle.** These cells do not differentiate, but remain meristematic, able to divide as needed.

Within the pericycle are the vascular tissues, the root's xylem and phloem, which transport substances to and from the shoot system. Conducting cells of the phloem are among the first cells of the young root to differentiate. They then bring in food for the growth and development of cells near the root tip. Trans-

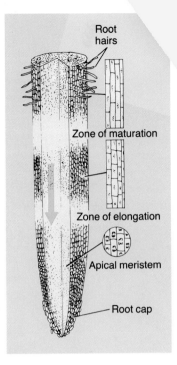

Figure 33-4
A growing root tip. The circle and boxes show what the cells in each area look like through the microscope. Cells in the apical meristem divide and produce new cells for the root and the root cap. The zone of elongation contains newly produced cells that are growing and pushing the root tip through the soil (arrow). Older cells are differentiating in the zone of maturation, to form the tissues shown in Figure 33-5.

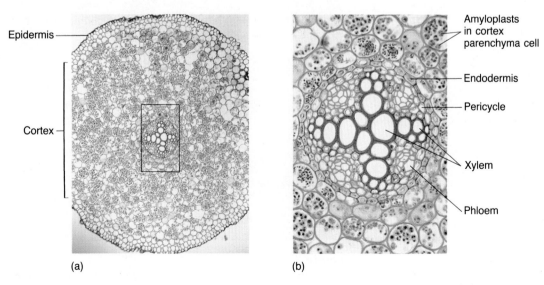

Figure 33-5
The mature primary tissues of a root. (a) Cross section of a buttercup root. (b) Close-up of the center of the root in (a). (Ed Reschke)

port cells in the xylem do not start to function until they have grown to their final size and died, leaving a hollow cell wall to conduct water upward. We shall examine the structure and function of vascular tissues in the next chapter. For now, note that they usually occupy the core of the root. Cross sections of dicot roots commonly show a central "star" of xylem, with the phloem in pockets between the arms of the xylem star.

Primary Growth of Roots: Production of Laterals

The first root of a bean seedling grows quickly and soon begins to produce side branches, or **laterals.** Cells in the pericycle, which remained meristematic when their neighbors differentiated, divide and form a new apical meristem. As the innermost cells elongate, they push the new meristem out through the endo-dermis, cortex, and epidermis, and on into the soil (Figure 33-6). This new branch root follows the same pattern of growth outlined before, and becomes a lateral root with the same kinds of primary tissues.

There are two basic types of root systems (Figure 33-7). A **taproot system** has one main root, the **taproot,** by far the plant's longest and thickest root. Its side branches are much shorter and thinner. By concentrating growth into one axis, a taproot may penetrate deep enough into the soil to reach a reliable supply of water. The taproot of an old grape vine may reach 15 metres down and obtain water far beneath the soil surface. Many kinds of trees have taproots as seedlings or saplings and develop a more branching root system later.

Fibrous root systems branch throughout a large volume of soil, from which they absorb water and minerals. Instead of one main root, they have several roots of about equal size, each with smaller lateral roots. Plants with fibrous root systems, such as grasses, are very good at holding soil and therefore controlling soil erosion.

Some roots do not arise from existing roots. **Adventitious roots** are formed by meristems growing from other parts of the plant. For example, corn plants have "prop roots" that arise adventitiously from the base of the stem (Figure 33-8). Many climbing plants form tiny roots on the undersides of their

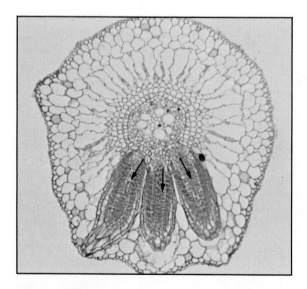

Figure 33-6
Origin of lateral roots. In this water hyacinth root, three branch roots (arrows) are growing out from the pericycle. (Carolina Biological Supply Company)

Figure 33-7
Two types of root systems. Left, a taproot system; right, a fibrous root system.

stems as they grow up tree trunks or brick walls. African violets can form roots from the undersides of leaves placed on moist soil. Cuttings of many plants form adventitious roots if the stems are placed in water or moist soil.

Functions of Roots

We can now list the four main functions of roots:

1. Roots anchor the plant in the soil. Taproots are especially good anchors, since they are long, and so thick and tough that it is hard to break them. Fibrous roots are good at holding soil in place.

2. Roots absorb water and minerals from the soil. Fibrous root systems often expose more absorptive surface area to the soil than do taproot systems, but taproots may grow deep enough to reach a more reliable supply of water.

3. Roots transport water and minerals up to the shoot system. The central core of xylem serves this function. The endodermal layer exerts some control over the passage of substances into or out of this central core of vascular tissue.

4. Roots store food. The root cortex is the primary food storage area in some plants. In many other kinds of plants, the root stores only a small supply of food for itself.

33-C Stems

Primary Growth of Stems: Growth in Length

Once the bean seedling has established roots, the shoot begins to grow. First, the lower part of the stem forms a loop that pushes up through the soil and then straightens, pulling the cotyledons free of the soil (Figure 33-9). Meanwhile, the apical meristem remains enclosed and protected between the cotyledons. Once the cotyledons are above ground, they spread apart. The first foliage leaves,

Figure 33-8
Prop roots of a corn plant.

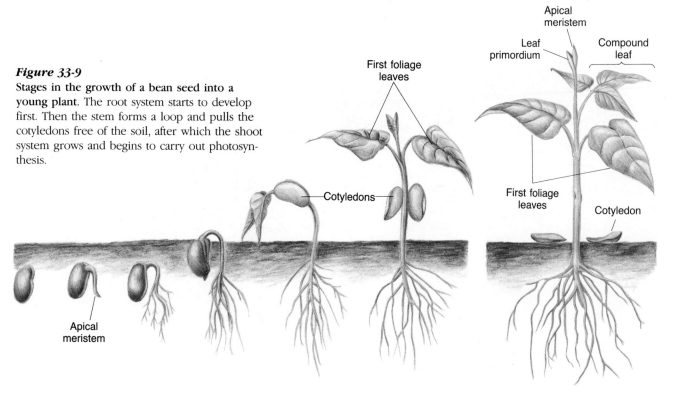

Figure 33-9
Stages in the growth of a bean seed into a young plant. The root system starts to develop first. Then the stem forms a loop and pulls the cotyledons free of the soil, after which the shoot system grows and begins to carry out photosynthesis.

First foliage leaves

Cotyledons

Apical meristem

Apical meristem

Leaf primordium

Compound leaf

First foliage leaves

Cotyledon

which were fully formed but tiny in the bean embryo, have already begun to expand. When they are exposed to sunlight they become green and expand more rapidly. Until these leaves can feed the plant by their photosynthesis, the bean seedling continues to use the food stored in the cotyledons. Eventually, the cotyledons shrivel as their food supply is used up, and they fall off.

The shoot's apical meristem, at the stem tip, consists of a little mound of rapidly dividing cells. The cells laid down by the apical meristem continue to divide, and this (rather than division in the apical meristem itself) produces most of the new cells during the primary growth of the stem. As in the root tip, the new cells in the stem eventually elongate and then mature. Growth in both length and diameter by cells in the shoot may last longer than in the root. Hence growth in the shoot can be detected quite a distance behind the growing tip of the apical meristem.

Primary Structure of Stems

Cells laid down by the shoot's apical meristem divide, enlarge, and differentiate to become the primary tissues of the stem. The epidermis of the stem is usually just one cell thick. Its cells fit tightly together, preventing both loss of moisture and damage by invading fungi or insects. These cells also conserve water by secreting a layer of waxy substances, the **cuticle,** which forms a waterproof covering over the stem surface. The cuticle of the stems is continuous with a layer of cuticle on the leaves.

Most of the primary tissue in the stem is ground tissue in the cortex and pith. The cortex is the next layer inside the epidermis (Figure 33-10). Most of the cortex is made up of parenchyma cells. Some of these cells contain chloroplasts and carry out photosynthesis, and some may store starch in amyloplasts. The **pith** in the center of the stem is also composed of parenchyma cells. In this respect, stems differ from roots, in which the center is usually occupied by the cylinder of vascular tissue. As the stem continues to grow, the cells in the outer parts may enlarge and expand so much that the pith is pulled apart, and the stem becomes hollow.

The stem's primary vascular tissue occurs as a ring of discrete vascular bundles rather than as the central core found in roots. Generally the xylem lies

■ *Cells in different parts of a root or stem tip are of different ages and so are at different stages of development: the farther from the apical meristem, the older the cells.*

Figure 33-10
Primary tissues of a dicot stem. (a) Cross section of a sunflower stem. Most of the stem consists of pith, which lies inside the ring of vascular bundles. (b) Close-up of part of the stem, pointing out the various tissues. (Ed Reschke)

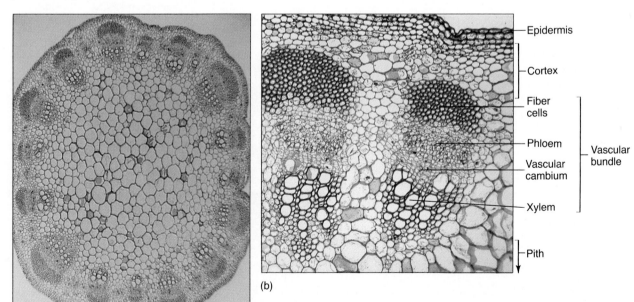

(a)

(b)

Epidermis

Cortex

Fiber cells

Phloem

Vascular cambium

Xylem

Vascular bundle

Pith

in the part of the bundle toward the inside of the stem, with the phloem toward the epidermis. Each vascular bundle is partly surrounded by thick-walled **fiber** cells, which provide extra support. The walls of the conducting cells in the xylem may also be especially thick. This extra strengthening material in the vascular bundles helps to support the stem in the air, which gives much less support than the soil that surrounds the roots.

Stems usually lack the endodermis and pericycle found in roots.

Primary Growth of Stems: Production of Laterals

As the shoot's apical meristem grows, it lays down not only the cells that divide and form the primary tissues of the stem but also **leaf primordia,** which develop into leaves (Figure 33-11). The angle between the base of the leaf and the stem is called the leaf's **axil** ("armpit"). Cell division in the axil produces a small group of meristematic cells called an **axillary** or **lateral bud,** which remains dormant for a time.

A new shoot branch arises when an axillary bud becomes active and begins to grow as the apical meristem of a new stem (Figure 33-11c). These branches produce leaves with axillary buds, just as the main stem does.

While some axillary buds produce new branches, others remain dormant. The growth of most of a plant's axillary buds is repressed by hormones from the apical meristems, a phenomenon called **apical dominance** (Section 35-B).

Functions of Stems

From our study, we can identify four main functions of stems:

1. Stems support structures of the shoot system. The thick cell walls in the vascular tissue of stems form a tough framework that is often difficult to break. In addition, the parenchyma cells in the pith and cortex contribute **turgor,** the internal pressure of the fluid inside them, which keeps the stem firm.

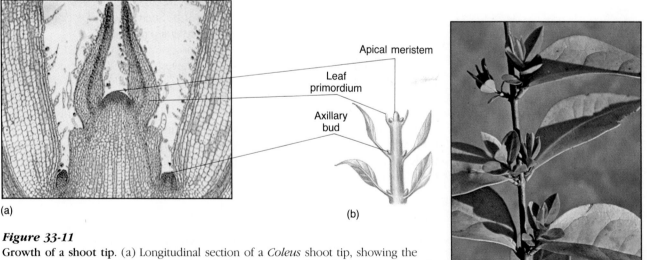

(a)

Apical meristem

Leaf primordium

Axillary bud

(b)

(c)

Figure 33-11

Growth of a shoot tip. (a) Longitudinal section of a *Coleus* shoot tip, showing the apical meristem flanked by new leaf primordia. Note the axillary buds in the axils of the next older pair of leaves, which are still differentiating. (b) Drawing to assist in identifying the structures shown in part (a). The leaf primordia are shown disproportionately large. (c) Axillary buds developing into new shoots on a privet twig. The bud on the left in the third set from the top is still dormant. (a, Carolina Biological Supply Company; c, Biophoto Associates)

Figure 33-12
Specialized stems. The stems of a cow's tongue cactus grow in the form of jointed, green pads, which carry out photosynthesis. They are also succulent (water-storing), swelling as the plant absorbs water after a rain. The leaves are modified as sharp spines that fend off animals who might attempt to consume the stems' food and water. The plant shown here also has a few yellow flowers and many red fruits, a favorite food of birds.

With this tough, resilient structure the stem can hold the leaves up to the sunlight they need for photosynthesis and can bend in the wind without breaking.

2. Stems transport substances between the roots and leaves. The vascular tissue of stems is continuous with that of the roots and leaves. The xylem transports water and minerals taken in by the roots up to the leaves and to the living cells of the stem, and the phloem carries food from the leaves or from storage areas to the living and growing parts of the plant, some of which cannot carry on photosynthesis.

3. Stems produce food. Some stems are green and photosynthetic. In most plants they supplement photosynthesis carried out in the leaves, but in plants such as cacti, stems are the main photosynthetic organs (Figure 33-12).

4. Stems store substances. Some stems contain many amyloplasts. A potato is an underground shoot with a large stem specialized for storage. Other stems store lesser amounts of food. Some stems, such as those of some cacti, also store a lot of water.

33-D Leaves

A leaf begins as a leaf primordium, a slender rod of meristematic cells laid down by the activity of an apical meristem. These cells divide and grow in various patterns, producing mature leaves with a shape typical of the species (Figure 33-13). If the cells near the edges of the developing leaf keep dividing in a uniform manner, the leaf grows in a smooth, rounded shape. However, if cells stop dividing in some areas but not in others, an irregular shape results, with lobes or teeth alternating with indentations. In extreme cases, the cells in some areas near the midrib do not divide at all, and the mature leaf is composed of several **leaflets,** connected only by the midrib. Such leaves are called **compound leaves,** with each leaflet looking like a separate leaf.

Our bean plant has both simple and compound leaves. The first foliage leaves are simple, somewhat heart-shaped leaves. The leaves formed by the plant later, however, are compound leaves, with three leaflets each (see Figure 33-9).

Most leaves have a broad, flat shape, which maximizes the amount of surface area per unit of volume. A large surface area is important both for capturing the energy of sunlight and for exchanging gases with the atmosphere. The comparatively small volume makes the leaves lightweight; heavy leaves would require more supporting tissue in the stems and roots.

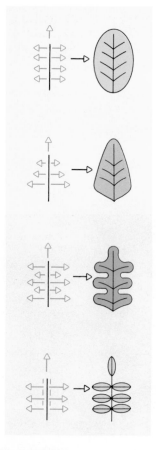

Figure 33-13
Leaf growth. The rate of growth and division of cells along different parts of the original leaf axis determines the leaf's final shape.

Structure of Leaves

As in the primary structure of the root and stem, a leaf's outer surface, both top and bottom, is covered by the epidermis. The epidermal cells of leaves, like those of stems, secrete a waxy, waterproof cuticle, which is especially thick in plants that must conserve water most strictly. Although the cuticle retards the loss of water to the atmosphere, it also impedes the passage of gases from the air into the plant. Since the leaves need carbon dioxide from the air for photosyn-

Figure 33-14
A stoma. The stoma is a slit between a pair of lip-like guard cells. Cell nuclei are stained red. (Carolina Biological Supply Company)

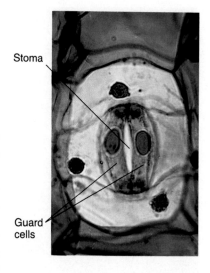

thesis, and must release the oxygen they produce, the secretion of a continuous layer of cuticle would solve one problem but create another.

In fact, the cuticle is not continuous but is interrupted at intervals by **stomata** ("mouths"; singular: **stoma**). Each stoma is a pore between a pair of lip-like **guard cells** (Figure 33-14). The size of the pore can be controlled by the guard cells. When the stoma is open, more carbon dioxide can enter the plant, but also more water vapor can escape into the atmosphere. Closing the stoma reduces the exchange of these gases between the leaf's interior and the atmosphere. Most dicot leaves have stomata only in the lower epidermis. However, plants such as water lilies have them in the upper epidermis, the only surface exposed to the air. Stomata occur not only in leaves, but also in the epidermal layers of stems and flower parts. All parts of a plant need oxygen for respiration, and some also need carbon dioxide for photosynthesis.

The ground tissue, which makes up the bulk of the leaf, consists of two layers of photosynthetic tissue (Figure 33-15). The cells in these layers are modified parenchyma, with many chloroplasts. Just beneath the upper epidermis is the **palisade mesophyll:** long, thin cells standing on end, perpendicular to the upper epidermis. Below the palisade mesophyll is the **spongy mesophyll,** its name suggested by the network of air spaces among the cells. The air spaces in both the palisade and spongy layers open to the atmosphere via the stomata, allowing gases to circulate to the photosynthetic cells inside the leaf.

The leaf also has vascular tissue, running through the mesophyll in strands called **veins.** The main vein, in the midrib, is a continuation of one of the stem's vascular bundles, and it in turn is continuous with the system of smaller veins in

■ *Leaves are the chief food-producing organs in most plants. The veins carry away food made in the leaf. They also provide the leaf with water, most of which escapes through the stomata into the air.*

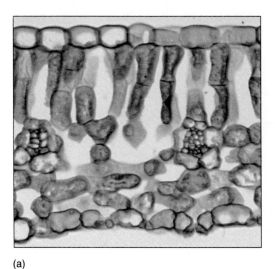

(a)

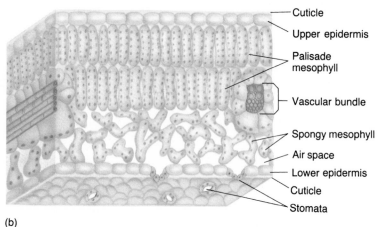

Cuticle
Upper epidermis
Palisade mesophyll
Vascular bundle
Spongy mesophyll
Air space
Lower epidermis
Cuticle
Stomata

(b)

Figure 33-15
Leaf tissues. (a) Cross section of a leaf as seen through a light microscope. (b) Drawing to assist in interpreting part (a). Both the upper and lower epidermis secrete cuticle. In dicots, most stomata occur in the lower epidermis. Sandwiched between the epidermal layers are the layers of photosynthetic mesophyll. Vascular bundles supply the photosynthetic cells with water and also transport food made in the mesophyll cells to other parts of the plant. (a, J. Robert Waaland, University of Washington/BPS)

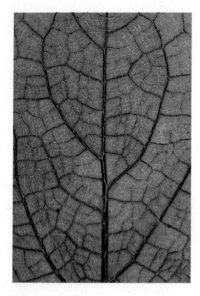

Figure 33-16
Leaf veins. The leaf's photosynthetic tissue is stretched out on a supporting framework of veins, composed of bundles of vascular tissue that branch out from the midrib (center) and carry water and minerals to all parts of the leaf.

■ *Wood produced during secondary growth thickens and strengthens stems and roots. New phloem and cork replace those tissues destroyed by this growth of secondary xylem.*

the leaf. The vascular tissue brings water to the leaves and carries away the products of photosynthesis. It also provides a supporting framework that stretches the delicate photosynthetic tissues out where they can intercept the rays of the sun and the gases in the atmosphere (Figure 33-16).

33-E Secondary Growth

Secondary growth is growth in the diameter of stems and roots by the addition of new cells. It is most familiar in woody, perennial (living for many years) dicots and gymnosperms. It is also found to some extent in nonwoody plants such as alfalfa, sunflowers, and some annuals (plants that live just one season).

The cells that form secondary tissues are produced by **lateral meristems.** The first of these is the **vascular cambium,** which lays down cells that become the **secondary vascular tissues.** In the stem, cells between the primary xylem and primary phloem in the vascular bundles become meristematic and form part of the vascular cambium. Additional cells between the vascular bundles also become meristematic. Hence the vascular cambium can be seen in a cross section of the stem as a continuous ring of tissue, with the xylem and pith on the inside and the phloem, cortex, and epidermis on the outside (Figure 33-17a).

Now cells in the vascular cambium divide, producing new cells. Those produced inside the ring of vascular cambium differentiate into **secondary xylem** tissue, also called **wood.** Most of these cells produce very thick cell walls and then die. As the vascular cambium produces new wood on the inside of the ring, the stem grows in diameter, and the phloem outside the vascular cambium becomes stretched and crushed. Meanwhile, though, cells produced just outside the vascular cambium have differentiated as **secondary phloem,** which transports food in place of the destroyed primary phloem. As more secondary xylem forms, the first secondary phloem is destroyed in its turn, and more secondary phloem is added just outside of the vascular cambium—that is, immediately *inside* the previously formed secondary phloem (Figures 33-17 and 33-18).

As the stem grows thicker with secondary xylem and phloem, the outer primary tissues—the cortex and epidermis—are also stretched and destroyed. The epidermis is replaced by a new protective outer layer of tissue. Cells in the

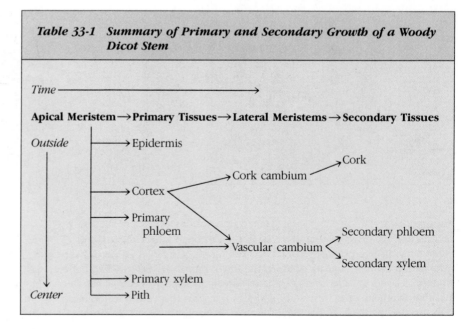

Table 33-1   Summary of Primary and Secondary Growth of a Woody Dicot Stem

Time ⟶

Apical Meristem → Primary Tissues → Lateral Meristems → Secondary Tissues

Outside ⟶ Epidermis

Cortex → Cork cambium → Cork

Primary phloem

Cortex → Vascular cambium → Secondary phloem / Secondary xylem

Primary xylem

Center ⟶ Pith

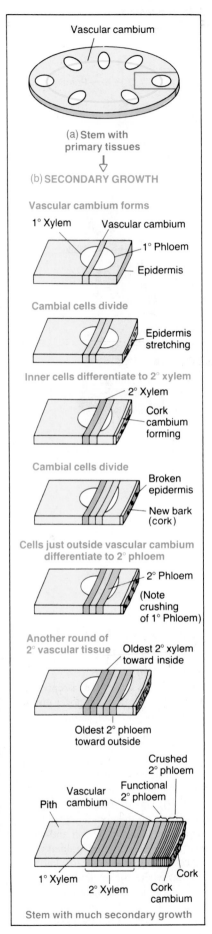

Figure 33-17

Secondary growth of a woody stem. (a) Cross section of a stem, showing the position of the vascular cambium among the primary tissues. The part cut out along the red lines is enlarged in part (b). (b) Sequence of events in the formation of secondary tissues: secondary (2°) xylem, secondary phloem, and cork.

cortex form another lateral meristem, the **cork cambium,** which divides and produces new cells to the outside. These cells become impregnated with a waterproof waxy material and then die, forming a protective layer of **cork,** or **outer bark,** on the outside of the tree.

In the cross section of a stem with well-developed secondary growth, we would find these tissues, starting from the center and working outward: pith, primary xylem, secondary xylem, vascular cambium, secondary phloem, the crushed remains of primary phloem and cortex (these would be present but perhaps not identifiable), cork cambium, and cork (bark) (Figure 33-17 and Table 33-1). The primary layer of epidermis has by this time ruptured and sloughed away. When the bark is peeled off a tree, it breaks at the layer of delicate undifferentiated cells in the region of the vascular cambium. The trunk of a tree is almost entirely secondary xylem, with a very slender column of pith and primary xylem in the center (Figure 33-19). The tissues outside the vascular cambium are all in the part of the trunk commonly called the bark.

From this discussion, we can see that most of the tree's wood and bark consist of dead cells. In essence, the living cells of the vascular and cork cambia

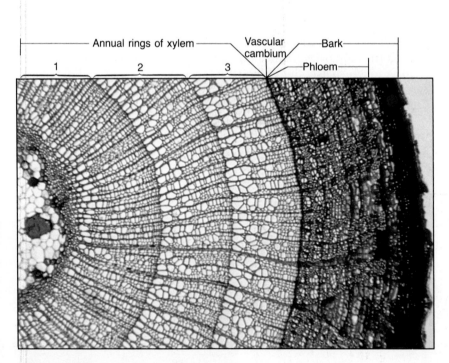

Figure 33-18

Cross section of a basswood twig after three years of secondary growth. The secondary xylem formed during each growing season forms distinct bands, the annual rings, which can be counted to determine the stem's age. Each ring has large cells on the inside, where the spring growth occurs. Cells laid down later in the season are progressively smaller. The phloem is part of the bark and can be removed by peeling the bark from the twig. (Carolina Biological Supply Company)

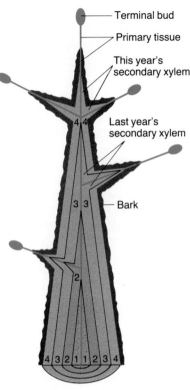

- Terminal bud
- Primary tissue
- This year's secondary xylem
- Last year's secondary xylem
- Bark

4 4

3 3

2

4 3 2 1 1 2 3 4

Figure 33-19

Relationship between primary and secondary tissues. This longitudinal section through a (fat) four-year old tree shows the primary plant body (green) and secondary growth for the first year (1), second year (2), and so on.

keep growing a new tree outside of the dead xylem core of its former self, and inside of its old, dead "skin" (Figure 33-19).

Secondary growth in roots is similar to that in stems. The main roots of a tree are large and woody. They provide support and serve as transport conduits. Encased in their tough bark, these roots cannot absorb water and minerals from the soil, a task performed by young roots at the far ends of the root system.

33-F Monocotyledons

Monocotyledons include grasses, palms, and flowering bulb plants. The large seeds of the grasses known as cereals (rice, wheat, corn, oats, barley, rye, and so forth) are the staples in the diets of most human societies. Monocotyledons differ from dicotyledons in the number of cotyledons (seed leaves) possessed by the embryo: monocotyledon embryos have only one cotyledon. The embryos of the two groups also have other structural differences.

The corn kernel is convenient for studying the structure of a monocot seed. A kernel of corn is actually a one-seeded fruit, and its "skin" contains both the seed coat and fruit parts, closely attached to each other. In the corn seed, the **endosperm,** a nutritive tissue, is separate from the embryo, whereas in beans the food-rich endosperm has been absorbed into the cotyledons. (The distribution of endosperm is not consistently different between monocots and dicots, however.) The corn embryo digests the food stored in the endosperm and uses it for growth until it becomes established as a self-sufficient, photosynthetic plant. The corn embryo has several developing leaves, wrapped above the apical meristem. A tough sheath, the **coleoptile,** protects the leaves and the apical meristem from injury as the seedling pushes up through the soil (Figure 33-20). Once the coleoptile has penetrated into the light above the ground, the leaves expand greatly, rupture the tip of the coleoptile, and grow out.

Monocots also differ from dicots in the structure of the primary plant body and the arrangement of its tissues. The xylem of a monocot root usually has many more "arms" (corn has 20 to 40), and most monocot roots have pith in the

Figure 33-20

Structure of a corn seed, a representative monocot. (a) A monocot embryo has only one cotyledon. Notice that the food supply of the endosperm occupies a great deal of the seed. The coleoptile is a sheath covering the leaves, which are rolled above the apical meristem of the shoot. (b) Growth of a corn seedling. Root hairs, on the root at the right, anchor the plant in the soil and absorb water. The pointed, golden tip of the coleoptile grows upward, into a position where the enclosed leaves will be able to reach light. Then the leaves will expand, rupture the coleoptile, and begin photosynthesis. (b, Biophoto Associates)

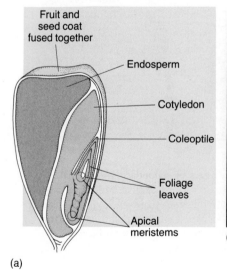

Fruit and seed coat fused together

- Endosperm
- Cotyledon
- Coleoptile
- Foliage leaves
- Apical meristems

(a)

(b)

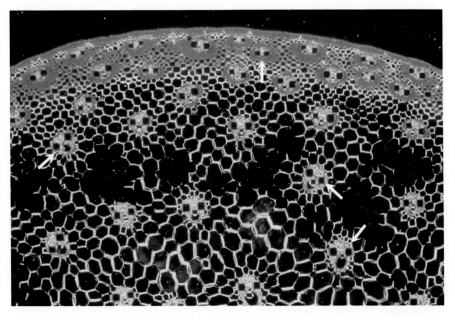

Figure 33-21
Monocot stem. This cross section shows the mature primary tissues of a corn plant stem. Note that the vascular bundles are scattered throughout the stem (white arrows), rather than being arranged in a single ring as in dicots (compare with Figure 33-10). (Phil Gates, Durham University, U.K./ BPS)

(a)

(b)

(c)

Figure 33-22
Monocots and dicots. Monocot features include (a) leaves with parallel venation, as in this banana plant; and (b) flower parts in multiples of three, as in the lilies. (c) The net-like leaf venation and flower parts in fives mark the primrose as a dicot.

Leaves to the Defense

A JOURNEY INTO LIFE

A plant's leaves are the parts most exposed to the environment. Here they spread out and absorb sunlight and carbon dioxide for photosynthesis. And here they also catch the attention of hungry herbivores. It is not surprising, therefore, that many plants have evolved leaves with protective mechanisms.

The most obvious defense is having sharp prickles. Leaves of many hollies have several sharp projections along the edges, and thistle leaves carry this motif over the whole leaf surface. In cacti, the entire leaf has become a sharp spine, with the stem taking over the role of photosynthesis completely (see Figure 33-12). Some grasses grow their weapons in miniature, producing saws rather than swords (Figure 33-A).

Many leaves have more subtle defenses, such as specialized epidermal hairs. The itchy juice secreted by the hairs of tomato plants, for instance, deters some animals from brushing against them, much less eating them. The hook-shaped hairs of beans entangle small insects, and the insect stops feeding on the plant while it struggles to free itself. The nettle leaf in Figure 33-B has hairs of both these types. Epidermal hairs of other kinds of plants release glue, which traps small insects and immobilizes the feet and mouthparts of larger ones.

Some leaves' defenses are not external but internal. For example, leaf cells may contain sharp, needle-like crystals of calcium oxalate, called raphides, which ensure that animals will not take a second mouthful of leaves from the plant.

Other plants produce toxic chemicals. For example, milkweed leaves contain cardiac glycosides, chemicals that cause the heart to beat faster, sometimes leading to death of the animal. However, this defense is not absolute: the caterpillars of monarch butterflies have evolved the ability to eat milkweed leaves without ill effects (Figure 33-C). In fact, these caterpillars store the milkweeds' cardiac glycosides in their own bodies, making themselves unpalatable to predators. These defensive chemicals are retained when the caterpillar metamorphoses into an adult butterfly. Hence the adult is protected from predators by chemicals from the food plant it ate as a larva (*A Journey Into Life*, Chapter 18). Other examples of

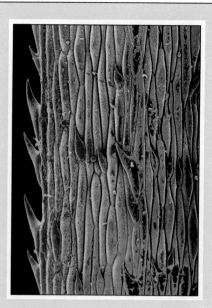

Figure 33-A
Armed grass. The scanning electron microscope shows that leaves of grass have saw-like teeth that can inflict a painful cut. (Biophoto Associates)

chemical toxins produced by leaves are discussed in *A Journey Into Life*, Chapter 14.

In some plants, leaves produce their defenses only when needed. For example, tannins are defensive chemicals that precipitate the digestive enzymes of many herbivores; unable to digest its food, the animal starves.

Table 33-2 Comparison of Monocotyledons and Dicotyledons

Characteristic	Monocots	Dicots
Seed leaf (cotyledon)	One	Two
Flower parts	In 3's or multiples of 3, or irregular	4, 5, their multiples, or irregular
Leaf venation	Parallel	Netted or fan-like
Vascular tissue	Bundles scattered through stem; usually no cambium	Single ring of bundles or continuous ring; cambium in woody forms
Form	Mostly herbs, few trees, e.g., palms	Herbs, shrubs, and trees
Examples	Grasses, cereals, palms, lilies, onions, orchids, irises, crocuses, daffodils	Oaks, maples, legumes, roses, mints, squashes, daisies, walnuts, cacti, violets, buttercups, poppies

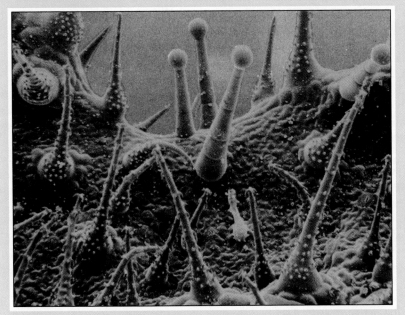

Figure 33-C
Chemical defense. Unlike most other animals, a monarch caterpillar can feed on milkweed, despite the toxic chemicals in the leaves.

Figure 33-B
Protective hairs. A South American nettle leaf has two kinds of epidermal hairs: one kind is pointed, with hooked ends, and the other kind secretes droplets of irritating fluid. (Biophoto Associates)

Sugar maples produce leaves with little tannin in the spring. In years when this first set of leaves is eaten by a heavy infestation of insects, the tree grows a set of replacement leaves containing more tannins.

A plant must fend off not only hungry herbivores but also other plants that compete with it for sunlight and water. Many plants produce chemicals that inhibit the growth of near neighbors, and what better way to deploy them into a "no trespassing" zone than by washing off the leaves? This is how shrubs of the chaparral in California inhibit the germination of seeds in the surrounding soil. Similarly, it is notoriously difficult to grow anything under a walnut tree because of a substance called juglone, which washes onto the ground and inhibits the growth of other plants.

center. Monocot stems have numerous small scattered vascular bundles (Figure 33-21), rather than the single ring of bundles found in dicots. Therefore the ground tissue has no distinct pith and cortex. In monocot leaves, the principal veins of vascular tissue run parallel to each other, rather than in the fan-like or net-like pattern common in dicots (Figure 33-22 and Table 33-2). Monocot leaves often stand more or less upright, and stomata are plentiful in both epidermal layers of the leaf.

Grasses and some other monocotyledons have growth areas besides the apical meristems. **Intercalary meristems** at the bases of leaves permit the plants to keep growing after their tips have been eaten by grazing animals or clipped by a lawn mower.

Most monocots are herbaceous (soft-bodied, non-woody) annuals, or perennials with leaves that die back to the ground after each growing season. Except for Joshua trees and other yuccas, and dragon trees (which have an anomalous type of secondary growth), most monocots lack secondary growth and secondary tissue. Even the tallest monocots, palm trees, grow in diameter without secondary growth: they have "primary thickening meristems," close behind the apical meristem of the trunk.

SUMMARY

A vascular plant grows in one place, making its own food and competing with other plants for the basic resources of sunlight, water, and minerals. The plant body consists of roots, stems, and leaves, all containing vascular tissue, which conducts water, minerals, and food rapidly from one part of the body to another. This body plan is adapted to obtaining water and sunlight efficiently. A plant grows and produces new organs throughout its life.

A seed consists of a protective covering, food supply, and plant embryo with a rudimentary root, stem, and leaves.

Early in life, all cells of the plant can divide. Later most cells mature, specialize, and lose their capacity to divide. Cells that retain the ability to divide are found in meristems at specific locations in the plant.

Primary growth is principally growth in length and production of new root and shoot branches. The tips of stems and roots grow longer through the activity of apical meristems. Stem branches arise from axillary buds, located at the bases of leaves. An axillary meri-stem thus becomes the apical meristem of a new branch. Lateral roots arise from the pericycle tissue inside the root. This produces a new apical meristem of a branch root.

Secondary growth is growth in girth by production of secondary tissues. Vascular cambium, a lateral meristem, produces secondary vascular tissues. Secondary xylem adds strength and increases the capacity for conducting water to the leaves; secondary phloem replaces primary phloem, or older secondary phloem, destroyed as interior tissues grow. The cork cambium, another lateral meristem, produces cork, which replaces the epidermis that has been destroyed by expansion of the interior tissues.

Monocotyledons and dicotyledons differ in the number of cotyledons found in the embryo (one versus two, respectively); in the arrangement of vascular tissue, especially in the stems and leaves; and often in the arrangement of leaf stomata. In addition, very few monocots have secondary growth, whereas secondary growth is found even in some annual dicots.

SELF-QUIZ

1. A fibrous root system is apt to perform which function better than a taproot system?
   a. absorption
   b. anchorage
   c. food storage
   d. transport
2. One difference between a bean seed and a kernel of corn is that:
   a. a bean seed has a seed coat but a kernel of corn does not
   b. only the bean has two cotyledons; the corn has one
   c. the bean contains stored food; the corn does not
   d. the bean embryo has leaves, while the corn embryo lacks leaves
   e. the bean embryo has two apical meristems; the corn embryo has only one
3. Primary growth of a tree:
   a. occurs through the activities of apical meristems
   b. occurs through the activity of a vascular cambium
   c. occurs through the activity of the root cap
   d. occurs only in the first year of the tree's life
   e. occurs in stems, but generally not in roots
4. Arrange the following events during the growth of a shoot in proper order:
   a. cell division
   b. cell maturation
   c. cell elongation
5. Lateral roots arise from:
   a. root hairs
   b. pericycle
   c. cork cambium
   d. endodermis
   e. axillary buds

6. Secondary xylem and phloem are laid down by:
   a. apical meristems
   b. axillary meristems
   c. vascular cambium
   d. cork cambium
   e. intercalary meristems
7. Which of the following would *not* secrete a cuticle?
   a. leaf epidermis
   b. stem epidermis
   c. root epidermis
8. For each characteristic listed below, tell whether it is characteristic of monocotyledons, dicotyledons, or both:
   ____ a. stomata on both leaf surfaces
   ____ b. roots with root caps
   ____ c. parallel venation in leaves
   ____ d. many vascular bundles scattered throughout stem cross section
   ____ e. vascular cambium
9. Horace and Hermione visited a forest on their honeymoon. Horace selected a tree 10 metres tall and 30 centimetres in diameter. He carved their initials into its bark 1.5 m above ground level. On their tenth anniversary, Horace and Hermione return to the forest; the tree is now 12 m tall and 33 cm in diameter. Their initials are now:
   a. 1.5 m above ground level
   b. 2 m above ground level
   c. 3.5 m above ground level

QUESTIONS FOR DISCUSSION

1. List specialized structures in a plant's dermal tissue that (a) exchange substances with the environment; (b) defend the plant.
2. What is the advantage of the arrangement of xylem in a star-shaped pattern in the root, rather than its being internal to the phloem as it is in the shoot?
3. Since the cotyledons of a bean seedling become photosynthetic, one might expect them to be valuable organs of the plant even after their stored food is used up. Why might it be adaptively advantageous to the plant to shed them soon after it becomes well established?

4. Leaf structure varies from one species to another, and it is often closely correlated with the habitat of the plant. In what type of habitat would you expect to find each of the following modifications of leaf structure?
   a. more than one cell layer in the epidermis
   b. extra thick layers of cuticle
   c. little or no cuticle
   d. little or no cuticle; little or no xylem; no stomata
   e. large air spaces in the mesophyll; stomata in the upper epidermis instead of in the lower epidermis

SUGGESTED READINGS

Blackmore, S. and E. Tootill. *The Facts on File Dictionary of Botany.* Aylesbury, U.K.: Market House Books Ltd., 1984. A small, very useful, dictionary of botanical terms.

Jensen, W. A., and F. B. Salisbury. *Botany: An Ecological Approach,* 2d ed. Belmont, CA: Wadsworth, 1984. A fresh, modern presentation of botany.

Mauseth, J. D. *Botany: An Introduction to Plant Biology.* Philadelphia: Saunders, 1991. An up-to-date, complete, introductory botany text, heavily illustrated.

Prance, G., and K. Sandved. *Leaves.* New York: Crown, 1985.

Ray, P. M. *The Living Plant,* 2d ed. New York: Holt, Rinehart and Winston, Inc., 1972. A clear, straightforward little book.

Shigo, A. L. "Compartmentalization of decay in trees." *Scientific American,* April 1985. How the structure of wood and activities of its cells protect against disease.

Weier, T. E., C. R. Stocking, and M. G. Barbour. *Botany: An Introduction to Plant Biology,* 6th ed. New York: John Wiley and Sons, 1982. An especially well-illustrated botany text.

Nutrition and Transport in Vascular Plants

CHAPTER

34

Agave

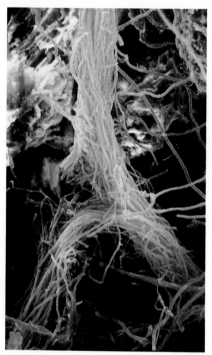

Root (SEM)

*P*lants take inorganic substances from the environment and use them to make their organic food molecules. All living organisms, including plants, need the same major nutrients. Land plants obtain carbon, from carbon dioxide in the air, through their stomata (Section 33-D). However, it is the roots that absorb the wide array of mineral nutrients and the huge amounts of water the plant needs. Taken all together, a plant's roots contain a vast total surface area of plasma membranes that absorb water and minerals from the soil.

Roots absorb water and minerals but cannot make food by photosynthesis in the darkness of the soil. The leaves rely on water and minerals from the roots, and in turn supply the roots with food. This complementary division of labor is possible because of the vascular tissues: xylem, which conducts water and minerals from the roots to the leaves; and phloem, which transports food from the leaves to the roots, as well as to growing buds, flowers, and fruits. Having an efficient transport system permits a plant to have parts specialized for different functions, such as water-gathering, energy-gathering, and reproduction. Vascular tissue also provides strength and support, as seen most clearly in the wood of roots and stems, and the veins of leaves. This support permits the plant to exploit a much larger volume of soil and air, and so increases its ability to compete with other plants.

How does a plant obtain nutrients from the soil? How does it move them up to the leaves, and move the food produced in the leaves to the roots and to growing buds and fruits? To answer these questions, we must study the plant's roots and the soil in which they grow, and the features of vascular tissue. We must also apply our knowledge of active transport, osmosis, and the other properties of water (Sections 4-D, 4-C, and 2-F). This is because transport in plants, as in animals, moves substances down a pressure gradient, from a higher to a lower pressure area. Phloem contents are pushed by a positive pressure generated by osmosis. A similar mechanism may move substances in xylem. However, most often xylem transport results from a negative pressure—a pull—generated by evaporation of water from the leaves.

- Roots absorb water and minerals from the soil solution through a large surface area of selectively permeable cell membranes.
- Transport in xylem and phloem follows a fluid pressure gradient.
- Water and minerals move upward in plants through xylem. The main force is the pull exerted when water evaporates from the leaves via the stomata.
- Food moves through phloem from leaves to sites of use or storage. Phloem transport follows a gradient of pressure built up by active transport of solutes and osmotic movement of water.

34-A *Nutritional Requirements of Plants*

Plants obtain carbon, hydrogen, and oxygen from carbon dioxide and water during photosynthesis. Plants also need the elements nitrogen (N), phosphorus (P), potassium (K), calcium (Ca), magnesium (Mg), sulfur (S), and iron (Fe). This list of plant nutrients is easy to remember using a phrase made up of the elements' chemical symbols: C HOPK'NS CaFe, Mighty good.

All of these nutrients except iron are needed in relatively large amounts, and so are called **macronutrients.** Calcium and magnesium are plentiful in many soils. Farmers have long fertilized their soil by adding the other elements in either organic or inorganic form. Nitrogen, phosphorus, and potassium are required in the greatest quantities, and commercial fertilizers are rated by the percentage of each of these elements they contain. For example, a "5-10-5" fertilizer contains 5% nitrogen, 10% phosphorus, and 5% potassium, by weight. (The other 80% is inert materials.)

Table 34-1 The Roles of Mineral Nutrients in Plants

Nutrient	Role in plant	Deficiency symptoms
Macronutrients		
Nitrogen	Component of proteins, nucleic acids, chlorophyll, some hormones	Mild: older leaves yellow; purplish leaf veins, stems. Severe: stunting
Potassium	Not well understood; important in membrane potentials and opening of stomata; activates enzymes in photosynthesis, respiration, and starch and protein synthesis	Various, general; mottled chlorosis and dead spots, starting in older leaves; weak stems
Phosphorus	Component of nucleic acids, ATP, phospholipids	Various, general; dark color; loss of older leaves; stunting; slow maturation
Sulfur	Component of some amino acids, coenzyme A (Section 7-D)	Chlorosis, poor roots
Calcium	Component of middle lamella of cell walls (Section 11-D); ties up waste products as insoluble salts; involved in membrane function	Meristem death; abnormal cell division; deformed tissues; breakdown of membrane structure
Magnesium	Component of chlorophyll; cofactor of many metabolic enzymes	Chlorosis, appearing first in older leaves
Micronutrients		
Iron	Needed for chlorophyll production; part of electron transport molecules and of some enzymes	Chlorosis, appearing first in youngest leaves
Boron	Mostly unknown; nucleic acid synthesis; pollen germination; carbohydrate transport	Various; thick dark leaves; malformations; cell division and elongation and flowering inhibited
Zinc	Synthesis of tryptophan (precursor of auxin, see Table 35-1); component of some enzymes	Small puckered leaves; reduced stem elongation
Manganese	Activates citric acid cycle enzymes; involved in evolution of O_2 during photosynthesis	Mottled chlorosis
Chlorine	Evolution of O_2 during photosynthesis	Small leaves, slow growth, thick stunted roots
Molybdenum	Part of enzymes for nitrate reduction and nitrogen fixation	Same as nitrogen deficiency
Copper	Component of some enzymes in respiration and photosynthesis	Dark misshapen leaves with dead spots

Plants also need various **micronutrients,** minerals used in small amounts: iron, boron, zinc, manganese, chlorine, molybdenum, and copper. Whereas nitrogen is applied to the soil at the rate of several hundred pounds per acre per year, the treatment for molybdenum-deficient soils in Australia is 2 ounces of MoO_3 per acre, applied once every 10 years.

Many minerals are vital components of important biological molecules (Table 34-1). An inadequate supply of any one of several minerals may result in

Figure 34-1
Sunflower seedlings grown in various nutrient solutions. Left to right: nutritionally complete medium, followed by media lacking only sulfur, nitrogen, phosphorus, and potassium, respectively.

Complete   −S   −N   −P   −K

rather general deficiency symptoms, such as **chlorosis** (paleness) and poor growth (Figure 34-1). On the other hand, deficiency symptoms of some nutrients may be quite specific.

34-B Soil

The minerals used by plants come ultimately from the soil around their roots. Soil contains particles formed as the underlying rock breaks down. The type of rock is largely responsible for the minerals present in the soil (except for nitrogen, discussed shortly).

The size of soil particles influences the soil's capacity to hold **soil water.** Soil can hold water by adhesion to the surface of each particle (Section 2-F) and by capillarity. **Capillarity** is the ability of water to fill narrow spaces, such as those between soil particles. This occurs because water adheres to hydrophilic ("water-loving") surfaces on the sides of the space and pulls other water molecules along by cohesion, so that water fills the space. The smaller the soil particles, the more water the soil can hold, since smaller particles have a greater collective surface area and more small-sized spaces between them than an equal volume of larger particles (Figure 34-2). In fact, clay particles, with the smallest particle size, tend to hold water so tightly that plants cannot withdraw much of it. The large particles of a sandy soil, on the other hand, often let so much water drain out that the soil is too dry for most plants. Most plants grow best in soil with a mixture of particle sizes.

Rainfall exceeding the soil's water-holding capacity drains away, carrying dissolved nutrients with it. This is called **leaching.**

Fertile soil contains a great deal of organic matter. Manure, dead leaves, and bits of wood act like sponges, soaking up water and swelling when it rains, and releasing the water slowly later, as the soil dries out. This alternate swelling and shrinking helps loosen the soil, allowing roots to grow through it easily. Organic matter on top of the soil (mulch) acts as a sunshield, reducing evaporation from the soil. More importantly, the organic matter in soil is a reservoir of nutrients, which are released slowly by decomposition.

Does organic matter make better fertilizer than inorganic products? Plants can take up nutrients only in certain chemical forms. Inorganic fertilizers usually provide these forms, bypassing the steps of microbial digestion needed to make organic fertilizers available to plants. Organic fertilizers do not provide better nutrition. They are superior because they help hold water in the soil and because they release their minerals gradually, giving plants a sustained source of nutrients. By contrast, nutrients in inorganic fertilizers may leach rapidly out of the soil after each treatment. This wastes money and may pollute water supplies (Section 38-G). Too much inorganic fertilizer may also make the soil solution so concentrated that plants lose water by osmosis. Such "fertilizer burn" may dehydrate and kill the plant.

An ongoing study is comparing the effects of inorganic and organic fertilizers on two adjacent farms in Washington state. One farm, first cultivated in 1908, has been conventionally managed since 1948, with inorganic fertilizer and pesticides applied as recommended by the state. Crops of winter wheat and spring peas are grown in succession in each field. The "organic" farm next door has received no inorganic fertilizer since it was first plowed in 1909. It too grows wheat and peas, but every third year, each field grows **green manure,** a crop that is not harvested but plowed back into the soil to add organic fertilizer to the soil. The organic farm now has 16 centimetres (6 inches) more topsoil, more soil organic matter, more water storage capacity, and less rain-shedding top crust.

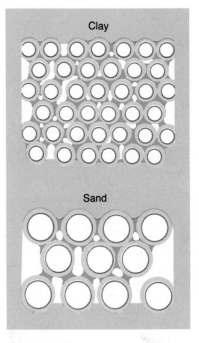

Figure 34-2
Soil particle size and water capacity. More water can be held by clay soils (small particles) than by sandy soils (larger particles). Light blue shows water held as a film on the surface of each particle; dark blue is water held by capillarity in the tiny crevices between particles.

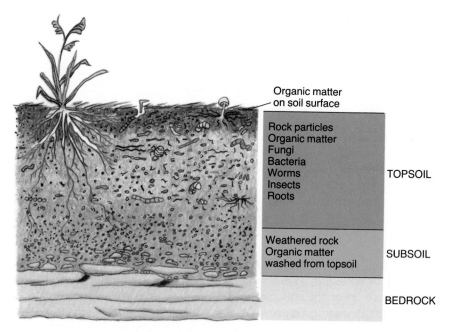

Organic matter
on soil surface

Rock particles
Organic matter
Fungi
Bacteria
Worms
Insects
Roots

TOPSOIL

Weathered rock
Organic matter
washed from topsoil

SUBSOIL

BEDROCK

Figure 34-3
Soil. Fertile topsoil is created by soil organisms from the rock particles and decaying vegetation that they feed on. The subsoil consists largely of rock particles and contains few organisms.

Living organisms are a vital component of soil (Figure 34-3). Bacteria and fungi digest organic matter, slowly releasing minerals that plants can absorb. Nitrogen-fixing bacteria convert nitrogen from the air (N_2), a form plants cannot use, to a form plants can use to make amino acids—ammonium ion (NH_4^+). If plants do not pick up ammonium quickly, other nitrifying bacteria oxidize it to nitrate (NO_3^-), which is easily leached out of the soil by rain. Plants of the legume family (peas, beans, clover, alfalfa) avert this problem by housing nitrogen-fixing bacteria in their root nodules (see Figure 19-5). Because legumes increase the soil's nitrogen content, they are planted as part of crop rotation programs.

The roots of a large majority of plants form **mycorrhizae,** symbiotic associations with soil fungi (Section 19-L). Fungi absorb soil minerals very efficiently. The hyphae of mycorrhizal fungi extend out into the soil around the root and also grow into the root itself. In this arrangement, they transfer minerals they have absorbed from the soil to the plant, especially phosphorus, which is not very mobile in the soil and is therefore hard to obtain. In return, these fungi receive organic molecules made by the plant. Most mycorrhizal fungi cannot

Figure 34-4
Which plants have fungi? The two orange seedlings on the left were grown for six months with a mycorrhizal fungus; the two on the right without the fungus. All the plants were given nitrogen-containing fertilizer (calcium nitrate to the left-hand plant and ammonium nitrate to the right-hand plant in each pair). However, only the plants with mycorrhizae absorbed the nitrogen and grew well. (Dr. J. Menge, University of California, Riverside)

grow without plant partners and are totally dependent on the plant for food. The plants also do much better with fungal partners (Figure 34-4).

Oxygen in the soil is important because most living things, including plant roots, require oxygen for respiration. The burrowing of organisms such as worms and insects mixes and breaks up the soil, allowing air to penetrate.

The Soil Solution

Soil water contains various dissolved molecules and ions. However, some minerals become tightly bound to soil particles. Bits of clay or organic matter have negatively charged surfaces and so bind most of the soil's positively charged ions. Only the ions that stay dissolved in the soil solution are actually available to the roots of plants.

Acid soil is low in nutrients because its H^+ displaces other positively charged ions from the surfaces of soil particles. The displaced mineral ions are then readily leached out of the soil. Soil pH determines the mix of nutrients that are dissolved and hence available to plants. Most plants do best at a soil pH of 6.0 to 7.5, although there are plants adapted to growing at higher or lower pH ranges.

Soil pH can be adjusted to suit the plants we wish to grow. Lime is applied to acid soils to raise the pH, whereas organic matter, sulfur, or ammonium sulfate is applied to alkaline soils. These treatments have other effects besides changing the pH: lime provides calcium, ammonium sulfate contributes nitrogen and sulfur, and organic matter provides many nutrients. If an acid soil's pH is raised too much, iron may precipitate out of the soil solution. One way around this is to apply **chelated iron**—iron bound to organic molecules so that it cannot be precipitated—along with lime.

Figure 34-5
Root hairs. The primary root of this germinating radish seedling has many root hairs, which provide anchorage among soil particles and increase the surface area available to take up water and minerals from the soil.

34-C Absorption by the Roots

A plant's small, young roots, near the ends of the root system, absorb water and minerals from the soil solution. The external surface area of these roots is increased by root hairs, epidermal cells with extensions that grow out between soil particles (Figure 34-5). But the epidermis is only part of a root's absorptive surface. The soil solution also moves freely into the root cortex via spaces between the cellulose fibers in cell walls (Figure 34-6a). Hence a root can take up water and minerals through the entire surface area of the plasma membranes of cells in its epidermis and cortex (Figure 34-6b).

At the inner edge of the cortex, the soil solution reaches a barricade: the root's endodermis. The cells of the endodermis form a layer like a brick wall, separating the epidermis and cortex from the root's interior. A **Casparian strip**, made of an impermeable waxy material, runs right around each cell in the endodermis, as shown in Figure 34-7. This material blocks the cell wall spaces and butts against the Casparian strips of the adjacent endodermal cells. Since the Casparian strips are impermeable to water and dissolved solutes, no substances can pass through the pores of the cell walls from the outside of the endodermis to the inside, or vice versa.

All substances that travel through the endodermis, therefore, must pass through the cytoplasm of the endodermal cells themselves. Thus the endodermis lives up to the translation of its name, "inner skin," by serving as a selective barrier. Substances in the cortex must pass through the living cells of the endodermis before they can enter the vascular tissue, in the center of the root, and be transported to the rest of the plant.

Some substances from the soil solution enter the plant through the plasma membrane of the endodermis, but most are taken up by the membranes of cells

■ ***The cell wall spaces in the root's epidermis and cortex are virtually continuous with the external soil solution.***

■ ***Minerals move into the root in the soil solution that bathes the epidermis and cortex. Here a large area of selectively permeable membrane absorbs minerals into the cytoplasm. All minerals must pass through the cytoplasm if they are to reach the xylem and, from it, the rest of the plant.***

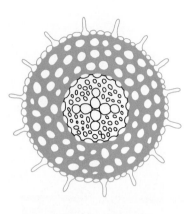

(a) Volume of root occupied by soil solution

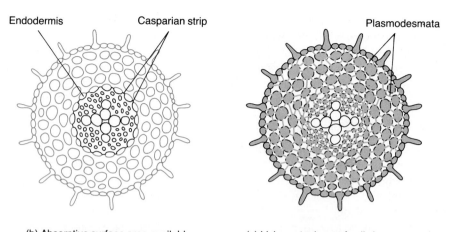

Endodermis   Casparian strip   Plasmodesmata

(b) Absorptive surface area available for uptake of water and minerals across plasma membranes

(c) Living cytoplasm of cells is connected throughout the root

Figure 34-6

Uptake of substances from soil solution by roots. (a) The soil solution (blue) freely penetrates the porous cell walls and the intercellular spaces in the outer part of the root, up to the outer face of the endodermis. (b) The root's absorptive surface consists of the plasma membranes of cells in the epidermis, cortex, and outer face of the endodermis (pink). (c) The living cytoplasm in the root cells (orange). Once water and minerals have been taken into a cell through its selectively permeable membrane, they can move throughout the root from cell to cell via plasmodesmata, eventually reaching the vascular tissues, which transport them to other parts of the plant.

in the epidermis and cortex. All of these living cells form a continuous system of cytoplasm extending throughout the root, connected by strands of cytoplasm called **plasmodesmata** (Figure 34-6c). Hence minerals absorbed by outer cells can be transported through the cytoplasm, which continues right through the endodermis, and into the living cells around the dead xylem conducting cells.

Some minerals, such as calcium and magnesium, enter root cells by diffusion. However, most are taken in by active transport, including potassium, nitrate, phosphate, and sulfate. Active transport uses the energy of ATP, and it stops in the absence of oxygen, which is needed to make ATP during cellular respiration.

Once minerals have passed through the endodermis, they must leave the cytoplasm of living cells, either by diffusion or by active transport, and move into the dead conducting cells of the xylem. From here they are transported upward to the rest of the plant.

In all of this movement of minerals into the root and then into the xylem, water follows the solutes by osmosis.

Although plants take up substances selectively, they do not totally exclude substances that are unnecessary, or even toxic. This can pose problems. For example, nutrient-rich sludge from sewage treatment plants is often recycled by using it as fertilizer. This sludge must be free of toxic substances, because plants will take them up along with nutrients. Wastes from industrial towns often contain high levels of toxic elements, such as antimony, cadmium, and tin, as well as

Figure 34-7

Structure of the endodermis. The cells are arranged like bricks fitting tightly against neighboring endodermal cells. The Casparian strip is a layer of watertight "mortar" impregnating the cell walls in a continuous strip, preventing passage of the soil solution from the exterior to the interior of the root by way of spaces in or between the endodermal cell walls. Any substance that reaches the interior of the root has passed through the plasma membrane and living cytoplasm of (at least) an endodermal cell (red arrows).

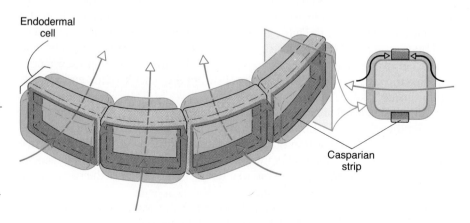

Endodermal cell

Casparian strip

toxic levels of micronutrients such as selenium. Plants fertilized with these wastes may take up so much of these substances that they become unsafe for human consumption.

Some plants inhabit soils with high concentrations of heavy metals, such as the slag heaps of mines. In Wales, a species of the bent grass *Agrostis* grows well on soil containing tailings from copper mines. This grass tolerates levels of copper in its body that would kill other plants. Not only does the plant thrive where other species would perish, so that it has little competition from other plants, but also the plant's copper content makes it toxic to herbivorous animals. The ability of some plants to absorb toxins can be used to clean up polluted soil. In California, workers are planting a wild mustard from Pakistan on soil polluted with excessive selenium. The plant removes selenium from the soil and can then be fed to cattle, replacing the selenium supplement often added to their fodder.

34-D *Functions of Xylem and Phloem*

The young roots that absorb water and minerals in a tall tree may be many metres from the leaves. The roots and leaves are connected by the vascular tissues, xylem and phloem, which form a transport system linking the various parts of the plant and moving substances between them.

In 1679, the Italian scientist Marcello Malpighi performed an experiment that showed the functions of xylem and phloem. He peeled off the bark in a complete ring around the trunk of a tree, a procedure known as **girdling.** This removes the phloem, which makes up the inner bark, but leaves the secondary xylem, or **wood,** intact. After this treatment, Malpighi found that a swelling appeared in the bark just above the stripped area. Fluid exuded from this swelling was sweet (we now know that it contained sucrose). The leaves showed no effects for days or months, but eventually they wilted and then died, and the entire tree was soon dead (Figure 34-8).

From these observations, Malpighi concluded that phloem transports food, such as the sugar in the liquid exuded from the bark, to the roots. Without this supply of food, the roots died after they had used the food already stored below the girdle. Leaves deprived of water wilt and die within hours. Since the leaves of the girdled tree remained healthy for much longer, Malpighi concluded that xylem was transporting water to the leaves. Other girdling experiments have since shown that phloem also conducts food to growing buds, flowers, and fruits.

■ *The vascular tissues divide the task of transport: xylem moves water from the roots to the shoot system, and phloem moves food from leaves to roots or other organs that need food.*

Figure 34-8
A girdling experiment. (a) Malpighi removed the bark, which includes the phloem, in a complete ring around a tree. (b) Following this treatment, a sugary fluid exuded out of the phloem above the girdle, but the leaves remained green for a time, supplied with water through the intact xylem. (c) The wilting and death of the leaves signalled that the roots had finally died of starvation, cut off from the food normally supplied by the phloem.

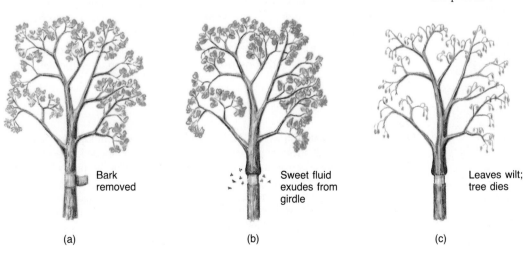

(a) Bark removed

(b) Sweet fluid exudes from girdle

(c) Leaves wilt; tree dies

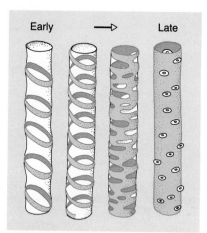

Figure 34-9
Secondary thickenings in xylem conducting cells. In a new length of stem, the first xylem cells lay down thickenings (pink) in disconnected rings or spirals, which permit the cell to elongate by stretching of the primary cell wall between them. Xylem cells formed after the stem reaches its final length lay down more extensive thickenings, which cannot be stretched, and the cell is fixed at this size.

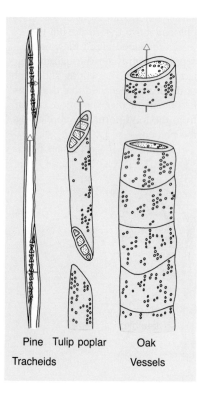

Pine Tulip poplar   Oak

Tracheids   Vessels

34-E Structure of Xylem

Xylem tissue contains many different types of cells. Only some of these conduct the mixture of water and solutes called **sap.** Both conducting and nonconducting cells may strengthen the plant body.

Xylem structure shows many adaptations that enhance its dual functions of transport and support. Right after being formed by cell division, most plant cells lay down **primary cell walls** of cellulose and other polysaccharides. Later, the cells that will become xylem conducting cells add more material, which strengthens the cell walls. These secondary thickenings vary from disconnected rings to extensive **secondary cell walls,** which cover the cell almost completely (Figure 34-9). The secondary thickenings consist of cellulose and **lignin,** a tough, complex organic compound that makes wood woody.

An important feature of xylem conducting cells is that they die once their cell walls are complete. The cell contents disintegrate, leaving a strong, hollow cylinder filled with sap. Since the conducting cells are stacked on top of one another, sap can travel in a more or less straight line up the plant.

The conducting cells of primitive vascular plants are **tracheids,** long, extremely thin cells with slanting end walls. The wood of a gymnosperm such as a pine is almost all tracheids (Figures 34-10, 34-11b). Heavy secondary cell walls slow the passage of sap from one tracheid to the next, except in thin areas called **pits,** where little or no secondary material has been added to the primary cell wall.

Wood also contains living parenchyma cells, in **rays** running out from the center of the tree. The ray cells conduct materials laterally, out to the living tissues in the vascular cambium and phloem. They may also serve as food storage depots. In pine wood, other living parenchyma cells secrete **resin,** a sticky, pungent fluid that repels wood-boring insects and inhibits the growth of certain pathogenic (disease-causing) organisms. Hence resin helps protect an injured tree from disease. Resin may be collected and distilled to make turpentine. It is also the raw material for pitch and tars, used to protect wood from rotting.

Angiosperms (flowering plants) have even more efficient xylem. Some conducting cells have literally made an evolutionary breakthrough: their last act in life is to digest parts of their end walls, forming real holes, not just thin pit areas (see Figure 34-10). This allows sap to flow more rapidly from one cell to the next. In some angiosperms, the end walls are entirely absent, and the cells are like sections of pipe.

Angiosperm conducting cells are also shorter and wider than tracheids. Their greater diameter reduces the cell walls' frictional drag and permits sap to travel faster. The hollow xylem tubes of angiosperms are called **vessels,** and the individual cells in the vessels are **vessel elements,** or **vessel members.**

The wood of an angiosperm such as an oak contains some vessels, but most of it consists of other types of cells (Figure 34-11c). Some are **fibers,** long, thin cells with thick cell walls that help support the tree long after the cell contents

Figure 34-10
Conducting cells found in xylem. Red arrows show the path of sap. Tracheids occur in all vascular plants, including angiosperms. Sap flows unimpeded through a tracheid's hollow interior. Gaps in the secondary cell wall occur in the pits, which occur at the ends of the tracheids (and in the side walls), but sap must still cross the primary cell wall to move from one tracheid to the next. The vessel elements of angiosperms are thought to have evolved from tracheids; they are generally shorter and wider than tracheids, with end walls perforated or absent. Sap moves freely from each cell to the one above.

have died (Figure 34-12). Angiosperm xylem also contains rays made up of living parenchyma cells, and many have tracheids as well as vessel elements.

Changes in the Xylem of Woody Plants

The xylem of a woody plant changes throughout its life. As we saw in Chapter 33, the vascular cambium adds new xylem outside the existing xylem. In climates where good growing conditions alternate with cold winters or dry seasons, the

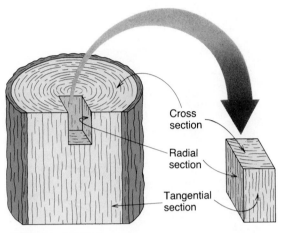

(a) Planes of section

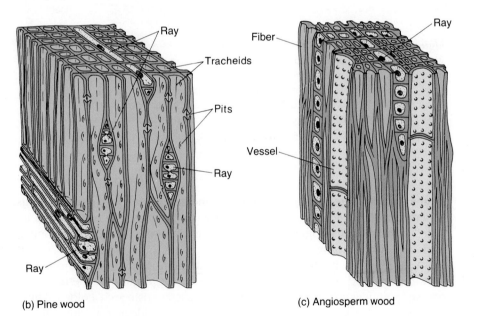

(b) Pine wood

(c) Angiosperm wood

Figure 34-11
Structure of secondary xylem. (a) The planes in which logs are cut to produce the sections shown in (b) and (c). Cutting straight across the log produces a cross section, with a circular outline. A radial section is made by a lengthwise cut that passes through the center of the log. A tangential section is made by a lengthwise cut that does not pass through the center. The colored section of the log is enlarged in parts (b) and (c). The top of each block shows a cross section, the front a tangential section, and the side a radial view. (b) The secondary xylem (wood) of a pine. The tracheids conduct sap. (c) The secondary xylem of an angiosperm. The vessels conduct sap, and the fibers provide strength. Note that ray cells in both kinds of wood are elongated for lateral conduction along a radius of the tree trunk.

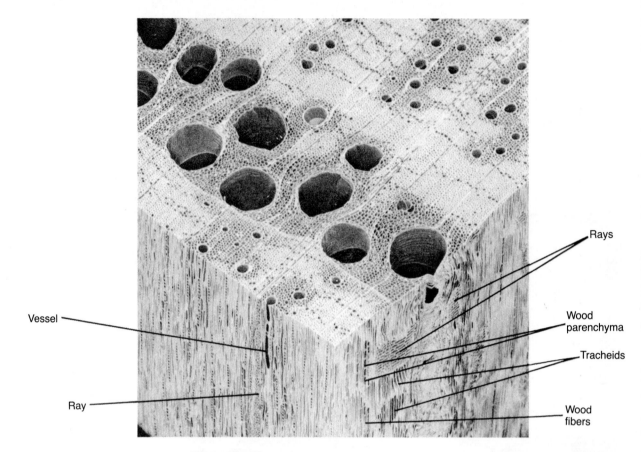

Vessel

Ray

Rays

Wood parenchyma

Tracheids

Wood fibers

Figure 34-12
Wood from a red oak tree (an angiosperm). This scanning electron micrograph shows many vessels, with large diameters and no end walls between vessel elements. Heavy secondary thickenings cover most of the insides of the vessels. Most of the wood consists of tracheids and fibers, much smaller in diameter. Magnification: 100×. (W. A. Cote, SUNY College of Environmental Science and Forestry, Center for Ultrastructure Studies)

cambium's activity follows the weather, and the new wood forms **growth rings.** In temperate climates, where a new ring is added each year, these rings are called **annual rings** (Figure 34-13a). Trees in uniform climates grow continuously and do not form distinct rings.

A tree "writes its diary" in its wood. Variations in cell size, thickness of growth rings, and so on, record rainfall, fire, insect plagues, or the death of a neighbor. Annual growth rings are sometimes used to date archaeological finds, or to determine the climate at some time in history. A wide growth ring reflects a good growing season, while a narrow band is formed in a poor year. By comparing the pattern of tree rings in building timbers with the rings in large, old trees in the same area, archaeologists can determine when the structures were built. Chemical analysis of tree rings from different years can also provide information on an area's history of air pollution and acid rain.

The latest several years' xylem is called **sapwood** because it actually conducts the sap in trees. Interior to the sapwood is the **heartwood,** older xylem that no longer conducts sap. The heartwood is often plugged by deposits of waste material. Sometimes it rots away, leaving a hollow tree that is still alive because its sapwood and phloem still function.

(a)

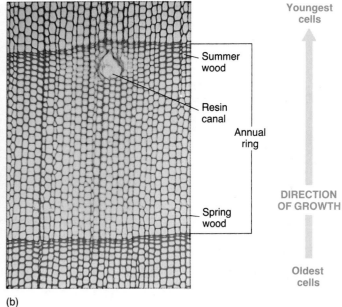

- Summer wood
- Resin canal
- Annual ring
- Spring wood

Youngest cells

DIRECTION OF GROWTH

Oldest cells

(b)

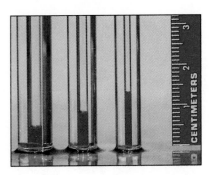

(c)

Figure 34-13
Annual rings. (a) A cross section of a pine tree trunk shows annual rings of alternating light and dark wood.
(b) Viewed with a light microscope, a cross section of pine wood shows the differences in tracheid cell diameter between spring and summer wood.
(c) This elm log shows a marked difference in color between the dark inner heartwood and the pale outer sapwood. (b, Biophoto Associates)

34-F Transport in the Xylem

Tracheids and vessels, the conducting cells of xylem, are stacked in columns, an advantage in the transport of sap. How does sap move through these pipelines of dead cell walls?

If you touch the tip of a thin glass tube to the surface of water, water rises into the tube quickly by capillarity (Section 34-B). The narrower the tube, the higher water rises (Figure 34-14). Water should also move in this way up the cell walls in the xylem tubes. However, calculations indicate that water moving by

Figure 34-14
Capillarity. When the ends of glass capillary tubes are touched to water (dyed red here), water molecules are attracted to the walls of the tube, and these molecules pull others up into the center of the tube. The smaller the bore of the tube, the higher water rises.

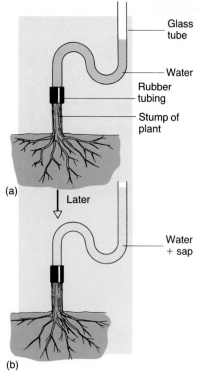

(a)

Later

(b)

Figure 34-15
Demonstration of root pressure. (a) A curved glass tube is sealed to a freshly decapitated plant and is then filled with water. (b) Sap rising into the tube from the roots raises the water level in the tube until the hydrostatic pressure in the tube equals the root pressure. The pressure can be calculated using the final height of the fluid in the tube.

Figure 34-16
Guttation in a strawberry plant. Guttation occurs when ample water and oxygen are available in the soil and the air is too humid to permit evaporation of water as fast as it is pushed up from the roots. Sap is exuded from the tips of xylem columns in the leaves.

capillarity in the slenderest tracheids can reach a maximum height of less than 2 metres, not high enough to account for the rise of sap in many plants.

Root Pressure

If sap cannot climb on its own, it must be either pushed up from the bottom of the plant, or pulled from the top. A push from below can be seen in some plants by cutting the plant off near the ground and sealing the stump into a glass tube: sap rises into the tube (Figure 34-15). The push is called **root pressure.**

Root pressure depends on active transport of ions from the soil solution into the xylem in the center of the root. Water follows by osmosis, building up hydrostatic pressure in the xylem vessels. The endodermis keeps the sap from flowing out of the root back to the soil. So, the sap has nowhere to go but up.

Because roots use a great deal of energy for active transport, they must have an adequate oxygen supply. When the soil contains ample water and oxygen, root pressure may be high. If, at the same time, the air is very humid, water does not evaporate readily from the leaves, and sap forced up from the roots may be exuded from the tips of the leaves. This **guttation** occurs only in rather short plants, where the leaf tips are relatively close to the root endings, the source of the root pressure (Figure 34-16).

However, the greatest root pressures measured can push sap less than a metre, not far enough to reach the tops of many plants. Thus, root pressure alone can account for xylem conduction only in certain short plants growing in certain environmental conditions.

Transpiration Pull

This evidence forces us to look at another mechanism for xylem transport, the pull from above. In 1727 Stephen Hales, an English clergyman, showed that **transpiration,** the evaporation of water from leaves, can pull sap up through the xylem of a plant. Hales cut sets of similar leafy branches and removed dif-

Start

Finish

Figure 34-17
Leaves and water movement. Stephen Hales measured the rates of water absorption by branches with differing amounts of leaf surface. He determined that water movement depends on the leaves.

Figure 34-18
Measuring transpiration pull. Hales attached a tube filled with water to the cut root of a pear tree. He then placed a vessel of mercury in contact with the water and measured how far the mercury was pulled up into the tube as the tree withdrew water.

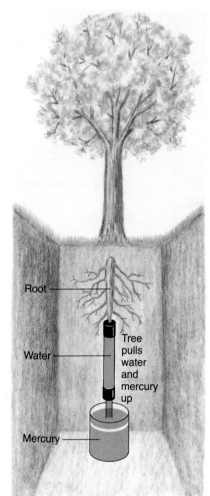

ferent amounts of leaves from some of them. He then set each branch in a container of water and observed that the amount of water removed from the container was roughly proportional to the area of leaf surface on the branch (Figure 34-17). Hales decided that some activity of the leaves caused sap to rise in the stem of a plant.

In another experiment, Hales found that the leaves had to be dry and exposed to the air in order for water to move through the branch. A branch with its leaves immersed in water could not "perspire" (or transpire, as we would say today), and almost no water moved through the branch.

In yet another experiment, Hales dug a hole to expose a root of a pear tree and measured how strongly the tree could pull on water (Figure 34-18). The pull was stronger on bright, sunny dry days than on cloudy or damp days, and it slackened at night. These were exactly the results expected if evaporation of water from the leaves were indeed the force pulling water up through the tree.

The chain of events that moves sap upward depends on the structure of plants and the properties of water:

1. Transpiration occurs: water evaporates from the walls of leaf cells into the air spaces inside the leaves and then diffuses out of the leaf into the atmosphere by way of the stomata (Figure 34-19).

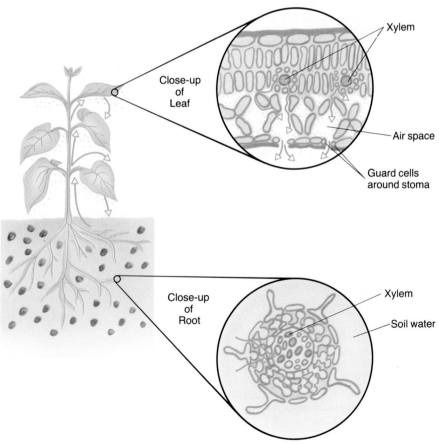

Figure 34-19
The "transpiration stream." Movement of water through a plant depends on the cohesiveness of water and its natural tendency to move into water-deficient areas. Movement begins as water evaporates into the air spaces of the leaves and then diffuses out through the stomata into the air. This starts a chain of movement of water from cell walls nearer xylem veinlets. Eventually this water is replaced by water from the xylem, which coheres to water behind it, and the whole column of sap in the xylem, all the way to the roots, creeps up. The roots replace water by taking in more from the soil (provided soil water is available).

The Kindest Cut

Figure 34-A
The importance of a continuous column of fluid in the xylem. Results of a demonstration described in the text.

After thanking the donor for a bouquet of flowers and sniffing them appreciatively, the next step is to find a vase and put them in water. The florist usually sends a message to cut off an inch or two of stem first; purists even instruct that this should be done while the stem is immersed in water.

Figure 34-A shows the importance of this minor surgery. Three stems were cut from the same plant at the same time. Stem 1 was placed in water immediately. Stem 2 was left for half an hour, and then a 2-inch length was cut off the bottom end. It was then placed in water. Stem 3 was also left for half an hour and then placed in water without further treatment. The photograph was taken one hour after the stems were cut from

the plant. Stem 3 is distinctly wilted, whereas the other two look normal.

Figure 34-B shows why Stems 1 and 2 have fared so much better: in both, the column of water in the xylem is continuous with the water in the container. As the leaves transpire, the water lost from the plant is replaced by pulling more water into the xylem from the container. While Stems 2 and 3 were left out of water, transpiration from the leaves pulled the water column in the xylem up into the stem, leaving room for air to enter the base of the xylem. Cutting

Stem 2 the second time removed this air-filled xylem, permitting the water still in the xylem to link up with the water in the container. However, Stem 3 was left with an air bubble at the base of its xylem, which blocks the entry of water from the container. As the leaves transpire, they lose water faster than the dwindling xylem contents can replace it. Water is pulled out of other cells in the leaves until they no longer fill their walls. Without this internal support the walls buckle and the plant wilts (Section 5-J).

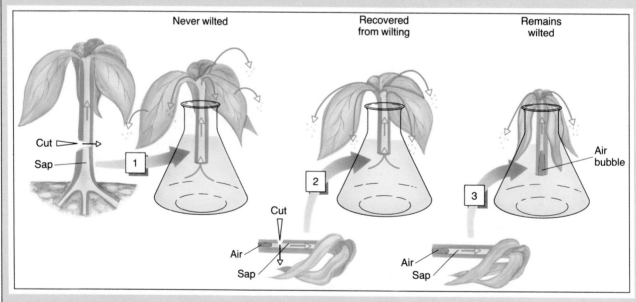

Figure 34-B
Diagram to explain the results shown in Figure 34-A. Xylem is represented by one conduit in the center of each stem, although in real life each stem has many vascular bundles, each containing several xylem vessels.

2. The loss of water by evaporation creates a water deficit in the cell walls next to the air spaces. Water is quickly replaced by movement of water in from the walls of neighboring cells. These cells get replacement water from their neighbors' cell walls, and so on. Water moves between the cellulose fibers of the porous cell walls by capillarity. Eventually, the water lost from the cell walls of the leaf cells is replaced by water from a conducting cell in the xylem at the tip of a veinlet in the leaf.

3. Since water molecules attract one another strongly, they stick together, or cohere. Pulling water out of the top of a xylem column is like pulling on a rope of water that extends all the way down the thin xylem pipeline to its ends in the roots. All the water in the xylem moves up a bit, and eventually the pull reaches root cells which take in soil water to replace that lost from the leaves.

Transpiration sets this process in motion, and the cohesion of water molecules allows it to continue. Hence this is called the **transpiration pull–water cohesion** mechanism of xylem transport.

Transpiration seems to be a necessary evil, the price plants pay for having leaves, which are very efficient structures for obtaining sunlight and carbon dioxide. The cuticle, guard cells, and internal air spaces seem to have evolved as a means of limiting water loss through the leaves.

Evaporation of water creates a negative water pressure in the leaves. So, when a plant is transpiring rapidly, the water in the xylem conducting cells is under tension. Cutting the xylem lets the water column snap apart, just as the two ends of a stretched elastic snap when it is cut. Air rushes into the xylem column, breaking the cohesion of the water and preventing further movement of sap by transpiration pull through the affected tracheids or vessels.

■ *A continuous "transpiration stream" moves from the plant's roots, up through the xylem, and out through the stomata. Water lost through the leaves is continuously replaced as soil water is taken up through the surfaces of root cells.*

Rate of Transpiration

Rate of Transpiration   Hales estimated that, weight for weight, a plant's daily water intake is 17 times that of a human being (see Table 26-1), because the plant transpires so much water to the atmosphere. On a warm, dry day Hales found that a 1-metre-tall sunflower plant lost 0.9 litre of water by transpiration in 12 hours. On dry nights, the plant lost less than 0.1 litre in 12 hours, and on nights with dew, the plant gained water by absorbing the dew through its leaves. A large tree may transpire more than 5 tons of water on a hot, sunny day.

A plant's transpiration rate depends on many factors. Structural features such as the density of stomata and the thickness of the cuticle have an effect. So do environmental factors. The availability of soil water determines how much water the plant can "afford" to lose by transpiration. Sunlight stimulates photosynthesis and causes the opening of the stomata and the faster uptake of carbon dioxide, and so the plant loses water more rapidly in bright sunlight. Low humidity increases transpiration because the concentration gradient between the leaf interior and the air is steeper: water diffuses from the leaf to the air faster. High temperatures increase evaporation. Furthermore, the higher the temperature, the more water vapor the air can gain before it becomes saturated. In still air, the water vapor transpired by a leaf remains nearby, forming a layer of saturated air that retards further evaporation from the leaf. However, if there is a breeze, this layer of still, saturated air blows away, and the leaf loses more water.

Transpiration can be regulated according to environmental conditions. This regulation is performed largely by the guard cells, which open and close the stomata.

■ *Transpiration is affected by the plant's anatomy and by environmental factors such as availability of soil water, intensity of sunlight, humidity, temperature, and wind.*

Operation of Guard Cells

Operation of Guard Cells   Each stoma is surrounded by two guard cells (Figure 34-20). Guard cells can control the size of the stomatal opening because of their peculiar structure: their cell walls are thicker next to the stomatal open-

(a)

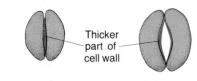

(b)

Figure 34-20
Guard cells and stomata. (a) The waxy lower surface of a rose leaf, showing stomata varying from closed to well opened. Each stoma is surrounded by a pair of lip-like guard cells. (b) Opening and closing of a stoma depends on turgor pressure in the guard cells. The guard cells have thicker walls on the side facing the stoma. Hence, when the cells swell, these inner parts of their cell walls can expand less than the ends and outer portions, and the stoma opens. (a, Biophoto Associates)

ing than elsewhere. Guard cells open the stoma by accumulating solutes. This causes water to enter the cells by osmosis, but because of the uneven thickness of their cell walls, the guard cells do not swell evenly all around. The thicker parts of the cell wall expand less than thinner parts, and so the guard cells assume a curved shape, with an opening between them.

Guard cells are the only epidermal cells with chloroplasts. The chloroplasts produce ATP that is used for active transport of solutes, mostly potassium ions, into the cell, and this transport leads to the opening of the stoma. The stomata open when the carbon dioxide level in the guard cells is low. This normally happens in the daytime, when photosynthesis is using up carbon dioxide, but it also occurs in the dark if carbon dioxide levels in the air are lowered. We do not know how low carbon dioxide levels stimulate transport of potassium.

Sometimes water is lost by transpiration faster than the xylem can replace it. When the leaves' water content becomes too low, the hormone abscisic acid is quickly released from nearby cells. This hormone causes the guard cells to release potassium ions and other solutes rapidly. As water follows by osmosis, the guard cells shrink, and the stomata close until the xylem's delivery of water catches up with the needs of the leaves.

34-G *Phloem Structure*

The other transport tissue, phloem, carries food and other organic compounds throughout the plant from sites of production to sites of use or storage. Most of the food moves in the form of sucrose (table sugar). Phloem may also contain minerals taken back into the plant from dying leaves. Phloem usually lies near the xylem (see Figure 33-2).

In angiosperms, the conducting cells of the phloem are called **sieve tube elements,** or **sieve tube members,** and they are stacked in long columns called **sieve tubes.** The walls of these cells have special **sieve areas,** where many pores permit exchange of substances with neighboring cells. Sieve areas in the end walls have evolved as **sieve plates** with rather large pores (Figure 34-21).

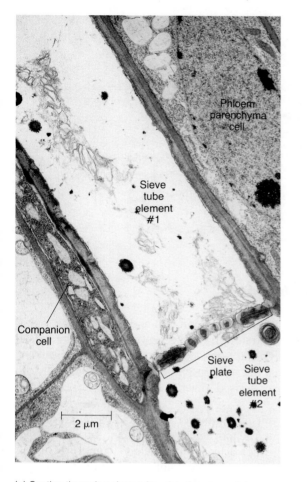

(a) Section through a sieve tube, showing sieve plate

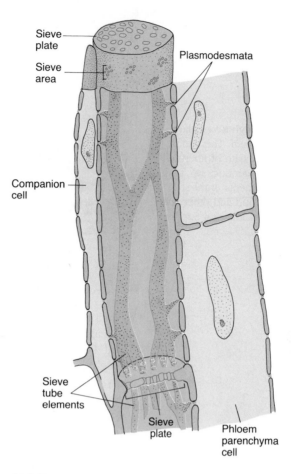

(b) Cell types found in phloem

Figure 34-21

Cell types found in phloem of angiosperms. (a) Electron micrograph of a longitudinal section showing two sieve tube elements, the sieve plate between them, a companion cell (of sieve tube element #1), and a phloem parenchyma cell. (b) Drawing of the kinds of cells in angiosperm phloem. Materials pass up and down from one sieve tube element to the next through sieve plates, and sideways to neighboring sieve tube elements through sieve areas. These conducting cells contain living cytoplasm, in which substances are transported, but they lack nuclei. "Life support" is provided by the nuclei of neighboring companion cells. Phloem parenchyma cells also contain nuclei and cytoplasm. Note the many plasmodesmata connecting the cells. (a, Biophoto Associates)

The conducting cells of phloem, unlike those of xylem, contain living cytoplasm. Before a cell becomes able to conduct, it loses its nucleus and most other organelles. In addition, there is a great enlargement of its plasmodesmata, which pass through the sieve areas and sieve plates and connect neighboring cells.

After a conducting cell has lost most of its own metabolic machinery, it is maintained by the metabolism of its next door neighbor, a **companion cell,** which remains intact. The sieve tube element and its companion cell are the offspring of a single cell that divided unequally (Figures 34-21 and 34-22).

As in xylem, phloem contains living parenchyma cells. Phloem also contains nonliving **phloem fibers,** similar to the strengthening fibers in xylem.

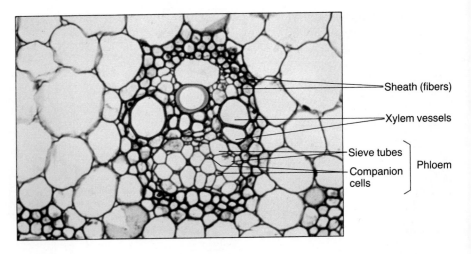

Sheath (fibers)

Xylem vessels

Sieve tubes
Companion cells

Phloem

Figure 34-22
Cross section through a vascular bundle of a corn stem. The bundle is surrounded by a sheath of fiber cells. The xylem vessels conduct sap, and the sieve tube elements of the phloem conduct food and other substances. (Carolina Biological Supply Company)

34-H *Transport in Phloem*

When a phloem sieve tube is injured, plugs of protein and polysaccharide rapidly seal the sieve plates and stop transport. This reaction, like the clotting of blood in an animal, keeps the plant from losing valuable food molecules through the wound. It also makes transport in phloem impossible to study directly, and so investigators have devised several indirect ways to observe phloem transport.

One much-used method is tracing the movement of radioactive materials applied to a plant's roots or leaves. The movement of a substance in a large tree may be measured by following the radioactivity with a Geiger counter. The usual method for small plants is autoradiography: the plant is given time to take up and transport the radioactive substance and is then pressed flat and placed on a sheet of photographic film. Particles emitted by radioactive decay expose (blacken) the film. The plant thus takes its own radioactivity portrait, which shows where the substance has been transported, by what route, and how fast (see Figure 34-25).

Phloem transport can also be studied by using aphids, tiny insects that feed on plant fluids (Figure 34-23). An aphid's mouthparts form a long tube, which the aphid can insert into a plant in such a way that it taps a single sieve tube element. Since the cell's contents are under pressure, fluid from the phloem oozes into the tube and then into the aphid's gut. After the aphid starts to feed, it is anesthetized with carbon dioxide and severed from its mouthparts. The fluid exuded from the phloem through the mouthparts can then be collected and analyzed. By using several aphids on different parts of the plant, an investigator can introduce substances at one point and study their speed and direction of travel.

Other methods of studying phloem include tracing dyes that move through the phloem, using chemicals that inhibit the activities of the phloem, applying heat to small areas, and studying tissue structure with microscopes.

Observations of phloem transport show that the contents of the conducting cells are under pressure. A huge volume of fluid passes through each cell in a short time, moving at speeds of 30 to 200 centimetres per hour. The direction of flow in a particular cell may reverse from time to time, and neighboring cells may conduct in opposite directions at the same time. Killing the cells stops transport in that part of the plant.

The most widely accepted model of phloem transport is the **mass flow** or **pressure flow** hypothesis, suggested by Ernst Munch in 1926. Munch built a physical model that behaved much like the phloem system in a living plant

Figure 34-23
Aphids. These insects suck the fluid from conducting cells in the phloem, and so are useful tools in studying how the phloem works. (Gill Renard)

(Figure 34-24). Two chambers were connected by a tube, and each chamber had a membrane permeable to water but not to sucrose. One chamber was filled with plain water, the other with a sucrose solution. When the two chambers were immersed in plain water, water entered the chamber with the sucrose solution by osmosis. This forced the solution to rise into the connecting arm and flow across to the other chamber, taking some of the sucrose along. As sucrose solution arrived at the second chamber, it forced some of the water there to flow out through the membrane into the water bath. The flow continued until the sucrose was equally concentrated in the two chambers, and then stopped.

If sucrose could be continuously added to the first chamber, and removed from the second chamber as it arrived, the osmotic gradient would be maintained, and the flow would continue indefinitely. Such conditions are indeed found in the phloem system. In a living plant, some areas are **sucrose sources,** continually making sugar (in leaves or green stems) or releasing it by breaking down stored starch (in roots or stems). Other areas are **sucrose sinks,** where sucrose is consumed as it arrives. This could be any cell that needs energy. Food storage tissues also act as sucrose sinks when sugar molecules are being converted into starch granules, which have less osmotic activity. Sucrose is the main solute in the phloem, and the phloem solution always moves down a sucrose gradient, regardless of the concentration gradients of other substances present.

In Munch's model, the water bath represents the xylem, the source of the water that enters the phloem solution by osmosis. The entrance of water would increase the pressure in the phloem, forcing the solution through the phloem until it arrived at a tissue that removed the dissolved nutrients. The water would then be forced out of the phloem by the pressure of the more concentrated fluid moving behind it and taken back into the xylem.

Sugar enters the phloem sieve tube elements in the smallest leaf veins. This is thought to occur by active transport, carried out by surrounding companion cells, which can build up sugar concentrations of 10 to 25%. There are no further membranes to cross, since strands of cytoplasm connect the cells of the sieve tube to each other through the sieve plates. Once sucrose arrives near a hungry cell, it is removed from the phloem and stored or metabolized.

This theory agrees well with the observations on the speed and pressure at which substances move through phloem. It can also account for observed changes in direction of flow. Flow proceeds from an area of high pressure to one of lower pressure. The pressure is lowest in tissues that withdraw nutrients most actively, which may be different parts of the plant at different times. At one time, the roots may be most active as they store food, but later growing buds or fruits may begin to withdraw nutrients faster, reversing the flow. In this case, the food storage tissues may even switch roles, from sucrose sinks to sucrose sources, as they break down starch reserves and release sucrose to support growth or reproduction in other parts of the plant.

■ *Transport in the phloem occurs down a sucrose gradient from "source" to "sink," moving from a high-sucrose, high-pressure area to a lower-sucrose, lower-pressure area.*

Figure 34-24
Munch's model of mass flow in the phloem. The selectively permeable membrane is permeable to water (blue) but not to sucrose (red dots). Sucrose solution is placed in the left-hand chamber, plain water in the right. The tube is then immersed with its ends in a water bath. Water enters the sucrose solution by osmosis through the membrane, and the solution rises into the neck of the tube and flows across to the right-hand chamber, where water is eventually forced out through the membrane. This system stops when the sucrose concentration becomes equal in the two chambers, but in a living plant some areas constantly produce sucrose as others remove it, and movement would continue.

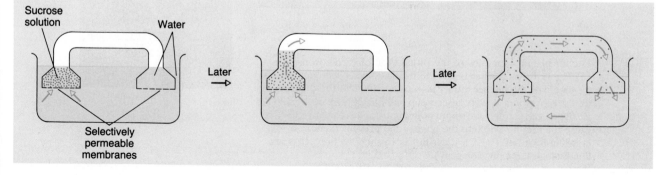

Sucrose solution | Water | Later → | Later → | Selectively permeable membranes

Carnivorous Plants

Many plants can take up nutrients through the leaf surfaces. This ability is carried to fascinating lengths by carnivorous plants, which not only absorb organic food molecules through the leaves, but first use the leaves to capture and digest animal prey.

Carnivorous plants inhabit acid soils. Acidity retards the growth of bacteria that release nutrients from dead organic matter, and also makes the nutrients that are released very soluble, so that they are easily leached away. Thus some plants, particularly in acid bogs, cannot obtain an adequate supply of nutrients through their roots. Some of these plants have "turned the tables" on some members of the animal kingdom by becoming carnivorous. The bodies of animals contain relatively high concentrations of protein, a good source of nitrogen, as well as minerals that plants need. Carnivorous plants, like most other plants, produce food by photosynthesis, and indeed they experience little competition for sunlight because few noncarnivorous plants can survive in their habitats.

There are three basic types of carnivorous plants: pitfall traps, such as pitcher plants; flypaper traps, such as sundews; and active traps, such as the Venus's flytrap. All three have certain features in common. The traps are modified leaves, supplied with nectar glands that exude substances attractive to insects. In most other plants, nectar glands are confined to the flowers (which also consist of modified leaves). The trapping leaves also have glands that produce digestive enzymes. A third feature in common is the modification of leaf hairs in ways that aid in capturing prey (Figures 34-C, 34-D, and 34-E).

Once the trapped insect has been digested, the leaves absorb nitrogen and minerals from its body. The plant absorbs most of its minerals through the leaves. The roots are small and serve mainly to anchor the plant and to absorb water.

Sundew and flytrap plants show growth adaptations as well as structural adaptations. When an insect is caught, rapid changes in hormonal levels make the cells enlarge in a manner that provides a more secure grip on the prey. After the insect has been digested, these plants "reset"

(a)

(b)

Figure 34-C
Pitcher plants, examples of the pitfall type of carnivorous plants. (a) The leaves of pitcher plants are tubular. Nectar-secreting glands on the lip of the pitcher attract insects. Just beyond these glands is a slick area where unwary insects slip and plunge to a pool of digestive juices below. Downward-pointing hairs inside the pitcher cause the prey to skid into the pitcher and prevent them from crawling back out. (b) A leaf cut in half to show its prey. (b, Carolina Biological Supply Company)

(a)

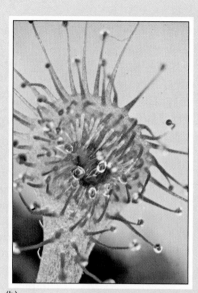

(b)

Figure 34-D

Sundew, a flypaper-type carnivorous plant. (a) Each
leaf is covered with hairs that secrete glistening,
sticky droplets attractive to insects. (b) Once an insect becomes entangled, nearby hairs grow toward it
and hold it more firmly. Again, digestive enzymes are
secreted and the digestion products absorbed. (Carolina Biological Supply Company)

their traps by another round of rapid
differential growth. Because cells can
grow only a limited amount, the
leaves of these plants can capture
only a few insects before their cells
have grown too much to be of further use. As some leaves become too
large, they are shed and new traps
are grown. Pitcher plants, however,
can reuse the same pitchers over and
over, and can even trap many insects
at once in each leaf.

How did carnivorous plants
evolve? In the preceding survey of

their adaptations, we did not find any
real innovations. Rather, we saw that
several common plant features have
been modified or exaggerated, forming a suite of adaptations that permit
carnivorous plants to live in acid
bogs, where few other plants can.
Having little competition for resources is a big advantage, and was a
major selective pressure for the evolution of the specializations found in
carnivorous plants.

In fact, this is just one example
of a general pattern seen among all

organisms adapted to living in unusual environments, such as acid bogs,
deserts, polar areas, mountaintops, or
the deep sea. In all of these cases,
lack of competition has been a dominant force. It has selected for modifications of existing features that made
organisms better equipped to meet
environmental challenges and so to
exploit the resources of sparsely settled habitats.

Figure 34-E

Venus's flytrap. This endangered carnivorous plant is
native only to the Carolinas. The red lining of the bi-
lobed leaves attracts insects. Sensitive hairs on the
inner surface of the leaves respond to an insect's jos-
tling by starting a wave of depolarization similar to a
nerve impulse. This causes rapid changes in the water
content of certain leaf cells; the leaf folds up quickly
and imprisons the prey. Hairs along the edges of the
leaf form a cage around the prey, as in the leaf at
bottom right. Glands on the trap's inner surface secrete enzymes that digest the insect's soft parts. The
rest of the insect blows away when the trap re-opens,
ready for another victim. (Carolina Biological Supply
Company)

34-I   *Distribution of Substances*

The two transport tissues, xylem and phloem, between them carry the various substances that must be moved from one part of the plant to another. Water and minerals taken up by the roots are transported by xylem and may be removed by living cells in any part of the plant. Normally, the vast bulk of water moved to the leaves is lost to the atmosphere. However, some water becomes part of the cytoplasm, some serves as a raw material for photosynthesis, and some moves into the phloem (this water eventually returns to the xylem).

Some minerals also eventually make their way into the phloem. As leaves grow old, their minerals may be moved back into the plant for use in younger leaves or for storage over the winter (Figure 34-25). Phosphorus, potassium, and nitrogen especially are reclaimed, and sometimes iron as well. Calcium is a mineral that does not move once the xylem has delivered it to the leaves.

Most plants do not store much food in the leaves, where it is made, but move it through the phloem to other parts of the plant. When sucrose or other sugars arrive in the roots, some of them are used by the root cells in respiration, some are stored as starch, and some are combined with ammonium ion (NH_4^+) or nitrate ion (NO_3^-), taken in through the roots, to form amino acids. These amino acids are then moved to other parts of the plant by the xylem or the phloem. Phloem also carries various other organic compounds.

The food storage areas of plants include roots, such as those of radishes, carrots, and rutabagas; underground stems, such as those of potatoes; and some-

(a)   (b)

Figure 34-25
Autoradiographs showing movement of phosphorus from older to younger plant parts. The roots of living bean plants were placed in a solution containing radioactive phosphorus (^{32}P). After plant (a) had had time to take up and transport the phosphorus, it was pressed and autoradiographed. The darkest areas represent the parts of the plant with the most concentrated radioactive substance: the roots and youngest leaves. Plant (b) was removed from the solution containing ^{32}P at the same time as plant (a), but was allowed to continue growing for a time in nonradioactive solution before it was autoradiographed. Note that the radioactive phosphorus has been moved to the younger leaves, which had not yet formed when the plant was in the radioactive solution. (Susann and Orlin Biddulph)

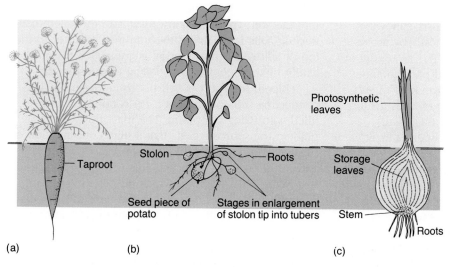

Figure 34-26
Underground food storage. (a) Many plants, such as carrots, store food in enlarged taproots. (b) A potato is an underground storage stem (tuber). Potato plants are grown from pieces of potatoes, each containing an "eye" (bud) that sprouts to form the stem and leaves. As food is produced it is stored in modified underground stems, the stolons, whose tips swell and form tubers. (c) An onion is composed of modified, fleshy food-storing leaves formed underground at the bases of the green photosynthetic leaves.

times even underground leaves, such as those of onions and other bulbs (Figure 34-26). One adaptive advantage of underground storage may be that it hides the food from hungry herbivores. In addition, underground storage organs are less vulnerable to freezing or drying out.

During reproduction, plants move their food reserves from storage areas to the developing seeds and fruits. The food in a seed nourishes the young embryo as it germinates, before it can make its own food. In some cases, storage of food in fruits is an adaptation to seed dispersal by animals: the animal is attracted by the nutritious fruit, and eats it, depositing the seeds elsewhere.

Practical Applications

Transport in the vascular tissues of plants has more than academic interest. Knowing how plants handle various substances can be helpful in the management of plants and plant pests. For example, leaves can absorb some substances placed on their surfaces in solution, and move them into the phloem, which transports them throughout the plant. This is the basis of **foliar feeding,** or fertilizing with a nutrient solution sprayed onto the leaves. Other substances must be worked into the soil because only the roots will absorb them.

Systemic pesticides, watered into the soil or sprayed onto the leaves, are absorbed and distributed throughout the plant. Systemic pesticides combat both sucking insects (such as aphids) and internal pests (such as leaf miners, insect larvae that live entirely inside leaves and never come into contact with sprays applied externally). **Topical pesticides** are not absorbed into the plant, but remain on the leaf surfaces and protect the plant by killing leaf-chewing insects or invading fungi.

SUMMARY

The minerals required by plants are divided into two groups, macronutrients and micronutrients, depending on the quantities needed.

Plants take up most of the water and minerals they need from the soil. The nature of the soil is determined by the organisms that live in it, by the type of rock from which it is derived and rock particle size, by rainfall, and by its content of organic matter and

oxygen. All of these components of soil interact and affect the availability of water and minerals in the soil solution.

The soil solution moves freely into the root cortex. The endodermis forms a living, selective barrier between the soil solution and the rest of the plant. The cells of the root epidermis and cortex absorb minerals and move them through the endodermis to the xylem, which transports them to the rest of the plant. Water enters these cells by osmosis, following the minerals that were accumulated mostly by active transport.

The conducting cells oˆ xylem and phloem show structural adaptations for their rapid transport of substances. Tracheids, found in the xylem of both gymnosperms and angiosperms, have pits in their walls, while vessel cells, found only in angiosperms, may have perforated end walls or no end walls. The sieve tube elements of angiosperm phloem have sieve areas in their cell walls and well-developed sieve plates in their end walls. Companion cells carry on life support for the living but organelle-depleted sieve tube elements. Both xylem and phloem also contain dead, supporting fiber cells and living parenchyma cells.

In xylem, a series of dead, tubular cells conducts water and dissolved minerals upward from the roots to the leaves, flowers, and fruits. Root pressure pushes sap up a short distance in the stems of some plants. In most cases, however, much more upward movement results from the transpiration of water vapor from the leaves. Transpiration creates a pull from the top of the water column down the xylem to the roots and out into the soil. The cohesion of water in the xylem conducting cells is crucial to the transport of water to the top of tall trees. Transpiration is controlled to some extent by the opening and closing of the stomata.

Phloem transport is still poorly understood. The conducting cells of phloem form continuous tubes filled with living cytoplasm. The mass flow theory of phloem transport suggests that a high concentration of sugar in the phloem cells of leaves causes water to enter these cells by osmosis. This uptake of water creates hydrostatic pressure, which pushes the phloem contents along from one cell to the next. Phloem distributes sugars, amino acids, hormones, and some minerals to the roots, fruits, and growing buds.

An understanding of transport mechanisms in plants, as well as a knowledge of the pathways followed by various substances within the plant body, is important in the planning of fertilization, pest control, and hormone treatment programs in modern agriculture.

Plants can store both organic and inorganic nutrients for future use in growth or reproduction.

SELF-QUIZ

Match the following components of the soil with their role in plant nutrition:

_____ 1. Living organism
_____ 2. Organic matter
_____ 3. Oxygen
_____ 4. Rock particles
_____ 5. Water

a. ultimate source of most soil minerals
b. used in breakdown of organic molecules to release energy and minerals
c. dissolves minerals and carries them into roots
d. provides food for fungi and bacteria in soil
e. releases minerals bound in organic molecules

6. Which of the following would *not* make minerals more available to plants?
a. increasing the rainfall in a wet, forested area
b. raising the pH of a very acid soil
c. spreading manure or other fertilizer
d. introducing mycorrhizal fungi into sterilized soil

7. Soil solution fills the cell wall spaces of the root's _____ and _____, and substances are absorbed by the living cells in these layers. The soil solution is prevented from penetrating the entire plant by the presence of the _____ in the _____ layer of root cells.

8. From the list below, pick the three nutrients needed by plants from the soil in highest amounts.

boron	magnesium	phosphorus
calcium	manganese	potassium
copper	molybdenum	sulfur
iron	nitrogen	zinc

9. The main solute transported by phloem is:
a. glucose
b. potassium
c. sucrose
d. starch
e. amino acids

10. The girdling experiments performed by Malpighi supported the theory that:
a. water moves in a tree by the root pressure mechanism
b. water moves in a tree by a transpiration-cohesion mechanism
c. xylem is primarily responsible for conducting water from the roots to the leaves
d. phloem is primarily responsible for conducting water from the roots to the leaves

11. List two functions of xylem tissue, and describe at least one adaptation of cells found in the xylem that contributes to the performance of each function.

12. Movement of water up through a tree trunk depends on:
a. the high boiling point of water
b. capillary movement of water
c. the vapor pressure of water
d. attraction between water molecules
e. osmotic movement of the sap

13. Would root pressure increase, decrease, or remain the same under each of the following conditions?
_____ a. high humidity
_____ b. watering dried-out soil
_____ c. darkness

14. Would the rate of transpiration increase, decrease, or remain the same under the following conditions?
    ___ a. high humidity
    ___ b. increased turgor pressure in the guard cells
    ___ c. increased light
    ___ d. increased wind
15. Xylem that is not conducting water is called:
    a. heartwood
    b. sapwood
    c. rays
    d. vessels
    e. tracheids

16. Which of the following is *not* necessary to the operation of the mass flow mechanism, according to the theory presented in this book?
    a. ATP
    b. root pressure
    c. intact membranes in conducting cells
    d. different osmotic concentrations in different parts of the plant
    e. constant production or release of sugar molecules

QUESTIONS FOR DISCUSSION

1. Explain why chemical analysis of the elements present in a plant is not a good indication of its nutritional needs.
2. Why is vegetation often sparse in soil with many pebbles and boulders?
3. Boron-deficient soils are improved by applying sodium tetraborate at a recommended rate of 20 to 50 pounds per acre, which contains about 2 to 5 pounds of boron. One study showed that wheat plants took up only 0.3 pound of boron per acre in one growing season. If these values are typical, what happens to the rest of the boron applied to the soil?
4. Design an experiment to determine how carnivorous plants obtain mineral nutrients other than nitrogen.
5. In 1936, Bruno Huber performed an experiment in which he inserted thin heating wires into the xylem of a tree. He placed a thermocouple (a sensitive heat-detecting device) farther up the stem and timed how long it took before the heated sap passed the thermocouple. Huber found that the sap moved slowly at night. In the morning, the sap movement speeded up first in the twigs; later, the sap began to rise more quickly in the trunk farther down the tree. Do these results support the root pressure or the transpiration pull theory of xylem transport? Justify your answer.
6. What is the best time of day to cut flowers? Justify your answer.
7. We have seen in this chapter that cohesion of water keeps a column of water travelling up through the xylem of a tall tree. How did the water reach the top in the first place?
8. Why should the ground around evergreen plants be watered thoroughly before the ground freezes for the winter?
9. Contrast the conditions affecting transpiration experienced by a houseplant in winter with those in summer.
10. What is the advantage of bean leaves' having more stomata in the lower epidermis than in the upper epidermis? Why are the stomata of a corn leaf located in both epidermal layers?
11. What does wilting indicate about the movement of water in a plant? What does it indicate about the water content of the soil?
12. The virus disease known as "beet yellow" is transmitted from plant to plant by aphids. Why does the disease spread through the plant rapidly and kill it quickly?
13. Large marine brown algae known as kelps have a system of "sieve filaments" whose cells look remarkably like the phloem conducting cells of vascular plants. These sieve filaments transport carbohydrates from the blades (photosynthetic parts) to the stipe (stalk) and holdfast of the plant. Why do these kelps have phloem-like tissue but no xylem-like tissue?

SUGGESTED READINGS

Cohen, I. B. "Stephen Hales." *Scientific American,* May 1976. A biographical sketch of the life and experimental work of an energetic pioneer in plant physiology.

Epstein, E. "Roots." *Scientific American,* May 1973. Compares the form of root systems in different types of plants as well as how roots absorb minerals from the soil.

Fritts, H. C. "Tree rings and climate." *Scientific American,* May 1972. How tree ring patterns are used to analyze the climate of bygone times.

Heslop-Harrison, Y. "Carnivorous plants." *Scientific American,* February 1978. Briefly describes the types of carnivorous plants and presents experimental and photographic evidence on how prey is captured and digested.

Reganold, J. P., L. F. Elliott, and Y. L. Unger. "Long-term effects of organic and conventional farming on soil erosion." *Nature* 330:370, 1987.

Wooding, F. B. P. "Phloem." *Carolina Biology Reader.* Burlington, NC: Carolina Biological Supply Company, 1978.

Zimmermann, M. H. "How sap moves in trees." *Scientific American,* March 1963. Outlines various methods used to discover how transport occurs in xylem and phloem.

Regulation and Response in Plants

CHAPTER

35

Shagbark hickory leaves opening

Sumac

Harebell

Field of corn poppies

A plant grows through the activity of its meristems. Meristematic cells divide and produce cells that grow, differentiate, and mature (Chapter 33). How is this growth controlled? Why does one cell differentiate into an epidermal cell while another becomes part of a xylem vessel? Why does the root grow down into the soil and the shoot grow up into the air? What determines whether a lateral bud forms a new branch this year, next year, or never? Why do all the cherry trees in an orchard bloom at the same time in the early spring, and why do their fruits ripen over a short season in early summer? And why do all the thistles in a field bloom at once in the late summer?

All of these events, and more, are under the control of plant hormones, chemicals produced in various parts of the plant and distributed throughout its body, affecting virtually every area in one way or another.

KEY CONCEPTS

- Every aspect of the production, differentiation, growth, and maturation of a plant is regulated by chemicals, the plant hormones.
- Environmental stimuli affect the production and distribution of hormones. The hormones in turn govern responses by which the plant adapts to these external factors.

35-A Plant Hormones

Hormones are chemical messengers produced in one part of an organism and transported to other parts, where they exert effects out of proportion to their very small concentrations.

Unlike animals, plants do not have distinct endocrine organs. Plant hormones are produced in various organs that have other functions as well (Figure 35-1). Another difference between plant and animal hormones is that plant hormones can affect virtually any tissue in the plant, rather than having specific target tissues. (In both plants and animals, however, different tissues may respond differently to a particular hormone, depending on the tissue and its stage of development, and on the amount of hormone.) Plant hormones do not usually maintain homeostasis, as do many animal hormones. Instead they affect differentiation and growth by causing permanent changes such as cell division, elongation, and death.

How do plant hormones influence differentiation and growth? Applying a hormone to part of a plant often results in changes in the activity of certain enzymes or in membrane permeability. In addition, the plant hormones all have long-term effects on protein synthesis. However, it is not yet clear whether the hormones themselves bring about these changes, or whether they act indirectly, via second messengers.

A cell's differentiation is determined by the hormones and other chemicals it contains, which in turn are affected by the rest of the plant. Hormonal messages passing between different parts of the plant regulate growth of new parts. Thus the root system remains in physiological balance with the shoot system, and the position of a branch or leaf or root determines where others are added.

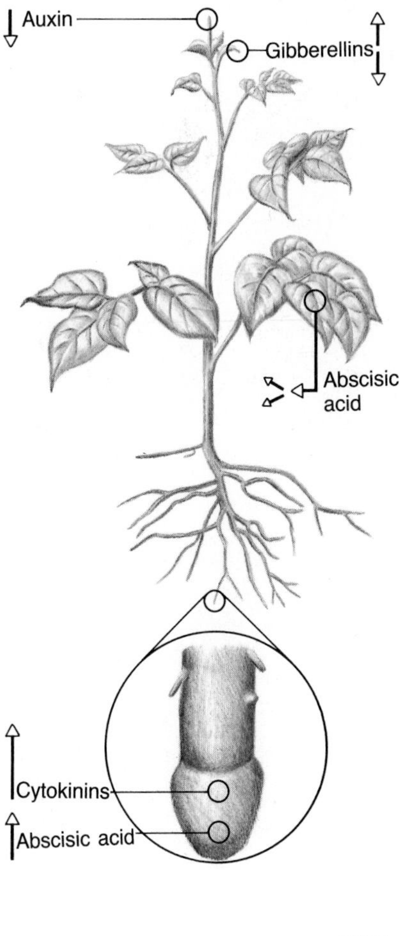

Figure 35-1
Sites of hormone production in a plant. Auxin is produced in growing regions of shoots, including apical meristems and young leaves. Gibberellins are produced in young leaves. Cytokinins are produced in the apical meristems of the roots. Abscisic acid is produced in older leaves and in the root caps. Ethylene (not shown) is produced in many areas when the concentration of auxin exceeds a certain threshold. Arrows indicate direction in which each hormone is transported.

707

Table 35-1   Summary of the Chemical Structure, Production, Transport, and Actions of the Major Plant Hormones

Hormone	Actions	Produced in	Transported by
Auxin	Meristem growth, cell division, enlargement Apical dominance Xylem production Sex differentiation in flower parts Fruit development Formation of new roots Inhibition of root cell elongation Stimulation of stem cell elongation Promotion of ethylene production Gravitropism: positive in roots, negative in shoots Positive phototropism in shoots	Young tissues: shoot apex, young leaves, seeds	Slow cell-by-cell movement toward roots
Gibberellins	Stem cell elongation Cell division in apical meristem Phloem differentiation Sex differentiation in flower parts Leaf growth Inhibition of root formation Stimulation of auxin production	Young leaves Embryo	Unknown
Cytokinins	Cell division Growth Production of fruits and seeds Inhibition of dormancy Inhibition of senescence	Roots	Xylem
Abscisic acid	Dormancy in shoots and seeds Formation of leaves into bud scales Closure of stomata during water stress	Mature leaves Root caps	Vascular tissue
Ethylene	Counteracts effects of auxin Ripening of fruit Senescence	Plant parts with high auxin concentration	Diffusion

A plant must also respond to stimuli in its physical environment, such as light, temperature, wind, water, the pull of gravity, and the change of seasons. These responses, too, are controlled by hormones.

The traditional list of plant hormones has five major entries: auxin, gibberellin, and cytokinin, which induce cell division or cell growth; and abscisic acid and ethylene. Each hormone plays a leading role in certain activities, but many plant responses are governed by the interaction of two or more hormones. Table 35-1 summarizes the production, transport, and activities of these hormones. Recently, investigators have identified a sixth group of regulatory molecules, the oligosaccharins, which act in their own unique roles and also may serve as intermediaries for some of the five major hormones.

Auxin

Auxin was the first plant hormone discovered. It affects many plant activities. Auxin is important in meristematic growth and cell division, but it inhibits the growth of lateral buds (Section 35-B). It also induces production of xylem, development of fruits, and differentiation of sexual parts in flowers. Although auxin stimulates formation of new roots, it inhibits elongation of root cells except at very low concentrations. High concentrations of auxin promote production of ethylene, another plant hormone.

Auxin also mediates many responses to the environment. It causes roots to grow downward and shoots to grow upward, and it is responsible for the bending of stems and leaves toward light. Auxin also induces trees to produce more secondary xylem in response to mechanical disturbance, such as wind. This additional xylem thickens and strengthens the tree trunk.

We have at least a partial picture of how auxin stimulates growth. Auxin causes the plasma membrane to transport hydrogen ions out of the cell. Here the increased acidity activates enzymes that loosen the crosslinks between the cellulose fibers of the cell wall. As water enters the cell from the plant's dilute sap, the cell expands and pushes its wall outward. This expansion starts in the first half hour after auxin is supplied. Afterward, auxin continues to promote growth by speeding the synthesis of proteins needed for growth.

Synthetic auxins, such as 2,4-D, are used as weed killers in lawns. Dicots, which include broadleaved weeds, are much more sensitive to low levels of auxins than are monocots, which include the grasses. Hence, applying a low dosage of synthetic auxins to a lawn can cause the dicots literally to grow themselves to death. To be most effective, these weed killers must be applied when the plants are actively growing and therefore sensitive to extra doses of hormones.

During the Vietnam conflict, American airplanes sprayed Agent Orange, a herbicide containing synthetic auxins, on forests and crops. This had a drastic effect on the ecology of the countryside. Experiments with laboratory animals showed that a contaminant in the spray (a dioxin) causes malformations of developing animal fetuses. After dioxin was detected in drinking water and in fish, one of the few sources of protein in the Vietnamese diet, the spraying was abandoned. Manufacturers later found ways to reduce the concentration of dioxin in the herbicide, and it was used in the United States to kill unwanted forest trees and to control weeds on grazing ranges. However, further research showed that dioxin is a powerful carcinogen, and probably a mutagen, even in minute concentrations. After dioxin was found in beef fat and in human milk, the herbicide was banned.

Other synthetic auxins are sprayed on pear and apple trees to keep them from dropping fruits before they ripen, or applied to cuttings of plant shoots to promote root formation.

Gibberellins

Gibberellins are an important group of plant hormones. The first one known, discovered in Japan, was actually produced by a fungus, *Gibberella fujikuroi,* which causes "foolish seedling" disease of rice. Rice seedlings with the disease grow abnormally quickly, but are spindly and unhealthy, and seldom yield fruit. In vascular plants, gibberellins are produced in young leaves.

One important role of gibberellins is to stimulate elongation in stem cells. Dwarf or miniature strains of plants are often genetic mutants that do not pro-

Figure 35-2
Effect of gibberellin on dwarf and normal strains of corn plants. From left to right: Dwarf plant, untreated; dwarf plant treated with gibberellin; normal plant treated with gibberellin; normal plant, untreated. Notice that the dwarf plant responds much more dramatically than the normal plant does to gibberellin. The dwarf plant is homozygous recessive for a gene needed for normal gibberellin production (Courtesy of B. O. Phinney)

duce gibberellins, but they will grow as tall as normal varieties if treated with gibberellins. Giving extra gibberellins to normal plants has little effect (Figure 35-2).

Gibberellins also stimulate cell division at the stem apex, leaf growth, development of sexual flower parts, and (with sugars) differentiation of phloem tissue. They also stimulate the production of auxin, as well as making plants more responsive to auxin treatment. On the other hand, they inhibit root formation. Gibberellins are also involved in the responses of plants to their environment. Some plants that require a period of cold, or exposure to light, before flowering or before seed germination will respond after gibberellin treatment even if they have not experienced the proper exposure to cold or light.

Cytokinins

The first known **cytokinin** was a component of coconut milk. This compound proved tricky to isolate, but whole coconut milk became a standard ingredient in plant tissue cultures in the laboratory because it stimulates cell division.

In intact plants, cytokinins are involved with growth and with production of fruits and seeds. Cytokinins also prevent the onset of dormancy and slow the aging of cut leaves or fruits. Holly that is cut early for the holiday season can be kept fresh and green by spraying it with cytokinins.

Cytokinins may interact in various ways with other hormones. This is often studied using plant cells grown in laboratory tissue cultures. For example, the relative concentrations of cytokinins and auxin in tissue cultures determine whether or not cells divide and differentiate, and whether they develop into roots or into shoots with leaves (Figure 35-3). To grow an entire plant from cultured cells, hormones must be applied at certain concentrations, in particular ratios to each other, and in the correct sequence. Presumably, similar hormonal interactions in intact plants govern the switching on and off of genes, resulting in the differentiation of the various tissues and organs during embryonic development and throughout the plant's life.

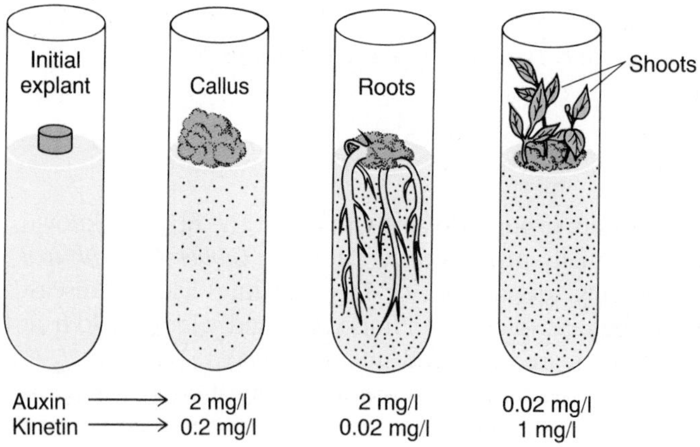

Figure 35-3
Response of cultured plant tissue to auxin and cytokinin. The initial explant is a plug of pith cut from the center of a tobacco stem and placed on nutrient medium. Depending on the concentrations of auxin and cytokinin (kinetin) in the medium, the tissue may develop as an undifferentiated lump of cells (callus) or develop roots or shoots.

	Initial explant	Callus	Roots	Shoots
Auxin		2 mg/l	2 mg/l	0.02 mg/l
Kinetin		0.2 mg/l	0.02 mg/l	1 mg/l

Abscisic Acid

Auxin, gibberellins, and cytokinins have many growth-stimulating roles. In contrast, **abscisic acid** is often called the "stress hormone" because of its roles in helping the plant cope with adverse environmental conditions. For example, during periods of water shortage abscisic acid makes the stomata close, reducing the plant's loss of water by transpiration (Section 34-F).

Abscisic acid is produced in mature leaves. At the end of the growing season, its concentration in twigs is high compared to the levels of gibberellins and cytokinins. This high level of abscisic acid induces the apical meristem to stop dividing, and the newest leaf primordia form into bud scales around the tip instead of becoming normal photosynthetic leaves (Figure 35-4). The bud scales protect the delicate apical meristem from freezing or drying out in the cold of winter. In spring, overwintering twigs are released from dormancy by destruction of abscisic acid, or production of substances that counteract it. Abscisic acid also keeps seeds dormant until it is leached away by water or overcome by a stimulatory hormone, usually gibberellin. Again, this keeps the tender seedling from being blighted by winter's cold. Abscisic acid does, however, promote the embryo's growth and synthesis of storage proteins, making it better prepared for life after it does germinate.

Ethylene

Unlike the other known plant hormones, **ethylene** is a gas, and so it travels in the air as well as within the plant. Hence it affects not only the plant that produced it, but others as well. For example, ethylene given off by apples inhibits the sprouting of potatoes stored in the same bin.

The existence of a gas that affects plant physiology was known long before it was shown to be ethylene. Nineteenth-century growers of greenhouse plants knew that something in the gas used in gas lighting caused blossoms to wither prematurely. Mango and pineapple growers commonly lit fires in their groves because something in the smoke synchronized flowering and fruit ripening.

The production and effects of ethylene are closely tied to auxin. After auxin exceeds a certain level, it stimulates production of ethylene, which in turn counteracts the effect of auxin. Thus production of ethylene by lateral buds as they receive auxin seems to play some part in repressing their growth.

Ethylene also has interesting roles in plant senescence (Section 35-F) and in the ripening of fruits. The development of a fruit is stimulated by auxin, gibberellins, and cytokinins. The level of auxin builds up and then drops; this is followed by production of ethylene. Ethylene stimulates the activity of various enzymes, which (1) convert starch and acids of the unripe fruit to sugars, and (2) soften the fruit by breaking down pectins in the cell walls.

The release of ethylene has a positive feedback effect: the more ethylene produced, the more the fruit is stimulated to produce additional ethylene. Hence the entire fruit ripens at once, and if there are many fruits in an area, the first to begin ripening stimulates ripening of its neighbors. By the same token, overripening is also contagious, and it is quite true that "one bad apple spoils all the good ones." Apples are now stored in refrigerated, airtight rooms in an atmosphere enriched with carbon dioxide, which inhibits the action of ethylene.

(a)

(b)

Figure 35-4

Leaves and bud scales. (a) Abscisic acid causes new leaves on this azalea bush to develop as bud scales, like those surrounding the bud shown, rather than as the more familiar foliage leaves nearby. (b) A hickory bud released from dormancy in the spring. New green foliage leaves are swelling and bursting through the layers of tan bud scales that protected them through the winter.

Figure 35-5
Fruit ripening. Fruits ripen in synchrony in response to the gaseous plant hormone ethylene, either produced by the fruits themselves or supplied artificially by humans.

Ethylene finds uses, however, in the production of tropical fruits for far-away markets. Rather than letting the fruit ripen on the trees and risk loss by over-ripening in transit, growers pick fruits such as bananas, citrus fruits, and pineapples when they are still green and ripen them by applying ethylene after they reach their destination (Figure 35-5).

Recent research also shows that ethylene induces plants to make certain enzymes, including chitinase, which defends the plant against fungi by breaking down chitin in the fungal walls.

Oligosaccharins

The newly discovered **oligosaccharins** (oligo = few) consist of sugars in short, branched chains, and they occur at extremely low concentrations. They are released when enzymes cleave larger, complex polysaccharides in the cell wall matrix (the material in which the wall's cellulose fibers are embedded). Each oligosaccharin is liberated by a different enzyme and mediates a particular response, unlike the major plant hormones, each of which plays several roles.

Some oligosaccharins help to defend the plant by prompting nearby cells to produce substances that kill pathogens or that make the plant less digestible to insects. Some may also act as second messengers, mediating a specific effect of one of the five major plant hormones. We still have much to learn about oligosaccharins.

35-B   Apical Dominance

■ *Auxin, produced in the apical meristems of shoots, tends to repress growth of lateral buds into new branches.*

Many plants exhibit **apical dominance,** in which the growing apical meristem represses the growth of lateral buds on the same stem, causing them to remain dormant. This is due primarily to auxin, which is produced in the apical meristem and transported toward the base of the plant. If the apical meristem is cut off, the lateral buds develop into new branches. This can be prevented by placing auxin on the cut stump immediately after the apical meristem is removed (Figure 35-6).

Other plant hormones may interact with auxin in apical dominance. Auxin stimulates lateral buds to produce ethylene, which in turn seems to play some

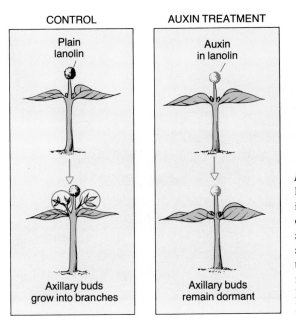

CONTROL   AUXIN TREATMENT

Plain lanolin

Auxin in lanolin

Axillary buds grow into branches

Axillary buds remain dormant

Figure 35-6
Demonstration that apical dominance is mediated by auxin. (Left) Removal of the apical bud permits growth of axillary buds, but if the apical meristem is replaced with an auxin solution (right), the axillary buds are still inhibited from growing. (Why is plain lanolin applied to the plant on the left?)

part in inhibiting the buds' growth. In contrast, cytokinins may help to overcome apical dominance. Virtually all of the cytokinins produced in a plant's roots are transported to growing areas such as apical and lateral meristems, and fruits. When the apical meristem is removed, the lateral buds receive more cytokinins. As a shoot grows, its lower buds break dormancy, and this may be due to the accumulation of more cytokinins from the roots and the reduction in the amount of auxin reaching them from the apical meristem. So it may be more cytokinin, perhaps combined with less auxin, that releases the lateral buds from inhibition.

Along similar lines, it is known that food and water are transported preferentially into areas of the plant where auxin concentration is high. Thus it is possible that the lateral buds fail to grow because of starvation rather than repression. Removal of the auxin-producing apical meristem would reduce auxin levels at the tip of the plant and permit the transport of more food to the lateral buds. These buds could then begin to grow and produce their own auxin, causing them, in turn, to receive more nutrients from the transport system.

It is hard to design experiments to show whether the inhibition of lateral buds comes from too much auxin, from too little cytokinin, from the wrong combination of auxin and cytokinin, from starvation, or from some combination of these.

A plant with strong apical dominance has a distinct main axis, with few side branches, whereas a plant with weak apical dominance has many well-developed branches and a bushy appearance. Gardeners often pinch out apical meristems to make their plants grow into compact, bushy shapes.

By looking at the plants around us, we can easily see that the degree of apical dominance varies a great deal (Figure 35-7). Some of this variation is due to genetic differences, but environment can also affect apical dominance. A beech tree in a forest is tall and slender, with short, thin branches, whereas another, growing in the open, is wide and spreading; a third, growing at the edge of the wood, has large branches on the open side, and small, thin branches on the shady side. These variations in shape suggest that light may somehow counteract apical dominance, allowing the tree to grow new branches into any openings in the forest canopy.

(a)

(b)

Figure 35-7
Tree shapes. (a) Conifers with strong apical dominance. The trunk grows much faster than the branches, and the terminal bud represses the nearest branches most strongly. (b) Very weak apical dominance results in a rounded, spreading shape, with many large branches.

Figure 35-8
Tropisms. (a) Positive gravitropism: these bean seeds were germinated in the dark in several positions, but the roots always grew downward.
(b) These houseplants show positive phototropism, growing sideways toward the window, the primary source of light, and positioning the leaves perpendicular to the sun's rays.

35-C Responses to the Environment

A plant's growth is governed by the interactions of its hormones, produced in various parts of the plant and transported throughout its body in varying concentrations. Factors in the environment influence the production and distribution of hormones. This results in adaptive adjustments in growth, as when light shining on a lateral bud helps to release it from apical dominance: the bud grows into a new branch, with leaves that can use the light for photosynthesis to make more food for the plant.

Many plants require a period of cold before they break dormancy and resume growth. Examples include germination of some seeds, blooming of spring bulbs, and breaking of dormancy in overwintering twigs of some woody plants. In all of these cases, the requirement for a cold period ensures that the plant delays growth until spring, when growing conditions are likely to be favorable. Many of these plant species do not grow well in climates where the winters are not cold enough to provide the necessary stimulus.

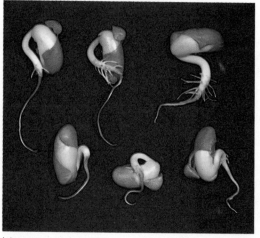

(a)

(b)

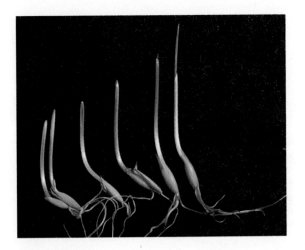

Figure 35-9
Growth of oat seedlings. The roots grow downward in response to the pull of gravity. The coleoptile, a white sheath over the leaves, grows toward light. Illumination of the coleoptile stimulates the leaves within it to expand, rupturing its tip and growing out into the light where the coleoptile has positioned them, as seen in the two seedlings at the right.

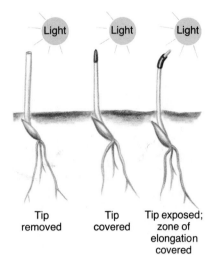

Figure 35-10
Phototropism in the coleoptile of an oat seedling. Removing or covering the coleoptile tip destroys the phototropic response, whereas covering the zone of elongation, where bending actually occurs, has no effect.

Tropisms: Growth in Response to Gradients

Growth responses along the direction of environmental gradients are called **tropisms.** The response of a plant to gravity, for example, is called **gravitropism.** In most plants, the roots are **positively gravitropic,** tending to grow downward, while the shoot is **negatively gravitropic,** growing away from the pull of gravity, or upward. **Thigmotropism** is a response to contact, as when climbing plants wrap themselves around a supporting object.

Tropisms orient the parts of a plant favorably with respect to the resources it needs. The first root of a germinating seed always grows downward, no matter what the position of the seed in the soil (Figure 35-8a). This positive gravitropism ensures that the root establishes a firm anchorage for the plant. It also gives the best possible chance of finding water and dissolved nutrients. Roots may also respond to water itself. For example, the roots of a willow may grow horizontally a dozen metres or more toward water sources. This strongly positive **hydrotropism** is responsible for clogging many a sewer line or septic system, an expensive botany lesson for unwary homeowners.

Unlike roots, most shoots show negative gravitropism, growing away from the pull of gravity. This ensures that a seedling's shoot grows upward and reaches the soil surface by the shortest route. Once the shoot reaches light, it also shows **positive phototropism,** growing toward light. If the light shines mainly from one side, the plant bends in that direction (Figure 35-8b).

Auxin was first detected because of its role in phototropism in oat seedlings. A sheath, the **coleoptile,** covers the first leaves of the oat seedling and other grasses and cereals (Figure 35-9). If the coleoptile of a seedling grown in the dark is exposed to light from one side, it bends toward the light. Charles Darwin studied this phototropic response. Although the growth curvature takes place in the region of elongation below the coleoptile tip, Darwin found that covering this area did not affect the phototropic response. However, when he removed the coleoptile tip, or covered it with an opaque cap, the phototropic response did not occur. He concluded that phototropic bending in the zone of elongation was controlled by the tip (Figure 35-10).

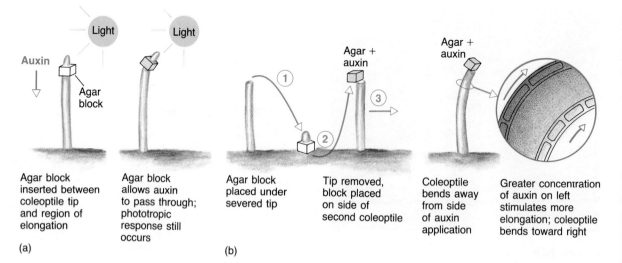

Agar block inserted between coleoptile tip and region of elongation

(a)

Agar block allows auxin to pass through; phototropic response still occurs

Agar block placed under severed tip

Tip removed, block placed on side of second coleoptile

(b)

Coleoptile bends away from side of auxin application

Greater concentration of auxin on left stimulates more elongation; coleoptile bends toward right

Figure 35-11

Chemical control of phototropism. (a) An agar block can transmit the agent responsible for phototropism from the coleoptile tip to the zone of elongation. (b) This agent can be collected from the tip and later transferred to a second plant, which then bends as though it too were lit from one side. Bending is due to greater elongation of cells on the side away from the light, that is, on the side with a higher concentration of auxin.

Later experiments showed that phototropism is caused by the chemical we now call auxin. A coleoptile tip was cut off and placed on a block of gelatin-like agar. When this block was later placed on one side of a freshly decapitated coleoptile, the coleoptile grew and bent away from that side (Figure 35-11). Furthermore, if two coleoptile tips instead of one were put on the agar block, the block caused the headless coleoptile to bend twice as much.

These results suggest that phototropism in the intact seedling is due to an auxin concentration that is higher on the dark than on the lighted side of the coleoptile. By applying radioactively labelled auxin to plants, it was found that auxin is transported from the lit to the shaded side. Here it causes increased cell elongation—hence the increased bending seen in the phototropic response. So plants actually grow toward light by growing away from the dark. Plants grown in bright light show a decreased sensitivity to light, and are often shorter (and sturdier) than plants grown in dimmer conditions. In this way the plant does not waste energy growing toward light when it already receives enough.

■ *Differences in auxin distribution result in positive phototropism of shoots and gravitropism of shoots and roots.*

Gravitropism is also mediated mainly by auxin. In roots, the pull of gravity is detected in the root cap, and the root bends farther back, in the zone of elongation (see Figure 33-4). Auxin is transported down through the center of the root tip, and when it reaches the tip of the root cap it makes a U-turn and goes back up the root through the outer cell layers. If the root is vertical, auxin is distributed evenly to all sides, and the root continues to grow straight down. However, if the root is horizontal, more auxin is distributed to the lower side of the root. Unlike stem cells, root cells are inhibited by all but the lowest levels of auxin. Hence cells on the lower side of the root grow less than the cells on the upper side, and the root bends in the zone of elongation until it is growing straight down.

A root's response to gravity depends on the starch-storing **amyloplasts** of root cap cells. The heavy amyloplasts fall to the bottoms of the cells (Figure 35-12). Here they apparently press on the endoplasmic reticulum and cause it to release stored calcium ions. This leads to activation of auxin pumps in the nearby areas of plasma membrane (on the lower sides of cells). Auxin is

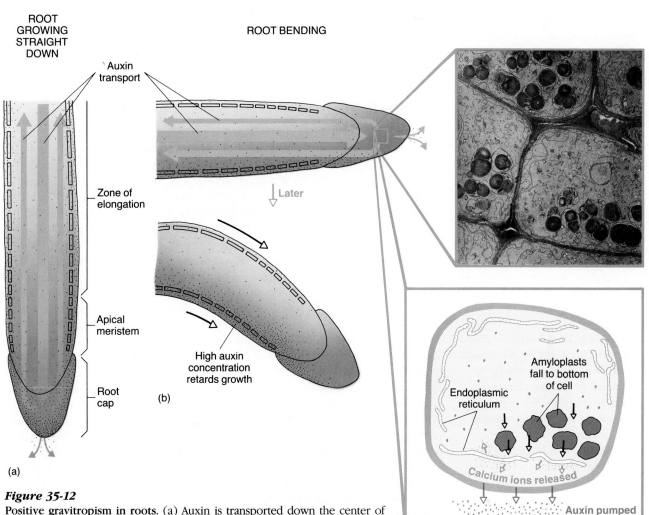

ROOT GROWING STRAIGHT DOWN

ROOT BENDING

Auxin transport

Zone of elongation

Apical meristem

Root cap

(a)

Later

High auxin concentration retards growth

(b)

Endoplasmic reticulum

Amyloplasts fall to bottom of cell

Calcium ions released

Auxin pumped out of cell

(c)

Figure 35-12
Positive gravitropism in roots. (a) Auxin is transported down the center of the root to the root cap, where it is distributed sideways and sent back toward the stem (pink arrows). (b) A root turned on its side bends in the zone of elongation and grows downward. (c) Bending occurs because amyloplasts (containing heavy starch grains) fall toward the bottom of the cell (that is, in the direction of gravity). This is thought to cause the underlying endoplasmic reticulum to release calcium ions, which activate nearby auxin pumps. As a result, auxin is transported mainly toward the bottom of the root. Here it inhibits elongation of cells on the lower side of the root. (c, Barrie Juniper)

pumped out of the lower sides of the cells and soon builds up more on the lower side of the root cap than on the upper side. This difference is maintained while the auxin is transported back up the outside of the root to the zone of elongation.

35-D *Flowering*

To many people, the parade of flowers blooming one after the other symbolizes the passing of the seasons. The fragrance of roses ushers in the summer, and the rich hues of goldenrod and purple asters herald the coming of autumn.

The predictable flowering of many plants depends on environmental cues. Some familiar plants, such as carrots and parsley, flower the second year after the seed began to grow—following a period of cold in the winter. Such plants can be forced to flower by placing them in the cold for several weeks, or by applying gibberellins. Evidently a cold period somehow triggers a hormonal change in these plants, switching them from the vegetative to the reproductive phase.

Photoperiodism: Responses to the Length of Night and Day

In higher latitudes, the period of daylight becomes longer in spring and shorter again in the fall. Many plants flower only when exposed to certain light and dark periods, a response called **photoperiodism.** Surprisingly, this was not recognized until 1920, when W. W. Garner and H. A. Allard of the U.S. Department of Agriculture studied two flowering problems. One was the behavior of "Maryland mammoth" tobacco, which reached prodigious heights in the field but did not flower. However, cuttings rooted and grown in greenhouses during the winter flowered even at much smaller sizes. The second problem concerned "Biloxi" soybeans, which flowered on the same date even when farmers staggered their plantings in order to try to stagger the harvest. Since all of the tobacco plants and all of the soybean plants flowered at the same time, regardless of size or age, it looked as though flowering must be determined by some cue in the environment.

Experiments with different temperatures and light intensities failed to link these factors to flowering. Garner and Allard then turned to another hypothesis, that plants could respond to varying lengths of daylight—that is, to different photoperiods. This seemed silly because it meant that plants would have to be able to measure time. But, sure enough, when the researchers artificially changed the photoperiod, they found that the tobacco and soybean plants would flower only if exposed to light for less than a certain maximum amount of time each day! Accordingly, these were dubbed **short-day plants.**

We now know that these plants should really be called "long-night" plants because what they need to flower is a certain minimum period of uninterrupted darkness, the **critical night length** (which occurs naturally if the period of light—the **critical daylength**—is short enough). Interrupting the dark period with a flash of light prevents the plants from changing to the flowering mode. That is why, if you own a Christmas cactus or poinsettia that you wish to flower during the holiday season, you must put the plant in a place kept dark from sunset to sunrise. It usually takes several days of such treatment to make the plant flower.

Later experiments showed that some plants are **long-day plants,** requiring a certain minimum length of light period (or less than a certain maximum length of dark) for flowering (Figures 35-13 and 35-14). Examples are spinach,

■ *Short-day and long-day plants are defined not by the actual length of light (or dark) required to induce flowering, but by whether the critical length of the photoperiod is a* maximum *(for short-day plants) or a* minimum *(for long-day plants).*

Figure 35-13
Plants with photoperiodic requirements for flowering. (a) A short-day variety of chrysanthemums, blooming in late October. (b) These petunias are long-day plants.

(a)

(b)

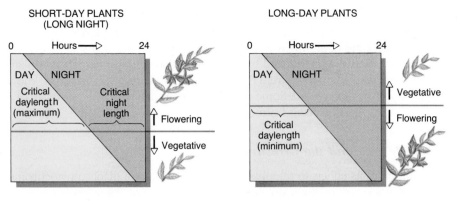

SHORT-DAY PLANTS
(LONG NIGHT)

LONG-DAY PLANTS

Figure 35-14
**Comparison of requirements for
flowering in short-day and long-day
plants.** A horizontal line drawn across
either box is 24 hours long. Each box
shows various possible combinations
of day and night length, from very
short days and long nights (top) to
very long days and short nights (bottom). The red line in each box indicates the critical daylength. At any
daylength above the red line in the
short-day plant box, the plant will
flower; at longer daylengths (below
the red line), the plant remains vegetative. The opposite is true for long-
day plants (right-hand box).

radish, barley, and black-eyed Susan. (In some long-day plants, treatment with gibberellins can substitute for the proper light/dark regimen.)

There are also many **day-neutral** plants, such as tomatoes and cucumbers, which begin to flower at a certain stage of growth regardless of daylength.

Many plants require other factors in addition to the proper daylength: stage of maturity, temperature, day/night temperature changes, moisture, or a se-quence of short days after long days or vice versa. The florist industry profits from the study of these factors. We can now obtain chrysanthemum plants in full bloom for Mother's Day and send carnations to our Valentines.

35-E Phytochrome

How do plants measure the length of photoperiods? Any timing mechanism must involve the biological clock found in all eukaryotes (Section 31-G). Mea-suring a period of light (or dark) must involve the interaction of this clock with pigment molecules, the means used by all organisms for detecting light.

To discover which pigment mediates photoperiodic control of flowering, we must find out which wavelengths of light affect flowering, and then look for a pigment that absorbs those wavelengths. By using different wavelengths of light to interrupt the dark period of short-day plants, researchers found that red light of about 660 nanometres inhibited flowering most effectively. Curiously, when far-red light (730 nanometres) was used, the plants acted as though no light were given in the dark period. Given alternate flashes of red and far-red light, the plants responded to the last wavelength, flowering when far-red was given last, and remaining vegetative when red was last (Figure 35-15). On the other hand, long-day plants held on short-day photoperiods could be made to flower if they were given a flash of red light during the dark period. This could be reversed by far-red light.

The pigment that absorbs red and far-red light is **phytochrome,** a blue-green molecule found in plasma membranes of plant cells. It has two different forms: one form, P_r, absorbs red light, which changes it to the other form, P_{fr}, which absorbs far-red light and is changed back to P_r. P_{fr} is the active form. Phytochrome works like a switch that is turned on and off by red and far-red

■ *Photoperiodism seems to in-*
volve interaction of phyto-
chrome with the metabolic
reactions controlled by the
plant's biological clock.

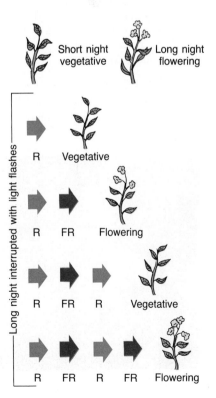

Figure 35-15
Response of a short-day plant when its critical dark period is interrupted with flashes of light. A flash of red light (R) will suppress the flowering response normally initiated by a regime of dark periods of critical length; following a flash of red light with a flash of far-red light (FR) reverses the effects of the red flash, and the plant enters the flowering phase. When a train of flashes is given, the plant always re-sponds to the last flash received.

Figure 35-16
Flowering of short-day plant exposed to long-day treatment. A single leaf of the plant is darkened to give it the short-day treatment normally required for flowering; hormones produced in this leaf are transported to the rest of the plant and induce flowering.

light. Flowering apparently depends on phytochrome's being in the proper form for a certain length of time during the right part of the plant's daily cycle.

The outcome of all this is a change in the plant's hormonal state. It can be shown that flowering is induced by hormones produced in one part of a plant and transported to other parts (Figure 35-16). However, we don't yet know which hormone is involved. Some researchers have proposed the existence of a yet undiscovered flowering hormone, already named "florigen"!

Phytochrome and the Growth of Seeds

In addition to its effects in the photoperiodic flowering of plants, phytochrome plays an important ecological role in the germination of many kinds of seeds. The seeds of plants such as "Grand Rapids" lettuce, and many kinds of weeds, require light for germination. As we expect from a response mediated by phytochrome, red light is most effective, whereas far-red light inhibits germination. In nature, this response keeps seeds from germinating when they are buried too deeply in the soil for light to penetrate. If a seed is buried too deep, the young seedling may exhaust its food supply before its shoot can reach light and begin photosynthesis. Such a seed may have a better chance of survival if it remains dormant until the soil is later disturbed, bringing the seed into the light, which then stimulates its germination. Overhanging plants, with their green leaves, also screen out most sunlight except far-red. Again, this keeps the seed dormant until the leaves above it die and the seed receives full sun, when it germinates.

Phytochrome plays a further role in the development of the young shoot. While the shoot is growing in the dark, it elongates rapidly, increasing its chances of reaching the light before its food supply runs out. The stem is long, spindly, and unpigmented (white or pale yellow), and the leaves remain small. This condition, known as **etiolation,** is most familiar in the bean sprouts used in Oriental and vegetarian cooking, and in forgotten potatoes sprouting in a cupboard. When the shoot reaches the light, its phytochrome is converted to the active form, initiating a series of changes: the stem's rate of elongation decreases, and the stem becomes thicker and sturdier; the leaves expand; and undifferentiated plastids develop into chloroplasts with chlorophyll, giving the stem and leaves a healthy green color.

35-F Senescence

Flowering, setting of seeds and fruit, and germination of seeds are events in a plant's life that occur predictably in response to environmental or genetic factors. **Senescence,** the process of aging that makes all or part of the plant more

Figure 35-17
Leaf senescence and abscission.
(a) Senescence occurs regularly each autumn. Chlorophyll is destroyed, revealing pigments of other colors, and the nutrients of chlorophyll are drawn back into the plant. (b) The attachment of the leaves weakens as the abscission zone forms, as seen in a line running down from the top center of this picture. The leaf eventually separates from the stem at this area of weakness and falls off the plant. (b, Carolina Biological Supply Company)

(a)

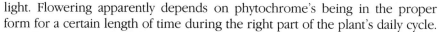

(b)

susceptible to death, is also an integral part of its life history. A wheat plant turns yellow, dries up, and dies after it has set seed; a plum tree drops its fruits during a short period in early summer and loses all its leaves during the fall.

Senescence is under hormonal control, probably by ethylene. In annual plants, the hormonal changes that initiate senescence of the whole plant often seem to be triggered by the setting of seed. In perennial plants, the leaves, fruits, and withered flower parts senesce and drop off each year, a process called **abscission.** Although its name suggests a link with abscisic acid, abscission is promoted mainly by ethylene. (Auxin and cytokinins inhibit senescence and abscission.)

Before abscission, several changes take place. Enzymes that degrade the leaf tissue become active, and nutrients may be withdrawn from the leaf. Cell divisions occur in the **abscission zone,** where the leaf joins the stem. In this region, ethylene stimulates the production and release of the enzyme cellulase, which breaks down the cell walls, thereby creating a zone of weakness (Figure 35-17). Finally, this zone breaks apart, and the leaf drifts away. The leaf scar is quickly healed by the deposition of a corky, waterproof seal that prevents the loss of water or entry of pathogens (disease-causing organisms).

Electrical Responses of Plants

A JOURNEY INTO LIFE

Figure 35-A
A sensitive *Mimosa* in the Big Thicket of Texas. As the finger strokes from left to right, the leaflets fold quickly, suddenly looking limp and wilted. The extent of the plant's response varies with the intensity of the stimulus.

Most responses of plants to environmental stimuli are mediated by hormones. However, some plants can make fairly rapid electrical responses similar to those that occur in the nerves and muscles of animals (Chapter 28). When a Venus's flytrap closes around an insect victim, the first, rapid movement begins as an electrical response that leads to swift osmotic changes in cells on the outside of the leaf's two lobes. The electrical signal causes these cells to pump out hydrogen ions (H^+), which activate enzymes that loosen the crosslinks between fibers in their cell walls. The cells then expand as water rushes into them by osmosis. This pushes the two lobes of the leaf toward each other and the trap snaps shut. Then slower, differential growth caused by hormones ensures a firm grip on the prey (Chapter 34, *A Journey Into Life,* "Carnivorous Plants").

Another notable example is the rapid closure and drooping of the sensitive leaves of mimosa plants (Figure 35-A). This response may protect the plant from being eaten. In these plants, certain cells in the phloem have highly negative membrane potentials. Various stimuli such as touch, light, or certain chemicals, when sufficiently strong, can elicit an all-or-nothing action potential. This electrical signal travels from cell to cell by way of plasmodesmata. It spreads rather slowly, at about one to two centimetres per second, rather than the tens of metres per second for animals' action potentials.

The effectors in these plants are particular cells known as motor cells, which respond to the electrical signals by sudden membrane and osmotic changes. Potassium ions rush out of the motor cells, followed by water, and the cells collapse as they lose their internal fluid pressure. The motor cells are located at key points called hinge regions, and their collapse results in rapid folding of the leaflets.

SUMMARY

Five groups of major plant hormones are known. Auxin, gibberellins, and cytokinins are generally growth-promoting; abscisic acid enables the plant to cope with stress, and ethylene promotes ripening of fruits and senescence of leaves. Probably no single effect can be attributed to just one hormone; rather, the interactions of these hormones govern a plant's growth. Hence, different parts, such as the leaves and roots, remain in anatomical and physiological balance with each other. Hormones are also involved in a plant's response to its environment. They enable the plant to respond appropriately to the direction of light, gravity, or prevailing winds, and to changes in daylength and temperature that signal changes in the seasons.

How a plant responds to a particular hormone depends on the tissue that receives the hormone, the concentration of the hormone and of other hormones also present, the age and physiological state of the tissue, and environmental factors such as temperature, light, or photoperiod. Since plant hormones can change a cell's synthesis of proteins, a plant's response to a particular hormone is greatly influenced by its genetic makeup, and different species, varieties, or even individual plants will respond to the same treatment in different ways.

Oligosaccharins make up a sixth group of chemical mediators; each one performs a specific function.

SELF-QUIZ

Match the hormones listed below to their effects.

_____ 1. Promotes ripening of fruits
_____ 2. Initiates cell division in tissue culture
_____ 3. High concentrations stimulate ethylene production
_____ 4. Substitutes for cold period in the flowering of some plants
_____ 5. Responsible for gravitropic response in roots
_____ 6. Counteracts the effects of auxin
_____ 7. Promotes onset of dormancy

a. abscisic acid
b. auxin
c. cytokinin
d. ethylene
e. gibberellin

8. In phototropism, auxin:
   a. promotes growth of cells
   b. stimulates differential growth of cells in different sides of the plant
   c. inhibits growth of cells
   d. inhibits cell division

e. absorbs stimuli and signals the direction of light or gravity to the plant

9. The flowering of certain plants only under "short-day" conditions is an example of:
   a. apical dominance
   b. positive phototropism
   c. negative phototropism
   d. photoperiodism

10. A long-day plant is one that:
   a. requires more than 12 hours of light in order to flower
   b. increases in height when it flowers
   c. needs a certain minimum length of photoperiod in order to flower
   d. is not affected by temperature in its flowering response
   e. will not flower if its dark period is interrupted by a flash of light

QUESTIONS FOR DISCUSSION

1. Propose a mechanism by which hydrotropism might work (see Section 35-C).
2. The fungus that causes the "foolish seedling" disease in rice appears not to need the gibberellin it produces for its own growth. Why might it be selectively advantageous for the fungus to secrete this substance?
3. Why are apples, oranges, and grapefruit sold in plastic bags with holes in them rather than in unperforated bags? Why does it often turn out that produce packaged in market trays with clear wrap is too soft on the underside, which you could not see when you picked it out in the store?
4. Many plants of the forest floor produce seeds that germinate only in the early spring, before the canopy leafs out. Propose an explanation for the timing of this germination, and explain its adaptive advantage.

5. In some species of trees, individuals growing near streetlights become dormant later in the fall than do other individuals. How could you account for this?
6. Exposure to a flash of light during the dark period can change the plant's subsequent flowering response (or lack of it). However, interrupting the light period with an interval of dark has no effect. How can you explain this?
7. Some plants use the end of a cold period as their cue that spring has come and it is time to flower, whereas others use increasing daylength. Which is a more reliable predictor of favorable conditions for flowering and seed production? Why do different plants use different cues?
8. Oligosaccharins originate from the cell wall rather than from structures inside the cell. What is the adaptive value to the plant of having regulatory molecules with such a source?

SUGGESTED READINGS

Albersheim, P., and A. G. Darvill. "Oligosaccharins." *Scientific American,* September 1985.

Evans, M. L., R. Moore, and K.-H. Hasenstein. "How roots respond to gravity." *Scientific American,* December 1986.

Galston, A. W., P. J. Davies, and R. L. Satter. *The Life of the Green Plant,* 3d ed. Englewood Cliffs, NJ: Prentice-Hall, 1980. A clear, well-illustrated plant physiology text.

Reproduction in Flowering Plants

CHAPTER

36

Hibiscus: pistil and stamens

Chestnut oak fruits, acorns

Salsify

Composite flower

724

*P*lants acquire resources from their environment as they grow (Chapters 33 and 34). Ultimately, many of these resources are channelled into the plant's reproduction.

Two kinds of reproduction are to be found among flowering plants. **Vegetative reproduction** is an extension of the kinds of growth we saw in Chapter 33. It gives rise to new individuals with genetic makeup identical to the parent's, and thus perpetuates gene combinations that are well adapted to the local environment. Individuals with the favorable genetic combination quickly spread through the area where the parent plant is growing.

Sexual reproduction involves more complex events: production and growth of a group of structures making up the flower, production and fertilization of gametes, and development of the embryo, seed, and fruit. Sexual reproduction has two main advantages. First, it forms new genetic combinations in each generation. It also produces seeds, which can disperse over a wide area, and which are protected against adverse environmental conditions that might kill the parent plant.

In Chapter 20, we studied the basic plant life history, in which the diploid (2N) sporophyte generation alternates with the haploid (N) gametophyte generation. We can diagram the life history for flowering plants:

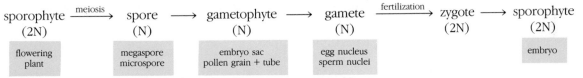

(You may wish to refer back to this diagram as you read.)

In flowering plants, the sporophyte generation is dominant, the gametophyte very much reduced. The familiar plants of garden or forest are sporophytes. The male gametophyte consists of a pollen grain and the tube that grows from it, and the female gametophyte is hidden within the female flower parts.

Modern human society depends on our ability to grow plants for food and for many other uses. In this we employ both vegetative and sexual reproduction. Plant breeders manipulate the sexual reproduction of economically important plants in order to produce individuals with more desirable combinations of genetic features. Once such a set of features is achieved, vegetative propagation can be used to increase the number of plants available to farmers and gardeners.

- Sexual reproduction in flowering plants has two important outcomes.
  1. It produces new genetic combinations, the raw material for evolution by natural selection.
  2. It produces seeds, units that disperse offspring to new areas some distance from the parent plant.
- Vegetative reproduction perpetuates combinations of genes suited to the local environment and allows plants with these combinations to spread widely within this local area.
- Both kinds of reproduction are important in human attempts to improve the strains of plants that we grow for our own use.

36-A *Flowers*

Flowers are sexual reproductive structures, produced in response to hormonal changes in the plant. Some plants flower at a certain age, whereas others are induced to flower by certain environmental stimuli (Section 35-D). The new hormone balance makes a meristem develop into an abbreviated shoot with a cluster of many highly modified leaves—the flower parts.

Figure 36-1
Flower structure. (a) *Trillium,* with its flower parts arranged in threes, is a monocot. (b) Marsh marigolds are dicots (note the five petals). A wreath of stamens surrounds a central group of pistils.

■ *Flowers typically have four kinds of flower parts: protective sepals, petals, male stamens, and female pistils.*

A typical flower has four types of modified leaves. Beginning at the outside, the first flower parts are the **sepals,** which are often green. The sepals develop first and protect the other parts maturing inside the flower bud. Just inside the sepals are the **petals,** which are often large and showy, with bright colors and patterns that attract animal pollinators. Next come the **stamens,** the male, or pollen-producing, parts. These bear **anthers,** chambers where pollen grains develop. In the center of the flower are one or more **pistils,** structures containing the flower's female parts (Figure 36-1). At the tip of the pistil is the sticky **stigma,** which traps pollen grains. Next is the long **style,** and at the base of the pistil lies the **ovary,** enclosing one or more **ovules** (Figure 36-2c).

Many variations on these typical flower parts occur among the 275,000 species of flowering plants. For example, in lilies the three sepals look almost exactly like the three petals (Figure 36-2a). In wind-pollinated plants, the sepals

Figure 36-2
Structure of a lily flower. (a) The petals and sepals look almost alike. Six stamens surround the pistil. (b) Some sepals and petals removed to show how the stamens attach to the flower stalk. The anthers produce pollen. (c) A closer view of the pistil.

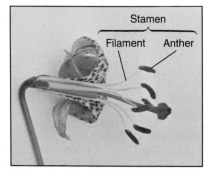
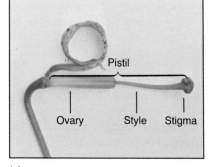

Figure 36-3
Wind pollination. The gray birch bears separate clusters of female and male flowers. The pendulous male catkins swing in the breeze and shed pollen, which is intercepted by the pink stigmas of female flowers in upright catkins. Nonreproductive parts of the flowers are small, allowing the stigmas and anthers free access to the air.

Female catkin   Male catkin

and petals are often tiny or absent, allowing greater exposure of the stamens as they shed pollen, and of the stigmas as they receive it (Figure 36-3). Some plants produce separate male and female flowers (for example, corn and members of the squash family, including cucumbers and pumpkins). Still others have separate "male" and "female" plants, as in spinach, willows, and some hollies. There are also many plants in which structures near the flowers act as parts of the "flower." The "petals" of poinsettias, dogwood, and bougainvillea, for example, are really modified leaves around clusters of small, inconspicuous flowers (Figure 36-4).

36-B Pollen

In plants meiosis gives rise to haploid cells called **spores,** rather than to gametes, which are the products of meiosis in animals. The haploid spores grow into haploid **gametophytes,** which in turn produce the gametes that take part in fertilization. Flowering plants produce spores of two sizes, microspores and megaspores, which give rise to male and female gametophytes, respectively.

Microspores are produced in the chambers of the anthers and begin to develop into male gametophytes. A **pollen grain** is an immature male gametophyte, enclosed in a protective wall. Before it can complete its development, the male gametophyte must be deposited on the stigma of a flower.

Just as leaf and flower structures vary among plants, so too do the shape and pattern of the pollen grain wall. In fact, experts can easily place a particular pollen grain into the proper genus (and sometimes species) by its distinctive

Figure 36-4
Non-flower "flowers." (a) Showy pink bracts (leaf-like structures below flowers) surround the inconspicuous whitish-yellow flowers of bougainvillea. (b) The "petals" of a dogwood are also bracts surrounding a cluster of inconspicuous flowers.

(a)   (b)

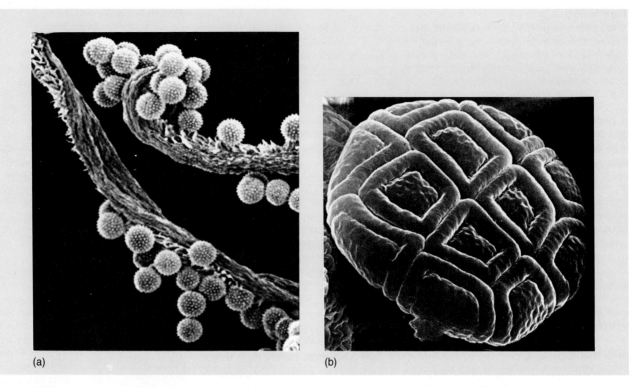

(a)　　　　　　　　　　　　　　　　　　　(b)

Figure 36-5
Pollen grains. (a) Pollen of hollyhock on strands of tissue from the anther. (b) A pollen grain of mimosa. (Biophoto Associates)

cell wall pattern (Figure 36-5). Pollen grains may last for millions of years, preserved in rock formations or peat deposits. The history of the vegetation in an area can be traced by examining this fossil pollen.

Pollination

Pollination is the transfer of pollen from the (male) anther, where it forms, to the stigma of the (female) pistil (Figure 36-6). Pollen may simply fall from the anther onto the stigma of the same flower, resulting in **self-pollination.** Some flowers, such as peas and their relatives (see Figure 12-B), are so constructed that their stamens and pistils are completely enclosed within the petals, resulting in a high percentage of self-pollination.

Cross-pollination, the transfer of pollen to another individual of the same species, gives more genetic variety. This is often an evolutionary advantage, and many plants have adaptations that ensure cross-pollination. For example, a flower's pistils may mature only after its anthers have shed their pollen. The existence of separate male and female plants or flowers is probably due to selective pressure for cross-pollination.

Pollen cannot move on its own power; plants rely on wind or animals as agents of pollination. From a plant's point of view, pollination by animals may have advantages over pollination by wind. First, wind pollination wastes a lot of the energy invested in pollen production because much of the pollen never reaches another flower. Second, wind pollination is very inefficient for a plant that does not live in dense populations. If the nearest neighbor of the same species is far away, there is a good chance that no pollen will reach its stigmas. By contrast, an animal that visits only one kind of plant carries pollen directly from one individual to another of the same species. Many flowers have evolved structures such that only one species of animal can pollinate them, and these flowers enjoy highly specific transfer of pollen from one individual to another.

Animal pollinators are attracted by some type of reward, usually a sweet nectar. The reward is made easier for the animal to find by an attention-catching "advertisement," such as the odor, shape, or color of a flower—preferably all

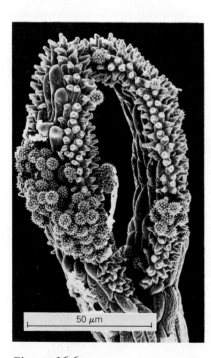

50 μm

Figure 36-6
Pollen on a stigma from a daisy. (Biophoto Associates)

three. The reward is so located that the animal cannot reach it without picking up pollen at the same time. All of this has a cost: the animal-pollinated flower must invest energy in making its nectar and its large, showy petals, even though it need not make the prodigious amounts of pollen required for wind pollination.

Animals that serve as pollinators include insects—bees, butterflies, moths, wasps, flies, and beetles—and vertebrates such as birds, bats, and even a South African mouse!

Pollen Maturation

A pollen grain completes development into a mature male gametophyte after it has landed on the stigma. The protective coat of the pollen grain ruptures and a **pollen tube** grows out (Figure 36-7). The pollen grain wall contains glycoproteins that must be compatible with proteins in the stigma if the pollen is to grow. Hence pollen will usually not germinate on the stigma of flowers of a different species. Some species of plants also prevent self-fertilization, by means of a system of compatibility genes: pollen cannot complete its development when it contains the same genetic alleles as the stigma. This assures that the flower is cross-fertilized and so maintains genetic diversity in the population.

If the pollen and stigma are compatible, the pollen tube grows down the style toward the ovule(s) in the ovary at the base of the pistil. Many pollen grains may land and produce pollen tubes in the same style. The pollen tube grows into the ovule through a tiny pore, the **micropyle,** and releases two **sperm nuclei** (see Figure 36-8b). The micropyle then closes, preventing the entry of any more pollen tubes. If there are more ovules, other pollen tubes enter through their micropyles. The traffic problem that must exist in the style of a cantaloupe flower is fearful to contemplate!

Figure 36-7
A lily pollen grain germinated to form a pollen tube. (Carolina Biological Supply Company)

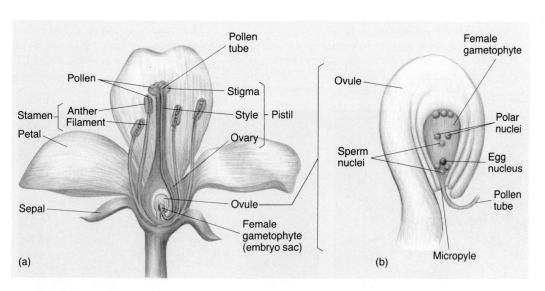

(a)

(b)

Figure 36-8
Parts of a flower. Beginning with the sepal (lower left), and going clockwise around the flower, the labels progress from the outermost to innermost floral parts. Pollen produced in the anthers of the stamens lands on the sticky stigma of the pistil and grows a pollen tube down through the style to the female gametophyte inside the ovule. (b) Close-up of the ovule, showing double fertilization. The pollen tube releases two sperm nuclei (red). One fertilizes the egg nucleus, forming a zygote (diploid); the other sperm nucleus fertilizes the two haploid polar nuclei, forming a triploid endosperm nucleus.

Coevolution of Flowers and Their Pollinators

A JOURNEY INTO LIFE

Much of the evolutionary success of flowering plants comes from the fact that they evolved after animal life was well established on land. Flowering plants and animals have exerted strong selective pressures on one another, and each has shaped the evolution of the other in many ways. Pollination systems offer many fascinating examples of such **coevolution.**

The most important pollinating animals are the bees (Figure 36-A). A flower enjoys several advantages in being pollinated by bees. Bees are widely distributed and numerous. Bees also work very hard at visiting flowers because many bees depend entirely on the food they obtain from flowers, both to nourish themselves and to feed the larvae. The behavior of bees also makes them highly desirable pollinators. Bees quickly learn to tell the different types of flowers

apart, and they are faithful to one kind of flower for long periods. Bees are also available throughout the growing season, and they can remain active even at very low air temperatures, which immobilize most other insects.

Various butterflies and moths are important flower pollinators in all parts of the world. Since these insects use nectar only as a supplementary food for their short-lived adult stage, however, they are not as effective pollinators as bees. Most moths fly at night, and the flowers that depend on them for pollination tend to have pale colors visible in dim light. Some flowers, such as *Nicotiana* (a member of the tobacco family), produce scent only at night, when the moths that pollinate them are active. Flowers pollinated by butterflies, on the other hand, are more likely to have bright colors that stand out by day (Figure 36-B).

Although many butterflies and moths feed on nectar from more than one species of flower, they concentrate on one species at a time. Thus, a hawk moth feeds only on, say, toadflax for as many as five days, and then switches to feeding on nothing but bedstraw. This faithfulness is plainly advantageous to both flowers and insects. The flower benefits because the insect is likely to convey pollen to another flower of the same species. The pollinator benefits by "keying in" on certain cues provided by the flower. The insect can then

find more flowers of that species efficiently and ignore the cues from competing "restaurants" (just as some people key in on the "Golden Arches"!).

Many species of birds feed on nectar and supplement their diet with insects. However, birds often pierce the sides of tubular flowers and so obtain the nectar without picking up pollen, a situation that is disadvantageous to the flower. This may be one of the selective pressures that led to the evolution of flowers shaped in such a way that birds can reach the nectar more easily from a position where they also brush against the pollen.

The 300 species of hummingbirds are the largest group of bird pollinators. They nearly always feed while in flight, hovering in front of a flower and using their long bills and tube-like tongues to suck up the nectar deep within the flower (Figure 36-C). Flowers pollinated by hummingbirds usually have long stigmas that pick up pollen from the bird's head. In tropical areas particularly, the length of the bird's bill and the depth of the flower trumpet dictate considerable specificity, making any one species of hummingbird able to feed only on certain species of flowers. Most flowers growing at high elevations in the tropics are bird-pollinated. The frequent rains at these altitudes hinder the flight of insects but not of birds.

Figure 36-A
A wild bee liberally dusted with pollen. (Biophoto Associates, N.H.P.A.)

Figure 36-B
A butterfly sucking nectar through its tubular proboscis. (Biophoto Associates)

Figure 36-C
A broadtail hummingbird visiting a thistle. (Wendy Shattil and Robert Rozinski/Tom Stack & Associates)

Plants need to ensure not only that their pollinators are faithful to their species but also that the pollinators visit the flowers frequently. Flowers produce small amounts of nectar, making it necessary for the pollinator to visit many flowers, and to revisit each flower again and again. Another adaptation that ensures frequent visits is the flower's distinctiveness: if the flower looks and smells different from other flowers nearby, the pollinator can find the flower easily and discover more flowers of the same species quickly, so that it does not waste time and energy hunting around.

There is also strong selection for different plant species to bloom at different times, so that each species receives the attentions of pollinators in its turn, rather than having all flowers competing for attention during a brief period. Such staggered blooming also provides pollinators with a steady food supply throughout the growing season.

Although most flower-pollinator relationships are mutually beneficial, some plants have evolved ways to secure animals' services without paying a reward. In many orchids, part of the flower resembles the rear end of a female bee (Figure 36-D). The flower may even emit the same chemicals used by the bee as her sex-attractant pheromone. Deluded male bees mistakenly copulate with the flower and pick up pollen, which they carry to other orchids of the same species. This adaptation ensures that the flowers are visited frequently and faithfully.

The hairy red petals and rotting-meat stench of carrion flowers mimic dead mammals well enough to fool a female blowfly looking for a place to lay her eggs (Figure 36-E). The fly travels from flower to flower, transferring pollen as she leave clusters of eggs. Since her larvae cannot survive without animal protein, this arrangement boosts the plant's evolutionary success but lowers the fly's.

Figure 36-D
A bee-mimic orchid flower. (Biophoto Associates, N.H.P.A.)

Figure 36-E
A fly on a carrion flower.

36-C   Preparation of the Ovule

Before the pollen tube arrives, a megaspore has formed inside the ovule and developed into an **embryo sac,** a mature female gametophyte. The embryo sac usually contains eight haploid nuclei: three at the end near the micropyle, one of which is the egg nucleus; three at the opposite end; and two **polar nuclei** in the center (Figure 36-8b). This female gametophyte is now ready to be fertilized.

36-D   Fertilization

Fertilization may take place as little as an hour after pollination, as in barley, or as much as several months later: witch hazel flowers and is pollinated in late fall but is not fertilized by the arrival of the sperm nuclei at the micropyle until spring.

During the sexual reproduction of flowering plants, both sperm nuclei take part in fertilization. In this **double fertilization,** one sperm nucleus fertilizes the egg nucleus, forming the zygote. The other sperm nucleus fuses with the two polar nuclei, forming an **endosperm** nucleus that is triploid ($3N$, where N is the number of chromosomes in a haploid nucleus). The adaptive value of this second fertilization is unclear, although the endosperm tissue that arises from this nucleus has a very important role, as we shall see shortly.

36-E   Development of the Seed and Fruit

In the next stage of development, the zygote develops into an embryonic plant, and the parent plant supplies it with nutrients that will help it to establish itself as an independent individual. In addition, the wall of the ovule develops into a protective **seed coat,** and the wall of the ovary develops into a **fruit.**

Right after fertilization, the zygote enters a period of dormancy. Meanwhile, the endosperm nucleus divides many times to form endosperm tissue, which enlarges and absorbs food from the parent plant. When the zygote breaks dormancy, the endosperm tissue has a food supply ready for it.

The first structure formed by division of the zygote is a line of cells called the **suspensor.** The suspensor cells near the micropyle elongate and push the cells at the far end into the nutrient-rich endosperm. Soon the embryo starts to develop at this end of the suspensor (Figure 36-9). The mature plant embryo has a tiny axis with an apical meristem at each end (one for the primary root, the other for the shoot system) and one or two **cotyledons,** depending on whether it is a monocotyledon or a dicotyledon (see Table 33-2).

As the embryo grows, the endosperm continues to absorb food from the parent plant. The endosperm may persist as a food supply for the embryo or may be completely absorbed into the embryo's cotyledons as the seed matures (as in the bean seed, Figure 33-3). The wall of the ovule, which is part of the parent plant, becomes larger as the embryo grows, and usually hardens to form the protective seed coat.

Outside the seed coat, the wall of the ovary also enlarges and absorbs more nutrients to form a fruit (Figure 36-10). Fruit growth begins when the pollen tube releases tiny amounts of the hormones auxin and gibberellin (Chapter 35). Soon the developing seed begins to produce its own hormones, which continue to stimulate growth of the fruit. In addition, there is enhanced transport of cytokinins, hormones that stimulate cell division, from the parent plant. These three hormones promote fruit growth by both cell division and cell enlargement.

■ *A seed consists of:*
1. *An embryonic plant, developed from the zygote.*
2. *A food supply, absorbed by the endosperm (or cotyledons).*
3. *A protective seed coat, developed from the ovule wall.*

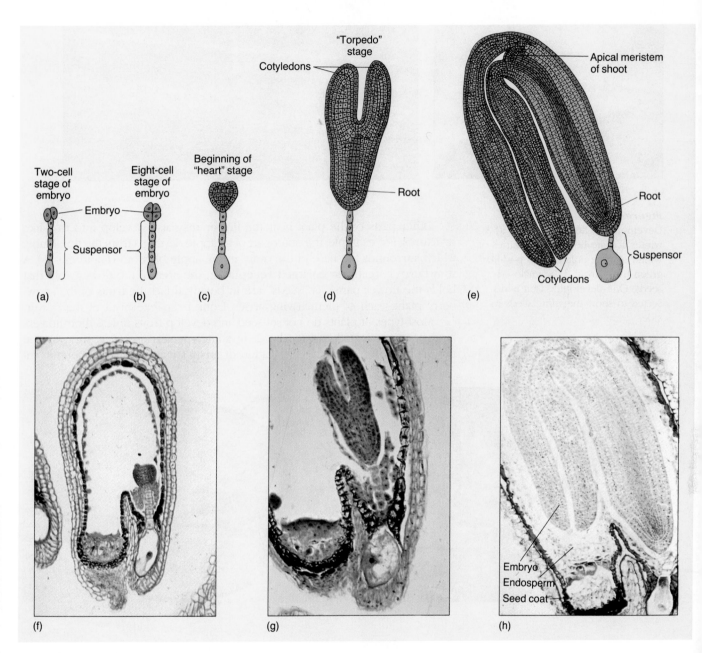

Figure 36-9
Early development of the embryo. (a) A series of cell divisions forms a string of cells (suspensor); one end (top) is pushed into the nutritive endosperm. The end cell divides to form an embryo (brown), which grows by cell division and absorption of food from the endosperm (b, c). (d) "Torpedo" stage: the cotyledons are forming (top); the end near the suspensor becomes the root. (e) The embryo has grown much larger and the cotyledons have folded over, with their ends beside the root. (f, g, h) Embryos of shepherd's purse *(Capsella bursa-pastoris)* at the stages shown in c, d, and e. Around the embryo lie patches of endosperm; outside both is the wall of the ovule, which develops into the seed coat. (f, g, h, Biophoto Associates)

(a)

(b)

Figure 36-10
Development: flower to fruit. (a) A winter aconite flower. (b) Ovaries of winter aconite enlarging as pod-like green fruits around the enclosed seeds. One developing fruit is dissected to show the white seeds inside.

Other parts of the plant near the flower may also develop into fruit-like structures. For example, the outer part of an apple develops from the floral tube, which surrounds the base of the ovary in the apple blossom (Figure 36-11). A strawberry is really an enlarged **receptacle,** the area of the flower stalk that holds the flower parts. Its "seeds" are in botanical fact the fruits of the strawberry plant, each of them having arisen from a separate pistil of the flower.

Most types of plants do not set seed and develop fruits unless their flowers have been pollinated and fertilized. In some species, however, spraying the flowers with the proper concentration of auxin or auxin plus gibberellin in-

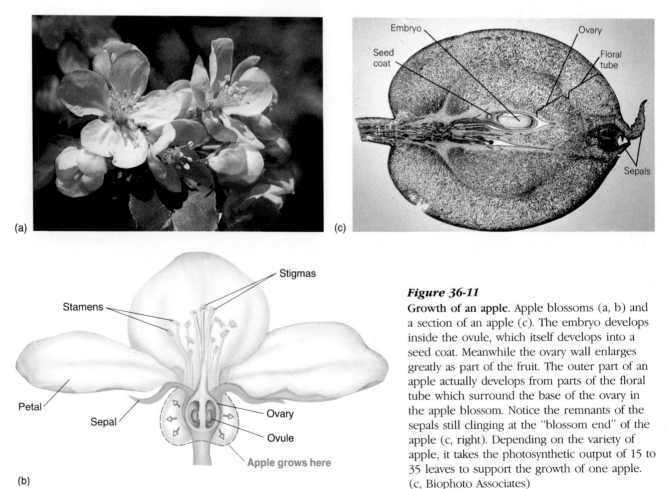

(a)

(c)

Figure 36-11
Growth of an apple. Apple blossoms (a, b) and a section of an apple (c). The embryo develops inside the ovule, which itself develops into a seed coat. Meanwhile the ovary wall enlarges greatly as part of the fruit. The outer part of an apple actually develops from parts of the floral tube which surround the base of the ovary in the apple blossom. Notice the remnants of the sepals still clinging at the "blossom end" of the apple (c, right). Depending on the variety of apple, it takes the photosynthetic output of 15 to 35 leaves to support the growth of one apple. (c, Biophoto Associates)

(b)

734

duces the production of seedless fruits. Some fruits, such as cultivated strains of bananas and pineapples, develop naturally without fertilization and are therefore seedless. Humans have increased the populations of these economically desirable plants by vegetative propagation, and the plants have thus become successful through artificial selection.

Not all fruits are fleshy and delicious. Many are rather minimal protective layers. When the seeds are ripe, the fruit may rupture along its lines of weakness and release the seeds. Pods of peas, beans, and milkweeds are examples of such fruits.

36-F   Dispersal of Seeds and Fruit

Once mature, the seeds are ready for dispersal. In most cases it is advantageous for the seeds to grow well away from the parent. First, this distributes the population of the plant's descendants over a wider area. Second, it avoids competition for light, water, and soil minerals between the parent plant and its offspring.

The usual dispersal agents are wind and animals. Small, lightweight milkweed seeds and dandelion fruits have parachute-like tufts of fiber that enable them to disperse by floating through the air (Figure 36-12). The thin, flat wings of maple fruits whirl in the breeze like the blades of a helicopter. Larger seeds will have a distinct advantage when they start growing because they contain more food for the embryo, but wind cannot waft a large seed far from the parent plant. An animal, on the other hand, is strong enough to carry even a large seed quite a distance.

Many plants invest a lot of energy in adaptations that promote dispersal by animals. Most commonly, the seeds are protected in indigestible seed coats and surrounded with a tasty, nutritious fruit that an animal will eat. The seeds then pass unharmed through the animal and are deposited with a small pile of organic fertilizer. In fact, the seeds of some plants will not grow unless their seed coats have been eroded somewhat by an animal's digestive enzymes.

Fruits are usually protected from being eaten before the seeds have matured. Unripe fruit is often distasteful and may even contain toxic chemicals. When the fruit is ripe, its chemical composition changes so that it tastes good.

Figure 36-12

Wind dispersal. (a) Milkweed seeds float off on parachute-like tufts of fibers after the wall of the fruit dries and splits open. (b) The "wings" of maple fruits, reminiscent of aircraft propellers, are enormous outgrowths of the ovary wall. The seeds are enclosed in the green swellings at their bases. Note the stigmas and style, still visible in the center of the "V," and the stamen, petals, and sepals at the base of the extremely enlarged ovary.

(a)

(b)

Figure 36-13
Ripening fruits. These porcelain-berries change from white to deep blue by way of several intermediate shades, a signal that they are ready to be eaten.

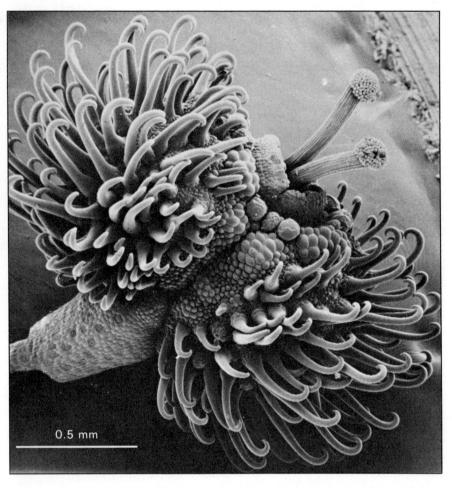

0.5 mm

Figure 36-14
A herby hooker. This scanning electron micrograph shows a goosegrass flower with a pair of developing fruits. These fruits disperse by hooking onto an animal or onto human clothing when we walk through the meadow in late summer. The stigmas can still be seen (top right). (Biophoto Associates)

The fruit may also change color, a visual signal that it is now ready to be eaten (Figure 36-13).

Some fruits or seeds have hook-like extensions that attach to the feathers, fur, or clothing of passing animals or people, which give the seed a free ride to a new home (Figure 36-14). These hitchhikers gain the use of the animal's mobility for a small energy investment in the production of hooks.

Seeds and Seed Predators

Because seeds contain the food supply for an embryonic plant, they also make good food for animals. Hence there is strong selection for adaptations that protect the seeds from predation (Figure 36-15). The types of adaptations that are effective depend on the main types of predators eating the seeds. Some seed predators gobble up every seed they find. Plants with this type of predator usually produce many small seeds. This gives a good chance for some of the seeds to escape notice, so not all are eaten.

Other seed predators maximize their food intake for the energy they spend. They may attack plants that have the most seeds in a fruit, or the seeds that are largest or easiest to chew. Plants attacked by such predators usually enclose their seeds in a hard covering, such as a nutshell, that discourages the predator (Figure 36-16). Producing smaller seeds works only if the seeds become smaller than those of another species to which the predator might switch.

Another adaptation is **seed masting,** the simultaneous release of seeds by all the plants of the same species in an area, at intervals of two years or more. This makes seeds available to predators for a minimum time period. Beech trees and some oaks do this, but the most impressive examples are bamboos. Part of a bamboo stand in India was collected and sent to botanical gardens in the United States and Britain at the beginning of the nineteenth century. The plants grew vegetatively for the next 130 years, and then the bamboo stands on all three continents produced seeds in the same year! The advantage of seed masting to the plant is that most of the time there are no seeds to support the growth of large populations of seed-eaters. When seeds are finally shed, there are so many that a small population of seed predators cannot eat them all, and some seeds escape to produce the next generation of plants.

Some large seeds, such as nuts and acorns, are actually dispersed by would-be predators. Squirrels and some birds collect these large, nutritious seeds and hide them for later use. However, they do not return to all of their caches. The forgotten seeds sprout the next spring and grow into new trees. In the long run, this predation behavior benefits the tree, which has gained some surviving offspring at the expense of others. Researchers attribute the wide distribution of oak trees to the industry of jays, which may fly several kilometres to bury acorns in soft earth or under moist leaves (squirrels have much smaller home ranges). In one study, 50 jays spent September hiding 150,000 acorns! The diligence of jays is thought to account for the rapid spread of oak trees north, following the retreating glaciers, after the last ice age—a feat the heavy acorns could never have accomplished themselves.

36-G   Germination

Many seeds become dormant after they have formed. As a seed enters dormancy, it dries out, and its final water content may be less than 5% of its total weight. In many kinds of seeds, dryness seems to be the main factor assuring the seed's **viability,** or ability to break dormancy and grow into a new plant after an extended period of time. Many commercial seed suppliers now dry their seeds thoroughly and wrap them in moisture-proof foil packets.

The viability of seeds varies among species, and among individuals within a species. The viability record is held by water lily seeds found in a peat bed in

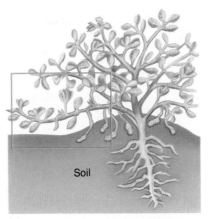

Soil

(a)

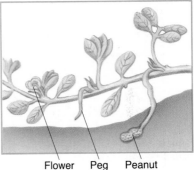

Flower   Peg   Peanut

(b)

Figure 36-15

An unusual adaptation. (a) Peanut plants bury their own seeds. (b) An enlargement of one branch. After the flower petals die, the female part of the flower elongates, forming a "peg" which pushes into the ground. Underground, the end of the peg enlarges into a peanut shell, enclosing the seed. This hides the nutritious peanut from hungry animals as well as planting it. Peanut plants have nitrogen-fixing bacteria in their roots. Keeping the seeds near the parent permits the young plants to benefit from the nitrogen that the parent plant contributes to the soil. (Thom Smith)

Figure 36-16

A defensive arsenal. A large, nutritious seed (left) with a tough shell, which originally was covered by a prickly husk (right), a deterrent to prospective seed predators. (Biophoto Associates)

Manchuria and dated at over 1000 years old by radioactive isotope methods (*A Journey Into Life,* Chapter 2). Seeds from even older deposits have been grown, but no dating was done to prove that the seeds were as old as the layer in which they were found. In contrast, sugar maple seeds live less than a week. These seeds do not have a dormant period, and their viability is better if they do not dry out much.

In order to **germinate,** or begin growing into a new plant, seeds must be supplied with water. Most seeds also require oxygen, and many require particular temperature and light conditions before they will germinate. Furthermore, germination requirements may vary not only from species to species, but also from individual to individual. Thus a plant does not have all of its seeds "in one basket": germination may spread over months or years, and at least some seedlings are likely to find conditions that favor their survival.

Germination of seeds is associated with an increase in the growth-stimulating gibberellin hormones. Applying gibberellins can often overcome a seed's special light or temperature requirements for germination. In some species, other plant hormones are needed for germination.

Germination has been best studied in barley. The barley embryo secretes gibberellin, which induces the secretion of various enzymes. These enzymes break down starch and other stored food in the endosperm, making it available for absorption by the developing embryo. As the embryo uses this food, the dry weight of the seed decreases until the embryo begins to make its own food through photosynthesis. The seedling then grows as described in Chapter 33.

36-H Breeding Programs

Probably since the beginning of agriculture, humans have been breeding plants selectively to improve the features that make them useful to us. This starts by choosing plants that produce well in the field and collecting their seeds to plant for the next generation. For thousands of years, this was the only available way to improve plants. Nevertheless it produced strains of cultivated plants strikingly different from their wild ancestors, especially in improved food yield. More control of the process can be gained by carefully cross-pollinating plants selected for desirable traits, in an attempt to get offspring with all of the desired features in the same plants. Our modern knowledge of genetics guides the selection of parental strains for such breeding programs.

Figure 36-17

Plant breeding. (a) Corn plants with striped bags over their anthers (top) to collect pollen and plain bags over the stigmas (lower on the plants) to prevent pollination until the breeder brings the desired pollen to the future ear of corn. (b) Almost 20 years of selective breeding went into developing "sugar bush," a type of watermelon with desirable flavor, color, and disease resistance that also grows on space-saving bushy vines ("normal" watermelon vines may sprawl over several metres). (b, W. Atlee Burpee Co.)

(a)

(b)

The genetics of some crop plants are fairly well understood, especially for major crops like tomatoes, wheat, and corn. The popular strains of hybrid corn are produced by carefully planned crosses between a number of strains, each bred for particular traits (Figure 36-17). It may take several generations of crosses to produce the seed used to grow hybrid corn, and these crosses must be made anew each year in a continuous breeding program, since the hybrids do not breed true. So seed for hybrid corn must be purchased from breeders each year.

Corn is a fairly easy crop to use in genetics programs. Because male and female flowers grow in large, separate clusters, it is easy to carry out a desired pollination. And, since corn is an annual plant, the success of a cross can be judged within a year.

Other plants pose difficulties. For example, a breeder of apples must wait four to ten years until seeds grow into mature trees and bear fruit. Furthermore, about 99% of such trees carry new genetic combinations that are inferior to old varieties. Any new tree that looks promising must be screened for another ten years before it is ready for marketing—or the woodpile.

36-I Vegetative Reproduction

Most flowering plants can reproduce sexually. Sexual reproduction increases the genetic variation in a population, and so provides many "trial" assortments of genes that natural selection can act upon. This has the potential to produce genetic combinations better suited to the environment than those of the parent plants. However, a large percentage of these offspring end up with less favorable genetic combinations. Many of the rest fall prey to animals or fungal disease, while still others end up in habitats unfavorable to their growth.

Asexual reproduction is also common among plants. Dandelions, hawk-weeds, and many grasses reproduce asexually by seeds that develop from unfertilized ovules. In some of these plants pollination occurs, but it is only a stimulus to the ovule to develop into a seed, and the pollen contributes no genetic material to the offspring.

Many plants reproduce asexually by **vegetative reproduction,** in which new individuals arise from the parent's roots, stems, or leaves (Figure 36-18). Asexual reproduction allows these plants to perpetuate combinations of genes that are well adapted to their environment. They spread and cover a wide area with a **clone** of plants, a population of individuals with identical genotypes.

Vegetative propagation is often desirable from the human as well as from the plant point of view. Home gardeners root cuttings of *Coleus,* geranium, or

Figure 36-18
Vegetative reproduction. (a) Stolons of strawberry plants begin as slender stems that arch away from the parent plant and form roots and leaves of a new plant where they touch the ground. (b) The leaves of *Kalanchoe* bear tiny plantlets, complete with roots, which eventually fall off and grow on their own.

(a) (b)

Figure 36-19
A successful graft in an apple tree. Scion (above) and stock (below) formed a bulging callus as they grew together and sealed themselves into a single unit. Eventually the callus will disappear as new tissue grows smoothly around it.

ivy shoots by placing them in water, or set leaves of African violets or jade plants on moist soil until they grow roots. Some plants can be rooted more successfully with the use of commercial plant hormone preparations. People also help spread plants such as daffodils and onions by digging up the bulbs when the tops die back and separating those that have multiplied, giving each more room to grow.

Potatoes are an important crop produced by vegetative means. A potato is an underground stem, or **tuber.** Farmers get many offspring from a good potato plant by digging up "seed potatoes" (tubers of good quality), cutting them into pieces, each with an "eye" (bud), and replanting them. The seeds produced by potato flowers usually have inferior genetic combinations. However, breeders do grow plants from these seeds in an attempt to produce new strains with desirable traits, such as resistance to certain diseases and pests and the ability to form tubers when grown in tropical climates.

Grafting is another means of artificial vegetative propagation. A **scion,** a twig or bud of a desirable plant, is attached to a **stock,** the root system or stem of another plant, from which a twig or bud similar to the scion has just been removed. Scion and stock are then wrapped closely together, and their cut areas soon produce new cells that merge the two parts into a functioning unit (Figure 36-19). Each part of the graft retains its genetic identity, however.

Grafts work only between plants of the same or closely related species. The method is used to produce fruit trees, grape vines, or rose bushes that combine the desirable fruits or flowers of the scion with a sturdy, pest-resistant rootstock. For example, all the Red Delicious, Golden Delicious, and McIntosh apple trees now in existence are derived, by grafting, from single fence-row "volunteer" trees that happened to have desirable fruits. Dwarf fruit trees are produced by grafting scions onto rootstocks of related species. Thus, dwarf pear trees consist of pear scions grafted onto quince roots. It is even possible to purchase small, manageable apple and pear trees with five or six grafted branches, each bearing fruits of a different variety.

An exciting development in vegetative propagation is the production of entire plants from meristematic cells grown in laboratory culture. The culture medium contains high levels of plant hormones, which produce high rates of mutation. Researchers start with cells from plants with many desirable qualities, allow the cells to multiply and mutate in the laboratory, and then grow plants from them. The resulting plants are very similar genetically—more so than sexually produced offspring—but they may differ in important traits such as resistance to drought or to particular diseases. This method allows plant breeders to "fine-tune" the genetic makeup of crop plants. It also gives a way to produce more replicas of a desirable plant quickly. Only a small lump of meristematic cells is needed to start each new plant, rather than a large, leafy cutting. Cells kept in a flask or two in the laboratory can substitute for acres of plants formerly kept as sources for cuttings. Many plants, such as chrysanthemums, are now grown commercially in laboratory culture from meristematic cells. Laboratory culture is also used to grow genetically engineered cells into complete plants (Figure 36-20).

■ *Vegetative propagation allows us to make many exact replicas of plants with desirable genetic combinations.*

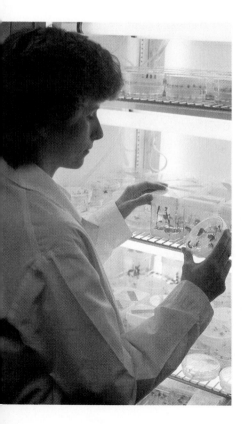

Figure 36-20
Plantlets grown in tissue culture from cells containing foreign genes. The cultures are kept in a growth chamber where the environment is carefully controlled. (Courtesy of Calgene)

Better Plants Through Engineering

Breeding better strains of plants by sexual crosses is restricted to genes that occur naturally in some member of the species (or of a closely related species able to hybridize with the one being developed). Often, breeders would like to develop strains of plants with genetic features not already found in any member of the species. To do this, they must first identify the gene of interest in an organism of another species. Next, they must isolate and clone this gene, and then transplant it into cells taken from the plant they are trying to develop, using methods described in *A Journey Into Life,* Chapter 9. These cells must be grown into whole plants in laboratory culture. The plants are then tested and bred to make sure that the transplanted gene is present, and active, in each plant and in the plants that grow from its seeds. Eventually, seed containing the transplanted gene will be ready for market.

Researchers have succeeded in producing stable gene transplants such as this in a number of plant species. However, there are still many species that have not yet been successfully grown from single cells in the laboratory. Finding the conditions needed to grow these plants is an area of ongoing research, and the list of species that can be cultured this way is steadily increasing.

Researchers have given some experimental crop plants genes for resistance to weed-killers, major diseases, or insect pests (Figure 36-F). They are also working to isolate and transplant the genes that enable mem-bers of the legume (pea and bean) family to house nitrogen-fixing bacteria in their roots. These bacteria convert nitrogen gas, abundant in the atmosphere, into a form the plants can use. If we could transplant the relevant legume genes into other crop plants and set them up with their own bacterial tenants, we might eliminate the need for nitrogen fertilizers. These fertilizers are expensive to produce and apply. They also contribute to water pollution in agricultural areas because rainwater often dissolves the nitrogen and carries it into lakes and streams.

In 1987, the first field trials of a plant containing a pesticide gene took place using tobacco plants. The pesticide gene comes from the bacterium *Bacillus thuringiensis,* best known as the active ingredient of sprays to combat gypsy moths. The bacterium produces a toxin that kills moth larvae but is not toxic to most other insects, nor to mammals or birds. The problem with using the spray is that it rapidly washes off plants, and it also degrades in sunlight. Both of these drawbacks are avoided by transplanting the gene for the bacterial toxin into the plant itself.

Field trials of other plant species, containing other transplanted genes, are also under way. Some companies hope to be able to market seed with useful transplanted genes to farmers by the mid-1990s. If these endeavors succeed, researchers hope that farmers will be able to reduce their use of expensive and environmentally dangerous pesticides and fertilizers.

Experimental plant: healthy

Control plant: diseased

Figure 36-F
Resistance of transformed tobacco plants to tobacco mosaic virus infection. The experimental plant (left) was transformed with tobacco mosaic virus genes, which were spliced to a vector to carry them into tobacco cells. The control plant (right) was transformed with the vector alone. Then, both of the plants shown were inoculated with tobacco mosaic virus. After 15 days, typical mosaic symptoms have developed only on the control plant. (Photo courtesy of Milton Zaitlin)

SUMMARY

Plants flower in response to specific cues that differ greatly among the various species, and each species of angiosperm has its own distinctive flower structure. In all this diversity, however, we can find a basic unity in the structure and function of flowers.

A flower is an abbreviated shoot, with modified leaves, the flower parts. The male parts, the stamens, produce the haploid male gametophytes, the pollen grains. The female parts, the pistils, produce the female gametophytes, the embryo sacs. Double fertilization forms a zygote and an endosperm mother cell in the embryo sac. The endosperm mother cell divides and develops into the endosperm, which absorbs food from the parent plant, and the zygote soon develops into the embryo of a new plant. The parent plant, besides contributing food to the new embryo, also protects it and its food supply in a seed coat derived from the wall of the ovule. This in turn is surrounded by the fruit, derived from the wall of the ovary, another parental structure.

Much of the evolutionary success of flowering plants is undoubtedly due to the fact that they have coevolved with animals. Many species rely on animals, rather than on wind and water, to pollinate their flowers and disperse their seeds. They devote considerable energy to attracting animals that will perform these services, and they are rewarded by pollination that is efficient and specific, and by seed dispersal that distributes even large seeds over a relatively wide range.

Many seeds enter a period of dormancy following their release from the parent plant. Eventually the seed germinates in response to environmental cues and establishes itself as a new plant.

Sexual reproduction results in individuals with new combinations of genetic characters. In many cases, these combinations are less desirable than those of the parents, from either the human or the plant point of view. Many plants have some means of asexual reproduction (which perpetuates a favorable combination of genes unchanged) in addition to, or instead of, sexual reproduction. Humans propagate many plants vegetatively by artificial means such as rooting and grafting.

SELF-QUIZ

Label the structures numbered in the diagram below:

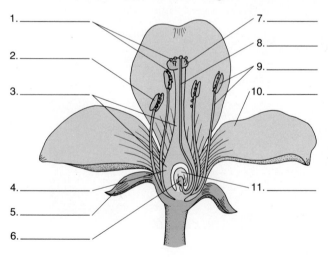

1. _____
2. _____
3. _____
4. _____
5. _____
6. _____
7. _____
8. _____
9. _____
10. _____
11. _____

For each of the following descriptions, give the number and name of the structure in the diagram:

____ 12. Site of pollen production
____ 13. Female gametophyte
____ 14. Protective flower part
____ 15. Develops into the seed coat
____ 16. Contains structure that gives rise to endosperm
____ 17. Develops into the fruit
18. True or False: The terms pollination and fertilization can be used interchangeably.
19. Which of the following is *never* required for seed germination?
    a. certain temperature conditions
    b. oxygen
    c. water
    d. light
    e. none of the above
20. Grafting is used to propagate plants because:
    a. it is faster than growing seeds
    b. it maintains a desired set of genetic characteristics
    c. it combines the genetic characteristics of two desirable strains of plants
    d. healthy plants will graft by themselves, so that they reproduce profusely
    e. a plant can produce many more scions than seeds

QUESTIONS FOR DISCUSSION

1. Why do banana plants put so much energy into producing fruits that contain no seeds?
2. Plants given large amounts of fertilizer, especially fertilizer containing much nitrogen, often flower poorly or not at all, and do not accumulate food reserves; instead they engage in vigorous vegetative growth. Is there an adaptive advantage to this?
3. Some plants, such as dandelions and hawkweeds, have

lost the ability to reproduce sexually but still produce flowers and set seed by development of the ovule without meiosis or fertilization. What is the advantage of this system over a more orthodox means of vegetative reproduction?

4. Pollen is produced at the tips of the stamens, whereas ovaries lie at the bases of the pistils. What is the adaptive advantage of these differences in position?

SUGGESTED READINGS

Barrett, S. C. H. "Mimicry in plants." *Scientific American,* September 1987. Some plants cheat pollinators or farmers by resembling other species so closely that the animals are duped into helping the plant without reaping the expected reward.

Echlin, P. "Pollen." *Scientific American,* April 1968. Many interesting facts and illustrations.

Faegri, K., and L. van der Pijl. *The Principles of Pollination Ecology,* 3d ed. New York: Pergamon Press, 1979.

Handel, S. N., and A. J. Beattie. "Seed dispersal by ants." *Scientific American,* August 1990.

Heinrich, B. "The energetics of the bumblebee." *Scientific American,* April 1973. The relationships between bumblebees and the flowers they pollinate, viewed in terms of the influence of energy expenditure on evolution of adaptations.

Koller, D. "Germination." *Scientific American,* April 1959.

Meeuse, B., and S. Morris. *The Sex Life of Flowers.* New York: Facts on File, 1984.

Miller, J. A. "Somaclonal variation." *Science News* 128:120, 1985. How new varieties of plants are grown from cells that mutate in tissue culture.

Proctor, M., and P. Yeo. *The Pollination of Flowers.* Glasgow: William Collins Sons, 1973.

Wickelgren, I. "Please pass the genes." *Science News* 136:120, August 19, 1989. A discussion of whether or not genetically engineered food crops pose novel hazards to human diners.

The World of Life: Ecology

PART 6

Organisms do not live in isolation. Their lives depend upon interacting with the world around them. **Ecology** is the study of these interactions. The word "ecology," coined in 1869, is based on the Greek word *oikos,* meaning "house" or, more loosely, "habitat." The term **ecosystem,** from the same root, is the habitat or environment where organisms live and interact. The ecosystem includes **abiotic** (nonliving) factors such as sunlight, temperature, water, and soil and **biotic** factors (all the other organisms in the ecosystem). Each ecosystem contains its own **community** of organisms, a collection of different species living together. Ecologists study the patterns of distribution and abundance of organisms in nature, how these patterns are maintained in the short run, and how they change during the course of evolution.

The science of ecology grew out of natural history—the observation and description of organisms in nature. As Western naturalists explored the world, they discovered two main patterns. First, every new area explored contains species not previously known to science. Second, no matter how many species are described, all live in only a small number of types of communities, which have similar characteristics wherever they are found. Thus shrubland is found on the coast of California and on the coast of the Mediterranean Sea. The species are different in the two places but the plants are of similar heights, distribution, and even chemistry, and animals in the two areas are of similar size, types, and habits. In Chapter 37 we examine the types of communities in which organisms live, why these are found where they are, and how they develop and maintain themselves.

Life on earth requires water, a source of energy, and various nutrients. These are the main factors that determine where organisms can live. Energy and nutrients (and water) are transferred between different members of a community of organisms. When an insect eats a plant, it takes in energy and nutrients for its own use. In Chapter 38, we examine transfers of materials and energy among organisms that live in the same ecosystem and the factors that determine an ecosystem's productivity—how much life it can support. In the twentieth century, the study of ecology gained impetus with the realization that human activities have a profound effect upon transfers of energy and material in the living world, and that this in turn affects us by altering our own environment. We consider the pollution of ecosystems as an example of how we alter our environment.

A community of organisms is made up of populations of many species. What determines how many oak trees live in a forest? Why has the population of right whales in the Atlantic shrunk so far that the species is in danger of extinction? Why is the human population of the earth growing so rapidly? The factors that determine the sizes of populations are examined in Chapter 39.

In Chapter 40, we consider the ecology of our own species, paying particular attention to the invention of agriculture and industry—events that changed our world dramatically. Like other organisms, human beings interact with their environment. Because of our numbers and our technology, however, we affect our environment more drastically and more permanently than do populations of any other organism. Ecologists studying human impact upon natural ecosystems warn us that we are changing the natural world beyond recognition, destroying other species and threatening our own existence.

Three more every second!

Tiger

Cacti, Apache Trail, AZ

Antarctic lichen

Home, sweet home

Distribution of Organisms

CHAPTER

37

Yuma, Arizona

White-faced capuchin

Antarctic lichen

Alpine pool

*L*ife on earth requires a moderate temperature, water, a source of energy, and various chemical nutrients found in the soil, water, and air. A hundred kilometres beneath our feet, the earth is white hot. Thirty kilometres above our heads the air is too thin and cold for organisms to survive. Suitable combinations of the things organisms need are found only in a narrow layer that forms a rough sphere around the surface of the earth. This layer is called the **biosphere** because it is, as far as we know, the only place where life can exist. The biosphere extends about 8 kilometres up into the atmosphere (where insects and the spores of bacteria and plants may be found) and as much as 8 kilometres down into the ocean.

Organisms do not live randomly scattered through the biosphere. They live in **communities,** collections of different species living in the same area at the same time. For instance, a particular community of organisms lives in an oak wood. A community of organisms together with its physical environment is known as an **ecosystem.**

The world contains only a limited number of types of community. Walking through a tropical forest in South America, we would find tall trees with large leaves and fruits, festooned with climbing vines, and we would see colorful butterflies and birds flitting through the gloomy shade. A tropical forest in Asia would look much the same, but the species of trees and vines, and of butterflies and birds, would be different. Forests in South America and Asia are similar because they occur in areas with similar climates, with high temperatures and heavy rainfall. Wherever a particular pattern of temperature and precipitation occurs, plants adapted to that climate will be found. The climate and kinds of plants, in turn, determine the community's animal life.

Tropical forest, found in several parts of the world, is known as a **biome,** a type of community defined by its climate. Other biomes—desert, grassland, or tundra—occur in parts of the world with particular climates and also look much the same wherever in the world they occur.

Aquatic communities are not classified into biomes, but their distribution is also determined by the physical environment. For instance, photosynthetic organisms live only near the water surface because deep in the water there is not enough light for photosynthesis.

In this chapter, we consider the question, "Why are organisms where they are?" And we also consider how organisms establish themselves in a particular area after a disturbance such as a fire or landslide.

- The organisms on earth make up only a small number of different types of communities, each of which is found in several parts of the world.
- On land, the biome in an area is determined largely by the pattern of temperature and precipitation.
- The community found in a particular aquatic ecosystem depends largely on light, temperature, salinity, water currents, and the type of bottom in the area.
- In places where vegetation is destroyed, organisms invade and replace each other in succession until species typical of the mature biome are reestablished.

37-A *Climate and Vegetation*

If we look at a map of the world showing the kinds of communities in different places, we find that areas with similar climates have communities of the same type (Figure 37-1). Climate is the main factor determining the type of soil and the types of plants in an area. The plant life and climate, in turn, determine the types of animals and microorganisms present.

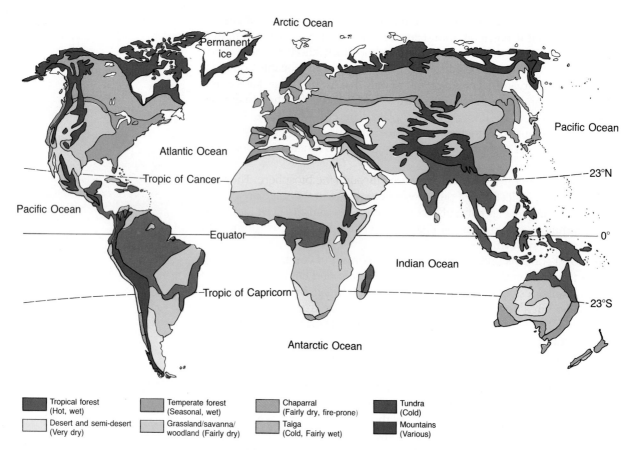

Figure 37-1

Biomes of the earth. This map shows the major biomes, simplified to emphasize the overall pattern. The order of biomes northward from the equator is mirrored by the same biomes at similar latitudes south of the equator. Oceans, land masses, and mountain ranges affect climate, and so vegetation, making the map more complicated than it would be otherwise.

Climate depends on the sun. Solar energy provides the heat that determines the average temperature, causes the winds, and powers precipitation. Tropical climates, receiving near-vertical sunlight throughout the year, have fairly steady, high temperatures (Figure 37-2). In other areas, the temperature varies roughly with the amount and intensity of sunlight at different seasons. Temperature varies with altitude (height above sea level) as well as with latitude (distance from the equator). As a result, the plant life on mountains shows changes from base to peak similar to those seen when travelling farther and farther north or south from the equator (Figure 37-3).

In addition to temperature, moisture is the other major factor determining where organisms can live, and this also depends on the sun. Warm air holds more moisture than cool air, and as air cools some of its moisture condenses as rain, snow, or dew. Air heated by the sun at the equator rises, expanding and cooling as it mounts higher into the atmosphere. This makes it release much of its moisture, producing the teeming rains of tropical jungles. The dry air that remains moves both north and south from the equator, and eventually sinks to earth again, becoming warmer as it does so. The descent of this dry air creates the world's great deserts. Still further north and south, in the temperate latitudes that include most of the United States and Europe, swirling winds pull masses of air, sometimes from warm tropical areas, sometimes from frigid polar regions, producing the varied weather found between the two.

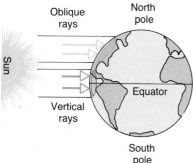

Figure 37-2

Vertical versus oblique rays of sunlight. A beam of sunshine striking the earth away from the equator is spread over a wide area. It is therefore less intense at any one point than a similar beam near the equator, which strikes the earth vertically.

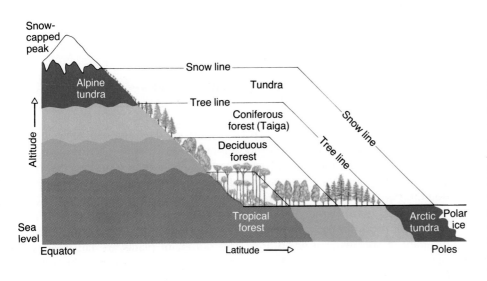

Figure 37-3
Similar effects of altitude and latitude on vegetation. As you go up a mountain, the vegetation changes much as it does when you travel north or south from the equator. This is because similar temperatures favor similar types of plants. Vegetation type is also influenced by moisture. This example shows a cross section through communities with abundant precipitation. (The horizontal axis is greatly compressed.)

Biomes

Each biome has a characteristic type of vegetation, but biomes seldom have sharp boundaries. Instead they gradually merge into one another, forming gradients of changing community type along gradients of changing climate.

Here we class the world's terrestrial communities into a small number of major biomes (see Figure 37-1). Starting at the equator, we shall discuss them in order of the increasing distance from the equator at which they occur. In tropical and temperate areas, the biomes can be arranged along gradients of increasing dryness (Figure 37-4). For instance, trees use a lot of water and can survive only where there is heavy rainfall. Progressively lighter rainfall supports communities dominated by small trees, shrubs, grasses, and finally scattered cacti or other desert plants. In extreme cases, there is so little rainfall that plants cannot grow at all.

■ ***The higher the rainfall and temperature in an area, the more and the larger the plants it can support.***

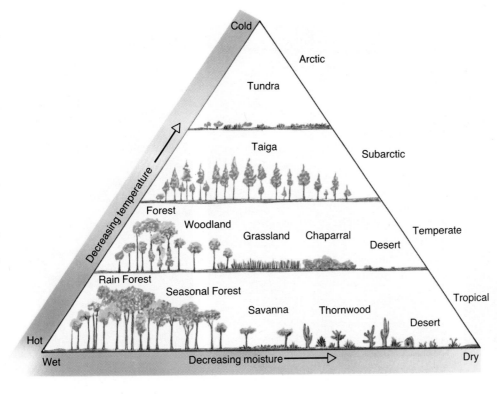

Figure 37-4
The relationship between climate and biome. This diagram shows that temperature and precipitation are the main factors determining what biome occurs where.

37-B   Tropical Biomes

Tropical Forest

Tropical rain forest occurs where the climate provides excellent growing conditions for plants throughout the year: intense sunlight and high, fairly constant temperature and rainfall.

The soil in most rain forests is waterlogged and thin. The high temperature and moisture are ideal for decomposer organisms that break down organic matter. Here, a fallen leaf may decompose in two months, a process that takes one to seven years in a temperate forest. The minerals released by decomposition are rapidly taken up again by the fast-growing plants, and so almost all the forest's nutrients are inside the bodies of living organisms instead of in the soil.

■ *Tropical rain forest, containing an enormous variety of plants and animals, is found where rainfall and temperature are high throughout the year.*

Tropical rain forest is the richest of all biomes, in that it has the greatest diversity of species in a given area. The dominant plants are tall trees with slender trunks that branch only near the top, covering the forest with a dense canopy of leathery evergreen leaves that shed water rapidly. The tall trees provide surfaces for the anchorage of many other plants called **epiphytes** (epi = upon; phyte = plant; also called air plants). Epiphytes include a great variety of orchids, bromeliads, and ferns.

The animal life of rain forests is also exceedingly rich. Birds, butterflies, beetles, and frogs exhibit an almost bewildering diversity of striking color patterns. Since most of the plant food is high in the canopy, most of the animals also live in the trees (Figure 37-5). The difficulty of working in the high canopy is one reason we know so little about the species that inhabit tropical rain forests.

Figure 37-5
Life in a tropical rain forest. (a) Tree ferns, tangled vines, and tall, skinny trees (top left) characterize the perennial wet twilight of this patch of rain forest. Some of the larger trees have been cut down, allowing understory plants to grow. (b) A South American parrot, member of a family common in rain forests throughout the world. (c) An orchid growing as an epiphyte on the trunk of a mahogany tree. (a, Biophoto Associates, N.H.P.A.; c, Paul Feeny)

(a)

(b)

(c)

Farther from the equator, climates have distinct seasons, with rainfall concentrated during part of the year and a definite dry season. These areas support **tropical seasonal forest.** As the length of the dry season increases, we find more and more **deciduous trees** (trees that lose their leaves for part of the year). Tropical seasonal forests include the monsoon forests of India and Southeast Asia.

All tropical forests are being cut down rapidly as a result of the enormous growth of human populations in these biomes. The loss of tropical forest is a pressing world problem (Section 40-D).

Tropical Savanna and Tropical Thornwood

Tropical savanna extends over large areas where rainfall is too low to support many trees, or where growth of forests is prevented by recurrent fires. Typical savanna consists of grassland dotted with scattered small trees or shrubs, such as acacias (Figure 37-6). Some savannas are entirely grassland, while others contain many trees.

Savannas are most extensive in Africa, where they support a rich variety of grazing mammals, such as zebras, wildebeest, and gazelles. The spectacular mi-

■ *Tropical savanna and thornwood, consisting of grassland with scattered trees, feed many native mammals.*

(a)

(b)

Figure 37-6
Savanna. (a) Cape buffalo kick up a dust in Kenya, in a typical savanna landscape of low vegetation dotted with trees. (b) In the dry season, thousands of square kilometres of savanna are parched and brown. (b, Paul Feeny)

grations of some of these species are related to shifting patterns of local rainfall that permit the growth of the young, nutritious foliage of grasses.

Tropical thornwood occurs in many regions too dry to support forest, but with at least a short rainy season each year. Spiny acacias and other trees of the pea family often dominate thornwoods of the Americas and Africa. Many of the plants in a tropical thornwood lose their small leaves during the long dry season, and their growth and reproduction take place entirely during the wet season.

37-C Desert

Deserts occur in regions having less than about 20 centimetres of rain each year. Typical hot deserts are found around latitudes 20 to 30° north and south, where dry air from the equator falls from the upper atmosphere, warming as it is compressed near the earth. Because it contains little water vapor, the air over a desert is a poor insulator, and although days can be very hot, nights are often cold because the ground radiates heat rapidly.

The Sahara desert, stretching across north Africa, is the largest hot desert in the world. Hot deserts also occur in southwestern North America, the west coast of South America, and central Australia. Desert areas with less than 2 centimetres of rain per year support little life of any kind, and the terrain is mainly rocks and sand. Less extreme areas have highly specialized plants, many of them annuals that grow, bloom, and set seed in the few days when water is available. Most desert perennials are small woody shrubs that shed their leaves during the dry season, or else **succulents,** plants that store water in their tissues, such as the American cacti (Figure 37-7). Desert animals have adaptations that restrict the loss of water through their skin and lungs and in their urine and feces. Many are nocturnal, avoiding the heat of the day, when they would lose water rapidly, by burrowing into the cooler soil.

37-D Temperate Biomes

Temperate Forest

North and south of the tropics and their adjacent deserts lie the world's temperate regions, so called because their climate typically has moderate temperatures (although you may not think so as you struggle to start a car on a February morning in Minnesota).

Figure 37-7
Desert in Arizona. (a) The woody shrubs are green because rain has fallen recently, inducing them to grow leaves. In dry periods, they lose their leaves and so lose little water by transpiration. The cacti in the foreground can remain green and photosynthesize throughout the year using water stored in their succulent, prickly stems. (b) A coati, native of American deserts.

(a)

(b)

The **temperate forest** biome occurs in temperate regions with plentiful rainfall. The canopy trees absorb about 40% of the sunlight reaching them. Below the canopy grows an **understory** of smaller trees. Less than 10% of the initial sunlight may reach the next level down, the shrubs. Beneath the shrubs there is usually a layer of low-growing, nonwoody **herbs** that receives less than 5% of the original sunlight striking the forest. Mosses and creeping herbs may provide yet another layer of vegetation close to the ground. Vertical structure continues down into the soil, where the roots of different plants extend to different depths.

Temperate forest falls into three major categories: deciduous, evergreen, and rain forests. **Temperate deciduous forests** occur in moderately humid inland climates where precipitation occurs throughout the year, but where winters are cold, restricting plant growth to the warm summers. Broad-leaved deciduous trees, such as beeches, oaks, hickories, and maples, dominate this kind of forest. There is also a well-developed understory of shrubs and herbaceous plants on the forest floor (Figure 37-8). The soil is rich in minerals and organic matter.

Mammals of North American deciduous forests include white-tailed deer, chipmunks, squirrels, and foxes. Wolves, black bears, bobcats, and mountain lions roamed widely until they were largely eliminated by human activities. As winter draws near, many of the birds migrate south, and many of the mammals hibernate. In the spring, plants such as trilliums, violets, and Solomon's seal produce their leaves and flower before the tree canopy leafs out and cuts off most of the light from the forest floor.

Temperate evergreen forests occur where poor soils, droughts, and forest fires favor gymnosperms or broad-leaved evergreens over deciduous trees (Section 37-H). In the United States, temperate evergreen forests include impressive stands of ponderosa and other pines in the west, as well as the pine forests of the southern states. These are now prime areas for commercial timber operations. Elsewhere in the world, temperate evergreen forests occur in eastern Asia, in southern Chile, in New Zealand, and in Australia, where forests are dominated by various species of *Eucalyptus*.

Figure 37-8
Temperate deciduous forest. (a) In spring, azaleas bloom in this young forest. (b) Deciduous forest in autumn. (c) A raccoon, an animal of temperate forests. (d) Many forest wildflowers, such as these dogtooth violets, bloom before the trees leaf out in spring.

(c)

(a)

(b)

(d)

■ *Temperate forests and woodlands are less productive and diverse than tropical forest because for part of the year the temperature (and sometimes rainfall) is too low for plant growth.*

Temperate rain forests occur in cool climates near the sea with abundant winter rainfall and summer cloudiness or fog. They include the forests of giant trees along the Pacific coast of North America, from the mixed coniferous forest of Washington's Olympic Peninsula to the coastal redwood forests of Oregon and northern California. Although there is little rainfall in California in summer, the foliage of redwoods can absorb water from the frequent fogs.

Temperate woodland occurs in climates too dry to support forests, yet with enough moisture to support more than grassland. Pygmy conifer woodlands of piñon pine and juniper cover extensive areas of the American west. Oak woodlands are common in central California, and extensive evergreen oak and oak-pine woodlands occur in the southwestern states and in Mexico.

Temperate Shrubland

The **temperate shrubland** biome is best represented by the **chaparral** communities in all five areas of the world with a Mediterranean climate: the Mediterranean region, southern Australia, the southern tip of Africa, and coastal Chile and California. These areas have moderately dry climates with little or no summer rain. The shrubs are mainly angiosperms (flowering plants). They are often distinctly aromatic, and their leaves contain volatile and flammable organic compounds. Fires are frequent and pose a constant threat to residents of Santa Barbara and other cities in this biome.

Temperate Grassland

■ *Temperate grassland has produced the world's deepest soils and, therefore, its best agricultural land.*

Early visitors to the American West were most impressed not by the forest but by the prairie with its burrowing prairie dogs and large grazing mammals, such as bison and pronghorn antelope. Prairie is **temperate grassland,** which covers extensive areas in the interiors of continents where there is not enough moisture to support forest or woodland. Scattered shrubs may occur, often in depressions or watercourses where extra water is available.

Although grassland vegetation forms only a single layer, many plant species may be present. Rich deep soil underlies much temperate grassland because dead vegetation is added to the soil faster than it decomposes. These regions of deep soil, including the midwestern United States, the Asian steppe, and the Ukraine in the Soviet Union, have become prime areas for farming. As a result of its agricultural value, prairie has suffered more complete destruction than any other biome in North America. Some types of prairie have been so completely

Figure 37-9
Prairie. (a) Short-grass prairie in South Dakota. (b) Researchers sampling the plants in a long-grass prairie preserved and restored by conservationists. (a, Paul Feeny; b, Richard Thom/VU).

(a)

(b)

Figure 37-10
Temperate semidesert in Idaho. This land has low rainfall, so it supports only a thin cover of vegetation. As a result, the soil is thin and easily washed or blown down the steep hills if the vegetation is removed. This land, like much of America's desert and semidesert, is used for cattle grazing and has suffered severe soil erosion as a result.

eliminated that ecologists are not even sure what plants and animals they contained. Conservation groups are undertaking the restoration of partly destroyed prairie in several parts of the United States (Figure 37-9).

Temperate Desert

Temperate desert or semidesert, sometimes called scrubland, occurs in regions too dry to support fertile grassland. Cool semidesert occupies much of the Great Basin east of the Cascade and northern Sierra Nevada mountain ranges in the western United States. Large areas are dominated by sagebrush, interspersed with perennial grasses (Figure 37-10). Typical animals include jack rabbits, sage grouse, pocket mice, and kangaroo rats. Cool temperate semideserts also occur in central Asia, South America, and Australia.

37-E Taiga

The **taiga** biome is dominated by conifers—spruces, pines, and firs—that can survive extreme cold in winter. Trees in the taiga tend to be farther apart than those in a deciduous forest, and light penetrating to the forest floor supports a ground cover of shrubs. The taiga, or **boreal forest,** as it is sometimes called, stretches in a giant circle through Canada and Siberia (Figure 37-11). The forest,

Figure 37-11
Taiga. (a) Trees of the taiga consist of a few coniferous species adapted to very cold or very dry soil. (b) Animals of the taiga include the snowshoe hare, whose coat turns completely white (matching the snow) in winter.

(a)

(b)

containing only a few tree species, is occasionally interrupted by extensive areas of bog, or "muskeg," in poorly drained areas.

Much of the precipitation in the taiga falls as snow, and in the winter many of the animals grow white fur or plumage that blends with the background. Animals of the North American taiga include moose, wolverines, wolves, lynx, spruce grouse, gray jays, crossbills, and (in summer) many species of warblers (birds).

37-F   Tundra

The **tundra,** a treeless biome, occurs far north in the arctic regions, where winters are too cold and dry to permit the growth of trees (Figure 37-12). In many areas the deeper layers of soil remain frozen, as **permafrost,** throughout the year, and only the surface thaws during the summer. Because the ground is so cold, decomposition is slow, so the soil is shallow, and plant growth slow. As a result, tundra takes a long time to recover when it is destroyed. This is why conservationists are so concerned about the effects of oil spills and oil industry traffic on the wildlife of the tundra.

Tundra vegetation is dominated by sedges, grasses, mosses, lichens, and dwarf woody shrubs. Bogs are common because the permafrost prevents water from draining away. The largest animals of the tundra are caribou in North America and reindeer in Europe and Asia. Hordes of mosquitoes, deerflies, and blackflies breed in the wet spots during the brief arctic summer. These insects contribute to the food available for a variety of birds, including various plovers, sandpipers, and horned larks, which nest in the tundra.

Neither taiga nor tundra occurs at sea level in the Southern Hemisphere because the continents do not extend far enough south. Antarctica harbors only a few forms of life around its edges.

■ *Tundra is a biome of low-growing plants found where there is enough light and water for plant growth for only a few months a year.*

(a)

(b)

(c)

Figure 37-12
Tundra and its inhabitants. (a) The treeless landscape; (b) a reindeer; (c) a dwarf willow, one of the small shrubs that can survive in the thin soil. (Biophoto Associates)

37-G *Aquatic Communities*

Strictly speaking, the term biome refers only to communities on land. However, there are also many different kinds of aquatic communities, both marine and freshwater, which, like biomes, exhibit similarities wherever in the world they occur. Here we shall consider temperate freshwater lakes and life in the sea.

As on land, environmental conditions influence the distribution of organisms in water. Temperature, nutrient supply, the intensity of sunlight, and salinity (salt concentration) determine what can grow where. Lack of water is not a problem, but in some areas a shortage of minerals dissolved in the water limits plant life. The types of organisms found in an aquatic ecosystem also depend upon the type of bottom (mud, sand, or rock), and upon wave action and water currents.

Lakes and Rivers

The world's large temperate region lakes, such as the Great Lakes of North America, and Lake Baikal in the Soviet Union, are vitally important sources of fresh water for drinking, agriculture, and industry.

Sunlight is a lake's source of energy. As it passes down through the water, some of it is used in photosynthesis by phytoplankton and some is absorbed by the water itself. So, as light passes deeper into the water, it becomes dimmer. In deep lakes there is a **compensation depth** where the available light is just bright enough for green plants to eke out a living: their photosynthesis (production of food and oxygen) exactly offsets their respiration (use of food and oxygen). Above the compensation depth plants produce more oxygen than they use, and so extra oxygen is available for the respiration of other organisms; below it there is not enough photosynthesis to offset respiration, and any available oxygen must come down from the water above.

Rooted aquatic plants such as water lilies and rushes grow in shallow water around the edge of the lake. In many lakes, most of the lake's productivity comes from photosynthesis in this **littoral zone.** Fish, amphibians, insects and other arthropods, snails, and worms live and feed among the plants. In the lake's open surface waters live floating plants, which need light, and animals that need abundant oxygen, such as fish and small arthropods.

The amount of oxygen dissolved in the lake's waters is one of its most important features. Oxygen affects nearly every aspect of the lake, including what animals and plants can live where and the solubility of many inorganic nutrients. Oxygen enters the water from the photosynthesis of aquatic plants and by dissolving into the water from the air. Oxygen leaves the water when it is used in respiration. The cooler the water, the more dissolved oxygen it can hold. A lot of oxygen is used by the bacteria that decompose **detritus** (dead organic matter). Many desirable fish, such as trout, can survive only in waters containing large amounts of oxygen.

Deep in a lake, where there is not enough light for photosynthesis, the main biological process is decomposition. Here the chief input of energy is detritus falling from above, which feeds decomposers, fish, and invertebrates.

Lakes can be divided into categories based on how much plant life they support. **Oligotrophic** ("few food") lakes are low in nutrients such as phosphorus, calcium, and nitrogen, so they support little plant growth and contain few organisms. Oligotrophic lakes are usually deep, with steep sides and narrow littoral zones. Their water is usually very clear, and the deep waters always contain oxygen (Figure 37-13). **Eutrophic** ("good food") lakes are rich in nutrients and organisms and are usually shallow. Such lakes contain little oxygen because decomposer organisms rapidly use it up metabolizing the organic matter produced by the lake's many other residents. In the normal course of events, a lake ages as it is steadily filled in with minerals and organic matter, becoming

Figure 37-13
An oligotrophic lake in the Rocky Mountains.

more eutrophic as it ages. Natural eutrophication takes thousands of years, but the process may be speeded up to take only a few years if the lake becomes polluted. When nutrients in sewage or minerals such as chemical fertilizers wash into a lake, they speed plant growth and hence eutrophication.

The main difference between rivers and lakes in the same climate is that rivers usually have stronger currents. Where the current is strong, only organisms that can anchor themselves or swim against it can survive. A shallow, rapidly flowing river is usually well oxygenated because it has a large surface area to absorb oxygen from the air.

The Edges of the Ocean

The edges of the sea are the hatcheries and nurseries of many important species of marine life. Coastal wetlands and estuaries (areas where rivers enter the ocean) also serve as nesting, feeding, and resting spots for migratory waterfowl, and they reduce erosion and flooding inland. Recognizing these important but indirect contributions to human food, fun, and safety, ecologists are alarmed by the draining and filling of these areas to build towns, marinas, and resorts (see *A Journey Into Life,* "Salt Marsh and Seafood," Chapter 38).

Along the seacoasts, many kinds of plants and animals thrive in the **intertidal zone,** the area between the high and low tide marks, where they are submerged for part of the day. There are three main types of intertidal zone: muddy, sandy, and rocky shores, which support very different communities.

Mudflats occur where the water moves slowly enough to deposit a sediment of small particles. Algae cover the particles and provide food for a multitude of burrowing molluscs, worms, and crustaceans.

Sandy beaches are less stable than mudflats, for sand shifts constantly and dries out faster than mud when the tide is out. Most of the tiny protists, worms, and crustaceans that live between the sand grains eat marine plankton stranded when the tide goes out, or algae attached to the sand grains. A wide variety of shore birds feeds on the invertebrate inhabitants.

■ *Intertidal and subtidal zones are among the most densely populated marine communities because they receive abundant light, and mineral nutrients are readily available to plants.*

(a)

(b)

Figure 37-14
Sea shores. (a) On a sandy beach, large populations of shore birds feed on molluscs and crustaceans in the sand. (b) In a rocky intertidal zone, snails, a sea star, barnacles, and other sessile organisms cling to the rocks as the tide goes down.

Neither muddy nor sandy shores provide much foothold for sessile animals or anchored seaweed. These are much more common on rocky shores, which support a wider variety of organisms. Since the water crashes onto the rocks, motile animals such as crustaceans anchor themselves firmly to rocks or seaweeds by their legs, or hide in crevices. Very few vertebrates live in the intertidal zone, although a number of birds come in at low tide to scavenge or to prey on invertebrates (Figure 37-14).

The **subtidal zone** occupies the **continental shelves**—the edges of continents—extending from the low tide mark to a depth of about 200 metres. Here, temperature fluctuates less, and wave movement is less violent, than in the intertidal zone. Mineral nutrients are also readily available, washed from the land by rivers. Continental shelves are among the most densely populated areas on earth.

■ *The distribution of life in the ocean depends on water temperature and the availability of sunlight, nutrients, and surfaces to which organisms can attach.*

Coral Reefs

Coral reefs are restricted to warm oceans, where the water temperature seldom falls below 21°C. Corals are cnidarians that live in symbiotic association with photosynthetic protists. The reef itself is made up of calcareous material, secreted by the coral animals themselves and by red and green algae. Since photosynthetic organisms are so important to their formation, coral reefs are found only in clear, shallow water where there is enough light for photosynthesis.

Most of a coral reef is submerged, although its top may be exposed at low tide. A reef acts physically like a rocky shore in providing anchorage for algae and sessile animals. A great variety of fish and swimming invertebrates find

(a)

(b)

Figure 37-15
A coral reef. (a) Snappers patrolling their territory on a Caribbean reef. (b) Hydroids (cnidarians) on a reef. (Steven Webster)

shelter within the reef's crevices (Figure 37-15). Today, many coral reefs are threatened by the dumping of wastes from nearby tourist areas and by drilling for underlying oil deposits. States such as the American and British Virgin Islands in the Caribbean enforce protective laws because a coral reef, once destroyed, takes many years to regrow.

The Open Ocean

Figure 37-16
Animals of the nekton. (a) A baby turtle struggles toward the sea soon after it has hatched. These turtles live for decades as members of the nekton, but they lay their eggs ashore. Humans have brought them to the brink of extinction by eating adults and eggs, killing adults to make "tortoiseshell," and drowning adults accidentally in fishing nets. (b) A school of fish. (a, Biophoto Associates, N.H.P.A.; b, Steven Webster)

The open ocean can be divided into the top 100 metres or so, where photosynthesis can occur, the deep ocean, and the ocean floor. In the surface waters live the **plankton,** drifting protists, plants, and animals not powerful enough to swim against currents. They can, however, control their vertical position in the water, thereby moving from one current to another, for currents flow in different directions at different depths in many parts of the ocean. Fish and similar large animals in the ocean make up the **nekton,** creatures that can swim independent of currents. These animals feed mainly on plankton or on each other (Figure 37-16). Because mineral nutrients are scarce in many areas of the open ocean, the world's major fisheries lie over continental shelves, which receive

(a)

(b)

(a)

(b)

Figure 37-17
Primary succession. (a) This island in the St. Lawrence River
formed several thousand years
ago when a glacier dropped its
load of granite boulders. Lichens
and moss grow on the rocks. In
cracks between the rocks,
enough debris has accumulated
to support the growth of shrubs
and trees. (b) Lichens and resur-
rection ferns grow on a dead
branch. The fern is an epiphyte.
It does not need soil to grow,
merely somewhere to anchor itself. Soil may form rapidly in this
situation as the log is broken
down by decomposer organisms
and fern roots hold the material
in place.

minerals washed down rivers, and in parts of the open ocean where currents
carry minerals up from the bottom.

Seventy-five percent of the ocean's water lies more than 1000 metres deep.
Diving techniques permit sampling at depths of more than 6000 metres and
reveal fascinating communities on the ocean floor—communities that contain
no plants. The **benthos** (depths of the sea), the community of the ocean floor,
includes decomposer bacteria that live on dead organisms falling from the surface layers above them or on dead members of the deep-sea community. Larger
members of the benthos also feed on falling carcasses or on the decomposers,
and filter-feeders strain food out of the water.

37-H Ecological Succession

Coastal California is predominantly covered, not by chaparral, but by farms,
roads, and buildings. Human civilization has disturbed the natural communities,
clearing the vegetation to make room for human affairs and their adjunct parking lots. So, when we say that climate determines the type of community in an
area, we mean the community that would exist if the area were left alone long
enough, rather than what actually exists there. The community that forms if the
land is left undisturbed, and that remains as long as no disturbances occur, is
called the **climax community.**

After a climax community is disturbed, either by human activities or by
natural means such as floods or fires, the area slowly returns to its original state.
Ecological succession is the progressive series of changes that ultimately produces a climax community.

Primary Succession

Primary succession occurs where there is initially no soil—when a new island rises out of the sea, or when a glacier retreats or a mountainside caves in,
leaving a pile of rocks. Consider an area of rock created by a landslide. Lichens
adapted to exposed conditions may spread over the rock surface. They produce
organic acids, which dissolve some of the rock. Dead lichens contribute organic
matter to the forming soil, and mosses may gain a hold in even a thin layer of
lichen remains and rock dust. As the mosses break up the rock further and add
their own dead bodies to the pile, the seeds of small, rooted plants can germinate and grow. The process continues along similar lines until the climax community becomes established. It may take thousands of years for the soil and the
climax vegetation to develop fully (Figure 37-17).

■ *Primary succession occurs
slowly because it takes time
for soil to form and for plant
life to move into an area.*

Primary succession also occurs as an oligotrophic lake or pond ages. It fills up with silt and fallen leaves, becoming more eutrophic as the shoreline creeps toward the center of the lake. Gradually the lake turns into a marsh and then into dry land, eventually colonized by plants of climax species from surrounding communities.

Secondary Succession

Secondary succession is the series of changes that occur in a community that has been disturbed but not totally stripped of its soil and vegetation. Although it may take a hundred years or more for the climax vegetation to return during secondary succession, the process is much faster than primary succession because soil already exists.

A familiar example of secondary succession in the New England area is "old field succession," by which abandoned farms return to the climax temperate forest (Figure 37-18). When a farmer stops cultivating the land, grasses and annual weeds quickly move in and clothe the earth with a carpet of wild carrot, black mustard, and dandelions. The **pioneers** of newly available habitats, these plants grow rapidly and produce seeds adapted to disperse over a wide area. Soon taller plants, such as goldenrod and perennial grasses, move in. Because these newcomers shade the ground and their long root systems monopolize the soil water, it is difficult for seedlings of the pioneer species to grow. But even as these tall weeds choke out the sun-loving pioneer species, they are in turn shaded and deprived of water by the seedlings of pioneer trees, such as pin cherries, dogwoods, sumac, and aspens, which take longer to become established but command the lion's share of the resources once they reach a respectable size.

Succession is still not complete, for the pioneer trees are not members of the climax forest. After 5 to 30 years, slower-growing oak, maple, and hickory trees will take over, shading out the saplings of the pioneer tree species. After perhaps a century or two, the land is covered with mature climax forest.

In any tract of land, we can always find at least small patches that are undergoing succession following disturbance—a spot where a large tree has fallen, leaving a light gap where pioneer weeds can move in, or a burned forest. The existence of various patches undergoing succession ensures that there is a

■ *In secondary succession, an area whose vegetation has been disturbed or destroyed undergoes a series of species replacements until the soil and organisms of the area's climax community are restored.*

Figure 37-18
Secondary succession in an old field in New York. (Paul Feeny)

Figure 37-19
A controlled burn in a southern pine forest. The fire was set to burn undergrowth and controlled to ensure that it did not endanger property.

steady supply of **fugitive** plants, the fast-growing, here-today-and-gone-tomorrow pioneers, which include many kinds of weeds.

Animals, as well as plants, may be fugitive species. Insects that specialize in eating a particular plant species may travel far and use their keen senses to smell out new patches of their food plant some distance away. Some of our agricultural pest problems stem from the fact that most crop plants originated as fugitive species. By planting fields exclusively to one crop year after year, farmers create a paradise for fugitive animals such as cabbage worms and cucumber beetles, which no longer have to spend energy to find food and have nothing to do but eat and multiply.

Succession occurs because of progressive changes that make the environment less favorable for the species that are present and more favorable for colonization by others. Some of the changes are purely physical, like the silting in of a lake or the weathering of rock, but many are caused by the organisms themselves. As succession proceeds, the supply of available nutrients in the soil declines as minerals become increasingly locked up in living organisms. Both the community's production of new organic matter (through photosynthesis) and the total weight of all the organisms in the community increase during succession, levelling off as the climax is approached.

Fire-Maintained Communities

Fire is one cause of disturbance and succession that has been particularly well studied. Fires, set by lightning or by human activities, occasionally sweep through large areas of taiga and temperate forest, burning trees and destroying entire communities of animals and plants. Burned areas undergo secondary succession.

In some communities, fire occurs often enough to determine the nature of the dominant vegetation. Such communities include some pine forests, chaparral, and temperate grassland. Grasses readily regenerate after fires that would kill trees; thus recurrent fires may prevent grassland or savanna from turning into woodland.

Seedlings and saplings of deciduous trees are especially susceptible to fire, whereas many pines are adapted to survive, and even to exploit, fires. In some pines, for example, the cones open and their seeds germinate only when exposed to temperatures of several hundred degrees. This ensures that the seeds germinate in areas that have just been burned. If fires are prevented in a fire-adapted pine forest, deciduous trees may become established. In addition, dead wood and litter build up on the ground, adding extra fuel. When a fire eventually does occur, it is more severe than usual, destroying not only any deciduous colonizers but also the pines and other species.

Odd though it may seem at first, frequent burning is essential for the preservation of many natural communities (Figure 37-19). This is the reason that the

The Disappearing Soil

A JOURNEY INTO LIFE

Every year, an average of 25 to 30 tons of soil erodes from each hectare (2.5 acres) of American agricultural land. **Soil erosion** is the movement of soil from one location, usually agri-cultural land, to somewhere else, usually a lake or river. Wind and water are the forces that move the soil. In the United States, about 70% of all farmland has had its productivity reduced by soil loss, and many areas have lost essentially all their soil. The soil of an area is considered destroyed when crop plants will no longer grow there.

Soil destruction threatens our ability to feed the world's growing population, and it also causes other problems. For instance, soil washing into streams and rivers carries fertilizer and pesticides with it and so contributes to water pollution. The soil particles may also block out light needed by photosynthetic organisms in streams and lakes (Figure 37-A).

Erosion is not the only cause of soil destruction. Soil in many parts of the world has become so salty that plants can no longer grow in it, especially in arid regions, including the semidesert and shrubland of the western United States. **Arid regions** are areas where more water leaves the soil by evaporation than reaches it through rainfall. (Water moves up from the water table underground, drawn by plant transpiration and by capillarity.) When water evaporates, the salts dissolved in it are left behind, so arid land soil contains high concentrations of salts. Because arid land receives less than 25 centimetres of rain a year, most agriculture in arid areas is irrigated. When irrigation water evaporates, even more salts are

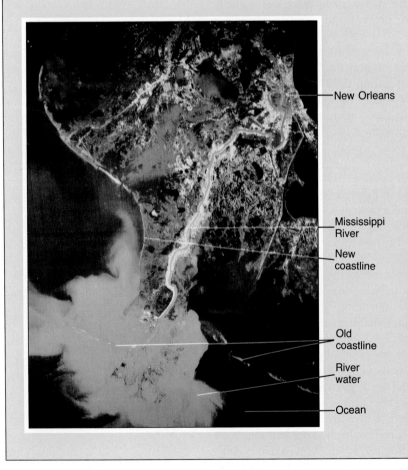

New Orleans

Mississippi River

New coastline

Old coastline

River water

Ocean

Figure 37-A
Soil erosion. This computer-enhanced photograph from a Landsat satellite shows the mouth of the Mississippi River. Pale green river water, loaded with soil eroded from farmland, forest, and construction sites, flows into the ocean in the Gulf of Mexico. Louisiana is losing land faster than any other state or country in the world (including rapidly eroding Bangladesh). About 130 square kilometres disappear from Louisiana every year. Much of this land is coastal wetland and farmland that sinks below sea level. The land sinks because soil erodes from the surface, the sea level is rising, and the land subsides as water and oil are pumped out of the ground beneath it. (VU/NASA)

added to the soil. Eventually, the soil may become so salty that plants cannot grow in it. The accumulation of salts in the soil is called **salinization.**

The usual way to grow crops on very salty soil is to add much more water than the plants can absorb. Then the water will run off the land, carrying dissolved salts with it. This salt-laden water is often toxic. Kesterton Wildlife Refuge in California receives such high concentrations of the element selenium from the irrigated land of the San Joaquin Valley that birds and plants in the refuge die. Selenium occurs naturally in the soil in low, harmless concentrations. Evaporation from arid land has concentrated it to a dangerous level. Scientists are attempting to clean up the toxic soil in Kesterton by growing plants and fungi that absorb selenium.

Saline soil can be restored to usefulness without intensive care by planting salt-tolerant trees. These absorb salts and hold the soil in place. The leaves and twigs that fall from the trees add organic matter to the soil. Australia has organized its Girl Scouts and other groups to plant more than a billion trees on land turned to saline desert by intensive

agriculture. Tree seedlings survive best if they are watered and protected from animals for the first few years, so soil reclamation projects are most successful if residents of the area undertake to care for the trees.

The problem of soil destruction is largely economic: an individual farmer profits for a short period by farming practices that destroy the soil. These practices include planting crops instead of windbreak trees, irrigating arid land, or leaving land without plant cover between crops (Figure 37-B). Because the problem is economic, the solution usually involves altering government regulations and subsidies. For instance, in 1985, the United States passed a soil conservation act that rewards farmers for planting trees and grass on previously plowed land that erodes easily (such as that on steep hillsides). This program greatly reduced the rate of soil erosion.

Soil destruction is sometimes called **desertification,** the formation of desert from previously productive land as a result of human activities. The name arose because many of the world's deserts have been produced or enlarged in this way, and each

year desertification adds to them an estimated million hectares (nearly 4000 square miles). In prehistoric times, the Sahara Desert was less than a fifth its present size.

The desert created by destroying soil can be made to bloom again, but this is a long, slow process. The basic solution is to keep the soil covered with plants at all times and let some of this plant material decay into the soil. If only rock or gravel remains when the soil is gone, soil will take thousands of years to form again by primary succession. If sand and clay remain, the outlook is brighter. Both can be made into fertile soil by the addition of organic matter. This can be done intensively, by planting **green manure,** plants that will grow in nutrient-poor soil and are then plowed into the soil to decompose. After only a few years of this treatment, soil will regain its fertility.

Loss of soil is probably the most devastating environmental problem that we face. Governments around the world are beginning to realize this and to design economic incentives that will encourage farmers to conserve and restore the soil.

Figure 37-B
Erosion by wind. This sandy semidesert soil in Idaho was left without vegetation cover. It has blown from the field into the neighboring road. Soil that is dry and contains little organic matter is particularly susceptible to erosion. Wind erosion can be prevented by planting windbreaks to slow the wind, by keeping the soil covered with plants, and by adding organic matter (which holds water) to the soil.

Figure 37-20
Adaptations to desert life. Spines and water-storing stems are characteristic of desert plants in all parts of the world. Here they are seen in the tree-like saguaro cactus. This attractive American native is a valued landscape plant in arid regions but is disappearing from its natural habitat. Saguaros grow very slowly, so thousands of mature specimens are collected illegally and sold to householders each year.

■ *A species' distribution depends on the type of climate it requires and on its ability to disperse (or be dispersed) to suitable new areas.*

U.S. Park Service adopted the policy of letting fires in National Forests burn if they do not endanger human life or property. This policy caused an outcry when fires burned millions of hectares in Yellowstone National Park in 1988 because most people did not understand the ecology of fire-adapted communities. These fires became an opportunity for visitors to learn about succession following a fire. In 1989, the park had more visitors than ever before—come to survey the damage and to watch pioneer species colonizing the blackened landscape.

37-I   Why Are Organisms Where They Are?

Fugitive species get around quickly, with efficient dispersal mechanisms, such as seeds that float on the wind or are carried by animals. However, there is a limit to how far an organism can travel over territory that is unsuitable for its survival. If we return to the question that began this chapter and ask again, "Why are organisms where they are?" we find that at least two factors determine the answer: climate and the organisms' ability to disperse to areas where they can survive.

When we look again at African and South American rain forests and ask, "Why are there different species in these two areas?" our answer is that, although each area has a climate suitable for growth of the plants and animals of the other area, they are separated by wide stretches of ocean, an impassable barrier to dispersal. There are exceptions: for example, small animals called rotifers form cysts that can be blown almost anywhere in the world, but the animals can live only in very restricted types of environments. As a result, the rotifer species in a marble cemetery urn in Pennsylvania may be the same as that in a marble urn in a South African cemetery, but different from that in the granite urn on the next grave!

Because most species cannot travel between continents (unless they are transported by humans), there are different species in the tropical rain forests of the different continents. The similarities of form and color of species in each place result from convergent evolution (Section 18-C). For instance, the advantage of being able to conserve water in desert habitats has led to the evolution of plants with thick, water-storing stems, as well as spiny leaves that deter animals from using the stems as the source of their own water, in deserts all over the world (Figure 37-20).

SUMMARY

A worldwide survey of the distribution of organisms reveals two main patterns:

1. Different areas of the world are inhabited by different species of plants and animals.
2. Terrestrial communities in different parts of the world can be divided into a fairly small number of categories, or biomes, each with a characteristic array of plant life. These biomes are worldwide, occurring wherever a suitable climate exists.

While the actual species found in an area depend on the area's evolutionary history, the biome depends mainly upon the annual pattern of rainfall and temperature—that is, climate. Similar changes in biomes occur with increasing altitude and increasing latitude.

The richest biome is tropical rain forest, where high temperatures and rainfall permit plants to grow throughout the year. Most of the plant and animal life is found in the canopy among the broad evergreen leaves of trees. The soil is poor because decomposition is so rapid that most of the nutrients are locked in the bodies of living organisms.

In temperate deciduous forest, the soil is much richer in nutrients because the trees lose their leaves in the fall, creating a litter layer that decomposes and releases the leaves' nutrients only slowly. Deciduous forest is an important biome of North America, Europe, and Asia in areas with warm, moist summers and cold winters.

Where the soil is poor or fire is frequent, temperate evergreen forest replaces temperate deciduous forest. Farther north, both are replaced by taiga, a biome dominated by conifers adapted to growing in sparse soil and to resisting extreme cold and water loss in winter.

North of the taiga lies the tundra, dominated by cold-resistant woody shrubs or by grasses.

Grasslands receive less rainfall than forests but more than deserts. Grasslands occur in the drier interiors of continents. Shrubs and trees may be scattered among the tall grasses. Grasses are replaced by small woody shrubs in areas where there is too little water for grasses to grow.

Deserts have hot days, cold nights, and very little rainfall. Their plant life is mainly annuals with very short growing seasons, succulent perennials adapted to the low rainfall, and small-leaved woody shrubs.

The distribution of aquatic organisms is determined by water temperature, depth, salinity, and motion, and by the availability of light, oxygen, and minerals. Shallow areas near the coast are well supplied with both light and minerals, and they support dense communities of life. Coral reefs are specialized communities found only in tropical ocean waters. In the open ocean, the availability of light for photosynthesis restricts plankton to the upper layers of the water, but scarcity of nutrients in these layers may limit the numbers of organisms. Larger nektonic organisms are found primarily where planktonic food is abundant. Dead organisms from the surface layers of the ocean supply food for a benthic community of bacteria and other organisms that live on the sea floor.

Although the climate of an area determines the composition of its climax community, patches of the area are always in various stages of ecological succession following disturbances. Organisms adapted to living in the unstable communities of early successional stages have effective dispersal mechanisms and perpetuate themselves by continuously colonizing new habitats as they arise.

Most organisms cannot travel far and so they do not colonize all possible habitats. Therefore, in different parts of the world we find similar communities inhabited by similar species, which have arisen by the convergent evolution of similar adaptations to similar environments.

SELF-QUIZ

1. Which of the following has a vegetation structure with only one level?
   a. tropical rain forest
   b. taiga
   c. grassland
   d. shrubland
   e. desert

2. Which of the following communities would have trees?
   a. taiga
   b. intertidal zone
   c. shrubland
   d. tundra
   e. plankton

3. A biome with high temperature, high rainfall, and poor soil is:
   a. shrubland
   b. coral reef
   c. semidesert scrub
   d. tropical rain forest
   e. temperate evergreen forest

4. Which of the following communities has no living green plants?
   a. a rocky shore
   b. the plankton
   c. a mud flat
   d. the deep ocean floor
   e. a coral reef

5. Compared with a eutrophic lake, an oligotrophic lake contains a greater concentration of:
   a. organic matter
   b. plants
   c. oxygen
   d. bacteria
   e. mineral nutrients

In questions 6 and 8, choose the correct term from each pair in parentheses.

6. Colonization of an abandoned stone quarry would be an example of (primary/secondary) succession.

7. A pond in a deciduous forest becomes filled in with rock particles and dead leaves, creating soil. List, in order, the types of vegetation that would be seen as this area undergoes ecological succession, and name the climax community that would eventually result.

8. A(n) (early successional/climax) community would have a high proportion of fugitive species.

9. The American prairies and the Asian steppes do not have the same species of grasses because ____. However, both are inhabited primarily by grasses because ____.

QUESTIONS FOR DISCUSSION

1. What biome do you live in?

2. The 30°N latitude line runs through southern Louisiana and northern Florida as well as through desert country in Mexico and Texas. Why is the area in Louisiana and Florida not desert like the area in Mexico and Texas?

3. Why is it proving difficult to carry out large-scale "agribusiness" farming in vast tracts of land cleared of their tropical rain forest vegetation?

4. Why is there less variation in size of vegetation in the tundra than in tropical regions?

5. Why does secondary succession slow down as it proceeds?

6. How can frequent fires increase the species diversity of a region?

7. Why does a light gap contain different species of animals and plants from those in surrounding climax forest?

SUGGESTED READINGS

BioScience 39 (10), 1989. "Yellowstone Fires." An entire issue devoted to the effect of the 1988 fires on the ecology of Yellowstone National Park and the controversy over the extent to which fires should be extinguished or left to burn naturally.

Childress, J. J., H. Felbeck, and G. N. Somero. "Symbiosis in the deep sea." *Scientific American,* May 1987. Tube worms and clams live on the energy from hydrothermal vents in the deep sea floor. They are supplied with nutrients by symbiotic bacteria that live on them.

Colinvaux, C. "The past and future Amazon." *Scientific American,* May 1989. Colinvaux argues that the enormous diversity of species in the Amazon Basin is partly a result of frequent disturbances (of climate and geology). We can, therefore, be reasonably optimistic that the Amazon will survive recent human disturbances—as long as these are not too destructive.

Perry, D. R. "The canopy of the tropical rain forest." *Scientific American,* November 1984. Its inventor tells the story of a climbing method that has made it possible to do research in the once-inaccessible rain forest canopy.

Ecosystems and Communities

CHAPTER

38

Bison

Tiger

Hermit thrush in flowering dogwood

Long-nose hawkfish

A tropical rain forest or a desert may cover a huge area. For convenience, ecologists usually study smaller units—for example, a hillside, a lake, or a field. Each of these is an **ecosystem,** consisting of the **community** of all the organisms living there, along with their physical environment. We usually treat an ecosystem as an isolated unit, but in fact things invariably move from one ecosystem to another, as when leaves blow from a forest into a lake, or birds migrate between their summer and winter homes.

Not all ecosystems are natural: a space station, an aquarium, and a pot of houseplants are artificial ecosystems. A farm is often considered as an ecosystem because farmers must recognize the interactions between crop plants, fertilizers, pesticides, soil, climate, and the natural plant and animal life in order to manage the farm effectively.

The organisms in an ecosystem interact with each other and with their physical environment. These interactions can be viewed on two different time scales. Ecologists study what is happening here and now: plant growth, animals eating plants, and so on. But over the long term, every environmental event affecting organisms may also be a selective force that shapes their evolution. Each time an owl catches a mouse, it not only feeds itself and reduces the number of mice but also selects against a set of mouse genes that are not effective at avoiding capture by owls.

KEY CONCEPTS

- The organisms in an ecosystem interact with one another and with their physical environment, influencing each others' lives and evolution.
- Nutrients may cycle indefinitely in an ecosystem, but energy is continuously lost.
- A self-sustaining ecosystem must contain nutrients, producers, and decomposers, and receive a continuous input of energy.

38-A The Basic Components of Ecosystems

To be **sustainable,** capable of lasting indefinitely, an ecosystem must contain the resources necessary to support its resident organisms and to dispose of their wastes. The necessary components of an ecosystem are water, various minerals,

(a)

(b)

Figure 38-1
Producers and decomposers. (a) The main producers on land are grasses, shrubs, and trees, here in oak woodland in California. (b) Mushrooms are the reproductive structures of decomposer fungi. (b, Biophoto Associates)

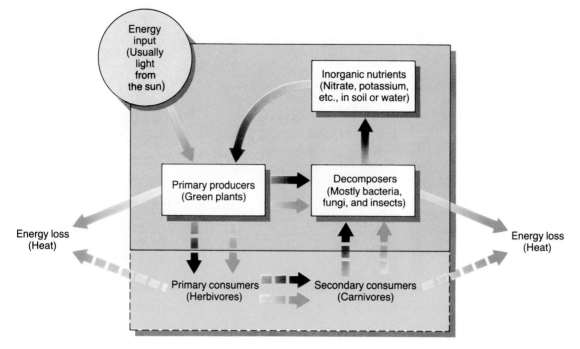

Figure 38-2
Energy flow and nutrient cycling. The general pattern of energy (blue arrows) and nutrient (gray arrows) exchange in an ecosystem. The solid line encloses the minimal components required for a self-sustaining ecosystem. In addition to these, most ecosystems contain primary, secondary, and even higher levels of consumers (dashed box and arrows). Energy is continuously lost from the ecosystem in the form of heat produced by the organisms' metabolism.

carbon dioxide, oxygen (in most cases), and various kinds of organisms. An ecosystem must also receive a continuous supply of energy.

The sun is the ultimate source of energy for almost all ecosystems. It drives photosynthesis by green plants. Because green plants are **autotrophs,** making their own food from inorganic substances, they are called the **producers** of all the food in the ecosystem (Figure 38-1).

Plants eventually die. Their remains are usually broken down by **decomposers,** organisms that acquire their food molecules from dead organic material. In the process of extracting energy and nutrients from this material, decomposers release some of the nutrients back into the ecosystem, where they are again available to producers. Nutrients are thus cycled through the ecosystem and may be used again and again in the same small area. Energy, by contrast, is not cycled but is continuously lost from an ecosystem. Most organisms would soon die if the sun's energy were cut off for any length of time.

The basic requirements for a self-sustaining ecosystem are inorganic nutrients, producers, decomposers, and a continuously renewed source of energy (Figure 38-2). However, most ecosystems also contain **consumers,** animals (and other organisms) that eat plants or each other. Plant-eating animals are collectively known as **herbivores** or **primary consumers.** These may die and pass directly to the decomposers, or they may be eaten by **carnivores,** also called **secondary consumers** (Figure 38-3). There may be tertiary or even quaternary consumers, carnivores that feed on the secondary and tertiary consumers, respectively. Consumers and decomposers are **heterotrophs,** feeding on organic matter produced by other organisms.

(a)   (b)

Figure 38-3
Consumers. (a) A herbivore: a Masai giraffe eating a prickly acacia. (b) Carnivores:
part of a pride of lions killing a cape buffalo. (a, Biophoto Associates, N.H.P.A.)

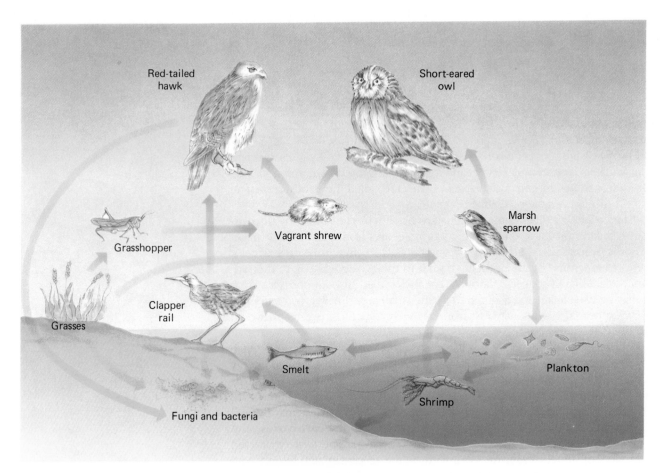

Figure 38-4
Part of the food web in a marsh. Arrows show the passage of energy and nutrients
as one organism eats another.

38-B   Food Webs

A **food chain** is a series of organisms, each of which is eaten by the next. Examples are the passage of nutrients and energy from leaves to caterpillars that eat them, to chickadees, to hawks, or from dead animals to fly maggots to parasitic wasps. Food chains are interconnected in a complex pattern called a **food web** (Figure 38-4).

■ *All the organisms in a community are connected in a complex food web.*

The **trophic level** to which an organism belongs is an indication of how far it is removed from plants in the food chain. Green plants make up the first (producer) trophic level. The second trophic level contains the plant-eating animals (primary consumers), and higher trophic levels are made up of carnivores (secondary consumers, and so forth). An organism cannot always be assigned to just one trophic level. Thus some plants, such as Venus's flytrap, are carnivores as well as producers. A frog tadpole eats diatoms or other plant life, but the adult frog is carnivorous. Many mammals, such as foxes, bears, and humans, are **omnivores,** organisms that belong to several trophic levels because they eat both plants and other animals.

38-C   Pyramids of Energy

The flow of energy through an ecosystem can be represented in the form of a **pyramid of energy,** which shows the total amount of incoming energy for successive trophic levels. These diagrams are smaller at the top because some energy is always lost in going from one trophic level to the next. This loss of energy also limits the number of trophic levels in the ecosystem: in terrestrial ecosystems, there are seldom more than five, although aquatic systems often have more.

■ *Energy from the sun enters a community during photosynthesis. Then it passes through one to five trophic levels in a terrestrial food web.*

Why So Few Trophic Levels?

There are several reasons ecosystems on land have so few trophic levels. First, not all of the food available at one trophic level is actually eaten by animals in the next level.

Second, not all of the food eaten is useful. For instance, the rate at which herbivores grow is often limited by the content of nutrients such as essential amino acids in their food plants. In the process of eating enough food to extract the amino acids they need, they may excrete and "waste" a lot of energy-rich plant material.

A third source of energy loss is respiration to drive the organism's metabolism. In Chapter 6 we saw that energy is lost in every energy transformation. Cellular respiration of glucose, the major body fuel, is only about 50% efficient. More energy is lost as the ATP formed during respiration is used in the maintenance and repair of body tissues; in functions such as feeding, excretion, and circulation; and in behavior. The energy that remains is stored as new **biomass,** material that makes up the bodies of organisms and is available for other organisms to eat (Figure 38-5).

Because of all these energy losses from one trophic level to the next, there is not enough energy left to support higher trophic levels on land. A wolf may have to travel 30 kilometres a day to find enough food to eat, and a tiger requires a home range of up to 240 square kilometres. An animal that fed on wolves or tigers would have to cover a wide hunting area to try to find enough of its widely scattered prey. It is not energetically feasible to try to harvest the small amount of food energy available in the highest trophic level. The organisms that do feast on these top predators are parasitic worms and fleas. They eat only part of the predator and get only a tiny crumb of the ecosystem's energy pie.

TROPHIC LEVEL

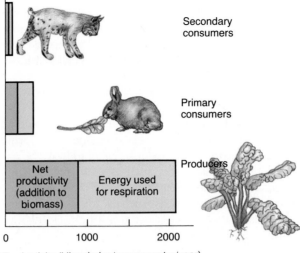

Figure 38-5
Energy flow for an actual ecosystem.
The graph shows total energy input at
each trophic level and its division
into new biomass (green) and energy
used during respiration. This type of
graph is often called a "pyramid of
energy." The pyramid is so much
wider at the bottom than at the top
that it would be much wider than the
page if more than three trophic levels
were shown. (Thom Smith)

38-D   Productivity

The flow of energy through an ecosystem can be measured at various points by
answering questions such as these: How much solar energy is trapped in the
food made during photosynthesis? How much of the energy in plant material
can a herbivore use? How much energy does a herbivore use before it is eaten
by a carnivore? Similar questions may be asked, proceeding from one trophic
level to the next. Let us consider some of these questions.

Primary Productivity

Primary productivity is the rate at which energy is stored in organic matter by
photosynthesis. It is usually measured as increases in energy stored or in biomass. Not all of the organic matter made during primary productivity accumulates as plant biomass: about half is used up in the plant's own respiration. What

Table 38-1 Net Primary Productivity of Some Major Community Types* (In Grams of Dry Plant Material per Square Metre Per Year)†	
Community	**Average net primary productivity**
Coral reef	2500
Tropical rain forest	2200
Temperate forest	1250
Savanna	900
Cultivated land	650
Open ocean	125
Semidesert	90

* See Chapter 37 for characteristics of the various communities mentioned in this table.

† After R. H. Whittaker, *Communities and Ecosystems,* 2d ed. New York: Macmillan, 1975.

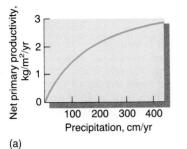

Figure 38-6

Productivity and precipitation. (a) A graph showing how net primary productivity increases with average annual precipitation. (b) The relationship between rainfall and productivity is striking in arid ecosystems, where productivity is limited by scarcity of water. In this scene from Death Valley, plants grow only along a dried-up stream bed. This is where water collects after the infrequent rains, providing enough moisture to sustain plants through the long dry spells. (c) Artificial irrigation supports the much higher productivity of this Death Valley golf course.

(a)

(b)

(c)

remains is the **net primary productivity,** which appears as plant growth (new biomass) and is available to heterotrophs.

Note that productivities are rates. The **yield** of an agricultural crop, by contrast, is the amount of the crop that can be harvested. It is part of the crop's productivity multiplied by the length of the growing season and the area of crop. Nor should productivity be confused with **standing crop,** which is the amount of biomass present at any one time. In some situations productivity is actually higher at a lower standing crop. For instance, a field or lawn that is mowed regularly or grazed by cattle (a low standing crop of grass) can have a higher net productivity than it would have if just left to grow (higher standing crop).

Productivity varies considerably from one type of community to another (Table 38-1). Why should this be? We have seen that plants require a number of resources for growth. If one of these is in short supply, it becomes the **limiting factor** for plant growth. As might be expected, energy is one of the most crucial factors: productivity generally increases from polar regions toward the tropics, because of the increasing sunlight and temperature. Where water is scarce, or nutrients are lacking, productivity is low no matter how much sunlight reaches the area. In dry areas, such as savannas and deserts, water is usually the limiting factor (Figure 38-6). Intensive agriculture can achieve net productivities as high as those of any natural vegetation on land (see *A Journey Into Life,* "Declining Productivity").

Secondary Productivity

Secondary productivity is the rate of formation of new organic matter by heterotrophs (their growth and production of offspring). Herbivores eat some of the plant biomass in any ecosystem, but not all of this becomes secondary productivity. A caterpillar's gut absorbs only about half of the leaf material eaten; the rest is egested as feces. Furthermore, of the food absorbed, about two thirds

Declining Productivity

The annual net primary productivity of the whole planet is about 170 billion tons of dried organic matter. Of this total, nearly 70% is produced on land and only 30% in the oceans (despite the fact that the oceans occupy about 70% of the earth's surface).

Scientists calculate that humans take over about 40% of the earth's primary productivity. We harvest plants for food, we harvest trees and other crops for purposes other than food, and we destroy sources of primary productivity by clearing the plant life from land and then building over it.

Three types of ecosystems—farmland, grazing land, and forest—support the world's economy. With seafood, they produce all the food we eat. With fossil fuels and minerals, they supply all the raw materials for industry. The decline in the total productivity of these ecosystems during the twentieth century threatens our ability to support a still-growing human population.

The amount of the earth's surface covered with these three types of ecosystem is shrinking, while the areas of desert and human settlement are increasing. The area of land used for crops increased steadily until 1981. Since then, new land has been converted to farmland, but just as much land has been lost from farm production by being built on and by soil destruction (see *A Journey Into Life,* Chapter 37). The area of grassland has also shrunk since 1970, much of it converted to desert by overgrazing. Forests in most parts of the world have been shrinking for centuries, but the rate has accelerated since 1980 (Figure 38-A).

To make matters worse, productivity is declining in the ecosystems that remain. One cause of declining productivity of forests and farmland is air pollution. Brown spots on tomato leaves these days do not necessarily result from disease. They may be a result of acid rain or excess amounts of ozone in the air. The U.S. Forest Service reports that the annual growth of yellow pines, an economically important species in the southeastern United States, declined by 30 to 50% between 1955 and 1985. More than 50% of the grazing land managed by the U.S. Bureau of Land Management has also suffered declines in productivity because it is overgrazed.

The earth's primary productivity ultimately limits the number of humans (and other species) it can support. At a time when the demand for that productivity is rising rapidly because of human population growth, the earth's productivity is falling. These two trends cannot continue together indefinitely. At some point, almost certainly within the next 50 years, we shall be asking more of the earth than it can produce. What will happen then?

Figure 38-A
Deforestation in the Himalayan Mountains in India. This photograph is taken from a forest protected for religious reasons. As late as 1970, these mountains were completely covered with forest. The population of the area has doubled since that time and the trees have been cut down for fuel and to make these narrow terraced fields for growing food. Without trees to absorb rainfall and hold the soil in place, much more water than usual runs down the steep hills. This water causes terrible floods downstream in Bangladesh. In the middle of the picture is a gully where a landslide has washed soil from a large area despite the terraces designed to hold the soil in place. Since 1970, soil washed from the deforested Himalayas has formed a new island in the ocean in the Bay of Bengal. (Paul Feeny)

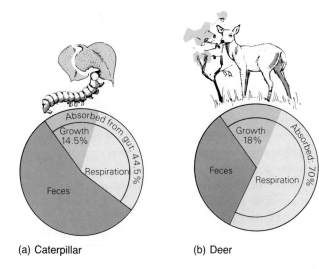

Figure 38-7
Food and energy budgets. The fate of energy in black swallowtail butterfly caterpillars and deer. Energy is expressed as percent of calories taken in as food. The deer is more efficient at absorbing food from the gut, but uses more energy in respiration than the caterpillar because it is warm-blooded. (From data courtesy of J. M. Scriber, P. Feeny, and R. L. Cowan)

(a) Caterpillar   (b) Deer

is used in respiration, so that only about 15% of the food eaten appears as secondary productivity (Figure 38-7). Grasshoppers in a Tennessee field were found to convert only about 4% of the food they ate into secondary productivity. An average figure for animals is about 10%. The average is lower for herbivores and higher for carnivores, because a given weight of meat contains more of the nutrients an animal needs than the same weight of plant food.

Even if the organisms at each trophic level could find, capture, and eat all of the net productivity from the previous level, the tertiary consumers would receive only about $\frac{1}{10} \times \frac{1}{10} \times \frac{1}{10} = \frac{1}{1000}$th of the energy present in the original producers in their food web. It is clear from this that in times of food scarcity, eating meat is a luxury for omnivores, including humans. By adopting a vegetarian diet, it is possible to skip the energy losses at one trophic level so that more people can be fed from a given area of land.

■ *Because energy is lost at each transfer, productivity decreases at each higher trophic level. On average, animals convert only about 10% of the energy they eat into new animal (growth or reproduction).*

38-E   Cycling of Mineral Nutrients

Although an ecosystem's productivity may be limited by the supply of sunlight or water, in many cases it is limited instead by the availability of nutrients.

Living organisms require eight elements in relatively large amounts: carbon, hydrogen, oxygen, nitrogen, potassium, calcium, phosphorus, and sulfur. These elements are present in the environment. Some occur in rocks and are released by erosion and weathering into soil, rivers, lakes, and the oceans. Some are also present in the atmosphere. The movements of nutrient elements through the biosphere, or through any particular ecosystem, by physical and biological processes, are called **biogeochemical cycles.** They are called cycles because nutrient elements, unlike energy, may be used over and over again by living systems.

Nutrients are sometimes recycled rapidly through ecosystems. In other cases, nutrients spend many years apart from the activities of the biological world. For example, remains of marine organisms may sink to the ocean bottom and be incorporated into sedimentary rocks that are lifted and exposed at the earth's surface only after millions of years. Every nutrient element has a somewhat different fate, depending on its physical and chemical properties and on its role in living organisms. We shall illustrate the concept of nutrient cycling with a few simplified examples.

The Carbon Cycle

Carbon is said to move in an **atmospheric cycle** because carbon occurs in the air, as well as in rocks and dissolved in water. Much of the carbon in terrestrial ecosystems travels rapidly between living organisms and the atmosphere as a result of photosynthesis and respiration (Figure 38-8). Most of the carbon fixed as organic matter by photosynthesis is rapidly broken down and released back into the air as carbon dioxide (CO_2) produced in respiration. In aquatic ecosystems, the exchange occurs between living organisms and CO_2 or bicarbonate (HCO_3^-) dissolved in the surrounding water. Another carbon compound, methane (CH_4), is released into the atmosphere by anaerobic organisms, such as those in sewage treatment plants, the bottom mud of wetlands, and the digestive systems of ruminants (such as cattle and sheep).

Some carbon enters a long-term cycle. For instance, deposits of coal, oil, and natural gas were part of the net productivity of ancient ecosystems. These fossil fuels are carbon compounds that were buried before they decomposed, and were then transformed by time and geological processes.

The Greenhouse Effect   The atmosphere is about 2% water vapor (by volume), about 0.03% carbon dioxide, and about 0.0002% methane. But these tiny quantities, together with other "greenhouse gases," are important because they absorb infrared radiation (heat) from the earth. They trap heat within the atmosphere, preventing it from escaping into space (Figure 38-9). Without the greenhouse effect, the earth would be too cold for life. The more of these gases the atmosphere contains, the more heat is retained.

The quantity of methane in the air is increasing, partly because the numbers of ruminants (mainly cattle and sheep) and of sewage plants on earth are in-

■ *Carbon is fixed in organic molecules during photosynthesis, and most of it is soon released by respiration. The carbon balance of the biosphere is moderated by the exchange of CO_2 between the atmosphere and the oceans.*

Figure 38-8
A **simplified carbon cycle.** This diagram shows how carbon moves through several ecosystems. Carbon passes through the processes indicated by dashed arrows more rapidly than through those indicated by solid arrows.

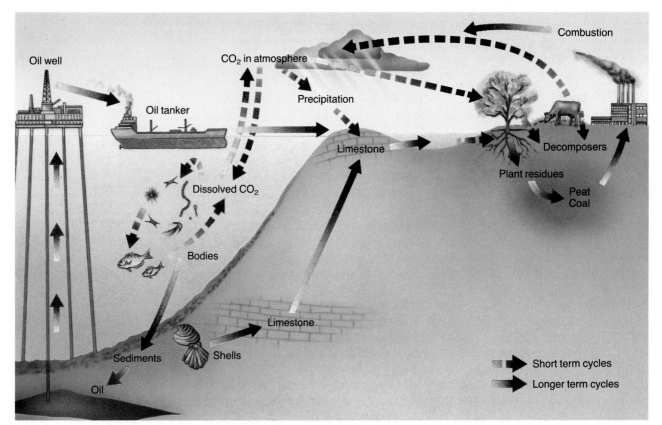

creasing rapidly. The carbon dioxide content of the atmosphere is also increasing. Our burning of fossil fuels and other organic matter releases CO_2 into the air. In addition, widespread **deforestation,** the cutting down of trees, has slowed the rate at which CO_2 is removed from the air. Most of the terrestrial vegetation that absorbs CO_2 during photosynthesis is found in wet parts of the tropics. Most of the world's tropical forests have already been destroyed. When tropical forest is cut down to make way for agriculture, the trees are usually burned to get them out of the way, releasing still more CO_2 into the air. However, this produces only about one-fifth as much CO_2 each year as does burning fossil fuel in cars, power stations, and factories.

The concentration of CO_2 in the atmosphere is increasing much more slowly than would be predicted from all these human activities. The rest of the CO_2 is almost certainly being absorbed by the oceans, which act as a global "sink" for CO_2. It is not clear how much more CO_2 the oceans can hold.

It is difficult to say precisely how much our addition of carbon to the atmosphere will heat up the earth. Both heat and CO_2 are stored in various ways, and many factors besides the greenhouse gases affect the average temperature of the atmosphere. Even with these uncertainties, however, most estimates predict a global warming of 1 to 5°C during the next century. This temperature rise would produce many effects that we are not prepared for. Climate zones and ocean currents would shift, agriculture would be displaced, and the world's major vegetation zones would alter. Another dramatic effect might be thermal expansion of the ocean and further melting of the world's great ice caps, which would raise the sea level.

Even if the extent of global warming cannot be predicted accurately, it would seem prudent to encourage moves that counteract the greenhouse effect as well as being useful in other ways. For instance, by using less fossil fuel, we can reduce air pollution, save money, and reduce emissions of carbon dioxide. Planting trees has the triple advantages of counteracting the greenhouse effect, moderating the local climate, and preventing soil erosion.

Figure 38-9
The fate of solar energy as it passes through the earth's atmosphere. Much of the energy that reaches the earth is reflected back into the atmosphere. Here it is trapped by "greenhouse gases." The more concentrated the greenhouse gases in the atmosphere, the greater the greenhouse effect. (The percentages are approximate, which is why they do not add to 100.)

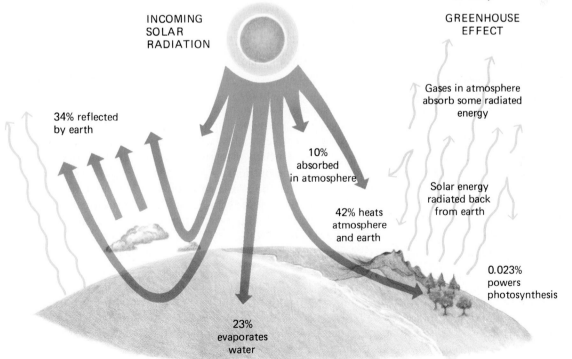

INCOMING SOLAR RADIATION

GREENHOUSE EFFECT

Gases in atmosphere absorb some radiated energy

34% reflected by earth

10% absorbed in atmosphere

42% heats atmosphere and earth

Solar energy radiated back from earth

0.023% powers photosynthesis

23% evaporates water

(a)

(b)

Figure 38-10

Reforestation. (a) Part of an "urban forest" in a parking lot. Many cities employ foresters to plant and care for trees. On a hot day, the temperature under a tree is up to 10°C cooler than in the open, reducing air conditioning costs. The trees also absorb pollutants, as well as rainwater that might otherwise flood streets and storm sewers (which may overload the sewage system). As a result of these benefits, urban reforestation is growing rapidly. (b) A tree nursery in Ghana where trees are bred and grown. Reforestation is vitally important to developing countries, where wood is the main fuel for cooking and heating. (b, Biophoto Associates)

 Nitrogen is abundant in the air, but in a form few organisms can use. The scarcity of fixed nitrogen compared to plants' ability to use this vital resource makes nitrogen a limiting nutrient in many ecosystems.

Humankind is not sitting idle while our forests burn. Reforestation is taking place on a massive scale (Figure 38-10). About 11 million hectares of tropical forest are felled each year, but by 2000 about 17 million hectares of land are expected to be reforested each year. Among the examples: China has doubled its forested area in less than 20 years, New England's forested area is 50% greater than it was a century ago, Chile has doubled its area of pine plantations in 15 years, and in Rwanda, trees planted by rural people now cover 200,000 hectares, more than the area of the country's remaining natural forest and artificial plantations put together.

The Nitrogen Cycle

Nitrogen is a vital part of many essential organic compounds, especially proteins and nucleic acids. About 80% of the atmosphere is molecular nitrogen gas (N_2), which is much more abundant than CO_2. However, whereas all producers use CO_2 directly, only certain species of prokaryotes can use N_2. These bacteria fix nitrogen by reducing it to ammonia (NH_3) (Section 19-B). Plants must get this vital nutrient either directly from nitrogen-fixing microorganisms or indirectly from the limited pool of nitrogen found as ammonium (NH_4^+) or nitrate (NO_3^-) ions in soil water, rivers, lakes, and oceans. Nitrogen fixation is slow compared with the rate at which plants can take up fixed nitrogen, and so nitrogen is often the limiting nutrient in plant growth. Nitrogen is one of the elements most commonly applied in crop fertilizers. This additional input of usable nitrogen into the biosphere is based on the industrial manufacture of ammonia-based fertilizers from atmospheric nitrogen, using catalysts and a lot of fossil fuel energy (Figure 38-11).

There is a second difference between the movements of nitrogen and carbon through an ecosystem. Most organisms release carbon dioxide during respiration, but the conversion of nitrogen from organic molecules into inorganic forms such as nitrates involves steps carried out by a series of different organisms. Certain bacteria and fungi use the proteins and amino acids in dead organic matter for their own dietary nitrogen. They release the excess as nitrogen compounds that can be absorbed by plants.

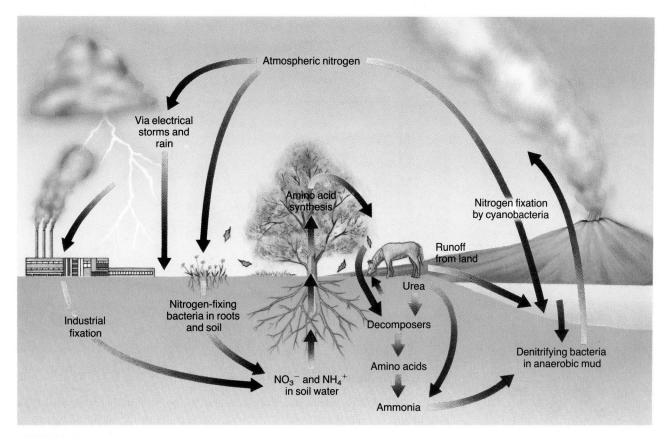

Figure 38-11
The nitrogen cycle.

Nitrogen is returned to the atmosphere by denitrifying bacteria living in the anaerobic mud of fertile lakes, bogs, estuaries, and parts of the ocean floor.

The Phosphorus Cycle

Phosphorus is a major constituent of nucleic acids and of membrane phospholipids, and many animals also need a lot of phosphorus for shells, bones, and teeth. Since phosphorus almost never occurs as a gas, its cycle, unlike the atmospheric cycles of nitrogen or carbon, is a **sedimentary cycle** (Figure 38-12). When rocks are eroded by weather, small amounts of phosphorus dissolve, usually as phosphate, and so become available to plants. Much of the phosphorus excreted by animals is in the form of phosphate, which plants can use immediately. Thus the cycling of phosphorus in terrestrial ecosystems is usually very efficient, although small amounts are continually lost downstream and to the oceans.

The phosphorus cycle is, in the short run, a one-way flow—from rocks to land ecosystems to the ocean, and finally to ocean sediments. The only natural way for phosphorus to return to land is by slow geological processes in which sea floor sediments may again become terrestrial rocks. However, terrestrial ecosystems are able to retain much of their phosphorus since soil particles absorb phosphate ions, helping to provide a steady supply of it for plant growth. Soil erosion robs an ecosystem of its phosphorus, and it may take thousands of years to recoup this loss through the weathering of rocks.

■ *Phosphorus moves through ecosystems in a one-way, sedimentary cycle. It is usually in short supply but tends to be retained efficiently.*

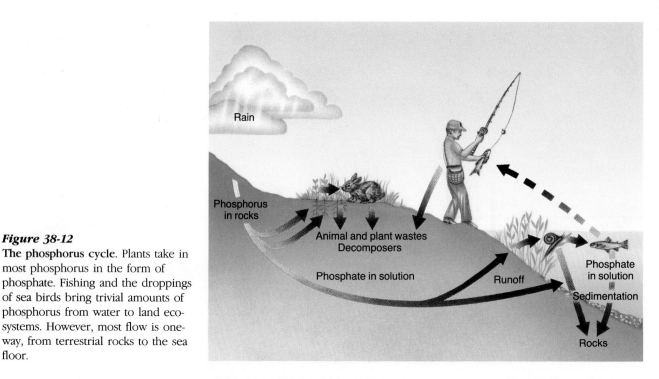

Figure 38-12
The phosphorus cycle. Plants take in most phosphorus in the form of phosphate. Fishing and the droppings of sea birds bring trivial amounts of phosphorus from water to land ecosystems. However, most flow is one-way, from terrestrial rocks to the sea floor.

Labels in figure:
Rain
Phosphorus in rocks
Animal and plant wastes
Decomposers
Phosphate in solution
Runoff
Phosphate in solution
Sedimentation
Rocks

An Experimental Ecosystem

Hubbard Brook Experimental Forest in the White Mountains of central New Hampshire is one of several areas where researchers study nutrient cycles and how they are affected by human activities. The forest consists of a group of valleys, each with its own creek running down the middle.

(a)

(b)

Figure 38-13
Hubbard Brook Experimental Forest. (a) Experiments on how methods of harvesting trees affect the flow of nutrients. In this winter photograph, the white area near the bottom has been clear cut. In the center of the picture is another area where trees have been cut down in horizontal strips. (b) A V-notch weir. All the water that runs off one of the valleys is funnelled through this weir so that its volume can be measured and its nutrient content analyzed. (Peter Marks)

Data from Hubbard Brook reveal that the forest is extremely efficient at retaining nutrients. Nutrients that reach the forest dissolved in rain and snow approximately balance those leaving the forest, washed away in its streams. Both quantities are small relative to the total amounts of nutrients present.

Investigators also examined what happens when a forest is cut down. One winter they cut down all of the trees and shrubs in one of the six valleys. Dramatic effects became obvious almost immediately. The rate at which nutrients left the forest dissolved in the stream water increased six- to eightfold. Other experiments revealed that nutrient losses are reduced if the forest is cut in horizontal strips, leaving strips of standing trees, rather than being clear-cut (Figure 38-13). The remaining trees absorb many of the nutrients that enter the soil water after their neighbors are felled.

■ *Nutrients recycle indefinitely in a natural ecosystem, with small losses into the air and in runoff and sedimentation. Disturbances can increase an ecosystem's loss of nutrients enormously.*

38-F Species Diversity

We turn from mineral nutrients to the types of organisms they sustain. Communities of organisms differ in their **species diversity,** the number of species they contain. The simplest measure of such diversity is the number of species in a given area. Often there are a few dominant species, which are especially abundant or obvious in the ecosystem, and a host of species that are less common.

We might expect that one kind of decomposer would become so efficient at digesting and absorbing dead matter that it would drive competing species to extinction. Competition between species for limited resources sometimes does lead to the extinction of one of them, but this is not always the case. What maintains species diversity? What prevents one or more species in a trophic level from eliminating the others through competition? One answer is that when species are in potential competition for a food supply, they may subdivide this resource in some way. Different insect-eating birds, for instance, search for food at different heights in a forest. An important reason for species diversity within a particular community, therefore, is specialization in the use of limited resources.

A second contribution to the overall species diversity of a region is the variety of habitats it contains, each characterized by its own diversity. Species diversity may vary with height above sea level, soil type, and so forth. Another cause of diversity is the creation of different habitats within a region by periodic disturbance. Light gaps in a forest may be inhabited by different species of birds and insects from those in the nearby climax forest.

Predation also helps to maintain species diversity. This was shown in a study of an intertidal community on the rocky coast of Washington, where mussels, barnacles, limpets, and so forth, were all fed on by the food web's top carnivore, a sea star. When the sea stars were removed from the rocks, barnacles settled and took up about 80% of the available space within three months. Later, the barnacles were gradually crowded out by two species of mussels. After a year or so, one mussel species dominated the experimental area, and the number of species present had dropped from 15 to 8. Thus, in the natural community, the sea star predator maintains diversity by preventing some of its prey species from excluding the others by competition for space. Grazing by herbivores sometimes has similar effects. In several experiments, herbivorous mammals were fenced out of areas of grassland. A few species of grasses and herbs took over more and more of the area and the species diversity was reduced.

Specialized herbivorous insects can contribute to diversity. In tropical forests, almost all of the seeds and seedlings of some plant species die if they remain near their parents, because they are attacked by insects. Only if seeds are dispersed some distance do they have a reasonable chance of escaping the

Figure 38-14
Species diversity. Plants at the edge of tropical rain forest in the Caribbean show some of the incredible diversity found in this species-rich community.

insects near the parent plant. This sort of predation probably contributes to the enormous diversity of species in tropical forests (Figure 38-14). Species that become common are most easily discovered by enemies, which quickly eat them so that they become rare again, leaving space for other species.

Species Turnover in Communities

A 1968 survey of the number of species of birds breeding on each of the nine Channel Islands off the coast of southern California showed that the total number of species on each island had changed little since a similar survey 50 years earlier. However, the species composition had changed markedly. On San Nicolas Island, for example, there were 11 species of birds in 1917 and the same number in 1968, but only five of these species were the same. Six species had disappeared from the island, but six other species had colonized it. It seems that each undisturbed habitat can always support the same number of species, although the particular species present may change from time to time.

Ecologists tested this hypothesis experimentally. First, they counted the number of species, mainly insects, on six small mangrove islands in Florida. Next, they completely exterminated all of the arthropods on these islands by enclosing the islands in enormous plastic tents and fumigating them (Figure 38-15). They then removed the tents and sampled the animals regularly. Within six months, the number of species on each island had returned to approximately the number present before fumigation, and remained at about the same level for as long as the islands were observed. As predicted, however, the species compositions were not identical to those before the experiment, and species turnover continued on each island, with losses approximately balanced by immigration.

Species turnover is not restricted to island communities. Turnover is slow in many communities on continents because populations are large and any losses are easily replaced by immigration from nearby areas. Small, isolated communities in places such as mountain tops, bogs, and lakes, however, may have appre-

Figure 38-15
Measuring species diversity. Scaffolding completely encloses a small mangrove island in the Florida Keys, in preparation for enclosing the island with plastic sheeting. After elimination of the island's insects by fumigation, tent and scaffolding were removed and the process of recolonization of the island was monitored. (Daniel Simberloff)

ciable species turnover. An understanding of island communities promises to be a valuable tool in the design of parks and nature reserves, artificial "islands" where decisions must be made as to what area of habitat should be preserved and whether more species will be saved if the park contains one large or several smaller pieces of the habitat.

38-G Air and Water Pollution

All organisms expel their wastes into the environment. However, in a balanced ecosystem, one organism's wastes are another's food and drink. Decomposers break down waste materials, and so keep them from accumulating and polluting the environment.

Pollution is an undesirable change in an ecosystem's physical, chemical, or biological characteristics. It results when wastes accumulate, making the environment less hospitable to the life of humans and other organisms. Various forms of pollution have been a problem for human societies for thousands of years. But the problems have accelerated in recent years as a result of increasing numbers of people and our increased use of fossil fuels. Pollution cannot be avoided entirely, but it can be minimized. This requires an understanding of its effects on ecosystems. Here we consider a few examples.

Ozone and the Ozone Layer

Ozone (O_3) is a gas that forms in the atmosphere both naturally and as a result of human activities. When it occurs in smog near the ground, it is a dangerous pollutant. When it occurs high in the atmosphere, it is a shield to be preserved. There is little mixing between layers in the atmosphere.

Cities where motor vehicles are the main source of pollution tend to suffer from yellowish **photochemical smog.** Nitric oxide from car and truck exhausts reacts with oxygen to form nitrogen dioxide, a yellow-brown gas that produces a colored haze and has a strong, choking smell. Ultraviolet rays from the sun release an atom of oxygen from nitrogen dioxide. Such oxygen atoms react with oxygen gas in the air to produce various compounds, including ozone. Smog also contains organic compounds, such as industrial solvents and spilled or unburned gasoline, which rise from the ground during the day.

Ozone in the atmosphere is also formed by photochemical reactions. High in the atmosphere, ultraviolet radiation from the sun is intense enough to form small amounts of ozone from oxygen. The resulting **ozone layer** lies more than 30 kilometres above the earth. Once formed, ozone itself is a good absorber of high-energy ultraviolet radiation, and so it reduces the amount of this radiation that reaches the earth's surface. Ultraviolet radiation is dangerous to living organisms because it damages DNA. Large amounts of it can cause skin cancer in humans and kill many microorganisms outright.

We are destroying the ozone layer, permitting increasing amounts of ultraviolet radiation to reach the earth. A large, and growing, hole in the ozone layer over Antarctica was discovered in the 1980s. Damage to the ozone layer comes mainly from chlorofluorocarbons (CFCs), gases used in the production of aerosol sprays, in refrigeration, and in air conditioners. CFCs are ideal aerosol propellants because they do not react chemically with whatever is being sprayed. But this same lack of chemical reactivity means that they are not broken down in the air. They make their way to the upper atmosphere, where they are broken down by the intense ultraviolet light, releasing free chlorine atoms that react with ozone and break it down.

A number of countries have endorsed agreements to cut world use of CFCs, but these plans are far from being complete solutions. Our best hope probably

■ **We are destroying the ozone layer in the atmosphere, and this permits increasing amounts of ultraviolet radiation to reach the earth.**

lies with large chemical companies, which are working to produce substitutes for CFCs. Industry officials are optimistic that these will be available soon.

Acid Rain

Since 1970, scientists have documented the death of more than 7 million hectares of trees in once-thriving forests in Europe and North America. In parts of Germany, more than 50% of the forest is affected and the timber lost is worth more than $1 billion a year.

These trees have died of exposure to **acid rain,** rain made highly acidic by pollutants in the air. The most important such pollutants are nitrogen and sulfur oxides, which form acids when they dissolve in rain. In both Europe and America, the most severely damaged trees are those on high slopes facing the prevailing winds, particularly on slopes often covered with clouds and fog. The condensed moisture of clouds and fog subjects trees to concentrated doses of the pollutants that kill them (Figure 38-16a)

The pH of "normal" rain is somewhat acidic, about 5.6, because carbon dioxide and other gases form acids as they dissolve in rainwater. Since the mid-1960s, however, rain throughout the world has become increasingly acidic, with a pH averaging 4.0 to 4.2, and sometimes reaching 3.0 or even lower (recall that the pH scale is logarithmic; *a drop of one unit means a tenfold increase in hydrogen ion concentration*).

At first, only areas to the east of industrial centers were affected by acid rain because prevailing winds blow from west to east in the industrialized northern hemisphere. Now, the rain everywhere on earth is more or less acidic.

Acid rain does untold damage. It dissolves paint and stone from buildings, kills trees, crops, and garden plants, and destroys natural ecosystems. When the pH of a lake falls below about 5, the fish die. Norway and some other countries spend millions of dollars every year spreading lime ($CaCO_3$) on their lakes to raise the pH to a level that will support animal life.

The sulfur and nitrogen oxides found in acid rain come mainly from industrial pollution, automobile exhaust, and natural sources such as volcanic eruptions and wetlands (where anaerobic bacteria produce sulfurous gases). Since proteins contain sulfur and nitrogen, burning any organic substance releases acid compounds into the air. In many states, wood-burning stoves are required to have catalytic combusters to reduce pollution from this source. The only cure for acid rain would be stringent pollution control on every form of combustion.

Figure 38-16
Damage caused by acid rain.
(a) Trees killed by acid rain in a North Carolina forest. Note that there are many more dead trees on one side of the mountain—the side that faces the prevailing wind and rain.
(b) Acid rain has dissolved part of the stone faces of this group of carved figures on a church in Paris. (a, Carolina Biological Supply Company)

(a)

(b)

Figure 38-17
The effect of phosphorus on the productivity of a lake. This lake in Manitoba was divided in two by plastic sheeting across the narrow neck in the middle. Phosphorus was added to the half of the lake in the upper part of the photograph. Several weeks later, the phosphorus-fertilized half of the lake was opaque as a result of massive plankton bloom. The lower part of the lake remained clear and oligotrophic. (David Schindler)

Water Pollution

Acid rain is only one of the substances that is polluting our water. For instance, toxic chemicals are a common cause of pollution in lakes. Chlorinated hydrocarbons, such as DDT and the polychlorinated biphenyls (PCBs), and heavy metals such as mercury and lead, are a danger to natural ecosystems, just as they are to humans.

Organisms have only limited ability to excrete these substances, so they accumulate and are passed on to predators higher in the food chain. In aquatic ecosystems, they are especially damaging to carnivores such as fish-eating birds. For example, DDT causes birds to lay eggs with thin shells prone to break when the bird sits on them. DDT persists in the environment for a long time. Although it is now banned in the United States, and is no longer found in high concentrations, it is still used in other countries. It has dispersed all over the world and can be detected even in the Antarctic, thousands of miles from where it was originally used.

Less obvious forms of pollution, at first sight, are nutrients and heat. Plant nutrients may enter the ecosystem in a form plants can use or may be released from organic pollutants by decomposers. Phosphorus is a good example. Phosphorus is one of the elements most likely to limit the productivity of natural ecosystems, including lakes (Figure 38-17). Many lakes and rivers receive phosphorus continuously from sewage, high-phosphate detergents, and fertilizer washed off farmland, lawns, and gardens. In these ecosystems, productivity remains high and eutrophication is hastened, changing the character of the lake (Section 37-G). Such changes have fundamentally altered Lake Erie, which is relatively shallow. Even deep oligotrophic lakes, such as Lake Tahoe, have become noticeably more eutrophic in the last 20 years because of nutrient pollution.

Oligotrophic lakes with clear water are much more appealing and useful than eutrophic lakes covered with algae, clogged with weeds, and stinking from the by-products of anaerobic bacteria. For this reason, there has been strong pressure to ban the use of detergents containing phosphorus and to speed the installation of tertiary sewage treatment plants, which remove nutrients from the water. (If you want to help, most liquid detergents contain no phosphorus.) In some cases, these measures have brought lakes back toward a more oligotrophic condition.

Thermal pollution results from adding excessive amounts of heat to a lake—for example, in the waste water from the cooling towers of an electrical power station. This increases the production of organic matter in the lake, which in turn means more work for the aerobic decomposers. Aerobic decomposers use up oxygen when they break down organic matter, leaving less oxygen for fish and other aerobic organisms. The lake's oxygen may all be used up, leading to the same kind of anaerobic conditions that result from nutrient enrichment.

■ *Pollution degrades our environment when we expel wastes faster than they can be broken down or diluted by ecosystems' natural processes.*

Salt Marsh and Seafood

A JOURNEY INTO LIFE

Revellers at clambakes, oyster roasts, or shrimp boils seldom pause to consider that their favorite seafood is on the menu only because the local salt marsh has not yet been converted to highways or condominiums.

Salt marsh dominates much of the flat shoreline of the Gulf of Mexico and Atlantic coasts of the United States. Here rivers, such as the Mississippi, divide into numerous channels and deposit their load of mud and sand as they wend slowly across huge marshy deltas to the sea. Smooth cordgrass *(Spartina alterniflora)* covers hundreds of thousands of hectares of the marsh, with widely scattered stands of other plants on higher, dryer ground (Figure 38-B).

Cordgrass is the mainstay of an ecosystem that is unusual in two ways. First, it contains few species of plants. Salt marshes are flushed twice a day by the tide, and so the soil is made up of wet, salty mud and sand. Few other flowering plants can tolerate these conditions, but cordgrass thrives in them. Second, few animals can eat cordgrass because its leaves contain quantities of glass-like silica.

So, although cordgrass is a major producer in the ecosystem, its energy and nutrients pass to consumers indirectly, by way of decomposers. Because the soil is constantly flushed by water full of nutrients and oxygen, the cordgrass grows rapidly, and salt marsh is a very productive ecosystem. Every year, the dead stalks of last year's cordgrass break off and wash up high into the marsh, where they are broken down by decomposers into **detritus,** molecules and small particles of organic matter. The tide then washes the detritus slowly down toward the many creeks that meander through the marsh. The organic matter provides nutrients for a large biomass of microscopic algae, and food for crabs, mussels, worms, snails, and clams that live in the soil and among the cordgrass stems. At the edges of creeks, oysters feed on detritus and algae that they filter out of the water. All these invertebrates, in turn, feed human collectors, raccoons, marsh rats, and a host of wading birds (Figure 38-C).

Figure 38-B

Springtime at low tide in a Georgia salt marsh. This year's young cordgrass is especially tall and green where nutrients are most plentiful at the edges of creeks in the foreground. The sandy patch in the middle is high marsh, which is seldom flushed by the tide and is, therefore, so salty that even cordgrass cannot grow on it. The decaying brown stalks of last year's cordgrass are visible, especially in the background. (Thom Smith)

Figure 38-C
A marsh resident. A blue heron wading in a Florida salt marsh strikes at one of the invertebrates or small fish on which it feeds.

The marsh is the nursery where a number of transient animals find food and protection while they are small. For instance, southern commercial shrimp live on the ocean floor as adults. They lay their eggs on the bottom, and the young move in with the tidal currents until they reach the marsh, where they grow to adulthood (Figure 38-D). Other commercially important species that spend part of their lives in the marsh include the young stages of several species of crabs, as well as fish such as flounders, menhaden, and sea trout. As they grow to maturity, these animals migrate down the creeks and out to sea, where they may end up as seafood on the table or as food for ocean fish such as swordfish, snapper, grouper, and tuna.

The economic importance of salt marsh was not widely recognized until recently, after large areas of this ecosystem had been destroyed. In Louisiana, which contains about 40% of the United States's remaining coastal wetland, one hectare of marsh is lost every *hour*. (It is built on,

washed away, or sinks beneath the rising sea.) Led by ecologist Eugene Odum, Georgia was among the first states to try to save its salt marshes. A 1970 act protects the marsh against all incursions except road-building and the Federal Government (which has destroyed vast areas of marsh with "flood control" projects and military bases).

Legislators are particularly concerned about protecting the marsh's roles in seafood production and pollution control. Salt marsh has a considerable ability to absorb and detoxify pollutants. A 1987 oil spill in the Savannah River caused considerable damage in the river itself, but closed the oyster beds for only a few days when it leaked into the nearby salt marsh. Large quantities of inadequately treated sewage, farm fertilizer runoff, and even heavy metal industrial waste, which kill fish and eutrophy lakes in other places, are absorbed by bottom mud, or broken down by organisms, and rendered harmless by the marsh.

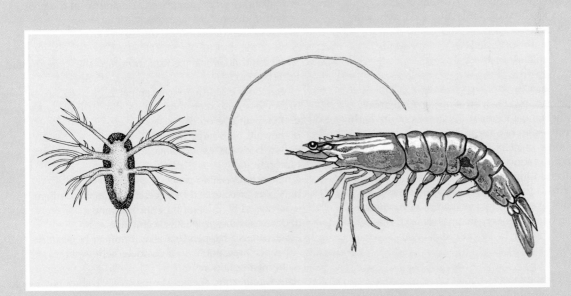

Figure 38-D
The southern pink shrimp *(Penaeus duorarum)*. Left, the nauplius larva; right, the adult. The nauplius hatches from an egg on the sea floor and makes the long journey up a creek into the marsh. Here it develops into a small shrimp, which returns to the sea. These shrimp grow to about 20 centimetres long. Every year, millions of dollars' worth of them are caught for the dinner table by trawlers. (Carol Johnson)

SUMMARY

An ecosystem consists of a community of organisms that depend on one another in various ways, plus their physical environment. Two of the most important factors determining the character of an ecosystem are its productivity and nutrient cycles.

The productivity of an ecosystem is determined by the temperature and by the availability of light, water, and minerals for photosynthesis. Energy flows one way, from one trophic level to the next, with only about 10% passing to each successive level. In consequence, the biomass that an ecosystem can support at each successive trophic level declines rapidly.

Nutrients cycle through an ecosystem. They are taken in by organisms as inorganic substances, many of which are incorporated into organic molecules. Nutrients may pass through the food web for a time, but eventually they are once again released into the environment as inorganic substances. The availability of nutrients often limits the productivity of an ecosystem, even though an ecosystem may be very efficient at conserving and recycling its nutrients.

Species diversity in a community is maintained by the ability of potentially competing species to subdivide limited resources (such as food) within habitats, and to specialize in using slightly different habitats. Species diversity may also be maintained by predation, which can prevent one prey species from eliminating others by competition.

The study of islands shows that there is species turnover in communities, with new species moving in and resident species becoming locally extinct. The number of species varies from one community to another depending on the size of the area the community occupies, diversity of habitats in this area, and availability of colonists from nearby areas.

The pollution of our water, air, and land results from the vast number of people on earth today, and from our use of fossil fuel to power modern technology. Pollution results when wastes are discharged into the environment faster than natural processes can disperse them or break them down.

Nutrient pollution may promote the growth of some organisms at the expense of others, leading to an imbalance that destroys the character of the ecosystem. Thermal pollution may increase growth rates of some organisms while making oxygen less available for respiration.

SELF-QUIZ

1. The role of decomposers in an ecosystem is ____.
2. Using the items listed below, diagram a food web; indicate the trophic level of each organism.

   deer   herbivorous insect
   soil bacteria   spider
   shrub   sparrow
   wolf   hawk

3. The annual primary productivity of any ecosystem is greater than the annual increase in biomass of the herbivores in that ecosystem because:
   a. plants are more efficient than animals in converting energy input to biomass
   b. energy is lost during each energy transformation
   c. there are always more plants than plant eaters
   d. woody plants live much longer than most herbivores
4. Of the total amount of energy that passes from one trophic level to another in a food chain, about 10% is:
   a. transpired
   b. "burned" in respiration
   c. stored as body tissue
   d. reradiated in the form of heat
   e. passed out in the feces
5. Nutrient cycles may involve:
   a. movement of the nutrient from the organism to the atmosphere
   b. movement of nutrients into the soil

   c. limitations on the number of organisms in the ecosystem due to shortage of some nutrients
   d. loss of the nutrient from the ecosystem
   e. all of the above
6. Wolves and lions may be said to occupy the same trophic level because:
   a. they both eat primary consumers
   b. they both use their food with about 10% efficiency
   c. they both live on land
   d. they are both large mammals
   e. they both eat a wide range of dietary items
7. Eutrophication:
   a. is a nonreversible process
   b. is often accelerated by excessive phosphorus input
   c. is caused by inhibition of algal blooms
   d. decreases the productivity of a lake
   e. need never happen to a lake if human beings take proper precautions to avoid detergent, sewage, and thermal pollution
8. State whether the species diversity of a community would be likely to increase or decrease under each of the following conditions:
   ____ a. increased frequency of disturbances
   ____ b. extermination of a predator that preys on many other species
   ____ c. partitioning of a shared resource among competing species

9. Which of the following is a likely result of deforestation?
   a. The amount of CO_2 removed from the atmosphere is reduced.
   b. Wind blows the soil away because its plant cover has been removed.
   c. Water washes off the land more rapidly, causing floods.
   d. Water washes soil into rivers and streams, causing them to silt up.
   e. All of the above.

QUESTIONS FOR DISCUSSION

1. Is more energy lost from an ecosystem when an herbivore eats a plant or when a carnivore eats an animal? Why?
2. How is the flow of energy in an ecosystem linked to the flow of nutrients? How do energy and nutrient flow differ?
3. The productivity of an ecosystem increases during the course of ecological succession (Section 37-H). Why is this so?
4. Robert MacArthur found that the number of bird species in forests is correlated not with plant species diversity but with the amount of layering of foliage at different heights. Can you account for these findings?
5. How can frequent fires increase the species diversity of a region?
6. Suppose you were a member of a congressional committee developing legislation to set aside 100,000 acres of land within a particular region for a biological preserve or national park. The committee must decide whether to recommend government purchase of one large area or several smaller areas. What would your advice be if the prime objective was to preserve (1) an endangered large mammal population (such as the Texas red wolf), (2) as many species as possible, (3) as many local habitats as possible, and (4) the best possible compromise of these?
7. What actions can you take to improve the quality of the environment, as an individual? As a member of an organization? As a voter?

SUGGESTED READINGS

Anderson, J. M. *Ecology for Environmental Sciences: Biosphere, Ecosystems and Man.* London: Edward Arnold (Publishers) Ltd., 1983. The effects of pollutants on the workings of ecosystems, populations, and communities.

Cohn, J. P. "Gauging the biological impacts of the greenhouse effect." *BioScience* 39(3):142, 1989. The high points of a World Wildlife Fund conference on the probable effects of global warming on individual plants and animals and upon the probable distribution of organisms.

Elkington, J., J. Hailes, and J. Makower. *The Green Consumer.* New York: Viking Penguin, 1990. A guide to doing your bit for the environment.

Hillary, E. *Ecology 2000: The Changing Face of Earth.* New York: Beaufort Books, Inc., 1980. Readable account of the world's major environmental problems.

Houghton, R. A., and G. M. Woodwell. "Global climatic change." *Scientific American,* April 1989. Presents the evidence that the greenhouse effect is already warming the earth and discusses how to control it.

MacEachern, D. *Save Our Planet: 750 Everyday Ways You Can Help Clean Up the Earth.* Washington, DC: Dell, 1990

Myers, N., ed. *The Gaia Atlas of Planet Management.* London and Sydney: Pan Books, 1985. Our environmental problems portrayed in the form of maps. Some excellent illustrations and loads of interesting information.

Populations

CHAPTER

39

Fungia coral

Cacti, Apache Trail, AZ

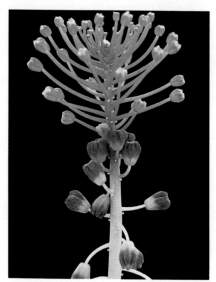

Liliaceae

Elephants drinking

Walruses

*B*iologists study the nonliving components of an ecosystem, such as climate, energy input, and nutrient flow, but they are most interested in the populations that make up the ecosystem's community of living organisms. A **population** consists of all the members of a species occupying an area at the same time. Examples are the bass population of a lake and the human population of the United States. In our study of evolution, we saw that each population evolves adaptations suited to its own locale. Hence the population is the unit that evolves. It is also an ecological unit playing a role in the community of organisms in an ecosystem.

A population has characteristics not found in its individual members. For example, each population has a gene pool and a certain pattern of distribution, density, and age structure. Population **density** is the number of individuals per unit area or volume—the number of bison per hectare, for example (Figure 39-1). The age structure of a population is the percentage of individuals of each age. These and other features can be used to describe populations and to predict their fates.

39-A Population Growth

A population gains individuals by birth and immigration, and loses them by death and emigration. Whether a population grows, declines, or remains the same size, depends upon the balance between these factors:

Change in population size = (Births + Immigration) − (Deaths + Emigration)

Figure 39-1
A population. A herd of bison grazing on a western stream bank.

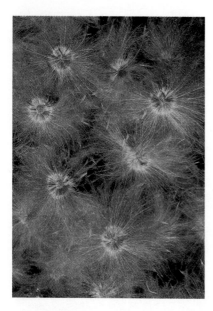

Figure 39-2
Each mountain avens flower produces many seeds. However, most of these seeds never grow into new plants, and so the population does not grow at its maximum possible rate.

■ *Age at first reproduction is the most important factor determining a species' biotic potential.*

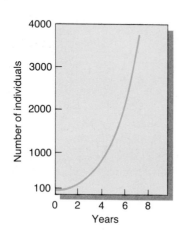

Figure 39-3
Exponential growth of a population. The population grows by adding an ever-increasing number of individuals per unit time. (The equation describing this graph is given in the box on page 799.)

The size of a population of mice in a field or of violets in a woodlot seems, at first sight, to vary little from year to year. Is this really the case? Surely organisms produce so many offspring that their populations could increase greatly from one year to the next (Figure 39-2). What limits the size of natural populations?

To answer these questions, it is convenient first to find out how rapidly populations could increase if nothing stopped their growth. (The mathematics of population growth is described in the box on page 799.)

A population's **biotic potential** is its fastest possible rate of growth, given ideal conditions. In nature, conditions are seldom perfect. Predators, disease, or food shortages nearly always prevent populations from growing as fast as their biotic potentials would permit. However, there are times when populations grow very fast.

The Russian ecologist G. F. Gause studied the growth of populations of the protist *Paramecium caudatum.* Every few hours a well-nourished paramecium divides into two new individuals. Gause set up tubes containing plenty of bacteria for food and introduced one paramecium into each. He observed that the population of *Paramecium* in each tube showed **exponential growth,** that is, growth in which an ever-increasing number of new individuals is added in each unit of time. Exponential growth, plotted as a graph, produces a J-shaped curve (Figure 39-3).

Populations sometimes grow exponentially when they have a superabundant supply of resources. History records many cases of exponential growth by species imported, by accident or on purpose, into areas where resources were available and natural enemies or competitors were lacking. For example, dandelions, starlings, and water hyacinths introduced into the United States underwent dramatic population explosions. Similar population explosions occur when bacteria invade the intestinal tract of a newborn animal or when decomposers invade a freshly dead animal or plant.

Exponential growth does not necessarily mean that the population is growing at its biotic potential. The human population started growing exponentially in the mid-eighteenth century, although women were not bearing infants as frequently as is biologically possible. In addition, many people who could have reproduced did not.

Biotic potential differs from one species to another. An individual's contribution to population growth can be increased in any or all of three ways:

1. By producing more offspring at a time (that is, a larger litter size).
2. By having a longer reproductive life, so that it reproduces more times.
3. By reproducing earlier in life.

Of these three factors, the last is by far the most important. A bacterium does not live for long, and it produces only two offspring at a time. Nevertheless, a population of bacteria has a higher biotic potential than does a population of dogs. This is because most bacteria can reproduce within an hour after being formed, whereas a dog cannot reproduce until it is about six months old. The shorter the generation time of a species (that is, the younger its members when they first reproduce), the higher its biotic potential (Figure 39-4).

Lamont Cole, who drew attention to the significance of age at first reproduction, calculated that a woman who bears three children, one a year starting at age 13, contributes as much to the growth rate of a human population as a woman who bears five children but starts at age 30.

Reproductive Strategies and Survivorship

In each species, the number of offspring produced and the age at first reproduction result from natural selection. These and other factors are part of the species' **reproductive strategy.** There are two extremes in reproductive strategies. At one extreme, parents may produce many small individuals, and provide each

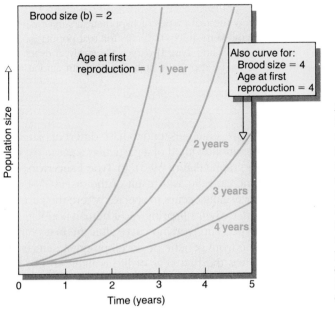

Figure 39-4
How (female) age at first reproduction affects population growth. In all these curves, females produce two offspring per year, but the age at which females first reproduce differs in each curve (first reproduction at 1, 2, 3, or 4 years of age). The boxed note indicates that changing the age at first reproduction from 4 to 3 (one year younger) has the same effect as *doubling* the brood size from 2 to 4. (After Cole, *Quarterly Review of Biology* 29:103, 1954)

with little food or parental care. At the other extreme, an organism may produce very few offspring but invest a lot of energy in each one (Figure 39-5). As we might expect, the rate of survival among the young tends to be higher in the second case. Where many small offspring are produced and receive no care, most of them die at a very young age. Because the parents have produced so many offspring, however, there is a good chance that some will survive to become parents of the next generation.

(a)   (b)

Figure 39-5
Reproductive strategies. (a) A fern produces thousands of tiny spores in the many sporangia on the undersides of its leaves. Some of the spores have caught in a spider's web as they were shed into the air. The spores are at the mercy of air currents to deposit them where they can grow. Needless to say, many do not survive. (b) By contrast, birds such as this warbler produce only a few young but invest a lot of energy in each: birds spend energy building the nest, the female invests a great deal of food energy in the production of each egg, and the parents feed the young after they hatch. (b, Biophoto Associates)

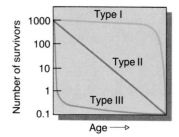

Figure 39-6
Survivorship curves. These graphs show the main types of hypothetical survivorship curves (note the logarithmic scale on the vertical axis). Curves for real populations may fall between those shown here. Type I and Type II curves are rarely, if ever, exactly as shown, because there is always an especially high rate of death among very young individuals.

Species with different reproductive strategies tend to have different patterns of **survivorship,** the length of time an individual of a particular age can expect to survive. This can be shown by a graph (Figure 39-6). In Type I survivorship, most individuals live for a long time, and die as a result of the diseases of old age. This type of survivorship curve usually occurs in species where the parents devote considerable energy and care to their offspring. Most human populations in developed nations approach a Type I survivorship curve after the first year of life (when there is a high death rate from genetic or developmental defects or birth accidents). A baby who survives the first year of life is likely to live for another 60-plus years (Figure 39-7).

In Type III survivorship, most individuals die young, as spores, eggs, or larvae. This type of survivorship is characteristic of many species of invertebrates, bony fishes, plants, and fungi, where large numbers of offspring are produced but not cared for.

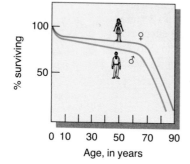

Figure 39-7
Survivorship curves for Americans, based on the 1980 census data. Human survivorship approximates a Type I curve.

Table 39-1   Life Table for the Total Population of the United States, 1986*

Age interval	Of 100,000 born alive		Average remaining lifetime
Period of life between two ages (in years)	Number living at beginning of age interval	Number dying during age interval	Average number of years of life remaining at beginning of age interval
0–1	100,000	1,036	74.8
1–5	98,964	234	74.6
5–10	98,730	106	70.7
10–15	98,624	185	65.8
15–20	98,439	480	60.9
20–25	97,959	590	56.2
25–30	97,369	616	51.5
30–35	96,753	752	46.8
35–40	96,001	961	42.1
40–45	95,040	1,392	37.4
45–50	93,648	2,137	32.9
50–55	91,511	3,438	28.5
55–60	88,073	5,361	24.3
60–65	82,712	8,297	20.4
65–70	74,415	9,004	16.8
70–75	65,411	12,200	13.6
75–80	53,211	14,432	10.7
80–85	38,779	14,473	8.1
85 and over	24,306	24,306	6.0

*Data courtesy of the National Center for Health Statistics.

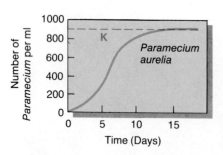

Figure 39-8
Growth of a *Paramecium aurelia* population in a culture medium to which Gause added a constant supply of bacteria as food each day. The population grew exponentially at first but then the growth rate decreased and eventually reached zero, so that the population size levelled out as its numbers approached the carrying capacity, K (red line). (The equation for these data is described in the box on page 799.)

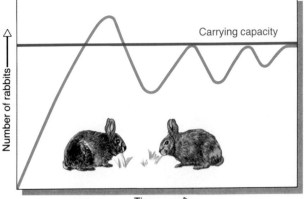

Figure 39-9
Carrying capacity. The red line shows how the size of the rabbit population changed after rabbits were introduced into Australia from Europe. At first the number of rabbits grew exponentially. Then the population crashed. Now, the population size oscillates around a value that represents the carrying capacity of the environment for rabbits.

The Type II curve falls between Types I and III. There is again an initial period of high mortality (death) due to defective genes or accidents during development, birth, or hatching. However, once past this critical period, an individual is just as likely to die at any age. The chances of dying or being killed are equal throughout life. This type of curve is typical of several birds and of human beings exposed to poor nutrition and hygiene.

Survivorship within a population can also be represented in the form of a life table, a summary of the likelihood of death in groups of individuals of each age (Table 39-1). Life tables for human populations are used by life insurance companies to predict how much longer people of a given age are likely to live. This determines the price of insurance for people of various ages. Life tables for populations of animals and plants are useful aids for summarizing and analyzing the effects of different causes of death acting on populations.

Carrying Capacity

No population can grow exponentially for long. Gause found that his *Paramecium* populations eventually stopped growing (Figure 39-8 and the box, "Changes in Population Size"). They had reached their environment's **carrying capacity,** the number of individuals that this particular environment could support indefinitely. When a population reaches this size it may stabilize, with fluctuations above and below the carrying capacity (Figure 39-9).

Carrying capacity is determined by many factors, including predation, competition, and climate. The factors that limit a population's growth may change, and so the carrying capacity of any area for a population of a given species also changes with time.

39-B   Regulation of Population Size

The number of individuals in a natural population varies, sometimes dramatically. In spite of fluctuations, however, the average size of most large populations changes relatively little over the years (Figure 39-10). This suggests that population sizes are usually regulated in such a way that small populations grow quickly, larger populations grow more slowly, and still larger populations decline.

One reason for this is that at least some of the **mortality factors** (causes of death) affecting populations are **density-dependent.** That is, as the population density increases, these factors kill a larger *proportion* (not just a larger number) of individuals. Several kinds of mortality can act in a density-dependent manner. For instance, predation and disease are density-dependent factors,

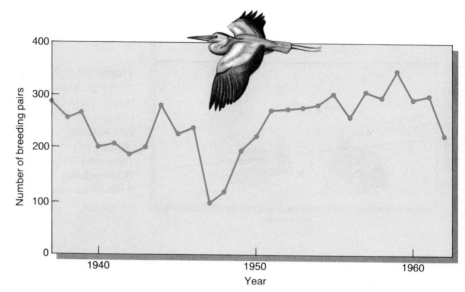

Figure 39-10
A stable population. The number of breeding pairs of gray herons in part of northwest England fluctuated little over 25 years. The population recovered rapidly from the severe winter of 1947. Fluctuations about the carrying capacity are small compared with the heron's reproductive potential (three new birds per breeding pair per year). (After Lack, D. *Population Studies of Birds.* New York: Clarendon Press, 1966)

Figure 39-11
Competition in a dense population. These northern fur seals compete for space on a limited number of breeding beaches. Lack of space on breeding beaches limits the rate at which the population can grow and the ultimate size of the seal population. It is also obvious that disease or a predator (such as a human hunter) would kill a higher proportion of this dense population than of a less dense one. (Carolina Biological Supply Company)

partly because a disease-causing organism is more likely to encounter a host, or a predator its prey, when there are more hosts or prey in the area.

In addition, studies on some wild animal populations have shown that individuals in dense populations have decreased health and vigor. This may decrease their immunity to disease and make them more susceptible to predation. In some extremely crowded populations, stress itself seems to lead to changes in body function that eventually prove fatal. It is not clear whether poor health is caused directly by overcrowding or by some underlying factor, such as too little of some vital nutrient to go around in a dense population. (Overcrowding does not appear to harm human health directly. In cities such as Hong Kong, New York, and Mexico City, people live at incredibly high densities with no apparent ill-effects.)

Other mortality factors are **density-independent,** killing a proportion of the population without regard to its density. For example, in several insect species, harsh winter weather kills about 90% of a population regardless of its density. Hurricanes, drought, or earthquakes may also kill a large proportion of individuals, no matter what the population density. However, bad weather may sometimes cause death in a density-dependent manner: if it is possible to survive by finding shelter, and if the number of shelters is limited, then all of the members of a sparse population may survive whereas only a fraction of a denser population will be protected.

It would be convenient if we could pinpoint one or two factors and say that they determine the size of a particular population. However, the sizes of natural populations are often affected by many different factors, whose interactions can be complex.

39-C   Competition

Competition is a density-dependent factor that helps to regulate the size of many populations. **Competition** occurs when two or more organisms attempt to exploit a limited resource, such as food or space (Figure 39-11). Competition between individuals of the same species is very common. Generally, members of the same species need the same resources and so they usually compete for them.

Changes in Population Size

Exponential Growth

When the number of individuals in a population is increasing exponentially, the growth of the population in any unit of time is equal to the size of the population at the beginning of the time interval multiplied by the population's growth rate. This growth is described by this formula:

$$\frac{dN}{dt} = rN$$

where $\frac{dN}{dt}$ is the increase of population size per unit time,

r is the rate of population growth, and

N is the number of individuals in the population at the start of the time period being considered. If we insert actual figures into this formula and plot the results in a graph, they look like this:

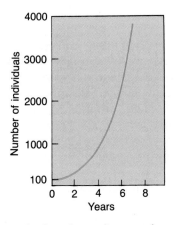

This is the graph plotted on a linear scale:

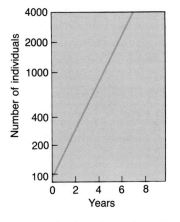

This is the same graph plotted on a logarithmic scale:

Carrying Capacity

Populations cannot continue to grow exponentially. Eventually, some vital resource runs short and growth slows down. In an actual experiment, Gause grew a population of *Paramecium aurelia*. He fed them by adding the same weight of bacteria to the culture every day. At first, the number of individuals in the culture grew exponentially. Eventually, however, the constant number of bacteria added each day was insufficient to feed the ever-growing number of *Paramecium* and the growth rate fell, eventually reaching zero. The population size levelled out as its numbers (N) approached the carrying capacity, K. The graph depicting this experiment looks like this:

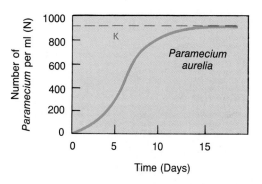

The data Gause obtained are described by the **logistic equation** for population growth, which is an extension of the equation given above for exponential growth:

$$\frac{dN}{dt} = rN\left(\frac{K - N}{K}\right)$$

where $\frac{K - N}{K}$ indicates how much of the resource that limits growth (in this case, food) is still available to the population. When N is much less than K, this term approximates 1 and the equation becomes $\frac{dN}{dt} = rN$ (the equation describing exponential growth). As N grows until it is almost equal to K, the value of the term $\frac{(K - N)}{K}$ approaches zero and $\frac{dN}{dt}$ (that is, growth rate) also becomes zero. The carrying capacity of the environment for the population has now been reached.

Figure 39-12
The effect of seed density on survival of white clover seedlings subjected to the stress of water shortage. When water was available, survival was about the same at all population densities (blue bars). Brown bars represent the survival of seedlings that were not watered after the 18th day. Survival for these seedlings was much lower in the dense population. (After J. L. Harper, *Society for Experimental Biology Symposium* 15:1, 1961, Cambridge University Press)

In one experiment, seeds of white clover were planted at three different densities. Half of the plants at each density were watered throughout the experiment, but the other half were watered only for the first 18 days. After seven weeks, the densities of the surviving seedlings were measured. Among the seedlings that were watered regularly, mortality was low regardless of density. Among the seedlings deprived of water, however, the proportion of seedlings killed was three times greater in the high density than in the intermediate density plots, dramatic evidence of density-dependent mortality when a resource (water) was scarce (Figure 39-12).

Instead of competing directly for a limiting resource, individuals of many animal species compete indirectly for social dominance or for a territory. A **territory** is an area occupied by one or more individuals and defended by its occupants against other members of the same species (and sometimes of other species). Possessing a territory may guarantee the territory-holder an adequate supply of some limiting resource such as food, space, or a nesting site. In one study, a particular area of marsh in Iowa always contained about 400 adult muskrats. Males compete for territories, and the marsh provided about 180 territories, each with food and a refuge from predators for a pair of muskrats and their young. Animals in their territories are relatively safe from predators because they know their territories well, and the territories have ample cover for escape. Muskrats unsuccessful in the annual competition for territories were forced to live in unfavorable areas at the edge of the marsh, where they and their offspring suffered a high rate of death from overcrowding, predation, inadequate food, and interference by other animals.

Niche

When members of different species compete, they are said to have overlapping niches. The **niche** of a population or species is its functional role in an ecosystem. To illustrate niche, consider a badger, an inhabitant of grassland or woodland. The badger is a carnivore; to live, it eats mainly rodents, and it eats some species more than others. It alters the soil by digging a burrow to live in,

Figure 39-13
A badger emerging from the entrance to its burrow under a tree. (E. R. Degginger)

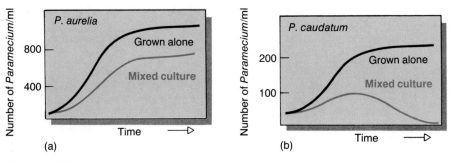

Figure 39-14

Gause's experiment with two species of *Paramecium.* Both species grew well by themselves in culture tubes with daily changes of water and inputs of bacterial food. When placed together under these conditions, *P. aurelia* (a) survived, while *P. caudatum* (b), the larger and slower-growing of the two species, always declined to extinction. If the water was not changed, waste built up, and the competitive outcome was invariably reversed.

and by digging its prey out of their burrows. It drinks water; it produces feces, little ecosystems in themselves; it attracts flies that feed on its feces; it may be eaten by wolves or other large predators, and so forth (Figure 39-13). Every interaction of the badger with its environment is part of its niche and determines where it can live and what other organisms can live in the same habitat with it.

The theory of **competitive exclusion** states that two species with very similar niches cannot survive together because they compete so intensely that one species eliminates the other. Gause set out to test this theory. He chose two species of *Paramecium* that have similar food requirements and hence similar niches. When he raised the two in the same culture, he found that one species always eliminated the other. The particular conditions in the culture determined which species survived (Figure 39-14).

In nature, it often looks as though several species coexist while competing strongly for the same resources. Whenever such cases have been examined in detail, however, it has always been found that the species divide the resources in some way. For instance, Robert MacArthur found that different species of wood warblers in northeastern forests forage for insects in different parts of the trees, reducing the competition between them (Figure 39-15).

An example of overlapping niches and division of resources comes from a study of two species of barnacles on the rocky coast of western Scotland, *Balanus* and *Chthamalus* (don't pronounce the "Ch"). *Chthamalus* occupies the upper part of the intertidal zone and *Balanus* a lower zone, with little overlap

■ *Competition among members of the same or different species for a limiting resource acts as a density-dependent control of population size.*

Figure 39-15

Coexistence by avoiding competition. Several species of warbler of the genus *Dendroica* hunt for insects in coniferous trees in the same New England forests. Each usually forages in a different part of the tree (green), thus reducing the competition for food.

BLACKBURNIAN WARBLER   BAY-BREASTED WARBLER   MYRTLE WARBLER

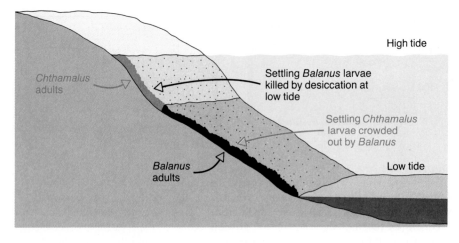

Figure 39-16
Division of space between two species of barnacles. On the rocky coast of Scotland, the vertical distribution of *Chthamalus* barnacles is limited by competitive exclusion. *Chthamalus* larvae settling from the plankton are crowded out by growing *Balanus* barnacles. The upper limit of *Balanus* is determined by its lesser tolerance of exposure at low tide.

between them. The planktonic larvae of both species settle on rocks in both zones. However, *Balanus* cannot survive in the upper zone because it is less tolerant than *Chthamalus* of exposure to the air at low tide. By removing *Balanus* larvae as they settled in the lower zone, Joseph Connell showed that *Chthamalus* survives in the lower zone when *Balanus* is absent. *Balanus* grows faster than *Chthamalus,* so that when the species compete for space on the rocks, *Balanus* grows over *Chthamalus* and crowds it out. Thus the lower limit of the zone occupied by *Chthamalus* is determined by whether or not *Balanus* is present (Figure 39-16). Competition of this sort, which restricts the area where a species can live, and hence limits the size of a population, is probably very common in nature.

39-D Predation and Pest Control

Predation causes considerable mortality in most species and is therefore an important factor in regulating population size. In this context, we can define **predation** broadly, as any case where individuals of one species exploit a living prey species for food. By this definition, **predators** include herbivores and parasites, and **prey** includes plants and the parasites' hosts.

Predators that specialize in eating only one species of prey may control the population size of their prey. For instance, in California, a small beetle controls the numbers of an attractive yellow-flowered plant called Klamath weed. This plant, introduced from Europe, spread throughout the United States. Klamath weed is toxic to cattle, and is also an aggressive competitor, displacing desirable plants from grazing land. By the late 1940s it covered 80% of the ground in many areas of California and Oregon, making it useless for cattle ranching (Figure 39-17).

In its European home, Klamath weed is attacked by several insects, including two species of beetles. Supplies of these beetles were let loose in California in 1945 and 1946. They flourished, and by 1959 Klamath weed had been reduced to less than 1% of its former abundance, chiefly due to the voracious appetites of the beetle larvae. Because the beetles are specialists, they cannot switch to other food plants when the Klamath weed population declines. (If they could, they might become pests themselves.) Both Klamath weed and its beetle predators persist in the United States, but at low densities. This is an example of **biological control** of a pest species by a natural enemy.

Today, the search is on for predators of more pest species as the most effective way to control them. For example, imported water hyacinth has clogged southern waterways and resisted all efforts at control. Finally, in 1989,

■ **Specialized predators may hold the population size of their prey species at very low levels. Generalized predators seldom do so because they switch from one prey species to another as the abundance of prey species changes.**

(a)   (b)

Figure 39-17
Biological control proved very effective in reclaiming range land overgrown with
Klamath weed. (a) A field with a large population of yellow-flowered Klamath weed
in California, June 1948. (b) The same field in June 1949, a year after the introduction of beetles adapted to feeding on Klamath weed. (c) Beetle larva eating Klamath
weed. (F. E. Skinner)

scientists sent to Pakistan for a species of beetle that eats this pest. The beetle
offers the first sign of hope that this weed may finally be controlled (Figure
39-18).

Both these examples show that specialist predators, which attack only one
or a few species, may limit the populations of their prey. It appears that generalized predators, those that can feed on many different species, seldom have this
effect. Such predators can exploit any prey species that becomes common in
their habitat. By concentrating on such a prey species for a while, they may slow
a population explosion, but as the prey becomes scarce they switch to other
foods.

(c)

39-E   Extinction and Endangered Species

A species becomes **extinct** when its last member dies. Local populations of
many species do become extinct quite often, but the population is reestablished
later by immigration from neighboring populations of the same species. This
happens, for example, when butterfly populations in some of the high mountain
valleys of Colorado are occasionally exterminated by freak midsummer snow-
storms.

When an entire species consists only of one small, restricted population, it
is especially prone to extinction. Such species are most common on islands or in
small isolated areas of restricted habitat. Here, a disaster may wipe out the
species in one blow. On the Caribbean island of Martinique, the Martinique rice
rat was exterminated by a volcanic explosion, which also killed every person in
a town of 30,000 except a prisoner in an underground jail cell.

Extinction is not new. About 65 million years ago, more than half of all
species on earth, including the dinosaurs, became extinct in a relatively short
period. Biologists are still arguing about the reasons. Some 50,000 years ago, the
rate of extinctions again started to increase. A large number of animals, including many African game species, the Irish elk, mammoth, steppe bison, woolly
rhinoceros, giant sloth, and mastodon, became extinct between 50,000 and
10,000 years ago. The dates of the extinctions coincide with the spread of human

Figure 39-18
Water hyacinths choke a waterway.

populations into new areas. A site at Solutré in France contains the remains of over 100,000 horses! Native Americans slaughtered bison by the thousands by driving them into pits. However, there were still about 60 million bison in North America when European settlers arrived. It took less than 200 years of hunting with firearms to reduce the bison to near-extinction in 1850, when only 250 remained in all of North America (see Figure 39-1).

The extinctions caused by humans since 1900 dwarf anything in history. Hunting, for food, fur, livestock protection, or pleasure, has exterminated many species of large mammals and left nearly all large carnivores endangered. But most species become extinct because humans destroy their habitat, converting it into farmland, highways, or suburbs. The fate of a species in the twentieth century depends largely on how compatible it is with human population growth.

As humans occupy more and more of the earth, animals that need a lot of space are particularly vulnerable. Many of these are large carnivores occupying the top trophic level of a food chain. Mountain lions, Florida panthers, and California condors are examples of animals that need large areas in order to find enough food. Their habitat has steadily disappeared as the human population has increased.

Many species cannot survive competition from the organisms that accompany human invasions: weeds such as dandelions and goldenrod, and animals such as goats, pigs, dogs, cats, rats, and mice. These animals eat eggs, young, and adults of many native species, and compete with them for space. For instance, when the Polynesian discoverers of Hawaii arrived there 1500 years ago, they brought with them dogs, pigs, rats, chickens, and some 30 species of plants. Centuries later, European settlers brought dozens more new species to the islands. Competition with these imports has caused the extinction of hundreds of species. An estimated half of Hawaiian native species are now extinct.

Experts believe that humans have exterminated about a million species in the twentieth century, mainly in tropical forests. Tropical forest contains millions of species that have never been described, most of which will be extinct before they are even named. The area of tropical forest has decreased by more than 50% in the last 200 years. Each year, an area of forest about the size of West Virginia is cut down, and soon little will be left.

Probably 99% of all the species that have ever lived are now extinct. Species form and then they become extinct, anywhere from a few weeks to millions of years later. So why worry about what is, after all, a natural process? The reason for concern is that species can become extinct much faster than new ones can form. Agricultural and medical researchers tell us that extinctions deprive us of irreplaceable drugs and genetic resources. They feel we should stop our heedless extermination of other species, at least until we have assessed their possible usefulness. Threatened forests have already yielded a wealth of medicines, foods, and new seed stocks for crops. In the 1970s, a strain of disease-resistant wild corn (maize) was discovered in the Mexican forest. The hardy hybrids produced from breeding this corn are already worth billions of dollars to farmers. Many of the prescription drugs sold in the United States are derived from substances discovered in tropical forests. For example, the drug vincristine, used to cure childhood leukemia, comes from the Madagascar periwinkle (Figure 39-19).

Ecologists point out that any species may be a vital link in a food web. Its extinction may disrupt an ecosystem or cause the extinction of other species. Others feel that there are more important considerations than the economics of extinction: the esthetic loss we suffer when a species disappears (Figure 39-20).

Most countries have enacted laws and regulations designed to prevent more species from becoming extinct. The United States' Endangered Species Act is considered by many to be the strongest of these laws, but this does not mean that it does the job it was designed to do. Since European settlers arrived, more

■ *Human activities have greatly increased the rate of extinction of species, mainly by destruction of habitat and competition or predation by species imported by humans.*

Figure 39-19
Madagascar periwinkle, from which vincristine is extracted. (This common ornamental is also known as vinca, its old generic name. The international commission on nomenclature for plants has changed its name to *Catharanthus roseus.*)

(a)

(b)

Figure 39-20

Endangered species. (a) Some species of giant tortoises from the Galapagos archipelago are already extinct, and those remaining have greatly declined in numbers. Early travellers carried tortoises off by the thousands because tortoises can live for months without food or water, providing fresh meat as needed on long sea voyages. More recently, competition from wild goats for food plants, and predation on eggs and young by wild pigs, dogs, and rats, have severely depleted the populations that remain. Tortoises are now being bred in captivity and the young are returned to the wild. (b) The black rhinoceros, well able to defend itself against most enemies with its huge bulk and sharp horns, has been brought to the brink of extinction by human poachers. Rhinoceros horns are carved to make dagger handles, treasured by wealthy men in the Middle East. Demand for these daggers provides the economic incentive for poaching. The rhinoceros is also used in folk medicine in various cultures.

than 500 species of plants and animals have disappeared from North America, and some 3900 species of American plants and animals are currently endangered, 700 of them in Florida alone.

39-F   Human Population Growth

Human population growth is responsible for the extinction of other species and for most of our own environmental problems, from traffic jams to pollution.

Declining Death Rates

The number of people on earth has increased steadily for at least 10,000 years. But the greatest population growth has taken place in the last 200 years, and the rate of growth is still increasing (Figure 39-21). During the 1990s nearly one billion people will be added to the human population, more than in any previous decade. They will increase the population by 20%, from about 5.3 billion in 1990 to about 6.3 billion in 2000.

The human population explosion is largely a result of reduced death rates, particularly among infants. Infant mortality is the most important factor determining **life expectancy,** the average number of years that a newborn baby can be expected to survive. In 1900, the infant mortality rate in North America, the U.S.S.R., and Western Europe was about 40 per 1000 babies born, and the life

■ *Modern increases in human life expectancy are due mostly to better nutrition and hygiene, which have boosted infant survival.*

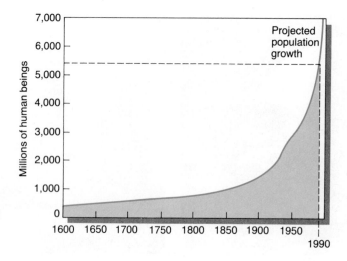

Figure 39-21
Growth of the world human population in the last 400 years. The dotted line shows the approximate world population in 1990.

expectancy was 46 years. In 1986, the infant mortality rate for these countries was about 10 per 1000 and life expectancy was more than 70 years.

In developing countries, most deaths are due to respiratory and digestive tract infections in infants, and such deaths are easily reduced without expensive medical care. The simple lack of clean water, nutritious food, and basic education in hygiene is the reason that life expectancy for most of Africa remains less than 45 years, while the average for Latin America and East Asia (including Japan and China) has risen to more than 65 years since 1950 (Figure 39-22).

The Demographic Transition

Demography is the study of (particularly human) populations.

The theory of the **demographic transition** states that economic and social progress affects population growth in three stages. In the first stage, in preindustrial societies, birth and death rates are both high and the population grows slowly, if at all. It is in the second stage that a population explosion occurs. Death rates fall as public health improves, but birth rates remain high. The population grows very fast, by about 3% per year, and so it doubles every 25 years. In the

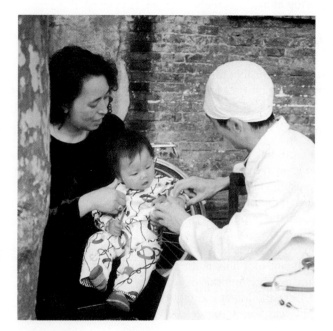

Figure 39-22
A doctor immunizes a baby at a roadside clinic in China. China is an example of a country that reduced its infant mortality rate very rapidly by making health care and education free and available to all. (Tricia Smith)

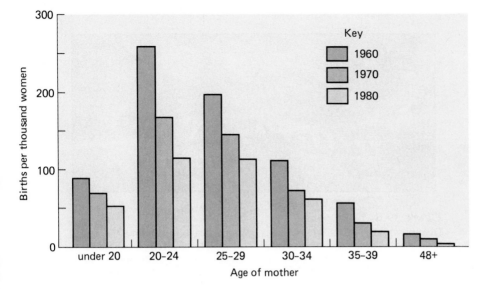

Figure 39-23
America's falling birth rate. This graph shows the number of births per thousand women in various age groups during 1960, 1970, and 1980. The number of births in all age groups fell from 1960 to 1980, while the average age of the mother increased (compare women in their twenties in 1960 and 1980). Both these trends slow population growth.

third stage, birth rates fall until they roughly equal death rates, and population growth slows down and stops (Figure 39-23).

The theory of the demographic transition was based largely on what had happened to European populations. The developed countries of Europe, North America, and parts of Asia have indeed passed through these stages. Populations in these areas have grown as much as five-fold since 1850, but today they are growing slowly or even declining.

The single factor most clearly correlated with a decline in birth rate during the demographic transition is increasing education and economic independence for women. The spread of knowledge that lowers the death rate is usually part of a general program to improve the educational level. Educated women find that they need not bear many children to ensure that a few survive, and they also learn contraceptive techniques. In addition, women find that they can contribute to the family's increasing prosperity by holding a job and by spending less time and energy on raising children. This is usually attractive to women, even in countries where religious doctrine and tradition dictate large families.

The demographic transition has taken from one to three generations to spread through the populations in most developed countries. During its progress, death rates are low, but birth rates remain high, and so the population grows enormously. The population of the United States will almost have quadrupled during the twentieth century.

■ *Low birth rates are found where women have at least some education and economic control over their own lives.*

The Demographically Divided World

Some countries are not proceeding through the demographic transition as predicted by the theory. They appear to have stopped in the second stage, unable to make the educational and economic gains that would reduce the birth rate. If the vast rate of population growth in the second stage is sustained, it begins to overwhelm food production, medical care, and education. When the demands of the population exceed the sustainable production of local farming and graz-

(text continues on p. 810)

Camouflage

A JOURNEY INTO LIFE

Figure 39-B
Disruptive coloration. When it rests against a black background, the black border of this butterfly becomes invisible, leaving only a serrated green area that does not remind a predator of a butterfly. (Biophoto Associates)

Most organisms are vulnerable to predation, and so there is strong selective pressure for adaptations that serve as defenses against predators. Many animals are **camouflaged**— disguised in such a way that they are difficult to perceive even when they are in plain sight. Experiments have shown that camouflage is in fact of selective advantage to its owner. We have studied one such case in this book, the evolution of melanism in moths: moths the same color as their background are less likely to be eaten by birds than moths that contrast with their background (see Figure 14-13).

Since vision is the dominant sense in humans, visual camouflage is very obvious to us. We recognize an animal by three main visible features: its silhouette, its eye, and its bulk, or appearance of being rounded. An animal's camouflage must disguise these three features.

Bulk is nearly always disguised by countershading (Figure 39-A). If an object is the same color all over, its underside appears darker when light falls on it from above and makes it appear rounded. The vast majority of animals have light-colored bellies and dark backs. Light falling from above makes them look uniformly colored and therefore flat. A camouflaged gun or plane is also painted with a pattern that countershades it.

Camouflage often involves coloration that disrupts the silhouette (Figure 39-B). Some parts of the body appear the same color and intensity as the normal background, while others contrast with it. Under these conditions, some parts of the object stand out whereas others seem to disappear. The result is a pattern of splotches rather than a recognizable animal.

Camouflaging the eye is important for two reasons. First, where there is an eye, there is an animal.

Second, an eye is always near the brain, one of an animal's most vital organs. The eyes may be disguised in various ways (Figure 39-C).

Camouflage is useless without appropriate behavior patterns. For instance, a "leaf" wandering up and down a twig is apt to be noticed by a predator. Most butterflies are not camouflaged; they fly by day and their motion makes them visible whatever their coloration. Camouflage is much more common in moths, most of which are active at night; during the day they rest motionless and camouflaged on some appropriate surface.

What is the advantage of visual camouflage? First, coloration that hides an animal against its background might have been selected for in animals with predators that hunt by sight. If this is true, we would expect that animals with few predators would be less likely to be cryptically colored. This is in fact the case. Large birds such as swans and gulls, which have few predators, are conspicuous, but their small, vulnerable young are cryptically colored (Figure 39-D).

Plants, too, can be camouflaged. Mistletoes in Australian forests, for example, mimic the leaf shapes of their plant hosts in a most remarkable fashion. This probably makes the mistletoes less apparent to the butterflies (and perhaps to some mammals) that feed on them.

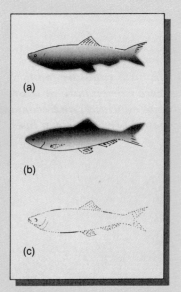

(a)

(b)

(c)

Figure 39-A
Countershading. (a) Appearance of a uniformly colored object lit from above. (b) Appearance of a countershaded object lit uniformly from all sides. (c) Appearance of a countershaded object lit from above.

(a)

(b)

Figure 39-C
Disguising the eye. (a) The eye is the most recognizable part of any animal. In many animals, the eye is made less obvious by such devices as a black stripe across it.
(b) This butterfly carries eye disguise to extremes by having a false head (yellow) on the hind wing tips. Predatory birds peck at the false head, which frequently saves the real head at the other end of the body from damage. (Biophoto Associates)

(a)

(b)

Figure 39-D
Cryptic coloration is common in animals that remain in sight of potential predators for any length of time.
(a) A turnstone sitting on her nest.
(b) A gull chick in its nest. (a, Biophoto Associates)

Figure 39-24
A street scene in Jaipur, India. India is a crowded country with an area less than half that of the United States but a population three times as large. It has made heroic efforts to slow its population growth. Nevertheless, the population is expected to almost double from its present 850 million before population growth ceases sometime in the twenty-first century. (Paul Feeny)

ing land, forests, and water supplies, people begin to consume the resource base itself. Vegetation disappears, soil erodes, wells run dry, and productivity declines. This in turn reduces food production and incomes in a downward spiral.

Population trends appear to be driving about half the world's people toward this kind of economic decline and half toward a better future (Table 39-2). The world is divided into countries that have completed the demographic transition and countries stuck in the second stage, with few countries still in the process of passing through the demographic transition.

In slowly growing areas, mainly Europe, North America, Australia, China, and Japan, populations are growing on average 0.8% per year, and living conditions are fairly stable. Some ecological resources may actually be increasing, as these countries invest in pollution control, reforestation, and restocking wild populations of plants and animals. In complete contrast, populations in high-growth areas (chiefly Southeast Asia, Latin America, India, the Middle East, and Africa) are growing more than three times as fast—on average by 2.5% per year. The slow-growth countries add 19 million people a year to the world's population; the fast-growth countries add 64 million (Figure 39-24).

TABLE 39-2   Income and Population Change for Selected Countries*

Country	Change in per capita† income, 1980–86	Predicted increase in population from 1986 until growth ceases	Change in per capita grain production 1972–85
World Average		+74%	+0.4%
Nigeria	−28%	+406%	−7%
Argentina	−21%	+74%	+12%
Philippines	−16%	+129%	+13%
Peru	−11%	+158%	−24%
Kenya	−8%	+455%	−19%
Sudan	−7%	+264%	−6%
Brazil	−6%	+113%	+12%
China	+58%	+50%	+20%
South Korea	+34%	+130%	+9%
Japan	+21%	+6%	−9%
India	+14%	+116%	+8%
W. Germany	+10%	−15%	+13%
U.S.A.	+9%	+20%	+8%
U.K.	+12%	+5%	+19%
France	+4%	+1%	+12%

* data from The World Bank.

† "per capita" means per person.

SUMMARY

Populations of various species of organisms are the basic biological units of ecosystems. Given ideal environmental conditions, the number of individuals in a population increases at its biotic potential. This biotic potential is determined mainly by the age of the (female) parent at first reproduction, but it is also influenced by the number of offspring produced at each reproductive event and by the parent's reproductive lifespan. A population seldom, if ever, reproduces at its biotic potential even when it is growing exponentially. Among the fastest-growing populations are those of organisms introduced into new, favorable environments.

Survivorship curves for members of a population reflect the population's reproductive strategy. At the two extremes, the members of a population may produce many small offspring and leave them to fend entirely for themselves, or they may produce a few, large offspring that are nourished and trained by the parents. Survivorship data are the basis for constructing life tables, useful in predicting future changes of population size.

Most populations remain about the same average size from year to year. Populations living in areas with

extreme climates may be kept in check by density-independent natural events (such as droughts or freezes). However, most populations are generally limited by density-dependent factors, such as predation, disease, or competition for resources. When growth of a population ceases, under the influence of one or more of these factors, the size of the population stabilizes at approximately the carrying capacity of the environment for that species.

Competition for resources may play an important part in regulating the size of a population. Competition between members of different species with similar niches leads either to the extinction of one species, the weaker competitor, within the area of overlap, or to selection for adaptations that reduce the competition. For example, each species may become specialized in such a way that it uses only part of a limited resource. Predation is another factor that may reduce the size of a population. Many specialized predators and parasites are known to keep their prey species at low density.

Many factors influence the size of a population, and the main factor actually controlling population size may vary from time to time and from place to place.

Extinction is the inevitable fate of populations and of whole species. Species consisting of only one or a few small populations, living in restricted habitats (such as islands), are particularly vulnerable to extinction. Humans have greatly increased the rate at which species become extinct by introducing predators or competitors into new areas, by hunting, and especially by destroying habitats. The resources that we are losing as a result of these extinctions are cause for concern.

Like all other organisms, humans inhabit an environment with limited resources. The number of peo-

ple in the world has grown exponentially in the last few hundred years as a result of improved nutrition and hygiene, which have reduced the death rate, especially among infants. Today the populations of many countries are growing faster than the supply of resources their people need to survive. The nations of the world can be divided into those that have passed through the demographic transition and show little population growth, and those that appear stuck in the middle of the demographic transition, whose populations are doubling about every 25 years.

SELF-QUIZ

1. Draw a graph showing the long-term growth of a population of bacteria on a nutrient medium in a laboratory culture.
2. A population can grow exponentially:
   a. when food is the only limiting resource
   b. when first invading a suitable and previously unoccupied habitat
   c. only if there is no predation
   d. only in the laboratory
3. Which of the following does *not* directly affect biotic potential?
   a. a female's age at first reproduction
   b. carrying capacity of the environment
   c. length of time a female is fertile
   d. average number of offspring per brood or litter
4. If a population exceeds the carrying capacity of the environment:
   a. it will evolve adaptations to avoid a population crash
   b. its numbers will probably decrease rapidly
   c. its food supply will increase in the next generation
   d. the average number of young per individual will increase
5. Which of the following would be *least* likely to act as a density-dependent factor limiting the size of a population of mice?

   a. parasitism
   b. buildup of waste products
   c. predation
   d. unfavorable climate
6. A female elephant bears one offspring every two to four years. Which type of survivorship curve would you expect elephant populations to show: I, II, or III?
7. Studies suggest that:
   a. two species may share the same resource only if both are strong competitors for it
   b. two species that appear to be sharing the same resource are probably specializing so that each uses only a particular part of the resource
   c. if two species are sharing the same resource, neither is capable of exploiting the part of the resource used by the other
   d. no resource can be used indefinitely by more than one species
8. The demographic transition in a country is correlated with:
   a. education for women
   b. improvements in agricultural production
   c. the industrial revolution
   d. improved hygiene and nutrition
   e. improved treatment for illnesses

QUESTIONS FOR DISCUSSION

1. Paul Ehrlich has said, "It is quite possible that the penalty for frantic attempts to feed burgeoning populations may be a lowering of the carrying capacity of the entire planet." Ecologist Lee Talbot has said, "We haven't inherited the earth from our parents. We've borrowed it from our children." What does each of these statements mean? Do you agree? Why?
2. What are some of the lines of evidence that indicate that the human population has already exceeded the earth's carrying capacity?

3. How does a project such as filling in a marsh for a housing development or building a four-lane highway affect the populations of organisms in an area?
4. The 1970s saw a decline in the birth rate in the United States. Why is the population of the United States still growing?
5. Consider two women born in the same year, each of whom will give birth to twin girls as her only children. However, one woman (A) will have her twins at age 18, the other (B) at age 36. Each daughter will have twin

daughters at the same age her mother gave birth, and so on. All mothers will die at age 72.

a. How many descendants does A have when she dies?

b. How many descendants does B have when she dies?

c. Construct a graph to show the growth of populations A and B.

d. How do the rates of increase compare in the two populations? (Find a numerical answer if you can!)

SUGGESTED READINGS

Barrett, S. C. H. "Waterweed invasions." *Scientific American,* October 1989. The story of the damage done by introduced water hyacinth and kariba weed, the failure of mechanical and chemical control, and the hope that natural predators will control the invasions.

Ehrlich, P. R., and A. E. Ehrlich. *The Population Explosion.* New York: Simon and Schuster, 1990. The causes and effects of human population growth.

Hoage, R. J., ed. *Animal Extinctions. What Everyone Should Know.* Washington, DC: Smithsonian Institution, 1985. The record of a symposium to acquaint the public with the major issues of species extinction and related habitat destruction; dozens of well-documented examples.

Wilson, E. O., ed. *Biodiversity.* Washington: National Academy Press, 1988. Essays by many authors on the crisis caused by twentieth-century extinction.

Human Ecology

Home, sweet home

Urbanization

Exploring a new frontier

Three more every second!

*L*ike other organisms, human beings interact with their environment. Because of our numbers and our technology, however, we affect our environment more dramatically and more permanently than do populations of any other organism. During the course of history, two changes in the way we live have been particularly important in altering our relationship with our environment. The first of these was the development of agriculture, when humans started to grow their own food, instead of finding it. The second was the industrial revolution, which also completely changed the way we live and our effects on the environment.

In this chapter we consider how human interactions with the environment have changed and how these changes have led to the environmental problems that confront us today.

- Ecologically, the most important developments in human history were the spread of agriculture and the industrial revolution.
- Modern problems that result from agriculture include population growth, habitat destruction, food simplification, and soil loss.
- Problems that result from the industrial revolution include population growth and pollution.

40-A *Hunter-Gatherer Populations*

For most of our evolutionary history, human beings were **hunter-gatherers,** people who obtain their food by collecting plants and killing wild animals. Our knowledge of hunter-gatherers comes from archeological finds and from studies of modern hunter-gatherer groups such as Eskimos, Australian aboriginals, and the !Kung bushmen of the Kalahari desert in southern Africa (Figure 40-1).

Cooperative efforts permit hunters to spear or trap large game animals. But the main source of food in hunter-gatherer groups is plant materials—seeds, fruits, leaves, nuts, roots, and berries—gathered as they are used, or preserved and stored for later. The use of fire opened up a new range of plant foods. Cooking can remove plant toxins and can also soften plants, making them more digestible.

Seafood is another source of food for hunter-gatherers. Native Americans in the southeastern United States left piles of shells up to 20 feet high from the

■ *The environmental impact of early hunter-gatherers included changing the distribution of plants, starting fires, and possibly also the extinction of game animal species.*

(a) (b)

Figure 40-1
Bushmen of the Kalahari desert in southern Africa. (a) Lightweight bows and arrows are used to shoot small game animals. (b) Bushmen have an encyclopedic knowledge of the plants in their environment and gather many different species for food. Children do not hunt or gather food for the group because both are skilled activities which take many years to learn. (In agricultural societies, in contrast, very young children usually participate in food production.) (Biophoto Associates, N.H.P.A.)

Figure 40-2
A large pot made by Stone Age Guale Indians of coastal Georgia. The pot, fashioned from the local clay, was decorated with intricate grooves while the clay was still soft and then baked to harden it. (Karen Roeder)

oysters that formed a major part of their diet. They also caught fish by spearing and trapping them. Although some fish are farmed today, most of our fish still comes from "hunting" these animals in their wild habitat.

Many early hunter-gatherer groups were nomadic, moving from place to place according to the availability of plant and animal food. Later, more or less permanent settlements developed and people began to build more elaborate dwellings and to accumulate possessions such as the decorated tools and pots that began to appear in Europe and Asia about 20,000 years ago, and in America at least 12,000 years ago (Figure 40-2). Human populations spread from Europe and Asia to the Americas and Australia some 25,000 years ago.

Settlements, permanent or temporary, usually lead to environmental problems such as pollution. Deformities in the skeletons of ancient hunter-gatherers suggest that some of these people suffered from diseases caused by polluting the water. A village well or a nearby stream is easily polluted by human waste or the runoff from a garbage pile.

Even the small populations of ancient hunter-gatherers affected their environment in many ways. As they travelled to new areas, early humans carried plant seeds with them, altering the distribution of plants (and possibly animals). We can only guess at the size of the Old Stone Age population (before the invention of agriculture), but estimates put it at around 5 million people. This would seem to be too few to have had very dramatic effects on the environment. Nevertheless, human activities such as starting or spreading fires do not need many people to produce major effects.

A large number of animal species, including many African game species, the Irish elk, the mammoth, steppe bison, woolly rhinoceros, and mastodon became extinct during the Old Stone Age. Many more species became extinct then than in earlier or later periods. The traditional explanation has been that these extinctions were caused by rapid changes in climate during the Ice Ages which occurred between about 2 million and 10,000 years ago. Some workers, however, believe that many of these extinctions were caused by human hunting and destruction of habitat (by starting fires to drive game animals into traps).

40-B The Agricultural Revolution

Agriculture is the practice of breeding and caring for animals and plants that are used for food, clothing, housing, and other purposes. Agriculture originated, probably independently, in many different places at about the same time some 10,000 years ago (Figure 40-3). Fossils of domesticated dogs dating from 11,000 years ago have been found in Iraq, and cultivated plants date back to 10,000 years ago in America.

At first, agriculture merely provided part of the food for a hunter-gatherer society. For instance, when Spanish missionaries landed in the southeastern United States in the sixteenth century, they found hunter-gatherer societies living largely on deer, seafood, and local plants. In addition, however, many tribes cultivated a few plants around their huts. They also cleared temporary plots in the forest, where they grew maize (corn) and beans, using seeds that were originally imported from Mexico.

Later, human societies became completely dependent on agriculture for food. Although it occurred slowly, the changeover from hunting and gathering to agriculture had such a dramatic impact on human societies that it is often called the **agricultural revolution.**

We are so used to thinking of agriculture as a superior way of life that the relative advantages of hunter-gatherer culture may come as a surprise. Hunter-gatherers do not face the constant battle with pests, droughts, and famine that beset all agricultural communities. Studies in southern Africa during a drought

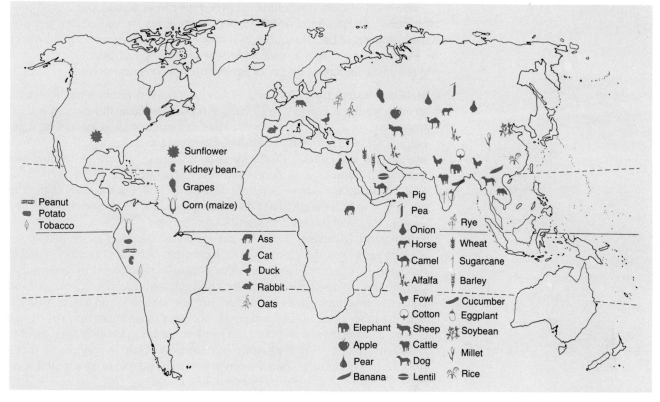

Figure 40-3
Areas where plants and animals we use today were probably first domesticated.
Some species are believed to have been domesticated independently in two or more
areas, and these are shown in both places.

showed that farmers starved while the population of hunter-gatherer bushmen
in the Kalahari desert remained stable in size and the people well-fed. This is
probably because hunter-gatherers keep their populations well below the size
that their territory can support. This is not a result of a high death rate. The
people make a conscious effort to keep their population size down by such
practices as abstention from sexual intercourse, abortion, infanticide, late marriage, and late weaning. Furthermore, these bushmen have a more balanced diet
than most farmers, and their life expectancies are comparable with those of
agricultural peoples in most of the world.

A striking consequence of agriculture is that a new type of division of labor
grows up within the group. In the more prosperous agricultural societies, a few
people can produce food for everyone. The rest are freed to become builders,
bakers, and merchants. Finally, the population may even be able to afford the
luxury of poets, scholars, and students, who contribute little to the group's physical well-being, but are the basis of its cultural life.

Once farming had begun anywhere on earth, it inevitably spread over the
face of the globe. Because farming supports larger populations, an agricultural
community always expands into the land of any nearby hunter-gatherer group,
fighting for the territory if necessary, and driving the hunter-gatherers to extinction or to become integrated into the farming community. Similarly, settled
agriculture overwhelms nomadic livestock herding as a way of life. For instance,
the battle for land between farmers and ranchers in the West is part of American
history. In Africa today, expanding agriculture is destroying the traditional lifestyles of Masai and Bedouin herders.

The roots of most environmental problems can be traced to the evolutionary success of agricultural societies. No one believes we can return to a pre-agricultural way of life. The question is how our destructive, agricultural way of life can be changed into a sustainable way of feeding our populations.

The most important effects of agriculture can be summarized:

1. **Habitat destruction.** The area of land devoted to farming has increased steadily, with the conversion of natural habitats to farmland causing the extinction of thousands of species of plants and animals. Clearing natural vegetation, particularly forests, to produce farmland has had many effects, including altering the climate, producing a widespread shortage of cooking and heating fuel, and causing floods and water shortages.
2. **Soil erosion.** Agriculture need not destroy the soil, but it usually does. The amount of soil on the surface of the earth has steadily decreased since the first digging tool was invented thousands of years ago.
3. **Population increase.** Not only can agriculture support more people on a given land area than hunting and gathering, but the population control practiced by hunter-gatherer societies is usually abandoned by agricultural communities. This is partly because children, who are not important food collectors in a hunter-gatherer society, are useful as labor on a farm. In addition, the inheritance of land and goods becomes more important. The desire to have children who will inherit the property and care for their aged parents is a recurrent theme in mythology and literature. Only in very recent times has effective population control been practiced by people in agricultural societies.

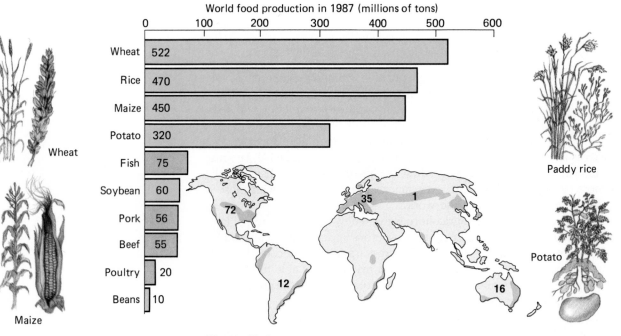

Figure 40-4
Annual world production of major foods. The four most important are shown at the top. The numbers on the map show 1987 grain exports in millions of tons. Several grain-growing areas have no numbers, meaning either that they produce just enough grain for their own use (India and China) or that they import grain (Eastern Europe, Latin America, Africa). Note that North America produces the lion's share of the grain that goes to food-importing nations. As a result, the food security of much of the world is threatened by falling grain yields in North America in recent years. Note the overwhelming importance of the major cereals (wheat, rice, and maize).

4. **Food simplification.** Studies of hunter-gatherers make it clear that humans once ate a much greater variety of plant and animal species than we do today. Several thousand species of plants and several hundred species of animals once formed part of the human diet. The Cherokee Indians of North America alone ate more than 400 different species of plants. Nowadays, few people eat more than 50 species of organisms, and the vast majority of human food comes from just four species of plants: wheat, rice, maize, and potatoes (Figure 40-4). This dependence on a few species lays societies open to famine on a scale unimaginable in earlier times: a new disease or pest of any of these crops can rapidly eliminate a large part of our food supply.

40-C   Feeding the Human Population

The enormous growth of the human population, particularly in the last hundred years, has resulted in widespread starvation. About 40,000 people die of starvation every day, and at least 10 million children in the world are so **malnourished** (poorly fed) that their lives are in danger. **Starvation** means death from lack of food. However, most people who are inadequately fed do not actually die because they eat too few calories to sustain life, but because their malnourished bodies have little resistance to diseases that would not be fatal to the properly fed.

Most malnourished people get about as many calories as they need, but their food is deficient in essential amino acids and vitamins. The problem of diet, then, is not merely the problem of obtaining enough calories, but of obtaining a balanced diet, one that contains the mixture of macronutrients and micronutrients needed to keep the body healthy.

Human starvation is not biologically necessary—yet. The world's farmers produce enough food calories, proteins, and vitamins to keep more than 6 billion people in good health. The trouble is that food, and the income to buy food, are unevenly distributed between the rich and the poor. The problem of hunger today is more economic and political than biological. But we are pushing the limits of the earth's productivity. Before the year 2000, food production per person will probably be falling in most parts of the world.

The Efficiency of Agriculture

Food production depends ultimately upon photosynthesis. However, lack of moisture, nutrients, and warmth usually limits the growth of crops to a fraction of their photosynthetic potential.

Agriculture can feed many more people from a given area of land when people eat plants or plant parts instead of meat. If we eat animals, the animals must eat plants first, and animals convert only about 10% of their food calories into calories that are available to those that eat them. The highest efficiency with which herbivorous animals convert calories from their food into calories in animal products is about 25%, attained in milk and egg production. In contrast, about one third of a plant's productivity can usually be harvested for food, and up to 80% of such a harvest can be digested by humans.

Most of the calories in plants are in the form of carbohydrates. The protein content of plants varies from about 6 to 20%. Since a plant must expend more energy to make proteins than to make carbohydrates, and since protein synthesis also reduces the amount of energy available for a plant to store, it is more efficient to grow and eat plants with a low protein content, as long as enough of the amino acids essential to human health are present. Plants such as wheat and rice produce around 10% protein. This is nutritionally adequate for an adult human, although children need a higher percentage of protein if they are to grow normally.

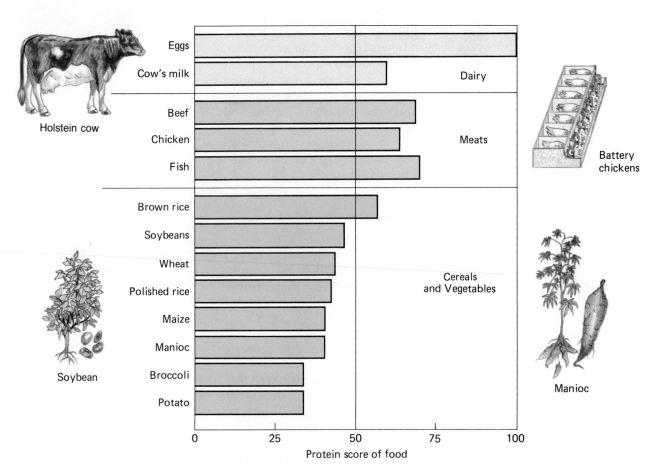

Figure 40-5
Essential amino acid content of various foods. Eggs are considered to have an almost ideal amino acid content for human nutrition, and the other foods are rated in comparison with egg to give a protein score (which is not the same as protein content).

■ *Wheat, rice, maize, and potatoes supply more than half of all human food. Cereals are the most important human food because they contain the minimum nutritionally necessary protein content (about 10%).*

More than 70% of the world's farmland is devoted to growing cereals (grains), such as wheat, rice, maize (corn), barley, rye, and oats, all members of the grass family. Grains are the fruits of these grass plants and they contain a lot of energy-rich stored starch and some fats, protein, minerals, and vitamins (provided by the parent plant for the nourishment of the young seedling). Cereals produce high yields, are easy to collect, and may be stored for long periods without spoiling. The cereals are excellent sources of human food except that they are deficient in some essential amino acids (Figure 40-5). Besides feeding humans directly, the grains and leaves of grasses are the main food of most domesticated animals.

Although plants are much more important than animals as human food, there are places in the world better suited to grazing animals than to growing crops. Livestock herds have long been the basis of life for humans in northern Africa, Lapland, and parts of South America. People who rely heavily on grazing livestock must have access to large areas of land. Thus, many North American ranchers graze their cattle partly on their own fenced fields and partly on federal land, where they have rights to graze a certain number of cattle. As more and more land is fenced and plowed for crops, the area of grazing land decreases. We shall probably reach the point where nearly all animals grown for food are not allowed to graze at all but are raised in feed lots and chicken-farming operations, with all their food delivered from elsewhere.

Only about 50 animal species have ever been **domesticated,** meaning that their reproduction is controlled by humans, and this figure includes honeybees, silkworms, and a few aquatic animals, such as oysters, carp, and trout. Aside from cats and dogs, chickens are the most numerous domesticated animals. Then come sheep, and then cattle. Goats, pigs, and water buffalo are also important in

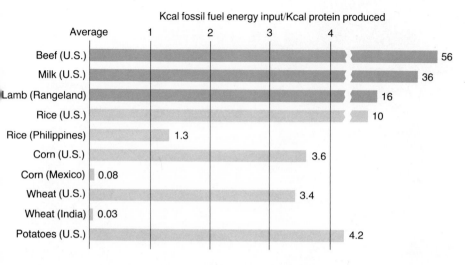

Kcal fossil fuel energy input/Kcal protein produced

Average	value
Beef (U.S.)	56
Milk (U.S.)	36
Lamb (Rangeland)	16
Rice (U.S.)	10
Rice (Philippines)	1.3
Corn (U.S.)	3.6
Corn (Mexico)	0.08
Wheat (U.S.)	3.4
Wheat (India)	0.03
Potatoes (U.S.)	4.2

Figure 40-6
Kilocalories of fossil fuel energy used to produce each kilocalorie of edible protein. The numbers vary for different foods in different parts of the world. Animal products are in red, plants in gold. The difference in energy cost reflects the degree to which agriculture is mechanized. Rice is cultivated largely by manual labor in the Philippines, but by machine in the United States. The enormous cost of producing meat and milk in the United States results from the fact that animal feed is produced by machinery and an animal converts only 10 to 15% of the plant food it consumes into meat or milk. (Data courtesy of David Pimentel)

many parts of the world. Fifteen of the 22 most important domesticated animals are artiodactyls, the even-toed ungulates. Many of these are ruminants (cud-chewing animals), including three of the most important as well as the first to be domesticated: cattle, sheep, and goats. Symbiotic microorganisms in ruminants' stomachs permit them to digest food, such as the cellulose cell walls of plants, that humans cannot. We can therefore use ruminants to convert plants that we cannot digest—such as grass stems and woody shrubs—into foods that we can digest—the meat or milk of cattle, sheep, or goats.

The value of cattle for food has led to their introduction into parts of the world to which they are not well adapted, such as the semi-arid scrub region of the western United States and the tropical forest biome of South America. Advances in breeding and feeding have made cattle more efficient meat producers than they used to be, but beef is still the most expensive meat to produce. More than half the maize (corn) grown in the United States is used to feed livestock.

Food takes a lot of energy to produce (Figure 40-6). However, in both traditional and industrial societies, the energy used to process, distribute, and cook food is greater than the energy used to produce the food in the first place. For instance, about twice as much energy goes into cooking a kilogram of rice in rural India as was invested in producing it. Energy shortages, especially for cooking, have caused problems, including deforestation, in many parts of the world. In the United States, it has been estimated that each calorie on our dinner tables has cost nine calories to put there. Half a calorie represents investment on the farm; the rest represents the cost of processing, packaging, distributing, and cooking. (Packaging was the single largest contributor to the increase in retail food prices in the United States between 1967 and 1977.)

The Green Revolution

The production of wheat in Mexico increased more than eightfold from 1950 to 1970. The land planted to crops doubled, while the yield per unit area quadrupled. In the same period, India doubled its production of grain and now has food reserves, a future that appeared impossible 50 years ago. These spectacular increases, hailed as the "green revolution," resulted from intensive efforts to breed new strains of plants (especially wheat and rice).

Unlike traditional crops, these new strains produce enormous yields if, and only if, they are supplied with large amounts of water, fertilizer, and pesticides. A vital part of the green revolution, therefore, was educating farmers in **high-input** agriculture, agriculture that uses chemicals, irrigation, and farm equip-

Figure 40-7
"Miracle" rice. Many new varieties of rice were bred in the paddies of this agricultural experiment station in the Philippines. Rice is irrigated by diverting water from a stream into the crop during part of the growing season. This is a common method of irrigation, especially in developing countries where the pipes and power sources for sprinkler irrigation systems are not available. (Biophoto Associates/Holt Studios)

ment. The green revolution has transformed the lives and prospects of hundreds of millions of people and is considered the most successful achievement of international development since World War II (Figure 40-7).

High-input agriculture is environmentally expensive. It uses and pollutes an enormous proportion of our fresh water. An estimated half of all chemical fertilizer used does not end up fertilizing plants but washes off the land, together with eroding soil, into waterways where it pollutes rivers, lakes, and underground water supplies.

We have access to only a tiny fraction of the water on earth, nearly all of it water that is purified by the water cycle (Figure 40-8). The water cycle is driven by the sun's heat, which causes water to evaporate. Water vapor, the gas that

Figure 40-8
The water cycle. Water that reaches the atmosphere by evaporation or transpiration is pure fresh water. (Carol Johnson)

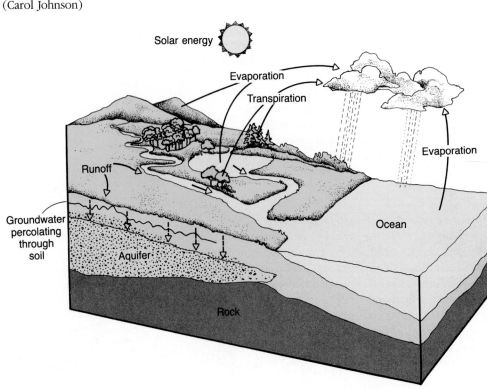

evaporates from a damp surface into the air, is pure water. Some of the water vapor in the air eventually descends on the land as rain or snow. This volume makes up our "water income," the recycling supply of purified water upon which life depends.

The water cycle replaces every body of water more or less rapidly: water leaves the area and "new" water from rain or runoff takes its place. For instance, the water in any part of a river is replaced every 10 to 20 days. The water in a deep lake may be completely replaced only once every 100 years. The rate of replacement is important to us because it influences how long natural processes will take to replace polluted water once the pollution stops.

More than 80% of all the fresh water used in the United States is used for agricultural irrigation. Many areas face water shortages as a result. The media tend to blame southern California's water shortage on lawns, golf courses, and domestic use, but this is not the main cause. More than twice as much water is used on the fields east of Los Angeles as in the urban area itself. This water is used to grow crops such as alfalfa, used to feed cattle.

Some of the gains produced by the green revolution are now in jeopardy. Worldwide grain production actually fell in 1988, the first year that the United States produced less grain than it consumed. The shortfall was only partly made up in 1989. One reason for the decline is that the United States has been forced to take land out of production because its soil is so badly eroded.

Food production is declining partly because of shortages of water for irrigation and partly because intensive farming is destroying the soil on existing farms. There is very little more land that can be converted into farmland. Modern farming has had other adverse effects. For instance, the intensive use of pesticides has exerted strong selective pressure for insects and fungi to evolve pesticide resistance. Pesticide use on American farms is now falling. The pesticides simply do not work well any more, so farmers do not spend the money on them. Pesticide resistance affects not only agricultural pests but also disease carriers, such as the mosquitoes that transmit malaria. The dramatic increase of malaria in many parts of the world in recent years is largely due to the use of pesticides and to the spread of irrigation ponds and ditches, which provide mosquitoes with breeding areas.

The land that the green revolution reached is now producing close to the maximum possible crop yields. Because of this, any further increases in worldwide agricultural productivity will have to come from small farmers who were barely touched by the green revolution (Figure 40-9). Let us first consider how these small farms compare with modern, mechanized farms.

Figure 40-9
Labor-intensive agriculture.
(a) Threshing wheat in Iran. (b) Picking tea in Java. Traditional agriculture like this is labor-intensive. People are needed to care for livestock, to plow, plant, weed, and harvest, among other tasks. Much of the work is unskilled and can be performed by young children who are assets to the farm and receive little education. This is in contrast to the situation in hunter-gatherer societies (Figure 40-1). (a, Paul Feeny; b, Biophoto Associates/Holt Studios)

(a)

(b)

The world's farms can be loosely divided into **subsistence farms,** which provide the family with food and sometimes a small crop that can be sold for cash, and modern farms, where crops are raised to be sold. Modern farms are much more productive in terms of output per person per year, but they usually produce less food per unit area, sometimes because the plants must be far enough apart to permit machinery to move between the rows. A subsistence farmer, or a vegetable gardener, usually produces much more per year from the same area of soil by planting one crop between the rows of another, by applying organic matter to the soil, and by planting another crop in places where the first is harvested. There is nothing inherently wrong with labor-intensive small farms. In China they make up much of the world's most efficient agricultural system. (China feeds over 20% of the world's population on less than 7% of the world's farmland.)

Because small farms produce more food per unit area than large farms, land reform that places more land in the hands of small farmers would boost food production. In most parts of the world, however, the economics of agriculture dictate the opposite trend: farms get larger and larger. The U.S. Department of Agriculture, for instance, predicts that almost one million American farmers will go out of business between 1985 and 2000 as they go bankrupt and their land is added to large farms (Figure 40-10).

Soviet farming could produce much more food if it were efficiently managed. But experts believe that the main increases that we can hope for in agricultural productivity will come from low-input farming (including organic farming). **Low-input farming** uses less pesticide, fertilizer, and irrigation than green revolution methods. It relies upon keeping the soil as fertile as possible by adding manure, compost, and other organic matter, keeping the soil planted at all times to prevent erosion, interspersing different crops to reduce pest infestations, and similar practices. Low-input farming is often more profitable than conventional farming because the farmer spends less on inputs such as irrigation, pesticides, and fertilizer. The sums that North American farmers spend on chemicals are now falling as the trend toward low-input farming gathers momentum.

Plant breeders are working to produce the crop varieties needed by low-input farmers—varieties that produce high yields without irrigation, pesticides,

Figure 40-10
The falling number of small farms in the United States. When a small farm goes out of business, the farmhouse is usually abandoned and the land added to a neighboring larger farm. (Note: ha is the abbreviation for hectare, about 2.5 acres)

Number of farms with less than 70 ha of land

2500
2000
1500
1000

1955   1965   1975   1984
Year

or fertilizers. Agriculture specialists have also realized that the most efficient sustainable agriculture is that tailored to suit a particular area. The agriculture of developed nations often cannot be exported efficiently to developing countries where the soil and climate are different. This realization has led to the establishment of agricultural research stations in many developing nations, and particularly in tropical areas, in attempts to find sustainable agricultural methods for the skyrocketing populations of tropical countries.

Toward Sustainable American Agriculture

The future of North American agriculture is particularly important to the world's food supply because the United States and Canada grow nearly one quarter of all the world's grain. They also export more grain than all other countries put together. Any decline in North American farm productivity threatens worldwide, not just local, food supplies.

Agriculture is America's biggest industry and the foundation of the country's economic success. The food industry is the country's biggest employer, and its productivity has grown enormously during the twentieth century.

Agriculture is threatened by diminishing resources. The main agricultural resources are soil, water, and energy (to fuel machinery and manufacture fertilizer and pesticides). These, or any other resources, become less available as a result of depletion or pollution. **Depletion** is the using up of a resource so that it becomes more and more expensive and, eventually, unavailable. Most Americans first became aware that agricultural resources were being depleted during the Dust Bowl of the 1930s. During several dry years, high winds blew millions of tons of soil from overworked farms into dust storms that darkened the sky and put thousands of farmers out of work. The next crisis occurred during the 1970s, when the rising cost of energy raised the cost of fuel and fertilizers. Energy conservation became a major goal of farmers. Also in the 1970s, the dangers of pesticide pollution and poisoning led to stronger laws to regulate pesticide use.

The federal government showed little awareness of the threats to American agriculture until the 1980s, and national policy still subsidizes production rather than sustainability. However, there are signs of progress. A bill providing increased subsidies for soil conservation was passed in 1985. Since 1985, U.S. food production has fallen for the first time in history, and this has shocked many into realizing the need for reform. American agriculture may never regain its former productivity. But further losses can be prevented if agriculture is made more sustainable. Money for research, pollution control, and conservation of soil and water are urgently needed to achieve this goal.

40-D   Tropical Deforestation

Hand in hand with the spread of agriculture goes the loss of forest as it is converted into farmland. Today, the world's forests and woodlands are being destroyed at an alarming rate. This destruction is changing patterns of rainfall, causing the extinction of forest-dwelling species and genetic resources we cannot replace, and producing a shortage of wood for building and fuel.

In Europe and North America, the main cause of tree death is acid rain resulting from pollutants emitted by industry and automobiles (Section 38-G). In other parts of the world, the usual problem is that trees are being cut down faster than they can regrow.

For most people in the world, wood is the main source of energy, with a consumption of one to two tons per person per year. Half of all wood used in the world is burned as fuel. The extent to which demand exceeds supply can be

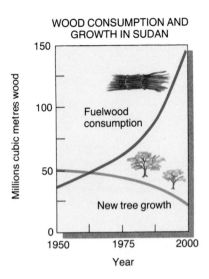

WOOD CONSUMPTION AND GROWTH IN SUDAN

Figure 40-11
The fuelwood shortage. The graph shows that fuelwood use has increased much faster than wood has grown in Sudan. Projections of the worsening problem by the year 2000 are included.

Figure 40-12
The deforestation problem. (a) Making charcoal for use as domestic fuel in Ghana. Newly cut wood is covered with soil and burned slowly to produce charcoal. (b) Cakes of cow dung, to be used as domestic fuel for cooking and heating, have been set out to dry on this riverbank in India. (a, Biophoto Associates/Holt Studios; b, Biophoto Associates)

gauged from some examples. Around Khartoum, the capital of Sudan, nearly every tree has been felled within a radius of 100 kilometres (Figure 40-11). In Upper Volta, women (who do this work) may have to walk for 6 hours three times a week to find enough wood to cook the evening meal, and urban families spend as much as a quarter of their incomes on fuel wood and charcoal.

The shortage of wood causes a number of other damaging consequences. For lack of wood, many people burn animal dung, which would otherwise have been used to fertilize the land (Figure 40-12). One expert estimates that burning dung reduces grain production by enough to feed 100 million people each year.

Because deforestation removes plants that hold soil in place, it permits floods and soil erosion. Sometimes, it permits the topsoil to wash away slowly; often it creates more spectacular landslides and avalanches. In India, 75% of the trees on the Himalaya Mountains have been felled. The few remaining trees cannot absorb the periodic torrential monsoon rains as the forests used to do. Floods now kill people and animals and wash away soil from farmland every year. The flood water deposits this soil as the water slows down behind irrigation and flood control(!) dams. In 1979, a silted-up dam in India burst, unleashing a 5-metre wall of water and mud that killed 15,000 people. India spends about $250 million each year merely to repair the immediate damage done by floods. The damage is not confined to India. North America also loses millions of tons of soil to erosion caused by cutting down trees (Table 40-1).

Agriculture in Tropical Forest Biomes

The traditional method of growing crops in tropical forest biomes is called "slash-and-burn," or swidden agriculture. A patch of forest is cut down and crops are grown in the area for a few years. Because most tropical soil is thin, rain quickly washes the minerals away and the soil becomes less fertile. Except in young, nutrient-rich volcanic soils, the cleared land rapidly loses its fertility, and agriculture becomes impossible. After farming the area for a few seasons, the tribe moves on, leaving the surrounding forest to fill in from the edges and return the area to forest once again.

Recently, however, developing nations have been clearing their rain forests on a massive scale, in attempts to provide farmland for their rapidly growing populations. (90% of the human population growth in the next 20 years will occur in moist tropical forest areas.) If this clearing continues at the present rate, there will soon be very little tropical rain forest left, and many species of organisms, including a large number that scientists have not discovered and described, will have become extinct. Loss of tropical forests will also decimate

(a)

(b)

Table 40-1 Soil erosion caused by cutting forest and building dirt roads through forested areas*

Site	Volume of soil lost (cu m/sq km/yr)
Alder Creek, Oregon	
Uncut forest	45
Clear-cut area†	117
Dirt road	15,565
Coast Mountains, SW British Columbia	
Uncut forest	11
Clear-cut area	25
Dirt road	283

*Data from Swanston, D. N., and F. J. Swanson, 1976. "Timber harvesting, mass erosion and steepland forest geomorphology in the Pacific northwest." In D. R. Coates, ed. *Geomorphology and Engineering.* Stroudburg: Dowden, Hutchinson and Ross.

† A forest is clear-cut when all the trees in an area are felled. The ground still has some vegetation cover, however, because herbs and shrubs in the undergrowth survive.

populations of North American birds that spend the winter feeding in these areas.

Ecologists are very concerned about the problem of forest destruction, and recently some nations have reassessed their policies. Brazil has found that about two thirds of its land is unsuitable for farming and has set aside 100,000 square kilometres of rain forest to be used for recreation and research. Costa Rica's policies have long been the model for conservation in the tropics (Figure 40-13).

Figure 40-13
Palo Verde national park in Costa Rica. (Paul Feeny)

The Environmental Movement and the Tragedy of the Commons

A JOURNEY INTO LIFE

In 1968, ecologist Garrett Hardin published an influential parable entitled *The Tragedy of the Commons.* In it, Hardin argued that the main difficulty in our attempts to solve problems such as overpopulation and pollution is the conflict between the short-term welfare of individuals and the long-term welfare of society.

Hardin illustrated this conflict with the case of commons, such as those of medieval Europe or colonial America. A common was grazing land that belonged to the whole village. Any member of the community could graze cows and sheep there. The tragedy of the commons is that it was in the interest of each individual to put as many animals on the common as possible, to take advantage of the free animal feed. However, if too many animals grazed on the common, they destroyed the grass. Then everyone suffered because no one could raise animals on it. For this reason, common land was eventually replaced by individually owned, enclosed fields. Before the era of subsidized agriculture, owners were careful not to put too many cows on one patch of grass, because overgrazing one year would

mean that fewer cows could be supported the next year (Figure 40-A).

The tragedy of the commons is a result of natural selection. During the course of evolution, some people gave priority to increasing the health and wealth of their own families. These people raised more children than those who put the long term welfare of their village or nation first.

The parable of the commons throws fresh light on many public problems. Congress as a whole deplores the budget deficit, but each member goes on increasing it with expenditures that benefit his or her own district. Each fur trapper or whaler contributes to the extinction of the species from which he or she makes a living. Hardin asserted that the commons (which we can think of as all natural resources) are limited, and that people will pursue their own self-interest, destroying the commons to the point of society's collapse.

Hardin was certainly right in thinking that individuals acting separately cannot be expected to solve all environmental problems. If we, as individuals or as corporations, knew we should pay directly for the overpopulation and the pollution each of us causes, we would each have fewer children and contribute less to pollution. However, it is not usually possible to assign responsibility for ecological problems directly to the people who cause them. Future generations will pay most of the price for the fact that we have too many children and cause soil erosion now, and our individual contributions to water pollution cannot be easily distinguished. Thus, although incentives for individual action are effective ways of getting most things done, many environmental problems can be solved only by effective action by groups such as

governments that, at least in theory, can look beyond the immediate interests of individuals and plan for the long-term welfare of society.

Hardin pursued his argument in a 1974 essay, *The Ethics of a Lifeboat,* which explores the way the wealthy countries of the world treat their impoverished neighbors. An allegorical lifeboat filled with the world's rich is surrounded by the struggling poor who attempt to clamber aboard. How should the rich behave? They cannot let everyone into the boat because it would sink. If they allow some into the boat, they remove the lifeboat's safety margin and are also presented with the further ethical dilemma of whom to save and whom to condemn. To Hardin, the only sensible course of action is to ignore the pleas for help and maintain the boat's safety margin.

The lifeboat argument proposes that it is counterproductive to supply economic aid to poor countries. The argument is particularly strong in the case of countries that do not attempt to curb their population growth. If a poor country can call on aid in times of need so that its people do not starve to death, its population, and its need, continue to grow indefinitely, reducing still further the resources that are left for future generations everywhere. Lifeboat philosophers argue that the most humane course, in the long run, is to withhold aid so that starving people die. The smaller population that remains will have more chance of developing sustainable agriculture that does not degrade the environment and can feed the population.

If Hardin's analysis of human nature were the whole story, it would make the task of solving environmental problems almost impossible. Act-

40-E   The Industrial Revolution

■ *Industrialization has accelerated every kind of human impact on the environment, including population growth, resource depletion, and pollution.*

For some 8000 years, most of the people on earth lived in agricultural societies. Most people lived on farms, and towns were mainly centers of trade and culture. The sources of energy were such things as wood, sunlight, and moving water. These sources are **renewable,** replaced by natural processes (Figure 40-14). Then, some 300 years ago, people began to make extensive use of **nonrenewable** energy sources, particularly fossil fuels. The enormous amounts of energy provided by fossil fuels changed society in ways that are collectively known as the **industrial revolution.**

Figure 40-A
An encampment of nomadic shepherds on overgrazed land in Iran. This part of the
Middle East was once the world's bread basket and was known as the "fertile crescent."
Soil destruction has reduced most of the area to grinding poverty. (Paul Feeny)

ing each with rational self-interest, we should continue to destroy the earth's resources until we became extinct. However, Hardin ignored important additional facts about the nature of human beings: we are intelligent, deeply social animals, and we are capable of altruistic behavior (Section 16-E).

Altruistic behavior is part of our makeup because we are social animals. We depend upon cooperation with other people for our very survival. This is, ultimately, why we can prevent ourselves and each other from destroying the environment. The chieftain who burns down the local forest to enlarge his own wheat field

runs the risk that he will be thrown out of the village, and if he is ejected from human society he will die. The many heads of nations who have abused their citizens and natural resources and then been deposed (and often killed) are obvious examples. So if our urge to work first for ourselves and our own families endangers the environment, our need to be accepted by society can save it. When enough people, in a village or a nation, want a particular reform, the rest must go along with it. You can make almost any change in society if you can convince enough people that it is the right thing to do.

As we would also predict from this, people are more inclined to work for the general welfare if their actions are effective and do not cost them very much. Public opinion polls show that, even during economic recessions, the vast majority of people will accept tax increases to pay for pollution control and the conservation of resources. Politicians are frequently surprised at the results of such polls, but they should not be. People are behaving logically in voting for effective environmental action that will not cost each of them very much.

Like the agricultural revolution, the industrial revolution has reduced the land space required to support each individual and has increased the depletion of natural resources. This accelerating use of resources has led to enormous increases in our standard of living, but it has also caused many environmental problems. These include the modern population explosion, pollution, and diminishing reserves of the fuel and minerals that are the basis of the industrial revolution itself.

The modern pollution crisis dates from the industrial revolution with its use of fossil fuels. Until about 1950, most air pollution was caused by burning coal. Modern "smogs" are caused largely by the combustion products of gasoline

(a)

(b)

Figure 40-14
Renewable energy. (a) This electricity-generating plant in New Zealand is powered by hot springs. (The clouds are steam, not smoke.) (b) Water-driven mills were once used to grind grain to flour; nowadays, many generate electricity. (a, Biophoto Associates)

Figure 40-15
An oil refinery, symbol of the industrial revolution. The refinery processes petroleum, fossil fuel from planktonic organisms that lived millions of years ago, into heating oil, gasoline, aviation fuel, the raw materials from which plastics and chemicals are made, and dozens of other products.

from cars. In cities such as Los Angeles, Mexico City, Denver, London, and Tokyo, smog frequently renders the air unfit for human consumption. Since it is impractical to filter air as people breathe it, air pollution must be controlled at its source.

Since about 1950, pollution from solid waste has also become an acute problem. Part of this is due to the population explosion, which has resulted in a greater total amount of solid waste. In addition, our higher standard of living in a "throw-away" society has increased the amount of waste produced per person. The use of fossil fuel, this time as the raw material for the manufacture of many artificial polymers collectively known as "plastics," has also contributed to solid waste pollution.

Modern artificial polymers are made from organic molecules in fossil fuel (Figure 40-15). Polyester, polyethylene, vinyls, and acetates are just some of the artificial polymers used in clothing, containers, skis, bottles, artificial arteries, and thousands of other items. Plastics have replaced animal and vegetable products in many areas of our lives. Most artificial polymers cannot be broken down by any microorganism, and so they are apparently with us essentially forever.

Biodegradable plastics are not really what their name implies. Most of them consist of short chains of artificial polymers interspersed with short chains

Figure 40-16
Quito, Ecuador. The city's population has increased six-fold in 20 years, a rate of growth matched by more than 50 cities in developing countries. By contrast, Paris took more than a century for its population to grow six-fold.

of starch (which is why they often feel powdery, like flour). Aerobic decomposers metabolize the starch, leaving flakes of undegraded plastic. Even most degradable plastics will not be decomposed by bacteria in the anaerobic conditions found in a landfill. Although degradable plastics help solve the problem of plastic litter, they do little to reduce the total volume of solid waste. They may also increase air pollution as bits of plastic are picked up by breezes.

We have used modern technology to clean up as well as to pollute our environment. Cities in places such as North America and Western Europe are more pleasant places to live than they have ever been. This is of prime importance because a major effect of the industrial revolution has been to hasten **urbanization**—the movement of people from the countryside, where relatively few people now find jobs on farms, into cities, where industries are established and most jobs are to be found (Figure 40-16).

SUMMARY

Early humans were hunter-gatherers, hunting animals and gathering plants for food. Some were nomadic, and others lived in settled villages or migrated between seasonal homes. These people permanently altered their environment in a number of ways, but agricultural societies have had a much greater impact.

Agriculture supports more people on a given area of land than does a hunter-gatherer way of life. Agricultural societies have steadily squeezed hunter-gatherer populations into less productive areas as agriculture has expanded. The most important environmental effects of agriculture are habitat destruction, soil erosion, population increase, and food simplification.

Humans feed themselves most cheaply when they eat plants that are nutritionally adequate but contain little protein. Most of the world's food supply comes from grains, which fulfill this requirement and are also easy to store. The new plant varieties of the green revolution permitted huge increases in grain production by means of high-input agriculture, dependent on chemicals and irrigation. The environmental damage caused by high-input agriculture includes disproportionate use of water, water pollution, and soil erosion. The current trend is toward low-input agriculture, which is cheaper and less environmentally destructive.

We are still producing enough food for everyone, but many people are too poor to buy their share. Today, population is increasing faster than grain production and we face food shortages during the 1990s. Unless we succeed in curbing our population growth, and the loss of farmland to soil erosion, desertification, and building projects, we shall soon exceed the capacity of the world's farmland to support us.

The industrial revolution came about when societies started to power technology, including agriculture, with the enormous quantities of energy available from fossil fuels. Industrial development accelerated population growth and raised standards of living, but contributed to modern problems, particularly air pollution, toxic wastes, and the huge volume of solid waste we produce today.

SELF-QUIZ

1. As a result of the agricultural revolution, all of the following increased *except:*
   a. human population size
   b. quality of human nutrition
   c. pollution
   d. rate of deforestation
   e. soil erosion

2. The agricultural revolution is believed to have been responsible for a dramatic increase in the human population. Which of the following was *not* a factor in this increase?
   a. Food became more concentrated and thus easier to obtain.
   b. Many methods of birth control were abandoned.
   c. Improved medical knowledge increased life expectancy.
   d. Larger amounts of food could be produced by fewer people.
   e. People could accumulate possessions and wanted more children to pass them to.

3. Humans can eat animals, or plants, or both. Which of the following statements about the feeding of human populations is *not* true?
   a. Because animals eat plants, less land is required to produce plant than animal food.
   b. A diet consisting only of meat contains a higher percentage of protein than a human adult needs.
   c. About 10% of the calories in a plant are lost when an animal eats the plant.
   d. An adult human eating nothing but wheat, rice, or potatoes would receive approximately enough essential amino acids.

4. Which of the following is a likely result of deforestation?
   a. The amount of carbon dioxide removed from the atmosphere is reduced.
   b. Wind blows soil away because its plant cover has been removed.
   c. Water washes off the land more rapidly, causing floods.
   d. Water washes soil into rivers and streams, causing them to silt up.
   e. All of the above.

5. The industrial revolution increased all of the following *except:*
   a. use of fossil fuel
   b. water pollution
   c. urbanization
   d. human birth rate
   e. air pollution

QUESTIONS FOR DISCUSSION

1. Do you think that *Homo sapiens* will be extinct within the next thousand years? Why, or why not?

2. The world contains a few remaining hunter-gatherer societies in places such as Australia, New Guinea, Brazil, and Peru. The governments that control the countries where they live generally believe these people should be left alone and perhaps given title to the land where they have lived for centuries. In some places this may actually happen. In others, hunter-gatherers are rapidly disappearing. What pressures determine one fate rather than the other? Is it worth trying to save modern hunter-gatherers?

3. This chapter stresses the environmental problems caused by agriculture and industry. But agriculture and industry have brought inestimable benefits to human societies. List as many of these benefits as you can.

4. It has been argued that advances in human civilization can be traced to exploitation of new sources of energy. Major steps included the addition of meat to the largely herbivorous diet of our ancestors, the taming of fire, and the use of fossil fuels. Other advances along the way have been domestication of beasts of burden and harnessing of wind and water power. How has each of these affected the ecology of human beings?

SUGGESTED READINGS

Brown, L. R., and others. *State of the World 1991, A Worldwatch Institute Report on Progress Toward a Sustainable Society.* New York: W. W. Norton, 1991. A volume of this book is produced each year by the Worldwatch Institute. It contains about a dozen chapters on environmental problems and progress.

Costanza, R. "Social traps and environmental policy." *BioScience* 37(6):407, 1987. Argues that "social traps" tempt people into situations from which environmental solutions are extremely difficult. Hazardous waste sites, the eroding coast of Louisiana, and nuclear war are the examples used. The tragedy of the commons is seen as an example of a social trap.

Ellis, D. *Environments at Risk: Case Histories of Impact Assessment.* New York: Springer-Verlag, 1989. A series of case studies, including the stories of Minamata, the Amoco Cadiz, Bhopal, Chernobyl, the Thames Estuary, and Hell's Gate. Includes chapters on reducing environmental risk.

Hardin, G. "The tragedy of the commons." *Science* 162:1243, 1968. The thought-provoking argument that ecologically ethical individuals are evolutionarily doomed.

Harlan, J. R. "The plants and animals that nourish man." *Scientific American,* September 1976. An interesting discussion of human food plants and animals: where they originated and how they have become tamed or cultivated.

McKay, B. J., and J. M. Acheson, eds. *The Culture and Ecology of Communal Resources.* Tucson: University of Arizona Press, 1987. A collection of papers evaluating Hardin's parable "The Tragedy of the Commons" in the light of studies of fisheries, grazing lands, and similar commons. The authors conclude that sustainable management of commons is possible if the users of the commons make up a single group of accountable people, and if the limits of the commons are under the group's control.

Reganold, J. P., R. I. Papendick, and J. F. Parr. "Sustainable agriculture." *Scientific American,* June 1990. How American farmers are turning to more environmentally sound practices.

Scientific American, "Managing Planet Earth." September 1989. An entire issue devoted to environmental affairs including management, pollution, biodiversity, population growth, sustainable agriculture, and energy.

Appendix A   GEOLOGIC TIME SCALE

Years Ago (millions)	Era	Period*	Epoch*	Climate and Physical Events
	CENOZOIC	Quaternary	Recent	4 ice ages; rise of Sierra Nevada
			Pleistocene	
2.5		Tertiary	Pliocene	Rise of Panama; cool; extinction of many species
7			Miocene	Rocky Mountains rise further
26			Oligocene	Rise of Alps and Himalayas
38			Eocene	Mild to tropical weather
53			Paleocene	Most continental seas disappear
65	MESOZOIC	Cretaceous		Rise of Rockies reduces rainfall to their east
135		Jurassic		Much of Europe covered by sea. Breakup of Pangaea
195		Triassic		Large areas arid and mountainous. Appalachians rising
225	PALEOZOIC	Permian		Appalachians rising; glaciation in Southern Hemisphere
280		Carboniferous	Pennsylvanian	Land low, covered by shallow seas and swamps; subtropical climate
			Mississippian	
345		Devonian		U.S. largely low and sea-covered; Europe mountainous
395		Silurian		Continents flat; mountains beginning to rise in Europe
430		Ordovician		Mild climate; shallow seas cover continents, which are flat
500		Cambrian		
600	PRECAMBRIAN	Precambrian		Planet cooling, shallow seas, many mountains rise

(The Earth is about 4½ billion years old)

*Remember the periods and epochs (beginning with the earliest) with a mnemonic such as: Penguins Can Only Swim Deeply, Carried Past The Jewel-Crusted Pyramids Ensconced On Mildly Pettish Palfreys.

Life

H. sapiens; extinction of many large mammals; many desert forms evolve.

Large carnivores; hominoid apes.

Forests dwindle; grassland spreads.

Anthropoid apes; ungulates, whales.

Earliest tiny horses.

First primates and carnivores.

Angiosperms originate and spread; extinction of many dinosaurs; marsupials present.

Origin of birds; reptiles dominant; cycads, ferns.

First dinosaurs and mammals, forests of gymnosperms and ferns.

First conifers, cycads, ginkgos.

Origin of reptiles; amphibians dominant.

Fungi, first insects, sphenopsids, lycopsids.

Fishes dominant, origin of modern vascular plants. Sharks.

First vascular plants, modern groups of algae and fungi.

First (agnathan) fish. Plants invade land. Invertebrates and marine plants.

Origin of eukaryotes. Cyanophytes and bacteria.

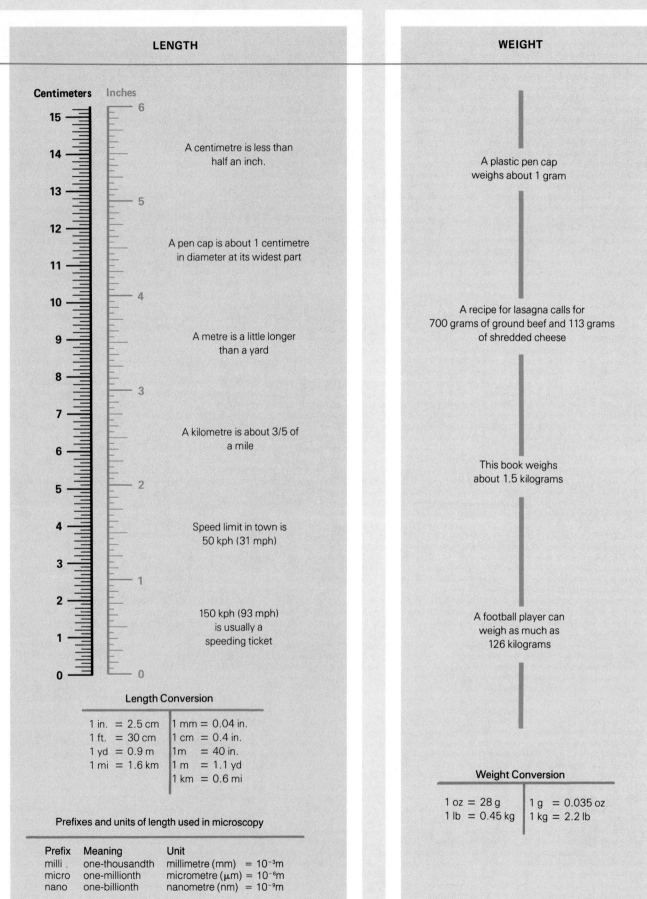

LENGTH

Centimeters **Inches**

A centimetre is less than half an inch.

A pen cap is about 1 centimetre in diameter at its widest part

A metre is a little longer than a yard

A kilometre is about 3/5 of a mile

Speed limit in town is 50 kph (31 mph)

150 kph (93 mph) is usually a speeding ticket

Length Conversion

1 in. = 2.5 cm	1 mm = 0.04 in.
1 ft. = 30 cm	1 cm = 0.4 in.
1 yd = 0.9 m	1 m = 40 in.
1 mi = 1.6 km	1 m = 1.1 yd
	1 km = 0.6 mi

Prefixes and units of length used in microscopy

Prefix	Meaning	Unit	
milli	one-thousandth	millimetre (mm)	$= 10^{-3}$m
micro	one-millionth	micrometre (μm)	$= 10^{-6}$m
nano	one-billionth	nanometre (nm)	$= 10^{-9}$m

WEIGHT

A plastic pen cap weighs about 1 gram

A recipe for lasagna calls for 700 grams of ground beef and 113 grams of shredded cheese

This book weighs about 1.5 kilograms

A football player can weigh as much as 126 kilograms

Weight Conversion

1 oz = 28 g	1 g = 0.035 oz
1 lb = 0.45 kg	1 kg = 2.2 lb

VOLUME

A millilitre is about one-fifth of a teaspoon

A litre is becoming the standard size for soft drink, wine, and liquor bottles in the U.S. If its price is the same as for the old quart bottle, you're in luck because a litre is equal to slightly more than a quart

The gas tank of a large car holds 64 to 75 litres of gasoline

Volume Conversion

1 tsp = 5 m*l*	1 m*l* = 0.03 fl oz
1 tbsp = 15 m*l*	1 *l* = 2.1 pt
1 fl oz = 30 m*l*	1 *l* = 1.06 qt
1 cup = 0.24 *l*	1 *l* = 0.26 gal
1 pt = 0.47 *l*	
1 qt = 0.95 *l*	
1 gal = 3.8 *l*	

TEMPERATURE

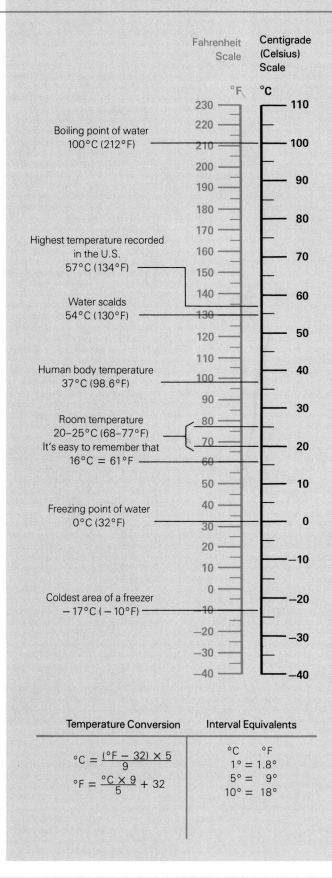

Fahrenheit Scale

Centigrade (Celsius) Scale

Boiling point of water 100°C (212°F)

Highest temperature recorded in the U.S. 57°C (134°F)

Water scalds 54°C (130°F)

Human body temperature 37°C (98.6°F)

Room temperature 20–25°C (68–77°F) It's easy to remember that 16°C = 61°F

Freezing point of water 0°C (32°F)

Coldest area of a freezer −17°C (−10°F)

Temperature Conversion

$$°C = \frac{(°F - 32) \times 5}{9}$$

$$°F = \frac{°C \times 9}{5} + 32$$

Interval Equivalents

°C	°F
1° =	1.8°
5° =	9°
10° =	18°

CHAPTER 2
Some Basic Chemistry

1. alkaline; decreased
2. a.
3. a. H_2O
   b. NaCl
   c. CO_2
   d. O_2
4. HCl: 36.5 grams
   CO_2: 44 grams
5. covalent
6. covalent
7. a.
8. h.
9. f.
10. c.
11. d.
12. g.

CHAPTER 3
Biological Chemistry

1. nucleic acids
2. nucleotides
3. *any two of the following:*
   carry genetic information
   direct the synthesis of proteins
   supply energy to enzyme reactions
   act as coenzymes
4. lipids
5. C, H, O (P, N)
6. energy storage
   formation of biological membranes
   hormones
7. C, H, O, N (S)
8. amino acids
9. *any of the following:*
   catalyze reactions (enzymes)
   form structural elements of body
   hormones
   muscle contraction
   defense against disease
   transport of substances
10. carbohydrates
11. C, H, O (N)
12. monosaccharides
13. energy storage
    structural support and protection
14. b.
15. c.
16. d.
17. b.
18. c.

CHAPTER 4
Cells and Their Membranes

1. d.
2. hypotonic; higher; from the environment into the animal; a.
3. a.
4. a.
5. a.
6. c., d.
7. a.
8. b.
9. d.
10. endocytosis
11. active transport, diffusion through channels, endocytosis

CHAPTER 5
Cell Structure and Function

1. p.
2. d., f.
3. a.
4. c.
5. g.
6. i.
7. o.
8. q.
9. n.
10. m.
11. d., f., k.
12. all
13. animals, some plants, some prokaryotes
14. plants, prokaryotes
15. animals, plants (prokaryote DNA is not organized into chromosomes)
16. animals, plants
17. e.

CHAPTER 6
Energy and Living Cells

1. photosynthesis
2. reduced
3. true
4. ADP; P_i; endergonic; energy
5. c.
6. e.

CHAPTER 7
Food as Fuel: Cellular Respiration and Fermentation

1. c.
2. c.
3. inner mitochondrial
4. a. pyruvate, NADH, ATP
   b. CO_2, ATP, NADH, $FADH_2$, oxaloacetic acid, CoA
   c. ATP, NAD^+, CO_2, ethanol
   d. NAD^+, FAD, H_2O
   e. ATP, NAD^+, lactate
5. a.
6. true
7. true

CHAPTER 8
Photosynthesis

1. c.
2. a.
3. c.
4. d.
5. c.
6. a., d.
7. d.
8. b.
9. c.
10. b., c.
11. a.
12. d.
13. c.

CHAPTER 9
DNA and Genetic Information

1. a.
2. c.
3. d.
4. T-A-G-A-C-A-T-A-C-T
5. c.
6. e.
7. *Escherichia coli*
8. both
9. your own cells
10. your own cells
11. both
12. your own cells

CHAPTER 10
RNA and Protein Synthesis

1. mRNA: A-U-G-U-U-C-A-U-G-A-A-C-A-A-A-G-A-A
   amino acids: methionine—phenylalanine—methionine—asparagine—lysine—glutamic acid
2. The second amino acid would be leucine instead of phenylalanine, and the fourth amino acid would be lysine instead of asparagine.
3. The second amino acid would be leucine instead of phenylalanine, and the chain would terminate after this amino acid, because the next codon is now a *STOP* codon.
4. e.
5. a. DNA double-stranded; RNA single-stranded
   b. DNA nucleotides contain deoxyribose; RNA nucleotides contain ribose
   c. DNA contains thymine nucleotides; RNA contains uracil nucleotides
6. a.
7. d.
8. b.

CHAPTER 11
Reproduction of Eukaryotic Cells

1. b.
2. N = 2; 2N = 4
3. b.
4. b.
5. d.
6. b.
7. nucleus; cytoplasm; eggs; sperm
8. b.

CHAPTER 12
Mendelian Genetics

1. a. both are *Tt*
   b. $\frac{3}{4}$ tasters: $\frac{1}{4}$ nontasters
   c. all tasters
   d. $\frac{1}{2}$ tasters: $\frac{1}{2}$ nontasters
2. a. dumpy recessive to normal, which is dominant
   b. both heterozygous for dumpy wings

3. 40
4. a. 250   b. 125
5. a. Sniffles: homozygous dominant (colored)
      Whiskers: heterozygous for albino
      Esmeralda: homozygous recessive (albino)
   b. $\frac{3}{4}$ colored: $\frac{1}{4}$ albino
   c. $\frac{1}{2}$ colored: $\frac{1}{2}$ albino
6. Mate his dog to bitches known to carry the trait. If any pups show it, the dog is heterozygous for the allele. If none of a large number of pups shows it, the dog is probably homozygous normal.
7. a. yes, if both are heterozygous
   b. no
8. a.

	BS	Bs	bS	bs
Bs	BBSs	BBss	BbSs	Bbss
bs	BbSs	Bbss	bbSs	bbss

   b.

	BS	Bs
Bs	BBSs	BBss
bs	BbSs	Bbss

   c.

	BS	Bs	bS	bs
bs	BbSs	Bbss	bbSs	bbss

   d. $\frac{1}{2}$
9. a. $\frac{3}{16}$ b. $\frac{3}{16}$ c. $\frac{9}{16}$
10. a. 480   b. 160   c. 40
11. a. $\frac{1}{8}$ b. $\frac{1}{8}$ c. $\frac{3}{8}$
12. a. stamens: straight dominant to incurved; petals (red vs. streaky): can't tell from information given
    b. stamens: both parents heterozygous; petals: one heterozygous, one homozygous recessive, but no indication which is which from information given
    c. red × red and streaky × streaky: if red is dominant, red × red will produce some streaky progeny, and vice versa
13. a. $\frac{1}{2}$ b. $\frac{1}{4}$ c. $\frac{1}{4}$
14. let T^A = crosswise stripes
    T^L = lengthwise stripes

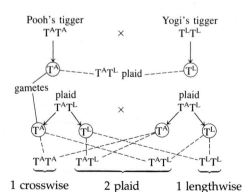

15. a. genotype: SsRR′
       phenotype: straight roan
    b. straight red ($\frac{1}{8}$)

straight roan ($\frac{1}{4}$)
straight white ($\frac{1}{8}$)
curly red ($\frac{1}{8}$)
curly roan ($\frac{1}{4}$)
curly white ($\frac{1}{8}$)

16. a. clover patch: $\frac{1}{2}$ roan, $\frac{1}{2}$ white
alfalfa field: $\frac{1}{2}$ red, $\frac{1}{2}$ roan
cornfield: $\frac{1}{4}$ red, $\frac{1}{2}$ roan, $\frac{1}{4}$ white
b. it doesn't matter; $\frac{1}{2}$ the calves will be roan in any case
17. a. $\frac{1}{4}$ b. $\frac{1}{2}$ c. $\frac{1}{4}$ d. $\frac{3}{16}$ e. $\frac{1}{8}$
18. a. The genes are linked.
b. In the female parent, the genes for sable body and normal wing are on one chromosome, and the genes for normal body and miniature wing are on its homologue. This arrangement is indicated by the large numbers of offspring with the combinations sable body + normal wing and normal body + miniature wings.
19. The genes are probably unlinked; this is shown by the almost-equal numbers of offspring in each phenotype category.

CHAPTER 13
Inheritance Patterns and Gene Expression

1. a. $\frac{2}{3}$
b. sell cows bearing these calves and try a new bull
2. a. $\frac{1}{4}$
b. $\frac{1}{2}$ normal: $\frac{1}{2}$ brachydactylic
3. a. Yellow mice are heterozygous.
b. "Yellow" allele is lethal in the homozygous condition, early in embryonic development (2:1 ratio).
c. Homozygous yellow individuals die as early embryos and are resorbed by the uterus.
d. He could carry out yellow × yellow matings and examine contents of females' uteri early in pregnancy to detect defective embryos.
4. a. $I^A i$ d. $I^B i$ g. $I^A i$ or $I^B i$ or ii
b. $I^B i$ e. $I^A I^A$ or $I^A i$ or $I^A I^B$ h. $I^A i$
c. $I^A i$ or $I^A I^B$ f. $I^A i$ i. $I^B i$
5. Yes, John Smith is really Tom Jones! A baby with blood type M cannot have a parent with blood type N (Ms. Smith). Also, a parent with blood type AB (Ms. Jones) cannot have a child with blood type O.
6. a. $\frac{1}{2}$ normal: $\frac{1}{4}$ chinchilla: $\frac{1}{4}$ Himalayan
b. $\frac{1}{2}$ chinchilla: $\frac{1}{4}$ Himalayan: $\frac{1}{4}$ albino
c. $\frac{3}{4}$ chinchilla: $\frac{1}{4}$ albino
7. $\frac{1}{2}$ barred males: $\frac{1}{2}$ non-barred females (remember, male birds are ZZ and females ZW [see Figure 13-8])

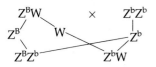

8. $\frac{1}{2}$ the sons hemophiliacs: $\frac{1}{2}$ the sons normal
$\frac{1}{2}$ the daughters heterozygous carriers: $\frac{1}{2}$ the daughters homozygous normal
9. when the mother is either a hemophiliac or a carrier
10. a. mother homozygous recessive colorblind; father normal allele + Y

b. $\frac{1}{2}$ of the sons are expected to be colorblind
c. $\frac{1}{4}$ ($\frac{1}{2}$ of the offspring are expected to be daughters; $\frac{1}{2}$ of these are expected to be colorblind; $\frac{1}{2} \times \frac{1}{2} = \frac{1}{4}$)
11. 2 female offspring: 1 male
12. No; since the baldness trait is carried on the autosomes, a man could inherit the trait from either parent who carried it, and from any of his four grandparents who passed it on to the appropriate parent.

CHAPTER 14
Evolution and Natural Selection

1. c.
2. d.
3. e.
4. b.
5. Individuals with longer necks could reach food higher on trees and therefore had a better chance to survive and reproduce when food was scarce nearer the ground. These individuals passed on the genes for longer necks to their offspring. Continued selection pressure of this sort over many generations led to evolution of longer and longer neck length.
6. e.

CHAPTER 15
Population Genetics and Speciation

Identification of dichotomous key specimens (Figure 15-11):
I. Diplopoda
II. Chilopoda
III. Crustacea
IV. Insecta
V. Arachnida

1. d.
2. d.
3. The normal allele would gradually increase, and the sickle allele would decrease, but not to zero; the sickle allele would remain at low levels in the heterozygous condition.
4. a.
5. c.
6. d.
7. a. increase
b. increase
c. increase
d. increase
8. c.
9. d.

CHAPTER 16
Evolution and Reproduction

1. b.
2. favorable; unfavorable
3. b.
4. c.
5. a.
6. c.
7. a.

CHAPTER 17
Origin of Life

1. oxygen: less
   carbon dioxide: more
   water vapor: more
2. e.
3. 1. organic monomers
   2. polymers
   3. fermentation
   4. water-splitting photosynthesis
   5. respiration
   6. intracellular organelles
4. c.
5. b.
6. 1. Addition of O_2 to the atmosphere made it oxidizing rather than mildly reducing.
   2. An ozone layer formed.
   Effects:
   1. Addition of O_2 acted as selection pressure for the evolution of respiration.
   2. Ozone layer permitted organisms to move onto land.
7. c.

CHAPTER 18
Classification of Organisms and the Problem of Viruses

1. b.
2. d.
3. d.
4. d.
5. a., b.
6. Protista
7. d.
8. c.
9. a. lytic (true of lysogenic when it has become lytic)
   b. lysogenic
   c. lysogenic
   d. both
10. true

CHAPTER 19
Bacteria, Protists, and Fungi

1. d.
2. a.
3. true
4. false
5. b.
6. eukaryotic, unicellular
7. a. pseudopodia
   b. cilia
   c. flagella
8. c.
9. b.
10. b.
11. a.
12. c.
13. a.

CHAPTER 20
The Plant Kingdom

1. d.
2. c.
3. a.
4. a.
5. a.
6. b., d.
7. b. zygote, d. sporophyte, e. spore, c. gametophyte, a. gamete
8. c.
9. d.
10. d.
11. d.
12. c.
13. a. growing on moist soil or rocks in shady areas
   b. growing on surface of moist soil in vicinity of fern plants
   c. male: immature is pollen grain, in pollen cone; mature is pollen tube that has grown to female gametophyte in seed cone.
   female: at base of seed cone scales
   d. in male and female flower parts, respectively

CHAPTER 21
The Animal Kingdom

1. bilateral
2. dispersal
3. a.
4. c.
5. d.
6. b.
7. a.
8. feathers
9. b.
10. 1. support
    2. dehydration of body and of eggs laid outside the body
    3. extraction of O_2 from (dry) air rather than water
11. Reptilia
12. Reptilia
13.

	Amphibians	*Reptiles*
Body structure:	no claws	claws on toes
	no scales	scales on skin
	legs out to sides	legs under body
	slender bones	robust bones
Reproduction:	eggs lack shells	eggs have water-proof shells
	eggs laid in water	eggs laid on land

14. f.
15. i.
16. b.
17. d.
18. d., h., i.
19. e.
20. a.
21. fire: permitted eating of more types of food and migration to colder climates; later, permitted metal ores to be smelted

tools: at first, probably more efficient hunting and gathering; later, many kinds of manipulation of environment: farming, felling trees, etc.

home base: familiarity with local resources, climate, etc.; ability to accumulate more stored food, tools, and other possessions than could be carried easily in a nomadic existence

CHAPTER 22
Animal Nutrition and Digestion

1. f.	6. b.	11. b.
2. a.	7. f.	12. c.
3. a., b., c.	8. a.	13. a.
4. a., c.	9. a.	
5. d.	10. e.	

CHAPTER 23
Gas Exchange in Animals

1. a.
2. c.; b.
3. b.
4. a.
5. a.
6. a.
7. d.

CHAPTER 24
Transport and Temperature Regulation

1.

cnidarian	earthworm	insect	fish	mammal
x	x	x	x	x
	x		x	x
				x
	x	x	x	x

2. c., d.
3. d.
4. b.
5. c. (flow to the skin and lungs decreases as exercise begins but later increases)
6. c.
7. a.
8. a. ectotherms
   b. both
   c. endotherms

CHAPTER 25
Defenses Against Disease

1. b.	6. d.
2. d.	7. b.
3. e.	8. a.
4. a.	9. c.
5. d.	

CHAPTER 26
Excretion

1. a.	8. a.
2. b. (it is a mammal)	9. b.
3. a.	10. d.
4. a.	11. c.
5. b.	12. b.
6. d.	13. decrease; c.
7. e.	

CHAPTER 27
Sexual Reproduction and Embryonic Development

1. j.
2. i.
3. b., e.
4. d.
5. a.
6. f.
7. h.
8. f. (testis) → j. (vas deferens) → g. (urethra) → i. (vagina) → a. (cervix) → h. (uterus) → d. (oviduct)
9. a.
10. f.
11. d.
12. b.
13. a.
14. c.
15. a.
16. c.
17. d., a.
18. g.
19. e.
20. c.
21. C
22. N
23. O
24. C
25. G

CHAPTER 28
Nervous Systems

1. d.
2. a.
3. a. A neurotransmitter substance carries information between two neurons in that it is released as a result of the arrival of an action potential at the presynaptic membrane of one neuron, and its arrival at the postsynaptic membrane stimulates activity in the postsynaptic cell.
   b. Vesicles store neurotransmitter molecules in the presynaptic terminal and release them when an action potential arrives at the terminal.
   c. On combining with neurotransmitter molecules, receptor molecules change the permeability of the postsynaptic membrane to ions, resulting in a local potential.
   d. Enzymes that destroy neurotransmitter molecules in

effect "turn off" the signal brought across the synapse by the neurotransmitter.

4. d.
5. b.
6. d.
7. d.
8. A. cell body of sensory neuron
   B. dendrite of sensory neuron
   C. axon of motor neuron
   D. axon of sensory neuron
   E. synapse between sensory and motor neuron
9. c.
10. norepinephrine; acetylcholine

CHAPTER 29
Sense Organs

1. a.
2. b.
3. e.
4. d.
5. d.
6. b., c.
7. a.
8. check your drawing against Figure 29-10; structures that modify incoming light: cornea, iris, lens; transmission of stimulus to brain is via the optic nerve; receptor types: rods (throughout retina except fovea) and cones (fovea and adjacent parts of retina)
9. adaptation
10. rods; dim; poor

CHAPTER 30
Muscles and Skeletons

1. b.
2. a., c.
3. a., c.
4. c.
5. a., c.
6. A. Z line
   B. actin, calcium (during contraction), tropomyosin, troponin
   C. mitochondria, calcium (during relaxation)
   D. myosin
7. During tetanic contraction, motor neurons are firing a continuous train of closely spaced nerve impulses.
8. A muscle does not relax between nerve impulses during tetanic contraction because each new impulse arrives before the muscle completes the sequence of chemical and mechanical events of contraction initiated by the previous impulse.
9. c.
10. e.

CHAPTER 31
Animal Hormones and Chemical Regulation

1. c.
2. c.
3. a.

4. b.
5. a. negative feedback
   b. target cells
6.

$$\text{Thyroxin} \downarrow \quad \xrightarrow{\text{TSH} \uparrow} \quad \text{Thyroxin} \uparrow$$
$$\xleftarrow{\text{TSH} \downarrow}$$

CHAPTER 32
Behavior

1. g.	5. f.	8. d.
2. a.	6. c.	9. c.
3. h.	7. b.	10. e.
4. j.		11. e.

CHAPTER 33
Plant Structure and Growth

1. a.
2. b.
3. a.
4. a., c., b.
5. b.
6. c.
7. c.
8. a. monocotyledons
   b. both
   c. monocotyledons
   d. monocotyledons
   e. dicotyledons
9. a.

CHAPTER 34
Nutrition and Transport in Vascular Plants

1. e.
2. d.
3. b.
4. a.
5. c.
6. a.
7. epidermis, cortex, Casparian strip, endodermal
8. nitrogen, phosphorus, potassium
9. c.
10. c.
11. function: transport of sap
    adaptation: tubular, dead and hollow, pits or open ends, stacked end to end
    function: support of plant body
    adaptation: secondary wall thickenings
12. d.
13. a. remain the same
    b. increase
    c. remain the same
14. a. decrease
    b. increase
    c. increase
    d. increase
15. a.
16. b.

CHAPTER 35
Regulation and Response in Plants

1. d.   6. d.
2. c.   7. a.
3. b.   8. b.
4. e.   9. d.
5. b.   10. c.

CHAPTER 36
Reproduction in Flowering Plants

1. pollen   12. #2: anther
2. anther   13. #6: embryo sac
3. pistil   14. #5: sepal
4. ovary   15. #11: ovule
5. sepal   16. #6: embryo sac
6. embryo sac   17. #4: ovary
7. stigma   18. false
8. pollen tube   19. e. (but not all seeds require all
9. stamen   factors listed)
10. petal   20. b.
11. ovule

CHAPTER 37
Distribution of Organisms

1. c.
2. a.
3. d.
4. d.
5. c.
6. primary
7. (1) marsh plants
   (2) short fugitive plants
   (3) taller, perennial plants
   (4) shrubs and pioneer trees
   (5) climax trees
8. early successional
9. they are separated by great distances; they have similar
   climates, which exert similar selective pressures (in this
   case, nothing much taller than grasses can grow with so
   little rainfall)

CHAPTER 38
Ecosystems and Communities

1. recycling of nutrients
2.

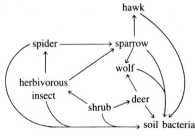

decomposers: soil bacteria
producer: shrub
primary consumers: deer, herbivorous insect
secondary consumers: wolf, spider, (sparrow)
tertiary consumers: hawk, (sparrow)
(sparrow feeds at several possible trophic levels)
3. b.
4. c.
5. e.
6. a.
7. b.
8. a. increase
   b. decrease
   c. increase

CHAPTER 39
Populations

1.

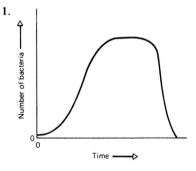

The population grows exponentially, levels off, and then
declines rapidly to extinction in this "closed" environment. Mark yourself $\frac{1}{2}$ off if you forgot to label the axes.
2. b.
3. b.
4. b.
5. d.
6. I.
7. b.
8. a.

CHAPTER 40
Human Ecology

1. b.
2. c.
3. c.
4. e.
5. d.

Glossary

abdomen In vertebrates and some arthropods, the rearmost part of the body, containing the end of the digestive tract, external genital organs, etc.

abiotic Nonbiological; occurring neither within living cells nor under their influence.

abortion Process whereby a mammalian embryo or fetus becomes detached from the uterine wall and expelled from the female's body either naturally ("miscarriage," spontaneous abortion) or by medical means (induced abortion).

Acacia Large genus of trees of the legume family. Most acacias are tropical. Common name: thornwood.

acetyl CoA Acetyl coenzyme A: a molecule that can donate its (two-carbon) acetyl group to production of fatty acids or to the citric acid cycle, where it is broken down to yield energy.

acetylcholine An important neurotransmitter in vertebrates and invertebrates; the transmitter at some synapses in the brain, many nerve-muscle junctions, etc.

acid 1. A substance that releases hydrogen ions (H^+) in aqueous solution. 2. A substance that accepts electrons.

acid rain Rain containing pollutants that give it an acid pH, usually of less than 5.0. The most common pollutants with this effect are oxides of nitrogen and sulfur which form acids when they dissolve in water.

Acrasiomycota Phylum containing the cellular slime molds.

actin A protein that forms the thin filaments involved in muscle contraction and the microfilaments involved in changes of shape in other kinds of cells.

actinomycetes Filamentous prokaryotic organisms.

action potential A fast-moving, all-or-nothing electrical disturbance in the membrane of a nerve cell; a nerve impulse. Also occurs in the membrane of a muscle when it is stimulated to contract.

activation energy Energy input required before a chemical reaction can proceed.

active transport Process in which energy is expended in moving a substance through a membrane against its concentration gradient.

acylation The attachment of an amino acid to a transfer RNA molecule.

adaptation 1. Process by which populations evolve to become suited to their environments over the course of generations. 2. Characteristic that increases an organism's evolutionary success. 3. Process by which a sense organ ceases to respond to a constant stimulus.

adaptive radiation Formation of two or more new species, macromolecules, or physiological pathways, adapted to different ways of life, from one ancestral species, molecule, or pathway.

adhesion Sticking together (of cells, molecules of different substances, etc.).

adipose tissue Tissue made up mainly of fat-storing cells.

adrenalin See epinephrine.

aerial In the air.

aerobic 1. Requiring molecular oxygen (O_2) to live. 2. In the presence of molecular oxygen.

affected individual In genetics, an individual homozygous for a deleterious recessive allele, with a phenotype expressing the allele.

air bladder Air-filled sac providing buoyancy in fish (= swimbladder) or algae.

albinism (adj. albino) White coloration due to lack of genes for normal pigmentation.

alcohol Organic compound containing an alcohol group (COH).

alcoholic fermentation Biochemical pathway in some yeasts (and other organisms) in which pyruvate is converted into ethyl alcohol (ethanol), releasing NAD; used to make alcoholic beverages.

alga (pl. algae) Photosynthetic organism with a one-celled or simple multicellular body plan and lacking a multicellular embryo protected by the female reproductive structure.

alkali (= base) A substance that releases hydroxide ions (OH^-) in water or that accepts hydrogen ions (H^+) or that gives up electrons.

alkaline Having a (basic) pH of more than 7.0.

allantois An embryonic membrane: sac that grows out of the embryonic gut in higher vertebrates; stores the embryo's nitrogenous waste until hatching in reptiles and birds; forms the main embryonic respiratory surface and part of the placenta in mammals.

allele One of two or more genes carrying the information for contrasting forms of the same genetic trait (e.g., blue eyes and brown eyes).

allopatric Living in different areas.

allosteric protein A protein capable of changing shape in a way that alters its binding sites and its activity.

alpine Of high mountain regions.

altitude Height above sea level.

altruistic behavior Behavior that favors the reproductive success of another member of the same species.

ambient temperature Temperature of the surroundings (equivalent to "room temperature").

amino acid Small organic molecule containing both a carboxyl group and an amino group bonded to the same carbon atom. Amino acids are the monomers from which polypeptides and proteins are made.

aminoacyl attachment site That part of a transfer RNA molecule to which an amino acid attaches.

ammonia NH_3.

ammonium NH_4^+.

amniocentesis Technique for obtaining a sample of fetal cells by inserting a syringe through the mother's abdominal wall and uterus, into the amniotic sac surrounding the fetus, which contains cells sloughed off the fetus.

amnion A fluid-filled sac that surrounds the embryo in reptiles, birds, and mammals.

amniotic egg Type of egg found in reptiles and their descendants (birds and mammals) in which the embryo is surrounded by an amnion and other membranes (allantois and chorion).

amphibian Member of the vertebrate class Amphibia, e.g., frogs, toads, salamanders, newts, and their relatives.

amylase An enzyme that breaks glucose molecules from starch molecules.

amyloplast Plastid that stores starch.

amylose, amylopectin Polymers of glucose that combine to make up starch.

anaerobic (n. anaerobe) 1. Without molecular oxygen (O_2). 2. Not requiring molecular oxygen for extraction of energy from food (respiration).

analogous In biology, of two or more organs with the same function but with different evolutionary origins.

anaphase The stage of nuclear division during which chromosomes move to the poles of the spindle apparatus.

anatomy The physical structure of an organism.

anemia A deficiency of hemoglobin or of red blood cells, resulting in a deficiency of oxygen supply to the tissues.

angiosperms Flowering plants.

anion A negatively charged ion.

annelid A segmented worm, member of the phylum Annelida; e.g., earthworm, polychaete, leech.

anterior At or toward the front end of an animal.

anther Organ in flowers that produces pollen.

antheridium Organ that produces male gametes in fungi, algae, mosses, and ferns.

anthocyanins Pigments of red, blue, or purple hue, commonly found in vacuoles of plant cells.

Anthophyta Plant division containing the flowering plants (= angiosperms).

anthropoids Monkeys, apes, and humans, which make up the suborder Anthropoidea of the order Primates.

anthropomorphism Attribution of human characteristics to other animals.

antibiotics Substances that kill microorganisms.

antibody (= immunoglobulin) One of a huge group of proteins, produced by B lymphocytes, which will react with a specific antigen as part of the body's defensive immune response to disease.

anticodon The row of three nucleotides on a tRNA molecule that base-pairs with the complementary codon on an mRNA molecule attached to a ribosome. This allows the amino acid carried by the tRNA to be placed correctly into the peptide chain that is being synthesized on the ribosome.

antigen A substance capable of stimulating an immune response by binding to an antibody or lymphocyte receptor; usually a protein or other substance that is not naturally part of the body.

Anura The order of amphibians containing frogs and toads.

aorta (pl. aortae) Large artery that carries blood from the heart toward the rest of the body.

apical Of the tip of something.

apical meristem Area of dividing cells at the root and stem tips of a plant.

Apicomplexa Phylum of parasitic protists with a distinctive, complex arrangement of components at one end of the cell.

aposematic coloration Conspicuous, "warning" coloration of an animal that is dangerous or distasteful to eat.

aquatic Of water (fresh or salt).

aqueous Containing, or composed largely of, water.

arboreal Living in trees.

Archaeobacteria Ancient group containing the methane-producing, salt-loving, and hot acid-loving bacteria.

arid Dry; of areas where more water leaves the ecosystem (by evaporation and transpiration) than enters it (as precipitation).

arteriole A small artery carrying blood from an artery to a capillary bed.

artery Vessel that carries blood away from the heart.

arthropods Members of the phylum Arthropoda: segmented animals with jointed appendages, and stiff chitin-containing external skeletons; e.g., crabs, lobsters, barnacles, insects, spiders.

ascidian A marine invertebrate chordate (sea squirt).

Ascomycota Division of fungi which reproduce sexually by means of ascospores produced inside asci; sac fungi.

ATP Adenosine triphosphate; most common energy-donating molecule in biochemical reactions.

atrium (pl. atria) A chamber; e.g., part of the heart that receives blood as it enters.

autonomic nervous system The division of the vertebrate nervous system over which the animal usually has no control; composed of the sympathetic and parasympathetic systems; regulates blood pressure, breathing, gut movements, etc.

autoradiography Technique for determining the position of a radioactive element introduced into a system; based on the fact that radiation emitted by radioactive substances exposes (blackens) photographic film that is placed near them.

autosome Any chromosome that is not a sex chromosome.

autotroph (adj. autotrophic) Organism that can make its own (organic) food molecules from inorganic constituents, using energy either from sunlight (green plants) or inorganic chemical reactions (chemosynthetic bacteria). Compare heterotroph.

avian Pertaining to birds, members of the class Aves.

axial Pertaining to an axis.

axil (adj. axillary) In plants, the angle between the stalk (petiole) of a leaf and the stem.

axon (adj. axonal) Extension of a neuron that carries nerve impulses to the next cell(s) in line.

bacterial transformation Uptake and incorporation of genetic material by a bacterium.

bacteriophage (abbr. phage) A virus that infects, and reproduces within, a bacterium.

bacterium Member of the phylum Monera, containing organisms without nuclear envelopes around the single, circular DNA genome; most are very small, unicellular or forming colonies of independent cells.

basal body Structure at the base of a cilium or flagellum, containing a circle of nine triple microtubules.

base (adj. basic) See alkali.

base pairing In nucleic acids, the attachment of one nucleotide base to another by hydrogen bonds, in unique pairs: adenine-thymine in DNA, adenine-uracil in RNA, or guanine-cytosine, in both).

bicarbonate ion HCO_3^-, a common buffer in biological systems.

binary fission Division (usually of a cell) into two equal parts.

biodiversity The number of different species of organisms found in an area.

biogeochemical cycle Description of the geological and biological processes that affect the movement of an element among different components of an ecosystem.

biological species A group containing all the members of one or more populations that interbreed or potentially interbreed with one another under natural conditions.

bioluminescence Production of light by a living organism (such as a firefly).

biomass Amount of material that is part of the bodies of living organisms.

biome Major type of terrestrial communities of organisms; e.g., tropical rain forest, desert.

biosphere Total of all areas on earth where living organisms are found; includes deep ocean and part of the atmosphere.

biota The organisms living in an area.

biotic Having to do with living organisms.

biotic potential The rate at which a population can increase in size by reproduction under ideal conditions.

bipedalism Habit of walking on two legs.

bloom Generally means flower; ecologists also use it to mean a rapid increase in population of microorganisms, especially cyanobacteria or unicellular algae.

bone marrow Soft tissue in the center of the long bones in the limbs.

brachiation Mode of locomotion using the forelimbs to swing from branch to branch.

bromeliad Member of a large family of mostly epiphytic plants.

bryophyte A member of the plant division Bryophyta: embryophyte plants lacking vascular tissue; mosses and liverworts.

bug Member of the insect order Hemiptera.

C₃ cycle ($=$ Calvin cycle) Metabolic pathway of photosynthesis in which NADPH and ATP are used to fix carbon dioxide into three-carbon organic molecules.

caecum A pouch leading off the digestive tract in some animals.

calcareous Composed of, or containing, calcium carbonate ($CaCO_3$).

calorie (adj. caloric) 1 Calorie $=$ 1000 calories $=$ 1 kilocalorie or kcal. A calorie is the amount of heat needed to raise one gram of water from 14.5°C to 15.5°C. The energy values of foods are expressed in Calories.

cambium (adj. cambial) Meristematic tissue that produces new cells which increase the diameter of a woody stem or root.

canine 1. (adj.) Of or pertaining to dogs. 2. (noun) The (usually) pointed tooth next to the incisors in mammals.

capillaries The smallest blood vessels in a circulatory system; exchange of substances between the blood and the extracellular fluid takes place across the thin walls of the capillaries.

capillarity Movement of fluid into a narrow tube or other space because of attraction of the walls of the space for the molecules of the fluid.

carbohydrates A class of compounds whose members have the general formula ($CH_2O)_n$ and contain at least one double-bonded oxygen.

carbon dioxide Gas (CO_2) produced during respiration by organisms, and by burning of organic matter or other carbon-containing substances.

carboxyl group A functional group (—COOH) consisting of a carbon atom double-bonded to an oxygen atom and single-bonded to another oxygen atom, which in turn is bonded to a hydrogen atom.

carcinogen Cancer-causing agent.

cardiac Of the heart.

cardiac muscle The type of muscle that makes up the vertebrate heart.

cardiovascular Of the heart and circulatory system.

carnivore 1. Animal that eats other animals. 2. Member of the mammalian order Carnivora; e.g., dogs, cats, weasels, bears, raccoons, skunks.

carotenoid Accessory photosynthetic pigment that usually appears yellow, orange, or brown.

carpals The bones in the wrist.

carpel A female flower part.

carrier In genetics, an individual heterozygous for a (usually damaging) recessive allele.

carrying capacity Number of individuals of a species that a particular environment can support indefinitely; represented in formulas by the letter K.

cartilage (adj. cartilaginous) 1. Tissue composed of scattered cells surrounded by tough, flexible intercellular protein fibers. 2. A skeletal element composed of cartilage.

cation A positively charged ion.

caudal Of the tail.

cell wall Stiff, fibrous layer that lies outside the plasma membrane of a cell. Occurs in plants, fungi, bacteria, and some protists.

cellular respiration The stepwise release of energy from food molecules, accompanied by storage of the energy in short-lived energy intermediates such as ATP.

cellulose Polysaccharide that makes up fibers that form a large part of the cell walls of plant cells.

central nervous system The brain and spinal cord.

centrioles Pair of structures containing nine triple microtubules, found in most animal cells and in cells of lower plants and fungi; usually close to the nucleus outside the nuclear membrane.

centromere The constricted area where chromatids of a replicated chromosome remain joined until anaphase.

cephalization Concentration of nervous tissue and sense organs in the head.

Cephalopoda Class of molluscs containing octopuses, squids, *Nautilus*.

cereals Those grass species of the plant family Gramineae which make up the bulk of human food.

cerebral cortex Outer layer of the cerebral hemispheres.

cerebral hemispheres The two halves of the cerebrum in the forebrain.

cerebrospinal fluid A clear fluid, derived from blood, that bathes and cushions the brain and spinal cord.

cerebrum A large, rounded area of the forebrain divided by a fissure into right and left halves ($=$ cerebral hemispheres).

Cetacea The order of mammals whose members are most highly adapted to aquatic life, including whales and porpoises.

chaparral Dry shrublands of temperate coastal regions such as California and the Mediterranean.

chelicerae Piercing and sucking mouthparts of spiders and their relatives; associated with poison glands in spiders.

Chelonia Order of reptiles including turtles and tortoises.

chemiosmosis (adj. chemiosmotic) A means of releasing useful energy by permitting hydrogen ions to pass through a membrane down their concentration gradient.

chemoreceptor A sense organ that responds to chemical stimuli.

chemosynthesis Production of organic molecules from inorganic molecules, using energy released by chemical reactions rather than light energy, which is used in the much more common process of photosynthesis. Also called chemoautotrophy or chemolithotrophy.

chitin (adj. chitinous) A stiff polysaccharide, a prominent component of the cuticle of arthropods (e.g., insects) and of the cell walls of many fungi.

chiton A member of the molluscan class Amphineura, having a body covered with eight crosswise skeletal plates.

chlorophyll Green pigment that traps light energy during photosynthesis.

Chlorophyta A division containing the green algae.

chloroplast Green plastid (containing chlorophyll) in which photosynthesis occurs.

Chondrichthyes Class of vertebrates containing cartilaginous fish with jaws; the sharks, skates, and rays (also called elasmobranchs).

chordate Member of the phylum Chordata; animal with a notochord and pharyngeal gill slits at some stage of its life.

chorion The embryonic membrane lying immediately under the egg shell in reptiles and birds; in mammals it forms part of the placenta.

chromatid One of the two replicated strands of a chromosome still held together at the centromere.

chromatin Combination of DNA and proteins, visible as a loosely arranged mass in the nucleus of an appropriately stained eukaryotic cell during interphase.

chromatography Separation of a mixture of substances into its components by differences in their electrical charges and particle sizes.

chromoplast Plastid that produces and stores yellow and orange pigments.

chromosome (adj. chromosomal) Thread-like structure in the nucleus of a eukaryotic cell, consisting of DNA and proteins and carrying genetic information.

Chrysophyta Phylum of protists: diatoms and their relatives. Most are photosynthetic.

ciliate Member of the protistan phylum Ciliophora.

Ciliophora Phylum of heterotrophic protists that move by means of cilia and have two types of nuclei in their cells.

cilium (pl. cilia; adj. ciliary, ciliated) Thread-like organelle containing microtubules, present on the surfaces of many eukaryotic cells and used in movement.

circadian rhythm Cycle of about 24 hours in the physiology and behavior of a eukaryotic organism.

clade The set of all species descended from a particular ancestral species, together with their common ancestor.

cleavage A series of cell divisions that convert the zygote into a multicellular blastula.

cleidoic egg (= amniotic egg) "Closed" egg of reptiles, birds, and mammals.

cline A series of interconnected populations extending from one area to another, with gradual, continuous changes from one population to the next.

cloaca The vestibule, found in most vertebrates, into which urine, feces, and sperm or eggs are discharged before they leave the body.

clone A population of cells or individuals descended from one original cell or individual by asexual propagation, and hence genetically identical.

Cnidaria Phylum of simple animals with only two well-developed layers of cells, only one opening into the gastrovascular cavity, tentacles, and stinging nematocysts (e.g., jellyfish, *Hydra,* corals, sea anemones).

coccus (pl. cocci) Sphere-shaped bacterium.

codominance Condition in which both alleles are expressed in the phenotype of a heterozygote.

codon Series of three messenger RNA nucleotides which, in the genetic code, specifies a particular amino acid.

Coelenterata Old name of a phylum whose members are now placed in the phyla Cnidaria and Ctenophora.

coelom (adj. coelomic) A body cavity lined with mesodermal tissue, in which the internal organs are suspended.

coelomates Animals with bodies containing coeloms.

coenocyte Structure in fungi or plants in which many nuclei occupy a mass of cytoplasm, without being separated into individual cells by plasma membranes.

coenzyme Organic molecule that must be present for an enzyme to function.

coevolution Evolution together of two or more species whose members exert selective pressures on one another.

cofactor Ion or molecule that must be associated with an enzyme for the enzyme's proper functioning.

cohesion The sticking together of molecules of the same substance.

coleoptile The sheath covering the embryonic shoot in a monocot seed.

collagen A structural protein that forms fibers; common intercellular component of connective tissue.

colony (adj. colonial) In animals, bacteria, or protists, a group of more or less independent individuals attached to one another.

commensalism Close association between members of different species in which one member benefits and the effect on the other is neutral or unknown.

community All the organisms living in a particular habitat.

competitive exclusion Name given to the idea that two species with identical niches cannot exist together in the same place and time.

concentration The proportion of one substance found in the total of a mixture of several substances; may be given in terms of weight, of proportion of molecules, and so on. Concentration is symbolized by square brackets, as in [sugar] = concentration of sugar.

concentration gradient The change in concentration of a substance over a distance.

condensation reaction Chemical reaction in which two molecules become covalently bonded by removing —H from one and —OH from the other, with the removed atoms joining to form a molecule of water.

conifer Member of the plant division Coniferophyta: cone-bearing gymnosperms, including pines and spruces (as well as junipers and yews, whose reproductive structures do not resemble cones).

conjugation Method of genetic recombination in which one cell passes DNA to another via a physical link formed between the two.

connective tissue Tissue composed of scattered cells and much intercellular material secreted by the cells.

conservative Of a character that has changed very little during the course of evolution.

consumer Organism that eats other organisms.

convergent evolution Evolution of similar features by unrelated organisms in response to similar selective pressures.

copulation (= mating, coitus) Linking of sexual organs of male and female that permits transfer of sperm (in semen, spermatophores, etc.) from male to female.

coronary Pertaining to the heart.

corpus luteum "Yellow body" that forms in an egg follicle of an ovary after the follicle has ruptured and released an egg.

cortex (adj. cortical) Layer lying just inside the outermost boundary (epidermal or epithelial) layer of a stem, root, kidney, brain, etc.

cotyledon A seed leaf of a plant embryo.

courtship Behavior that precedes mating.

Crocodilia Order of reptiles including crocodiles, alligators, etc.

cross section View of an organism from a cut made perpendicular to the long axis of the body.

cross-fertilization Fertilization of one plant by sperm nuclei from another plant (not to be confused with cross-pollination).

cross-pollination Transfer of pollen from male flower parts of one plant to female flower parts of another.

crossing over In genetics, the process by which homologous chromosomes exchange pieces of DNA and form new assortments of genes.

crustacean Member of the arthropod class Crustacea: lobsters, shrimp, crabs, barnacles, pill bugs, copepods, ostracods, etc.

cubic centimetre (abbr. cc, cm^3) A volume equal to one millilitre.

cutaneous Of the skin.

cuticle Layer of waxy waterproof substance secreted on the outer surface of an organism.

cyanobacteria Prokaryotic organisms that use chlorophyll *a* in their photosynthesis and produce O_2 as a byproduct (also called blue-green bacteria or cyanophytes).

cycad Palm-like gymnosperm; member of the division Cycadophyta.

cyst 1. A dormant organism within a resistant covering; stage in which some organisms pass through adverse conditions. 2. Sac-like nonmalignant tumor.

cytochrome An electron carrier molecule consisting of a protein and a porphyrin ring containing a metal ion.

cytokinesis Division of a eukaryotic cell in two, following nuclear division.

cytokinins Plant hormones that stimulate cell division.

cytoplasm The fluid that makes up all of a cell except its nucleus and organelles.

cytoplasmic streaming Flow of cytoplasm within a cell or between adjacent cells.

cytoskeleton The "skeleton" of a cell, composed of thin tubules and contractile filaments in the cytoplasm; much of the cytoskeleton can be assembled from subunits and disassembled as needed.

cytosol The soluble portion of cytoplasm.

cytotoxic T cell Lymphocyte that destroys a cell displaying both self and foreign antigens on its surface.

deamination Removal of an amino (—NH$_2$) group.

deciduous Of plants that lose their leaves during one season of the year; not evergreen.

decomposer In ecology, an organism that feeds on the dead bodies, body parts, or wastes of other organisms, thereby breaking down and recycling the nutrients they contain.

dendrite Extension of a neuron that receives stimuli or impulses from other neurons.

depolarization Decrease in the electrical potential difference across a membrane.

desiccation Drying out.

desmosome Attachment between two cells in which the cells' membranes are fastened together in areas of high stress.

detritus Molecules and larger particles of dead organic matter.

Deuteromycota Division of fungi whose members do not reproduce sexually.

diaphragm 1. In mammals, a muscular partition between the thorax and abdomen, used in breathing. 2. A disk-like rubber device inserted into the vagina to bar sperm from reaching an egg, used for birth control. 3. A disk with a central opening that controls the amount of light entering a microscope.

dichotomous Forking into two.

dicotyledon (abbr. dicot) Member of the group of flowering plants whose embryos have two cotyledons.

dikaryon Fungus or part of fungus in which each cell contains two haploid nuclei.

dinoflagellate One-celled organism with two flagella belonging to the phylum Pyrrophyta; most are photosynthetic.

diploid Containing twice the number of chromosomes found in a gamete; having paired homologous chromosomes.

disaccharide A molecule formed by a condensation reaction between two simple sugars (monosaccharides).

distal In a position or direction away from the point of an appendage's attachment to the body.

disulfide bond In protein structure, a covalent bond between sulfur atoms that are parts of two different cysteine monomers; the bond attaches parts of the same or different polypeptides to each other.

diurnal 1. Active during the daytime. 2. Daily.

division A taxon of plants or fungi equivalent to a phylum in the animal kingdom.

dizygotic twins Twins arising from two different zygotes (fertilized eggs); nonidentical twins.

DNA Deoxyribonucleic acid, the genetic material of organisms and of many viruses.

dominant allele Allele expressed in the heterozygote.

dominant generation Larger, more conspicuous stage in the life history of a species with alternation of generations.

dorsal Toward the back, or uppermost surface, of an animal.

echinoderm ("spiny-skinned") Member of the invertebrate phylum Echinodermata, e.g., sea stars, sea urchins, sand dollars, sea cucumbers.

ecology Study of the relationships of organisms with other organisms and with their physical environment.

ecosystem All of the organisms present in a particular area, together with their physical environment.

ecto- Prefix meaning external or outermost.

ectoderm Outermost of the three germ layers of the embryonic gastrula, giving rise to skin, nervous system, and associated structures.

ectothermy Regulation of body temperature by behavioral means (obtaining heat from outside the body).

EEG Electroencephalogram; recording of electrical activity in the brain.

effectors Structures that carry out an animal's response to a stimulus; e.g., muscles, glands.

EKG Electrocardiogram; recording of electrical activity in heart muscle.

elasmobranchs Cartilaginous fish with jaws; the sharks, skates, and rays. Another name for Chondrichthyes.

electrical potential Difference in concentration of electrically charged particles on two sides of a membrane.

electron Fundamental negatively charged particle occupying space outside the nucleus of an atom.

electron micrograph Photograph taken using an electron microscope.

electron transport chain (= electron transport system) A series of proteins found in certain biological membranes that splits hydrogen atoms into electrons and hydrogen ions, and removes the electrons.

electrophoresis A technique that separates substances, using the fact that they move at different rates (depending on their size and electrical charge) when subjected to an electric current.

embryo A multicellular developing plant or animal still enclosed inside the parent's body or in a seed or egg.

embryophyte A plant in which the zygote remains within the parent plant as it develops into an embryo; a land plant.

emigration Leaving an area of residence for some other place.

endemic (adj. sometimes used as a noun) Peculiar to a particular population or locality, where it originated.

endergonic reaction Reaction in which the energy of the products is greater than the energy of the reactants.

endo-, end- Prefix meaning inside or within.

endocrine gland A gland whose hormone secretion enters the body fluids directly rather than being transported to the site of action through a duct.

endocytosis Engulfing of a particle by a cell.

endoderm The innermost of the three germ layers in the early embryo of higher animals; gives rise to lining of the digestive tract.

endodermis Layer of cells between the cortex and the pericycle in a root.

endoplasmic reticulum System of membranous sacs and tunnels found in most eukaryotic cells.

endoskeleton Internal skeleton.

endosperm Nutritive tissue in an angiosperm seed.

endothermy Maintenance of a particular body temperature by physiological regulation of the loss of metabolically generated heat.

energy The capacity to do work.

entomology The study of insects.

environment An organism's physical and biological surroundings (often includes conditions within the organism's own body).

enzyme Protein that catalyzes a particular biochemical reaction involving specific substrate (reactant) molecules.

epi- Prefix meaning on, above, or around.

epidermis Layer of cells covering the outside of the body.

epinephrine A hormone released by the medulla of the adrenal glands; one of its effects is to bring about the physiological changes associated with stress. Also called adrenalin.

epiphyte Plant that grows on the surface of larger plants, which are not harmed by having the smaller plant growing there; e.g., most orchids.

epithelial tissue Animal tissue that forms a sheet covering an external or internal surface.

equator An imaginary circle round the earth's surface formed by the intersection of a plane passing through the earth's center perpendicular to its axis of rotation.

equilibrium 1. Point in a chemical reaction at which the rates of forward and reverse reactions are equal and energy change during the reaction is zero. 2. Sense of balance.

erythrocyte Red blood cell.

essential amino acid An amino acid that an animal must obtain in its diet because it cannot make enough of the molecule to survive.

estivation (also aestivation) Period of dormancy during the summer.

estrus (adj. estrous) Sexually receptive state of female mammals; usually called rut in males.

Eubacteria ("true bacteria") Prokaryote group containing most bacteria, including actinomycetes, mycoplasmas, spirochetes, and cyanobacteria. (Archaeobacteria are the only other prokaryotes.)

Euglenophyta Phylum of photosynthetic protists containing *Euglena* and its relatives.

eukaryotic Having a nuclear membrane surrounding the genetic material, and with other membrane-bounded organelles in the cytoplasm.

eutrophic Of a body of fresh water rich in nutrients and hence in living organisms.

eutrophication Process in which debris accumulates in a body of water, making it richer in nutrients and hence in organisms, until eventually it fills in and becomes dry land.

evaporative cooling Reduction of temperature by escape of the fastest-moving (that is, warmest) water molecules as water vapor.

evolution 1. Descent of modern species of organisms from related, but different, species that lived in previous times. 2. Change in the gene pool of a population from generation to generation.

ex-, exo- Prefix meaning outside or proceeding from.

excitatory Of nerve or muscle, tending to cause depolarization of the postsynaptic membrane, enhancing the chances that the postsynaptic cell will fire an action potential.

exergonic reaction Reaction that releases energy, that is, one in which the energy of the products is less than the energy of the reactants.

exocytosis Method of expelling substances contained in a membrane-bounded vesicle to the exterior of the cell by fusion of the vesicle membrane with the plasma membrane.

exon The part of a structural gene that is translated into protein (as opposed to an intron, which is not).

exoskeleton External skeleton, outside the rest of the body.

exothermic Heat-releasing.

extinction (of a species) Disappearance of a species from earth when its last surviving member dies.

extracellular Outside a cell.

extracellular fluid (ECF) Fluid surrounding cells.

FAD Flavin adenine dinucleotide, a hydrogen carrier molecule; a coenzyme in cellular respiration.

Fallopian tube Oviduct, in mammals.

fat A lipid made up of three fatty acids and glycerol; solid at room temperature.

fauna Animal life of an area.

fermentation Anaerobic breakdown of food molecules to release energy, in which the final electron acceptors (and end products) are organic molecules.

fertility 1. Of a female, the ability to reproduce. 2. Of soil, the ability to supply plants with the nutrients they need. 3. (rate) Of a population, number of offspring produced per female per unit time.

fertilization Union of an egg with a sperm.

filter feeders Animals that eat smaller organisms or particles of organic matter, which they strain out of the surrounding water.

filtrate Substance that has passed through a filter.

fission Division into (usually two) parts.

fitness In population genetics, a measure of the ability of an organism to produce surviving offspring.

fix 1. In chemistry, to incorporate into a less volatile compound. 2. In genetics, to establish one allele in place of all alternate alleles in a population.

flagellate 1. (noun) Protist that moves by means of one or more flagella. 2. (adj.) Bearing one or more flagella.

flagellum (pl. flagella) Long thin projection from a cell, which moves and propels the cell.

flatworm Member of the invertebrate phylum Platyhelminthes; e.g., planarians, flukes, tapeworms.

flavin adenine dinucleotide See FAD.

flora Plant life of an area (often includes bacteria and fungi, reflecting the old two-kingdom system).

fossil fuel A fuel created by decomposition and geological processes from the remains of dead organisms; includes oil, coal, natural gas, and peat.

fossils Remains of organisms, or other evidence of once-living organisms, preserved in rocks.

frameshift mutation Mutation that adds or deletes nucleotides, thereby altering the reading frame of the genetic message.

free energy Energy that is available or usable to do work.

frond A leaf, usually highly divided (usually applied to ferns or palms).

fruit Structure that develops from the ovary of a flower, surrounding one or more seeds.

fruiting body Rather large, prominent reproductive structure formed by some fungi.

fugitive species In ecology, species that occur in an area for only a short time, during succession.

Fungi Kingdom of organisms containing eukaryotic, unicellular (yeasts) or multicellular heterotrophs feeding by absorbing nutrients through their cell walls.

gamete Sexual reproductive cell: egg or sperm.

gametophyte Haploid plant that produces haploid gametes by mitosis.

ganglion (pl. ganglia) A group of neuron cell bodies.

gap junction An area where ions can move directly between the cytoplasm of two animal cells by way of many tiny tunnels through their plasma membranes.

gastrocoel The hollow in the embryonic gastrula that becomes the lumen of the gut.

gastropod Member of the molluscan class Gastropoda, including snails, slugs, nudibranchs, and limpets.

gastrovascular cavity A cavity in the body that serves for both digestion and distribution of food.

gated channel Channel through a membrane that can be opened or closed by chemical or electrical events.

gene A length of DNA that functions as a unit.

gene bank An institution where plant material is stored in a viable condition. Usually seeds are dried and frozen in a sealed container. Some plants have seeds that will not survive this treatment and they must be maintained as growing plants or in tissue culture.

gene flow Transfer of genes between one more-or-less isolated population and another.

gene pool All of the genes present in a population of organisms.

genera Plural of genus.

generation time Time elapsed from production (birth) of a new individual to production of its first off-spring.

genetic drift Changes in the gene pool of a population caused by random events (as opposed to natural selection).

genetic engineering The isolation of useful genes from a donor organism or tissue and their incorporation into an organism that does not normally possess them.

genetic reassortment Production of new gene combinations in sexually reproducing organisms by crossing over and independent assortment during meiosis and by combination of gametes at fertilization.

genetic recombination Formation of new combinations of genes by joining DNA from two different molecules into one.

genital Of the reproductive system.

genome All the genetic material contained by an individual (or by a representative member of a population).

genotype The particular genes present in an individual, some of which may not be expressed in the phenotype; usually refers to only one or a few gene pairs.

genus (pl. genera; adj. generic) The taxon above species in the hierarchical classification of organisms. The genus name is the first word of the Latin binomial for a species; e.g., *Ursus* is the generic name of the grizzly bear, *Ursus horribilis.*

germ cells Cells that give rise to reproductive cells (eggs, sperm, spores).

germ layers The three layers of cells in the embryonic gastrula (ectoderm, mesoderm, endoderm).

gestation Period during which an embryo is carried within the mother's body before birth.

gibberellins Plant hormones that stimulate cell enlargement.

gills Thin extensions of the body surfaces of many aquatic animals, used for gas exchange and/or feeding.

Ginkgophyta Gymnosperm division containing *Ginkgo* trees.

gizzard Part of the stomach or gut modified as a heavy-walled grinding chamber.

glaciation (= Ice Age) One of four cold periods during the Pleistocene era when ice and glaciers extended farther south from the North Pole than they do now.

glucose A six-carbon monosaccharide.

glycogen A storage polysaccharide made from glucose monomers, commonly found in animals.

glycolipid Molecule made up of carbohydrate and lipid subunits.

glycolysis Biochemical pathway found in most organisms, in which glucose is broken down into pyruvate and hydrogen atoms, with the storage of energy in ATP.

glycoprotein Protein with carbohydrate attached.

Golgi complex Stack of membrane-enclosed sacs that modify and package molecules produced in a eukaryotic cell.

gonad In animals, an organ that produces gametes.

grafting A procedure in which a tissue or organ of one plant or animal is attached to and incorporated into the body of another.

grain 1. The fruit of members of the monocotyledonous plant family Gramineae from which most human food comes. 2. A cluster of molecules of starch.

gram molecular weight See mole.

gray matter Nervous tissue made up of unmyelinated neuron cell bodies and extensions.

green algae Members of the division Chlorophyta: plants with unicellular or multicellular body plans, using chlorophylls *a* and *b* for photosynthesis and storing food as starch.

green plants Photosynthetic organisms that give off oxygen during their photosynthesis, including prokaryotic cyanobacteria and all photosynthetic eukaryotes (protists and plants).

greenhouse effect Heating of the earth caused by gases in the atmosphere that trap infrared radiation from the earth and prevent it escaping into space.

guard cells Two cells surrounding a stoma (pore) in the epidermis of a plant that are able to regulate the pore's opening and closing.

gustation Tasting.

guttation Exudation of water from tips of xylem veinlets in leaves, due to root pressure.

gymnosperm Nonflowering plant that produces seeds, e.g., pines, redwoods, cycads.

habitat The physical area where an organism lives.

haploid Containing the number of (unpaired) chromosomes found in a gamete; equal to half the number of chromosomes found in a body cell of most higher plants and animals.

haustorium (pl. haustoria) An extension of a fungus inside the cell wall of a living plant cell; the fungus lies against the plant cell's plasma membrane and absorbs nutrients from the plant cell.

hectare 10,000 square metres, or about 2.5 acres.

heme Iron-containing group in hemoglobin and some cytochromes.

hemocoel Body cavity containing a fluid that acts as the transport medium in many invertebrates.

hemoglobin Respiratory pigment that carries oxygen in the blood of vertebrates and various invertebrates.

hemolymph Fluid that fills the body cavity and acts as blood in animals with open circulatory systems.

hemolysis The bursting of blood cells.

hemophilia Condition in which blood fails to clot, resulting in excessive bleeding, caused by a defect in a gene for clotting factor.

hemorrhage Bleeding, e.g., from a ruptured blood vessel.

herbaceous Having nonwoody stems.

herbicide A chemical used to kill plants.

herbivore An animal that eats plants or parts of plants.

hermaphroditic Containing both male and female organs.

hetero- Prefix meaning "other" or "different."

heterospory In plants, production of spores of two distinct sizes, which give rise to gametophytes of the two sexes: microspores to male gametophytes, megaspores to female.

heterotroph (adj. heterotrophic) Organism dependent on other organisms to produce its organic (food) molecules.

heterozygote (adj. heterozygous) Individual with two different alleles in a given gene pair.

heterozygote advantage (= heterosis) Selective advantage of a heterozygote over either homozygote.

histone Type of protein with an alkaline pH, characteristic of eukaryotic chromosomes and playing a role in the packing of DNA and regulation of transcription.

homeostasis (adj. homeostatic) The maintenance of conditions inside the body within the narrow limits required for life.

homeothermy Maintenance of a constant, high body temperature and metabolic rate.

hominids Humans and their direct ancestors: members of the primate family Hominidae, with large brains, small teeth, and bipedal locomotion.

hominoids Humans and apes: large tail-less primates.

homologous 1. Of chromosomes, of the same origin and containing the same kinds of genetic information. 2. Of structures or organs, originating from the same structure in ancestral forms (e.g., a bird's wing and a seal's front flipper are homologous structures).

homozygote (adj. homozygous) Individual having same allele in both members of a given gene pair.

homozygous recessive Having two copies of a recessive allele in a given gene pair.

hormone Chemical messenger produced in one part of the body and specifically influencing certain activities of cells in another part of the body.

hunter-gatherers Members of human society in which food is obtained by collecting plants and hunting wild animals.

hybrid Offspring of a mating between genetically different individuals.

hydrogen bond Weak link between two molecules, or two parts of the same macromolecule, due to the attraction of a hydrogen atom with a partial positive charge to an oxygen or nitrogen atom with a partial negative charge.

hydrolysis The breaking apart of a molecule into its monomer subunits by addition of the components of a water molecule into each of the covalent bonds linking the monomers.

hydrophilic "Water-loving"; able to dissolve in water; polar or ionic.

hydrophobic "Water-hating"; unable to dissolve in water; nonpolar.

hydrostatic pressure Pressure exerted by confined fluid.

Hymenoptera The order of insects that includes bees, wasps, ants, etc.

hyperpolarization Increase in the membrane potential.

hypertonic Of a solution, tending to gain water from a solution to which it is being compared, when the two are separated by a membrane; usually this means having a higher concentration of dissolved particles (lower osmotic potential) than the other solution.

hypha (pl. hyphae) One of the thread-like structures that make up the body of a fungus.

hypothalamus A part of the brain, responsible for monitoring internal conditions in the body and initiating behaviors that tend to maintain physiological homeostasis.

hypothesis (pl. hypotheses) A proposed answer to a question.

hypotonic Of a solution, tending to lose water to a solution to which it is being compared, when the two are separated by a membrane; usually this means having a lower concentration of dissolved particles (higher osmotic potential) than the other solution.

immigration Movement of new individuals into an area.

immunoglobulin See antibody.

immunology Study of the reactions that provide protection specifically against individual diseases.

inbreeding Mating of related individuals that share many of the same alleles, and therefore produce offspring with a high degree of homozygosity.

incisors Chisel-shaped cutting teeth found in the center of the lower and upper jaws in mammals.

incomplete dominance Condition in which neither of a pair of alleles hides the presence of the other in the phenotype.

independent assortment Creation of new haploid combinations of chromosomes due to random lining up of members of homologous chromosome pairs on either side of the spindle equator during meiosis.

inducer molecule Molecule (often a food molecule) that causes the genetic machinery of a cell to transcribe the genes for and to produce particular metabolic enzymes.

induction 1. The initiation of protein synthesis because a particular inducer substance is present. 2. In development, conversion of one type of tissue into another as a result of the tissue's contact with a particular stimulus.

infant mortality rate The death rate for humans in the first year of life.

ingest To take into the body through the mouth.

innervate Supply nervous connections to.

insect Arthropod with three distinct body areas (head, thorax, abdomen); adults with three pairs of legs, and usually two pairs of wings, attached to the thorax.

insectivore (adj. insectivorous) Insect-eating.

insulin A small protein hormone, one of whose functions is to regulate cellular uptake of glucose from the blood; it is produced by the pancreas.

interferon Protein produced by mammalian cells that interferes with replication of viruses.

intermediary metabolism Total of all of a cell's enzyme-mediated reactions involved with extracting energy from food molecules and using it to synthesize the cell's own molecules.

interneuron Type of neuron in the central nervous system that is neither sensory nor motor, but transmits information between other neurons.

internode Portion of a stem between sites of leaf attachment.

intertidal Between tidemarks; covered by water at high tide and exposed to the air at low tide.

intervening sequence See intron.

intracellular Inside a cell.

intraspecific Within one species.

intron Part of a structural gene that is transcribed into messenger RNA but not translated into protein.

invertebrate An animal that lacks a backbone; e.g., earthworm, snail.

ion (adj. ionic) Particle carrying one or more positive or negative electrical charges.

iso- Prefix meaning "same."

isomers Molecules containing identical numbers and types of atoms, but with these atoms arranged differently.

isotonic Of a solution, having osmotic properties such that it neither gains nor loses water across a membrane separating it from a solution to which it is being compared.

isotopes Two or more forms of an element, differing by the numbers of neutrons in the nuclei of their atoms.

K 1. Chemical symbol for potassium, which usually occurs as an ion (K^+). 2. Symbol for carrying capacity.

kelps Large marine brown algae with considerable differentiation of tissues.

keratin A structural protein that makes up hair and fingernails.

killer cells Poorly understood lymphocyte-like immune cells, which are apparently specialized to destroy abnormal body cells.

kin selection Natural selection that causes an individual to act in such a way as to enhance the survival or reproduction of other, genetically related individuals (kin).

kinetochore Structure on a chromosome to which microtubules attach; usually located on or near the centromere.

krill Marine planktonic arthropods, up to about 15 centimetres in length.

labelled Prepared in such a way that it can be traced; e.g. containing a high proportion of a rare isotope, allowing the fate of the atoms to be traced by methods that can distinguish between the isotopes of an element.

lactation Secretion of milk from mammary glands.

larva (pl. larvae) Immature stage of an animal with different appearance and way of life from the adult.

larynx Voice box ("Adam's apple") located at the top of the trachea (windpipe).

lateral 1. (adj.) Pertaining to the side. 2. (noun) A side branch or branch root of a plant.

lateral line organ Pressure-sensitive sense organ found in fish and larval amphibians, used to detect water currents, etc.

laterite A hard crust that may develop when vegetation is removed from the surface of soil containing metals such as aluminum and iron in tropical regions with wet and dry seasons. In the dry season, soil solution rises to the surface by capillarity, and aluminum and iron oxides accumulate and combine to form the crust. Laterization results in an infertile soil called latosol.

latitude Distance north or south of the equator.

leach To dissolve and carry out of; leached soil has lost mineral nutrients that have dissolved in rainwater running through the soil and have been carried away into streams.

leaf nodes Areas of stem at which leaves are attached.

legumes Members of the plant family Leguminosae, including beans, peas, peanuts, alfalfa, clover, *Acacia*.

Lepidoptera The order of insects that includes moths and butterflies.

lethal allele Allele whose expression causes premature death of the individual that carries it (usually, in the homozygous recessive condition).

leucoplast Colorless plastid.

leukocyte White blood cell.

life expectancy The number of years a particular person can expect to live, calculated from actuarial statistics.

linkage group All of the genes on the same chromosome, which are therefore all inherited together (except for crossing over).

lipid One of a large class of organic molecules not soluble in water; includes fats, waxes, oils, steroids, carotenes.

lipopolysaccharide Molecule composed of lipid joined to polysaccharide.

lipoprotein Molecule made up of lipid and polypeptide subunits.

locus (pl. loci) The position on a chromosome that is occupied by an allele for a particular gene; e.g., the hemoglobin beta chain locus may be occupied by an allele for normal or sickle hemoglobin. (This book uses the less technical term, gene location.)

lumen Space in the center of a tube.

Lycophyta A division of primitive vascular plants: club mosses and ground pines.

lymphocyte (= T cell or B cell) One of a group of white blood cells responsible for the specificity of immune responses: production of antibodies, recognition of antigens, and immunological memory.

lysis (adj. lytic) Bursting of a cell.

lysosome Membrane-bounded organelle filled with hydrolytic (digestive) enzymes.

macromolecule Large molecule. Usually refers to a polymer with a molecular weight of many thousands, made up of many (identical or different) monomers.

macronutrient Nutrient needed in relatively large amounts.

macroscopic Visible to the unaided eye.

maize *Zea mays*, corn, important cereal crop.

major histocompatibility complex See MHC.

mammal Warm-blooded vertebrate with lower jaw consisting of only one bone (the mandible) on each side, with fur or hair, with young nourished by milk from the mammary glands of the female parent; e.g., humans, rabbits, cattle.

mantle Sheet of tissue covering the visceral mass of a mollusc and secreting the shell, if present.

map In genetics, the plan of a genome showing the relative positions of particular genes on the organism's chromosomes.

marine Of the sea.

marsupials Mammals whose young are born quite early in development and complete their development attached to a nipple in the mother's marsupium, or pouch.

mating types The equivalent of

sexes in fungi, some bacteria, and protists.

median The value that falls between the lowest and the highest 50% of individual measurements.

medusa (pl. medusae) One of the two possible forms of the cnidarian body (the other is the polyp); often the reproductive stage in the life history; e.g., the body of a jellyfish is a medusa.

megaspores Large spores produced by meiosis in some plants and giving rise to female gametophytes.

meiosis A series of nuclear divisions that produces four new nuclei, each with half the number of chromosomes contained in the original nucleus.

melanin Black pigment that gives dark color to organisms.

melanism Dark coloration.

membrane potential The difference in electrical charge across a membrane; the inside of a cell is negatively charged with respect to the outside of the cell so the membrane usually has a potential of about −40 millivolts (about −70 millivolts for a neuron).

meristem (adj. meristematic) Region of dividing cells in a growing area of a plant.

mesentery A sheet of mesodermal tissue that suspends the internal organs in the coelom.

mesoderm The middle one of the three embryonic germ layers of the gastrula; gives rise to most of the muscles, heart, kidneys, gonads, etc.

mesoglea The "middle glue" layer, containing few, scattered cells, between the outer and inner layers of a cnidarian.

messenger RNA The molecule that carries genetic information from DNA to ribosomes, where the information is used as a code to direct the order in which amino acids are joined to form a polypeptide.

metabolic heat Heat released by chemical reactions in the body.

metabolic rate The rate of the total of an organism's biochemical reactions; usually measured as the rate of oxygen consumption by respiration, since respiration produces the energy needed for the other biochemical processes.

metabolism All the chemical reactions taking place within an organism.

metamorphosis The radical change in shape, physiology, and behavior that occurs when a larva becomes a very different-looking adult.

methane Gas (CH_4) produced by the metabolic processes of anaerobic methanogen bacteria.

mg (milligram) A thousandth of a gram.

MHC (major histocompatibility complex) The group of genes that determines many mammalian cell surface antigens, including those that cause rejection of transplanted organs.

microfilament Filament assembled from actin subunits, found near the plasma membrane; part of the cytoskeleton; responsible for much of the movement within a eukaryotic cell.

micrograph Photograph taken using a microscope.

micrometre (μm) 10^{-3} mm = 10^{-6} m

micron Outmoded name for micrometre.

micronutrient Nutrient needed by an organism in relatively small amounts.

microorganisms Unicellular or simple many-celled organisms: bacteria, fungi, protists, or small algae.

microspores Small spores produced by meiosis in some plants and giving rise to male gametophytes.

microtubule Thickest type of filament in the cytoskeleton of a eukaryotic cell; present in eukaryotic centrioles and in cilia, flagella, and their basal bodies, and in mitotic and meiotic spindles; assembled from protein subunits.

microvillus Tiny, finger-like projection that increases the surface area of a cell.

middle lamella The shared partition between the cell walls of adjacent plant cells.

mimicry Resemblance of one organism to another, or to a nonliving object, providing an offensive or defensive advantage to the mimic.

mineral Any naturally occurring inorganic substance having a definite chemical composition and a particular crystalline structure, color, and hardness.

mites Small arthropods in the class Arachnida. Mites have eight legs, and the body is not divided into two parts as it is in other arachnids, such as spiders.

mitochondrion (pl. mitochondria)

Large, self-replicating membrane-bounded organelle where most of a eukaryotic cell's ATP is produced.

mitosis (adj. mitotic) Series of events that results in the division of one cell nucleus into two nuclei containing sets of chromosomes identical to that in the original nucleus.

mole Gram molecular weight; e.g., 1 mole of water = 18 grams because the molecular weight of H_2O is 18.

molecular biology Study of the molecular basis of inheritance.

molecular weight Sum of the atomic weights of the atoms in a compound.

Mollusca Phylum of soft-bodied invertebrate animals with a muscular head-foot and a mantle, which usually secretes a shell; e.g., snails, clams, squids.

molting Shedding of skin, exoskeleton, or feathers.

Monera Kingdom containing all prokaryotic organisms (bacteria).

monocotyledon (abbr. monocot) Member of the group of flowering plants whose embryos have only one cotyledon.

monogamy The mating of one male with one female, either for life or for the duration of one breeding season.

monomers Small molecules that may become joined together to form large (macro) molecules; e.g., amino acids are the monomers that make up polypeptides.

monophyletic Of a clade, i.e., a taxon that contains a common ancestor and all the species descended from it.

monosaccharide Simple sugar, with formula given by $(CH_2O)_n$; e.g., glucose, ribose.

monotremes Egg-laying mammals.

monozygotic twins Twins originating from the same fertilized egg; identical (genetically) twins.

morph Form, variety.

morphogenesis (adj. morphogenetic) The formation of shape or structures during development.

morphology (adj. morphological) Structure, anatomy.

mortality Death.

motile Able to move itself from place to place.

multicellular Composed of more than one cell.

mutagen Agent (e.g., chemicals, certain kinds of radiation) that causes mutation.

mutation Inheritable change in the genetic material (DNA).

mutualism Close association between members of different species that benefits both.

mycelium (pl. mycelia) The body of a fungus.

mycology Study of fungi.

mycorrhiza (pl. mycorrhizae) Mutualistic association between a fungus and the roots of a higher plant; the fungus takes up mineral nutrients from the soil and passes them to the plant, receiving some organic (food) molecules made by the plant in return.

myelin sheath Layers of fatty, insulating wrapping around the axons of some neurons.

myoglobin Oxygen-storing molecule in muscles.

myosin Protein that interacts with actin to produce movement in cells, such as contraction in muscle cells.

myotome A block of muscle that forms one of a series of such blocks along the back of an animal.

Myxomycota Acellular slime molds.

NAD (nicotinamide adenine dinucleotide) A hydrogen-carrying coenzyme in cellular respiration.

nanometre (abbr. nm) 10^{-9} metre.

natural selection Differential reproduction among the variety of genotypes in a population; leads to different genetic contributions to future generations.

negative feedback Mechanism whereby the change detected in some condition stimulates compensating physiological activity that brings the condition back toward its average value.

nematocyst Stinging structure characteristic of cnidarians.

nematode A roundworm, member of the phylum Nematoda.

neuron (adj. neuronal) Nerve cell.

neurotransmitter Chemical that transmits information in the nervous system; it travels across the synaptic cleft from a neuron that has just fired an action potential to

another cell, whose electrical or chemical activity it affects.

niche The way of life of a species; includes the habitat, food, nest sites, etc., that it needs in order to survive.

nitrogen fixation Conversion of gaseous nitrogen (N_2) to ammonia (NH_3).

nocturnal Active at night.

nomadic Moving from place to place, with no single fixed home.

nondisjunction Failure of homologous chromosomes or of sister chromatids to separate; if this occurs during meiosis, one of the resulting nuclei will have an extra copy of the chromosome, while another will have one chromosome too few.

notochord Elastic rod dorsal to the gut in all chordate embryos; in most adult chordates (i.e., vertebrates), the notochord is replaced by vertebrae, which form around it.

nucleic acids Class of macromolecules, made up of nucleotide monomers, that contain the genetic information of organisms; DNA and RNA.

nucleolus (pl. nucleoli) Area of the cell nucleus where the nucleolar organizer (part of one or two particular chromosomes) lies and where ribosomes are made; often appears denser than the rest of the nucleus in micrographs.

nucleosome Particle-like structure formed by part of a chromosome in which DNA is wound around a cluster of histones.

nucleotide Monomer unit that makes up nucleic acids; consists of a single- or double-ring nitrogenous base, a pentose sugar (ribose or deoxyribose), and one to three phosphate groups.

nucleus (pl. nuclei) 1. That part of a eukaryotic cell surrounded by the nuclear envelope and containing the genetic material. 2. The more-or-less central part of an atom, consisting of one or more protons and (except in most hydrogen atoms) neutrons. 3. A cluster of neuron cell bodies in the brain.

nutrient Any chemical that an organism must take in from its environment because it cannot produce it (or cannot produce it as fast as it needs it).

olfactory Pertaining to the sense of smell.

oligotrophic Of a body of fresh water that contains few nutrients and few organisms.

omnivore Animal that eats both plants and animals.

oncogene A gene capable of taking part in the transformation of a normal cell into a cancerous cell.

oncogenic Of something that causes cancer.

oocyte Cell that undergoes meiosis during oogenesis.

oogamy Form of sexual reproduction involving an egg, which is large and nonmotile, and a sperm, which is small and motile.

oogenesis Formation of eggs or ova, the female gametes.

Oomycota Division of fungi that produce egg-like structures during sexual reproduction; "water molds."

operculum Common covering of all the gills on each side in bony fish.

operon A cluster of genes with related functions: usually one or more structural genes and the regulatory genes that control their activity.

organ Group of tissues assembled in such a way that the entire structure (organ) performs a particular function; e.g. liver, kidney, heart.

organelle Structure within a cell that takes part in carrying out the cell's life functions.

organism An individual living thing, made up of one or more cells.

osmoregulation Regulation of the body's salt and water content.

osmosis Movement of water through a membrane down a water potential gradient.

osmotic potential Tendency of a solution to gain water when separated from pure water by an ideal selectively permeable membrane; the negative of osmotic pressure.

osmotic pressure Pressure that must be exerted on a solution to keep it from gaining water when separated from pure water by an ideal selectively permeable membrane.

Osteichthyes Class of vertebrates containing the bony fishes.

oviduct Tube through which eggs pass from the ovary toward the exterior of the body.

oviparous Reproducing by laying

eggs which develop outside the female's body.

oviposition Laying of eggs.

ovoviviparous Retaining eggs within the mother's body until hatching; e.g., most sharks and some reptiles.

ovule Structure inside the ovary of a flower enclosing a female gametophyte.

ovum (pl. ova) Female gamete, egg.

oxidation A chemical reaction involving removal of electrons or hydrogen atoms, or addition of oxygen; always paired with a reduction.

ozone O_3. A poisonous gas, a common pollutant in smog; also formed by the action of sunlight on oxygen in the ozone layer of the atmosphere.

ozone layer Layer in the upper stratosphere where solar radiation converts some O_2 atoms into ozone molecules. The ozone layer absorbs much ultraviolet radiation and prevents it from reaching the earth.

pancreas In vertebrates, an organ near the stomach that produces hormones and digestive enzymes.

parasite Organism that feeds on another living organism without killing it.

parasympathetic nervous system Part of the autonomic nervous system; its activities promote digestion and elimination, slowing of heart rate, etc.

parthenogenesis Production of young from unfertilized eggs.

pathogenic Disease-causing.

pectoral Pertaining to the anterior fins or limbs of vertebrates.

pelvic Pertaining to the rear pair of appendages (limbs or fins) of vertebrates.

peptide bond Covalent bond joining the carboxyl carbon of one amino acid to the amino nitrogen of the next.

perennial Living for many years and surviving normal seasonal changes.

peripheral nervous system The part of the nervous system outside the brain and spinal cord; it consists of nerves to and from the muscles, sense organs, and internal organs.

permafrost Permanently frozen layer in the soil, found in arctic and antarctic regions.

pest Any organism that is undesirable at the time and in the place where it exists. Includes plant-eating insects, molluscs, and nematodes on crop plants, parasites on animals, weeds in fields of crops.

pesticide Substance used to kill undesirable organisms; includes insecticides, herbicides, nematocides, etc.

pH Logarithm to the base 10 of the hydrogen ion concentration of a solution; a measure of how acidic or basic a solution is, on a scale of 0 to 14 (0 = very acidic, 14 = very basic, 7 = neutral).

Phaeophyta Taxonomic division containing the brown algae.

phage See bacteriophage.

phagocyte Scavenger cell that ingests and destroys pathogens, dead body cells, and other debris.

pharynx Part of the gut just behind the mouth in many animals.

phenotype The characteristic produced by expression of an organism's genes.

pheromone Chemical released by one member of a species that influences the behavior of another member of the species.

phloem Tissue in plants that conducts food from sites of synthesis or storage to sites where food is used or stored.

phospholipids Group of structural lipids that are the main components of biological membranes; made up of fatty acid(s), phosphate, and (usually) nitrogen-containing choline.

phosphorylation Addition of a phosphate group.

photochemical reaction Chemical reaction powered by light energy.

photoperiodism Production of a physiological response (such as flowering or coming into breeding condition) in response to length of period of daylight each day.

photosynthesis Process whereby plants, and some bacteria and protists, capture solar energy and store it as chemical bonds in carbohydrate molecules, using CO_2 to build the carbohydrate.

photosystem Cluster of pigment molecules in which light-absorbing reactions of photosynthesis occur.

phototropism Growth toward or away from light.

phycobilins Accessory photosynthetic pigments found in Cyanobacteria and Rhodophyta; major ones are phycocyanin and phycoerythrin.

phycology Study of algae.

phylogeny (adj. phylogenetic) Line of evolutionary descent.

physiology The processes by which an organism carries out its various biological functions; how an organism works.

phytoplankton Plants floating in the upper layers of a body of water.

P_i Abbreviation for an inorganic phosphate group.

pigment Molecule that differentially absorbs particular wavelengths of visible light and so appears colored.

placenta Organ in mammals in which blood capillaries from mother and fetus lie close together and exchange substances via the extracellular fluid between the two bloodstreams.

placentals Mammals that undergo part of their development in the mother's uterus, where they receive food and oxygen via the placenta, until a fairly advanced stage of development.

plankton Organisms that drift around in water because they are not capable of swimming against currents in the water.

plasma membrane Phospholipid and protein membrane surrounding a cell. Also called plasmalemma, cell membrane.

plasmid Small, circular DNA molecule, found in some bacteria and fungi in addition to the organism's own genome.

plasmodesma (pl. plasmodesmata) Strand of cytoplasm that directly links the cytoplasms of two neighboring plant cells.

plastid Organelle found only in plant cells; depending on the cell's specialized function, its plastids may develop as chloroplasts, chromoplasts, amyloplasts, etc.

Platyhelminthes Phylum of animals containing the flatworms, e.g., planarians, tapeworms, flukes.

poikilothermy Condition of having a body temperature that changes with that of the environment (opposite of homeothermy).

polar 1. Electrically asymmetrical.

2. Pertaining to the poles (ends) of the mitotic or meiotic spindle.

pollen grain Immature male gametophyte of a gymnosperm or flowering plant; it will produce the sperm nuclei that fertilize the egg.

pollination Deposition of pollen on or near the female parts of a gymnosperm or angiosperm.

pollutant Substance that causes pollution.

pollution A change in the physical, chemical, or biological properties of air, water, or soil that can adversely affect the health, survival, or activities of humans and other living organisms.

poly- Prefix meaning many.

polyandry Mating system in which a female mates with more than one male.

polygamy Mating system in which an individual may have more than one mate.

polygenic character Trait whose phenotypic expression is governed by many pairs of alleles (e.g., human skin or hair color).

polygyny Mating system in which one male mates with more than one female.

polymer Large molecule made up of many subunits that are smaller molecules similar or identical to one another.

polymorphism Simultaneous presence in a population of two genetically different forms of a trait at frequencies higher than could be maintained by recurrent mutation.

polyp 1. One of the two alternate body forms found in members of the phylum Cnidaria. 2. A small tumor, often attached by a stalk.

polypeptide A polymer composed of amino acid monomers joined by peptide bonds.

polyphyletic Of a taxon containing members of several evolutionary lines but not including their common ancestor (if there is one).

polyploidy Possession of three or more haploid sets of chromosomes.

polysaccharide A macromolecule made up of many subunits which are simple sugars.

polytene chromosome A chromosome that is exceptionally large because it contains multiple copies of its DNA side by side.

population All members of a species living in a particular area and making up one breeding group.

portal system A group of blood vessels that carry blood from one capillary bed to another without passing through the heart.

posterior At or toward the rear end of an animal.

postsynaptic Receiving stimuli from a nerve cell across a small gap, or synapse.

prebiotic Before life arose.

Precambrian Before the Cambrian period, which began about 600 million years ago.

predator Animal that captures other organisms (usually animals) for food.

presynaptic Pertaining to part of the neuron that sends a signal across a synapse; the receiving cell is postsynaptic.

primary growth Growth in length and production of new stem and root branches in plants.

primary productivity The rate at which food is made from inorganic substances by photosynthetic and chemosynthetic organisms.

primary structure Of a protein, the order in which amino acids are joined together to form a polypeptide.

primary tissues Plant tissues that differentiate from cells laid down by the apical meristem.

Primates The order of mammals that contains monkeys, apes, humans, etc.

primitive Showing features believed to have arisen early in evolution.

producers Photosynthetic and chemosynthetic organisms.

productivity Amount of organic matter produced by members of a given trophic level during a given period of time.

progeny Offspring.

prokaryotes Bacteria: organisms that lack both a nuclear envelope separating the DNA from the cytoplasm and membrane-bounded organelles.

promoter site A section of DNA to which RNA polymerase must bind before transcription can occur.

protein A functional unit made up of one or more polypeptides.

Protista Kingdom (under the five-kingdom system of classification) of unicellular or colonial eukaryotes (e.g., *Amoeba, Paramecium, Euglena*).

protozoa Heterotrophic unicellular eukaryotes.

proximal Toward the center of the body or point of origin of a limb or other structure.

pseudopodium (pl. pseudopodia) Flowing extension of the plasma membrane and cytoplasm of a cell, used for locomotion in organisms such as amoebas.

Pterophyta Division of plants containing the ferns.

pulmonary Of the lungs.

punctuated equilibrium Idea that evolution sometimes proceeds in fits and starts, with new species forming or changing rapidly and then remaining unchanged for long periods of time.

pupa Stage between larva and adult in insects with complete metamorphosis, e.g., butterflies, flies, beetles.

pupation 1. Time of entry into the pupal stage. 2. State of being in the pupal stage.

quadrupeds Animals that walk on four legs.

receptor 1. A structure whose function is to recognize and bind a particular molecule; e.g., a T-cell receptor on the surface of a lymphocyte binds a specific antigen, and receptor proteins in the cytoplasm bind specific steroid hormones. 2. A cell or part of a cell in a sensory system that intercepts a stimulus and causes an effect such as generation of a nerve impulse.

recessive allele An allele not expressed in the heterozygote's phenotype.

recombinant DNA DNA produced by combining genes from more than one organism.

recombination In genetics, the formation of DNA molecules with new combinations of genes in an individual as a result of crossing over during sexual reproduction in eukaryotes or as a result of transduction, transformation, or conjugation in prokaryotes.

red blood cells Cells in the blood that contain hemoglobin, a red, oxygen-carrying pigment. Also called erythrocytes.

reduction A chemical reaction involving addition of electrons; often takes the form of adding entire hydrogen atoms.

reflex Unit of automatic action of the nervous system, controlled by a reflex arc, which consists of a sensory neuron, usually one or more interneurons, and one or more motor neurons.

regeneration The process by which an animal regrows an amputated organ.

renal Of the kidney.

replication (= duplication) The making of an exact copy.

repressor In genetics, a molecule that binds to part of an operon and prevents its transcription.

reproductive potential See biotic potential.

reptile Vertebrate with dry, scaly skin and eggs laid on land, e.g., snakes, lizards, alligators, turtles.

residue In molecular structure, what is left of a molecule when it has reacted with other molecules; e.g., a molecule of glucose and a molecule of fructose may condense together to form sucrose, which consists of a glucose residue and a fructose residue.

resorption Absorption back into the body.

respiration Series of oxidation-reduction reactions by which organisms break the chemical bonds in food molecules to release energy, using an inorganic substance (usually O_2) as the final electron receptor.

respiratory pigment Colored molecule that transports oxygen in an animal's blood.

restriction enzyme Bacterial enzyme that breaks DNA between specific nucleotides in a particular sequence; used in genetic engineering.

retina Layer of receptor cells in the eye that responds to light.

retrovirus A virus with an RNA genome that is replicated by the action of reverse transcriptase.

reverse transcriptase Viral enzyme that synthesizes DNA on an RNA template.

rhizoid Root-like structure that anchors bryophytes, some fungi, etc.

rhizome Underground stem of a vascular plant.

Rhodophyta Division containing the red algae.

rodents Members of the mammalian order Rodentia: rats, mice, and their relatives.

Rotifera Phylum of small freshwater animals with a mouth surrounded by cilia.

ruminants Mammalian herbivores in which the stomach contains fluids of an alkaline (basic) pH and is divided into fermentation chambers housing microorganisms that digest the food.

salinity Saltiness.

salt A substance composed of a positively charged ion other than H^+ and a negatively charged ion other than OH^-; an ionic compound whose cation comes from a base and whose anion comes from an acid.

sampling error Statistical error in scientific experiment resulting from sampling too few subjects.

sap Mixture of water, minerals, etc., conducted in xylem tissue of plants.

saprobe Organism using nonliving organic matter for food.

scavenger Animal that eats dead organisms or organic matter.

Schwann cell Type of cell forming the fatty myelin sheath that surrounds some peripheral neurons.

seaweed Multicellular algae in marine habitats; some members of the Rhodophyta, Phaeophyta, and Chlorophyta.

secondary growth Growth in girth in plants.

secretion 1. Expulsion of a product of a gland or cell. 2. Movement of substances from the blood into forming urine inside the nephron tube.

seed Dispersal unit of gymnosperms and angiosperms, consisting of a seed coat, embryonic plant, and food supply.

seed coat Outer covering of a seed, developed from the outer layers of the ovule.

selective permeability Property of allowing some substances to pass through more easily than others.

self-pollination The transfer of pollen from male reproductive structures such as cones or flower parts to female structures on the same plant.

semen Fluid containing sperm and attendant secretions.

sensory Having to do with detection of changes in the external or internal environment of an organism.

septum (pl. septa) Partition.

sessile (= sitting) Not moving from place to place.

sex chromosomes A pair of chromosomes that cause their carrier to develop as a member of one sex if homozygous for the chromosome pair and the other sex if heterozygous (e.g., X and Y chromosomes in most mammals).

sexual dimorphism Dimorphism = "two forms"; in sexual dimorphism, the two sexes differ in appearance or behavior.

sexual selection Type of natural selection in which females choose to mate with males with particular hereditary traits.

shoot system Part of a plant consisting of the stems, leaves, and any reproductive structures borne thereon.

siliceous Containing silica (SiO_2).

skeletal muscle (= striated muscle) The type of muscle that forms the muscles of the limbs, back, etc.; contains multinucleated fibers.

smooth muscle The type of muscle that lines the walls of internal organs, e.g., digestive tract, blood vessels; consists of single cells.

sociobiology The study of social behavior, the cooperative interactions of animals of the same species.

somatic cells Body cells (as opposed to germ cells, which give rise to gametes).

sorus (pl. sori) Cluster of sporangia in ferns.

speciation Formation of a new species.

species Group of organisms whose members share a common gene pool. See also biological species.

spermatocyte Cell that undergoes meiosis during spermatogenesis.

spermatogenesis Formation of spermatozoa (sperm), male gametes.

spermicidal Sperm-killing.

Sphenophyta A division of lower vascular plants: "horsetails."

sphincter A circular muscle whose contraction closes a tube.

spindle Structure made up of microtubules upon which chromosomes move during mitosis and meiosis.

spirillum (pl. spirilla) Spiral-shaped bacterium.

spontaneous generation The ancient belief that living things routinely arise from nonliving matter.

sporangium (pl. sporangia) Structure in which spores develop.

spore Reproductive cell that can grow into a new individual without fertilization; produced by meiosis in plants, by meiosis or mitosis in fungi. Bacterial spores form when an individual cell encases itself in a protective covering when conditions are unfavorable for growth.

sporophyte Plant that produces haploid spores following meiosis.

Sporozoa Old phylum of protists; most of its members are now placed in the phylum Apicomplexa, the rest in a small phylum, Microspora.

Squamata Order of reptiles containing lizards and snakes.

standing crop Mass of organisms actually present at any one time.

starch Storage polysaccharide made up of glucose monomers, commonly found in plants.

sterilization 1. Any more-or-less permanent change that prevents an animal from reproducing sexually. 2. Cleaning of an object by destroying all the organisms on it.

sternum Breastbone.

steroid A lipid containing four contiguous carbon rings; e.g., cholesterol, estrogen, testosterone.

stigma Tip of female flower part, usually sticky, allowing pollen to adhere to it easily.

stimulus (pl. stimuli) Energy (chemical, electrical, thermal, light, mechanical, etc.) in the external or internal environment of an organism, to which the organism may respond.

stoma (pl. stomata) In a plant, a pore between two guard cells through which gases are exchanged between the plant and the air.

striated muscle See skeletal muscle.

stroma In chloroplasts, the material surrounding the thylakoid membrane and containing the chloroplast's ribosomes, DNA, and enzymes of carbon fixation.

structural gene A section of chromosome that carries information determining the sequence of amino acids in a polypeptide or protein.

substrate 1. Reactant in an enzyme-mediated chemical reaction. 2. Underlying surface, e.g., a rock in the ocean floor.

subtropical (or semitropical) Lying near, but not within, the tropics.

succession In ecology, process in which the community of species inhabiting an area that has been disturbed changes with time in a regular sequence; succession finishes when the organisms of the climax community of the area have become established.

succulents Plants that store water in fleshy stems or leaves.

sucrose A disaccharide consisting of a glucose monomer and a fructose monomer; table sugar.

sustainable Capable of continuing indefinitely in approximately its present form.

symbiont Organism that lives in close association with a member of another species (often refers to a microorganism living in relationship with a larger host organism).

symbiosis (adj. symbiotic) Close association between members of two or more species (see mutualism, commensalism, parasite.)

sympathetic nervous system The part of the nervous system that prepares the body to meet stressful or dangerous situations.

sympatric Living in the same area.

synapse Tiny gap across which information is transferred from one neuron to an adjacent one.

synapsis The lining up of sister chromatids with their homologues in the early stages of meiosis.

syncytium An animal "cell" that contains more than one nucleus within its plasma membrane.

tactile Of the sense of touch.

taxon (pl. taxa) Any one of the hierarchical categories into which organisms are classified; e.g., species, order, class.

taxonomy Study of the classification and identification of living organisms.

TCA cycle Abbreviation for tricarboxylic acid cycle, old name for the citric acid cycle.

template A pattern or mold.

temporal Of time.

terrestrial Of land (as opposed to water or air).

territory Area defended by one or more animals against intruders.

tetrad Foursome consisting of two sets of two (= 4 altogether) linked sister chromatids lined up at synapsis.

tetrapods "Four-footed" vertebrates; amphibians, reptiles, birds, and mammals.

thermochemical reaction Chemical reaction whose rate varies with the temperature.

thorax (adj. thoracic) Part of the body between the head and the abdomen.

thylakoid Flattened membranous sac in which the light-energy capturing reactions of photosynthesis take place.

tissues Groups of cells that perform a particular task in an organism; e.g., blood, cartilage, xylem.

tonoplast Membrane enclosing the central vacuole of a plant cell.

toxin (adj. toxic) Poison.

trachea 1. In tetrapod vertebrates, the windpipe: tube that conducts air from the pharynx to the lungs. 2. In insects, one of the tubes that conduct air from the outside throughout the interior of the body.

tracheophyte Vascular plant.

transcription Synthesis of RNA using a DNA template.

transduction 1. Conversion of the energy of a stimulus into electrical energy that can be transmitted by the nervous system. 2. Transfer of genetic material from one bacterium to another by a bacteriophage.

transfer RNA RNA molecule that transports amino acids to ribosomes during protein synthesis.

translation The assembly of amino acids to form a polypeptide, in a sequence specified by the order of nucleotides in a molecule of messenger RNA.

translocation 1. In genetics, the movement of a segment of nucleic acid from one part of a chromosome to another part of the same chromosome, or to a different chromosome. 2. In protein synthesis, the movement of messenger RNA and a transfer RNA molecule attached to a growing polypeptide, from the A site to the P site along a ribosome.

transpiration Loss of water by evaporation through pores (stomata) in the shoot system of a plant.

transposable element A length of DNA that can move from its position on one chromosome to another position on the same chromosome or to another chromosome.

trophic level The level in the food chain at which an organism functions; e.g., herbivores, members of the second trophic level, eat autotrophs, members of the first trophic level.

tropics That part of the earth lying between the tropic of Cancer (at latitude 23 degrees 27 minutes north of the equator) and the tropic of Capricorn (at the same latitude south of the equator). These latitudes mark the limits of the sun's apparent movement north and south during the year.

tuber Underground storage stem, e.g., potato.

tumor Abnormal growth of tissue; sometimes malignant (cancerous), but may be benign.

turgid Swollen with fluid.

turgor Internal pressure that results from being filled with fluid.

type specimen Individual specimen preserved by a person who describes a new species, as the standard for comparison in determining whether other individuals are members of the same species.

ungulates Hoofed mammals.

unicellular With a body consisting of only one cell.

ureter Tube through which urine passes from the kidney to the urinary bladder.

urethra Tube from the urinary bladder to the exterior.

Urodela Order of amphibians including newts, salamanders, mud puppies, etc.

uterus (adj. uterine) The organ in females of most species of mammals where the young develop; situated between the oviducts and the vagina.

vacuole A membrane-enclosed sac filled with fluid.

variety A subdivision of the species in the hierarchy of taxonomic classification; the variety (sometimes called subspecies) name follows the species epithet.

vascular Of tissues that transport fluids around the body; e.g., veins, arteries, xylem, phloem.

vascular cambium Meristematic tissue that produces secondary xylem and phloem.

vascular tissue Tissue that conducts water, minerals, and food from one part of the plant to another.

vegetative Carrying out the basic life activities of photosynthesis, metabolism, etc., as opposed to reproduction.

vegetative reproduction Reproduction by growth of an individual's body or fragments of its body; reproduction without production of gametes or spores.

vein 1. Vessel carrying blood toward the heart. 2. Thickened ridge in the wing of an insect. 3. Bundle of vascular tissue in a leaf or flower part.

vena cava A large vein that delivers blood to the (right) atrium (of the heart).

venation The arrangement of veins in leaves or in the wings of insects.

venereal disease (VD) Traditional name for a disease, such as syphilis or AIDS, transmitted by sexual activity. These diseases are now generally known as sexually transmitted diseases (STDs).

ventral Pertaining to the undersurface of a bilaterally symmetrical animal; e.g., underside of a worm, belly of a dog.

ventricle (adj. ventricular) 1. A heart chamber that pumps blood into one or more arteries. 2. Fluid-filled cavity in the brain.

venule A small vein in the circulatory system.

vertebrate Animal with a backbone; e.g., fish, human.

vesicle Membrane-enclosed sac, holding secretory products, enzymes, etc., in a cell.

vestigial Small and nonfunctional.

virus Particle composed of nucleic acid and protein that can be reproduced only in a living cell.

vitamin An organic micronutrient.

viviparity (adj. viviparous) Condition of giving birth to young rather than laying eggs.

warbler Member of a group of small insect-eating birds.

wavelength Light and sound may be considered as travelling in wavy lines. The distance between adjacent peaks of the line is the wavelength, symbolized by λ (lambda).

white blood cells Cells in the blood that are involved in defending the body against foreign organisms and substances; also called leukocytes.

white matter Nervous tissue consisting mainly of myelinated axons.

wild type In genetics, showing the normal phenotype of wild members of the species for the trait in question.

wood Secondary xylem.

xylem Plant tissue that conducts sap from the roots to the leaves.

Zoomastigina Phylum of heterotrophic protozoa that move by means of flagella.

zooplankton Animals and protozoa floating in the surface layers of a body of water.

Zygomycota Division of fungi whose members form zygospores during sexual reproduction.

zygote Fertilized egg.

Photo Credits

CHAPTER *11*

Paramecium Mating, Carolina Biological Supply Company
Fission in Amoeba Proteus, Carolina Biological Supply Company
Roquefort penicillium, CNRI, Science Photo Library/Photo Researchers

CHAPTER *12*

Normal Male Chromosomes, Biophoto Associates/Photo Researhers
Silver Leaf Monkey and Baby, Fletcher & Baylis/Photo Researchers
Tulips, Jon Feingersh/Tom Stack & Associates
Mitosis, Michael Abbey/Photo Researchers

CHAPTER *13*

Male Mallard Head, Mary Clay/Tom Stack & Associates
Ocelot, Chip and Jill Eisenhart/Tom Stack & Associates

CHAPTER *14*

Chimpanzee, Tom McHugh/Photo Researchers
King Vulture, Gerlad and Buff Corsi/Tom Stack & Associates
Allosaurus, Brian Parker/Tom Stack & Associates
Fossil Fish, Joyce Photographics/Photo Researchers

CHAPTER *15*

Boga, Larry Lipsky/Tom Stack & Associates
Green Turtle, © 1990 Frans Lanting/Minden Pictures
Daffodils, Bonnie Rauch/Photo Researchers

CHAPTER *16*

Lowland Gorillas, Tom McHugh/Photo Researchers
Peacock, Thomas Kitchin/Tom Stack & Associates
Male Stag Beetle, Hans Pfletschinger/Peter Arnold, Inc.
Honey Bee Queen & Attendant Worker Bees, 1978 Scott Camazine/Photo Researchers

CHAPTER *17*

Amargosa River Entering Death Valley, Mojave Desert, © David Muench
Chambered Nautilus, Todd Gipstein/Photo Researchers

CHAPTER *18*

Kangaroo, Gérard Lacz/Peter Arnold, Inc.
Amoebas, Carolina Biological Supply Company
Horn Shark, Fred Bavendam/Peter Arnold, Inc.

CHAPTER *19*

Pretzel Slime Mold, Nemitrichia serpula, R. M. Meadows/Peter Arnold, Inc.
Bacteria–Staphylococcus epidermidis, David Scharf/Peter Arnold, Inc.

CHAPTER *20*

Kelp (Brown Algae), Robert B. Evans/Peter Arnold, Inc.
Pollen and Seed Cones of Pine, Michel Viard/Peter Arnold, Inc.

CHAPTER *21*

Frilled Lizard, David G. Barker/Tom Stack & Associates
Nudibranch, Dave B. Fleetham/Tom Stack & Associates
Eclectus Parrot, Female, John Cancalosi/Peter Arnold, Inc.

CHAPTER *22*

Blue and Red Sea Cucumber, Brian Parker/Tom Stack & Associates
Jaguar (Panthera onca), Kevin Schafer/Tom Stack & Associates
Feeding Spoonbill, © 1990 Frans Lanting/Minden Pictures
Grasshopper, C. Allan Morgan/Peter Arnold, Inc.

CHAPTER *23*

Angel Fish, Jeff Rotman/Peter Arnold, Inc.
Versicolored Emerald, Luiz Claudio Marigo/Peter Arnold, Inc.

CHAPTER *24*

Adelie Penguins, Antarctica, Luiz C. Marigo/Peter Arnold, Inc.
Mountain Lion, S. J. Krasemann/Peter Arnold, Inc.
Jerusalem Cricket, © 1986 Frans Lanting/Minden Pictures

CHAPTER *25*

Innoculation of Baby Boy, Jim Selby, Science Photo Library/Photo Researchers
T4 Lymphocyte Releasing AIDS viruses, Prof. Luc Montagnier, Institut Pasteur, Science Photo Library/Photo Researchers

CHAPTER *26*

Soft Coral, Ed Robinson/Tom Stack & Associates
Banded Gila Monster, J. Cancalosi/Peter Arnold, Inc.

CHAPTER *27*

Five-month-old Fetus, Alex Bartel, Science Photo Library/Photo Researchers

Egg with Sperm, SIU, School of Medicine/Peter Arnold, Inc.
Leatherback Turtle Hatching, © 1986 Frans Lanting/Minden Pictures
Egg Arrangement, © 1986 Frans Lanting/Minden Pictures

CHAPTER *28*

Three Neurons of the Human Cerebral Cortex, Secchi-Lecaque, Roussel-DCLAF, CNRI, Science Source
X-ray Human Head, Density Slicing, Manfred Kage/Peter Arnold, Inc.

CHAPTER *29*

Platypus, Steve Kaufman/Peter Arnold, Inc.
Ommatidia in Fly Eye, Paulette Brunner/Tom Stack & Associates
Rana Catesbeiana Frog, Carolina Biological Supply Company
Blue Shark, Norbert Wu/Peter Arnold, Inc.

CHAPTER *30*

San José Relays, California, Keith H. Murakami/Tom Stack & Associates
Skull Being X-rayed onto Xerox Sheet, Manfred Kage/Peter Arnold, Inc.
Human Bone, Haversian Canal, Michael Abbey, Science Source/Photo Researchers

CHAPTER *31*

Young Bicyclists, Laurent Dardelet
Impala Bucks, © 1990 Frans Lanting/Minden Pictures
Canada Goose, S. J. Krasemann/Peter Arnold, Inc.

CHAPTER *32*

Polar Bear Males Get Acquainted, Fred Bruemmer/Peter Arnold, Inc.
Tiger (Panthera tigris) Gérard Lacz/Peter Arnold, Inc.
Mountain Gorilla, V. Arthus-Bertrand/Peter Arnold, Inc.

CHAPTER *33*

Rainbow Cactus, Michel Viard/Peter Arnold, Inc.
Redwood Sorrel, © 1990 Frans Lanting/Minden Pictures

CHAPTER *34*

Root, Manfred Kage/Peter Arnold, Inc.
Beach Primrose Sand Dune Vegetation, © 1983 Frans Lanting/Minden Pictures
Agave, John Gerlach/Tom Stack & Associates

CHAPTER *35*

Field of Corn-poppies, Photo Aubert/Photo Researchers

Harebell, Rod Planck/Photo Researchers
Shagbark Hickory Leaves Opening, John Shaw/Tom Stack & Associates

CHAPTER 36

Red Flower, Brian Parker/Tom Stack & Associates
Hibiscus: Pistil and Stames, Paulette Brunner/Tom Stack & Associates
Salsify at Sunrise, Wendy Shattil and Bob Rozinski/Tom Stack & Associates
Chestnut Oak Fruits, Acorns, W. H. Hodge/Peter Arnold, Inc.

CHAPTER 37

Alpine Pool, © David Muench
Yuma, Arizona, Ken Biggs/Photo Researchers
Antarctic Lichen, © 1986 Frans Lanting/Minden Pictures
White-faced Capuchin Costa Rica, Kevin Schafer/Peter Arnold, Inc.

CHAPTER 38

Hermit Thrush in Flowering Dogwood, Stephen Collin/Photo Researchers
Tiger (Pathera tigris), Gérard Lacz/Peter Arnold, Inc.
Long-nose Hawkfish, Ed Robinson/Tom Stack & Associates
Bison, © 1986 Frans Lanting/Minden Pictures

CHAPTER 39

Walruses, Mark Newman/Tom Stack & Associates
Cacti, Apache Trail, AZ, © David Muench
Elephant Drinking, Leonard Lee Rue/Photo Researchers
Liliaceae, Michel Viard/Peter Arnold, Inc.
Fungia Coral, Larry Tackett/Tom Stack & Associates

CHAPTER 40

Baby Smiling, Tom Stack/Tom Stack & Associates
Lightning, Kent Wood/Photo Researchers
Diver and Icilogorgia, Dr. Steven K. Webster
Full Moon and Surf, Chris Luneski/Photo Researchers

PART 1

Mountain Gorilla, Arthus-Bertrand/Peter Arnold, Inc.
Water, David Muench
Protein (Cytochrome c), Irving Geis, Science Source/Photo Researchers
Amoeba, Ed Reschke/Peter Arnold, Inc.
Microtubules, Dr. Gopal Murti, Science Photo Library/Photo Researchers
Human Brain Cell, Manfred Kage/Peter Arnold, Inc.
Energy Transfer, John Vucci/Peter Arnold, Inc.
Arrow Crab Feeding, Brian Parker/Tom Stack & Associates
Aspen Leaves, D. Cavagnaro/Peter Arnold, Inc.

PART 2

DNA Double Helix, Nelson Max, Lawrence Livermore Lab/Peter Arnold, Inc.
Sumatran Orang-utans, Brian Parker/Tom Stack & Associates
Red-Humped Caterpillar, D. Wilder/Tom Stack & Associates
Paramecium, Carolina Biological Supply Company
Chromosomes, Biophoto Associates/Photo Researchers

PART 3

King Vultures, Gerlad and Buff Corsi/Tom Stack & Associates
Boga, Larry Lipsky/Tom Stack & Associates
Peacock, Thomas Kitchin/Tom Stack & Associates
Amoebas, Carolina Biological Supply Company

Fungus, E. Guerko, CNRI, Science Photo Library/Photo Researchers
Micro-Orchids, Luiz Claudio Marigo/Peter Arnold, Inc.
Frilled Lizard, David G. Barker/Tom Stack & Associates

PART 4

Sea Cucumber, Brian Parker/Tom Stack & Associates
Jellyfish, Sea Studios, Inc./Peter Arnold, Inc.
Adelie Penguins, Luiz C. Marigo/Peter Arnold, Inc.
Macrophage, Manfred Kage/Peter Arnold, Inc.
Zebras, 1989 Frans Lanting/Minden Pictures
Fetus, Alex Bartel, Science Photo Library/Photo Researchers
Polar Bear, S.J. Kraseman/Peter Arnold, Inc.
Galago, Gary Milburn/Tom Stack & Associates
Haversian Canal, Michael Abbey, Science Source/Photo Researchers
Dingo Nursing, Gérard Lacz/Peter Arnold, Inc.
Learning, Suzanne Szasz/Photo Researchers

PART 5

Rainbow Cactus, Michel Viard/Peter Arnold, Inc.
Lilac Stem, Manfred Kage/Peter Arnold, Inc.
Beach Primrose, Frans Lanting/Minden Pictures
Corn Poppies, Photo Aubert/Photo Researchers
Harebell, Rod Planck/Photo Researchers
Composite Flower, Brian Parker/Tom Stack & Associates

PART 6

Lichen, Frans Lanting/Minden Pictures
Tiger, Gérard Lacz/Peter Arnold, Inc.
Cacti, David Muench
Baby, Tom Stack/Tom Stack & Associates

Index†